R. Gupta's®

Popular Master Guide

B.Tech/B.E./B.Sc./B.Sc. (Hon.)

Biotechnology

Entrance Exams

CONDUCTED BY

AIIMS, New Delhi/GGS Indraprastha Univ. Delhi/Jamia Millia Islamia Univ./Amity Institute of Biotechnology, New Delhi/IIT's (New Delhi, Chennai & Kharagpur)/ UPTU, Lucknow/Univ. of Rajasthan, Jaipur/Osmania Univ., Hyderabad/APECET/ Gulbarga Univ. Gulbarga/Panjab Univ., Chandigarh/Guru Nanak Dev Univ., Amritsar/ Bharati Vidyapeeth, Pune/Univ. of Kerala/Vellore Institute of Technology, Tamil Nadu and Other Prestigious Indian Institutes/Universities.

2027
EDITION

RAMESH PUBLISHING HOUSE, NEW DELHI

Published by

O.P. Gupta *for* Ramesh Publishing House

Admin. Office

12-H, New Daryaganj Road, Opp. Officers' Mess,
New Delhi-110002 ✆ 23275224, 23245124

E-mail: info@rameshpublishinghouse.com
For Online Shopping: www.rameshpublishinghouse.com

Showroom

- Balaji Market, Nai Sarak, Delhi-110006 ✆ 23282525 📱 9354373464
- 4457, Nai Sarak, Delhi-110006

Book Code: R-1441

ISBN: 978-93-5012-152-8

Price: ₹ 680

Printed at: Deepak Offset, Delhi

CONTENTS

Previous Years' Paper

B.Sc. BIOTECHNOLOGY

Entrance Exam, 2025*

PHYSICS

1. If the protons and electrons are the only basic charges in the universe, all the observable charges have to be integral multiples of *e*. Thus, if an object contains *x* electrons and *y* protons, the net charge on the object will be:

1. $-(x + y)e$
2. $(x + y)e$
3. $(x - y)e$
4. $(y - x)e$

2. A charge of magnitude 3×10^{-7} C is located at a distance of 0.09 m from a point P. Obtain the work done in bringing a charge of 2×10^{-9} C from infinity to the point P.

1. 6×10^{4} J
2. 6×10^{-2} J
3. 6×10^{-5} J
4. 6×10^{5} J

3. In a series combination of capacitors connected across a battery:

1. each capacitor has an equal charge for certain values of capacitances only.
2. each capacitor has different charge for a certain value of its capacitance.
3. each capacitor has equal charge for any value of its capacitance.
4. each capacitor has different charge for any value of its capacitance.

4. Two point charges, 4 μC and –3 μC (with no external field) are placed at (–6 cm, 0, 0) and (6 cm, 0, 0), respectively. The amount of work required to separate the two charges infinitely away from each other will be:

1. 0.9 J
2. 0.18 J
3. –0.9 J
4. –0.018 J

5. If the net flux through a cube is 1.05 N m^2 C^{-1}, what will be the total charge inside the cube? (Given : The permittivity of free space is 8.85×10^{-12} C^2 N^{-1} m^{-2}).

1. 9.29×10^{-11} C
2. 9.27×10^{-10} C
3. 9.22×10^{-6} C
4. 9.29×10^{-12} C

6. A parallel plate capacitor having plate area 200 cm^2 and separation 2.0 mm holds a charge of 0.06 μC on applying a potential difference of 60 V. The dielectric constant of the material filled in between the plates is:

1. 0.113
2. 1.13
3. 11.3
4. 113

1. 4	**2.** 3	**3.** 3	**4.** 1	**5.** 4	**6.** 3

* *Conducted by NTA for CUET (UG).*

7. The electric potential due to an electric dipole:

A. depends on r, where r is the magnitude of position vector $\vec{r}$

B. depends on the angle between the position vector $\vec{r}$ and the dipole moment vector $\vec{p}$

C. falls off at long distances, as $1/r^2$

D. does not depend upon the distance separating the charges

Choose the ***correct*** answer from the options given below:

1. A, B and D only
2. A, B and C only
3. A, B, C and D
4. B, C and D only

8. Two point charges placed a distance d apart in vacuum exert a force of magnitude F on each other. One of the two charges is doubled. To keep the magnitude of force same the separation between the charges should be changed to:

1. $2d$ 2. $d/2$
3. $\sqrt{2}d$ 4. $d/\sqrt{2}$

9. Conductors develop electric currents in them:

1. on applying electric field
2. on placing the conductor in the magnetic field
3. on applying gravitational field only
4. on applying the magnetic field and gravitational field

10. Resistivity of a conductor depends on:

1. its material and the dimensions of the conductor
2. its material and temperature of the conductor
3. the dimensions of the conductor only
4. the temperature of the conductor only

11. In the (*i*) absence of electric field, and in the (*ii*) presence of electric field, the paths of electrons between successive collisions with the positive ions of the metal, are:

1. (*i*) Straight line, (*ii*) straight line
2. (*i*) Curved, (*ii*) straight line
3. (*i*) Curved, (*ii*) curved
4. (*i*) Straight line, (*ii*) Curved in general

12. A resistor develops 800 J of thermal energy in 20 s on applying a potential difference of 20 V. Its resistance is:

1. 20 Ω 2. 10 Ω
3. 40 Ω 4. 0.5 Ω

13. A wire of resistance 4 Ω is used to make a coil of radius 7 cm. The wire has a diameter of 1.4 mm and the resistivity of its material is 2×10^{-7} Ω m. The number of turns in the coil will be:

1. 70 2. 40
3. 140 4. 20

14. A battery of emf 12 V and internal resistance 3 Ω is connected to an external resistor. If the current in the circuit is 0.6 A, the voltage across the external resistor will be:

1. 10.2 V 2. 17.0 V
3. 12.0 V 4. 13.8 V

15. A uniform wire of resistance 12 Ω is cut into three pieces in the ratio of length 1 : 2 : 3. Now the three pieces are connected to form a triangle. A cell of emf 8 V and internal resistance 5 Ω is connected across the highest of the three resistors. The current through the circuit is:

1. 4 A 2. 2 A
3. 0.5 A 4. 1 A

16. A bar magnet of magnetic moment 5.0 A m^2 has poles 20 cm apart. The pole strength would be:

1. 20 A m 2. 30 A m
3. 1 A m 4. 25 A m

7. 2	8. 3	9. 1	10. 2	11. 4	12. 2	13. 1	14. 1	15. 4	16. 4

17. Match List-I with List-II.

List-I (Physical Quantity)	List-II (Units)
A. Magnetic field	I. $J\ T^{-1}$
B. Magnetic moment	II. $T\ m\ A^{-1}$
C. Pole strength	III. $J\ T^{-1}\ m^{-1}$
D. Permeability of free space	IV. $Wb\ m^{-2}$

Choose the ***correct*** answer from the options given below:

1. A-IV, B-II, C-III, D-I
2. A-IV, B-I, C-III, D-II
3. A-II, B-I, C-IV, D-III
4. A-IV, B-III, C-I, D-II

18. A long wire with a small current element of length 1 cm is placed at the origin and carries a current of 10 A along the *x*-axis. The magnitude of the magnetic field, due to the element, on the *y*-axis at a distance 0.5 m from it, would be:

1. 4×10^{-8} T
2. 5×10^{-8} T
3. 6×10^{-8} T
4. 2×10^{-8} T

19. A conductor is placed along *z*-axis carrying current in *z*-direction in uniform magnetic field directed along *y*-axis. The magnetic force acting on the conductor is directed along:

1. positive *x*-axis
2. positive *y*-axis
3. positive *z*-axis
4. negative *x*-axis

20. A galvanometer of resistance 520 Ω is shunted with 20 Ω resistance to convert it into an ammeter. The resistance of the ammeter will be:

1. 16.8 Ω
2. 540 Ω
3. 19.3 Ω
4. 18 Ω

21. The magnitude of magnetic field inside a solenoid of length 0.3 m having 800 turns carrying a current of 6 A is:

1. 2.03 T
2. 60.3 mT
3. 20 mT
4. 6.03 T

22. A charged particle accelerated through a potential difference of V volts acquires a speed *u*. The particle is then made to enter perpendicularly in a uniform magnetic field B. The radius of the circular path followed by the charged particle will be proportional to:

1. V/u
2. u/V
3. V^2/u^2
4. u^2/V^2

23.

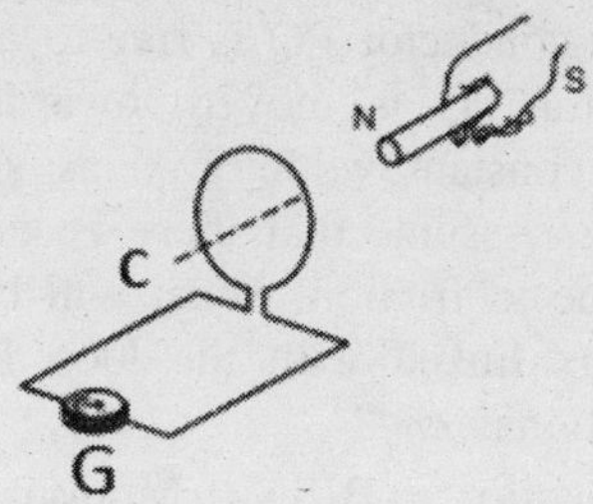

Figure shows a coil C connected to a galvanometer G. When the North-pole of a bar magnet is pushed towards the coil, the pointer in the galvanometer deflects. Regarding this set up, the following statements are given:

A. It indicates the presence of electric current in the coil.
B. The deflection is found to be smaller when the magnet is pushed towards the coil faster.
C. There is repulsion in the moving magnet and the magnetic pole induced in the coil facing towards the N pole of the magnet.
D. If the bar magnet does not move, there is no induced current in the coil.

Choose the ***correct*** answer from the options given below:

1. A, C and D only
2. A, B and C only
3. A, B, C and D
4. B, C and D only

17. 2	18. 1	19. 4	20. 3	21. 3	22. 1	23. 1

24. 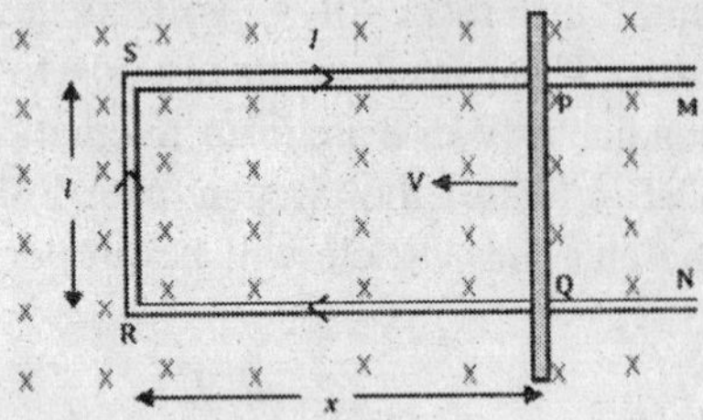

Figure shows a rectangular conductor PQRS in which the conductor PQ is free to move. The conductor PQ is moved towards the left with a constant velocity V as shown in the figure. Assume that there is no loss of energy due to friction. What will be the magnetic flux linked with the loop PQRS and the motional *emf*?

1. Magnetic *flux* = $Bl(xV)$; Motional *emf* = Blx
2. Magnetic *flux* = Blx; Motional *emf* = $(BlV)/t$
3. Magnetic *flux* = Blx; Motional *emf* = BlV
4. Magnetic *flux* = $B/(lx)$; Motional *emf* = BlV

25. The average emf induced in a coil is 2 V when current is changed in 0.4 s:

A. from 5 A to 2 A and the self-inductance of the coil is 0.266 mH.
B. from 4 A to 4 A in the opposite direction, the self-inductance of the coil is 0.10 mH.

1. (A) is correct, (B) is incorrect
2. Both (A) and (B) are correct
3. (A) is incorrect (B) is correct
4. Both (A) and (B) are incorrect

26. An inductor of 500 mH is in series with a resistance and a variable capacitor connected to a source of frequency 0.4 kHz. The value of capacitance of the capacitor to get a maximum current will be:

1. 2.3 μF
2. 0.32 μF
3. 63 μF
4. 0.62 μF

27. A 12 V battery connected to a 6 Ω, 10 mH coil through a switch drives a constant current in the circuit. The switch is suddenly opened. Assuming that it took 1 ms to open the switch, the average emf induced across the coil would be:

1. 10 V
2. 20 V
3. 200 V
4. 12 V

28. For an ac source rated at 220 V, 50 Hz, which of the following statements is correct?

1. The peak value over a period of (1/50) s is 220 V.
2. The average value over a period of (1/50) s is 220 V.
3. The average value over a period of (1/50) s is 0 V.
4. The average value over a period of (1/50) s is $200\sqrt{2}$ V.

29. Which of the following statements is **not** correct for electromagnetic induction?

1. The magnitude of induced emf in a circuit is equal to the time rate of change of magnetic flux through the circuit.
2. The magnitude of induced emf in a circuit is equal to the total change of magnetic flux through the circuit.
3. The induced emf can be increased by increasing the number of turns N of a closed coil.
4. The polarity of induced emf is such that it tends to produce a current which opposes the change in magnetic flux that produced it.

30. Maxwell's displacement current is:

1. due to flow of charges
2. due to changing gravitational field
3. due to changing electric field
4. ε_0 times the rate of change of magnetic flux

24. 3	**25.** 4	**26.** 2	**27.** 2	**28.** 3	**29.** 2	**30.** 3

31. Match List-I with List-II.

List-I (Electromagnetic waves)	List-II (Wavelength range)
A. X-rays	I. 780 nm to 1 mm
B. Radio waves	II. 0.1 m to 1 mm
C. Infrared waves	III. 1 nm to 10^{-3} nm
D. Microwaves	IV. > 0.1 m

Choose the ***correct*** answer from the options given below:

1. A-IV, B-II, C-III, D-I
2. A-I, B-III, C-II, D-IV
3. A-II, B-I, C-IV, D-III
4. A-III, B-IV, C-I, D-II

32. The electric field E associated with a progressive electromagnetic wave is given by $E = E_0 \sin(kx - \omega t)$. If B_0 is the amplitude of the magnetic field associated with the wave, then:

1. $\frac{E_0}{B_0} = \frac{\omega}{k}$
2. $\frac{E_0}{B_0} = \frac{\omega^2}{k^2}$
3. $\frac{E_0}{B_0} = \frac{k}{\omega}$
4. $\frac{E_0}{B_0} = \frac{k^2}{\omega^2}$

33. Which of the following statements are correct?

A. All mirrors follow the laws of reflection.
B. The angle between the ray of incidence and the plane surface of the mirror is equal to the angle between the plane surface of mirror and the ray of reflection for plane mirror.
C. The rays coming parallel to the principal axis will go after reflection through the focus of the curved mirror.
D. The rays coming to the pole of a curved mirror making an angle with axis will be reflected making the equal angle with the axis on the other side of the axis.

Choose the ***correct*** answer from the options given below:

1. A, B, C and D
2. A, C and D only
3. A, B and C only
4. B, C and D only

34. Which of the following statement(s) is/are correct?

A. The power of a lens is the ability of the lens to converge or diverge the incident rays.
B. S.I. unit of the power of a lens is dioptre while focal length is in centimetres.
C. For a lens of larger focal length, power is smaller.
D. In any combination of lenses, the power of combination is not algebraic addition of power of combined lenses.

Choose the ***correct*** answer from the options given below:

1. A and C only
2. B, C and D only
3. A and B only
4. A only

35. In an experiment with a convex lens, the length of an image is 1 cm, and the object length is 5 cm. If the object is placed at a distance of 40 cm from the lens, then the focal length of the lens is:

1. 6.67 cm
2. 13.5 cm
3. 5.6 cm
4. 3.6 cm

36. In a Young's double-slit experiment, two slits are 1.5 mm apart while the screen is 1.2 m away. When a light of wavelength 600 nm is incident on slits, the fringe width will be:

1. 0.48 mm
2. 4.5 mm
3. 4.8 mm
4. 4.2 mm

31. 4	**32.** 1	**33.** 1	**34.** 1	**35.** 1	**36.** 1

37. The critical angle of incidence i_c for a ray incident from a denser to rarer medium, is that angle for which:

1. the angle of reflection is 90°.
2. the angle of refraction is 90°.
3. the angle of refraction is 0°.
4. the angle of reflection is 0°.

38. A polaroid sheet is rotated between two crossed polarizers. The intensity of transmitted light would be maximum, when the angle between the axes of the first polarizer and the polaroid sheet is:

1. $\pi/2$ 2. $\pi/4$
3. π 4. $\pi/3$

39. Which of the following statements are correct?

A. Total internal reflection occurs when a ray of light travels from a rarer transparent medium to a denser medium.
B. In total internal reflection, the incident ray of light remains in the same medium after reflection.
C. In total internal reflection, the angle of incidence inside the denser transparent medium is equal to the angle of reflection in the same medium.
D. In total internal reflection inside a denser medium there is no angle of refraction.

Choose the ***correct*** answer from the options given below:

1. B and C only
2. A and B only
3. B, C and D only
4. C and D only

40. If focal length of a concave lens is 50 cm, then the power of the lens would be:

1. +5 D 2. –5 D
3. +2 D 4. –2 D

41. The photoelectric current is directly proportional to the number of photo electrons emitted per second. This implies that:

1. the number of photoelectrons emitted per second is equal to the frequency of incident radiation.
2. the number of photoelectrons emitted per second is inversely proportional to the intensity of incident radiation.
3. the number of photoelectrons emitted per second is directly proportional to the intensity of incident radiation.
4. the number of photoelectrons emitted per second is not related to the intensity of incident radiation.

42. The de-Broglie wavelength associated with a ball of mass 150 g traveling at 30.0 m/s would be:

1. 1.47×10^{-34} m
2. 14.7×10^{-34} m
3. 0.147×10^{-34} m
4. 7.14×10^{-34} m

43. Which of the following statements are correct?

A. A nucleus of mass number A has a radius R given by the expression $R = R_o A^{1/3}$
B. Volume of nucleus is proportional to mass number A
C. The density of nucleus increases with the radius of nucleus
D. Density of nuclear matter does not depend on its mass number A

Choose the ***correct*** answer from the options given below:

1. A, B and C only
2. A, B, C and D
3. A B and D only
4. B and C only

37. 2	**38.** 2	**39.** 3	**40.** 4	**41.** 3	**42.** 1	**43.** 3

44. An electron in the ground state of a hydrogen atom absorbs 12.09 eV energy. The angular momentum of the electron increases by:

1. $(h/2\pi)$ 2. $2(h/2\pi)$
3. $3(h/2\pi)$ 4. $4(h/2\pi)$

45. Which of the following statements are correct?

A. The electrostatic repulsive force between the protons can be greater than the nuclear force to bind the nucleons together inside a nucleus.
B. The repulsive electrostatic force between protons in smaller nuclei is much smaller than the nuclear force between nucleons inside a nucleus.
C. The gravitational force between nucleons is much smaller than the nuclear force between the nucleons inside a nucleus.
D. The binding energy per nucleon between nucleons is almost constant because the nuclear force is a long range force.

Choose the ***correct*** answer from the options given below:

1. A and D only 2. B and C only
3. C and D only 4. A only

46. Below given are some statements about electronic devices:

A. Diodes can be used for rectifying an ac voltage.
B. For semiconductors, band gap energy $E_g > 3$ eV.
C. By changing the external applied voltage, junction barriers can be changed.
D. *p-n* junction is the 'key' to all semiconductor devices.

Choose the ***correct*** answer from the options given below:

1. A, B and D only
2. A, B and C only
3. B, C and D only
4. A, C and D only

47. On connecting a device X in a series circuit with a battery and a resistor, a current passes through the circuit. On reversing the polarity of the battery, the current in the circuit drops to almost zero. The device X may be a:

1. *p*-type semiconductor
2. *n*-type semiconductor
3. *p-n* junction diode
4. capacitor

48. Which of the following statement(s) is/are true for *p*-type semiconductors?

A. Holes are minority carriers and pentavalent atoms are the dopants.
B. Electrons are majority carriers and trivalent atoms are the dopants.
C. Holes are majority carriers and trivalent atoms are the dopants.
D. Electrons are minority carriers and pentavalent atoms are the dopants.

Choose the ***correct*** answer from the options given below:

1. B only 2. C only
3. C and D only 4. A and C only

49. In the following nuclear reaction,
$^{1}_{0}n + {}^{235}_{92}\text{U} \rightarrow {}^{140}_{54}\text{Xe} + {}^{a}_{b}\text{Sr} + 2^{1}_{0}n$ we have:

1. $a = 38,\ b = 94$
2. $a = 94,\ b = 38$
3. $a = 94,\ b = 40$
4. $a = 96,\ b = 38$

50. A proton accelerated through a potential difference of V volts has a de-Broglie wavelength λ associated with it. In order to get the same wavelength associated with an α-particle, the required accelerating potential is:

1. V/16 2. V/8
3. 4V 4. 8V

44. 2	**45.** 2	**46.** 4	**47.** 3	**48.** 2	**49.** 2	**50.** 2

CHEMISTRY

51. The value of van't Hoff factor, *i*, for CH_3COOH solution in water will be:

1. Between 1 and 2
2. Less than 1
3. 2
4. 1

52. The following solutions were prepared by dissolving 1 g of solute in 1 L of the solution. Arrange the following solutions in **decreasing** order of their molarity:

A. Glucose (molar mass = 180 g mol^{-1})
B. NaOH (molar mass = 40 g mol^{-1})
C. NaCl (molar mass = 58.5 g mol^{-1})
D. KCl (molar mass = 74.5 g mol^{-1})

Choose the ***correct*** answer from the options given below:

1. A, D, C, B
2. A, B, D, C
3. B, C, D, A
4. D, C, A, B

53. Match List-I with List-II.

List-I (Solutions)	List-II (Explanation)
A. Saturated solution	I. Solution having two components.
B. Isotonic solutions	II. A solution whose osmotic pressure is more than that of another.
C. Binary solution	III. A solution which contains the maximum amount of solute that can be dissolved in a given amount of solvent at a given temperature.
D. Hypertonic solution	IV. The solutions having same osmotic pressure at a given temperature.

Choose the ***correct*** answer from the options given below:

1. A-I, B-II, C-III, D-IV
2. A-I, B-III, C-II, D-IV
3. A-I, B-II, C-IV, D-III
4. A-III, B-IV, C-I, D-II

54. Give the reason for low concentration of oxygen in the blood and tissues of people living at high altitude:

1. Both low temperature and high atmospheric pressure
2. Low temperature
3. Low atmospheric pressure
4. High atmospheric pressure

55. Calculate the molality of KI if the density of 20% (mass/mass) aqueous solution of KI is 1.202 g mL^{-1}.

(Molar mass of KI is 166 g mol^{-1})

1. 1.5 mol kg^{-1}
2. 1.2 mol kg^{-1}
3. 15 mol kg^{-1}
4. 0.12 mol kg^{-1}

56. The unit of E_{cell} is:

1. V m^{-1}
2. S cm^{-1}
3. V
4. S cm^{-2} mol^{-1}

57. Match List-I with List-II.

List-I (Property)	List-II (Unit)
A. Cell constant	I. cm^{-1}
B. Molar conductance	II. ohm^{-1} cm^2 mol^{-1}
C. Specific conductance	III. ohm^{-1} cm^{-1}
D. Conductance	IV. ohm^{-1}

Choose the ***correct*** answer from the options given below:

1. A-I, B-II, C-III, D-IV
2. A-I, B-III, C-II, D-IV
3. A-I, B-II, C-IV, D-III
4. A-III, B-IV, C-I, D-II

51. 1	**52.** 3	**53.** 4	**54.** 3	**55.** 1	**56.** 3	**57.** 1

58. The following statements describe various properties of a Mercury cell:

A. It converts energy of combustion into electrical energy
B. It is rechargeable
C. The cell reaction involved is $Zn(Hg) + HgO(s) \rightarrow ZnO(s) + Hg(l)$
D. It is a low current device used in hearing aids

Choose the ***correct*** answer from the options given below:

1. A, B and D only 2. A and C only
3. A, B, C and D 4. C and D only

59. Which cell is used in automobiles and inverters?

1. Mercury cell 2. Dry cell
3. Lead storage cell 4. Fuel cell

60. A galvanic cell behaves as electrolytic cell when?

1. $E_{cell} = E_{ext}$ 2. $E_{cell} = 0$
3. $E_{ext} > E_{cell}$ 4. $E_{cell} > E_{ext}$

61. Which of the following **does not represent** a correct application of the coordination compound?

1. *cis*-platin effectively inhibit the growth of tumours.
2. chlorophyll help in photosynthesis
3. desferrioxime B is used in treatment of lead poisoning
4. cyanocobalamine, antipernicious anemia factor is a coordination compound of Co

62. Arrange the following ions in **increasing** order of number of 3*d*-electrons:

A. Cr^{2+} B. Cu^{+}
C. Ti^{3+} D. Mn^{+}

Choose the ***correct*** answer from the options given below:

1. B, A, C, D 2. C, A, D, B
3. C, D, A, B 4. D, B, C, A

63. The atomic number of Lanthanum is 57. Its electronic configuration will be _____.

1. $[Xe]5d^1 6s^2$ 2. $[Xe]4f^1 5d^2$
3. $[Xe]4f^3$ 4. $[Xe]4f^1 5d^1 6s^1$

64. Match List-I with List-II.

List-I (Compound/Elements)	List-II (Uses)
A. Magnesium based alloy is constituent of	I. Bullets
B. Lanthanoid oxide	II. Petroleum cracking
C. Mixed oxides of Lanthanoids are employed in	III. Television screen
D. Misch metal	IV. Lanthanoid metal and iron

Choose the ***correct*** answer from the options given below:

1. A-I, B-II, C-III, D-IV
2. A-I, B-III, C-II, D-IV
3. A-I, B-II, C-IV, D-III
4. A-III, B-IV, C-I, D-II

65. What is the colour of Fe^{3+} (aqs) ion?

1. Yellow 2. Violet
3. Colourless 4. Bluepink

66. Why is HCl not used to make the medium acidic in oxidation reactions of $KMnO_4$ in acidic medium?

1. $KMnO_4$ is weaker oxidizing agent than HCl
2. $KMnO_4$ oxidises HCl into Cl_2 which is also an oxidizing agent
3. Both HCl and $KMnO_4$ act as oxidizing agent
4. $KMnO_4$ act as a reducing agent in the presence of HCl

67. Acidified potassium dichromate oxidizes sulphides (S^{2-}) to _____.

1. SO_4^{2-} 2. SO_3^{2-}
3. Sulphur (S) 4. SO_2

58. 4	59. 3	60. 3	61. 3	62. 2	63. 1	64. 2	65. 1	66. 2	67. 3

68. What is the **decreasing** order of field strength of given ligands?

A. S^{2-} B. Ethylenediamine
C. NCS^- D. CN^-

Choose the ***correct*** answer from the options given below:

1. A, C, B, D 2. A, B, C, D
3. D, B, C, A 4. D, C, B, A

69. What is the IUPAC name of $[Pt(NH_3)_2Cl_2]$?

1. Diamminedichloridoplatinum (IV)
2. Diamminedichloridoplatinum (II)
3. Diamminedichloridoplatinum (0)
4. Dichloridodiammineplatinum (IV)

70. Select the correct statements for $[Fe(CN)_6]^{3-}$ complex:

A. Paramagnetic
B. sp^3d^2 hybridization
C. Magnetic moment = 5.92 BM
D. d^2sp^3 hybridization

Choose the ***correct*** answer from the options given below:

1. A and D only
2. A, B and C only
3. B and C only
4. B, C and D only

71. Match List-I with List-II.

List-I	List-II
A. Ambident nucleophiles	I. Symmetrical object
B. Plane polarized light	II. Saytzeff rule
C. Superimposable mirror image	III. Cyanides and nitrites
D. β-elimination reaction	IV. Nicol Prism

Choose the ***correct*** answer from the options given below:

1. A-III, B-IV, C-I, D-II
2. A-I, B-III, C-II, D-IV
3. A-I, B-II, C-IV, D-III
4. A-III, B-IV, C-II, D-I

72. Optically active alkyl halide undergoing S_N^2 substitution involves:

1. retention of configuration
2. recemic mixture
3. inversion of configuration
4. formation of carbocation

73. Gabriel phthalimide synthesis is used for the preparation of:

1. tertiary amine
2. acid synthesis
3. primary amine
4. secondary amine

74.

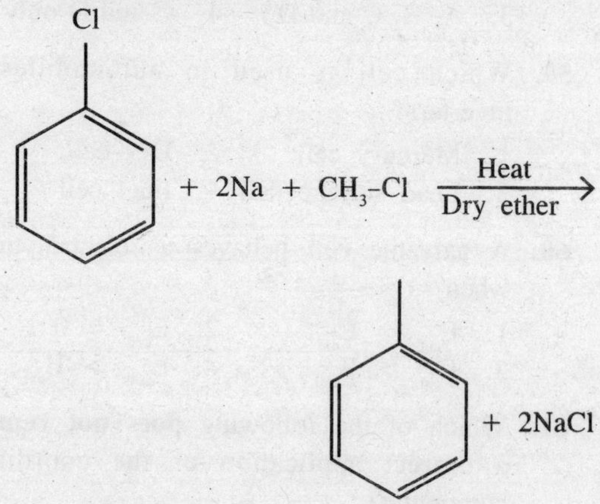

The above reaction is an example of:

1. Sandmeyer's reaction 2. Wurtz reaction
3. Wurtz Fittig reaction 4. Kolbe reaction

75. The reagent(s) used in hydroboration oxidation of propene are

A. B_2H_6 B. H_2O
C. H_2O_2 D. OH^-

Choose the ***correct*** answer from the options given below:

1. A, B and D only 2. A, B and C only
3. A, B, C and D 4. B, C and D only

76. The carbohydrate used as storage molecule in plants is:

1. Starch 2. Glycogen
3. Cellulose 4. Glucose

68. 3 **69.** 2 **70.** 1 **71.** 1 **72.** 3 **73.** 3 **74.** 3 **75.** 3 **76.** 1

77. Hell-Volhard Zelinsky reaction is used for the formation of:

1. alcohols
2. aldehydes
3. ketones
4. α-halocarboxylic acids

78. What is the correct sequence of increasing reactivity of the following compounds towards nucleophilic addition reaction?

A. Ethanal B. Propanone
C. Propanal D. Butanone

Choose the ***correct*** answer from the options given below:

1. A, B, C, D 2. D, B, C, A
3. A, C, B, D 4. C, B, D, A

79. Which of the following reagent(s) is required for the conversion of Benzene to methyl benzoate?

A. $Br_2/FeBr_3$
B. Mg, dry Ether
C. CO_2, H_3O^+
D. Methanol, Conc. H_2SO_4

Choose the ***correct*** answer from the options given below:

1. A, B and C only
2. A, B and D only
3. A, B, C and D
4. B, C and D only

80. Which reagent will distinguish Benzophenone from acetone?

1. Fehling's reagent 2. Tollen's reagent
3. 2,4-DNP reagent 4. $I_2/NaOH$

81. The structural feature in carbonyl compound for Aldol condensation:

1. presence of at least one β-hydrogen
2. presence of at least one α-hydrogen
3. Concentrated base
4. Lack of α-hydrogen

82. The nitrogen atom in amines is trivalent and possess an unshared pair of electrons. The geometry of trimethyl amine is _____.

1. Tetrahedral 2. Pyramidal
3. Square planar 4. Triangular

83. What happens when C_6H_5-O-R is treated with HX?

1. RX and C_6H_5OH are formed
2. ROH and C_6H_5X are formed
3. $C_6H_4X_2$ and ROH are formed
4. RX and C_6H_5X are formed

84. In the nitration of benzene using a mixture of conc. H_2SO_4 and conc. HNO_3, the nitrating species is:

1. NO_2^- 2. NO_2^+
3. NO^+ 4. NO_2^- and NO_2^+

85. Which of the following compounds will not give azo coupling reaction with benzene diazonium chloride?

1. Nitrobenzene 2. Aniline
3. *o*-Toluidine 4. Phenol

86. Amylose is a water-soluble part of starch. What is the percentage solubility of it?

1. 20 to 30% 2. 15 to 20%
3. 30 to 60% 4. 50 to 70%

87. What is an example of globular protein?

A. Insulin B. Keratin
C. Albumin D. Myosin

Choose the ***correct*** answer from the options given below:

1. A, B and D only
2. A and C only
3. A, B, C and D
4. B, C and D only

88. Which types of bonds or interactions are found in the β-helix of protein?

1. Ionic bond 2. Covalent interaction
3. H-bond 4. Banana bond

77. 4	**78.** 2	**79.** 3	**80.** 4	**81.** 2	**82.** 2
83. 1	**84.** 2	**85.** 1	**86.** 2	**87.** 2	**88.** 3

89. Match List-I with List-II.

List-I (Name of amino acid)	List-II (One letter code)
A. Lysine	I. W
B. Tryptophan	II. Q
C. Tyrøsine	III. K
D. Glutamine	IV. Y

Choose the *correct* answer from the options given below:

1. A-II, B-I, C-III, D-IV
2. A-III, B-I, C-IV, D-II
3. A-I, B-II, C-IV, D-III
4. A-III, B-IV, C-I, D-II

90. Arrange the following compounds in **increasing** order of their acidic strength:

A. 3-Nitrophenol
B. 3,5-Dinitrophenol
C. 2,4,6 -Trinitrophenol
D. Phenol

Choose the *correct* answer from the options given below:

1. D, C, B, A 2. C, A, B, D
3. D, A, B, C 4. A, B, C, D

Directions (Q.No. 91-95): *Read the passage carefully and answer the questions:*

The speed at which a chemical reaction takes place is called the rate of reaction. The rate of reaction depends on various factors like concentration of the reactants, temperature, etc. The relation between the rate of reaction and the concentration of reacting species is represented by the equation $r = k[A]^x[B]^y$, where x and y are the order of the reaction with respect to the reactants A and B, respectively. The overall order of the reaction is $x + y$. The rate of reaction can also be increased by the use of a catalyst which provides an alternate pathway of lower activation energy. It increases the rate of forward and backward reaction to an equal extent. It does not alter the Gibbs energy of the reaction.

91. The rate of a gaseous reaction is given by $r = k[A][B]$. If the volume of the reaction vessel is suddenly reduced to (1/4)th of its initial value, the reaction rate relating to the initial rate will become:

1. 2 times 2. 1/4 times
3. 16 times 4. 1/16 times

92. Calculate the order of a reaction whose Rate = $k\ [A]^{1/2}\ [B]^{3/2}$.

1. second order 2. half order
3. first order 4. zero order

93. The rate law of a reaction is given by $r = k\ [CH_3OCH_3]^{3/2}$. If the pressure is measured in bar and time in minutes, then the unit of rate constant will be:

1. $bar^{1/2}\ min^{-1}$. 2. $bar^{-3/2}\ min^{-1}$.
3. $bar^{-1/2}\ min^{-1}$. 4. $bar^{3/2}\ min^{-1}$.

94. If the rate of reaction becomes twenty-seven times upon increasing the concentration of reactant by three times, the order of this reaction is:

1. 0 2. 1
3. 3 4. 2

95. The role of a catalyst is to change:

1. Gibbs energy of the reaction.
2. Enthalpy of a reaction.
3. Activation energy of a reaction.
4. Equilibrium constant.

Directions (Q.No. 96-100): *Read the passage carefully and answer the questions:*

Replacement of a hydrogen atom in a hydrocarbon by an alkoxy or carboxyl group yields a class of compounds known as ethers. Ethers are classified as symmetrical or unsymmetrical on the basis of groups attached to the oxygen atoms. Diethyl ether, a symmetrical ether, has been widely used as an inhalation anesthetic. Ethers can be prepared by acid catalyzed intermolecular dehydration of alcohols and Williamson's synthesis. Acid catalyzed dehydration of alcohols is not generally preferred

89. 2	**90.** 3	**91.** 3	**92.** 1	**93.** 3	**94.** 3	**95.** 3

as it gives a mixture of elimination and substitution products. In Williamson's synthesis, an alkyl halide is allowed to react with sodium alkoxide. Ethers containing substituted Alkyl groups may also be prepared by this method. The C—O bond in ether is weakly polar and is cleaved under drastic conditions with excess of hydrogen halides. In electrophilic substitution, the alkoxy group deactivates the aromatic ring and directs the incoming group to ortho and para positions.

96. When ethanol is dehydrated in the presence of H_2SO_4 at 443 K and 413 K respectively the products formed are:

1. Ethane and ethoxyethane
2. Ethylmethyl ether and butene
3. Ethylmethyl ether and propene
4. Ethene and ethoxyethane

97. The major product in the reaction of anisole with bromine in ethanoic acid, is:

1. *o*-bromoanisole
2. *p*-bromoanisole
3. *m*-bromoanisole
4. *o*-bromoanisole and *p*-bromoanisole

98. In Williamson synthesis, the alkoxide ion attacks the alkyl halide via which pathway?

1. S_N2
2. S_N1
3. Depends on the nature of alkoxide ion
4. Depends on the nature of Alkyl halide

99. Which is most reactive hydrogen halide for cleavage of ethers?

1. HF　　2. HCl
3. HBr　　4. HI

100. Which type of ether is anisole?

1. Dialkyl ether
2. Diaryl ether
3. Phenyl Alkyl ether
4. Alkoxy Alkyl ether

BIOLOGY

101. Which one of the following guides the entry of the pollen tube into the embryo sac?

1. Antipodal Cells　　2. Filiform apparatus
3. Micropyle　　4. Synergids

102. Antibody molecule has:

1. Two peptide chain
2. Three peptide chain
3. Four peptide chain
4. One polypeptide chain

103. Choose the correct equation from the following options to show correct relationship between Gross Primary Productivity (GPP), Respirational Losses (R) and Net Primary Productivity (NPP).

1. $GPP = R/(NPP)^2$
2. $GPP = 2NPP + R$
3. $GPP = NPP - R$
4. $GPP - R = NPP$

104. Match List-I with List-II.

List-I (Disease)	List-II (Characteristic symptoms)
A. Malaria	I. Inflammation of lower limb
B. Filariasis	II. Cycles of fever
C. Ringworms	III. Blood clots and excess mucus in stools
D. Amoebiasis	IV. Scaly lesions on nails

Choose the ***correct*** answer from the options given below:

1. A-IV, B-II, C-III, D-I
2. A-I, B-III, C-II, D-IV
3. A-II, B-I, C-IV, D-III
4. A-III, B-IV, C-I, D-II

96. 4	**97.** 2	**98.** 1	**99.** 4	**100.** 3	**101.** 2	**102.** 3	**103.** 4	**104.** 3

105. Match List-I with List-II.

List-I (Producer)	List-II (Acid)
A. *Clostridium butylicum*	I. Lactic acid
B. *Aspergillus niger*	II. Butyric acid
C. *Acetobacter aceti*	III. Citric acid
D. *Lactobacillus*	IV. Acetic acid

Choose the ***correct*** options:

1. A-I, B-II, C-III, D-IV
2. A-II, B-III, C-IV, D-I
3. A-III, B-II, C-IV, D-I
4. A-III, B-IV, C-I, D-II

106. Which one of the following is opposite part of micropylar end of ovule?

1. Hilum
2. Chalaza
3. Funicle
4. Nucellus

107. Match List-I with List-II.

List-I (Technique/term)	List-II (Features)
A. Amniocentesis	I. Ligation of fallopian tube
B. Maternal mortality rate	II. Ligation of vas deferens
C. Vasectomy	III. Death rate
D. Tubectomy	IV. Testing of sex

Choose the ***correct*** options:

1. A-I, B-II, C-III, D-IV
2. A-IV, B-III, C-II, D-I
3. A-III, B-IV, C-I, D-II
4. A-II, B-I, C-III, D-IV

108. Which type of innate Immunity is exhibited by interferons secreted by virus infected cells to protect non infected cells?

1. Physical barriers
2. Cytokine barriers
3. Physiological barriers
4. Cellular barriers

109. The process of copying genetic information from one strand of DNA into RNA is termed as:

1. Replication
2. Translation
3. Transcription
4. Regulation

110. Arrange the following events in the correct order pertaining to fertilization in the human reproductive system.

A. The blastocyst becomes embedded in the endometrium.

B. Finger-like projections appear on the trophoblast called chorionic villi.

C. The blastomeres are arranged into trophoblast and the inner cell mass.

D. The zygote divides miotically and transforms into an embryo with 8-16 blastomeres, called a morula.

Choose the ***correct*** answer from the options given below:

1. D, C, A, B
2. A, C, B, D
3. B, A, D, C
4. C, B, D, A

111. Which of the following are **not** involved in intra uterine devices?

A. Lippes Loop
B. LNG-20
C. Saheli
D. Implants

Choose the ***correct*** answer from the options given below:

1. A, B and D only
2. C and D only
3. A, B, C and D
4. C, B and D only

112. Match List-I with List-II.

List-I (Character)	List-II (Recessive Trait)
A. Flower colour	I. Green
B. Seed colour	II. Yellow
C. Pod colour	III. Constricted
D. Pod shape	IV. White

Choose the ***correct*** answer from the options given below:

1. A-II, B-IV, C-I, D-III
2. A-I, B-IV, C-II, D-III
3. A-I, B-II, C-IV, D-III
4. A-IV, B-I, C-II, D-III

105. 2	**106.** 2	**107.** 2	**108.** 2	**109.** 3	**110.** 1	**111.** 2	**112.** 4

113. Which one of the following is **not** associated with the process of transcription in bacteria?

1. Rho factor
2. Methyl guanosine triphosphate
3. Sigma factor
4. DNA dependent RNA polymerase

114. Which of the following disorders are the results of aneuploidy?

A. Haemophilia B. Down's Syndrome
C. Thalassemia D. Turner's Syndrome

Choose the ***correct*** answer from the options given below:

1. B and D only
2. A, B and C only
3. A, B, C and D
4. A and C only

115. Which one of the following options will express intermediate skin colour in an individual?

1. AABBCC 2. aabbcc
3. AaBbCc 4. aaBbcc

116. Match List-I with List-II.

List-I (Name of the gene)	List-II (Encodes)
A. 'i'	I. Permease
B. 'z'	II. Repressor
C. 'y'	III. Transacetylase
D. 'a'	IV. β-Galactosidase

Choose the ***correct*** answer from the options given below:

1. A-IV, B-II, C-III, D-I
2. A-I, B-III, C-II, D-IV
3. A-II, B-IV, C-I, D-III
4. A-III, B-IV, C-I, D-II

117. Which of the following is **incorrect** with reference to drug abuse?

1. Cannabinoids affect cardiovascular system of the body.
2. Heroin is extracted from the latex of *Papaver somniferum*.
3. Nicotine is a very effective sedative and pain killer.
4. Excessive dosage of coca alkaloid causes hallucinations.

118. Arrange the given steps involved in gel electrophoresis used for separation of DNA fragments?

A. Exposure to UV light
B. Staining with ethidium bromide
C. Moving of DNA fragments towards anode
D. Elution

Choose the ***correct*** options given below:

1. A, B, C, D 2. B, A, D, C
3. A, D, B, C 4. C, B, A, D

119. Match List-I with List-II.

List-I (Product)	List-II (Producer)
A. Citric Acid	I. *Trichoderma polysporum*
B. Ethanol	II. *Monascus purpureus*
C. Statins	III. *Saccharomyces cerevisiae*
D. Cyclosporin A	IV. *Aspergillus niger*

Choose the ***correct*** answer from the options given below:

1. A-IV, B-III, C-II, D-I
2. A-I, B-III, C-II, D-IV
3. A-I, B-II, C-IV, D-III
4. A-III, B-IV, C-I, D-II

120. *EcoRI*, a significant tool in rDNA technology is:

1. Bacteria 2. Plasmid
3. Enzyme 4. Purine

113. 2 **114.** 1 **115.** 3 **116.** 3 **117.** 3 **118.** 4 **119.** 1 **120.** 3

121. Arrange the following geological periods in their occurrence from latest to oldest order.

A. Triassic   B. Carboniferous
C. Tertiary   D. Jurassic

Choose the *correct* answer from the options given below:

1. D, B, C, A   2. A, B, D, C
3. B, A, D, C   4. C, D, A, B

122. Which of the following steps are related to polymerase chain reaction (PCR)?

A. Extension   B. Annealing
C. Propagation   D. Denaturation

Choose the *correct* answer from the options given below:

1. A, B and D only
2. A, B and C only
3. A, B, C and D
4. B, C and D only

123. Baculoviruses are pathogens that attack:

1. Insects   2. Roundworms
3. Molluscs   4. Birds

124. The number of individuals in the reproductive age group is more than the number of individuals in the pre-reproductive age group, the shape of its age pyramid would reflect the growth status of the population as:

1. Expanding   2. Stable
3. Declining   4. Homeostasis

125. Which of the following statements are true with reference to homology or homologous organs?

A. Homology indicates common ancestry.
B. Whale and Cheetah share similarities in the pattern of the bones of the forelimbs.
C. Vertebrate heart is an example of homologous organs.
D. Thorn of *Bougainvillea* and tendrils of cucurbita represent homology.

Choose the *correct* answer from the options given below:

1. A, B and D only
2. A, B and C only
3. A, B, C and D
4. B, C and D only

126. Arrange the following groups of plants according to their appearance on earth.

A. Angiosperms
B. Seed ferns
C. Rhynia type plants
D. Psilophyton

Choose the *correct* answer from the options given below:

1. C, D, B, A
2. A, C, B, D
3. B, A, D, C
4. C, B, D, A

127. If a man is eating fish as his food, which trophic level is occupied by him in the food chain?

1. First   2. Second
3. Third   4. Fourth

128. Thorn of *Bougainvillea* and tendril of *Cucurbita* are:

1. Homologous structures
2. Analogous structures
3. Vestigial structures
4. Developing structures

129. A small standing crop of phytoplankton supports a large standing crop of zooplankton. This can be correctly represented by:

1. Inverted pyramid of energy
2. Inverted pyramid of biomass
3. Upright pyramid of biomass
4. Sometimes upright and sometimes inverted pyramid of energy

121. 4	122. 1	123. 1	124. 3	125. 3	126. 1	127. 4	128. 1	129. 2

130. The first isolated restriction endonuclease was:
1. *Hind II* 2. *EcoR I*
3. *BamH I* 4. *Pvu II*

131. Arrange the given products formed during sewage treatment in correct sequence:
A. Biogas B. Activated sludge
C. Flocs D. Primary sludge

Choose the ***correct*** options given below:
1. A, B, C, D 2. D, C, B, A
3. C, D, B, A 4. B, A, D, C

132. Wildlife safari parks are example of:
1. *Ex situ* conservation
2. *In situ* conservation
3. Biodiversity hot spots
4. Sacred grooves

133. Humification leads to:
1. Soil erosion
2. Soil conservation
3. Accumulation of humus
4. Accumulation of salts

134. If a double-stranded DNA has 15% of adenine, find out the percent of cytosine in the DNA:
1. 15% 2. 30%
3. 35% 4. 85%

135. Transgenic animals are used to understand the contribution of genes in the development of diseases such as:
1. Cholera and typhoid
2. Elephantiasis and ringworm
3. Cancer and cystic fibrosis
4. Pneumonia and kala-azar

136. Genetically modified plants have been useful in:
1. Increasing post harvest losses
2. Decreasing crop yield
3. Making crops tolerant to stresses
4. Decreasing efficiency of mineral usage by plants

137. Which of the following is the primary female sex organ?
1. Mammary glands
2. Uterus
3. Ovaries
4. Cervix

138. Which one of the following structure is haploid (*n*) in relation to male reproductive system?
1. Secondary spermatocytes
2. Primary spermatocytes
3. Leydig cells
4. Sertoli cells

139. Of the incident solar radiation, what is the percentage of photosynthetically active radiation (PAR)?
1. 100% 2. Less than 50%
3. 1-5% 4. 2-10%

140. ELISA is based on the principle of:
1. Antigen-antibody interaction
2. PCR
3. Radioactive molecule
4. Amount of DNA

Directions (Q.No. 141-145): *Read the passage carefully and answer the questions based on the passage:*

Species diversity on earth is not uniformly distributed. It is generally highest in the tropics and decreases towards the poles. Earth's fossil history reveals the incidences of mass extinctions in the past. Earth's rich biodiversity is vital for the very survival of mankind. It is believed that communities with high diversity tend to be less variable and more productive. The reasons of conserving biodiversity are narrowly utilitarian, broadly utilitarian and ethical. Biodiversity conservation may be *in situ* as well as *ex-situ*.

130. 1	**131.** 2	**132.** 1	**133.** 3	**134.** 3	**135.** 3
136. 3	**137.** 3	**138.** 1	**139.** 2	**140.** 1	

141. Which of the following is not included in *in-situ* conservation?

1. Zoological park
2. National park
3. Wild life sanctuary
4. Biosphere reserves

142. Which one of the following does not exhibit narrowly utilitarian argument for conserving biodiversity?

1. Construction materials
2. Pollination
3. Industrial products
4. Medicines

143. Which of the following might **not** account for the greater biological diversity in the tropic region?

1. Frequent glaciations in the past
2. More solar energy available
3. Less seasonal and more constant and predictable
4. Undisturbed for million of years

144. How many episodes of mass extinction of species have occurred since the origin and diversification of life on earth?

1. Two 2. Three
3. Five 4. Seven

145. Which of the following hot spots does not cover India's biodiversity regions?

1. Western Ghats-Sri Lanka
2. Amazon forests
3. Indo-Burma
4. Himalaya

Directions (Q.No. 146-150): *Read the passage carefully and answer the questions based on the passage:*

Pollen pistil interaction involves all events from the landing of pollen grains on the stigma until the pollen tube enters the embryo sac (when the pollen is compatible). When a pollen tube grows through the style and enters into the ovules, it finally discharges two male gametes in one of the synergids. Syngamy and triple fusion are two fusion events occur in angiosperms. Thus, angiosperms exhibit double fertilization. The products of these fusions are the diploid zygote and triploid primary endosperm nucleus. Zygote develops into embryo and primary endosperm cell forms the endosperm tissue. The developing embryo passes through different stages before maturation.

146. With reference to reproduction in flowering plants, which one of the following is **incorrect**?

1. Endosperm develops into seed
2. Ovary develops into fruit
3. Ploidy of PEN is $3n$
4. Syngamy is the fusion of male and female gamete

147. Which of the following is not a stage of growing embryo in dicotyledon plants?

1. Heart-shaped 2. Globular
3. Proembryo 4. Epiblast

148. The coconut water from tender coconut, a good source of nutrition is nothing but:

1. Free-nuclear endosperm
2. Synergids
3. Antipodal cells
4. Scutellum

149. Formation of diploid zygot is a result of:

1. Emasculation 2. Triple fusion
3. Syngamy 4. Bagging

150. If there are 38 chromosomes in the zygote, how many chromosomes will be there in its haploid egg cell?

1. 38 2. 57
3. 19 4. 76

141. 1	**142.** 2	**143.** 1	**144.** 3	**145.** 2
146. 1	**147.** 4	**148.** 1	**149.** 3	**150.** 3

EXPLANATORY ANSWERS

1. Electrons carry charge $-e$ and protons carry charge $+e$. If an object contains x electrons and y protons, the net charge is:
$q = (+y)e + (-x)e = (y - x)e.$

2. Potential at point P due to charge $q_1 = 3 \times 10^{-7}$ C at distance $r = 0.09$ m;

$$V = \frac{kq_1}{r} = \frac{9\times10^9 \cdot 3\times10^{-7}}{0.09}$$
$$= 3.0 \times 10^4 \text{ V}$$

Work done in brining $q_2 = 2 \times 10^{-9}$ C from infinity:

$$W = q_2V$$
$$= (2 \times 10^{-9})(3.0 \times 10^4)$$
$$= 6 \times 10^{-5} \text{ J}.$$

3. In a series combination of capacitors, the charge on each capacitor is the same because the same current flows through all. This is true for any value of capacitances.

4. Distance between charges:

$$r = 6 + 6$$
$$= 12 \text{ cm}$$
$$= 0.12 \text{ m}$$

Potential energy initially:

$$U_i = \frac{kq_1q_2}{r}$$
$$= \frac{9\times10^9\cdot(4\times10^{-6})(-3\times10^{-6})}{0.12}$$
$$= -0.9 \text{ J}$$

At infnity, $U_f = 0$.
Work required:

$$W = U_f - U_i = 0 - (-0.9)$$
$$= 0.9 \text{ J}.$$

5. By Gauss's law:

$$\phi = \frac{Q_{in}}{\varepsilon_0} = Q_{in} = \varepsilon_0\phi$$

$$Q_{in} = (8.85 \times 10^{-12})(1.05)$$
$$= 9.29 \times 10^{-12} \text{ C}.$$

6. We are given:

$$\text{Plate area A} = 200 \text{ cm}^2$$
$$= 200 \times 10^{-4} \text{ m}^2$$
$$= 0.02 \text{ m}^2$$
$$\text{Separation } d = 2.0 \text{ mm}$$
$$= 2.0 \times 10^{-3} \text{ m}$$
$$\text{Charge Q} = \mu_C$$
$$= 0.06 \times 10^{-6} \text{ C}$$
$$= 6.0 \times 10^{-8} \text{ C}$$

Potential difference V = 60 V
Step 1: Capacitance from Q = CV:

$$C = \frac{Q}{V} = \frac{6.0\times10^{-8}}{60}$$
$$= 1.0 \times 10^{-9} \text{ F}$$

Step 2: Relation of capacitance with dielectric constant:

$$C = \kappa\varepsilon_0\frac{A}{d}$$

where, $\varepsilon_0 = 8.85 \times 10^{-12}$ F/m.
Step 3: Solve of κ:

$$\kappa = \frac{Cd}{\varepsilon_0 A}$$

Substitute values:

$$\kappa = \frac{(1.0\times10^{-9})(2.0\times10^{-3})}{(8.85\times10^{-12})(0.02)}$$

Step 4: Simplify numerator and denominator:

$$\text{Numerator} = 2.0 \times 10^{-12}$$
$$\text{Denominator} = 1.77 \times 10^{-13}$$
$$\kappa = \frac{2.0\times10^{-12}}{1.77\times10^{-13}} \approx 11.3.$$

7. For an electric dipole with moment $\vec{p}$ at a point with position vector $\vec{r}$, the potential is

$$V = \frac{1}{4\pi\varepsilon_0}\frac{\vec{p}\cdot\vec{r}}{r^3} = \frac{1}{4\pi\varepsilon_0}\frac{p\cos\theta}{r^2},$$

where $r = |\vec{r}|$ and θ is the angle between $\vec{p}$ and $\vec{r}$.

A. True — $V \propto 1/r^2$, so it depends on r.

B. True — $V \propto \cos\theta$, so it depends on the angle between $\vec{p}$ and $\vec{r}$.

C. True — at large distance, $V \sim 1/r^2$.

D. False — V depends on the dipole moment $p = qd$ (with d the charge separation), so it does depends on the separation through p.

8. According to Coulomb's law, the electrostatic force between two changes is:

$$F = \frac{1}{4\pi\varepsilon_0}\frac{q_1q_2}{d^2}$$

Initially, the force is:

$$F = k\frac{q_1q_2}{d^2}$$

When one charge is doubled ($q_1 \to 2q_1$), the new force becomes:

$$F' = k\frac{(2q_1)q_2}{r^2} = \frac{2kq_1q_2}{r^2}$$

To keep the force the same ($F' = F$):

$$\frac{2kq_1q_2}{r^2} = \frac{kq_1q_2}{d^2}$$

Canceling common terms:

$$\frac{2}{r^2} = \frac{1}{d^2}$$

$$r^2 = 2d^2$$

$$r = \sqrt{2}d.$$

9. Electric current in a conductor arises when free electrons drift under the influence of an electric field. Magnetic or gravitational fields alone do not set up a continuous drift current.

10. Resistivity ρ is a material property, independent of length or area. It varies only with the nature of the material and the temperature. Dimensions do not affect resistivity (they affect resistance R, not ρ).

11. • In the absence of electric field, electrons move randomly between collisions, so their path is a straight line until the next collision.

• In the presence of an electric field, electrons experience a force and hence a continuous acceleration, making the trajectory generally curved before collision.

12. Thermal energy produced = electrical energy consumed:

$$H = \frac{V^2}{R}\cdot t$$

Given: $H = 800$ J, $V = 20$V, $t = 20$s:

$$800 = \frac{(20)^2}{R}\times 20$$

$$800 = \frac{400}{R}\times 20 = \frac{8000}{R}$$

$$R = \frac{8000}{800} = 10\ \Omega.$$

13. Given:

Wire resistance $R = 4\ \Omega$;

Coil radius $r = 7$ cm $= 0.07$ m;

Wire diameter $d = 1.4$ mm $= 1.4 \times 10^{-3}$ m;

Resistivity $\rho = 2 \times 10^{-7}\ \Omega$m.

Cross-sectional area:

$$A = \pi\left(\frac{d}{2}\right)^2 = \pi(0.7 \times 10^{-3})^2$$

$$= \pi \times 0.49 \times 10^{-6}$$

$$\approx 1.539 \times 10^{-6}\ \text{m}^2.$$

Wire length from $R = \rho L/A$:

$$L = \frac{RA}{\rho}$$

$$= \frac{4\times 1.539\times 10^{-6}}{2\times 10^{-7}} \approx 30.78\ \text{m}$$

Length per turn:

Circumference $= 2\pi r$

$= 2\pi(0.07)$

$= 0.14\pi \approx 0.4398$ m.

Number of turns: $N = \dfrac{L}{2\pi r} \approx \dfrac{30.78}{0.4398} \approx 70.$

14. Battery emf E = 12 V,
Internal resistance $r = 3\ \Omega$, Current I = 0.6 A
Terminal voltage across external resistor:

$$\begin{aligned} V &= E - Ir \\ &= 12 - (0.6)(3) \\ &= 12 - 1.8 \\ &= 10.2\ \text{V}. \end{aligned}$$

15. A uniform 12 Ω wire is cut in length ratio 1 : 2 : 3
⇒ resistances $R_1 = 2\ \Omega$, $R_2 = 4\ \Omega$; $R_3 = 6\ \Omega$ (since 12 split over 6 equal parts).
Forming a triangle, connecting the cell across the largest side 6 Ω. The alternate path between the same nodes is 2 Ω + 4 Ω = 6 Ω.
Equivalent external resistance:

$$R_{ext} = 6\ \Omega \parallel 6\ \Omega = 3\ \Omega.$$

Including internal resistance $r = 5\ \Omega$:

$$R_{tot} = 3 + 5 = 8\ \Omega.$$

Circuit current:

$$I = \frac{E}{R_{tot}} = \frac{8}{8} = 1\ \text{A}.$$

16. Magnetic moment of a bar magnet: $M = p(2\ell)$, where p is pole strenght and 2ℓ is pole separation.
Given: $M = 5.0\ \text{A m}^2$,
Separation $2\ell = 20\ \text{cm} = 0.20\ \text{m}$.

Pole strength: $p = \dfrac{M}{2\ell} = \dfrac{5.0}{0.20} = 25\ \text{A m}$.

17. A. Magnetic field has unit Tesla (T), which is equivalent to Wb m^{-2}. → (IV)
B. Magnetic moment has unit A · m^2 = J T^{-1}. → (I)
C. Pole strength has unit A · m, so moment/pole separation = J T^{-1} m^{-1}. → (III)
D. Permeability of free space has unit T m A^{-1}. → (II)

18. Magnetic field due to a small current element (Biot–Savart law):

$$dB = \frac{\mu_0}{4\pi}\frac{I\ dl \sin\theta}{r^2}$$

Given:

$$I = 10\ \text{A},\ dl = 0.01\ \text{m},$$
$$r = 0.5\ \text{m},\ \theta = 90°$$

$$\begin{aligned} dB &= \frac{10^{-7}\cdot(10)(0.01)}{(0.5)^2} \\ &= \frac{10^{-7}\cdot 0.1}{0.25} = \frac{1.0\times10^{-8}}{0.25} \\ &= 4\times10^{-8}\ \text{T}. \end{aligned}$$

19. Force on a current element: $\vec{F} = I(\vec{L}\times\vec{B})$.
Current along z-axis $(\vec{L}in + z)$.
Magnetic field along $+y$.
$\vec{L}\times B = \hat{z}\times\hat{y} = -\hat{x}$.

20. Equivalent resistance of galvanometer (G) shunted with shunt (S):

$$R_{ammeter} = \frac{G\cdot S}{G+S}$$
$$= \frac{(520)(20)}{520+20} = \frac{10400}{540} \approx 19.26\ \Omega.$$

21. Magnetic field inside a solenoid is given by:

$$B = \mu_0 nI$$

where, $n = \dfrac{N}{L}$

$$N = 800,$$
$$L = 0.3\ \text{m},$$

So, $n = \dfrac{800}{0.3} = 2666.7$ turns/m

Current I = 6A

$$\begin{aligned} \mu_0 &= 4\pi\times10^{-7}\ \text{H/m} \\ B &= (4\pi\times10^{-7})(2666.7)(6) \\ B &\approx (1.256\times10^{-6})(16000) \\ &\approx 2.01\times10^{-2}\ \text{T} \\ &= 20\ \text{mT}. \end{aligned}$$

22. Kinetic energy gained by charge q after acceleration through potential V:

$$\frac{1}{2}mu^2 = qV$$

$$\Rightarrow \quad u = \sqrt{\frac{2qV}{m}}$$

In magnetic field, radius of circular path:

$$r = \frac{mu}{qB}$$

Substituting for u:

$$r = \frac{m}{qB}\sqrt{\frac{2qV}{m}} = \frac{1}{B}\sqrt{\frac{2mV}{q}}$$

Thus, $r \propto \sqrt{V}$. Since $u \propto \sqrt{V}$, we have $r \propto \frac{V}{u}$.

23. The setup is based on Faraday's law of electromagnetic induction: a changing magnetic flux through the coil induces an emf, producing current detected by the galvanometer.

A. True. Deflection of the galvanometer shows induced current in the coil.

B. False. Faster motion of the magnet increases rate of change of flux, hence deflection is larger, not smaller.

C. True. By Lenz's law, induced current produces a magnetic field opposing the change, so the face of the coil becomes a north pole to repel the incoming north pole.

D. True. Without motion of the magnet, flux is constant → no induced current.

Correct statements: A, C, D.

24. The loop area at any instant is the rectangle swept by the sliding rod PQ:

$$A = l \times x,$$

where l is the separation between the rails (length of PQ) and x is the distance of PQ from the left side. With uniform field B into the page, the magnetic flux linked with the loop is:

$$\Phi = BA = Blx$$

As PQ moves left with constant speed V (perpendicular to PQ and B), the motional emf induced across the rod (and hence around the loop) is

$$\varepsilon = BlV.$$

25. Average induced emf in a coil:

$$|\varepsilon| = L\frac{|\Delta I|}{\Delta t}.$$

Given: $|\varepsilon| = 2V$ and $\Delta t = 0.4$ s.

A. Current changes 5 → 2A

$$\Rightarrow \quad |\Delta I| = 3 \text{ A}.$$

$$L = \frac{\varepsilon \Delta t}{\Delta I} = \frac{2 \times 0.4}{3}$$

$$= 0.266\overline{6} \text{ H} = 266.7 \text{ mH}.$$

Given value 0.266 mH is 10^3 times smaller ⇒ (A) incorrect.

B. Current changes +4 → −4 A

$$\Rightarrow \quad |\Delta I| = 8A.$$

$$L = \frac{2 \times 0.4}{8} = 0.10 \text{ H} = 100 \text{ } m\text{H}.$$

Given value 0.10 mH again off by 10^3 ⇒ (B) incorrect.

26. Maximum current in a series RLC occurs at resonance:

$$\omega = \frac{1}{\sqrt{LC}}$$

Given: $L = 500$ mH $= 0.5$ H, $f = 0.4$ kHz $= 400$ Hz

$$\Rightarrow \quad \omega = 2\pi f = 2\pi(400).$$

$$C = \frac{1}{\omega^2 L} = \frac{1}{(2\pi \cdot 400)^2 \cdot 0.5}$$

$$\approx 3.2 \times 10^{-7} \text{ F} = 0.32 \text{ }\mu\text{F}.$$

27. Steady current before opening:

$$I_0 = \frac{V}{R} = \frac{12}{6} = 2 \text{ A}.$$

On opening, current drops from 2 → 0 A in

$$\Delta t = 1 \text{ ms} = 10^{-3} \text{ s}.$$

Induced emf across inductor:

$$|\varepsilon| = L\frac{|\Delta I|}{\Delta t}.$$

With L = 10 mH = 0.01 H:

$$|\varepsilon| = 0.01 \cdot \frac{2}{10^{-3}}$$
$$= 0.01 \cdot 2000$$
$$= 20 \text{ V}.$$

28. An AC source rated at 220 V, 50 Hz means 220 V is the RMS value.

Peak Value:

$$V_0 = V_{rms}\sqrt{2} = 200\sqrt{2} \text{ V} \approx 311 \text{ V}.$$

So, option 1 is incorrect.

Average value of AC voltage over one full cycle:

Since positive and negative half cycles cancel,

$$\langle V \rangle = 0.$$

So, statement 3 is correct.

Statement 2 is wrong (average is not 220 V).

Statement 4 is wrong (average is not $200\sqrt{2}$).

29. Faraday's law: the induced emf magnitude in a circuit is the time rate of change of magnetic flux:

$$|\varepsilon| = \left|\frac{d\phi}{dt}\right| \left(\text{for N turns, } |\varepsilon| = N\left|\frac{d\phi}{dt}\right|\right).$$

Hence, (1) is correct. Increasing the number of turns N increases the induced emf ⇒ 3 is correct. Lenz's law gives the polarity that opposes the change in flux ⇒ (4) is correct. Statement (2) says emf equals the total change of flux (not the rate), which is incorrect.

30. Maxwell introduced displacement current to preserve continuity in Ampere's law for time varying fields. Its density is $J_d = \varepsilon_0 \frac{\partial E}{\partial t}$, *i.e.*, it arises from a changing electric field, not form real charge flow or changing magnetic flux directly.

31. Typical wavelength bands:

X-rays ~ 1 nm – 10^{-3} nm ⇒ A–III.

Radio waves > 0.1 m ⇒ B–IV.

Infrared 780 nm – 1 mm ⇒ C–I.

Microwaves 0.1 m – 1 mm ⇒ D–II

32. For an electromagnetic wave, the relation between electric and magnetic field amplitudes is:

$$\frac{E_0}{B_0} = c$$

where c is the speed of light.

Also, from the wave equation,

$$c = \frac{\omega}{k}$$

where ω is angular frequency and k is wave number.

Thus,

$$\frac{E_0}{B_0} = \frac{\omega}{k}.$$

33. Laws of reflection hold for all mirrors (plane or curved) → A true.

For a plane mirror, if θ_i and θ_r are angles w.r.t. the normal, then the angles w.r.t. the plane are 90° – θ_i and 90° – θ_r; since $\theta_i = \theta_r$, these are equal → B true.

Rays parallel to the principal axis reflect through the focus of a (paraxial) curved mirror → C true.

A ray hitting the pole obeys the law of reflection with the axis as the normal there; it reflects making an equal angle on the other side of the axis → D true.

34. Power P is the lens's ability to converge/diverge: $P = \frac{1}{f}$ (with in metres) → A true.

S.I. unit of power is dioptre (D), but focal length must be in metres, not centimetres → B false.

Larger f ⇒ smaller P → C true.

For thin lenses in contact, powers add algebraically $P_{eq} = P_1 + P_2 + \ldots$ → D false.

35. Linear magnification $|m| = \frac{\text{image length}}{\text{object length}} = \frac{1}{5}$.

With object distance u = 40 cm

(Cartesian : u = –40 cm)

$$|m| = \left|\frac{v}{u}\right| = \frac{1}{5} \Rightarrow v = 8 \text{ cm}.$$

Lens formula:

$$\frac{1}{f} = \frac{1}{v} - \frac{1}{u} = \frac{1}{8} - \frac{1}{-40}$$

$$= \frac{1}{8} + \frac{1}{40} = \frac{6}{40} = \frac{6}{40} = \frac{3}{20}.$$

$$f = \frac{20}{3} = 6.67 \text{ cm.}$$

36. Young's fringe width: $\beta = \frac{\lambda D}{d}$.

$\lambda = 600$ nm $= 6 \times 10^{-7}$ m,

D = 1.2 m,

d = 1.5 mm

$= 1.5 \times 10^{-3}$ m.

$$\beta = \frac{6 \times 10^{-7} \times 1.2}{1.5 \times 10^{-3}}$$

$$= \frac{7.2 \times 10^{-7}}{1.5 \times 10^{-3}}$$

$= 4.8 \times 10^{-4}$ m

= 0.48 mm.

37. Critical angle i_c (from denser to rarer medium) is defined for the refracted ray just grazing the interface. Hence, the refracted angle becomes 90°.

38. Unpolarized light → first polarizer gives intensity $I_0/2$ polarized along its axis. A Polaroid sheet at angle θ transmits $(I_0/2)$ and sets the polarization along its own axis. The second polarizer is crossed (at 90° to the first), so its angle to the sheet is 90° − θ; transmitted factor $\cos^2 (90° - \theta) = \sin^2 \theta$.

$$I = \frac{I_0}{2} \cos^2 \theta \sin^2 \theta = \frac{I_0}{8} \sin^2(2\theta)$$

Maximum when $\sin^2 (2\theta) = 1$

⇒ 2θ = 90°

⇒ θ = 45° = π/4.

39. A. False — TIR occurs from denser → rarer medium, not rarer → denser.

B. True — the ray remains in the same (denser) medium after reflecting.

C. True — law of reflection holds: angle of incidence angle of reflection.

D. True — there is no refracted ray (hence no refracted angle).

Correct set: B, C and D

40. Concave lens has negaive focal length:

f = −50 cm = −0.50 m.

Power $P = \frac{1}{f} = -2D$.

41. In the photoelectric effect (for a given metal and for incident frequency $v > v_0$), the photoelectric current is proportional to the rate of electron emission. Since higher intensity means more photons per second on the surface (at the same frequency), it ejects more photoelectrons per second, thus larger current.

42. de Broglie wavelength:

$$\lambda = \frac{h}{mv}.$$

With m = 0.150 kg,

v = 30.0 m/s,

$p = mv$ = 4.50 kgm/s,

$h = 6.626 \times 10^{-34}$ J/s:

$$\lambda = \frac{6.626 \times 10^{-34}}{4.50} \approx 1.47 \times 10^{-34} m.$$

43. Nuclear radius: $R = R_0 A^{1/3}$ ⇒ A true.

Nuclear volume $V \propto R^3 \propto A$ ⇒ B true.

Nuclear density $\rho \sim \frac{\text{mass}}{\text{volume}} \propto \frac{A}{A}$ constant (independent of A)

⇒ C false, D true.

Correct set: A, B and D.

44. Hydrogen energy levels: $E_n = -\frac{13.6}{n^2} eV$.

Ground state E_1 = −13.6 eV.

Absorbing gives 12.09 eV,

E_f = −13.6 + 12.09

= −1.51 eV

⇒ n = 3 (since $-13.6/3^2 = -1.51$).

Angular momentum $L_n = n\hbar$.

Increase: $\Delta L = (3 - 1)\hbar = 2\hbar = 2(h/2\pi)$.

45. Inside nuclei, the strong nuclear force is short-range, attractive, and far stronger than Coulomb at nucleon separations; this dominance is why nuclei bind.

A. False — if electrostatic repulsion were greater, nucleons wouldn't bind.

B. True — in light (smaller) nuclei, Coulomb repulsion is much smaller than the strong force.

C. True — gravitational force between nucleons is utterly negligible vs. the strong force.

D. False — binding energy per nucleon is roughly constant due to the short-range (saturation) nature of the strong force, not because it is long-range.

46. A. True -- a diode rectifies AC by allowing current mainly in one direction.

B. False — E_g > 3 eV corresponds to insulators; semiconductors typically have E_g ~ 0.2 - 2 eV (Si ≈ 1.1 eV).

C. True — external bias changes the depletion region/barrier height.

D. True — the p–n junction underlies most semiconductor devices.

47. Current flows for one polarity but drops to almost zero when the battery is reversed — hallmark of a *p–n* junction diode (forward vs. reverse bias). Pure *p*- or *n*-type pieces conduct in both polarities; a capacitor blocks steady DC regardless of polarity (after a transient).

48. In a *p*-type semiconductor, the dopants are trivalent atoms (like B, Al, Ga) which create holes as majority carriers.

Therefore, C is true.

Electrons are minority carriers in *p*-type, but the dopants are not pentavalent (so D is false).

A is false (holes are majority, not minority).

B is false (that describes n-type).

49. Nuclear reaction:

$${}_0^1n + {}_{92}^{235}\text{U} \rightarrow {}_{54}^{140}Xe + {}_b^aSr + 2{}_0^1n$$

Mass number balance:

On the left-hand side: 1 + 235 = 236

On the right-hand side: 140 + a + 2 = 236.

So, a = 94.

Atomic number balance:

On the left-hand side: 0 + 92 = 92.

On the right-hand side: 54 + b = 92.

So, b = 38.

Therefore, the missing fragment is Strontium-94 with atomic number 38.

50. De Broglie wavelength:

$$\lambda = \frac{h}{\sqrt{2mqV}}$$

For a proton: m_p, $p_q = e$.

For an α-particle: $m_\alpha = 4m_p$, $q_\alpha = 2e$.

Condition for same λ:

$$2m_p eV = 2m_\alpha q_\alpha V_\alpha$$

$$m_p eV = m_\alpha q_\alpha V_\alpha$$

$$V_\alpha = \frac{m_p eV}{m_\alpha q_\alpha} = \frac{m_p eV}{(4m_p)(2e)} = \frac{V}{8}.$$

51. CH_3COOH (acetic acid) in water behaves as a weak electrolyte. When it dissolves, it undergoes partial ionization according to the reaction:

$$CH_3COOH \rightleftharpoons CH_3COO^- + H^+.$$

The van't Hoff factor (*i*) is defined as the ratio of the actual number of particles in solution after dissociation or association to the number of formula units initially dissolved. For a non-electrolyte that does not dissociate, *i* = 1. For a strong electrolyte that completely dissociates into two ions, *i* = 2. Since acetic acid dissociates only partially, the actual number of particles is greater than in the undissociated state but less than in complete dissociation. Therefore, the value of i for CH_3COOH solution in water will be greater than 1 but less than 2. This intermediate value reflects the extent of dissociation (α) being between 0 and 1.

52. The molarity (M) of a solution is given by: M = (mass of solute in *g*) ÷ (molar mass in *g* mol^{-1} × volume of solution in L). Here, for each solute, the mass is 1 *g*, the volume is 1 L, so the molarity depends solely on the molar mass.

- Glucose (molar mass = 180 *g* mol^{-1}):
  M = 1 ÷ 180 = 0.00556 mol L^{-1}.
- NaOH (molar mass = 40 g mol^{-1}):
  M = 1 ÷ 40 = 0.0250 mol L^{-1}.
- NaCl (molar mass = 58.5 g mol^{-1}):
  M = 1 ÷ 58.5 ≈ 0.01709 mol L^{-1}.
- KCl (molar mass = 74.5 g mol^{-1}):
  M = 1 ÷ 74.5 ≈ 0.01342 mol L^{-1}.

Since molarity is inversely proportional to molar mass for the same mass of solute and same volume of solution, the decreasing order of molarity is: NaOH (B) > NaCl (C) > KCl (D) > Glucose (A).

53. Matching each term with its correct explanation:

- **Saturated solution:** A solution that contains the maximum amount of solute that can be dissolved in a given amount of solvent at a specific temperature; any additional solute will remain undissolved. This corresponds to III.
- **Isotonic solutions:** Two solutions having the same osmotic pressure at a given temperature; no net movement of solvent occurs between them through a semipermeable membrane. This corresponds to IV.
- **Binary solution:** A solution made up of two components — one solute and one solvent. This corresponds to I.
- **Hypertonic solution:** A solution whose osmotic pressure is greater than that of another solution, causing solvent movement toward it if separated by a semipermeable membrane. This corresponds to II.

Thus, the correct matching is: A–III, B–IV, C–I, D–II.

54. At high altitudes, the atmospheric pressure decreases significantly compared to sea level. The partial pressure of oxygen in the air is directly proportional to the total atmospheric pressure. At low atmospheric pressure, the partial pressure of oxygen also drops, which reduces the oxygen gradient between the air in the lungs and the blood. This weakens the diffusion of oxygen into the bloodstream, leading to lower oxygen saturation in both blood and tissues. This physiological effect is known as hypoxia and is the main cause of altitude sickness. The factor responsible here is low atmospheric pressure, not temperature.

55. Given:

Density of solution = 1.202 g mL^{-1}

concentration = 20% (m/m),

molar mass of KI = 166 g mol^{-1}.

Step 1: Mass of 1 L solution
= 1000 mL × 1.202 g mL^{-1}
= 1202 g.

Step 2: Mass of KI in solution
= 20% of 1202 g
= (20/100) × 1202
= 240.4 g.

Step 3: Moles of KI
= 240.4 g ÷ 166 g mol^{-1}
= 1.448 mol.

Step 4: Mass of water (solvent)
= 1202 − 240.4
= 961.6 g
= 0.9616 kg.

Step 5: Molality (m) = moles of solute ÷ mass of solvent in kg
= 1.448 ÷ 0.9616
≈ 1.505 mol kg^{-1}.

Hence, molality ≈ 1.5 mol kg^{-1}.

56. The EMF of a cell (E_{cell}) is a measure of the potential difference between two electrodes when no current is drawn. It is expressed in volts (V), which is equivalent to joules per coulomb (J C^{-1}). It is not expressed in units of conductivity (S cm^{-1}) or cell constant (cm^{-1}) because it deals with electrical potential, not conductance or resistance.

57. Matching each property with its correct unit:

A. Cell constant: Defined as the ratio of the distance between electrodes to the cross-sectional area of the electrodes (l/A). Its unit is cm^{-1} because length is in cm and area in cm^2. → I.

B. Molar conductance: Conductance of all ions from 1 mole of electrolyte when placed between two electrodes 1 cm apart. Its unit is ohm^{-1} cm^2 mol^{-1}. → II.

C. Specific conductance (conductivity): Conductance of a solution of 1 cm length and 1 cm^2 cross-section. Its unit is ohm^{-1} cm^{-1} (S cm^{-1}). → III.

D. Conductance: Reciprocal of resistance; unit is ohm^{-1} (siemens, S). → IV.

Thus, A–I, B–II, C–III, D–IV.

58. A Mercury cell is a primary cell (non-rechargeable) with a cell reaction:

$$Zn(Hg) + HgO(s) \rightarrow ZnO(s) + Hg(l).$$

It produces constant voltage until exhaustion and is commonly used in hearing aids and watches, which require low current. It does not convert combustion energy to electrical energy (that is the principle of fuel cells) and is not rechargeable. Therefore, correct points are C and D only.

59. Automobiles and inverters require high current supply and rechargeable storage. The lead storage cell (lead-acid battery) fulfills this requirement. It consists of lead dioxide cathode, sponge lead anode, and sulfuric acid electrolyte, and can be recharged multiple times. Mercury cells and dry cells are not suitable for high current, and fuel cells are not conventionally used in standard cars and inverters.

60. In a galvanic cell, spontaneous reactions occur when $E_{cell} > E_{ext}$. However, if an external potential (E_{ext}) applied in reverse exceeds the EMF of the cell ($E_{ext} > E_{cell}$), the spontaneous reaction is suppressed and the cell begins to operate as an electrolytic cell, forcing a non-spontaneous reaction.

61. Coordination compounds are widely used in medicine, biology, and industry.

- cis-Platin ($[Pt(NH_3)_2Cl_2]$) is a well-known anticancer drug that effectively inhibits tumour growth by binding to DNA and disrupting replication → correct.
- Chlorophyll is a magnesium-containing coordination complex that captures light energy for photosynthesis → correct.
- Desferrioxamine B is a chelating agent primarily used for removing excess iron (Fe^{3+}) from the body in iron poisoning, not lead poisoning → incorrect statement. Lead poisoning is treated with EDTA or dimercaprol.
- Cyanocobalamin (vitamin B_{12}) is a cobalt-containing coordination compound used to treat pernicious anemia → correct.

The incorrect application here is statement 3.

62. The number of 3d electrons depends on the electronic configuration after removing electrons according to the rule: remove from 4s before 3d for transition metals.

- Cr^{2+}: Cr (Z = 24)
  = [Ar] $3d^5 4s^1$ → Cr^{2+}
  = remove 2 electrons
  → [Ar] $3d^{10}$ → 4 electrons.
- Cu^+: Cu (Z = 29)
  = [Ar] $3d^{10} 4s^1$ → Cu^+
  = remove 1 electron
  → [Ar] $3d^{10}$ → 10 electrons.
- Ti^{3+}: Ti (Z = 22)
  = [Ar] $3d^2 4s^2$ → Ti^{3+}
  = remove 3 electrons
  → [Ar] $3d^1$ → 1 electron.

- Mn^+: Assuming Mn^+ means Mn^{1+}, Mn (Z = 25)
  = [Ar] $3d^54s^2$ → Mn^+
  = remove 1 electron
  → [Ar] $3d^54s^1$ → 5 electrons in 3d.

Increasing order: Ti^{3+} (1) < Cr^{2+} (4) < Mn^+ (5) < Cu^+ (10). That is C, A, D, B.

63. Lanthanum (Z = 57) has the electron configuration after xenon of $5d^16s^2$. It does not have electrons in the 4f subshell in its ground state; the lanthanide series begins from cerium (Z = 58) where 4f starts filling.

64. Matching uses:

- Magnesium-based alloy is constituent of bullets because it is light and strong → I.
- Lanthanoid oxide is used in television screens as phosphors → III.
- Mixed oxides of lanthanoids are employed in petroleum cracking catalysts → II.
- Misch metal (an alloy of lanthanoid metals and iron) is used in flints for lighters and other ignition devices, matching "lanthanoid metal and iron" → IV.

Thus: A–I, B–III, C–II, D–IV.

65. Fe^{3+} (aq) ions in solution are typically yellow to pale brown due to d–d transitions and charge-transfer processes involving water ligands. The colour arises from electronic transitions in the octahedral $[Fe(H_2O)_6]^{3+}$ complex.

66. In oxidation reactions of $KMnO_4$ in acidic medium, an acid is required to provide H^+ ions. H_2SO_4 is usually preferred instead of HCl. This is because $KMnO_4$ is a very strong oxidizing agent and readily oxidizes Cl^- ions from HCl into Cl_2 gas:

$$2MnO_4^- + 16H^+ + 10Cl^- \rightarrow 2Mn^{2+} + 5Cl_2 + 8H_2O.$$

This side reaction consumes both $KMnO_4$ and HCl, leading to waste of oxidant and liberation of toxic chlorine gas. The correct reason is that $KMnO_4$ oxidizes HCl to Cl_2.

67. Acidified potassium dichromate ($K_2Cr_2O_7$ in dilute H_2SO_4) is a strong oxidizing agent. When it reacts with sulphides (S^{2-}), such as in hydrogen sulphide (H_2S), it oxidizes sulphur from oxidation state –2 to 0, producing elemental sulphur. The reaction is:

$$K_2Cr_2O_7 + 4H_2SO_4 + 3H_2S \rightarrow K_2SO_4 + Cr_2(SO_4)_3 + 7H_2O + 3S.$$

During the process, the orange colour of dichromate ion ($Cr_2O_7^{2-}$) changes to green due to the formation of chromium (III) sulphate $[Cr_2(SO_4)_3]$. This is a classic test for sulphides, where the precipitated sulphur appears as a yellow solid.

68. The ligand field strength is given by the spectro-chemical series:

CN^- > en (ethylenediamine) > NCS^- > S^{2-}.

Here, CN^- is a strong field ligand (causes large Δ_0), en is strong due to chelation, NCS^- is weaker, and S^{2-} is the weakest among these due to large size and low electronegativity.

Thus, decreasing order: D (CN^-) > B (en) > C (NCS^-) > A (S^{2-}).

69. In $[Pt(NH_3)_2Cl_2]$, Pt has oxidation state:

$x + 0\ (NH_3) + 2(-1)\ (Cl^-) = 0 \rightarrow x = +2.$

The ligands are named in alphabetical order: "ammine" before "chlorido." The oxidation state is written in Roman numerals. Hence the name is diammine dichloridoplatinum (II).

70. In $[Fe(CN)_6]^{3-}$, Fe^{3+} has configuration [Ar] $3d^5$. CN^- is a strong field ligand (high Δ_0) causing pairing of electrons in the 3d orbitals, giving a low-spin complex. Thus, all five *d*-electrons pair into t_2g orbitals, leaving only one unpaired electron → paramagnetic. The hybridization is d^2sp^3 (inner orbital complex). Magnetic moment $\mu = \sqrt{(n(n + 2))} = \sqrt{(1 \times 3)} \approx 1.73$ BM, not 5.92 BM. Therefore, only A and D are correct.

71. • Ambident nucleophiles: Nucleophiles that can attack through two different atoms. Examples include cyanides (CN^-) and nitrites (NO_2^-) which can bond via carbon or nitrogen → III.

• Plane polarized light: Produced by polarizing devices such as Nicol Prism → IV.

• Superimposable mirror image: Refers to an object that is identical to its mirror image, i.e., a symmetrical object → I.

• β-elimination reaction: Follows Saytzeff's rule, where the more substituted alkene is the major product → II.

Thus, matching is A–III, B–IV, C–I, D–II.

72. In an S_N2 reaction, the nucleophile attacks from the side opposite to the leaving group in a single-step mechanism, leading to a complete inversion of configuration at the stereogenic centre (Walden inversion). Hence, an optically active alkyl halide undergoing S_N2 substitution will result in inversion of configuration.

73. Gabriel phthalimide synthesis uses potassium phthalimide to prepare primary amines. This method avoids formation of secondary or tertiary amines because phthalimide anion selectively alkylates and after hydrolysis gives only primary amine.

74. The reaction shown is between an aryl halide (chlorobenzene) and an alkyl halide (CH_3Cl) in the presence of sodium metal and dry ether, resulting in the formation of an alkyl-substituted benzene (toluene). This type of coupling reaction, when sodium metal couples an aryl halide with an alkyl halide, is known as the Wurtz–Fittig reaction.

• Wurtz reaction involves coupling of two alkyl halides.

• Fittig reaction involves coupling of two aryl halides.

• Wurtz–Fittig reaction is a combination where an aryl halide couples with an alkyl halide.

Hence, the given example matches Wurtz–Fittig reaction.

75. Hydroboration-oxidation of propene is a two-step reaction:

Step 1: Hydroboration with diborane (B_2H_6) adds boron and hydrogen across the double bond in an anti-Markovnikov manner.

Step 2: Oxidation with hydrogen peroxide (H_2O_2) in the presence of hydroxide ions (OH^-) replaces boron with OH group, giving alcohol.

Water (H_2O) is not directly used as a reagent in this process. The reagents are B_2H_6, H_2O_2, and OH^-.

76. In plants, the main carbohydrate used for long-term storage of energy is starch. It is a polysaccharide composed of amylose (linear) and amylopectin (branched) units of α-D-glucose. Starch is stored in plastids such as amyloplasts and is broken down into glucose when energy is required.

77. The Hell–Volhard–Zelinsky (HVZ) reaction involves halogenation of the α-carbon atom of carboxylic acids in the presence of red phosphorus and halogen, producing α-halocarboxylic acids. It works because the α-position is activated via formation of the acyl halide intermediate.

78. Reactivity towards nucleophilic addition is influenced by both inductive and steric effects. Aldehydes are generally more reactive than ketones due to less steric hindrance and stronger partial positive charge on the carbonyl carbon. Among aldehydes, smaller alkyl groups make the compound more reactive. Thus:

Ethanal (A) > Propanal (C) > Propanone (B) > Butanone (D).

Increasing order (least to most reactive): D < B < C < A.

79. Conversion of benzene to methyl benzoate involves:

Step 1: Bromination of benzene with Br_2/$FeBr_3$ to form bromobenzene (A).

Step 2: Formation of phenylmagnesium bromide using Mg in dry ether (B).

Step 3: Reaction with CO_2 followed by acid hydrolysis to give benzoic acid (C).

Step 4: Esterification with methanol in presence of conc. H_2SO_4 to produce methyl benzoate (D).

All four reagents are required.

80. I_2/NaOH (iodoform test) distinguishes methyl ketones from other ketones. Acetone (a methyl ketone) gives a positive iodoform test (yellow precipitate of CHI_3), whereas benzophenone (no methyl group attached to carbonyl carbon) gives a negative test.

81. Aldol condensation occurs in carbonyl compounds (aldehydes or ketones) that have at least one α-hydrogen atom adjacent to the carbonyl carbon. This β-hydrogen is abstracted by a base to form an enolate ion, which then attacks another carbonyl group to form a β-hydroxy carbonyl compound, which can dehydrate to give an α, β-unsaturated carbonyl. Without α-hydrogen, the reaction does not proceed.

82. In trimethylamine ($N(CH_3)_3$), nitrogen is sp^3 hybridized, forming three sigma bonds with carbon atoms and having one lone pair. The lone pair distorts the tetrahedral arrangement, resulting in a trigonal pyramidal geometry, not perfectly tetrahedral.

83. When phenyl alkyl ethers (C_6H_5–O–R) are treated with hydrogen halides (HX), the bond between oxygen and alkyl group breaks. The alkyl group forms the alkyl halide (R–X), and phenol (C_6H_5OH) is produced. The aryl–oxygen bond remains intact because it is stronger and involves partial double bond character due to resonance.

84. In nitration of benzene, conc. H_2SO_4 protonates nitric acid to form nitronium ion (NO_2^+), which is the electrophile that attacks the benzene ring in the electrophilic substitution mechanism. The reaction:

$$HNO_3 + 2H_2SO_4 \rightarrow NO_2^+ + H_3O^+ + 2HSO_4^-.$$

85. Azo coupling occurs with activated aromatic compounds, usually containing electron-donating groups like –OH or $-NH_2$. Nitrobenzene has a strongly electron-withdrawing $-NO_2$ group, which deactivates the ring toward electrophilic substitution and thus does not undergo azo coupling with benzene diazonium chloride.

86. Starch is made up of two main components — amylose and amylopectin. Amylose is the water-soluble fraction of starch, generally constituting about 15–20% of the total starch content. This range is the most commonly reported, though some references mention 15–30% or even 10–30%. Amylopectin, the water-insoluble branched component, makes up the remaining 70–85% of starch. Thus, the percentage solubility of amylose in starch is generally accepted as 15–20%.

87. Globular proteins are generally soluble in water and have compact, spherical shapes, functioning mostly as enzymes, hormones, or transport proteins. Examples:

- Insulin: peptide hormone, globular.
- Albumin: abundant plasma protein, globular.
- Keratin: fibrous protein in hair and nails, not globular.
- Myosin: motor protein in muscles, globular domains with fibrous tail.

Globular examples here are insulin and albumin.

88. The β-helix (β-sheet structure) in proteins is stabilized by hydrogen bonds between the carbonyl oxygen of one peptide bond and the amide hydrogen of another on adjacent strands. These H-bonds are responsible for the stability of β-structures in proteins.

89. Matching amino acids with one-letter codes:
- Lysine → K (A–III)
- Tryptophan → W (B–I)
- Tyrosine → Y (C–IV)
- Glutamine → Q (D–II)

Thus, A–III, B–I, C–IV, D–II.

90. Acidic strength of phenols increases with the presence and position of electron-withdrawing groups like $-NO_2$. Ortho and para NO_2 groups have a stronger effect than meta, and more $-NO_2$ groups increase acidity further.

Phenol (D) < 3-Nitrophenol (A) < 3,5-Dinitrophenol (B) < 2,4,6-Trinitrophenol (C).

91. The rate law is $r = k[A][B]$. Reducing the reaction vessel's volume to 1/4 of the original increases the concentration of each reactant fourfold (since concentration $\propto$ 1/volume). Substituting into the rate law: $r' = k(4[A]) \times (4[B]) = 16 \times k[A][B]$. Thus, the new rate is 16 times the initial rate.

92. The given rate equation is Rate $= k[A]^{1/2}[B]^{3/2}$. The order of the reaction is the sum of the exponents:

$$\text{Order} = 1/2 + 3/2 = 4/2 = 2.$$

Hence, the reaction is of second order.

93. For $r = k[CH_3OCH_3]^{3/2}$, the rate has units of bar min^{-1}. From the rate law:

$$\text{bar min}^{-1} = k \times (\text{bar})^{3/2}$$

$$k = \text{bar}^1 \times \text{min}^{-1} \times \text{bar}^{-3/2}$$

$$= \text{bar}^{(1 - 3/2)} \text{ min}^{-1}$$

$$= \text{bar}^{(-1/2)} \text{ min}^{-1}.$$

94. If the rate increases 27 times when the concentration increases 3 times:

$$\text{Rate ratio} = (3)^n$$

$$= 27 \rightarrow 3^n$$

$$= 3^3 \rightarrow n$$

$$= 3.$$

Thus, the order of the reaction is 3.

95. A catalyst works by lowering the activation energy of the reaction, providing an alternative pathway, but does not change the Gibbs energy, enthalpy, or equilibrium constant.

96. Ethanol undergoes acid-catalyzed dehydration at different temperatures:
- At 443 K (high temperature), elimination predominates, producing ethene.
- At 413 K (lower temperature), substitution occurs between two ethanol molecules to form ethoxyethane (diethyl ether).

Thus, at 443 K → ethene; at 413 K → ethoxyethane.

97. Anisole (methoxybenzene) contains a methoxy ($-OCH_3$) group, which is an ortho–para directing group in electrophilic substitution. The lone pairs on the oxygen atom are delocalized into the aromatic ring by resonance, increasing electron density at ortho and para positions. When anisole undergoes bromination, the electrophile (Br^+) preferentially attacks these positions. The para position, being less sterically hindered than the ortho position, becomes the major site of substitution, leading predominantly to p-bromoanisole.

98. Williamson's synthesis involves the nucleophilic substitution of an alkyl halide by an alkoxide ion. The alkoxide ion is a strong nucleophile, and the reaction proceeds via the bimolecular S_N2 pathway, involving backside attack and inversion of configuration in primary halides.

99. For ether cleavage with hydrogen halides, reactivity increases with the halide's ability to act as a nucleophile and the strength of the acid: HI > HBr > HCl > HF. HI is the strongest acid and has the most nucleophilic iodide ion, making it the most reactive.

100. Anisole (methoxybenzene) has one phenyl group and one alkyl group attached to the oxygen atom. Such compounds are classified as phenyl alkyl ethers.

101. The filiform apparatus is a finger-like wall thickening present at the micropylar end of synergids. It secretes chemical signals that attract and guide the pollen tube into the embryo sac toward the egg apparatus. This ensures precise delivery of male gametes for fertilization.

102. An antibody (immunoglobulin) molecule is made up of four polypeptide chains – two heavy chains and two light chains, linked by disulfide bonds. This gives the characteristic Y-shaped structure.

103. The correct equation representing the relationship is:

$$GPP - R = NPP$$

Here, Gross Primary Productivity (GPP) is the total energy fixed, R is the energy lost in respiration, and Net Primary Productivity (NPP) is what remains available for biomass.

104. A. Malaria → (II) Cycles of fever (caused by Plasmodium)

B. Filariasis → (I)Inflammation of lower limb (elephantiasis, caused by Wuchereria)

C. Ringworm → (IV)Scaly lesions on nails/skin (caused by fungi)

D. Amoebiasis → (III)Blood clots and excess mucus in stools (caused by Entamoeba histolytica)

So, the correct match is: A–II, B–I, C–IV, D–III

105. This is about microbes as industrial producers of acids:

A. Clostridium butylicum → (II) Butyric acid

This anaerobic bacterium ferments sugars to produce butyric acid, which has industrial uses in plastics, pharmaceuticals, and flavorings.

B. Aspergillus niger → (III) Citric acid

This fungus is widely used in biotechnology for the production of citric acid, an important preservative, flavoring agent in food, and in pharmaceuticals.

C. Acetobacter aceti → (IV) Acetic acid

This bacterium is responsible for the conversion of alcohol (ethanol) into acetic acid, which is used in making vinegar.

D. Lactobacillus → (I) Lactic acid

These bacteria ferment lactose and other sugars to form lactic acid, which is important in curd formation, food preservation, and also used in the dairy industry.

Thus the correct match is: A–II, B–III, C–IV, D–I.

106. The chalaza is the part of the ovule situated opposite to the micropyle. It is the basal region where the nucellus, integuments, and funicle meet. It plays a key role in the supply of nutrients from the parent plant to the developing embryo sac.

107.
- Amniocentesis → Testing of sex (A-IV)
- Maternal mortality rate → Death rate (B-III)
- Vasectomy → Ligation of vas deferens (C-II)
- Tubectomy → Ligation of fallopian tube (D-I)

Correct set: A-IV, B-III, C-II, D-I.

108. Interferons are signaling proteins released by virus-infected cells. They do not kill the virus directly but warn nearby uninfected cells, making them resistant to viral infection. This defense mechanism is part of cytokine barriers, which are a form of innate immunity.

109. The process of transcription is the synthesis of RNA from a DNA template. Only one DNA strand (template strand) is used by RNA polymerase to produce a complementary RNA molecule. This RNA may function as mRNA, tRNA, or rRNA, depending on the need of the cell.

110. Correct sequence of early developmental events in human fertilization:

D. Zygote divides → Morula (8–16 blastomeres)

C. Blastomeres arrange into trophoblast & inner cell mass

A. Blastocyst embeds in endometrium

B. Chorionic villi appear from trophoblast

Order: D → C → A → B.

111. Intrauterine devices (IUDs) include copper-releasing devices (like Cu-T, Lippes Loop) and hormone-releasing devices (like LNG-20). Saheli (C) is an oral contraceptive pill and implants (D) are subdermal devices, so they are not IUDs.

112. A. Flower colour → Dominant = Violet/Purple, Recessive = (IV) White

B. Seed colour → Dominant = Yellow, Recessive = (I) Green

C. Pod colour → Dominant = Green, Recessive = (II) Yellow

D. Pod shape → Dominant = Inflated, Recessive = (III) Constricted

113. In bacteria, transcription requires DNA-dependent RNA polymerase, sigma factor (initiation), and rho factor (termination). Methyl guanosine triphosphate (m7G cap) is a eukaryotic RNA modification, not seen in bacteria.

114. Aneuploidy results from the gain or loss of chromosomes.

B. Down's Syndrome (trisomy-21) → due to extra chromosome 21.

D. Turner's Syndrome (XO) → due to monosomy of sex chromosome.

(A) Haemophilia and (C) Thalassemia are genetic disorders but not due to aneuploidy.

115. Skin colour inheritance is polygenic. Intermediate phenotype occurs in heterozygous condition (AaBbCc), as multiple alleles contribute additively. Homozygous dominants (AABBCC) → darkest, homozygous recessives (aabbcc) → lightest.

116. This is about the Lac Operon in E. coli, which regulates lactose metabolism:

A. 'i' gene → (II) produces repressor protein, which binds to the operator and prevents transcription.

B. 'z' gene → (IV) codes for β-galactosidase, an enzyme that breaks down lactose into glucose and galactose.

C. 'y' gene → (I) codes for permease, a membrane protein that facilitates lactose entry into the cell.

D. 'a' gene → (III) codes for transacetylase, which transfers acetyl groups to lactose analogs.

Correct match: A-II, B-IV, C-I, D-III.

117.
- Cannabinoids affect the cardiovascular system → True.
- Heroin is obtained from the latex of Papaver somniferum → True.
- Nicotine is actually a stimulant, not a sedative or painkiller → Incorrect.
- Coca alkaloids in excess cause hallucinations → True.

118. Steps of gel electrophoresis:

C. DNA fragments move towards anode

B. Staining with ethidium bromide

A. Exposure to UV light to visualize bands

D. Elution – desired DNA fragment is cut out

Correct sequence: C → B → A → D.

119. A. Citric acid → (IV) Aspergillus niger, used in food industry as a preservative and flavoring agent.

B. Ethanol → (III) Saccharomyces cerevisiae (yeast), produced during fermentation of sugars, widely used in beverages and biofuel.

C. Statins → (II) Monascus purpureus, used as cholesterol-lowering agents by inhibiting cholesterol synthesis.

D. Cyclosporin A → (I) Trichoderma polysporum, an immunosuppressive drug given during organ transplantation to prevent rejection.

Correct match: A-IV, B-III, C-II, D-I

120. EcoRI is a restriction endonuclease enzyme isolated from Escherichia coli. It cuts DNA at a specific recognition site (GAATTC), producing sticky ends useful in genetic engineering.

121. The correct order of geological periods from latest to oldest is:

C. Tertiary → Latest (after Cretaceous)

D. Jurassic → Mesozoic era

A. Triassic → Older than Jurassic

B. Carboniferous → Much older, in Paleozoic era

So, the sequence is C → D → A → B.

122. The three steps in Polymerase Chain Reaction (PCR) are:

D. Denaturation → Separation of DNA strands at high temperature.

B. Annealing → Primers bind to single-stranded DNA.

A. Extension → DNA polymerase synthesizes new strands.

C. Propagation is not a step in PCR.

123. Baculoviruses are pathogens of insects, mainly used as biological control agents for harmful pests. They are considered eco-friendly because they do not harm plants, animals, or humans.

124. If the number of individuals in the reproductive age group is greater than the pre-reproductive group, the base of the age pyramid is narrow. This means fewer young individuals are available to replace the reproductive generation in the future, leading to a gradual fall in population size. Hence, the population is declining.

125. All four statements (A, B, C, and D) are correct regarding homology and homologous organs.

A. Homology shows common ancestry and evolutionary relationships.

B. Whale flippers and cheetah forelimbs share the same skeletal plan but different functions.

C. The vertebrate heart has a common basic structure in all vertebrates, despite variation in chambers.

D. Thorn of Bougainvillea and tendril of Cucurbita are both stem modifications, serving different functions but having the same origin.

126. In plant evolution:

C. Rhynia-type plants → earliest simple vascular plants (Silurian/Devonian).

D. Psilophyton → early vascular land plants (Devonian).

B. Seed ferns → gymnosperm-like plants (Palaeozoic).

A. Angiosperms → most recent, flowering plants (Cretaceous).

Correct order: C → D → B → A.

127. If a man is eating fish, he occupies the fourth trophic level in the food chain.

- First trophic level: Phytoplankton (producers) make food by photosynthesis.
- Second trophic level: Zooplankton (herbivores) feed on phytoplankton.
- Third trophic level: Fish (secondary consumers) eat zooplankton.
- Fourth trophic level: Man, by eating fish, becomes a tertiary consumer.

128. The thorn of Bougainvillea is a stem modification for protection, while the tendril of Cucurbita is a stem modification for climbing. Both develop from the same origin (stem) but perform different functions. This makes them homologous structures, showing divergent evolution.

10. A point source causing photoelectric emission from a metallic plate is moved away from the plate. The variation of photoelectric current with distance from the source is correctly represented by the graph _______ .
Fill in the blank with the **correct** answer from the options given below.

A.
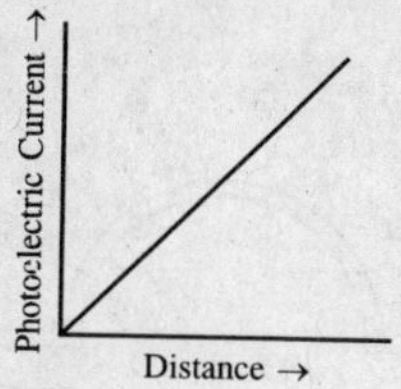

B.
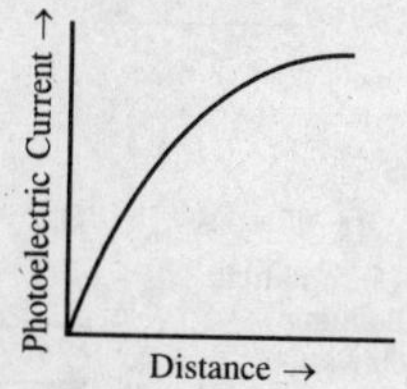

C.
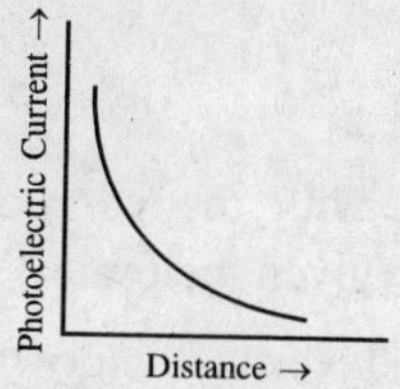

D.
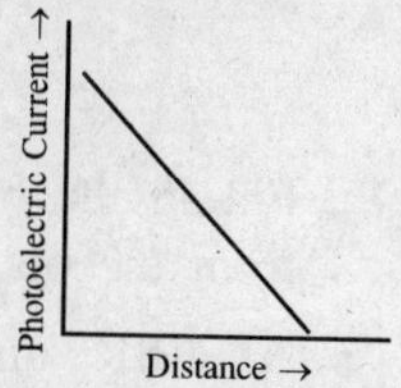

11. A proton accelerated through a potential difference V has a de Broglie wavelength λ. On doubling the accelerating potential, de Broglie wavelength of the proton _________ .
Fill in the blank with the **correct** answer from the options given below.
A. remains unchanged
B. becomes double
C. becomes four times
D. decreases

12. The kinetic energy of an electron in ground level in hydrogen atom is K units. The values of its potential energy and total energy respectively are _______ .
Fill in the blank with the **correct** answer from the options given below.
A. –2K, –K
B. +2K, –K
C. –K, +2K
D. +K, +2K

13. Two nuclei have mass numbers A and B respectively. The density ratio of the nuclei is _______ .
Fill in the blank with the **correct** answer from the options given below.
A. A : B
B. $\sqrt{A} : \sqrt{B}$
C. $A^2 : B^2$
D. 1 : 1

14. The shortest wavelengths emitted in hydrogen spectrum corresponding to different spectral series are as under:
(*a*) Pfund series
(*b*) Balmer series
(*c*) Brackett series
(*d*) Lyman series

The wavelengths arranged correctly in decreasing order are _________ .
Fill in the blank with the **correct** answer from the options given below.
A. (*a*), (*b*), (*c*), (*d*)
B. (*a*), (*c*), (*b*), (*d*)
C. (*b*), (*a*), (*d*), (*c*)
D. (*a*), (*c*), (*d*), (*b*)

15. Silicon can be doped using one of the following elements as dopant:
(*a*) Arsenic
(*b*) Indium
(*c*) Phosphorus
(*d*) Boron

To get *n*-type semiconductor, the dopants that can be used are _______ .
Fill in the blank with the **correct** answer from the options given below.
A. (*a*) and (*c*) only
B. (*b*) and (*c*) only
C. (*a*), (*b*), (*c*) and (*d*)
D. (*c*) and (*d*) only

10. C **11.** D **12.** A **13.** D **14.** B **15.** A

16. Given below are V versus I graphs for different types of *p-n*-junction diodes marked A, B, C and D.

(*a*) 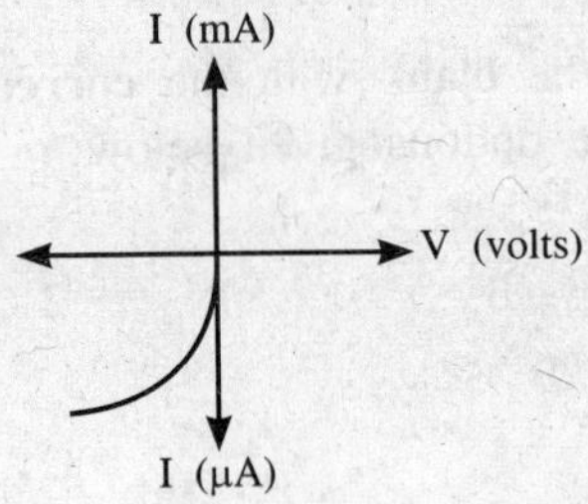

(*b*) 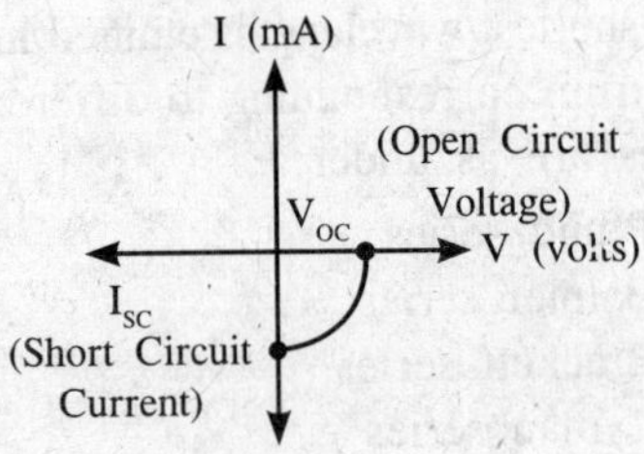

(*c*) 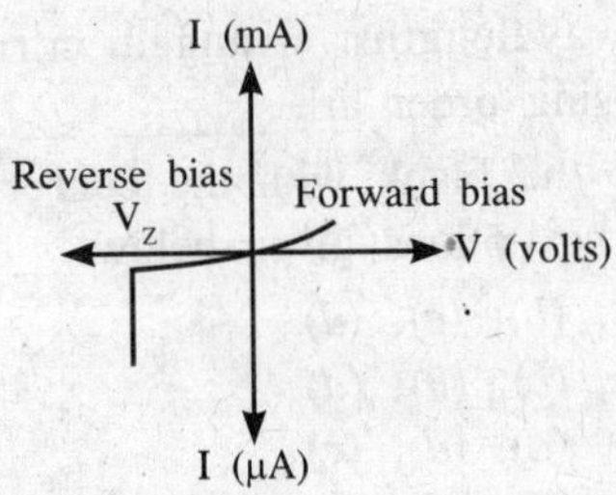

(*d*) 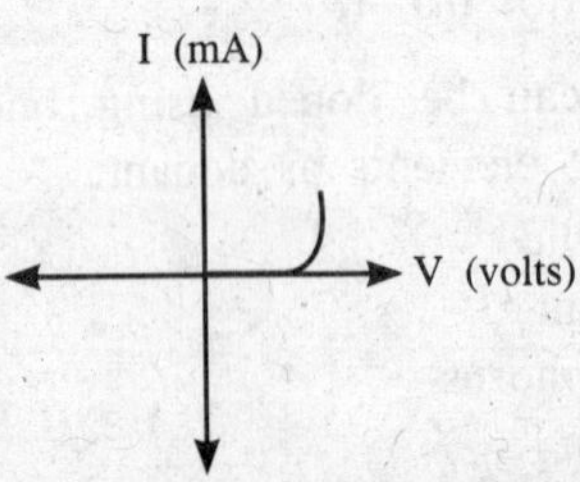

The correct sequence of graphs corresponding to forward biased *p-n* junction; Zener diode; Photo diode and Solar cell in order is ______.

Fill in the blank with the **correct** answer from the options given below.

A. (*d*), (*c*), (*a*), (*b*)

B. (*a*), (*c*), (*b*), (*d*)

C. (*b*), (*a*), (*d*), (*c*)

D. (*c*), (*b*), (*d*), (*a*)

17. A wire carrying current I, bent as shown in the figure, is placed in a uniform field B that emerges normally out from the plane of the figure. The force on this wire is ______.

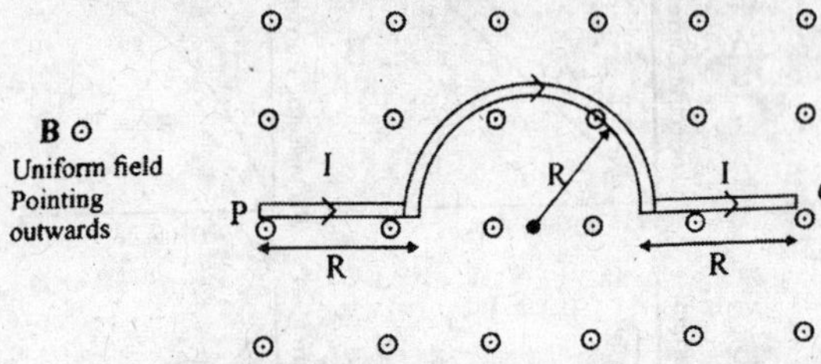

Fill in the blank with the **correct** answer from the options given below:

A. 4BIR, directed vertically downward

B. 3BIR, directed vertically upward

C. BI(2R + πR), vertically downward

D. 2πBIR, from P to Q

18. The refractive index of the material of an equilateral prism is $\sqrt{2}$. The angle of minimum deviation of that prism is ______.

Fill in the blank with the **correct** answer from the options given below.

A. 60° B. 75°

C. 30° D. 90°

19. The transfer of integral number of ______ is one of the evidence of quantization of electric charge.

Fill in the blank with the **correct** answer from the options given below.

A. photons B. nuclei

C. electrons D. neutrons

16. A **17.** A **18.** C **19.** C

129. In aquatic ecosystems, the standing crop biomass of phytoplankton is very small because they multiply and are consumed rapidly. Yet, they support a larger biomass of zooplankton and higher consumers. This is represented by an inverted pyramid of biomass (not energy, as energy pyramids are always upright).

130. The first restriction endonuclease enzyme discovered was Hind II in 1970, from Haemophilus influenzae. It was the first enzyme proven to cut DNA at a specific nucleotide sequence. This discovery revolutionized molecular biology and laid the foundation of recombinant DNA technology.

131. During sewage treatment, the sequence is:

D. Primary sludge: solids settle after primary treatment.

C. Flocs: masses of bacteria + fungi form during secondary treatment.

B. Activated sludge: flocs settle after aeration.

A. Biogas: produced from anaerobic digestion of sludge.

Correct order: D $\rightarrow$ C $\rightarrow$ B $\rightarrow$ A.

132. Wildlife safari parks are an example of *ex situ* conservation.

Ex situ conservation means protecting biodiversity outside its natural habitat. Safari parks provide animals with a semi-natural but controlled environment, ensuring their safety, breeding, and care. Since it is not their original natural home, it is classified as *ex situ*.

133. Humification is the process during decomposition where organic matter in soil is broken down into a dark, amorphous, and colloidal substance called humus. Humus is resistant to further decomposition, improves soil fertility, increases water-holding capacity, and provides essential nutrients for plants.

134. According to Chargaff's rule:

A = T and G = C.

If A = 15%,

then T = 15% $\rightarrow$ A + T = 30%.

Remaining = 70% $\rightarrow$ equally divided between G and C $\rightarrow$ each = 35%.

135. Transgenic animals are genetically modified by introducing human genes to study their role in health and disease. They are used to understand complex disorders such as cancer (uncontrolled cell growth) and cystic fibrosis (a genetic disorder affecting lungs and pancreas). These models help in developing new drugs and therapies.

136. Genetically modified (GM) plants are developed to increase tolerance to stresses such as drought, salinity, pests, and pathogens. They can also improve nutritional quality and reduce chemical pesticide use.

137. The primary female sex organs are the ovaries, as they produce female gametes (ova) and secrete sex hormones like estrogen and progesterone. Structures like uterus and mammary glands are secondary sex organs.

138. In male gametogenesis:

- Primary spermatocytes ($2n$) $\rightarrow$ undergo meiosis I $\rightarrow$ Secondary spermatocytes (n)
- Leydig and Sertoli cells are diploid somatic cells.

Thus, secondary spermatocytes are haploid (n).

139. Out of the total incident solar radiation, about 50% is Photosynthetically Active Radiation (PAR), used for photosynthesis. Hence, the correct option is less than 50%.

140. ELISA (Enzyme-Linked Immunosorbent Assay) is a diagnostic test based on the principle of antigen–antibody interaction, where enzyme-linked antibodies produce a color change for detection.

141. In-situ conservation means conserving biodiversity within natural habitats, e.g., National Parks, Wildlife Sanctuaries, and Biosphere Reserves. Zoological parks are man-made and belong to ex-situ conservation.

142. Narrowly utilitarian values include direct benefits like food, medicines, fuel, industrial products, and construction materials. Pollination is an ecosystem service, which falls under broadly utilitarian values, not narrow.

143. Tropical regions have greater biodiversity because they have been undisturbed for millions of years, receive more solar energy, and have stable, less seasonal climates. Frequent glaciations (which caused species loss) happened in temperate/polar regions, not in the tropics.

144. Since the origin and diversification of life, Earth has witnessed five episodes of mass extinction, including the famous Cretaceous–Paleogene extinction (dinosaurs).

145. India has three biodiversity hotspots:

- Indo-Burma
- Himalaya
- Indo-Malayan (including Western Ghats–Sri Lanka)

The Amazon forests are in South America and not part of India's biodiversity.

146.
- Zygote → Embryo
- Primary endosperm nucleus ($3n$) → Endosperm
- Ovary → Fruit

Thus, the incorrect statement is "Endosperm develops into seed", because it nourishes the embryo, not forms the seed.

147. Stages of dicot embryo development are: Proembryo → Globular → Heart-shaped → Torpedo stage. Epiblast is part of monocot embryo, not dicot.

148. Coconut water is the free-nuclear endosperm formed before cellularization, rich in nutrients to support the developing embryo.

149. Syngamy is the fusion of one male gamete (n) with the egg cell (n) to form a diploid zygote ($2n$).

150. If the zygote ($2n$) has 38 chromosomes, then the haploid gametes (n) will have half = 19 chromosomes.

Previous Years' Paper

B.Sc. BIOTECHNOLOGY

Entrance Exam, 2024*

PHYSICS

1. In an electromagnetic wave, the ratio of energy densities of electric and magnetic fields is ________.

Fill in the blank with the **correct** answer from the options given below.

A. 1 : 1 B. 1 : c

C. c : 1 D. 1 : c^2

2. Match List-I has four graphs showing variation of opposition to flow of ac versus frequency with circuit characteristic in List-II.

List-I **List-II**

(*a*) 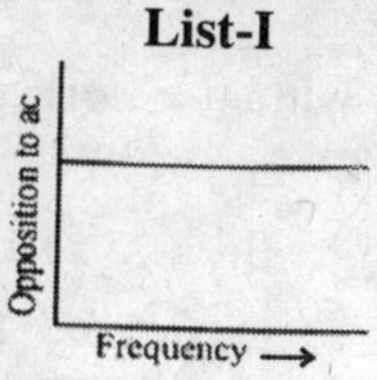

I. Impedance

(*b*) 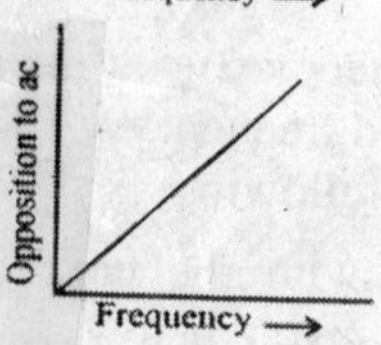

II. Capacitive reactance

(*c*) 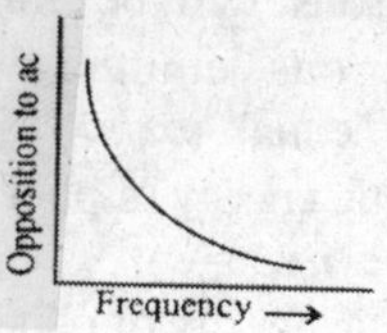

III. Inductive reactance

(*d*) 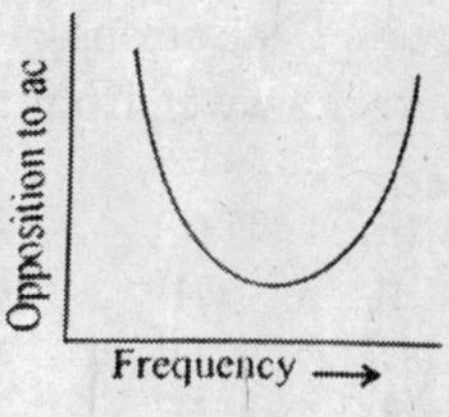

IV. Resistance

Choose the **correct** answer from the options given below:

	(*a*)	(*b*)	(*c*)	(*d*)
A.	I	II	III	IV
B.	IV	III	II	I
C.	I	II	IV	III
D.	III	IV	I	II

3. Of the following, the **correct** arrangement of electro-magnetic spectrum in decreasing order of wavelength is ________.

Fill in the blank with the correct answer from the options given below.

A. Radio waves, X-rays, Infrared waves, Microwaves, Visible waves

B. Infrared waves, Microwaves, Radio waves, X-rays, Visible waves

C. Radio waves, Microwaves, Infrared waves, Visible waves, X-rays

D. X-rays, Visible waves, Infrared waves, Microwaves, Radio waves

1. A **2.** B **3.** C

Conducted by NTA for CUET (UG).

4. Match Electromagnetic waves listed in Column-I with Production method/device in Column-II.

Column-I (Electromagnetic waves)	**Column-II (Production method/ device)**
(*a*) Microwaves	I. LC oscillator
(*b*) Infrared	II. Magnetron
(*c*) X-rays	III. Vibration of atoms/ molecules
(*d*) Radio waves	IV. Bombarding large atomic number metal target with fast moving electrons

Choose the **correct** answer from the options given below:

	(*a*)	(*b*)	(*c*)	(*d*)
A.	I	II	III	IV
B.	II	III	IV	I
C.	II	I	IV	III
D.	III	IV	I	II

5. In the figure given below, APB is a curved surface of radius of curvature 10 cm separating air and a transparent material (μ = 4/3). A point object O is placed in air on the principal axis of the surface 20 cm from P. The distance of the image of O from P will be ________.

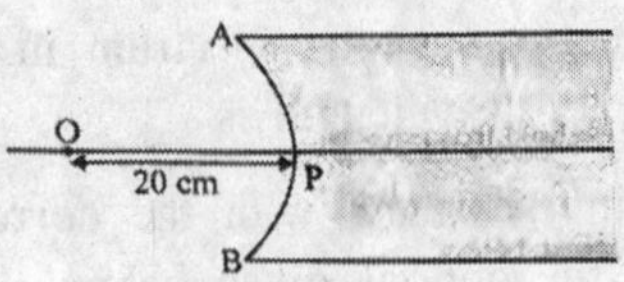

Fill in the blank with the **correct** answer from the options given below.

A. 16 cm left of P in air

B. 16 cm right of P in water

C. 20 cm right of P in water

D. 20 cm left of P in air

6. For fixed values of radii of curvature of lens, power of the lens will be ________.

Fill in the blank with the **correct** answer from the options given below.

A. $P \propto (\mu - 1)$ B. $P \propto \mu^2$

C. $P \propto 1/\mu$ D. $P \propto \mu^{-2}$

7. The graph correctly representing the variation of image distance '*v*' for a convex lens of focal length '*f*' versus object distance '*u*' is ________.

Fill in the blank with the **correct** answer from the options given below.

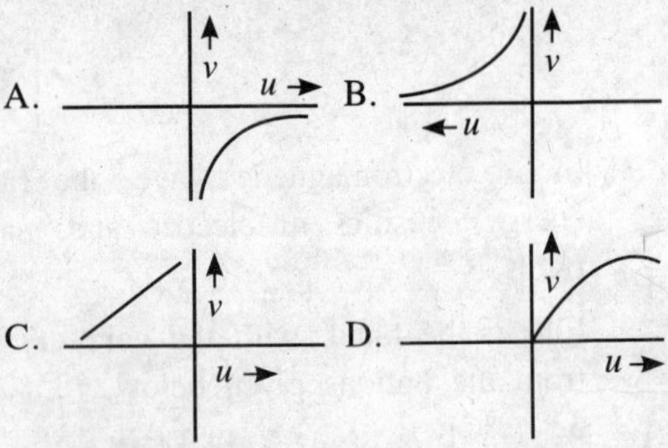

8. Using light from a monochromatic source to study diffraction in a single slit of width 0.1 mm, the linear width of central maxima is measured to be 5 mm on a screen held 50 cm away. The wavelength of light used is ________.

Fill in the blank with the **correct** answer from the options given below.

A. 2.5×10^{-7} m B. 4×10^{-7} m

C. 5×10^{-7} m D. 7.5×10^{-7} m

9. Radiation of frequency $2\nu_0$ is incident on a metal with threshold frequency ν_0. The correct statement of the following is ________.

Fill in the blank with the **correct** answer from the options given below.

A. No photoelectrons will be emitted

B. All photoelectrons emitted will have kinetic energy equal to $h\nu_0$

C. Maximum kinetic energy of photoelectrons emitted can be $h\nu_0$

D. Maximum kinetic energy of photoelectrons emitted will be $2h\nu_0$

4. B **5.** A **6.** A **7.** B **8.** C **9.** C

20. When a slab of insulating material 4 mm thick is introduced between the plates of a parallel plate capacitor of separation 4 mm, it is found that the distance between the plates has to be increased by 3.2 mm to restore the capacity to its original value. The dielectric constant of the material is ________.
Fill in the blank with the **correct** answer from the options given below.

A. 2 B. 5
C. 3 D. 7

21. A copper ball of density 8.0 g/cc and 1 cm in diameter is immersed in oil of density 0.8 g/cc. The charge on the ball if it remains just suspended in oil in an electric field of intensity 600π V/m acting in the upward direction is ________.
Fill in the blank with the correct answer from the options given below. (Take g = 10 m/s^2)

A. 2×10^{-6} C B. 2×10^{-5} C
C. 1×10^{-5} C D. 1×10^{-6} C

22. A metal wire is subjected to a constant potential difference. When the temperature of the metal wire increases, the drift velocity of the electron in it ______.
Fill in the blank with the **correct** answer from the options given below.

A. increases, thermal velocity of the electrons decreases
B. decreases, thermal velocity of the electrons decreases
C. increases, thermal velocity of the electrons increases
D. decreases, thermal velocity of the electrons increases

23. For the given mixed combination of resistors calculate the total resistance between points A and B.

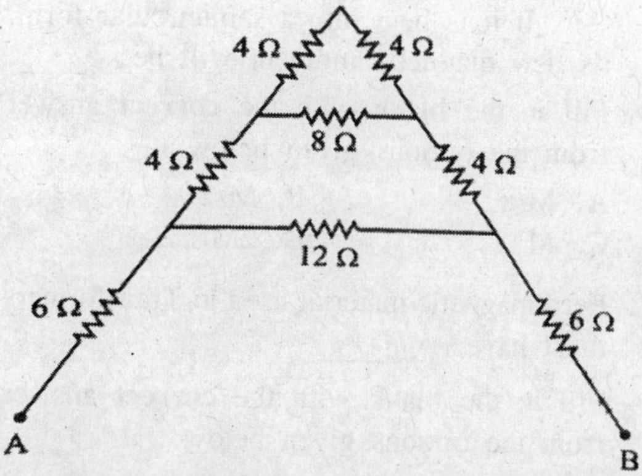

Choose the **correct** answer from the options given below.

A. 9 Ω B. 18 Ω
C. 4 Ω D. 14 Ω

24. A cell of emf 1.1 V and internal resistance 0.5 Ω is connected to a wire of resistance 0.5 Ω. Another cell of the same emf is now connected in series with the intention of increasing the current but the current in the wire remains the same. The internal resistance of the second cell is ______.
Fill in the blank with the **correct** answer from the options given below.

A. 1 Ω
B. 2.5 Ω
C. 1.5 Ω
D. 2 Ω

25. P, Q, R and S are four wires of resistances 3, 3, 3 and 4 Ω respectively. They are connected to form the four arms of a wheatstone bridge circuit. The resistance with which S must be shunted in order that the bridge may be balanced is ______.
Fill in the blank with the **correct** answer from the options given below.

A. 14 Ω B. 12 Ω
C. 15 Ω D. 7 Ω

20. B **21.** B **22.** D **23.** B **24.** A **25.** B

26. Magnetic moment of a thin bar magnet is 'M'. If it is bent into a semicircular form, its new magnetic moment will be ______.

Fill in the blank with the **correct** answer from the options given below.

A. M/π B. M/2
C. M D. 2M/π

27. Ferromagnetic material used in Transformers must have ________.

Fill in the blank with the **correct** answer from the options given below.

A. Low permeability and High hysteresis loss
B. High permeability and Low hysteresis loss
C. High permeability and High hysteresis loss
D. Low permeability and Low hysteresis loss

28. A conducting ring of radius r is placed in a varying magnetic field perpendicular to the plane of the ring. If the rate at which the magnetic field varies is x, the electric field intensity at any point of the ring is ______.

Fill in the blank with the **correct** answer from the options given below.

A. rx B. $rx/2$
C. $2rx$ D. $4r/x$

29. A 50 Hz ac current of crest value 1 A flows through the primary of a transformer. If the mutual inductance between the primary and secondary be 0.5 H, the crest voltage induced in the secondary is ________.

Fill in the blank with the **correct** answer from the options given below.

A. 75 V B. 150 V
C. 100 V D. 200 V

30. A long solenoid of diameter 0.1 m has 2×10^4 turns per meter. At the centre of the solenoid a coil of 100 turns and radius 0.01 m is placed with its axis coinciding with the solenoid axis. The current in the solenoid reduces at a constant rate to 0 A from 4 A in 0.05 s. If the resistance of the coil is $10\pi^2$ Ω, then the total charge flowing through the coil during this time is ______.

Fill in the blank with the **correct** answer from the options given below.

A. 16 μC B. 32 μC
C. 16π μC D. 32π μC

31. Lower half of a convex lens is made opaque. Which of the following statement describes the image of the object placed in front of the lens?

(*a*) No change in image
(*b*) Image will show only half of the object
(*c*) Intensity of image gets reduced

Choose the **correct** answer from the options given below:

A. (*a*) only B. (*b*) only
C. (*c*) only D. (*b*) and (*c*) only

32. Two slits are made 0.1 mm apart and the screen is placed 2 m away. The fringe separation when a light of wavelength 500 nm is used is ______.

Fill in the blank with the **correct** answer from the options given below.

A. 1 cm B. 0.15 cm
C. 1.5 cm D. 0.1 cm

33. For an astronomical telescope having objective lens of focal length 10 m and eyepiece lens of focal length 10 cm, telescope's tube length and magnification respectively are ______.

Fill in the blank with the **correct** answer from the options given below.

A. 20 cm, 1 B. 1000 cm, 1
C. 1010 cm, 1 D. 1010 cm, 100

26. D **27.** B **28.** B **29.** C **30.** B **31.** C **32.** A **33.** D

34. According to Bohr's Model:

(*a*) The radius of the orbiting electron is directly proportional to '*n*'.

(*b*) The speed of the orbiting electron is directly proportional to '1/*n*'.

(*c*) The magnitude of the total energy of the orbiting electron is directly proportional to '$1/n^2$'.

(*d*) The radius of the orbiting electron is directly proportional to 'n^2'.

Choose the **correct** answer from the options given below:

A. (*a*), (*b*) and (*c*) only
B. (*a*), (*b*) and (*d*) only
C. (*a*), (*b*), (*c*) and (*d*)
D. (*b*), (*c*) and (*d*) only

35. For a full wave rectifier, if the input frequency is 50 Hz, the output frequency will be ______ .

Fill in the blank with the **correct** answer from the options given below.

A. 50 Hz B. 100 Hz
C. 25 Hz D. 0 Hz

36. For an electric dipole in a non-uniform electric field with dipole moment parallel to direction of the field, the force F and torque τ on the dipole respectively are ______ .

Fill in the blank with the **correct** answer from the options given below.

A. $F = 0, \tau = 0$ B. $F \neq 0, \tau = 0$
C. $F = 0, \tau \neq 0$ D. $F \neq 0, \tau \neq 0$

37. Two large plane parallel sheets shown in the figure have equal but opposite surface charge densities $+\sigma$ and $-\sigma$. A point charge *q* placed at points P_1, P_2 and P_3 experiences forces F_1, F_2 and F_3 respectively. Then

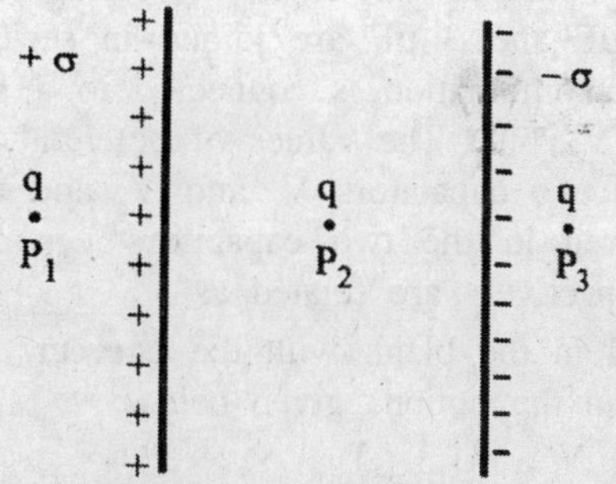

Choose the **correct** answer from the options given below.

A. $\vec{F}_1 = 0, \vec{F}_2 = 0, \vec{F}_3 = 0$
B. $\vec{F}_1 = 0, \vec{F}_2 \neq 0, \vec{F}_3 = 0$
C. $\vec{F}_1 \neq 0, \vec{F}_2 \neq 0, \vec{F}_3 \neq 0$
D. $\vec{F}_1 = 0, \vec{F}_3 \neq 0, \vec{F}_2 = 0$

38. Two charged metallic spheres with radii R_1 and R_2 are brought in contact and then separated. The ratio of final charges Q_1 and Q_2 on the two spheres respectively will be ______ .

Fill in the blank with the **correct** answer from the options given below.

A. $\frac{Q_1}{Q_2} = \frac{R_2}{R_1}$ B. $\frac{Q_1}{Q_2} < \frac{R_1}{R_2}$
C. $\frac{Q_1}{Q_2} > \frac{R_1}{R_2}$ D. $\frac{Q_1}{Q_2} = \frac{R_1}{R_2}$

39. Two charged particles, placed at a distance *d* apart in vacuum, exert a force F on each other. Now, each of the charges is doubled. To keep the force unchanged, the distance between the charges should be changed to ______ .

Fill in the blank with the **correct** answer from the options given below.

A. 4*d* B. 2*d*
C. *d* D. *d*/2

34. D	**35.** B	**36.** B	**37.** B	**38.** D	**39.** B

40. Two parallel plate capacitors of capacitances 2 μF and 3 μF are joined in series and the combination is connected to a battery of V volts. The values of potential across the two capacitors V_1 and V_2 and energy stored in the two capacitors U_1 and U_2 respectively are related as ________ .

Fill in the blank with the **correct** answer from the options given below.

A. $\frac{V_1}{V_2} = \frac{U_1}{U_2} = \frac{3}{2}$

B. $\frac{V_1}{V_2} = \frac{U_1}{U_2} = \frac{2}{3}$

C. $\frac{V_1}{V_2} = \frac{3}{2}$ and $\frac{U_1}{U_2} = \frac{2}{3}$

D. $\frac{V_1}{V_2} = \frac{2}{3}$ and $\frac{U_1}{U_2} = \frac{3}{2}$

41. Two resistances of 100 Ω and 200 Ω are connected in series across a 20 V battery as shown in figure below. The reading in a 200 Ω voltmeter connected across the 200 Ω resistance is ______ .

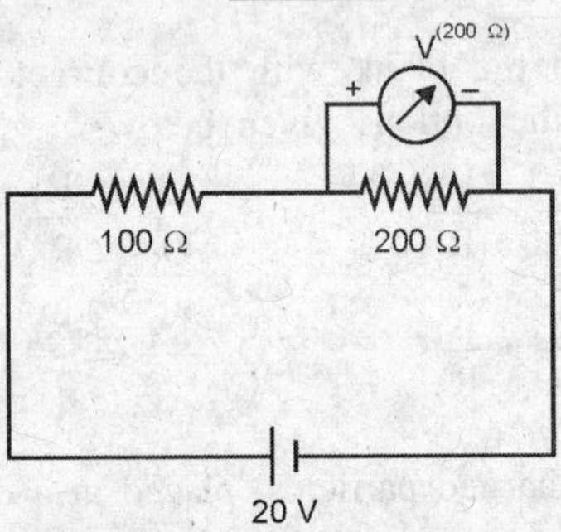

Fill in the blank with the **correct** answer from the options given below.

A. 4 V B. 20/3V
C. 10 V D. 16 V

42. The current through a 4/3 Ω external resistance connected to a parallel combination of two cells of 2 V and 1 V emf and internal resistances of 1 Ω and 2 Ω respectively is ___ .

Fill in the blank with the **correct** answer from the options given below.

A. 1 A B. 2/3 A
C. 3/4 A D. 5/6 A

43. A metallic wire of uniform area of cross section has a resistance R, resistivity ρ and power rating P at V volts. The wire is uniformly stretched to reduce the radius to half the original radius. The values of resistance, resistivity and power rating at V volts are now denoted by R′, ρ′ and P′ respectively. The corresponding values are correctly related as ________ .

Fill in the blank with the **correct** answer from the options given below.

A. ρ′ = 2ρ, R′ = 2R, P′ = 2P
B. ρ′ = (1/2)ρ, R′ = (1/2)R, P′ = (1/2)P
C. ρ′ = ρ, R′ = 16R, P′ = (1/16)P
D. ρ′ = ρ, R′ = (1/16)R, P′ = 16P

44. Three magnetic materials are listed below:

(*a*) paramagnetics (*b*) diamagnetics
(*c*) ferromagnetics

Choose the **correct** order of the materials in increasing order of magnetic susceptibility.

A. (*a*), (*b*), (*c*) B. (*c*), (*a*), (*b*)
C. (*b*), (*a*), (*c*) D. (*b*), (*c*), (*a*)

45. Two infinitely long straight parallel conductors carrying currents I_1 and I_2 are held at a distance *d* apart in vacuum. The force F on a length L of one of the conductors due to the other is ______ .

Fill in the blank with the **correct** answer from the options given below.

A. Proportional to L but independent of $I_1 \times I_2$
B. Proportional to $I_1 \times I_2$ but independent of length L
C. Proportional to $I_1 \times I_2 \times L$
D. Proportional to $\frac{L}{I_1 \times I_2}$

40. A	**41.** C	**42.** D	**43.** C	**44.** C	**45.** C

46. In the circuit shown below, a current 3I enters at A. The semicircular parts ABC and ADC have equal radii 'r' but resistances 2R and R respectively. The magnetic field at the center of the circular loop ABCD is ______ .

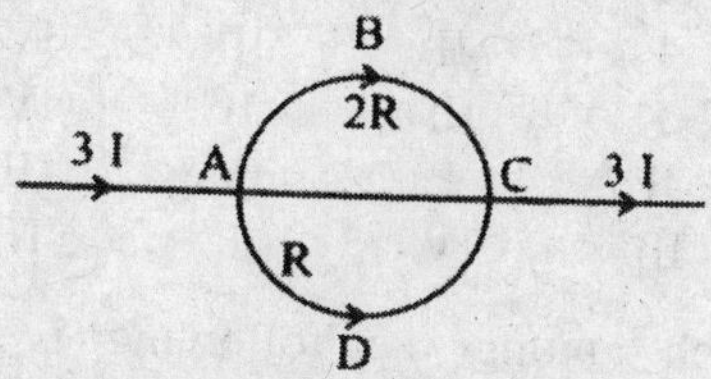

Fill in the blank with the **correct** answer from the options given below.

A. $\frac{\mu_0 I}{4r}$ out of the plane

B. $\frac{\mu_0 I}{4r}$ into the plane

C. $\frac{\mu_0 3I}{4r}$ out of the plane

D. $\frac{\mu_0 3I}{4r}$ into the plane

47. A square loop with each side 1 cm, carrying a current of 10 A, is placed in a magnetic field of 0.2 T. The direction of magnetic field is parallel to the plane of the loop. The torque experienced by the loop is ______ .

Fill in the blank with the **correct** answer from the options given below.

A. zero B. 2×10^{-4} Nm

C. 2×10^{-2} Nm D. 2 Nm

48. In an ac circuit, the current leads the voltage by $\pi/2$. The circuit is ______ .

Fill in the blank with the **correct** answer from the options given below.

A. purely resistive

B. should have circuit elements with resistance equal to reactance

C. purely inductive

D. purely capacitive

49. In a pair of adjacent coils, for a change of current in one of the coils from 0 A to 10 A in 0.25 s, the magnetic flux in the adjacent coil changes by 15 Wb. The mutual inductance of the coils is ______.

Fill in the blank with the **correct** answer from the options given below.

A. 120 H B. 12 H

C. 1.5 H D. 0.75 H

50. A wire of irregular shape in figure (*a*) and a circular loop of wire in figure (*b*) are placed in different uniform magnetic fields as shown in the figures below. In figure (*a*), the magnetic field is perpendicular into the plane. In figure (*b*), the magnetic field is perpendicular out of the plane.

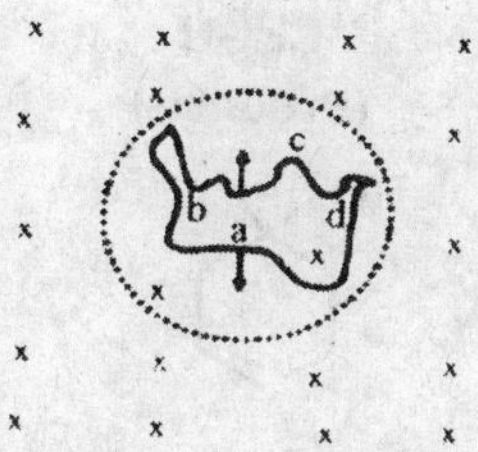

(*a*)

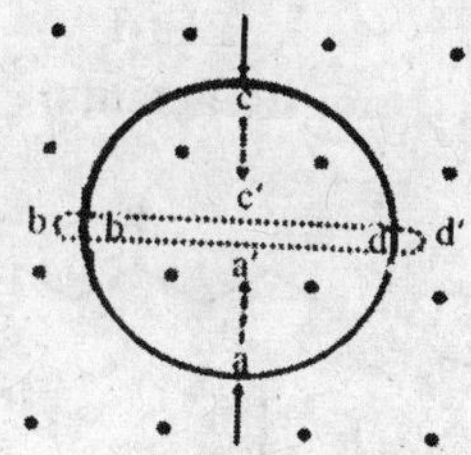

(*b*)

The wire in figure (*a*) is turning into a circular loop and that in figure (*b*) into a narrow straight wire. The direction of induced current will be ______.

Fill in the blank with the **correct** answer from the options given below.

A. clockwise in both (*a*) and (*b*)

B. anticlockwise in both (*a*) and (*b*)

C. clockwise in (*a*) and anticlockwise in (*b*)

D. anticlockwise in (*a*) and clockwise in (*b*)

46. A **47.** B **48.** D **49.** C **50.** B

CHEMISTRY

51. The increasing order of acidity of the following compounds based on pKa values is:

(*a*) $BrCH_2COOH$
(*b*) $ClCH_2COOH$
(*c*) FCH_2COOH
(*d*) HCOOH

Choose the **correct** answer from the options given below:

A. (*d*) < (*a*) < (*b*) < (*c*)
B. (*a*) < (*d*) < (*c*) < (*b*)
C. (*b*) < (*a*) < (*d*) < (*c*)
D. (*c*) < (*b*) < (*d*) < (*a*)

52. In the following table, match the reactants given in List-I with the correct product in List-II as the reaction of hydration of alkene under acidic condition.

List-I (Reactants)		**List-II (Products)**	
(*a*)	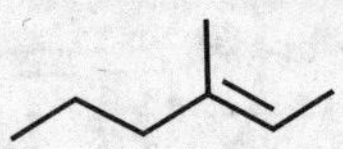	I.	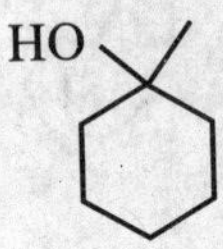
(*b*)	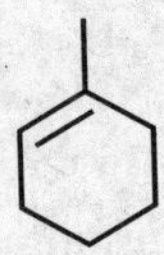	II.	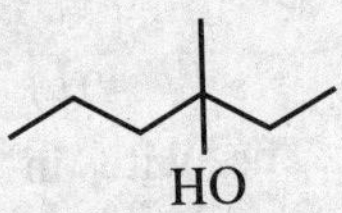
(*c*)	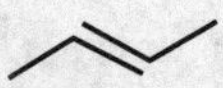	III.	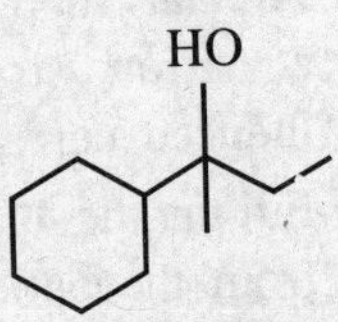
(*d*)	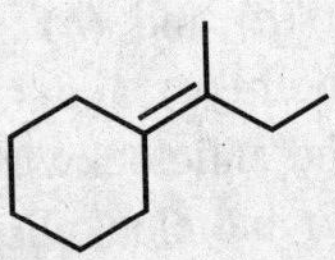	IV.	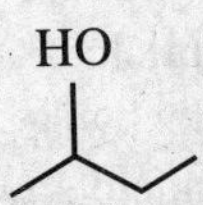

Choose the **correct** answer from the options given below:

	(*a*)	(*b*)	(*c*)	(*d*)
A.	I	II	III	IV
B.	I	III	II	IV
C.	II	I	IV	III
D.	III	IV	I	II

53. Which among the following is **not** an Analgesic?

A. Morphene B. Heroin
C. Codeine D. Ranitidine

54. For S_N2 reaction, the increasing order of the reactivity of the following alkyl halides is:

(*a*) $CH_3CH_2CH_2CH_2Br$
(*b*) $CH_3CH_2CH(Br)CH_3$
(*c*) $(CH_3)_3CBr$
(*d*) $(CH_3)_2CHCH_2Br$

Choose the **correct** answer from the options given below:

A. (*a*) < (*b*) < (*c*) < (*d*)
B. (*a*) < (*c*) < (*b*) < (*d*)
C. (*b*) < (*a*) < (*d*) < (*c*)
D. (*c*) < (*b*) < (*d*) < (*a*)

Directions (Qs. No. 55-59): *Read the following passage and answer the next five questions based on it.*

Battery or cell converts chemical energy of the redox reaction to electrical energy. In fuel cell (a galvanic cell), the chemical energy of combustion of fuels like H_2, ethanol, etc. are directly converted to electrical energy. In a fuel cell, H_2 and O_2 react to produce electricity, where H_2 gas is oxidised at anode and oxygen is reduced at cathode and the reactions involved are:

51. A **52.** C **53.** D **54.** D

Anode reaction : $H_2 + 2OH^- \rightarrow 2H_2O + 2e^-$

Cathode reaction : $O_2 + 2H_2O + 4e^- \rightarrow 4OH^-$

67.2 L of H_2 at STP reacts in 15 minutes.

55. The number of moles of hydrogen oxidized is:

A. 0.33 moles
B. 33.3 moles
C. 3.0 moles
D. 1.33 moles

56. The number of moles of electrons produced in the oxidation of 67.2 L of H_2 at STP is:

A. 2 moles
B. 4 moles
C. 1 mole
D. 6 moles

57. The quantity of electricity produced in the oxidation of 67.2 L of H_2 at STP is:

A. 96500 C
B. 579000 C
C. 193000 C
D. 48250 C

58. If the entire current produced is used for the electrodeposition of Silver (at.wt. 108 g mol^{-1}) from Silver (I) solution, the amount of silver deposited will be:

A. 324 g
B. 648 g
C. 108 g
D. 216 g

59. The source of electrical energy on the Apollo moon flight was:

A. Lead storage battery
B. A generator set
C. Ni-Cd cells
D. H_2-O_2 Fuel cell

Directions (Qs. No. 60-64): *Read the following passage and answer the next five questions based on it.*

Sc Ti V Cr Mn Fe Co Ni Cu Zn

Y Zr Nb Mo Tc Ru Rh Pd Ag Cd

La Hf Ta W Re Os Ir Pt Au Hg

In any transition series, as we move from left to right the d-orbitals are progressively filled and their properties vary accordingly.

Ce Pr Nd Pm Sm Eu Gd Tb Dy Ho Er Tm Yb Lu

Th Pa U Np Pu Am Cm Bk Cf Es Fm Md No Lr

The above are the two series of f-block elements in which the chemial properties won't change much. The 5f-series elements are radioactive in nature and mostly are artificially synthesized in laboratories and thus much is not known about their chemical properties.

60. Identify the **incorrect** statement.

A. Second ionisation enthalpy of Ag is greater than second ionisation enthalpy of Pd
B. Zr and Hf shares almost identical nuclear properties.
C. Melting point of Mn is lower than that of Cr.
D. Interstitial compounds are non-stoichiometric and neither ionic nor covalent in nature.

61. Which of the following is the correct order of second ionisation enthalpy?

A. V > Cr > Mn
B. V < Cr < Mn
C. V < Cr > Mn
D. V > Cr < Mn

62. Which of the following pair of compounds exhibits same colour in aqueous solution?

A. $FeCl_2$, $CuCl_2$
B. $VOCl_2$, $CuCl_2$
C. $VOCl_2$, $FeCl_2$
D. $VOCl_2$, $MnCl_2$

63. Which metal has the highest oxidation state in the first row transition series?

A. Cr
B. Fe
C. Mn
D. V

55. C **56.** D **57.** B **58.** B **59.** D **60.** B **61.** C **62.** B **63.** C

64. Why do the actinoids exhibit higher number of oxidation states than lanthanoids?

A. 4f orbitals are more diffused than the 5f orbitals
B. Energy difference between 5f and 6d is less with respect to the energy difference between 4f and 5d
C. Energy difference between 5f and 6d is more with respect to the energy difference between 4f and 5d
D. Actinoids are more reactive in nature than the lanthanoids.

65. Camphor in nitrogen gas is a type of solution:

A. Gas – Gas B. Solid – Gas
C. Liquid – Gas D. Solid – Liquid

66. Identify the correct order of organic compounds in the following chemical reaction:

$$\underline{?} + \text{Mg} \xrightarrow{\text{Dry Ether}} \underline{?} \xrightarrow{H_2O} \underline{?} \xrightarrow{Cl_2,\ \Delta} \underline{?}$$

(*a*) CH_3MgBr (*b*) CH_3Br
(*c*) CH_3Cl (*d*) CH_4

Choose the **correct** answer from the options given below:

A. (*b*), (*a*), (*d*), (*c*) B. (*a*), (*c*), (*b*), (*d*)
C. (*b*), (*a*), (*c*), (*d*) D. (*c*), (*b*), (*d*), (*a*)

67. Consider the following statements regarding osmotic pressure:

(*a*) Molar mass of a protein can be determined using osmotic pressure method.
(*b*) The osmotic pressure is proportional to the molarity.
(*c*) Reverse osmosis occurs when a pressure larger than osmotic pressure is applied to the concentrated solution side.
(*d*) Edema occurs due to retention of water in tissue cells as a result of osmosis.

Choose the **correct** statements with reference to osmotic pressure.

A. (*a*), (*b*) and (*d*) only
B. (*a*), (*b*) and (*c*) only
C. (*a*), (*b*), (*c*) and (*d*)
D. (*b*), (*c*) and (*d*) only

68. Vapour pressures of pure liquids 'A' and 'D' at 50°C are 500 mm Hg and 800 mm Hg respectively. The binary solution of 'A' and 'D' boils at 50°C and 700 mm Hg pressure. The mole percentage of 'D' in the solution is:

A. 33.33 mole per cent
B. 66.67 mole per cent
C. 25.75 mole per cent
D. 75.25 mole per cent

69. For the following reaction:

$$2A_2(g) + \frac{1}{4}X(g) \rightarrow 2A_2X(g)$$

volume is increased to double its value by decreasing the pressure on it. If the reaction is first order with respect to X and second order with respect to A_2, the rate of reaction will:

A. Decrease by eight times of its initial value
B. Increase by eight times of its initial value
C. Increase by four times of its initial value
D. Remain unchanged

70. The total number of sigma bonds present in P_4O_{10} are:

A. 6 B. 7
C. 16 D. 17

71. In the electrolysis of alumina to obtain Aluminium metal, the cryolite is added mainly to:

A. lower the melting point of alumina.
B. dissolve the alumina in the molten cryolite.
C. remove the impurities of alumina.
D. increase the electrical conductivity.

64. B	**65.** B	**66.** A	**67.** C	**68.** B	**69.** A	**70.** C	**71.** A

72. Identify the order of reaction if its rate constant is $k = 2 \times 10^{-2}\ s^{-1}$.

A. Zero order B. First order
C. Second order D. Half order

73. For a complex reaction, the order of reaction is equal to:

A. Sum of stoichiometric coefficients in balanced chemical reaction.
B. The molecularity of overall reaction.
C. Order of fastest step of the reaction.
D. The molecularity of slowest step of reaction.

74. A molecule X associates in a given solvent as per the following equation:

$$X \rightleftharpoons (X)_n$$

For a given concentration of X, the van't Hoff factor was found to be 0.80 and the fraction of associate molecules was 0.3. The correct value of '*n*' is:

A. 2 B. 3
C. 1 D. 5

75. The oxidation number of Co in complex $[Co(H_2NCH_2CH_2NH_2)_3]_2(SO_4)_3$ is:

A. 3 B. 4
C. 2 D. 5

76. The correct structure of dipeptide, Gly-Ala (glycyl alanine) is:

A. $H_2N-CH_2-CO-NH-CH(CH_3)-COOH$
B. $HOOC-CH_2-NH-CO-CH(CH_3)-NH_2$
C. $HOOC-CH(CH_3)-NH-CO-CH_2-NH_2$
D. $H_2N-CH(CH_3)-CO-NH-CH_2-COOH$

77. The total number of ions produced from the complex $[Cr(NH_3)_6]Cl_3$ in aqueous solution will be ______.

A. 2 B. 3
C. 4 D. 5

78. Arrange the following in decreasing order of number of molecules contained in:

(*a*) 16 g of O_2 (*b*) 16 g of CO_2
(*c*) 16 g of CO (*d*) 16 g of H_2

Choose the **correct** answer from the options given below:

A. (*a*), (*b*), (*c*), (*d*) B. (*d*), (*c*), (*a*), (*b*)
C. (*b*), (*a*), (*d*), (*c*) D. (*c*), (*b*), (*d*), (*a*)

79. The Cu metal crystallises into *fcc* lattice with a unit cell edge length of 361 pm. The radius of Cu atom is:

A. 127 pm B. 181 pm
C. 157 pm D. 108 pm

80. If 75% of a first order reaction gets completed in 32 minutes, time taken for 50% completion of this reaction is:

A. 16 minutes B. 78 minutes
C. 8 minutes D. 4 minutes

81. Which of the following compounds will be repelled when placed in an external magnetic field?

A. $Na_2[CuCl_4]$ B. $Na_2[CdCl_4]$
C. $K_4[Fe(CN)_6]$ D. $K_3[Fe(CN)_6]$

82. The spin only magnetic moment of Hexa-cyanido-manganate(II) ion is ______ BM.

A. 5.90 B. 1.73
C. 4.90 D. 3.87

83. The correct order of increasing boiling points of the following compounds is:

Pentan-1-ol, n-Butane, Pentanal, Ethoxyethane

A. Ethoxyethane, Pentanal, n-Butane, Pentan-1-ol
B. Pentanal, n-Butane, Ethoxyethane, Pentan-1-ol
C. n-Butane, Pentanal, Ethoxyethane, Pentan-1-ol
D. n-Butane, Ethoxyethane, Pentanal, Pentan-1-ol

72. B **73.** D **74.** B **75.** A **76.** A **77.** C **78.** B **79.** A **80.** A **81.** B **82.** B **83.** C

84. In the following reaction, identify the product D.

$$C_6H_5-OH \xrightarrow{\text{Zn dust}} A \xrightarrow{CH_3Cl\ +\ \text{anhy. } AlCl_3} B$$
$$\xrightarrow{K_2Cr_2O_7\ +\ H_2SO_4} C \xrightarrow{H_2SO_4\ +\ HNO_3} D$$

A. o-Nitrobenzoic acid

B. p-Nitrobenzoic acid

C. o,p-Dinitrobenzoic acid

D. m-Nitrobenzoic acid

85. The gold number range of some of the lyophilic colloids is given below:

A : 0.005 – 0.01,

B : 0.15 – 0.25,

C : 0.04 – 1.0 and

D : 15 – 25.

Which among these can be used as a better protective colloid?

A. A B. B

C. C D. D

86. Reaction of aniline with conc. HNO_3 and conc. H_2SO_4 at 298 K will produce 47% of:

A. p-Nitroaniline

B. o-Nitroaniline

C. m-Nitroaniline

D. 2,4-Dinitroaniline

87. What will be increasing order of basic strength of the following compounds?

$C_2H_5NH_2$, $(C_2H_5)_2NH$, $(C_2H_5)_3N$, $C_6H_5NH_2$

A. $C_2H_5NH_2 < (C_2H_5)_2NH < (C_2H_5)_3N < C_6H_5NH_2$

B. $C_6H_5NH_2 < C_2H_5NH_2 < (C_2H_5)_3N < (C_2H_5)_2NH$

C. $(C_2H_5)_3N < (C_2H_5)_2NH < C_6H_5NH_2 < C_2H_5NH_2$

D. $(C_2H_5)_2NH < (C_2H_5)_3N < C_2H_5NH_2 < C_6H_5NH_2$

88. Which of the following compounds will give Hell-Volhard-Zelinsky reaction?

A. $R-CH_2-COOH$

B. R_3C-CHO

C. R_2CO

D. $H-COOH$

89. Arrange the following acids in increasing order of their acidic strengths:

$HCOOH$, FCH_2COOH, NO_2CH_2COOH, $ClCH_2COOH$

A. $HCOOH < FCH_2COOH < NO_2CH_2COOH < ClCH_2COOH$

B. $HCOOH < NO_2CH_2COOH < ClCH_2COOH < FCH_2COOH$

C. $NO_2CH_2COOH < HCOOH < ClCH_2COOH < FCH_2COOH$

D. $HCOOH < ClCH_2COOH < FCH_2COOH < NO_2CH_2COOH$

90. In the following compounds, what is the increasing order of their reactivity towards nucleophilic addition reactions?

Benzaldehyde, p-Tolualdehyde, p-Nitrobenzaldehyde, Acetophenone

A. Benzaldehyde < p-Tolualdehyde < p-Nitro-benzaldehyde < Acetophenone

B. Acetophenone < Benzaldehyde < p-Tolualdehyde < p-Nitrobenzaldehyde

C. Acetophenone < p-Tolualdehyde < Benzaldehyde < p-Nitrobenzaldehyde

D. Benzaldehyde < Acetophenone < p-Tolualdehyde < p-Nitrobenzaldehyde

84. D **85.** A **86.** C **87.** B **88.** A **89.** D **90.** C

91. The Gatterman-Koch reaction is used in the industrial preparation of benzaldehyde. The electrophile involved in this reaction is:

A. CO^+
B. $HCl + CO_2 +$ anhydrous $AlCl_3$
C. HCO^+
D. $CO +$ anhydrous $AlCl_3$

92. Formaldehyde undergoes Cannizzaro reaction because:

(*a*) It has alpha-hydrogen atom.
(*b*) It does not have alpha-hydrogen atom.
(*c*) It does not undergo self-oxidation and reduction on heating with conentrated alkali.
(*d*) It undergo self-oxidation and reduction on heating with concentrated alkali.

Choose the **correct** answer from the options given below:

A. (*b*) and (*d*) only B. (*a*) and (*c*) only
C. (*b*) and (*c*) only D. (*a*) and (*d*) only

93. In the reaction, $(CH_3)_3C - O - CH_3 + HI \rightarrow$ Products

CH_3OH and $(CH_3)_3CI$ are the products and not CH_3I and $(CH_3)_3C - OH$. It is because:

(*a*) in step 2 of the reaction the departure of leaving group ($HO - CH_3$) creates less stable carbocation.
(*b*) in step 2 of the reaction the departure of leaving group ($HO - CH_3$) creates more stable carbocation.
(*c*) the reaction follows S_N1 mechanism.
(*d*) the reaction follows S_N2 mechanism.

Choose the **correct** answer from the options given below:

A. (*b*) and (*d*) only
B. (*b*) and (*c*) only
C. (*a*) and (*d*) only
D. (*a*) and (*c*) only

94. Aniline does not undergo Friedel-Crafts reaction because:

(*a*) It forms salt with the Lewis acid catalyst, $AlCl_3$.
(*b*) Nitrogen of aniline acquires negative charge.
(*c*) Nitrogen of aniline acquires positive charge.
(*d*) Nitrogen acts as a strong deactivating group in the further reaction.

Choose the **correct** answer from the options given below:

A. (*a*), (*b*) and (*d*) only
B. (*a*), (*b*) and (*c*) only
C. (*a*), (*c*) and (*d*) only
D. (*b*), (*c*) and (*d*) only

95. Although chlorine is an electron withdrawing group, yet it is ortho- and para-directing in electrophilic aromatic substitution reaction because:

(*a*) Chlorine withdraws electrons through inductive effect.
(*b*) Chlorine destabilises the intermediate carbocation formed during electrophilic substitution.
(*c*) Chlorine accepts electrons through resonance.
(*d*) Chlorine releases electrons through resonance.

Choose the **correct** answer from the options given below:

A. (*a*), (*b*) and (*d*) only
B. (*a*), (*b*) and (*c*) only
C. (*a*), (*c*) and (*d*) only
D. (*b*), (*c*) and (*d*) only

96. In Etard reaction, the final product is:

A. Aromatic aldehyde
B. Aromatic chloride
C. Aromatic amine
D. Aromatic alcohol

91. C **92.** A **93.** B **94.** C **95.** A **96.** A

97. Match List-I with List-II.

List-I	List-II
(*a*) Amino acids linked in a specific sequence	I. Primary structure of proteins
(*b*) Regular folding of a specific sequence of amino acids due to H-bonding	II. Secondary structure of proteins
(*c*) Fibrous proteins	III. Quaternary structure of proteins
(*d*) Spatial arrangement of two or more polypeptide chains	IV. Tertiary structure of proteins

Choose the **correct** answer from the options given below:

	(*a*)	(*b*)	(*c*)	(*d*)
A.	I	II	III	IV
B.	I	III	II	IV
C.	I	II	IV	III
D.	III	IV	I	II

98. Match List-I with List-II.

List-I	List-II
(*a*) Tollen's reagent	I. Rochelle salt
(*b*) Jones reagent	II. Conc. HCl and $ZnCl_2$
(*c*) Lucas reagent	III. Ammoniacal silver nitrate
(*d*) Fehling solution	IV. Chromium trioxide-sulphuric acid

Choose the **correct** answer from the options given below:

	(*a*)	(*b*)	(*c*)	(*d*)
A.	III	IV	II	I
B.	IV	III	I	II
C.	I	IV	II	III
D.	III	I	IV	II

99. Match List-I with List-II.

List-I	List-II
(*a*) Swarts Reaction	I. $C_6H_5NH_2 + NaNO_2 + HX + Cu_2X_2 \rightarrow C_6H_5X + N_2$
(*b*) Finkelstein reaction	II. $2RX + 2Na \rightarrow R - R + 2NaX$
(*c*) Sandmeyer's reaction	III. $RX + AgF \rightarrow R - F + Agx$
(*d*) Wurtz reaction	IV. $RX + NaI \rightarrow R - I + NaX$

Choose the **correct** answer from the options given below:

	(*a*)	(*b*)	(*c*)	(*d*)
A.	I	II	III	IV
B.	I	III	II	IV
C.	I	II	IV	III
D.	III	IV	I	II

100. Match List-I with List-II.

List-I (Biomolecule)	List-II (Function/ Diseases)
(*a*) Vitamin A	I. Menstrual cycle
(*b*) Thiamine	II. Xerophthalmia
(*c*) Glucocorticoids	III. Beri-Beri
(*d*) Estradiol	IV. Addison's Disease

Choose the **correct** answer from the options given below:

	(*a*)	(*b*)	(*c*)	(*d*)
A.	III	II	I	IV
B.	II	III	I	IV
C.	III	II	IV	I
D.	II	III	IV	I

97. C **98.** A **99.** D **100.** D

BIOLOGY

101. Analogous structures are a result of:

A. Convergent evolution
B. Divergent evolution
C. Parallel evolution
D. Retrogressive evolution

102. Which of the following does **not** affect the Hardy-Weinberg equilibrium?

A. Natural selection
B. Genetic drift
C. Gene pool
D. Gene migration

103. Which of the following primates was more like an ape?

A. *Homo erectus*
B. *Dryopithecus*
C. *Australopithecines*
D. *Ramapithecus*

104. Nucleosome is associated with ______ molecules of histones.

A. Four
B. Nine
C. Two
D. Eight

105. Select the observations drawn from the human genome project which are correct.

(*a*) The human genome contains 3164.7 million bp.
(*b*) The average gene consists of 3000 bases.
(*c*) Total number of genes is estimated at 30,000.
(*d*) The functions are unknown for over 50% of discovered genes.
(*e*) Less than 2% of the genome codes for proteins.

Choose the **correct** answer from the options given below:

A. (*a*), (*b*), (*c*) and (*d*) only
B. (*a*), (*c*), (*d*) and (*e*) only
C. (*a*), (*c*) and (*e*) only
D. (*a*), (*b*), (*c*), (*d*) and (*e*)

106. Match List-I with List-II:

List-I (Placental Mammals)	**List-II (Counterpart Marsupials)**
(*a*) Anteater	I. Spotted cuscus
(*b*) Bobcat	II. Numbat
(*c*) Lemur	III. Flying Phalanger
(*d*) Flying squirrel	IV. Tasmanian tiger cat

Choose the **correct** answer from the options given below:

	(*a*)	(*b*)	(*c*)	(*d*)
A.	II	IV	I	III
B.	II	I	IV	III
C.	IV	I	II	III
D.	IV	I	III	II

107. Identify the **incorrect** statement/s:

(*a*) Intestinal perforation and death may occur in severe cases of typhoid infection.
(*b*) Common cold is caused by Rhinoviruses
(*c*) Lips and fingernails may turn grey to bluish colour in severe cases of pneumonia.
(*d*) Pneumonia is caused by *Salmonella*.
(*e*) Typhoid fever could be confirmed by Widal test.

Choose the answer from the options given below:

A. (*a*), (*c*) and (*d*) only
B. (*b*) and (*e*) only
C. (*d*) only
D. (*a*) and (*d*) only

101. A **102.** C **103.** B **104.** B **105.** D **106.** A **107.** C

108. Match List-I with List-II:

List-I (Types of barriers)	List-II (Examples)
(*a*) Cytokine barriers	I. Mucus coating
(*b*) Physical barriers	II. Tears from eyes
(*c*) Cellular barriers	III. Phagocytosis
(*d*) Physiological barriers	IV. Interferons

Choose the **correct** answer from the options given below:

	(*a*)	(*b*)	(*c*)	(*d*)
A.	IV	III	I	II
B.	II	IV	III	I
C.	II	I	IV	III
D.	IV	I	III	II

109. Smack is chemically:

A. Diacetyl morphine
B. Cocaine
C. Benzodiazepine
D. Amphetamine

110. Antibodies are secreted by:

A. T-Cells
B. B-Cells
C. α-Cells
D. β-Cells

111. In sewage treatment, flocs are:

A. the solids that settle during sedimentation.
B. the supernatant that is formed above the primary sludge.
C. the masses of bacteria associated with fungal filaments.
D. the bacteria which grow anaerobically and are also called anaerobic sludge digesters.

112. Match List-I with List-II:

List-I (Products)	List-II (Organisms)
(*a*) Statin	I. *Streptococcus*
(*b*) Clot buster	II. *Trihoderma*
(*c*) Swiss cheese	III. *Monascus*
(*d*) Cyclosporin-A	IV. *Propionibacterium*

Choose the **correct** answer from the options given below:

	(*a*)	(*b*)	(*c*)	(*d*)
A.	II	I	IV	III
B.	III	I	IV	II
C.	III	IV	II	I
D.	II	III	I	IV

113. The beetle used as a biocontrol agent for aphids and mosquitoes is:

A. *Trichoderma*
B. Dragonflies
C. Ladybird
D. Silver fish

114. Downstream processing method involves:

A. Identification
B. Amplification
C. Fermentation
D. Purification

115. Which of the following is **not** the correctly matched pair of organism and its respective cell wall degrading enzyme?

A. Fungi – Chitinase
B. Algae – Methylase
C. Plant cells – Cellulase
D. Bacteria – Lysozyme

116. Arrange the following steps involved in transformation of bacteria in a sequence from initiation to end.

(*a*) Incubation of rDNA with bacterial cell on ice
(*b*) Treatment with divalent cations
(*c*) Heat shock treatment
(*d*) Selection on antibiotic containing agar plate
(*e*) Placed them again on ice

Choose the **correct** answer from the options given below:

A. (*a*), (*b*), (*d*), (*c*), (*e*)
B. (*b*), (*a*), (*c*), (*e*), (*d*)
C. (*b*), (*c*), (*d*), (*a*), (*e*)
D. (*a*), (*c*), (*b*), (*d*), (*e*)

108. D **109.** A **110.** B **111.** C **112.** B **113.** C **114.** D **115.** B **116.** B

117. Which of the following statements are **incorrect**?

(*a*) Fragments of DNA can be separated by ELISA.

(*b*) Transformation is a procedure through which a piece of DNA is introduced in a host bacterium.

(*c*) Recombinant DNA technology does not involve isolation of a desired DNA fragment.

(*d*) DNA ligases are used for stitching DNA fragments into a vector.

Choose the answer from the options given below:

A. (*a*) and (*c*) only B. (*a*) and (*b*) only
C. (*b*) and (*c*) only D. (*a*), (*c*) and (*d*) only

118. Which of the following statements are **true**?

(*a*) Milk obtained from 'Rosie' is nutritionally more balanced for human babies than natural human milk.

(*b*) Biopiracy refers to the use of bioresources without proper authorisation from MNCs.

(*c*) GEAC is the decisive body for safety and validity of GMOs and GM research respectively.

(*d*) Transgenic animals help us to understand the contribution of genes in the development ot disease.

Choose the **correct** answer from the options given below:

A. (*a*) and (*c*) only B. (*c*) and (*d*) only
C. (*a*) and (*d*) only D. (*b*) and (*c*) only

119. Match List-I with List-II:

List-I (Transgene)	**List-II (Used for Products)**
(*a*) α-l-antitrypsin	I. Meloidegyne incognitia
(*b*) crylAc	II. Corn borer
(c) Antisense RNA	III. Treat emphysema
(d) crylAb	IV. Cotton bollworms

Choose the **correct** answer from the options given below:

	(*a*)	(*b*)	(*c*)	(*d*)
A.	III	IV	I	II
B.	I	II	III	IV
C.	III	II	I	IV
D.	I	IV	III	II

120. Expand "GEAC"

A. Genetic and Enviroment Advisory Committee

B. Gene Establisment Approval Committee

C. Genetic Engineering Advisory Committee

D. Genetic Engineering Approval Committee

121. When an insect feeds on the Bt plant, the insect dies due to the conversion of inactive protein to active protein in:

A. Alkaline pH of the gut

B. Acidic pH of the gut

C. Acidic pH of the saliva

D. Alkaline pH of saliva

122. Match List-I with List-II:

List-I (Interspecies Relationships)	**List-II (Features)**
(*a*) Commensalism	I. One species is benefitted at the expense of the other
(*b*) Mutualism	II. One species is harmed and the other is unaffected
(*c*) Amensalism	III. Both the species are benefitted
(*d*) Parasitism	IV. One species benefits and other remains unaffected

Choose the **correct** answer from the options given below:

	(*a*)	(*b*)	(*c*)	(*d*)
A.	I	II	III	IV
B.	IV	III	II	I
C.	II	I	III	IV
D.	III	IV	I	II

117. A **118.** B **119.** A **120.** D **121.** A **122.** B

123. In a country at any time, the population has the same number of youngs and mature ones. What type of growth does, it reflect?

A. Expanding B. Declining
C. Stable D. S-shaped

124. Two closely related species can coexist indefinitely and violate the Gause's '*Competitive Exclusion Principle*' by:

A. eliminating the inferior species.
B. resource partitioning.
C. interacting with each other symbiotically.
D. changing the area of grazing.

125. The process of mineralisation by microorganisms helps in the release of:

A. inorganic nutrients from detritus and formation of humus
B. organic nutrients from humus.
C. inorganic nutrients from humus.
D. organic and inorganic nutrients from detritus.

126. In which ecosystem is the biomass of primary consumers greater than producers?

A. Forests B. Grassland
C. Desert D. Sea

127. Choose the correct statements with respect to decomposition from the following:

(*a*) Decomposition is an anaerobic process.
(*b*) Decomposition rate of detritus depends upon the chemical nature of it.
(*c*) Water-soluble organic nutrients go into the soil and get precipitated in the process of leaching.
(*d*) Humification follows mineralisation.

Choose the **correct** answer from the options given below:

A. (*b*) and (*d*) only
B. (*a*) and (*c*) only
C. (*b*) and (*c*) only
D. (*a*) and (*d*) only

128. Match List-I with List-II:

List-I (Concepts)	**List-II (Explanation)**
(*a*) Standing state	I. Available biomass for the consumption of heterotrophs
(*b*) Secondary productivity	II. Rate of formation of organic matter by consumers
(*c*) Standing crop	III. Mass of living matter in a trophic level at a given time
(*d*) Net primary productivity	IV. Amount of mineral nutrients in the soil at a given time

Choose the **correct** answer from the options given below:

	(*a*)	(*b*)	(*c*)	(*d*)
A.	IV	III	II	I
B.	I	II	III	IV
C.	IV	II	III	I
D.	I	IV	II	III

Directions (Qs. No. 129-133): *Read the passage carefully and give the answer to the next five questions:*

India was amongst the first countries in the world to initiate action plans and programmes at a national level to attain total reproductive health as a social goal. These programmes called 'family planning' were initiated in 1951 and were periodically assessed over the past decades. Improved programmes covering wider reproduction-related areas are currently in operation. Creating awareness among the people about various reproduction-related aspects and providing facilities and support for building up a reproductively healthy society are the major tasks under these programmes.

129. Which of the following is **not** a Sexually Transmitted Disease?

A. Chlamydiasis B. Filariasis
C. Genital herpes D. Trichomoniasis

123. C	**124.** B	**125.** C	**126.** D	**127.** C	**128.** C	**129.** B

130. Which of the following statements is incorrect with respect to Medical Termination of Pregnancy?

A. They are considered safe during the first, trimester
B. It is legalised in India from 1971.
C. MTPs can be performed even after 24 weeks, but with the opinion of 2 registered medical practitioners on specific grounds.
D. About 20% of the total number of conceived pregancies undergo MTP in a year globally.

131. Match List-I with List-II:

List-I (Various Assisted Reproductive Technologies)	**List-II (Process Involved)**
(*a*) ZIFT	I. Formation of embryo in vitro by injecting sperm directly into ovum
(*b*) ICSI	II. Transferring of embryo with more than 8 blastomeres into the uterus
(*c*) IUI	III. Transferring of fertilised egg up to 8 blastomeres into fallopian tube
(*d*) IUT	IV. Transfer of semen from a healthy donor into the uterus Artificially

Choose the **correct** answer from the options given below:

	(*a*)	(*b*)	(*c*)	(*d*)
A.	III	I	II	IV
B.	III	I	IV	II
C.	II	III	IV	I
D.	IV	III	I	II

132. Which of the following methods of contraception is not meant for females?

A. IUDs
B. Lactational amenorrhea
C. Vasectomy
D. Condoms

133. 'Saheli' an oral contraceptive pill, also known as the 'Once a week' pill, was developed by:

A. AIIMS
B. NBRl
C. CDRI
D. NBPG

Directions (Qs. No. 134-138): *Read the passage carefully and give the answer to the next five questions:*

Does the number of species in a community really matter to the functioning of the ecosystem? This is a question for which ecologists have not been able to give a definitive answer. For many decades, ecologists believed that communities with more species, generally, tend to be more stable than those with less species. According to the International Union for Conservation of Nature and Natural Resources (IUCN) (2004), the total number of plant and animal species described so far is more than 1.5 million.

134. Which of the following is **not** a characterstic of a stable biological community?

A. It must be resistant to invasions by alien species.
B. It should not show too much variation in productivity from year to year.
C. All the species are equally important in a stable community and absence of any one leads to its unstability.
D. It is resilient to occasional disturbances, whether natural or man-made.

135. In 'rivet popper hypothesis' the 'rivet' signifies:

A. Key species
B. Endemic species
C. Community
D. Species

130. C **131.** B **132.** C **133.** C **134.** C **135.** D

136. The scientist who proved that species richness directly correlates with the stability of a community, was _____ .

A. Paul Ehrlich B. David Tilman
C. Robert May D. Edward Wilson

137. Among the vertebrates, which of the following is the most species-rich group?

A. Reptiles B. Fishes
C. Insects D. Mammals

138. The following are the various hypotheses proposed in explaining the greatest biological diversity in tropics *except*:

A. Temperate regions are subjected to glaciations, but tropical latitudes have remained relatively undisturbed.
B. Tropical environments have more humidity/moisture which helps the diversity to flourish.
C. Tropical environments are less seasonal and more constant.
D. There is more solar energy available in the tropics which contributes to higher productivity and hence, biodiversity.

139. Cells present in the mature pollen grains are _______.

A. Central cell and generative cell
B. Antipodal cell and vegetative cell
C. Vegetative cell and generative cell
D. Filiform cell and micropylar cell

140. Match List-I with List-II:

List-I (Structures)	List-II (Functions)
(*a*) Filiform apparatus	I. Made up of sporopollenin
(*b*) Tapetum	II. Attachment of ovule to the placenta
(*c*) Exine	III. Guides pollen tube into the synergid
(*d*) Funicle	IV. Nourishes the pollen grain

Choose the **correct** answer from the options given below:

	(*a*)	(*b*)	(*c*)	(*d*)
A.	IV	I	II	III
B.	III	IV	I	II
C.	II	I	III	IV
D.	I	III	IV	II

141. Primary Endosperm Nucleus is the product of:

A. Double fusion B. Triple fusion
C. Parthenogenesis D. Apomixis

142. In humans, mammary gland is divided into _____ lobes.

A. 10 - 12 B. 25 - 30
C. 30 - 35 D. 15 - 20

143. Sex in human embryo is determined by:

A. 'X' chromosome of egg
B. 'X' or 'Y' chromosome of sperm
C. Only 'Y' chromosome of sperm
D. Health of mother

144. Arrange the following stages of oogenesis in order of their occurrence.

(*a*) Ovum
(*b*) Oogonia
(*c*) Primary oocyte
(*d*) Secondary oocyte

Choose the **correct** answer from the options given below:

A. (*c*), (*b*), (*d*), (*a*)
B. (*b*), (*c*), (*d*), (*a*)
C. (*d*), (*c*), (*a*), (*b*)
D. (*a*), (*d*), (*c*), (*b*)

145. Which of the following pair of contrasting traits was not studied by Mendel?

A. Pink and white flowers
B. Inflated and constricted pods
C. Axial and terminal flowers
D. Green and yellow pods

136. B **137.** B **138.** B **139.** C **140.** B **141.** B **142.** D **143.** B **144.** B **145.** A

146. Failure of chromatids to segregate during cell division cycle results in:

A. Polyploidy B. Euploidy
C. Aneuploidy D. Autopolyploidy

147. Select the correctly matched pair about sickle cell anaemia:

Genotype	:	**Phenotype**
(*a*) Hb^A Hb^A	:	Diseased phenotype
(*b*) Hb^A Hb^S	:	Diseased phenotype
(*c*) Hb^S Hb^S	:	Diseased phenotype
(*d*) Hb^S Hb^A	:	Carrier of disease

Choose the **correct** answer from the options given below:

A. (*c*) and (*d*) only
B. (*a*) and (*c*) only
C. (*b*), (*c*) and (*d*) only
D. (*a*), (*b*) and (*c*) only

148. Match List-I with List-II:

List-I (Scientists)	**List-II (Discovery)**
(*a*) Sutton and Boveri	I. X-Body
(*b*) Sturtevant	II. Chromosomal Theory of Inheritance
(*c*) Henking	III. Transforation in bacteria
(*d*) Griffith	IV. Genetic maps

Choose the **correct** answer from the options given below:

	(*a*)	(*b*)	(*c*)	(*d*)
A.	II	IV	I	III
B.	II	I	IV	III
C.	I	III	II	IV
D.	IV	I	III	II

149. Which of the following statements are incorrect with respect to nucleotides?

(*a*) Purines and pyrimidines are nitrogenous bases.
(*b*) Nucleotides are non-enzymatic molecules.
(*c*) Phosphate group is linked to –OH of 5′ C of a nucleoside through phosphoester linkage.
(*d*) In RNA, every nucleotide residue has an additional – OH group present at 2′ position in the ribose.
(*e*) Thymine is an example of Pyrimidine.

Choose the **correct** answer from the options given below:

A. (*a*), (*b*) and (*e*) only
B. (*d*) and (*e*) only
C. (*b*) and (*d*) only
D. (*b*) and (*e*) only

150. Arrange the given steps of DNA fingerprinting in the sequence from initiation to end.

(*a*) Digestion of DNA by restriction endonuclease
(*b*) Isolation of DNA
(*c*) Hybridisation using labelled VNTR probe
(*d*) Transferring (blotting) of separated DNA fragments to synthetic membrane

Choose the **correct** answer from the options given below:

A. (*a*), (*b*), (*c*), (*d*)
B. (*a*), (*d*), (*c*), (*b*)
C. (*b*), (*a*), (*d*), (*c*)
D. (*c*), (*a*), (*b*), (*d*)

---

146. C **147.** A **148.** A **149.** * **150.** D

Previous Years' Paper

B.Sc. BIOTECHNOLOGY

Entrance Exam, 2023*

PHYSICS

1. A converging lens is used to form image on a screen. When the lower half of the lens is covered by an opaque screen then, the correct statements are:

(*a*) Intensity of Image will decrease.
(*b*) Intensity of Image will increase.
(*c*) Complete Image will be formed.
(*d*) Half the Image will disappear.

Choose the **correct** answer from the options given below:

A. (*a*), (*b*) and (*c*) only
B. (*a*), (*c*) and (*d*) only
C. (*a*) and (*c*) only
D. (*b*) and (*c*) only

2. In Young's double slit experiment, the ratio of slit widths is 4 : 1. The intensity ratio in the interference pattern would be:

A. 1 : 2 B. 1 : 3
C. 9 : 1 D. 1 : 9

3. The instantaneous magnetic flux linked with a coil is given by $\phi = (15t^3 - 100t^2 + 40)$ Wb. The emf induced in the coil at time $t = 1$ s is:

A. +150 V
B. –155 V
C. –150 V
D. +155 V

4. Two antennas A and B have maximum line of sight range $R_{AB} = 45.5$ km. If the height of antenna A is 32 m then the height of antenna B is: (Given radius of earth is = 6400 km)

A. 25 m
B. 100 m
C. 50 m
D. 75 m

5. When strong magnetic field is applied on a stationary electron, the electron:

A. moves in the direction of the field.
B. moves in opposite directions of the field.
C. starts spinning.
D. remains stationary.

6. The current I flowing in the circuit is:

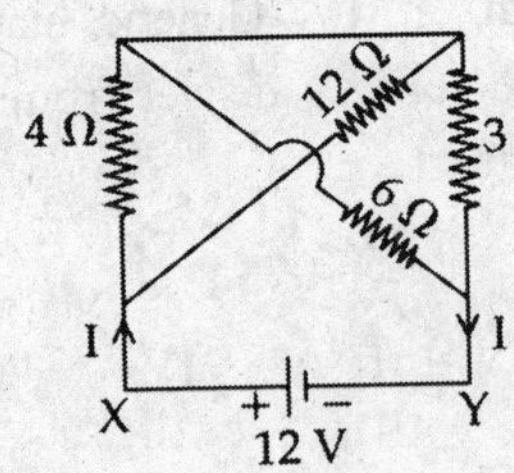

A. 2.5 A
B. 2.4 A
C. 2.2 A
D. 2.0 A

1. C	2. C	3. D	4. C	5. D	6. B

* *Conducted by NTA for CUET (UG).*

7. Match List-I with List-II:

List-I (Scientist)	List-II (Phenomenon)
(*a*) Rutherford	(*i*) Nuclear Model of atom
(*b*) A.H. Becquerel	(*ii*) Plum-Pudding Model of atom
(*c*) Niels Bohr	(*iii*) Radioactivity
(*d*) J.J. Thomson	(*iv*) Hydrogen atom model

Choose the **correct** answer from the options given below:

	(*a*)	(*b*)	(*c*)	(*d*)
A.	(*i*)	(*iii*)	(*iv*)	(*ii*)
B.	(*iii*)	(*i*)	(*ii*)	(*iv*)
C.	(*iv*)	(*ii*)	(*iii*)	(*i*)
D.	(*i*)	(*ii*)	(*iii*)	(*iv*)

8. A carrier wave of peak voltage 24 V is used to transmit a message signal. What should be the peak Voltage of the modulating signal in order to have a modulation index of 70%?

A. 16.8 V B. 2.9 V
C. 18 V D. 16.8×10^2 V

9. When a potential difference is applied across the ends of a conductor, the electric field exists:

A. inside the conductor.
B. outside the conductor.
C. both inside and outside the conductor.
D. no where.

10. Huygen's wave theory of light could not explain the phenomenon of:

A. interference
B. diffraction
C. rectilinear propagation of light
D. photoelectric effect

11. The de Broglie wavelength of an electron having kinetic energy 56 eV is:

A. 0.022 nm B. 16.4 Å
C. 0.164 nm D. 0.0164 nm

12. The following that is **not** transported by Electromagnetic waves is:

A. Momentum B. Energy
C. Charge D. Information

13. The current in a circuit having constant resistance is tripled. The power dissipation will become:

A. 6 times B. 9 times
C. 3/2 times D. 1/2 times

14. Match List-I with List-II:

List-I (EM waves)	List-II (Wavelength Range)
(*a*) Light	(*i*) 1 mm to 10^{-3} nm
(*b*) Infra-red	(*ii*) 700 nm to 400 nm
(*c*) X-rays	(*iii*) 400 nm to 1 nm
(*d*) UV rays	(*iv*) 1 mm to 700 nm

Choose the **correct** answer from the options given below:

	(*a*)	(*b*)	(*c*)	(*d*)
A.	(*i*)	(*ii*)	(*iii*)	(*iv*)
B.	(*ii*)	(*iv*)	(*i*)	(*iii*)
C.	(*iii*)	(*ii*)	(*iv*)	(*i*)
D.	(*iv*)	(*i*)	(*iii*)	(*ii*)

15. For a photodiode, the I-V characteristics for different illumination intensities I_1, I_2, I_3 and I_4 were plotted. Which of the following is correct?

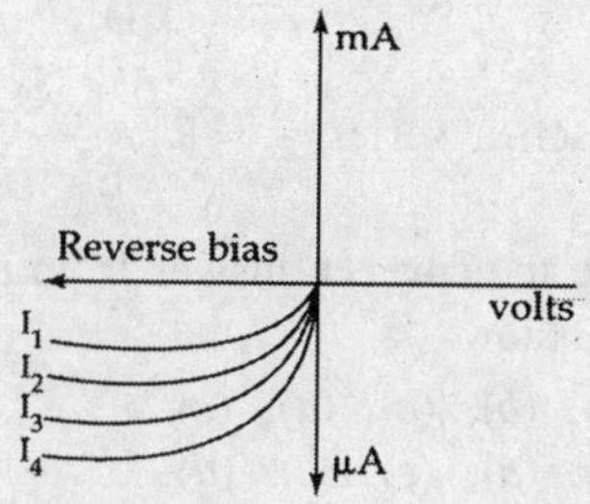

A. $I_1 > I_2 > I_3 > I_4$
B. $I_4 > I_3 > I_2 > I_1$
C. $I_2 > I_1 > I_3 > I_4$
D. $I_4 > I_3 > I_1 > I_2$

7. A	**8.** A	**9.** A	**10.** D	**11.** C	**12.** C	**13.** B	**14.** B	**15.** B

16. A positively charged glass rod attracts an object. The object must be:

A. negatively charged
B. neutral
C. either negatively charged or neutral
D. a magnet

17. When a current of 5 A flows through an ammeter, it gives full deflection. The resistance of ammeter is 20 Ω. If the same ammeter is to be used for measuring a maximum current of 10 A, then the ammeter must be connected to a resistance of:

A. 40 Ω in series
B. 40 Ω in parallel
C. 20 Ω in series
D. 20 Ω in parallel

18. The correct sequence (*a*), (*b*), (*c*), (*d*), (*e*) with reference to atomic spectral series is:

(*a*) Lyman series $\frac{1}{\lambda} = R\left(\frac{1}{A^2} - \frac{1}{n^2}\right)$

(*b*) Balmer series $\frac{1}{\lambda} = R\left(\frac{1}{B^2} - \frac{1}{n^2}\right)$

(*c*) P fund series $\frac{1}{\lambda} = R\left(\frac{1}{C^2} - \frac{1}{n^2}\right)$

(*d*) Bracket series $\frac{1}{\lambda} = R\left(\frac{1}{D^2} - \frac{1}{n^2}\right)$

(*e*) Paschen series $\frac{1}{\lambda} = R\left(\frac{1}{E^2} - \frac{1}{n^2}\right)$

Choose the **correct** answer from the options given below:

A. (*a*), (*b*), (*c*), (*d*), (*e*)
B. (*b*), (*a*), (*e*), (*c*), (*d*)
C. (*a*), (*b*), (*e*), (*d*), (*c*)
D. (*e*), (*d*), (*c*), (*b*), (*a*)

19. A radioactive isotope has half-life of 'K' years. How long will it take, so that activity reduces to 6.25% of its original value?

A. 5 K
B. 4 K
C. 3 K
D. 2 K

20. The linear magnifications of the object in the conjugate positions of a convex lens are 4 and 2, respectively. The distance between the conjugate positions is 40 cm. This convex lens of glass (μ = 1.5) is immersed in water (μ = 1.33). The new focal length of convex lens in water is:

A. 40 cm
B. 20 cm
C. 60 cm
D. 80 cm

21. The angle of dip at a location in Southern India is about 18°. Which of the following is correct for dip angle in Britain?

(*a*) Dip angle will be greater in Britain than in India.
(*b*) Dip angle will be smaller in Britain than in India.
(*c*) Nearer the place to the poles, lesser is the dip angle.
(*d*) Nearer the place to the poles, more is the dip angle.
(*e*) Dip angle is also called the angle of declination.

Choose the **correct** answer from the options given below:

A. (*a*) and (*d*) only
B. (*b*) and (*c*) only
C. (*a*), (*c*) and (*e*) only
D. (*d*) and (*e*) only

22. Wavelength of matter wave is independent of:

A. momentum
B. charge
C. velocity
D. mass

16. C	**17.** D	**18.** C	**19.** B	**20.** D	**21.** A	**22.** B

23. Match List-I with List-II:

List-I	List-II
(*a*) Electric potential (V) due to a single charge	(*i*) $\propto r^0$
(*b*) Electric flux due to a positive point charge through sphere of radius *r*	(*ii*) $\propto \frac{1}{r^2}$
(*c*) Electric intensity (E) due to a point charge	(*iii*) $\propto \frac{1}{r}$
(*d*) Electric intensity (E) due to dipole	(*iv*) $\propto \frac{1}{r^3}$

Choose the **correct** answer from the options given below:

	(*a*)	(*b*)	(*c*)	(*d*)
A.	(*iv*)	(*ii*)	(*iii*)	(*i*)
B.	(*iii*)	(*i*)	(*ii*)	(*iv*)
C.	(*iii*)	(*iv*)	(*ii*)	(*i*)
D.	(*iii*)	(*iv*)	(*i*)	(*ii*)

24. According to Einstein's photoelectric equation:

A. $K_{max} = h\nu - \phi_o$

B. $h\nu = K_{max} - \phi_o$

C. $K_{max} = h\nu + \phi_o$

D. $\phi_o = K_{max} + h\nu$

(Where all symbols have their usual meaning)

25. An uncharged capacitor is connected to a battery. How much amount of energy supplied by the battery is lost as heat while charging the capacitor?

A. $\frac{1}{4}QV$ B. zero

C. $\frac{1}{2}QV$ D. $\frac{1}{3}QV$

26. A wire of length 4 m carrying current I A is bent in the form of a circle. Its magnetic moment will be:

A. $\frac{IL}{4\pi}$ B. $\frac{IL^2}{4\pi}$

C. $\frac{I^2L^2}{4\pi}$ D. $\frac{LI^2}{4\pi}$

27. The graph showing the distribution of kinetic energy of electrons emitted during beta decay would be:

A. 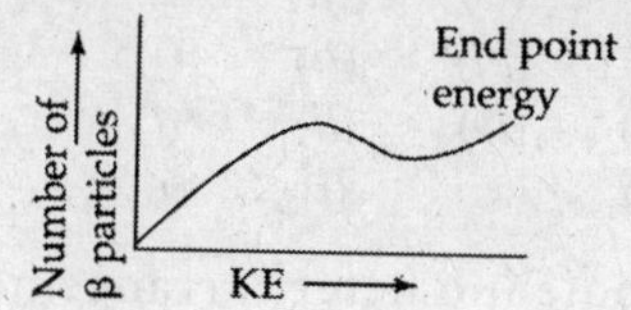

B. 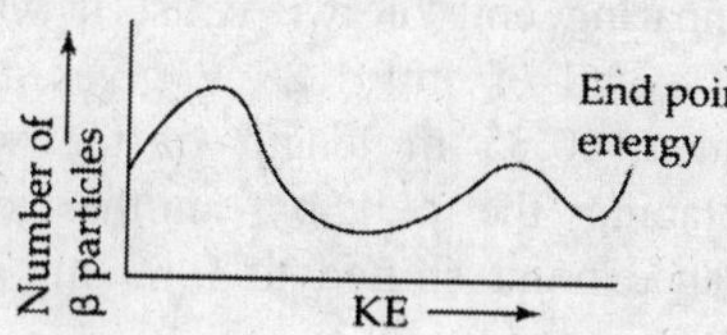

C. 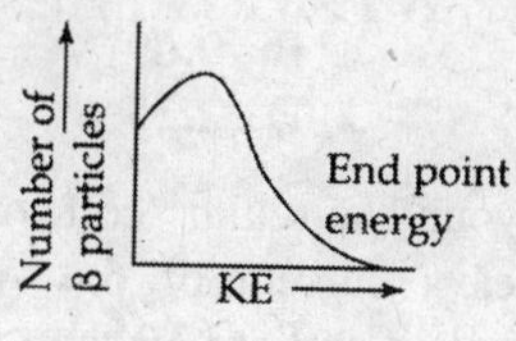

D. 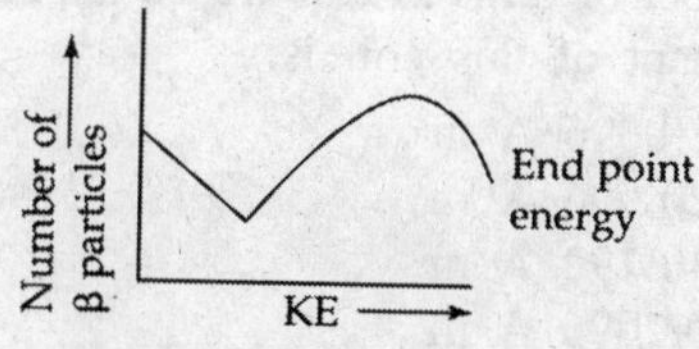

28. The number of electrons in one coulomb of electric charge is:

A. 6.25×10^{-18}

B. 6.25×10^{18}

C. 1.6×10^{-19}

D. $1.6 \times 10^{+19}$

23. B **24.** A **25.** C **26.** B **27.** C **28.** B

29. Match List-I with List-II.

List-I (Electronic Device)	List-II (Application)
(*a*) Zener diode	(*i*) Rectifier
(*b*) Capacitor	(*ii*) Amplifier
(*c*) Transistor	(*iii*) Voltage regulator
(*d*) p-n junction diode	(*iv*) Filter circuit

Choose the **correct** answer from the options given below:

	(*a*)	(*b*)	(*c*)	(*d*)
A.	(*i*)	(*ii*)	(*iii*)	(*iv*)
B.	(*iv*)	(*iii*)	(*ii*)	(*i*)
C.	(*iii*)	(*iv*)	(*ii*)	(*i*)
D.	(*iv*)	(*iii*)	(*i*)	(*ii*)

30. In a potentiometer arrangement for comparing emf of two cells, it was found that a cell of emf 1.25 V gives a balance point at 0.35 m length of the wire. On replacing the cell by another cell the balance point shifted to 0.63 m. The emf of the second cell is:

A. 1.5 V B. 0.69 V
C. 1.75 V D. 2.25 V

31. A closely wound circular coil of radius 10 cm carries a current of 6.4 A. If the number of turns in coil are 10 then magnetic moment of this coil is:

A. 2.0096 A m^2
B. 20.096 A m^2
C. 40.192 A m^2
D. 0.2096 A m^2

32. A transformer has 4000 turns in primary and 100 turns in secondary coils. The primary coil is connected to a 240 V and secondary to a bulb of 1.2 Ω resistance. The current in the bulb is:

A. 0.50 A B. 0.25 A
C. 0.20 A D. 5.00 A

33. Arrange the following elements of a generalised communication system in the sequencial order:

(*a*) Channel
(*b*) Information source
(*c*) Transmitter
(*d*) Receiver
(*e*) User of information

Choose the correct answer from the options given below:

A. (*b*), (*d*), (*a*), (*c*), (*e*)
B. (*e*), (*d*), (*c*), (*b*), (*a*)
C. (*a*), (*b*), (*c*), (*d*), (*e*)
D. (*b*), (*c*), (*a*), (*d*), (*e*)

34. Six point charges are kept at the vertices of a regular hexagon of side *a* and centre O is shown. The correct statement(s) are:

$$\left(\text{Given } K = \frac{1}{4\pi \epsilon_0} \frac{q}{a^2}\right).$$

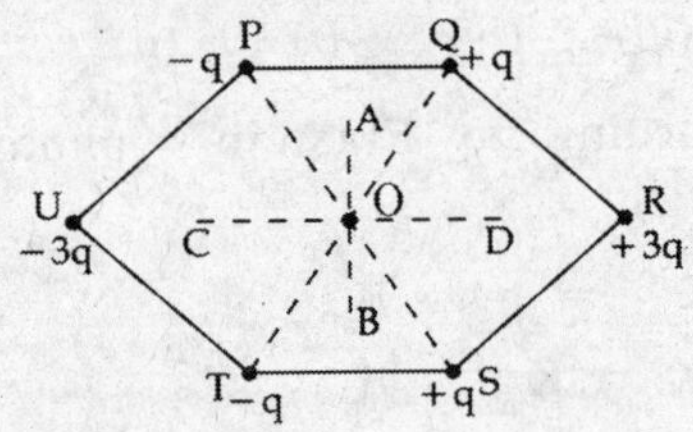

(*a*) Net electric field at O is 8K along OU.
(*b*) Electric field at O will be zero if all charges are of same magnitude and sign.
(*c*) Electric potential at all points on the line AB is same.
(*d*) Electric potential at all points on the line CD is same.
(*e*) Electric potential at centre O is zero.

Choose the **correct** answer from the options given below:

A. (*a*) and (*b*) only
B. (*a*), (*b*) and (*e*) only
C. (*b*), (*c*) and (*e*) only
D. (*a*), (*b*), (*c*) and (*e*) only

29. C **30.** D **31.** A **32.** D **33.** D **34.** D

35. If ν is the frequency of ac mains, then the frequency of sonometer wire at resonance is:

A. $\frac{\nu}{2}$ B. ν

C. 2ν D. 4ν

36. Two sources are producing waves as given in the options below. Which two sources are called coherent?

A. of same velocity
B. of equal wavelength
C. having a constant phase difference
D. having wavefront of same shape

37. A battery with potential V across it is connected to a combination of two identical resistors, and then has current i through it. What are the potential difference across and the current through either resistor if the resistors are connected in series?

A. V, $\frac{i}{2}$ B. $\frac{V}{2}, \frac{i}{2}$

C. $\frac{V}{2}, i$ D. V, i

38. The speed of an electromagnetic wave in a medium of dielectric constant 2.25 and relative permeability 4 is:

A. 3×10^8 m s^{-1}
B. 9×10^8 m s^{-1}
C. 10^8 m s^{-1}
D. 1.33×10^8 m s^{-1}

39. The radii of curvature of the faces of a double convex lens of focal length 12 cm made up of glass (μ = 1.5) are 10 cm and 'p' cm respectively. The value of p is:

A. 10 cm B. 15 cm
C. 30 cm D. 20 cm

40. For an LCR circuit:

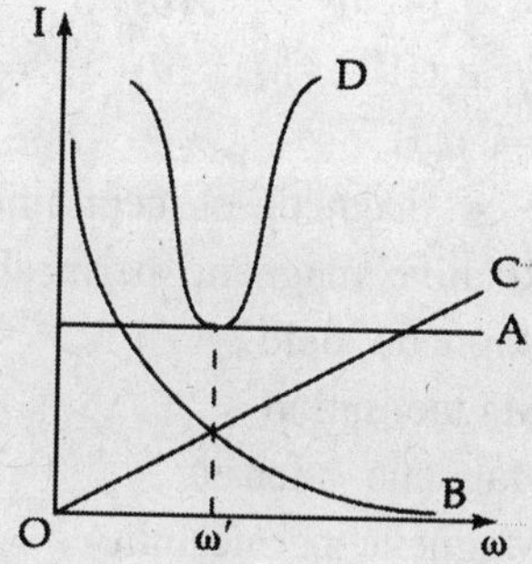

(*i*) C and D represent R and Z, respectively.
(*ii*) A, B, C, D represent R, X_C, X_L and Z, respectively.
(*iii*) C and D represent Z and R, respectively.
(*iv*) A and B represent R and X_L, respectively.
(*v*) For ω = ω′, the phase difference between current and voltage is zero.

(R, Z, X_C, X_L, ω, ω′ have their usual meaning)

Choose the **correct** answer from the options given below:

A. (*i*), (*ii*) only
B. (*ii*), (*iv*) only
C. (*ii*), (*v*) only
D. (*iii*), (*v*) only

41. A metal foil of negligible thickness is introduced between two parallel plates of a capacitor at the centre. The new capacitance of the capacitor would be:

A. C B. 4C

C. $\frac{1}{2}$C D. 2C

42. The essential condition for total internal reflection to occur is:

A. angle of incidence i > critical angle i_c
B. angle of incidence i < critical angle i_c
C. angle of incidence i = critical angle i_c
D. angle of incidence i = 90°

35. C	**36.** C	**37.** C	**38.** C	**39.** B	**40.** C	**41.** A	**42.** A

43. The correct statement given below are:

(*a*) $X = 1 + \mu_r$ (*b*) $\mu_r = 1 + X$

(*c*) $B = \mu_0(H + M)$ (*d*) $H = B_0\mu_0$

(*e*) $B_0 = \mu_0 H$

Here X = magnetic susceptibility

μ_r = Relative magnetic permeability

B = Magnetic field

M = Magnetisation

H = Magnetic intensity

μ_0 = Magnetic permeability

Choose the **correct** answer from the options given below:

A. (*b*) only

B. (*a*), (*b*) and (*c*) only

C. (*b*), (*c*) and (*e*) only

D. (*b*), (*c*) and (*d*) only

44. What will be the half life of a radioactive isotope whose $\left(\frac{1}{16}\right)^{th}$ of initial amount remains unchanged after 2 hours?

A. 60 min B. 45 min

C. 30 min D. 15 min

45. If the forward voltage in a p-n junction diode is increased, the width of the depletion region:

A. increases B. decreases

C. fluctuates D. no change

46. The current flowing through an inductor of self inductance L is continuously increasing. The correct graph, showing the variation of induced emf(e) versus $\frac{dI}{dt}$?

A. 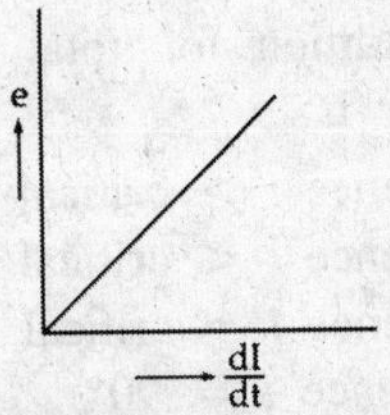

B. 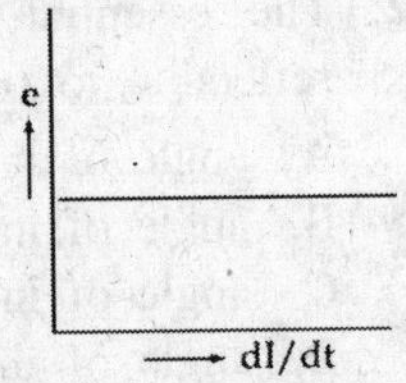

C. 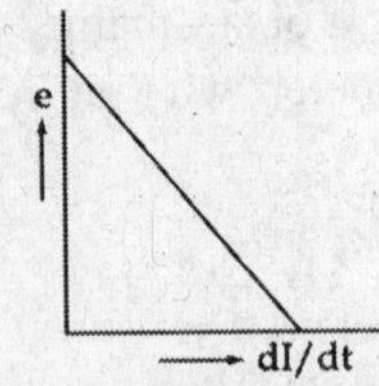

D. 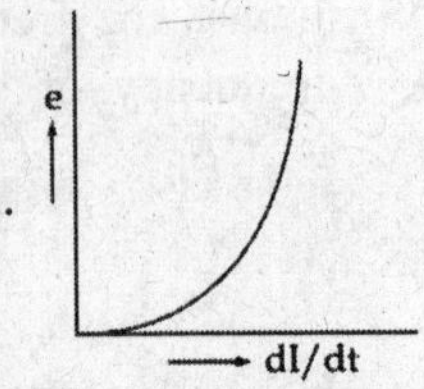

47. Which of the following statements best describe the temperature dependence of resistances of Cu and Si in the temperature range 27° to 127°C.

A. Linear decrease for both Cu and Si

B. Linear increase for both Cu and Si

C. Linear decrease for Cu and exponential increase for Si

D. Linear increase for Cu and exponential decrease for Si

48. A student has drawn the following courses of rays through a glass prism. The one which represents the position of minimum deviation is:

A. 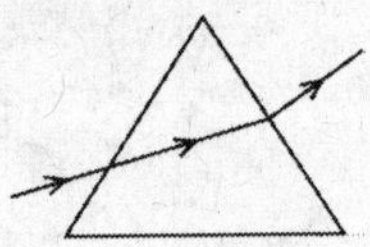

B. 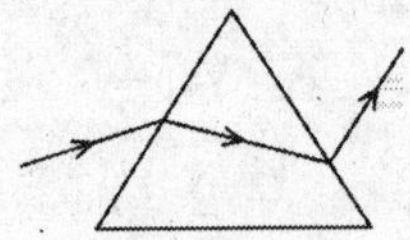

C. 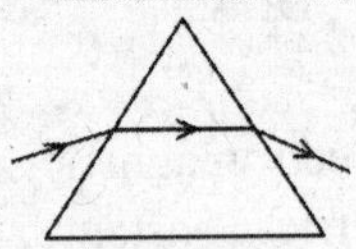

D. 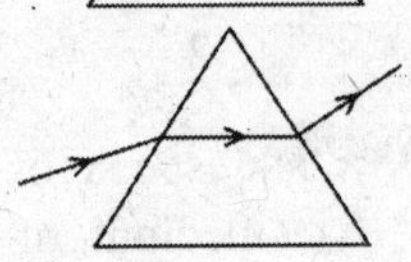

49. Three charges $+q$, $-2q$ and $4q$ are placed on the vertices of an equilateral triangle of side 0.2 m. The electrostatic potential energy of the system (Take $q = 10\ \mu C$ is:

A. -6.3×10^{-3} J B. -9.0×10^{-3} J

C. -27 J D. $+27$ J

50. Two solenoids of equal number of turns having their lengths and radii in the same ratio 1 : 3. The ratio of their self inductances $L_1 : L_2$ will be:

A. 1 : 27 B. 3 : 1

C. 1 : 9 D. 9 : 1

43. C **44.** C **45.** B **46.** B **47.** D **48.** C **49.** C **50.** A

CHEMISTRY

51. Match List-I with List-II.

List-I	List-II
(*a*) Williamsons Synthesis	I. $C_6H_5Cl + NaOH \xrightarrow[\text{High pressure}]{350°C}$
(*b*) Reimer Tiemann reaction	II. $C_6H_5OH + C_6H_5COCl \xrightarrow{\text{aq NaOH}}$
(*c*) Dow's process	III. $CH_3CH_2ONa + CH_3Cl \xrightarrow{\text{heat}}$
(*d*) Schotten Baumann reaction	IV. $C_6H_5OH + CHCl_3 + NaOH \xrightarrow{\text{heat}}$

Choose the **correct** answer from the options given below:

	(*a*)	(*b*)	(*c*)	(*d*)
A.	III	IV	I	II
B.	I	II	III	IV
C.	IV	III	II	I
D.	II	III	I	IV

52. A + B → Nylon 6, 6

Predict A and B:

A. Ethylene glycol + adipic acid

B. Pentamethylene diamine + adipic acid

C. Hexamethylene amine + tartaric acid

D. Hexamethylene diamine + adipic acid

53. Which of the following compound is the most acidic?

A. 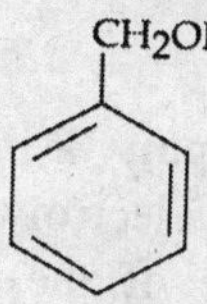

B. 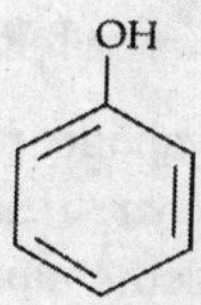

C. 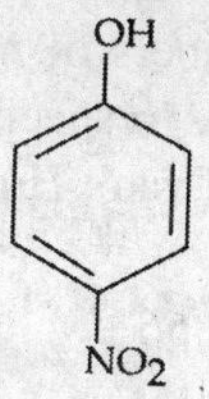

D. 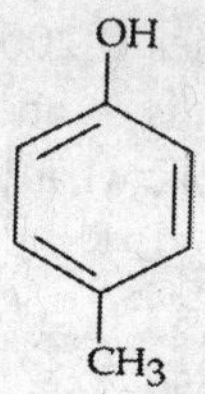

54. $Ph-\underset{\underset{(+)}{Cl}}{CH}-CH_3 \xrightarrow[\Delta]{\text{alc.KCN}} ?$

Product formed is:

A. $Ph-\underset{\underset{(\pm)}{CN}}{CH}-CH_3$

B. $Ph-\underset{\underset{(-)}{CN}}{CH}-CH_3$

C. $Ph-CH=CH_2$

D. $Ph-\underset{\underset{(+)}{CN}}{CH}-CH_3$

55. Which of the following hormone is also known as flight or fight hormone?

A. Adrenaline B. Androgens

C. Thyroxine D. Insulin

56. Match List-I with List-II.

List-I	List-II
(*a*) Adenine	I. Purine
(*b*) Thymine	II. Pyrimidine
(*c*) Uracil	III. DNA
(*d*) Deoxyribose	IV. RNA

Choose the **correct** answer from the options given below:

	(*a*)	(*b*)	(*c*)	(*d*)
A.	II	III	IV	I
B.	I	II	IV	III
C.	III	IV	I	II
D.	II	III	I	IV

57. Which of the following is the correct range of critical micelle concentrations (CMC) for soaps?

A. 10^2 to 10^4 mol/L

B. 10^{-5} to 10^{-8} mol/L

C. 10^{-1} to 10^{-3} mol/L

D. 10^{-4} to 10^{-3} mol/L

51. A **52.** D **53.** C **54.** C **55.** A **56.** B **57.** D

58. Which of the following is an example of negatively charged sol?

A. $Al_2O_3.xH_2O$ B. As_2S_3
C. TiO_2 D. Blood

59. Identify incorrect statement about $H_2 - O_2$ fuel cell from the following:

A. At cathode O_2 is reduced to OH^-
B. At anode H_2 is oxidised to form water
C. The cell runs continuously even if reactants are not supplied
D. Water is formed as a by product

60. The correct statement about 4f-series elements is:

A. Paramagnetism falls to minimum in Nd.
B. Lu^{3+} ions are bright green in aqueous solution.
C. Promethium is artificially synthesised 4f series element.
D. Absorption bands of Lanthanoids are broad.

61. Which of the following has the highest flocculating power?

A. $[Fe(CN)_6]^{4-}$ B. PO_4^{3-}
C. SO_4^{2-} D. Cl^-

62. Match List-I with List-II.

List-I (Complex ion)	List-II (Hybridisation)
(*a*) $[CoF_6]^{3-}$	I. d^2sp^3
(*b*) $[Co(CN)_4]^{2-}$	II. sp^3d^2
(*c*) $[FeCl_4]^-$	III. sp^3
(*d*) $[Co(C_2O_4)_3]^{3-}$	IV. dsp^2

Choose the **correct** answer from the options given below:

	(*a*)	(*b*)	(*c*)	(*d*)
A.	II	III	IV	I
B.	I	IV	III	II
C.	II	IV	III	I
D.	I	III	IV	II

63. Match List-I with List-II.

List-I	List-II
(*a*) Cheese	I. Emulsions
(*b*) Cell fluids	II. Aerosol
(*c*) Hair cream	III. Gel
(*d*) Clouds	IV. Sol

Choose the **correct** answer from the options given below:

	(*a*)	(*b*)	(*c*)	(*d*)
A.	I	III	IV	II
B.	III	IV	I	II
C.	IV	I	III	II
D.	I	IV	II	III

64. Which of the following reagents, are used for denaturation of alcohol?

A. $CuSO_4$ and pyridine
B. $FeSO_4$ and $CHCl_3$
C. $FeCl_3$ and pyridine
D. $CuSO_4$ and freon

65. Which of the following is the correct formula of a colligative property?

A. $\dfrac{P_1^o - P_1}{P_1^o} = x_1$

B. $\Delta T_b = \dfrac{K_b \times 1000 \times w_2}{M_1 \times w_1}$

C. $\Delta T_f = \dfrac{K_f \times 1000 \times w_2}{M_2 \times w_1}$

D. $\pi = \Delta T_f RT$

66. The vapour pressure of pure Benzene is 0.900 bar. If a nonvolatile electrolytic solid is added, vapour pressure of the solution becomes 0.200 bar. If numbers of moles of solute is 1 and solvent is 2 respectively in solution, what is the value of van't Hoff factor (i)?

A. 2.33 B. 0.56
C. 1.56 D. 2.00

58. B	59. C	60. C	61. A	62. C	63. B	64. A	65. C	66. A

67. Which of the following alkyl halides will undergo S_N2 reaction most readily?

A. 1-Bromo-2,2-dimethylpropane
B. 1-Bromo-2-methylbutane
C. 1-Bromobutane
D. 1-Bromo-3-methylbutane

68. Match List-I with List-II.

List-I	List-II
(*a*) Tetracarbonylnickel(0)	I. $[CoCl_2(en)_2]^+$
(*b*) Tetraammineaquachloridocobalt(III) chloride	II. $K_2[Zn(OH)_4]$
(*c*) Potassium tetrahydroxidozincate(II)	III. $[Co(NH_3)_4(H_2O)Cl]Cl_2$
(*d*) Dichloridobis(ethane-1, 2-diamine)cobalt(III)	IV. $[Ni(CO)_4]$

Choose the **correct** answer from the options given below:

	(*a*)	(*b*)	(*c*)	(*d*)
A.	I	II	III	IV
B.	IV	III	II	I
C.	III	II	I	IV
D.	III	IV	II	I

69. For the reaction, R → P, the $\log([R]_o/[R])$ verses time (t) plot is given below. What is the order of reaction and what are the units of rate constant?

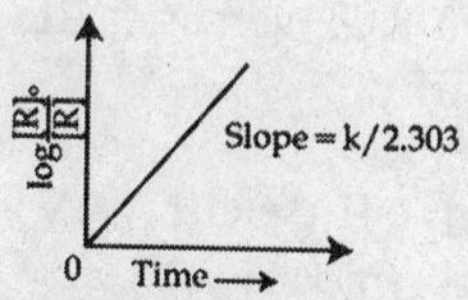

A. Zero order, s^{-1}
B. First order, s^{-1}
C. Second order, $mol^{-1}\ L\ s^{-1}$
D. Zero order, $mol\ L^{-1}\ s^{-1}$

70. An analysis of iron oxide shows that its formula is $Fe_{1.9}O_3$. What percentages of iron exists as Fe^{3+} and Fe^{4+} ions:

A. Fe^{3+} = 82.98% and Fe^{4+} = 17.02%
B. Fe^{3+} = 84.21% and Fe^{4+} = 15.79%
C. Fe^{3+} = 15.79% and Fe^{4+} = 84.21%
D. Fe^{3+} = 17.02% and Fe^{4+} = 82.98%

71. 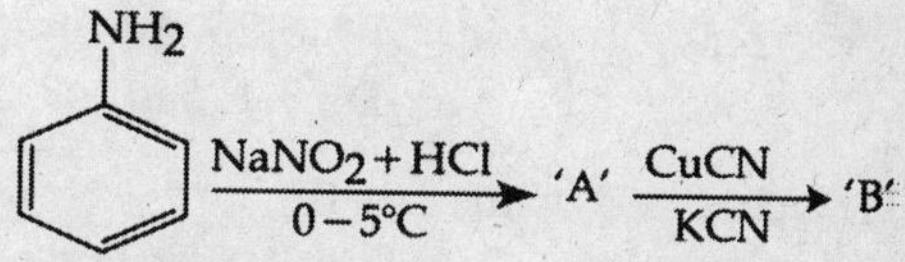

The compound B is:

A. Acetanilide B. Benzonitrile
C. Benzamide D. Benzene

72. Match List-I with List-II.

List-I

(*a*) 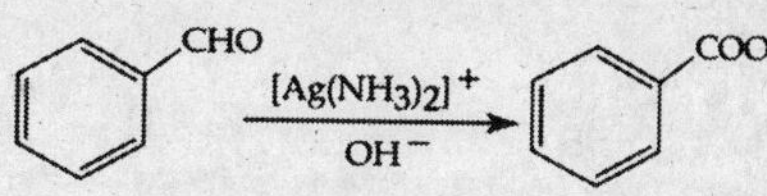

(*b*) 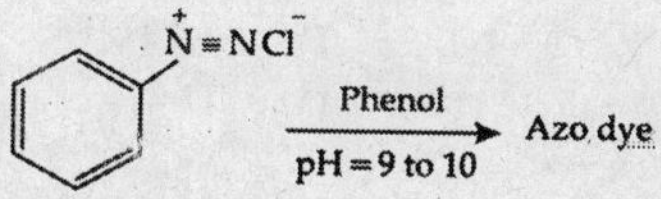

(*c*) 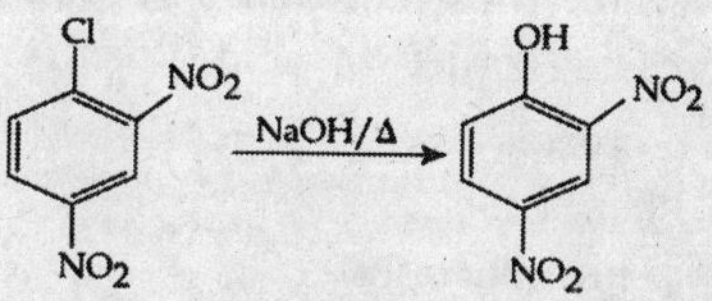

(*d*) 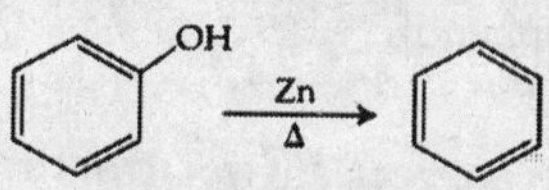

List-II

I. Nucleophillic substitution Reaction
II. Elimination Reaction
III. Electrophilic substitution Reaction
IV. Redox Reaction

Choose the **correct** answer from the options given below:

	(*a*)	(*b*)	(*c*)	(*d*)
A.	IV	III	I	II
B.	IV	III	II	I
C.	III	I	IV	II
D.	III	I	II	IV

67. C **68.** B **69.** B **70.** B **71.** B **72.** A

73. The amount of radioactive substance left after 75 hrs, when 100g of it is taken initially, will be:

(half life of radioactive substance is 15 hrs)

A. 12.5 g B. 6.25 g

C. 1.5625 g D. 3.125 g

74. Match List-I with List-II.

List-I	List-II
(*a*) Cubic	I. $a = b \neq c$
(*b*) Tetragonal	II. $a \neq b \neq c$
(*c*) Orthorhombic	III. $a = b \neq c$
(*d*) Hexagonal	IV. $a = b = c$

Choose the **correct** answer from the options given below:

	(*a*)	(*b*)	(*c*)	(*d*)
A.	IV	I	III	II
B.	IV	III	II	I
C.	I	III	IV	II
D.	I	II	IV	III

75. Arrange the following compounds in decreasing order of acidity:

(*a*) *o*-cresol

(*b*) Phenol

(*c*) *m*-nitrophenol

(*d*) *o*-nitrophenol

(*e*) Ethanol

Choose the **correct** answer from the options given below:

A. D > B > C > E > A

B. E > A > B > C > D

C. C > D > A > B > E

D. D > C > B > A > E

76. Match List-I with List-II.

List-I	List-II
(*a*) PbS → PbO	I. Roasting
(*b*) $CaCO_3$ → CaO	II. Calcination
(*c*) ZnS → Zn	III. Carbon reduction
(*d*) Cu_2S → Cu	IV. Self reduction

Choose the **correct** answer from the options given below:

	(*a*)	(*b*)	(*c*)	(*d*)
A.	IV	III	I	II
B.	I	II	III	IV
C.	III	IV	II	I
D.	II	III	IV	I

77. In which of the following cases there is a change in oxidation number?

A. Aqueous solution of CrO_4^{2-} is acidified

B. SO_2 gas is passed through acidified $Cr_2O_7^{2-}$ solution

C. $Cr_2O_7^{2-}$ solution is made alkaline

D. CrO_2Cl_2 is dissolved in NaOH

78. The cell $Zn|Zn^{2+}$ (lM)$||Cu^{2+}$ (1 M Cu. (E°_{Cell} = 1.10 V) was allowed to be completely discharged at 298 K. The relative concentration of Zn^{2+}/Cu^{2+} is:

A. 9.65×10^4 B. Antilog 24.8

C. 37.7 D. $10^{37.3}$

79. Match List-I with List-II.

List-I	List-II
(*a*) Λ_m of KC1 increases on dilution	I. is a linear graph
(*b*) Λ_m of CH_3COOH increases on dilution	II. is a non linear
(*c*) Λ_m of KCl versus $(\text{molarity})^{1/2}$	III. due to increased degree of ionisation
(*d*) Λ_m of CH_3COOH versus $(\text{molarity})^{1/2}$	IV. due to weakening of interionic attractions

Choose the correct answer from the options given below:

	(*a*)	(*b*)	(*c*)	(*d*)
A.	I	II	III	IV
B.	IV	III	I	II
C.	III	IV	II	I
D.	II	III	IV	I

73. D **74.** B **75.** D **76.** B **77.** B **78.** D **79.** B

80. Which of the following reaction has fractional order with respect to one of the reactant?

A. $2NO(g) + O_2(g) \rightarrow 2NO_2(g)$

B. $CHCl_3 + Cl_2 \rightarrow CCl_4 + HCl$

C. $CH_3COOC_2H_5 + H_2O \rightarrow CH_3COOH + C_2H_5OH$

D. $CH_3COOC_2H_5 + NaOH \rightarrow CH_3COONa + C_2H_5OH$

81. Identify the name reaction given below:

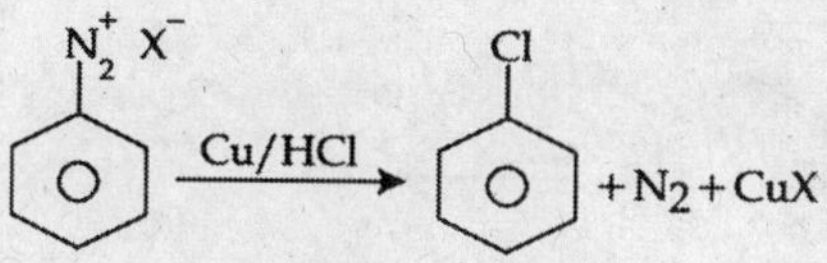

A. Gatterman Reaction

B. Sandmeyer Reaction

C. Gatterman - Koch Reaction

D. Dow's Process

82. Out of the following which will not undergo S_N^1 reaction pathway:

A. 1-Chloro-1,1,1-triphenylmethane

B. 2-Iodo-2-methylpropane

C. 1-Bromo-3,3-dimethylbutane

D. 3-Bromoprop-1-ene

83. Which of the following is an example of minimum boiling azeotrope?

A. Phenol and aniline

B. Chloroform and acetone

C. Nitric acid and water

D. Ethanol and acetone

84. Match List-I with List-II.

List-I	List-II
(*a*) Copper pyrites	I. Cu_2O
(*b*) Malachite	II. Cu_2S
(*c*) Cuprite	III. $CuCO_3.Cu(OH)_2$
(*d*) Copper glance	IV. $CuFeS_2$

Choose the **correct** answer from the options given below:

	(*a*)	(*b*)	(*c*)	(*d*)
A.	I	III	IV	II
B.	I	III	II	IV
C.	IV	III	I	II
D.	III	II	IV	I

85. Glucose on reaction with bromine water forms gluconic acid which confirms the presence of:

A. –COOH group B. –CHO group

C. Five –OH groups D. Ketonic group

86. The correct bond order for carbon-oxygen bond in following metal carbonyls $[Cr(CO)_6]$, $[V(CO)_6]^-$, $[Mn(CO)_6]^+$ is:

A. $[V(CO)_6]^- < [Cr(CO)_6] < [Mn(CO)_6]^+$

B. $[V(CO)_6]^- = [Cr(CO)_6] = [Mn(CO)_6]^+$

C. $[Mn(CO)_6]^+ < [Cr(CO)_6] < [V(CO)_6]^-$

D. $[Cr(CO)_6] < [V(CO)_6]^- < [Mn(CO)_6]^+$

87. The magnetic nature of elements depends on the presence of unpaired electrons. Identify the configuration of transition elements which shows highest magnetic moment:

A. $3d^7$ B. $3d^5$

C. $3d^8$ D. $3d^2$

88. Match List-I with List-II.

List-I (Drug)	List-II (Use)
(*a*) Phenelzine	I. Antibiotic
(*b*) Terfenadine	II. Antihistamine
(*c*) Codeine	III. Antidepressant
(*d*) Salvarsan	IV. Analgesics

Choose the **correct** answer from the options given below:

	(*a*)	(*b*)	(*c*)	(*d*)
A.	III	II	IV	I
B.	III	I	IV	II
C.	II	III	IV	I
D.	IV	III	II	I

80. B **81.** A **82.** C **83.** D **84.** C **85.** B **86.** A **87.** B **88.** A

89. Match List-I with List-II.

List-I	List-II
(*a*) Element with highest second ionisation enthalpy	I. Co
(*b*) Element with highest third ionisation enthalpy	II. Cr
(*c*) M in $M(CO)_6$ is	III. Cu
(*d*) Element with highest heat of atomisation	IV. Zn

Choose the **correct** answer from the options given below:

	(*a*)	(*b*)	(*c*)	(*d*)
A.	I	II	III	IV
B.	IV	III	II	I
C.	II	III	I	IV
D.	III	IV	II	I

90.

Sol		Charge Present
(*a*) Blood	—	positive
(*b*) Congo red	—	positive
(*c*) Gum	—	negative
(*d*) Gold sol	—	positive
(*e*) $Fe(OH)_3$ sol	—	negative

Choose the **correct** answer from the options given below :

A. (*a*) and (*c*) only
B. (*b*) and (*d*) only
C. (*c*) and (*e*) only
D. (*b*) and (*e*) only

Directions (Qs. No. 91-95): *Answer the question on basis of passage given below:*

Organic compounds containing carbon-oxygen double bonds $\left(\rangle C=O\right)$ are called carbonyl compounds. These are highly polar molecules. Hence, their boiling points are higher than hydrocarbons and ethers. Aldehydes can be prepared from dehydrogenation or oxidation of alcohols and by reduction of acylhalides. Ketones are prepared by oxidation of secondary alcohols or by hydration of alkynes etc. Aldehydes and ketones undergo nucleophilic addition reaction with the number of nucleophiles such as HCN, $NaHSO_3$ ammonia derivatives. Carboxylic acids are prepared by the oxidation of primary alcohols, aldehydes and alkenes, by hydrolysis of nitriles. Carboxylic acids are more acidic than alcohols and phenols.

91. The product (A) is:

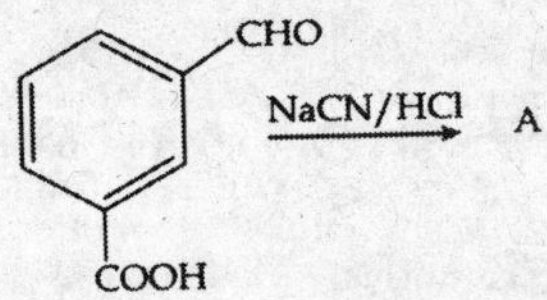

A. 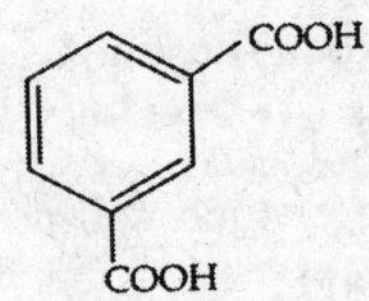

B. 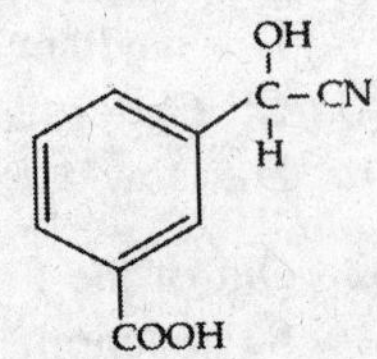

C. 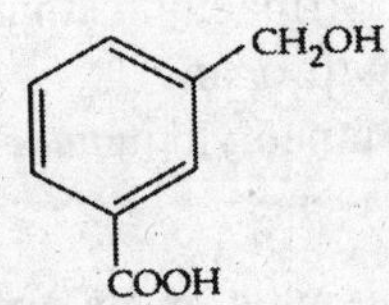

D. 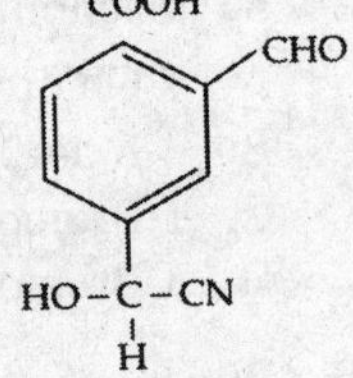

92. The correct increasing order of acidic strength is:

A. $C_6H_5COOH < C_6H_5CH_2COOH < NO_2CH_2COOH < FCH_2COOH$
B. $C_6H_5CH_2COOH < C_6H_5COOH < NO_2CH_2COOH < FCH_2COOH$
C. $C_6H_5CH_2COOH < C_6H_5COOH < FCH_2COOH < NO_2CH_2COOH$
D. $C_6H_5COOH < FCH_2COOH < C_6H_5CH_2COOH < NO_2CH_2COOH$

89. D	**90.** A	**91.** B	**92.** C

93. The correct sequence of reagents for the conversion of 3-nitrobromobenzene to 3-nitrobenzoic acid is:

A. Mg/ether, CO_2, H_3O^+

B. KCN, $LiAlH_4$, NaOH

C. $KMnO_4/OH^-$, dil.H_2SO_4, CO_2/H^+

D. $AgNO_3$ (Ammonical), CO_2, H_3O^+

94. Which of the following is most reactive towards nucleophilic addition reaction?

A. Butanone B. Propanone

C. Ethanal D. Propanal

95. The product (X) is:

$$R-CH = CH-CHO + NH_2-\overset{\overset{O}{||}}{C}-NH-NH_2 \xrightarrow{H^+} X$$

A. $R-CH = CH-CH = \overset{\overset{NH_2}{|}}{CH}-NH-NH_2$

B. $R-CH = CH-CH = N-\overset{\overset{O}{||}}{C}-NH-NH_2$

C. $R-CH_2-\overset{\overset{O}{||}}{C}-CH=N-NH-\overset{\overset{O}{||}}{C}-NH_2$

D. $R-CH = CH-CH = N-NH-\overset{\overset{O}{||}}{C}-NH_2$

Directions (Qs. No. 96-100): *Answer the question on basis of passage given below:*

Nitrogen differs from the rest of the members of its group due to its small size, high electronegativity, high ionisation enthalpy and non-availability of d-orbitals. Nitrogen has unique ability to form pπ-pπ multiple bonds with itself and with other elements having small size and high electronegativity (e.g. C, O) Heavier elements of this group do not form pπ-pπ bonds as their atomic orbitals are so large and diffused that they cannot have effective overlapping.

96. Single bond is not formed by two atoms of:

A. P – P B. N – N

C. As – As D. Sb – Sb

97. Maximum covalency of Nitrogen is:

A. 2 B. 3

C. 4 D. 5

98. Nitrogen family show reducing characters. Least basic amongst the following is:

(*a*) SbH_3 (*b*) BiH_3

(*c*) NH_3 (*d*) PH_3

Choose the **correct** answer from the options given below:

A. (*c*) Only

B. (*a*) Only

C. (*b*) Only

D. (*d*) Only

99. Most stable hydride amongst the following is:

(*a*) NH_3 (*b*) PH_3

(*c*) AsH_3 (*d*) SbH_3

(*e*) BiH_3

Choose the **correct** answer from the options given below:

A. (*a*) only

B. (*b*) only

C. (*c*) & (*d*) only

D. (*d*) & (*e*) only

100. Brown ring in nitrates test is formed when:

A. Fe^{2+} oxidises

B. Fe^{3+} reduces

C. NO_3^- oxidises

D. HNO_3 oxidises

93. A	**94.** C	**95.** D	**96.** B	**97.** C	**98.** C	**99.** A	**100.** A

BIOLOGY

101. The unit of Productivity in terms of energy is:

A. Kcal m^{-2}

B. [Kcal m^{-2}] yr^{-1}

C. gm^{-2} yr^{-1}

D. gm^{-2}

102. *Saccharum barberi* was originally grown in:

A. North America

B. North India

C. South America

D. South India

103. Arrange the following steps of PCR in correct sequence:

(*a*) Temperature induced denaturation of double stranded DNA.

(*b*) Isolation of gene of interest

(*c*) Amplification of gene by using *Taq* polymerase

(*d*) Primer annealing on the separated templates

Choose the **correct** answer from the options given below:

A. (*a*), (*c*), (*b*), (*d*)

B. (*b*), (*a*), (*d*), (*c*)

C. (*a*), (*b*), (*d*), (*c*)

D. (*b*), (*a*), (*c*), (*d*)

104. Which of the following ecosystem is most productive in terms of net primary productivity?

A. Deserts

B. Estuaries

C. Tropical rain forests

D. Oceans

105. Match List-I with List-II:

List-I	List-II
(*a*) Dodo	I. Africa
(*b*) Quagga	II. Russia
(*c*) Thylacine	III. Mauritius
(*d*) Stellar Sea Cow	IV. Australia

Choose the **correct** answer from the options given below:

	(*a*)	(*b*)	(*c*)	(*d*)
A.	III	I	IV	II
B.	I	IV	II	III
C.	IV	II	III	I
D.	II	I	IV	III

106. A DNA fragment contains 18% Thymine (T). The combined content of Cytosine (C) and Guanine (G) in it would be:

A. 36%

B. 18%

C. 64%

D. 32%

107. Match List-I with List-II:

List-I	List-II
(*a*) Pneumonia	I. *Wuchereria bancrofti*
(*b*) Typhoid	II. *Streptococcus pneumoniae*
(*c*) Filariasis	III. *Entamoeba histolytica*
(*d*) Amoebiasis	IV. *Salmonella typhi*

Choose the **correct** answer from the options given below:

	(*a*)	(*b*)	(*c*)	(*d*)
A.	II	III	IV	I
B.	II	IV	III	I
C.	III	IV	II	I
D.	II	IV	I	III

101. B **102.** B **103.** B **104.** C **105.** A **106.** C **107.** D

108. Identify the statements which hold true for apiculture:

(*a*) *Apis dorsata* is the most common honey bee species used for apiculture

(*b*) Honey is the food of high nutritive value and finds use in indigenous system of medicine

(*c*) Bee keeping is not labour intensive

(*d*) Honeybees also produce beewax

Choose the **correct** answer from the options given below:

A. (*a*), (*b*) and (*d*) only

B. (*a*), (*b*) and (*c*) only

C. (*a*), (*b*), (*c*) and (*d*)

D. (*b*), (*c*) and (*d*) only

109. Fossils are generally found in:

A. Igneous and Metamorphic rocks

B. Sedimentary rocks

C. Metamorphic rocks

D. Sedimentary and Igneous rocks

110. Thalassaemia is a:

A. Bone disease B. Blood disease

C. Eye disease D. Respiratory disease

111. In *Vallisneria*, the pollination is carried by:

A. Wind B. Water

C. Insects D. Animals

112. Leydig cells secrete:

A. Gonadotropin Releasing Hormone (GnRH)

B. Follicle Stimulating Hormone (FSH)

C. Androgens

D. Estrogen

113. Gene therapy is a collection of methods for:

A. Correcting physiological defects

B. Correcting morphological defects

C. Treating blood related disorders

D. Correcting a genetic defect diagnosed in a child or embryo

114. Which is not an example of convergent evolution?

A. Eye of an octopus and eye of man

B. Sweet potato and Potato

C. Thorns of *Bougainvillea* and tendrils of *Cucurbita*

D. Wings of butterfly and wings of an Eagle

115. Brief exposure to the sound of minimum .... can damage ear drums of humans.

A. 90 dB B. 110 dB

C. 150 dB D. 100 dB

116. Match List-I with List-II:

List-I	List-II
(*a*) ZIFT	I. Transfer of an ovum from a donor into the fallopian tube of another female
(*b*) GIFT	II. Semen collection from husband/ healthy donor is artificially introduced into the uterus of the female
(*c*) IUT	III. Transfer of zygote or embryo of upto 8 blastomeres into fallopian tube
(*d*) IUI	IV. Transfer of embryo with more than 8 blastomeres into uterus

Choose the correct answer from the options given below:

	(*a*)	(*b*)	(*c*)	(*d*)
A.	III	II	IV	I
B.	I	II	III	IV
C.	III	I	IV	II
D.	IV	III	II	I

117. Which of the following microbes is not used as a biocontrol agent?

A. *Lactobacillus*

B. *Bacillus thuringiensis*

C. *Trichoderma*

D. *Nucleopolyhedrovirus*

108. D **109.** B **110.** B **111.** B **112.** C **113.** D **114.** C **115.** C **116.** C **117.** A

118. The following are the natural methods of contraception except:

A. Lactational amenorrhea
B. Coitus interruptus
C. Intra Uterine Devices (IUD)
D. Periodic Abstinence

119. The organism that breeds only once in their life time:

A. Oysters
B. Banyan tree
C. Prawn
D. Pacific salmon fish

120. Identify the example of Australian Marsupial from the following:

A. Flying squirrel B. Flying phalanger
C. Bobcat D. Mole

121. In the exponential growth model equation

$$\frac{dN}{dt} = rN$$

'N' stands for:

A. Population size
B. Number of births
C. Number of deaths
D. Difference between births and deaths per unit time

122. Arrange the steps of biological sewage treatment process in a sequence.

(*a*) Vigorous growth of flocs and significant reduction in BOD
(*b*) Activated sludge pumped into anaerobic sludge digester
(*c*) Effluent passed into an aeration tank and agitated mechanically
(*d*) Effluent transfer to settling tank followed by sedimentation of bacterial flocs

Choose the **correct** answer from the options given below:

A. (*a*), (*b*), (*c*), (*d*) B. (*b*), (*c*), (*d*), (*a*)
C. (*c*), (*a*), (*d*), (*b*) D. (*d*), (*c*), (*b*), (*a*)

123. What is the full form of ELISA?

A. Enzyme Linked Immunodeficient-sorbent Assay
B. Enzyme Linked Immune-sorbent Assessment
C. Enzyme Linked Immuno-sorbent Assay
D. Enzyme Linked Immuno-sorbent Assessment

124. Arrange the following in correct sequence with respect to the events of menstrual cycle:

(*a*) Development of primary follicle to graafian follicle
(*b*) Attainment of peak level in luteinising hormone
(*c*) Release of ovum by graafian follicle
(*d*) Formation of corpus luteum

Choose the **correct** answer from the options given below:

A. (*a*), (*b*), (*d*), (*c*) B. (*b*), (*d*), (*a*), (*c*)
C. (*a*), (*c*), (*d*), (*b*) D. (*a*), (*b*), (*c*), (*d*)

125. Which of the following organism helps the plants in the absorption of phosphorus from soil?

A. *Anabaena* B. *Nostoc*
C. *Glomus* D. *Oscillatoria*

126. Identify the statements which hold true for cancer:

(*a*) Cancerous cells do not show contact inhibition
(*b*) Chemotherapeutic drugs are used to kill cancerous cells
(*c*) Benign tumors can reach distant sites through blood
(*d*) Non-ionising radiations can also cause cancer

Choose the **correct** answer from the options given below:

A. (*a*), (*b*) and (*d*) only
B. (*a*), (*b*) and (*c*) only
C. (*a*), (*b*), (*c*) and (*d*)
D. (*b*), (*c*) and (*d*) only

118. C **119.** D **120.** B **121.** A **122.** C **123.** C **124.** D **125.** C **126.** A

127. Heroin, commonly called as *smack* is a/an:

A. Cannabinoid
B. Opioid
C. Phylloid
D. Metalloid

128. In angiosperms, the outer wall of ovary develops into:

A. Perisperm
B. Pericarp
C. Periderm
D. Sporoderm

129. Which of the following device is used to culture large volumes of living cells to get products on large scale?

A. PCR machine
B. Agitator
C. Bioreactor
D. Assimilator

130. Identify the statements which are **not** true with respect to Heterochromatin:

(*a*) Loosely packed
(*b*) Densely packed
(*c*) Stains light
(*d*) Transcriptionally active

Choose the **correct** answer from the options given below:

A. (*a*), (*b*) and (*d*) only
B. (*a*), (*c*) and (*d*) only
C. (*a*), (*b*), (*c*) and (*d*)
D. (*c*) and (*d*) only

131. Identify the set of cells which have haploid number of chromosomes?

A. Egg cell and Synergid cell
B. Egg cell and Nucellar cell
C. Zygote and Egg cell
D. Zygote and Antipodal cell

132. Match List-I with List-II:

List-I	List-II
(*a*) First Restriction Endonuclease	I. GAATTC
(*b*) Disarmed Plasmid	II. Selectable markers
(*c*) Antibiotic resistant genes	III. *Hind II*
(*d*) *EcoR1*	IV. Vector

Choose the correct answer from the options given below:

	(*a*)	(*b*)	(*c*)	(*d*)
A.	III	I	II	IV
B.	I	III	II	IV
C.	III	IV	II	I
D.	I	IV	II	III

133. Match List-I with List-II:

List-I	List-II
(*a*) Buds	I. *Chlamydomonas*
(*b*) Conidia	II. Sponge
(*c*) Gemmules	III. *Hydra*
(*d*) Zoospores	IV. *Penicillium*

Choose the **correct** answer from the options given below:

	(*a*)	(*b*)	(*c*)	(*d*)
A.	IV	III	II	I
B.	I	III	IV	II
C.	III	IV	II	I
D.	III	I	II	IV

134. Ex-situ conservation method does not include:

(*a*) Cryopreservation
(*b*) National park
(*c*) Zoological park
(*d*) Sacred groves

Choose the **correct** answer from the options given below:

A. (*a*) and (*d*) only
B. (*b*) and (*c*) only
C. (*b*) and (*d*) only
D. (*a*) and (*c*) only

127. B **128.** B **129.** C **130.** B **131.** A **132.** C **133.** C **134.** C

135. ABO blood group is an example of:

A. Single allele
B. Multiple alleles
C. Single gene
D. Two genes

136. If an *E.coli* with heavy DNA (N^{15}) is put in a medium with $NH_4Cl[N^{14}]$ then, what type of DNA will be obtained after two generations/replications?

A. $N^{14}N^{14}$ type
B. $N^{14}N^{15}$ and $N^{14}N^{14}$ type
C. $N^{14}N^{15}$ and $N^{15}N^{15}$ type
D. $N^{15}N^{15}$ and $N^{14}N^{14}$ type

137. Match List-I with List-II:

List-I (Animal Breeding)	**List-II (Type)**
(*a*) Crossing tall F_2 progeny with recessive parent to determine its genotype	I. Cross-breeding
(*b*) Breeding within the same breed with no common ancestor	II. Inter-specific hybridization
(*c*) Breeding of superior male of one breed with the superior female of another breed	III. Test-crossing
(*d*) Breeding of animals of two different but related species	IV. Out-crossing

Choose the **correct** answer from the options given below:

	(*a*)	(*b*)	(*c*)	(*d*)
A.	I	II	III	IV
B.	II	I	III	IV
C.	I	II	IV	III
D.	III	IV	I	II

138. Identify the metal which is NOT used in catalytic converters.

A. Gold
B. Platinum
C. Rhodium
D. Palladium

139. Match List-I with List-II:

List-I	**List-II**
(*a*) Down's syndrome	I. 44 + XX/XY
(*b*) Klinefelter's syndrome	II. 44 + XO
(*c*) Turner's syndrome	III. 44 + XXY
(*d*) Normal human	IV. 45 + XX/XY

Choose the **correct** answer from the options given below:

	(*a*)	(*b*)	(*c*)	(*d*)
A.	I	III	II	IV
B.	II	IV	III	I
C.	IV	III	II	I
D.	III	II	IV	I

140. Which of the following statements are NOT true for biopiracy?

(*a*) It is the use of bio-resources by multinational companies with proper authorisation by the concerned country
(*b*) Developing countries are rich in biodiversity
(*c*) Developed countries are rich in traditional knowledge related to bio-resources
(*d*) Biopiracy involves inadequate compensation and benefit sharing between developed and developing countries, while using latter's bioresources by the former

Choose the **correct** answer from the options given below:

A. (*a*) and (*c*) only
B. (*a*) and (*b*) only
C. (*b*) and (*d*) only
D. (*c*) and (*b*) only

135. B **136.** B **137.** D **138.** A **139.** C **140.** A

Directions (Qs. No. 141-145): *Observe the diagram and answer the following question.*

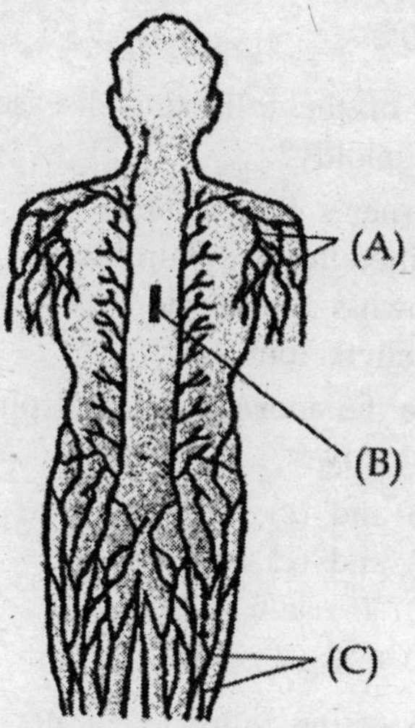

Diagrammatic representation of Lymphatic system

141. A, B and C labelled in the diagram are:

(*a*) Thymus
(*b*) Lymph nodes
(*c*) Lymphatic vessels
(*d*) Bone marrow

Choose the **correct** answer from the options given below:

A. (*a*), (*b*), (*d*) respectively
B. (*b*), (*a*), (*c*) respectively
C. (*a*), (*b*), (*c*) respectively
D. (*c*), (*b*), (*a*) respectively

142. Identify the primary lymphoid organs:

(*a*) Lymph nodes (*b*) Bone marrow
(*c*) Thymus (*d*) Tonsils

Choose the **correct** answer from the options given below:

A. (*a*) and (*b*) only
B. (*c*) and (*d*) only
C. (*b*) and (*c*) only
D. (*a*) and (*c*) only

143. Which of the following organ reduces in size with age?

A. Thymus B. Spleen
C. Lymph nodes D. Bone marrow

144. Match List-I with List-II:

List-I	List-II
(*a*) MALT	I. Largest reservoir of erythrocytes
(*b*) Spleen	II. Provides microenvironment for development and maturation of T-lymphocytes
(*c*) Thymus	III. Peyer's patches
(*d*) Small intestine	IV. Lining of urogenital tract

Choose the **correct** answer from the options given below:

	(*a*)	(*b*)	(*c*)	(*d*)
A.	I	II	IV	III
B.	II	III	I	IV
C.	IV	I	II	III
D.	III	IV	I	II

145. What is the full form of MALT?

A. Mucosa Associated Lymphatic Tissue
B. Mucous Associated Lymphoid Tissue
C. Mucous Associated Lymphatic Tissue
D. Mucosa Associated Lymphoid Tissue

Directions (Qs. No. 146-150): *Read the paragraph and answer the questions.*

Broadly, genetic disorders may be grouped into two categories—Mendelian disorders and Chromosomal disorders. Mendelian disorders are mainly determined by alteration or mutation in the single gene. These disorders are transmitted to the offspring on the same lines as we have studied in the principle of inheritance. The trait may also be linked to the sex chromosome as in case of haemophilia. It is evident that this X-linked recessive trait shows transmission from carrier female of male progeny.

141. B **142.** C **143.** A **144.** C **145.** D

146. The disease which is transmitted to the offsprings following the principles of inheritance is:

A. Sickle cell anaemia
B. Down's Syndrome
C. Klinefelter's Syndrome
D. Turner's Syndrome

147. Match List-I with List-II:

List-I	List-II
(*a*) Autosomal dominant trait	I. Klinefelter's Syndrome
(*b*) Autosomal recessive trait	II. Sickle Cell Anaemia
(*c*) Sex linked recessive trait	III. Myotonic Dystrophy
(*d*) Gain of chromosomes	IV. Haemophilia

Choose the **correct** answer from the options given below:

	(*a*)	(*b*)	(*c*)	(*d*)
A.	III	IV	I	II
B.	II	I	III	IV
C.	III	II	IV	I
D.	I	II	IV	III

148. What is the probability of having a haemophilic child to a carrier mother and a normal father?

A. 50%
B. 75%
C. 100%
D. 25%

149. Which of the following disease is a result of aneuploidy?

(*a*) Turner's Syndrome
(*b*) Klinefelter's Syndrome
(*c*) Down's Syndrome
(*d*) Phenylketonuria

Choose the **correct** answer from the options given below:

A. (*a*) and (*c*) only
B. (*b*) and (*c*) only
C. (*a*), (*b*) and (*c*) only
D. (*a*) and (*d*) only

150. Which of the following is not a Mendelian disorder?

A. Cystic Fibrosis
B. Turner's Syndrome
C. Sickle cell anaemia
D. Thalassaemia

---

146. A **147.** C **148.** D **149.** C **150.** B

Previous Paper (Solved)

B.Sc. BIOTECHNOLOGY

Entrance Test, 2021

PHYSICS

1. A charge of 1500 μC is distributed over a very large sheet having a surface area of 300 m^2. The electric field at a distance of 25 cm will be:

A. 2.8×10^5 N/C

B. 2.24×10^5 N/C

C. 2.08×10^5 N/C

D. None of these

2. A uniformly charged sphere has a total charge of 300 μC and radius 8 cm. The electric field intensity at a point 4 cm inside from the centre of sphere will be:

A. 4.2×10^7 N/C

B. 10.5×10^7 N/C

C. Zero

D. None of these

3. On the surface of the moon a person weighs:

A. Six times his weight on the earth

B. Equal to his weight on the earth

C. 0.67 times his weight on the earth

D. 0.167 times his weight on the earth

4. In CB transistor amplifier, the current gain is:

A. 1000 B. 10

C. 1 D. Less than 1

5. D.C. can be converted to A.C. by using:

A. Multimeter

B. Rectifier

C. Cathode ray tube

D. Oscillator

6. If blue light can cause photoelectric emission from a metal, but yellow light cannot. If red light is incident on the metal, then:

A. Photoelectric current will increase.

B. Rate of emission of photoelectrons will decrease

C. No photoelectric emission will occur

D. None of these

7. A hydrogen atom of mass 1.7×10^{-27} kg moving with a velocity of 1000 m/sec. has waves associated with it. Their wavelength is:

A. 3.8 m B. 3.8 Å

C. 1.7×10^{-30} m D. 1.7×10^{-24} m

8. The angle of prism is 60° and R.I. of the prism is $\sqrt{2}$. The angle of minimum deviation will be:

A. 24° B. 35°

C. 30° D. None of these

9. The fringe width in a diffraction due to single slit is:

A. Independent of wavelength of incident light

B. Directly proportional to the wavelength of incident light

C. Directly proportional to the square of wavelength of incident light

D. Inversely proportional to the wavelength of incident light

10. A plano-concave lens is made of glass of refractive index 1.5 and the radius of curvature of the curved face is 50 cm. The power of the lens is:

A. –1.0 D B. –0.5 D

C. +1.0 D D. +0.5 D

11. In vacuum, the speed of light depends on:

A. Wavelength

B. Frequency

C. Speed of source

D. None of these

12. Power is generally transmitted over long distances on high voltage A.C. The reason for doing so is:

A. Power can be transmitted more quickly on higher voltage

B. There is a less wastage of power in transmission at higher voltages

C. The high voltage of transmission line prevents theft of the line wire

D. Large amount of power cannot be generated at low voltages

13. 5 ampere current becomes zero in 0.1 second and average e.m.f. is 200 volts. The self-inductance will be:

A. 0.4 henry B. 4 mH

C. 4 henry D. 0.04 henry

14. Two parallel conductors carrying current in the same direction experience a force which is:

A. Repulsive

B. Attractive

C. May be any depending on the length of the conductors

D. None of the above

15. The specific resistance of a wire 1.1 m long, 0.4 mm diameter, having a resistance of 4.2 ohms will be:

A. 48×10^{-8} ohm m

B. 40×10^{8} ohm m

C. 4×10^{-8} ohm m

D. None of these

16. A spherical liquid drop has a diameter of 2 cm and is given a charge of 1 μC. The potential at the surface of the drop is :

A. 9×10^{8} volt

B. 9×10^{9} volt

C. 1×10^{8} volt

D. 1×10^{-4} volt

17. The number of lines of force that radiate outward from one coulomb of charge is:

A. 8.85×10^{-12}

B. 1.13×10^{11}

C. Infinite

D. Negligibly small

18. The bandwidth required for transmission of pictures is:

A. 6.5 MHz B. 3.8 MHz

C. 4.2 MHz D. None of these

19. How many AND gates are required to realize Z = AB + DC + F ?

A. 3 B. 4

C. 1 D. 2

20. If you connect the inputs of a NAND gate, the resulting gate will be:

A. AND B. NAND

C. NOT D. None of these

21. A particle of mass *m* is hanging vertically by an ideal spring of force constant *k*. If the mass is made to oscillate vertically, its total energy is:

A. Maximum at the extreme position

B. Maximum at the near equilibrium

C. Minimum at the near equilibrium

D. None of the above

22. The speed of sound in a gas at N.T.P. is 300 m/s. If the pressure is increased to four times without change in temperature, the velocity of sound will be:

A. 150 m/s

B. 300 m/s

C. 600 m/s

D. 1200 m/s

23. The specific heats at constant pressure and volume for 1 g of a gas are 0.21 and 0.16 respectively. If J 4.2 × 10^7 ergs/cal, then the density of gas at NTP will be:
A. 1.76 g/litre
B. 1.89 g/litre
C. 1.81 g/litre
D. None of these

24. According to the kinetic theory, the inter-molecular forces between the molecules of a gas are:
A. Zero
B. Very large
C. Very small
D. Infinitely large

25. Hydraulic brakes make use of:
A. Bernoulli's principle
B. Archemedes' principle
C. Newton's law of viscous drag
D. None of the above

26. The graph between the applied force and change in length of a wire within elastic limit is:
A. Straight line with negative slope
B. Straight line with positive slope
C. Smooth curve
D. None of the above

27. Two wires, made of the same material and of same area of cross-section, are 1 m and 2 m long respectively. Force required to change the length of one metre wire by 1 cm is F. The force required to change the length of the 2 m wire by 1 cm will be:

A. $\frac{F}{4}$ B. $\frac{F}{2}$
C. F D. 2F

28. The escape velocity of a projectile on the earth's surface is 11.2 km/s. A body is projected out with twice this speed. The speed of the body far away from the earth will be:
A. 18.8 km/s
B. 18.3 km/s
C. 19.4 km/s
D. Can't be determined

29. A person standing on a turn table with dumb-bells in his hands suddenly withdraws his hands to his chest. The angular velocity of the table will:
A. Decrease
B. Increase
C. Remain the same
D. Become double

30. A body of mass 10 kg moving with a speed of 2 m/s on a frictionless table impinges on a mounted spring of force constant 4 × 10^5 N/m. The compression of the spring when the body comes to rest is:
A. 1.5 cm
B. 2.3 cm
C. 1.2 cm
D. 1 cm

31. The angle between frictional force and instantaneous velocity of a body moving over a rough surface is:

A. π B. $\frac{\pi}{2}$
C. zero D. None of these

32. A cricketer can throw a ball to a maximum horizontal distance of 100 m. How much high above the ground can be cricketer throw the same ball?
A. 25 m
B. 26.5 m
C. 28 m
D. None of these

33. Tripling the speed of a car multiplies the distance needed for stopping it by:
A. 3
B. 6
C. 9
D. Some other number

CHEMISTRY

34. The enthalpy of neutralization of HCl and NaOH is:
A. – 57.3 kJ
B. Zero
C. + 57.3 kJ
D. Cannot be predicted

35. The number of g-atom of oxygen in 6.02×10^{24} CO molecules is:
A. 10 B. 5
C. 0.5 D. 01

36. For principal quantum number $n = 4$, the total number of orbitals having $l = 3$ is:
A. 9 B. 7
C. 3 D. 5

37. Which of the following has the highest boiling point?
A. Kr B. Ar
C. Xe D. Ne

38. The ONO angle is maximum in:
A. NO_3^- B. NO_2^-
C. NO_2^+ D. NO_2

39. Two gases A and B having same volume diffuse through a porous partition in 10 and 20 seconds respectively. The molecular mass of A is 49 u. Molecular mass of B is:
A. 12.25 u B. 13.25 u
C. 14.25 u D. 15.25 u

40. In a reversible reaction, two substances are in equilibrium. If the concentration of each one is doubled the equilibrium constant will be:
A. constant
B. doubled
C. reduced to one-fourth of its original value
D. reduced to half of its original value

41. Solubility product of silver bromide is 5.0×10^{-13}. The quantity of potassium bromide molar mass taken as 120 g mole^{-1} to be added to 1 L of 0.05 M solution of silver nitrate to start the precipitation of AgBr is:
A. 5.0×10^{-12} g
B. 6.0×10^{-12} g
C. 2.0×10^{-12} g
D. 1.2×10^{-9} g

42. For the feasibility of redox reaction in a cell the EMF should be:
A. positive B. fixed
C. negative D. zero

43. Polyphosphates are used as water softening agents because they:
A. form soluble complexes with anionic species
B. precipitate anionic species
C. form soluble complexes with cationic species
D. precipitate cationic species

44. The byproduct of Solvay ammonia process is:
A. carbon dioxide
B. ammonia
C. calcium chloride
D. calcium carbonate

45. Amorphous silica is:
A. Tridymite
B. Fumed silica
C. Cristobalite
D. Asbestos

46. The IUPAC name of acrolein is:
A. Prop-2-enal
B. Cyanomethane
C. Methanenitrile
D. Cyanoethane

47. The compound that *does not* give blue colour in Lassaigne's test is:
A. aniline
B. glycine
C. hydrazine
D. urea

48. Which of the following *cannot* be used for the one-step preparation of ethene?
A. ethyl bromide
B. ethanol
C. ethylene dibromide
D. ethane

49. 1-Chlorobutane on reaction with alcoholic potash gives:
A. 1-butene
B. 1-butanol
C. 2-butene
D. 2-butanol

50. Which of the following is responsible for depletion of ozone layer in the upper strata of the atmosphere?
A. polyhalogen
B. ferrocene
C. fullerenes
D. freons

51. When benzene sulphonic acid and paranitrophenol are treated with aqueous $NaHCO_3$, the gases released respectively are:
A. CO_2, CO_2
B. SO_2, CO_2
C. SO_2, NO
D. SO_2, NO_2

52. The order of ease of the following leaving groups would be:
1. CH_3COO^-
2. MeO^-
3. $CH_3SO_3^-$
4. $CF_3SO_3^-$
A. 1 > 2 > 3 > 4
B. 3 > 4 > 1 > 2
C. 4 > 3 > 1 > 2
D. 2 > 1 > 3 > 4

53. The electronegativity difference is highest for the pair:
A. K, F B. Li, F
C. Li, Cl D. Na, Cl

54. The shape of BrF_3 is:
A. trigonal pyramidal
B. trigonal bipyramidal
C. trigonal planar
D. T shape

55. The paramagnetic species is:
A. Na_2O B. KO_2
C. N_2 D. CO

56. The electrode potential E° for Fe^{3+}/Fe = –0.036 and Fe^{2+}/Fe = –0.439 respectively. The electrode potential of $E^{\circ}_{Fe^{3+}/Fe^{2+}}$ will be:
A. 0.770 V B. – 0.072 V
C. 0.385 V D. – 0.270 V

57. The reaction of Zn with dilute and concentrated nitric acid respectively produces:
A. N_2O, NO_2 B. NO_2, NO
C. NO, NO_2 D. NO_2, NO_2

58. Purple colour of $KMnO_4$ is due to:
A. *d-d* transition
B. ligand field transition
C. charge transfer transition
D. sigma to pi antibonding transition

59. The metal ion and the macrocyclic skeleton present in the green pigment of plants respectively, are:
A. Mg and chlorine
B. N and corrinaz
C. Mn and chlorine
D. Mg and porphine

60. Which of the following polymers is used for making electrical switches and computer discs?
A. Glyptal B. PVC
C. Bakelite D. Polystyrene

61. On complete hydrolysis of starch we get:
A. Glucose
B. Fructose
C. Glucose and fructose
D. Sucrose

62. Bithional is added to soap as an additive to function as:

A. softener B. hardener

C. dryer D. antiseptics

63. In the pinacol-pinacolone rearrangement the reactive species undergoes rearrangement is:

A. free radical

B. carbanion

C. carbocation

D. carbene

64. Which among the following will be most reactive towards nucleophilic substitution?

A. CH_3F B. CH_3Cl

C. CH_3Br D. CH_3I

65. Which among the following is most basic?

A. NH_3 B. CH_3NH_2

C. $(CH_3)_2NH_2$ D. $(CH_3)_3N$

66. Which of the following is *not* a necessary requirement for a compound to show liquid crystal behaviour?

A. the molecule should possess elongated shape

B. the compound should have long alkyl chain

C. the molecule should be rigid with possible flexible ends

D. in the aromatic portion, the subsituents should be at para position

BIOLOGY

67. Which of the following is *not* a correct match?

A. Protoplasm – Huxley

B. Cell – Robert Hooke

C. Vitamin – Funk

D. Chromosome – Mendel

68. The walls of the cork cells are thick due to the presence of:

A. Cuticle B. Suberin

C. Lignin D. None of these

69. If there are 999 bases in an RNA that codes for a protein with 333 amino acids and the base at position 901 is deleted such that the length of the RNA becomes 998 bases, how many codons will be altered?

A. 1 B. 11

C. 33 D. 333

70. Removal of RNA polymerase-III from nucleoplasm will affect the synthesis of:

A. *t* RNA B. *hn* RNA

C. *m* RNA D. *r* RNA

71. Which of the following is *not* related to connective tissues?

A. Mast cells B. Immunocytes

C. Sarcolemma D. Adipose cells

72. When two species of different geneology come to resemble each other as a result of adaptation, the phenomenon is termed as:

A. Divergent evolution

B. Micro-evolution

C. Co-evolution

D. Convergent evolution

73. C-peptide of human insulin is:

A. A part of mature insulin molecule

B. Responsible for formation of disulphide bridges

C. Removed during maturation of pro-insulin to insulin

D. Responsible for its biological activity

74. A DNA with unequal nitrogen bases would most probably be:

A. Single stranded

B. Double stranded

C. Triple stranded

D. Four stranded

75. The first clinical gene therapy was done for the treatment of:

A. AIDS
B. Cancer
C. Cystic fibrosis
D. SCID resulting from deficiency of ADA

76. Distance between the genes and percentage of recombination shows:

A. A direct relationship
B. An inverse relationship
C. A parallel relationship
D. No relationship

77. Which one of the following is *not* a living fossil?

A. King crab B. Sphenodon
C. Peripatus D. Archaeopteryx

78. A cross between two tall plants resulted in offspring having only a few dwarf plants. What would be the genotypes of both the parents?

A. TT and Tt B. Tt and Tt
C. TT and TT D. Tt and tt

79. Which of the following is used as an atmospheric pollution indicator?

A. Lepidoptera
B. Lichens
C. Lycopersicon
D. Lycopodium

80. Mycorrhiza *does not* help the host plant in:

A. Enhancing its phosphorous uptake capacity
B. Increasing its tolerance to drought
C. Enhancing its resistance to root pathogens
D. Increasing its resistance to insects

81. *E. coli* about to replicate was placed in a medium containing radioactive thymidine for five minutes. Then it was made to replicate in a normal medium. Which of the following observations shall be correct?

A. Both the strands of DNA will be radioactive
B. One strand will be radioactive
C. Each strand will be half radioactive
D. None of the strand will be radioactive

82. In which of the following pairs of parts of a flowering plant is epidermis absent?

A. Root tip and shoot tip
B. Shoot bud and floral bud
C. Ovule and seed
D. Petiole and pedicel

83. In virus-infected plants, the meristematic tissues in both apical and axillary buds are free of virus because:

A. The dividing cells are virus resistant
B. Meristems have anti-viral compounds
C. The cell division of meristems is faster than the rate of virus multiplication
D. Viruses cannot multiply within meristem cells

84. DNA is a polymer of nucleotides which are linked to each other by 3'—5' phosphodiester bond. To prevent polymerization of nucleotides, which of the following modifications would you choose?

A. Replace purines with pyrimidines
B. Remove/Replace 3'OH group in deoxyribose
C. Remove/Replace 2'OH group with some other group in deoxyribose
D. Both (B) and (C)

85. The vitamin whose content increases following the conversion of milk into curd by lactic acid bacteria is:

A. Vitamin C
B. Vitamin D
C. Vitamin B_{12}
D. Vitamin E

86. BOD of waste water is estimated by measuring the amount of:

A. Total organic matter
B. Biodegradable organic matter
C. Oxygen evolution
D. Oxygen consumption

87. While isolating DNA from bacteria, which of the following enzymes is *not* required?
A. Lysozyme
B. Ribonuclease
C. Deoxyribonuclease
D. Protease

88. Amongst the animal groups given below, which one appears to be more vulnerable to extinction?
A. Insects B. Mammals
C. Amphibians D. Reptiles

89. Edaphic factor refers to:
A. Soil
B. Water
C. Relative humidity
D. Altitude

90. The human chromosome with the highest and least number of genes in them are respectively:
A. Chromosome 21 and Y
B. Chromosome 1 and X
C. Chromosome X and Y
D. Chromosome 1 and Y

91. A bacterial cell was transformed with a recombinant DNA molecule that was generated using a human gene. However, the transformed cells did not produce the desired protein. Reasons could be:
A. Human gene may have intron which bacteria cannot possess
B. Amino acid codons for humans and bacteria are different
C. Human protein is formed but degraded by bacteria
D. All of the above

92. Which one of the following sets of animals share a four-chambered heart?
A. Amphibians, Reptiles, Birds
B. Crocodiles, Birds, Mammals
C. Crocodiles, Lizards, Turtles
D. Lizards, Mammals, Birds

93. The most abundant chemical in living organisms could be:
A. Protein B. Water
C. Sugar D. Nucleic acid

94. At which stage of meiosis is the genetic constitution of gametes finally decided?
A. Metaphase I B. Anaphase II
C. Metaphase II D. Anaphase I

95. During light reaction in photosynthesis, the following are formed:
A. ATP and Sugar
B. Hydrogen, O_2 and Sugar
C. ATP, Hydrogen and O_2 donor
D. ATP, Hydrogen donor and O_2

96. Mycorrhiza is a symbiotic association of fungus with root system which helps in:
I. Absorption of water
II. Mineral nutrition
III. Translocation
IV. Gaseous exchange
A. Only I B. Only II
C. Both I and II D. Both II and III

97. What is common to the techniques?
(*i*) in-vitro fertilization,
(*ii*) cryo-preservation, and
(*iii*) tissue culture?
A. All are *in situ* conservation methods
B. All are *ex situ* conservation methods
C. All require ultra modern equipment and large space
D. All are methods of conservation of extinct organisms

98. Which of the following will exhibit the highest rate of respiration?
A. Germinating seed
B. Growing shoot apex
C. Root tip
D. Leaf bud

99. Which of the following statements is *incorrect*?
A. A person of "O" blood group has anti-A and anti-B antibodies in his plasma.
B. A person of "B" blood group can't donate blood to a person of "A" blood group.

C. Blood group is designated on the basis of the presence of antibodies in the blood plasma.
D. A person of AB blood group is a universal recipient.

100. The correct sequence of flow of electrons in the light reaction is:

A. PS II, plastoquinone, cytochromes, PS I, ferredoxin
B. PS I, plastoquinone, cytochromes, PS II, ferredoxin
C. PS I, ferredoxin, PS II, cytochromes, plastoquinone
D. PS I, plastoquinone, cytochromes, PS II, ferredoxin

GENERAL APTITUDE

101. Study the following information carefully and answer the question given below:

Amongst six friends, P, Q, R, S, T and U, each got different marks in the examination. T scores more marks than U and P. R scores more marks than U but less marks than S. P does not score marks that is an odd number. S scores marks that is an odd number. S scores more marks than T and does not score highest marks. R does not score less than T. The one who scored the minimum marks scored 29 marks and the one, who scored the highest, scored 80 marks.

Who among the following persons score more than P and less than S?

A. R B. Q
C. T D. Both (A) and (C)

102. Find out the missing number in the following number series:

3, 8, 13, 24, 41, ?

A. 70 B. 75
C. 80 D. 85

103. In this question, a group of digits/letters is given followed by four combinations of symbols numbered A, B, C and D. You have to find out which of the following four combinations correctly represent the group of digits/letters based on the symbol codes and the conditions given below:

A	E	I	O	U
@	#	$	%	&

And the rest of the letters will be coded as the sum of the digits of the numerical position of them in the alphabetical series.

Condition 1:

If the sum of the numbers in the code is a multiple of 3 then the last two elements of the code are to be interchanged.

Condition 2:

If the product of the first and last element in the code is a multiple of 5 then the code will be written in the reversed order.

Condition 3 :

If there are more than two symbols in the code then the code will be written in the reversed order except for first and the last element.

What would be the code of the word 'JESUS'?

A. 1#1&1 B. 1&1#1
C. 11&#1 D. 1#11&

104. Pointing to a photograph, a woman says, "This man's son's sister is my mother-in-law." How is the woman's husband related to the man in the photograph?

A. Grandson B. Son
C. Son-in-law D. Nephew

105. In a city, Fruit lane was to the north-west of Jewellery lane. Grocery lane was in the east of Fruit lane which was towards the south of Vegetable lane. Toy lane was towards the south-west of Grocery lane such that it was situated towards the east of Jewellery lane. Clothes lane was 20 km towards the north of Jewellery lane, which was 4 times the distance between Jewellery lane and Toy lane. Inner circle was the midpoint between Fruit lane and Grocery lane and also Jewellery lane and Clothes lane. The distance between Vegetable lane and Fruit lane was half of the distance between Clothes lane and Inner circle.

Vegetable lane is in which direction with respect to the Clothes lane?

A. West
B. East
C. South-West
D. North-East

106. Read the information carefully and answer the question given below:

In a certain code language,
'speak nicely to all' is coded as "ka cu ma he"
'all are like us' is coded as "si fo he to"
'teach us lesson nicely' is coded as "po ma fo re"
'lesson like all humans' is coded as "he re gu si"

What would be the code for 'nicely'?

A. he
B. ma
C. si
D. fo

107. Four positions of a dice are given below. Identify the number on the face opposite to the face showing 4.

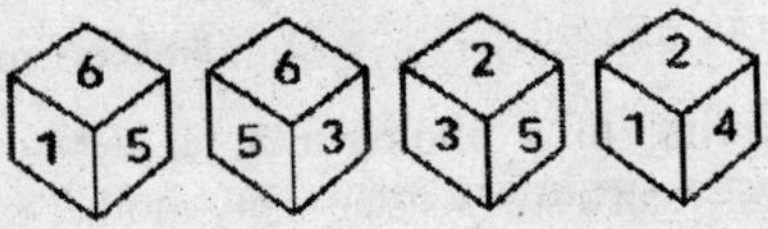

A. 5
B. 6
C. 3
D. 1

108. From the given options, select a figure in which the Question Figure is hidden/ embedded:

A. 

B. 

C. 

D. 

109. In this question is given a statement followed by three courses of action numbered I, II and III. A course of action is a step or administrative decision to be taken for improvement, follow up or further action in regard to the problem, policy, etc. On the basis of information given in the statement, you have to assume everything in the statement to be true. Then decide which of the three given suggested course(s) of action logically follow(s) worth pursuing:

Statement: A large number of tribal inhabitants are found to have been suffering from various diseases due to exposure to radioactive waste near the uranium mine.

Courses of Action:

I. The Government should immediately close down the mine.
II. The Government should immediately take steps to save local people from exposure to radioactive waste.
III. The tribal people should be rehabilitated at a safer place.

A. Only I follows
B. Only III follows
C. Only II follows
D. Only II and III follow

110. Acme : Climax :: Pathetic : ?

A. Moving
B. Disagreeable
C. Designed
D. Lucid

GENERAL ENGLISH

111. Find the synonym of THRONG:
A. Excitement
B. Crowded mass of people
C. Danger
D. Suspense

112. In the following question, out of the four alternatives, choose the word opposite in meaning to the given word:

SUCCINCT
A. Failure
B. Inelegance
C. Verbosity
D. Nomenclature

113. We cannot .... such a/an .... act of violence.
A. tolerate, insipid
B. consider, important
C. commit, magnificent
D. pardon, egregious

114. In this question, out of the four alternatives, choose the one which can be substituted for the given sentence, group of words or clauses:

A round or cylindrical container used for storing things such as food, chemicals or rolls of film
A. Tanker B. Canister
C. Vessel D. Casket

115. Idiom and Phrase : Wrangle over an ass' shadow
A. Draw the picture of an ass
B. To be alert
C. To quarrel over trifles
D. To watch a light and shadow play

116. Which one of the following words is incorrectly spelt?
A. Sapphire B. Scurrilous
C. Secratariat D. Seethe

117. "Sarojini was not .... by the criticism and paid no .... even when her best friend talked against her." Which pair of words make sentence meaningful?
A. bothered, attention
B. troubled, brained
C. threatened, warning
D. shaken, indication

118. Improve the sentence:

Education is a strong instrument for moulding the character of the young.
A. striking
B. powerful
C. potent
D. no improvement

119. Fill in the blanks:

I called .... his house yesterday.
A. at B. on
C. of D. to

120. Rearrange the following sentences in the proper sequences so as to form a meaningful paragraph, and then answer the question given below them:
(*a*) The interest subsidy it pays to banks.
(*b*) Farm loans in the budget.
(*c*) Had suggested that farm loans be made available at four percent.
(*d*) The government plans to.
(*e*) The working group on agricultural production.
(*f*) Provide incentives on.
(*g*) Will have to increase.
(*h*) This would mean that the government.

Which of the following should be the LAST sentence after the rearrangement?
A. (*a*) B. (*d*)
C. (*e*) D. (*c*)

ANSWERS

1	2	3	4	5	6	7	8	9	10
A	C	D	D	D	C	B	C	B	A
11	**12**	**13**	**14**	**15**	**16**	**17**	**18**	**19**	**20**
D	B	C	B	A	A	B	C	D	C
21	**22**	**23**	**24**	**25**	**26**	**27**	**28**	**29**	**30**
D	B	A	A	C	B	B	C	B	D
31	**32**	**33**	**34**	**35**	**36**	**37**	**38**	**39**	**40**
A	A	C	A	A	B	C	C	A	A
41	**42**	**43**	**44**	**45**	**46**	**47**	**48**	**49**	**50**
D	A	C	C	B	A	C	D	A	D
51	**52**	**53**	**54**	**55**	**56**	**57**	**58**	**59**	**60**
A	C	A	D	B	A	A	C	A	C
61	**62**	**63**	**64**	**65**	**66**	**67**	**68**	**69**	**70**
A	D	C	D	C	B	D	B	C	A
71	**72**	**73**	**74**	**75**	**76**	**77**	**78**	**79**	**80**
C	D	C	A	D	A	D	B	B	D
81	**82**	**83**	**84**	**85**	**86**	**87**	**88**	**89**	**90**
B	A	C	B	C	D	C	C	A	D
91	**92**	**93**	**94**	**95**	**96**	**97**	**98**	**99**	**100**
A	B	B	D	D	C	B	A	C	A
101	**102**	**103**	**104**	**105**	**106**	**107**	**108**	**109**	**110**
D	A	D	A	C	B	A	B	D	A
111	**112**	**113**	**114**	**115**	**116**	**117**	**118**	**119**	**120**
B	C	D	B	C	C	A	B	A	A

Previous Paper (Solved)

B.Sc. BIOTECHNOLOGY

Entrance Test, 2020*

1. Genetically Modified (GM) plants have been useful in many ways. Genetic modification has:
   A. Decreased efficiency of mineral usage by plants
   B. Genetically Modified (GM) plants helped to reduce post-harvest losses
   C. Genetically Modified (GM) plants reduced reliance on chemical pesticides (pest-resistant crops)
   D. Genetically Modified (GM) plants made crops more tolerant to abiotic stresses (cold, drought, salt, heat).

2. In which year, the Air (Prevention and Control of Pollution) Act was amended to include noise as an air pollutant?
   A. 1981 B. 1978
   C. 1987 D. 1997

3. Which of the following groups has recorded the maximum no. of species?
   A. Algae B. Fungi
   C. Mammalia D. Aves

4. Succession of plants in wet areas is called:
   A. Hydrarch succession
   B. Xerarch succession
   C. Hydroponic
   D. Hydrophilic succession

5. Who formulated the law of "Competitive Exclusion Principle" to explain the possibility of co-existence of two closely related species?
   A. Georgy Gause B. Alfred J. Lotka
   C. Vito Volterra D. Richard Goodwin

6. Which of the following gene codes a toxin protein that controls corn borer?
   A. cryIIAb B. cryIAb
   C. cryIAc D. cryIIIAbc

7. Who is called as the Darwin of the 20th century?
   A. Ernst Mayr
   B. Greg Mendel
   C. H.G. Khorana
   D. G.N. Ramachandran

8. What is the biological name of Men?
   A. Primate B. Homo sapiens
   C. Adam D. Hominidae

9. Which of the following famous botanical garden is located at Lucknow?
   A. Royal Botanic Gardens
   B. National Botanical Research Institute
   C. Indian Botanical Garden
   D. Botanical Gardens of the Indian Republic

10. Identify the correct order of taxonomical sequences?
    A. Species, Genus, Order, Phylum, Kingdom
    B. Genus, Order, Species, Phylum, Kingdom
    C. Species, Order, Genus, Kingdom, Phylum
    D. Species, Order, Genus, Phylum, Kingdom

11. The HIV genome consists of:
    A. Two identical single-stranded DNA molecules that are enclosed within the core of the virus particle.
    B. Two identical single-stranded RNA molecules that are enclosed within the core of the virus particle.

* *Exam conducted by Jamia Millia Islamia University*

C. One identical single-stranded RNA molecules that are enclosed within the core of the virus particle.
D. None of these

12. Which organism was used for proving the hypothesis of semi-conservative mode of replication of DNA by Matthew Meselson and Franklin Stahl in 1957?
A. *E. coli* B. *Oryzasativa*
C. *Mirabillisjalapa* D. *Lathyrusodaratus*

13. How many base pairs are there in E. coli whose DNA is 1.36 mm in length?
A. 4000 bp B. 1,36,000 bp
C. 4,00,000 bp D. 40,000 bp

14. Mendel's experimental material was:
A. *Mirabillisjalapa*
B. *Oryzasativa*
C. *Lathyrusodaratus*
D. *Pisumsativum*

15. Mendel presented his work in:
A. Natural History Society in America
B. Natural History Society in Germany
C. Natural History Society in Brunn
D. Natural History Society in Russia

16. The innate tendency of offspring to resemble their parents is called:
A. Resemblance
B. Variation
C. Heredity
D. Inheritance

17. The reproductive cycle or the menstrual cycle occurs in the female primates of:
A. Monkeys
B. Apes
C. Human beings
D. All of the above

18. Which of the following is a part of the hind brain?
A. Cerebellum and medulla
B. Cerebrum
C. Corpora quadrigemina
D. Cerebral aqueduct

19. Which one of the following is not a layer of cranial meninges in brain?
A. Dura mater
B. Pia mater
C. Meningitis
D. Arachnoid

20. The no. of bones in human cranium is:
A. 14 B. 8
C. 206 D. 33

21. Proteins involved in muscle contraction are:
A. Opsin and rhodopsin
B. Haemoglobin and myoglobin
C. Carbonic anhydrase and glucose oxidase
D. Actin and myosin

22. How many chromosomes are there in a fruit fly?
A. 8 B. 46
C. 23 D. 42

23. Which one of the following is a steroid hormone?
A. Insulin B. Epinephrine
C. Cortisol D. Glucagon

24. What are the two main types of cells present in the Islet of Langerhans?
A. α-cells and β-cells
B. B cells
C. White blood cells
D. Red blood cells

25. Melatonin is secreted by:
A. Parathyroid gland
B. Adrenal gland
C. Thyroid gland
D. Pineal gland

26. Auxin was first isolated from:
A. Sunflower
B. Bovine serum
C. Lotus root
D. Human urine

27. Which one of the following statements is incorrect?
A. Carotenoids is blue in colour
B. Xanthophyll is yellow in colour

C. Chlorophyll b is yellow green in colour
D. Chlorophyll a is bright or blue green in colour

28. During photosynthesis, in which part of cell, is sugar synthesised?
A. Stroma
B. Lysosome
C. Ribosome
D. Mitochondria

29. Older dying leaves export much of their mineral content to younger leaves through a process called remobilisation. Which of the following mineral ion is generally not remobilised.
A. Potassium
B. Nitrogen
C. Calcium
D. Sulphur

30. During the cell division in apical meristem, the nuclear membrane appears in:
A. Cytokinesis B. Anaphase
C. Telophase D. Metaphase

31. Which of the following is not the key feature of metaphase during cell division?
A. Chromosomes are seen to be composed of two chromatids attached together at the centromere.
B. Chromosomes get aligned along metaphase plate through spindle fibres.
C. Chromosomes are moved to spindle equator.
D. Spindle fibres attach to kinetochores of chromosomes.

32. In a polysaccharide, a bond called .... links the individual monosaccharides.
A. Ionic bond
B. Phosphodiester bond
C. Glycosidic bond
D. Peptide bond

33. Major component of plant cell wall is ....
A. DNA B. RNA
C. Proteins D. Cellulose

34. If the external solution balances the osmotic pressure of the cytoplasm, it is said to be....
A. Exothermic
B. Hypertonic
C. Hypotonic
D. Isotonic

35. Which protein is present in membrane and help in transport of molecules and small proteins to pass through the membrane?
A. Tubulin
B. Collagen
C. Transferrin
D. Porin

36. The best bio material for the study of mitosis in laboratory is:
A. Ovary B. Leaf tip
C. Anther D. Root tip

37. In a somatic cell cycle, DNA synthesis take place in:
A. Prophase of mitosis
B. S phase
C. G 2 prophase
D. G 1 phase

38. Who discovered triple helical structure of collagen?
A. James Watson
B. Srinivasa Ramanujan
C. G.N. Ramachandran
D. H.G. Khorana

39. Ureotelic animal excretes:
A. Urea
B. Uric acid
C. Ammonia
D. Nitrogen

40. Which one of the following is incorrect for an isobilateral leaf?
A. Reticulate venation
B. Parallel venation
C. The mesophyll is not differentiated into palisade and spongy parenchyma
D. The stomata are present on both the surfaces of the epidermis

41. A staminode is:
A. A fertile stamen
B. A sterile carpel
C. A stamina with ovules
D. A sterile stamen

42. The most notable disease caused by prions is:
A. Bovine spongiform encephalopathy (BSE)
B. Parkinson's disease
C. Alzheimer's disease
D. Meningitis

43. Identify the incorrect habitat of archaebacteria.
A. Saliva
B. Marshy areas
C. Hot springs
D. Extreme salty areas

44. Which of the following is correct as per the Five Kingdom Classification proposed by R.H. Whittaker in 1969.
A. Monera, Protista, Fungi, Plantae and Animalia.
B. Prokaryotica, Protista, Fungi, Plantae and Animalia.
C. Prokaryotica, Fungi, Plantae, Animalia and Eukaryotica
D. Algae, Bryophytes, Pteridophytes, Gymnosperms and Angiosperms

45. Which of the following is the smallest living cell that can survive without oxygen?
A. E. coli
B. Mycoplasma
C. HIV
D. Corona virus

46. Which one of the following is not an example of class Aves?
A. Pavo
B. Camelus
C. Columba
D. Corvus

47. The Anatomy of Seed Plants was written by:
A. Katherine Esau
B. Marx Albert
C. Lewin Gene
D. Albert Lehringer

48. Chlorophyll d is present in:
A. Rhodophyceae
B. Phaeophyceae
C. Fucoxanthin
D. Chlorophyceae

49. Carrageen is a
A. Fungi
B. Virus
C. Brown algae
D. Red algae

50. Bohr's orbitals are called stationary because:
A. Electrons in Bohr's orbitals are stationary
B. The protons remains in nuclei and are stationary.
C. The electrons in Bohr's orbitals have a fixed energy.
D. Bohr's orbitals have a fixed radii

51. BF_3 acts as an acid according to a concept of:
A. Bronsted
B. Arrhenius
C. Lewis
D. Michaelis and Menten

52. The isomers which can be converted into another form by rotation of the molecule around single bond are called:
A. Geometrical isomers
B. Enantiomers
C. Diasteoreomers
D. Conformers

53. In an aqueous solution, hydrogen will not reduce:
A. Fe^{3+} B. Cu^{2+}
C. Ag^{+} D. Zn^{2+}

54. Dehydration of alcohol is an example of:
A. Addition reaction
B. Substitution reaction
C. Redox reaction
D. Elimination reaction

55. Which of the following has least polarity in bond?
A. H-S B. H-O
C. H-Cl D. H-F

56. Tyndall is effect that can be observed in:
A. Precipitate
B. Unsaturated solute
C. Good solvent
D. Colloidal solution

57. If gas is expended at constant temperature:
A. The number of molecules of gas increases
B. The kinetic energy of the molecules decreases
C. The kinetic energy of the molecules remains the same
D. The pressure increases

58. The process of separation of a racemic mixture in to d- and I-enantiomers is called:
A. Dehydrohalogenation
B. Revolution
C. Dehydration
D. Resolution

59. How many chain isomers could be obtained from the C_6H_{14}?
A. 4 B. 5
C. 6 D. 7

60. The molar concentration of 20 gram of NaOH present in 5 litre of solution is:
A. 1.0 moles/litre
B. 0.5 moles/litre
C. 0.2 moles/litre
D. 0.1 moles/litre

61. One of the following molecules is not a polymer:
A. Cellulose B. Protein
C. DNA D. Amino acid

62. Electrons in the outer orbit are called:
A. Orbitals B. Valences
C. Shells D. Nuclei

63. Tree of 21st Century is:
A. Rubber B. Neem
C. Mango D. Acacia

64. Most stable ecosystem is:
A. Ocean B. Mountain
C. Desert D. Forest

65. The pH of a solution is determined by:
A. Concentration of salt
B. Dielectric constant of the medium
C. Relative concentration of acids and bases
D. Environmental effect

66. One nanometre is equal to
A. 10^{-6} mm B. 10^{-6} m
C. 10^{9} m D. 10^{-9} mm

67. When a planet comes nearer to sun, it moves:
A. Standstill
B. Constant at every point
C. Slower
D. Faster

68. In p-type semiconductor germanium is doped with:
A. Aluminium
B. Gallium
C. Boron
D. All of the above

69. Which one of the following is a bad conductor of electricity?
A. Copper
B. Plastic
C. Silver
D. Gold

70. Among the following which is used for the reductive cleavage of disulphide bonds:
A. Urea
B. SDS
C. Performic acid
D. β-Mercaptoethanol

71. According to Human Genome Project, genetic similarity among all humans is:
A. 90% B. 99.9%
C. 99.5% D. 95%

72. The period of geostationary artificial satellite is:
A. 24 hours
B. 30 days
C. 7 days
D. 365 days

73. For a spontaneous chemical process, the free energy change is:
A. No change
B. Either positive or negative
C. Negative
D. Positive

74. Water falls from a height of 500 m. The rise in temperature at bottom if whole of the energy remains in water, will be:
(Specific heat of water = 4.2 kJ/Kg)
A. 1.02 °C
B. 0.96 °C
C. 1.16 °C
D. 0.23 °C

75. The SI unit of power is:
A. Joule
B. Ampere
B. Ohm
D. Watt

76. The value of 'g' at a particular point is 9.8 m/sec^2, suppose the earth suddenly shrink uniformly to half its present size without losing any mass. The value of 'g' at the same point (assuming that the distance of the point from the centre of the earth does not shrink) will become:
A. 9.8 m^2/s^2
B. 19.6 m/s^2
C. 4.9 m/s^2
D. 9.8 m/s^2

77. A point object is 24 cm above the surface of water (m = 4/3) in lake. A fish inside the water will observe the image to be at a point.
A. 18 cm below the surface of water
B. 32 cm above the surface of water
C. 6 cm below the surface of water
D. 32 cm below the surface of water

78. Sticky ends are overhanging pieces of single stranded:
A. DNA
B. RNA
C. Proteins
D. Messenger RNA

79. Which of the following viral vectors has the largest capacity for foreign DNA with respect to gene transfer to animal cells?
A. Adenovirus
B. Retrovirus
C. Baculovirus
D. Corona Virus

80. The term Single Cell Protein (SCP) was coined by:
A. Ian Wilmot
B. Louis Pasteur
C. Gregor Mendel
D. Carroll L. Wilson

81. Which one of the following statement about Angiotensin II is incorrect?
A. Angiotensin II is a powerful vasodilator.
B. Angiotensin II activates the adrenal cortex to release Aldosterone.
C. Angiotensin II increases the glomerular blood pressure and thereby GFR.
D. Angiotensin II is powerful vasoconstrictor.

82. Ammonotelism is:
A. The process of excreting uric acid
B. The process of excreting urea and uric acid
C. The process of excreting urea
D. The process of excreting ammonia

83. Who discovered blood groups?
A. Ernst Haeckel
B. Camillo Golgi
C. Karl Landsteiner
D. Thomas Cooley

84. What is the normal platelet count in 1 mm^{-3} of blood?
A. 1,50,000-3,50,000
B. 1-8
C. 1,500-3,000
D. 10,000-80,000

85. Which is the most abundant protein in the whole of the biosphere?
A. Collagen
B. Ribulosebisphosphate Carboxy lase-Oxygenase (RuBisCO)

C. Hemoglobin
D. Serum albumin

86. Which of the following is not correct?
A. Robert Brown discovered the cell.
B. A unicellular organism carries out its life activities within a single cell.
C. Virchow explained that cells are formed from pre-existing cells.
D. Schleiden and Schwann formulated the cell theory.

87. Which type of chromosome has centromere slightly away from the middle of it resulting into one shorter arm and one longer arm?
A. Telocentric chromosome
B. Acrocentric chromosome
C. Sub-metacentric chromosome
D. Metacentric chromosome

88. Which of the following statement is correct for ribosome?
A. Ribosome have only cell membrane, not cell wall.
B. Ribosomes are found in nucleus of a cell.
C. Ribosome is found only in animal cell.
D. Ribosome is a non-membrane bound organelles found in a cell.

89. Who discovered Viroids?
A. Ananda Mohan Chakraborty
B. Charles Darwin
C. Kari Mulis
D. T.O. Diener

90. Chipko movement started to conserve:
A. Grasslands
B. Forests
C. Deserts
D. Soil

91. Which of the following will easily dissolve in a polar solvent?
A. Sodium chloride
B. Argon
C. Methane
D. Benzene

92. The dissociation constant of H_2O at 25°C is:
A. 10^7 M
B. 10^{-7} M
C. 10^{14} M
D. 10^{-14} M

93. Absorption maxima of DNA is around:
A. 260 nm
B. 280 nm
C. 409 nm
D. 180 nm

94. Which of the following alcohols would be most soluble in water?
A. Octanol
B. Butanol
C. Ethanol
D. Methanol

95. The resistance of a wire varies inversely as:
A. Length
B. Temperature
C. Resistivity
D. Area of cross section

96. Voltage is a kind of:
A. Solar energy
B. Fuel energy
C. Kinetic energy
D. Potential energy

97. The velocity of an electron in the innermost orbit of an atom is:
A. 0
B. Average of all
C. Lowest
D. Highest

98. If the momentum of a particle is doubled, then its de-Broglie wave length will become:
A. Unchanged
B. 4 Times
C. 2 Times
D. ½ Times

99. When ultraviolet radiation is incident on a surface, no photoelectrons are emitted. If a second beam causes photoelectrons to be ejected, it may consist of:
A. Radio waves
B. Visible light rays
C. X-rays
D. Infra-red waves

100. 16 gram of oxygen at 37 °C is mixed with 14 gram of nitrogen at 27 °C, the temperature of the mixture is:
A. 32.0 °C
B. 27.0 °C
C. 37.0 °C
D. 30.5 °C

ANSWERS

1	2	3	4	5	6	7	8	9	10
D	C	B	A	A	B	A	B	A	A
11	**12**	**13**	**14**	**15**	**16**	**17**	**18**	**19**	**20**
B	A	C	D	C	C	D	A	C	B
21	**22**	**23**	**24**	**25**	**26**	**27**	**28**	**29**	**30**
D	A	C	A	D	D	A	A	C	B
31	**32**	**33**	**34**	**35**	**36**	**37**	**38**	**39**	**40**
A	C	D	D	D	D	B	C	A	A
41	**42**	**43**	**44**	**45**	**46**	**47**	**48**	**49**	**50**
D	A	A	A	B	B	A	A	D	C
51	**52**	**53**	**54**	**55**	**56**	**57**	**58**	**59**	**60**
C	D	D	D	A	D	D	D	B	D
61	**62**	**63**	**64**	**65**	**66**	**67**	**68**	**69**	**70**
D	B	B	A	C	A	D	D	B	D
71	**72**	**73**	**74**	**75**	**76**	**77**	**78**	**79**	**80**
B	A	C	C	D	D	B	A	A	D
81	**82**	**83**	**84**	**85**	**86**	**87**	**88**	**89**	**90**
A	D	C	A	A	A	A	A	D	B
91	**92**	**93**	**94**	**95**	**96**	**97**	**98**	**99**	**100**
A	D	A	D	D	D	D	D	C	A

Previous Paper (Solved)

B.Sc. BIOTECHNOLOGY

Common Entrance Test (CET), 2019*

MATHEMATICS

1. The adjoint of $A = \begin{pmatrix} 1 & 4 & -2 \\ -2 & -5 & 4 \\ 1 & -2 & 1 \end{pmatrix}$ is:

A. $\begin{pmatrix} 1 & 4 & 2 \\ -2 & -5 & 4 \\ 1 & -2 & 1 \end{pmatrix}$ B. $\begin{pmatrix} 1 & 4 & -2 \\ 2 & -5 & 4 \\ 1 & -2 & 1 \end{pmatrix}$

C. $\begin{pmatrix} 3 & 0 & 6 \\ 6 & 3 & 0 \\ 9 & 6 & 3 \end{pmatrix}$ D. $\begin{pmatrix} 3 & 2 & 1 \\ 4 & 1 & -1 \\ 0 & 3 & 4 \end{pmatrix}$

2. If A is a square matrix of order 3, then (adj A). A =

A. A. (adj A) B. A × (adj A)

C. A – (adj A) D. A + (adj A)

3. The inverse of $A = \begin{pmatrix} 2 & 3 \\ 2 & 5 \end{pmatrix}$ is:

A. $\begin{pmatrix} 5/4 & -3/4 \\ 1/2 & 1/2 \end{pmatrix}$ B. $\begin{pmatrix} 5/4 & 3/4 \\ -1/2 & 1/2 \end{pmatrix}$

C. $\begin{pmatrix} 5/4 & -5/4 \\ -1/2 & 1/2 \end{pmatrix}$ D. $\begin{pmatrix} 5/4 & -3/4 \\ -1/2 & 1/2 \end{pmatrix}$

4. If $A = \begin{pmatrix} 3 & 2 & x \\ 4 & 1 & -1 \\ 0 & 3 & 4 \end{pmatrix}$ is a singular matrix, then the value of x is:

A. $11/12$ B. $-11/12$

C. $3/12$ D. $5/4$

5. If $A = \begin{pmatrix} 3 & 1 \\ -1 & 2 \end{pmatrix}$, then $A^2 - 5A + 7I$ is:

A. $\begin{pmatrix} 1 & 0 \\ 0 & 1 \end{pmatrix}$ B. $\begin{pmatrix} 0 & 3 \\ 2 & 0 \end{pmatrix}$

C. $\begin{pmatrix} 0 & 0 \\ 0 & 0 \end{pmatrix}$ D. $\begin{pmatrix} 2 & 3 \\ 2 & 5 \end{pmatrix}$

6. Resolve $\frac{3x+7}{(x-1)(x-2)}$ into partial fractions:

A. $\frac{12}{(x-2)} - \frac{10}{(x-1)}$ B. $\frac{13}{(x-2)} - \frac{10}{(x-1)}$

C. $\frac{13}{(x-5)} - \frac{10}{(x-1)}$ D. $\frac{13}{(x-2)} - \frac{10}{(x-7)}$

7. Resolve $\frac{5x^2+1}{x^2-3}$ into partial fractions:

A. $\frac{12}{(x-2)} - \frac{10}{(x-1)}$ B. $\frac{13}{(x-2)} - \frac{10}{(x-1)}$

C. $\frac{13}{(x-5)} - \frac{10}{(x-1)}$ D. $\frac{2}{(x-1)} + \frac{3x+1}{x^2+x+1}$

8. If $\tan^2\theta + \sec\theta = 5$, then the value of $\cos\theta$ is:

A. $-1/3$ or $1/2$ B. $-11/12$ or $1/2$

C. $13/12$ or $-1/3$ D. $5/4$ or $1/2$

* *Exam Held on 30 April, 2019 (AP-ECET).*

9. The value of 16 $\sin^3\theta$ + 8 $\cos^3\theta$ is:

A. 3 B. 1

C. –3 D. 0

10. If $\sin\alpha = \frac{15}{17}$, $\cos\beta = \frac{12}{13}$, then the value of $\sin(\alpha + \beta)$ is:

A. $\frac{110}{105}$ B. $-\frac{121}{152}$

C. $\frac{220}{221}$ D. $\frac{5}{4}$

11. The value of cos 20° cos 40° cos 60° cos 80° is:

A. $\frac{11}{12}$ B. $\frac{1}{16}$

C. $\frac{13}{12}$ D. $\frac{5}{4}$

12. The value of $\frac{\cos 17° + \sin 17°}{\cos 17° - \sin 17°}$ is:

A. cos 20° B. tan 65°

C. tan 60° D. tan 62°

13. The value of $\sin\frac{\pi}{5}\sin\frac{2\pi}{5}\sin\frac{3\pi}{5}\sin\frac{4\pi}{5} =$

A. $\frac{4}{15}$ B. $\frac{5}{16}$

C. $\frac{-5}{16}$ D. $\frac{7}{15}$

14. If $\tan^{-1}x + \tan^{-1}y + \tan^{-1}z = \frac{\pi}{2}$, then the value of $xy + yz + zx$ is:

A. –1 B. 3

C. 5 D. 1

15. The general solution of 4 $\cos^2 x - 3 = 0$ is:

A. $2n\pi \pm \frac{\pi}{6}$ B. $2n\pi \pm \frac{7\pi}{6}$

C. $3n\pi \pm \frac{5\pi}{6}$ D. $2n\pi \pm \frac{11\pi}{6}$

16. The modulus of a complex number $\sqrt{3} + i$ is:

A. –2 B. 3

C. 2 D. 5

17. The value of

$(a-b)^2\cos^2\left(\frac{C}{2}\right) + (a+b)^2\sin^2\left(\frac{C}{2}\right)$ is:

A. c^3 B. c

C. c^5 D. c^2

18. If $x + \frac{1}{x} = 2\cos\theta$, then the value of $x^n + \frac{1}{x^n}$ is:

A. 2 cos $n\theta$ B. –2 cos $n\theta$

C. 3 cos θ D. 2 sin $n\theta$

19. The value of $2\tan^{-1}\left(\frac{1}{3}\right) + \tan^{-1}\left(\frac{1}{7}\right)$ is:

A. $\frac{\pi}{4}$ B. $-\frac{\pi}{4}$

C. $\frac{\pi}{6}$ D. $\frac{\pi}{3}$

20. The length of the major axis of the ellipse: $4x^2 + 3y^2 = 48$ is:

A. 10 B. 11

C. 8 D. 13

21. The Centre of the ellipse:

$9x^2 + 25y^2 - 18x + 100y - 116 = 0$ is:

A. (2, –1) B. (–1, –2)

C. (1, –2) D. (1, 2)

22. The equation of the parabola with vertex (2, –1) and focus (2, –3) is:

A. $x^2 - 4x + 8y + 12 = 0$

B. $x^2 - 4x - 8y - 12 = 0$

C. $x^2 + 4x - 8y - 12 = 0$

D. $x^2 + 5x - 8y - 11 = 0$

23. The length of the latus rectum of the hyperbola: $\frac{x^2}{9} - \frac{y^2}{16} = 1$ is:

A. 32/3 units B. 5 units

C. 6 units D. 13 units

24. If the length of latus rectum is $\frac{9}{2}$ and the distance between its foci is 10, then the equation of hyperbola is:

A. $\frac{x^2}{16}+\frac{y^2}{9}=1$ B. $\frac{x^2}{18}-\frac{y^2}{9}=1$

C. $\frac{x^2}{16}-\frac{y^2}{6}=1$ D. $\frac{x^2}{16}-\frac{y^2}{9}=1$

25. The equation of the parabola with focus at (–3, 2) and vertex (–2, 2) is:

A. $x^2 - 4x + 8y + 12 = 0$

B. $x^2 + 5x - 8y - 11 = 0$

C. $y^2 + 4x - 4y + 12 = 0$

D. $x^2 - 4x - 8y - 12 = 0$

26. If $y=\frac{a+bx}{b-ax}$, then the derivative of y with respect to x is:

A. $\frac{a^2+b^2}{(b-ax)^2}$ B. $\frac{a^2+b^2}{(b+ax)^2}$

C. $\frac{a^2-b^2}{(b-ax)^2}$ D. $\frac{a+b}{(b-ax)^2}$

27. If $y=\frac{2+3\sin hx}{3+2\sin hx}$, then the derivative of y with respect to x is:

A. $\frac{5\cos hx}{(3+2\sin hx)^2}$ B. $\frac{5\sin hx}{(3+2\sin hx)^2}$

C. $\frac{5\sin x}{(3-2\cos hx)^2}$ D. $\frac{\sin h^2x}{(2-3\sin hx)^2}$

28. The range of x for which the function $x^3 - 3x^2 - 45x + 2$ is increasing with x is:

A. (3, –5) B. (–3, –5)

C. (3, 5) D. (–3, 5)

29. If u is a homogeneous function of x and y with degree n, then $x\frac{\partial u}{\partial x}+y\frac{\partial u}{\partial y}=$

A. $-nu$ B. n^2u

C. nu D. $nu^2 + u$

30. The angle between the curves $y = x^2 + 3x - 7$ and $y^2 = 2x + 5$ at (2, 3) is:

A. $\tan\theta = 2$ B. $\sec\theta = 2$

C. $\cos\theta = 1$ D. $\sin\theta = 3$

31. The maximum value of the function $2x^3 - 12x^2 + 18x + 5$ is:

A. 13 B. 12

C. 10 D. 15

32. The three sides of a trapezium are equal each being 6" long then the area of the trapezium when it is maximum is:

A. 27 square units

B. 33 square units

C. $27\sqrt{3}$ square units

D. $29\sqrt{3}$ square units

33. The interval in which the function $f(x) = x^2 \log x$ is an increasing function is:

A. $(1, e^{-1/2})$ B. $(2, e^{-1/2})$

C. $(0, e^{1/2})$ D. $(0, e^{-1/2})$

34. The stationary points and the corresponding values of the function $f(x) - x^3 - 9x^2 + 15x - 1$ is:

A. 6, –26 B. 3, –26

C. 6, 26 D. –6, –26

35. If $u=\log\left(\frac{x^2+y^2}{x+y}\right)$, then $x\frac{\partial u}{\partial x}+y\frac{\partial u}{\partial x}=$

A. 2 B. 4

C. 5 D. 1

36. The value of $\int \log x\,dx$ is:

A. $x\log x + x + c$ B. $x^2\log x - x + c$

C. $x\log x - x + c$ D. $x\log x-\frac{x^2}{2}+c$

37. The value of $\lim_{n\to\infty}\left[\frac{1}{n+1}+\frac{1}{n+2}+....+\frac{1}{n+n}\right]$ is:

A. log 2 B. log 3

C. –log 2 D. log n

38. The value of $\int\frac{\cos\sqrt{x}}{\sqrt{x}}dx$ is:

A. $2\sin\sqrt{x}+c$ B. $3\sin\sqrt{x}+c$

C. $2\sin x+c$ D. $\sin\sqrt{x}+c$

39. The area enclosed between the curve $y^2 = 4ax$ and the line $x = 2y$ is:

A. $\frac{64}{5}$ square units B. $\frac{64}{3}$ square units

C. $\frac{65}{4}$ square units D. $\frac{63}{4}$ square units

40. The value of $\int_1^{\frac{\pi}{2}} \sin^2 x\, dx$ is:

A. $\frac{\pi}{2}$ B. $-\frac{\pi}{4}$

C. $\frac{\pi}{6}$ D. $\frac{\pi}{4}$

41. The value of $\int_1^4 \left(\sqrt{x}+\frac{1}{\sqrt{x}}\right) dx$ is:

A. $\frac{20}{3}$ B. $-\frac{20}{3}$

C. $\frac{10}{3}$ D. $\frac{15}{3}$

42. The value of $\int_0^{\pi/4} \sqrt{1+\sin 2x}\, dx =$

A. −1 B. −3

C. 3 D. 1

43. The value of $\int_0^{\pi/2} \frac{\sin x}{1+\cos^2 x} dx =$

A. $\frac{\pi}{4}$ B. $-\frac{\pi}{4}$

C. $\frac{\pi}{3}$ D. $\frac{\pi}{2}$

44. The particular integral of $(D^2 + 5D + 6)y = e^x$ is:

A. $\frac{-e^{-x}}{12}$ B. $\frac{e^{2x}}{12}$

C. $\frac{e^x}{12}$ D. $\frac{e^x}{6}$

45. Form the differential equation by eliminating the arbitrary constant a from $ay^2 = x^3$:

A. $\frac{dy}{dx} = \frac{3y}{2x}$ B. $\frac{dy}{dx} = \frac{2x}{3y}$

C. $\frac{dy}{dx} = \frac{x}{y}$ D. $\frac{dy}{dx} = \frac{2y}{x}$

46. The solution of $\frac{dy}{dx} + y = e^{-x}$ is:

A. $(x + c)e^{-x}$ B. $(x - c)e^x$

C. $(x + c)e^x$ D. $(x + c)e^{-2x}$

47. The complementary function of

$(D^2 + 3D + 2)y = 8 \sin 5x$ is:

A. $c_1e^{-x} + c_2e^{-2x}$ B. $c_1e^x + c_2e^{2x}$

C. $c_1e^{-x} + c_2e^{2x}$ D. $c_1e^{2x} + c_2e^{3x}$

48. The solution of exact differential equation $2xy\, dx + x^2\, dy = 0$ is:

A. $x^2y^2 = c$ B. $x^2y = c$

C. $x^3y = c$ D. $x^2y^3 = c$

49. Form the differential equation representing the family of curves $x^2 - 4ay$, where a is any arbitrary constant:

A. $x\frac{dy}{dx} - 2y = 0$ B. $x\frac{dy}{dx} + 2y = 0$

C. $x\frac{dy}{dx} - 6y = 0$ D. $x\frac{dy}{dx} - y = 0$

50. The solution of $\frac{dy}{dx} + y\cot x = \cos x$ is:

A. $y\sin x = \frac{-\cos 2x}{4} + c$

B. $y\sin x = \frac{\cos 2x}{4} + c$

C. $y\sin x = \frac{-\cos 5x}{4} + c$

D. $y\cos x = \frac{-\cos 2x}{4} + c$

PHYSICS

51. In the equation $\frac{\alpha}{t^2} = Fv + \frac{\beta}{x^2}$ the dimensional formula for [α], [β] is (here t = time, F = force, v = velocity, x = distance)

A. MLT^{-1}, MLT^{-3} B. ML^2T, ML^4T^2
C. ML^2T^{-1}, ML^4T^{-3} D. ML^3T^{-1}, MLT^{-3}

52. Which of the following quantities has not been expressed in proper units?
A. Young's modulus = N/m^2
B. Surface tension = N/m
C. Pressure = N/m^2
D. Energy = kg m/s

53. Three vectors A, B and C satisfy the relation A.B = 0 and A.C = 0. The vector A is parallel to:
A. B B. C
C. B . C D. B × C

54. If three vectors A, B and C are 12, 5 and 13 in magnitude such that C = A + B, then the angle between A and B is:
A. 60° B. 90°
C. 120° D. 30°

55. A stone dropped from a certain height, can reach the ground in 5s. It is stopped after 3 seconds of its fall and then allowed to fall again. The time taken by the stone to reach the ground for the remaining distance is:
A. 2 s B. 6 s
C. 4 s D. 1 s

56. The range of projectile fired at an angle of 15° is 50 m. If it is fired with the same speed at an angle of 45°, its range will be:
A. 25 m B. 37 m
C. 50 m D. 100 m

57. A freely falling body acquires a velocity 'v' m/s in falling through a distance of 80 m. How much further distance should it fall, so as to acquire a velocity of '2v' m/s? (Take g = 10 m/s^2)
A. 240 m B. 200 m
C. 400 m D. 280 m

58. A block is projected along a rough horizontal road with a speed of 10 m/s. If the coefficient of kinetic friction is 0.10, how far will it travel before coming to rest?
A. 50 m B. 60 m
C. 40 m D. 10 m

59. What force is required to push a 200 N body up a 30° smooth incline with an acceleration of 2 m/s^2? The force is to be applied along the plane is (Take g = 10 m/s^2)
A. 40 N B. 60 N
C. 80 N D. 140 N

60. A block of mass 2 kg rests on a rough inclined plane making an angle of 30° with the horizontal. The coefficient of static friction between the block and the plane is 0.7. The frictional force on the block is:
A. 9.8 N B. 0.78 × 9.8 N
C. $9.8 \times \sqrt{3}$ N D. $0.7 \times 9.8\sqrt{3}$ N

61. A man moves on a straight horizontal road with a block of mass 2 kg in his hand. If he coveres a distance of 40 m with an acceleration of 0.5 m/s^2, the work done by the man on the block during the motion is: (Take g = 10 m/s^2)
A. 40 J B. 1 J
C. 80 J D. 20 J

62. In a factory it is desired to lift 2000 kg of metal through a distance of 12 m in 1 minute. The minimum horse power of the engine to be used is:
A. 3.5 B. 5.3
C. 4.3 D. 5.8

63. Energy harnessed from flowing water is called _____ energy.
A. Hydel B. Solar
C. Tidal D. Geothermal

64. When a particle executing simple harmonic motion passes through the mean position, it has:
A. minimum K.E. and maximum P.E.

B. maximum K.E. and maximum P.E.
C. maximum K.E. and minimum P.E.
D. minimum K.E. and minimum P.E.

65. A particle of mass 200 g executes a simple harmonic motion. The restoring force is provided by a spring of spring constant 80 N/m. The time period is:
A. 0.2 s B. 0.41 s
C. 0.31 s D. 0.5 s

66. The temperature at which the speed of sound will be double of its value at 10°C is:
A. 819°C B. 850°C
C. 919°C D. 900°C

67. If the source of sound moves towards an observer, then:
A. The frequency of the source is increased
B. The velocity of sound in the medium is increased
C. The wavelength of sound in the medium towards the observer is decreased
D. The amplitude of vibration of the particles is increased

68. A cinema hall has a volume of 7500 m^3. The total absorption in the hall if the reverberation time of 1.5 s is to be maintained is:
A. 800 OWU B. 925 OWU
C. 950 OWU D. 825 OWU

69. One mole of oxygen is heated at constant pressure starting at 0°C. The heat energy that must be supplied to the gas to double its volume is:
A. 2.5 × 273 × R B. 3.5 × 273 × R
C. 2.5 × 546 × R D. 3.5 × 546 × R

70. A vessel contains a gas at a temperature of 27°C and a pressure of 20 atm. If one half of the gas is released and the temperature of the remaining gas is raised by 50°C, the new pressure will be:
A. 12.24 atm B. 11.67 atm
C. 13.79 atm D. 11 atm

71. The temperature of 5 gm of air is raised from 0°C to 1°C. The increase in the internal energy of air is: (C_V = 0.172 cal/gm/°C and J = 4.18 × 10^7 erg/cal)
A. 3.595 × 10^7 erg B. 3 × 10^7 erg
C. 4.5 × 10^7 erg D. 2.595 × 10^7 erg

72. In all reversible processes entropy of the system:
A. decreases B. increases
C. remains constant D. remains zero

73. If one mole of a monoatomic gas ('Y' = 5/3) is mixed with one mole of a diatomic gas ('Y' = 7/5), the value of 'Y' for the mixture is:
A. 1.40 B. 1.50
C. 1.53 D. 3.07

74. Electrons are emitted with zero velocity from a certain metal surface when it is exposed to radiations of wavelength 7000 A°. The work function of the metal is:
A. 1 eV B. 1.52 eV
C. 2.52 eV D. 1.77 eV

75. A superconducting material exhibits:
A. zero conductivity and complete diamagnetism
B. zero resistivity and complete paramagnetism
C. infinite conductivity and complete paramagnetism
D. zero resistivity and complete diamagnetism

CHEMISTRY

76. The splitting of spectral lines in a strong magnetic field is called:
A. Stark effect
B. Pauli Exclusion Principle
C. Zeeman effect
D. Aufbau Principle

77. Bohr's model can explain:
A. The spectrum of hydrogen atom only
B. The spectrum of hydrogen molecule
C. The solar spectrum
D. Spectrum of an atom or ion containing one electron only

78. The maximum number of electrons that a d-orbital can accommodate is:
A. 2 B. 6
C. 10 D. 14

79. Magnesium Atomic number is 12, which of the following is the electronic configuration:
A. $1S^2\ 2S^1\ 2P^6\ 3S^2$
B. $1S^2\ 2S^2\ 2P^5\ 3S^2$
C. $1S^2\ 2S^2\ 2P^6\ 3S^2$
D. $1S^2\ 2S^2\ 2P^6\ 3S^1\ 3D^1$

80. N_2 molecule contains:
A. Covalent bond B. Ionic bond
C. Hydrogen bond D. Metalic bond

81. One mole of any of the particles contains:
A. 6.023×10^{-23} B. 6.022×10^{23}
C. 60.23×10^{23} D. 6.023×10^{25}

82. The normality of the solution obtained by dissolving 4 gm of NaOH in 1 litre is:
A. 1 N B. 0.1 N
C. 0.5 N D. 0.02 N

83. Molecular weight of H_2SO_4 is:
A. 92 B. 96
C. 98 D. 99

84. A Lewis acid is a substance which:
A. Accept protons
B. Accept a lone pair of electrons
C. Donate protons
D. Donate a lone pair of electrons

85. P^H of a solution is 9.5, the solution is:
A. Basic B. Acidic
C. Neutral D. Amphoteric

86. Laws of electrolysis were given by:
A. Ostwald B. Faraday
C. Arrhenius D. Volta

87. Common electrolyte used in the salt bridge is:
A. NaOH B. $NaCO_3$
C. KCl D. KOH

88. Standard Reduction Potential of an element is equal to:
A. 1 X Its reduction potential
B. –1 X Its standard oxidation potential
C. –1 X Its reducton potential
D. 1 X Its standard oxidation potential

89. The standard emf for the cell reaction, $Zn + Cu^{+2} \rightarrow Cu + Zn^{2+}$ is 1.10 V at 25°C. The emf of the cell reaction when 0.1 M Cu^{+2} and 0.1 M Zn^{+2} solutions are used at 25°C is:
A. 1.10 V B. 0.11 V
C. –1.10 V D. –0.11 V

90. Which chemical is responsible for permanent hardness of water?
A. KCl B. $MgCl_2$
C. NaCl D. AgCl

91. Permutit is chemically:
A. Sodium Silicate
B. Aluminium Silicate
C. Hydrated Sodium Alumino Silicate
D. Calcium Silicate

92. The cation exchange resin possesses:
A. Acidic group
B. Basic group
C. Amphoteric group
D. Benzo group

93. Chemically the rust is:
A. Fe_2O_3 B. $Fe_2O_3 . FeO$
C. $Fe_2O_3 . XH_2O$ D. $Fe_2O_3 . NH_3$

94. Galvanizing is the process of coating iron with:
A. Mg B. Cu
C. Au D. Zn

95. Which of the following is not a thermoplastic?
A. Bakelite B. Polystyrene
C. Polythene D. Nylon

96. Isoprene is a monomer of:
A. Starch B. Cellulose
C. Natural rubber D. Lignin

97. Buna-S is a copolymer of:
A. Butadiene and Styrene
B. Butadiene and Acrylonitrile
C. Butadiene and Isoprene
D. Formaldehyde and Styrene

98. Main constituent of natural gas is:
A. Ethane B. Methane
C. Butane D. Carbon Monoxide

99. Ozone layer is present at:
A. Stratosphere B. Inosphere
C. Thermosphere D. Atmosphere

100. The amount of DO required to aerobically decompose biodegradable organic matter of a given volume of water is:
A. Biochemical Oxygen Demand
B. Biological Oxygen Demand
C. Chemical Oxygen Demand
D. Biomagnification

BIOTECHNOLOGY

101. Which of the following is an upstream process?
A. Product recovery
B. Product purification
C. Media formulation
D. Screening

102. Fungal strain used for the large scale production of citric acid is:
A. *Penicillium chrysogenum*
B. *Aspergillus niger*
C. *Saccharomyces cerevisae*
D. *Lactobacillus*

103. Which of the following components is a rich source of Vitamin B?
A. Peptone B. Yeast extract
C. Beef extract D. Agar

104. The isolation of Cholera causing bacterium, *Vibrio cholorae* by using antibiotics in media is an example of:
A. Selective media
B. Differential media
C. Enriched media
D. Assay media

105. Which of the following carbon sources is not used in media preparation?
A. Carbohydrates B. Oils and fats
C. Hydrocarbons D. Peptones

106. Mutant strain of a bacteria with an additional nutritional requirement that is not seen in the wild type strain is called:
A. Autotroph B. Heterotroph
C. Mixotroph D. Auxotroph

107. MacConkey agar medium inhibits the growth of Gram positive bacteria due to the presence of:
A. Antibiotic B. Phenolethanol
C. Bile salts D. Lactose

108. Azolla is used as a biofertilizer as it contains:
A. Mycorrhiza B. Cyanobacteria
C. Rhizobium D. Humus

109. Bacteria used in a Bio gas plant are:
A. Methanogens
B. Nitrifying Bacteria
C. Denitrifying Bacteria
D. Ammonifying Bacteria

110. Biofertilizer which improves the phosphorous uptake of plants is:
A. Rhizobium B. Azospirillum
C. Nostoc D. Anabaena

111. Covalent bonding between two molecules requires:
A. Electrons with opposite spins
B. Electrons of the same spins
C. Electrons of the same orbital
D. Electrons with different orbital

112. Instrument used for the measurement of optical activity is:
A. Spectrophotometer B. Polarimeter
C. Infantometer D. Calorimeter

113. When voltage 'V' is applied across a pair of electrode (cathode and anode), a potential gradient 'E' is created between the electrodes. 'E' can be calculated as:
A. $E = V/d$ B. $E = (1/V) \times Q$
C. $E = (Vd)/Q$ D. $E = V + d$

114. The radio velocity 'v' of a biomolecule in a medium under constant electric field is 'E' is called electrophoretic mobility denoted as 'μ'. Mathematical expression of μ is:

A. $\mu = E/v$ B. $\mu = v/E$
C. $\mu = 1/Ev$ D. $\mu = vE$

115. A cell without a cell wall is called:

A. Tonoplast B. Amyloplast
C. Protoplast D. Cytoplast

116. If the power of an ocular lens is $10x$ and objective lens is $20x$, the magnification is:

A. 30 times B. 20 times
C. 200 times D. 2000 times

117. Resolving power of a microscope can further be enhanced by:

A. Using an illumination of longer wavelength and by decreasing the numerical aperture
B. Using an illumination of longer wavelength and by increasing the numerical aperture
C. Using an illumination of shorter wavelength and by decreasing the numerical aperture
D. Using an illumination of shorter wavelength and by increasing the numerical aperture

118. Microscopy technique that relies on the specimen interfering with the wavelength of light to produce a high contrast image without the need for dyes or any damage to the sample is:

A. Bright field light microscopy
B. Electron microscopy
C. Fluorescence microscopy
D. Phase contrast microscopy

119. Plasma membrane is impermeable to all molecules except:

A. Glucose B. ATP
C. Urea D. K^+

120. Type of transport that induce a conformational change in proteins?

A. Simple diffusion
B. Facilitated diffusion
C. Active transport
D. Ion driven transport

121. The geometrical device that helps in finding out all possible combinations of male and female gametes is:

A. Bateson square B. Mendel square
C. Punnete square D. Morgan square

122. Test cross is performed to determine the:

A. genotype of a plant
B. phenotype of a plant
C. inbreeding of a plant
D. Allellomorphs

123. Genes in Drosophila males are completely linked. The reason behind this is:

A. Genes are very closely located on chromosomes
B. All genes are coupled
C. No formation of synaptonemal complex
D. No crossing over

124. You cross a p+/v+ p+/v+ male drosophila to a p-/v- p-/v- and obtain the F1 hybrid. Now you cross the F1 male with double recessive female. What will be the recombination phenotype in F2?

A. p+/v+ p-/v- only
B. p+/v+ p-/v- and p-/v- p-/v-
C. p+/v+ p-/v- and p-/v- p-/v- also p+/v- p-/v- and p-/v- p-/v-
D. p+/v+ p-/v- and p-/v- p-/v- also p-/v+ p-/v- and p-/v- p-/v-

125. Chromatin fibres are observed only during:

A. Interphase B. Prophase
C. Metaphase D. Anaphase

126. Chromosomal replication occurs during which phase of cell cycle:

A. G1 B. G2
C. S D. M

127. A bivalent consists of:

A. Two chromatids and one centromere
B. Four chromatids and two centromeres
C. Two chromatids and two centromeres
D. Four chromatids and four centromeres

128. During which stage of meiosis chaismata are first seen:
A. Leptotene B. Zygotene
C. Pachytene D. Diplotene

129. Cellular structures that disappear during mitosis and meiosis:
A. Plasma membrane B. Nucleolus
C. Centromere D. Mitochondria

130. Balbiani rings occur in:
A. Polytene chromosomes
B. Polygenic chromosomes
C. Lampbrush chromosomes
D. B-Chromosomes

131. A Mutation which brings about structural changes in a DNA molecule is known as:
A. Somatic mutation
B. Spontaneous mutation
C. Point mutation
D. Duplication mutation

132. Haemophilia occurs more commonly in males because it is a:
A. Y linked recessive
B. Y linked dominant
C. X linked dominant
D. X linked recessive

133. A colour blind girl will be born when:
A. Her mother and maternal grandfather are colour blind
B. Her father and maternal grandfather are colour blind
C. Her mother is colour blind and father has normal vision
D. Both mother and father have normal vision but grandparents are colour blind

134. In a nucleotide, the nitrogen base is joined to the sugar molecule by:
A. Phosphodiester bond
B. Glycosidic bond
C. Hydrogen bond
D. Tripple bond

135. Which phase of bacterial growth produces secondary metabolites?
A. Lag phase B. Log phase
C. Stationary phase D. Death phase

136. _______ method is used for the enumeration of bacteria in vaccines and cultures.
A. Microscopic Count
B. Membrane filter
C. Plate count
D. Dry weight determination

137. At what temperature the medium is maintained for pour plate method of bacterial culture?
A. 37°C B. 67°C
C. 45°C D. 4°C

138. Method applied for preserving the microbes in an active metabolic state is:
A. Overlaying culture with mineral oil
B. Vacuum drying
C. Lyophilization
D. Cryopreservation

139. The term facultative anaerobe refers to an organism that:
A. Does not use oxygen but tolerates it
B. Requires less oxygen than present in air
C. Is killed by oxygen
D. Uses oxygen when present and grows without oxygen when absent

140. Disinfectants that act by disrupting the microbial membranes are:
A. Halogens
B. Heavy metals
C. Cationic detergents
D. Aldehydes

141. When a substance is added to a liquid medium for inhibiting the growth of unwanted bacteria and favouring the growth of wanted bacteria, it is known as:
A. Differential medium
B. Enrichment medium
C. Selective medium
D. Basal medium

142. Micro-organisms that produce antibiotics are detected by:
A. Wilkins agar plate method
B. Crowded plate method
C. Enrichment method
D. Agar plug method

143. Process of killing and removal of micro-organism is known as:
A. Sterilization B. Pasteurization
C. Disinfection D. Destruction

144. Method applied for calculating the percentage similarity (%S) of each strain to every other strain is:
A. Intuitive Method
B. Numerical Taxonomy
C. Genetic Relatedness
D. DNA homology

145. Bacteria that lives in the root nodules of legumes are:
A. Azotobacter B. Penicillium
C. Rhizobium D. Mycobacterium

146. What is the unit of influent flow rate?
A. m d B. m/d
C. m^2/d D. m^3/d

147. In which type of reactor, aeration is generally accomplished in a separate vessel?
A. Fluidised bed
B. Trickle bed
C. Packed bed
D. Stirred and air-driven reactors

148. Penetration theory assumes that turbulent eddies travel from the bulk of the phase to the interface where they remain constant for a constant exposure time (t_e). The model correlating K_L, mass transfer coefficient and D_{AB}, diffusivity can be expressed as:
A. $K_L = 2(D_{AB}/\pi t_e)^{0.25}$
B. $K_L = 2(D_{AB}/\pi t_e)^{0.5}$
C. $K_L = 2(D_{AB}/\pi_T t_e)^{0.75}$
D. $K_L = 2(D_{AB}/\pi t_e)$

149. The oxygen uptake requirements of a microbial population is characterized by the following parameters: $\mu_m = 0.2\ h^{-1}$, $K_0 = 0.2$ mg $O_2.l^{-1}$, $Y_0 = 0.5$ mg dry weight/mg O_2 and $C_{0,crit} = 0.8\ mg.l^{-1}$. The required concentration of cells is 1000 $mg.l^{-1}$ and the saturation oxygen concentration of the medium is 5.8 $mg.l^{-1}$. The required K_L a must be greater than:
A. 64 h^{-1} B. 32 h^{-1}
C. 16 h^{-1} D. 8 h^{-1}

150. If density and viscosity of the liquid remains constant, then the Reynolds number in a stirred tank reactor will vary with the:
A. Diameter of the impeller
B. Square of the impeller diameter
C. Square root of the impeller diameter
D. Cube of the impeller diameter

151. A double spiral heat-exchanger is a:
A. Direct heat exchanger
B. Indirect heat exchanger
C. Temperature controlled device
D. Thermostat

152. The process of ethanol production by yeast cells under high glucose concentrations rather than biomass production by TCA cycle is known as:
A. Warburg effect B. Sympson's effect
C. Crabtree effect D. Olivosky's effect

153. What is the ratio of diameter of impeller to the diameter of tank ($D_a : D_t$)?
A. 1 : 3 B. 2 : 1
C. 4 : 1 D. 1 : 5

154. At 25°C, the saturation concentration of oxygen in water is 1.26 mmol/l and its partial pressure is 1 atm. The Henry's law constant will be:
A. 0.793 atml/mmol
B. 0.207 atml/mmol
C. 1.26 atml/mmol
D. 8.74 atml/mmol

155. The volume of liquid (V_L) in a cylindrical reactor can be calculated from the liquid height (H_L) and tank diameter (D_t) using the following equation:
A. $V_L = 4/3 \times \pi \times H_L \times D_t^3/8$
B. $V_L = H_L \times \pi \times D_t^2/4$
C. $V_L = H_L \times \pi \times D_t^2$
D. $V_L = 4 \times \pi \times D_t^2$

156. Addition of detergents in an aerated bioreactor will increase the rate of oxygen transfer due to:
A. Increase in the bubble coalescence
B. Increase in the bubbles expansion
C. Decrease in the bubble coalescence
D. Increase in the surface tension of the liquid

157. If a DNA fragment consists of 80 thymine and 80 guanine bases, what is the total number of bases in that fragment:
A. 40 B. 160
C. 320 D. 640

158. Type of sugar present in the DNA is:
A. Triose B. Tetrose
C. Pentose D. Hexose

159. Anticodon region is an important structural component of:
A. DNA B. mRNA
C. tRNA D. SnRNA

160. Which ratio is constant for DNA?
A. A + G / T + C B. A + T / G + C
C. A + C / U + G D. A + U / G + C

161. Segment of DNA that reads the same from forward and backward is called:
A. Palindromic DNA
B. Complementary DNA
C. Plasmid DNA
D. Copy DNA

162. Which of the following is not a nonsense codon?
A. UAA B. UUU
C. UAG D. UGA

163. Enzyme responsible for the unwinding of the DNA during replication is:
A. Polymerase B. Topoisomerase
C. Helicase D. Primase

164. Mode of DNA replication in *Escherichia coli* is:
A. Conservative and unidirectional
B. Semiconservative and unidirectional
C. Conservative and bidirectional
D. Semiconservative and bidirectional

165. DNA and histone proteins in the nucleus of a cell are associated by:
A. Covalent bonding
B. Hydrogen bonding
C. Hydrophobic bonding
D. Vander Waal's interation

166. Peptide bond formation between amino acids of growing polypeptide chain is catalyzed by:
A. Peptidyl transferase
B. Amino acyl t-RNA synthetase
C. Peptide polymerase
D. Peptidyl synthetase

167. Aminoacids are assembled into polypeptide chain on:
A. Nucleus
B. Ribosome
C. Endoplasmic reticulum
D. Glogi

168. The sex complement of a male child suffering with Down's syndrome is:
A. XO B. XY
C. XX D. XXY

169. Albinism is an example of:
A. Dominance
B. Recessiveness
C. Incomplete dominance
D. Incomplete recessiveness

170. Which one of the following methodology can be used for the gene silencing?
A. Transposon insertion
B. PCR
C. Antisense RNA
D. Southern blot hybridization

171. Hormone pair required for the differentiation of callus are:
A. Auxin and Cytokinin
B. Auxin and Gibberellin
C. Auxin and ABA
D. Cytokinin and Gibberellin

172. Which of the following plant cells will show totipotency?
A. Xylem vessels B. Seive tubes
C. Meristem D. Cork cells

173. Protoplasts are the cells devoid of:
A. Cell membrane B. Cell wall
C. Nucleus D. Cytoplasm

174. Cybrids are produced by:
A. Fusion of two different nuclei of two different plants
B. Fusion of two same nuclei from two same plants
C. Fusion of nuclei of one species and cytoplasm of both the species
D. Fusion of two different cytoplasm of from two different plants

175. _______ is required to distinguish between the cell that have taken the vector and that have not.
A. Multiple cloning site
B. Origin of replication
C. High copy number
D. Selection marker

176. *Agrobacterium tumefaciens* is a:
A. Gram (+) Bacteria B. Gram (–) Bacteria
C. Fungi D. Virus

177. Biological nitrogen fixation is the conversion of:
A. N_2 to NO_3^- and NH_3
B. N_2 to N
C. N_2 to Urea
D. N_2 to NH_3

178. Which of the following organisms fix the nitrogen in water logged soil?
A. Azotobacter B. Nitrobacter
C. Nostoc D. Clostridium

179. Insecticides that act by permeating the entire plant are called as:
A. Penetrating pesticides
B. Porous insecticides
C. Contact poisons
D. Systemic insecticides

180. Pesticides that are used to control plant pest are:
A. Toxicols B. Weed agents
C. Herbicides D. Fungicides

181. The first vaccine developed from the animal cell culture was:
A. Hepatitis B Vaccine B. Influenza vaccine
C. Polio vaccine D. Pox vaccine

182. Which of the following is not a source of energy in active muscle cells?
A. Lactic acid B. ATP
C. Creatine Phosphate D. Glucose

183. What are the main constituents needed in the culture for the animal cell Growth?
A. Glucose and glutamine
B. Lactate
C. Cytokines
D. Growth factors

184. Application of Hybridoma cells is to produce:
A. Antigens B. Antibodies
C. Cancer cells D. Cell lines

185. Gene therapy is a technique for _____ the defective genes responsible for disease development.
A. Altering B. Replacing
C. Correcting D. Analysing

186. The onco mouse is also known as:
A. Phidelphia mouse B. Ohio mouse
C. Smart mouse D. Harvard mouse

187. Eggs in mammals are:
A. Alecithal B. Microlecithal
C. Mesolecithal D. Macrolecithal

188. Gene knockout means ______ of a specific gene.
A. Removal B. Purification
C. Inactivation D. Transfection

189. BLAST programme is used for:
A. DNA sequencing B. DNA identification
C. DNA barcoding D. DNA alignment

190. GeneBank and SWISSPORT are examples of:
A. Primary Databases
B. Secondary Databases
C. Composite Databases
D. Tertiary Databases

191. BLOSUM matrix is used in:
A. Single sequence alignment
B. Pairwise sequence alignment
C. Multiple sequence alignment
D. Phylogenetic analysis

192. Enzyme which helps in changing shape of molecule is called:
A. Ligases B. Dehydrogenases
C. Hydrolases D. Isomerases

193. Proteases are commonly rich in:
A. Pineapple B. Papaya
C. Paprika D. Pomegranate

194. Which of the following is produced with the combination of apoenzyme and coenzyme?
A. Holoenzyme
B. Enzyme substrate complex
C. Prosthetic group
D. Enzyme product complex

195. Zymogen is a:
A. Hormone B. Modulator
C. Vitamin D. Enzyme precursor

196. Immobilized enzyme produced by micro encapsulation technique provides:
A. Large surface area
B. Small surface area
C. High amount of solvent
D. Low amount of solvent

197. The rate of substrate transfer during the enzymatic reaction of an immobilized enzyme is:
A. equal to that of substrate consumption
B. more than that of substrate consumption
C. lesser than that of substrate consumption
D. is nothing to do with the substrate consumption

198. For a steady state condition, the change of substrate concentration (dC_s/dt) is:
A. Zero B. 1
C. > 1 D. < 1

199. The catalytic efficiency of two different enzymes can be compared by:
A. Formation of a product
B. Km value
C. Molecular size of the enzymes
D. pH of the optimum value

200. Which of the following factors is responsible for the inhibition of the enzymatic process during feedback inhibition?
A. Enzymes B. End Product
C. Temperature D. Substrate

ANSWERS

1	2	3	4	5	6	7	8	9	10
C	A	D	A	C	B	D	A	A	C
11	**12**	**13**	**14**	**15**	**16**	**17**	**18**	**19**	**20**
B	D	B	D	A	C	D	A	A	C
21	**22**	**23**	**24**	**25**	**26**	**27**	**28**	**29**	**30**
C	A	A	D	C	A	A	D	C	A
31	**32**	**33**	**34**	**35**	**36**	**37**	**38**	**39**	**40**
A	C	D	A	D	C	A	A	B	D
41	**42**	**43**	**44**	**45**	**46**	**47**	**48**	**49**	**50**
A	D	A	C	A	A	A	B	A	A
51	**52**	**53**	**54**	**55**	**56**	**57**	**58**	**59**	**60**
C	D	D	B	C	D	A	A	D	A
61	**62**	**63**	**64**	**65**	**66**	**67**	**68**	**69**	**70**
A	B	A	C	C	A	C	D	B	B
71	**72**	**73**	**74**	**75**	**76**	**77**	**78**	**79**	**80**
A	C	B	D	D	C	D	C	C	A
81	**82**	**83**	**84**	**85**	**86**	**87**	**88**	**89**	**90**
B	B	C	B	A	B	C	B	A	B

91	**92**	**93**	**94**	**95**	**96**	**97**	**98**	**99**	**100**
C	A	C	D	A	C	A	B	A	B
101	**102**	**103**	**104**	**105**	**106**	**107**	**108**	**109**	**110**
C	B	B	A	D	D	C	B	A	B
111	**112**	**113**	**114**	**115**	**116**	**117**	**118**	**119**	**120**
A	B	A	B	C	C	A	D	C	B
121	**122**	**123**	**124**	**125**	**126**	**127**	**128**	**129**	**130**
C	A	C	B	A	C	B	D	B	A
131	**132**	**133**	**134**	**135**	**136**	**137**	**138**	**139**	**140**
C	D	B	B	C	A	C	A	D	C
141	**142**	**143**	**144**	**145**	**146**	**147**	**148**	**149**	**150**
B	B	C	B	C	D	C	B	A	B
151	**152**	**153**	**154**	**155**	**156**	**157**	**158**	**159**	**160**
B	C	A	A	B	C	C	C	C	A
161	**162**	**163**	**164**	**165**	**166**	**167**	**168**	**169**	**170**
A	B	C	D	B	A	B	B	B	C
171	**172**	**173**	**174**	**175**	**176**	**177**	**178**	**179**	**180**
A	C	B	C	D	B	D	D	D	C
181	**182**	**183**	**184**	**185**	**186**	**187**	**188**	**189**	**190**
C	A	A	B	C	D	A	C	D	A
191	**192**	**193**	**194**	**195**	**196**	**197**	**198**	**199**	**200**
B	D	B	A	D	A	A	A	B	B

EXPLANATORY ANSWERS

1. We have, $A = \begin{pmatrix} 1 & 4 & -2 \\ -2 & -5 & 4 \\ 1 & -2 & 1 \end{pmatrix}$

Co-factors of A

$$C_{11} = \begin{vmatrix} -5 & 4 \\ -2 & 1 \end{vmatrix} = (-5)(1) - (4)(-2)$$
$$= -5 + 8 = 3$$

$$C_{12} = \begin{vmatrix} -2 & 4 \\ 1 & 1 \end{vmatrix} = (-2)(1) - (4)(1)$$
$$= -2 - 4 = -6$$

$$C_{13} = \begin{vmatrix} -2 & -5 \\ 1 & -2 \end{vmatrix} = (-2)(-2) - (-5)(1)$$
$$= 4 + 5 = -9$$

$$C_{21} = \begin{vmatrix} 4 & -2 \\ -2 & 1 \end{vmatrix} = 4 - (-2)(-2)$$
$$= 4 - 4 = 0$$

$$C_{22} = \begin{vmatrix} 1 & -2 \\ 1 & 1 \end{vmatrix} = 1 - (-2)(1)$$
$$= 1 + 2 = 3$$

$$C_{23} = \begin{vmatrix} 1 & 4 \\ 1 & -2 \end{vmatrix} = 1(-2) - 4 \times 1$$
$$= -2 - 4 = -6$$

$$C_{31} = \begin{vmatrix} 4 & -2 \\ -5 & 4 \end{vmatrix} = 4 \times 4 - (-2)(-5)$$
$$= 16 - 10 = 6$$

$$C_{32} = \begin{vmatrix} 1 & -2 \\ -2 & 4 \end{vmatrix} = 4 - (-2)(-2)$$
$$= 4 - 4 = 0$$

$$C_{33} = \begin{vmatrix} 1 & 4 \\ -2 & -5 \end{vmatrix} = -5 \times 1 - 4(-2)$$
$$= -5 + 8 = 3$$

$$A = \begin{pmatrix} 3 & -6 & 9 \\ 0 & 3 & -6 \\ 6 & 0 & 3 \end{pmatrix}$$

$\because$ Sign of cofactor is $\begin{pmatrix} + & - & + \\ - & + & - \\ + & - & + \end{pmatrix}$

$$A = \begin{pmatrix} 3 & 6 & 9 \\ 0 & 3 & 6 \\ 6 & 0 & 3 \end{pmatrix}$$

$$\text{Adj } A = \begin{pmatrix} 3 & 6 & 9 \\ 0 & 3 & 6 \\ 6 & 0 & 3 \end{pmatrix}^T = \begin{pmatrix} 3 & 0 & 6 \\ 6 & 3 & 0 \\ 9 & 6 & 3 \end{pmatrix}.$$

3. We have, $A = \begin{pmatrix} 2 & 3 \\ 2 & 5 \end{pmatrix}$

$$|A| = 2 \times 5 - 3 \times 2 = 10 - 6 = 4$$

Cofactors of A are

$$C_{11} = 5, C_{12} = 2$$
$$C_{21} = 3, C_{22} = 2$$
$$= \begin{pmatrix} 5 & -2 \\ -3 & 2 \end{pmatrix}$$

$$\text{Adj } A = \begin{pmatrix} 5 & -2 \\ -3 & 2 \end{pmatrix}^T = \begin{pmatrix} 5 & -3 \\ -2 & 2 \end{pmatrix}$$

$$A^{-1} = \frac{1}{|A|}\text{Adj } A$$

$$= \frac{1}{4}\begin{pmatrix} 5 & -3 \\ -2 & 2 \end{pmatrix} = \begin{pmatrix} \frac{5}{4} & -\frac{3}{4} \\ -\frac{1}{2} & \frac{1}{2} \end{pmatrix}.$$

4. Matrix A is singular iff $|A| = 0$

$$\begin{vmatrix} 3 & 2 & x \\ 4 & 1 & -1 \\ 0 & 3 & 4 \end{vmatrix} = 0$$

$$3(4 - (-1)(3)) - 2(4 \times 4 - 0) + x(4 \times 3 - 0) = 0$$
$$3(4 + 3) - 2(16) + x(12) = 0$$
$$21 - 32 + 12x = 0$$
$$12x = 11$$
$$x = \frac{11}{12}.$$

5. We have,

$$A = \begin{pmatrix} 3 & 1 \\ -1 & 2 \end{pmatrix}$$

$$A^2 = A.A. = \begin{pmatrix} 3 & 1 \\ -1 & 2 \end{pmatrix}\begin{pmatrix} 3 & 1 \\ -1 & 2 \end{pmatrix}$$

$$= \begin{pmatrix} 3\times 3 + (1)(-1) & 3 \times 1 + (1)(2) \\ (-1)(3) + (2)\times(-1) & (-1)\times 1 + 2(2) \end{pmatrix}$$

$$= \begin{pmatrix} 9-1 & 3+2 \\ -3-2 & -1+4 \end{pmatrix} = \begin{pmatrix} 8 & 5 \\ -5 & 3 \end{pmatrix}$$

$A^2 - 5A + 7I$

$$= \begin{pmatrix} 8 & 5 \\ -5 & 3 \end{pmatrix} - 5\begin{pmatrix} 3 & 1 \\ -1 & 2 \end{pmatrix} + 7\begin{pmatrix} 1 & 0 \\ 0 & 1 \end{pmatrix}$$

$$= \begin{pmatrix} 8 & 5 \\ -5 & 3 \end{pmatrix} - \begin{pmatrix} 15 & 5 \\ -5 & 10 \end{pmatrix} + \begin{pmatrix} 7 & 0 \\ 0 & 7 \end{pmatrix}$$

$$= \begin{pmatrix} 8-15+7 & 5-5+0 \\ -5+5+0 & 3-10+7 \end{pmatrix} = \begin{pmatrix} 0 & 0 \\ 0 & 0 \end{pmatrix}.$$

6. $\dfrac{3x+7}{(x-1)(x-2)}$

For partial fractions, we can written this as

$$\frac{3x+7}{(x-1)(x-2)} = \frac{A}{(x-1)} + \frac{B}{(x-2)} \quad ...(i)$$

$$\frac{3x+7}{(x-1)(x-2)} = \frac{A(x-2)+B(x-1)}{(x-1)(x-2)}$$

Compare Numerators

$$3x + 7 = A(x - 2) + B(x - 1)$$
$$3x + 7 = (A + B)x + (-2A - B)$$
$$A + B = 3 \quad ...(a)$$
$$-2A - B = 7 \quad ...(b)$$

Add equation (*a*) and (*b*) we get

$-A = 10$, $A = -10$, $B = 13$

Put value A and B in equation (*i*), we get,

$$= \frac{-10}{(x-1)} + \frac{13}{(x-2)} = \frac{13}{x-2} - \frac{10}{x-1}.$$

7. We have, $\dfrac{5x^2+1}{x^3-1} = \dfrac{5x^2+1}{(x-1)(x^2+x+1)}$

For Partial fractions,

$$\frac{5x^2+1}{(x-1)(x^2+x+1)} = \frac{A}{x-1} + \frac{Bx+C}{x^2+x+1} \quad ...(i)$$

$$\frac{5x^2+1}{(x-1)(x^2+x+1)}$$

$$= \frac{A(x^2+x+1)+(Bx+C)(x-1)}{(x-1)(x^2+x+1)}$$

After Comparing Numerator we get,

$$A + B = 5 \quad ...(a)$$

$$A - B + C = 0 \quad ...(b)$$

$$A - C = 1 \quad ...(c)$$

Put B = 5 – A from equation (*a*) to equation (*b*)

$$A - (5 - A) + C = 0$$

$$2A + C = 5 \quad ...(d)$$

From equation (*c*) and (*d*)

$$A = 2, C = 1, B = 3$$

Put A, B and C value in equation (*i*), we get

$$= \frac{2}{(x-1)} + \frac{3x+1}{(x^2+x+1)}.$$

8. We have, $\tan^2\theta + \sec\theta = 5$

$$\frac{\sin^2\theta}{\cos^2\theta} + \frac{1}{\cos\theta} = 5$$

$$\frac{\sin^2\theta + \cos\theta}{\cos^2\theta} = 5$$

$$\sin^2\theta + \cos\theta = 5\cos^2\theta$$

$$(1 - \cos^2\theta) + \cos\theta = 5\cos^2\theta$$

$$6\cos^2\theta - \cos\theta - 1 = 0$$

$$6\cos^2\theta - 3\cos\theta + 2\cos\theta - 1 = 0$$

$$3\cos\theta(2\cos\theta - 1) + 1(2\cos\theta - 1) = 0$$

$$(2\cos\theta - 1)(3\cos\theta + 1) = 0$$

$$\cos\theta = \frac{1}{2}, -\frac{1}{3}$$

$\because \cos\theta\ \varepsilon\ [-1, 1]$ both value valid.

10. We have, $\sin\alpha = \frac{15}{17}$, $\cos\beta = \frac{12}{13}$

$$\sin(\alpha+\beta) = \sin\alpha\cos\beta + \cos\alpha.\sin\beta$$

$$\cos\alpha = \sqrt{1-\sin^2\alpha} = \sqrt{1-\left(\frac{15}{17}\right)^2}$$

$$= \sqrt{\frac{289-225}{289}} = \sqrt{\frac{64}{289}} = \frac{8}{17}$$

$$\sin\beta = \sqrt{1-\cos^2\beta} = \sqrt{1-\left(\frac{12}{13}\right)^2}$$

$$= \sqrt{\frac{169-144}{169}} = \sqrt{\frac{25}{169}} = \frac{5}{13}$$

$\therefore \sin(\alpha+\beta)$

$$= \frac{15}{17}\times\frac{12}{13} + \frac{8}{17}\times\frac{5}{13} = \frac{180}{221} + \frac{40}{221} = \frac{220}{221}.$$

11. We have,

$$= \cos 20^\circ \cos 40^\circ \cos 60^\circ \cos 80^\circ$$

$$= \cos 20^\circ \cos(60-20)\cos(60+20)\cos 60^\circ$$

$$= \frac{1}{4}\cos(3\times 20^\circ).\cos 60^\circ$$

$$\left[\because \cos\theta.\cos(60-\theta)\cos(60+\theta) = \frac{1}{4}\cos(3\theta)\right]$$

$$= \frac{1}{4}\cos 60^\circ.\cos 60^\circ$$

$$= \frac{1}{4}\cos^2 60^\circ = \frac{1}{4}\times\left(\frac{1}{2}\right)^2 = \frac{1}{16}.$$

12. We have, $\frac{\cos 17^\circ + \sin 17^\circ}{\cos 17^\circ - \sin 17^\circ}$

$$= \frac{\cos 17^\circ + \sin(90-73)}{\cos 17^\circ - \sin(90-73)} = \frac{\cos 17^\circ + \cos 73^\circ}{\cos 17^\circ - \cos 73^\circ}$$

$$= \frac{2\cos\left(\frac{17+73}{2}\right)\cos\left(\frac{17-73}{2}\right)}{2\sin\left(\frac{17+73}{2}\right)\sin\left(\frac{73-17}{2}\right)}$$

$$= \frac{2\cos\left(\frac{90}{2}\right).\cos\frac{56}{2}}{2\sin\left(\frac{90}{2}\right)\sin\frac{56}{2}}$$

$$= \frac{2\cos 45^\circ\cos 28^\circ}{2\sin 45^\circ\sin 28^\circ} = \frac{2\times\frac{1}{\sqrt{2}}\cos 28^\circ}{2\times\frac{1}{\sqrt{2}}\sin 28^\circ}$$

$$= \cot 28^\circ = \cot(90^\circ - 62^\circ) = \tan 62^\circ.$$

14. $\tan^{-1} x + \tan^{-1} y + \tan^{-1} z = \frac{\pi}{2}$

$$\tan^{-1}\left(\frac{x+y+z}{1-xy-yz-zx}\right) = \frac{\pi}{2}$$

$$\frac{x+y+z}{1-xy-yz-zx} = \tan\frac{\pi}{2}$$

$$\frac{x+y+z}{1-xy-yz-zx} = \frac{1}{0}$$

$$1 - xy - yz - zx = 0$$

$$xy + yz + zx = 1.$$

15. $4\cos^2 x - 3 = 0$

$$\cos^2 x = \frac{3}{4}$$

$$\cos x = \frac{\sqrt{3}}{2}$$

$$\cos x = \cos\frac{\pi}{6}$$

$$x = 2n\pi \pm \frac{\pi}{6}.$$

16. We have, $z = \sqrt{3} + i$

$$|z| = \sqrt{(\text{Re}(z))^2 + (\text{Im}(z))^2}$$

$$= \sqrt{\left(\sqrt{3}\right)^2 + (1)^2} = \sqrt{3+1} = \sqrt{4} = 2.$$

17. $(a-b)^2 \cos^2\left(\frac{C}{2}\right) + (a+b)^2 \sin^2\left(\frac{C}{2}\right)$

$$= (a^2+b^2-2ab)\cos^2\frac{C}{2} + (a^2+b^2+2ab)\left(1-\cos^2\frac{C}{2}\right)$$

$$= a^2\cos^2\frac{C}{2} + b^2\cos^2\frac{C}{2} - 2ab\cos^2\frac{C}{2} + a^2 + b^2 + 2ab$$

$$-a^2\cos^2\frac{C}{2} - b^2\cos^2\frac{C}{2} - 2ab\cos^2\frac{C}{2}$$

$$= a^2 + b^2 + 2ab - 4ab\cos^2\frac{C}{2}$$

$$= a^2 + b^2 + 2ab\left(1 - 2\cos^2\frac{C}{2}\right)$$

$$= a^2 + b^2 - 2ab\left(2\cos^2\frac{C}{2} - 1\right)$$

$$= a^2 + b^2 - 2ab\,(\cos C)$$

$$= a^2 + b^2 - 2ab\cos C = c^2.$$

18. We have, $x + \frac{1}{x} = 2\cos\theta$

$$x^2 - 2\cos\theta\, x + 1 = 0$$

$$x = \frac{2\cos\theta \pm \sqrt{4\cos^2\theta - 4}}{2}$$

$$x = \cos\theta \pm \sqrt{\cos^2\theta - 1}$$

$$x = \cos\theta \pm i\sin\theta$$

$\because$ $x + \frac{1}{x} \geq 2$

$\therefore$ x min value $= 1$

$\therefore$ $x = \cos\theta + i\sin\theta$

$$x^n = (\cos\theta + i\sin\theta)^n$$

(By De-moiver's Theorem)

$$= \cos n\theta + i\sin n\theta$$

$$x^{-n} = (\cos\theta + i\sin\theta)^{-n}$$

$$= \cos(-n\theta) + i\sin(-n\theta)$$

$$= \cos n\theta - i\sin n\theta$$

$$x^n + x^{-n} = 2\cos n\theta.$$

19. $2\tan^{-1}\left(\frac{1}{3}\right) + \tan^{-1}\left(\frac{1}{7}\right)$

$$= \tan^{-1}\frac{\left(2\times\frac{1}{3}\right)}{1-\left(\frac{1}{3}\right)^2} + \tan^{-1}\left(\frac{1}{7}\right)$$

$$\left[2\tan^{-1} x = \tan^{-1}\frac{2x}{1-x^2}\right]$$

$$= \tan^{-1}\left(\frac{\frac{2}{3}}{\frac{8}{9}}\right) + \tan^{-1}\left(\frac{1}{7}\right) = \tan^{-1}\left(\frac{3}{4}\right) + \tan^{-1}\left(\frac{1}{7}\right)$$

$$\left[\because \tan^{-1} x + \tan^{-1} y = \tan^{-1}\left(\frac{x+y}{1-xy}\right)\right]$$

$$= \tan^{-1}\left(\frac{\frac{3}{4}+\frac{1}{7}}{1-\frac{3}{4}\times\frac{1}{7}}\right) = \tan^{-1}\left(\frac{\frac{21+4}{28}}{\frac{28-3}{28}}\right)$$

$$= \tan^{-1}\left(\frac{25}{25}\right) = \tan^{-1}(1) = \frac{\pi}{4}.$$

20. We have,

$$4x^2 + 3y^2 = 48$$

$$\frac{x^2}{12} + \frac{y^2}{16} = 1$$

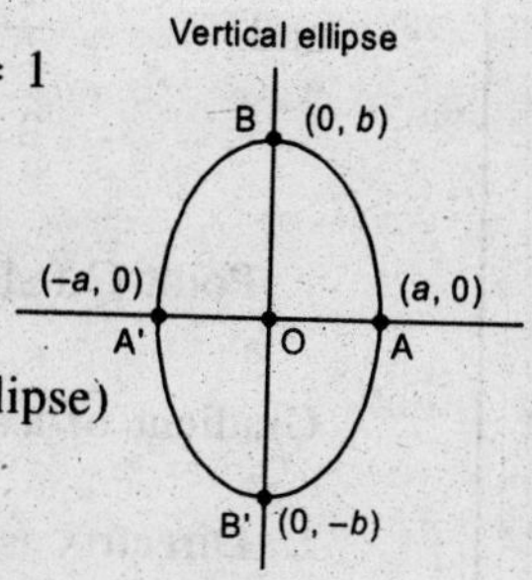

Compare with

$$\frac{x^2}{a^2} + \frac{y^2}{b^2} = 1$$

$\because$ $b > a$, (vertical ellipse)

$a^2 = 12,\ b^2 = 16$

$b = 4$

Major axis of ellipse

$BB' = 2b = 2 \times 4 = 8.$

21. We have,

$9x^2 + 25y^2 - 18x + 100y - 116 = 0$

$= (3x)^2 + (5y)^2 - 2 \times 3x \times 3 + 9 + 2 \times (5y) \times (10) + 100 - 116 - 9 - 100 = 0$

$= [(3x)^2 - 2 \times (3x) \times 3 + (3)^2] + [(5y)^2 + (10)^2 + 2(5y)(10)] = 225$

$\Rightarrow \quad (3x - 3)^2 + (5y + 10)^2 = 225$

$\Rightarrow \quad 9(x - 1)^2 + 25(y + 2)^2 = 225$

$$\Rightarrow \quad \frac{9(x-1)^2}{225} + \frac{25(y+2)^2}{225} = 1$$

$$\Rightarrow \quad \frac{(x-1)^2}{25} + \frac{(y+2)^2}{9} = 1$$

The Centre of the ellipse are $(1, -2)$.

22. Let the point of intersection of Axis and D irectrix is B $(x_1\ y_1)$

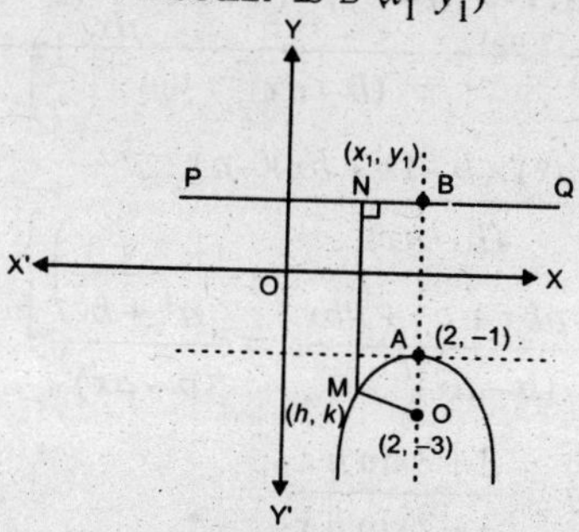

$$\frac{x_1 + 2}{2} = 2$$

$$x_1 = 2$$

$$\frac{y_1 - 3}{2} = -1$$

$$\Rightarrow y_1 - 3 = -2$$

$$y_1 = 1$$

$\therefore$ Point of B $(2, 1)$

$$\text{Gradient of Axis} = \frac{y_2 - y_1}{x_2 - x_1} = \frac{-3+1}{2-2} = \frac{1}{0} = \infty$$

$\because$ Directrix is perpendicular to Axis, so the Gradient of Directrix = 0

$\therefore$ Equation of Directrix

$\Rightarrow \quad y - 1 = 0\ (x - 2)$

$y - 1 = 0$

Let a point $m(h, k)$ in the parabola

$$OM^2 = MN^2$$

$$(k + 3)^2 + (h - 2)^2 = \left(\frac{k-1}{\sqrt{1^2 + 0}}\right)^2$$

$k^2 + 9 + 6k + h^2 + 4 - 4h = k^2 + 1 - 2k$

$h^2 - 4h + 8k + 12 = 0$

Replace h, k by x and y respectively

$x^2 - 4x + 8y + 12 = 0.$

23. We have, $\frac{x^2}{9} - \frac{y^2}{16} = 1$ Compare with

$\frac{x^2}{a^2} - \frac{y^2}{b^2} = 1$, We get

$a^2 = 9,\ b^2 = 16,\ a = 3$

$\because$ Length of latus rectum

$$LL_1 = L'L_1'$$

$$= \frac{2b^2}{a} = \frac{2 \times 16}{3} = \frac{32}{3}.$$

24. We have,

$$\text{length of latus rectum} = \frac{9}{2}$$

$$\frac{2b^2}{a} = \frac{9}{2} \Rightarrow b^2 = \frac{9}{4}a \quad ...(i)$$

Distance between foci = 10

$2ae = 10$

$ae = 5$

$$e = \frac{5}{a} \qquad ...(ii)$$

$$\because \quad \text{eccentricity, } e = \sqrt{1 + \frac{b^2}{a^2}}$$

$$e^2 = 1 + \frac{b^2}{a^2}$$

From equation (*i*) and (*ii*)

$$\left(\frac{5}{a}\right)^2 = 1+\frac{\frac{9}{4}a}{a^2}$$

$$\frac{25}{a^2} = 1+\frac{9}{4a}$$

$$\frac{25}{a^2} = \frac{4a+9}{4a}$$

$$4a^3 + 9a^2 - 100a = 0$$

$$4a^2 + 9a - 100 = 0$$

$$4a^2 + 25a - 16a - 100 = 0$$

$$a(4a + 25) - 4(4a + 25) = 0$$

$$(a - 4)(4a + 25) = 0$$

$$4a + 25 \neq 0,\ a = 4$$

$a = -\dfrac{25}{4}$ (*a* will not be negative)

$\therefore \quad a = 4$

Put in equation (*i*)

$$b^2 = \frac{9}{4}\times 4 \Rightarrow b^2 = 9$$

$\because$ Equation of hyperbola

$$\frac{x^2}{a^2}-\frac{y^2}{b^2} = 1$$

$$\frac{x^2}{(4)^2}-\frac{y^2}{9} = 1 \Rightarrow \frac{x^2}{16}-\frac{y^2}{9} = 1.$$

25. We have focus (–3, 2) and Vertex (–2, 2)

Let the point of intersection parabola axis and directrix is $C(x_1, y_1)$

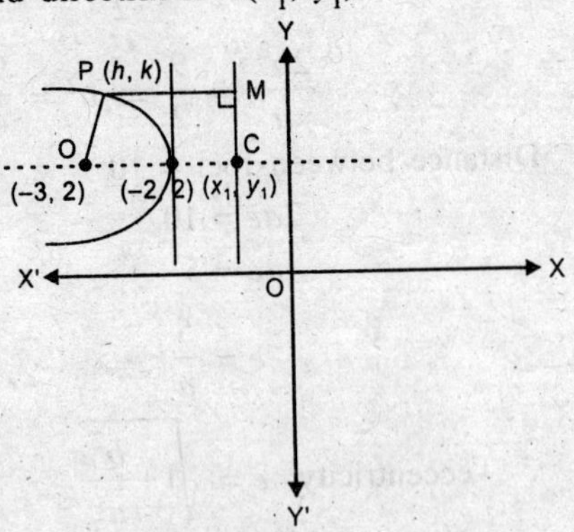

Then, $\dfrac{-3+x_1}{2} = -2$

$$-3 + x_1 = -4$$

$$x_1 = -1$$

and $\dfrac{2+y_1}{2} = 2$

$$2 + y_1 = 4$$

$$y_1 = 2$$

$\therefore$ Point C (–1, 2)

Gradient of axis line $= \dfrac{y_2-y_1}{x_2-x_1} = \dfrac{2-2}{-3+2} = 0$

$\because$ Directrix is perpendicular to axis then its gradient = 0

Equation of directrix

$$y - 2 = \frac{1}{0}(x+1)$$

$$x + 1 = 0$$

Let A point P(*h*, *k*) on the parabola then, $PO^2 = PM^2$

$$(2 - k)^2 + (-3 - h)^2 = \left(\frac{h+1}{\sqrt{1^2+0^2}}\right)^2$$

$$4 + k^2 - 4k + 9 + h^2 + 6h = h^2 + 1 + 2h$$

$$k^2 - 4k + 4h + 12 = 0$$

Replace *h*, *k* by *x* and *y* respectively

$$y^2 - 4y + 4x + 12 = 0.$$

26. We have,

$$y = \frac{a+bx}{b-ax}$$

$$\frac{dy}{dx} = \frac{(b-ax)\frac{d}{dx}(a+bx)-(a+bx)\frac{d}{dx}(b-ax)}{(b-ax)^2}$$

$$= \frac{(b-ax)\times b-(a+bx)(-a)}{(b-ax)^2}$$

$$= \frac{b^2-abx+a^2+abx}{(b-ax)^2} = \frac{a^2+b^2}{(b-ax)^2}.$$

27. We have, $y = \dfrac{2+3\sin h\,x}{3+2\sin h\,x}$

$$\frac{dy}{dx} = \frac{(3+2\sin h\,x)\frac{d}{dx}(2+3\sin h\,x)-(2+3\sin h\,x)\frac{d}{dx}(3+2\sin h\,x)}{(3+2\sin h\,x)^2}$$

$$= \frac{(3+2\sin h\, x)(3\cos h\, x) - (2+3\sin h\, x)(2\cos h\, x)}{(3+2\sin h\, x)^2}$$

$$= \frac{9\cos h\, x + 6\sin h\, x\cos h\, x - 4\cos h\, x - 6\sin h\, x\cos h\, x}{(3+2\sin h\, x)^2}$$

$$= \frac{5\cos h\, x}{(3+2\sin h\, x)^2}.$$

28. We have, $f(x) = x^3 - 3x^2 - 45x + 2$

$f'(x) = 3x^2 - 6x - 45$

For increasing function,

$$f'(x) > 0$$
$$3x^2 - 6x - 45 > 0$$
$$x^2 - 2x - 15 > 0$$
$$x^2 - 5x + 3x - 15 > 0$$
$$x(x-5) + 3(x-5) > 0$$
$$(x+3)(x-5) > 0$$
$$x \in (-3, 5).$$

30. We have, $y = x^2 + 3x - 7$

$$m_1 = \left(\frac{dy}{dx}\right)_{(2,3)}$$

$$= 2x + 3 = 2 \times 2 + 3 = 7$$

and $y^2 = 2x + 5$

$$m_2 = \left(\frac{dy}{dx}\right)_{(2,3)} = \frac{2}{2y} = \frac{1}{y} = \frac{1}{3}$$

$$\because \quad \tan\theta = \frac{m_1 - m_2}{1 + m_1 m_2}$$

$$\tan\theta = \frac{7 - \frac{1}{3}}{1 + 7 \times \frac{1}{3}} = \frac{\frac{20}{3}}{\frac{3+7}{3}} \Rightarrow \tan\theta = 2.$$

31. We have, $f(x) = 2x^3 - 12x^2 + 18x + 5$

$f'(x) = 6x^2 - 24x + 18$

For maximа and minima

$$f'(x) = 0$$
$$6x^2 - 24x + 18 = 0$$
$$x^2 - 4x + 3 = 0$$
$$(x-3)(x-1) = 0$$
$$x = 1, 3$$
$$f''(x) = 12x - 24$$
$$f''(x) < 0 \text{ at } x = 1$$

$\therefore$ Function will give maximum value at $x = 1$

Maximum value of the function

$= 2(1)^3 - 12(1)^2 + 18(1) + 5$

$= 2 - 12 + 18 + 5 = 13.$

32. In ΔDBC, $(6)^2 = h^2 + \left(\frac{b-6}{2}\right)^2$

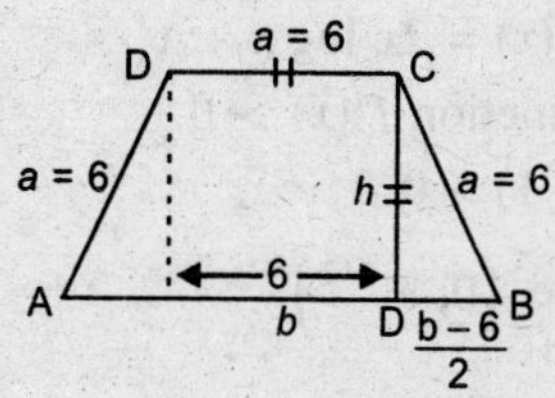

$$36 = h^2 + \frac{(b-6)^2}{4}$$

$$h^2 = 36 - \frac{(b-6)^2}{4}$$

$$h = \frac{\sqrt{144-(b-6)^2}}{2}$$

$\because$ Area of Trapezium

$$= \frac{1}{2} \times \text{sum of || side} \times h$$

$$A = \frac{1}{2} \times (6+b) \times \frac{\sqrt{144-(b-6)^2}}{2}$$

$$A = \frac{(6+b)\sqrt{144-(b-6)^2}}{4} \quad ...(i)$$

For maxima minima

$$\frac{dA}{db} = \frac{\sqrt{144-(b-6)^2} + (6+b)\frac{1}{2}(144-(b-6))^{-\frac{1}{2}}(-2(b-6))}{4}$$

$$\frac{dA}{db} = 0$$

$$\frac{1}{4}\left(\sqrt{144-(b-6)^2} - \frac{(6+b)(b-6)}{\sqrt{144-(b-6)}}\right) = 0$$

$$144 - (b-6)^2 - (b^2 - 36) = 0$$
$$144 - b^2 - 36 + 12b - b^2 + 36 = 0$$
$$2b^2 - 12b - 144 = 0$$
$$b^2 - 6b - 72 = 0$$
$$b^2 - 12b + 6b - 72 = 0$$
$$b(b-12) + 6(b-12) = 0$$
$$(b+6)(b-12) = 0$$

$b = -6,\ b = 12$

Put b value in equation (i), we get,

$$A = \frac{(6+12)\sqrt{144-36}}{4}$$

$$A = \frac{18\times\sqrt{108}}{4} \Rightarrow A = \frac{18\times 6\sqrt{3}}{4}$$

$$A = 27\sqrt{3} \text{ square unit.}$$

33. We have, $f'(x) = x^2 \log x$

$$f'(x) = 2x\log x + x^2 \times \frac{1}{x}$$

$$f'(x) = 2x \log x + x$$

for increasing function $f'(x) > 0$

$x(2 \log x + 1) > 0$

Function increase $\left(0, e^{-1/2}\right)$.

35. We have, $u = \log\left(\frac{x^2+y^2}{x+y}\right)$

We can written this as

$$e^u = \frac{x^2+y^2}{x+y}$$

$$e^u.\frac{\partial u}{\partial x} = \frac{(x+y).2x-(x^2+y^2).1}{(x+y)^2}$$

$$e^u.\frac{\partial u}{\partial x} = \frac{2x^2+2xy-x^2-y^2}{(x+y)^2}$$

$$\frac{(x^2+y^2)}{x+y}.\frac{\partial u}{\partial x} = \frac{x^2y^2+2xy}{(x+y)^2}$$

$$\frac{\partial u}{\partial x} = \frac{x^2-y^2+2xy}{(x^2+y^2)(x+y)}$$

Similarly,

$$e^u.\frac{\partial u}{\partial y} = \frac{(x+y).(2y)-(x^2+y^2).1}{(x+y)^2}$$

$$\frac{x^2+y^2}{x+y}.\frac{\partial u}{\partial y} = \frac{2xy+y^2-x^2}{(x+y)^2}$$

$$\frac{\partial u}{\partial y} = \frac{2xy+y^2-x^2}{(x+y)(x^2+y^2)}$$

Then, $x\frac{\partial u}{\partial x}+y\frac{\partial u}{\partial y}$

$$= \frac{x^3-xy^2+2x^2y}{(x+y)(x^2+y^2)}+\frac{2xy^2+y^3-x^2y}{(x+y)(x^2+y^2)}$$

$$= \frac{x^3+y^3-xy^2+2x^2y-x^2y+2xy^2}{(x+y)(x^2+y^2)}$$

$$= \frac{(x+y)(x^2+y^2-xy)-xy(x+y)+2xy(x+y)}{(x+y)(x^2+y^2)}$$

$$= \frac{(x+y)(x^2+y^2-xy-xy+2xy)}{(x+y)(x^2+y^2)}$$

$$= \frac{(x+y)(x^2+y^2)}{(x+y)(x^2+y^2)} = 1.$$

36. We have, $\int \log x\, dx = \int \log x.1\, dx$ (I II)

$$= \log x\int dx - \int\left\{\left(\frac{d}{dx}.\log x\right)\int 1.dx\right\}dx$$

$$= \log x\times x - \int\frac{1}{x}\times x\,dx = x\log x - \int dx$$

$$= x \log x - x + c.$$

37. $\lim_{n\to\infty}\left[\frac{1}{n+1}+\frac{1}{n+2}+\ldots\frac{1}{n+n}\right]$

$$= \lim_{n\to\infty}\sum_{r=1}^{r}\frac{1}{n+r}$$

$$= \lim_{n\to\infty}\sum_{r=1}^{\infty}\frac{1}{n\left(1+\frac{r}{n}\right)}$$

$$= \int_0^1\frac{1}{1+x}.dx = \left[\log(1+x)\right]_0^1$$

$$= \left[\log(1+1)-\log 1\right] = \log 2 - 0 = \log 2.$$

38. $\int\frac{\cos\sqrt{x}}{\sqrt{x}}dx$

Let, $\sqrt{x} = t$

$$\frac{1}{2\sqrt{x}}dx = dt \Rightarrow \frac{dx}{\sqrt{x}} = 2dt$$

Then, $\int\cos t\times 2dt = 2\int\cos t\, dt$

$$= 2 \sin t + c$$

$$= 2\sin\sqrt{x}+c.$$

40. $\int_0^{\pi/2} \sin^2 x\,dx$

We can written,

$$\sin^2 x = \frac{1-\cos 2x}{2}$$

Then,

$$= \int_0^{\pi/2}\left(\frac{1}{2}-\frac{\cos 2x}{2}\right)dx$$

$$= \left[\frac{1}{2}x-\frac{1}{2}\times\sin 2x\times 2\right]_0^{\pi/2} = \left[\frac{x}{2}-\sin 2x\right]_0^{\pi/2}$$

$$= \left[\left(\frac{\pi}{4}-\sin\pi\right)-(0-\sin 0)\right] = \frac{\pi}{4}-0=\frac{\pi}{4}.$$

41. $\int_1^4\left(\sqrt{x}+\frac{1}{\sqrt{x}}\right)dx = \int_1^4\left(x^{1/2}+x^{-1/2}\right)dx$

$$= \left[\frac{x^{\frac{1}{2}+1}}{\frac{1}{2}+1}+\frac{x^{-\frac{1}{2}+1}}{-\frac{1}{2}+1}\right]_1^4 = \left[\frac{2}{3}x^{\frac{3}{2}}+2x^{\frac{1}{2}}\right]_1^4$$

$$= \left[\left(\frac{2}{3}(4)^{3/2}+2(4)^{1/2}\right)-\left(\frac{2}{3}(1)^{3/2}+2(1)^{1/2}\right)\right]$$

$$= \left[\left(\frac{2}{3}\times 8+4\right)-\left(\frac{2}{3}+2\right)\right] = \left(\frac{28}{3}-\frac{8}{3}\right)=\frac{20}{3}.$$

42. $\int_0^{\pi/4}\sqrt{1+\sin 2x}\,dx$

$$= \int_0^{\pi/4}\sqrt{\sin^2 x+\cos^2 x+2\sin x.\cos x}\,dx$$

$$= \int_0^{\pi/4}\sqrt{(\sin x+\cos x)^2}\,dx$$

$$= \int_0^{\pi/4}(\sin x+\cos x)\,dx = \left[-\cos x+\sin x\right]_0^{\pi/4}$$

$$= \left[\left(-\cos\frac{\pi}{4}+\sin\frac{\pi}{4}\right)-(-\cos 0+\sin 0)\right]$$

$$= \left[\left(-\frac{1}{\sqrt{2}}+\frac{1}{\sqrt{2}}\right)-(-1+0)\right]$$

$= 0 + 1 = 1.$

43. $\int_0^{\pi/2}\frac{\sin x}{1+\cos^2 x}dx$

Let $\cos x = t$

$\Rightarrow$ $x = 0,\ t = 1$ (Lower limit)

$x = \frac{\pi}{2},\ t = 0$ (Upper limit)

$-\sin x\,.\,dx = dt$

$\sin x\,dx = -dt$

$$\int_1^0\frac{-dt}{1+t^2} = -\left[\tan^{-1}t\right]_1^0 = -\left[\tan^{-1}0-\tan^{-1}1\right]$$

$$= -\left[0-\frac{\pi}{4}\right]=\frac{\pi}{4}.$$

44. $(D^2 + 5D + 6)y = e^x$ when $f(x) = e^{ax}$

$$y_p = \left(\frac{1}{F(D)}\right)e^{ax}$$

Put $D = a$ in the denominator

$$y_p = \frac{1}{D^2+5D+6}e^x$$

$$y_p = \left(\frac{1}{F(a)}\right)e^{ax} \quad \text{if } F(a) \neq 0$$

$$y_p = \frac{1}{(1)^2+5\times 1+6} \Rightarrow y_p = \frac{e^x}{12}.$$

45. We have, $ay^2 = x^3$ ...(*i*)

$$a\times 2y\times\frac{dy}{dx} = 3x^2 \Rightarrow \frac{dy}{dx} = \frac{3x^2}{2ay} \quad ...(ii)$$

Put $a = \frac{x^3}{y^2}$ from equation (*i*) to equation (*ii*)

$$\frac{dy}{dx} = \frac{3x^2}{2\times\frac{x^3}{y^2}\times y} \Rightarrow \frac{dy}{dx} = \frac{3y}{2x}.$$

46. We have, $\frac{dy}{dx}+y = e^{-x}$

This is a linear differential equation.

Compare with $\frac{dy}{dx}+py=Q$

$P = 1,\ Q = e^{-x}$

Now integrating factor (IF)

$= e^{\int P\,dx} = e^{\int dx} = e^{x}$

Solution, $y(\text{IF}) = \int Q(\text{IF})dx + C$

$y \times e^{x} = \int e^{-x}e^{x}.dx + c$

$y \times e^{x} = x + c$

$y = (x + c)e^{-x}.$

47. We have, $(D^2 + 3D + 2)y = 8 \sin 5x$

Auxiliary equation of the given function is

$m^2 + 3m + 2 = 0$

$m^2 + 2m + m + 2 = 0$

$m(m + 2) + 1(m + 2) = 0$

$(m + 1)(m + 2) = 0 \Rightarrow m = -1, -2$

Roots real and distinct

$\therefore$ Complementary function

$= c_1e^{-x} + c_2e^{-2x}.$

48. We have, $2xy.dx + x^2 . dy = 0$

$\Rightarrow \quad \frac{dy}{dx} = -\frac{2xy}{x^2}$

This is a homogeneous function of same degree put $y = vx$

$\frac{dy}{dx} = v + x\frac{dv}{dx}$

$\therefore \quad v + x\frac{dv}{dx} = -\frac{2 \times x \times vx}{x^2}$

$v + x\frac{dv}{dx} = -2v$

$x\frac{dv}{dx} = -3v$

$\frac{dv}{v} = -3\frac{1}{x}dx$

$\int\frac{dv}{v} = -3\int\frac{1}{x}dx$

$\log v = -3 \log x + c$

$\log v + \log x^3 = c$

$\log vx^3 = c \Rightarrow vx^3 = e^c$

Where e^c is a constant

$\frac{y}{x} \times x^3 = c \Rightarrow x^2y = c.$

49. We have, $x^2 - 4ay = 0$ ...(*i*)

Differentiate w.r.t. x

$2x - 4a\frac{dy}{dx} = 0 \Rightarrow x - 2a\frac{dy}{dx} = 0$ ...(*ii*)

Put a value from equation (*i*) $a = \frac{x^2}{4y}$

$x - 2 \times \frac{x^2}{4y}.\frac{dy}{dx} = 0 \Rightarrow x - \frac{x^2}{2y}.\frac{dy}{dx} = 0$

$1 - \frac{x}{2y}.\frac{dy}{dx} = 0 \Rightarrow 2y - x\frac{dy}{dx} = 0$

$x\frac{dy}{dx} - 2y = 0.$

50. We have, $\frac{dy}{dx} + y\cot x = \cos x$

This is a linear differential equation

Compare with, $\frac{dy}{dx} + P.y = Q$

We get, $P = \cot x,\ Q = \cos x$

Now, Integrating Factor (IF) $= e^{\int P\,dx}$

$= e^{\int \cot x.dx} = e^{\log \sin x} = \sin x$

Solution, $y(\text{IF}) = \int Q(\text{IF})dx + c$

$y \times \sin x = \int \cos x.\sin x.dx + c$

$= \frac{1}{2}\int 2\sin x.\cos x.dx + c = \frac{1}{2}\int \sin 2x.dx + c$

$= \frac{1}{2}\left(-\frac{\cos 2x}{2}\right) + c = -\frac{\cos 2x}{4} + c.$

Previous Paper (Solved)

B.Sc. BIOTECHNOLOGY

Common Entrance Test (CET), 2017*

MATHEMATICS

1. If the traces of A and B are 20 and –8 then the trace of (A + B) is:

A. 12 B. – 12
C. 28 D. – 28

2. If $A = \begin{bmatrix} x & 1 \\ 1 & 0 \end{bmatrix}$ is an involutory matrix then x =

A. 0 B. – 2
C. – 1 D. 2

3. The determinant of $\begin{bmatrix} \log e & \log e^2 & \log e^3 \\ \log e^2 & \log e^3 & \log e^4 \\ \log e^3 & \log e^4 & \log e^5 \end{bmatrix}$ is ....

A. 0 B. 1
C. 4 $\log e$ D. 5 $\log e$

4. If $A = \begin{bmatrix} 1 & 1 & 0 \\ 2 & 1 & 3 \\ 0 & 1 & 2 \end{bmatrix}$ then det(*adj* A) = ...

A. det A B. det A^2
C. – det A D. $(\det A)^2$

5. If A, B are two matrices and AB = B, BA = A then $A^2 + B^2$ =

A. A + B B. A – B
C. AB D. 0

6. If $\dfrac{3x+2}{(x+1)(2x^2+3)} = \dfrac{A}{x+1} + \dfrac{Bx+C}{2x^2+3}$, then A + C – B = ...

A. 0 B. 2
C. 3 D. 5

7. If $\dfrac{3x}{(x-a)(x-b)} = \dfrac{2}{x-a} + \dfrac{1}{x-b}$ then $a : b$ = ...

A. – 2 : 1 B. 2 : 1
C. 1 : 2 D. 3 : 1

8. The value of tan 855° = ...

A. 1 B. $\dfrac{1}{\sqrt{2}}$
C. – 1 D. $-\dfrac{1}{\sqrt{2}}$

9. If $\tan\alpha = \dfrac{m}{m+1}$ and $\tan\beta = \dfrac{1}{2m+1}$ then $\tan(\alpha+\beta)$ = ...

A. – 1 B. 0
C. 1 D. 2

10. The value of $6\sin 20° - 8\sin^3 20°$ =

A. 2 B. $\dfrac{1}{\sqrt{2}}$
C. $\sqrt{3}$ D. $\dfrac{1}{\sqrt{3}}$

11. If $3\sin\theta + 4\cos\theta = 5$ then the value of $4\sin\theta - 3\cos\theta$ =

A. 0 B. – 1
C. 1 D. 2

12. The sine function with period 3 is:

A. $\sin\dfrac{2\pi x}{3}$ B. $\sin\dfrac{\pi x}{3}$
C. $\sin 3\pi x$ D. $\sin\dfrac{3\pi x}{2}$

*Held on 03/05/2017 (AP–ECET)

13. The maximum value of $3 \sin^2 x + 5 \cos^2 x$ is ....

A. 8 B. 3
C. 5 D. 34

14. The equation $\sqrt{3}\sin x + \cos x = 4$ has ....

A. only one solution
B. two solutions
C. infinite solutions
D. no solution

15. The solution of $\cos^{-1}\left(\sqrt{3x}\right) + \cos^{-1} x = \frac{\pi}{2}$ is ....

A. $\frac{1}{2}$ B. $\frac{1}{5}$
C. $-\frac{1}{2}$ D. $-\frac{1}{5}$

16. The value of $\sin\theta + \sin(\theta + 120°) - \sin(120° - \theta) =$ ....

A. 0 B. $\sin \theta$
C. 1 D. $-\sin \theta$

17. The principal solution of 3 cosec A = 4 sin A is ....

A. $\frac{\pi}{4}$ B. $\pm\frac{\pi}{3}$
C. $\pm\frac{\pi}{6}$ D. $\pm 2\pi$

18. If $|z^2 - 1| = |z|^2 + 1$, then z lies in ....

A. The real axis
B. A circle
C. The imaginary axis
D. A parabola

19. If $\left(\frac{1+i}{1-i}\right)^3 - \left(\frac{1-i}{1+i}\right)^3 = a + ib$, then a and b are ....

A. 1, 1 B. 2, – 2
C. 0, – 2 D. 0, – 1

20. If the line $y = 2x + c$ is a tangent to $x^2 + y^2 = 5$, then the value of c is ....

A. 2 B. 3
C. 4 D. 5

21. The vertex of the parabola $x^2 + 8x + 12y + 4 = 0$ is:

A. (– 4, 1) B. (4, – 1)
C. (– 4, – 1) D. (4, 1)

22. The number of tangents of the ellipse $\frac{x^2}{4} + \frac{y^2}{2} = 1$ through (2, 1) is ....

A. 0 B. 1
C. 2 D. 3

23. The length of the latus rectum of the hyperbola $x^2 - 4y^2 = 4$ is ....

A. 2 B. 1
C. 4 D. 3

24. The length of the diameter of the circle $x^2 + y^2 - 6x - 8y = 0$ is ....

A. 10 B. 15
C. 5 D. 20

25. If the line $2y = 5x + k$ touches the parabola $y^2 = 6x$ then $k =$ ....

A. $\frac{2}{3}$ B. $\frac{4}{3}$
C. $\frac{3}{5}$ D. $\frac{6}{5}$

26. $\lim_{x \to 2+} \frac{x|x-2|}{x-2} =$ ....

A. 1 B. – 1
C. 2 D. – 2

27. If $f(x) = (1+x)^{\frac{2}{x}}$ is continuous at $x = 0$, then $f(0) =$ ....

A. e B. e^2
C. e^3 D. e^4

28. If $x = a \sec \theta$, $y = b \tan \theta$ then $\frac{dy}{dx} =$ ....

A. $\frac{b}{a}\sec\theta$ B. $\frac{b}{a}\operatorname{cosec}\theta$
C. $\frac{a}{b}\sec\theta$ D. $\frac{a}{b}\operatorname{cosec}\theta$

29. If $x^y = e^{x-y}$ then $\frac{dy}{dx} =$ ....

A. $\frac{\log x}{(1+\log x)^2}$ B. $\frac{\log x}{(1-\log x)^2}$

C. $\frac{-\log x}{(1+\log x)^2}$ D. $\frac{-1}{(1+\log x)^2}$

30. If $y=\sin^{-1}\left(\frac{x}{\sqrt{1+x^2}}\right)$ then $\frac{dy}{dx}$ = ...

A. $-\frac{1}{1+x^2}$ B. $\frac{1}{1+x^2}$

C. $\frac{2}{1+x^2}$ D. $-\frac{2}{1+x^2}$

31. The slope of the normal to the curve $x = a \sec\theta$, $y = a\tan\theta$ at $\theta=\frac{\pi}{6}$ is ....

A. 2 B. 0

C. $-\frac{1}{2}$ D. 1

32. The rate of change of area of a circle with respect to radius when $r = 5$ cm is:

A. 2π sq. cm/sec B. 10π sq. cm/sec

C. 100π sq. cm/sec D. 20π sq. cm/sec

33. Which of the following function has maxima or minima?

A. e^x B. $\log x$

C. $x^3 + x^2 + x + 1$ D. $\sin x$

34. In the increase in the side of a square is 2% then the approximate percentage increase in the area of the square is ....

A. 2 B. 4

C. 6 D. 8

35. For the function $f(x) = \log(x^2 + y^2)$. Which of the following is true?

A. $f_x + f_y = 0$ B. $f_{xx} + f_{yy} = 0$

C. $f_x - f_y = 0$ D. $f_{xx} - f_{yy} = 0$

36. $\int \text{cosec}^5\theta\cot\theta d\theta$ = ...

A. $\frac{\cot^2\theta}{2}$ B. $\frac{-\text{cosec}^5\theta}{5}$

C. $\frac{\text{cosec}^6\theta}{6}$ D. $\frac{-\text{cosec}^6\theta}{6}$

37. $\int_2^3\frac{dx}{x^2-x}$ = ...

A. $\log\frac{2}{3}$ B. $\log\frac{4}{3}$

C. $\log\frac{8}{3}$ D. $\log\frac{1}{4}$

38. If $a < 0 < b$ then $\int_a^b\frac{|x|}{x}dx$ = ...

A. $b - a$ B. $a - b$

C. $a + b$ D. 0

39. $\int_0^1 x\tan^{-1}x dx$ = ...

A. $\frac{\pi}{4}-\frac{1}{2}$ B. $\frac{\pi}{8}-\frac{1}{2}$

C. $\frac{\pi}{4}+\frac{1}{2}$ D. $\frac{\pi}{8}+\frac{1}{2}$

40. $\lim_{n\to\infty}\sum_{r=1}^{n}\frac{1}{n}e^{\frac{r}{n}}$ = ...

A. e B. $(1 + e)$

C. $(1 - e)$ D. $(e - 1)$

41. $\int_0^{\pi/4}\sec^6 x dx$ = ...

A. $\frac{8}{3}$ B. $\frac{28}{15}$

C. $-\frac{28}{15}$ D. $\frac{4}{5}$

42. The area bounded by the curve $y = \log x$, x-axis and the straight line $x - e = 0$ is ..... square units.

A. e B. $(e - 1)$

C. 0 D. $(1 - e)$

43. The volume of the solid generated by rotating one arch of the curve $y = \sin 3x$ about the x-axis is ....

A. π^2 B. $\frac{\pi^2}{2}$

C. $\frac{\pi^2}{4}$ D. $\frac{\pi^2}{6}$

44. $y = cx - c^2$ is the general solution of the differential equation:

A. $\left(\frac{dy}{dx}\right)^2 - x\left(\frac{dy}{dx}\right) + y = 0$

B. $\frac{d^2y}{dx^2} = 0$

C. $\frac{dy}{dx} = c$

D. $\left(\frac{dy}{dx}\right)^2 + x\left(\frac{dy}{dx}\right) + y = 0$

45. The general solution of the differential equation $\frac{dy}{dx} + \frac{y}{3} = 1$ is:

A. $y = 3 + ce^{\frac{x}{3}}$ B. $y = 3 + ce^{-\frac{x}{3}}$

C. $3y = c + e^{\frac{x}{3}}$ D. $3y = c + e^{-\frac{x}{3}}$

46. The differential equation corresponding to the family of curves $y = ae^{bx}$, where a and b are arbitrary constants, is ....

A. $\frac{d^2y}{dx^2} = y\frac{dy}{dx}$

B. $y\frac{d^2y}{dx^2} - \frac{dy}{dx} = 0$

C. $y\frac{d^2y}{dx^2} = \left(\frac{dy}{dx}\right)^2$

D. $\frac{dy}{dx} - y^2 = 0$

47. An integrating factor of the differential equation $(x^2y + y + 1)dx + (x + x^3)dy = 0$ is:

A. e^x B. x^2

C. $\frac{1}{x}$ D. x

48. The differential equation whose solution is $Ax^2 + By^2$, where A, B are arbitrary constants are of ....

A. 1st order and 1st degree

B. 2nd order and 1st degree

C. 2nd order and 2nd degree

D. 1st order and 2nd degree

49. The general solution of the differential equation $\frac{d^2x}{dt^2} - 4\frac{dx}{dt} + 5x = 0$ is:

A. $x = (c_1 \cos t + c_2 \sin t)e^{2t}$

B. $t = (c_1 \cos x + c_2 \sin x)e^{2x}$

C. $x = (c_1 \cos 2t + c_2 \sin 2t)e^t$

D. $t = (c_1 \cos 2x + c_2 \sin 2x)e^x$

50. The particular integral of $(D - 2)^2 y = \sin 2x$ is:

A. $\frac{\cos 2x}{8}$ B. $\frac{\sin 2x}{8}$

C. $\frac{-\cos 2x}{2}$ D. $\frac{-\sin 2x}{2}$

PHYSICS

51. The unit of impulse is the same as that of:

A. moment of force B. linear momentum

C. force D. pressure

52. If the force is given by $F = at + bt^2$, where t is the time. The dimensions of a and b are:

A. MLT^{-4}, MLT^{-2} B. MLT^{-3}, MLT^{-4}

C. ML^2T^{-3}, ML^2T^{-2} D. ML^2T^{-3}, ML^3T^{-4}

53. Vector parallel to $6\hat{i} + 8\hat{j}$ and having a magnitude of 5 is:

A. $4\hat{i} + 3\hat{j}$ B. $12\hat{i} + 16\hat{j}$

C. $16\hat{i} + 8\hat{j}$ D. $3\hat{i} + 4\hat{j}$

54. If $|\vec{A} \times \vec{B}| = K(AB)$ then angle between $\vec{A}$ and $\vec{B}$ is:

A. $\cos^{-1} K$ B. $\cos^{-1}(1/K)$

C. $\sin^{-1} K$ D. $\sin^{-1}(1/K)$

55. A cricket ball is thrown at a speed of 28 m/s in a direction 30° above the horizontal. The maximum height reached by the ball is:

A. 10 m B. 20 m

C. 30 m D. 40 m

56. Two bodies are projected at angles of 45° and 60° with the horizontal with same velocity simultaneously. Ratio of their horizontal ranges is:

A. $\sqrt{3}:2$ B. $2:\sqrt{3}$
C. 1 : 2 D. 2 : 1

57. A ball thrown by a boy is caught 2 seconds later by another at some distance away on the same level. If the angle of projection is 30°, the velocity of projection is:

A. 19.6 m/sec B. 9.8 m/sec
C. 4.9 m/sec D. 5.2 m/sec

58. A 200 m wide river flows with a velocity of 5 m/sec. A man crosses the river in the shortest time of 25 sec. If there is no flow and he swims with the same velocity, the time taken to cross the river is:

A. $\frac{200}{5\sqrt{3}}$ sec B. 20 sec
C. 25 sec D. $25\sqrt{2}$ sec

59. A body of mass 1 kg lies on an inclined plane of angle 60° to the horizontal. If the coefficient of friction is 0.4, the frictional force along the inclined plane is:

A. 1.96 N B. 0.98 N
C. 0.49 N D. 0.245 N

60. A force of 20 kg weight is required to just slide a wooden box weighing 50 kg over ice. Then coefficient of static friction between the surfaces in contact is:

A. 0.2 B. 0.4
C. 0.8 D. 0.1

61. A cyclist comes to a skidding stop in 10 m. During this process, the force on the cycle due to the road is 200N and is directly opposed to the motion. The work done by the road on the cycle is:

A. 1000 J B. 2000 J
C. – 1000 J D. – 2000 J

62. A sphere of mass 4 kg is dropped from a certain height. After 5 s, its kinetic energy is (g = 10 m/s^2):

A. 5 J B. 50 J
C. 5 KJ D. 50 KJ

63. An elevator weighing 500 kg is to be lifted up at a constant velocity of 0.20 m/s. What would be the minimum power of the motor to be used?

A. 100 W B. 500 W
C. 980 W D. 900 W

64. At t = 0, the displacement of a particle in SHM is half its amplitude. Its initial phase is (referring to mean position):

A. $\frac{\pi}{6}$ B. $\frac{\pi}{3}$
C. $\frac{2\pi}{3}$ D. $\frac{\pi}{2}$

65. The length of seconds pendulum is 100 cm. To have a period half of this value, the length is to be reduced by:

A. 25 cm B. 75 cm
C. 50 cm D. 100 cm

66. Inside a big hall, the reverberation time is:

A. directly proportional to volume
B. inversely proportional to sound absorption
C. both directly proportional to volume and inversely proportional to sound absorption
D. depends on temperature

67. The voice of lion is different from that of a mosquito because:

A. the sounds have different pitch
B. they are of different size
C. the two voices travel with different velocities
D. the sounds have different phases

68. A car is travelling at $\frac{V}{10}$ m/s and sounds horn of frequency 990 Hz. The apparent frequency heard by a police chasing the car at $\frac{V}{9}$ m/s (V is the velocity of sound) is:

A. 990 Hz B. 900 Hz
C. 100 Hz D. 1000 Hz

69. When ice cube melts and becomes water, the ice-water system undergoes a change such that:

A. entropy of the system decreases and internal energy decreases
B. entropy of the system decreases and internal energy increases
C. entropy of the system increases and internal energy increases
D. entropy of the system increases and internal energy decreases

70. A mass of 300 gm falls from a height of 3 m (g = 9.8 m/s^2). Assuming that the whole energy is converted into heat, the amount of heat produced is:

A. 2 cal B. 2.1 cal
C. 4 cal D. 4.2 cal

71. During an adiabatic expansion of 2 moles of a gas, the change in internal energy was found to be equal to 100 J. The work done during the process will be equal to:

A. zero B. – 100 J
C. 200 J D. 100 J

72. The pressure and density of a diatomic gas $\left(Y = \frac{7}{5}\right)$ change adiabatically from (P, d) to (P^1, d^1). If $\frac{d^1}{d} = 32$, then $\frac{P^1}{P}$ is:

A. 128 B. 32
C. 256 D. 64

73. Boyle's law holds good for an ideal gas during:

A. isobaric changes B. isothermal changes
C. isochoric changes D. isotopic changes

74. The threshold frequency of metal is v_0. When a light of frequency 4 v_0 is incident on metal then the K.E$_{max}$ of emitted electrons is:

A. 2 v_0h B. 3 v_0h
C. 4 v_0h D. v_0h

75. Superconductors are .... materials.

A. dielectric B. paramagnetic
C. ferromagnetic D. diamagnetic

CHEMISTRY

76. The Pauli exclusion principle is concerned with:

A. Energy of orbital
B. Spin of electron
C. Energy of electron
D. Angular momentum of electron

77. According to Bohr's model of hydrogen atom, the following is quantized:

A. Linear momentum
B. Linear velocity
C. Angular momentum
D. Angular velocity

78. How many 'd'-orbitals have two perpendicular nodal planes:

A. Two B. Three
C. Four D. Five

79. In sodium chloride crystal, each Na^+ ion is surrounded by:

A. Two Cl^- ions B. Four Cl^- ions
C. Six Cl^- ions D. Eight Cl^- ions

80. Which among the following molecule contains a π-bond?

A. H_2 B. O_2
C. F_2 D. HCl

81. Which among the following is insoluble in water?

A. Alcohol B. Ammonia
C. Benzene D. Acetone

82. The normality of 2.3 M H_2SO_4 solution is:

A. 0.46 N B. 0.23 N
C. 2.3 N D. 4.6 N

83. 8 grams of substance of molecular weight 40 is dissolved in 250 g of water. Then the molality of the solution is:

A. 0.4 B. 0.8
C. 0.2 D. 0.6

84. The pH value of 0.05M $Ba(OH)_2$ solution is:

A. 10 B. 12
C. 13 D. 11

85. Which of the following molecule is not a Lewis Base?
A. H_2O B. BF_3
C. NH_3 D. CO

86. During the electrolysis of brine, 710 g of Cl_2 was liberated at anode. The weight of NaOH formed:
A. 800 g B. 400 g
C. 80 g D. 40 g

87. In the Danniel cell, which electrode acts as anode?
A. Cu B. Hg
C. Zn D. Pt

88. The molar conductance of HCl is more than that of NaCl because:
A. NaCl is more polar than KCl
B. NaCl is ionic while HCl is covalent
C. Ionic mobility of H^+ is more than that of Na^+
D. H^+ get hydrated.

89. The units for electrochemical equivalent are:
A. grams B. grams ampere
C. Coulomb D. grams per coulomb

90. Zeolite softening process removes:
A. Only permanent hardness of water
B. Only temporary hardness of water
C. Both temporary and permanent hardness of water
D. The dissolved gases in permanent hard water

91. The permanent hardness of water is caused by the presence of:
A. Bicarbonates of Ca and Mg
B. Carbonates of Na and K
C. Chlorides and Sulphates of Ca and Mg.
D. Phosphates of Na and K

92. The secondary treatment of water uses .... to consume wastes in water.
A. Filtration B. Sedimentation
C. Chemicals D. Microorganisms

93. Difficult to monitor and very dangerous form of corrosion is:
A. Galvanic B. Pitting
C. Crevice D. Stress

94. When Pt and Co are electrically connected, which one gets corroded?
A. Co B. Pt
C. None D. Both

95. What rubber was invented when Dr. Joseph C. Patrick tried to make antifreeze?
A. Methyl rubber B. Chloroprene
C. Bruna N D. Thiokol

96. The first plastic ever synthesized was called ....
A. Bakelite B. Nylon
C. Dacron D. Cellulose

97. .... is a brand of polyester textile fiber that is wrinkle resistant and strong.
A. Cellulose B. Dacron
C. Bakelite D. Nylon

98. Water gas is a mixture of:
A. $H_2 + CO$ B. $N_2 + CO$
C. $H_2 + CO_2$ D. $H_2 + CH_4$

99. Which of the following is not a greenhouse gas?
A. CO B. CO_2
C. water vapour D. CH_4

100. Burning of fossil fuels causes:
A. Global warming
B. Ozone depletion
C. Acid rain
D. Eutrophication

BIOTECHNOLOGY

101. Which of the following is not a source of inorganic nitrogen for industrial microbes?
A. Ammonium sulphate
B. Di-ammonium hydrogen phosphate
C. Ammonia
D. Protein

102. Sulphite waste liquor is a by-product of:
A. Sugar industry

B. Dairy industry
C. Leather industry
D. Paper industry

103. Which of the following is used as an antifoaming agent in fermenters?
A. Methanol
B. Silicone compounds
C. Ethanol
D. PEG

104. Which among the following is not a part of fermentation process?
A. Propagation step
B. Downstream processing
C. Pilot scale fermentation
D. Main production fermentation

105. Which among the following is the cheapest source of carbon for Industrial fermentations?
A. Sucrose B. Molasses
C. Glucose D. Fructose

106. Peptones are prepared by acid or enzyme hydrolysis of:
A. High lipid material
B. High carbohydrate material
C. High sugar material
D. High protein material

107. Which is a limitation of liquid biofertilizers over carrier based fertilizers?
A. Longer shelf life
B. No contamination
C. Better survival on seeds and soil
D. Low export potential

108. Which of the following is not a bio-pesticide?
A. Chitosan B. Insect pheromones
C. Azolla D. Bt toxin

109. Which of the following does not come in the category of biopesticides?
A. Microbial pesticides
B. Plant incorporated protectants
C. Biochemical pesticides
D. Chemical pesticides

110. Which of the following is not a biofertilizer?
A. Rhizobium B. Azobactor
C. Agrobacterium D. Azospirilum

111. According to the cell theory:
A. All the organism are composed of several cells
B. Life continues to evolve a new with each new cell
C. The smallest living thing is a cell
D. New cells arise spontaneously when conditions are right

112. Which among the following is the most appropriate definition of Biophysics?
A. Using and developing tools from physics to study biological systems
B. Study of physics
C. Study of physical chemistry
D. Study of chemical properties

113. Which among the following is not an area of biophysics?
A. Medical imaging B. Structural biology
C. Systems biology D. Microbiology

114. Who proposed cell theory?
A. Joseph Lister
B. Robert Koach
C. Nitch
D. Schleiden and Schwain

115. What correctly describes of atomic theory?
A. Matter is made up of protons
B. Cells are basic unit of life
C. It is the characteristics and properties of atoms that make up matter
D. Matter is made up of neutrons

116. What is incorrect about significance of biological membranes being selectively permeable?
A. It permits selective uptake of nutrients and elimination of waste
B. Allows cells to concentrate particular ions on either side of the membrane
C. Prevents toxic material from entering the cells
D. Hydrophilic heads are not immerged in water

117. Which among the following is used for visualization of internal structure by projection of an electron beam on to the sample?

A. SEM
B. TEM
C. Stereo microscope
D. Compound microscope

118. Movement of molecules across the membrane against a concentration gradient at the expense of ATP is called:
A. Active transport
B. Passive transport
C. Osmosis
D. Facilitated diffusion

119. The biological membranes are made up of:
A. Lipids B. Proteins
C. Carbohydrates D. Lipoproteins

120. Motile bacterium can be seen with the help of:
A. Bright field microscope
B. Electron microscopy
C. Darkfield microscopy
D. Fluorescence microscopy

121. Alternate form of gene governing the expression of same trait that occur at the same position on homologous chromosomes are known as:
A. Chromatids B. Autosomes
C. Alleles D. mRNA

122. For Mendel's pea plants tall was dominant over dwarf (*f*) and red (R) was dominant over white flowers (P). A plant with genotype ttPp would have the phenotype:
A. tall with white flowers
B. dwarf with white flowers
C. dwarf with red flowers
D. tall with red flowers

123. The condition that results from one allele affecting more than one trait is called:
A. Incomplete dominance
B. Epistasis
C. Partial dominance
D. Pleiotropy

124. Which among the following is not true about meiosis?
A. It occurs in reproductive cells
B. It results in four haploid daughter cells
C. Homologous pair are pulled apart
D. Occurs only in plants

125. Which is the correct order of organisation of genetic material from largest to smallest?
A. Genome, chromosome, gene, nucleotide
B. Gene, chromosome, nucleotide, genome
C. Chromosome, gene, genome, nucleotide
D. Chromosome, genome, nucleotide, gene

126. What is a source of genetic recombination during gamete production?
A. Mutation
B. Crossing over
C. Controlled assortment
D. Nondisjunction

127. Males who have XXY combination due to nondisjunction have the condition called:
A. Turner syndrome
B. Haemophilia
C. Klinefelter syndrome
D. Down syndrome

128. The mutations which express differently depending on the sex of the parent the gene is passed through are known as:
A. Sex linked inheritance
B. Imprinting
C. Penetrance
D. Epistasis

129. When there are two breaks in a chromosome and the detached segment becomes reinserted in the reversed order, it is called a:
A. Inversion B. Translocation
C. Transversion D. Deletion

130. Which is true about holandric genes?
A. They are carried on X chromosome
B. They are carried on Y chromosome and can only be passed by males to their sons
C. They code for femaleness
D. They are transferred from mothers to daughters

131. Which of the following is not a sterilization method?
A. Steam B. Dry heat
C. Ethylene oxide D. PEG treatment

132. Which of the following is used for ionizing radiation sterilization?
A. UV radiation
B. Gamma radiation
C. Formaldehyde
D. NaOCl treatment

133. Which among the following is not an authentic Microbial Culture collection centre?
A. ATCC B. NCL
C. NBPGR D. MTCC

134. The number of surviving microbial cells in the exponential death phase are given in terms of:
A. Decimal death time
B. Decimal reduction time
C. Thermal death time
D. Thermal death point

135. Which among the following is not a preservation method?
A. Freezing
B. Mineral oil Overlay
C. Lyophilisation storage in silica gel
D. Autoclaving

136. Chemotherapeutic agent:
A. Kills the surface
B. Is used for surface sterilization
C. Is used internally to kill microbe with selective toxicity towards microbes
D. Is used for disinfection

137. The growth phase when doubling time is constant and shortest is called:
A. Log phase
B. Exponential phase
C. Stationary phase
D. Decline phase

138. General purpose media supplemented by special nutrients to encourage the growth of fastidious heterotrophs is known as:
A. Enriched media
B. Selective media
C. Differential media
D. Complex media

139. Media which favours the growth of particular microbe and inhibits others is known as:
A. Enriched media
B. Selective media
C. Differential media
D. Complex media

140. Media that distinguishes between different groups of bacteria on the basis of their biological characteristics and causes observable changes in media when biochemical reactions occur is called:
A. Enriched media B. Selective media
C. Differential media D. Complex media

141. In a plug flow reactor, the highest concentration of substrates is exposed to the cells that:
A. Are near the effluent or exit of the reactor
B. Are in the mid way along the reactor
C. Are located near entrance of feed
D. Are near the wall of the reactor

142. In fed batch bioreactor modelling, the rate of change in the bioreactor volume is assumed to be equal to:
A. Flow rate
B. Initial volume
C. Volume of solids in the reactor
D. Volume of liquid

143. Production of organic acids in batch culture is generally, growth associated phase followed by non-growth associated production. The probable reason for this is that:
A. Organic acids uncouple Catabolism from Anabolism
B. Biomass yields increase as the fermentation proceeds
C. Organic acids are secondary metabolites
D. High concentrations of organic acid promote the growth of the microbial population

144. The ideal tubular-flow fermenter without radial variations is called a:
A. Plug flow fermenter

B. Continuous stirred tank fermenter (CSTF)
C. Column fermenter
D. Rotating drum fermenter

145. When the product formation is approximately equal to the rate of cell growth, the pattern of product formation is termed as:
A. uncoupled
B. growth associated
C. non-growth associated
D. metabolically uncoupled

146. For continuous mode of operation the specific mass balance equation for chemostat is:
A. $Y_{S_x} = \mu / r_s$
B. $\mu = D$
C. $\mu_{max} = Cs/Cs - ks$
D. F/V

147. The vast majority of closed loop controllers used in bioprocess engineering are:
A. Proportional control
B. Integral control
C. Adaptive control
D. Proportional integral derivative (PID) control

148. Saturated clean steam is used for sterilization in place operations in bioreactors. The optimal steam pressure required is:
A. 10.2 – 10.5 bar gauge
B. 5.6 – 10.2 bar gauge
C. 1.1 – 1.4 bar gauge
D. 12.3 – 16.0 bar gauge

149. Sterilization in-place operation in a bioreactor mainly requires a complex arrangement of pipe work, valves and:
A. Foam Breaker B. Filters
C. Sight Glass D. Cold Water

150. In a 50 L Stirred Tank Bioreactor for animal or plant cell culture applications the impeller speed should not exceed:
A. 180 rpm B. 120 rpm
C. 30 rpm D. 200 rpm

151. Polymerase chain reaction is used to:
A. Amplify small amount of DNA
B. Cleave the DNA
C. Seal the sticky end
D. Identification of plasmids

152. During DNA replication what is the first process to occur?
A. Synthesis of lagging strand
B. Unwinding of parental strand
C. Synthesis of leading strand
D. Sealing of nicks between short DNA sections

153. The process where DNA is copied into messenger RNA is called:
A. Translation
B. Translocation
C. Transversion
D. Transcription

154. Which is not true about alkatonuria?
A. It is called black wine disease
B. It is inherited genetic disorder of phenylalanine and tyrosine metabolism
C. Affects glucuronidase
D. Homogenistic acid build up in the body.

155. DNA is called double helix. What does it mean?
A. Two X shaped strands
B. Two loops like figure eight
C. Two spirals, like a twisty ladder
D. Two Y shaped strand

156. Agarose gel, electrophoresis separate nucleic acids based on their:
A. G + C/A + T content
B. Ratio of mass / charge
C. Length
D. Origin of organism

157. An operon is a transcriptional unit in bacteria that contains:
A. RNA polymerase loading zones
B. A promoter site, an operator site one or more regulatory genes
C. A promoter site, an operator site and two or more structural genes

D. Two acting elements with transacting factors

158. Eukaryotic DNA replication is:
A. Conservative
B. Semiconservative
C. Dispersive
D. Semi-dispersive

159. Which is an incorrect statement about essential genes?
A. Essential genes are relatively preserved throughout bacterial kingdom than nonessential genes
B. The have higher proportion of large and small proteins
C. They are thought to be critical for the survival of the organism
D. They are not necessary for survival

160. DNA is made from RNA with the help of:
A. Restriction enzyme
B. Polymerase
C. Reverse transcriptase
D. Ligase

161. Callus can be defined as:
A. A differentiated mass of cells
B. An undifferentiated mass of cells
C. Shoot cultures
D. Organized cultures

162. Media most widely used for plant tissue culture:
A. LB media B. Nutrient agar
C. MS media D. YEB media

163. Which among the following is not an application of plant tissue culture?
A. Micropropagation
B. Haploid production
C. Protoplast fusion
D. Wide hybridization in field

164. The enzyme combination used for protoplast isolation is:
A. Cellulase, pectinase, hemicellulase
B. Protease, lipase, xylanase
C. Ligase, endonucleases
D. Polymerases, proteases

165. Major component of plant cell wall is:
A. Peptidoglycan
B. Chitin
C. Cellulose
D. Phospholipids

166. Which part of the Ti plasmid gets transferred to plant during transformation?
A. T-DNA
B. Virulence genes
C. Origin of replication
D. Overdrive

167. Which among the following is not a direct gene transfer technique?
A. Microinjection
B. PEG mediated transformation
C. Gene gun
D. *Agrobacterium* mediated transformation

168. Which of the following gene is not a part of nitrogen fixation?
A. Nod gene B. Nif gene
C. Hup gene D. Rol gene

169. Which of the following plant does not have an association with *Rhizobium*?
A. Lentil B. Peas
C. Alfalfa D. Pumpkin

170. Which of the following is not applied for sustained crop protection by preventing losses due to pests and diseases?
A. Insecticide act
B. Integrated pest management
C. Plant quarantine
D. DNA banking

171. Primary cell cultures are composed of cells taken directly from:
A. Cells from dead animals
B. Cells taken directly from living animals
C. Stem cells
D. Tissues from accident sites

172. Enzymes required for establishment of primary cell lines are:
A. Trypsin, Collagenases
B. Pectinase, Cellulose

C. Xylanase
D. Lipase

173. Which among the following is not related to purification of pharmaceutical products from cell cultures?
A. The purity must fulfil the specifications
B. Structure and activity should be stable
C. Product should be free from viruses
D. Modifications

174. The parameters maintained in CO_2 incubator used in animal cell culture laboratory are:
A. 5% CO_2, 37°C, 100% humidity
B. 100% CO_2, 37°C, 5% humidity
C. 0% CO_2, 37°C, 5% humidity
D. 0% CO_2, 37°C, 100% humidity

175. Advantage of using animal cells rather than microbial cells for the production of recombinant proteins is:
A. Post transcriptional modification
B. DNA replication
C. Transcription
D. Post translational modification

176. Which one of the following is used as a cryoprotectant?
A. Dimethyl sulphoxide
B. Sodium dodecyl sulphate
C. Ethylene diamine tetra acetic acid
D. Serum

177. Which among the following is not an application of organ culture?
A. Study of pattern of growth
B. Bioassays for action of drugs and carcinogenic agents
C. Production of tissues for implantations
D. Production of phytochemicals

178. In microinjection transgenic animals can be produced by transfer of:
A. DNA into somatic cell
B. DNA into pronucleus of reproductive cell
C. RNA into somatic cell
D. RNA into pronucleus of reproductive cell

179. Which of the following is used for maintaining the pH of the media of mammalian cell culture?
A. Phenol red
B. Neutral red
C. Trypan blue
D. Bromophenol blue

180. The technique not used for organ culture is:
A. Plasma clot
B. Raft methods
C. Grid method
D. Electroporation

181. Which among the following is a DNA sequence database?
A. Genbank B. PIR
C. Swiss prot D. TREMBL

182. Which among the following is a protein sequence database?
A. EMBL B. GenBank
C. DDBJ D. Swiss prot

183. Which is not an application of bioinformatics in Biotechnology?
A. Sequences and structure of genes and proteins
B. 3D molecular structure
C. Genome structure and functions
D. Antisense technology

184. Which among the following BLAST search programme deals with amino acid sequences?
A. BLASTN B. BLASTX
C. TBLASTN D. BLASTP

185. Which statement is not correct about proteomics?
A. Analyses all the proteins in a cell with then individual functions
B. Interaction of specific proteins with other cellular components
C. Study of genome
D. Protein interactions

186. NCBI is:
A. National Center for Biotechnology Information

B. National Center for Biological Information
C. National Center for Bioinformatics Information
D. National Center for Biochemical Information

187. The field of study involving the sequencing of the genomes of organisms is:
A. Proteomics
B. Genomics
C. Bioinformatics
D. Molecular genetics

188. Orthologues mean:
A. They are related by descent from a common ancestor. Orthologous genes normally have the same cellular function.
B. They are related by descent from a common ancestor. Orthologous genes normally have the different cellular function.
C. They are related by descent from a different ancestor. Orthologous genes normally have the different cellular function.
D. They are related by descent from a different ancestor. Orthologous genes normally have the same cellular function.

189. Phylogenetic analysis deals with:
A. Finding out repetitive DNA
B. Finding out evolutionary relationship among Biomolecules (DNA, Proteins)
C. Sorting out structural and non-structural genes
D. Finding out promoter regions

190. Which is not a application of proteonomics technology?
A. Protein expression mapping
B. Annotation of the genome
C. Protein complex identification
D. Sequence and structure of genes

191. Which of the following statements is correct about Coenzymes?
A. Protein component of enzyme
B. Non-protein component that bind with an enzyme to catalyze a reaction
C. Apoenzyme by non-covalent bond
D. Can function alone

192. Which is not a type of enzyme specificity?
A. Absolute B. Group
C. Linkage D. Single

193. Which does not affect enzyme activity?
A. Environment conditions
B. Relative humidity
C. Cofactors and coenzymes
D. Enzyme inhibitors

194. Each enzyme has classification number consisting of:
A. Four digits
B. Three digits
C. Two digits
D. One digit

195. Tenderization of meat can be done using pectolytic enzymes such as:
A. Amylase
B. Lipolase
C. Papain
D. Esperase

196. In confectionary industry which enzyme is used for inversion of sucrose to a mixture of glucose and fructose?
A. Carboxylase
B. Invertase
C. Celluzyme
D. Alcalase

197. Which among the following is not applied for enzyme immobilization?
A. Lyophilization
B. Covalent bonding
C. Entrapment
D. Cross linking

198. The two main types of secondary structure of proteins are:
A. Amino acid sequence
B. α-helix and the β-sheet
C. Aggregate protein complex
D. 3-D structure

199. Which is not a property of enzymes?

A. Enzymes are proteins that increase the rate of reaction by lowering energy of activation

B. They catalyse nearly all the chemical reactions occurring in the body

C. No altered or consumed during reaction

D. Non reusable

200. Which is incorrect about the active site of an enzyme?

A. The area where substrate attaches to

B. It is a small region of enzyme

C. It is a large region of enzyme

D. It is a special pocket on cleft

ANSWERS

1	**2**	**3**	**4**	**5**	**6**	**7**	**8**	**9**	**10**
A	A	A	D	A	B	A	C	C	C
11	**12**	**13**	**14**	**15**	**16**	**17**	**18**	**19**	**20**
A	A	C	D	A	A	B	C	C	D
21	**22**	**23**	**24**	**25**	**26**	**27**	**28**	**29**	**30**
A	C	B	A	D	C	B	B	A	B
31	**32**	**33**	**34**	**35**	**36**	**37**	**38**	**39**	**40**
A	B	D	B	B	B	B	C	C	D
41	**42**	**43**	**44**	**45**	**46**	**47**	**48**	**49**	**50**
B	C	D	A	B	C	D	B	A	A
51	**52**	**53**	**54**	**55**	**56**	**57**	**58**	**59**	**60**
B	B	D	C	A	B	A	C	A	B
61	**62**	**63**	**64**	**65**	**66**	**67**	**68**	**69**	**70**
D	D	C	A	B	C	A	D	C	B
71	**72**	**73**	**74**	**75**	**76**	**77**	**78**	**79**	**80**
B	A	B	B	D	B	C	C	C	B
81	**82**	**83**	**84**	**85**	**86**	**87**	**88**	**89**	**90**
C	D	B	C	B	A	C	C	D	C
91	**92**	**93**	**94**	**95**	**96**	**97**	**98**	**99**	**100**
C	D	B	A	D	A	B	A	A	C
101	**102**	**103**	**104**	**105**	**106**	**107**	**108**	**109**	**110**
D	D	B	B	B	D	D	C	D	C
111	**112**	**113**	**114**	**115**	**116**	**117**	**118**	**119**	**120**
C	A	D	D	C	D	B	A	D	C
121	**122**	**123**	**124**	**125**	**126**	**127**	**128**	**129**	**130**
C	C	D	D	A	B	C	A	A	B
131	**132**	**133**	**134**	**135**	**136**	**137**	**138**	**139**	**140**
D	B	C	B	D	C	B	A	B	C
141	**142**	**143**	**144**	**145**	**146**	**147**	**148**	**149**	**150**
C	A	A	A	B	B	D	C	B	B

151	**152**	**153**	**154**	**155**	**156**	**157**	**158**	**159**	**160**
A	B	D	C	C	C	C	B	D	C
161	**162**	**163**	**164**	**165**	**166**	**167**	**168**	**169**	**170**
B	C	D	A	C	A	D	D	D	D
171	**172**	**173**	**174**	**175**	**176**	**177**	**178**	**179**	**180**
B	A	D	A	D	A	D	B	B	D
181	**182**	**183**	**184**	**185**	**186**	**187**	**188**	**189**	**190**
A	D	D	D	C	A	B	A	B	D
191	**192**	**193**	**194**	**195**	**196**	**197**	**198**	**199**	**200**
B	D	B	A	C	B	A	B	D	C

Previous Paper (Solved)

B.Sc. BIOTECHNOLOGY
Common Entrance Test (CET), 2016*

PHYSICS

1. Two point charges of 25 μC and 100 μC are placed 30 cm apart. The electric field intensity will be zero at a point on the line joining the two charges:

A. 10 cm from the 100 μC charge
B. 12 cm from the 25 μC charge
C. 18 cm from the 25 μC charge
D. 10 cm from the 25 μC charge

2. The longest wavelength in Lyman series is 1216 Å. The shortest wavelength of this series is:

A. 1621 Å
B. 412 Å
C. 912 Å
D. 632 Å

3. A dipole is consisting of two charges $+q$ and $-q$ separated by distance d. The dipole is placed at angle θ_1 in the uniform electric field $\vec{E}$. The work done in rotating the electric dipole from angle θ_1 to θ_2 in the electric field is:

A. $qdE[\cos\theta_1 - \cos\theta_2]$
B. $q(d/2)E[\cos\theta_1 - \cos\theta_2]$
C. $qdE[\cos\theta_1 + \cos\theta_2]$
D. $qdE[\sin\theta_1 - \sin\theta_2]$

4. The electric flux through a surface of area 20 units lying in the $y - z$ plane due to the electric field $\vec{E} = 6\hat{i} + 3\hat{j} + 4\vec{k}$ is:

A. 120 units
B. 100 units
C. 60 units
D. 80 units

5. During a negative beta decay:

A. an atomic electron is ejected
B. an electron which is already present inside the nucleus is ejected
C. a neutron in the nucleus decays emitting an electron
D. a proton in the nucleus decays emitting a positron

6. A charged particle of mass 'm' and charge 'q' is released from rest in a uniform electric field of intensity 'E'. The kinetic energy acquired by the particle in time 't' will be given by:

A. $\frac{q^2 E^2 t^2}{2m}$ B. $\frac{mq^2 E^2 t^2}{2m}$

C. $\frac{mq^2 E^2 t^2}{2}$ D. $\frac{qE^2 t^2}{2m}$

7. A potential difference of 30 V is applied across a colour coded carbon resistor with rings of blue, black and yellow colours. The number of electrons (electron charge = 1.6×10^{-19} C) flowing through the resistor per second will be:

A. 8.0×10^{12} B. 3.1×10^{14}
C. 8.0×10^{15} D. 3.1×10^{13}

8. The kinetic energy of a particle varies with time according to the relation $E_k = (3t + 4)$. The force acting on the particle:

A. is constant
B. varies inversely with velocity
C. varies directly with velocity
D. none of the above

* *Exam Held on 22nd May, 2016.*

9. In the *Laclanche* cell:
A. MnO_3 acts as depolarizer and NH_4Cl solution as electrolyte. Zinc and carbon rods act as anode and cathode, respectively
B. NH_4Cl acts as depolarizer and MnO_2 solution as electrolyte. Zinc and carbon rods act as anode and cathode, respectively
C. NH_4Cl acts as depolarizer and MnO_2 solution as electrolyte. Carbon and zinc rods act as anode and cathode, respectively
D. MnO_2 acts as depolarizer and NH_4Cl solution as electrolyte. Carbon and zinc rods as anode and cathode, respectively

10. An electric power station transmits 100 MW power to a distant load through long and thin cables. The ratio of power loss in the transmission of power at 200 V and 20 kV will be:
A. 2×10^4 B. 10^4
C. 10^2 D. 10^6

11. Four point masses, each of mass M, are placed at the corners of a square ABCD of side L. The moment of inertia of this system about an axis passing through A and parallel to BD is:
A. $3ML^2$ B. $2ML^2$
C. ML^2 D. $5ML^2$

12. The magnetic field at a point is 0.4 Gauss and angle of dip is 30°. The horizontal and vertical components of the field are:
A. 0.2 and $0.2\sqrt{3}$ Gauss, respectively
B. $0.5\sqrt{2}$ and 0.5 Gauss, respectively
C. $0.2/\sqrt{3}$ and 0.2 Gauss, respectively
D. $0.2\sqrt{3}$ and 0.2 Gauss, respectively

13. An A.C. generator consists of a coil of 50 turns and area 2.5 m^2 rotating at angular speed of $(1800/\pi)$ rotations/minute in a uniform field B = 0.2 T between two fixed pole pieces. The resistance of circuit including that of coil is 500 ohm. The maximum current will be:
A. 3 Amperes
B. 3π Amperes
C. 18 Amperes
D. $40/\pi$ Amperes

14. A hydrogen $\left({}^1_1H\right)$ nucleus and tritium $\left({}^3_1H\right)$ nucleus enter a uniform magnetic field at 90° with same velocity. The period of rotation of tritium nucleus inside the magnetic field is:
A. same as that of hydrogen nucleus
B. 9 times that of hydrogen nucleus
C. 3 times that of hydrogen nucleus
D. 4 times that of hydrogen nucleus

15. One mole of oxygen at 0°C temperature and 1 atmosphere pressure is compressed adiabatically, till its pressure is 8 atmosphere. The ratio of specific heat at constant pressure to that at constant volume is given to be 1.5. The final temperature of the gas will be:
A. 273 K B. 546 °C
C. 273 °C D. 200 °C

16. Given coefficient for linear expansion for brass $= 1.9 \times 10^{-5}\,°C^{-1}$. At the room temperature of 25 °C, the volume of a thin brass vessel and volume of a solid brass cube are equal to one litre each. The change in the volume of the vessel and that of the cube upon being heated to 50 °C will be:
A. 1,425 cm^3 in both cases
B. 1,425 cm^3 and 2,850 cm^3, respectively
C. 1,425 cm^3 and 4,275 cm^3, respectively
D. 4,275 cm^3 in both cases

17. Diameter of a plano-convex lens is 6 cm and its thickness at the centre is 3 mm. The radius of curvature for the convex surface is:
A. 3 cm B. 12 cm
C. 15 cm D. 5.7 cm

18. A reversible engine converts one-sixth of heat input into work. When the temperature of the sink is reduced by 62°C, its efficiency is doubled. The temperature of the source will be:
A. 99 °C B. 370 °C
C. 172 °C D. 199 °C

19. A potential difference of 20 kV is applied across a Cu-anode X-ray tube. Given value of Planck's constant $= 6.62 \times 10^{-34}$ J-s. The minimum wavelength of continuous spectrum of photons generated will be:

A. 10^{-9} m B. 6.2×10^{-13} m
C. 1.32 nm D. 6.2 Å

20. The number of air molecules in a room of capacity 30 m^3 at a temperature of 27°C and one atmosphere pressure. Given Boltzmann constant $k = 1.38 \times 10^{-23}$ J K^{-1}:

A. 6.1×10^{26} B. 7.3×10^{26}
C. 6.1×10^{23} D. 7.3×10^{23}

21. The degree of freedom of a monoatomic gas molecule, a diatomic gas molecule and a triatomic gas molecule are:

A. 1, 2 and 3 respectively
B. 3, 6 and 9 respectively
C. 2, 4 and 6 respectively
D. 3, 5 and 6 respectively

22. At a given temperature, the ratio of r.m.s. speed of oxygen molecules to that of ozone molecules will be:

A. 1.5 B. 1.22
C. 0.66 D. 0.81

23. A point bulb is placed at the bottom of tank containing liquid (refractive index = 1.5) to a depth of 100 cm. The approximate area of the surface of water through which light from the bulb can emerge out is approximately equal to:

A. 22,500 cm^2 B. 1,500 cm^2
C. 25,100 cm^2 D. 27,500 cm^2

24. For the Glass-Air system, the value of critical angle is in increasing order for the:

A. Blue, Yellow, Green, Red colours of light
B. Green, Yellow, Red, Blue colours of light
C. Red, Yellow, Green, Blue colours of light
D. Blue, Green, Yellow, Red colours of light

25. If the two slits in Young's double slit experiment have width ratio 4 : 1, the ratio of intensity at maxima and minima in the interference pattern will be:

A. 1 : 4 B. 4 : 1
C. 9 : 1 D. 16 : 1

26. A body of mass 100 kg falls on the earth from infinity. Given radius of earth = 6400 km and g = 9.8 m/s^2. Air friction may be neglected. Its velocity and energy on reaching the earth are:

A. 8 km/s and 3.2×10^9 J, respectively
B. Both infinity
C. 11.2 km/s and 6.27×10^8 J, respectively
D. 11.2 km/s and 6272 J, respectively

27. A sample of milk diluted with pure water has density of 1032 kg/m^3. If pure milk has density of 1080 kg/m^3. The percentage of water by volume added in the milk is:

A. 50% B. 60%
C. 40% D. 30%

28. A water drop of radius 'R' falling on the earth achieves terminal velocity due to viscous force on the drop due to air. The terminal velocity will be proportional to:

A. R B. $1/R^2$
C. R^2 D. $1/R^3$

29. If 10% of a radioactive material decays in 5 days, then the amount of the original material left after 20 days is approximately:

A. 60% B. 65%
C. 70% D. 75%

30. A geostationary satellite is orbiting the earth at a height 6 R above the surface of earth; where R is the radius of the earth. The time period of another satellite orbiting at the height 2.5 R from the surface of the earth will be approximately:

A. 6.5 hours B. 8.5 hours
C. 7.5 hours D. 6 hours

31. The ratio of the respective resolving powers of an optical microscope corresponding to the wavelengths of light λ_1 = 400 nm and λ_2 = 500 nm used is:

A. 5 : 4 B. 4 : 5
C. 16 : 25 D. 25 : 16

32. Rainbow is formed due to combination of:

A. Dispersion and total internal reflection
B. Refraction and polarization
C. Dispersion and focusing
D. Reflection and polarization

33. Light of wavelength 500 nm is made to pass from air to glass, its wavelength and frequency in the glass (refractive index = 1.5) will be:

A. 500 nm and 6×10^{14} Hz

B. 333 nm and 6×10^{14} Hz

C. 500 nm and 4×10^{14} Hz

D. 333 nm and 4×10^{14} Hz

34. The electromagnetic waves spectrum in the correct increasing order of wavelengths is:

A. Microwaves, X-rays, Ultraviolet, Visible, Infrared

B. Ultraviolet, Visible, Infrared, Microwaves, X-rays

C. X-rays, Ultraviolet, Visible, Infrared, Microwaves

D. X-rays, Microwaves, Ultraviolet, Visible, Infrared

35. Two air columns (closed at one end) 100 cm and 101 cm long give 17 beats in 20 s, when each is sounding its fundamental note. The velocity of sound is:

A. 34340 cm/s

B. 30340 cm/s

C. 33240 cm/s

D. 32000 cm/s

36. The moment of force $F = \hat{i} + \hat{j} + \hat{k}$ acting at point (–2, 3, 4) about the point (1, 2, 3) is equal to:

A. $4(\hat{i} + 2\hat{j} - 3\hat{k})$ B. $4(2\hat{i} + 3\hat{j} - 4\hat{k})$

C. $3(\hat{j} - \hat{k})$ D. $4(\hat{j} - \hat{k})$

37. If the linear density of a rod varies as $\lambda = Bx$ along its length L kept along the *x*-axis, its centre of mass will be at:

A. $X_{CM} = 4L/5$ B. $X_{CM} = L/2$

C. $X_{CM} = 2L/3$ D. $X_{CM} = 5L/3$

38. The dimensions of $\varepsilon_0 E^2$ (ε_0 is permittivity of free space and E is electric field) is:

A. $M L T^{-1}$ B. $M L^2 T^{-2}$

C. $M L^{-1} T^{-2}$ D. $M L^2 T^{-1}$

39. A neutron travelling with a velocity V and kinetic energy *E* collides elastically head on with the nucleus of an atom of mass number A at rest. The fraction of total energy retained by the neutron is approximately equal to:

A. $\left[\frac{A+1}{A-1}\right]^2$ B. $\left[\frac{A-1}{A+1}\right]^2$

C. $\left[\frac{A}{A-1}\right]^2$ D. $\left[\frac{A+1}{A}\right]^2$

40. Energy levels A, B and C of a certain atom correspond to increasing values of energy, *i.e.*, $E_A < E_B < E_C$. If λ_1, λ_2 and λ_3 are the wavelengths of radiations for the transitions C → B, B → A and C → A, respectively, which of the following is correct?

A. $1/\lambda_3 = 1/\lambda_1 + 1/\lambda_2$ B. $(\lambda_3)^2 = (\lambda_1)^2 + (\lambda_2)^2$

C. $\lambda_1 + \lambda_2 + \lambda_3 = 0$ D. $\lambda_3 = \lambda_1 + \lambda_2$

41. The minimum distance between an object and its real image formed by a convex lens of focal length '*f*' is:

A. 2 *f* B. 4 *f*

C. *f* D. zero

42. The height y (in metres) and distance *x* (in metres) along a horizontal plane of a projectile of earth are given by $x = 6t$ and $y = (8t - 5t^2)$, where *t* represents time in seconds. The velocity with which the projectile is projected is:

A. 8 m/s making angle $\tan^{-1}$ (3/4) with *x*-axis

B. 6 m/s making angle $\tan^{-1}$ (4/3) with *x*-axis

C. 10 m/s making angle $\tan^{-1}$ (4/3) with *x*-axis

D. 12 m/s making angle $\tan^{-1}$ (4/3) with *x*-axis

43. If the relation between distance *x* and time *t* is of the form $t = ax^2 + bx$, then acceleration of the particle is:

A. $-2av^3$, where $v = dx/dt$

B. $2bv^3$, where $v = dx/dt$

C. $-2av^{3/2}$, where $v = dx/dt$

D. $-2bv^3 + 2av$, v here $v = dx/dt$

44. A particle of mass m is projected with velocity V making an angle of 45° with the horizontals. The magnitude of the angular momentum of the projectile about the point of projection when the particle is at maximum height is:

A. Zero

B. $\frac{mv^3}{2\sqrt{2}\,g}$

C. $\frac{mv^2}{4\sqrt{2}\,g}$

D. $\frac{mv^3}{4\sqrt{2}\,g}$

45. Two linear harmonic motions of equal amplitude and frequencies ω and 2ω are impressed by a particle along x and y-axes, respectively. If the initial phase difference between them is 90, the resultant path followed by the particle will be:

A. Parabola
B. Ellipse
C. Circle
D. Hyperbola

46. Limbs of a manometer consist of uniform capillary tubes of radii 'r_1' and 'r_2', respectively, are connected to pressures p_1 and p_2, respectively. Given $r_2 < r_1$. The level of the liquid (density = d, surface tension = T) in the narrower tube stands at height 'h' above the broader tube. Then the correct pressure difference ($p_1 - p_2$) between the limbs of manometer is given by:

A. $hdg - 2T\left[\frac{1}{r_2} - \frac{1}{r_1}\right]$

B. hdg

C. $hdg - T\left[\frac{1}{r_2} - \frac{1}{r_1}\right]$

D. $hdg - 2T\left[\frac{1}{r_1} - \frac{1}{r_2}\right]$

47. A barking dog delivers about 1 mW of power, which is uniformly distributed over a hemispherical area. The intensity of sound at a distance of 5 m is:

A. 0.04 mW/m^2
B. 3.19 μW/m^2
C. 12.74 μW/m^2
D. 6.37 μW/m^2

48. A progressive wave of frequency 500 Hz is travelling with a velocity of 360 m/s. The distance between the two points 60° out of phase is:

A. 0.05 m
B. 0.06 m
C. 0.12 m
D. 0.24 m

49. A message signal of 12 kHz and peak voltage 20 V is used to amplitude modulate a carrier wave of frequency 12 MHz and peak voltage 30 V. The value of modulation index (m_a) and the side band frequencies produced will be:

A. m_a = 1.5 and LSB = 11988 kHz, HSB = 12012 kHz

B. m_a = 0.67 and LSB = 11.88 kHz, HSB = 12.12 kHz

C. m_a = 0.67 and LSB = 11988 kHz, HSB = 12012 kHz

D. m_a = 1.5 and LSB = 11.88 kHz, HSB = 12.12 kHz

50. A TV transmitting antenna in the plane area is 81 m tall. For the receiving antenna height to be at near ground level, the service area covered by the transmitting antenna will be about:

A. 81 km^2
B. 6400 km^2
C. 5100 km^2
D. 3200 km^2

51. A full wave rectifier consists of two identical junction diodes having negligible forward resistance. When a sinusoidal voltage supply of peak value 100 V is fed to the rectifier, the RMS value of the voltage at its output will be:

A. 100 V

B. $100\sqrt{2}$

C. 70.7 V

D. $\frac{100}{(2\sqrt{2})}$

52. An n-type semiconductor is having electron concentration of 8 × 10^{15} m^{-3} and hole concentration of 5 × 10^{16} m^{-3}. Given electron mobility = 2.3 m^2 V^{-1} s^{-1} and hole mobility = 0.01 m^2 V^{-1} s^{-1}. Electron charge = 1.6 × 10^{-19} C. The resistivity of the sample is:

A. 3.4 × 10^{-2} Ω-m
B. 29.4 Ω-m
C. 2.94 Ω-m
D. 3.3 Ω-m

53. A train moves towards a stationary observer with speed 34 m/s. The train sounds a whistle and its frequency registered by an observer is f_1. If the train's speed is reduced to 17 m/s, the frequency registered is f_2. If the speed of the sound is 340 m/s then the ratio $\frac{f_1}{f_2}$ is:

A. $\frac{19}{17}$ B. $\frac{1}{2}$

C. 2 D. $\frac{19}{18}$

54. An electron is released at rest in an electric field of intensity 2×10^4 N/C. Given mass of electron = 9×10^{-31} kg. The time taken for the electron to travel 1.6 cm in the electric field is:

A. 28.8 *ns* B. 9 *ns*

C. 3 *ns* D. 3×10^{-8} *s*

55. There are two 500 pF capacitors A and B. The capacitor A is charged by connecting it to a 200 V battery. Then it is disconnected from the battery and connected in parallel to capacitor B. The electrostatic energy stored in the final system will be:

A. 10 μJ B. 20 μJ

C. 5 μJ D. 2.5 μJ

56. N identical cells each of emf 'E' and internal resistance 'r' are put in parallel across an external resistance 'R'. The current drawn from each of the cell will be:

A. $\frac{(n-1)\mathrm{E}}{n\mathrm{R}+r}$ B. $\frac{n\mathrm{E}}{nr+\mathrm{R}}$

C. $\frac{n\mathrm{E}+r}{n\mathrm{R}}$ D. $\frac{\mathrm{E}}{n\mathrm{R}+r}$

57. A ray of light passes through an equilateral prism such that the angle of incidence is equal to the angle of emergence and the latter is equal to 3/4 times the angle of prism. The angle of deviation is:

A. 20° B. 60°

C. 30° D. 45°

58. Given Density of silver = 10.0 g/c.c and E.C.E. of silver = 0.00112 g/Coulomb. A current of one ampere is passed through a silver voltameter in order to deposit a layer of 0.14 mm thick on a metal plate of 140 cm^2 area. The time taken for the deposit is:

A. 17500 sec

B. 175 sec

C. 20000 sec

D. 1750 sec

59. A hall is dropped from a height *h* on to a floor and undergoes multiple bounces. If in each collision its speed becomes '*e*' times of its striking value, the time taken by the ball between first bounce and second bounce is:

A. $\frac{e}{g}\sqrt{2gh}$ B. $2e\sqrt{\frac{2h}{g}}$

C. $\frac{2e\sqrt{h}}{g}$ D. $\frac{2e\sqrt{g}}{h}$

60. The truth table for the following logic circuit is:

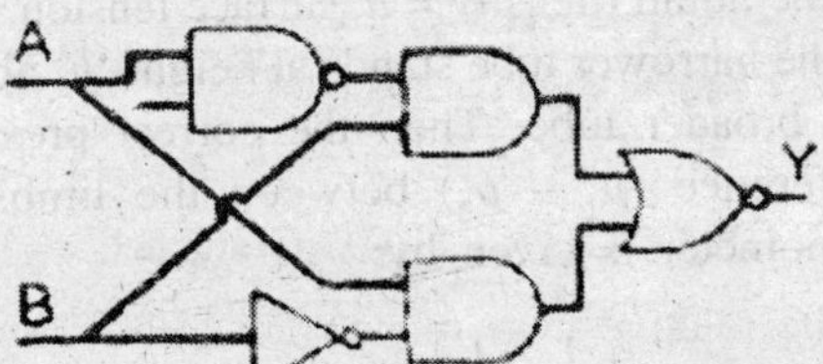

A.

A	B	Y
0	0	1
0	1	1
1	0	0
1	1	1

B.

A	B	Y
0	0	0
0	1	0
1	0	0
1	1	1

C.

A	B	Y
0	0	1
0	1	0
1	0	0
1	1	0

D.

A	B	Y
0	0	1
0	1	0
1	0	0
1	1	1

CHEMISTRY

1. Under conditions of fixed temperature and amount of gas, Boyle's law requires that:

I. $P_1V_1 = P_2V_2$

II. PV = constant

III. $P_1/P_2 = V_2/V_1$

A. I, II and III
B. II only
C. III only
D. I only

2. The type of hybridization of each carbon in CH_3CN:

A. sp^2, sp^3 B. sp^1, sp
C. sp^3, sp^2 D. sp, sp^3

3. The correct order for ACIDITY values among the following is:

A. $CH_3CH_2COOH < CH_3COOH < C_6H_5CH_2COOH < C_6H_5COOH$

B. $CH_3CH_2COOH < CH_3COOH < C_6H_5COOH < C_6H_5CH_2COOH$

C. $C_6H_5COOH < C_6H_5CH_2COOH < CH_3CH_2COOH < CH_3COOH$

D. $C_6H_5COOH < C_6H_5CH_2COOH < CH_3COOH < CH_3CH_2COOH$

4. The relative case of dehydration of alcohols follows the order:

A. Tertiary > Primary > Secondary
B. Tertiary > Secondary > Primary
C. Secondary > Primary > Tertiary
D. Secondary > Tertiary > Primary

5. The product formed when ethanol is treated with sulphuric acid at 413 K:

A. Ethoxyethane B. Ethene
C. Methoxyethane D. Ethanoic acid

6. The conversion of nitriles to primary amines can be achieved with the help of following reagent:

A. Cu/HCl B. Br_2/NaOH
C. Br_2/P D. $Na(Hg)C_2H_5OH$

7. Which of the following is NOT a monosaccharide?

A. Glucose B. Sucrose
C. Ribose D. Fructose

8. The polymer obtained by interaction of caprolactum with water is:

A. Nylon 6 B. Bakelite
C. Dacron D. Glyptal

9. Heroin is an example of:

A. Antiseptics B. Tranquilizers
C. Analgesics D. Antibiotics

10. The reaction between toluene and halogen proceeds in the presence of Lewis acid to furnish halotoluenes. This is an example of:

A. Electrophilic substitution
B. Electrophilic addition
C. Nucleophilic addition
D. Free radical substitution

11. The correct IUPAC name of the given chemical compound is:

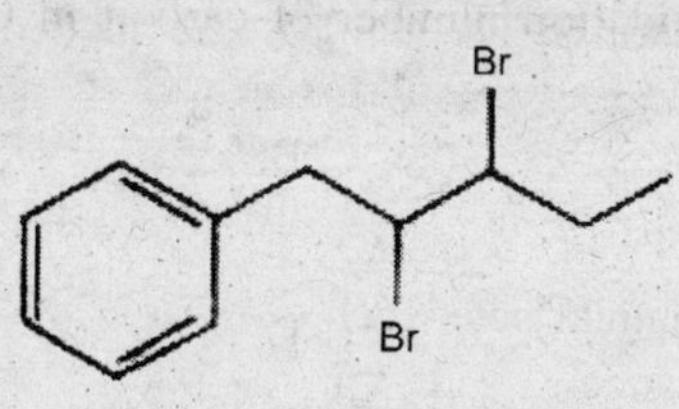

A. 3,4 Dibromo-5-phenyl pentane
B. 2,3 Dibromo-1-phenyl pentane
C. 3,4 Dibromo-1-phenyl pentane
D. 3,4 Dibromo-6-phenyl pentane

12. During the estimation of halogen in Carius method, 0.8 g of an organic compound gave 0.188 g of AgBr. The percentage of bromine in the compound is:

A. 0.1 B. 1.0
C. 10 D. 100

13. The number of pi (π) and sigma (σ) bonds in 1,3,5,7 octatetraene are:

A. $\pi = 4, \sigma = 17$ B. $\pi = 8, \sigma = 8$
C. $\pi = 17, \sigma = 4$ D. $\pi = 8, \sigma = 17$

14. The maximum prescribed concentration (ppm) of iron in drinking water is:
A. 0.002 B. 0.02
C. 0.2 D. 2

15. Ozone is present in:
A. Exosphere B. Stratosphere
C. Troposphere D. Thermosphere

16. Sodium salt of which acid will be needed for the preparation of propane?
A. Propionic acid
B. Butanoic acid
C. Hexanoic acid
D. Pentanoic acid

17. HBr added to propene gives:
A. 1-Bromopropane
B. 2-Bromopropane
C. 1,1-dibromopropane
D. 1,3-dibromopropane

18. Which of the following substituent groups give-R effect?
A. OH B. NH_2
C. COOH D. NHR

19. The oxidation number of carbon in C_3O_2 is:
A. 2 B. $\frac{4}{3}$
C. $\frac{3}{4}$ D. 4

20. The correct order for decreasing reducing strength among the following is:
A. $Li > Na > Ag > Cl^- > I^-$
B. $I^- > Cl^- > Na > Ag > Li$
C. $I^- > Na > Ag > Cl^- > Li$
D. $Li > Na > I^- > Ag > Cl^-$

21. Dihydrogen is reduced by sodium to form:
A. NaH_2 B. Na_2H
C. NaH D. Na_2H_2

22. The strength of 100 volume solution of hydrogen peroxide is:
A. 3% B. 30%
C. 300% D. 0.3%

23. The energy released per mole (in kJ) on combustion would be highest for:
A. LPG B. Dihydrogen
C. CH_4 D. Octane

24. The radioactive isotope of water is:
A. Deuterium
B. Protium
C. Tritium
D. Deuterium and Protium

25. The correct order for decrease in ionic sizes among the following is:
A. $Li^+ > Na^+ > K^+ > Rb^+ > Cs^+$
B. $Cs^+ > Rb^+ > K^+ > Na^+ > Li^+$
C. $Li^+ > K^+ > Rb^+ > Na^+ > Cs^+$
D. $Cs^+ > K^+ > Rb^+ > Na^+ > Li^+$

26. In group 14, the most acidic dioxide is formed by:
A. Lead B. Silicon
C. Carbon D. Germanium

27. The correct order of catenation is:
A. $C >> Si > Ge \approx Sn$
B. $C >> Si > Ge > Sn$
C. $Si > Ge \approx Sn >> C$
D. $Ge \approx Sn >> C > Si$

28. $B(OH)_3$ is:
A. Tetra basic acid
B. Tribasic acid
C. Dibasic acid
D. Monobasic acid

29. In the transitions of the electron in hydrogen atom, Lyman series belong to which spectral region?
A. Infra red B. Visible
C. Ultra violet D. Far infra red

30. The mass of photon of 3.6 Å wavelength is:
A. 6.315×10^{-29} kg
B. 3.157×10^{-29} kg
C. 6.315×10^{-29} gm
D. 3.157×10^{-29} gm

31. The uncertainty principle was given by:
A. Bohr B. E. Schrodinger
C. W. Heinsenberg D. Rutherford

32. Which of the following is correct?

A. $n = 0, \ell = 0, m_1 = 0, m_s = +\frac{1}{2}$

B. $n = 1, \ell = 0, m_1 = 0, m_s = -\frac{1}{2}$

C. $n = 1, \ell = 1, m_1 = 0, m_s = +\frac{1}{2}$

D. $n = 3, \ell = 3, m_1 = -3, m_s = +\frac{1}{2}$

33. Which of the following oxidation states is most common for all lanthanides?

A. +4 B. +3

C. +2 D. +1

34. The most negative electron gain enthalpy among the following is of:

A. P B. S

C. Cl D. F

35. The covalency of Al in $[AlCl(H_2O)_5]^{2+}$ is:

A. 5 B. 2

C. 6 D. 3

36. The correct geometry for BF_3 is:

A. Trigonal pyramidal

B. Trigonal planar

C. Tetrahedral

D. Bent

37. The correct hybridization of SF_6 is:

A. sp B. sp^3d^1

C. sp^3d^3 D. sp^3d^2

38. The bond order of C_2 is:

A. 1 B. 2

C. 3 D. 4

39. Which of the following is paramagnetic?

A. H_2 B. C_2

C. Li_2 D. O_2

40. The enthalpics of all elements in their standard states are:

A. Unity

B. < 0

C. Different for each element

D. Zero

41. The value of K_C for the reaction:

$2A \rightleftharpoons B + C$ is 2×10^{-3}.

At a given time the concentration of reaction mixture is $[A] = [B] = [C] = 3 \times 10^{-4}$ M. In which direction the reaction will proceed?

A. Forward direction

B. Reverse direction

C. Steady state

D. Static equilibrium

42. A Bronsted-Lowry base is defined as a substance that:

A. acts as a proton donor

B. increases $[H^+]$ when placed in water

C. acts as a proton acceptor

D. decreases $[H^+]$ when placed in water

43. The pH of a 1.0×10^{-8} M solution of HCl:

A. 8 B. 6

C. 6.98 D. 7.02

44. The packing efficiency in *Cubic Close Packing* structure is:

A. 74%

B. 68%

C. 52.4%

D. 45.2%

45. Raoult's Law describes:

A. How the partial pressure of solvent vapour varies with solute molecular mass

B. How the partial pressure of solvent vapour varies with solute concentration

C. How the partial pressure of a gas varies with temperature

D. How the solubility of a gas varies with pressure

46. 200 cm^3 of an aqueous solution of a protein contains 1.26 g of the protein. The osmotic pressure of such a solution at 27°C is found to be 2.57×10^{-3} bar. The molar mass of the protein is:

A. 1,22,078 g mol^{-1}

B. 1,00,786 g mol^{-1}

C. 86,437 g mol^{-1}

D. 61,022 g mol^{-1}

47. The units of specific conductance are:

A. ohm^{-1}

B. ohm^{-1} cm^{-1}

C. ohm^{-1} cm^3 mol^{-1}

D. ohm^{-1} cm^2 $(g\ eq.)^{-1}$

48. The standard electrode potential for Daniell cell is 1.1 V. The standard Gibbs energy for the reaction following is:
$Zn(s) + Cu^{+2}(aq) \rightarrow Zn^{+2}(aq) + Cu(s)$

A. -212.27 kJ mol^{-1}

B. -313.17 kJ mol^{-1}

C. -142.54 kJ mol^{-1}

D. -116.22 kJ mol^{-1}

49. If the molar conductance at infinite dilution of NaCl, HCl and CH_3COONa are 126.4, 425.9 and 91.0 S cm^2 mol^{-1} respectively. The molar conductance at infinite dilution for Acetic acid is:

A. 390.5 S cm^2 mol^{-1}

B. 300.9 S cm^2 mol^{-1}

C. 290.5 S cm^2 mol^{-1}

D. 200.9 S cm^2 mol^{-1}

50. The half-life of a first order reaction is found 1.26×10^{13} sec. The rate constant of the reaction is:

A. $k = 1.5 \times 10^{-15}$ sec^{-1}

B. $k = 5.5 \times 10^{-14}$ sec^{-1}

C. $k = 5.1 \times 10^{-15}$ sec^{-1}

D. $k = 1.5 \times 10^{-11}$ sec^{-1}

51. The rate of a chemical reaction doubles for every 10°C rise of temperature. If the temperature is raised by 50°C, the rate of the reaction increases by about:

A. 10 times B. 24 times
C. 32 times D. 64 times

52. The rate constant of a first order reaction depends on the:

A. Time

B. Concentration of the product

C. Concentration of the reactant

D. Temperature

53. The rate constant of a reaction at 500 K and 700 K are 0.02 s^{-1} and 0.07 s^{-1} respectively. The energy of activation is:

A. 18.231 kJ B. 24.342 kJ
C. 28.321 kJ D. 31.801 kJ

54. In Freundlich Adsorption isotherm, the value of $1/n$ is:

A. Between 2 and 4 in all cases

B. Between 0 and 1 in all cases

C. 1 in case of chemisorption

D. 1 in case of physical adsorption

55. Which one of the following is an example for homogenous catalysis?

A. Hydrogenation of oil

B. Manufacture of ammonia by Haber's process

C. Manufacture of sulphuric acid by Contact process

D. Hydrolysis of sucrose in presence of dilute hydrochloric acid

56. Which one of the following forms micelles in aqueous solution above certain concentration?

A. Sodium Chloride

B. Glucose

C. Dodecyl trimethyl ammonium chloride

D. Urea

57. Hypochlorous acid is:

A. HOCl

B. HOClO

C. $HOClO_2$

D. $HOClO_3$

58. Which one of the following has an optical isomer? (en = ethylenediamine)

A. $[Zn(en)(NH_3)_2]^{2+}$

B. $[Co(H_2O)_4(en)]^{2+}$

C. $[Co(en)_3]^{3+}$

D. $[Zn(en)_2]^{2+}$

59. Magnetite is:

A. Fe_2O_3

B. Fe_3O_4

C. $CuFeS_2$

D. $CuCO_3.Cu(OH)_2$

60. P_4O_{10} is the anhydride of:

A. H_3PO_2 B. H_3PO_3
C. H_3PO_4 D. $H_4P_2O_7$

BIOLOGY

1. Which one of the following regarding embryonic development in humans is correct?
A. Cleavage brings considerable increase in the mass of protoplasm
B. In the second cleavage division, one of the two blastomeres usually divides a little sooner than the second
C. With more cleavage divisions, the resultant blastomeres become larger and larger
D. Cleavage division results in a solid ball of cells called morula

2. The correct sequence of events during fertilization in mammals is:
A. Capacitation, activation of eggs, acrosomal reaction, and entry of sperm in egg
B. Capacitation, acrosomal reaction, activation of egg and entry of sperm in egg
C. Aerosomal reaction, activation of egg, capacitation and entry of sperm in egg
D. Activation of egg, acrosomal reaction, and entry of sperm in egg and capacitation

3. Hypersensitivity to an allergen is associated with:
A. Aberrant functioning of the immune mechanism
B. Increase in umbient temperature
C. Age of the individual
D. Food habits

4. In gastrula, the formation of three germinal layers is the beginning of:
A. Differentiation
B. Involution
C. Neurulation
D. Gastrulation

5. In a hybridization experiment a plant shows phenotypic ratio of 15:1. How many genes control the trait for observed phenotypic ratio?
A. One B. Two
C. Three D. Polygene

6. When a diploid cell undergoes normal meiotic division, then all:
A. Dominant genes segregate together
B. Homologous chromosomes segregate
C. Daughter cells will have different homologous chromosome pairs
D. Linkage groups will be disrupted

7. Extrusion of second polar body from egg nucleus occurs:
A. After entry of sperm before completion of fertilization
B. After completion of fertilization
C. Before entry of sperm
D. Without any relation of sperm entry

8. Genetic drift is most likely to occur:
A. Among migrating populations
B. In populations which show no mutations
C. In very large sized populations
D. In populations which are geographically isolated

9. The chromosome analysis of a newly born male revealed that non-disjunction of sex chromosomes at mitosis has occurred once. The most likely sex chromosome composition of the individual is:
A. XY B. XO
C. XXY D. XYY

10. The enzyme peptidyl transferase which helps in the formation of peptide bond is present in the:
A. t-RNA
B. Larger subunit of ribosome
C. Smaller subunit of the ribosome
D. m-RNA

11. The insect responsible for the spread of kala-azar is:
A. *Musca domestica*
B. *Phlebotomus argentipus*
C. *Glossina palpalis*
D. *Cimex lectularis*

12. Which vector can clone only a small fragment of DNA?

A. Bacterial artificial chromosome
B. Yeast artificial chromosome
C. Plasmid
D. Cosmid

13. The reason why Antarctic penguins are much larger than the penguins living in more temperate regions is:

A. Cold air contains a larger percentage of oxygen than the warm air
B. The fishes on which penguins feed are larger in cold water than in warm water
C. In Antarctica penguins lay fewer and larger eggs than penguins in warm latitudes
D. Large penguins have a smaller surface to volume ratio than small penguins

14. Clitoris is an outgrowth which is:

A. Non functional penis
B. Analogous to penis
C. Functional penis in females
D. Homologous to penis

15. Bacteria cannot be classified as species by the biological species concept because they:

A. Reproduce asexually
B. Have high growth rate
C. Exhibit little morphological variation
D. Are enucleated

16. Which of the following statements is incorrect about histone proteins?

A. They are highly basic because they contain many positively charged amino acid side chains
B. They are extensively modified after their translation
C. In combination with DNA, they are primary constituents of chromatin
D. They account for approximately one-fifth of the mass of a chromosome

17. Ecotone is best described as:

A. State of equilibrium among various trophic levels in an ecosystem
B. Potentiality of an animal to adjust to new circumstances
C. Transition zone between two or more diverse communities
D. Maximum biomass an ecosystem can support

18. Members of which trophic level would suffer the most from biological magnification of harmful substances?

A. First trophic level
B. Second trophic level
C. Third trophic level
D. Fourth trophic level

19. Which of the following characteristics is undesirable in cloning vectors?

A. Readily isolated from cells
B. High copy number
C. Vulnerable at several sites to a restriction enzyme
D. Small in size

20. Transgenic plants are easier to produce than transgenic animals because:

A. Plants can more easily be grown from single cultured cell into which foreign DNA has been introduced
B. Plant DNA is easier to clone
C. Plant cells can be transformed by bacterial infection
D. DNA passes more readily through the plant cells wall than through the animal cell membrane

21. Which of the following types of vectors would be most suitable for introducing DNA into a human cell?

A. Plasmid
B. Bacteriophage
C. Cosmid
D. Adenovirus

22. Human insulin is commercially produced by:

A. Transgenic species of *Saccharomyces*
B. Transgenic species of *Escherichia*
C. Transgenic species of *Mycobacterium*
D. Transgenic species of *Rhizobium*

23. Which cells are involved in antigen clearance non-specifically by engulfing them?

A. B cells
B. T cells
C. Phagocytic cells
D. Cytotoxic T cells

24. Agar is used as a medium for bacteria cultures. It is obtained from:
A. bones of dead animals
B. starch
C. banana
D. seaweed

25. Marine pollution with oil discharged from damaged tankers is now dealt with by spraying a suspension of bacterium known as:
A. *Lactobacillus* B. *Streptococcus*
C. *Xanthomonas* D. *Pseudomonas*

26. In biogas plant, which bacteria is utilized?
A. Methanogens
B. Nitrifying bacteria
C. Ammonifying bacteria
D. Denitrifying bacteria

27. In which ecosystem would you find the highest biodiversity?
A. Freshwater marsh
B. Tundra
C. Prairie grassland
D. Boreal forest

28. There is more competition for survival between:
A. Different animals of same niche
B. Same animals of same niche
C. Different animals of different niche
D. Same animals of different niche

29. Vaccines which are result of recombinant DNA technology are called:
A. First generation vaccines
B. Second generation vaccines
C. Third generation vaccines
D. Fourth generation vaccines

30. Tissue macrophages are mature:
A. B cells B. T cells
C. NK cells D. Monocytes

31. Fungi reproduce by only asexual methods in:
A. Myxomycetes
B. Ascomycetes
C. Basidiomycetes
D. Deuteromycetes

32. Nutrition in *Chlamydomonas* is:
A. Heterotrophic B. Parasitic
C. Autotrophic D. Saprophytic

33. Gemmae are vegetative reproductive structures found in:
A. Angiosperms B. Bryophytes
C. Algae D. Gymnosperms

34. The function of ligule is:
A. Hygroscopic
B. Perennation
C. Reproduction
D. All of the above

35. Largest ovules in plant kingdom are found in:
A. *Cycas revoluta* B. *Cycas thousarsii*
C. *Cycas circinalis* D. *Pinus longifolia*

36. Ray-florets and disc-florets are the characteristic features of family:
A. Brassicaceae B. Solanaceae
C. Asteraceae D. Malvaceae

37. Which type of leaf is that of *Mimosa pudica*?
A. Simple pinnate
B. Simple palmate
C. Compound bipinnate
D. Decompound

38. Abnormal secondary growth due to accessory cambium is found in:
A. *Helianthus* B. *Cucurbita*
C. *Dracaena* D. Maize

39. The closing and opening of stomata is due to influx of:
A. K ions B. Na ions
C. Fe ions D. Mg ions

40. Discovery of Emerson effect has shown the existence of:
A. Light and dark reaction
B. Photorespiration
C. Photophosphorylation
D. Two distinct pigment systems

41. Which one of the following hormones is not found in plants?
A. 2, 4-D B. GA_3
C. Gibberellin D. IAA

42. Pappus is a modification of:
A. Calyx B. Corolla
C. Stamens D. Gynoecium

43. Phenomenon of opening of flower buds is called:
A. Gametogenesis
B. Sporogenesis
C. Anthesis
D. None of the above

44. Development of embryo sac from any cell of nucellus is called:
A. Apogamy B. Apospory
C. Parthenogenesis D. Sporogenesis

45. To which category of fruits, maize grain can be assigned?
A. Pyxidium B. Caryopsis
C. Capsule D. Sorosis

46. How many ascospores are found in *Saccharomyces cerevisiae*?
A. 2 B. 4
C. 8 D. 16

47. Palmella stage in *Chlamydomonas* is produced:
A. In rainy season
B. During unfavourable conditions
C. During favourable conditions
D. None of the above

48. Which one of the following is absent in sporophyte of Funaria?
A. Foot B. Seta
C. Elaters D. Columella

49. The gametophytic generation in pteridophytes is commonly called:
A. Thallus B. Plant body
C. Prothallus D. Protonema

50. Mucilage canals of *Cycas* stem are:
A. Lysigenous B. Schizogenous
C. Schizolysogenous D. Filled with tannin

51. Monocotyledonous leaf showing reticulate venation is found in:
A. *Smilax* B. *Zea mays*
C. *Cocas* D. *Calophyllum*

52. Epiphytic roots are found in:
A. Indian Rubber
B. *Orchids*
C. *Tinospora*
D. *Cuscuta*

53. The apical meristem, in root is sub-terminal because it:
A. Is covered by root hairs
B. Is covered by root cap
C. Has many corpus cells
D. Is covered by tunica cells

54. The bulliform cells of monocot leaves show:
A. Tropic movement
B. Autonomic movement
C. Nastic movement
D. Turgor movement

55. Energy molecule is produced during light reaction and transferred to the dark reaction by:
A. Chlorophyll B. ADP
C. ATP D. RuBP

56. Presence of which of the following is associated with seed dormancy?
A. Starch B. Ethylene
C. IAA D. Abscisic acid

57. Which technical term is used for a sterile stamen?
A. Spur B. Staminode
C. Stallion D. Staminate

58. Myrmecophily is a beneficial association between some flowering plants and:
A. Bacteria B. Ants
C. Virus D. Mycoplasma

59. First androgenic haploid was raised by:
A. J.P. Nitsch
B. Guha and Maheshwari
C. Haberlandt
D. F.C. Steward

60. Which one of the following fruits is derived from a monocarpellary pistil?
A. Pome B. Samara
C. Capsule D. Legume

BIOTECHNOLOGY

1. Which is the longest phase of cell cycle?
A. Mitotic phase B. Cytokinesis
C. S phase D. Interphase

2. IPR is defined as ....
A. Important Property Rights
B. Intellectual Property Rights
C. Indian Patent Rights
D. Intellectual Provisional Rights

3. Which of the following is an Archaebacteria?
A. Methanogens
B. Cyanobacteria
C. Helicobacter
D. Proteobacteria

4. Enzyme used in Polymerase Chain reaction is ....
A. DNA Pol III B. Ligase
C. Taq Polymerase D. Hind III

5. Aneuploidy refers to ....
A. Loss of a part of a chromosome
B. Gain of a part of chromosome
C. Presence of an abnormal number of chromosomes
D. Both A and B

6. The uptake of naked DNA by the bacteria is called ....
A. Electrophoresis B. Transformation
C. Conjugation D. Recombination

7. Net yield of ATP's in complete oxidation of one molecule of glucose is ....
A. 38 B. 32
C. 33 D. 12

8. Study of interaction among organism and their environment is called:
A. Synecology B. Ecology
C. Autecology D. Diversity

9. A test cross is carried between an individual with a phenotypically recessive individual to determine its zygosity, will have phenotypic ratio ....
A. 3:1 B. 3:4
C. 1:2:1 D. 1:1

10. Which antibody is responsible for response during allergic and hypersensitive reaction?
A. IgM B. IgE
C. IgG D. IgA

11. In an experiment you have to use two reducing sugars which of the following combination you will choose?
A. Lactose and Maltose
B. Trehalose and Sucrose
C. Maltose and Trehalose
D. Lactose and Sucrose

12. The enzyme that catalyses the reduction of nitrogen to ammonia is ....
A. Nitrogenase B. Nitrate reductase
C. Nitrite reductase D. Deaminase

13. Restriction Enzyme which recognizes the same sequence are called ....
A. Neoisochizomer B. Isochizomer
C. Isocaudomer D. Neocaudomer

14. Choose the right combination:

	Stop codon	Start codon
A.	AUG/UGA/AGA	AUG
B.	AAG/UAA/AGU	GAG
C.	UAA/UAG/UGA	AUG
D.	UAA/UAG/UGA	UAG

15. Which is not an auto immune disease?
A. Grave's disease
B. Hashimoto disease
C. Herpes disease
D. Myasthenia gravis

16. Isoelectric focusing separate molecules by their ....
A. Conductivity
B. Molecular size
C. Positive or negative charge
D. Isoelectric point

17. First stable product in C3 plant is ....
A. 3-phosphoglycerate
B. 3-phosphoglycolate
C. Oxaloacetate
D. Phosphoenol pyruvate

18. What are the basic building blocks of cellulose and what type of linkage they have?

A. Glucose β 1-4
B. Maltose α 1-4
C. Glucose α 1-4
D. Maltose β 1-4

19. Which of the following is not a fat soluble vitamin?

A. D B. B
C. E D. A

20. Apart from protein which molecule can act as an enzyme?

A. RNA B. DNA
C. Carbohydrate D. Lipids

21. The only antibody capable of crossing the placenta in humans is ....

A. IgA B. IgM
C. IgG D. IgE

22. Which of the following is not present in the eukaryotic cell?

A. Microsome B. Mesosome
C. Centrosome D. Glycosome

23. The most abundant protein in nature:

A. Collagen
B. Chlorophyll
C. RUBISCO
D. Haemoglobin

24. A Single mRNA molecule that codes for more than one protein:

A. Cistronic mRNA
B. Intronic mRNA
C. Polycistronic mRNA
D. Both B and C

25. Allosteric enzymes functions through ....

A. Reversible non-covalent binding of a regulatory metabolite
B. Irreversible non-covalent binding of a regulatory metabolite
C. Reversible covalent binding of a regulatory metabolite
D. Irreversible covalent binding of a regulatory metabolite

26. Which of the following enzyme complex converts pyruvate obtained from the glucose breakdown into the acetyl Co-A to enter the citric acid cycle?

A. Cytochrome c oxidase
B. NADH-Q oxidoreductase
C. Pyruvate dehydrogenase
D. Pyruvate oxidase

27. Which microbiologist disproved the doctrine of spontaneous generation?

A. Francesco Redi
B. Robert Koch
C. Elie Metchnikoff
D. Louis Pasteur

28. Which of the following is not a vector?

A. pSC101 B. PAC
C. EcoRI D. Lambda phage

29. Glycolysis results in the production of two pyruvate molecules, before entering into the citric acid cycle this pyruvate are transported into which cell organelle?

A. Endoplasmic Reticulum
B. Golgi body
C. Lysosome
D. Mitochondria

30. Which of the following is an example of negative interaction?

A. Epiphyte growing on some woody plant
B. Rhizobacterium and leguminous plant
C. Antibiosis
D. Lichen

31. The five different classes of antibodies are defined on the basis of ....

A. Structure of constant region of the heavy chain
B. Structure of variable region of the heavy chain
C. Structure of constant region of the light chain
D. Structure of variable region of the light chain

32. Operon has:

A. A group of closely linked structural genes regulated by the same operator

B. Single promoter for all structural genes
C. Produces polycistronic mRNA
D. All of the above

33. The repetitive stretches of DNA located at the ends of linear chromosomes:
A. Kinetochore B. Acromere
C. Telomere D. Centromere

34. First genetically modified plant was:
A. An antibiotic resistant tobacco plant
B. *Flavr Savr* tomato which has longer shelf life
C. Bt cotton
D. Bt tobacco plant

35. Considering meiosis which of the following is true statement?
A. Recombination occurs in Prophase I
B. Meiosis I is called the equational division and meiosis II is the reductional division
C. Creates germ cells only
D. Both A and C

36. Plant tissue culture is possible because:
A. Plant cells exhibit plasticity
B. Plant cells exhibit totipotency
C. Plant cells have higher regeneration power
D. All of the above

37. Each chromosome is made up of:
A. Tightly coiled DNA many times around proteins called histones
B. Condensation of chromatins
C. DNA only
D. Both A and B

38. Match the column:

(*a*) Lysosome	1. Makes lipid, breakdown drugs and pack proteins
(*b*) Golgi Body	2. Power house of the cell
(*c*) Mitochondria	3. Digest food particle and foreign invaders
(*d*) Endoplasmic Reticulum	4. Processes and transport proteins out of the cell

A. (*a*)-4/(*b*)-3/(*c*)-2/(*d*)-1
B. (*a*)-1/(*b*)-2/(*c*)-3/(*d*)-4
C. (*a*)-2/(*b*)-3/(*c*)-4/(*d*)-1
D. (*a*)-3/(*b*)-4/(*c*)-2/(*d*)-1

39. Which of the following is defined as the law of independent assortment?
A. Allele pairs separate independent of each other during the formation of gametes. Hence, traits are transmitted to offspring independently of one another.
B. During the production of gametes the allele of each hereditary factor segregate so that offspring acquire one factor from each parent.
C. One factor in a pair of traits dominates the other inheritance unless both factors in the pair are recessive.
D. All of the above

40. The phenomenon in which a single gene affects a number of phenotypic traits in the same organism is known as:
A. Polytropism
B. Co-dominance
C. Pleitropism
D. Imprinting

41. Haemoglobin is an example of .... protein.
A. Storage B. Protective
C. Regulatory D. Transport

42. Subunits of prokaryotic and eukaryotic ribosomes are:
A. 60S, 40S and 50S, 30S respectively
B. 50S, 40S and 60S, 30S respectively
C. 50S, 30S and 60S, 40S respectively
D. 60S, 30S and 50S, 40S respectively

43. In bacterial conjugation F plasmid is called:
A. Fertility factor B. Female factor
C. Facilitating factor D. Functional factor

44. Activated Sludge process is ....
A. Aerobic B. Anaerobic
C. Ionic D. Non-ionic

45. Bioremediation is:
A. Use of plants for the degradation of pollutant
B. Use of microorganisms for the degradation of pollutant
C. Both A and B
D. Use of chemicals for the degradation of pollutant

46. pUC stands for:
A. Plasmid Universal Coding
B. Plasmid University of California
C. Plasmid University of Cambridge
D. Plasmid Unknown Coding

47. Which of the following is not a Green House Gas?
A. Methane
B. Carbon dioxide
C. Argon
D. Carbon Monoxide

48. First amino acid incorporated to make protein chain during translation is:
A. Lysine
B. Arginine
C. Methionine
D. Valine

49. Na^+/K^+ pump in a cell helps in transportation across the membrane by:
A. Passive Transport
B. Active Transport
C. Facilitated Diffusion
D. Osmosis

50. What do you mean by degeneracy of genetic code?
A. One amino acid can be coded by more than one codon
B. One codon can code for more than one amino acid
C. Gene sequence degenerate with time
D. Genetic code varies with the cell type

51. If protein of interest forms 10% of total protein present in crude enzyme, how many maximum folds of it can be purified?
A. 90
B. 10
C. 100
D. 0

52. Specific activity is:
A. enzyme per milligram of total protein
B. enzyme per microgram of total protein
C. enzyme per nanogram of total protein
D. enzyme per milliliter of total protein

53. ELISA is defined as:
A. Enzyme Linked Immuno Soluble Assay
B. Enzyme Linked Immuno Sorbent Assay
C. Enzyme Linked Instant Soluble Assay
D. Enzyme Linked Immuno Sandwich Assay

54. Which of the following are the essential components of the plasmid vector?
A. Origin of replication
B. Multiple Cloning Site
C. Antibiotic resistant gene
D. All of the above

55. The light reaction of photosynthesis:
A. Occurs in the stroma of chloroplast
B. Involves only photosystem II
C. Produce ATP and NADH
D. Also called Calvin Benson Cycle

56. Stem cells are:
A. Present only in embryo
B. The cells which are differentiated
C. Can develop into any kind of cell
D. Examples are nerve cell, muscle cell, RBC

57. Which of the following statements are true regarding PS(Photosystem) II?
A. In PS II, Photocenter is p680
B. Participates only in non-cyclic photophosphorylation
C. Located at the inner surface of thylakoid membrane
D. All of the above

58. Hybridoma is used for:
A. Hybridization
B. Gene amplification
C. Monoclonal antibody production
D. DNA segments separation

59. Mostly used soil bacteria in plant biotechnology for transformation is:
A. *Bacillus thuringiensis*
B. *Agrobacterium tumefaciens*
C. *Rhizobacterium*
D. *Escherichia coli*

60. Enzyme used for stitching two fragments of DNA is:
A. Alkaline phosphatase
B. DNA ligase
C. DNA topoisomerase
D. DNA gyrase

MATHEMATICS

1. If $f(x)=\dfrac{\cos^2 x+\sin^4 x}{\cos^4 x+\sin^2 x}$, then $f(2016)$ is equal to:

A. 1 B. 2
C. 3 D. 4

2. If $f(x)=\begin{Bmatrix} x & \text{if } x \text{ is rational} \\ -x & \text{if } x \text{ is irrational}\end{Bmatrix}$,

then $f:\mathbb{R}\to\mathbb{R}$ is

A. one-one and into
B. neither one-one nor onto
C. many-one and onto
D. one-one and onto

3. Let R be the relation from {11, 12, 13} to {8, 10, 12}, defined by $y = x - 3$, then R^{-1} is

A. {(8, 11), (10, 13)}
B. {(11, 8), (13, 10)}
C. {(11, 8), (12, 9), (13, 10)}
D. {(8, 11), (9, 12), (10, 13)}

4. Let A = $\mathbb{R}$ – {3}, B = $\mathbb{R}$ – {1} and let $f: A \to B$ be defined by $f(x)=\dfrac{x-2}{x-3}$. Then which of the following is not true?

A. f is bijective
B. f is one-one
C. f is onto
D. f is one-one but not onto

5. The period of the function

$f(x)=\dfrac{\sin x+\sin 3x+\sin 5x+\sin 7x}{\cos x+\cos 3x+\cos 5x+\cos 7x}$ is:

A. π B. $\dfrac{\pi}{2}$

C. $\dfrac{\pi}{3}$ D. $\dfrac{\pi}{4}$

6. Let A = {1, 2, 3}. Then the number of equivalence relations containing (1, 2) is

A. 1 B. 2
C. 3 D. 4

7. Which of the following is true?

A. $\tan 1 \,.\, \tan^{-1} 1 = 1$ B. $\tan 1 < \tan^{-1} 1$
C. $\tan 1 > \tan^{-1} 1$ D. $\tan 1 = \tan^{-1} 1$

8. If $[\sin^{-1} \cos^{-1} \sin^{-1} \tan^{-1} x] = 1$, where [] denotes the greatest integer value function, then x lies in the interval

A. [tan sin cos 1, tan sin cos sin 1]
B. (tan sin cos 1, tan sin cos sin 1)
C. [–1, 1]
D. [sin cos tan 1, sin cos sin tan 1]

9. If $\tan\theta=\sqrt{n}$ for some non-square natural number n, then sec 2θ is:

A. a rational number
B. an irrational number
C. a +ve real number
D. none of the above

10. cot 15° + cot 75° + cot 135° – cosec 30° equals

A. –1 B. 0
C. 1 D. 1/2

11. Two rays are drawn through a Point A at an angle of 30°. A point B is taken on one of them at a distance 'a' from the Point A. A perpendicular is drawn from the point B to the other ray, and another perpendicular is drawn from its foot to AB to meet AB at another point from where the similar process is repeated indefinitely. The length of the resulting polygonal line is

A. a B. $\left(2+\sqrt{3}\right)a$

C. $\left(\sqrt{2}+3\right)a$ D. $\left(\sqrt{2}+\sqrt{3}\right)a$

12. Let $S(n) = 1 + 3 + 5 + ... + (2n - 1) = 3 + n^2$. Then which of the following is true?

A. S(1) is correct
B. $S(n) \Rightarrow S(n + 1)$
C. $S(n) \not\Rightarrow S(n+1)$
D. Principle of mathematical induction can be used to prove the formula

13. The locus of the inequality $\log_{(1/3)} |z + 1| > \log_{(1/3)} |z - 1|$ is

A. Re $z < 0$ B. Re $z > 0$

C. Im $z < 0$ D. Im $z > 0$

14. Let $z_a = 2 - 2i$ and $a = 2 - i$. If $z_a + (a - z_n)e^{i\theta}$ is the complex number on $|z - z_0| = 1$ with least absolute value then θ equals

A. $-\frac{\pi}{4}$ B. 0

C. $\frac{\pi}{4}$ D. $\frac{\pi}{2}$

15. The number of real solutions of $\sin(e^x) = e^x + e^{-x}$ is

A. 0 B. 1

C. 2 D. infinite

16. The value of x satisfying the equation $x + \log_{10} (1 + 2^x) = x \log_{10}5 + \log_{10} 6$ is

A. −1 B. 0

C. 1 D. 2

17. If $a < b < c < d$ then $f(x) = (x - a)(x - c) + \lambda(x - b)(x - d)$ has real roots

A. For all λ

B. Only when $\lambda > 0$

C. Only when $\lambda < 0$

D. For no λ

18. If the coefficients of second, third and fourth terms in the expansion of $(1 + x)^n$ are in AP then n equals

A. 2 B. 3

C. 5 D. 7

19. $\lim_{n\to\infty} \sum_{r=1}^{n} \tan^{-1}\left(\frac{1}{2r^2}\right)$ equals

A. $\frac{\pi}{2}$ B. $\frac{\pi}{3}$

C. $\frac{\pi}{4}$ D. $\frac{\pi}{6}$

20. The sum to $(n + 1)$ terms of the series

$3C_0 - 8C_1 + 13C_2 - 18C_3 + \ldots\ldots\ldots$ is

A. −1 B. 0

C. 1 D. 5.2^{n-1}

21. The point of intersection of the lines $\frac{x}{a} + \frac{y}{b} = 1$ and $\frac{x}{b} + \frac{y}{a} = 1$ does not lie on the line

A. $x - y = 0$

B. $(x + y)(a + b) = 2ab$

C. $(2x + 3y)(a + b) = 5ab$

D. $(2x + 3y)(a + b) = -ab$

22. Area of an equilateral triangle inscribed in the circle $x^2 + y^2 - 4x + 6y - 3 = 0$

A. 12 B. $\frac{12}{\sqrt{3}}$

C. $12\sqrt{3}$ D. $12\sqrt{2}$

23. Equation of the hyperbola with foci $\left(0, \pm\sqrt{10}\right)$ and passing through (2, 3) is

A. $\frac{x^2}{18} - \frac{y^2}{8} = 1$ B. $\frac{y^2}{18} + \frac{x^2}{8} = 1$

C. $\frac{x^2}{1} - \frac{y^2}{3} = 1$ D. $\frac{x^2}{5} - \frac{y^2}{5} = -1$

24. If a line segment AM = a moves in the plane XOY remaining parallel to OX so that the left end point A slides along the circle $x^2 + y^2 = a^2$, the locus of M is

A. $x^2 + y^2 = 4a^2$

B. $x^2 + y^2 = 2ax$

C. $x^2 + y^2 = 2ay$

D. $x^2 + y^2 - 2ax - 2ay = 0$

25. Shortest distance between the lines $\frac{x+1}{7} = \frac{y+1}{-6} = \frac{z+1}{1}$ and $\frac{x-3}{1} = \frac{y-5}{-2} = \frac{z-7}{1}$ is

A. $2\sqrt{29}$ B. $\frac{3\sqrt{2}}{2}$

C. $\frac{3}{\sqrt{19}}$ D. $\frac{8}{\sqrt{29}}$

26. The centre of the circle given by $\vec{r}-(\hat{i}+2\hat{j}+2\hat{k})=15$ and $|\vec{r}-(\hat{j}+2\hat{k})|=4$ is:

A. (0, 1, 2) B. (1, 3, 4)
C. (–1, 3, 4) D. (1, 0, 2)

27. The angle between a diagonal and an edge of the cube intersecting the diagonal is:

A. $\cos^{-1}\frac{1}{3}$ B. $\cos^{-1}\sqrt{\frac{2}{3}}$
C. $\tan^{-1}\sqrt{2}$ D. $\frac{\pi}{4}$

28. The range of the function

$$f(x)=\sin^{-1}\left[x^2+\frac{1}{2}\right]+\cos^{-1}\left[x^2-\frac{1}{2}\right],$$

where [.] is the greatest integer function is:

A. $\left\{\frac{\pi}{2},\pi\right\}$ B. $\left\{0,\frac{\pi}{2}\right\}$
C. $\{\pi\}$ D. $\left(0,\frac{\pi}{2}\right)$

29. The domain of the function $f(x) = \cos^{-1}\log_2(x^2 + 5x - 8)$ is:

A. [2, 3] B. [–3, –2]
C. [–2, 2] D. [–3, 1]

30. The number of points at which $f(x) = \frac{1}{\log|x|}$ is discontinuous in its domain is

A. 1 B. 2
C. 3 D. 4

31. If $f(x)=\begin{vmatrix}\sin x & \cos x & \tan x\\ x^3 & x^2 & x\\ 2x & 1 & 1\end{vmatrix}$

then $\lim_{x\to 0}\frac{f(x)}{x^2}$ equals

A. 3 B. –1
C. 0 D. 1

32. Which of the following is true for any two statements p and q?

A. $\sim[p \vee (-q)] = (\sim p) \wedge q$
B. $(p \vee q) \vee (\sim q)$ is a tautology
C. $(p \wedge q) \wedge (\sim q)$ is a contradiction
D. $\sim[p \wedge (\sim p)]$ is a tautology

33. If a variable x takes values x_j such that $a \le x_j \le b$ for j = 1, 2, ... n then

A. $a^2 \le \text{var}(x) \le b^2$ B. $a \le \text{var}(x) \le b$
C. $\frac{a^2}{4} \le \text{var}(x)$ D. $(b-a)^2 \ge \text{var}(x)$

34. If a variable takes values 0, 1, 2, ... n with frequencies proportional to the binomial coefficients $^nc_0$, $^nc_1$, ... ^nc_n respectively then mean of the distribution is:

A. $\frac{n}{2}$ B. $\frac{n(n+1)}{2}$
C. $\frac{n(n-1)}{2}$ D. $\frac{n+1}{2}$

35. If A and B are two events such that P(A) + P(B) – P(A and B) = P(A) then

A. P(B | A) = 1 B. P(A | B) = 1
C. P(B | A) = 0 D. P(A | B) = 0

36. The value of $\sin\left(2\tan^{-1}\frac{1}{3}\right)+\cos\left(\tan^{-1}2\sqrt{2}\right)$ is:

A. $\frac{11}{12}$ B. $\frac{12}{13}$
C. $\frac{13}{14}$ D. $\frac{14}{15}$

37. A man in a boat rowing away from a cliff 150 m high takes 2 minutes to change the angle of elevation of the top of the cliff from 60° to 45°. The speed of the boat is:

A. 1 km/hr B. $\frac{3}{2}(3-\sqrt{3})$ km/hr
C. $\frac{3}{2}(\sqrt{3}+3)$ km/hr D. 2 km/hr

38. If $f(x) = (1 + x)^n$ then the value of

$f(0)+f^n(0)+\frac{f^n(0)}{2!}+.....+\frac{f^n(0)}{n!}$ is

A. n B. 2^n
C. 2^{n-1} D. 2^{n+1}

39. If $e^{f(x)} = \log x$ then $(f^{-1})^1\ (0)$ equals

A. e B. e^{-1}
C. 1 D. e^2

40. If $f'(a) = 2$ and $f(a) = 4$

then $\lim_{x\to a}\frac{x\,f(a)-a\,f(x)}{x-a}$ equals

A. 4
B. $2a$
C. $4 + 2a$
D. $4 - 2a$

41. A function $f(x)$ is defined for all $x > 0$ and satisfies $f(x^2) = x^3$. Then f is differentiable at $x = 0$ with $f'(4)$ equal to

A. 1 B. 2
C. 3 D. 4

42. The number of real roots of the equation $e^{x-1} + x - 2 = 0$ is

A. 1 B. 2
C. 3 D. 4

43. Equation of normal to the curve $y = (1 + x)^2 + \sin^{-1}(\sin^2 x)$ at $x = 0$ *i.e.*, at (0, 1) is:

A. $x + y = 1$
B. $x + y = -1$
C. $x - y = 1$
D. $x - y = -1$

44. $\int\frac{\cos 2x}{(\sin x+\cos x)^2}dx$ equals

A. $\frac{-1}{\sin x+\cos x}$

B. $\log|\sin x + \cos x|$

C. $\log|\sin x - \cos x|$

D. $\frac{1}{(\sin x+\cos x)^2}$

45. If $I_n = \int\cot^n x\,dx$ then $\frac{\cot^{n-1} x}{n-1}$ equals

A. $I_n + I_{n-2}$
B. $-(I_n + I_{n-2})$
C. $I_n - I_{n-2}$
D. $-I_n + I_{n-2}$ $(n \geq 4)$

46. The value of $\int_0^{1000} e^{x-[x]}\,dx$ (where [.] is the greatest integer function) equals

A. $\frac{e^{1000}-1}{1000}$ B. $\frac{e^{1000}-1}{e-1}$

C. $1000\,(e - 1)$ D. $\frac{e-1}{1000}$

47. The value of the integral

$\int_0^{\infty} e^{-2x}(\sin 2x+\cos 2x)dx$ is

A. 0 B. 1

C. 2 D. $\frac{1}{2}$

48. General solution to the differential equation $y\,dx + (x + x^2y)\,dy = 0$ is

A. $-\frac{1}{xy}+\log y=c$ B. $\frac{1}{xy}-\log y=cx$

C. $\frac{1}{xy}+\log y=c$ D. $\frac{1}{xy}+\log y=\frac{c}{x}$

49. If A is a square matrix such that $A^2 = A$, then $(I + A)^3 - 7A$ is equal to

A. A B. I – A
C. I D. 3A

50. The function $\Delta=\begin{vmatrix} x & a & a \\ b & x & a \\ b & b & x \end{vmatrix}$ (*a*, *b* are +ve constants) has

A. Local maxima at both $-\sqrt{ab}$ and $\sqrt{ab}$

B. Local minima at both $-\sqrt{ab}$ and $\sqrt{ab}$

C. Local maxima at $-\sqrt{ab}$ and local minima at $\sqrt{ab}$

D. Local minima at $-\sqrt{ab}$ and local maxima at $\sqrt{ab}$

51. If a^{-1}, b^{-1}, c^{-1} are pth, qth and rth terms of an AP then the value of $\begin{vmatrix} bc & ca & ab \\ p & q & r \\ 1 & 1 & 1 \end{vmatrix}$ is

A. –1 B. 0
C. 1 D. 2

52. If $A = \begin{bmatrix} 3 & 1 \\ -1 & 2 \end{bmatrix}$ then the matrix $A^2 - 5A + 8I$ is

A. 0 B. I
C. 2I D. 3I

53. If $a + b + c \neq 0$ then the number of solutions (x, y, z) to the system

$(b + c)(y + z) - ax = b - c$
$(c + a)(z + x) - by = c - a$
$(a + b)(x + y) - cz = a - b$ is

A. 0 B. 1
C. 3 D. infinite

54. Inverse of $f(x) = \begin{bmatrix} \cos x & \sin x & 0 \\ -\sin x & \cos x & 0 \\ 0 & 0 & 1 \end{bmatrix}$ is:

A. $f(x)$ B. $-f(x)$
C. $f(-x)$ D. $-f(-x)$

55. If $\vec{a}$ and $\vec{b}$ are vectors such that $|\vec{a}| = 3$, $|\vec{b}| = \frac{\sqrt{2}}{3}$ and $|\vec{a} \times \vec{b}| = 1$ then an angle between $\vec{a}$ and $\vec{b}$ could be

A. $\frac{\pi}{6}$ B. $\frac{\pi}{4}$

C. $\frac{\pi}{3}$ D. $\frac{\pi}{2}$

56. Three vectors $\hat{i} + \hat{j}$, $\hat{j} + \hat{k}$ and $\hat{k} + \hat{i}$ taken two at a time form three planes. The three unit vectors drawn perpendicular to these three planes form a parallelopiped of volume

A. $\frac{1}{3}$ B. 4

C. $\frac{3\sqrt{3}}{4}$ D. $\frac{4}{3\sqrt{3}}$

57. The random variable X has a probability distribution P(X) of the form

$$P(X) = \begin{cases} k & \text{if } X = 0 \\ 2k & \text{if } X = 1 \\ 3k & \text{if } X = 2 \\ 0 & \text{otherwise} \end{cases}$$ with k being some positive constant. The P $(X \leq 2)$ is

A. $\frac{1}{2}$ B. $\frac{1}{3}$

C. $\frac{1}{4}$ D. $\frac{1}{6}$

58. The variance of the number obtained on a throw of an unbiased die is

A. $\frac{21}{6}$ B. $\frac{91}{6}$

C. $\frac{35}{12}$ D. $\frac{12}{35}$

59. Maximum value of the objective function $x + 2y$ subject to the constraints $2x + 3y \leq 6$, $x + 4y \leq 4$, $x \geq 0$, $y \geq 0$ is

A. 3.2 B. 4.2
C. 2.3 D. 2.4

60. Consider the feasible region R as given by $x + y \geq 1$, $7x + 9y \leq 63$, $x \leq 6$, $y \leq 5$, $x \geq 0$, $y \geq 0$. The number of points in R at which the objective function $3.5x + 4.5y$ obtains maximum value is:

A. 0 B. 1
C. 2 D. infinite

ANSWERS

Physics

1	2	3	4	5	6	7	8	9	10
D	C	A	A	C	A	B	B	D	B
11	**12**	**13**	**14**	**15**	**16**	**17**	**18**	**19**	**20**
A	D	A	C	C	A	C	A	B	B
21	**22**	**23**	**24**	**25**	**26**	**27**	**28**	**29**	**30**
D	B	C	D	C	C	B	C	B	B
31	**32**	**33**	**34**	**35**	**36**	**37**	**38**	**39**	**40**
A	A	B	C	A	D	C	C	B	A
41	**42**	**43**	**44**	**45**	**46**	**47**	**48**	**49**	**50**
B	C	A	D	A	A	D	C	C	D
51	**52**	**53**	**54**	**55**	**56**	**57**	**58**	**59**	**60**
C	A	D	C	C	D	C	A	B	D

Chemistry

1	2	3	4	5	6	7	8	9	10
A	B	A	B	A	D	B	A	C	A
11	**12**	**13**	**14**	**15**	**16**	**17**	**18**	**19**	**20**
B	C	A	C	B	B	B	C	B	D
21	**22**	**23**	**24**	**25**	**26**	**27**	**28**	**29**	**30**
C	B	D	C	B	C	A	D	C	A
31	**32**	**33**	**34**	**35**	**36**	**37**	**38**	**39**	**40**
C	B	B	C	C	B	D	B	D	D
41	**42**	**43**	**44**	**45**	**46**	**47**	**48**	**49**	**50**
B	C	C	A	B	D	B	A	A	B
51	**52**	**53**	**54**	**55**	**56**	**57**	**58**	**59**	**60**
C	D	A	B	D	C	A	C	B	C

Biology

1	2	3	4	5	6	7	8	9	10
D	B	A	A	B	B	A	D	C	B
11	**12**	**13**	**14**	**15**	**16**	**17**	**18**	**19**	**20**
B	C	D	D	A	D	C	D	C	A

21	22	23	24	25	26	27	28	29	30
D	B	C	D	D	A	A	B	C	D
31	**32**	**33**	**34**	**35**	**36**	**37**	**38**	**39**	**40**
D	C	B	A	C	C	C	C	A	D
41	**42**	**43**	**44**	**45**	**46**	**47**	**48**	**49**	**50**
A	A	C	B	B	B	B	C	C	B
51	**52**	**53**	**54**	**55**	**56**	**57**	**58**	**59**	**60**
A	B	B	D	C	D	B	B	B	D

Biotechnology

1	2	3	4	5	6	7	8	9	10
D	B	A	C	C	B	A	B	D	B
11	**12**	**13**	**14**	**15**	**16**	**17**	**18**	**19**	**20**
A	A	B	C	C	D	A	A	B	A
21	**22**	**23**	**24**	**25**	**26**	**27**	**28**	**29**	**30**
C	B	C	C	A	C	D	C	D	C
31	**32**	**33**	**34**	**35**	**36**	**37**	**38**	**39**	**40**
A	D	C	A	D	D	D	A	A	C
41	**42**	**43**	**44**	**45**	**46**	**47**	**48**	**49**	**50**
D	C	A	A	C	B	C	C	B	A
51	**52**	**53**	**54**	**55**	**56**	**57**	**58**	**59**	**60**
B	A	B	D	C	C	D	C	B	B

Mathematics

1	2	3	4	5	6	7	8	9	10
A	D	A	A	D	B	C	A	A	C
11	**12**	**13**	**14**	**15**	**16**	**17**	**18**	**19**	**20**
B	B	A	C	A	C	A	D	C	B
21	**22**	**23**	**24**	**25**	**26**	**27**	**28**	**29**	**30**
D	C	D	B	A	B	X	C	B	X
31	**32**	**33**	**34**	**35**	**36**	**37**	**38**	**39**	**40**
D	X	D	A	B	D	B	B	A	D
41	**42**	**43**	**44**	**45**	**46**	**47**	**48**	**49**	**50**
C	A	A	B	B	C	D	A	C	C
51	**52**	**53**	**54**	**55**	**56**	**57**	**58**	**59**	**60**
B	B	B	C	B	D	X	C	A	D

Note : *An 'X' is the answer key indicated that either the question is antiguous or it has printing mistake. All candidates will be given credit for this question.*

EXPLANATORY ANSWERS

Mathematics

1. $f(x) = \dfrac{\cos^2 x + \sin^4 x}{\cos^4 x + \sin^2 x}$

$= \dfrac{\cos^2 x + (\sin^2 x)^2}{\cos^4 x + \sin^2 x}$

$= \dfrac{\cos^2 x + (1 - \cos^2 x)^2}{\cos^4 x + \sin^2 x}$

$= \dfrac{\cos^2 x + 1 + \cos^4 x - 2\cos^2 x}{\cos^4 x + \sin^2 x}$

$= \dfrac{\cos^4 x + (1 - \cos^2 x)}{\cos^4 x + \sin^2 x}$

$= \dfrac{\cos^4 x + \sin^2 x}{\cos^4 x + \sin^2 x}$

$= 1.$

$\therefore f(2016) = 1.$

3. $f(x) = y = x - 3$

$\Rightarrow \quad x = y + 3$

$f'(x) = x + 3$

Then, $R^{-1} = \{(8, 11), (10, 13)\}$

4. $f(x_1) = f(x_2) = \dfrac{x_1 - 2}{x_1 - 3}$

$= \dfrac{x_2 - 2}{x_2 - 3}$

$\Rightarrow \quad x_1 = x_2$

Hence, $f(x)$ is one-one *i.e.* injective.

Again, $y = \dfrac{x-2}{x-3}$

$\Rightarrow \quad x = \dfrac{2-3y}{1-y}$

$\Rightarrow \quad f^{-1}(y) = \dfrac{2-3y}{1-y}$

Hence, f is onto *i.e.,* surjective

Thus, $f(x)$ is bijective.

5. $f(x) = \dfrac{\sin x + \sin 3x + \sin 5x + \sin 7x}{\cos x + \cos 3x + \cos 5x + \cos 7x}$

$= \dfrac{(\sin x + \sin 7x) + (\sin 3x + \sin 5x)}{(\cos x + \cos 7x) + (\cos 3x + \cos 5x)}$

$= \dfrac{2\sin 4x.\cos 3x + 2\sin 4x.\cos x}{2\cos 4x.\cos 3x + 2\cos 4x.\cos x}$

$= \dfrac{\sin 4x(\cos 3x + \cos x)}{\cos 4x(\cos 3x + \cos x)}$

$= \tan 4x$

Now, period of $\tan x = \pi$

$\therefore$ Period of $\tan 4x = \dfrac{\pi}{4}$

$\therefore$ Period of $f(x) = \dfrac{\pi}{4}$.

9. $\tan\theta = \sqrt{n}$ {given}

$\sec 2\theta = \dfrac{1}{\cos 2\theta}$

$= \dfrac{1 + \tan^2\theta}{1 - \tan^2\theta}$

$= \dfrac{1 + (\sqrt{n})^2}{1 - (\sqrt{n})^2}$

$= \dfrac{1+n}{1-n}$

Here, n is non-square of natural number.

Thus, $\sec 2\theta = \dfrac{1+n}{1-n}$ is a rational number.

10. $\cot 15° + \cot 75° + \cot 135° - \text{cosec } 30°$

$$= \frac{\cos 15°}{\sin 15°} + \frac{\cos 75°}{\sin 75°} + \cot(180° - 45°) - 2$$

$$= \frac{\sin 75°.\cos 15° + \cos 75°.\sin 15°}{\sin 75°.\sin 15°} - \cot 45° - 2$$

$$= \frac{\sin(75° + 15°)}{\frac{1}{2}\left(\cos(75° - 15°) - \cos(75° + 15°)\right)} - 1 - 2$$

$$= \frac{2}{\cos 60° - \cos 90°} - 3$$

$$= \frac{2}{\frac{1}{2} - 0} - 3 = 4 - 3 = 1.$$

11.

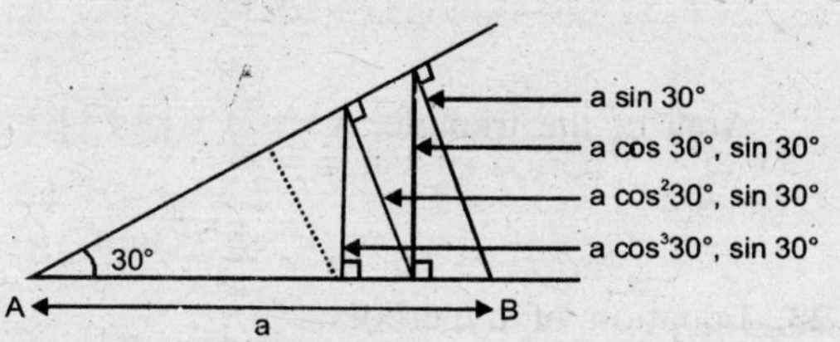

Sum of polygonal line

$= a \sin 30° + a \cos 30° \,.\, \sin 30° + a \sin 30° \,.\, \cos^2 30° + a \sin 30° \,.\, \cos^3 30° + \ldots\ldots \infty$

$= a \sin 30° \,[1 + \cos 30° + \cos^2 30° + \cos^3 30° + \ldots\ldots \infty]$

$$= \frac{a \sin 30°}{1 - \cos 30°}$$

$$= \frac{\frac{a}{2}}{1 - \frac{\sqrt{3}}{2}}$$

$$= \frac{a}{2 - \sqrt{3}}$$

$$= \left(2 + \sqrt{3}\right)a.$$

15. Let, $e^x = y$, here $y > 0$

$$\sin y = \frac{y^2 + 1}{y}$$

For, $0 \le \sin y \le 1$,

$$\Rightarrow \quad 0 \le \frac{y^2 + 1}{y} \le 1$$

$$0 \le y^2 + 1 \le y$$

$$0 \le e^{2x} + 1 \le e^x$$

This is not true for any real values of x.

16. $x + \log_{10}(1 + 2^x)$

$$= x \log_{10} 5 + \log_{10} 6$$

$$x = \log_{10} 5^x + \log_{10} 6 - \log_{10}(1 + 2^x)$$

$$x = \log_{10}\left(\frac{6.5^x}{1 + 2^x}\right)$$

$$10^x = \frac{6.5^x}{1 + 2^x}$$

This is satisfied for $x = 1$

$$10^1 = \frac{6 \times 5^1}{1 + 2^1} = \frac{6 \times 5}{3} = 10.$$

17. $f(x) = (x - a)(x - c) + 1\,(x - b)(x - d)$

$= x^2 - (a + c)x + ac + \lambda\,\{x^2 - (b + d)x + bd\}$

For real root, $D \ge 0$

$(a + c)^2 - 4ac \ge 0$

and $\{(b + d)^2 - 4bd\} \ge 0$

$(c - a)^2 \ge 0$

and $(d - b)^2 \ge 0$

This is true for all λ.

18. Expansion of $(1 + x)^n$

$= {}^nc_0 + {}^nc_1 x + {}^nc_2 x^2 + {}^nc_3 x^3 + \ldots\ldots {}^nc_n x^n$

From question, ${}^nc_1$, ${}^nc_2$ and ${}^nc_3$ are in A.P.

$$2({}^nc_2) = {}^nc_1 + {}^nc_3$$

$$n(n-1) = n+\frac{n(n-1)(n-2)}{6}$$

$$6n(n-1) = 6n + n^3 - 3n^2 + 2n$$

$$6n^2 - 6n = 8n + n^3 - 3n^2$$

$$n^3 - 9n^2 + 14n = 0$$

$$n(n^2 - 9n + 14) = 0$$

as $n \neq 0$.

$\therefore \quad n^2 - 9n + 14 = 0$

$$n = 2 \text{ and } 7$$

Hence, $n = 7$.

19. $\tan^{-1}\left(\frac{1}{2r^2}\right) = \tan^{-1}(2r+1) - \tan^{-1}(2r-1)$

Now, $\sum_{r=1}^{\infty}\tan^{-1}\left(\frac{1}{2r^2}\right)$

$= \sum_{r=1}^{\infty}\left\{\tan^{-1}(2r+1)-\tan^{-1}(2r-1)\right\}$

$= \tan^{-1}(3) - \tan^{-1}(1) + \tan^{-1}(5) - \tan^{-1}(3) + \tan^{-1}(7) - \tan^{-1}(5) + ... + \tan^{-1}(\infty)$ ...

$= \tan^{-1}(\infty) - \tan^{-1}(1)$

$= \frac{\pi}{2}-\frac{\pi}{4}$

$= \frac{\pi}{4}$.

21. $\frac{x}{a}+\frac{y}{b} = 1$ ...(*i*)

$\frac{x}{b}+\frac{y}{a} = 1$ ...(*ii*)

From (*i*) and (*ii*), we have,

$$\frac{x}{a^2}-\frac{x}{b^2} = \frac{1}{a}-\frac{1}{b}$$

$$\frac{x[b^2-a^2]}{a^2b^2} = \frac{(b-a)}{ab}$$

$\Rightarrow \quad x = \frac{ab}{a+b}$

Similarly, $\quad y = \frac{ab}{a+b}$

Clearly,

Point $(x, y) = \left(\frac{ab}{a+b},\frac{ab}{a+b}\right)$

doesn't lies on $(2x + 3y)(a + b) = -ab$.

22. Circle : $x^2 + y^2 - 4x + 6y - 3 = 0$

Centre of the circle $= (2, -3)$

radius of the circle $= \sqrt{(2)^2+(-3)^2-(-3)}$

$= 4$

Side length of equilateral triangle

$= 2 \times 4 \times \sin 60°$

$= 8\times\frac{\sqrt{3}}{2}$

$= 4\sqrt{3}$ unit

Area of the triangle, $A = \frac{\sqrt{3}}{4}\times\left(4\sqrt{3}\right)^2$

$= 12\sqrt{3}$ unit.

23. Equation of hyperbola :

$$\frac{y^2}{b^2}-\frac{x^2}{a^2} = 1$$

As this passes through Point (2, 3)

$$\frac{3^2}{b^2}-\frac{2^2}{a^2} = 1$$

$\frac{9}{b^2}-\frac{4}{a^2} = 1$ ...(*i*)

Focii of hyperbola :

$$\left(0,\pm\sqrt{a^2+b^2}\right) = \left(0,\pm\sqrt{10}\right)$$

$$\sqrt{a^2+b^2} = \sqrt{10}$$

$\Rightarrow \quad a^2 + b^2 = 10$ ...(*ii*)

From (*i*) and (*ii*), we have

$$\frac{9}{b^2} - \frac{4}{10 - b^2} = 1$$

$$9(10 - b^2) - 4b^2 = b^2 (10 - b^2)$$

$$\Rightarrow \quad b^4 - 23b^2 + 90 = 0$$

$$\Rightarrow \quad b^2 = 5$$

and $\quad a^2 = 5$

$\therefore$ Equation of hyperbola :

$$\frac{y^2}{5} - \frac{x^2}{5} = 1$$

or $\quad \frac{x^2}{5} - \frac{y^2}{5} = -1.$

35. $P(A) + P(B) - P(A \cap B) = P(A)$

$$P(B) - P(A \cap B) = 0$$

$$P(B) = P(A \cap B)$$

Now, $\quad P(A \mid B) = \frac{P(A \cap B)}{P(B)}$

$$= \frac{P(A \cap B)}{P(A \cap B)} = 1.$$

36. $\sin\left(2\tan^{-1}\frac{1}{3}\right) + \cos\left(\tan^{-1} 2\sqrt{2}\right)$

$$= \sin\left(\tan^{-1}\left(\frac{\frac{2}{3}}{1-\left(\frac{1}{9}\right)^2}\right)\right) + \cos\left(\tan^{-1} 2\sqrt{2}\right)$$

$$= \sin\left(\tan^{-1}\frac{3}{4}\right) + \cos\left(\tan^{-1} 2\sqrt{2}\right)$$

$$= \sin\left(\sin^{-1}\frac{3}{5}\right) + \cos\left(\cos^{-1}\frac{1}{3}\right)$$

$$= \frac{3}{5} + \frac{1}{3}$$

$$= \frac{14}{15}.$$

37.

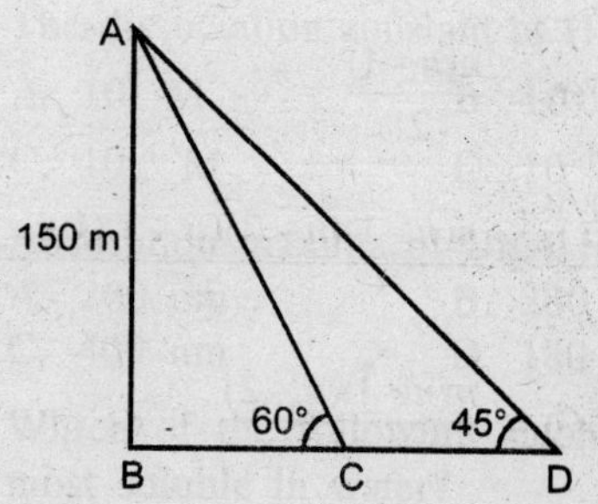

From ΔABC, $\quad BC = AB.\cot 60°$

$$= 150 \times \frac{1}{\sqrt{3}}$$

$$= 50\sqrt{3} \text{ m}$$

In ΔABD, $\quad BD = AB.\cot 45°$

$$= 150 \times 1$$

$$= 150 \text{ m}$$

$$CD = 150 - 50\sqrt{3}$$

$$= 50\left(3 - \sqrt{3}\right) \text{m}$$

From question, man row distance CD in 2 minutes

Speed of the boat

$$= \frac{50\left(3-\sqrt{3}\right)}{2 \times 60}$$

$$= \frac{5\left(3-\sqrt{3}\right)}{12} \text{ m/sec}$$

$$= \frac{5\left(3-\sqrt{3}\right)}{12} \times \frac{18}{5}$$

$$= \frac{3}{2}\left(3-\sqrt{3}\right) \text{ km/h}.$$

38. $f(x) = (1 + x)^n,\ f(0) = 1 = {}^nc_0$

$$f'(x) = n(1 + x)^{n-1}$$

$$f'(0) = n = {}^nc_1$$

$$\frac{f''(x)}{2!} = \frac{n(n-1)(1+x)^{n-1}}{2!}$$

$$f''(0) = \frac{n(n-1)}{2!} = {}^nc_2$$

$$\frac{f'''(x)}{3!} = \frac{n(n-1)(n-2)(n-3)(1+x)^{n-3}}{3!}$$

$$f'''(0) = \frac{n(n-1)(n-2)}{3!} = {}^nc_3$$

... ...

$$\frac{f^n(x)}{n!} = \frac{(n(n-1)(n-2)....1)}{n!}$$

$$\frac{f^n(0)}{n!} = 1$$

$$\therefore\ f(0)+f'(0)+\frac{f''(0)}{2!}+\frac{f'''(0)}{3!}+\ ...\ \frac{f^n(0)}{n!}$$

$$= {}^nc_0 + {}^nc_1 + {}^nc_2 + ...{}^nc_n$$

$$= 2^n.$$

39. $e^{f(x)} = \log x$

$f(x) = \log(\log x)$

Let $y = f(x)$

Then, $e^y = \log x$

$\Rightarrow$ $x = e^{e^y}$

$f^{-1}(x) = e^{e^x}$

Again, $z = f^{-1}(x)$ {say}

$z = e^{e^x}$

$\log z = e^x$

Differentiating both sides, we have

$$\frac{1}{z}.z' = e^x$$

$$z' = z.e^x$$

$$\left(f^{-1}(x)\right)' = \left(e^{e^x}\right).e^x$$

$$\left(f^{-1}(0)\right)' = \left(e^{e^0}\right).e^0 = e.$$

44. $\int \frac{\cos 2x}{(\sin x + \cos x)^2}.dx$

$$= \int \frac{\cos 2x}{(1+\sin 2x)}.dx$$

Let $1 + \sin 2x = z$

$dz = 2\cos 2x.\ dx$

or, $\cos 2x\ .\ dx = \frac{dz}{2}$

$$\int \frac{\cos 2x}{(1+\sin 2x)}.dx = \frac{1}{2}\int \frac{dz}{z}$$

$$= \frac{1}{2}\log z + c$$

where, c = constant

$$= \frac{1}{2}\log(1+\sin 2x) + c$$

$$= \frac{1}{2}\log|\sin x + \cos x|^2 + c$$

$$= \log|\sin x + \cos x| + c.$$

47. Let, $e^{-2x}.\cos 2x = z$

then, $dz = \left(-2.e^{-2x}.\cos 2x - 2e^{-2x}.\sin 2x\right).dx$

$\therefore\ e^{-2x}.(\cos 2x + \sin 2x)\ dx = -\frac{dz}{2}$

Now, $\int e^{-2x}(\sin 2x + \cos 2x).dx$

$$= -\frac{1}{2}\int dz = \frac{-z}{2} + c$$

$$= -\frac{e^{-2x}.\cos 2x}{2}$$

Applying the limit from $x = 0$ to $x = \infty$

$$I = -\frac{1}{2}\left[e^{-2x}.\cos 2x\right]_0^\infty$$

$$= -\frac{1}{2}[0-1] = \frac{1}{2}.$$

51. Let first term of the A.P. is x and common difference is d.

p^{th} term : $x + (p-1)d = \frac{1}{a}$ ...(*i*)

q^{th} term : $x + (q-1)d = \frac{1}{b}$ ...(*ii*)

r^{th} term : $x + (r-1)d = \frac{1}{c}$ ...(*iii*)

from (*i*), (*ii*) and (*iii*), we have

$$(p-q)d = \frac{1}{a} - \frac{1}{b}$$

$$\Rightarrow \quad (p-q)ab = \frac{b-a}{d}$$

$$(q-r)d = \frac{1}{b} - \frac{1}{c}$$

$$\Rightarrow \quad (q-r)bc = \frac{c-b}{d}$$

$$(r-p)d = \frac{1}{c} - \frac{1}{a}$$

$$\Rightarrow \quad (r-p)ac = \frac{a-c}{d}$$

Now, $\begin{vmatrix} bc & ca & ab \\ p & q & r \\ 1 & 1 & 1 \end{vmatrix}$

$= (q-r)bc + (r-p)ca + (p-q)ab$

$= \frac{c-b}{d} + \frac{a-c}{d} + \frac{b-a}{d} = 0.$

52. $A^2 = \begin{bmatrix} 3 & 1 \\ -1 & 2 \end{bmatrix}\begin{bmatrix} 3 & 1 \\ -1 & 2 \end{bmatrix}$

$= \begin{bmatrix} 9-1 & 3+2 \\ -3-2 & -1+4 \end{bmatrix} = \begin{bmatrix} 8 & 5 \\ -5 & 3 \end{bmatrix}$

Now, $A^2 - 5A + 8I$

$$\begin{bmatrix} 8 & 5 \\ -5 & 3 \end{bmatrix} - 5\begin{bmatrix} 3 & 1 \\ -1 & 2 \end{bmatrix} + 8\begin{bmatrix} 1 & 0 \\ 0 & 1 \end{bmatrix}$$

$$= \begin{bmatrix} 8-15+8 & 5-5+0 \\ -5+5+0 & 3-10+8 \end{bmatrix}$$

$$= \begin{bmatrix} 1 & 0 \\ 0 & 1 \end{bmatrix} = I.$$

54. $|f(x)| = \begin{vmatrix} \cos x & \sin x & 0 \\ -\sin x & \cos x & 0 \\ 0 & 0 & 1 \end{vmatrix}$

$= |\cos^2 x + \sin^2 x| = 1$

$\text{Adj}(f(x)) = \begin{vmatrix} \cos x & -\sin x & 0 \\ \sin x & \cos x & 0 \\ 0 & 0 & 1 \end{vmatrix}$

$f^{-1}(x) = \frac{\text{Adj}(f(x))}{|f(x)|}$

$= \begin{vmatrix} \cos x & -\sin x & 0 \\ \sin x & \cos x & 0 \\ 0 & 0 & 1 \end{vmatrix}$

and, $f(-x) = \begin{vmatrix} \cos(-x) & \sin(-x) & 0 \\ -\sin(-x) & \cos(-x) & 0 \\ 0 & 0 & 1 \end{vmatrix}$

$= \begin{vmatrix} \cos x & -\sin x & 0 \\ \sin x & \cos x & 0 \\ 0 & 0 & 1 \end{vmatrix}$

$= f^{-1}(x)$

$\therefore \quad f^{-1}(x) = f(-x).$

55. $|\vec{a} \times \vec{b}| = |\vec{a}|.|\vec{b}|.\sin\theta$

where, θ is the angle between vectors $\vec{a}$ and $\vec{b}$.

$$1 = 3 \times \frac{\sqrt{2}}{3}.\sin\theta$$

$$\sin\theta = \frac{1}{\sqrt{2}} \Rightarrow \theta = \frac{\pi}{4}.$$

59.

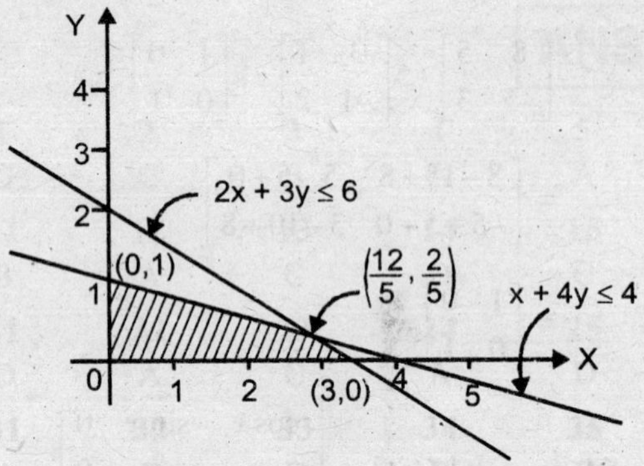

Point of intersections are (3, 0), $\left(\frac{12}{5}, \frac{2}{5}\right)$ and (0, 1)

Given function $f(x) = x + 2y$

At Point (3, 0), $f(x) = 3$

At Point $\left(\frac{12}{5}, \frac{2}{5}\right)$,

$$f(x) = \frac{16}{5} = 3.2$$

At Point (0, 1),

$$f(x) = 2$$

$\therefore f(x)$ is maximum at $\left(\frac{12}{5}, \frac{2}{5}\right)$ and

$$f_{max}(x) = 3.2.$$

60.

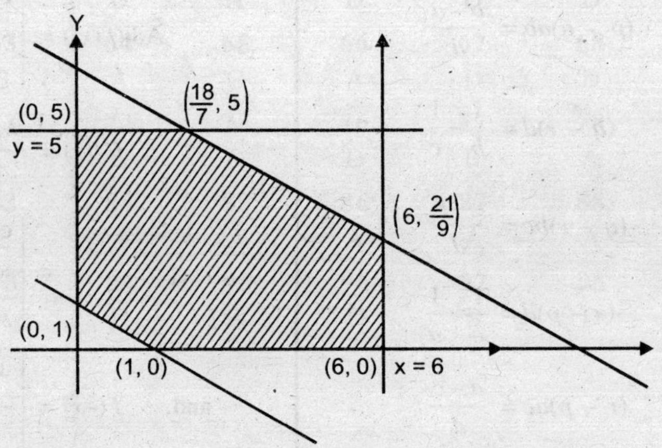

Point of intersections are (1, 0), (6, 0), $\left(6, \frac{21}{9}\right), \left(\frac{18}{7}, 5\right)$, (0, 5) and (0, 1)

Objective function

$$f(x) = 3.5x + 4.5y.$$

	(1, 0)	(6, 0)	$\left(6, \frac{21}{9}\right)$	$\left(\frac{18}{7}, 5\right)$	(0, 5)	(0, 1)
$f(x) = 3.5x + 4.5y$	3.5	21	31.5	31.5	22.5	4.5

Thus, f(x) is maximum at two points i.e., $\left(6, \frac{21}{9}\right)$ and $\left(\frac{18}{7}, 5\right)$, and $f_{max}(x) = 31.5$.

PHYSICS

1 UNITS AND MEASUREMENT

Units for Measurement

(i) Fundamental units are independent of each other and can not be further resolved into any other units. All other units can be expressed in terms of these fundamental units. The quantities mass, length and time are called fundamental quantities. For measuring these quantities, there are independent units such as kilogram, metre and second.

(ii) It is found that units of the other physical quantities can be expressed in terms of the fundamental units of mass, length and time. The units of all such physical quantities which can be expressed in terms of the fundamental units of mass, length and time are called derived units.

Example: The unit of area is a derived unit. The unit of area is area of a square having its length and breadth each equal to the unit length. Let us consider the defining equation of speed.

$$\text{Speed} = \frac{\text{Distance travelled}}{\text{time taken}}$$

$$\therefore \text{ Unit of speed} = \frac{\text{unit of distance, i.e., length}}{\text{unit of time}}$$

$$= \frac{\text{metre}}{\text{second}} = \text{ms}^{-1}.$$

Systems of Units

These are mainly four in number:

1. **CGS system:** This system was established in France. It is based on centimetre, gram and second as the fundamental units of length, mass and time respectively.
2. **FPS system:** This system is called as British system of units. It is based on foot, pound and second as the fundamental units of length, mass and time respectively.
3. **MKS system:** This system of units was also established in France. It makes use of metre, kilogram and second as the fundamental units of mass, length and time.
4. **SI system:** The units of mass, length and time can be used to find the units of physical quantities in mechanics only. It is abbreviated as SI from the French name Le Systeme International d'Unites. It is based on the following seven fundamental (or basic) and two supplementary units:

	Basic physical quantity	Name of the unit	Symbol
1.	Length	metre	m
2.	Mass	kilogram	kg
3.	Time	second	s
4.	Electric current	ampere	A
5.	Temperature	kelvin	K
6.	Luminous intensity	candela	Cd
7.	Amount of substance	mole	mol

Basic and Supplementary Units of SI System

1. **Kelvin:** It was adopted as the unit of temperature.
2. **Ampere:** It was adopted as the unit of current.

3. **Candela:** It was adopted as the unit of luminous intensity.
4. **Mole:** It was adopted as the unit of amount of substance.
5. **Radian:** It was adopted as the unit of plane angle.
6. **Steradian:** It was adopted as the unit of solid angle with its apex at the centre of sphere.
7. **Parallactic second (Parsec):** It is the largest practical unit of distance.

   1 parsec = 3.08×10^{16} m = 4.2 light year
8. **Light Year:** It is also a practical unit of distance used in astronomy to measure the distance of nearer stars.
9. **Astronomical unit:** It is defined as the average distance between sun and earth, i.e., radius of earth's orbit.

   1 A.U. = 1.496×10^{11} m
10. **Microns:** It is a unit of distance defined as micrometre.

    1 microns = 1 μm = 10^{-6} m
11. **Angstrom.** It is also a practical unit of length used in atomic physics.

    1 Å = 10^{-10} m = 10^{-8} cm
12. **Fermi:** It is the smallest practical unit of distance used in nuclear physics.

    1 Fermi = 10^{-15} m = 10^{-13} cm

Dimensions of a Physical Quantity

(i) In all the systems of units, the derived units of all the physical quantities can be expressed in terms of the fundamental units of mass, length and time. raised to some power. The powers to which fundamental units must be raised in order to express a physical quantity, are called its dimensions. To make it clear, consider the physical quantity density which is defined as mass per unit volume.

$$\text{Hence, density} = \frac{\text{mass}}{\text{volume}} = \frac{M}{L^3} = \left[M^1L^{-3}T^0\right]$$

(ii) *(a)* Above expression shows that dimensions of density are 1 in mass, –3 in length and 0 in time.

(b) Dimensional formula for density is $[ML^{-3}]$ or $[ML^{-3}T^0]$

(c) Dimensional equation for density is $[d] = [ML^{-3}T^0]$

Theory of Dimensions Applied to Heat

1. In heat, temperature is assumed to be a fundamental quantity with dimension θ and unit *K*.
2. **Heat:** As *Q* represents energy, hence

   $$[Q] = [ML^2T^{-2}]$$

   and unit is calories.
3. **Sp. heat:** $Q = ms/\Delta\theta$

   $\therefore \quad L = Q/m\Delta\theta$

   Hence $\quad [s] = [ML^2T^{-2}]/[M\theta]$

   $= [L^2T^{-2}\theta^{-1}]$

   and unit is cal/gm °C
4. **Latent heat:** $Q = mL$

   $\therefore \quad L = Q/m$

   i.e. $\quad [L] = [Q]/[m]$

   $= [ML^2T^{-2}]/[M]$

   $= [L^2T^{-2}]$

   Its unit is cal/gm.
5. **Coefficient of thermal conductivity**

   $$Q = K.A\,(\theta_1 - \theta_2)\,t/d$$

   or $\quad K = Qd\,/\,[A(\theta_1 - \theta_2)\,t]$

   $$\therefore \quad K = \frac{(dQ/dt)}{[A(d\theta/dx)]}$$

   $$= \frac{[ML^2T^{-2}]/[T]}{[L^2][\theta/L]}$$

   $$= [MLT^{-3}\theta^{-1}]$$

Theory of Dimensions Applied to Electrostatics

1. While dealing with electrostatics or electricity normally current is treated as fundamental quantity with dimensions [A] and unit ampere.
2. **Charge:** $\quad I = q/t$

   or $\quad q = It$

$\therefore \quad [q] = [AT]$

Unit of charge = Ampere × sec = coulomb.

3. **Potential:** $V = W/q$,

$$\therefore \quad [V] = \frac{[W]}{[q]} = \frac{[ML^2T^{-2}]}{[AT]}$$

$$= [ML^2T^{-3}A^{-1}]$$

Its unit is joule/coulomb or Volt.

4. **Intensity of electric field:**

$$E = \frac{F}{q};$$

$$\therefore \quad [E] = \frac{[F]}{[q]} = \frac{[MLT^{-2}]}{[AT]}$$

$$= [MLT^{-3}A^{-1}]$$

Its unit is $\frac{\text{newton}}{\text{coulomb}}$ or $\frac{\text{Volt}}{\text{metre}}$.

5. **Capacity:** $q = CV$

or $$C = \frac{q}{V} = \frac{q}{Q/q} = \frac{q^2}{W}$$

$$\therefore \quad [C] = \frac{[A^2T^2]}{[ML^2T^{-2}]}$$

$$= [M^{-1}L^{-2}T^4A^2]$$

Its unit is coulomb/volt or farad.

Theory of Dimensions Applied to Electricity

1. **Resistance:** $V = IR$

or $R = V/I = W/qI$

$$\therefore \quad [R] = \frac{[W]}{[q][I]}$$

$$= \frac{[ML^2T^{-2}]}{[AT][A]}$$

$$= [ML^2T^{-3}A^{-2}]$$

Its unit is volt/amp. or ohm.

2. **Conductance:** $C = \frac{1}{R}$

Hence, $$[C] = \frac{1}{[R]}$$

$$= [M^{-1}L^{-2}T^3A^2]$$

Its unit is $(\text{ohm})^{-1}$ or mho.

3. **Specific resistance (*r*) or resistivity:**

$$R = \rho\,(l/A)$$

or $$\rho = RA/l$$

$$\therefore \quad [\rho] = \frac{[ML^2T^{-3}A^{-2}]\,[L^2]}{[L]}$$

$$= [ML^3T^{-3}A^{-2}]$$

Its unit is ohm × metre or ohm × cm.

4. **Conductivity or specific conductance (σ):**

It is defined as

(i) $\sigma = 1/\rho$

or $[\sigma] = 1/[\rho] = [M^{-1}L^{-3}T^3A^2]$

Its unit is $(\text{ohms} \times \text{metre})^{-1}$ or mhos/ metre

(ii) σ can also be defined as

$$J = \sigma E$$

or $$\sigma = \frac{J}{E} = \frac{I/A}{V/l} = \frac{Il}{VA}$$

As $[V] = ML^2T^{-2}]/[AT]$

$$[\sigma] = \frac{[AL][AT]}{[ML^2T^{-2}][L^2]}$$

$$= [M^{-1}L^{-3}T^3A^2]$$

MULTIPLE CHOICE QUESTIONS

1. The ratio of one kWh to one MeV is
   (*a*) 2.25×10^{17} (*b*) 2.25×10^{19}
   (*c*) 2.25×10^{23} (*d*) 4.4×10^{25}

2. The product (PV) has the dimensions
   (*a*) $M^1L^{-1}T^{-2}$ (*b*) $M^1L^2T^4$
   (*c*) $M^1L^2T^{-2}$ (*d*) $M^1L^2T^{-3}$

3. 1 kg f/cm^2 is equal to
(*a*) 760 mm of Hg (*b*) 735.6 mm of Hg
(*c*) 1 mm of Hg (*d*) 100 mm of Hg

4. The unit of co-efficient of thermal conductivity is
(*a*) watt-kelvin/meter (*b*) watt/m-k
(*c*) Joule/sec (*d*) Joule

5. Dimension of $\frac{1}{\epsilon_0}\frac{e^2}{hc}$; where e = electronic charge, h = planck constant, c = velocity of light
(*a*) $A^2T^4L^{-3}M^{-1}$ (*b*) $A^{-2}T^{-4}L^3M$
(*c*) $A^0T^0L^0M^0$ (*d*) $AT^2L^3M^{-1}$

6. If *E, G, J* and *m* represent energy, universal gravitational constant, angular momentum and mass respectively, then dimensions of $\frac{EJ^2}{m^5G^2}$ are those of:
(*a*) $[\text{mass}]^{-2}$ (*b*) $[\text{length}]^{-4}$
(*c*) $[\text{time}]^{-6}$ (*d*) angle

7. When Bernoulli's theorem is expressed as $\frac{P}{\rho g}+\frac{1}{2}\frac{v^2}{g}+h$ = constant the dimensions of the constant on the right hand side of the equation are given as
(*a*) $M^0L^0T^0$ (*b*) $M^0L^1T^0$
(*c*) $M^1L^0T^0$ (*d*) $M^1L^2T^{-2}$

8. The dimensional formula of Hall coefficient *h* is
(*a*) M^0L^3AT (*b*) $M^0L^3A^{-1}T^{-1}$
(*c*) M^0L^0AT (*d*) $M^0L^0A^{-2}T^{-1}$

9. The physical quantity which has the dimensional formula $M^{-1}L^1T^{+2}$ is
(*a*) coefficient of viscosity
(*b*) modulus of rigidity
(*c*) compressibility
(*d*) surface tension

10. If frequency (F), velocity (V) and density (D) are considered as fundamental units, the dimensional formula for momentum will be
(*a*) DVF^2 (*b*) DV^2F^{-1}
(*c*) $D^2V^2F^2$ (*d*) DV^4F^{-3}

11. The dimensional formula for Joule's mechanical equivalent 'J' is
(*a*) $M^1L^1T^1$ (*b*) $M^1L^0T^1$
(*c*) $M^0L^0T^1$ (*d*) $M^0L^0T^0$

12. If x is the distance covered in 4 seconds, t represents time measured in seconds, then what are α and β in the following dimensionally correct relationship,

$$x = \alpha t + \frac{1}{2}\beta t^2$$

(*a*) LT^{-1} and LT^{-2} (*b*) LT^{-2} and LT^{-4}
(*c*) LT^{-2} and LT^{-3} (*d*) None of these

13. The dimensions of coefficient of permittivity for free space (ϵ_0) in a relation $F = \frac{1}{4\pi\epsilon_0}\frac{q_1q_2}{r^2}$, where symbols have their usual meanings is
(*a*) $M^{-1}L^{-3}A^2T^4$ (*b*) $M^1L^3A^{-2}T^{-4}$
(*c*) $M^1L^3A^2A^{-4}$ (*d*) $M^{-1}L^{-3}A^{-2}T^{-4}$

14. The dimensional formula for bulk modulus of elasticity is
(*a*) $M^1L^{-2}T^2$ (*b*) $M^1L^{-3}T^{-2}$
(*c*) $M^1L^2T^{-2}$ (*d*) $M^1L^{-1}T^{-2}$

15. If force (F), acceleration (A) and time (T) be taken as fundamental physical quantities, the dimensions of length on this system of units are
(*a*) FT (*b*) FAT
(*c*) AT^2 (*d*) FAT^2

16. If L = 36 ± 4 mm, B = 14 ± 2 mm, what is the maximum error possible in the determination of x = 2 (L – B)
(*a*) ± 4 mm (*b*) ± 6 mm
(*c*) ± 12 mm (*d*) None of these

17. If length were measured in metres and time in seconds but a new unit of mass were choosen in order to make G = 1. How many kilograms would there be in this new unit of mass?
(*a*) 6.67×10^{-8} kg (*b*) 6.67×10^{-11} kg
(*c*) 1.5×10^{10} kg (*d*) 1.5×10^7 kg

18. Which one of the following has not been expressed in proper units

(a) $\frac{\text{Stress}}{\text{Strain}} = \text{newton/m}^2$
(b) Surface tension = newton/m
(c) Energy = kg – m/sec
(d) pressure = newton/m^2

19. A pressure of 10^6 dyne/cm^2 is equivalent to
(a) 10^5 n/m^2 (b) 10^4 n/m^2
(c) 10^6 n/m^2 (d) 10^7 n/m^2.

20. Error in the measurement of radius of a sphere is 1%. Then error in the measurement of volume is
(a) 1% (b) 5%
(c) 3% (d) 8%

21. The measured mass and volume of a body are 22.42 gm and 4.7 cm^3 with possible error 0.01 gm and 0.1 cc. The maximum error in density is about
(a) 0.2% (b) 2%
(c) 5% (d) 10%

22. The specific resistance σ of a thin circular wire of radius *r* cm, resistance R ohms and length L is given by $\sigma = \pi r^2 \frac{R}{L}$. If *r* = 0.26 ± 0.02 cm, R = 32 ± 1 ohm and L = 7.8 ± 0.01 cm, find the percentage error in σ
(a) 9% (b) 13%
(c) 19% (d) 7%

23. The length of a string of simple pendulum is 102.3 cm and the diameter of the bob is 2.56 cm. What is the effective length of simple pendulum (i.e. the distance between the point of suspension and the centre of gravity of the bob) upto required significant figures.
(a) 104 cm (b) 103.58 cm
(c) 103.6 cm (d) 104.86 cm

24. The length, breadth and thickness of a block are given by *l* = 12 cm, *b* = 6 cm, *t* = 2.45 cm. What is the volume of the block according to the idea of significant figures.
(a) 1.764×10^2 cm^3 (b) 1.77×10^2 cm^3
(c) 2×10^2 cm^3 (d) 1.8×10^2 cm^3

25. A resistor of 6KΩ with tolerance 10% and another of 4KΩ with tolerance 10% are connected in series. The tolerance of combination is about
(a) 5% (b) 10%
(c) 12% (d) 15%

26. In the above question, the tolerance when resistors are connected in parallels
(a) 10% (b) 20%
(c) 30% (d) 40%

27. On the main scale of vernier callipers, each centimetre length is divided into 10 divisions. If the vernier has to have a least count of 0.005 cm, the vernier scale should have
(a) 10 divisions (b) 20 divisions
(c) 25 divisions (d) 50 divisions

28. A spherometer has 250 equal divisions marked along the periphery of its disc; and the full rotation of the disc advances on the main scale by 0.0625 cm. The least count of the system is
(a) 2.5×10^{-2} cm (b) 2.5×10^{-3} cm
(c) 2.5×10^{-4} cm (d) None of these

29. A vernier calliper has 20 divisions on the vernier scale which coincide with 19 on the main scale. The least count of the instrument is 0.1 mm. The main scale division has value
(a) 0.25 mm (b) 0.5 mm
(c) 1 mm (d) 2 mm

30. If the area of a square is (100 ± 0.2) cm^2, then the side of the square is
(a) 10 ± 0.1 cm (b) 10.0 ± 0.02 cm
(c) (10 ± 0.2) cm (d) None of these

31. Neutron was discovered by
(a) Yukawa (b) Thomson
(c) Chadwick (d) Roentgen

32. Order of magnitude of light year expressed in metres is
(a) one (b) 10^{15}
(c) 16 (d) 10^{16}

33. The approximate distance of the nearest galaxy (Andromeda) is
(a) 10^{-12} m (b) 10^{22} m
(c) 10^{26} m (d) 10^{10} m

34. Hubble describes
(*a*) Superconductivity
(*b*) Expansion of universe
(*c*) Propagation of electromagnetic waves
(*d*) Unification of weak and electromagnetic interaction

35. Who discovered photoelectric effect?
(*a*) Albert Einstein
(*b*) Halwaches
(*c*) Hertz
(*d*) Max Planck

Directions Qs. 36-35: *In each of the following questions there are two statements Assertion (A) is given followed by a corresponding statement of Reason (R) just below it. Of the statements, mark the correct answer:*
(*a*) If both assertion and reason are true and reason is the correct explanation of assertion.
(*b*) If both assertion and reason are true but reason is not the correct explanation of assertion
(*c*) If assertion is true but reason is false
(*d*) If assertion is false but reason is true.

36. Assertion (A): In mechanics, we treat length, mass and time as the three basic or fundamental quantities.
Reason (R): Length, mass and time cannot be obtained from one another.

37. Assertion (A): Light year is a unit of time.
Reason (R): Light year is the distance travelled by light in vacuum in one year.

38. Assertion (A): Radian is the unit of distance.
Reason (R): One radian is the angle subtended at the centre of a circle by an arc equal in length to the radius of the circle.

39. Assertion (A): A.U. is much bigger than Å.
Reason (R): A.U. stands for astronomical unit and Å stands for angstrom.

40. Assertion (A): When the change the unit of measurement of a quantity, its numerical value changes.
Reason (R): Smaller the unit of measurement smaller is its numerical value.

41. Assertion (A): Impulse has the dimensions of momentum.
Reason (R): Impulse is directly proportional to force and time.

42. Assertion (A): Dimension of coefficient of viscosity is $[ML^{-1}T^{-1}]$.
Reason (R): Coefficient of viscosity is force acting per unit area per unit velocity gradient.

43. Assertion (A): Pressure can be subtracted from pressure gradient.
Reason (R): Only like quantities can be added or subtracted from each other.

44. Assertion (A): Dimensional constants are the quantities whose value are constant.
Reason (R): Dimensional constants are dimensionless.

ANSWERS

1	2	3	4	5	6	7	8	9	10
(*b*)	(*c*)	(*b*)	(*b*)	(*c*)	(*d*)	(*b*)	(*b*)	(*c*)	(*d*)
11	**12**	**13**	**14**	**15**	**16**	**17**	**18**	**19**	**20**
(*d*)	(*c*)	(*a*)	(*d*)	(*c*)	(*c*)	(*b*)	(*c*)	(*a*)	(*c*)
21	**22**	**23**	**24**	**25**	**26**	**27**	**28**	**29**	**30**
(*b*)	(*c*)	(*c*)	(*c*)	(*b*)	(*c*)	(*b*)	(*c*)	(*d*)	(*d*)
31	**32**	**33**	**34**	**35**	**36**	**37**	**38**	**39**	**40**
(*c*)	(*c*)	(*b*)	(*b*)	(*a*)	(*a*)	(*d*)	(*d*)	(*b*)	(*c*)
41	**42**	**43**	**44**						
(*a*)	(*a*)	(*d*)	(*c*)						

SOME SELECTED EXPLANATORY ANSWERS

1. $1\text{kWh} = 1000 \times 3600 \text{ J}$

$$= \frac{3.6\times10^{6}}{1.6\times10^{-13}} \text{ MeV} = 2.25 \times 10^{19} \text{ MeV}$$

2. $$[PV] = \frac{[\text{Force}]}{[\text{Area}]}\times[\text{Volume}]$$

$$= \frac{[M^1L^1T^{-2}]\,L^3}{L^2} = [M^1L^2T^{-2}]$$

3. $$\frac{1\text{kgf}}{\text{cm}^2} = \frac{9.8\text{N}}{\text{cm}^2} = 9.8\times10^5\frac{\text{dyne}}{\text{cm}^2}$$

$= 735.6$ mm of Hg

(760 mm of Hg = 1×10^6 dyne/cm^2)

4. $$Q_1 = KA\,(\theta_1 - \theta_2)\frac{t}{d},\ K = \frac{Q_1 d}{A\theta t} = \text{watt / mK}$$

5. $$\frac{e^2}{\epsilon_0\,hc},\ \text{use } F = \frac{q_1q_2}{4\pi\epsilon_0\,r^2},\ \left[\frac{e^2}{\epsilon_0}\right] = [Fr^2] \text{ etc.}$$

Further, answer cannot have any A. Hence, (C) alone can be the correct answer.

6. $$\left[\frac{EJ^2}{m^5G^2}\right] = \frac{ML^2T^{-2}\left(ML^2T^{-1}\right)^2}{M^5\left(M^{-1}L^3T^{-2}\right)^2} = M^0L^0T^0 = \text{angle}$$

7. $$[\text{Constant}] = [h] = \left[\frac{v^2}{g}\right] = \left[\frac{p}{\rho g}\right] = L^1$$

8. Hall coefficient $$[h] = \frac{1}{(ne)} = \frac{1}{L^{-3}AT}$$

$$= L^3A^{-1}T^{-1}$$

10. $M^1L^1T^{-1} = M^1L^{-3}L^4T^{-1}$

$= DL^4T^{-4}T^3 = DV^4F^{-3}$

11. $$[J] = \frac{[W]}{[H]} = \frac{M^1L^2T^{-2}}{M^1L^2T^{-2}} = M^0L^0T^0$$

19. 10^6 dyne/cm^2

$$= \frac{10^6}{10^5}\frac{N}{\left(\frac{1}{100}\text{m}\right)^2} = 10^5\text{N / m}^2$$

20. $$\text{Vol. } V = \frac{4}{3}\pi r^3;\ \frac{\Delta V}{V} = \frac{3\Delta r}{r} = 3\times\frac{1}{100}\times100$$

$= 3\%$

21. $$\text{Density} = \frac{\text{Mass}}{\text{Volume}} \text{ or } D = \frac{M}{V}$$

$$\frac{\Delta D}{D} = \frac{\Delta M}{M}+\frac{\Delta V}{V} = \left(\frac{0.01}{22.42}+\frac{0.1}{4.7}\right)\times100$$

$= 2$

22. $$\sigma = \pi r^2\frac{R}{L}$$

$$= \frac{\Delta\sigma}{\sigma} = \frac{2\Delta r}{r}+\frac{\Delta R}{R}+\frac{\Delta L}{L}$$

$$= \left(2\times\frac{0.2}{0.26}+\frac{1}{32}+\frac{0.01}{7.8}\right)\times100\%$$

23. $$l = l_0+\frac{D}{2} = l_0 + r$$

$= 102.3 + 1.28 = 103.58 = 103.6$ m

24. $V = lbt = 12 \times 6 \times 2.45 \text{ cm}^3$

$= 176.4 \text{ cm}^3 = 2 \times 10^2 \text{ cm}^3$

Retain only one significant figure.

25. $R_s = R_1 + R_2 = 6 + 4 = 10\ \Omega$

$\Delta R_s = \Delta R_1 + \Delta R_2$

$$= \frac{10}{100}\times6+\frac{10}{100}\times4$$

$= 0.6 + 0.4 = 1$

$$\frac{\Delta R_s}{R_s} = \frac{1}{10} = 10\%$$

29. $$\text{Vernier constant} = \frac{1}{N}\times\text{M.S.D.}$$

$$0.1 \text{ mm} = \frac{1}{20}\times1 \text{ main scale division}$$

2 MOTION IN A STRAIGHT LINE

Motion

Motion is the change in position of the object with respect to time.

(i) Absolute motion (ii) Relative motion

Absolute Motion : Absolute motion is the motion with respect to a body which is at absolute rest. This type of motion is impossible.

Relative Motion : Relative motion is the motion with respect to a body which is at relative rest.

Various Terms Regarding Motion

(i) Displacement : Displacement is the change in position of the body in a certain direction. It is a vector quantity.

(ii) Distance : Distance is the total actual path covered by the body between initial and final states. It is a scalar quantity.

(iii) Velocity : Rate of change of displacement with respect to time is called velocity.

$$\text{Velocity } \vec{v} = \frac{\text{displacement}}{\text{time}} = \frac{d\vec{x}}{dt}$$

(iv) Speed : Speed is the scalar form of velocity.

$$\text{speed } v = \frac{d}{t} \text{ m/sec.}$$

(v) Acceleration : Rate of change of velocity with respect of time is known as acceleration. It is a vector quantity.

$$\vec{a} = \frac{\Delta \vec{V}}{\Delta t} \quad \text{or} \quad a = \frac{d^2x}{dt^2}.$$

Equations of Motion

The equations of motion are as follows :

(i) $v = u + at$

(ii) $s = ut + (1/2)\, at^2$

(iii) $v^2 = u^2 + 2as$

where u = initial velocity, v = final velocity after t seconds, a = acceleration, s = distance travelled in t seconds.

Note : The distance travelled in n^{th} second is given by

$$S_n = u + (2n - 1)\, a/2$$

Motion Under Gravity

(a) Downward Motion (↓)

(i) $v = u + gt$

(ii) $h = ut + \frac{1}{2}gt^2$

(iii) $v^2 = u^2 + 2gh$

(b) Upward Motion (↑)

(i) $v = u - gt$

(ii) $h = ut - \frac{1}{2}gt^2$

(iii) $v^2 = u^2 - 2gh$

Relative Velocity

If the distance between the two bodies is changing either in magnitude or direction or both then each is said to have a relative velocity with respect to other.

The relative velocity of first body with respect to the second body is obtained by the vector addition of the negative velocity (velocity in opposite direction) of second body to the velocity of first body.

Particulars Cases

(a) If the two bodies A and B are moving with velocities u and v in the same direction, then

Relative velocity of A with respect to $B = u - v$

Relative velocity of B with respect to $A = v - u$

(b) If the two bodies A and B are moving in opposite directions with velocities u and v then, relative velocity is given by

Relative velocity of A w.r.t. $B = u + v$

Relative velocity of B w.r.t. $A = -(u + v)$

MULTIPLE CHOICE QUESTIONS

1. If velocity of a particle is given by $v = \sqrt{180-16x}$. Its acceleration will be
(a) $-16\ m/s^2$ (b) $-8\ ms^{-2}$
(c) $4\ ms^{-2}$ (d) $32\ m/s^2$

2. The velocity of a body depends on time according to the equation $v = 20 + 0.1\ t^2$. The body is undergoing.
(a) uniform acceleration
(b) uniform retardation
(c) non-uniform acceleration
(d) zero acceleration

3. A particle moves in xy = plane according to equations $x = 4t^2 + 5t + 16$, $y = 5t$. The acceleration of the particle must be
(a) 8 m/sec^2 (b) 13 m/sec^2
(c) 14 m/sec^2 (d) None of these

4. A body is projected vertically with a speed of 20 m/s in a region where g = 10 m/s^2. The distance travelled in 3 seconds is
(a) 15 m (b) 25 m
(c) 20 m (d) 30 m

5. A car moving at 12 m/s due east attains a speed of 16 m/s towards north in 4 seconds. The magnitude of its average acceleration in m/s^2 is
(a) 1 (b) 3
(c) 5 (d) 7

6. A car covers the first half distance at 40 km per hour and the second half distance at 60 km per hour along a straight road. The average speed is
(a) 48 km per hour (b) 50 km per hour
(c) 52 km per hour (d) 60 km per hour

7. A tiger chases a deer 30 m ahead of it and gains 3m in first 5s after the chase started. After 10s, the distance between them is (consider both at rest initially)
(a) 6 m (b) 14 m
(c) 18 m (d) 24 m

8. A stone is allowed to fall from a top of a tower and covers half the height of tower in the last second of its journey, the approximate height of tower is (g = 9.8 ms^{-2})
(a) 78.4 m (b) 57.1 m
(c) 37.6 m (d) 29.4 m

9. A stone is allowed to fall from the top of a tower and; covers half the height of tower in the last second of its journey. The time taken by the stone to reach the foot of the tower is
(a) $(2-\sqrt{2})s$ (b) $(2+\sqrt{2})s$
(c) 4 s (d) $(2\pm\sqrt{2})s$

10. A particle moves along x-axis in such a way that its coordinate (x) varies with time (t) according to the expression $x = 2 - 5t + 6t^2$. The initial velocity of the particle is
(a) –5m/s (b) –3m/s
(c) 3 m/s (d) 6 m/s

11. A body covers 200 cm in the first 2 seconds and 220 cm in the next 4 seconds. The velocity at the end of 7th second is
(a) 20 cms^{-1} (b) 10 cms^{-1}
(c) 115 cms^{-1} (d) –10 cms^{-1}

12. The speed of the car is reduced to one third of its original value in travelling a distance S. Then the car is brought to rest in a minimum distance.

(*a*) 3 S (*b*) 9 S

(*c*) $\frac{8}{9}$S (*d*) $\frac{9}{8}$S

13. Two balls of the same mass are thrown upwards at an interval of 2 seconds, along the same vertical line with the same initial velocity of 39.2 ms^{-1}. They will collide at a height

(*a*) 44.1 m (*b*) 58.8 m

(*c*) 73.5 m (*d*) 78.4 m

14. A ship of mass 3×10^7 kg, initially at rest, is pulled by force of 5×10^4 N through a distance of 3m. Assuming that the resistance due to water is negligible the speed of the ship is.

(*a*) 1.5 ms^{-1} (*b*) 60 ms^{-1}

(*c*) 0.1 ms^{-1} (*d*) 5 ms^{-1}

15. The initial velocity of a particle is *u* (at, $t = 0$) and the acceleration *f* is given by *at*, where "*a*" is a constant. Which of the following relations is valid.

(*a*) $v = u + at^2$ (*b*) $v = u + \frac{at^2}{2}$

(*c*) $v = u + at^2/4$ (*d*) $v = u$

16. A body falls freely from top of a tower and falls $\frac{7}{16}$th of the total distance in its last second. The time taken by the body to complete motion is

(*a*) 3 seconds (*b*) 4 seconds

(*c*) 5 seconds (*d*) 6 seconds

17. A balloon is at a height of 81 metres and is ascending upwards with a velocity of 12 m/s. A body of 2 kg weight is dropped from it. If $g = 10$ m/s^{-2}, the body will reach the surface of the earth in

(*a*) 1.5 s (*b*) 4.025 s

(*c*) 5.4 s (*d*) 6.75 s

18. A stone thrown vertically upward flies past a window one second after it was thrown upward and after three seconds on its way downward the initial velocity of stone is

(*a*) 10 ms^{-1} (*b*) 20 ms^{-1}

(*c*) 15 ms^{-1} (*d*) 5 ms^{-1}

19. A body falls freely without an initial velocity from a height of 20 metres. At what height will it reach a velocity equal to half the velocity with which it reaches the ground

(*a*) 10 m (*b*) 15 m

(*c*) 5 m (*d*) 17.5 m

20. The distance travelled by a body falling from rest in the first, second and third seconds are in the ratio

(*a*) 1 : 2 : 3 (*b*) 1 : 3 : 5

(*c*) 1 : 4 : 9 (*d*) None of these

21. A particle is dropped under gravity from rest from a height *h* ($g = 9.8$ ms^{-2}) and it travels a distance $\frac{9h}{25}$ in the last second, the height *h* is

(*a*) 100 m (*b*) 122.5 m

(*c*) 145 m (*d*) 167.5 m

22. The velocity of a particle at an instant is 10 m/s, after 5 seconds, the velocity of the particle is 20 m/s. The velocity at 3 seconds before is (assume constant acceleration)

(*a*) 8 m/s (*b*) 4 m/s

(*c*) 6 m/s (*d*) 7 m/s

23. A stone weighing 3kg falls from top of tower 100 m high and burries itself 2 m deep in sand. The time of penetration is

(*a*) 0.09 second (*b*) 0.9 second

(*c*) 2.1 second (*d*) 1.3 second

Directions Qs. 24-33: *In each of the following questions there are two statements Assertion (A) is given followed by a corresponding statement of Reason (R) just below it. Of the statements, mark the correct answer:*

(*a*) If both assertion and reason are true and reason is the correct explanation of assertion.

(*b*) If both assertion and reason are true but reason is not the correct explanation of assertion

(*c*) If assertion is true but reason is false

(*d*) If assertion is false but reason is true.

24. **Assertion (A):** A body having non-zero acceleration can have a constant velocity.

Reason (R): Acceleration is the rate of change of velocity.

25. **Assertion (A):** A body can have acceleration even if its velocity is zero at a given instant of time.

Reason (R): A body is momentarily at rest when it reverses its direction of motion.

26. **Assertion (A):** A body, whatever its motion is always at rest in a frame of reference which is fixed to the body itself.
Reason (R): The relative velocity of a body with respect to itself is zero.

27. **Assertion (A):** Displacement of a body may be zero when distance travelled by it is not zero.
Reason (R): The displacement is the longest distance between initial and final position.

28. **Assertion (A):** The equations of motion in scalar form can be applied only if acceleration is along the direction of velocity.
Reason (R): If the acceleration of a body is constant then its motion is known as uniform motion.

29. **Assertion (A):** A bus moving due north takes turn and start moving towards east with same speed. There will be no change in the velocity of bus.
Reason (R): Velocity is a vector quantity.

30. **Assertion (A):** The relative velocity between any two bodies moving in opposite direction is equal to sum of the velocities of two bodies.
Reason (R): Sometimes relative velocity between two bodies is equal to difference in velocities of the two bodies.

31. **Assertion (A):** The displacement - time graph of a body moving with uniform acceleration is a parabola.
Reason (R): The displacement is proportional to time.

32. **Assertion (A):** Velocity-time graph for an object in uniform motion along straight path is a straight line parallel to the time axis.
Reason (R): In uniform motion of an object velocity increases as the square of time elapsed.

33. **Assertion (A):** A body may be accelerated even when it is moving at uniform speed.
Reason (R): When direction of motion of the body changing then body may have acceleration.

ANSWERS

1	2	3	4	5	6	7	8	9	10
(*b*)	(*c*)	(*a*)	(*b*)	(*c*)	(*a*)	(*c*)	(*b*)	(*b*)	(*a*)
11	**12**	**13**	**14**	**15**	**16**	**17**	**18**	**19**	**20**
(*b*)	(*d*)	(*c*)	(*c*)	(*b*)	(*b*)	(*c*)	(*b*)	(*b*)	(*b*)
21	**22**	**23**	**24**	**25**	**26**	**27**	**28**	**29**	**30**
(*b*)	(*b*)	(*a*)	(*d*)	(*a*)	(*a*)	(*c*)	(*b*)	(*d*)	(*b*)
31	**32**	**33**							
(*c*)	(*c*)	(*a*)							

SOME SELECTED EXPLANATORY ANSWERS

1. $v = \sqrt{180-16x}$

$v^2 = 180 - 16x$

compare it $v^2 = u^2 + 2ax$

$a = -8 \text{ ms}^{-2}$

2. $v = 20 + 0.1t^2$

$a = \dfrac{dv}{dt} = 0.1\times 2t;\ a = .2t$

Time dependent acceleration, *i.e.*, non-uniform acceleration

3. $x = 4t^2 + 5t + 16;\quad y = 5t$

Differentiating w.r.t t

$\dfrac{dx}{dt} = v_x = 8t+5$ and $v_y = \dfrac{dy}{dt} = 5$

Differentiating w.r.t. t

$a_x = \dfrac{dv_x}{dt} = 8,\ \dfrac{dv_y}{dt} = 0$

$a = \sqrt{a_x^2 + a_y^2} = \sqrt{8^2+0^2} = 8\text{ms}^{-2}$

4. Time of ascent $= \frac{u}{g} = \frac{20}{10} = 2s$

Maximum height $= \frac{u^2}{2g} = \frac{20^2}{2\times 10} = 20\,\text{m}$

During fall, distance covered in one second

$= \frac{1}{2}\times 10\times 1^2 = 5\text{m}$

Total distance covered in 3 second

$= 20 + 5 = 25$ m

8. $n = 2+\sqrt{2} = 2+1.41 = 3.41\text{s}$

$h = \frac{1}{2}\times 9.8(3.41)^2 = 57.1\,\text{m}$

9. $\frac{h}{2} = \frac{g}{2}(2n-1)$

$h = \frac{1}{2}gn^2,\ n = (2+\sqrt{2})\text{s}$

discard negative sign as $n > 1$

so $n = (2+\sqrt{2})\text{s}$

10. $x = 2 - 5t + 6t^2$,

$v = \frac{dx}{dt} = -5+12t$

Put $t = 0$ $v_0 = -5$ m/s

11. $200 = u\times 2+\frac{1}{2}a\times 2^2$

$200+220 = u(2+4)+\frac{1}{2}a\,(2+4)^2$

$v = u + at = 0.1\ \text{ms}^{-1} = 10\ \text{cms}^{-1}$

12. $v^2 = u^2 - 2as = u^2 - \left(\frac{u}{3}\right)^2 = 2as$ ...(*i*)

$= u^2 - 0^2 = 2\ as'$ ...(*ii*)

Divide (*i*) by (*ii*) solves $s' = \frac{9}{8}s$

13. $h = 39.2t-\frac{1}{2}gt^2$ ...(*i*)

$= 39.2\,(t-2)-\frac{1}{2}g(t-2)^2 = t = 5\text{s}$

Putting values in (*i*) gives $h = 73.5$ m

14. $u^2 = u^2 + 2as$

$v^2 = 0+2\left(\frac{5\times 10^4}{3\times 10^7}\right)\times 3$, using $a = \frac{\text{F}}{\text{m}}$

$v = 10^{-1} = 0.1\ \text{ms}^{-1}$

15. Acceleration, $f = \frac{dv}{dt}$

$\therefore\ dv = fdt$

$\int_u^v dv = \int_0^t fdt = \int_0^t at\ dt\ (\because f = at)$

$\therefore\ v = u+\frac{at^2}{2}$

16. $\frac{\frac{7}{16}h}{h} = \frac{\frac{g}{2}(2n-1)}{\frac{g}{2}n^2}$

$\frac{(1)}{(2)};\ \frac{7}{16} = \frac{2n-1}{n^2},\ n = 4\text{s}$

17. $h = ut+\frac{1}{2}at^2$

$-81 = +12t-\frac{1}{2}\times 10\times t^2$

or $5t^2 - 12t - 81 = 0$

$t = \frac{12\pm\sqrt{144+1620}}{10} = \frac{12\pm 42}{10} = 5.4$ sec

18. Sum of roots, $t_1+t_2 = \frac{2u}{g} = 1+3$

$mu = \frac{40}{2} = 20\,\text{ms}^{-1}$

19. $\frac{1}{2}\times\sqrt{2g\text{H}} = \sqrt{(2gh)}$

H = 20 m, $h = 5$ m

H − h = 15 m required

20. $s_1 = \frac{1}{2}g\,(2\times 1-1) = \frac{1}{2}g$

$s_2 = \frac{1}{2}g\,(2\times 2-1) = 3\times\frac{1}{2}g$

$s_3 = \frac{1}{2}g\,(2\times 3-1) = 5\times\frac{1}{2}g$

ratio 1 : 3 : 5

$s_1 : s_2 : s_3 :: 1 : 3 : 5$

21. $S_{nth} = \frac{9h}{25} = \frac{g}{2}(2n-1)$ ...(*i*)

$S_n = h = \frac{g}{2}n^2$ ...(*ii*)

divide (*i*) by (*ii*)

$\frac{9}{25} = \frac{2n-1}{n^2}$

gives $n = 5s$

$h = \frac{9.8}{2}\times 5^2 = 122.5\,\text{m}$

3 LAWS OF MOTION

Newton's First Law of Motion

(i) According to this law, **every body continues in its state of rest or of uniform motion in a straight line, unless it is compelled by some external force to change the state.**

(ii) Newton's first law of motion can be divided into two parts. The first part is concerned with a basic property of matter, called **Inertia.**

(iii) The second part of the first law of motion gives a definition for force. It is a push or pull which either changes or tends to change the state of rest or uniform motion of a body.

Mass and Momentum

(i) The physical quantity which is a measure of the inertia of a body is called its inertial mass.

(ii) Mass is a scalar quantity and is measured in Kilograms.

(iii) **Momentum:**

(a) It is defined as the total quantity of motion contained in a body and is measured as the product of the mass of body and its velocity.

(b) The momentum of a body of mass m moving with velocity $\vec{v}$ is given by

$$\vec{p} = m\vec{v}$$

(e) If two objects of different masses have same momentum, the lighter body possesses greater velocity.

Newton's Second Law of Motion

(i) According to this law, the time rate of change of momentum of a body is directly proportional to the external force applied on it and the change in momentum takes place in the direction of force.

(ii) If $\vec{p}$ be the momentum of a body and $\vec{F}$ be the external force acting on it, then according to Newton's second law of motion

$$\vec{F} \propto \frac{d\vec{p}}{dt}$$

or $$F = K\frac{d\vec{p}}{dt}$$

Here, K is a constant of proportionality. Its value depends on the unit adopted for measuring the force. Both in SI and CGS system, the unit of force is selected in such a manner so that $K = 1$. Hence above equation becomes:

$$\vec{F} = \frac{d\vec{p}}{dt}$$

i.e., force is equal to the time rate of change of momentum and it leads to the quantitative measurement of the force.

(iii) A body of mass m moving with a velocity $\vec{v}$ has got a momentum $\vec{p} = m\vec{v}$. Hence

$$\vec{F} = \frac{d}{dt}(m\vec{v}) = m\frac{d\vec{v}}{dt} = m\vec{a}$$

Newton's Third Law of Motion

(i) According to this law to every action, there is equal and opposite reaction.

(ii) In general, if a body A exerts force $\vec{F}_{AB}$ (which may be gravitational, electrical or magnetic etc.)

on a body B, then the body B will exert a force $\vec{F}_{BA}$ on body A, such that

$$\vec{F}_{AB} = -\vec{F}_{BA}$$

Motion of Two Bodies Connected by a String Passing Over a Smooth Pulley

Case (a): Suppose A and B are two bodies of masses m_1 and m_2 respectively suspended by means of a light string passing over a smooth pulley P. Let $m_2 > m_1$.then Tension

$$T = \frac{2m_1m_2}{m_1 + m_2} g$$

Motion of a Man in a Lift

Case (i): When the lift is accelerated upwards: Suppose R be the upward thrust of the floor on the man (normal Reaction) and mg is the weight of the man acting downwards.

Hence,

unbalanced force = mass × acceleration

or $R - mg = ma$

or $R = mg + ma = m(g + a)$

Thus, if the man is standing on weighing machine, it will show a larger weight than mg.

Case (ii): When the lift is accelerated downward: In this case, inertial force $F = ma$, is to be applied vertically upward. Now

$R + F = mg$

or $R + ma = mg$

or $R = m(g - a)$

Now the weighing machine will show a weight smaller than mg.

Case (iii): When the lift moves with uniform velocity (or is at rest)

In this case $a = 0$ and $R = mg$.

In the case of free fall of the lift, $a = g$, then $R = m(g - g) = 0$, i.e., the man will feel weightlessness.

FRICTION

Kinetic Friction

(i) When two bodies in contact move with respect to each other, rubbing the surfaces in contact, the friction between them is called Kinetic friction. The directions of the frictional forces are such that the relative motion is opposed by the friction.

(ii) The magnitude of the kinetic friction is proportional to the normal force acting between the bodies.

i.e. $F_k = \mu_k N$

where N is the normal force. The proportionality constant μ_k is called the coefficient of Kinetic Friction.

Static Friction

Frictional forces can also act between two bodies which are in contact but are not sliding with respect to each other. The friction in such cases is called static friction. For example, suppose many persons are trying to push a heavy box on the floor to take it out of a room.

Angle of Friction

(i) Suppose a body of mass m is lying in rest position on a rough table. Its weight mg is acting downward and normal reaction R is acting in opposite direction. Suppose we pull the body by some horizontal force P and F is the frictional force preventing the motion of the body.

Suppose S represents the resultant of R and F which makes an angle θ with R. Resolving S along R and F, we get

$S \cos\theta = R$

and $S \sin\theta = F$

i.e. $\tan\theta = F/R$

For equilibrium,

$R = W = mg$

and $F = P$

(ii) The coefficient of friction μ is defined as the ratio of limiting frictional force F to the normal reaction R between two surfaces.

i.e. $\mu = \dfrac{F}{R}$

$\tan\lambda = F/R$

From equations (1) and (2), we get

$$\mu = \tan\lambda$$

Applications of The Principle of Conservation of Linear Momentum

(i) When a man jumps from a boat to the shore, the boat slightly moves away from the shore

Initially, the total momentum of the boat and the man is zero. When a man jumps from the boat, total momentum can be zero, only if the boat moves in a direction opposite to the direction of jumping of man.

(ii) Rocket Launching

As the fuel in rocket undergoes combustion, the gases so produced leave the body of the rocketwith large velocity and give upthrust to the rocket. If we assume that the fuel is burnt at a constant rate, then the rate of change of momentum of the rocket will be constant. As more and more fuel gets burnt, the mass of the rocket goes on decreasing and it leads to increase of the velocity of rocket more and more rapidly.

MULTIPLE CHOICE QUESTIONS

1. The velocity time graph of a lift moving downwards is a straight line inclined to the time axis at 45°. If mass of the lift is M kg, what is the effective weight (in newton) of the lift? Take $g = 10\ ms^{-2}$.

(*a*) 10 *M* (*b*) 9 *M*
(*c*) *M* (*d*) none of the above

2. A monkey of mass 20 kg is holding a vertical rope. The rope can break when a mass of 25 kg is suspended from it. What is the maximum acceleration with which the monkey can climb up along the rope?

(*a*) 2.5 ms^{-2} (*b*) 5 ms^{-2}
(*c*) 7 ms^{-2} (*d*) 10 ms^{-2}

3. A mass when dropped vertically from the top of a smooth inclined plane reached the ground with a velocity of 50 ms^{-1}. If it is allowed to move along the inclined plane, what will be the horizontal component of the velocity on reaching the ground? Angle of inclination of the plane is 30°.

(*a*) 50 ms^{-1} (*b*) $25\sqrt{3}\ ms^{-1}$
(*c*) 10 ms^{-1} (*d*) zero

4. In which of the following cases the net force is not zero?

(*a*) a kite skilfully held stationary in the sky.
(*b*) a ball freely falling from a height.
(*c*) an aeroplane rising upwards at an angle of 45° with the horizontal with a constant speed.
(*d*) a cork floating on the surface of water.

5. Two masses of 10 kg and 5 kg are suspended from a rigid support as shown in figure. The system is pulled down with a force of 150 N attached to the lower mass. The string attached to the support breaks and the system accelerates downwards.

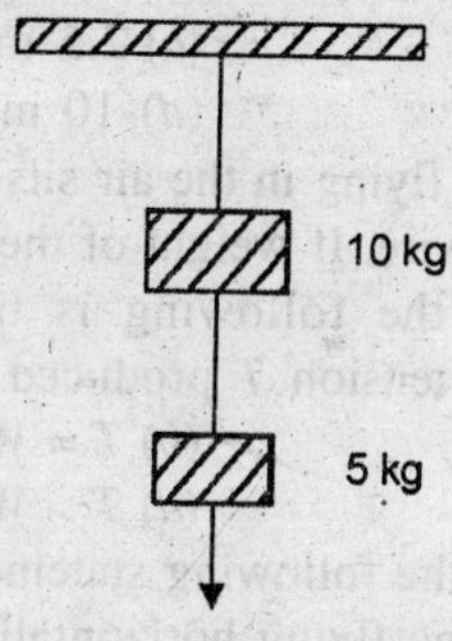

If the downward force continues to act, what is the acceleration of system?

(*a*) 20 ms^{-2} (*b*) 10 ms^{-2}
(*c*) 5 ms^{-2} (*d*) zero

6. Two masses M_1 and M_2 are attached to the ends of a string which passes over a pulley attached to the top of an inclined plane. The angle of inclination of the plane in θ. Take $g = 10\ ms^{-2}$

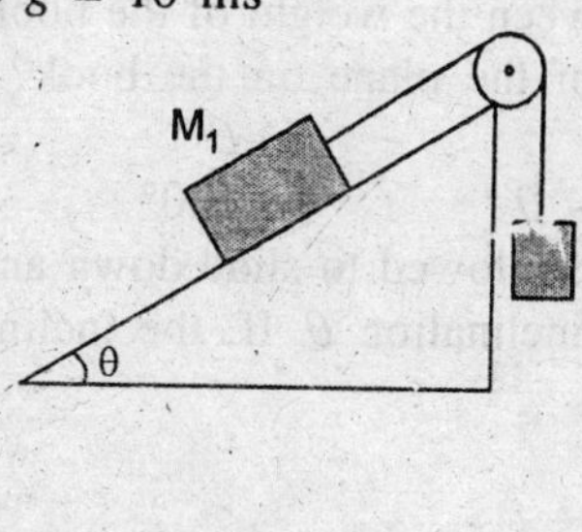

Given that $M_2 = 2M_1$ and M_2 moves vertically downwards with acceleration a. If the position of the masses are reversed the acceleration of M_2 down the inclined plane will be

(a) 2 a (b) a

(c) a/2 (d) none of the above

7. Two masses M_1 and M_2 are attached to the ends of string which passes over the pulley attached to the top of a double inclined plane. The angles of inclination of the inclined planes are α and β. See figure and answer the following questions. Take $g = 10\ ms^{-2}$.

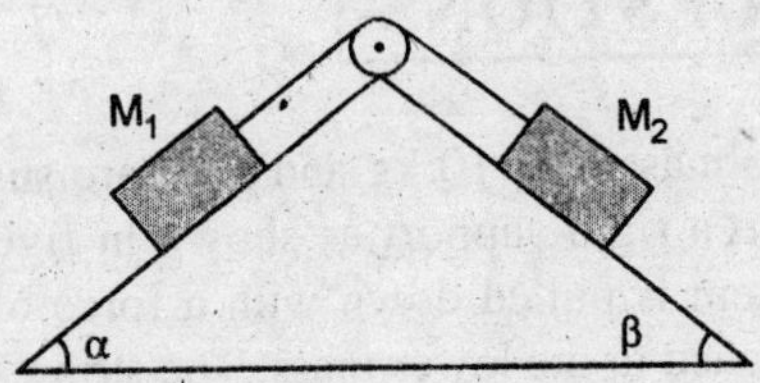

If $M_1 = M_2$ and $\alpha = \beta$, what is the acceleration of the system?

(a) zero (b) 2.5 ms^{-2}

(c) 5 ms^{-2} (d) 10 ms^{-2}

8. A sparrow flying in the air sits on a stretched telegraph wire. If weight of the sparrow is W, which of the following is true about the additional tension T produced in the wire

(a) $T = 0$ (b) $T = W$

(c) $T < W$ (d) $T > W$

9. Which of the following statements is true for the jet plane flying horizontally at very high speed?

(a) gravity does not act on it.

(b) upward thrust of the air balances gravity.

(c) thrust due to the gas ejected balances gravity.

(d) the flow of air around the plane causes lift that balances the gravity.

10. A book is lying on an inclined plane having inclination to the horizontal θ°. What is the angle between the weight of the book and the reaction of the plane on the book?

(a) 0° (b) θ°

(c) 180° – θ° (d) 180°

11. A block is allowed to slide down an inclined plane of inclination θ. If, the inclined plane is lying on the floor of a lift which is falling down with a retardation a, what will be the acceleration of the block with which it will slide down?

(a) $(g + a) \sin \theta$ (b) $(g - a) \sin \theta$

(c) $g \sin \theta + a$ (d) $g \sin \theta - a$

12. A rope of mass M is held vertically by fixing its upper end to a rigid support. What will be the tension in the rope at a distance x from the rigid support? Given that the length of the rope is L, and acceleration due to gravity is g

(a) Mg (b) $\frac{L-x}{L} \times Mg$

(c) $\frac{L}{L-x} \times Mg$ (d) $\frac{x}{L} \times Mg$

13. Three masses M_1, M_2 and M_3 are attached with strings passing over a fixed and frictionless pulley as shown in the figure.

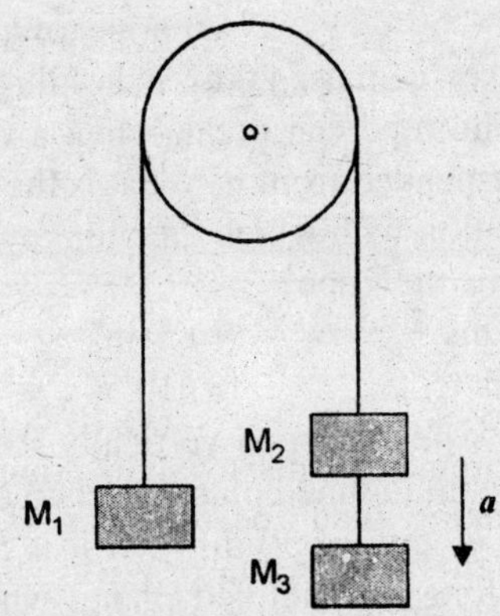

What is the acceleration of M_1?

(a) $\frac{M_3 + M_2 - M_1}{M_3 + M_2 + M_1}$ (b) $\frac{M_3 - M_2 + M_1}{M_3 + M_2 + M_1}$

(c) $\frac{M_3 - M_2 - M_1}{M_3 + M_2 + M_1}$ (d) $(M_2 + M_3 - M_1)g$

14. A body is at rest on the surface of the earth. Which of the following statements is correct?

(a) no force is acting on the body

(b) only weight of the body acts on it

(c) net downward force is equal to the net upward force

(d) none of the above

15. Two blocks of mass 4 kg and 6 kg are placed in contact with each other on a frictionless horizontal surface. See figure. The acceleration of the lighter mass will be

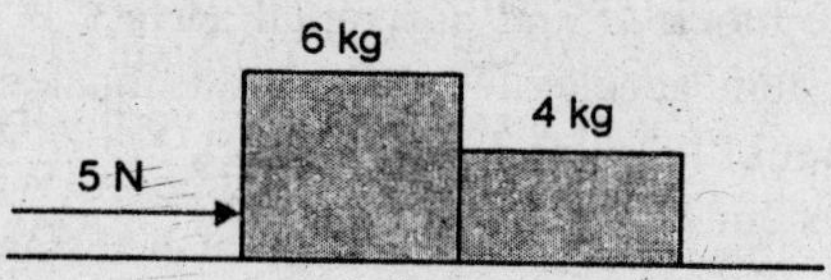

(*a*) 0.5 ms^{-2} (*b*) $\frac{5}{4}$ ms^{-2}

(*c*) $\frac{5}{6}$ ms^{-2} (*d*) none of the above

16. Three blocks masses 2 kg, 3 kg and 5 kg are connected to each other with light strings and are then placed on a smooth frictionless surface. See figure below. Let the system be pulled with a force F from the side of lighter mass so that it moves with an acceleration of 1 ms^{-2}. T_1 and T_2 denote the tensions in the other strings.

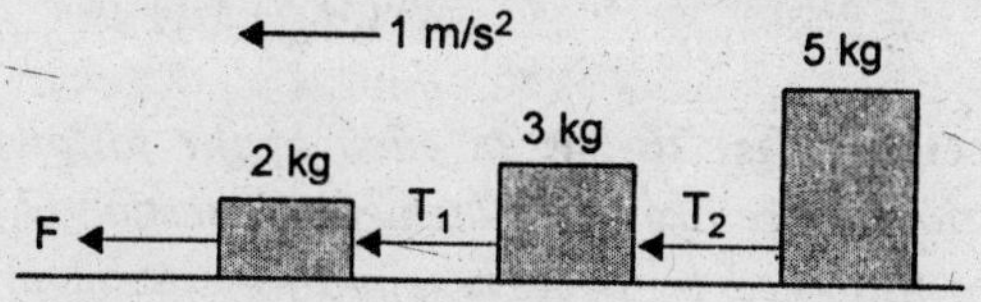

Then value of F is

(*a*) 2 N (*b*) 3 N

(*c*) 5 N (*d*) 10 N

17. A fireman wants to slide down a rope. The breaking load for the rope is 3/4th of the weight of the man. With what minimum acceleration should the fireman slide down? Acceleration due to gravity is g.

(*a*) $\frac{1}{4}g$ (*b*) $\frac{1}{2}g$

(*c*) $\frac{3}{4}g$ (*d*) zero

18. Two masses of 10 kg and 20 kg respectively are connected by a massless spring as shown in the figure. A pull of 100 N acts on the mass of 20 kg.

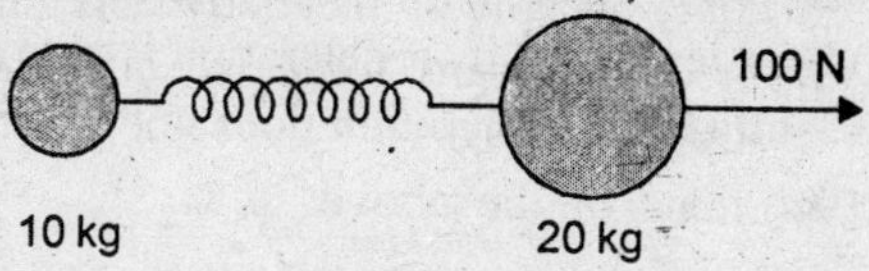

If the instantaneous acceleration of 10 kg mass be 2 ms^{-2}, the acceleration of 20 kg mass is

(*a*) 4 ms^{-2} (*b*) 3 ms^{-2}

(*c*) 2 ms^{-2} (*d*) 1 ms^{-2}

19. A bullet of mass m moving with a velocity v is fired into a large wooden block of masses M. If the bullet remains embeded on the wooden block, the velocity of the system will be

(*a*) $\frac{M}{M+m}v$ (*b*) $\frac{m}{M+m}v$

(*c*) $\frac{M}{M-m}v$ (*d*) $\frac{m}{M-m}v$

20. A particle of mass 2 kg is moving along a circular path of radius 1 m. If its angular speed is 2π rad s^{-1}, the centripetal force on it is

(*a*) 4π N (*b*) 8π N

(*c*) $4\pi^4$ N (*d*) $8\pi^2$ N

21. A mass M is suspended by a rope from a rigid support at P as shown in the figure. Another rope is tied at the end Q, and it is pulled horizontally with a force F. If the rope PQ makes angle θ with the vertical then the tension in the string PQ is

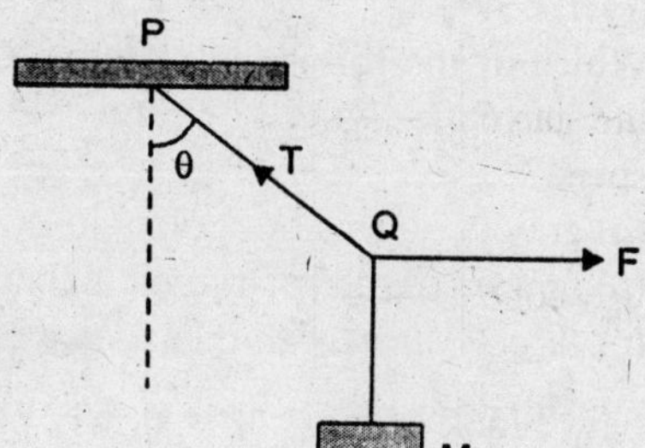

(*a*) $F \sin\theta$ (*b*) $F/\sin\theta$

(*c*) $F \cos\theta$ (*d*) $F/\cos\theta$

22. A block is suspended from a spring balance B, as shown in figure and it reads M_1. A beaker partially filled with water is placed on another spring balance B_2 as shown in figure (*b*) which reads M_2. If, we immerse the M_1 in the water in the beaker as shown in figure (*c*). What will be reading of balance B_1.

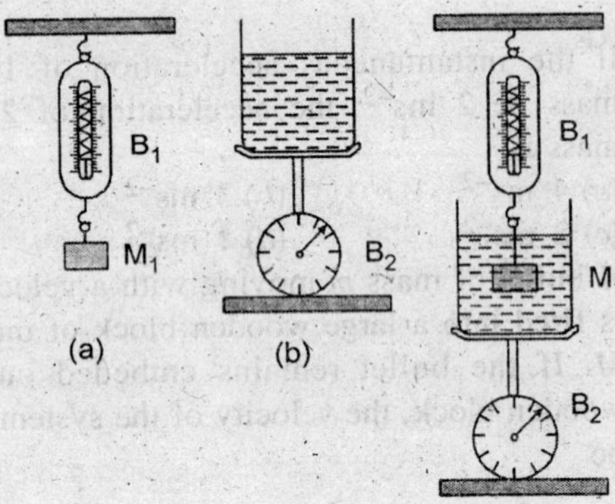

(a) M_1 (b) $\frac{M_1 + M_2}{2}$

(c) Less than M_1 (d) More than M_1

23. A block moves down a smooth inclined plane of inclination θ. Its velocity on reaching the bottom is v. If it slides down a rough inclined plane of same inclination its velocity on reaching the bottom is v/n, where n is a number greater than 0. The coefficient of friction μ is given by

(a) $\mu = \tan\theta\left[1 - \frac{1}{n^2}\right]$

(b) $\mu = \cot\theta\left[1 - \frac{1}{n^2}\right]$

(c) $\mu = \tan\theta\left[1 - \frac{1}{n^2}\right]^{\frac{1}{2}}$

(d) $\mu = \cot\theta\left[1 - \frac{1}{n^2}\right]^{\frac{1}{2}}$

24. A heavy block of mass M is slowly placed on a conveyer belt moving with a speed v. The coefficient of friction between the block and the belt is μ. Through what distance will the block slide on the belt?

(a) $\frac{v}{\mu g}$ (b) $\frac{v^2}{\mu g}$

(c) $\frac{v}{2\mu g}$ (d) $\frac{v^2}{2\mu g}$

25. Two blocks of mass M_1 and M_2 are connected with a string passing over a pulley as shown in the figure. The block M_1 lies on a horizontal surface. The coefficient of friction between the block M and horizontal surface is μ. The system accelerates. What additional mass m should be placed on the block M_1 so that the system does not accelerate?

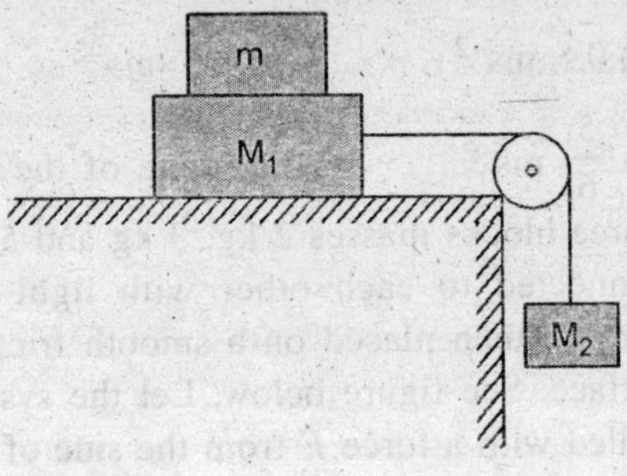

(a) $\frac{M_2 + M_1}{\mu}$ (b) $\frac{M_2}{\mu} - M_1$

(c) $M_2 - \frac{M_1}{\mu}$ (d) $(M_2 - M_1)\,\mu$

Directions Qs. 26-34: *In each of the following questions there are two statements Assertion (A) is given followed by a corresponding statement of Reason (R) just below it. Of the statements, mark the correct answer:*

(a) If both assertion and reason are true and reason is the correct explanation of assertion.

(b) If both assertion and reason are true but reason is not the correct explanation of assertion

(c) If assertion is true but reason is false

(d) If assertion is false but reason is true.

26. **Assertion (A):** Inertia is the property by virtue of which the body is unable to change by itself the state of rest only.

Reason (R): The bodies do not change their state unless acted upon by an unbalanced external force.

27. **Assertion (A):** If the net external force on the body is zero, then its acceleration is zero.

Reason (R): Acceleration does not depend on force.

28. **Assertion (A):** Newton's second law of motion gives the measurement of force.
Reason (R): According to Newton's second law of motion, force is directly proportional to the rate of change of momentum.

29. **Assertion (A):** Force is required to move a body uniformly along a circle.
Reason (R): When the motion is uniform, acceleration is zero.

30. **Assertion (A):** If two object of different masses have same momentum, the lighter body possess greater velocity.
Reason (R): For all bodies momentum always remains same.

31. **Assertion (A):** Aeroplanes always fly at low altitudes.
Reason (R): According to newton's third law of motion, for every action there is an equal and opposite reaction.

32. **Assertion (A):** No force is required by the body to remain in a state of uniform motion or at rest.
Reason (R): In uniform linear motion, acceleration has a finite value.

33. **Assertion (A):** Mass is a measure of inertia of the body in linear motion.
Reason (R): Greater the mass, greater is the force required to change its state of rest or of uniform motion.

34. **Assertion (A):** The slope of momentum versus time curve give us the force.
Reason (R): Acceleration is given by the rate of change of momentum.

ANSWERS

1	2	3	4	5	6	7	8	9	10
(b)	(a)	(b)	(b)	(a)	(d)	(a)	(d)	(d)	(c)
11	**12**	**13**	**14**	**15**	**16**	**17**	**18**	**19**	**20**
(a)	(b)	(a)	(c)	(a)	(d)	(a)	(a)	(b)	(d)
21	**22**	**23**	**24**	**25**	**26**	**27**	**28**	**29**	**30**
(b)	(c)	(a)	(d)	(b)	(d)	(c)	(a)	(c)	(c)
31	**32**	**33**	**34**						
(a)	(c)	(a)	(c)						

SOME SELECTED EXPLANATORY ANSWERS

1. The slope of the graph is tan 45° = 1. Hence, the lift is falling with acceleration $a = 1\ ms^{-2}$. Taking $g = 10\ ms^{-2}$, effective weight of the lift is $M(g - a) = M(10 - 1)$ newton = 9 M newton.

2. Let the monkey climb up with acceleration a. Then tension caused in the rope will be $T = M(g + a)$ where M is the mass of the monkey. Maximum value of T is $25 \times g$. Hence $25 \times g = 20(g + a)$.
Taking $g = 10\ ms^{-2}$, it gives $a = 2.5\ ms^{-2}$.

3. Let l be the length of the inclined plane and h be its height. Using $\upsilon^2 - \upsilon_0^2 = 2ax$, we find $h = 125$ m. And $l = h/\sin\theta = 250$ m. Acceleration down the inclined plane = $g \sin\theta$. So, velocity at the bottom will be 50 ms^{-1} directed along the plane. Hence, its horizontal component will be
$(50 \cos 30°)\ ms^{-1} = 25\sqrt{3}\ ms^{-1}$

4. When the body falls freely, it is accelerated downwards. Hence, a force acts on it.

5. Total force on the system will be weight of the system + 150 N = 300 N.
Hence acceleration of the system when force continues to act will be 300 N/15 kg = 20 ms^{-2} . = a (say).
Tension in the string will be $10 \times a = T + 10 \times g$. Which gives $T = 100$ N.
When the force stops acting, the system falls freely.

8. Let the angle subtended by the wire at the point where the sparrow sits be 2θ. Then $2T \cos \theta = w$. Since θ is very near to 90°, as the wire will sag by a small amount. Hence $T >> w$.

10. See figure. Angle between action and reaction is 180°. And that between reaction and weight is $180° - \theta$ °.

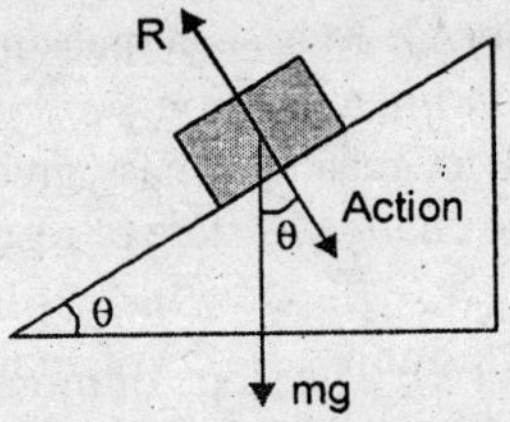

11. The effective value of the downward acceleration is $[g - (-a)] \sin \theta = (g + a) \sin\theta$.

12. The tension at a distance x from the support will be the weight of the rest of the rope having length $(L - x)$. Which comes out to be $(L - x)\, Mg/L$.

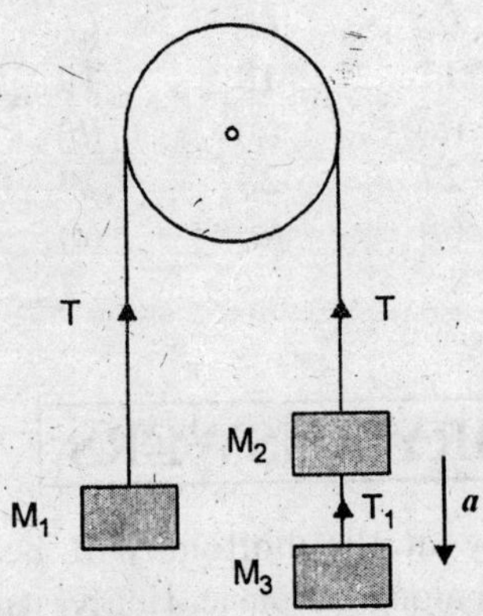

14. Impulse = change in momentum = 200 Ns. Hence the recoil velocity v = 200 Ns/10 kg = 20 ms^{-1}.

15. The net force on the body is zero. Weight of the body is balanced by the reaction of the ground.

16. Here $F = (2 + 3 + 5)$ kg × 1 ms^{-2} = 10 N.

17. Let the acceleration of the fireman be a. Then $Mg - R = M a$.

But maximum value of $R = \frac{3}{4}Mg$.

Hence $Mg - \frac{3}{4}Mg = M\, a$.

That is $a = g/4$.

18. Force acting on 10 kg mass = 10 × 2N = 20 N. The mass of 10 kg will pull the mass of 20 kg backwards with a force of 20 N. Hence net force on the mass of 20 kg = 100 N – 20 N = 80 N. Acceleration 80 N / 20 kg = 4 ms^{-2}.

19. Here $mv = (M + m) V$.

20. $F_c = MR\omega^2 = 2 \times 1 \times (2\pi)^2$ N

22. The reading in B_1 will decrease due to the up thrust caused by the buoyancy.

The component of the acceleration due to gravity along the plane will add up to the acceleration of the cabin.

23. Acceleration down the smooth inclined plane will be $g \sin \theta$. That on the rough inclined plane will be $(g \sin \theta - \mu g \cos \theta)$. If l be the length of the inclined plane, then

$$l = \frac{1}{2} g \sin \theta\, (t)^2$$

And $l = \frac{1}{2}\; g\, (\sin \theta - \mu \cos \theta)\, (n\, t)^2$

On solving we find $\mu = \tan \theta\, (1 - \frac{1}{n^2})$

We are using $v^2 - v^2_0 = 2ax$.

$v^2 = 2g \sin \theta \times 1$

$(v/n)^2 = 2g\, (\sin \theta - \mu \cos \theta) \times 1$.

On solving we again find the same value of μ.

24. The block will be acted by the force friction $= \mu\;\; Mg$. It will be causing acceleration $a = \mu g$. The block slides on the belt till its velocity becomes v. Using $v^2 - v^2_0 = 2ax$ we find $x = v^2/2\mu\, g$.

25. The motion will stop when the force of friction is equal to the weight of M_2. That is, when $M_2\, g = \mu\, (M_1 + m)\, g$.

4 WORK, POWER AND ENERGY

Work Done by a Constant Force

(i) The work done by a force is measured by the product of the applied force and the displacement of the body in the direction of the applied force.

If a force $\vec{F}$ acts on a body and produces a displacement $\vec{s}$ in the body in the direction of the force, then work done by the force is given by

$$W = Fs \qquad ...(1)$$

where F and s are the magnitudes of the force vector and the displacement vector in the direction of the applied force respectively.

(ii) If force $\vec{F}$ does not act along the direction of displacement $\vec{s}$, then work is calculated by resolving the force $\vec{F}$ into two mutually perpendicular components:

Hence, work is done only by component $\vec{F}_x$, given by

$$W = |\vec{F}_x|\,|\vec{s}| = (F\cos\theta)\,s = F\,s\cos\theta$$

(iii) Also, if $\vec{F}$ and $\vec{s}$ are perpendicular to each other, then

$$W = \vec{F}\cdot\vec{s} = F\,s\cos 90° = 0$$

Thus if the displacement of the body is perpendicular to the force, no work is done, e.g. when a satellite revolves around the earth, the direction of force applied by the earth is always perpendicular to the direction of motion of the satellite. Hence work done on the satellite by the centripetal force is zero.

Nature for Work Done

(i) Work done is given by

$$W = \vec{F}\cdot\vec{s}$$

According to this equation three possible situations arise regarding the sign or nature of work, which are given below:

(a) $W > 0$, if angle between $\vec{F}$ and $\vec{s}$ is **acute** or $\vec{F}$ and $\vec{s}$ are in the same direction.

(i) **Zero work done:**

(a) When a body is moved along a circular path of a string, the work done by the tension in the string is zero.

(b) When a person does not move from his position but he may be holding any amount of heavy load, the work done is zero.

(ii) **Negative work done:**

(a) When a body is made to slide over a rough surface, the work done by the frictional force is negative (as angle between $\vec{F}$ and $\vec{s}$ is 180°).

(b) When a gas filled in a cylinder fitted with a piston is compressed by applying some external force, then work done by external force is **negative**.

Power

(i) **The time rate of doing work is called power.** If W is the amount of work done in time, t, then average power, is

$$P_{av} = W/t. \quad \ldots(1)$$

(ii) Since power is the ratio of two scalars, the average power is a scalar. Power is defined as

$$P = \lim_{\Delta t \to 0} \frac{\Delta W}{\Delta t} = \frac{dW}{dt}$$

Now, $dW = \vec{F} \cdot d\vec{s}$

Hence $P = \vec{F} \cdot \frac{d\vec{s}}{dt}$

But, $\frac{d\vec{s}}{dt} = \vec{v}$, the instantaneous velocity of the particle.

Hence $P = \vec{F} \cdot \vec{v} \quad \ldots(2)$

Energy

(i) The energy of a body is defined as its capacity for doing work.

(a) it is measured in the same units as the work i.e. **joule** and **erg**

Kinetic Energy

(i) The energy possessed by a body by virtue of its motion is called kinetic energy.

Kinetic energy = ½ mv^2

Work-energy theorem

(i) If a body is initially at rest and a force $\vec{F}$ is applied on the body to move it through a distance $d\vec{s}$ along its own direction, then, work done

$$dW = \vec{F} \cdot d\vec{s} = F\,ds$$

$$W = \int_{v_1}^{v_2} mv\,dv = m\int_{v_1}^{v_2} v\,dv$$

$$= m\left[\frac{v^2}{2}\right]_{v_1}^{v_2} = \frac{1}{2}mv_2^2 - \frac{1}{2}mv_1^2$$

Potential Energy

Potential energy is the energy which a body has by virtue of its position in a conservative field and is represented by U.

Collision

Thus a collision will take place if either of the two bodies come in physical contact with each other or even when path of one body is affected by the force exerted due to the other.

(A) Elastic Collision: The collision, in which both the momentum and kinetic energy of the system remain conserved, are called elastic collision. Collision between two glass balls etc.

(B) Inelastic Collisions: The collisions in which only the momentum of the system is conserved but kinetic energy is not conserved are called inelastic collisions. Most of the collision in our day to day life are inelastic collisions. For example mud thrown on the wall is an example of perfectly inelastic collision.

Perfectly Inelastic Collision in One Dimension

(i) **Final Velocity:** When perfectly inelastic bodies moving along the same line collide, they stick to each other. Let m_1 and m_2 be the masses and v_1 and v_2 be their velocities before the collision and V be the common velocities of the bodies after the collision. According to law of conservation of linear momentum

$$m_1v_1 + m_2v_2 = m_1V + m_2V$$

$$\therefore \quad V = \frac{m_1v_1 + m_2v_2}{m_1 + m_2}$$

(ii) **Loss of Kinetic Energy:** The kinetic energy before the collision

$$= \frac{1}{2}m_1v_1^2 + \frac{1}{2}m_2v_2^2$$

Kinetic Energy after the collision

$$= \frac{1}{2}(m_1 + m_2)\,V^2$$

Loss of Kinetic Energy due to the collision

$$= \frac{1}{2}m_1v_1^2 + \frac{1}{2}m_2v_2^2 - \frac{1}{2}(m_1+m_2)V^2$$

$$= \frac{1}{2}\left[m_1v_1^2 + m_2v_2^2 - \frac{(m_1v_1+m_2v_2)^2}{(m_1+m_2)}\right]$$

$$= \frac{1}{2}\left[\frac{m_1m_2(v_1^2+v_2^2-2v_1v_2)}{(m_1+m_2)}\right]$$

$$= \frac{m_1m_2(v_1-v_2)^2}{2(m_1+m_2)}.$$

Obviously loss in kinetic energy is positive.

MULTIPLE CHOICE QUESTIONS

1. When a body is moving in a circular orbit, work done will be
(*a*) positive (*b*) negative
(*c*) zero (*d*) None of these

2. When a body moves with a constant speed along a circle
(*a*) no acceleration is produced in the body
(*b*) its velocity remains constant
(*c*) no work is done on it
(*d*) no force acts on the body

3. What is **F**. d**s**?
(*a*) Torque (*b*) Impulse
(*c*) Momentum (*d*) Work

4. A man lifts 20 kg mass upto 2m. The work done by him is
(*a*) 2 joule (*b*) zero
(*c*) none of these (*d*) 200 joule

5. A 60 kg weight is dragged on a horizontal surface by a rope. If coefficient of friction is μ = 0.5, the angle of rope with surface is 60° and g = 9.8 m/sec^2, then work done is
(*a*) 294 joule (*b*) 15 joule
(*c*) 588 joule (*d*) 197 joule

6. A horizontal force of 5 N is required to maintain a velocity of 2 m/s for a block of 10 kg mass sliding over a rough surface. The work done by this force in one minute is
(*a*) 600 J (*b*) 60 J
(*c*) 6 J (*d*) 6000 J

7. A body is acted upon by a force $\vec{F} = -12\hat{i} + 5\hat{j}$. Work done by the force in moving the body from the origin to a point, coordinates of which are (5, 4) is
(*a*) 153 (*b*) 80
(*c*) 73 (*d*) 7

8. Work done by a frictional force is
(*a*) negative (*b*) positive
(*c*) zero (*d*) none of these

9. A uniform chain of mass M and length L is lying on a smooth table with half of its length hanging vertically down the edge of the table. The work done in pulling the hanging part of the chain over the table is
(*a*) MgL / 2 (*b*) MgL / 4
(*c*) MgL / 8 (*d*) MgL / 16

10. A body moves a distance of 10 m along a straight line under the action of a 5 N force. If the work done is 25 J, then angle between the force and direction of motion of the body is
(*a*) 30° (*b*) 45°
(*c*) 60° (*d*) 75°

11. A body of mass 3 kg is under a force which causes a displacement in it, given by s = t^2/3 (in m). Find the work done by the force in 2 seconds
(*a*) 2 J (*b*) 3.8 J
(*c*) 5.2 J (*d*) 2.6 J

12. The work done in pulling up a block of wood weighing 2 kN for a length of 10 m on a smooth plane inclined at an angle of 15° with the horizontal is
(*a*) 4.36 kJ (*b*) 5.17 kJ
(*c*) 8.91 kJ (*d*) 9.82 kJ

13. A string of mass m is stretched by a length ℓ. What is the work done where g is the acceleration due to gravity?

(*a*) 1/2 $mg\ell$ (*b*) $mg\ell$
(*c*) 1/4 $mg\ell$ (*d*) 4 $mg\ell$

14. A mass is hanging on a wire, due to which its length increases by *l*, then work done is.
(*a*) 1/2 mgl (*b*) mgl
(*c*) zero (*d*) 2 mgl

15. If a particle is rotating in a circle, which of the following is true?
(*a*) No work is done
(*b*) Particle has no acceleration
(*c*) Velocity of particle is constant
(*d*) No force is acting on the particle

16. A car weighing 500 kg working against a resistance of 500 N, accelerates from rest to 20 ms^{-1} in 100 m. (g = 10 ms^{-2}). The work done by the engine of car is
(*a*) 1.0×15^5 J (*b*) 1.5×15^5 J
(*c*) 1.05×15^5 J (*d*) None of these

17. A force of $(3\hat{i}+4\hat{j})$ N acts on a body and displaces it by, $(3\hat{i}+4\hat{j})$ metres. The work done by the force is
(*a*) 10 J (*b*) 12 J
(*c*) 16 J (*d*) 25 J

18. Calculate the work done when a force $\vec{F}=2\hat{i}+3\hat{j}-5\hat{k}$ units acts on a body producing a displacement $\vec{s}=2\hat{i}+4\hat{j}+3\hat{k}$ units
(*a*) 30 units (*b*) 45 units
(*c*) 1 units (*d*) zero unit
(e) 15 units

19. A force $\vec{F}= 4\hat{i}+5\hat{j}$ newton displaces a particle through a distance S = $(3\vec{i}-\vec{j})$ metre, then find the work done
(*a*) – 7 joule (*b*) + 7 joule
(*c*) + 12 joule (*d*) – 12 joule

20. A mass M is lowered with the help of a string by a distance x at a constant acceleration g/2. The work done by the string will be
(*a*) Mgx (*b*) ½ Mgx
(*c*) ½ Mgx^2 (*d*) Mgx^2

21. A body moves a distance of 10 m under the action of force F = 10 N. If the work done is 25 J, the angle which the force makes with the direction of motion is
(*a*) 0° (*b*) 30°
(*c*) 60° (*d*) none of these

22. A uniform chain of length *l* and mass M lying on a smooth table and one-third of its length is hanging vertically down over the edge of the table. If g is the acceleration due to gravity, the work required to pull the hanging part on to the table is
(*a*) Mg*l* (*b*) Mg*l*/9
(*c*) Mg*l*/3 (*d*) Mg*l*/18

23. A force $\vec{F}=6\hat{i}+2\hat{j}-3\hat{k}$ acts on a particle and produces a displacement of $\vec{S}=2\hat{i}-3\hat{j}-x\hat{k}$. If the work done is zero, the value of x is
(*a*) – 2 (*b*) 1/2
(*c*) 2 (*d*) 3

24. A lorry and a car, moving with the same kinetic energy, are brought to rest by applying the same retarding force, then
(*a*) lorry will come to rest in a shorter distance
(*b*) both will come to rest in the same distance
(*c*) car will come to rest in a shorter distance
(*d*) none of these

25. A force $\vec{F}= 6\hat{i}+5\hat{j}$ newton acts and displaces the body through $\hat{S}=3\hat{i}-2\hat{k}$ meter. Then the work done will be
(*a*) 8 J (*b*) 28 J
(*c*) 18 J (*d*) 10 J

26. A force $\vec{F}=(5\hat{i}+4\hat{j})$ N acts on a body and produces a displacement $\vec{S}= 6\hat{i}-5\hat{j}+3\hat{k}$ m. The work done will be
(*a*) 10 J (*b*) 20 J
(*c*) 30 J (*d*) 40 J

27. A ball of mass M moves with velocity v and strikes with a wall having infinite mass and return with same velocity. Work done by the ball on the wall is
(*a*) zero (*b*) Mv J
(*c*) $\frac{M}{v}$J (*d*) $\frac{v}{M}$J

28. The work done by the centripetal force F when the body completes one rotation around the circle of radius R is

(*a*) 2πRF (*b*) 2 RF
(*c*) RF (*d*) Zero

29. A force acts on a 3 gm particle in such a way that the position of the particle as a function of time is given by $x = 3t - 4t^2 + t^3$, where x is in metres and t is in seconds. The work done during the first 4 second is

(*a*) 490 mJ (*b*) 450 mJ
(*c*) 912 mJ (*d*) 530 mJ

30. Natural length of a spring is 60 cm, and its spring constant is 4000 N/m. A mass of 20 kg is hung from it. The extension produced in the spring is, (Take g = 9.8 m/s^2)

(*a*) 4.9 cm (*b*) 0.49 cm
(*c*) 9.4 cm (*d*) 0.94 cm

Directions Qs. 31-37: *In each of the following questions there are two statements Assertion (A) is given followed by a corresponding statement of Reason (R) just below it. Of the statements, mark the correct answer:*

(*a*) If both assertion and reason are true and reason is the correct explanation of assertion.
(*b*) If both assertion and reason are true but reason is not the correct explanation of assertion
(*c*) If assertion is true but reason is false
(*d*) If assertion is false but reason is true.

31. **Assertion (A):** A person working on a horizontal road with a load on his head does no work.
Reason (R): No work is said to be done, if directions of force and displacement of load are perpendicular to each other.

32. **Assertion (A):** The work done during a round trip is always zero.
Reason (R): No force is required to move a body in its round trip.

33. **Assertion (A):** Work done by friction on a body in its round trip.
Reason (R): Work done is greater than zero, if angle between force and displacement is acute or both are in same direction.

34. **Assertion (A):** When a gas is allowed to expand, work done by gas is positive.
Reason (R): Force due to gaseous pressure and displacement (of piston) are in the same direction.

35. **Assertion (A):** A light body and heavy body have same momentum. Then they also have same kinetic energy.
Reason (R): Kinetic energy depend on mass of the body.

36. **Assertion (A):** The instantaneous power of an agent is measured as the dot product of instantaneous velocity and the force acting on it at that instant.
Reason (R): The unit of instantaneous power is watt.

37. **Assertion (A):** The change in kinetic energy of a particle is equal to the work done on it by the net force.
Reason (R): Change in kinetic energy of particle is equal to the work done only in case of a system of one particle.

ANSWERS

1	2	3	4	5	6	7	8	9	10
(*c*)	(*c*)	(*d*)	(*c*)	(*b*)	(*a*)	(*c*)	(*a*)	(*d*)	(*c*)
11	**12**	**13**	**14**	**15**	**16**	**17**	**18**	**19**	**20**
(*d*)	(*b*)	(*a*)	(*a*)	(*a*)	(*b*)	(*d*)	(*c*)	(*b*)	(*b*)
21	**22**	**23**	**24**	**25**	**26**	**27**	**28**	**29**	**30**
(*d*)	(*d*)	(*c*)	(*c*)	(*c*)	(*a*)	(*a*)	(*d*)	(*c*)	(*a*)
31	**32**	**33**	**34**	**35**	**36**	**37**			
(*a*)	(*c*)	(*d*)	(*a*)	(*d*)	(*b*)	(*b*)			

SOME SELECTED EXPLANATORY ANSWERS

1. Work done W = $\vec{F}.\vec{S}$. In this case, F and S are ⊥ to each other. This however, will not be zero if it is the case of non-uniform circular motion.

10. The work done W= $\vec{F}.\vec{d}$

W =Fd cos θ

⇒ 25 = 5 × 10 × cos θ

⇒ cos θ = 1/2

⇒ θ = 60°.

11. Work done = $\int F.ds$

$$= \int m\left(\frac{d^2s}{dt^2}\right)ds$$

Now, $s = \frac{t^2}{3}$

$$\therefore \frac{ds}{dt} = \frac{2t}{3} \text{ and } \frac{d^2s}{dt^2} = \frac{2}{3}$$

$$\therefore W = \int_{t=0}^{t=2} 3\times\frac{2}{3}\left(\frac{2t}{3}\right)dt$$

$$\text{or} \quad = \frac{4}{3}\left(\frac{t^2}{2}\right)_0^2 = \frac{4}{3}\times\frac{4}{2} = \frac{8}{3} = 2.6\text{ J}.$$

13. Work done = 1/2 × stress × strain
= 1/2 × mg × ℓ

17. Force (F) = $(3\hat{i}+4\hat{j})$N and displacement *(d)*

$= (3\hat{i}+4\hat{j})$metres

∴ Work done (W) $= \vec{F}.\vec{d} = (3\hat{i}+4\hat{j}).(3\hat{i}+4\hat{j})$

= [(3 × 3) + 16] = 9 + 16 = 25 J.

19. Work done = F.d

$= (4\hat{i}+5\hat{j})(3\hat{i}-\hat{j})$ = 12 − 5 = 7 joule

21. Fs cos θ = W ⇒ cos θ = 25/100 = 1/4.

25. W = F.S = $(6\hat{i}+5\hat{j}).(3\hat{i}-2\hat{k})$ = 18

26. Work done, W = $\vec{F}.\vec{S}$

$= (5\hat{i}+4\hat{j}).(6\hat{i}-5\hat{j}+3\hat{k})$

= 30 − 20= 10 J

27. Work done = F.S.
Here S = 0 ∴ Work done = 0

28. Work = force × displacement
for circular round, displacement = 0
∴ W = 0.

29. $x = 3t - 4t^2 + t^3$

$$\therefore \frac{dx}{dt} = 3 - 8t + 3t^2 \quad \therefore \frac{d^2x}{dt^2} = -8 + 6t$$

∴ Force, F = mass × acceleration

$= 3 \times 10^{-3} (-8 + 6t)$ N

∴ Work = $[3 \times 10^{-3} (-8 + 6t)] \times [3 - 8t + 3t^2]$

$= [3 \times 10^{-3} (-8 + 6 \times 4)] \times [3 - 8 \times 4 + 3 \times 4^2]$

$= 912 \times 10^{-3}$ J = 912 mJ.

30. Length of the spring = ℓ = 60 cm
spring constant of the spring = *k* = 4000 N/m
Load mass attached to the spring = 20 kg
Total load attached to the spring = 20 × 9.8 = 196 N
Let the extension produced = Δ*x*
Now, force applied by the spring = Downward force applied on the spring by the weight

$$K\Delta x = mg, \ \Delta x = \frac{mg}{K} = \frac{20\times 9.8}{4000}$$

Δ*x* = 0.049 *m*, Δ*x* = 4.9 cm

5 GRAVITATION

Kepler's Laws of Planetary Motion

(i) Every planet revolves round the sun in an elliptical orbit with sun at one of the foci.

(ii) The line joining the sun to the planet (radius vector) sweeps out equal area in equal interval of time, i.e., the areal velocity is constant

i.e., dA/dt = constant

Newtons's Universal Law of Gravitation

On the basis of Kepler's laws of planetary motion, Newton stated his famous law of gravitation. According to Newton's universal law of gravitation every two objects in the universe attract each other. The force of attraction is directly proportional to the product of masses and inversely proportional to the square of distance between the two masses, i.e., if two masses, m_1 and m_2 are separated from each other by a distance r then

$$F \propto \frac{m_1 m_2}{r^2} \quad \text{or} \quad F = G.\frac{m_1 m_2}{r^2}$$

where G = universal gravitational constant

$G = 6.67 \times 10^{-11}$ Newton m^{-2} kg^{-2}

$G = 6.67 \times 10^{-8}$ dyne cm^{-2} gm^{-2}

Acceleration Due to Gravity 'g'

Acceleration due to gravity is the acceleration produced in a body due to the force of gravity. Its value in S.I. system is 9.8 m/sec^2 and in F.P.S. system is 32 ft/sec^2.

Relation between g, G and M_e

Consider a body of mass m placed on the surface of earth of mass M_e. According to Newton's law of gravitation, the gravitational force acting on the body is

$$F = \frac{GM_e m}{R_e^2}$$

But $F = mg_e \quad \therefore \quad mg_e = \dfrac{GM_e m}{R_e^2}$

or $\quad g_e = \dfrac{GM_e}{R_e^2} \quad$ or $\quad GM_e = g_e R_e^2$

Variation in Value of 'g'

A. Variation of 'g' on the surface of earth. It is due to the following two reasons :

(a) Due to shape of earth. The earth is elliptical in shape. It is flatter at the poles and bulged out at the equator. Now, we know that $g \propto 1/R^2$, therefore the value of g at the equator is minimum and the value of g at the poles is maximum (Radius at poles is < Radius at equator).

(b) Due to rotation of earth. Since earth rotates about its axis hence every object placed on the surface of earth also rotates about the same axis. A centripetal force is required for circular motion. A component of true weight of the body provides this force and the other component is responsible for the observed value of g.

B. Variation of 'g' above the surface of earth

The value of 'g' decreases as we go above the surface of the earth.

(i) Consider a body of mass m at a height 'h' above the surface of earth. Thus, the gravitational force on the body is

$$F = \frac{GM_e m}{(R_e + h)^2}$$

If g' is the value of gravity at a height 'h' above the surface of earth, then

$$mg' = \frac{GM_e m}{(R_e + h)^2}$$

$$\Rightarrow \quad g' = \frac{GM_e}{(R_e + h)^2} = \frac{g_e R_e^2}{(R_e + h)^2}$$

or $$g' = \frac{g_e}{(1 + h/R_e)^2} = g_e \left[1 + \frac{h}{R_e}\right]^{-2}$$

where R_e is the radius of earth and M_e is the mass of earth, if $h << R_e$, then expanding binomially and neglecting higher powers of h/R_e, we get

$$g' = g_e\left(1 - \frac{2h}{R_e}\right)$$

$\therefore$ the decrease in the value of g on going up a height 'h' above the surface of earth,

$$\Delta g_e = \frac{2g_e h}{R_e}.$$

C. Variation of 'g' below the surface of earth

The value of 'g' decreases as we go below the surface of earth the effective mass of earth attracting the body decreases. Consider a body of mass m situated at a depth x from the surface of earth. Then the distance of the body from the centre of earth is $(R_e - x)$. Hence, there will be gravitational attraction between the mass m and a spherical region of radius $(R_e - x)$ of earth. If d_e is the density of earth, then the gravitational force on the body is

$$F = \frac{G\dfrac{4\pi(R_e - x)^3}{3} d_e . m}{(R_e - x)^2}$$

But, $$F = mg'$$

$$\therefore \quad mg' = \frac{4\pi G d_e (R_e - x)m}{3}$$

or $$g' = \frac{4\pi G d_e R_e}{3}\left(1 - \frac{x}{R_e}\right)$$

But $$\frac{4\pi G d_e R_e}{3} = g_e$$

hence, the value of 'g' at a depth x from the surface of earth is $g' = g\{1 - (x/R)\}$.

At the centre of earth $x = R$ hence $g = 0$.

The decrease in the value of 'g' at a depth 'x' below the surface of earth is $\Delta g = (gx/R)$.

Planets and Satellite

Planets. Planets are the heavenly bodies revolving round the sun. They are nine in number, e.g. Mercury, earth, venus, jupiter, mars, saturn, uranus and neptune.

Satellites. Satellites are the heavenly bodies revolving round the planets. They are of two types

(i) Natural satellites such as moon.

(ii) Artificial satellites launched by man such as Rohini, Aryabhatt etc.

Escape Velocity

The minimum velocity required to the body to enable it to escape away from the gravitational field of earth is called escape velocity.

Let the mass of the planet be M and its radius R, then the escape velocity from its surface will be

$$V_e = \sqrt{(2GM/R)} = \sqrt{(8\pi R^2 dG/3)}$$

or $$V_e = \sqrt{(2gR)}.$$

Escape velocity from the surface of earth is 11.2 km/sec.

MULTIPLE CHOICE QUESTIONS

1. The acceleration due to gravity at a height 1/20th of the radius of the earth above the earth surface is 9 ms^{-2}. Its value at a point at an equal distance below the surface of the earth in ms^{-2} is about
   (a) 8.5 (b) 9.5
   (c) 9.8 (d) 11.5
2. Mars has about 1/10th as much mass as the earth and half as great as diameter. The acceleration of falling body on Mars in ms^{-2} is about (acceleration due to gravity on earth = 9.8 ms^{-2})
   (a) 1.96 (b) 3.92
   (c) 4.9 (d) 9.8
3. Sun is about 330 times heavier and 100 times bigger in radius than earth. The ratio of mean density of the sun to that of earth is
   (a) 3.3×10^{-6} (b) 3.3×10^{-4}
   (c) 3.3×10^{-2} (d) 1.3
4. What will happen to the weight of the body at the south pole, if the earth stops rotating about its polar axis ?
   (a) no change
   (b) increases
   (c) decreases but does not become zero
   (d) reduces to zero
5. The radii of two planets are respectively R_1 and R_2 and their densities are respectively ρ_1 and ρ_2. The ratio of the accelerations due to gravity (g_1/g_2) at their surface is
   (a) $\frac{R_1\rho_2}{R_2\rho_1}$ (b) $\frac{R_1\rho_1}{R_2\rho_2}$
   (c) $\frac{\rho_1 R_2^2}{\rho_2 R_1^2}$ (d) $\frac{R_1 R_2}{\rho_2\, \rho_2}$
6. If one moves from the surface of earth to moon, what will be the effect on its weight ?
   (a) weight of a person decreases continuously with height from the surface of earth
   (b) weight of a person increases with height from the surface of earth
   (c) weight of a person first decreases with height and then increases with height from the surface of earth
   (d) weight of a person first increases with height and then decreases with height from the surface of earth
7. One goes from the centre of the earth to a distance two third the radius of the earth, where will the acceleration due to gravity be the greatest ?
   (a) at the centre of the earth
   (b) at a depth half the radius of the earth
   (c) at a depth one third the radius of the earth
   (d) at a depth two third the radius of the earth
8. Two spherical planets A and B have same mass but densities in the ratio 8 : 1. For these planets, the acceleration due to gravity at the surface of A to its value at the surface of B is
   (a) 1 : 4 (b) 1 : 2
   (c) 4 : 1 (d) 8 : 1
9. If both the masses and radius of the earth, each decrease by 50%, the acceleration due to gravity would
   (a) remain same
   (b) decrease by 50%
   (c) decrease by 100%
   (d) increase by 100%
10. A pendulum is taken to a place where the acceleration due to gravity is 1/6th that of earth's surface. How would its length be changed to keep its time period same ?
   (a) 1/12th (b) 1/6th
   (c) 1/3rd (d) none of the above
11. If R is the radius of the earth and g is acceleration due to gravity on the earth's surface, the mean density of earth is
   (a) $\frac{4\pi G}{3gR}$ (b) $\frac{3\pi G}{4gG}$
   (c) $\frac{3g}{4\pi RG}$ (d) $\frac{\pi Rg}{12G}$

12. At a distance 320 km above the surface of earth, the value of acceleration due to gravity will be lower than its value on the surface of the earth by nearly (radius of earth = 6400 km)
(a) 2% (b) 6%
(c) 10% (d) 14%

13. Gravitational intensity at a place is given by
(a) G (b) g
(c) mg (d) none of the above.

14. Intensity of gravitational field of earth is maximum at
(a) poles (b) equator
(c) centre of earth (d) same every where

15. The gravitational intensity at any point due to earth inside the earth is
(a) same as at surface
(b) increased as we go down
(c) same as at the centre of earth
(d) decreased as we go down.

16. If g is the acceleration due to gravity on the earth's surface, the gain in potential energy of the earth at a height equal to three times the radius R of the earth will be
(a) mgR (b) $\frac{1}{2}mgR$
(c) $\frac{1}{3}mgR$ (d) $\frac{3}{4}mgR$

17. Weight of a body is defined as the product of its mass and acceleration due to gravity i.e. $W = mg$. If weight of a body is 600 N and g is taken as 10 ms^{-2}, then the mass of the body will be
(a) 6 kg (b) 60 kg
(c) 600 kg (d) none of these

18. A mass m is placed at a point B in the gravitational field of mass M. When the mass m is brought from B to near point A, its gravitational potential energy will
(a) remain unchanged
(b) increase
(c) decrease
(d) become zero

19. Two bodies of masses 100 kg and 1000 kg are separated by a distance of 1 m. What is the intensity of gravitational field at the mid point of the line joining them ?
($G = 6.6 \times 10^{-11}$ Nm^2 kg^{-2})
(a) 2.4×10^{-9} N/kg (b) 2.4×10^{-8} N/kg
(c) 2.4×10^{-7} N/kg (d) 2.4×10^{-6} N/kg

20. A body of mass m rises to a height $h = R/5$ from the surface of earth, where R is the radius of earth. If g is the acceleration due to gravity at the surface of earth, the increase in potential energy is
(a) (4/5) mgh (b) (5/6) mgh
(c) (6/7) mgh (d) mgh

21. The gravitational potential difference between the surface of a planet and a point 20 m above it is 14 J kg^{-1}. The work done in moving a 2.0 kg mass by 8.0 m on a slope of 60° from the horizontal is equal to
(a) 7 J (b) 9.6 J
(c) 16 J (d) 32 J

22. In a gravitational field, if a body is bound, then total energy has
(a) positive value (b) negative value
(c) zero value (d) KE < PE

23. The gravitational field due to a mass distribution is $I = \frac{C}{x^2}$ in x direction. Here C is constant. Taking the gravitational potential to be zero at infinity, potential at x is
(a) $\frac{2C}{x}$ (b) $\frac{C}{x}$
(c) $\frac{2C}{x^2}$ (d) $\frac{C}{2x^2}$

24. A uniform ring of mass M and radius R is placed directly above a uniform sphere of mass 8 M and of same radius R. The centre of the ring is at a distance of $d = \sqrt{3}R$ from the centre of the sphere. The gravitational attraction between the sphere and the ring is
(a) $\frac{GM^2}{R^2}$ (b) $\frac{3GM^2}{2R^2}$
(c) $\frac{2GM^2}{2R^2}$ (d) $\frac{\sqrt{3}GM^2}{R^2}$

25. At what height from the surface of the earth, the total energy of the satellite is equal to its potential energy at a height of $2R$ from the surface of earth (R = radius of earth)

(a) $\frac{R}{4}$ (b) $\frac{R}{2}$

(c) $2R$ (d) $4R$

26. The self gravitational potential energy of a spherical shell of mass M and radius R is

(a) $-\frac{GM^2}{R}$ (b) $-\frac{GM^2}{2R}$

(c) $-\frac{GM^2}{4R}$ (d) $-\frac{3}{5}\frac{GM^2}{R}$

27. The period of a satellite in a circular orbit around a planet is independent of

(a) the mass of the planet

(b) the radius of the planet

(c) the mass of the satellite

(d) all the three parameters (a), (b) and (c)

28. The gravitational field due to a mass distribution is $E = K/x^3$ in the x-direction (K is a constant). Taking the gravitational potential to be zero at infinity, its value at a distance $x/\sqrt{2}$ is

(a) K/x (b) $K/2x$

(c) K/x^2 (d) $K/2x^2$

Directions Qs. 29-33: *In each of the following questions there are two statements Assertion (A) is given followed by a corresponding statement of Reason (R) just below it. Of the statements, mark the correct answer:*

(a) If both assertion and reason are true and reason is the correct explanation of assertion.

(b) If both assertion and reason are true but reason is not the correct explanation of assertion

(c) If assertion is true but reason is false

(d) If assertion is false but reason is true.

29. Assertion (A): Smaller the orbit of the planet around the sun, shorter is the time it takes to complete one revolution.
Reason (R): According to Kepler's third law of planetary motion, square of time period is proportional to cube of mean distance from sun.

30. Assertion (A): Gravitational force between two particles is negligibly small compared to the electrical force.
Reason (R): The electrical force is experienced by charged particles only.

31. Assertion (A): The universal gravitational constant is same as acceleration due to gravity.
Reason (R): Gravitational constant and acceleration due to gravity have same dimensional formula.

32. Assertion (A): The value of acceleration due to gravity does not depend upon mass of the body.
Reason (R): Acceleration due to gravity is a constant quantity.

33. Assertion (A): If a pendulum is suspended in a lift and lift is falling freely, then its time period becomes infinite.
Reason (R): Free falling body has acceleration equal to acceleration due to gravity.

ANSWERS

1	2	3	4	5	6	7	8	9	10
(b)	(b)	(b)	(a)	(b)	(c)	(c)	(c)	(d)	(b)
11	12	13	14	15	16	17	18	19	20
(c)	(c)	(b)	(a)	(d)	(d)	(b)	(c)	(c)	(b)
21	22	23	24	25	26	27	28	29	30
(b)	(b)	(b)	(d)	(b)	(b)	(c)	(c)	(a)	(b)
31	32	33							
(d)	(c)	(b)							

SOME SELECTED EXPLANATORY ANSWERS

1. Given $g_h = 9 = \frac{gR^2}{(R+R/20)^2} = \frac{20\times 20}{21\times 21}g$

or $g = \frac{9\times 21\times 21}{20\times 20}$

Now, $g_d = g\left(1-\frac{d}{R}\right)$

$= \frac{9\times 21\times 21}{20\times 20}\left[1-\frac{(R/20)}{R}\right]$

$= 9.5\ \text{ms}^{-2}$.

2. $g_m = GM_m/R^2_m$ and $g_e = GM_e/R^2_e$

$\therefore \frac{g_m}{g_e} = \frac{M_m}{M_e}\times\left(\frac{R_e}{R_m}\right)^2$

$= \frac{1}{10}\times(2)^2 = \frac{4}{10} = 0.4$

or $g_m = 0.4\ g_e$

$= 0.4\times 9.8 = 3.92\ \text{ms}^{-2}$

3. As mass, $M = \frac{4}{3}\pi R^3\rho$

or $\rho = \frac{3M}{4\pi R^3}$

$\therefore \frac{\rho_s}{\rho_e} = \frac{M_s}{M_e}\times\frac{R_e^3}{R_s^3} = 330\times\left(\frac{1}{100}\right)^3$

$= 3.3\times 10^{-4}$

4. As weight of body on pole = mg and g does not change at pole due to rotation of earth, so there is no change in the weight of body.

5. $g = \frac{GM}{R^2} = \frac{G}{R^2}\frac{4}{3}\pi R^3\rho = \frac{4}{3}\pi GR\rho$

So, $\frac{g_1}{g_2} = \frac{R_1\rho_1}{R_2\rho_2}$.

6. The gravitational attraction on a body due to earth decreases with height and increases due to moon. At a certain height, it becomes zero and with further increase in height, the gravitational attraction of moon becomes more than that of earth.

7. The acceleration due to gravity at a depth d inside the earth is

$$g' = g\left(1-\frac{d}{R}\right) = g\left(\frac{R-d}{R}\right) = g\cdot\frac{r}{R}$$

Where, $R-d = y$ = distance of a place from the centre of earth. Therefore $g' \propto r$.

8. Mass of two planets is same, so

$\frac{4}{3}\pi R_1^3\rho_1 = \frac{4}{3}\pi R_2^3\rho_2$

or $\frac{R_1}{R_2} = \left(\frac{\rho_2}{\rho_1}\right)^{1/3} = \left(\frac{1}{8}\right)^{1/3} = \frac{1}{2}$

$\frac{g_1}{g_2} = \frac{GM/R_1^2}{GM/R_2^2} = \left(\frac{R_2}{R_1}\right)^2 = (2)^2 = 4$

9. Here, $g = GM/R^2$ and

$g' = \frac{G(M/2)}{(R/2)^2} = \frac{2GM}{R^2} = 2g$

$\therefore$ % increase in $g = \left(\frac{g'-g}{g}\right)\times 100$

$= \left(\frac{2g-g}{g}\right)\times 100 = 100\%$.

10. $T = 2\pi\sqrt{\frac{l}{g}}$; for constant value of T, l/g is constant. If g becomes $g/6$ then l should be $l/6$ for the same value of T.

11. $g = \frac{GM}{R^2} = \frac{G}{R^2}\times\frac{4}{3}\pi R^3\rho = \frac{4}{3}\pi GR\rho$

or $\rho = \frac{3g}{4\pi RG}$

12. $\frac{g'}{g} = 1-\frac{2h}{R} = 1-\frac{2\times 320}{6400}$

$= 1-\frac{1}{10} = \frac{9}{10}$

$\therefore$ % decrease in $g = \left(\frac{g-g'}{g}\right)\times 100$

$= \frac{1}{10}\times 100 = 10\%$

13. Gravitational intensity at a point is equal to acceleration due to gravity at that point.

15. Gravitational intensity at a point inside the solid sphere is directly proportional to the distance of point from the centre of sphere *i.e.*, $I \propto r$.

16. Gravitational potential energy at any point at distance x from centre of earth is $E = -GMm/x$; At the surface of earth $x = R$, so,

$$E_1 = \frac{-GMm}{R} = -mgR$$

At height $3R$, from surface of earth. $x = 4R$

So, $$E_2 = \frac{-GMm}{4R} = -mgR/4.$$

Increase in potential energy

$$= \frac{-mgR}{4} + mgR = \frac{3}{4}mgR.$$

17. Weight $= 60 \times 10$ N $= mg = m \times 10$;

so $m = 60$ kg.

18. Gravitational potential energy of a body in the gravitational field, $E = \frac{-GMm}{r}$. When r decreases $-$ve value of E increases *i.e.* E decreases.

19. Resultant gravitational intensity at a mid point on the line joining the two mass bodies is

$$I = \frac{Gm_2}{(r/2)^2} - \frac{Gm_1}{(r/2)^2} = \frac{4G}{r^2}(m_2 - m_1)$$

$$= \frac{4 \times 6.6 \times 10^{-11}}{1^2}(1000 - 100)$$

$= 2.4 \times 10^{-7}$ N/kg.

20. Gravitational force on a body at a distance x from the centre of earth $F = \frac{GMm}{x^2}$. Work done,

$$W = \int_R^{R+h} F\,dx = \int_R^{R+h} \frac{GMm}{x^2}\,dx$$

$$= GMm\left[-\frac{1}{x}\right]_R^{R+h}$$

$$= mgR^2\left[\frac{1}{R} - \frac{1}{R+h}\right]$$

This work done appears as increase in potential energy ΔE_p. So

$$\Delta E_p = mgR^2\left[\frac{1}{R} - \frac{1}{R+h}\right]$$

$$= mg\,(5h)^2\left[\frac{1}{5h} - \frac{1}{6h}\right] = \frac{5}{6}mgh.$$

21. Gravitational intensity,

$$I = \frac{dV}{dx} = \frac{14}{20} = 0.7 \text{ N kg}^{-1}$$

Acceleration due to gravity,

$$g = I = 0.7 \text{ N kg}^{-1}$$

Work done under this field in displacing a body on a slope of 60° through a distances

$$= m\,(g \sin 60°)\,s$$

$$= 2 \times (0.7 \times \sqrt{3}/2) \times 8 = 9.6 \text{ J}$$

22. For a bound body, total energy of the body is negative. It is due to the fact that the energy is required to free the body from the gravitational attraction.

23. $$V = \int_\infty^x I\,dx = -\int_\infty^x \frac{C}{x^2}\,dx = \frac{C}{x}.$$

25. Total energy of a satellite at a height h

$$= \frac{GMm}{2(R+h)}$$

Potential energy at a height 2R from surface of

$$\text{Earth} = -\frac{GMm}{2R+R} = -\frac{GMm}{3R}$$

As per question, $-\frac{GMm}{2(R+h)} = -\frac{GMm}{3R}$.

or $h = \frac{R}{2}$.

26. When mass of the shell is m, its potential is

$$= -\frac{Gm}{R}$$

Work done in increasing the elementary mass dm is

$$dW = -\frac{Gm}{R}\,dm$$

Total work done = self potential energy

$$= \int_0^M -\frac{Gm}{R}\,dm = -\frac{GM^2}{2R}.$$

6 OSCILLATION

Oscillatory or Vibratory Motion

The movement of a body on either side of a point in definite time interval, is known as oscillatory or vibratory motion. Time taken by the body to complete one oscillation is known as time-period e.g., motion of a mass suspended from a spring, motion of simple pendulum etc.

Simple Harmonic Motion and its Characteristics

A motion in which the acceleration of the body is proportional to its displacement from the mean position and is always directed towards the mean position is known as the simple harmonic motion.

(i) Motion is on both sides of mean position. The maximum displacement on one side of mean position is called amplitude.

(ii) The body repeats its motion in a definite interval of time.

(iii) Acceleration is always proportional to the displacement and is directed opposite to it.

Acceleration $\propto -y$

(iv) The motion of foot of perpendicular dropped from the particle moving in a circle on the horizontal and vertical diameters is called S.H.M.

Equation of simple harmonic motion

The equation of S.H.M. is $y = A \sin \omega t$.

Velocity of a particle executing S.H.M.

We know $\bar{y} = A \sin \omega t$

$dy/dt = A\omega \cos \omega t$ or $v = \omega\sqrt{(A^2 - y^2)}$

At $y = 0$ (i.e. at mean position) velocity is maximum and $v_{max} = A\omega$.

At $y = A$ (i.e., at extreme position) velocity is minimum and $v_{min} = 0$.

Kinetic energy of a particle executing S.H.M.

Kinetic energy, $K = (1/2)\ mv^2 = (1/2)\ m\omega^2 (A^2 - y^2)$.

The kinetic energy is maximum at $y = 0$ i.e. at mean position and $K_{max} = (1/2)\ m\omega^2 A^2$, similarly the kinetic energy is minimum at $y = A$, i.e. at extreme position and $K_{min} = 0$.

Acceleration of a particle executing S.H.M.

$$\text{Acceleration} = \frac{d^2y}{dt^2} = -\omega^2 A \sin \omega t.$$

$\therefore$ Acceleration $= -\omega^2 y$

The negative sign indicate that the acceleration is directed towards the mean position.

Acceleration is maximum at extreme position (i.e., at $y = A$) and the maximum acceleration is given by

Maximum acceleration $= \omega^2 A$

Acceleration is minimum at $y = 0$, i.e. at mean position and the minimum acceleration is zero.

Potential energy of a particle executing S.H.M.

Maximum potential energy is at $y = \pm A$ and is given by $U_{max} = (½)\ m\omega^2 A^2$

Total energy of a particle executing S.H.M.

Total energy, $E =$ K.E. + P.E.

or $E = 2\pi^2 m A^2 n^2$.

Simple Pendulum

An ideal simple pendulum consists of a heavy point mass suspended from a rigid support by means of an elastic inextensible string.

The time-period of the simple pendulum is $T = 2\pi\sqrt{(l/g)}$.

Simple Harmonic Oscillations of Springs

Springs have a property that when they are stretched or compressed then a restoring force is immediately developed in them while try to bring them back to their initial state. If the expression produced in the spring is 'x', then

$$F = Kx$$

Where K is a constant known as spring constant or force constant or stiffness constant. K is numerically equal to $K = F/x = mg/x$.

Its units are Newton/metre. The time period of a spring loaded by mass 'm' is

$$T = 2\pi/\omega = 2\pi\sqrt{(I/C)}$$

MULTIPLE CHOICE QUESTIONS

1. The S.H.M. of a particle is given by the equation $y = 3 \sin \omega t + 4 \cos \omega t$. The amplitude is

(*a*) 7 (*b*) 12
(*c*) 1 (*d*) 5

2. Two mutually perpendicular simple harmonic vibrations have same amplitude, frequency and phase. When they superimpose, the resultant form of vibration will be

(*a*) a straight line (*b*) a parabola
(*c*) a circle (*d*) an ellipsoid.

3. A spring having a spring constant 'K' is loaded with a mass 'm'. The spring is cut into two equal parts and one of these is loaded again with the same mass. The new spring constant is

(*a*) K (*b*) $2K$
(*c*) $K/2$ (*d*) K^2

4. A mass m suspended from a massless spring stretches it by a distance. If pulled slightly more and let go, it will oscillate with a time period T where

(*a*) $T = 2\pi\sqrt{g/y}$ (*b*) $T = 2\pi\sqrt{my/g}$
(*c*) $T = 2\pi\sqrt{y/g}$ (*d*) $T = \sqrt{y/g}$

5. Two masses m_1 and m_2 are suspended together by a massless spring of force constant k. When the masses are in equilibrium, m_1 is removed without disturbing the system. The amplitude of oscillation is

(*a*) $\frac{m_1 g}{k}$ (*b*) $\frac{m_2 g}{k}$
(*c*) $\frac{(m_1 + m_2)g}{k}$ (*d*) $\frac{(m_1 - m_2)g}{k}$

6. For a particle executing simple harmonic motion, the kinetic energy K at an instant t is given by

$$K = K_0 \cos^2 \omega t$$

The maximum value of potential energy is

(*a*) K_0 (*b*) zero
(*c*) $\frac{K_0}{2}$ (*d*) not obtainable

7. A body executes S.H.M. with an amplitude a. The displacement of the body when its potential energy is one-third of its kinetic energy is

(*a*) $a/3$ (*b*) $a/2$
(*c*) $a/\sqrt{2}$ (*d*) $a/\sqrt{3}$

8. A particle is vibrating in S.H.M. with an amplitude of 4 cm. At what displacement from the equilibrium position is its energy half potential and half kinetic

(*a*) 1cm (*b*) $\sqrt{2}$ cm
(*c*) 2cm (*d*) $2\sqrt{2}$ cm

9. The work done by the string of a simple pendulum during one complete oscillation is equal to
   (*a*) total energy of the pendulum
   (*b*) kinetic energy of the pendulum
   (*c*) potential energy of the pendulum
   (*d*) zero

10. If a simple pendulum of length l has maximum angular displacement θ, then the maximum kinetic energy of bob of mass m is
   (*a*) $\frac{1}{2} m\left(\frac{l}{g}\right)$
   (*b*) $\frac{1}{2} \times \frac{mg}{l}$
   (*c*) $mgl \times (1 - \cos \theta)$
   (*d*) $\frac{1}{2} \times mgl \sin \theta$

11. A particle executes S.H.M. having time period T. Then the time period with which the potential energy changes is
   (*a*) T (*b*) $2T$
   (*c*) $T/2$ (*d*) ∞

12. For a simple Harmonic Oscillator, the potential energy is equal to kinetic energy
   (*a*) once during each cycle
   (*b*) twice during each cycle
   (*c*) when $x = a/2$
   (*d*) when $x = a$.

13. The displacement of a particle is represented by $y = 10 \sin (5t + \theta)$. The period of oscillation is
   (*a*) $2\pi/5s$ (*b*) $5/2\pi s$
   (*c*) $\pi/5s$ (*d*) $5/\pi s$

14. A particle is executing S.H.M. when its acceleration increases, its time period
   (*a*) decreases (*b*) increases
   (*c*) remains same (*d*) becomes half

15. For a particle in S.H.M. if the amplitude of displacement is a and the amplitude of velocity is v, the amplitude of acceleration is
   (*a*) va (*b*) $\frac{v^2}{a}$
   (*c*) $\frac{v^2}{2a}$ (*d*) $\frac{v}{a}$

16. A tunnel has been dug through the centre of the earth and a ball is released in it. It executes S.H.M. with time period
   (*a*) 42 minutes (*b*) 1 day
   (*c*) 1 hour (*d*) 84.6 minutes

Directions Qs. 17-20: *In each of the following questions there are two statements Assertion (A) is given followed by a corresponding statement of Reason (R) just below it. Of the statements, mark the correct answer:*
   (*a*) If both assertion and reason are true and reason is the correct explanation of assertion.
   (*b*) If both assertion and reason are true but reason is not the correct explanation of assertion
   (*c*) If assertion is true but reason is false
   (*d*) If assertion is false but reason is true.

17. **Assertion (A):** All oscillatory motions are necessarily periodic motion but all periodic motion are not oscillatory.
   Reason (R): Simple pendulum is an example of oscillatory motion.

18. **Assertion (A):** Simple harmonic motion is a uniform motion.
   Reason (R): Simple harmonic motion is the projection of uniform circular motion.

19. **Assertion (A):** Acceleration is proportional to the displacement. This condition is sufficient for simple harmonic motion.
   Reason (R): In simple harmonic motion direction of displacement is also considered.

20. **Assertion (A):** Sine and cosine functions are periodic functions.
   Reason (R): Sinusoidal functions repeat it values after a definite interval of time.

ANSWERS

1	2	3	4	5	6	7	8	9	10
(*d*)	(*a*)	(*b*)	(*c*)	(*a*)	(*a*)	(*b*)	(*d*)	(*d*)	(*c*)
11	**12**	**13**	**14**	**15**	**16**	**17**	**18**	**19**	**20**
(*c*)	(*b*)	(*a*)	(*c*)	(*b*)	(*d*)	(*b*)	(*d*)	(*a*)	(*a*)

SOME SELECTED EXPLANATORY ANSWERS

1. $y = 3 \sin \omega t + 4 \cos \omega t$

Let $3 = r \cos \theta$ and $4 = r \sin \theta$

then $y = r \cos \theta \sin \omega t + r \sin \theta \cos \omega t$

$= r \sin (\omega t + \theta)$

where $r = \sqrt{3^2 + 4^2} = 5.$

2. $x = a \sin \omega t$ and $y = a \sin \omega t$

So $x = y$. It will be a straight line equally inclined from X and Y-axis.

3. If k is the spring constant of each splitted part of spring, then $K = \dfrac{k \times k}{k + k} = \dfrac{k}{2}$ or $k = 2K$.

4. Restoring force $= Ky = mg$ or $K = mg/y$

$$T = 2\pi \sqrt{\frac{m}{K}} = 2\pi \sqrt{\frac{m}{mg/y}} = 2\pi \sqrt{\frac{y}{g}}.$$

5. For equilibrium of $(m_1 + m_2)$, $x_1 = \dfrac{(m_1 + m_2) g}{k}$

and for equilibrium of m_2, $x_2 = \dfrac{m_2 g}{k}$

$\therefore$ Amplitude of oscillation will be

$$A = x_1 - x_2 = \frac{(m_1 + m_2)g}{k} - \frac{m_2 g}{k} = \frac{m_1 g}{k}.$$

6. The maximum value of K.E. $= K_0 =$ maximum value of P.E.

7. $\frac{1}{2} m\omega^2 y^2 = \frac{1}{3} \times \frac{1}{2} m\omega^2 (a^2 - y^2)$

8. As per question,

$\frac{1}{2} m\omega^2 y^2 = \frac{1}{2} m\omega^2 (a^2 - y^2)$ or $2y^2 = a^2$

or $y = \dfrac{a}{\sqrt{2}}$ or $y = \dfrac{4}{\sqrt{2}} = 2\sqrt{2}$ cm.

10. In Figure. $AC = l \cos \theta$

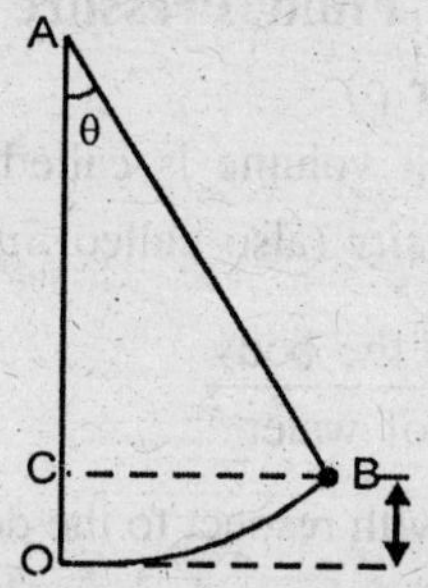

$\therefore$ $OC = OA - AC$

$= l - l \cos \theta = l(1 - \cos \theta)$

Max. K.E. of bob at O = Max. P.E. of bob at B

$= mg \times OC = mg\, l(1 - \cos \theta)$

12. Total energy = P.E. + K.E. When a particle executes S.H.M., there will be two position in each cycle where the P.E. is equal to K.E. of the body in S.H.M.

14. In S.H.M. $T = 2\pi \sqrt{\dfrac{\text{inertia factor}}{\text{spring factor}}}$, which is independent of acceleration.

15. $v_{max} = a\omega$ and (Acceleration)max $= \omega^2 a$

$$= \left(\frac{v}{a}\right)^2 a = \frac{v^2}{a}.$$

16. $T = 2\pi \sqrt{\dfrac{R}{g}} = 2\pi \sqrt{\dfrac{6.4 \times 10^6}{9.8}}$

$$= 2 \times \frac{22}{7} \times \frac{8 \times 10^3}{7 \times \sqrt{2}}$$

$$= \frac{\sqrt{2} \times 22 \times 8 \times 1000}{49 \times 60} \text{ min}$$

$= 84.6$ min.

7 FLUID MECHANICS

Buoyancy and Fluid Pressure

1. Density (d or ρ)

Mass per unit volume is called density.

Relative Density (also called *Specific Gravity*)

$$= \frac{\text{Density of the body}}{\text{Density of water}}$$

i.e., density with respect to the density of water.

2. Thrust (i.e., Total Downward Force) on a Plane Lamina Immersed in a Liquid

$= h\ d\ g\ a$

where

h = depth of the centroid of the plane lamina from liquid surface

d = density of liquid

a = Surface Area (of one face only) of the plane lamina

The centroid of the immersed portion of a floating body is called *Centre of Buoyancy*. It is same as the centre of gravity of the displaced liquid.

∴ The thrust on an immersed plane is independent of its inclination from free surface of the liquid.

3. Archimedes Principle

** If a body is immersed in a fluid, partly or fully, it experiences a loss in its weight. This loss of weight is equal to the weight of the liquid displaced, and is called 'Upthrust' or 'Buoyant Force'.

* Apparent weight of a body immersed in a fluid
= Actual weight of the body – upthrust
$= mg - (m/d)\ \rho g = mg\ (1 - \rho/d)$
where ρ = liquid density
d = density of the body

* If 2 fluids of densities d_1 and d_2 are mixed together, then the density d of the resultant mixture is
$= (d_1 + d_2)/\ (2)$
(when equal volumes are mixed)
$= 2d_1\ d_2/d_1 + d_2$
(when equal masses are mixed)

* Relative Density of a body =
wt. of body in air/wt. of liquid displaced.

4. Pascal's Law

In a confined fluid, externally applied pressure is transmitted uniformly in all directions. In a static fluid, force is transmitted at the speed of sound throughout the fluid and acts at right angles to any surface in or bounding the fluid. This principle is used in hydraulic jacks, hydraulic press, pneumatic tyres etc.

Motion in and of Fluids

Pressure Energy:

Pressure Energy per unit volume = p
Pressure Energy per unit mass = p/ρ

1. Bernoulli's Theorem

It is applicable to ideal fluid flowing in stream line in a tube of uniform or non-uniform cross-section. Ideal fluid means incompressible and non-viscous. It states that sum of P.E., K.E. and pressure energy *per unit volume* is constant over any cross-section, i.e.,

$$p+\frac{1}{2}\rho v^2+\rho gh = \text{constant} = \text{Energy per unit volume}$$

or $$\frac{p}{\rho g}+\frac{v^2}{2g}+h = \text{constant}$$

$$\frac{\rho}{\rho g} = \text{pressure head}$$

$$\frac{v^2}{2g} = \text{velocity head}$$

$$h = \text{gravitational head.}$$

If the tube is horizontal, then

$$p+\frac{1}{2}\rho v^2 = \text{constant}$$

Thus at places where velocity is high, pressure is low and vice-versa.

Thus, Bernoulli's Theorem is a consequence of conservation of energy.

2. Viscosity

Viscosity is the property of fluids because of which they offer resistance to any relative motion between their own respective layers or to the motion of any foreign body through them.

Coefficient of Viscosity: It is the force required per unit area to maintain unit velocity gradient between different layers of a flowing liquid.

According to *Newton's Formula for Viscous Flow*

$$F \propto A.\frac{dv}{dx}$$

F = Area of each layer

V = Velocity of any particular layer

x = Distance between layers

$\therefore$ $\frac{dv}{dx}$ = Velocity gradient, i.e., rate of change v as you go down (i.e., x changes)

$$\therefore \quad F = \eta . A.\frac{dv}{dx}$$

where η = constant of proportionality called "coefficient of viscosity".

Kinematic Viscosity (v) : It is the ratio of the coefficient of viscosity and the density of the fluid.

Stoke's Law : If a spherical ball of radius r moves with velocity v through a liquid of viscosity (i.e. coefficient of viscosity) η, then viscous force F acting on the body, opposing its motion is given by $F = 6\pi\eta rv$.

MULTIPLE CHOICE QUESTIONS

1. Soap helps in better cleaning of clothes because:
   (*a*) it reduces the surface tension of solution
   (*b*) it does not reduces the surface tension of solution
   (*c*) it absorbs the dirt
   (*d*) chemical of soaps change

2. If two soap bubbles of different radii are in communication with each other:
   (*a*) air flows from the larger bubble into the smaller one until the two bubbles are of equal size
   (*b*) the sizes of the bubbles remain unchanged
   (*c*) air flows from the smaller bubble into the larger one and larger bubble grows at the expense of the smaller one
   (*d*) air flows from the larger into the smaller bubble until the radius of the smaller one becomes equal to that of the larger one

3. More liquid rises in a thin tube because of:
   (*a*) larger value of radius
   (*b*) larger value of surface tension
   (*c*) small value of S.T.
   (*d*) small value of radius

4. When a soap bubble is charged:

(*a*) it contracts

(*b*) it expands

(*c*) it does not undergo any change in size

(*d*) None of these

5. A drop of oil is placed on the surface of water. Which of the following statement is correct?

(*a*) It will remain on it as a sphere

(*b*) It will spread as a thin layer

(*c*) It will partly be as spherical droplets and partly as thin film

(*d*) It will float as a distorted drop on the water surface

6. Small droplets of liquid are usually more spherical in shape than larger drops of the same liquid because:

(*a*) force of surface tension is equal and opposite to the force of gravity

(*b*) force of surface tension predominates the force of gravity

(*c*) force of gravity predominates the force of surface tension

(*d*) force of gravity and force of surface tension act in the same direction and are equal

7. It is easier to spray water to which soap is added, because

(*a*) addition of soap to water decreases surface tension of water

(*b*) addition of soap to water increases surface tension of water

(*c*) addition of soap to water makes surface tension of water zero

(*d*) addition of soap to water increases its density

8. A capillary tube is dipped in a water container, so that loss in weight of the capillary tube is:

(*a*) equal to the upward buoyant force

(*b*) less than the upward buoyant force

(*c*) more than the upward buoyant force

(*d*) half of the buoyant force

9. Soldering of two metals is possible because of the property of:

(*a*) viscosity (*b*) surface tension

(*c*) osmosis (*d*) cohesion

10. In order that a floating object be in a stable equilibrium, its centre of buoyancy should be:

(*a*) vertically above its centre of gravity

(*b*) vertically below its centre of gravity

(*c*) horizontally in line with its centre of gravity

(*d*) may be anywhere

11. A body is just floating in a liquid (their densities are euqal). If the body is slightly pressed down and released, it will:

(*a*) start oscillating

(*b*) sink to the bottom

(*c*) come back to the same position immediately

(*d*) will remain where it is left

12. A boat carrying a number of stones is floating in a water tank. If the stones are unloaded into water, the water level in the tank will:

(*a*) remain unchanged

(*b*) rise

(*c*) fall

(*d*) rise or fall depending on the number of stones unloaded

13. A bird, resting on the floor of an airtight box, which is being carried by a boy, starts flying. The boy will feel that the box is now:

(*a*) heavier

(*b*) lighter

(*c*) shows no change in weight

(*d*) lighter in the beginning and heavier afterwards

14. Pressure applied to enclosed fluid is:

(*a*) increased and applied to every part of the fluid

(*b*) diminished and transmitted to wall of container

(*c*) increased in proportion to the mass of the fluid and then transmitted

(*d*) transmitted unchanged to every portion of the fluid and walls of containing vessel.

15. A parrot is in a wire case, which is hanging from a spring balance. Initially the parrot sits in the cage and in the second instant the parrot flies inside the cage:

(*a*) the reading of the balance will be greater when the parrot flies in the cage

(*b*) the reading of the balance will be lesser when the parrot flies in the cage

(*c*) the reading will remain unchanged

(*d*) None of the above

16. If there were smaller gravitational effect which of the following forces do you think will alter in some respect?

(*a*) Viscous force (*b*) Archimedes' uplift

(*c*) Nuclear forces (*d*) Electrostatic forces

17. We live in:

(*a*) Stratosphere (*b*) Troposphere

(*c*) Ionosphere (*d*) Exosphere

18. A vessel containing water is in equilibrium on a beam balance when a man puts his finger into the water without touching the bottom or the sides of the vessel, the scale pan on which the vessel stands:

(*a*) will sink

(*b*) will rise

(*c*) will neither rise nor sink

(*d*) will be unpredictable

19. If you float on your back on water, your weight is:

(*a*) zero

(*b*) equal to your normal weight

(*c*) half to your normal weight

(*d*) greater than the weight of water you displace

20. A piece of ice is floating in a jar containing water, when the ice melts, the temperature of water falls from 4°C to 1°C. Then the level of water:

(*a*) rises (*b*) falls

(*c*) unchanged (*d*) None of these

21. A raft of wood (density 600 kg/m^3) of mass 120 kg floats in water. How much weight can be put on the raft to make it just sink?

(*a*) 120 kg (*b*) 200 kg

(*c*) 40 kg (*d*) 80 kg

22. Find the lifting force of a 4 kg cork life but in sea water if the density of cork and sea water are 0.2×10^3 kg/m^3 and 1.03×10^3 kg/m^3 respectively

(*a*) 163 N (*b*) 273 N

(*c*) 119 N (*d*) 289 N

23. Three liquids of densities *d*, 2*d* and 3*d* are mixed in equal volumes. Then the density of mixture is

(*a*) *d* (*b*) 2*d*

(*c*) 3*d* (*d*) 5*d*

24. Three liquids of densities *d*, 2*d* and 3*d* are mixed in equal proportions of weights. The relative density of the mixture is

(*a*) 11/7*d* (*b*) 18/11*d*

(*c*) 13/9*d* (*d*) 23/18*d*

25. Magnus effect is very near to the

(*a*) magnetic field

(*b*) electric field

(*c*) bernoulli's theorem

(*d*) magnetic effect of current

Directions Qs. 26-30: *In each of the following questions there are two statements Assertion (A) is given followed by a corresponding statement of Reason (R) just below it. Of the statements, mark the correct answer:*

(*a*) If both assertion and reason are true and reason is the correct explanation of assertion.

(*b*) If both assertion and reason are true but reason is not the correct explanation of assertion

(*c*) If assertion is true but reason is false

(*d*) If assertion is false but reason is true.

26. **Assertion (A):** Pascal law is the working principle of a hydraulic lift.

Reason (R): Pressure is equal to thrust per area.

27. **Assertion (A):** To float, a body must displace liquid whose weight is equal to the actual weight of the body.

Reason (R): The body will experiences no net downward force, in the case of floating.

28. **Assertion (A):** A man sitting in a boat which is floating on a pond. If the man drinks some water from the pond, the level of the water in the pond remain unchanged.
Reason (R): According to Archimede's principle the weight of water displaced by body is equal to the weight of the body, if the body is floating.

29. **Assertion (A):** It is better to wash the clothes in cold soap solution.
Reason (R): The surface tension of cold solution is more than the surface tension of hot solution.

30. **Assertion (A):** The velocity of flow of a liquid is smaller when pressure is larger and vice versa.
Reason (R): According to Bernoulli's theorem, for the steam line flow of an ideal liquid, the total energy per unit mass remains constant.

ANSWERS

1	2	3	4	5	6	7	8	9	10
(*a*)	(*c*)	(*d*)	(*b*)	(*b*)	(*b*)	(*b*)	(*b*)	(*d*)	(*a*)
11	**12**	**13**	**14**	**15**	**16**	**17**	**18**	**19**	**20**
(*b*)	(*c*)	(*c*)	(*d*)	(*b*)	(*b*)	(*b*)	(*a*)	(*a*)	(*a*)
21	**22**	**23**	**24**	**25**	**26**	**27**	**28**	**29**	**30**
(*d*)	(*a*)	(*b*)	(*b*)	(*c*)	(*b*)	(*b*)	(*a*)	(*d*)	(*a*)

SOME SELECTED EXPLANATORY ANSWERS

21. Volume of raft $= v =$ Mass/Density

$= 120/600 = 1/5\ \text{m}^3$

When raft is fully immersed in water, it displaces

$1/5\ \text{m}^3$ water so upthrust $= \frac{1}{5} \times 100 \times g\text{N}$

It x kg is the extra weight put on the raft to fully immerse in water then

$$(x + 120)y = \frac{1}{5} \times 1000 \times g$$

$$\therefore \quad x = 80\ \text{kg}$$

22. Volume of cork $= \dfrac{4}{0.2 \times 10^3}\ \text{m}^3$

weight of liquid displaced when cork is fully immersed in water

$$= \frac{4}{200} \times 1.03 \times 10^3 \times 9.8\ \text{N}$$

$$\text{Lifting force} = \frac{4}{200} \times 1.03 \times 10^3 \times 9.8 - 4 \times 9.8$$

$$= 163\ \text{N}$$

23. Let V be volume of each of three liquids of density d, $2d$ and $3d$. Then density of mixture

$$= \frac{\text{mass of mixture}}{\text{volume of mixture}}$$

$$= \frac{Vd + V(2d) + V \times 3d}{3V}$$

$$= \frac{6d}{3} = 2d.$$

24. Let M be mass of each of three liquids of density d, $2d$, $3d$, then density of mixture

$$= \frac{\text{mass of mixture}}{\text{volume of mixture}}$$

$$= \frac{3M}{\frac{M}{d} + \frac{M}{2d} + \frac{M}{3d}} = \frac{18d}{11}.$$

8 HEAT AND THERMODYNAMICS

Expansion of Solids

(i) There is expansion on heating and contraction while cooling in all solid substances. The expansion of solids on heating is found to be very small. The extent of expansion depends on the nature of the material and the temperature rise. The thermal expansion of solids is of three types namely linear expansion, superficial expansion and cubical expansion.

(a) Linear expansion: The expansion in the length of a solid is called linear expansion. Let L be the length of a solid in the form of a rod and ΔL be the increase in its length due to rise in temperature by ΔT, then ΔL is found to be proportional to

(i) its original length L and

(ii) rise in temperature ΔT.

i.e. $\Delta L \propto L\Delta T$

or $\Delta L = \alpha L\Delta T$

where α is called the coefficient of linear expansion.

(b) Superficial expansion: The expansion in area of a surface is called superficial expansion. Let A be the initial surface area of a solid and ΔA be the expansion or increase in surface area due to a temperature change ΔT, then increase in surface area

$$\Delta A \propto A\Delta T$$

or $\Delta A = \beta A\Delta T$

where β is called the coefficient of superficial expansion.

(c) Cubical expansion: The expansion in the volume of a solid is called cubical expansion. Let V be the initial volume of a solid and ΔV its expansion due to a temperature change ΔT, then the increase in volume

$$\Delta V \propto V\Delta T$$

or $\Delta V = \gamma V\Delta T$

where γ is called the coefficient of cubical expansion.

(ii) **Relation between α, β and γ :**

$$\alpha = (\beta/2) = (\gamma / 3)$$

Expansion of Liquids

(i) Liquids also expand on heating like solids. As a liquid has no definite shape it experiences volume expansion only.

(ii) A liquid is always taken in a vessel. If the liquid is heated, the vessel also gets heated equally and hence both the liquid and the vessel expand. But the expansion of liquid is much greater than that of the vessel.

(iii) The coefficient of apparent expansion of a liquid is defined as the ratio of the observed increase in volume per degree rise of temperature to the original volume of the liquid. It is represented by γ_a

$$\therefore \gamma_a = \frac{\text{Real increase in volume}}{\text{Original volume} \times \text{Rise in temperature}}$$

(iv) The coefficient of real expansion of a liquid is defined as the ratio of real increase in volume per degree rise of temperature to the original volume of the liquid. This is represented by γ_r.

$$\therefore \gamma_r = \frac{\text{Real increase in volume}}{\text{Original volume} \times \text{Rise in temperature}}$$

(v) Here, $\gamma_r = \gamma_a + \gamma_g$

where γ_g is the coefficient of cubical expansion of the material of the containing vessel.

Expansion of Gases

(i) **Boyle's Law:** At constant temperature the volume of a given mass of gas varies inversely as its pressure. Let P and V be the pressure and volume of a given mass of gas at a given temperature, then

$$P \propto (1/V) \quad \text{(when } T \text{ is constant)}$$

or $$PV = \text{constant}$$

or $$P_1V_1 = P_2V_2$$

(ii) **Charle's Law:** The volume of given mass of gas at constant pressure is directly proportional to the absolute temperature of the gas i.e.

$$V \propto T \quad \text{(when } P \text{ is constant)}$$

or $$\frac{V}{T} = \text{constant}$$

or $$\frac{V_1}{T_1} = \frac{V_2}{T_2}$$

Let V and V_0 be the volumes of a certain mass of gas at T K (corresponding to t °C) and 273 K (corresponding to 0 °C) respectively, then

$$\frac{V}{T} = \frac{V_0}{273}$$

or $$V = V_0\left(\frac{T}{273}\right)$$

$$= V_0\left(\frac{273+t}{273}\right)$$

$$= V_0\left(1+\frac{t}{273}\right)$$

Important conversion relations between different temperature scales:

(a) $$\frac{C}{100} = \frac{F-32}{180} = \frac{R}{80} = \frac{Ra-460}{212} = \frac{K-273}{100}$$

(b) $F = (9/5)\ C + 32$

(c) $C = (5/9)\ (F - 32) = K - 273$

(d) $K = C + 273 = (5/9)\ (F + 460)$

Heat and Mechanical Equivalent of Heat

As heat is just the energy in transit, its unit in SI is joule. Another unit of heat which is in wide use, is called calorie. **The amount of heat needed to increase the temperature of 1 g of water from 14.5°C to 15.5°C at a pressure of 1 atmosphere is called 1 calorie.**

Specific Heat

(i) **The specific heat of the material of a solid (or a liquid) is defined as the amount of heat required to raise the temperature of the unit mass of the solid (or the liquid) through 1°C or 1 K. It is represented by *c*.**

(ii) If an amount of heat ΔQ is required to raise the temperature of mass m of the solid through ΔT, then

$$c = \frac{\Delta Q}{m\Delta T}$$

Specific Latent Heat of Fusion and Vaporization

(i) Apart from increasing the temperature, heat supplied to a body may also cause a change in its phase such as solid to liquid or liquid to vapour.

(ii) During this process of melting or vaporization, the temperature remains constant. The quantity of heat required to melt a solid of mass m can be expressed as:

$$Q = mL_F$$

The constant L_F is known as specific latent heat of fusion.

Specific Heats of a Gas

Specific heat of a gas is defined in the following two ways:

(a) **Specific heat of a gas at constant pressure:** It is defined as the amount of heat required to raise the temperature of 1 g of a gas through

1°C at constant pressure. It is represented by c_p.

Molar specific heat of a gas at constant pressure: It is defined as the amount of heat required to raise the temperature of 1 mole of a gas through 1°C at constant pressure. It is represented by C_p.

Thus, $C_p = Mc_p$ ...(10)

where M = Molecular weight of the gas.

(b) Specific heat of a gas at constant volume: It is defined as the amount of heat required to raise the temperature of 1 g of a gas through 1°C at constant volume. It is represented by c_v.

Molar specific heat of a gas at constant volume: It is defined as the amount of heat required to raise the temperature of 1 mole of a gas through 1°C at constant volume. It is represented by C_v.

Thus, $C_v = Mc_v$

Definition of Some Terms Used in Thermodynamics

Thermodynamic process: A thermodynamic process takes place, if the thermodynamic variables of the system change with time. Thermodynamic process are of following types:

(a) Isothermal process: A thermodynamic process that takes place at constant temperature is called isothermal process.

(b) Adiabatic process: A thermodynamic process in which system is not allowed to exchange heat with surroundings is called adiabatic process.

(c) Isochoric process: A thermodynamic process that takes place at constant volume is called isochoric process.

(d) Isobaric process: A thermodynamic process that takes place at constant pressure is called isobaric process.

(e) Cyclic process: A thermodynamic process in which state of the system remains unchanged is called cyclic process.

First Law of Thermodynamics

Suppose, in a thermodynamic process, an amount of heat ΔQ is given to the gas and an amount of work ΔW is done by it i.e. the total energy of the gas must increase by an amount $(\Delta Q - \Delta W)$. As a result of this

(a) either the entire gas together with its cylinder may start moving or

(b) the internal energy of the gas molecules may increase.

If the energy does not appear in form *(a)* then this net energy $(\Delta Q - \Delta W)$ must be responsible for the increase in internal energy of the gas molecules. Hence, ΔU represents the change in internal energy of the gas, then we have

$$\Delta U = \Delta Q - \Delta W$$

or

$$\Delta Q = \Delta U + \Delta W$$

Transference of Heat

(i) Conduction : The process of heat transference from a place of higher temperature to a place of lower temperature by the aid of particles of the body without their migration (or their actual motion) is called conduction.

(ii) Convection : The process of heat transference from a place of higher temperature to a place of lower temperature by means of particles with their migrations from one place to another is callled convection.

(iii) Radiation : The process of heat transference from one place to another directly without the aid of intervening medium is called radiation. Heat from the sun reaches earth due to radiation without affecting the intervening medium.

Conduction

Coefficient of thermal conductivity

Let us consider a metal bar of cross-section area A in steady state and two sections at a distance d maintained at temperature θ_1 and θ_2 $(\theta_1 > \theta_2)$, then the quantity of heat Q passing across the two faces is given by

$$Q \propto \frac{A(\theta_1 - \theta_2)t}{d}$$

or

$$Q = \frac{KA(\theta_1 - \theta_2)t}{d} \quad \text{...(i)}$$

where the constant of proportionality K is called the coefficient of thermal conductivity while the quantity $(\theta_1 - \theta_2)/d$ the fall of temperature per unit distance along the direction of heat flow is called the temperature gradient and is denoted as $d\theta/dx$.

$$\therefore \quad Q = KA\frac{d\theta}{dx}t$$

or Heat flowing per second,

$$H = \frac{Q}{t} = KA\frac{d\theta}{dx} \quad \text{...(ii)}$$

From (ii),

$$K = \frac{H}{A(d\theta/dx)} \quad \text{...(iii)}$$

MULTIPLE CHOICE QUESTIONS

1. An iron ball is heated. The percentage increase will be the largest in
(a) diameter (b) surface area
(c) volume (d) density

2. Two rods of length L_1 and L_2 are made of materials having coefficients of linear expansions as α_1 and α_2 respectively. If $L_1 - L_2$ is independent of temperature, then which of the following relations is correct?
(a) $L_1\alpha_1 = L_2\alpha_2$ (b) $L_1\alpha_2 = L_2\alpha_1$
(c) $L_1L_2 = \alpha_1\alpha_2$ (d) none of the above

3. A process in which the heat content of the system remains constant is called
(a) isobaric (b) isochoric
(c) isothermal (d) none of the above

4. A process in which the volume remains constant is called
(a) isobaric (b) isochoric
(c) isothermal (d) none of the above

5. Which of the following is not the property of both heat and work.
(a) transient phenomenon
(b) boundary phenomenon
(c) path function
(d) exact differential

6. An ideal gas expands freely in a perfectly rigid and insulated cylinder. Which of the following parameters connected with it varies?
(a) temperature alone
(b) internal energy alone
(c) both temperature and internal energy
(d) neither temperature nor internal energy

7. When heat is added to a system, which of the following is not possible?
(a) internal energy of the system increases
(b) work is done by the system
(c) neither internal energy increases nor work is done by the system
(d) internal energy increases and also work is done by the system

8. The first law of thermodynamics is based on the law of conservation of
(a) energy (b) mass
(c) momentum (d) none of the above

9. Thermodynamics is concerned with
(a) measurement of heat
(b) transfer of heat
(c) change of state
(d) none of the above

10. Which of the following is the best container for gas during isothermal process?
(a) glass vessel (b) copper vessel
(c) wood vessel (d) thermos flask

11. In which of the following process all the three thermodynamic variables, that is pressure, volume and temperature can change?
(a) isobaric (b) isothermal
(c) isochoric (d) adiabatic

12. A sample of perfect gas is compressed isothermally to half its volume. If it is compressed adiabatically to the same volume, the final pressure of the gas will be
(a) more
(b) less
(c) same
(d) more or less depending on the initial temperature of the gas

13. Which of the following states of matter have two specific heats?
(a) solid (b) gas
(c) liquid (d) fluid

14. The internal energy of a perfect gas is
(*a*) wholly kinetic
(*b*) wholly potential
(*c*) sum of potential and kinetic energy of the molecules
(*d*) difference of kinetic and potential energies of the molecules

15. A sink, that is a system where heat is rejected is essential for the conversion of heat into work. From which law the above inference follows?
(*a*) zeroth (*b*) first
(*c*) second (*d*) third

16. What is the name of the device used to transfer heat from a subsystem at lower temperature to the surroundings at higher temperature?
(*a*) heat engine
(*b*) refrigerator
(*c*) Carnot's engine
(*d*) no such device exists

17. A Carnot engine has an efficiency 50% when its sink is at a temperature of 27°C. The temperature of the source is
(*a*) 273°C (*b*) 300°C
(*c*) 327°C (*d*) 373°C

18. The efficiency of the heat engine working between the freezing point and boiling point of water is very near to
(*a*) 50% (*b*) 25%
(*c*) 12.5% (*d*) 6.25%

19. The temperature determines the direction of net change of
(*a*) gross kinetic energy
(*b*) intermolecular kinetic energy
(*c*) gross potential energy
(*d*) intermolecular potential energy

20. What is heat?
(*a*) kinetic energy of molecules
(*b*) potential and kinetic energy of molecules
(*c*) energy in transit
(*d*) work done on the system.

21. Thermal equilibrium implies equality of
(*a*) energy (*b*) internal energy
(*c*) kinetic energy (*d*) none of the above

22. Heat required to melt 1 g of ice is 80 cal. A man melts 60 g of ice by chewing in one minute. His power is
(*a*) 4800 W (*b*) 336 W
(*c*) 1.33 W (*d*) 0.75 W

23. Which type of molecular motion does not contribute towards internal energy?
(*a*) translational (*b*) rotational
(*c*) vibrational (*d*) none of the above

24. A liquid with coefficient of cubical expansion γ is contained in a vessel having coefficient of linear expansion $\gamma/3$. When heated, what will happen to the level of the liquid in the vessel?
(*a*) it falls
(*b*) it rises
(*c*) it remains unchanged
(*d*) it may rise or fall depending upon the nature of container

25. A metallic bar is heated from 0°C to 100°C. The co-efficient of linear expansion is 10^{-5} K^{-1}. What will be the percentage increase in length
(*a*) 0.01% (*b*) 0.1 %
(*c*) 1% (*d*) 10 %

26. Coefficient of cubical expansion of water is zero at
(*a*) 0°C (*b*) 4°C
(*c*) 15.5°C (*d*) 100°C

27. If pressure on a gas is increased from P to $2P$, then its heat conductivity
(*a*) increases (*b*) decreases
(*c*) becomes zero (*d*) remains unchanged

28. In a room containing air, heat can go from one place to another by
(*a*) conduction (*b*) convection
(*c*) radiation (*d*) all these.

29. A piece of paper wrapped tightly on a wooden rod is found to get charred quickly when held over a flame compared to a similar piece of paper wrapped on brass rod. This is because
(*a*) wood contains large no. of free electrons
(*b*) wood is a good conductor of heat
(*c*) wood is a bad conductor of heat
(*d*) brass is a good conductor and wood is a bad conductor of heat

30. If l is length, A, the area of cross section and K, thermal conductvity, then the thermal resistance of the block is given by

(a) KlA

(b) l/KlA

(c) $l + KA$

(d) l/KA.

31. While measuring thermal conductivity of a liquid, we keep the upper part hot and lower part cold, so that

(a) radiation may start

(b) radiation may stop

(c) convection may start

(d) convection may be stopped

32. The phenomenon of black body radiation was explained statisfactorily by

(a) Kinetic theory

(b) Classical theory

(c) Theory of relativity

(d) Quantum theory

33. The ideal black body is

(a) lump of charcoal heated to a high temperature

(b) metal coated with a black dye

(c) glass surface coated with coal tar

(d) hollow enclosure blackened inside and having a small hole.

34. If temperature of a hot black body is raised by 5%, heat energy radiated would increase by

(a) 25%

(b) 15%

(c) 12.5%

(d) 21.55%

35. The maximum energy emitted by a body for a particular wavelength varies directly as

(a) square of temperature

(b) temperature

(c) fourth power of temperature

(d) reciprocal of temperature

Directions Qs. 36-40: *In each of the following questions there are two statements Assertion (A) is given followed by a corresponding statement of Reason (R) just below it. Of the statements, mark the correct answer:*

(a) If both assertion and reason are true and reason is the correct explanation of assertion.

(b) If both assertion and reason are true but reason is not the correct explanation of assertion

(c) If assertion is true but reason is false

(d) If assertion is false but reason is true.

36. **Assertion (A):** A brass disc is just fitted in a hole in a steel plate. The system must be cooled to loosen the disc from the hole.

Reason (R): The coefficient of linear expansion for brass is greater than the coefficient of linear expansion for steel.

37. **Assertion (A):** The coefficient of volume expansion has dimension K^{-1}.

Reason (R): The coefficient of volume expansion is defined as the change in volume per unit volume per unit change in temperature.

38. **Assertion (A):** We can not change the temperature of a body without giving (or taking) heat to (or from) it.

Reason (R): According to principle of conservation of energy, total energy of a system should remain conserved.

39. **Assertion (A):** A gas has a unique value of specific heat.

Reason (R): Specific heat is defined as the amount of heat required to raise the temperature of unit mass of the substance through unit degree.

40. **Assertion (A):** The specific heat of a gas in an adiabatic process is zero and in an isothermal process is infinite.

Reason (R): Specific heat of gas is directly proportional to change of heat in system and inversely proportional to change in temperature.

ANSWERS

1	2	3	4	5	6	7	8	9	10
(c)	(a)	(d)	(b)	(b)	(c)	(c)	(a)	(b)	(b)
11	12	13	14	15	16	17	18	19	20
(d)	(a)	(b)	(a)	(c)	(b)	(c)	(b)	(b)	(c)
21	22	23	24	25	26	27	28	29	30
(d)	(b)	(d)	(c)	(b)	(b)	(d)	(d)	(d)	(d)
31	32	33	34	35	36	37	38	39	40
(d)	(d)	(d)	(d)	(b)	(a)	(a)	(d)	(d)	(a)

SOME SELECTED EXPLANATORY ANSWERS

1. Volume coefficient is largest. Density decreases instead of increasing, when temperature increases.

2. Here $\Delta L_1 = \Delta L_2$.
That is $L_1 (1 + \alpha_1 T) = L_2 (1 + \alpha_2 T)$.
Which gives $L_1 - L_2 = (L_2\alpha_2 - L_1\alpha_1) T$.
Since $L_1 - L_2$ is independent of temperature.
Therefore: $L_2\alpha_2 - L_1\alpha_1 = 0$.

5. Work is exact differential but heat is not an exact differential.

6. It is adiabatic expansion. As the temperature falls, so does the internal energy.

7. $\Delta Q = \Delta W + \Delta U$. Hence when heat is added either ΔW or ΔU or both are non zero.

10. For isothermal process, heat generated should go out immediately. But in the adiabatic process, it should not go out at all.

12. For isothermal compression $P_2 = \dfrac{P_1 V_1}{V_2}$. For adiabatic compression $P_2 = P_1\left(\dfrac{V_1}{V_2}\right)^{\gamma}$. Since $\gamma > 1$, hence P_2 will be more as compared to that in case of isothermal compression.

13. Gases have C_p and C_v.

14. There is no inter molecular force for perfect gas. So, intermolecular potential energy does not exist.

17. $\eta = \dfrac{T_1 - T_2}{T_1} = 1 - \dfrac{T_2}{T_1}$. Hence

$$T_1 = \frac{T_2}{1-\eta} = \frac{300K}{\frac{1}{2}} = 600K = 327°C$$

18. $\eta = \dfrac{373-273}{373} = 0.268 = 26.8\%$

20. Heat appears as the transfer of energy from one body to the other due to the temperature difference.

21. Thermal equilibrium implies equality of temperature which depends on the average molecular kinetic energy.

22. $Q = ML = 60 \times 80 \text{ cal} = 4800 \text{ cal}$
$= 4800 \times 4.2 \text{ J}$.

$$P = \frac{Q}{t} = \frac{4800 \times 4.2 \text{ J}}{60 \text{ s}} = 336 \text{ W}$$

23. All types of molecular motions contribute to internal energy.

24. The coefficient of cubical expansion of the vessel is also y. Hence, change in volume of liquid is equal to the change on volume of the vessel.

25. $\Delta l / l = \alpha \Delta T = 10^{-5} \times 100 = 10^{-3}$
$= 10^{-3} \times 100\% = 0.1\%$

26. Density of water is maximum at 4°C. Volume decreases between 0°C and 4°C. Because water contracts within this range of temperature.

31. To measure conductivity, convection in liquid must be stopped.

32. Quantum theory alone could explain black body radiations.

9 WAVE MOTION

Progressive Wave

(i) If we generate the wave in a medium continuously, the particles of the medium oscillate continuously. In this situation, the disturbance produced in the medium is called a progressive wave.

(ii) Thus, **when a progressive wave propagates in a medium, then, at any instant, all the particles of the medium oscillate in the same way but the phase of oscillation changes from particle to particle.**

Two Types of Wave Motion or Mechanical Waves

The wave motion is classified into two catogories namely transverse and longitudinal wave motion, depending upon the direction of oscillation of the particle during the propagation of progressive mechanical waves in a medium.

(A) Transverse wave motion:

(i) If during the propagation of a mechanical wave in a medium, the particles of the medium vibrate about their mean positions in a direction perpendicular to the direction of propagation of wave, the wave motion is called the transverse wave motion.

(ii) All electromagnetic waves are transverse in character.

(iii) A transverse wave travels in the form of **Crests** and **Troughs**. A crest is a portion of the medium, which is highly raised above the normal positions of rest of the particles of the medium while a trough is a portion of the medium, which is highly depressed below the normal positions of rest of the particles of the medium.

(iv) The velocity of transverse wave motion is given by

$$v = \frac{\text{distance travelled}}{\text{time taken}}$$

Since, wavelength (λ) is the distance travelled by the wave in time T, in which the particle of the medium completes one vibration, we get

$$v = \frac{\lambda}{T} = \left(\frac{1}{T}\right)\lambda = n\lambda$$

where $n = 1/T$, is the frequency of the vibrating particle or the frequency of the wave motion i.e. the number of waves produced per second in the medium.

(B) Longitudinal wave motion

(i) If during the propagation of a mechanical wave in a medium, the particles of the medium oscillate parallel to the direction of propagation of the wave, the wave motion is called the longitudinal wave motion.

(ii) For example, when one end of a long spring is tied to a hook in a wall and the other end is moved forward and backward, then every turn of the spring osciilates parallel to the length of the spring and longitudinal waves propagate through the spring.

(iii) The states of compression and rarefaction continue to move forward along the length of spring. The distance travelled by the disturbance in the time, the particle of the medium completes one vibration is called wavelength. In case of a longitudinal wave it is found that it is equal to the distance between two successive compressions or rarefactions.

(iv) The velocity of longitudinal waves is also given by

$$v = n\lambda$$

(v) Longitudinal waves can be produced in all types of material medium (solid, liquid and gas). The waves produced in the air are always longitudinal. The wave produced in the interior of liquids are longitudinal, although transverse waves are possible only on the surface of liquids.

Equation of a Plane Progressive Wave

(i) If time is counted from the instant when the particle 1 at the origin just passes through mean position in positive direction, then the displacement of the particle after t seconds is given by

$$y = a \sin \omega t \quad \ldots(1)$$

where a is the amplitude of the oscillations of the particle and ω is the angular frequency ($\omega = 2\pi n$). We shall find here the displacement of particle 6 situated at a distance x from the origin at any time t.

(ii) But $\omega = 2\pi n$, hence we can also write equation (2) as

$$y = a \sin 2\pi n\, [t - (x/v)] \quad \ldots(3)$$

Now, $n = v/\lambda$, where λ is the wavelength. Therefore,

$$y = a \sin (2\pi v/\lambda)\, [t - (x/v)]$$

or $$y = a \sin (2\pi/\lambda)\, (vt - x) \quad \ldots(4)$$

Again, $v = n\lambda = \lambda/T$

where T is the time period.

Therefore,

$$y = a \sin (2\pi/\lambda)\, [\{(\lambda/T)t\} - x],$$

$$y = a \sin 2\pi\, [(t/T) - (x/\lambda)] \quad \ldots(5)$$

Refraction of Sound

When sound wave passes from one homogeneous medium to another homogeneous medium, it deviates from its path. This is called as phenomenon of refraction. If i and r are the angles of incidence and refraction, then according to Snell's law

$$\frac{\sin i}{\sin r} = \frac{v_1}{v_2} = \text{constant}$$

where v_1 and v_2 are the velocities of sound in first and second medium respectively.

Longitudinal Stationary Waves and Vibrations of air Columns

(i) Some musical instruments, in which sound is produced through vibrations of air columns are flute, whistle, clarinet etc. When air is blown at the mouth of a bottle, sound is produced due to vibrations of air column inside the bottle. Here we shall study the vibrations of air-columns in the cylindrical tubes of uniform diameter. Such sound-producing tubes are called organ pipes. The tube which is closed at one end and open at the other is called a closed organ pipe; and the tube which is open at both ends is called an open organ pipe.

(ii) **Vibrations of air column in closed organ pipe:** Only odd harmonics are produced in a closed organ pipe.

(iii) **Vibrations of air column in open organ pipe:** Both the even and odd harmonic are produced in an open pipe.

Doppler's Effect in Sound

If an observer is situated at a fixed distance from a sound-source, the frequency of sound heard by him is the same as produced by the source. But if the sound-source, or the observer, or both, are in state of motion; then the frequency of the sound appears to be changed to the observer. **This phenomenon of the apparent change in the frequency of the source due to a relative motion between the source and the observer is called as Doppler's effect.**

MULTIPLE CHOICE QUESTIONS

1. Under similar conditions of temperature and pressure, in which of the following gases the velocity of sound will be the largest?
(*a*) H_2 (*b*) N_2
(*c*) O_2 (*d*) CO_2

2. The velocity of sound at 0°C is 332 ms^{-1}. At what temperature will it be 664 ms^{-1}?
(*a*) 1092°C (*b*) 819°C
(*c*) 546°C (*d*) 273°C

3. Velocity of sound in air at the given temperature
(*a*) decreases with increase in pressure
(*b*) may increase on decrease with pressure
(*c*) is independent of the variation in pressure
(*d*) varies directly as the square root of pressure

4. A person hears the sound of jet aeroplane after it has passed over his head. The angle of the jet plane with the horizontal when the sound appears to be coming vertically downwards is 60°. If the velocity of the sound is *c*, then the velocity of the jet plane should be
(*a*) 2 *c* (*b*) $\frac{c}{\sqrt{3}}$
(*c*) $\sqrt{3}\,c$ (*d*) *c*

5. A student sees a jet plane flying from east to west. When the jet is seen just above his head, the sound of jet appears to reach him making angle 60° with the horizontal from east. If the velocity of sound is c, then that of the jet plane is
(*a*) 2 *c* (*b*) $\frac{\sqrt{3}}{2}c$
(*c*) $\frac{2}{\sqrt{3}}c$ (*d*) $\frac{c}{2}$

6. The velocity of sound in oxygen at NTP is *v*. The velocity of sound in hydrogen at NTP will be
(*a*) 4 *v* (*b*) $2\sqrt{2}$
(*c*) 2 *v* (*d*) none of the above

7. The velocity of sound in oxygen at NTP is *v*. The velocity of sound in helium at NTP should be
(*a*) 4 *v* (*b*) $2\sqrt{2}$
(*c*) 2 *v* (*d*) none of the above

8. The velocity of sound in air is 330 ms^{-1}. To increase the apparent frequency of the sound by 50%, the source should move towards the stationary observer with a velocity equal to
(*a*) 330 ms^{-1} (*b*) 220 ms^{-1}
(*c*) 165 ms^{-1} (*d*) 110 ms^{-1}

9. When both source and listener approach each other with a velocity equal to half the velocity of sound, the change in frequency of the sound as detected by the listener is
(*a*) zero (*b*) 25%
(*c*) 50% (*d*) none of the above

10. When both source and listener move in the same direction with a velocity equal to half the velocity of sound, the change in frequency of the sound as detected by the listener is
(*a*) zero (*b*) 25%
(*c*) 50% (*d*) none of the above

11. In which one of the following cases the jet aeroplane is flying at supersonic speed?
(*a*) its sound makes 30° with the vertical, when the jet is passing over the head of the listener
(*b*) its line of sight makes 30° with the vertical when sound appears to be coming vertically downwards
(*c*) its sound makes 60° with the vertical, when sound appears to be coming vertically downwards
(*d*) its line of sight makes 60° with the vertical, when sound appears to be coming vertically downwards

12. The wavelength of the sound produced by a source is 0.8 m. If the source moves towards the stationary listener at 32 ms^{-1}, what will be apparent wavelength of the sound? The velocity of sound is 320 ms^{-1}?
(*a*) 0.32 m (*b*) 0.40 m
(*c*) 0.72 m (*d*) 0.80 m

13. The difference between the apparent frequency of sound that is heard by the stationary listener, when the source approaches and then recedes with the same velocity is 5% of the natural frequency. Given that the velocity of sound is 320 ms^{-1}, the velocity of the source is

(*a*) 4 ms^{-1} (*b*) 8 ms^{-1}
(*c*) 12 ms^{-1} (*d*) 10 ms^{-1}

14. The pitch of the sound as detected by the observer is independent of

(*a*) original frequency
(*b*) the velocity of sound in the medium
(*c*) relative velocity of source and observer
(*d*) none of the above factors

15. The velocity of sound is 330 ms^{-1}. To hear a sound with 50% higher frequency, a listener should move towards the stationary source with a velocity equal to

(*a*) 330 ms^{-1} (*b*) 220 ms^{-1}
(*c*) 165 ms^{-1} (*d*) 110 ms^{-1}

16. Two factories are sounding their sirens at 800 Hz. A man goes from one factory to the other at a speed of 2 ms^{-1}. The velocity of sound is 320 ms^{-1}. The number of beats heard by the person in one second will be

(*a*) 2 (*b*) 4
(*c*) 8 (*d*) 10

17. A train is approaching the platform with a speed of 4 ms^{-1}. Another train is leaving the platform with the same speed. The velocity of sound is 320 ms^{-1}. If both the trains sound their whistles at frequency 280 hertz, the number of beats heard per second will be

(*a*) 6 (*b*) 7
(*c*) 8 (*d*) 10

18. A source of sound moves towards a stationary listener. The apparent pitch of the sound is found to be higher than its actual value. This happens because

(*a*) wavelength of sound waves decreases
(*b*) wavelength of sound waves increases
(*c*) the number of waves received by the listener increases
(*d*) the number of waves received by the listener decreases

19. Radio waves of wavelength λ are sent from a radar towards an aeroplane. If the aeroplane is moving towards the radar station, the wavelength of the radiowaves, received after reflection from the aeroplane will be

(*a*) λ
(*b*) more than λ
(*c*) less than λ
(*d*) more or less than λ, depending on the speed of aeroplane

20. In case of vibrating string, the frequency of the first overtone is equal to frequency of the

(*a*) fundamental note (*b*) first harmonic
(*c*) second harmonic (*d*) none of the above

21. A source of sound is moving towards a high wall with a speed of 20 ms^{-1}. The frequency of the sound produced by the source is 400 Hz. If the sound travels at 340 ms^{-1}, what will be the beat frequency heard by a man standing near the wall?

(*a*) zero (*b*) 2
(*c*) 5 (*d*) 10

22. The wavelength of light received from a galaxy is 10% greater than that received from identical source on the earth. The velocity of the galaxy relative to the earth is

(*a*) 3×10^8 ms^{-1} (*b*) 3×10^7 ms^{-1}
(*c*) 3×10^6 ms^{-1} (*d*) 3×10^5 ms^{-1}

23. A car sounding its horn at 480 Hz moves toward a high wall at a speed of 20 ms^{-1}. If the speed of sound is 340 ms^{-1}, the frequency of the reflected sound heard by the passenger sitting in the car will be nearest to

(*a*) 480 Hz (*b*) 510 Hz
(*c*) 540 Hz (*d*) 570 Hz

24. The sound of minimum frequency emitted by a vibrating string is not termed as

(*a*) first overtone
(*b*) first harmonic
(*c*) fundamental tone
(*d*) none of the above

25. If tension in the string is increased from 1 kN to 4 kN, other factors remaining unchanged, the frequency of the second harmonic will

(*a*) be halved
(*b*) remain unchanged
(*c*) be doubled
(*d*) become four times

Directions Qs. 26-30: *In each of the following questions there are two statements Assertion (A) is given followed by a corresponding statement of Reason (R) just below it. Of the statements, mark the correct answer:*

(*a*) If both assertion and reason are true and reason is the correct explanation of assertion.

(*b*) If both assertion and reason are true but reason is not the correct explanation of assertion

(*c*) If assertion is true but reason is false

(*d*) If assertion is false but reason is true.

26. **Assertion (A):** Two persons on the surface of moon cannot talk to each other.
Reason (R): There is no atmosphere on moon.

27. **Assertion (A):** Transverse waves are not produced in liquids and gases.
Reason (R): Light waves are transverse waves.

28. **Assertion (A):** Sound waves cannot propagate through vacuum but light waves can.
Reason (R): Sound waves cannot be polarised but light waves can be polarised.

29. **Assertion (A):** Sound passes through air in the form of longitudinal waves.
Reason (R): Longitudinal waves are easier to propagate.

30. **Assertion (A):** The velocity of sound increases with increass in humidity.
Reason (R): Velocity of sound does not depends upon the medium.

ANSWERS

1	2	3	4	5	6	7	8	9	10
(*a*)	(*b*)	(*c*)	(*b*)	(*d*)	(*a*)	(*b*)	(*d*)	(*d*)	(*a*)
11	**12**	**13**	**14**	**15**	**16**	**17**	**18**	**19**	**20**
(*d*)	(*c*)	(*b*)	(*d*)	(*c*)	(*d*)	(*b*)	(*a*)	(*c*)	(*c*)
21	**22**	**23**	**24**	**25**	**26**	**27**	**28**	**29**	**30**
(*a*)	(*b*)	(*c*)	(*a*)	(*c*)	(*a*)	(*d*)	(*b*)	(*c*)	(*c*)

SOME SELECTED EXPLANATORY ANSWERS

1. Velocity of sound $\propto \left[\dfrac{1}{\text{density}}\right]^{\frac{1}{2}}$.

2. $\dfrac{c_1}{c_2} = -\left[\dfrac{T_1}{T_2}\right]^{\frac{1}{2}}$.

Here $c_1 = 332\ \text{ms}^{-1}$, $T_1 = 0°\text{C} = 273\ \text{K}$,

$c_2 = 664\ \text{ms}^{-1}$. Hence $T_2 = 273 \times \left(\dfrac{c_2}{c_1}\right)^2$

$= 273 \times 4 = 1092\ \text{K}. = 819°\text{C}$.

5. Distance covered by sound to reach the student = (distance covered by jet) × cos 60°.

Hence velocity of jet $= \dfrac{\text{Velocity of sound}}{2}$

6. Density of oxygen is 16 times that of hydrogen. Hence velocity of sound in hydrogen $= [16]^{1/2}\ v = 4\ v$.

7. $\dfrac{c_1}{c_2} = \left[\dfrac{\rho_2}{\rho_1}\right]^{1/2}$.

Density of helium = 8 × density of oxygen.

8. $v' = 1.5\ v$. Hence $\dfrac{c}{c - u_s} v = 1.5\ v$.

This gives $u_s = c/3$.

9. Here $u_L = u_s = c/2$. Positive sign is to be taken with u_s and –ve sign with u_L. That is $u_L = -c/2$ and $u_s = +c/2$. Therefore $v' = \dfrac{c + c/2}{c - c/2} v = 3v$.

Hence $\Delta v = 2v = 200\%$.

10. $\nu' = \dfrac{c - u_L}{c - u_s}\nu$

Where +ve sign is taken with u_L and u_s if they are in the same direction as c, otherwise –ve sign is taken. Here $u_L = c/2$ and $u_s = c/2$. Positive sign is to be taken with both. Hence $\nu' = \nu$. So, there is no change in frequency.

11. Distance covered by sound to reach the person = (distance covered by jet) × cot 60°. Hence velocity of jet = velocity of sound/$\sqrt{3}$.

12. $\nu' = \dfrac{c}{c - u_s}\nu = \dfrac{320}{320-32}\nu = \dfrac{10}{9}\nu$.

Hence $\lambda' = \dfrac{c}{\nu'} = \dfrac{c}{\nu} \times \dfrac{\nu}{\nu'}$

$= \lambda \dfrac{\nu}{\nu'} = 0.8\,\text{m} \times \dfrac{9}{10} = 0.72$ m.

13. When source approaches $\nu'_a = \dfrac{c}{c - u_s}\nu$.

When source recedes $\nu'_r = \dfrac{c}{c + u_r}\nu$.

Also $(\nu'_a - \nu'_r) = \dfrac{2cu_s}{c^2 - u_s^2}\nu = \dfrac{5}{100}\nu$.

For 5% change in frequency $u_s < c$.

Hence u_s^2 may be neglected as compared to c^2.

Therefore we can write $\dfrac{2cu_s}{c^2} = \dfrac{1}{20}$.

15. $\nu' = 1.5\,\nu$. Hence $\dfrac{c + u_L}{c}\nu = 1.5\,\nu$.

This gives $u_L = c/2$.

16. $\nu'_a = \dfrac{c+u}{c}\nu$

$= \dfrac{320+2}{320} \times 800\text{ Hz} = \dfrac{322}{320} \times 800\text{ Hz}$.

$\nu'_r = \dfrac{c-u}{c}\nu$

$= \dfrac{320-2}{320} \times 800\text{ Hz} = \dfrac{318}{320} \times 800\text{ Hz}$

$\nu'_a - \nu'_r = \dfrac{322-318}{320} \times 800\text{ Hz}$

$= \dfrac{4 \times 800}{320}\text{ Hz} = 10\text{ Hz}$.

17. $\nu'_a = \dfrac{c}{c-u}\nu = \dfrac{320}{320-4} \times 280 = \dfrac{320}{316} \times 280\text{ Hz}$

$\nu'_r = \dfrac{c}{c+u}\nu = \dfrac{320}{320+4} \times 280 = \dfrac{320}{324} \times 280\text{ Hz}$

$\nu'_a - \nu'_r = 320 \times 280 \left[\dfrac{1}{316} - \dfrac{1}{324}\right]\text{ Hz}$

$= \dfrac{320 \times 280}{316 \times 324} \times 8\text{ Hz} = 7\text{ Hz}$.

19. Here $\nu' = \dfrac{c}{c-2u}\nu$. When the aeroplane approaches with velocity u. Hence ν' increases and so λ decreases.

21. Here the frequency of the direct waves received by the man near the wall = frequency of the reflected wave. Hence beat frequency is zero.

22. $\dfrac{\Delta\lambda}{\lambda} = \dfrac{v}{c}$. Here $\dfrac{\Delta\lambda}{\lambda} = 10\% = \dfrac{10}{100}$, and $c = 3 \times 10^8\text{ ms}^{-1}$, hence $v = 3 \times 10^8\text{ ms}^{-1} \times \dfrac{10}{100}$ $= 3 \times 10^7\text{ ms}^{-1}$.

23. $\nu' = \dfrac{c+u}{c-u}\nu = \dfrac{360}{320} \times 480 = 540$.

24. First over tone is the second harmonic.

25. $\nu \propto \sqrt{F}$. When tension F increases from 1 kN to 4 kN the frequency is doubled.

10 OPTICS

Introduction

Light falling on a surface, is divided into three parts: a part is reflected, a part is refracted and the remaining is absorbed.

Reflection. Depending on the nature of surface reflection is of two types.

(I) **Regular Reflection.** When the surface is smooth, the reflection is regular and follows the two laws:

(i) The incident ray, the reflected ray and normal to surface at the point of incidence all lie in the same plane.

(ii) The angle of incidence, i is equal to the angle of reflection, r

(II) **Diffuse Reflection.** When the surface is rough, the irregularities of surface act as reflecting centres and reflect waves irregularly in any direction. This phenmenon is called diffuse reflection or scattering. The objects in a room are visible due to duffuse reflection.

Mirrors. Generally reflecting surface is called mirror. Mirrors are of three types

(i) Plane mirror

(ii) Concave mirror

(iii) Convex mirror

(A) **Properties of a plane mirror :**

(i) The size of image is equal to the size of object.

(ii) The separation of image from mirror formed behind the mirror is equal to the separation of object from the mirror i.e. $OM = MI$.

(iii) The image is virtual, erect and laterally reversed.

(B) **Concave Mirror.** The mirror at which the reflection takes place at inner surface and whose outer surface is polished is called a concave mirror.

(C) **Convex Mirror.** The mirror at which reflection takes place from outer surface and whose inner surface is polished is called a convex mirror. The image formed is erect, virtual and behind the mirror.

Mirror Formula is

$$\frac{1}{f} = \frac{1}{v} + \frac{1}{u}$$

where u = distance of object from the pole of mirror

v = distance of image from the pole of mirror

$f = \frac{r}{2}$ focal length of mirror, r being radius of curvature of the mirror.

Lateral magnification produced by mirror,

$$m = \frac{I}{O} = -\frac{v}{u} = -\frac{f}{u-f}$$

where I = size of image and O = size of object.

Longitudinal magnification

$$mx = \frac{x_2}{x_1} = -\left(\frac{v}{u}\right)^2 = -\left(\frac{f}{u-f}\right)^2$$

The position and nature of image of an object placed at various positions from the mirror are given in the following table.

Mirror	Position of Object	Position and Nature of image
Concave	(i) At infinity	At principal focus in front of mirror, inverted, real very small.
	(ii) Beyond centre of curvature	Between focus and centre of curvature in front of mirror inverted, real, small.
	(iii) At centre of curvature	At centre of curvature in front of mirror, inverted, real, equal in size.
	(iv) Between focus and centre of curvature	Beyond centre of curvature in front of mirror, real, magnified and inverted.
	(v) At focus	At infinity in front of mirror, real highly magnified and inverted.
	(vi) Between pole and focus	Behind the mirror virtual, magnified and erect.
Convex	(i) At infinity	At principal focus, erect point image.
	(ii) Between pole and infinity	Between pole and focus behind the mirror, errect and small.

Refraction

There is a change in direction of ray when a ray of light falls on the boundary separating the two media. This penomenon is called *refraction*.

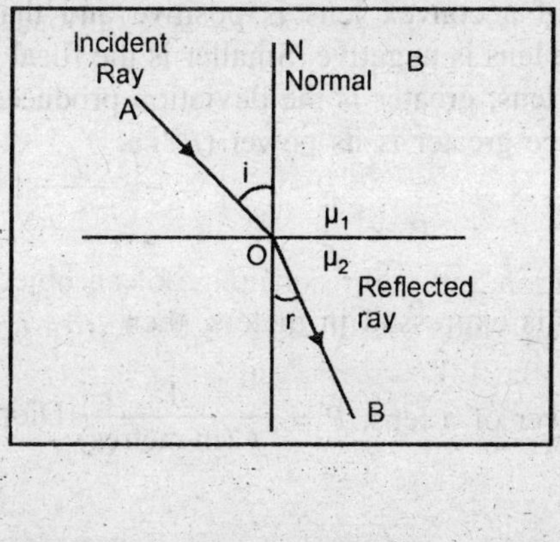

Laws of refraction

(i) The incident ray, the refracted ray and normal to the surface separating the two media all lie in the same plane.

(ii) Snell's law. For two media, the ratio of sine of angle of incidence of the sine of the angle of refraction is constant for a beam of particular wavelength i.e.

$$\frac{\sin i}{\sin r} = \text{Constant} = \frac{\mu_2}{\mu_1} = {}_1\mu_2$$

$$\Rightarrow \quad \mu_1 \sin i = \mu_2 \sin r \quad \text{...(i)}$$

where μ_1 and μ_2 are absolute refractive indices of media *I* and II respectively and ${}_1\mu_2$ is the refractive index of second medium with respect to *I* medium.

As light follows reversible path, we have

$$\frac{\sin r}{\sin i} = {}_2\mu_1 \quad \text{...(ii)}$$

Multiplying (i) and (ii), we get

$${}_2\mu_1 \times {}_1\mu_2 = 1 \text{ or } {}_2\mu_1 = \frac{1}{{}_1\mu_2} \quad \text{...(iii)}$$

Also the frequency of light, ν remains unchanged when passing from one medium to other and

$$\mu = \frac{c}{v} = \frac{\text{speed of light in vacuum or air}}{\text{Speed of light in air}}$$

$$= \frac{\nu\lambda_{\text{air}}}{\nu\lambda_{\text{medium}}} = \frac{\lambda_{\text{air}}}{\lambda_{\text{medium}}} \quad \text{...(iv)}$$

Where λ_{air} and λ_{medium} are wavelength of light in air and medium respectively.

$$\therefore \quad \frac{\sin i}{\sin r} = \frac{\mu_2}{\mu_1} = \frac{c/V_2}{c/V_1} = \frac{V_1}{V_2} = \frac{\lambda_1}{\lambda_2} \quad \text{...(v)}$$

Refraction Through a Prism

Suppose a monochromatic ray *EF* is incident on the face *PQ* of prism *PQR* of refracting angle *A* at angle of incidence i_1. The ray is refracted along *FG*, r_1 being angle of refraction. The ray *FG* is incident on the face *PR* at angle of incidence r_2 and is refracted in air along *GH*. Thus, *GH* is the emergent ray and i_2 the angle of emergence. The angle between incident ray *EF* and emergent ray *GH* is called *angle of deviation* δ.

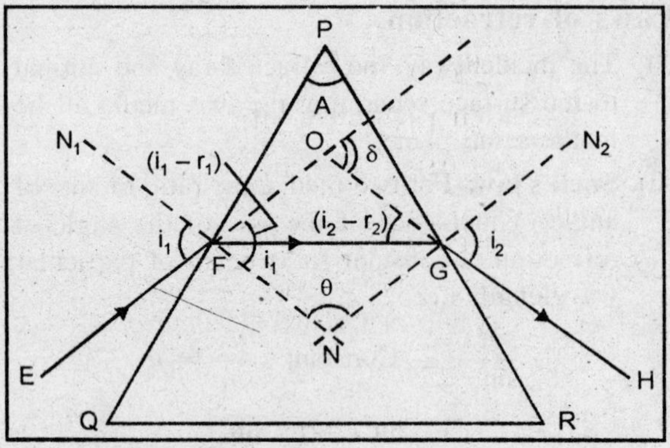

From triangle *OFG*

$$\delta = (i_1 - r_1) + (i_2 - r_2) = (i_1 + i_2) - (r_1 + r_2) \quad ...(i)$$

In quadrilateral *PFNG*,

$$A + 90^\circ + \theta + 90^\circ = 360^\circ$$

or $$A + \theta = 180^\circ \quad ...(ii)$$

In triangle *FGN*

$$r_1 + r_2 + \theta = 180^\circ \quad ...(iii)$$

Comparing equations, (ii) and (iii)

$$A = r_1 + r_2. \quad ...(iv)$$

∴ From (i) $\delta = i_1 + i_2 - A$

or $$i_1 + i_2 = A + \delta \quad ...(v)$$

Let μ be the refractive index of material of prism, then from Snell's law

$$\mu = \frac{\sin i_1}{\sin r_1} = \frac{\sin i_2}{\sin r_2}. \quad ...(vi)$$

Dispersion

Dispersion is the splitting of white light into constituent colours (VIBGYOR). White light falling on a prism is broken into constituent colours within the prism. Hence the emergent light has a number of coloured beams, the violet being deviated most and red the least in the visible region.

Thus the prism causes deviation as well as dispersion. If δv, δ_r and δ_y are the deviation caused by prism in violet, red and mean yellow colour, then angular dispersion = $\delta_v - \delta_r = (\mu_v - \mu_r) A$ for small angled prism.

Dispersive power,

$$\omega = \frac{\text{angular dispersion}}{\text{mean deviation}} = \frac{\delta_v - \delta_r}{\delta_y}$$

$$= \frac{(\mu_v - \mu_r)A}{(\mu_y - 1)A} = \frac{(\mu_v - \mu_r)}{\mu_y - 1}.$$

Refraction Through a Thin Lens

A transparent medium enclosed by two refracting surfaces, one of which at least must be spherical is called lens.

$$\frac{1}{f} = \frac{1}{v} - \frac{1}{u}$$

This is usual lens formula.

Ray diagram

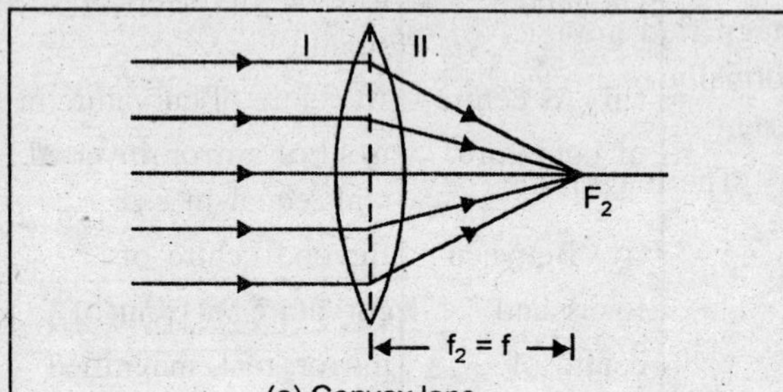

(a) Convex lens
$R_1 = (+)$, $R_2 = (-)$

(b) Concave lens
$R_1 = (-)$, $R_2 = (+)$

Power of Lens

The ability of a lens to deviate the rays towards the axis is called the power of a lens. Accordingly the power of a convex lens is positive and that of a concave lens is negative. Smaller is the focal length (*f*) of a lens, greater is the deviation produced by it and hence greater is its power (*P*) i.e.

$$P \propto \frac{1}{f}$$

If *f* is expressed in meters, then

$$\text{Power of a lens, } P = \frac{1}{f\text{ (in metres)}} \text{ Diopters.}$$

Magnification produced by a Lens

The ratio of size of image formed by a lens to that of the object is called magnification and is given by

$$m = \frac{\text{size of image } (I)}{\text{size of object } (O)} = \frac{v}{u} = \frac{f}{f+u}$$

Compound Microscope. To see distinct and magnified images of nearer tiny objects, not normally seen clearly by the eye a microscope is used.

The magnifying power of eye lens,

$$m_o = \frac{D}{v_e}\left(1+\frac{v_e}{f_e}\right) = \frac{D}{v_e} + \frac{D}{f_e}$$

∴ Magnifying power of microscope,

$$M = \frac{\text{angle substended by final image at eye } (\beta)}{\text{angle substended by the object at eye, when object placed at a distance of distinct vision } (\alpha)}$$

$$(= m_o \times m_e) = \frac{v_0}{u_0}\left(\frac{D}{v_e} + \frac{D}{f_e}\right)$$

The length of microscope,

L = separation between lenses = $v_0 + u_e$

MULTIPLE CHOICE QUESTIONS

1. Ray optics is valid when the characteristic dimensions are

(*a*) much larger than wavelength of light
(*b*) much smaller than wavelength of light
(*c*) of the order of one millimetre
(*d*) of same order as wavelength of light

2. The rectilinear propagation of light in a medium is due to

(*a*) its short wavelength
(*b*) its high frequency
(*c*) its high velocity
(*d*) the refractive index of medium.

3. A man who is 1.40 m tall stands in front of a mirror and can just see himself from head to toe. Assuming that his eyes are 0.14 m below the top of his head, the minimum height of the mirror is

(*a*) 1.40 m (*b*) 1.26 m
(*c*) 0.70 m (*d*) 0.63 m

4. A man runs towards a mirror at a rate of 6 m/s. If we assume the mirror to be at rest, his image will have a velocity

(*a*) + 12 m/s (*b*) + 6 m/s
(*c*) – 6 m/s (*d*) – 12 m/s

5. To get three images of a single object, one should have two plane mirrors at an angle of

(*a*) 50° (*b*) 60°
(*c*) 30° (*d*) 90°

6. A clock shows hours with the help of numbers. If it shows 8–20 in the mirror placed in front of it, then the real time is

(*a*) 4 – 40 (*b*) 3 – 40
(*c*) 4 – 20 (*d*) 4 –10

7. A man is 180 cm tall and his eyes are 10 cm below the top of his head. In order to see his entire height-right from toe to head he uses a plane mirror kept at a distance of 1 metre from him. The minimum height of the plane mirror required is

(*a*) 180 cm (*b*) 90 cm
(*c*) 85 cm (*d*) 170 cm

8. Plane mirrors A and B are kept at an angle θ with respect to each other. Light falls on A, is reflected, then falls on B and is reflected. The emergent ray is opposite to the incident direction. Then the angle θ is equal to

(*a*) 45° (*b*) 30°
(*c*) 60° (*d*) 90°

9. How many images will be formed if two mirrors are fitted on adjacent walls and one mirror on ceiling?

(*a*) 5 (*b*) 7
(*c*) 11 (*d*) 2

10. A thick mirror produces a number of images of an object. Which of the image is the brightest?

(*a*) first (*b*) second
(*c*) third (*d*) last one

11. Two plane mirrors are at right angle to each other. A man stands between them and combs his hair with his right hand. In how many the images will be seen using his right hand?
(*a*) none (*b*) 1
(*c*) 2 (*d*) 3

12. If an object is placed unsymmetrically between 2 plane mirrors, inclined at an angle of the 60°, then the total no. of images formed is
(*a*) 5 (*b*) 4
(*c*) 2 (*d*) infinite

13. A man moves towards a mirror with velocity 4 m/s. With what velocity its image moves towards him
(*a*) 8 m/s (*b*) 4 m/s
(*c*) 16 m/s (*d*) 2 m/s

14. Two mirrors are kept at 60° to each other and a body is placed at middle. The total number of images formed is
(*a*) six (*b*) four
(*c*) five (*d*) three

15. A pencil of light is incident on a plane mirror and after being reflected from it, forms a real image. Then the pencil, incident on the mirror is
(*a*) parallel (*b*) divergent
(*c*) convergent (*d*) statement is wrong

16. When a plane mirror is placed horizontally on level ground at a distance of 60 metres from the foot of a tower. The top of the tower and its image in the mirror subtend, at the eye close to mirror, an angle of 90°. The height of the tower is
(*a*) 30 m (*b*) 60 m
(*c*) 90 m (*d*) 120 m

17. Rays of light strike a horizontal plane mirror at an angle 45°. A second plane mirror is arranged at an angle θ with it. If the rays after reflection from the second mirror runs horizontally, i.e. parallel to the first mirror, then θ is
(*a*) 135° (*b*) 67° 30′
(*c*) 60° (*d*) 45° 30′

18. A plane mirror produces a magnification of
(*a*) – 1 (*b*) + 1
(*c*) zero (*d*) between 0 and + ∞

19. A man runs towards mirror with a speed of 5 m/s. What is the speed of his image?
(*a*) 5 m/s (*b*) 15 m/s
(*c*) 10 m/s (*d*) 20 m/s

20. A point source of light B is placed at a distance L in front of the centre of a mirror of width d hung vertically on a wall. A man walks in front of the mirror along a line parallel to the mirror at a distance 2L from it as shown. The greatest distance over which he can see the image of the light source in the mirror is

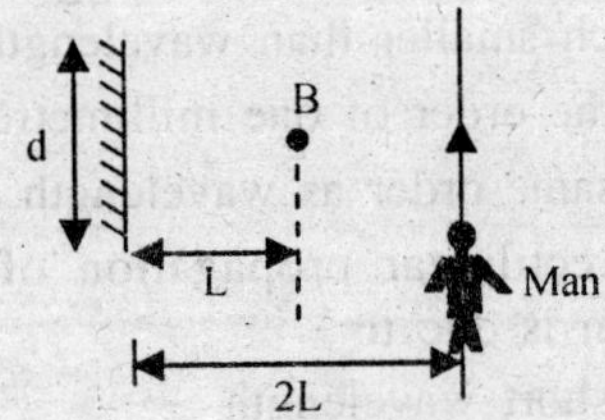

(*a*) d/2 (*b*) d
(*c*) 2d (*d*) 3d

21. The height of a man is 6 m. To see his full image the size of mirror is approximately
(*a*) 2.5 m (*b*) 3 m
(*c*) 6 m (*d*) 12 m

22. A light ray is incident normally on a plane mirror. The angle of reflection will be
(*a*) 135° (*b*) 90°
(*c*) 45° (*d*) 0°

23. A tall man of height 6 feet, want to see his full image. The required minimum length of the mirror will be
(*a*) 12 feet (*b*) 3 feet
(*c*) 6 feet (*d*) any length

24. A beam of light rays falls on a plane mirror and form a real image, so the incident rays are

(*a*) parallel (*b*) diverging
(*c*) converging (*d*) statement is false

25. A small object is placed 10 cm in front of a plane mirror. If you stand behind the object 30 cm from the mirror and look at its image, for what distance must you focus your eyes?

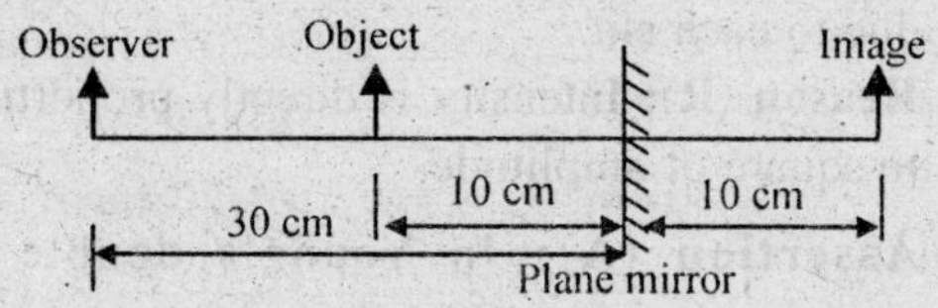

(*a*) 80 cm (*b*) 40 cm
(*c*) 20 cm (*d*) 60 cm

26. Evidence of the wave nature of light cannot be obtained from
(*a*) reflection (*b*) Doppler effect
(*c*) interference (*d*) diffraction

27. Light behave as
(*a*) particle
(*b*) wave
(*c*) both wave & particle
(*d*) sometimes particle & sometimes wave

28. Newton proposed his corpuscular theory of light on the basis of
(*a*) Planck's quantum theory
(*b*) rectilinear propagation of light
(*c*) refraction effect of photons
(*d*) dispersion effect of light

29. Which one of the following phenomena is not explained by Huygen's construction of wavefront?
(*a*) Refraction (*b*) Reflection
(*c*) Diffraction (*d*) Origin of spectra

30. Huygen's wave theory of light cannot explain
(*a*) diffraction (*b*) interference
(*c*) polarization (*d*) photoelectric effect.

31. According to Huygen's principle light is a form of
(*a*) particle (*b*) rays
(*c*) wave (*d*) none of these

32. Huygen's concept of secondary waves
(*a*) allows us to find the focal length of a thick lens
(*b*) is a geometrical method to find a wave front
(*c*) is used to determine the velocity of light
(*d*) is used to explain polarization.

33. According to Huygen's wave theory point on any wavefront may be regarded as
(*a*) a photon
(*b*) an electron
(*c*) a new source of wave
(*d*) neutron

34. Frequency of visible light is of the order of
(*a*) 10^{15} Hz (*b*) 10^{12} Hz
(*c*) 10^{8} Hz (*d*) 10^{5} Hz

35. Huygens's principle of secondary wavelets may be used to
(*a*) find the velocity of light in vacuum
(*b*) explain the particle behavior of light
(*c*) find the new position of the wavefront
(*d*) explain the photoelectric effect
(e) explain the scattering light

36. Light propagates rectilinearly because of its
(*a*) frequency
(*b*) velocity
(*c*) wavelength
(*d*) wave nature

37. The theory associated with secondary wavelets is
(*a*) Doppler's effect
(*b*) special theory of relativity
(*c*) Huygen's wave theory
(*d*) none of above

38. Huygen wave theory allows us to know
(*a*) the wavelength of the wave
(*b*) the velocity of the wave
(*c*) the amplitude of the wave
(*d*) the propagation of wave fronts

39. One important similarity between sound and light waves is that both

(*a*) can pass through even in the absence of any medium

(*b*) are transverse waves

(*c*) travel at the same speed in air

(*d*) can show interference effects

40. Two monochromatic light sources are said to be coherent if they have

(*a*) same frequency.

(*b*) constant relative phase difference.

(*c*) a phase difference changing with time.

(*d*) difference amplitude for their wave motion.

Directions Qs. 41-45: *In each of the following questions there are two statements Assertion (A) is given followed by a corresponding statement of Reason (R) just below it. Of the statements, mark the correct answer:*

(*a*) If both assertion and reason are true and reason is the correct explanation of assertion.

(*b*) If both assertion and reason are true but reason is not the correct explanation of assertion

(*c*) If assertion is true but reason is false

(*d*) If assertion is false but reason is true.

41. Assertion (A): When a light wave travels from a rarer to a denser medium, it loses speed. The reduction in speed does not imply a reduction in energy carried by the light wave.

Reason (R): The energy of a wave is proportional to velocity of wave.

42. Assertion (A): The maximum intensity in interference pattern is four times the intensity due to each slit.

Reason (R): Intensity is directly proportional to square of amplitude.

43. Assertion (A): In Young's double slit experiment, we observe an interference pattern on the screen if both the slits are illuminated by two bulbs of same power.

Reason (R): The interference pattern is observed when source is monochromatic and coherent.

44. Assertion (A): A narrow pulse of light is sent through a medium. The pulse will retain its shape as it travels through the medium.

Reason (R): A narrow pulse is made of harmonic waves with a large range of wavelengths.

45. Assertion (A): When monochromatic light is incident on a surface separating two media, the reflected and refracted light both have the same frequency as the incident frequency.

Reason (R): The frequency of monochromatic light depends on media.

ANSWERS

1	2	3	4	5	6	7	8	9	10
(*a*)	(*d*)	(*d*)	(*c*)	(*d*)	(*a*)	(*b*)	(*d*)	(*b*)	(*b*)
11	**12**	**13**	**14**	**15**	**16**	**17**	**18**	**19**	**20**
(*d*)	(*a*)	(*b*)	(*c*)	(*c*)	(*b*)	(*b*)	(*b*)	(*a*)	(*d*)
21	**22**	**23**	**24**	**25**	**26**	**27**	**28**	**29**	**30**
(*b*)	(*d*)	(*b*)	(*c*)	(*b*)	(*a*)	(*c*)	(*b*)	(*d*)	(*c*)
31	**32**	**33**	**34**	**35**	**36**	**37**	**38**	**39**	**40**
(*c*)	(*b*)	(*c*)	(*a*)	(*c*)	(*d*)	(*c*)	(*d*)	(*d*)	(*b*)
41	**42**	**43**	**44**	**45**					
(*c*)	(*b*)	(*d*)	(*d*)	(*c*)					

SOME SELECTED EXPLANATORY ANSWERS

1. We know that wave optics is valid, when the size of the objects is of the order of wavelength of light. And the ray optics is valid when the size of the object is much larger than the wavelength of light.

2. It is actually due to uniformity of refractive index of the medium. In case of a optically heterogenous medium, the light ray will not go straight.

4. In one second, the man moves towards the mirror through a distance of 6 m, and his image moves in the opposite direction through an equal distance.

5. $n = (360 / \theta) - 1;\quad \therefore \theta = 90°$

6. Plane mirror gives lateral inversion.

8. Ray is rotated through 90° at each reflection i.e. angle of incidence is 45°.

10. We know that first image is produced due to at the front surface and other images are produced due to multiple reflection at the front and rear surfaces. Since second image is, produced due to reflection from rear and silvered point. Therefore it is brightest image.

16. 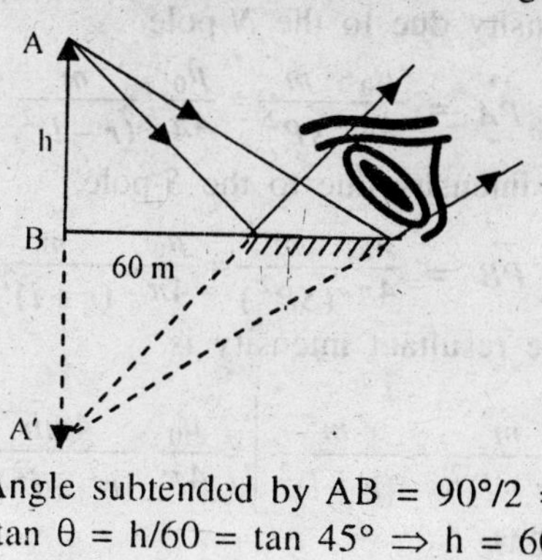

Angle subtended by AB = 90°/2 = 45°

$\tan\theta = h/60 = \tan 45° \Rightarrow h = 60$ m.

17. 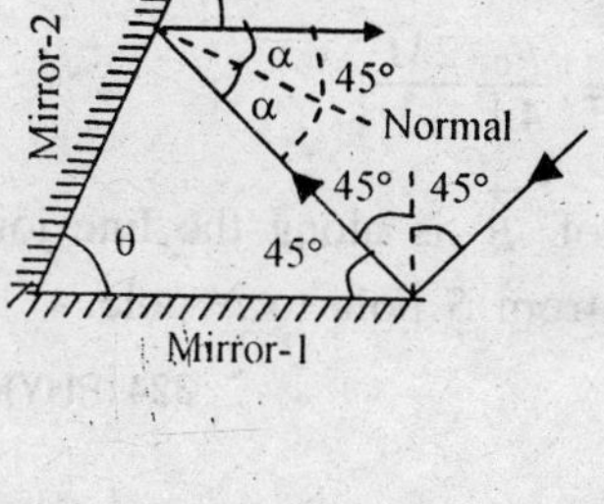

Let angle of incidence at mirror–2 be α.

So angle of reflection at mirror–2 is also α.

From the geometry of the figure

$\alpha + \alpha = 45° \Rightarrow \alpha = 22.5°$

Also $\alpha + \theta = 90° \Rightarrow \theta = 90° - \alpha$

$= 90° - 22.5° = 67.5° = 67°30'$.

19. In plane mirror, image distance = object distance

∴ Object speed = image speed

Speed of the image = 5 m/s.

23. The minimum mirror length should be half of the height of man.

25. Distance between the observer and the image = 30 + 10 = 40 cm.

26. Reflection is explained by ray optics.

28. On the basis of the rectilinear propagation of light Newton formulated his corpuscular theory of light because on its basis he could explain three things

(i) rectilinear propagation of light

(ii) reflection from a surface

(iii) refraction of light in an other medium.

30. According to Huygen's principle every point on a wavefront may itself be regarded as a source of secondary waves.

36. Light propagates rectilinearly. But it can take a slight turn at obstacles this property of light is due to its wave nature.

38. According to Huygen principle every point on primary wavefront acts as a source of spherical wavelets or secondary waves such that the primary wavefront at some later time is the envelope of these wavelets. The wavelets advance with a speed and frequency equal to those of primary wave at each point in space.

40. Coherent sources only can produce interference.

11 MAGNETISM

Magnetic Poles

A magnet has two poles. When a magnet is brought near a heap of iron fillings the ends of the magnet show the maximum attraction. These ends where the magnetic attraction is maximum are called poles of the magnet.

A magnet suspended freely, always points in $N - S$ direction. The pole which points towards north is called **north pole** (N) and the pole that points towards the south is called **south pole** (S).

The SI unit of strength of a magnetic pole is ampere metre (A m).

Magnetic Moment

The product of pole strength (m) and effective length ($2l$) of the magnet is called the Magnetic Moment. Thus $M = m \times 2l = 2ml$.

Its units are Ampere-m^2 and it is a vector quantity. Its direction is along the axis of the magnet from north pole of the magnet to the south pole.

Magnetic Field

The region around a pole in which magnetic effect can be experienced is called magnetic field.

Intensity of Magnetic Field

The number of lines of force crossing unit area of the surface, normally is called intensity of magnetic field.

OR

The force acting on unit pole placed in the magnetic field is called intensity of magnetic field.

i.e. $H = (\mu_0/4\pi) \cdot (m/r^2)$

Field Due to a Small Bar Magnet (Magnetic Dipole)

(i) In End-on position

In End-on position the point lies on the axis of the magnet. Let r be the distance of the point P from the centre O of the magnet. Let $2l$ be the length of the magnet and m be its pole strength. Then $NP = (r - l)$, $SP = (r + l)$.

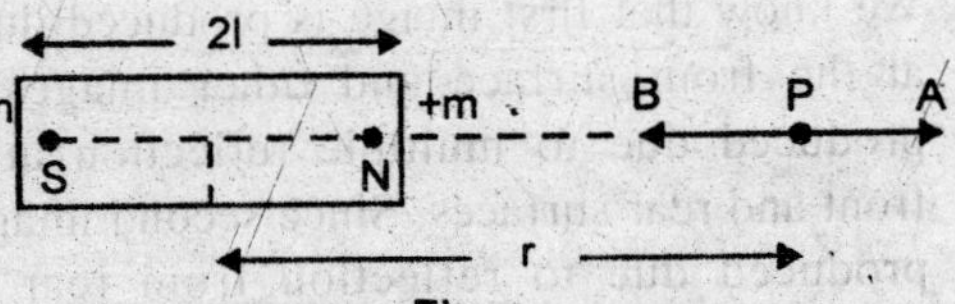

Fig.

The intensity due to the N-pole

$$\vec{PA} = \frac{\mu_0}{4\pi}\frac{m}{NP^2} = \frac{\mu_0}{4\pi}\cdot\frac{m}{(r-l)^2}$$

and, the intensity due to the S-pole,

$$\vec{PB} = \frac{\mu_0}{4\pi}\frac{m}{(SP^2)} = \frac{\mu_0}{4\pi}\cdot\frac{m}{(r+l)^2}$$

Thus, the resultant intensity is

$$B = \frac{\mu_0}{4\pi}\left[\frac{m}{(r-l)^2} - \frac{m}{(r+l)^2}\right] = \frac{\mu_0}{4\pi}\cdot\frac{4mlr}{(r^2-l^2)^2}.$$

$$B = \frac{\mu_0}{4\pi}\frac{2Mr}{(r^2-l^2)^2}$$

for a small magnet $r^2 >> l^2$, hence

$$B = \frac{\mu_0}{4\pi}\frac{2M}{r^3}$$

The direction of $\vec{B}$ is along the line joining the two poles from S pole to N pole.

(ii) Broad side-on position

In broad side-on the point lies on the magnetic equator of the magnet i.e. on the right bisector of the magnet.

Let us consider a magnet NS of pole strength m and length $2l$. P is the point on the right bisector of this magnet such that $OP = r$. Then,

$$SP = NP = (r^2 + l^2)^{1/2}$$

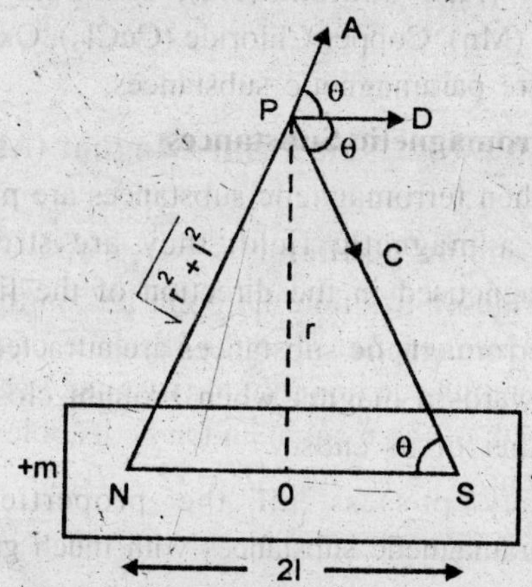

Fig.

Hence intensity of field at P due to N-pole $= \dfrac{\mu_0}{4\pi}\cdot\dfrac{m}{(r^2+l^2)}$ along $\overrightarrow{PA}$. The intensity of field at P due to S-pole $= \dfrac{\mu_0}{4\pi}\cdot\dfrac{m}{(r^2+l^2)}$ along $\overrightarrow{PC}$.

Each of these intensities can be resolved into two components : (i) along $\overrightarrow{PD}$ and (ii) perpendicular to $\overrightarrow{PD}$. The components perpendicular to $\overrightarrow{PD}$ being equal and opposite neutralise each other while the components parallel to $\overrightarrow{PD}$ add up.

Thus the resultant intensity at P is

$$B = \frac{\mu_0}{4\pi}\cdot\frac{m}{(r^2+l^2)}\cos\theta + \frac{\mu_0}{4\pi}\frac{m}{(r^2+l^2)}\cos\theta$$

or $$B = \frac{\mu_0}{4\pi}\left[\frac{2m}{(r^2+l^2)}\cdot\frac{l}{(r^2+l^2)^{1/2}}\right] = \frac{\mu_0}{4\pi}\cdot\frac{M}{(r^2+l^2)^{3/2}}$$

For a small magnet $r^2 >> l^2$, hence $B = \dfrac{\mu_0}{4\pi}\cdot\dfrac{M}{r^3}$

The direction of $\vec{B}$ is parallel to the axis from N-pole to S-pole.

Classification of Magnetic Substances

Curie and Faraday observed that almost all substances have certain magnetic properties. Placing many substances in magnetic fields he studied their behaviour. On the basis of magnetic behaviour of different materials, he divided them into three categories : (1) diamagnetic substances, (2) paramagnetic substances and (3) ferromagnetic substances.

1. Diamagnetic Substances

(i) When these substances are placed in a magnetic field they acquire feeble magnetism opposite to the direction of the magnetic field.

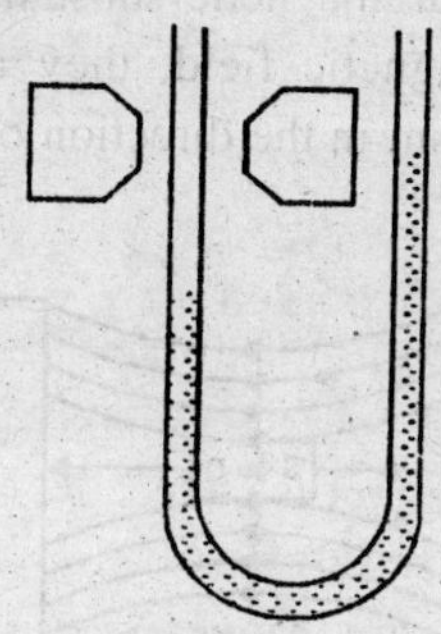
Fig.

(ii) When these substances are brought near the ends of a powerful magnet they are somewhat repelled. They are called 'diamagnetic' substances, and their magnetism is called the 'diamagnetism.'

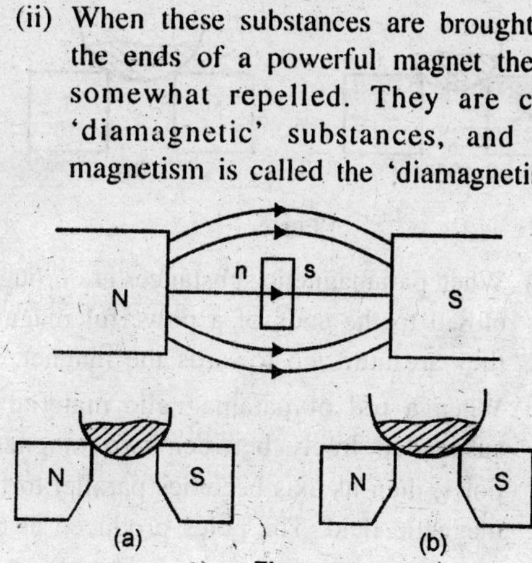

Fig.

(iii) If a diamagnetic solution is poured into a U-tube and one arm of this U-tube is placed between the poles of a strong magnet, the level of the solution in that arm is depressed (Fig. b).

(iv) Permeability of diamagnetic substances is less than 1 (i.e. $\mu < 1$).

(v) Susceptibility of diamagnetic substances is small and negative.

(vi) Susceptibility of diamagnetic substances is independent of temperature.

Examples of Diamagnetic Substance: Bismuth (Bi), Zinc (Zn), Copper (Cu), Silver (Ag), Gold (Au), (Diamond) (*C*), Salt (NaCl), Water (H_2O), Mercury (Hg), Nitrogen (N_2), Hydrogen (H_2), etc. are diamagnetic substances.

2. Paramagnetic Substances

(i) When paramagnetic substances are placed in a magnetic field, they acquire feeble magnetism in the direction of the magnetic field.

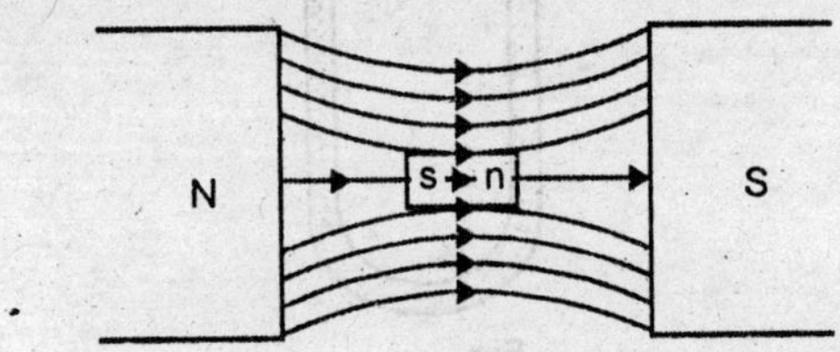

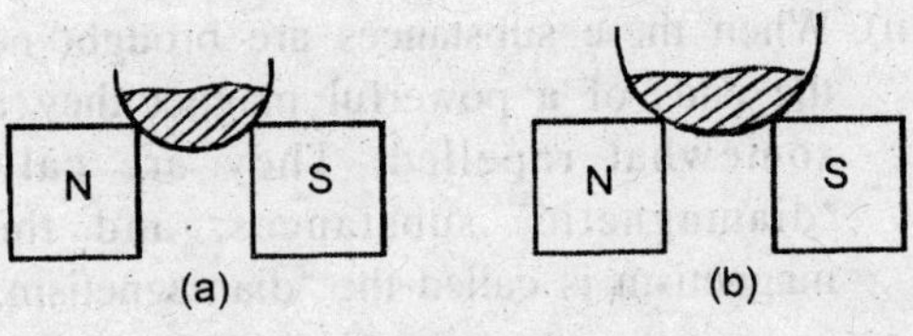

Fig.

(ii) When paramagnetic substances are brought closed to the ends of a powerful magnet, they are attracted towards the magnet.

(iii) When a rod of paramagnetic material is suspended freely between two magnetic poles, then its axis becomes parallel to the magnetic field. The poles produced at the ends of the rod are opposite to the nearer magnetic poles.

(iv) Permeability of paramagnetic substances is slightly greater than 1 i.e. $\mu > 1$.

(v) Susceptibility of paramagnetic substances is small and positive.

Examples of Paramagnetic substances Aluminium (Al), Sodium (Na), Platinum (Pt), Manganese (Mn), Copper Chloride ($CuCl_2$), Oxygen (O_2), etc. are paramagnetic substances.

3. Ferromagnetic Substances

(i) When ferromagnetic substances are placed in a magnetic field, they are strongly magnetised in the direction of the field.

(ii) Ferromagnetic substances are attracted fast towards a magnet when brought closer to either of its ends.

(iii) They possess all the properties of paramagnetic substances with much greater intensity.

(iv) Permeability of ferromagnetic substances is much greater than 1, ($\mu >> 1$).

(v) Ferromagnetic substances follow Curie law i.e. their susceptibility $\chi \propto (1/T)$. At Curie temperature ferromagnetic substances change into paramagnetic substances.

Examples of Ferromagnetic Substances. Iron (Fe), Nickel (Ni), Cobalt (Co), Magnetic (Fe_3O_4), etc. are ferromagnetic substances.

Elements of Earth's Magnetism

There are three elements of earth's magnetism (i) Angle of declination, (ii) Angle of dip, (iii) Horizontal component of earth's field.

Geographic Meridian : A vertical plane passing through the axis of rotation of the earth is called the geographical meridian.

Magnetic Meridian : A vertical plane passing through the axis of a freely suspended magnet is called the magnetic meridian.

MULTIPLE CHOICE QUESTIONS

1. S.I. unit of magnetic permeability is

(*a*) A-m (*b*) Am^2

(*c*) H-m (*d*) H/m

2. Potential at any point on equatorial line of dipole is

(*a*) $\mu_0\ M/4\pi d^2$ (*b*) $\mu_0 M/4\pi d^3$

(*c*) zero (*d*) none of these

3. The line joining a point to the centre of a short magnet makes angle θ with the axis. Potential at a point distant d from the centre of magnet, on this line is

(*a*) $\dfrac{\mu_0\ M \sin\theta}{4\pi d^2}$ (*b*) $\dfrac{\mu_0\ M \cos\theta}{4\pi d^2}$

(*c*) $\dfrac{\mu_0\ M}{4\pi d^3}$ (*d*) none of these

4. Magnetic potential at a point distant d from a magnetic pole of strength m is

(*a*) $\dfrac{\mu_0}{4\pi}\dfrac{m}{d}$ (*b*) $\dfrac{\mu_0}{4\pi}\dfrac{m}{d^2}$

(*c*) $\dfrac{\mu_0}{4\pi}\dfrac{2m}{d}$ (*d*) none of these

5. Which of the following is not a magnetic substances ?

(*a*) brass (*b*) iron

(*c*) cobalt (*d*) nickel

6. A magnet of magnetic moment M is freely suspended in a constant uniform magnetic field of intensity H. If magnet is delfected by an angle θ from direction of H, the work done is

(*a*) $MH\cos\theta$ (*b*) $MH(1-\cos\theta)$

(*c*) $MH\sin\theta$ (*d*) $MH(1-\sin\theta)$

7. A bar magnet is released into a copper ring directly below it. The acceleration of the magnet will be

(*a*) equal to the acceleration due to gravity at that place

(*b*) less than the acceleration due to gravity at that place

(*c*) greater than the acceleration due to gravity at that place

(*d*) twice the acceleration due to gravity at that place

8. Force between two magnetic poles depends on

(*a*) pole strength only

(*b*) distance only

(*c*) medium only

(*d*) all the three above

9. What is the resultant magnetic moment of two magnets, each of magnetic moment M in the following figures :

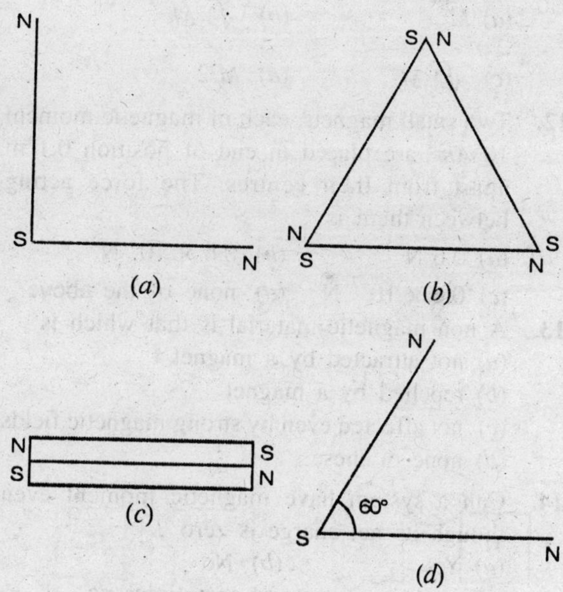

(*a*) (*b*) (*c*) (*d*)

(*a*) $\sqrt{2}\ M$; zero $\sqrt{3}\ M$; zero

(*b*) $\sqrt{2}\ M$; zero ; zero; $\sqrt{3}\ M$

(*c*) $\sqrt{2}\ M$; $\sqrt{3}\ M$; zero ; zero;

(*d*) none of the above

10. A bar magnet is held at right angles to a uniform magnetic field. The couple acting on the magnet is to be halved by rotating it from this position. The angle of rotation is

(a) 60° (b) 45°
(c) 30° (d) 75°

11. Two magnets of equal magnetic moments M each are placed as shown in Figure. The resultant magnetic moment is

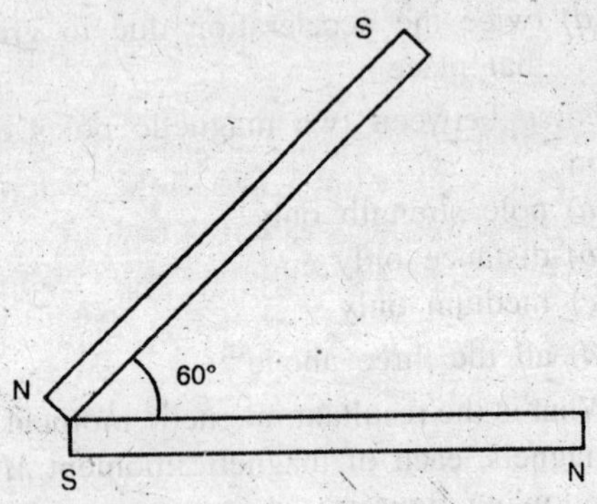

(a) M (b) $\sqrt{3}\,M$
(c) $\sqrt{2}\,M$ (d) $M/2$

12. Two small magnets, each of magnetic moment $10\ Am^2$ are placed in end of position 0.1 m apart from their centres. The force acting between them is
(a) 0.6 N (b) 0.6×10^7 N
(c) 0.6×10^{-7} N (d) none of the above

13. A non magnetic material is that which is
(a) not attracted by a magnet
(b) repelled by a magnet
(c) not affected even by strong magnetic fields
(d) none of these

14. Can a system have magnetic moment even though its net charge is zero ?
(a) Yes (b) No
(c) sometimes yes and sometimes no
(d) cannot say

15. The source of magnetic field is
(a) isolated magnetic pole
(b) static electric charge
(c) current loop
(d) none of the above

16. Two magnets have the same length and the same pole strength. But one of the magnets has a small hole at its centre. Then
(a) both have equal magnetic moment
(b) one with hole has smaller magnetic moment
(c) one with hole has larger magnetic moment
(d) one with hole loses magnetism through the hole

17. A thin magnet is cut into two equal parts by cutting it parallel to its length. If the original time period of vibration is 4 sec, the time period of each part in the same field will be
(a) 4s (b) 2s
(c) $4\sqrt{2}s$ (d) none of these

18. Two identical coils carrying same current have common centre but have their planes at right angles to each other. If the field due to each coil is B, the resultant field will be
(a) $\sqrt{2}B$ (b) $\frac{1}{\sqrt{2}}B$
(c) 2B (d) zero

19. A steel wire of length l has a magnetic moment M. It is bent into L shape from the middle. The new magnetic moment is
(a) M (b) $M/\sqrt{2}$
(c) $M/2$ (d) $2M$

20. The magnetic moment of a short magnet is $8\ Am^2$. The magnetic induction 20 cm away from its mid point on the axial line is
(a) 1×10^{-4}T (b) 2×10^{-4}T
(c) 4×10^{-4}T (d) 8×10^{-4}T

21. The moment of a magnet is $0.1\ Am^2$ and the force acting on each pole in a uniform magnetic field of 0.36 oersted is 1.44×10^{-4} N. The distance between the poles of magnet is
(a) 2.5 cm (b) 5.0 cm
(c) 1.25 cm (d) 1.17 cm

22. A magnet can be completely demagnetised by
(a) breaking the magnet into small pieces
(b) heating it slightly
(c) droppping into cold water
(d) applying a reverse field of appropriate strength

Directions Qs. 23-27: *In each of the following questions there are two statements Assertion (A) is given followed by a corresponding statement of Reason (R) just below it. Of the statements, mark the correct answer:*

(*a*) If both assertion and reason are true and reason is the correct explanation of assertion.

(*b*) If both assertion and reason are true but reason is not the correct explanation of assertion

(*c*) If assertion is true but reason is false

(*d*) If assertion is false but reason is true.

23. **Assertion (A):** Iron behaves as magnet.
Reason (R): In magnet, the molecular magnets are aligned in same direction.

24. **Assertion (A):** Basic difference between an electric line and magnetic line of force is that former is discontinous and the latter is continuous or endless
Reason (R): No electric lines of forces exist inside a charged body but magnetic lines do exist inside a magnet.

25. **Assertion (A):** Magnetic moment of an atom is due to both, the orbital motion and spin motion of every electron.
Reason (R): A charged particle produces a magnetic field.

26. **Assertion (A):** Torque experience by the bar magnet is maximum when field is applied perpendicular to magnetic moment.
Reason (R): Torque on a bar magnet depends on the angle between applied magnetic field and magnetic dipole moment.

27. **Assertion (A):** When radius of cirular wire carrying current is doubled, its magnetic moment becomes four times.
Reason (R): Magnetic moment depends on area of the loop.

ANSWERS

1	2	3	4	5	6	7	8	9	10
(*d*)	(*c*)	(*b*)	(*a*)	(*a*)	(*b*)	(*b*)	(*d*)	(*b*)	(*a*)
11	**12**	**13**	**14**	**15**	**16**	**17**	**18**	**19**	**20**
(*a*)	(*a*)	(*c*)	(*a*)	(*a, c*)	(*b*)	(*a*)	(*a*)	(*b*)	(*b*)
21	**22**	**23**	**24**	**25**	**26**	**27**			
(*a*)	(*d*)	(*d*)	(*a*)	(*c*)	(*a*)	(*b*)			

SOME SELECTED EXPLANATORY ANSWERS

1. $\mu_0 = \frac{B}{H} = \frac{\text{T}}{\text{Am}}$

$= TA^{-1}\, m^{-1} = Hm^{-1}$. [$\because TA^{-1}$ is henry (H)]

3. Mag. potential at any point due to a short magnet is

$$V = \frac{\mu_0}{4\pi}\frac{M\cos\theta}{d^2}.$$

4. Mag. potential $V = \frac{\mu_0}{4\pi}\frac{m}{d}$.

6. $\theta_1 = 0°, \theta_2 = \theta$

$W = -MH(\cos\theta_2 - \cos\theta_1)$

$= -MH(\cos\theta - \cos 0°)$

$= MH(1 - \cos\theta)$

7. A bar magnet falling freely induces current in copper ring, that opposes the motion of the bar magnet. Therefore, acceleration of bar magnet is less than accleration due to gravity at that place.

8. Force between two magnetic poles depends on pole strength, distance and medium between the poles.

9. $M' = \sqrt{M^2 + M^2} = \sqrt{2}M.$

As magnetic moments are in a closed loop in Figure (*b*)

$\therefore \quad M = 0$

In Figure (c) $M' = M - M = 0$

In Figure (d),

$$M' = \sqrt{M^2 + M^2 + 2MM\cos 60^\circ} = \sqrt{3}M$$

10. $\tau_1 = MB\sin 90^\circ = MB$

As $\tau_2 = \frac{1}{2}\tau_1$

$\therefore MB\sin\theta = \frac{1}{2}MB \quad \therefore \sin\theta = \frac{1}{2}, \theta = 30^\circ$

Angle of rotation $= 90^\circ - 30^\circ = 60^\circ$.

11. As magnetic moments are directed along SN, angle between $\vec{M}$ and $\vec{M}$ is $\theta = 120^\circ$ (from Figure)

$\therefore$ Resultant magnetic moment

$$= \sqrt{M^2 + M^2 + 2MM\cos 120^\circ}$$

$$= \sqrt{M^2 + M^2 + 2M^2(-1/2)} = M$$

12. Force between two magnetic dipoles is given by

$$F = \frac{\mu_0}{4\pi}\frac{6\,M_1\,M_2}{r^4}$$

$$= \frac{10^{-7}\times 6\times 10\times 10}{(0.1)^4} = 0.6\text{ N}$$

14. Yes, for example, a neutron has some magnetic moment. Atoms of para and ferro magnetic materials are neutral and still they have some net magnetic moment.

15. Magnetic field is associated with an isolated magnetic pole and also with a current loop.

17. Pole strength of each pole is halved, so is magnetic moment. As mass of each part is halved, moment of inertia I is halved.

As $T = 2\pi\sqrt{\frac{I}{MB}}$

$\therefore T' = \sqrt{\frac{(1/2)\,I}{(1/2)\,MB}} = T$

Hence T remains the same *i.e.* 4 seconds.

18. As planes of two coils are at right angle to each other, so must be their fields.

$\therefore$ Resultant field $= \sqrt{B^2 + B^2} = B\sqrt{2}$.

19. Strength of each pole $= M/l$

Distance between the poles on bending the wire into L shape

$$= \sqrt{(l/2)^2 + (l/2)^2}$$

$$= \frac{l}{2}\sqrt{2} = \frac{l}{\sqrt{2}}$$

$\therefore M' = \frac{M}{l}\times\frac{l}{\sqrt{2}} = \frac{M}{\sqrt{2}}$

20. $B = \frac{\mu_0}{4\pi}\frac{2M}{d^3} = \frac{10^{-7}\times 2\times 8}{(20\times 10^{-2})^3}$

$= 2\times 10^{-4}\ T$.

21. Here, $M = 0, 1\ Am^2$, $B = 0.36$ oersted

$= 0.36\times 10^{-4}$ T

$F = 1.44\times 10^{-4}$ N, $2l = ?$

As $F = mB$

$\therefore m = \frac{F}{B} = \frac{1.44\times 10^{-4}}{0.36\times 10^{-4}} = 4\text{A/m}$

As $M = m\times 2l$

$\therefore 2l = \frac{M}{m} = \frac{0.1}{4}\text{ m} = 2.5\text{ cm}$

12 ELECTROSTATICS AND CURRENT ELECTRICITY

Introduction : Electric Charge

1. Electric charge is a fundamental property like Length, Mass, and Time. Electric charges are of two types- Positive and Negative.
2. *(i)* A positive ion is an atom from which one or more electrons has been removed, i.e. **positive charge implies the deficiency of electrons.**

   (ii) A negative ion is an atom to which one more electrons has been added, i.e. **negative charge implies excess of electrons.**
3. **Units of Charges:**

   SI system – coulomb

Coulomb's Law

According to this law the electrostatic force F (either attractive or repulsive) between two stationary point charges q_1 and q_2 in free space (i.e. air or vacuum)

(i) is directly proportional to the product of the two charges.

(ii) is inversely proportional to square of the distance between the charges and

(iii) acts along the line joining the two charges.

Let r be the distance of charges q_2 from q_1 then

$$F \propto \frac{q_1 q_2}{r^2}$$

or $$F = C' \frac{q_1 q_2}{r^2}$$

where C' represents constant of proportionality.

Value of constant C depends on two factors.

(a) System of units.

(b) Medium in which charges are placed.

Electric Field

The region in which a charged particle placed at any point experiences a force, is called electric field. Electric field is measured in terms of electric field intensity $(\vec{E})$. The electric field intensity $\vec{E}$ at any point in an electric field is equal to force experienced by unit positive test charge q_0 placed at that point i.e.

$$\vec{E} = \frac{\vec{F}}{q_0}$$

Electric Potential

The electric potential at any point in an electric field is defined as the works done on the system against the electric force in moving an unit positive test charge from infinity (i.e. from beyond the electric field) to that point. If W is the work done in bringing the charge q_0 from infinity to given point, then potential at that point:

$$V = \frac{W}{q_0}$$

Electric Current

Electric current is the rate of flow of electric charge through any point of the circuit per unit time. It is denoted by I.

If a charge 'q' flows through the circuit for time 't', then current $I = q/t$.

Current is a *scalar* quantity and its unit is **ampere.**

Electric Resistance

The opposition offered by the substance to the flow of charge is known as electrical resistance. If the potential difference between the ends of the conductor is V and the current in the conductor is I, then the electric resistance of the conductor will be given by $R = V/I$.

Electric Resistance

The opposition offered by the substance to the flow of charge is known as electrical resistance. If the potential difference between the ends of the conductor is V and the current in the conductor is I, then the electric resistance of the conductor will be given by $R = V/I$.

The resistance of the conductor

(i) $R \propto$ length of conductor

(ii) $R \propto \dfrac{1}{\text{area of cross-section}}$

The Unit of resistance R in the MKS system is 'Ohm'.

Ohm's Law

According to Ohm's law, if physical state of a conductor (such as temperature), are unchanged then the ratio of the potential difference applied at its ends and the current flowing through it is constant.

Effect of Temperature on Resistivity

1. Resistivity of metals

Electric resistance of the metallic wire increases when its temperature is raised. Since the resistance of the wire is given by $R = ml/n\rho^2 A\tau$, where n is the number of free electrons per unit volume of the wire, m is the mass and e is the charge of the electrons.

2. Resistivity of Alloys

With rise in temperature the resistivity of alloys also increases. But this increase is much smaller compared to pure metals.

Combinations of Resistances

There are following two combinations of resistances :

(i) Resistance in Series

The equivalent resistance of combinations of three resistances in series

$$IR = IR_1 + IR_2 + IR_3$$

or $$R = R_1 + R_2 + R_3$$

(ii) Resistances in Parallel

The equivalent resistance of combination of three resistances R_1, R_2, R_3 in parallel

$$\frac{1}{R} = \frac{1}{R_1} + \frac{1}{R_2} + \frac{1}{R_3}$$

Electric Cell

A device which maintains the flow of charges continuously in a circuit is called the electric cell.

E.M.F. of Cell

It is defined in two ways :

(i) Qualitative Definition. The potential difference across the terminals of the cell in an open circuit is called the e.m.f.

(ii) Quantitative definition. The energy given by the cell in the flow of unit charge in the whole circuit (including the cell) is called the 'electromotive force (e.m.f.) of the cell. The e.m.f. is a characteristic of the cell which depends upon the nature of the plates and the electrolyte used in the cell. It is not affected by the quantity of the electrolyte or the size of the plates or the distance between the plates.

Suppose during the flow of a charge of q coulomb in an electric circuit, the energy supplied (or the work done) by a cell is W joule then the E.M.F. of the cell is

$$E = (W/q) \text{ joule/coulomb or volt.}$$

Potential Difference

Suppose during the flow of a charge of q coulomb in a part of an electrical circuit, W' joule work is done or W' joule energy is consumed, then the potential difference across that part will be $V = W'/q$ volt. It is measured by a voltmeter.

Combinations of Cells

As a cell is a source of electric current, a single cell cannot give a strong current, hence two or more cells are to be combined to get strong current. The combination of cells is called a 'battery'. Cells can be combined in three ways : (1) In series, (2) In parallel and (3) In mixed grouping.

1. In series.

Total resistance of the circuit = $(nr + R)$

Let i be the current in the circuit then $i = \dfrac{nE}{nr+R}$

Two cases arise here :

(i) If $nr << R$, then $i = \dfrac{nE}{R}$ (approx), i.e. if the internal resistance of the connected cells is much smaller then the external resistance, then the current given by these cells will be nearly n times, the current given by one cell. Hence, when the internal resistance of the connected cells is much smaller than the external resistance, then the cells should be connected in the series to obtain a strong current.

(ii) If $nr >> R$, then $i = \dfrac{nE}{nr} = \dfrac{E}{r}$ (approx), i.e. if the internal resistance of the connected cells is much greater than the external resistance, then nearly the same current is obtained by n cells as by a single cell. Hence, there is no advantage of connecting cells in series.

2. In Parallel

Total resistance of the circuit = $\left(\dfrac{r}{n} + R\right)$. If the current in the external circuit be i,

then $$i = \frac{E}{(r/n) + R} = \frac{nE}{r + nR}$$

Two cases arise here :

(i) If $\dfrac{r}{n} << R$, i.e. if the internal resistance of the cells is much smaller than the external resistance then r can be neglected in comparison to nR. Then from equation (i), $i = \dfrac{E}{R}$ (approx), i.e. the tōtal current will be equal to the current given by a single cell. Hence connecting the cells of small internal resistance in parallel has no advantage.

(ii) If $\dfrac{r}{n} >> R$, i.e. if the internal resistance of the cells is larger than the external resistance, then the current will be $i = n\dfrac{E}{r}$ (approx.) This current is nearly n times the current given by a single cell. Hence, when the internal resistance of the cells is much larger than the external resistance, then the cells should be connected in parallel.

3. Mixed Grouping

Suppose n cells are connected in each series, and such m rows are connected in parallel. Let E be the e.m.f. of each cell and r be the internal resistance. This battery of cells is sending current in an external resistance R.

∴ Total resistance of the circuit = $\{(nr/m) + R\}$.
Let the current in the external circuit be i, then

$$i = \frac{nE}{(nr/m) + R} = \frac{mnE}{nr + mR}$$

Wheatstone's Bridge

A bridge-arrangement of four resistances devised by Wheatstone by which resistance of a given conductor can be determined is called 'Wheatstone bridge'.

Four resistance P, Q, R and S are connected in the form of four arms of a parallelogram $ABCD$.

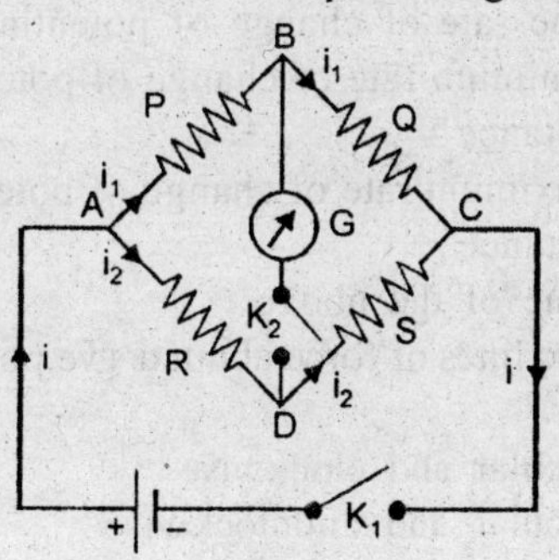

Fig.

$$\frac{P}{Q} = \frac{R}{S}$$

From this formula it is clear that if the ratio of the resistances P and Q, and the resistance R are known, then the unknown resistance S can be calculated.

MULTIPLE CHOICE QUESTIONS

1. Maximum value of electric intensity due to a charged sphere is at
(*a*) centre (*b*) surface
(*c*) infinity (*d*) none of the above

2. A positively charged rod is brought near an uncharged conductor. If the rod is then suddenly withdrawn the charge left on the conductor will be
(*a*) positive (*b*) zero
(*c*) negative (*d*) none of the above

3. Work done in carrying 2 C charge in a circular path of radius 3 m around a charge of 10 C is
(*a*) zero (*b*) 6.66 J
(*c*) 15 J (*d*) 60 J

4. The work done in displacing a charge of 2 C through 0.5 m on an equipotential surface is
(*a*) zero (*b*) 1 J
(*c*) 4 J (*d*) none of the above

5. The potential of a spherical conductor of radius 3 m is 6 V. The potential at its centre is
(*a*) zero (*b*) 2 V
(*c*) 6 V (*d*) 18 V

6. Electric intensity is equal to
(*a*) time rate of change of potential
(*b*) minimum rate of change of potential with distance
(*c*) maximum rate of change of potential with distance
(*d*) none of the above

7. Electric lines of force about a +ve point charge are
(*a*) circular and clockwise
(*b*) circular and anticlockwise
(*c*) radial outwards
(*d*) radial inwards

8. A charge is moved against electric repulsion in an electric field. Which of the following is correct?
(*a*) work is done by the electric field
(*b*) the potential energy of the charge decreases
(*c*) the strength of electric field decreases
(*d*) none of the above

9. What is the electric potential at the centre of a charged shell of radius 0.1 m if the potential at its surface is 10 V?
(*a*) 10 V (*b*) 1 V
(*c*) 0.1 V (*d*) zero

10. Electric potential is
(*a*) scalar and dimensionless quantity
(*b*) vector and dimensionless quantity
(*c*) scalar and dimensional quantity
(*d*) vector and dimensional quantity

11. Four charges each equal to q are placed at the corners of a square of side l. The electric potential at the centre of the square is
(*a*) $\frac{1}{4\pi\varepsilon_0}\frac{4q}{l}$ (*b*) $\frac{1}{4\pi\varepsilon_0}\frac{4q}{\sqrt{2}\,l}$
(*c*) $\frac{1}{\pi\varepsilon_0}\frac{\sqrt{2}\,q}{l}$ (*d*) $\frac{1}{\pi\varepsilon_0}\frac{2\,q}{l}$

12. Where is the electric potential due to a charged shell constant?
(*a*) outside it
(*b*) inside it
(*c*) both inside and outside it
(*d*) neither inside nor outside it

13. Where does the electric potential due to a charged conducting shell vary inversely as the distance from its centre?
(*a*) both inside and outside it
(*b*) only inside it
(*c*) only outside it
(*d*) neither inside nor outside it

14. Electric potential V due to a dipole is related to the distance r of the observation point as
(*a*) $V \propto r$ (*b*) $V \propto r^{-1}$
(*c*) $V \propto r^2$ (*d*) $V \propto r^{-2}$

15. A charge of 5 C is placed at the centre of a spherical gaussian surface of radius 5 cm. The electric flux through the surface is
(*a*) 0.1 C (*b*) 0.5 C
(*c*) 1 C (*d*) 5 C

16. A cubical gaussian surfaces encloses 30 C of charge. The electric flux through each surface of the cube is
(*a*) 30 C (*b*) 15 C
(*c*) 10 C (*d*) 5 C

17. For an irregularly shaped charged conductor, the potential is
(*a*) more at the flat parts
(*b*) more at the spherical parts
(*c*) more at the sharp edges
(*d*) same every where

18. A positively charged glass rod is brought near the disc of uncharged gold leaf electroscope. The leaves diverge. Which of the following statement is correct?
(*a*) no charge is present on the leaves
(*b*) positive charge is induced on the leaves
(*c*) negative charge is induced on the leaves
(*d*) positive charge is induced on one leaf and negative charge is induced on the other leaf

19. Two charges are placed a certain distance apart. A metallic sheet is placed between them. What will happen to the force between the charges?
(*a*) increases
(*b*) decreases
(*c*) remains unchanged
(*d*) may increase or decrease depending on the nature of the metal

20. Two balls carrying charges + 8 μ C and – 5 μ C attract each other with a force *F*. If a charge –3 μ C is added to both, the force between the balls will be
(*a*) 2 *F* (*b*) *F*
(*c*) *F*/2 (*d*) zero

21. How does the electric field strength vary when we enter a uniformly charged spherical cloud?
(*a*) decreases inversely as the square of the distance from the surface.
(*b*) decreases directly as the square of the distance from the surface
(*c*) increases directly as the square of the distance from the centre
(*d*) increases directly as the distance from the centre

22. From where do the electric lines of force are assumed to originate?
(*a*) positive charge
(*b*) negative charge
(*c*) both from +ve and –ve charges
(*d*) neither from +ve charge nor from –ve charge.

23. The electric field strength due to a circular loop of charge of radius *R* and linear density of charge λ at its centre is proportional to
(*a*) λR (*b*) $\frac{\lambda}{R}$
(*c*) $\frac{\lambda}{R^2}$ (*d*) none of the above

24. Which of the following does not allow the electric field lines to cross each other?
(*a*) they originate from +ve charge and end at –ve charge
(*b*) they do not pass through the conductor
(*c*) they exhibit longitudinal tension and lateral pressure
(*d*) superposition of electric fields

25. Which of the following increases, when the distance between the plates of the charged capacitor is increased?
(*a*) charge (*b*) potential difference
(*c*) capacitance (*d*) none of the above

26. The length of a conductor is doubled and its radius is halved, its specific resistance is
(*a*) unchanged (*b*) halved
(*c*) doubled (*d*) quadrupled

27. The resistance of a wire is *R*Ω. The wire is stretched to double its length keeping volume constant. Now the resistance of the wire will become
(*a*) 4*R*Ω (*b*) 2*R*Ω
(*c*) *R*/2Ω (*d*) *R*/4Ω

28. Conductivity of a conductor depends upon
(*a*) length (*b*) area of cross-section
(*c*) volume (*d*) temperature

29. Identify the set in which all the three materials are good conductors of electricity
(*a*) Cu, Ag and Au
(*b*) Cu, Si and diamond
(*c*) Cu, Hg and NaCl
(*d*) Cu, Ge and Hg

30. If a certain piece of copper is to be shaped into a conductor of minimum resistance, its length (L) and cross-sectional area A shall be respectively
(a) $L/3$ and $4A$ (b) $L/2$ and $2A$
(c) $2L$ and $A/2$ (d) L and A

31. A steady current is set up in a metallic wire of non-uniform cross-section. How is the rate of flow of electrons (R) related to the area of cross-section (A) ?
(a) $R \propto A^{-1}$ (b) $R \propto A$
(c) $R \propto A^2$ (d) R is independent of A

32. Two unequal resistances are connected in parallel across a cell. Which of the following statement is true ?
(a) current through smaller resistance is more
(b) current through larger resistance is more
(c) current is same through both the resistances
(d) nothing is definite, as it depends upon the e.m.f. of the cell and the circuit conditions

33. The resistance of human body is about
(a) 12 Ω (b) 120 Ω
(c) 12 kΩ (d) 120 kΩ

34. Constantan wire is used for making standard resistance, because it has
(a) high melting point
(b) low specific resistance
(c) high specific resistance
(d) negligible temperature coefficient of resistance

35. Which of the following has negative temperature coefficient of resistance ?
(a) platinum (b) tungsten
(c) carbon (d) nichrome

36. The resistance of a metal conductor increases with temperature due to
(a) change in current carriers
(b) change in the dimensions of the conductor
(c) increase in the number of collisions among the current carriers
(d) increase in the rate of collisions between the current carriers and the vibrating atoms of the conductor

37. On increasing the temperature of a conductor, its resistance increases because
(a) relaxation time decreases
(b) mass of the electron increases
(c) electron density decreases
(d) none of the above

38. In which of the following substances the resistance decreases with an increase of temperature ?
(a) Copper (b) Germanium
(c) Silver (d) Constantan

39. What length of a copper wire of cross-sectional area 0.01 mm^2, resistivity 1.7×10^{-8} Ωm, will be needed to prepare a resistance of 1.7 kΩ ?
(a) 0.1 km (b) 0.7 km
(c) 1 km (d) 10 km

40. A metallic resistor is connected across a battery. If the number of collisions of the free electrons with the lattice is some how decreased in the resistor (for example by cooling it,) the current will
(a) remain constant (b) increase
(c) decrease (d) become zero

41. When no current is passed through a conductor;
(a) the free electrons do not move
(b) the average speed of a free electron over a large period of time is not zero.
(c) the average velocity of a free electron over a large period of time is zero.
(d) the average of the velocities of all the free electrons at an instant is zero

42. A potential difference V is applied to a copper wire of length l and thickness d. If thickness is doubled, the drift velocity
(a) is doubled (b) is halved
(c) remains same (d) becomes zero.

43. A potential difference V is applied to a copper wire of length l and thickness d. If length is doubled, the drift velocity
(a) is doubled (b) is halved
(c) remains same (d) becomes zero.

44. A potential difference V is applied to a copper wire of length l and thickness d. If V is doubled, the drift velocity
(a) is doubled (b) is halved
(c) remains same (d) becomes zero.

45. A metallic block has no potential difference applied across it, if potential difference is applied, the drift velocity at temperature T is
(*a*) inversely proportional to T
(*b*) proportional to $\sqrt{T}$
(*c*) zero
(*d*) finite but independent of T

46. Three resistance of magnitude 2, 3, and 5 ohms are connected in parallel to a battery of 10 V and of negligible internal resistance. The potential drop across the 3 ohm resistance will be
(*a*) 10 volt (*b*) 2 volt
(*c*) 5 volt (*d*) 3 volt

Directions Qs. 47-50: *In each of the following questions there are two statements Assertion (A) is given followed by a corresponding statement of Reason (R) just below it. Of the statements, mark the correct answer:*
(*a*) If both assertion and reason are true and reason is the correct explanation of assertion.
(*b*) If both assertion and reason are true but reason is not the correct explanation of assertion
(*c*) If assertion is true but reason is false
(*d*) If assertion is false but reason is true.

47. **Assertion (A):** If there exists coulomb attraction between two bodies, both of them may not be charged.
Reason (R): In coulomb attraction two bodies are oppositely charged.

48. **Assertion (A):** No two electric lines of force can intersect each other.
Reason (R): Tangent at any point of electric line of force gives the direction of electric field.

49. **Assertion (A):** Electric force acting on a proton and on electron, moving in a uniform electric field is same, where as acceleration of electron is 1836 times that of a proton.
Reason (R): Electron is lighter than proton.

50. **Assertion (A):** As force is a vector quantity, hence electric field intensity is also a vector quantity.
Reason (R): The unit of electric field intensity is newton per coulomb.

ANSWERS

1	2	3	4	5	6	7	8	9	10
(*b*)	(*b*)	(*a*)	(*a*)	(*c*)	(*c*)	(*c*)	(*d*)	(*a*)	(*c*)
11	**12**	**13**	**14**	**15**	**16**	**17**	**18**	**19**	**20**
(*c*)	(*b*)	(*c*)	(*d*)	(*d*)	(*d*)	(*d*)	(*b*)	(*b*)	(*b*)
21	**22**	**23**	**24**	**25**	**26**	**27**	**28**	**29**	**30**
(*d*)	(*a*)	(*d*)	(*d*)	(*b*)	(*a*)	(*a*)	(*d*)	(*a*)	(*a*)
31	**32**	**33**	**34**	**35**	**36**	**37**	**38**	**39**	**40**
(*a*)	(*a*)	(*c*)	(*d*)	(*c*)	(*d*)	(*a*)	(*b*)	(*c*)	(*b*)
41	**42**	**43**	**44**	**45**	**46**	**47**	**48**	**49**	**50**
(c,d)	(*c*)	(*b*)	(*a*)	(*a*)	(*a*)	(*b*)	(*a*)	(*a*)	(*b*)

SOME SELECTED EXPLANATORY ANSWERS

3. Electrostatic force is conservative. So, when the charge returns to the starting point, the work done is zero.

4. Work done in displacing a charge = $q \times \Delta V$. On the equipotential surface $\Delta V = 0$. Hence work done is also zero.

9. Electric potential inside the charged shell is everywhere same as on the surface.

11. Electric potential due to each charge at the centre of the square is $\frac{1}{4\pi\varepsilon_0}\frac{\sqrt{2}q}{l}$.

Hence total potential is

$$4\times\frac{1}{4\pi\varepsilon_0}\frac{\sqrt{2}q}{l}=\frac{1}{\pi\varepsilon_0}\frac{\sqrt{2}q}{l}$$

15. Electric flux = charge enclosed.

16. Total number of surface = 6. Total flux = 30 C. So, flux through each surface = 5 C.

19. The electrostatic force varies inversely as the dielectric constant. The dielectric constant of the metals is very large. So, the electrostatic force will decrease.

20. The net charge will now be 5 μ C and –8 μ C. So, the magnitude of the force will remain unchanged.

21. Electric field inside uniformly cloud of charge is

$$E=\frac{1}{4\pi\varepsilon_0}\frac{\left(\frac{4}{3}\pi r^3\rho\right)}{r^2}=\frac{1}{4\pi\varepsilon_0}\left[\frac{4}{3}\pi r\rho\right]$$

So, $E \propto r$.

25. $V = q/C$. On increasing the separation, the C decreases, q remains unchanged, hence V increases.

26. When wire is stretched to double its length, its resistance becomes four times.

30. $R = \dfrac{\rho l}{A}$ or $R \propto \dfrac{l}{A}$

31. Rate of flow of electrons will depend upon drift velocity and drift velocity is inversely proportional to area of cross-section, i.e. $v_d \propto 1/A$. Therefore

$R \propto 1/A$

32. When resistance are connected in parallel to a cell, the potential difference across each resistance is the same.

Current = Pot. diff./resistance.

35. Carbon is a semiconductor ; its temp. coefficient of resistance is negative.

37. With increase in temperature, resistance of metal conductor increases. But $R \propto 1/\tau$, therefore time of relaxation decreases.

39. $R = \dfrac{\rho l}{A}$ or $l = \dfrac{RA}{\rho}$

$$= \frac{1.7\times10^3\times(1\times10^{-8})}{1.7\times10^{-8}}$$

$= 10^3$ m = 1 km.

40. If no. of collisions of the free electrons with the lattice is decreased, the time of relaxation of electrons will increase. Due to which drift velocity of electrons will increase and hence current will increase.

42. v_d is independent of area of cross-section of the conductor.

43. $v_d \propto 1/l$

44. $v_d = \dfrac{eE}{m}\tau = \dfrac{e}{m}\tau\left(\dfrac{V}{l}\right)$ or $v_d \propto V$

45. When temperature is raised, the ions/atoms of the conductor start vibrating with increased amplitude of vibration and greater frequency. Due to which the electrons moving towards the positive end of conductor will suffer more rapid collisions and hence time of relaxation (τ) decreases. As $v_d \propto \tau$, thus drift velocity decreases. Therefore $v_d \propto 1/T$.

46. Since the resistances are connected in parallel to the battery of negligible resistance, hence pot. difference across each resistance is same as that of E.M.F. of battery.

13 CHEMICAL EFFECT OF ELECTRIC CURRENT AND ITS APPLICATIONS

Chemical Effects of Electric Current

The phenomenon of the conduction of electricity through liquids was first studied by Faraday in about 1833. Pure liquids, with the exception of mercury, are, in general, bad conductors of electricity at ordinary temperatures, but the conductivity of a liquid increases if the solution contains a metallic salt or an acid. *The process of decompositions of a compound by the application of an electric field is called electrolysis.* The liquid containing the compound is called an electrolyte and the vessel containing the electrolyte is called a *voltmeter.*

Faraday's Laws of Electrolysis

The two laws of electrolysis are stated as follows:

(i) The mass of ion liberated from an electrolyte at the respective electrodes is directly proportional to the quantity of electricity which passes through it,

$$W \propto Q$$

But, $Q = It$

where Q is the charge, I is current and t is the time. If I is expressed in amperes and t in seconds, then Q will be in coulombs.

$\therefore$ $W \propto It$ or $W = ZIt$

where Z is a constant called the electro-chemical equivalent of the substance. The units of Z will be g/coulombs, if W is expressed in grams.

$$\therefore \quad Z = \frac{W}{I \times t} g/C$$

The electro-chemical equivalent of a substance is defined as the mass of the ion in grams liberated at the electrode when when one coulomb of electricity is passed through the electrolyte.

(ii) If the same quantity of electricity passes through different electrolytes, the masses of the ions liberated at the respective electrodes are proportional to their chemical equivalents.

Let W be the mass of ion liberated, then,

$$W \propto \text{chemical equivalent}$$

or $$\frac{W}{\text{Chemical equivalent}} = \text{constant}$$

provided the same quantity of electricity passes through different electrolytes.

Electromagnetic Induction

"Whenever the magnetic lines of force linked with a closed circuit change an induced e.m.f. is always produced in the circuit and lasts only so long as the change lasts." The induced e.m.f. giving rise to such currents is called the induced electromotive force and the phenomenon is called **electromagnetic induction**.

Faraday's Laws of Electromagnetic Induction

(i) The induced emf, e is equal to the negative rate of change of magnetic flux : If $\Delta\phi$ be the change in magnetic flux in a time interval Δt, then the induced emf in the circuit is

$$e = -\frac{\Delta\phi}{\Delta t}$$

In the limit $\Delta t \to 0$, $e = -\frac{d\phi}{dt}$

The negative sign indicated that the induced e.m.f. opposes the change in magnetic flux (Lenz's law).

Transformer

Transformer is a device used to obtain suitable A.C. voltage. It never works on D.C. There are of two types of transformer—

(a) Step-up Transformer. It is used to convert low voltage high current to high voltage low current.

(b) Step-down transformer. It is used to convert high voltage low current to low voltage high current.

Principle. Transformer works on the principle of mutual induction i.e. when a magnetic flux linked with one coil changes, an induced e.m.f. is produced in the other coil.

Generator or Dynamo

A device by which we can convert mechanical energy to electrical energy is called Dynamo.

Alternating Current Generator (A.C. Dynamo)

Principle. Alternating current generator is based on the principle of the electromagnetic induction. When a coil is rotated about an axis perpendicular to the direction of uniform magnetic field, an induced e.m.f. is produced across it.

Construction. The A.C. generator consists of the following parts :

(1) Armature. The armature is a rectangular coil *ABCD* consisting of a large number of turns wound over a soft iron core. The soft iron core is used to increase the magnetic flux.

(2) Field magnet. Field magnet is a strong magnet having concave poles. The armature is rotated between the two poles of this magnet so that axis of the armature is perpendicular to magnetic field lines.

(3) Slip ring. The leads from the arms of the armature are connected to the two rings R_1 and R_2 separately. As the armature and hence the leads rotate, the rings R_1 and R_2 also rotate about the central axis.

(4) Brushes. Brushes (B_1, B_2) are used to pass on the current from armature to the external load *R*. As the rings rotate, the brushes remain in constant touch with the rings and always keep on pressing against them.

Working. The working of the a.c. generator is illustrated with the help of five different positions of the armature *ABCD* at time $t = 0, \frac{T}{4}, \frac{T}{2}, \frac{3T}{4}$ and *T* respectively as shown in Fig. below.

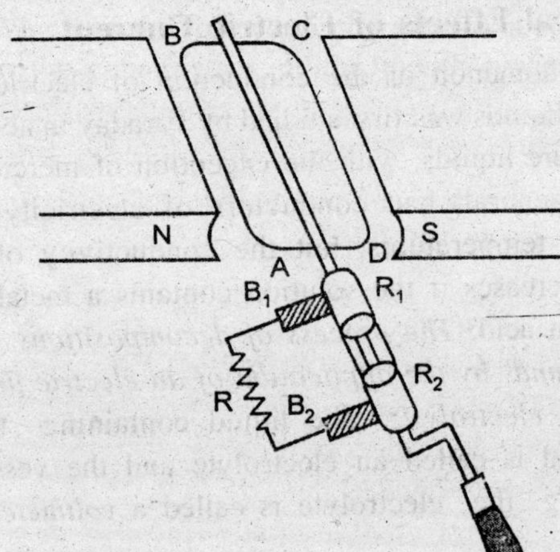

At $t = 0$, the armature *ABCD* is vertical with arm *AB* up and *CD* down. During the interval $t = 0$ to $t = \frac{T}{2}$, arm *AB* moves down and *CD* moves up. The direction of current can be found with the help of Fleming's right hand rule. The current in the armature will flow in the direction *DCBA*.

During the interval $t = \frac{T}{2}$ to $t = T$, just the arm *AB* moves up and *CD* moves down. Hence during this interval, the current in the armature will flow in the direction *ABCD*.

The arms *AB* and *CD* of armature move momentarily parallel to the field. whenever, the armature is vertical, hence the rate of change of magnetic flux is zero. So, the induced e.m.f. $\left(e = -\frac{d\phi}{dt}\right)$ is also zero. Thus at time $t = 0$, $\frac{T}{2}$ and *T*, there is no induced e.m.f. in the coil.

In the horizontal position of armature, the arms cut the magnetic lines of force at once, so the rate of change of magnetic flux is fastest and maximum e.m.f. is induced in this position. Thus at time $t = \frac{T}{4}$ and $\frac{3T}{4}$, the induced e.m.f. produced is maximum.

Hence output e.m.f. across R during a complete rotation will vary sinusoidally.

MULTIPLE CHOICE QUESTIONS

1. Mass of the ions liberated during electrolysis by 1 A current in one second is called
(*a*) equivalent weight
(*b*) chemical equivalent
(*c*) electrochemical equivalent
(*d*) none of the above

2. Voltameter cannot be used to measure
(*a*) current
(*b*) electrochemical equivalent
(*c*) potential difference
(*d*) charge

3. What determines the emf between the two metals placed in an electrolyte?
(*a*) relative position of metals in the electrochemical series
(*b*) distance between them
(*c*) strength of electrolyte
(*d*) nature of electrolyte

4. How much electric charge should pass through acidulated water to release 22.4 litre of hydrogen gas at NTP?
(*a*) 6.02×10^{23} C (*b*) 2×96500 C
(*c*) 22.4 C (*d*) 1 C

5. What carries current in an electrolyte?
(*a*) electrons only
(*b*) –ve ions only
(*c*) +ve ions only
(*d*) both +ve and –ve ions

6. Silver and zinc voltameter are connected in series and same current is passed through both of them for same time. If *x* kg of silver is liberated then the amount of zinc liberated will be very near to
(*a*) *x* (*b*) *x*/5
(*c*) *x*/3 (*d*) *x*/2

7. If 1 A of current is passed through $CuSO_4$ solution for 10 seconds, then the number of copper ions deposited at the cathode will be about
(*a*) 1.6×10^{19} (*b*) 3.1×10^{19}
(*c*) 4.8×10^{19} (*d*) 6.2×10^{19}

8. Same current is being passed through a copper voltameter and a silver voltameter. The rate of increase in weights of the cathodes in the two voltameters will be proportional to
(*a*) relative densities (*b*) atomic masses
(*c*) atomic numbers (*d*) none of the above

9. A cell sends a current through a resistance R_1 for one second and then sends current through another resistance R_2 for one second. Same heat is produced in both the resistance. What is the internal resistance of the cell?
(*a*) $R_1 - R_2$ (*b*) $R_1 + R_2$
(*c*) $\sqrt{R_1 R_2}$ (*d*) $\sqrt{R_1^2 - R_2^2}$

10. The resistance of a wire is 10 Ω. It is stretched so as to reduce its diameter to half. The resistance of the wire will now be
(*a*) 40 Ω (*b*) 80 Ω
(*c*) 160 Ω (*d*) 320 Ω

11. How many lamps each of 50 W and 100 V can be connected in parallel across a 120 V battery of internal resistance 10 Ω, so that each glows to full power?
(*a*) 16 (*b*) 8
(*c*) 4 (*d*) 2

12. Two similar head light lamps are connected in parallel to each other. Together, they consume 48 W from a 6 V battery. What is the resistance of each filament?
(*a*) 6 Ω (*b*) 4 Ω
(*c*) 3 Ω (*d*) 1.5 Ω

13. A lamp is marked 60 W, 220 V. If it operates at 200 V, the rate of consumption of energy will

(*a*) increase
(*b*) decrease
(*c*) remain unchanged.
(*d*) None of these

14. The potential difference across a conductor is doubled, the rate of generation of heat will
(*a*) become one fourth
(*b*) be halved
(*c*) be doubled
(*d*) become four times.

15. Resistance of conductor is doubled keeping the potential difference across it constant. The rate of generation of heat will
(*a*) become one fourth
(*b*) be halved
(*c*) be doubled
(*d*) become four times.

16. A constant potential difference is applied across the ends of a wire. Which one of the following operations will reduce the rate of heat generation to half ?
(*a*) both length and diameter are halved
(*b*) both length and diameter are doubled
(*c*) diameter is doubled and length is doubled
(*d*) diameter is halved and length is doubled

17. If the current in a conductor is doubled, the rate of rise of temperature in the conductor will be assuming that there is no loss of heat from the conductor
(*a*) halved (*b*) remain unchanged
(*c*) doubled (*d*) four times

18. What are the essential requirements for the heater filament?
(*a*) high resistivity, low melting point
(*b*) low resistivity, low melting point
(*c*) high resistivity, high melting point
(*d*) low resistivity, high melting point

19. What is converted into heat when current is set up through a conductor?
(*a*) electric current (*b*) electric potential
(*c*) electric energy (*d*) electric resistance

20. When a 40 watt lamp is connected in series with a 100 watt lamp across a 220 V supply, which one glows brighter?
(*a*) 100 Watt lamp
(*b*) 40 Watt lamp
(*c*) both will glow with equal brightness
(*d*) it cannot be predicted

21. Magnitude of *e*.m.f produced in a coil, when a magnet is inserted in the coil does not depend upon
(*a*) number of turns in the coil
(*b*) magnetic moment of the magnet
(*c*) speed of the magnet
(*d*) resistance of the coil.

22. A motor starter is a
(*a*) fixed resistance (*b*) variable resistance
(*c*) fixed capacitor (*d*) variable capacitor

23. Which of the following phenomena is utilised in the construction of the mouth piece of a telephone nowadays
(*a*) thermo electric effect
(*b*) photo electric effect
(*c*) change of resistance with pressure
(*d*) electromagnetic induction

24. The *e*.m.f produced in a wire by its motion across a magnetic field does not depend upon
(*a*) the length of the wire
(*b*) the composition of the wire
(*c*) the diameter of the wire
(*d*) the orientation of the wire

25. The equivalent quantity of mass in electricity is
(*a*) charge (*b*) current
(*c*) co-efficient of self inductance
(*d*) potential

26. The couple on an armature in an electric motor depend upon
(*a*) the number of turns in the coil
(*b*) the area of the coil
(*c*) the strength of field
(*d*) all of these

27. A cylinderical bar magnet is kept along the axis of circular coil. The magnet is rotated about its axis such that north pole faces the coil. The induced current in the coil
(*a*) is zero
(*b*) is clock-wise from magnet side
(*c*) may be clock-wise or anticlockwise
(*d*) is anticlockwise from magnet side

28. The direction of *e*.m.f induced in a coil is given by
(*a*) Faraday's laws
(*b*) Lenz's law
(*c*) Fleming's left hand rule
(*d*) Fleming's right hand rule

29. An inductor may store energy in
(*a*) an electric field
(*b*) a magnetic field
(*c*) in both (*a*) and (*b*)
(*d*) nothing can be said

30. When the rate of change of current is unity, induced *e*.m.f is equal to
(*a*) thickness of coil
(*b*) no. of turns in coil
(*c*) coefficient of self-induction
(*d*) total flux linked with coil

31. Two solenoids of same cross-sectional area have their lengths and number of turns in ratio of 1: 2. The ratio of self-inductance of two solenoids is
(*a*) 1 : 1 (*b*) 1 : 2
(*c*) 2 : 1 (*d*) 1 : 4

32. The back *e*.m.f. in a d.c. motor is maximum, when
(*a*) the motor has picked up max. speed
(*b*) the motor has just started moving
(*c*) the speed of motor is still on the increase
(*d*) the motor has just been switched off

Directions Qs. 33-37: *In each of the following questions there are two statements Assertion (A) is given followed by a corresponding statement of Reason (R) just below it. Of the statements, mark the correct answer:*
(*a*) If both assertion and reason are true and reason is the correct explanation of assertion.
(*b*) If both assertion and reason are true but reason is not the correct explanation of assertion
(*c*) If assertion is true but reason is false
(*d*) If assertion is false but reason is true.

33. Assertion (A): Magnetic field interacts with a moving charge and not with a stationary charge.
Reason (R): A moving charge produces a magentic field.

34. Assertion (A): The magnetic field intensity at the centre of a circular coil carrying current changes, if the current through the coil is doubled.
Reason (R): The magnetic field intensity is dependent on current in conductor.

35. Assertion (A): A linear solenoid carrying current is equivalent to a bar magnet.
Reason (R): The magnetic field lines of both are same.

36. Assertion (A): Force experienced by moving charge will be maximum if direction of velocity of charge is perpendicular to applied magnetic field.
Reason (R): Force on moving charge is independent of direction of applied magnetic field.

37. Assertion (A): If an electron is not deflected while passing through a certain region of space, then only possibility is that there is no magnetic field in this region.
Reason (R): Force is directly proportional to magnetic field applied.

ANSWERS

1	2	3	4	5	6	7	8	9	10
(*c*)	(*c*)	(*a*)	(*b*)	(*d*)	(*c*)	(*b*)	(*d*)	(*c*)	(*c*)
11	**12**	**13**	**14**	**15**	**16**	**17**	**18**	**19**	**20**
(*c*)	(*d*)	(*b*)	(*d*)	(*b*)	(*a*)	(*d*)	(*c*)	(*c*)	(*b*)
21	**22**	**23**	**24**	**25**	**26**	**27**	**28**	**29**	**30**
(*d*)	(*b*)	(*c*)	(*b*)	(*c*)	(*d*)	(*a*)	(b,d)	(*b*)	(*c*)
31	**32**	**33**	**34**	**35**	**36**	**37**			
(*b*)	(*a*)	(*a*)	(*a*)	(*a*)	(*c*)	(*d*)			

SOME SELECTED EXPLANATORY ANSWERS

6. Equivalent weight of silver and zinc respectively are 108 and 33. Same is the ratio of their chemical equivalents.

7. No. of copper ions

$$= \frac{\text{current} \times \text{time}}{\text{charge on electron} \times \text{valency}}$$

$$= \frac{1 \times 10}{1.6 \times 10^{-19} \times 2} = 3.125 \times 10^{19}$$

9. $I_1 = \frac{E}{R_1 + r}$ and $I_2 = \frac{E}{R_2 + r}$.

Also $I_1^2 R_1 = I_2^2 R_1$

That is $\frac{R_1}{(R_1 + r)^2} = \frac{R_2}{(R_2 + r)^2}$

This gives $r = \sqrt{R_1 R_2}$

10. $r = \frac{1}{\pi D^2/4}$. When diameter is halved, the length becomes L, such that

$$\pi \frac{D^2}{4} \times l = \pi \left(\frac{D}{2}\right)^2 \times \frac{1}{4} \times L.$$

That is $L = 4l$.

Hence $R = \rho \frac{4l}{\pi\left(\frac{D}{2}\right)^2 \times \frac{1}{4}} = 16\, r.$

11. Each glows to full power, if the terminal voltage of the battery is 100 V. That is maximum drop of potential in the battery should be 20 V. If external resistance be R, then $20 = I \times 10 = \frac{120}{R+10} \times 10$. This gives $R = 50$, which will be the case with 4 lamps in parallel.

12. Power consumed by each = 24 W. Hence using $R = V^2/P$, we find $R = 36/24$ $= 1.5\ \Omega$

13. $Q = \frac{V^2}{R} t$. When V decreases, Q also decreases.

14. Heat produced $Q = I^2 Rt = \frac{V^2}{R} t = VIt$.

When V is doubled, Q becomes 4 times, because resistance does not changed.

15. Heat produced $Q = I^2$ Rt $= \frac{V^2}{R} t = VIt$. When R is doubled, keeping V constant Q is halved.

16. $Q = \frac{V^2}{R} t$. So, Q will be halved when R is doubled.

Since $R = \rho\, l/A = \rho\, l/(\pi d^2/4)$. So, halving both l and d will double the resistance.

20. Current through each lamp is same. Since resistance of 40 W lamp is more, so greater heat is produced in the 40 W lamp.

22. A motor starter is a variable resistance which reduces the current as desired.

23. In the construction of mouth piece of a telephone, we use the pheonomenon of change of resistance with pressure (of sound waves).

24. e.m.f. does not depend upon composition of the wire.

25. Mass is measure of inertia, which corresponds to coefficient of self inductance L.

26. As τ = NABI, hence τ depends upon all the three N, A and B.

27. On rotating the magnet, no change in flux is linked with the coil. Therefore, induced e.m.f./current is zero.

28. Lenz's law and Fleming's right hand rule, both give the direction of e.m.f./current induced.

30. From $e = \frac{L di}{dt}$, when $\frac{di}{dt} = 1$, $e = L$.

32. The back e.m.f. in a motor is induced e.m.f. which is maximum, when speed of rotation of the coil is maximum.

14 ATOMIC STRUCTURE

Cathode Rays

(i) When a potential difference of about 10,000 volts is applied across the two electrodes of a discharge tube and pressure is reduced to about 10^{-2} mm of mercury, the whole of the discharge tube is filled with darkness and wall of the tube opposite to cathode is illuminated due to fluorescence whose colour depends upon the composition of glass of the tube. The fluorescence is produced due to string of particular type of invisible rays with the glass. These rays are emitted from the cathode and are called as Cathode Rays. These rays are independent of the nature of the gas in the discharge tube and their direction of propagation is not affected by the position of anode.

(ii) Thomson found following conclusions on the basis of his experiments:

(a) Cathode rays are negatively-charged particles which were called by him as electrons.

(b) Charge to mass ratio (*e*/*m*) of these particles is a constant quantity. Its value does not depend either on the gas filled in the discharge tube or the material of the cathode and the anode.

Properties of Cathode Rays

(i) **Cathode rays travel in straight line–** When an object is placed in front of the cathode rays a sharp shadow of the object is seen on the walls of the tube. This shows that cathode rays travel in straight line.

(ii) **Cathode rays are shot out normally from the surface of the cathode and not affected by the position of the anode–** If the surface of the cathode is plane, the rays emerge in the form of a parallel beam. If the tube has the concave cathode, the rays are found to be focussed at the centre of curvature. This shows that cathode rays emerge normally from the surface of cathode.

(iii) **Cathode rays exert mechanical pressure–** When a light paddle wheel of mica is placed in the path of cathode rays such that rays fall on half portion of the wheel, the wheel starts rotating. This is due to high momentum of cathode rays so that they exert pressure on striking a surface. This observation also confirms the fact that these rays are the moving particles of mass.

(iv) **Cathode rays produce heat when they fall on some material–** If a piece of platinum is placed at the centre of curvature of a concave cathode of the discharge tube, it becomes red hot. This is due to very high kinetic energy possessed by cathode rays because of their high velocity. When these highly energetic rays fall on platinum, their kinetic energy is converted into heat energy.

(v) **Cathode rays are deflected by electric and magnetic fields–** Both the electric and magnetic fields deflect the cathode rays in the direction, in which a negatively charged particle would be deflected.

(vi) Cathode rays can produce physical and chemical change– A photographic plate exposed to cathode rays gets blackened i.e. the cathode rays have a reducing effect on the salts of silver. These rays turn the colour of lithium chloride into violet.

(vii) Cathode rays can ionise gases– When cathode rays collide with atoms of a gas, they are able to eject electrons from them.

(viii) Cathode rays can produce X-rays– When cathode rays strike a solid target of high atomic weight and high melting point such as tungsten, molybdenum etc, they produce a highly penetrating radiation called the X-Rays.

(ix) Cathode rays produce fluorescence– When cathode rays fall on zinc-sulphide or Barium-platinocynide, coloured light is emitted from these substances, the colour depends upon the nature of substance.

Positive Rays

(i) In 1886 Goldstein discovered a new type of rays, called as positive rays. He observed that if in a cathode ray discharge tube cathode is perforated, then glass of the tube behind the cathode is found to be illuminated. The colour of this illumination is different from that produced by the cathode rays. He predicted that a new type of rays are emitted from the anode. He named these rays as positive rays.

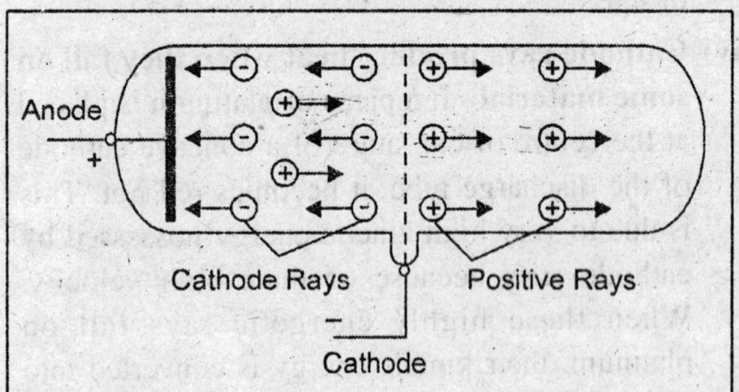

Fig.

(ii) Explanation of the production of positive rays: Since the gas molecules between the cathode and the anode are ionised by collision with the cathode rays positive rays are produced. Each of the atoms from which an electron is detached acquires a positive charge equal to the charge on the electron. These are called positive ions. The electrons detached from the atoms rush towards the anode alongwith the cathode rays and positive ions pass through the holes in the cathode and reach the other side. The moving positive ions constitute the positive rays.

Properties of Positive Rays

(i) These rays consist of positively charged particles which are actually the positive ions of the gas. The mass of these particles is nearly equal to the atoms of the gas filled in the discharge tube.

(ii) These rays are also deflected in electric and magnetic fields like cathode rays, but the direction of the deflection is opposite to that of cathode rays and the magnitude of their deflection is less than that of cathode rays.

(iii) These rays also affect the photographic plate and can penetrate through thin aluminium strips like cathode rays.

(iv) These rays also travel in straight line just like cathode rays but their velocity is far less than that of cathode rays. Moreover, these rays have velocity-distribution also, whereas it is not in the cathode rays.

(v) These rays are capable of producing both fluorescence and phosphorescence.

(vi) The value of specific charge (e/M) for positive rays is far less than the specific charge for electron. The value of e/M for positive rays produced by different gases is different because the masses (M) of atoms or positive ions of different gases are different. Hence e/M of positive rays is not a universal constant.

Determination of Charge to Mass Ratio (e/M) for Positive Rays

(i) Experimental arrangement: Thomson found the value of e/M for the positive rays from the different gases. The apparatus used by him is made of two parts. The first part is a spherical bulb V of very big size in which electric discharge takes place. The second part is a vacuum tube G in which ions are deflected (Fig.). At the end of this tube there exists a photographic plate. The bulb V is fitted with an anode A and a cathode K. The cathode K is a hollow copper tube of diameter 0.1 mm through which the positive ions pass through in the form of a fine beam and enter in the vacuum tube G. The gas is filled in the bulb V through a tube attached to the bulb and the pressure of the gas is so adjusted that on applying a potential difference of 30 kV to 50 kV, positive ions are emitted at very high speed.

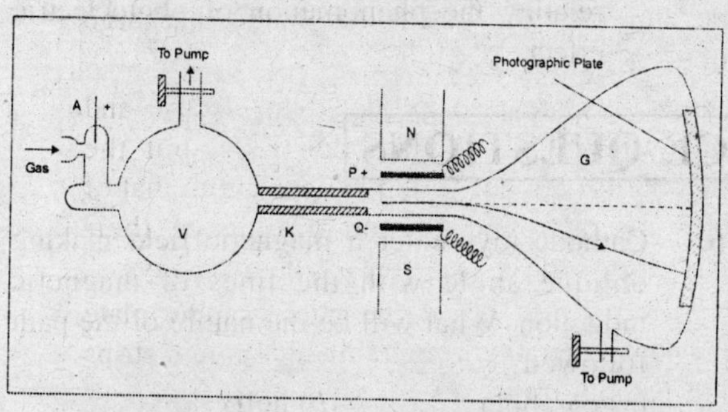

The positive ions emerging from the cathode K enter a region where electric and magnetic fields are applied simultaneously in the plane of paper in the same direction. The magnetic field is produced by keeping the tube between two poles N and S of a powerful electromagnet, while electric field is generated by applying a high potential difference between two plates P and Q. Thus the two fields are produced in the same direction i.e. directed downward in the plane of paper. Electric field deflects the positive ions in the plane of paper (in downward direction) while magnetic field deflects perpendicular to the plane of paper. The deflected ions travel in straight line in the vacuum tube G and fall on the photographic plate. A parabola is obtained on this plate.

(ii) Theory: Let the charge and mass of the positive ion be e and M respectively. Let it enters the region of two fields with some velocity v along the direction of x-axis. Let E and B represent the electric and magnetic field respectively, both being applied along y-axis.

Photoelectric Effect

(i) Hertz in 1887 observed that when ultraviolet rays are allowed to fall on negative plate of an electric discharge tube, then conduction takes place more easily. This shows that electrons are ejected from a metal surface when illuminated by light of suitable wavelength.

(ii) Hallwach confirmed Hertz observation through his following convincing experiment:

His apparatus consists of two zinc plates enclosed in an evacuated quartz tube. The plates are connected to a battery through a galvanometer (Fig.). He noted the following observations:

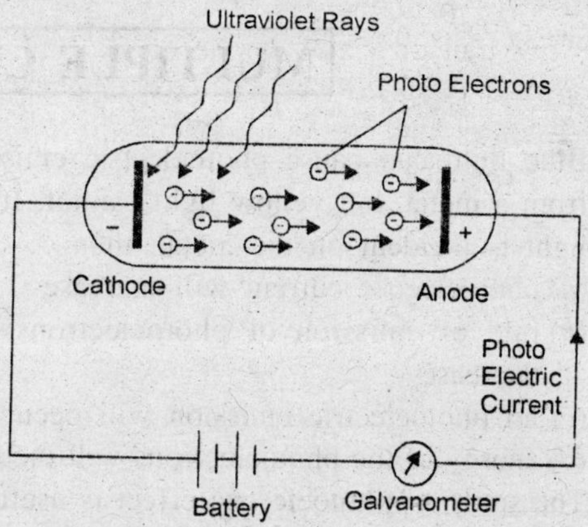

(a) When U.V. rays are allowed to fall on the cathode, a deflection is produced in the galvanometer i.e. a current flows in the circuit.

(b) As soon as rays are stopped, the deflection in the galvanometer becomes zero or current stops.

(c) If rays are made to fall on anode then either no current or a very small current flows in the circuit.

(iii) Lenard's explanation:

(a) He explained that when ultraviolet rays fall on cathode, electrons are ejected from it which are attracted towards anode or positively charged plate. Hence, the circuit which was incomplete till now, due to air gap between two plates in the tube, gets completed due to flow of electrons and a current starts flowing in the circuit.

(b) However, when rays fall on anode, the electrons are again emitted from the plate in the same way as earlier but due to being negatively charged do not reach the cathode i.e. circuit again remains incomplete and current does not flow.

(iv) This phenomenon of emission of electrons from a metallic surface when illuminated by light of appropriate wavelength or frequency is known as photoelectric effect. The electrons emitted in this process are called as photoelectrons and the current produced in the circuit is called as photoelectric current.

Experimental Study of Photoelectric Effect

Lenard and Millikan studied the photoelectric effect experimentally. They used plates of different materials and illuminated them with light rays of different intensities and different frequencies and measured in each case.

(a) maximum kinetic energy of emitted photoelectrons, and

(b) the strength of photoelectric current. On the basis of the results of these experiments they found many interesting conclusions relating the phenomenon of photoelectric effect.

MULTIPLE CHOICE QUESTIONS

1. Blue light can cause photoelectric emission from a metal, but yellow light cannot. If red light is incident on the metal, then

(*a*) photoelectric current will increase

(*b*) rate of emission of photoelectrons will decrease

(*c*) no photoelectric emission will occur

(*d*) energy of the photoelectrons will increase

2. The study of photoelectric effect is useful in understanding

(*a*) quantization of energy

(*b*) quantization of charge

(*c*) conservation of charge

(*d*) conservation of kinetic energy

3. Which of the following is not property of the photons?

(*a*) momentum (*b*) energy

(*c*) frequency (*d*) rest mass

4. The photoelectric effect is a phenomenon which can be used to convert

(*a*) electric energy into mechanical energy

(*b*) light energy into electric energy

(*c*) photon into electron

(*d*) light energy into mechanical energy

5. What is the order of pressure of the gas in cathode ray tube?

(*a*) 1 atmosphere (*b*) 1 mm of Hg

(*c*) 1 μ m of Hg (*d*) 1 nm of Hg

6. Cathode rays enter a magnetic field making oblique angle with the lines of magnetic induction. What will be the nature of the path followed?

(*a*) parabola (*b*) helix

(*c*) circle (*d*) straight line

7. The mass of electron is M, its charge is e and it is accelerated from rest through a potential V. The energy acquired by the particle in electron volt will be

(*a*) V (*b*) eV

(*c*) MeV (*d*) none of the above

8. The mass of electron is M, its charge is e and it is accelerated from rest through a potential V. The velocity required by the electron will be

(*a*) V (*b*) eV

(*c*) MeV (*d*) none of the above

9. The e/m is not constant for

(*a*) cathode rays (*b*) positive rays

(*c*) α-rays (*d*) β-rays

10. An electron and a proton are fired through uniform magnetic fields perpendicular to it with the same kinetic energy. What is the nature of trajectory?

(*a*) electron trajectory is more curved than the proton trajectory

(*b*) electron trajectory is less curved than the proton trajectory

(c) both trajectories are equally curved
(d) both particles move along straight line path

11. Given that the mass of neutron or proton is approximately 1840 times the mass of electron. What is the ratio of specific charge of electron to α-particle?
(a) $\frac{1}{1840}$ (b) $\frac{1}{3680}$
(c) 1840 (d) 3680

12. If an electron has an initial velocity perpendicular to direction of electric field, the path of the electron is
(a) a straight line (b) a parabola
(c) a circle (d) an ellipse

13. The magnitude of saturation photoelectric current depends upon
(a) frequency of radiations
(b) intensity of radiations
(c) work function
(d) stopping potential

14. Which electrons are emitted in the photoelectric effect?
(a) electrons in the inner orbits of the atom
(b) electrons in the outer most orbit of the atom
(c) electrons from within the nucleus
(d) electrons freely roaming about in the inter-atomic space

15. The photo electric effect supports a theory of light proposed by
(a) Newton (b) Huygen
(c) Einstein (d) none of the above

16. Which of the following cannot be found at rest
(a) photon. (b) electron
(c) proton (d) neutron

17. Which of the following achieves conversion of electromagnetic wave energy into electrical energy?
(a) Coolidge tube (b) Cathode ray tube
(c) Thermo couple (d) Photocell

18. Which of the following support the quantum nature of the em radiations?
(I) photoelectric effect
(II) Compton effect
(III) Doppler's effect
(IV) field effect
(a) I and II (b) II and III
(c) III and IV (d) IV and I

19. What determines the work function of a metal?
(a) frequency of the incident light
(b) energy required to ionise the atom
(c) energy required to eject the electron from the orbit
(d) none of the above

20. Why does the photoelectric effect reveal the particle nature of radiations?
(a) energy of the emitted electrons depends on the intensity of radiations
(b) energy of the emitted electrons depends on the frequency of radiations
(c) radiations eject electrons
(d) radiations are absorbed by the electrons

21. How does the photoelectric current vary with the intensity of the incident light?
(a) decreases with increase in intensity
(b) decreases with the decrease in intensity
(c) independent of the intensity
(d) decreases with intensity below the threshold frequency and increases above it

22. Which of the following is not the property of the laser beam?
(a) coherence
(b) very narrow angular spread
(c) can be focused to a tiny point
(d) consists of electrons

Directions Qs. 23-27: *In each of the following questions there are two statements Assertion (A) is given followed by a corresponding statement of Reason (R) just below it. Of the statements, mark the correct answer:*

(a) If both assertion and reason are true and reason is the correct explanation of assertion.
(b) If both assertion and reason are true but reason is not the correct explanation of assertion
(c) If assertion is true but reason is false
(d) If assertion is false but reason is true.

23. **Assertion (A):** Cathode rays get deflected by electric or magnetic fields but light rays remain unaffected by these fields.
Reason (R): Light rays are electromagnetic waves and not charged particles.

24. Assertion (A): An electron is not deflected on passing through certain region of space. This observation confirms that there is no magnetic field in that region.

Reason (R): The deflection of electron depends on angle between velocity of electron and direction of magnetic field.

25. Assertion (A): Electric conduction in gases is possible at normal pressure.

Reason (R): The electric conduction in gases depends only upon the potential difference between the electrodes and pressure.

26. Assertion (A): Gases become conducting only when their pressure is lowered.

Reason (R): At low pressure, the discharge current is high.

27. Assertion (A): Light is produced in gases in the process of electric discharge through them at low pressure.

Reason (R): At high pressure electrons of gaseous atoms collide and reach at excited state.

ANSWERS

1	2	3	4	5	6	7	8	9	10
(*c*)	(*a*)	(*b*)	(*b*)	(*c*)	(*b*)	(*a*)	(*d*)	(*b*)	(*a*)
11	**12**	**13**	**14**	**15**	**16**	**17**	**18**	**19**	**20**
(*d*)	(*b*)	(*b*)	(*d*)	(*d*)	(*a*)	(*d*)	(*a*)	(*d*)	(*b*)
21	**22**	**23**	**24**	**25**	**26**	**27**			
(*b*)	(*d*)	(*a*)	(*d*)	(*d*)	(*b*)	(*c*)			

SOME SELECTED EXPLANATORY ANSWERS

1. Threshold wave length is less than that of yellow light. So, the red light cannot cause photoelectric emission.

5. The pressure is very very low in CRO.

6. When the charged particle enters the magnetic field making angle other than 90° with it, the path is a helix.

9. Positive rays may consist of ions of different elements.

11. $M_\alpha = 4 \times M_e$ and $Q_\alpha = 2Q_e$.

12. It is similar to the throwing of stone horizontally from the top of a tower.

14. Free electrons are ejected in photoelectric effect.

15. Einstein explained photoelectric effect on the basis of quantum theory, which was proposed by Max Planck.

16. Rest mass of photon is zero.

17. Photocell produces electric current with the help of light.

18. Both photoelectric effect and Compton effect support quantum theory of light.

19. It is the energy required to free the electron from the metallic surface against Coulomb attraction.

20. Frequency is always an integral number.

15 MODERN PHYSICS

Bohr's Model of Atom

Bohr's model of atom is based on the following three postulates:

First Postulate: The electrons revolve around the nucleus in certain allowed circular orbits only and while in these orbits they do not radiate energy i.e. electronic orbits are stable. The necessary centripetal force for their circular motion is provided by the electrostatic attractive force between the positively charged nucleus and negatively and charged electrons. If Z is the atomic number of the nucleus and r is the radius of any allowed electrons orbit, then

$$\frac{mv^2}{r} = \frac{Ze \times e}{4\pi\varepsilon_0 r^2} \qquad \ldots(1)$$

where m is the mass of electron and v, the velocity of electron in the allowed orbit.

Second Postulate: The allowed orbits in which electron can revolve are those in which the angular momentum of an electron is an integer multiple of $h/2\pi$, where h is Planck's constant having value 6.63×10^{-34} Js i.e. if m and v are the mass and velocity of the electron in a permitted orbit of radius r, then

$$mvr = \frac{nh}{2\pi} \qquad \ldots(2)$$

where n is called principal quantum number and has the integer values 1, 2, 3, ... The equation (2) is also called Bohr's quantisation condition.

Third Postulate: According to Bohr's model, the electrons do not radiate any energy till they revolve in their own orbits i.e. orbits of electrons are stable. The energy is radiated only and only when an electron jumps from higher energy orbit to lower energy orbit and the energy is absorbed, when it jumps from lower energy orbit to higher energy orbit. **Both the radiation and absorption of energy takes place in the form of photons.**

If E_i and E_f are the energies associated with the orbits of principal quantum numbers n_i and n_f respectively ($n_i > n_f$), then the frequency of the emitted radiation is given by

$$hn = E_i - E_f \qquad \ldots(3)$$

The equation (3) is called Bohr's frequency condition.

Radioactivity

(i) **Substance capable of emitting radiations are called radioactive substances and the phenomenon of spontaneous emission of radiations from the substance is called Radioactivity.**

(ii) After the discovery of radioactivity in uranium, it was found that some other elements like – thorium, polonium, actinium, etc. are also radioactive. In 1898, Piere Curie and Madame Curie discovered a new radioactive element called radium which is found to be 10^6 times more radioactive than uranium.

Radiations Emitted By Radioactive Elements

A large number of experiments were conducted to determine the nature of radiations emitted by the radioactive substances immediate after the discovery of natural radioactivity. On the basis of these experiments, it was found that radiations emitted by a radioactive substance were of three different types namely; α, β and γ-rays. It should be noted here that all the radioactive elements do not necessarily emit all these three types of radiations.

Laws of Radioactive Decay

(i) The following two laws are obeyed by radioactive decay.

(a) The radioactive decay is spontaneous with the emission of α, β and γ-rays. It is not influenced by external condition such as temperature, pressure, electric and magnetic field (The nuclear forces are extremely strong and hence nuclear disintegration is unaffected by normal changes in external conditions).

(b) The rate of disintegration (i.e. number of atoms that disintegrate per second) is directly proportional to the number of radioactive atoms present. This is known as decay law.

(ii) Mathematical treatment of decay law:

Let at the beginning of disintegration (i.e. at $t = 0$), the number of radioactive atoms present be N_0. As the time passes, the number of original radioactive atoms decreases due to continuous disintegration. Suppose after time t, the number of atoms left is N. Let us suppose that dN atoms now disintegrate in small time dt.

$$\therefore \text{ Rate of disintegration } = \frac{dN}{dt}$$

According to decay law, the rate of disintegration is directly proportional to the number of radioactive atoms present i.e.

$$\frac{dN}{dt} \propto N$$

$$\text{or} \quad \frac{dN}{dt} = -\lambda N \qquad \ldots(1)$$

When λ is a constant of proportionality and is called decay constant or disintegration constant. The minus sign indicates that N is decreasing with respect to time

Equation (1) can be written as

$$\frac{dN}{N} = -\lambda\, dt$$

Integrating on both sides, we get

$$\log_e N = -\lambda t + K \qquad \ldots(2)$$

Where K is a constant of integration whose value can be found from the initial conditions.

When $\quad t = 0, N = N_0$

Putting these values in equation (2), we get

$$\log_e N_0 = -\lambda \times 0 + K$$

$$\therefore \quad K = \log_e N_0$$

Hence equation (2) becomes

$$\log_e N = -\lambda t + \log_e N_0$$

$$\text{or} \log_e N - \log_e N_0 = -\lambda t$$

$$\text{or} \quad \log_e \frac{N}{N_0} = -\lambda t$$

$$\text{or} \quad \frac{N}{N_0} = e^{-\lambda t}$$

$$N = N_0 e^{-\lambda t} \qquad \ldots(3)$$

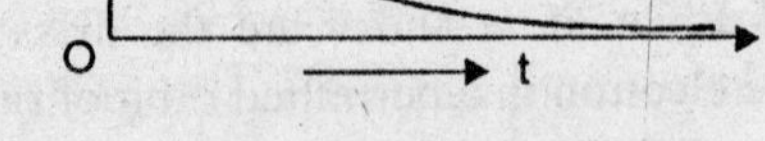

Fig.

Equation (3) follows exponential law and is known as decay equation. It can be used to find the number of radioactive atoms present at any time t if we know decay constant λ and the original number of radioactive atoms N_0. Fig. shows the curve for the decay of a radioactive substance. The curve shows that the number of atoms of a given substance decrease exponentially with respect to time. Since

N becomes zero only when t approaches infinite, therefore a radioactive substance will never disintegrate completely.

Nuclear Characteristics

(i) Nuclear mass: From Rutherford's α–particle scattering experiment we have seen that the mass of an atom is concentrated within a very small, positively charged region at the centre, called as nucleus. The total mass of nucleons in the nucleus is called as **nuclear mass** i.e.

Nuclear mass,

A = mass of protons + mass of neutrons

Since the mass of a proton or neutron is very small, therefore, nuclear masses are extremely small.

(ii) Size and Shape of the nucleus: The nucleus is believed to be very nearly spherical. Hence, its size is usually given in terms of radius. The radius of nucleus was measured by Rutherford in his α–particle scattering experiment. It has been found that nuclear radius R is given by:

$$R = R_0\, A^{1/3}$$

where $R_0 = 1.1$ fm (fermi) $= 1.1 \times 10^{-15}$ m and A is the mass number of the particular element. Which shows that the nuclear radius is of the order of 10^{-15} metre. Further, R_0 is the distance of closest approach to the nucleus and is also known as nuclear unit radius.

(iii) Nuclear charge: It is the total positive charge carried by the nucleus and is numerically equal to its atomic number times the electronic charge i.e. Nuclear charge = Ze

(iv) Nuclear density: The ratio of the mass of the nucleus to its volume is called nuclear density. Consider the nucleus of an atom having mass number A. If we neglect the mass of the orbital electrons then mass of the nucleus of the atom = A. a. m. u.= $A \times 1.660565 \times 10^{-27}$Kg

If R is the radius of the nucleus then volume of the nucleus = (4/3) ρR^3 = (4/3) $\rho R_0^{\,3}$ A.

∴ Density of the nucleus

$$\rho = \frac{\text{mass of nucleus}}{\text{Volume of nucleus}}$$

$$= \frac{A \times 1.660565 \times 10^{-27}\ \text{Kg}}{(4/3)\,\pi\,(1.1 \times 10^{-15})^3 \times A\text{m}^3}$$

$$= \mathbf{2.97 \times 10^{-17}\ Kg/m^3}$$

Einstein's Mass Energy Equivalence Principle

On the basis of his theory of Relativity, Einstein showed that the mass of a body is not constant but increases with velocity. In other words, when a body is given Kinetic energy, its mass is increased. This shows that mass depends upon energy and vice–versa. Therefore, mass and energy are inter–convertible i.e. they can be changed into each other. According to Einstein, if a substance loses an amount Δm of its mass, an equivalent amount ΔE of energy is produced, where

$$\Delta E = \Delta mc^2$$

where c, is the speed of light. This is called Einstein's mass-energy equivalence principle.

Atomic Mass Unit (A.M.U.)

(i) The masses of atoms, nuclei, sub atomic particles are very small. Hence, a small unit is used to express these masses. This unit is called as atomic mass unit (a.m.u.) **1 a.m.u. is equal to one twelfth part of the mass of carbon ($_6C^{12}$) atom.** Mass of $_6C^{12}$ is 12.00 a.m.u.

(ii) Now, the mass of 1 gm-atom of carbon is 12 gm and according to Avogadro's Hypothesis it has N (Avogadro's Number) atoms. Thus, the mass of one atom of carbon is (12/N) gms. According to the definition,

$$1\ \text{amu} = \frac{1}{12} \times (\text{mass of one carbon atom})$$

$$= \frac{1}{12} \times \frac{12}{N} = \frac{1}{N}\ \text{grams}$$

$$= \frac{1}{6.02 \times 10^{23}}\ \text{grams}$$

$$= 1.66 \times 10^{-24}\ \text{grams}$$

$$= 1.66 \times 10^{-27}\ \text{Kg}$$

(iii) We can now find energy equivalent to 1 a.m.u. mass with the help of Einstein's mass energy relation. If this energy be ΔE, then

$$\Delta E = \Delta mc^2$$
$$= (1.66 \times 10^{-27}) \times (3 \times 10^8)^2$$
$$= 1.49 \times 10^{-10} \text{ Joules}$$

But 1.6×10^{-19} Joules = 1 eV

$$\therefore \quad \Delta E = \frac{1.46 \times 10^{-10}}{1.6 \times 10^{19}}$$
$$= 0.931 \times 10^9 \text{ eV}$$
$$= \mathbf{931\ MeV}$$

i.e. **1 a.m.u. of mass is equivalent of 931 MeV energy.**

Binding Energy of Nucleus

(i) Binding energy of nucleus is the total energy required to liberate all the nucleons from the nucleus (i.e. to disintegrate the nucleus completely into its constituent particles).

(ii) Obviously, this is the same energy with which the nucleons are held together within the nucleus. The origin of binding energy results from strong nuclear exchange forces. In other words, we may think of the existence of binding energy in other useful way also. A nucleus is made by the coming together of various nucleons. It has been observed experimentally that mass of the nucleus always less than the sum of the masses of its constituents then measured in free state. For example, Deutron ($_1H^2$) is composed of one proton and one neutron. The mass of $_1H^2$ nucleus is less than the sum of masses of 1 proton and 1 neutron. The question arises where the difference in mass has gone. The answer is that this decrease in mass has been converted into energy binding the nucleons together according to the following relation:

$$\Delta E = \Delta mc^2$$

Where, ΔE = Binding energy of nucleus, Δm = decrease in mass, called mass defect and c = velocity of light.

Hence in the formation of stable nucleus, the following equation holds good

Mass of protons + Mass of neutrons
= Mass of Nucleus $+mc^2$ (B.E.)

Expression for binding energy per nucleon: We calculate binding energy per nucleon (*BEN*) in order to compare the stability of various nucleus. Higher is the binding energy per nucleon, more stable is the nucleus.

We have seen that mass defect during the formation of a nucleus

$$\Delta m = Zm_p + (A - Z)\, m_n - M$$

Total binding energy of nucleus

$$\Delta E = \Delta mc^2 = [Zmp + (A - Z)m_n - M] \times c^2$$

Mean binding energy per nucleon.

$$BEN = \frac{\Delta E}{A}$$
$$= \frac{\Delta mc^2}{A}$$
$$= \left[\frac{Z}{A}(m_p - m_n) + m_n - \frac{M}{A}\right] \times c^2$$

If the mass M of the nucleus is found experimentally, we can find the mean binding energy per nucleon since all other factors are known to us.

Nuclear Fission

(i) Nuclear Fission is the phenomenon of breaking a heavy nucleus into two light nuclei of almost equal masses alongwith the release of huge amount of energy.

(ii) The process of nuclear fission was first discovered by German Scientists Otto Hahn and Strassman in 1939. They bombarded uranium nucleus ($_{92}U^{235}$) with slow neutrons and found that intermediate product ($_{92}U^{236}$) was split into two medium weight part with the release of enormous energy. These fragments had atomic numbers far less than the target nucleus ($_{92}U^{235}$). The nuclear fission of $_{92}U^{235}$ is given by the following nuclear reaction:

$$_{92}U^{235} + {_0n^1} \rightarrow [_{92}U^{236}]$$
$$\rightarrow {_{56}Ba^{144}} + {_{36}Kr^{89}} + 3\,{_0n^1} + \text{energy}$$

(iii) When a neutron strikes $_{92}U^{235}$ nucleus, it is absorbed by it, producing a highly unstable $_{92}U^{236}$ nucleus. Instead of emitting α or β particles or γ-rays, this unstable nucleus is split into two middle weight parts viz $_{56}Ba^{144}$ and krypton ($_{36}Kr^{89}$). During this fission, three neutrons are given out and a mass defect occurs which is converted into enormous amount of energy.

Nuclear Fusion

(i) **Nuclear fusion is the process of combining two light nuclei to form a heavy nucleus. Clearly, the process of fusion is just the reverse of fission.** An important feature of fusion is that there is a release of huge amount of energy in the process. This can be easily understood. When two light nuclei are combined to form a heavy nucleus there occurs a small mass defect. In other words, the mass of the heavy nucleus turns out to be less than the sum of the masses of two light nuclei. This small mass defect results in the release of a huge amount of energy according to the relation

$$\Delta E = \Delta mc^2$$

(ii) *Example:* The following reaction is possible by the fusion of two nuclei of heavy hydrogen or deuterium ($_1H^2$).

$$_1H^2 + {_1H^2} \rightarrow {_1H^3} + {_1H^1} + 4.0 \text{ MeV}$$

The nucleus of tritium ($_1H^3$) so formed can again fuse with a deuterium nucleus

$$_1H^3 + {_1H^2} \rightarrow {_2He^4} + {_0n^1} + 17.6 \text{ MeV}$$

The net result of these two reactions is that then deuterium nuclei fuse together to form a helium nucleus and release 21.6 MeV energy which is obtained in the form of Kinetic energy of proton ($_1H^1$) and neutron ($_0n^1$).

Alternatively, following reactions are also possible for the fusion of three deuterium nuclei

$$_1H^2 + {_1H^2} \rightarrow {_2He^3} + {_0n^1} + 3.3 \text{ MeV}$$

$$_2He^3 + {_1H^2} \rightarrow {_2He^4} + {_1H^1} + 18.3 \text{ MeV}$$

(iii) The energy output in the process of Nuclear fusion (21.6 MeV) is quite less than the energy liberated in the fission of a U^{235} nucleus (200 MeV), but this does not imply that fusion is a weaker liberated in the fission of a U^{235} nuclei in 1 gram uranium. Therefore, **the energy liberated by the fusion of a certain mass of heavy hydrogen is much more than the energy released by the fission of equal mass of uranium.**

(iv) Nuclear fusion is quite a difficult process as compared to nuclear fission process. This is due to the repulsive force between them becomes very strong. For fusion against this force, they require very high energy ($\cong$ 0.1 MeV). To provide so much high energy to them very high temperature ($\cong 10^8$ K) and very high pressure is required. **Since high temperature is required to cause fusion, therefore fusion is sometimes called thermonuclear reaction.**

(v) The conditions of high temperature and high pressure are not available on earth. Such conditions are available either in the sun or are produced when a nuclear - fission bomb is exploded. **Thus nuclear fusion is possible on earth only by the explosion of a nuclear fission bomb. Hydrogen bomb is a nuclear fusion bomb.**

(vi) **Source of Solar Energy :** We knows that hydrogen and helium constitute about 90% of the mass of the sun and the rest 10% are other elements. The sun has been radiating huge amount of energy for billions of years. Chemical reactions cannot possibly be the source of energy because the energy released in such a reaction is very small. Similarly, it cannot be due to fission as the amount of heavy elements present in the sun is very small.

It is now believed that fusion reactions in the sun are responsible for its enormous heat. The following two sets of thermonuclear reactions account for the energy produced in the Sun and Stars:

(a) Carbon – Nitrogen cycles,

(b) Proton – Proton cycle.

Both these fusion reactions take place in the sun. Although these two reactions have different sequence of intermediate reactions, the net effect is the combination of four hydrogen nuclei (proton, $_1H^1$) to form a helium nucleus and two positrons alongwith the release of energy.

$$4\,_1H^1 \rightarrow {}_2He^4 + 2\,_{+1}\beta^0 + \text{Energy} + 2\nu \text{ (Neutrino)}$$
(Positron)

MULTIPLE CHOICE QUESTIONS

1. A metallic surface has a threshold wavelength 5200 Å. This surface is irradiated by monochromatic light of wavelength 4500 Å. Which of the following statements is true?
(*a*) the electrons are emitted from the surface having energy between 0 and infinity
(*b*) the electrons are emitted from the surface having energy between 0 and certain finite maximum value
(*c*) the electrons are emitted from the surface, all having certain finite energy
(*d*) no electron is emitted from the surface

2. In photoelectric effect when photons of energy $h\nu$ fall on a photosensitive surface (work function $h\nu_0$) electrons are emitted from the metallic surface with a kinetic energy. It is possible to say that
(*a*) all ejected electrons have same kinetic energy equal to $h\nu - h\nu_0$
(*b*) the ejected electrons have a distribution of kinetic energy from zero to $(h\nu - h\nu_0)$
(*c*) the most energetic electrons have kinetic energy equal to $h\nu_0$
(*d*) all ejected electrons have kinetic energy $h\nu_0$

3. Photoelectric effect can be explained by assuming that light
(*a*) is a form of transverse waves
(*b*) is a form of longitudinal waves
(*c*) can be polarised
(*d*) consists of quanta

4. In photoelectric effect, the photoelectric current
(*a*) does not depend on photon frequency, but only on intensity of incident beam
(*b*) depends both on intensity and frequency of incident beam
(*c*) increases when frequency of incident photons increases
(*d*) decreases when frequency of incident photons increases

5. Photo electric effect supports the quantum nature of light because
(*a*) there is minimum frequency of light below which no photoelectrons are emitted
(*b*) the maximum kinetic energy of photoelectrons depends only on the frequency of light and not on intensity
(*c*) even when a metal surface is faintly illuminated, the photoelectrons leave the surface immediately
(*d*) electric charge of the photoelectrons is quantised

6. A photo cell is illuminated by a small bright source placed 1 m away. When the same source of light is placed 2 m away. Which of the following in true about the electrons emitted by the photo cathode?
(*a*) each carry one quarter of their previous energy
(*b*) each carry one quarter of their previous momenta
(*c*) are half as numerous
(*d*) are one-quarter as numerous

7. Which one of the following is incorrect statement about a photon?
(*a*) photon's rest mass is zero
(*b*) photon's momentum is $h\nu/c$
(*c*) photon's energy is $h\nu$
(*d*) photons exert no pressure

8. In photoelectric effect, the current
(*a*) increases with increase of frequency of incident photon

(*b*) decreases with increase of frequency of incident photon
(*c*) does not depend on the frequency of photon but depends only on intensity of incident light
(*d*) depends both on intensity and frequency of incident beam

9. If the distance of 100 watt lamp is increased from a photo cell, the saturation current '*i*' in the photo cell varies with distance '*d*' as
(*a*) $i \propto d^2$ (*b*) $i \propto d$
(*c*) $i \propto 1/d$ (*d*) $i \propto 1/d^2$

10. Blue light can cause photoelectric emission from a metal, but yellow light cannot. If red light is incident on the metal, then
(*a*) photoelectric current will increase
(*b*) rate of emission of photoelectrons will decrease
(*c*) no photoelectric emission will occur
(*d*) energy of the photoelectrons will increase

11. The graph between, which of the following two factors for photoelectric effect, is a straight line?
(*a*) intensity of radiation and photoelectric current
(*b*) potential of anode and photoelectric current
(*c*) threshold frequency and velocity of photoelectrons
(*d*) intensity of radiations and stopping potential

12. A photosensitive material would emit electrons if excited by photons beyond a threshold. To cross the threshold you would increase
(*a*) intensity of light
(*b*) wavelength of light
(*c*) frequency of light
(*d*) the voltage applied to light source

13. The study of photoelectric effect is useful in understanding
(*a*) quantization of energy
(*b*) quantization of charge
(*c*) conservation of charge
(*d*) conservation of kinetic energy

14. Which of the following is not property of the photons?
(*a*) momentum (*b*) energy
(*c*) frequency (*d*) rest mass

15. Photoelectrons are being obtained by irradiating zinc by a radiation of 3100 Å. In order to increase the kinetic energy of ejected photoelectrons
(*a*) the intensity of radiation should be increased
(*b*) the wavelength of radiation should be increased
(*c*) the wavelength of radiation should be decreased
(*d*) both wavelength and intensity of radiation should be increased

16. When a photon collides with an electron which of the following characteristic of the photon increases?
(*a*) energy
(*b*) frequency
(*c*) wavelength
(*d*) none of the above

17. A photon stimulates the emission of another photon of the same energy. The two are
(*a*) in phase and travel in the same direction
(*b*) in phase and travel in the opposite direction
(*c*) out of phase and travel in the same direction
(*d*) out of phase and travel in the opposite direction

18. Which of the following makes use of photoelectric effect?
(*a*) television receiver
(*b*) television camera
(*c*) cathode Ray oscillograph
(*d*) radar

19. The momentum of the photon is given by
(*a*) $\frac{h}{\lambda}$ (*b*) $\frac{h}{c\lambda}$
(*c*) $\frac{hc}{\lambda}$ (*d*) $\frac{h\lambda}{c}$

20. The dynamic mass of the photon is given by

(a) $\frac{h\nu}{c}$ (b) $\frac{h\lambda}{c}$

(c) $\frac{h}{c\lambda}$ (d) $\frac{h}{c\nu}$

21. Which of the following characteristics of photoelectric effect supports the particle nature of radiations.

(a) threshold frequency.

(b) dependence of the velocity of photoelectron on frequency.

(c) independence of velocity of photoelectrons on intensity of radiations.

(d) instantaneous photoelectric emission.

22. A photo sensitive metal is not emitting photoelectrons when irradiated. It will do so when threshold is crossed. To cross the threshold we need to increase

(a) intensity (b) frequency

(c) wavelength (d) none of the above

23. The frequency and intensity of the incident beam of light falling on the surface of photoelectric material is increased by a factor of two. This will

(a) increase the maximum kinetic energy of the photoelectrons as well as photoelectrons current by a factor of two

(b) increase the maximum kinetic energy of photoelectric and would increase the photoelectric current by a factor of two

(c) increase the maximum kinetic energy of photoelectrons by a factor of two and will have no effect on photo electric current

(d) increase the photoelectric current by a factor of two but will have no effect in kinetic energy of emitted electrons

24. The maximum energy of the electrons released in photocell is independent of

(a) frequency of incident light

(b) intensity of incident light

(c) nature of cathode surface

(d) none of these

25. Ultraviolet radiation of 6.2 eV falls on an aluminium surface (work function 4.2 eV). The kinetic energy in joule of the fastest electron emitted is approximately

(a) 3×10^{-21} (b) 3×10^{-19}

(c) 4×10^{-17} (d) 3×10^{-15}

26. The ionisation potential of hydrogen atom is 13.6 V. How much energy need to be supplied to ionise the hydrogen atom in the first excited state?

(a) 13.6 eV (b) 27.2 eV

(c) 3.4 eV (d) 6.8 eV

27. Ionisation energy for hydrogen atom in the ground state is E. What is the ionisation energy of Li^{++} atom in the 2nd excited state?

(a) E (b) $3E$

(c) $6E$ (d) $9E$

28. The energy difference between the first two levels of hydrogen atom is 10.2 eV. For another element of atomic number 10 and mass number 20, this will be

(a) 1020 eV (b) 2040 eV

(c) 0.51 eV (d) 0.102 eV

29. For the first member of Balmer series of hydrogen spectrum, the wavelength is λ. What is the wavelength of the second member?

(a) $\frac{5}{30}\lambda$ (b) $\frac{3}{16}\lambda$

(c) $\frac{4}{9}\lambda$ (d) $\frac{20}{27}\lambda$

30. If the ionisation energy for hydrogen atom is 13.6 eV. The energy required to excite it from the ground state to the next higher state is nearly

(a) 3.4 eV (b) 10.2 eV

(c) 12.1 eV (d) 1.5 eV

31. The ionisation energy of hydrogen atom is 13.6 eV. Hydrogen atoms in the ground state are excited by monochromatic radiation of photon energy 12.1 eV. The spectral lines emitted by hydrogen atom according to Bohr's theory will be

(a) one (b) two

(c) three (d) four

32. The transition of the electron takes place from $n = 2$ orbit to $n = 1$ orbit. Which of the following gives the shortest wavelength?
(*a*) Hydrogen atom
(*b*) Deutrium atom
(*c*) singly ionised Helium
(*d*) doubly ionised Helium

33. The wavelength of K_α-line in copper is 1.54 Å. The ionisation energy of K electron in copper in joule is
(*a*) 11.2×10^{-17} (*b*) 12.9×10^{-16}
(*c*) 17×10^{-15} (*d*) 10×10^{-16}

34. The following figure indicates the energy levels of a certain atom. When the system moves from $2E$ level to E, a photon of wavelength λ is emitted. The wavelength of photon produced during its transition from level $4E/3$ to level E is

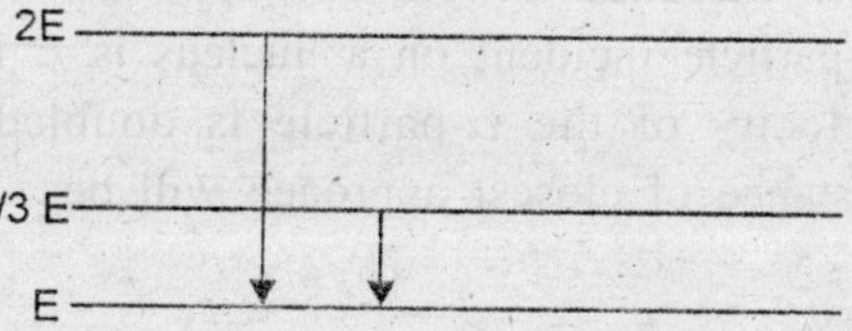

(*a*) $\frac{\lambda}{3}$ (*b*) $\frac{3\lambda}{4}$
(*c*) $\frac{4\lambda}{3}$ (*d*) 3λ

35. The ionisation energy of hydrogen atom is 13.6 eV. The ionisation energy of helium atom would be
(*a*) 1.6 eV (*b*) 27.2 eV
(*c*) 6.8 eV (*d*) 54.4 eV

36. The ionisation energy of 10 times ionised sodium atom is
(*a*) 13.6 eV (*b*) 13.6 × 11 eV
(*c*) $\frac{13.6}{11}$ (*d*) $13.6 \times (11)^2$ eV

37. The energy of hydrogen atom in its ground state is –13.6 eV. The energy of the level corresponding to $n = 5$ is
(*a*) – 0.54 eV (*b*) – 5.40 eV
(*c*) – 0.85 eV (*d*) – 2.72 eV

38. Which of the following sources give discrete emission spectrum?
(*a*) incandescent electric bulb
(*b*) sun
(*c*) mercury vapour lamp
(*d*) candle

39. Band spectrum is produced by
(*a*) H (*b*) He
(*c*) H_2 (*d*) Na

40. Energy levels *A*, *B*, *C* of a certain atom correspond to increasing values of energy i.e., $E_A < E_B < E_C$. If λ_1, λ_2, λ_3 are the wavelengths of radiation corresponding to the transitions *C* to *B*, *B* to *A* and *C* to *A* respectively, which of the following statement is correct?

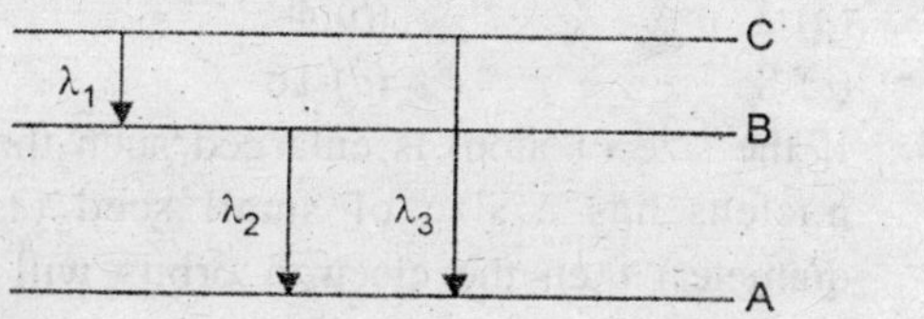

(*a*) $\lambda_3 = \lambda_1 + \lambda_2$
(*b*) $\lambda_3 = \frac{\lambda_1 \lambda_2}{\lambda_1 + \lambda_2}$
(*c*) $\lambda_1 + \lambda_2 + \lambda_3 = 0$
(*d*) $\lambda_3^2 = \lambda_1^2 + \lambda_2^2$

41. In the following figure the energy levels of hydrogen atom have been shown along with some transitions marked *A*, *B*, *C*, *D* and *E*. The transitions *A*, *B* and *C* respectively represent

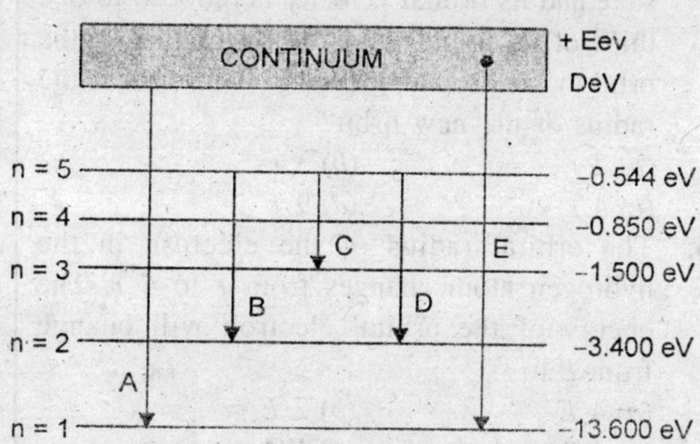

(*a*) the first member of Lyman series, third member of Balmer series and second member of Paschen series
(*b*) the ionisation potential of hydrogen, second member of Balmer series and third member of Paschen series
(*c*) the series limit of Lyman series, second member of Balmer series and second member of Paschen series
(*d*) the series limit of Lyman series, second member of Balmer series and third member of Paschen series

42. In an atom, the two electrons revolve around the nucleus. What is the ratio of their time periods in the first excited state and the ground state.
(*a*) 2 (*b*) 4
(*c*) 8 (*d*) 16

43. If the size of atom is enlarged such that the nucleus has a size of small seed (1 mm diameter) then the electron orbits will have the size of
(*a*) a big seed (1 cm diameter)
(*b*) a cricket ball (10 cm diameter)
(*c*) a cricket playground (100 m diameter)
(*d*) the earth

44. The radius of Bohr's first orbit is *r*. What is the radius of the first orbit in the singly ionised helium atom?
(*a*) 4 *r* (*b*) 2 *r*
(*c*) *r*/2 (*d*) *r*/4

45. The orbital electron of the hydrogen atom jumps from the ground state to a higher energy state and its orbital velocity is reduced to one third of its initial value. If the radius of the orbit in the ground state is *r*, then what is the radius of the new orbit?
(*a*) 2 *r* (*b*) 3 *r*
(*c*) 4 *r* (*d*) 9 *r*

46. The orbital radius of the electron in the hydrogen atom changes from *r* to 4 *r*. The energy of the orbital electron will change from *E* to
(*a*) 4 *E* (*b*) 2 *E*
(*c*) *E*/2 (*d*) *E*/4

47. The orbital velocity of the electron in the ground state of hydrogen atom is *v*. What will be its orbital velocity when excited to the energy state – 1.51 eV?
(*a*) *v*/9 (*b*) *v*/4
(*c*) *v*/3 (*d*) *v*/2

48. Who proposed that the orbits of the electron in the atom are elliptical?
(*a*) Thomson (*b*) Bohr
(*c*) Rutherford (*d*) Sommerfield

49. Suppose the elements with principal quantum number $n > 4$ where not allowed in nature, what would have been the total number of elements in nature?
(*a*) 10 (*b*) 28
(*c*) 60 (*d*) 110

50. The distance of closest approach of an α-particle incident on a nucleus is *r*. If the velocity of the α-particle is doubled, the distance of closest approach will be
(*a*) 2 *r* (*b*) 4 *r*
(*c*) *r*/2 (*d*) *r*/4

Directions Qs. 51-55: *In each of the following questions there are two statements Assertion (A) is given followed by a corresponding statement of Reason (R) just below it. Of the statements, mark the correct answer:*

(*a*) If both assertion and reason are true and reason is the correct explanation of assertion.
(*b*) If both assertion and reason are true but reason is not the correct explanation of assertion
(*c*) If assertion is true but reason is false
(*d*) If assertion is false but reason is true.

51. Assertion (A): Anode of Coolidge tube gets heated up at the time of emission of X-rays.
Reason (R): The anode of Coolidge tube is made of a material of high melting point.

52. Assertion (A): Penetrating power of X-rays increases with increasing wavelength.
Reason (R): The penetrating power of X-rays increases with the frequency of X-rays.

53. **Assertion (A):** X-rays are used for studying the structure of crystals.
Reason (R): The distance between the atoms of crystals is of the order of wavelength of X-rays.

54. **Assertion (A):** The phenomenon of X-ray production is basically inverse of photoelectric effect.
Reason (R): X-rays are electromagnetic waves.

55. **Assertion (A):** Soft and hard X-rays differ in frequency as well as velocity.
Reason (R): The penetrating power of hard X-rays is more than the penetrating power of soft X-rays.

ANSWERS

1	2	3	4	5	6	7	8	9	10
(*b*)	(*b*)	(*d*)	(*a*)	(a,b,c)	(*d*)	(*d*)	(*c*)	(*d*)	(*c*)
11	**12**	**13**	**14**	**15**	**16**	**17**	**18**	**19**	**20**
(*a*)	(*c*)	(*a*)	(*d*)	(*c*)	(*c*)	(*a*)	(*b*)	(*a*)	(*c*)
21	**22**	**23**	**24**	**25**	**26**	**27**	**28**	**29**	**30**
(*a*)	(*b*)	(*b*)	(*b*)	(*b*)	(*c*)	(*a*)	(*a*)	(*d*)	(*b*)
31	**32**	**33**	**34**	**35**	**36**	**37**	**38**	**39**	**40**
(*b*)	(*a*)	(*b*)	(*d*)	(*d*)	(*d*)	(*a*)	(*c*)	(*c*)	(*b*)
41	**42**	**43**	**44**	**45**	**46**	**47**	**48**	**49**	**50**
(*d*)	(*c*)	(*c*)	(*c*)	(*d*)	(*d*)	(*c*)	(*d*)	(*c*)	(*d*)
41	**42**	**43**	**44**	**45**	**46**	**47**	**48**	**49**	**50**
(*d*)	(*c*)	(*c*)	(*c*)	(*d*)	(*d*)	(*c*)	(*d*)	(*c*)	(*d*)
51	**52**	**53**	**54**	**55**					
(*b*)	(*d*)	(*a*)	(*b*)	(*d*)					

SOME SELECTED EXPLANATORY ANSWERS

1. Work function is the minimum amount of energy lost to get the electron ejected. More energy can also be lost due to the collision etc.
2. $h\nu_0$ is the minimum energy used to eject the electron. More energy of the electron can be lost due to collision etc.
3. Photoelectric effect, in fact, confirms the quantum nature of light.
4. Larger intensity means more incident photons which ejects larger number of electrons.
5. This shows that, there is a minimum amount of energy required to eject the electron. Also, according to quantum theory (i) max, KE depends only on frequency and is independent of intensity. (ii) photoelectric emission is instantaneous.
6. Intensity of light varies inversely as the square of the distance.
7. Photons exert pressure on the surface on which they fall.
8. It is the statement of fact.
9. The saturation current $\propto$ intensity $\propto$ $(1/d^2)$.
10. Threshold wave length is less than that of yellow light. So, the red light cannot cause photoelectric emission.
12. $h\nu = h\nu_0 + E_k$. Here, $\nu > \nu_0$.
14. Photons always travel with constant speed be decreased.
15. $\frac{hc}{\lambda} = \frac{hc}{\lambda_0} + E_k$.

To increase E_k, λ should be decreased.

20. $mc^2 = h\nu$. Hence $m = h\nu/c^2 = h/c\lambda$ and $p = h/\lambda$.

22. Current $\propto$ Intensity.
Kinetic energy increases with frequency.

24. $E_k = h\nu - h\nu_0$.

25. $E_k = 6.2 - 4.2 = 2$ eV
$= 2 \times 1.6 \times 10^{-19}$ J $= 3.2 \times 10^{-19}$ J.

26. Ionisation energy of nth state $= -E_n$.
For the first excited state $n = 2$ and energy in that state is $\dfrac{13.6}{2^2}$ eV $= 3.4$ eV.

27. $E_i = hcR_H Z^2$. Here $Z = 3$.
E_i (ground) for $Li^{++} = 9\,E$.
E_i (3rd ex. st.) $= 9E/(3)^2 = E$.

28. $\Delta E \propto Z^2$. Hence it will be $10.2 \times 10^2 = 1020$.

29. For Balmar series $\dfrac{1}{\lambda} = R_H\left[\dfrac{1}{2^2} - \dfrac{1}{n^2}\right]$

where $n = 3, 4, 5, 6$.
$\lambda_1 = 36/5\ R_H$ and $\lambda_2 = 16/3\ R_H$.

Hence $\lambda_2 = \dfrac{5 \times 16}{3 \times 36}\lambda_1 = \dfrac{20}{27}\lambda_1$

30. Energy in the next higher orbit
$= 13.6/(2)^2 = 3.4$ eV
Excitation energy $= 13.6 - 3.4 = 10.2$ eV.

31. Electron jumps to the energy level $13.6 - 12.1 = 1.5$ eV. It is the 3rd higher level. So, two lines corresponding to the transitions $3 \to 2$ and $3 \to 1$ will be obtained.

32. $\lambda \propto (1/Z^2)$ for hydrogen like atoms.

33. $E = \dfrac{hc}{\lambda}$.

34. $\dfrac{hc}{\lambda} = 2\,E - E = E$.

Hence $\dfrac{4}{3}E - E = \dfrac{1}{3}E$ that is $\lambda' = 3\lambda$.

35. $E \propto Z^2$ and Z for helium is 2.

36. It becomes hydrogen like atom with $Z = 11$.

37. $E_n = \dfrac{-13.6}{(5)^2} = -0.544$ eV.

40. $\dfrac{1}{\lambda} = R\left[\dfrac{1}{n_i} - \dfrac{1}{n_0^2}\right]$

Here $\dfrac{1}{\lambda_1} = R\left[\dfrac{1}{n_2^2} - \dfrac{1}{n_3^2}\right]$, $\dfrac{1}{\lambda_2} = R\left[\dfrac{1}{n_1^2} - \dfrac{1}{n_2^2}\right]$

and $\dfrac{1}{\lambda_3} = R\left[\dfrac{1}{n_1^2} - \dfrac{1}{n_3^2}\right]$

Hence $\dfrac{1}{\lambda_3} = \dfrac{1}{\lambda_1} + \dfrac{1}{\lambda_2}$.

41. The electron jump to $n = 1$ level in the Lyman series, $n = 2$ in the Balmer series and $n = 3$ level in the Paschen series.

42. $T^2 \propto r^3$ and $r \propto n^2$.
Hence $r_2/r_1 = 2^2/1^2 = 4$.
Hence $T_2/T_1 = [4^3]^{1/2} = 8$.

43. The radius of electron orbit is about 10^5 times the size of nucleus. Hence 1 mm $\times\ 10^5 = 10^2$m.

44. $r \propto (1/Z)$.

45. $v \propto (1/n)$ and $r \propto n^2$.

46. $r \propto n^2$ and $E \propto (1/n^2)$.

47. Energy state -1.51 eV correspond to $n = 3$.
And $v \propto (1/n)$

49. Total number of elements
$= 2\,[1^2 + 2^2 + 3^2 + 4^2]$.

50. $r \propto \dfrac{2Ze^2}{\frac{1}{2}Mv^2}$.

CHEMISTRY

A. PHYSICAL CHEMISTRY
B. INORGANIC CHEMISTRY
C. ORGANIC CHEMISTRY

A PHYSICAL CHEMISTRY

1 ATOMIC STRUCTURE

Dalton's Atomic Theory

The basic postulates of Dalton's atomic theory are as under:

1. All matters is made up of very tiny particles which cannot be further divided. These tiny particles are called atoms.
2. Atoms can neither be created nor destroyed during chemical reactions.
3. Atoms of a particular element are identical in all respects *i.e.*, same size, mass and other properties.
4. Atoms of different elements are different.
5. Atoms combine to form molecules.

Limitations of Dalton's Atomic Theory:

(*i*) It could not explain the laws of chemical combination by mass but failed to explain the law of gaseous volumes.

(*ii*) It could not explain by atoms of different elements have different masses, valencies, sizes etc.

(*iii*) It could not explain how and why atoms of different elements combine with each other to form compound atoms or molecules.

(*iv*) It failed to explain the nature of forces that bind together atoms in a molecule.

(*v*) It gave no satisfactory explanation between the ultimate particle of an element and that of a compound.

Rutherford's Model of Atom

Rutherford (1911) proposed the following atomic model :

1. An atom consists of a tiny nucleus which contains practically the whole mass of the atom. The atom is largely empty.

2. The nucleus is positively charged, and is surrounded by negatively charged electrons. The atom as a whole is electrically neutral.

3. The electrons are not stationary, but are revolving around the nucleus in orbits.

Drawbacks of Rutherford Model **:** The electrons orbiting around the nucleus are constantly changing their direction of motion. This cause the electron to continuously emit its energy as electromagnetic radiation. As a consequence of its continuously decreasing energy, the electron continuously decreasing radius, spiralling ultimately fall on to the nucleus. This, however, does "not happen—atoms are stable structures. The Rutherford model had sound basis of experimentally discovered nucleus. Although it failed to give a stable atom.

Postulates of Bohr's Theory

1. According to the laws of classical mechanics, an electron can revolve in an orbit of any radius

provided its velocity is of a suitable magnitude to make the centripetal force equal to electrostatic force of attraction. But Bohr postulated that an electron can revolve in a few widely separated permitted orbits. While moving along these orbits the electron does not radiate (loose) energy. This lead to the idea that each orbit is associated with definite energy. These orbits are therefore also known as energy levels. Bohr gave them number 1, 2, 3, 4 etc. These are termed as Principal quantum numbers and the various levels are designated by *K, L, M, N* etc.

2. The permissible orbit of an electron revolving round a positive nucleus are those for which the angular momentum of the electron is an integral multiple of $nh/2\pi$ where h is the Plank's constant.

3. The emission or absorption of energy takes place when an electron jumps from one permitted orbit to another.

Defects of Bohr's Atom Model:

1. It could not give the electronic structure of multi-electron atoms. Such structures with precise distribution of electrons in distinct shells and subshells, was made available later as a result of the precise study of various line spectra of atoms.

2. Even for hydrogen atom, Bohr's model could not provide the detailed structure which could explain the known effects and finer details of its line spectra.

3. Bohr's model could give only the frequency of emitted radiation but failed to give any structural basis to estimate the relative intensities of hydrogen spectral lines.

Orbital

The quantum theory describes the motion of electrons around the nucleus in terms of a set of mathematical expressions. And these expressions are called orbital wave function which are denoted by the symbol Ψ. The orbital wave functions are also commonly known as orbitals.

Shapes of Atomic Orbitals : Since an *s* orbital can have only one possible orientation, it should be spherical in shape because a sphere can be defined completely by a single value.

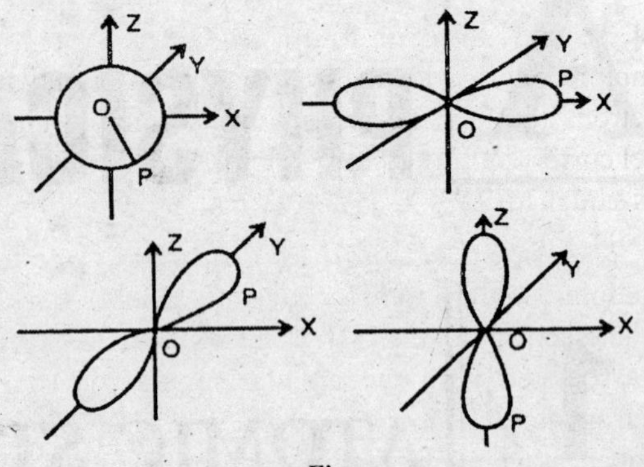

Fig.

For *d*-sublevel *m* has five values. This means *d*-sublevel can have five possible orientations. Thus there would be five *d*-orbitals. Whatever be the value of the principle quantum number, these five *d*-orbitals possess equal energies and differ only in orientations in three-dimensional spaces.

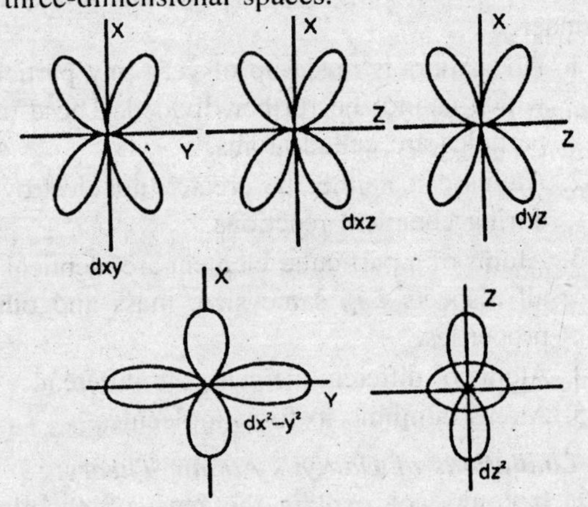

Fig.

Quantum Numbers

In an atom a large number of electron orbitals are permissible. These orbitals are designated by a set of numbers known as quantum numbers. In order to specify energy, size, shape and orientation of the electron orbital, three quantum numbers are required. These quantum number are;

1. ***Principal quantum number (n) :*** This is the most important quantum number as it gives the principle energy level to which the electron belongs. It is denoted by the letter '*n*'. This may have any integral value except zero. Thus *n* will be equal to 1, 2, 3 or 4 ... depending whether the electron is in

first, second, third or fourth (*K, L, M, N* or ...)... principle energy level. The maximum number of electrons that can be present in a principle energy level corresponding to the principle quantum number *n* is equal to $2n^2$.

2. ***Azimuthal quantum number*** (***l***) **:** The principle quantum number gives the energy position of the electron in general. A part of this energy must be associated with the angular momentum of the electron moving about the nucleons. The quantum number, as its name implies, determines the energy associated with the angular momentum of the electron. It is designated by letter *l*. The higher value of *l* evidently corresponds to greater angular momentum. The various integral values of *l* that are possible vary between 0 and $n - 1$.

3. ***Magnetic quantum number*** (***m***) **:** Since an electron have angular momentum, its motion may be compared to the flow of an electric current through a loop. Such a flow of current, as is well known, creats magnetic field. It is to be expected, therefore the angular momentum of electron should create a magnetic field. This magnetism is determined by a quantum number known as magnetic quantum number. It is denoted by '*m*'. Under the influence of magnetic field the electrons in a given energy sublevel prefers to orient themselves in certain specific regions of space around nucleons. These specific regions where the maximum probability of finding an electron is high are called orbitals: And the number of orientations of' space are given by possible values of *m*. The values of *m* depends upon the value of *l* and have equal to +1, +2, 0, −1, −2.

4. ***Spin quantum number*** (***s***) **:** A charged particle which spins about its own axis behaves as a small magnet. Hence the spinning of an electron adds to the angular momentum of the electron and to its energy relationship. The electron can spin clockwise or anticlockwise. Thus *s* can have either the value +1/2 or −1/2, depending upon the direction of spin.

Rules for filling of orbitals in atoms

The filling of electrons into the various orbitals of different atoms takes place according to the following rules:

1. Aufbau Principle : According to this principle "In the ground state of the atom, the electrons are added one by one to the various orbitals in order of their increasing energy starting with the orbital of lowest energy.

The order in which the energies of the orbitals increase and hence the order in which the orbitals are filled is as follows:

1*s*, 2*s*, 2*p*, 3*s*, 4*s*, 3*d*, 4*p*, 5*s*, 4*d*, 5*p*, 6*s*, 4*f*, 5*d*, 6*p*, 7*s*, 5*f*, 6*d*, 7*p*.

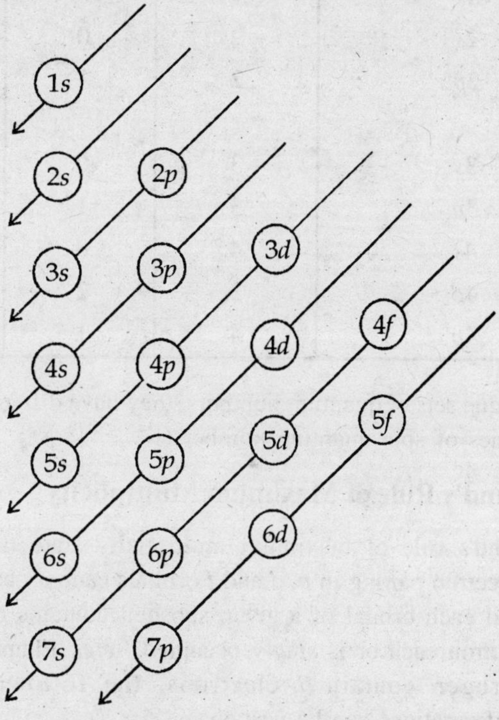

Fig. Order of filling of orbitals

Bohr-Bury Rule

The energy of an orbital depends upon the sum of values of the principal quantum number (*n*) and the azimuthal quantum number (*l*). This is called ($n + l$) rule. According to this rule,

"In neutral isolated atom, the lower the value of ($n + 1$) for an orbital, lower is its energy. However, if the two different type of orbitals have the same value of ($n + l$), the orbital with lower value of '*n*' has lower energy". To illustrate the ($n + l$) rule, different examples are given in the following table.

Pauli Exclusion Principle

This principle was discovered by Wolfgang Pauli in 1925. Pauli exclusion principle states that: "No two electrons in an atom can have same values for all the four quantum numbers."

All the electrons in particular orbital have same values of principal quantum numbers (n), azimuthal quantum number (l) and magnetic quantum number (m). For example, all the electrons in 3s orbital have $n = 3$, $l = 0$ and $m = 0$. Therefore, in order to have unique sets of quantum numbers, they have different values of spin quantum number (s).

Table : Illustration of ($n + l$) rule

Type of orbital	Value of 'n'	Value of l	Value of $n + l$	Relative energy
1s	1	0	$1 + 0 = 1$	Lowest energy
2s	2	0	$2 + 0 = 2$	Higher energy than 1s orbital
2p	2	1	$2 + 1 = 3$	2p orbital ($n = 2$) have lower energy than 3s ($n = 3$) obital
3s	3	0	$3 + 0 = 3$	3p orbitals ($n = 3$) have lower
3p	3	1	$3 + 1 = 4$	energy than 4s ($n = 4$) orbital
4s	4	0	$4 + 0 = 4$	3d orbitals ($n = 3$) have lower
4d	3	2	$3 + 2 = 5$	energy than 4p ($n = 4$) orbital
4p	4	1	$4 + 1 = 5$	

Hund's Rule of Maximum Multiplicity

Hund's rule of maximum multiplicity state that, "Electron pairing in p, d and f orbitals cannot occur until each orbital of a given subshell contains one electron each or is singly occupied." *e.g.*, Element nitrogen contain 7 electrons, the following configurations can be written N (Z = 7).

N (Z = 7) $1s^2\ 2s^2\ 2px^1\ 2py^1\ 2pz^1$

1s	2s	2px	2py	2pz
↑↓	↑↓	↑	↑	↑

In accordance with the Hund's rule the configuration in which the three unpaired electrons occupying 2px, 2py and 2pz orbitals have parallel spin (either all clockwise or anticlockwise).

The term maximum multiplicity *i.e.,* total spin of unpaired electrons is maximum. For example, the total spin of the unpaired electrons in-configuration. Thus, in accordance with Hund's rule, configuration with maximum multiplicity is correct.

MULTIPLE CHOICE QUESTIONS

1. Charge on fundamental particle neutrino is
   (*a*) 0 (*b*) +1
   (*c*) -1 (*d*) None of these

2. The fundamental particle that has least mass is
   (*a*) meson (*b*) α-particle
   (*c*) electron (*d*) neutron

3. Which of the following has highest mass?
   (*a*) Neutron (*b*) Alpha particle
   (*c*) Electron (*d*) Deuterium

4. Cathode rays contain a stream of
   (*a*) n (*b*) r
   (*c*) $+1^{e^0}$ (*d*) -1^{e^0}

5. Mass of positron is the same to that of
(*a*) proton (*b*) meson
(*c*) electron (*d*) neutron

6. The exchange of particles considered responsible for keeping the nucleons together are
(*a*) meson (*b*) electron
(*c*) positron (*d*) neutron

7. Charge on an electron was shown by
(*a*) J.J. Thomson (*b*) Kirchoff
(*c*) Ohm (*d*) M.Planck

8. Which of the following atoms has a non-spherical outermost orbital.
(*a*) H (*b*) Li
(*c*) Be (*d*) B

9. The number of neutrons present in $^{39}_{19}K$ are
(*a*) 28 (*b*) 39
(*c*) 20 (*d*) 19

10. In Lyman series an electron jumps from higher energy level to
(*a*) *K* energy level
(*b*) *M* energy level
(*c*) *N* energy *level*
(*d*) *L* energy level

11. The wavelength of an electron
(*a*) is equal to that of light
(*b*) remains constant with velocity
(*c*) decreases with an increasing velocity
(*d*) increases with an decreasing velocity

12. No two electrons in an atom can have the same values of all four quantum numbers according to
(*a*) Hund's rules
(*b*) Flemming rule
(*c*) Pauli's exclusion principle
(*d*) Bohr theory

13. The charge on an atom of $^{27}_{13}Al$ is
(*a*) + 40 (*b*) + 27
(*c*) + 14 (*d*) Zero

14. Bohr's concept of the orbit in an atom was contradicted by
(*a*) de Broglie relationship
(*b*) Uncertainty principle
(*c*) Planck's hypothesis
(*d*) Hund's rule

15. The velocity of a photon is
(*a*) dependent on its wavelength
(*b*) dependent on its source
(*c*) equal to cube of its amplitude
(*d*) independent of its wavelength

16. The number of unpaired electrons in fluorine atoms is/are
(*a*) 4 (*b*) 2
(*c*) 1 (*d*) 0

17. Among the following ions, which are has the highest paramagnetism?
(*a*) $[Cr(H_2O)_6]^{+3}$ (*b*) $[Fe(H_2O)_6]^{+2}$
(*c*) $[Cu(H_2O)_6]^{+2}$ (*d*) $[Zn(H_2O)_6]^{+2}$

18. In the case of atomic spectrum of hydrogen which series of lines lie in the visible region?
(*a*) Balmer (*b*) Paschen
(*c*) Pfund (*d*) None of these

19. When the electron is excited from *K* level to *M* level we get
(*a*) γ-rays
(*b*) cathode rays
(*c*) continuous spectra
(*d*) absorption spectra

20. Isotopes have different
(*a*) arrangement of electrons
(*b*) no. of *P* and e^-
(*e*) no. of neutrons
(*d*) no. of electrons

Directions Qs. 21-35: *In each of the following questions there are two statements Assertion (A) is given followed by a corresponding statement of Reason (R) just below it. Of the statements, mark the correct answer:*

(*a*) If both assertion and reason are true and reason is the correct explanation of assertion.
(*b*) If both assertion and reason are true but reason is not the correct explanation of assertion
(*c*) If assertion is true but reason is false
(*d*) If assertion is false but reason is true.

21. Assertion (A): Atom is electrically neutral.
Reason (R): A neutral particle, neutron is present in the nucleus of atom.

22. **Assertion (A):** Cathode rays get deflected towards the positive plate of electric field.

Reason (R): Cathode rays consist of negatively charged particles known as electrons.

23. **Assertion (A):** Electromagnetic radiations around 10^{15} Hz are called as visible light.

Reason (R): This is the only part of electro-magnetic radiation which is visible to eyes.

24. **Assertion (A):** VIBGYOR signifies the seven colour of visible light.

Reason (R): Red colour corresponds to higher frequency and blue colour to lower frequency region.

25. **Assertion (A):** The energy of quantum is given by $E = h\nu$.

Reason (R): Quantum in the energy equation signifies the principal quantum number.

26. **Assertion (A):** The kinetic energy of the photoelectron ejected increases with increase in intensity of incident light.

Reason (R): Increase in intensity of incident light increases the rate of emission.

27. **Assertion (A):** Threshold frequency is a characteristic for a metal.

Reason (R): Threshold frequency is a maximum frequency required for the ejection of electron from the metal surface.

28. **Assertion (A):** de-Broglie equation has significance for any microscopic or sub-microscopic particles.

Reason (R): de-Broglie wavelength is inversely proportional to the mass of the object.

29. **Assertion (A):** The radius of the first orbit of hydrogen atom is 0.529 Å.

Reason (R): Radius for each circular orbit (r_n) = 0.529 Å (n^2/Z), where n = 1, 2, 3 and Z = atomic number.

30. **Assertion (A):** The Mo (atomic number = 42) atom has six unpaired electrons.

Reason (R): Half-filled *s* orbital has greater stability.

31. **Assertion (A):** Ten distinct set of four quantum numbers are possible for *d* - subshell.

Reason (R): *d* - subshell splits into five orbitals.

32. **Assertion (A):** Energy of radiation is large if its wavelength is large.

Reason (R): Energy = $h\upsilon$ (υ = frequency, $\upsilon = \frac{c}{\lambda}$)

33. **Assertion (A):** Electronic configuration of an element is $1s^2\ 2s^1$.

Reason (R): In 2*s*, 2 signifies the maximum capacity of *s*-subshell.

34. **Assertion (A):** $3d_{z^2}$ orbital is spherically symmetrical.

Reason (R): $3d_{z^2}$ orbital is the only *d*-orbital which has two lobes in shape.

35. **Assertion (A):** Orbitals form the basis of the electronic structure of atoms.

Reason (R): An atomic orbital is the wave function ψ for an electron in an atom.

ANSWERS

1	2	3	4	5	6	7	8	9	10
(*a*)	(*d*)	(*b*)	(*d*)	(*c*)	(*a*)	(*a*)	(*d*)	(*c*)	(*a*)
11	12	13	14	15	16	17	18	19	20
(*c*)	(*c*)	(*d*)	(*b*)	(*d*)	(*c*)	(*b*)	(*b*)	(*d*)	(*c*)
21	22	23	24	25	26	27	28	29	30
(*b*)	(*a*)	(*a*)	(*c*)	(*c*)	(*d*)	(*c*)	(*a*)	(*a*)	(*c*)
31	32	33	34	35					
(*a*)	(*d*)	(*c*)	(*d*)	(*b*)					

2 PERIODIC CLASSIFICATION OF ELEMENTS

Modern Periodic Law : Since properties of an element depend on the number of electrons and their arrangement in the atom so their classification should be based on the number of these electrons or their arrangement in different orbitals. This led Moseley to change the basis of classification of elements from atomic weight to atomic number.

"The properties of elements are periodic function of their atomic number, *i.e.,* if the elements are arranged in the order of their atomic number, the similar elements are repeated after regular intervals."

Modern Periodic Table

In the modern periodic table (also known as **long form periodic table),** the elements are arraged according to increasing order of their atomic numbers in the horizontal rows known as **periods** and in vertical columns known as **groups.** There are seven periods and nine groups.

1st period, known as the 'shortest period', has 2 elements ($_1$H and $_2$He).

2nd period, known as 'short period' has 8 elements ($_3$Li to $_{10}$Ne).

3rd period, known as 'short period', has 8 elements ($_{11}$Na to $_{18}$Ar).

4th, period, known as 'long period', has 18 elements ($_{19}$K to $_{36}$Kr).

5th period, known as 'long period', has 18 elements ($_{37}$Rb to $_{54}$Xe).

6th period, known as the 'longest period', has 32 elements ($_{53}$Cs to $_{86}$Rn).

7th period, starting with $_{87}$Fr is incomplete. Elements upto $_{92}$U are naturally occurring and the remaining elements from $_{93}$Np to $_{106}$Ha are readymade.

Table : ***The Electronic Configuration of the Atoms***

Elements	*At. No.*	*1s*	*2s*	*2p*	*3s*	*3p*	*4s*
H	1	1					
He	2	2					
Li	3	2	1				
Be	4	2	2				
B	5	2	2	1			
C	6	2	2	2			
N	7	2	2	3			
O	8	2	2	4			
F	·9	2	2	5			
Ne	10	2	2	6			
Na	11	2	2	6	1		
Mg	12	2	2	6	2		
Al	13	2	2	6	2	1	
Si	14	2	2	6	2	2	
P	15	2	2	6	2	3	
S	16	2	2	6	2	4	
Cl	17	2	2	6	2	5	
Ar	18	2	2	6	2	6	
K	19	2	2	6	2	6	1
Ca	20	2	2	6	2	6	2

Modern Periodic Table (Extended Form)

Group	IA	IIA											IIIA	IVA	VA	VIA	VIIA	O
Period	H 1												Metalloids and Non-metals					
1	Active metals																	He 2
			Transition Metals								Cu-Zn Group							
2	Li 3	Be 4											B 5	C 6	N 7	O 8	F 9	Ne 10
3	Na 11	Mg 12	IIIB	IVB	VB	VIB	VIIB	VIII			IB	IIB	Al 13	Si 14	P 15	S 16	Cl 17	Ar 18
4	K 19	Ca 20	Sc 21	Ti 22	V 23	Cr 24	Mn 25	Fe 26	Co 27	Ni 28	Cu 29	Zn 30	Ga 31	Ge 32	As 33	Se 34	Br 35	Kr 36
5	Rb 37	Sr 38	Y 39	Zr 40	Np 41	Mo 42	Tc 43	Ru 44	Rh 45	Pd 46	Ag 47	Cd 48	In 49	Sn 50	Sb 51	Te 52	I 53	Xe 54
6	Cs 55	Ba 56	La* 57	Hf 72	Ta 73	W 74	Re 75	Oa 76	Ir 77	Pt 78	Au 79	Hg 80	Ti 81	Pb 82	Bi 83	Po 84	At 85	Rn 86
7	Fr 87	Ra 88	Ac** 89	Ku 104	Ha 105							Eka mercury 112						

*Lanthanide Elements	Ce 58	Pr 59	Nd 60	Pm 61	Sm 62	Eu 63	Gd 64	Tb 65	Dy 66	Ho 67	Er 68	Tm 69	Yb 70	Lu 71
**Actinide series	Th 90	Pa 91	U 92	Np 93	Pu 94	Am 95	Cm 96	Bk 97	Cf 98	Es 99	Fm 100	Md 101	No 102	Lr 103

The nine groups are denoted as groups I to VIII are zero (0) group. Groups I to VII are further divided into A and B subgroups. The elements in which the last added electron goes to *s*- or *p*-subshells are placed in A subgroups, while those in which it goes to *d*- or *f*-subshells are placed in B subgroups.

The long form of Periodic Table

With the shift of basis from atomic to atomic number a cause of periodicity, a large number of modified periodic tables were available. Most useful and at the same time easiest is the Bohr's Model of long form of periodic table. The mains features are:

1. In this table there are 16 vertical columns called subgroups. These are numbered IA IB to VIIB /VIII.
2. Element of the subgroups IA, IIA... VIIA have only their outermost shell incomplete. Each one of inner shell is complete. Hence, they are called **Normal elements.**
3. Elements of the subgroups IB IIB...VIIB and VIII have their outermost shell as well as their penultimate shell incomplete. These are termed as **Transition elements.**
4. Elements of subgroup O have all their shells complete. These show little reactivity and are termed as **Noble gases** and are grouped at the extreme right of the periodic table.
5. The lanthanides and the actinides have grouped separately and placed at the bottom. These elements are called as the inner transition elements as their penultimate as well as antipenultimate shell is incomplete.

s, p, d, f block elements

Elements with atom in which all but the outermost shells are complete. The electronic configuration in the outer shell of these atom varies from ns^1 to $ns^2 p^5$. They have been classified as :

***(i)* s-block elements :** Elements of Group IA and IIA have one electron only in their outer *s*-subshell and are termed as *s* block elements. They do not contain any electron in $(n-1)$ *s* orbital which $(n-1)$ *s* and *p* orbitals are completely filled. The differentiating electron enters *ns* orbitals.

***(ii)* p-block elements :** Elements of Group IIIA to VIIA have 1 to 5 electrons in their outer *p*-subshells and are called *p*-block elements. Their $(n-1)d$ orbitals along with $(n-1)s$ and *ns* are completely filled. The differentiating electron enters in *np* orbitals.

Both *s* and *p*-block elements enter into chemical combination to acquire ns^2p^6 configuration in the outermost shell by loosing or gaining or sharing electron.

***(iii)* d-block elements :** Elements with atoms in which two outermost shells are incomplete. These one commonly known as transition or d-block elements. They generally contain two electrons in their *ns* orbitals while the differentiating electron enters an available $(n-1)d$ orbitals. Strictly speaking the term transition metals is used only for those *d*-block elements which have one or more incompletely filled $(n-1)$ *d* orbitals.

Transition elements are so called because their properties are intermediate between those of *s* and *p* block elements. The transition elements resembles in a number of physical and chemical properties.

***(iv)* f-block elements :** Elements having three outermost shells incomplete. They are commonly known as *Inner transition elements* or the *f*-block elements. These are called inner transition elements because they have not only incomplete *d*-subshell but an incomplete *f*-shell also. They are tenned as *f*-block elements because their anti-penultimate shell of electrons is expanded due to addition of *f*-electrons. These are rare earth elements or lanthanides and actinides. The differentiating electron of these atoms are in antipenultimate shell.

Properties of Elements

1. Electronegativities : When two atoms are bonded by a covalent bond, both of them share a pair of electrons. The attraction that one atom exerts on this shared pair of electron is termed as electronegativity.

In general the smaller atoms attract electrons more than that of the larger one and are therefore more electronegative atoms with nearly filled shell of electrons will tend to have more electronegativity than those with sparsely occupied shells.

With moving from left to right in the periodic table, the electronegativity increases with the increase of nuclear charge and decreasing with moving down the group. Thus foi example in C — N → Nitrogen is more electronegative thus this is shown as $C^{\delta+} - N^{\delta-}$

2. Atomic radius (crystal radius) : This is used for metal atoms which are closely packed spheres and are solid at the ordinary temperature and exist as crystalline states. The distance between the neighbouring atoms can be found by the X-ray diffraction analysis. Half of the distance between the nucleus of neighbouring atom is taken as the atomic or crystal radius.

3. Covalent radius : These are used for the non-metals. It is half the distance of separation of nuclei in a covalent bonded molecule. The covalent radii for two different atoms can be added to give inter-nuclear distances.

4. Van der Waals' Radius : It is one half of the internuclear distance between two adjacent atoms belonging to two nearest neighbouring molecules of the same substance in the solid state. In genera!,

Covalent radius < Metallic radius < van der Waal's radius.

MULTIPLE CHOICE QUESTIONS

1. Which of the following has the maximum size?
(a) Al^- (b) Ne
(c) Na^+ (d) F^-

2. Elements belonging to the same period are
(a) Si, N, O (b) Li, Be, Mg
(c) Sc, Ti, V (d) Cu, Zn, Cd

3. Atomic number of an element X is 34, it belongs to
(a) IV period, VI A group
(b) IV period, II A group
(c) V period, III B group
(d) IV period, VII A group

4. An element A has atomic number 7. It will have properties similar to the element with atomic number
(a) 85 (b) 57
(c) 26 (d) 51

5. On moving across a period from left to right the ionisation energy increases because
(a) value of principal quantum number increases
(b) effective nuclear charge increases
(c) atomic size increases
(d) nuclear charge increases

6. IV Period has .... elements.
(a) 18 (b) 36
(e) 8 (d) 9

7. An element has atomic number 2. It belongs to
(a) O group (b) III A group
(c) VI group (d) II B group

8. Elements in the same vertical column has same
(a) valence shell electronic configuration
(b) electronic configuration
(c) atomic number of isotopes
(d) atomic size

9. Which of the following has the largest size?
(a) Br^- (b) I
(c) I^- (d) I^+

10. Arrange the following set of atoms in the order of increasing atomic radius: F, Cl, C,O
(a) F, Cl, O, C (b) C, O, F, Cl
(c) O, C, F, Cl (d) F, O, C, Cl

11. Which of the following atoms is paramagnetic?
(a) Zn (b) N
(c) Ca (d) Sr

12. Which of the following elements have maximum electronegativity?
(a) Cl (b) Br
(c) F (d) C

13. The first atom with maximum number of *d*-electrons is with atomic number
(a) 27 (b) 46
(c) 36 (d) 29

14. Which of the following has minimum ionisation energy?
(a) F (b) Cl
(c) Br (d) I

15. With the increase in atomic number in a period of the periodic table
(a) electron affinity decreases
(b) metallic character decreases
(c) ionisation energy decreases
(d) atomic mass decreases

16. The tendency towards complex formation is maximum in
(a) *s*-block elements
(b) *p*-block elements
(c) *d*-block elements
(d) None of these

17. In which of the following, the elements belong to the same period?
(a) Ca, Cr, Zn (b) Cr, Na, Pb
(c) H, He, F (d) Li, Na, K

18. The ionisation energy of nitrogen is more than that of oxygen because of
(a) the extra stability of half filled *p*-orbitals
(b) smaller size of the nitrogen atom

(c) more penetrating effect
(d) greater attraction of nucleus for the electrons

19. Long form of a periodic table is based upon
(a) electron affinity
(b) atomic size
(c) atomic number
(d) electronegativity

20. Among the following groups of elements, the one whose elements can have positive as well as negative oxidation states are
(a) He, Li, Be (b) Na, Mg, Al
(c) Kr, Ne, O (d) H, Cl, Br

Directions Qs. 21-30: *In each of the following questions there are two statements Assertion (A) is given followed by a corresponding statement of Reason (R) just below it. Of the statements, mark the correct answer:*
(a) If both assertion and reason are true and reason is the correct explanation of assertion.
(b) If both assertion and reason are true but reason is not the correct explanation of assertion
(c) If assertion is true but reason is false
(d) If assertion is false but reason is true.

21. Assertion (A): According to Mendeleef, periodic properties of elements is a function of their atomic masses.
Reason (R): Atomic number is equal to the number of protons.

22. Assertion (A): Atomic number of the element unubium is 112.
Reason (R): Name for digits 1 and 2 is un- an bi respectively in latin words.

23. Assertion (A): Second period consists of 8 elements.
Reason (R): Number of elements in each period is four times the number of atomic orbitals available in the energy level that is being filled.

24. Assertion (A): Element in the same vertical column have similar properties.
Reason (R): Elements have periodic dependence upon the atomic number.

25. Assertion (A): Helium is placed in group 18 along with p-block elements.
Reason (R): It shows properties similar to p-block elements.

26. Assertion (A): Chemistry of Actinoids is more complicated than Lanthanoids.
Reason (R): Actinoid elements are radioactive.

27. Assertion (A): Ionization enthalpy is the energy released to remove an electron from an isolated gaseous atom in its ground state.
Reason (R): Element has a tendency to lose or gain the electron to attain the stable configuration.

28. Assertion (A): Ionization enthalpy is always positive.
Reason (R): Energy is always released when electrons are removed.

29. Assertion (A): Second ionization enthalpy will be higher than the first ionization enthalpy.
Reason (R): Ionization enthalpy is a quantitiative measure of the tendency of an element to lose electron.

30. Assertion (A): First ionization enthalpy of boron is slightly less than that of beryllium.
Reason (R): Ionization enthalpy increases as we go across the period.

ANSWERS

1	2	3	4	5	6	7	8	9	10
(b)	(c)	(a)	(d)	(a)	(a)	(a)	(a)	(c)	(d)
11	**12**	**13**	**14**	**15**	**16**	**17**	**18**	**19**	**20**
(b)	(c)	(d)	(d)	(b)	(c)	(a)	(a)	(b)	(d)
21	**22**	**23**	**24**	**25**	**26**	**27**	**28**	**29**	**30**
(b)	(b)	(c)	(b)	(c)	(b)	(d)	(c)	(b)	(b)

3 CHEMICAL BONDING

The attractive force which holds together the constituent particles (*i.e.*, atoms, ions or molecules) in a chemical compound is known as chemical bond.

Ionic Lond

An Ionic bond is formed by the complete transference of one or more electrons from the outer energy shell of one atom to the energy shell of another atom. The atoms which looses an electron to acquire a positive charge and becomes a positive ion. The atom which gains a electron acquires a negative charge.

The crystals of sodium chloride obtained actually consist of a large number of Na^+ and Cl^- ions. The electrostatic attraction between the oppositely charged Na^+ and Cl^- is called an Ionic bond. In sodium chloride

$$Na^+ \, :\ddot{\underset{\cdot\cdot}{Cl}}: \rightarrow Na^+ + [:\ddot{\underset{\cdot\cdot}{Cl}}:]$$

Characteristics

1. Ionic compounds do not exhibit isomerism.
2. Ionic compounds are crystalline solids.
3. Ionic compounds have high mp and bp.
4. Ionic compounds are generally soluble in polar solvents like water and undergo ionic dissociation which makes the solution good conductor of electricity.
5. In molten state, the ionic compounds are good conductors of electricity.

Covalent bond

G.N. Lewis, an American chemist, was first to suggest that atoms may combine by sharing electrons in their outer shell with one another. The shared electrons contribute towards the stability of both atoms. This type of linkage is called the covalent linkage or the **covalent bond.** The compounds formed by this mechanism are called covalent compounds.

Formation of covalent bond in terms of decrease of energy **:** The bond formed by mutual sharing of a pair of electrons between the two atoms ɔf the same or different elements is known as a single covalent bond and is represented by a line (–). The compound so formed is known as a covalent compound. *Examples:*

$$H\cdot + \cdot H \rightarrow \quad H:H \quad H-H$$

Two hydrogen atoms — Hydrogen molecule

(1) (Single covalent bond)

$$:\ddot{\underset{\cdot\cdot}{Cl}}: \; + :\ddot{\underset{\cdot\cdot}{Cl}}: \; \rightarrow :Cl:Cl: \; Cl-Cl \text{ or } Cl_2$$

Two Chlorine atoms (2,8,7) — Chlorine molecule (Single covalent bond)

Each line (–) represents a pair of electrons, contributed and shared by tne two atoms. When two pairs of electrons are shared, it is known as a double bond. *Examples*:

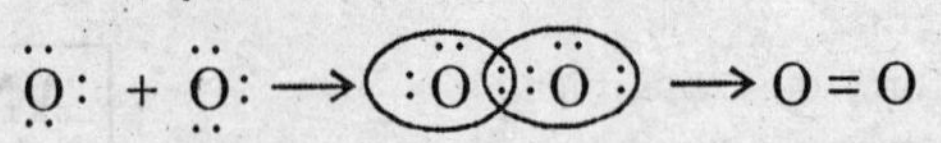

Two oxygen atoms (2, 6) — Oxygen molecule (Double bond)

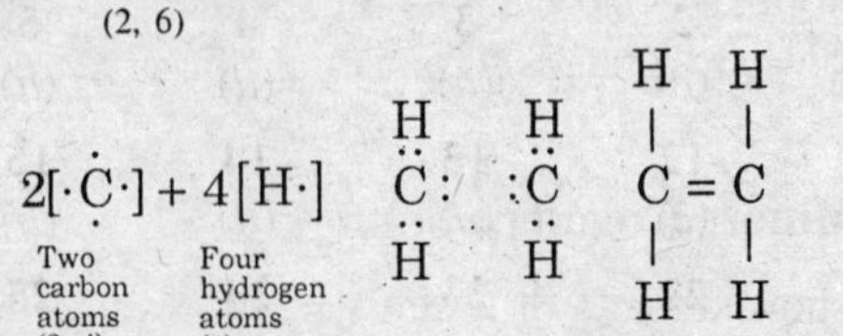

Two carbon atoms (2, 4) — Four hydrogen atoms (1) — Ethylene molecule (Double bond)

Similarly, when three pairs of electrons are shared between the two atoms, it is known as a triple bond. *Examples:*

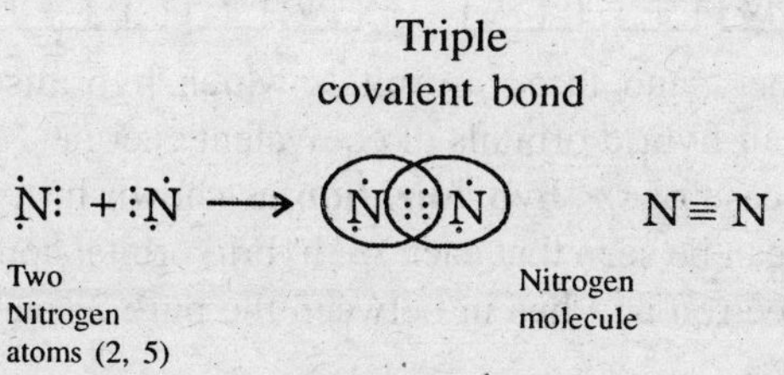

Triple covalent bond

A triple bond is usually stronger than a double bond, and a double bond is stronger than a single bond. Covalent compounds have the following characteristics :

1. Covalent bonds being directional, a covalent compound may exhibit some type of isomerism.

2. A covalent bond is weaker as compared to an ionic bond. The strength of a covalent bond is measured by its length. The greater the length, the lesser the strength. For example, the bond strength in halogen acids decreases as HF > HCl > HBr > HI.

3. As the inter-molecular forces in covalent compounds are weak, they have low mp and bp.

4. Covalent compounds are, in general, soluble in non-polar solvents like carbon tetrachloride, benzene, *etc.* In such solutions they do not ionise and hence do not conduct electricity. Also in molten and solid states they are bad conductors of electricity.

Polar Character of Covalent bonds **:** When two unlike atoms form a covalent bond, the electrons may not be shared equally by both the atoms. In such a case, the electron cloud is closer to the one atom than to the other. One end of the bond, therefore, develops partial positive charge and the other partial negative charge. Such a bond is called a partial polar bond. This is represented by symbol δ^+ and δ where δ stands for a partial charge. *Example:* H—F.

Co-ordinate Covalent bond

Lewis showed that a different type of covalent bond can be formed when both the electrons for sharing between the two atoms are contributed by only one atom. This type of bond is known as **Co-ordinate covalent bond simply dative bond.**

According to orbital theory a co-ordinate covalent bond results when an orbital containing a lone pair of electrons in one atom overlaps with an empty orbital present in the other atom. It is important to note that co-ordinate covalent bond once formed is indistinguishable from covalent bond. *For Example:*

Formation of Ammonium ion $NH_4^+ \rightarrow$:

The formation of co-ordinate covalent bond is generally and more conveniently represented by the dot method. The formation of hydronium ion by dot method is represented below :

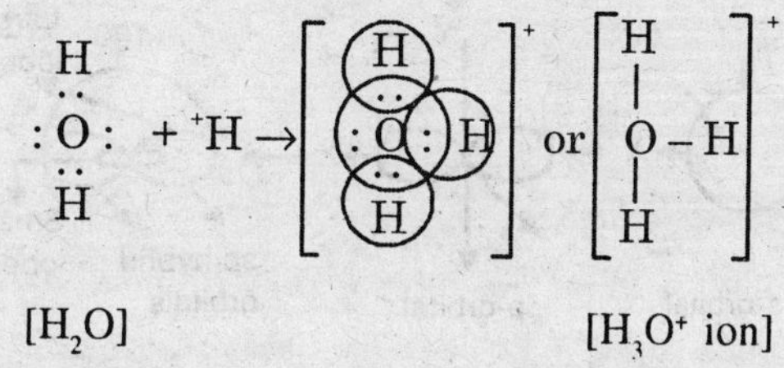

Hydrogen bond

The attractive force which binds hydrogen of one molecule with electronegative atom of another molecule of the same substance is known as Hydrogen bond.

In hydrogen fluoride, it may be noted that hydrogen atom is bonded to fluorine atom by a covalent bond in one molecule and by hydrogen bond to the fluorine atom in the adjacent molecule.

Hybridization

According to simple valence bond theory, the capacity of an atom to form covalent bond depends upon the number of unpaired electrons in its bonding orbitals. On this basis beryllium, boron and carbon are expected to form 0, 1 and 2 single covalent bonds. However, experimental results have indicated that bonding capacity of beryllium, boron and carbon is 2, 3 and 4 respectively. Moreover, the covalent bonds are equivalent in all respects. In order to account for this behaviour, the concept of hybridization was introduced.

Thus, hybridization is the process of intermixing of orbitals of the same atom having slightly different energies and the redistribution of their energies to form the new orbitals which have the identical shapes and equivalent energies. The new orbitals thus formed are called hybrid orbitals and the bond formed in each orbital are called **hybrid bonds.**

In the combination of beryllium and fluorine one *s*-and one *p* orbital is involved, it is called ***sp*-hybridization**. The hybridization is represented in figure and it can be seen that each hybrid orbital contain one electron and lies in between a pure *s* and *p* orbitals.

The electronic configuration of boron in the ground state is $1s^2$ $2s^2$ $2px$. When a

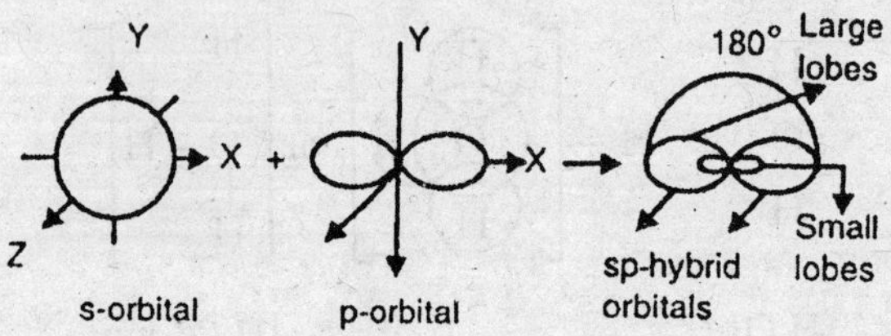

reacting atom approach to form a covalent bond, the energy released in the process promotes one of the 2*s* electron into the vacant 2*p*-orbitals. Now there are three half filled orbitals, *i.e.,* one *s* and two *p* orbitals which hybridise to give three orbitals of equivalent energies.

This type of hybridisation is known as ***sp*^{2} hybridization** as represented in figure and it can be seen that each hybrid orbital contain one electron and lies in between the pure 2*s* and pure 2*p*-orbitals.

The Ground State Excited State

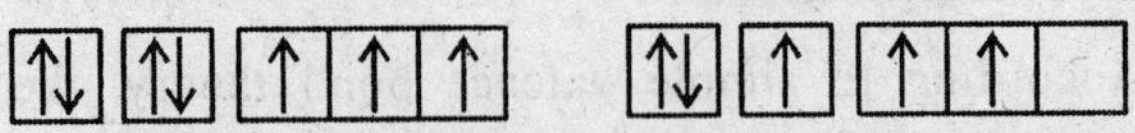

The electronic configuration of carbon is $1s^2\,2s^2\,2p_x^1\,2p_y^1$. When a reacting atom

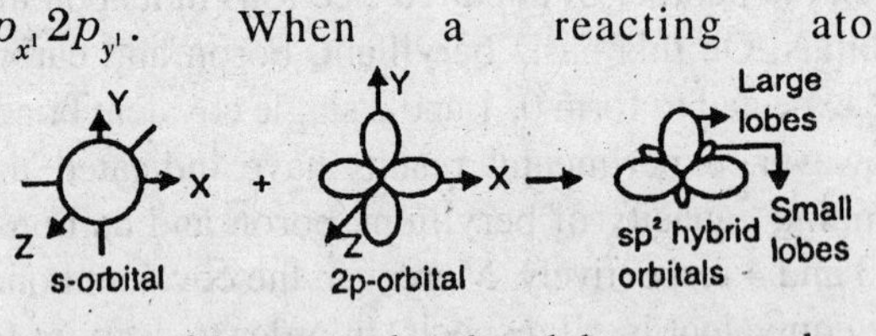

approach to form a covalent bond the energy released in the process to promote one 2*s*-electron to the vacant 2*p*-orbitals.

Thus, we have now four half filled orbitals

The Ground State Excited State

i.e., one *s* and three *p*-orbitals which hybridise to give four hybrid orbitals of equivalent energies. This is the case of ***sp*^{3} hybridization** as shown in figure and it can be seen that each *sp*^{3} hybrid orbital contain one electron and lies in between the pure 2*s* and 2*p* orbitals.

Sigma (σ) bond

Let us consider the formation of ethane molecule. It will be seen that one of the *sp*^{3} hybrid bond of each

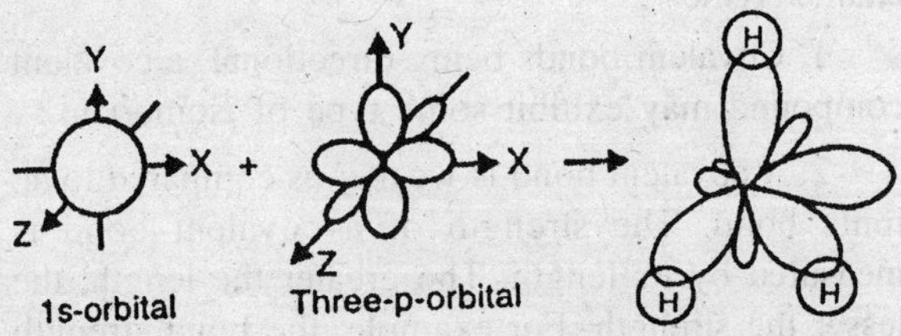

carbon atom used up in giving C–C bond. The remaining three *sp*^{3} hybrid bonds are linked to hydrogen atom. Evidently each bond in ethane is formed by the overlapping of orbitals along their axes. Such bonds are called as the **Sigma bond.**

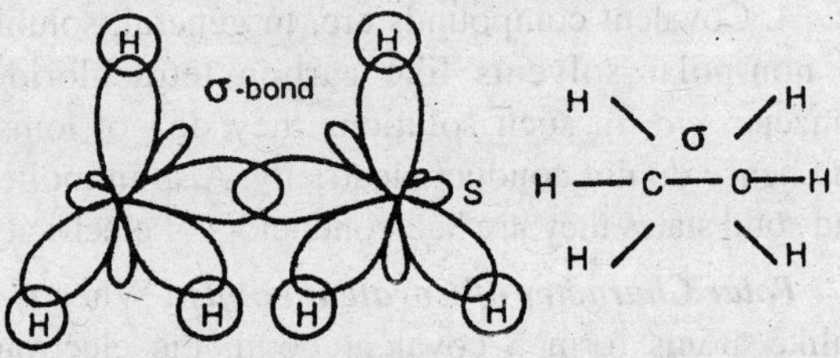

A single bond formed between the two atoms by overlapping of orbitals along their axis is known as a **sigma bond.**

Pi (π) bonds

The bond which results from the sidewise overlapping of *p*-orbitals are designated as Pi (π) bonds.

In the formation of ethylene three orbitals are used in forming bonds with two hydrogen atoms and one other carbon atom while *p* orbital remains half filled. If the two carbon atoms bonded to two hydrogen atoms and to one another are sufficiently closed, the half filled *p*-orbital in each carbon atom lying perpendicular to the plane of four hydrogen and two carbon atoms will overlap sidewise - and the sidewise overlapping of *p*-orbitals is called as the Pi (π) bonds.

Valence bond approach

In 1927, Heitler and London propounded a modified approach to the chemical combination called as the valence bond approach. The main features of that approach are given below :

(i) A covalent bond is formed by the overlapping of atomic orbitals of the individual atoms.

(ii) The overlapping orbitals should have electrons with opposite spin in accordance with the Pauli's Exclusion Principle.

(iii) The inner atomic orbitals from each atom forming the bond are undisturbed.

(iv) The valency can be determined from the number of unpaired electrons in bonding orbitals.

Molecular Orbital Methods

In 1932, Hund and Mulliken propounded a new theory known as molecular orbital theory to explain the formation of chemical bond. The main features are:

(i) This theory treats the distribution of electrons in the molecules in the same way as the distribution of electrons in the atomic orbitals of the atoms.

(ii) Molecular orbitals are postulated around all the nuclei present in the molecule. These molecular orbitals are the regions in space in which the electron is most likely to found.

(iii) The valence electrons are assumed as moving over the molecule as a whole.

Types of molecular orbitals : There are two types of molecular orbitals given as:

(i) Bonding molecular orbitals.

(ii) Anti-bonding molecular orbitals.

(i) The molecular orbitals result from the addition of the parts of atomic orbitals which overlaps, is known as bonding orbitals (molecular).

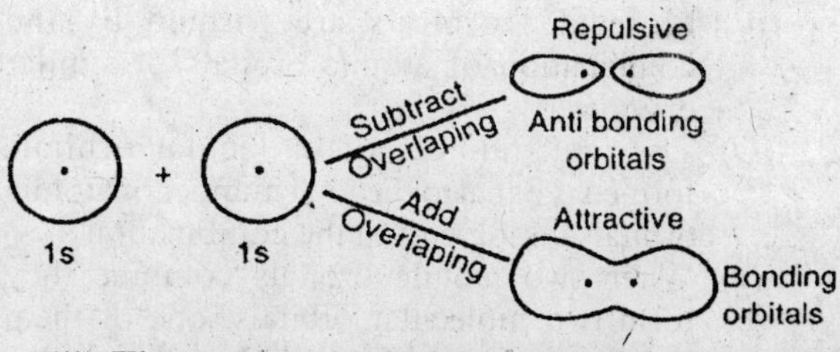

(ii) The molecular orbitals results from the subtraction of the parts of atomic orbitals which overlaps, is known as the antibonding orbitals (molecular).

Atomic Orbital (A.O.) **:** It is defined as the region in space where there is a probability of finding an electron. An atomic orbital has the following characteristics:

(i) An atomic orbital is distributed around the nucleus of a single atom.

(ii) An atomic orbital can accomodate maximum of the electrons and they should have opposite spins.

(iii) Atomic orbitals are of different shapes and sizes. *s*-orbital is spherically symmetrical while the *p*-orbital is dumb-bell shaped.

(iv) The four quantum numbers *n, l, m* and *s* characterise the atomic orbital.

Molecular Orbital (M.O.) **:** It is the region in space containing the nuclei of all the bonded atoms where the probability of finding an electron is a maximum. A M.O. has the following characteristics :

(i) A Molecular orbital is formed by the combination of atomic orbitals.

(ii) The number of molecular orbitals formed is the same as the number of atomic orbitals taking part in chemical combination. For example, the combination of the two atomic orbitals give rise to two molecular orbitals.

(iii) There are two types of M.O.'s, i.e., *bonding* and *antibonding*.

Important features of the M.O. theory :

(i) M.O.'s are composed of the nuclei of all the bonded atoms in a molecule.

(ii) In the formation of M.O's, the electrons of the bonding atoms are taken into consideration.

(*iii*) Molecular orbitals are formed by the combination of atomic orbitals of similar energies.

(*iv*) The total number of molecular orbitals formed is equal to the total number of atomic orbitals taking part in the combination.

(*v*) When two atomic orbitals combine they form two molecular orbitals, one of them is the bonding M.O. and the other is the antibonding M.O. The antibonding M.O. has always greater energy than the bonding M.O.

$$\text{Bond Order} = \frac{1}{2}(N_b - N_{ab})$$

(*xi*) A negative or zero bond order means an unstable molecule. A positive B.O. (Bond Order) means a stable molecule.

Geometry of Molecules

Hybridizations	*Shape of Molecule*	*Bond angle*	*Examples*
sp	Linear	180°	$HgCl_2$, C_2H_2, HCN
sp^2	Trigonal	120°	BF_3, BCl_3, C_2H_4,
	Triangular planar		CO_3^{2-}, NO_3^-, SO_3,
sp^3	Tetra hedral	109° 28'	CH_4, CCl_4

MULTIPLE CHOICE QUESTIONS

1. In which case no hydrogen bond is possible?
(*a*) HF (*b*) H_2O
(*c*) HCl (*d*) NH_3

2. The number of electrons that are shared between two atoms having a triple bond are
(*a*) 7 (*b*) 6
(*c*) 5 (*d*) 3

3. In which of the following central metal atom is sp^3 hybridised?
(*a*) $[Cu(H_2O)_4]^{2+}$ (*b*) $[Cu(CN)_4]^{2-}$
(*c*) $[Cu(CN)_4]^{3-}$ (*d*) $CuCl_2$

4. In which of the following octet rule is not applicable?
(*a*) Alkali earth metal
(*b*) Alkaline earth metal
(*c*) Transition elements
(*d*) *p*-block

5. The number of unpaired electrons in Ni^{2+} are
(*a*) 1 (*b*) 2
(*c*) 3 (*d*) 5

6. The molecules that contain both ionic and covalent bonds are
(*a*) KCN (*b*) H_2O
(*c*) H_2O_2 (*d*) NaCl

7. Tick mark the compounds that contain electron deficient central atom.
(*a*) NCl_3 (*b*) $FeCl_3$
(*c*) BCl_3 (*d*) $CuCl_2$

8. Which of the following is not tetrahedral?
(*a*) SO_4^{2-} (b) CCl_4
(c) NH_4^+ (d), CO_3^{2-}

9. Geometry of PCl_5 molecule is
(*a*) bipyramidal
(*b*) triangular bipyramidal
(*c*) hexagonal
(*d*) square planar

10. Which of the following has maximum ionic character?
(*a*) LiCl (*b*) CuCl
(*c*) NaCl (*d*) CsCl

11. The hydrogen halide which has maximum boiling point is
(*a*) HI (*b*) HBr
(*c*) HF (*d*) HCl

12. The increase in bond order results in
(*a*) increase in bond length and bond energy
(*b*) decrease in bond length and bond energy both
(*c*) decrease in bond length and increase in bond energy
(*d*) None of the above

13. Covalent character of an ionic compound increases with
(*a*) decrease in anion size
(*b*) decrease in cation size
(*c*) increase in cation size
(*d*) decrease in both cation and anion size

14. In the case of PCl_5 molecule, P is
(*a*) sp^2 (*b*) sp^3d
(c) dsp^2 (*d*) sp^3

15. Higher the bond order greater is
(*a*) bond energy (*b*) bond angle
(*c*) bond length (*d*) paramagnetism

16. Which of the following is most stable at room temperature?
(*a*) H_2 (*b*) N_2
(*c*) O_2 (*d*) Cl_2

17. Which of the following exists as dimers?
(*a*) $AlCl_3$ (*b*) $CaCl_2$
(*c*) HCl_3 (*d*) NCl_3

18. Between two elements A and B, the electronegativity difference is 2.0. The per cent ionic character of AB is
(*a*) 86 (*b*) 35
(*c*) 46 (*d*) 15

19. C—C bond length is minimum in
(*a*) C_6H_6 (*b*) C_2H_6
(*c*) CH_4 (*d*) C_2H_2

20. Which of the following have maximum bond order?
(*a*) H_2 (*b*) Br_2
(*c*) N_2 (*d*) Cl_2

Directions Qs. 21-30: *In each of the following questions there are two statements Assertion (A) is given followed by a corresponding statement of Reason (R) just below it. Of the statements, mark the correct answer:*

(*a*) If both assertion and reason are true and reason is the correct explanation of assertion.
(*b*) If both assertion and reason are true but reason is not the correct explanation of assertion
(*c*) If assertion is true but reason is false
(*d*) If assertion is false but reason is true.

21. **Assertion (A):** SF_6 and PF_5 are examples of expanded octet molecules.
Reason (R): P in PF_5 and S in SF_6 are in sp^3d and sp^3d^2 hybridization respectively.

22. **Assertion (A):** Octet theory cannot account for the shape of the molecule.
Reason (R): Octet theory can predict relative stability and energy of a molecule.

23. **Assertion (A):** The crystal structure gets stabilized even though the sum of electron gain enthalpy and ionization enthalpy is positive.
Reason (R): Energy is absorbed during the formation of crystal lattice.

24. **Assertion (A):** Ionic bond is non-directional.
Reason (R): Each ion is surrounded by a uniformly distributed electric field.

25. **Assertion (A):** Ionic compounds conduct electricity in solution as well as in molten form.
Reason (R): Ionic compounds are formed by sharing of electrons.

26. **Assertion (A):** Order of lattice energy for some halides are as LiX > NaX > KX.
Reason (R): Size of alkaline earth metal increases from Li to K.

27. **Assertion (A):** The lesser the lattice enthalpy more stable is the ionic compound.
Reason (R): The lattice enthalpy is greater for ions of higher charge and smaller radii.

28. **Assertion (A):** As lattice energy increases, melting and boiling points of ionic compound increases.
Reason (R): As lattice energy increases, stability of ionic compound increases.

29. **Assertion (A):** Water is one of the best solvent.
Reason (R): H-bonding is present in water molecules.

30. **Assertion (A):** Born-Haber cycle is based on Hess's law.
Reason (R): Lattice enthalpy can be calculated by Born - Haber cycle.

ANSWERS

1	2	3	4	5	6	7	8	9	10
(*c*)	(*b*)	(*c*)	(*c*)	(*b*)	(*a*)	(*c*)	(*a*)	(*b*)	(*d*)
11	12	13	14	15	16	17	18	19	20
(*c*)	(*c*)	(*b*)	(*b*)	(*c*)	(*b*)	(*a*)	(*c*)	(*a*)	(*c*)
21	22	23	24	25	26	27	28	29	30
(*b*)	(*c*)	(*c*)	(*a*)	(*c*)	(*c*)	(*d*)	(*a*)	(*b*)	(*b*)

4 CHEMICAL ENERGETICS

System : A system may be defined as any specified portion of the matter set aside in the imagination or on which whole attention is focused and have a direct bearing from its surroundings and the rest of the universe which might be in a position to exchange energy and matter with the system is called its surroundings.

Isolated System : A system which can exchange neither energy nor matter from the surroundings is called the isolated system.

Closed System : A system which can exchange energy but not matter from the surroundings is called the closed system.

Open System : A system which can exchange matter as well as the energy with its surroundings is called on open system.

Zeroth Law of Thermodynamics: When two systems are in thermal equilibrium with a third system then they are also in thermal equilibrium with each other. This statement is called the Zeroth law of Thermodynamics.

Work : In thermodynamics, the work done on a system is considered as positive while the work done by the system is taken as negative.

Heat : Thermodynamically the energy that flows from one body to another body or from the system to the surroundings, on account of the temperature difference between the two states is called **Heat**.

Reversible Process : From the thermodynamical point of view, a reversible process is one in which an infinitesimally small change in the external conditions will result in all changes taking place in the reverse order and in opposite sense.

Irreversible Process : From the thermodynamical point of view, a process is said to be **irreversible** if it cannot be retraced back exactly in the opposite direction.

First Law of Thermodynamics

First law of thermodynamics is only a mere case of law of conservation of energy. According to this, *"The energy of a system in definite state is fixed and is independent of the method of formation of the system."*

If heat is supplied to *a system capable of doing external work, then the quantity of heat absorbed by the system is equal to the amount of work done plus the change in internal energy.*

Internal Energy : In chemical thermodynamics we shall be concerned with the energy possessed by the system which is a function of temperature, pressure and chemical composition. This is called internal energy denoted by E.

Concept of Enthalpy

Enthalpy is an extensive thermodynamic property and is given by the symbol H. It is defined as the sum of internal energy and the product of pressure and volume as

$$H = E + PV$$

Likewise internal energy, enthalpy is the property of a system, determined by the state of the system.

Enthalpy of formation is the enthalpy change when one mole of the compound is formed from its constituent elements in their standard (commonly occurring) states, e.g.,

$$C(s) + O_2(g) \rightarrow CO_2(g);\ \Delta H = -393.7\ \text{kJ mol}^{-1}$$

Enthalpy of combustion is the enthalpy change when one mole of the substance is completely burnt in an excess of oxygen or air, *e.g.*,

$$C_6H_6\ (1) + 7\frac{1}{2}\,O_2(g) \rightarrow 6CO_2(g) + 3H_2O(l)$$

$$\Delta H = -3264\ \text{kJ mol}^{-1}$$

Enthalpy of neutralisation is the enthalpy change occurring when one gram equivalent of an acid and that of a base react in dilute solution, *e.g.*,

$$HCl\ (aq) + NaOH(aq) \longrightarrow NaCl\ (aq) + H_2O(l);\ \Delta = -57\ \text{kJ.}$$

The enthalpy of neutralisation of a strong acid with a strong base is always equal to— 57 kJ. However, the enthalpy of neutralisation of a weak acid (or a weak base) depends on the nature of the acid or the base.

Enthalpy of fusion is the change in enthalpy which occurs when one mole of a substance changes from the solid state to the liquid state at a fixed temperature, *e.g.*,

$$H_2(s) \text{ at } 273\ K \longrightarrow H_2O(l) \text{ at } 273\ K;\ \Delta H = 6.0\ \text{kJ.}$$

Enthalpy of vaporisation is the change in enthalpy occurring when one mole of liquid is converted into vapour at a fixed temperature, *e.g.*,

$$H_2O(l) \text{ at } 373\ K \rightarrow H_2O(g) \text{ at } 373\ K;\ \Delta H = 14.6\ \text{kJ.}$$

Enthalpy of sublimation is the enthalpy change accompanying the conversion of one mole of solid directly into vapour at a temperature below its m.p., *e.g.*,

$$I_2\ (s) \longrightarrow I_2(g);\ \Delta H = 62.4\ \text{kJ}$$

Enthalpy of sublimation of a solid is equal to the sum of the enthalpies of fusion and vaporisation.

Hess's Law of Constant Heat Summation (1840)

"The amount of heat involved in a chemical reoction is constant whether the reaction takes place in one step or in several steps." For a reaction A → B,

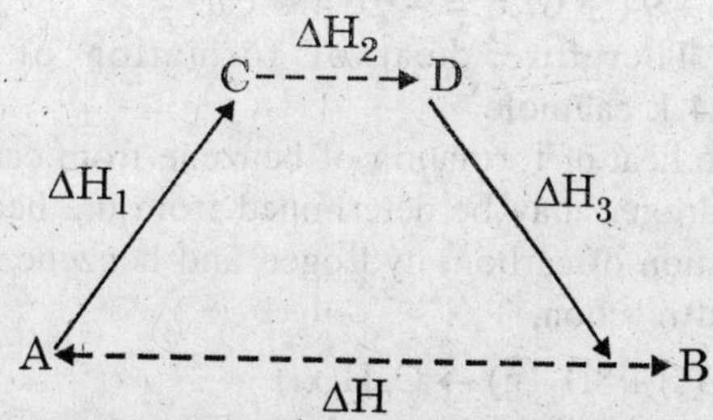

$$\Delta H = \Delta H_1 + \Delta H_2 + \Delta H_3$$

Heat of Combustion : The heat of combustion of an element or a compound is the change in enthalpy ΔH when 1 g molecule of the compound is completely oxidised.

Since combustion reactions are always exthermic the heat of combustion is negative.

$$CH_4(g) + 2O_2(g) \rightarrow CO_2(g) + 2H_2O(g) \qquad \Delta H = -19.0\ \text{kcal}$$

$$H_2 + O_2 \rightarrow H_2O\ \Delta H = -\ 286.2\ \text{kJ.}$$

Heat of Formation : The standard heat of formation $(\Delta H°)$ of a compound is the enthalpy change for the reaction in which 1 mole of that compound is formed from its element at 25°C and 1 atm.

$$C(g) + O_2(g) \rightarrow CO_2(g) \qquad \Delta H = -\ 393\text{kJ.}$$

Heat of Neutralization : The heat of neutralization of an acid is the enthalpy change for the reaction in which 1 g equivalent of the acid in dilute solution is being neutralized by one gram equivalent of a base in dilute sol.

$$HCl(aq) + NaOH(aq) \rightarrow NaCl(aq) + H_2O \qquad \Delta H = -\ 13.700/\text{kcal}$$

Application of Hess's Law

Heat of Formation : With the help of Hess's law, heat of reaction of such reactions can be determined which proceed with many side reactions. For example, heat of formation of CO can be determined by knowing the heat of formation of CO_2 from C and CO. Let the heat of formation of CO from C is Q, then we know that

Q = Heat of formation of CO_2 from C – Heat of formation of CO_2 from CO

= – 94 – (– 67.6)

$= -94 + 67.6 = -26.4$ k cal.

∴ Therefore, heat of formation of CO $= -26.4$ k cal/mole

The heat of formation of benzene from carbon and hydrogen may be determined from the heat of combustion of carbon, hydrogen and benzene. The required reaction,

$$6C(s) + 3H_2(g) \rightarrow C_6H_6(l)$$

can be obtained according to the following three steps:

Step 1 : Writing the combustion reactions of the reactants and products.

(i) $C(s) + O_2(g) \rightarrow CO_2(g); \quad \Delta H = -94 \text{ k cal}$

(ii) $H_2(g) + \frac{1}{2}O_2(g) \rightarrow H_2O(l); \ \Delta H = -68.4 \text{ k cal}$

(iii) $C_6H_6(l) + 7\frac{1}{2}O_2(g) \rightarrow 6CO_2(g) + 3H_2O(l);$

$$\Delta H = -781 \text{ k cal}$$

Step 2 : Since 6 carbon and 6 hydrogen atoms are present in benzene, equation *(i)* and *(ii)* must be multiplied by 6 and 3 respectively.

(iv) $6C(s) + 6O_2(g) \rightarrow 6CO_2(g);$

$$\Delta H = -564.0 \text{ k cal}$$

(v) $3H_2(g) + 1\frac{1}{2}O_2(g) \rightarrow 3H_2O(l);$

$$\Delta H = -205.2 \text{ k cal}$$

Step 3 : Add equations *(iv)* and *(v)* and subtract *(iii)* to get the required net reaction.

$$6C(s) + 3H_2(g) \rightarrow C_6H_6(l); \ \Delta H = -11.8 \text{ k cal}$$

Hence, the heat of formation of benzene $= -11.8$ k cal.

(2) Determination of heat changes of slow reactions, e.g., the transformation of rhombic sulphur into monoclinic sulphur.

$S(\text{rhombic}) + O_2(g) \rightarrow SO_2(g); \Delta H = -71.1 \text{ k cal}$

$S(\text{monoclinic}) + O_2(g) \rightarrow SO_2(g);$

$$\Delta H = -71.7 \text{ k cal}$$

Subtracting *(i)* from *(ii)* and rearranging.

$S(\text{monoclinic}) \rightarrow S(\text{rhombic}); \ \Delta H = -0.6 \text{ k cal}$

$S(\text{rhombic}) \rightarrow S(\text{monoclinic}); \ \Delta H = +0.6 \text{ k cal}$

Bond Energy

When a bond is formed between the two free atoms in a gaseous state to form a molecular product in a gaseous state, some heat is always evolved which is known as the *bond formation energy* or *bond energy*. The bond energy may be referred to as heat of formation of bond.

$$\Delta H_{reaction} = \begin{bmatrix} \text{Bond energy data} \\ \text{used for formation} \\ \text{of bond, to be} \\ \text{taken as -ve} \end{bmatrix} + \begin{bmatrix} \text{Bond energy data} \\ \text{used for dissocia-} \\ \text{tion of bond, to be} \\ \text{taken as +ve} \end{bmatrix}$$

Bond energy is always a positive quantity. The concept of average bond energy is based on the idea that a given type of bond, such as N—H or C=O, or $C \equiv C$, has about the same strength in any compound.

The bond energy of H—Cl bond is defined by ΔH for the process :

$$HCl(g) \rightarrow H(g) + Cl(g); \quad \Delta H = 432 \text{ kJ/mol}$$

Entropy (S)

It is a thermodynamic quantity which is a measure of the degree of disorder within any system. The greater the degree of disorder, the higher the entropy. Thus, for a given substance,

$$S_{gas} > S_{liquid} > S_{solid}.$$

The entropy of the universe always increase in the course of every natural change.

MULTIPLE CHOICE QUESTIONS

1. When a liquid is converted into vapours there is
   (a) no change in enthalpy
   (b) no change in entropy
   (c) decrease in enthalpy
   (d) increase in enthalpy
2. The total heat content of a system is known as

(*a*) Enthalpy (*b*) Thermal Energy
(*c*) Entropy (*d*) Activation Energy

3. In the reaction :
$CO_2(g) + H_2(g) \rightarrow CO(g) + H_2O(g)$
$\Delta H = 2.8$ kJ
then ΔH represents
(*a*) heat of formation
(*b*) heat of reaction
(*c*) heat of solution
(*d*) heat of combustion

4. The heat of neutralisation of a strong acid and a strong base is equal to .... kcal gm eq^{-1}.
(*a*) 27.3 (*b*) 31.7
(*c*) 15.7 (*d*) 13.7

5. Internal energy of the elements is as signed a value of .... cal per mole at standard states.
(*a*) 5 (*b*) 9
(*c*) 99 (*d*) 0

6. The apparatus used for measuring the heat changes is known as
(*a*) Thermometer (*b*) Ammeter
(*c*) Voltameter (*d*) Calorimeter

7. The highest temperature at which vapour pressure of a liquid can be measured is
(*a*) the boiling point of the liquid
(*b*) critical solution temperature
(*c*) ionisation temperature
(*d*) inversion temperature

8. If temperature is kept constant during the reaction, the process is said to be......................
(*a*) Adiabatic (*b*) Isobaric
(*c*) Isothermal (*d*) None of the above

9. The factor that does not influence the heat of reaction is
(*a*) method by which reactants change to products
(*b*) chemical state of the reactants and products
(*c*) whether it is taking place at constant temperature
(*d*) All of the above

10. The solution energy of Al^{+++} ion is more than that of Mg^{++} due to
(*a*) high ionisation energy of Al^{+++} ion
(*b*) greater size of Al^{+++} ion
(*c*) greater charge on Al^{+++} ion
(*d*) small size of Al^{+++} ion

11. 1 Joule is equal to
(*a*) 2.39 Cal (*b*) 0.04184 Cal
(*c*) 1.390 Cal (*d*) 0.239 Cal

12. Heat of combustion of methane and ethane are –291.7 kcal & –441.2 kcal respectively
(*a*) Methane is a better fuel because ethane is poisonous
(*b*) Methane is a better fuel than C_2H_6
(*c*) Ethane is a better fuel than CH_4
(*d*) Both are equally efficient as a fuel

13. Heat of fusion of a molecular solid is
(*a*) high (*b*) low
(*c*) average (*d*) very high

14. Heat of Reaction between sodium hydroxide and hydrochloric acid is 57.32 kJ $mole^{-1}$. So heat of dissociation of water is
(*a*) –57.32 kJ $mole^{-1}$
(*b*) +23.7 kJ $mole^{-1}$
(*c*) +57.32 kJ $mole^{-1}$
(*d*) –13.7 kJ $mole^{-1}$

15. For the reaction
$C + O_2 \rightarrow CO_2$, $\Delta H_1 = -394$ kJ mol^{-1} 2CO + $O_2 \rightarrow 2CO_2$, $\Delta H_2 = -569$ kJ
ΔH for 1 mole formation of CO is
(*a*) +265.0 kJ mol^{-1}
(*b*) –165.0 kJ mol^{-1}
(*c*) –109.5 kJ mol^{-1}
(*d*) +109.5 kJ mol^{-1}

16. The heat change taking place in the reaction
$H_2O(l) \rightarrow H_2O(g)$, ΔH_f $H_2O(g) = -57$ kcal
$\Delta H_f H_2O(l) = -68.3$ kcal is
(*a*) –11.3 kcal (*b*) +11.3 kcal
(*c*) –17.3 kcal (*d*) +125.5 kcal

17. The enthalpy of the elements in their standard states is taken as zero. So the enthalpy of formation of a compound may be
(*a*) negative (*b*) positive
(*c*) zero (*d*) negative or positive

18. Heat of neutralisation of HCl and NaOH is –57.32 kJ per gram equivalent and that of acetic acid and NaOH is –55.23 kJ per gram equivalent. The heat of dissociation of acetic acid per gram equivalent should be

(*a*) –2.09 (*b*) 2.09
(*c*) –2.196 (*d*) 2.196

19. Heat of reaction
(*a*) depends on the rate of reaction
(*b*) depends on pressure
(*c*) is independent of temperature
(*d*) may or may not depend on temperature

20. A person requires 2870 kcal of energy for leading a normal life. If heat of combustion of cane sugar is -1349 kcal then he must consume .... kg of cane sugar daily.
(*a*) 728 (*b*) 448
(*c*) 142 (*d*) 224

Directions Qs. 21-30: *In each of the following questions there are two statements Assertion (A) is given followed by a corresponding statement of Reason (R) just below it. Of the statements, mark the correct answer:*

(*a*) If both assertion and reason are true and reason is the correct explanation of assertion.
(*b*) If both assertion and reason are true but reason is not the correct explanation of assertion
(*c*) If assertion is true but reason is false
(*d*) If assertion is false but reason is true.

21. Assertion (A): Zeroth law can also be termed as law of thermal equilibrium.
Reason (R): Two objects in thermal equilibrium with the third one, are in thermal equilibrium with each other.

22. Assertion (A): Coffee in a stoppered thermos flask is an example of closed system.
Reason (R): A system is said to be closed if it can exchange energy but not matter.

23. Assertion (A): A cup of tea placed in a room is an example of open system.
Reason (R): Transfer of heat and water vapour can take place with the surrounding.

24. Assertion (A): All living systems are the examples of open system.
Reason (R): They cannot exchange matter and energy with the surrounding.

25. Assertion (A): T, P and V are state variables or state functions.
Reason (R): Their values depend on the state of the system and how it is reached.

26. Assertion (A): Thermodynamic process means change in atleast one of the state variables of the system.
Reason (R): Process is accompanied by an exchange of matter and energy between the system and the surroundings.

27. Assertion (A): A process is called adiabatic if the system does not exchange heat with the surroundings.
Reason (R): It does not involve increase or decrease in temperature of the system.

28. Assertion (A): In irreversible process changes are carried-out through quasi equilibrium state.
Reason (R): In reversible process changes are carried out so slowly that the system and surrounding are always in near equilibrium.

29. Assertion (A): First law of thermodynamics is identical to the law conservation of energy.
Reason (R): It states that the total energy of the universe is constant.

30. Assertion (A): Internal energy is an extensive property.
Reason (R): Internal energy depends upon the amount of the system.

ANSWERS

1	2	3	4	5	6	7	8	9	10
(*d*)	(*b*)	(*a*)	(*d*)	(*d*)	(*d*)	(*a*)	(*c*)	(*a*)	(*d*)
11	**12**	**13**	**14**	**15**	**16**	**17**	**18**	**19**	**20**
(*d*)	(*b*)	(*b*)	(*c*)	(*c*)	(*b*)	(*d*)	(*b*)	(*c*)	(*a*)
21	**22**	**23**	**24**	**25**	**26**	**27**	**28**	**29**	**30**
(*a*)	(*d*)	(*a*)	(*c*)	(*c*)	(*a*)	(*c*)	(*d*)	(*a*)	(*a*)

5 CHEMICAL EQUILIBRIUM

This state of the reaction at which the concentration of the reactants/products do not change with time is called **Chemical Equilibrium.**

(a) Reaction rate *vs* time

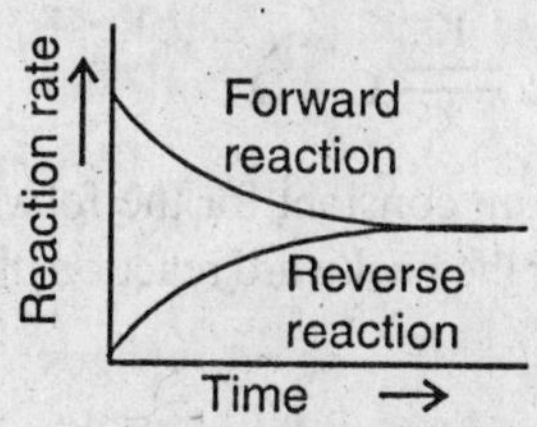

(b) Concentration *vs* time of reactants and products:

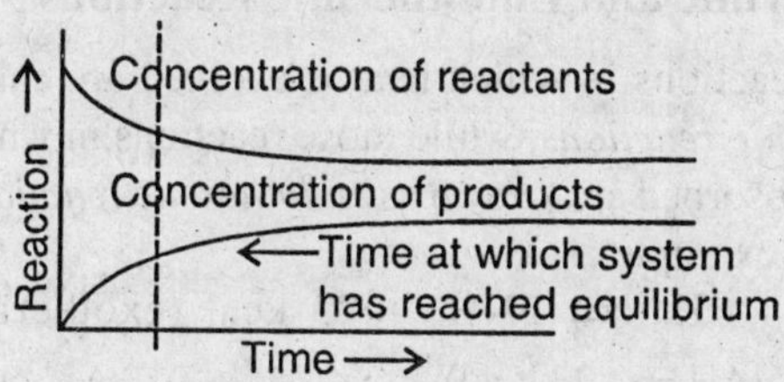

Properties of Chemical Equilibrium

1. This equilibrium can be attained from either direction.
2. We can attain this equilibrium only on the closure of system.
3. At this equilibrium, it seems the reaction stops, but both the forward and backward reactions continued and their rates of reactions become equal.

Law of Mass Action

It was first stated by Norwegian Chemist Cato Guldberg and Peter Waage in 1863. "At constant temperature, the rate of a homogeneous reaction at a given instant is usually proportional to some power of the concentration of each reactant at that instant" or in other words, under a given set of conditions, the rate of chemical reaction is directly proportional to the product of concentration (active mass) of the reacting substances.

Active mass (Molar concentration)

$$= \frac{\text{No. of moles}}{\text{Vol. in litres}}$$

Active mass of a gas or liquid means it molar concentration, while active mass of a solid is always taken as unity; irrespective of the quantity. Law of mass action is applicable only to reversible reactions.

Equilibrium Constant Expression **:** Let us consider a homogeneous reversible reaction

$$A + B \rightleftharpoons C + D$$

By applying law of mass action in this reversible reaction, we get

$$r_f \propto [A]^1[B]^1$$

$$r_f = K_1[A]^1[B]^1 \quad ...(i)$$

Similarly, $r_b = K_2[C]^1[D]^1 \quad ...(ii)$

where K_1 and K_2 are called velocity constants for forward and backward reactions.

Now, at equilibrium

$$r_f = r_b$$

$$\frac{K_1}{K_2} = \frac{[C]^1[D]^1}{[A]^1[B]^1}$$

$$K_c = \frac{[C][D]}{[A][B]} = \frac{K_1}{K_2}$$

where K_c is called equilibrium constant

Illustration-Reactions

$$N_2(g) + 3H_2(g) \rightleftharpoons 2NH_3(g)$$

$$K_c = \frac{[NH_3]^2}{[N_2]^1[H_2]^3}$$

Characteristics of Equilibrium Constant

(1) The numerical value of the equilibrium constant does not depend on :

(a) Concentration of the reactants

(b) Pressure change

(c) Presence of a catalyst

(d) Volume

(e) The direction from which equilibrium is attained

(f) Mechanism of reversible reactions.

(2) The value of K for a particular reaction is always the same, and is different for different reactions.

(3) The value of K does depend on temperature and has a definite value of a given temperature and changes with temperature.

For ***endothermic reversible reactions***, the equilibrium constant increases with rise in temperature and decrease with decrease in temperature. For ***endothermic* reactions**, equilibrium constant decreases with rise in temperature and increases with fall in temperature. The variation of equilibrium constant with temperature is given by Van't Hoff equation

$$\log\frac{K_2}{K_1} = \frac{\Delta H}{2.303R}\left(\frac{T_1 - T_1}{T_1T_2}\right)$$

where,

K_1 = Eq. constant at temperature T_1

K_2 = Eq. constant at temperature T_2

ΔH = Heat of reaction

(4) The *larger* the value of equilibrium constant, greater is the reaction *towards the product side*.

(5) The magnitude equilebrium constant also depends on how the balanced chemical equation of a given reaction is written.

The equilibrium constant is raised to a factor by which the balanced equation is multiplied.

(6) The equilibrium constant of a net reaction is the product of the values of equilibrium constants of individual reactions.

Relation between equilibrium constant of forward and backward reactions

Consider the following general reaction

$$A + B \underset{K_b}{\overset{K_f}{\rightleftharpoons}} C + D$$

If K_f is the equilibrium constant for the forward reaction and K_b is that for the backward reaction, then they are related as

$$K_f = \frac{1}{K_b} \text{ or } K_b = \frac{1}{K_f} \text{ or } K_fK_b = 1$$

Exothermic and Endothermic Reactions

Those reactions in which heat is evolved are called *exothermic reactions*, while those reactions in which heat is absorbed are known as *endothermic reactions*.

For example :

$N_2 + 3H_2 \rightleftharpoons 2NH_3 + 23$ kcal (exothermic reaction, $\Delta H = -23$ kcal)

$N_2 + O_2 \rightleftharpoons 2NO - 43.2$ kcal (endothermic reaction, $\Delta H = 43.2$ kcal)

Le-Chatelier's Principle

It was first of all stated in 1884. The effect of pressures, temperature and concentration on the equilibrium condition of system may be predicted by the application of a general rule known as Le-Chatelier's Principle.

According to this principle "If a system at equilibrium is subjected to a stress by changing pressure, temperature or concentration, the reaction proceeds in such a way as to tend to minimise the effect of that change. The presence of catalyst has no effect on the position of equilibrium."

(A) Effect of Pressure **:** Change in pressure or volume affect the equilibrium system in which at least one of the components is gas and also in which the no. of moles of gaseous reactants and gaseous products are different.

(B) Effect of Concentration **:** The increase in concentrations of one of the substances in the equilibrium mixture displaces the equilibrium in that direction which uses up some of the materials added. Conversely, by decreasing the concentration of a substance causes the production of more of that substance.

Example : $N_2 + 3H_2 \rightleftharpoons 2NH_3$

(C) Effect of Temperature **:** Increase in temperature causes reaction to occur in that direction which results in the absorption of heat, that is endothermic reactions are favoured at high temperature. Exothermic reactions give high yield at low temperature.

(i) K increases with increase in temperature for an endothermic reaction $(\Delta H = \text{positive})$.

(ii) K decreases with increase in temperature for an exothermic reaction $(\Delta H = \text{negative})$.

(iii) When $\Delta H = 0$, the temperature change has no effect on the equilibrium composition, because K is independent of temperature for which $\Delta H = 0$.

(D) Effect of Catalyst **:** The addition of a catalyst has no effect on position of equilibrium in a reversible reaction, since it accelerates the forward and reverse reactions to the same extent and equilibrium is achieved more rapidly. Consequently composition of the equilibrium mixture is unchanged by the presence of a catalyst.

MULTIPLE CHOICE QUESTIONS

1. For the reversible reaction

$N_2(g) + 3H_2(g) \rightleftharpoons 2NH_3(g)$ ΔH is -93.7 kJ. The equilibrium will shift in the forward direction, if conc. of

(a) NH_3 is increased
(b) N_2 is decreased
(c) N_2 is increased
(d) temperature is kept variable

2. According to Law of Mass action, the rate of reaction is directly proportional to

(a) molar concentration of reactants
(b) equilibrium constant
(c) nature of recants
(d) volume of the container

3. In the equilibrium,

$H_3PO_4 + H_2O \rightleftharpoons H_2PO_4^- + H_3O^+$

the equilibrium constant is most likely to be changed by

(a) adding a catalyst
(b) adding H_3PO_4
(c) heating the mixture
(d) adding $H_2PO_4^-$ ion

4. For the reaction, $2HI \rightleftharpoons H_2 + I_2$, the equilibrium constant K at 490°C is 0.022. The equilibrium constant for the reverse reaction will be

(a) 45.45 *(b)* 35.35
(c) 0.11 *(d)* None of these

5. The decomposition of PCl_5 at 323 K proceeds as $PCl_5(s) \rightleftharpoons PCl_3(g) + Cl_2(g)$. If the degree of dissociation of PCl_5 be α and the equilibrium pressure be P, then the partial pressure due to chlorine will be

(a) αP *(b)* $\frac{2-\alpha}{2+\alpha}P$

(c) $\frac{\alpha}{1+\alpha}P$ *(d)* None of these

6. One gm mole of CH_3COOH were made to react with one gm mole of C_2H_5OH till the equilibrium was established. The equilibrium mixture was found to contain 0.333 gm mole of unused acid. The equilibrium constant for the reaction is

(*a*) 8.0 (*b*) 4.0
(*c*) 3.0 (*d*) 0.15

7. A cylinder provided with a piston has some PCl_5 in equilibrium with PCl_3 and Cl_2 formed by the dissociation of PCl_5. If the system is compressed, then indicate the correct statement :

(*a*) The system will remain unaffected
(*b*) Some PCl_3 and Cl_2 will combine to give PCl_5
(*c*) Some more of PCl_5 will decompose
(*d*) None of the above

8. In the reaction $PCl_5(g) \rightleftharpoons PCl_3(g) + Cl_2(g)$ at 760 K, 4.0 moles of PCl_5 are taken in a 2-litre flask to contain 0.80 mole of Cl_2. The value of equilibrium constant will be

(*a*) 1.0 (*b*) 0.1
(*c*) 3×10^5 (*d*) 1×10^{-5}

9. The equilibrium constant for the reaction, $N_2(g) + 3H_2(g) \rightleftharpoons 2NH_3(g)$ is K The value of the constant for the reaction $NH_3(g) \rightleftharpoons \frac{1}{2} N_2(g) + \frac{3}{2} H_2(g)$ will be

(*a*) $\frac{1}{K}$ (*b*) $2\sqrt{K}$
(*c*) $\frac{1}{\sqrt{K}}$ (*d*) K/2

10. 0.2 mole of an inert gas were added to the system $2HI(g) \rightleftharpoons H_2(g) + I_2(g)$ at equilibrium. The dissociation of HI will

(*a*) increase
(*b*) be suppressed
(*c*) remain unchanged
(*d*) None of the above

11. When pressure is applied to equilibrium system of ice $\rightleftharpoons$ water, which of the following phenomenon will occur?

(*a*) Equilibrium will not be disturbed
(*b*) More water will be formed
(*c*) More ice will be formed
(*d*) Water will not evaporate

12. The decomposition of N_2O_4 to NO_2 is carried at 280 k in chloroform. When the equilibrium has been established, 0.2 mole of N_2O_4 and 2×10^{-3} mole of NO_2 are present in a two-litre solution. The equilibrium constant for the reaction $N_2O_4 \rightleftharpoons 2NO_2$ is

(*a*) 2×10^{-5} (*b*) 2×10^{-3}
(*c*) 1×10^{-5} (*d*) 1×10^{-2}

13. For an equilibrium reaction, the rate constant for the forward reaction is 2.38×10^{-4} and that for the reverse reaction is 8.15×10^{-4}. The equilibrium constant for the reaction is

(*a*) 0.531 (*b*) 1.420
(*c*) 55 (*d*) 0.292

14. One mole of A and one of mole B are mixed in a vessel to produce C and (*d*) At the equilibrium, 0.2 mole of B are left unreacted. The value of the equilibrium constant for the reaction will be

(*a*) 32 (*b*) 16
(*c*) 0.8 (*d*) None of these

15. An increase in reaction velocity is due to

(*a*) increase in volume
(*b*) increase in energy of nucleus
(*c*) increase in number of collisions
(*d*) increase in number of effective collisions

16. A relatively fast reaction is characterised by a

(*a*) rate constant equal to 1
(*b*) rate constant equal to Infinity
(*c*) large rate constant
(*d*) small rate constant

17. Given the reaction:

2A (gas) + B (gas) $\rightleftharpoons$ C (gas) + 100 kJ

Which combination of pressure and temperature would give the highest yield of C at equilibrium?

(*a*) 1000 atms. and 370 K
(*b*) 2000 atms. and 370 K
(*c*) 1000 atms. 773 K
(*d*) 500 atms. and 773 K

18. Which of the following is the correct statement for the manufacture of ammonia by Haber's process?

(*a*) High temperature, and high pressure
(*b*) Low temperature, high pressure and catalyst
(*c*) Low temperature and low pressure
(*d*) High temperature and low pressure .

19. When pressure is applied to the equilibrium system. Ice(s) $\rightleftharpoons$ $H_2O(l)$, which of the following will happen?
(*a*) More ice will be formed
(*b*) Water will be evaporated
(*c*) More water will be formed
(*d*) Equilibrium will not be reached

20. At constant pressure, upon the addition of helium at the equilibrium point in the reaction, $PCl_5(g) \rightleftharpoons PCl_3(g) + Cl_2(g)$, the degree of dissociation of
(*a*) PCl_5 will decrease
(*b*) PCl_5 will increase
(*c*) PCl_3 will increase
(*d*) Cl_2 will increase

Directions Qs. 21-27: *In each of the following questions there are two statements Assertion (A) is given followed by a corresponding statement of Reason (R) just below it. Of the statements, mark the correct answer:*
(*a*) If both assertion and reason are true and reason is the correct explanation of assertion.
(*b*) If both assertion and reason are true but reason is not the correct explanation of assertion
(*c*) If assertion is true but reason is false
(*d*) If assertion is false but reason is true.

21. **Assertion (A):** Equilibrium constant has meaning only when the corresponding balanced chemical equation is given.
Reason (R): Its value changes for the nev equation obtained by multiplying or dividing the original equation by a number.

22. **Assertion (A):** Equilibrium constant for the reverse reaction is the inverse of the equilibrium constant for the reaction in the forward direction.
Reason (R): Equilibrium constant depends upon the way in which the reaction is written.

23. **Assertion (A):** $K_p = K_c$ for all reaction.
Reason (R): At constant temperature, the pressure of the gas is proportional to its concentration.

24. **Assertion (A):** The value of K gives us a relative idea about the extent to which a reaction proceeds.
Reason (R): The value of K is independent of the stoichiometry of reactants and products at the point of equilibrium.

25. **Assertion (A):** Catalyst does not affects the final state of the equilibrium.
Reason (R): It enables the system to attain a new equilibrium state by complexing with the reagents.

26. **Assertion (A):** The equilibrium constant for the reaction $CaSO_4.\ 5H_2O_{(s)} \rightleftharpoons CaSO_4.\ 3H_2O_{(s)} + 2H_2O_{(g)}$ is $K_c = \dfrac{[CaSO_4.3H_2O][H_2O]^2}{[CaSO_4.5H_2O]}$.
Reason (R): Equilibrium constant is the ratio of the product of molar concentrations of the substances produced to the product of the molar concentrations of reactants with each concentration term raised to the power equal to the respective stoichiometric constant.

27. **Assertion (A):** For the reaction $2NH_3(g) \rightleftharpoons N_2(g) + 3H_2(g)$, the unit of K_p will be atm.
Reason (R): Unit of K_p is $(atm)^{\Delta n}$.

ANSWERS

1	2	3	4	5	6	7	8	9	10
(*c*)	(*a*)	(*c*)	(*a*)	(*c*)	(*b*)	(*b*)	(*b*)	(*c*)	(*c*)
11	**12**	**13**	**14**	**15**	**16**	**17**	**18**	**19**	**20**
(*b*)	(*c*)	(*d*)	(*b*)	(*d*)	(*c*)	(*a*)	(*b*)	(*c*)	(*b*)
21	**22**	**23**	**24**	**25**	**26**	**27**			
(*a*)	(*a*)	(*d*)	(*c*)	(*c*)	(*d*)	(*d*)			

6 CHEMICAL KINETICS

It is the branch of Physical Chemistry that deals with the study of the rates of chemical reactions and the factors influencing them. It also provides information about the paths by which the chemical reaction occurs. These paths are called the mechanism of the reaction.

Rate of Reaction

It may be defined as the change in concentration of anyone of the reactants or products per unit time. One of important factors governing the rate of reaction is the molar concentration, *i.e.*, number of moies per litre, of the reactants with the increase in the concentrations of the reacting substances, the rate of reaction increases and vice, versa. Thus, the rate of reaction is directly proportional to the concentration of the reactants.

Let us consider a general reaction as

$$\underset{\text{Reactants}}{A+B} \longrightarrow \underset{\text{Products}}{G+H}$$

$$\text{Then, rate} = \frac{-d[A]}{dt}, \frac{-d[B]}{dt}, \frac{+d[G]}{dt}, \frac{+d[H]}{dt}$$

Molecularity of a Reaction

The total number of molecules taking part in the rate determination, as represented by the simple chemical equation of that reaction is known as *Molecularity of the reaction. e.g.* the hydrolysis of methyl acetate is a reaction having the molecularity equal to two, thus, it is Bimolecular reaction.

$$CH_3COOCH_3 + H_2O \rightleftharpoons CH_3COOH + CH_3OH$$

The reactions are classified as

(i) Unimolecular, *(ii)* Bimolecular,
(iii) Tri-molecular and so on.

Pseudo-Unimolecular Reactions : A reaction of the type,

$$A + B \rightarrow \text{Products}$$

is theoretically bimolecular and is often overall of sound order with a rate law

$$-d[C_A]/dt = K[C_A][C_B]$$

If in such a case, the concentration of one of the reactants (say, B) is much larger than that of the other (say, A), then during the reaction, concentration of B does not alter appreciably, *i.e.*, remains constant.

Examples :

1. Hydrolysis of ethyl acetate:

$$CH_3COOC_2H_5 + H_2O\,[\text{excess}] \xrightarrow{Ni^{+2}(\text{catalyst})} CH_3COOH + C_2H_5OH$$

2. Inversion of cane sugar:

$$C_{12}H_{22}O_{11} + H_2O\,(\text{excess}) \rightarrow C_6H_{12}O_6 + C_6H_{12}O_6$$

Order of a Reaction

The number of reacting molecules whose concentration alter as result of chemical change is termed as the order of reaction. It can be classified as : *(i)* 1st order, *(ii)* 2nd order, *(iii)* 3rd order and so on.

Reactions of Zero Order

A reaction is said to be zero order if the rate of reaction is proportional to the zero power of the concentration of the reactants *i.e.*, independent upon zero power. For example:

1. **Combination of hydrogen and chlorine :** The combination of hydrogen and chlorine in region of light is the best example of zero order reaction.

$$H_2 + Cl_2 \longrightarrow 2HCl.$$

2. **Bromination of acetone :** It is the zero order reaction with respect to bromine taking place as

$$CH_3COCH_3 + Br_2 \rightarrow CH_3COCH_2\,Br + HBr$$

Reactions of First Order

A reaction is said to be of first order if the rate of reaction is directly proportional to the concentration of the reactants. The reaction of Ist order is given by

$$A \rightarrow \text{Product}$$

Example :

Decomposition of H_2O_2 : The reaction takes place as

$$H_2O_2 \xrightarrow{Pt} H_2O + \frac{1}{2}O_2$$

The decomposition is carried out in presence of platinum. The Kinetics of Reaction can be measured by measuring the volume of oxygen evolved after different intervals of time.

at $t = 0$ $\quad V = V_0$

$t = t$ $\quad V = V_t$

Then $V_0 \propto a_1$ and $V_t \propto (a - x)$ thus

$$K = \frac{2.303}{t}\log\frac{V_0}{V_t}$$

Reactions of 2nd Order

A reaction is said to be of 2nd order if the rate of reaction is directly proportional to the square of concentration. The second order reaction is given by

$$A + B \rightarrow \text{Products.}$$

When the concentration of reactant is equal, *i.e.*, *a*.

$$A + B \rightarrow \text{Product}$$

$(a) \quad (a)$

$t = 0 \rightarrow a \rightarrow$ Concentration of reactant.

$t = t \rightarrow a - x \rightarrow$ Concentration of reactant.

then $\frac{dx}{dt} \propto (a-x)(a-x)$

$$K = \frac{1}{t}\,\frac{x}{a(a-x)}$$

Example :

1. **Decomposition of ozone :** The reaction takes place as

$$2O_3 \xrightarrow{100°} 3O_2$$

2. **Thermal decomposition of N_2O :** The Reaction takes places as

$$2N_2O \rightarrow 2N_2 + O_2$$

Factors Affecting the Rate of Reaction

1. **Nature of Reactant :** The reaction between ionic or polar molecules, such as neutralisation are very fast. On the other hand, the oxidation or reduction reactions are very slow as there bonds are arranged or electrons are transferred. Thus, the nature of the reactant may increase or decrease the rate of reaction.

2. **Concentration of Reactant :** The rate of the reaction be determined by the no. of collisions between the reacting molecules. When we increase the concentration of the reactants the no. of collisions between the molecules increases and thus the rate of reaction increases and vice versa.

3. **Effect of Temperature :** The rate of a reaction increases considerably with the increase in temperature.

4. **Catalyst :** The catalyst considerably enhance the rate of reaction. A catalyst lowers the activation energy of a reaction by enabling the reaction to take place in a different way then the uncatalysed process.

Activation Energy

Activation energy is the energy which the reactants should overcome to convert into products. In other words, activation energy is the least amount of energy needed to initiate a chemical reaction. The symbol used to denote the Activation energy is E_a and the units of activation energy is Kilo joules per mole.

The formula, which Arrhenius introduces for the calculation of activation energy, is:

$$E_a = -RT\ \ln\ (K / A)$$

where, R = gas constant
T = Temperature (in Kelvin)
K = Coefficient of reaction rate
A = Frequency factor of the reaction

Facts Related to Activation Energy Rate

1. As per the equation, Activation energy depends upon Temperature

2. Activation energy can be calculated by reaction rate at any temperature.

3. Both Activation energy and rate constant are experimentally determined and represents macroscopic reaction-specific parameters.

4. To increase the rate of reaction, one should lower the activation energy.

5. It has been found that addition of a catalyst lowers the activation energy. As it provides an alternative pathway for the reaction and this path way has lower activation energy.

Catalyst

A catalyst is a substance which by its mere presence increases the rate of a reaction without being used up in the end. A catalyst actually participates in the reaction another step it is reproduced by lowering its activation energy.

In a reversible reaction a catalyst speeds both the forward and the backward reactions.

Types of Catalysis

1. **Homogeneous Catalysis :** In this, the catalyst is present in the same phase as the reacting substances, *e.g.*, nitric oxide (NO) in the conversion of SO_2 into SO_3 in lead chamber process for the manufacture of sulphuric acid.

2. **Heterogeneous Catalysis :** In this, the catalyst is present in a different phase, *e.g.*, platinum catalyst in the conversion of SO_2 into SO_3 in Contact Process for the manufacture of sulphuric acid.

Criteria of Catalysts

1. They increase the rate of a reaction. In case the rate is decreased, it is a negative catalyst.

2. A catalyst remains unchanged in quantity and in chemical composition.

3. Only a small quantity of a catalyst is required.

4. Catalysts do not change the position of equilibrium, it only speeds up its attainment.

Theory of Catalysis

There are two theories to explain the rule of catalysts:

1. **Intermediate Compound Theory :** The role of catalyst is understood on the basis of the formation of an intermediate compound. This is usually applicable to homogeneous catalysis. For example,

$A + B \rightarrow AB$
(slow process in the absence of catalyst)
$A + K \rightarrow AK$
(intermediate compound)
$AK + B \rightarrow AB + K$
(rapid process in the presence of K as catalyst)

2. **Contact Catalyst :** Adsorption play an important part in the chemical mechanism of heterogeneous catalysis. Adsorption increases the concentration of reacting particles on the catalytic surface and this itself increases the rate of reaction. Hydrogenation of oil in the presence of nickel as catalyst is a good example of contact catalysis.

Application of Catalyst in Manufacturing Process

Process	*Catalyst (type)*
1. Manufacture of ammonia (Haber's Process)	Finely divided iron (heterogeneous) (Potassium and aluminum oxide used as promoter)
2. Manufacture of nitric acid (Ostwald's process)	Platinum gauze (Heterogeneous)
3. Manufacture of sulphuric acid (Contact process)	Vanadium pentoxide, V_2O_5 (heterogeneous)
4. Hydrogenation of vegetable oils	Finely divided nickel (heterogeneous)
5. Manufacture of methanol	$ZnO + Cr_2O_3$ (heterogeneous)
6. Manufacture of chlorine (Deacon's process)	Copper (II) chloride

MULTIPLE CHOICE QUESTIONS

1. The reaction A + B $\rightleftharpoons$ C + D is an exothermic reaction. Activation energy of forward reaction is .... that of the backward reaction.
(*a*) lesser than (*b*) greater than
(*c*) equal to (*d*) twice

2. If there is 100°C rise in temperature then the rate constant will be increased by .... times.
(*a*) 50 (*b*) 100
(*c*) 150 (*d*) 125

3. In homogeneous catalytic reaction, rate of reaction depends on
(*a*) the concentration of catalyst
(*b*) physical state of catalyst
(*c*) the free energy change
(*d*) None of the above

4. All the radioactive changes are of
(*a*) Third order (*b*) Second order
(*c*) First order (*d*) Zero order

5. If half-life period is inversely proportional to the initial conc. then order of reaction is
(*a*) One (*b*) Two
(*c*) four (*d*) None of these

6. According to collision theory rate of reaction is equal to
(*a*) no. of collisions between the reactant molecules
(*b*) no. of effective collisions
(*c*) no. of collisions
(*d*) concentration of reactants

7. Cobalt present along with iron in the synthesis of ammonia acts as
(*a*) autocatalyst (*b*) induced catalyst
(*c*) promoter (*d*) inhibitor

8. For a first order reaction the graph will be a straight line in case the value........... are plotted.
(*a*) conc. inverse Vs. time
(*b*) log (conc.) Vs. time inverse
(*c*) log (conc.) Vs. time
(*d*) log (conc.) inverse Vs. time

9. The catalyst that is used in the manufacture of nitric add by Ostwald process
(*a*) Pt (*b*) Mo
(*c*) V_2O_5 (*d*) Rb

10. The rate of chemical reaction depends on the nature of chemical reactants because
(*a*) energy of activation differs from one reaction to another
(*b*) some are solid at room temperature
(*c*) some are coloured
(*d*) All of the above

11. During the manufacture of NH_3 by Haber's process, catalyst used is
(*a*) finely divide Fe (*b*) Rb
(*c*) Pt (*d*) V_2O_5

12. The minimum energy necessary to permit a reaction to occur is
(*a*) Threshold energy
(*b*) Activation energy
(*c*) Molar energy
(*d*) Free energy

13. The effect of temperature on the rate of reaction is given by
(*a*) Gibb's equation
(*b*) van der Waals' equation
(*c*) Freundlich equation
(*d*) Arrhenius equation

14. According to collision theory
(*a*) all collisions are ineffective
(*b*) all collisions cause energy loss
(*c*) collisions are violent
(*d*) only highly energetic molecules have enough energy to react

15. In the reaction between hydrogen and nitric oxide, when the initial concentra- tion of NO is kept constant and that of hydrogen is doubled, the reaction rate gets doubled. When the conc. of hydrogen is kept constant and that of NO is doubled, the reaction rate gets quadrupled. So order of this reaction is
(*a*) 2 (*b*) 6
(*c*) 0 (*d*) 3

16. In the case of unimolecular reaction, the time required for 99.9% of the reaction to take place is .... times that required for half of the reaction.

(*a*) 2 (*b*) 10
(*c*) 4 (*d*) 9

17. 75% of a reaction of a first order reaction is completed in 32 minutes. Its half-life is

(*a*) 16 min (*b*) 48 min
(*c*) 64 min (*d*) 96 min

18. What is the order of a chemical reaction whose rate is determined by the variation of one concentration term only?

(*a*) Second (*b*) Four
(*c*) First (*d*) Zero

19. The reaction though bimolecular but following the kinetics of the first order is called

(*a*) pseudo unimolecular reaction
(*b*) pseudo first order reaction
(*c*) pseudo molecular reaction
(*d*) None of the above

20. Active mass represents

(*a*) concentration in unit volume
(*b*) number of collisions in one min.
(*c*) number of moles per ml.
(*d*) molar concentration per litre

Directions Qs. 21-27: *In each of the following questions there are two statements Assertion (A) is given followed by a corresponding statement of Reason (R) just below it. Of the statements, mark the correct answer:*

(*a*) If both assertion and reason are true and reason is the correct explanation of assertion.
(*b*) If both assertion and reason are true but reason is not the correct explanation of assertion
(*c*) If assertion is true but reason is false
(*d*) If assertion is false but reason is true.

21. Assertion (A): Chemical kinetics deals only with the rate of a reaction.
Reason (R): Nature of reactants, concentration of reactants, products and catalyst affect the rate of reaction.

22. Assertion (A): Kinetics explains the reaction mechanism.
Reason (R): Kinetics explains the formation of products.

23. Assertion (A): For the reaction $PCl_5 \rightarrow PCl_3 + Cl_2$, the concentration of PCl_5 decreases with the increase in concentration of PCl_3 and Cl_2.
Reason (R): During the course of a reaction the concentration of the reactants decreases while that of products increases.

24. Assertion (A): The rate of the reaction is the rate of change of concentration of a reactant or a product.
Reason (R): Rate of reaction remains constant during the complete reaction.

25. Assertion (A): Instantaneous rate of reaction is equal to *dx/dt*.
Reason (R): It is the rate of reaction at any particular instant of time.

26. Assertion (A): The mass of the products formed in a reaction depends upon the limiting reactant.
Reason (R): Limiting reactant reacts completely in the reaction.

27. Assertion (A): Rate of reaction can be expressed as rate of change in partial pressure of the reactant or product.
Reason (R): Partial pressure of a gas is equal to its concentration.

ANSWERS

1	2	3	4	5	6	7	8	9	10
(*a*)	(*b*)	(*a*)	(*c*)	(*b*)	(*b*)	(*d*)	(*c*)	(*a*)	(*a*)
11	**12**	**13**	**14**	**15**	**16**	**17**	**18**	**19**	**20**
(*a*)	(*b*)	(*d*)	(*d*)	(*d*)	(*b*)	(*a*)	(*c*)	(*c*)	(*a*)
21	**22**	**23**	**24**	**25**	**26**	**27**			
(*d*)	(*a*)	(*a*)	(*c*)	(*b*)	(*a*)	(*c*)			

7 SOLUTIONS

A solution may be defined as a homogeneous mixture of two or more substances whose composition can be varied.

Solute and Solvent **:** In a solution the dissolved substance is called the solute. The substance in which the solute is dissolved is called the solvent.

Types of Solutions**:** Any state of matter *viz,* gas, liquid or solid may become a solute or a solvent in a solution. Thus, there are nine possible combinations of solute–solvent pairs for a binary solution.

True Solution

A true solution is a homogeneous solution in which the solute particles have diameters less than 10^{-7}cm. *i.e.*, the solute particles are of molecular dimensions. The particles are invisible even under powerful microscopes. For example, sodium chloride in water is a true solution. Most ionic compounds form true solutions in water. Organic compounds like sugar and urea also form true solutions in water.

Suspensions

A suspension is a heterogeneous mixture in which one or more components have a particle size greater than 10^{-5}cm. Very often these particles are visible to the naked eye.

For example, sand in water, sodium chloride in benzene, turmeric in water etc.

Colloids

A colloid is a solution in which the particle size ranges between 10^{-7} and 10^{-5} cm.

For example, milk, blood, honey, smoke, ink, gum, starch solution etc.

Table: Difference between true solutions, suspensions and colloids

Property	True solutions	Colloidal solutions	Suspensions
Particle size	Less than 10^{-7} cm	Between 10^{-5} and 10^{-7}cm	Greater than 10^{-5}cm
Visibility of particles	Invisible to naked eye not visible under powerful microscope	Invisible to naked eye. Visible under powerful microscope	Easily visible
Sedimentation of particles	Do not settle down	Settle down under high centrifugation	Settle down due to gravity
Filtration through filter paper	No residue is formed	No residue is formed	Residue is formed

Raoult's Law

When a liquid is allowed to evaporate in a closed container, the vapour from the liquid exerts a pressure. As the amount of vapour increases, so does the amount of molecules returning to the liquid until finally an equilibrium is reached. This final pressure of vapour is known as the **equilibrium vapour pressure** of the liquid. Two major factors that determine the vapour pressure are: The strength of the attractive forces between the molecules and the temperature. In a solution of two miscible liquids A and B, the partial vapour pressure P_A of the liquid A will be proportional to its mole fraction X_A. Similarly vapour pressure of B, P_B, will be proportional to its mole fraction X_B. Thus,

$$P_A \propto X_A \quad \text{and} \quad P_B \propto X_B$$

$$P_A^o = P_A X_A \quad \text{and} \quad P_B = P_B^o X_B$$

where P_A^o and P_B^o are proportionality constants and refer to vapour pressure of pure solvents A and B respectively. This relationship was given by Francois Raoult in 1886 and is known as **Raoult's law**. It states: *For a solution of volatile liquids, the partial vapour pressure of each component in solution is directly proportional to its mole fraction.*

Let us consider a solution which is obtained by adding a non-volatile solute to a volatile solvent (*e.g.*, glucose in water). In such a case there is no contribution of solute and the vapour pressure is only due to the solvent. The total pressure is only due to contribution of the solvent in solution. Therefore, if mole fraction of the solvent in a solution decreases, the vapour pressure of the solution will also decrease. Thus,

$$P = P_A = P_A^o X_A \qquad \text{....(1)}$$

An alternative statement of Raoult's law for non-volatile solute can be given as follows:

Relative lowering of vapour pressure for a solution is equal to the mole fraction of the solute when solvent alone is volatile.

Ideal and Non-ideal Solutions

An ideal solution is one that behaves in accordance with Raoult's law throughout the entire range of concentration and at all temperatures and pressures. The properties of such a solution are the average of the properties of its components in proportion to their mole fractions. On mixing the components in an ideal solution there will neither be a volume change (*i.e.*, $\Delta V = 0$) nor any change in enthalpy (*i.e.*, $\Delta H = 0$).

A non-ideal solution is that which does not obey Raoult's law and whose formation is accompanied by changes in volume and enthalpy.

Most of the real solutions are non-ideal and show considerable deviations, both positive and negative, from the ideal behaviour.

Colligative Properties

Certain properties of such solutions depend on the mole fraction of the solutes and not on their nature. Such properties which depend only on the mole fraction of the solute are called **Colligative properties**. These properties can be used to determine the molecular masses of substances. Some important colligative properties are:

1. *Relative Lowering of Vapour Pressure* : The relative lowering of vapor pressure is a colligative property as it depends only on the concentration of the solute and it is independent of its nature.

Consider a binary solution consisting of non volatile solute in a volatile solvent. The vapor pressure of the solution (P) is equal to vapor pressure of solvent (P_A) in the solution, which in turn is directly proportional to its mole fraction in solution.

$$P = P_A \, \alpha \, \chi_A$$

$$= P^0_A c_A \qquad \text{...}(i)$$

where c_A = mole fraction of solvent.

If mole fraction of solute is c_B then

$$c_A + c_B = 1 \text{ or } c_A = 1 - c_B \qquad \text{...}(ii)$$

From eq (*i*) and (*ii*)

$$P_A = P^0_A (1 - c_B) = P^0_A - P^0_A c_B$$

$$P^0_A - P_A = P^0_A c_B$$

$$\frac{P_A^0 - P_A}{P_A^0} = \chi_B$$

$$\frac{P_A^0 - P_A}{P_A^0}$$

is called the relative lowering of vapor pressure and is equal to the mole fraction of the solute.

2. *Boiling Point Elevation* : The normal boiling point of a liquid is the temperature at which its vapour pressure equals one atmosphere. It follows from Raoult's law that the boiling points of a solution will be higher than the boiling point of pure solvent. The extent of the boiling point elevation is given by the equation

$$\Delta T_b = K_b \times m$$

where ΔT_b is the amount of temperature increase, K_b is **molal boiling point elevation constant** which is characteristic for each solvent and m is the molality of the solution. From the above relationship the following expression can be worked out for the molecular weight determination :

$$M = \frac{1000 \times K_b \times w}{W \times \Delta T_p}$$

where M is the molecular weight of the solute, w is the weight of solute and W is the weight of the solvent.

3. *Depression of Freezing Point* : Solutions have lower freezing points than pure solvents. The extent of freezing point depression depends upon the molal concentration of the solute. The extent of freezing point depression is given by the expression

$$\Delta T_f = K_f \times m$$

where ΔT_f is the depression in freezing point and K_f is the **molal freezing point depression constant** of the solvent (characteristic of the solvent) and m is the molal concentration. K_f is also called **cryoscopic constant.** For the determination of molecular weight, the above relationship can be expressed as

$$M = \frac{100 \times K_f \times w}{W \times \Delta T_f}$$

where M is the molecular weight of the solute, w and W are the weights of solute and solvent respectively.

TABLE : Molal Boiling Point Elevation

Solvent	*Boiling Point (K)*	*K_b(K/m)*
Water	373.0	0.52
Ethyl alcohol	351.5	1.20
Benzene	353.3	2.53
Carbon tetrachloride	350.0	5.03
Ether	307.8	2.02
Acetic acid	391.3	3.07

TABLE : Molal Freezing Point Depression

Solvent	*Boiling Point (K)*	*K_b(K/m)*
Water	273.0	1.86
Ethyl alcohol	155.7	1.99
Benzene	278.6	5.12
Carbon tetrachloride	250.5	31.80
Ether	156.9	1.79
Acetic acid	289.7	3.90

Osmosis and Osmotic Pressure

When a solution and pure solvent are separated by a semi-permeable membrane, there is a net flow of the solvent into the solution through the membrane. This *phenomenon of the passage of pure solvent into a solution through a semi-permeable membrane is called* **osmosis.** Osmosis can also take place between solutions of different concentrations. In such cases, the solvent flows from the solution of low solute concentration to that of higher concentration.

This equilibrium hydrostatic pressure on the solution required to prevent the passage of solvent molecules into solution through a semi-permeable membrane is known as **osmotic pressure.**

It can be shown that for a solution containing n_B moles of solute in V litres of solution, the osmotic pressure π is given by

$$\pi = \frac{n_B}{V} RT$$

$$\pi = n_B\ RT$$

where R is the gas constant and T is the absolute temperature.

MULTIPLE CHOICE QUESTIONS

1. A standard solution is
(*a*) two molar
(*b*) one normal
(*c*) one of known concentration
(*d*) one in which no more solute can dissociate

2. 1 Normal solution of H_2SO_4 is also
(*a*) 1 molar (*b*) 1 molal
(*c*) 1/3 molar (*d*) 0.5 M

3. Molarity and Molality of a solution are related by

(*a*) $\frac{d}{M} = \frac{W}{1000} + \sqrt{\frac{1}{m}}$

(*b*) $M\left(\frac{W}{1000} + \frac{1}{M}\right) = d$

(*c*) $\frac{M}{d} = \frac{W}{1000} + \frac{1}{M}$

(*d*) $\sqrt{m\left(\frac{W}{2000} + \frac{1}{M}\right)} = d$

4. Equivalent weight of ammonia is
(*a*) 14/3 (*b*) 17
(*c*) 2.7 (*d*) 31

5. Amount of the substance required to prepare one litre of 2N solution is
(*a*) 3 g eq. (*b*) 2 g
(*c*) 2 g eq. (*d*) 2 g mole

6. Which of the following is temperature independent?
(*a*) N (*b*) m
(*c*) M (*d*) F

7. The weight of silver nitrate required to prepare 60 ml of 0.03 g per ml of $AgNO_3$ will be
(*a*) 1.8 g (*b*) 0.18 g
(*c*) 2.5 g (*d*) 9 g

8. For an ideal solution
(*a*) $\Delta H_{min} = 0$ (*b*) $\Delta G_{min} = 0$
(*c*) $\Delta F_{min} = 0$ (*d*) $\Delta S_{min} = 0$

9. 100 g of caustic soda is dissolved in 1 kg of water, the molality of the solution is
(*a*) 3.3 (*b*) 2
(*c*) 25 (*d*) 2.5

10. Equivalent weight of anhydrous oxalic acid is
(*a*) 45 (*b*) 54
(*c*) 36 (*d*) 72

11. 1 mole of sugar (non-volatile) is dissolved in two moles of water, the vapour pressure of the solution relative to that of pure solvent is
(*a*) 1/3 (*b*) 1/2
(*c*) 2/3 (*d*) 3/2

12. An azeotrope is one that
(*a*) can be separated by distillation below critical solution temperature
(*b*) can be separated by distillation above critical solution temperature
(*c*) can be separated by distillation at critical solution temperature
(*d*) cannot be separated by distillation

13. The mole fraction of sodium chloride in a solution containing 1 mole of sodium chloride in 1000 g of water is
(*a*) 0.3 (*b*) 0.0077
(*c*) 1.777 (*d*) 0.0177

14. In an aqueous solution of sulphuric acid, mole fraction of water is 0.85. The molality of the solution is
(*a*) 14 m (*b*) 9.8 m
(*c*) 7.9 m (*d*) 8.9 m

15. Equivalent weight of an oxidising agent is the molecular weight of the substance divided by
(*a*) No. of hydroxide ions that its formula contain
(*b*) No. of H^+ ion furnished per molecule
(*c*) No. of e^- lost
(*d*) No. of e^- gained

16. In the case of non-ideal solutions the enthalpy change taking place
(*a*) is equal to five
(*b*) is less than zero
(*c*) is more than zero
(*d*) may be less or more than zero

17. Which of the following shows positive deviation from Raoult's law?
(*a*) Acetone and ethyl alcohol
(*b*) Ethyl alcohol and cyclohexane
(*c*) Acetone and chloroform
(*d*) Benzene and toluene

18. The molarity of a solution containing 49 g of HPO_4 in 2 litres of solution is
(*a*) 0.37 (*b*) 0.25
(*c*) 1.00 (*d*) 0.50

19. Molality of a Solution is the number of moles of solute in
(*a*) 1 kg of solvent
(*b*) 1 litre of solvent
(*c*) 1 litre of solute
(*d*) 1 mole of solute

Directions Qs. 21-30: *In each of the following questions there are two statements Assertion (A) is given followed by a corresponding statement of Reason (R) just below it. Of the statements, mark the correct answer:*

(*a*) If both assertion and reason are true and reason is the correct explanation of assertion.
(*b*) If both assertion and reason are true but reason is not the correct explanation of assertion
(*c*) If assertion is true but reason is false
(*d*) If assertion is false but reason is true.

21. **Assertion (A):** A solution is a homogeneous mixture of two or more chemically non-reacting substances.
Reason (R): Solutions can be made between any two states of matter.

22. **Assertion (A):** If hydration energy is greater than lattice energy the solid dissolves in liquid.
Reason (R): The solubility of a solid in a liquid depends upon lattice energy and hydration energy.

23. **Assertion (A):** The solubility of the gas in a liquid increases with increase of pressure.
Reason (R): The solubility of a gas in a liquid is directly proportional to the pressure of the gas.

24. **Assertion (A):** Strength and Normality are two different methods of expressing the concentration of a solution.
Reason (R): Strength is equal to normality of a solution.

25. **Assertion (A):** One molal aqueous solution of urea contains 60 g of urea in 1 kg (1000 g) water.
Reason (R): Solution containing one mole of solute in 1000 g solvent is called as one molal solution.

26. **Assertion (A):** Normality and molarity can be calculated from each other.
Reason (R): Normality is equal to the product of molarity and *n*.

27. **Assertion (A):** One molar aqueous solution has always higher concentration than one molal.
Reason (R): The molality of a solution depends upon the density of the solution whereas molarity does not.

28. **Assertion (A):** If 100 cc of 0.1 N HCI is mixed with 100 cc of 0.2 N HCI, the normality of the final solution will be 0.15.
Reason (R): Normalities of similar solutions like HCI can be added.

29. **Assertion (A):** The vapour pressure of a liquid decreases if some nonvolatile solute is dissolved in it.
Reason (R): The relative lowering of vapour pressure of a solution containing a non-volatile solute is equal to the mole fraction of the solute in the solution.

30. **Assertion (A):** The vapour pressure of 0.1 M sugar solution is less than that of 0.1 M potassium chloride solution.
Reason (R): Lowering of vapour pressure is directly proportional to the number of species present in the solution.

ANSWERS

1	2	3	4	5	6	7	8	9	10
(*c*)	(*d*)	(*b*)	(*b*)	(*c*)	(*b*)	(*a*)	(*a*)	(*d*)	(*a*)
11	**12**	**13**	**14**	**15**	**16**	**17**	**18**	**19**	**20**
(*c*)	(*d*)	(*d*)	(*b*)	(*d*)	(*d*)	(*b*)	(*b*)	(*a*)	(*b*)
21	**22**	**23**	**24**	**25**	**26**	**27**	**28**	**29**	**30**
(*b*)	(*a*)	(*a*)	(*c*)	(*a*)	(*a*)	(*b*)	(*c*)	(*b*)	(*d*)

8 ELECTROCHEMISTRY

Electrolytes and Non-electrolytes

The substances which conduct electricity in their molten state or in aqueous solutions are known as electrolytes whereas the substances which do not conduct electricity in both these conditions ions are known as non-electrolytes. Based upon their ionisation the electrolytes are further divided as strong and weak electrolytes.

Strong Electrolytes : The substances which ionises completely into ions in an aqueous solution are called strong electrolytes H_2SO_4, HNO_3, HCl, KOH etc. are their examples

$$HCl + H_2O \rightarrow H_3O^+ + Cl^-$$

Weak Electrolytes : The substances which do not ionise completely in aqueous solution are known as weak electrolytes. HCN, $(NH_4)_2CO_3$, CH_3COOH are their examples.

$$CH_3COOH + H_2O \rightleftharpoons H_3O^+ + CH_3COO^-$$

Ionization of weak Electrotyles

In aqueous solutions, the weak electrolytes are completely ionised and an equilibrium is maintained ionised and unionised molecules. The equilibrium is known as Ionic Equilibrium.

The fraction of total number of molecules of electrolyte which ionise is called degree of ionisation. Thus,

α = Degree of ionisation

$$= \frac{\text{No. of molecules which ionises}}{\text{Total no. of molecules}}.$$

Ostwald's dilution Law : As the dilution of electrolyte increases, the degree of dissociation as well as the population of ion increases. And also the electrical conductivity of the electrolyte increases with accordance to the Arrhenius theory of weak electrolyte. An electrolyte BA ionizes as

$$BA + H_2O \rightarrow B^+(H_2O) + A^-$$

If we have C be concentration of the electrolyte originally and α be the degree of dissociation then we have $C\alpha$ be the concentration of each of B^+ and A^- ions and $C(1-\alpha)$ be the concentration of the undissociated electrolyte and the concentration of the water be 100/18 mole/litre be 55.5 mole /litre.

From the law of mass action

$$K_{aq} = \frac{[B^+][A^-]}{[BA][H_2O]}$$

$$= \frac{C\alpha\ C\alpha}{C(1-\alpha)\,55.5}$$

$$K_{aq} \times 55.5 = K$$

$$= \frac{C\alpha^2}{(1-\alpha)}$$

$$K = \frac{\alpha^2}{(1-\alpha)} \left(\because C = \frac{1}{V}\right)$$

This equation is known as Ostwald's dilution law. Thus, if $(1-\alpha) = 1.01$ for weak electrolyte we have $K = \frac{\alpha^2}{V} \Rightarrow \alpha = \sqrt{KV}$

$$\Rightarrow \quad \alpha = \sqrt{\frac{K}{C}}(V = 1/C)$$

Concept Acid-Base

1. *Bronsted-Lowry Concepts* : With accordance to this theory the substance which donates a proton (H^+) is called an acid while the substance which accepts a proton (H^+) is called a base. Thus, an acid is a proton donor while a base is a proton acceptor. Thus,

$$HCl + NH_3 \rightleftharpoons NH_4^+ + Cl^-$$

Here, HCl is an acid while NH_3 is a base

$$\underset{\text{(Acid)}}{NH_4^+(aq)} + \underset{\text{(Base)}}{H_2O(l)} \rightleftharpoons H_3O^+(aq) + NH_3(aq)$$

2. *Lewis-Concept* : With accordance to Lewis concept when a substance accepts a pair of electrons, it is known as an acid while when a substance donates a pair of electrons, it is known as base. Thus,

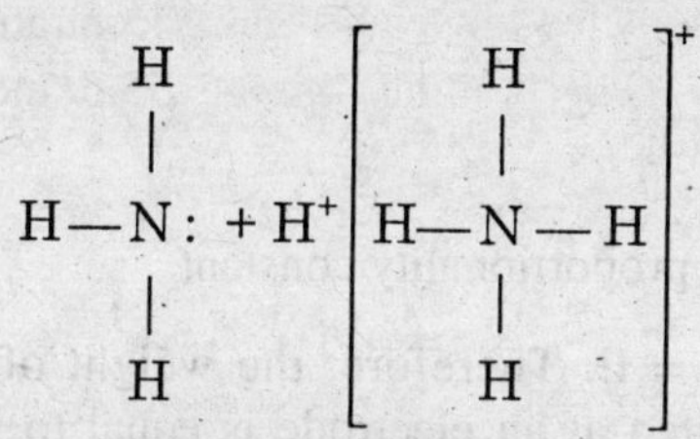

Here, NH_3 is a Lewis base, while H^+ is a Lewis acid.

pH Scale

The concentration of the [H^+] and the [OH^-] is 10^{-14} = Kw = Ionic product of water. Thus, it has been found that their concentration is of very wide range. In fact the term *pH* was introduced in 1909 by Sorensen, a Danish biochemist, used a logarithmic scale for expressing the H^+ concentration.

The pH scale can be defined as the log to the base 10 *of the effective concentration of the hydrogen ion or the activity of the hydrogen as such pH* = Δ – $\log_{10}$ *[H^+] and the concentration of hydrogen ion* = *Concentration of hydrogen ion + concentration of coefficient*

$$= CH^+ + rH$$

For the dilute solutions we have

$$rH = 1.0$$

and as the concentration increases the values of rH becomes less than one thus

$$[OH^+] = CH^+ \text{ therefore}$$

$$pH = -\log_{10}[CH^+]$$

For example concentration of [H^+] in water be 10^{-7}

then pH $= -\log 10\ [10^-]^{-7}$

$= [+ 7.0] = 7.0$

pOH : A relationship similar to the relationship of the pH and the hydrogen ion concentration is used to give the concentration of the [OH^-] other ion. Thus,

$$pOH = \log_{10}[OH^-]$$

It is defined as the logarithm to the base 10 of the concentration of the hydroxyl ion.

Buffer Solution

A solution whose pH remains unaltered by the addition of the small quantities of the acids and the base is defined to as the buffer solution and the buffer solution is considered to be composed of

(i) A solution of strong base and weak acid.
(ii) A solution of weak base and strong acid.
(iii) A solution of ampholite, which is sub- stance that can act as an acid as well as base.
(iv) A concentrated solution of strong acid.
(v) A concentrated solution of strong base.

The pH of the buffer solution be

$$pH = pka + \log\frac{[Salt]}{[Acid]}$$

And then the process in which the salt is completely or partially decomposed into water due to the interaction with water is called as the hydrolysis. Due to the hydrolysis the concentration of H^+ will not remain the same and will be greater or lesser than 10^{-7} mole/litre.

Ionic Product of water

We have the water molecule dissociates in the following manners:

$$H_2O \rightarrow [H^+] + [OH^-]$$

and from the law of mass action

$$K_{aq} = \frac{[H^+][OH^-]}{[H_2O^-]}$$

and we have the equilibrium constant and the concentration of water gives another constant K_W

$$K_{aq} \times [H_2O] = K_W$$

$$K_W = [H^+]\,[OH^-]$$

And this constant is defined to as the ionic product of water.

Faraday's Law of Electrolysis

Michael Faraday (1833) discovered certain relationships between the quantity of current passed through a solution of an electrolyte and the amount of substances liberated (or deposited) at the electrodes during electrolysis. He gave the results in the form of two laws of electrolysis.

First Law of Electrolysis **:** "The amount of any substance deposited (or dissolved) at an electrode during electrolysis is directly proportional to the quantity of electricity passed." The quantity of electricity is expressed in coulombs, one coulomb being the quantity of current flowing when a current of 1 ampere is passed for t seconds, the quantity of current Q = I × t coulombs.

∴ the first law may be given in the form

$$W \propto It \quad \text{or}\ W = Ite$$

where 'e' is a constant for a element and is called its 'electrochemical equivalent'. When It = 1 coulomb, W = e. So "the electrochemical equivalent may be defined as the weight in grams of an element liberated by the passage of 1 coulomb of current."

(e for hydrogen = 0.00001044, Cu = 0.000329, Ag = 0.001118)

This law may be used for measuring the quantity of electric current flowing in a circuit by determining the weight (W) of the substance deposited during electrolysis. If the electrochemical equivalent (e) is known, Q (= It) can be calculated.

Second Law of Electrolysis **:** The amount of different substances deposited (or dissolved) at electrodes by the passage of the same quantity of electricity are proportional to their chemical equivalents (E), *i.e.*,

$$W \propto E$$

For example, when the same quantity of current is passed through aqueous solutions of silver nitrate and copper sulphate.

$$\frac{\text{Wt. of Ag deposited}}{\text{Wt. of Cu deposited}} = \frac{\text{Chemical equivalent of Ag}}{\text{Chemical equivalent of Cu}}$$

In the weights of Ag and Cu deposited are found out and the chemical equivalent of Ag (108) is known, the chemical equivalent of copper can be calculated.

Combination of the two laws

$W \propto It$ (Ist law)

$W \propto E$ (IInd law)

$\therefore\ W \propto ItE$

$$\therefore W = \frac{ItE}{F}$$

Where $\frac{1}{F}$ is proportionality constant

when It = F, W = E. Therefore, the weight of a substance deposited at an electrode is equal to its chemical equivalent when It = F. The constant F is called *Faraday* and it is defined as the quantity of current that must be passed in order to deposit a gram equivalent of any substance at an electrode during electrolysis. The experimentally determined value of F is 96,500 coulombs. So we may write

$$W = \frac{ItE}{96{,}500}$$

Note : *(i)* One Faraday is the charge carried by a gram-equivalent of any ion.

(ii) The charge carried by an electron = $\frac{F}{N}$

$$= \frac{96500}{6.02 \times 10^{23}} = 1.602 \times 10^{-19} \text{ coulombs}$$

Solubility Product

When silver chloride is dissolved in water, an equilibrium exists between the undissolved solid silver chloride and silver ions and chloride ions, i.e., AgCl(s) $\rightleftharpoons$ Ag^+(aq) + Cl^-(aq)

Thus equilibrium constant expression for this reaction is :

$$K = \frac{[Ag^+][Cl^-]}{[AgCl]}$$

As the active mass or molar concentrations of a solid is taken to be constant, therefore [AgCl] is constant.

$\therefore$ K × [AgCl] = $[Ag^+][Cl^-]$

or K_{sp} = $[Ag^+][Cl^-]$

Where K_{sp} is a new constant called 'Solubility Product Constant'. Its value is constant at a given temperature. It is equal to the product of ionic concentrations in a saturated solution. As solubility increases with temperature, K_{sp} also increases. For a sparingly soluble salt that dissolves to give unequal number of cations and anions, the concentration of each ions is raised to a power given by the exponent in the balanced chemical equation. In general, for

$A_xB_y(s) \rightleftharpoons xA^{y+}(aq) + yB^{x-}(aq)$

K_{sp} is given by

$$K_{sp} = [A^{y+}]^x [B^{x-}]^y$$

Thus the solubility product of a sparingly soluble salt at a given temperature may be defined as the product of the concentration of its ions in a saturated solution with each concentration term raised to the power equal to the number of times each ion occurs in the balanced equation for the dissociation of the salt at that temperature.

MULTIPLE CHOICE QUESTIONS

1. One mole of electrons is same as
(*a*) faraday (*b*) coulomb
(*c*) ampere (*d*) None of these

2. For reducing 1 mole of ferrous ions to iron the number of faradays of electricity required will be
(*a*) 4 (*b*) 1.5
(*c*) 0.5 (*d*) 1

3. The number of faradays required to liberate 1 mole of any element tells about
(*a*) charge on the ion of that element
(*b*) power of that electrolyte
(*c*) stability of that element
(*d*) weight of that element

4. On passing electricity through copper sulphate solution, when platinum anode is used
(*a*) copper is discharged from anode
(*b*) H_2 is liberated from anode
(*c*) SO_4^{2-} is liberated at anode
(*d*) oxygen is formed at anode

5. Aqueous solution of HCl conducts electricity because it
(*a*) dissociates
(*b*) dimerises
(*c*) ionises
(*d*) forms hydrogen bond

6. To obtain 2.70 g of silver (at. mass 108) from an aqueous solution of $AgNO_3$ it is required to pass
(*a*) 4750 coulomb (*b*) 12062 coulomb
(*c*) 2412.5 coulomb (*d*) 241.5 coulomb

7. Voltaic cell stops functioning after sometimes because there is
(*a*) the decrease in oxidation potential outweigh the increase in reduction potential
(*b*) the decrease in oxidation potential value equalises the increase in reduction potential value
(*c*) decrease in oxidation potential value
(*d*) increase in oxidation potential and decrease in reduction potential

8. At infinite dilution of an electrolyte, the equivalent conductance of cations and anions are

(*a*) dependent on the nature of solvent
(*b*) independent of nature of ions
(*c*) independent of each other
(*d*) dependent on each other

9. Salt Bridge is used because it
(*a*) forms insoluble precipitates with the electrolytes in the two half cells
(*b*) enhances the flow of electrons by overcoming liquid junction potential
(*c*) maintains the electrical neutrality of the two half cells
(*d*) provides the extra ions

10. Four elements A, B, C and D have the standard oxidation potential values are +2.82 V ; +2.71 V ; +1.67 V and –2.87V. The strongest reducing agent will be
(*a*) A (*b*) B
(*c*) C (*d*) D

11. Is the reaction $2Al + 3Fe^{2+}\ 2Al^{3+} + 3Fe$ possible?
(*a*) No, because S.O.P value of Fe > Al
(*b*) Yes, because S.O.P. value of Al > Fe
(*c*) Experimental fact
(*d*) None of the above

12. More electropositive elements have more
(*a*) negative reduction potential
(*b*) positive oxidation potential
(*c*) tendency to loose electrons
(*d*) positive reduction potential

13. The SRP values of three elements A, B and C are +2.37V; –1.85V and –1.36V respectively. So the order of their reducing power will be
(*a*) B > C > A (*b*) C > B > A
(*c*) A > C > B (*d*) A > B > C

14. Will tin displace lead from an aqueous solution of lead bromide?
(*a*) No, because E° cell comes out to be negative
(*b*) Experimental fact, so cannot be predicted
(*c*) Yes, because SRP value of Pb > Sn
(*d*) Yes, because SRP value of Sn < Pb

15. In the cell,

$Cu\ ;\ Cu^{++}\ ||\ Ag^{+}\ ;\ Ag$

(*a*) Cu rod dissolves but Ag rod grows
(*b*) Cu^{++} ions are consumed and Ag^{+} ions are formed
(*c*) Ag dissolves
(*d*) Cu grows

16. It is impossible to measure the actual voltage of any half cell by itself because
(*a*) of resistance of wire
(*b*) electrical neutrality is maintained
(*c*) a reaction does not take place of its own
(*d*) both half cell reactions occur simul taneously

17. Which of the following metals when coupled will give maximum e.m.f. for a voltaic cell.
(*a*) Pb and Sn (*b*) Cu and Ag
(*c*) Ca and Cu (*d*) Fe and Cu

18. In the cell Cu; Cu^{++} || Ag^{+} ; Ag electrons will flow from
(*a*) Ag rod to Cu rod
(*b*) copper rod to Ag rod
(*c*) Cu^{++} to Ag^{+}
(*d*) neither direction

19. The compound commonly used as electrolyte in lead storage batteries is
(*a*) distilled water (*b*) H_3PO_4
(*c*) H_2SO_4 (*d*) $CuSO_4$

20. During discharge in the case of lead storage batteries density of sulphuric acid
(*a*) increases
(*b*) decreases
(*c*) may decrease or increase
(*d*) remains constant

Directions Qs. 21-30: *In each of the following questions there are two statements Assertion (A) is given followed by a corresponding statement of Reason (R) just below it. Of the statements, mark the correct answer:*

(*a*) If both assertion and reason are true and reason is the correct explanation of assertion.
(*b*) If both assertion and reason are true but reason is not the correct explanation of assertion

(*c*) If assertion is true but reason is false

(*d*) If assertion is false but reason is true.

21. **Assertion (A):** The conductivity of solution is greater than pure solvent.
Reason (R): Conductivity depends upon number of the ions present in solution.

22. **Assertion (A):** The electrical resistance of any object increases with increase in its length.
Reason (R): The electrical resistance of any object increases with increase in its area of cross-section.

23. **Assertion (A):** The resistivity for a substance is its resistance when it is one metre long and its area of cross section is one square meter.
Reason (R): The SI units of resistivity are ohm metre (Ω m) and ohm centimeter (Ω cm).

24. **Assertion (A):** Conductance of a substance increases with decrease in resistance.
Reason (R): The inverse of resistance is called conductance.

25. **Assertion (A):** Substances like glass, ceramics etc. having very low conductivity are known as insulators.
Reason (R): They do not allow the passage of electric current through them.

26. **Assertion (A):** Specific conductance decreases with dilution whereas equivalent conductance increases.
Reason (R): On dilution, number of ions per milliliter decreases but total number of ions increases considerably.

27. **Assertion (A):** The cell constant of a cell depends upon the nature of the material of the electrodes.
Reason (R): The observed conductance depends upon the nature of the electrolyte and the concentration of the solution.

28. **Assertion (A):** When acidified zinc sulphate solution is electrolysed between zinc electrodes, it is zinc that is deposited at the cathode and hydrogen evolution does not take place.
Reason (R): The electrode potential of zinc is more negative than hydrogen as the overvoltage for the hydrogen evolution on zinc is quite large.

29. **Assertion (A):** At the end of electrolysis using platinum electrodes, an aqueous solution of copper sulphate turns colourless.
Reason (R): Copper in $CuSO_4$ is converted to $Cu(OH)_2$ during the electrolysis.

30. **Assertion (A):** The conductivity of solutions of different electrolytes in the same solvent and at a given temperature is same.
Reason (R): The conductivity depends on the charge and size of the ions in which they dissociate, the concentration of ions or ease with which the ions move under potential gradient.

ANSWERS

1	2	3	4	5	6	7	8	9	10
(*b*)	(*d*)	(*a*)	(*d*)	(*c*)	(*c*)	(*b*)	(*c*)	(*c*)	(*a*)
11	**12**	**13**	**14**	**15**	**16**	**17**	**18**	**19**	**20**
(*b*)	(*a*)	(*a*)	(*d*)	(*a*)	(*d*)	(*c*)	(*b*)	(*c*)	(*b*)
21	**22**	**23**	**24**	**25**	**26**	**27**	**28**	**29**	**30**
(*a*)	(*c*)	(*c*)	(*a*)	(*a*)	(*c*)	(*d*)	(*a*)	(*c*)	(*d*)

9 OXIDATION-REDUCTION

Oxidation Process

(i) When an element or compound combines with oxygen or any other electronegative atom, the element or the compound is said to be oxidize

(ii) The removal of hydrogen or any other electropositive element from a substance is known as oxidation of that substance

(iii) According to electronic definition, when any atom, molecule or ions loses electrons, they are said to be oxidise Hence, *de-electronation* of any species is called oxidation of that species :

(iv) When oxidation number of any chemical species is increased, the species is said to be oxidise An increase in oxidation number is a tendency of losing electrons.

Reduction Process

(i) The attachment of hydrogen or any electropositive atom to any substance is called reduction of that substance :

(ii) The removal of oxygen or any other electronegative atom from a substance is called reduction of that substance :

(iii) According to electronic transition any atom, molecule or ion is said to be reduced when accepts electrons

(iv) Decrease in oxidation number of any species is called the reduction of that species. The decrease in oxidation number is the tendency of gaining electrons.

Redox Reactions

It is evident that if one substance loses the electrons, another substance at the same time must gain electrons because the electrons cannot be the products of any chemical change. This means that in any chemical process, oxidation can occur only if reduction is also taking place side by side and *vice versa*. This is the reason that reactions involving oxidation and reduction are called redox reactions.

During redox reaction, there is transference of electrons from the *reducing agent* to the oxidising agent.

$$\underset{\text{(A)}}{\text{Oxidising agent}} \xleftarrow{C^-} \underset{\text{(B)}}{\text{Reducing agent}} \longrightarrow \underset{\text{Reduced product}}{A^+} + \underset{\text{Oxidized product}}{B^-}$$

Oxidation $\rightarrow$ Loss of electrons

Reduction $\rightarrow$ Gains of electrons

Oxidising Agent $\rightarrow$ Species which gains electrons and itself is reduced

Reducing Agent $\rightarrow$ Species which loses electrons and itself is oxidized

Common Oxidising and Reducing Agents

(i) Atoms, ions and molecules that have an unusually *large affinity* for electrons tend to be good *oxidising agents*. eg. fluorine.

(ii) These compounds are strong oxidising agents because elements become more electronegative as the oxidation state of their atom increases. MnO_4^-, CrO_4^{2-}.

(iii) Metal hydrides, such as NaH, CaH_2 and $LiAlH_4$, which formally contains H^- ions are also good reducing agents.

(iv) Some compounds can act as either oxidising or reducing agents. Hydrogen gas, which acts as an oxidising agent when it combines with metals and reducing agent when it reacts with non-metals.

(v) Reactions in which a single reagent undergoes both oxidation and reduction are called 'disproportionation reactions! eg. Bromine.

Calculation of Oxidation Numbers

(i) The oxidation number of all the atoms in their respective elementary states is taken to be zero. eg. N_2, Cl_2, H_2, O_2, O_3, Na, P, Zn etc., the oxidation number of each atom is zero.

(ii) The oxidation number of hydrogen is + 1 in all its compounds except metallic hydrides where it is – 1.

(iii) The oxidation number of oxygen is – 2 in all compounds except peroxides where it is – 1, superoxides oxygen fluoride where it is + 2 and O_2F_2 where it is +1.

(iv) In halides, the oxidation number of halogens is –1.

(v) In sulphides, the oxidation number of sulphur is –2.

(vi) The oxidation number of an ion is the same as its charge. Thus the oxidation number Li^+, Al^{+3}, SO_4^{2-}, and PO_4^{3-} are + 1, + 3, – 2 and – 3 respectively.

(vii) The sum of oxidation numbers of all atoms in the formula of a compound is always zero.

(viii) Alkali metals and alkaline earth metals have + 1 and + 2 oxidation states respectively.

(ix) Oxidation number of a metal in carboxyl complex is always zero, *e.g.*, Ni in $[Ni(Co)_4]$.

Elements of group 4, 5, 6 and 7 show a variety of oxidation numbers. Variable oxidation number is more common in transition elements, *e.g.*, managenese can have five different oxidation numbers in different compounds.

MULTIPLE CHOICE QUESTIONS

1. Which of the following groups of iodine compounds shows increasing order of oxidation number of iodine?

(a) HIO_4, ICl, I_2, HI
(b) HI, I_2, ICl, HIO_4
(c) I_2, HI, HIO_4, ICl
(d) ICl, HIO_4, HI, I_2

2. Which of the following is a metathesis reactions?

(a) $Hg^{2+} + 2OH^- \rightarrow Hg_2O + H_2O$
(b) $Hg_2^{2+} + Sn^{2+} \rightarrow 2Hg + Sn^{4+}$
(c) $Hg_2^{2+} + H_2S \rightarrow Hg + HgS + 2H^+$
(d) None of the above

3. Which of the following reaction is a redox reaction?

(a) $PCl_3 + 3H_2O \rightarrow H_3PO_3 + 3HCl$
(b) $Co^{3+} + 6NO_2^- \rightarrow Co(NO_2)_6^{3-}$
(c) $HgCrO_4 + 2OH^- \rightarrow Hg_2O + CrO_4^{2-} + H_2O$
(d) $Br_2 + OH^- \rightarrow Br^- + BrO_3^-$

4. Which of the following pairs of ions cannot coexist in aqueous solution?

(a) Cr^{2+} and MnO_4^- *(b)* Fe^{3+} and $Cr_2O_7^{2-}$
(c) Cr^{2+} and I_3^- *(d)* Mn^{2+} and Cl^-

5. Which of the following transition metals is strongest reducing agent?

(a) Cr *(b)* Mn
(c) Ni *(d)* Co

6. Which of the following solutions is strongest oxidising agent?

(a) MnO_4^- in acid *(b)* MnO_4^- in base
(c) MnO_2 in base *(d)* CrO_4^{2-} in base

7. Which of the following cannot be a reducing agent?

(*a*) Cl^- (*b*) CaH_2
(*c*) Fe^{2+} (*d*) Br_2

8. Which of the following species does not contain hydrogen in negative oxidation state?
(*a*) $LiAlH_4$ (*b*) CaH_2
(*c*) NaH (*d*) N_3H

9. Which of the following compounds have lowest negative oxidation number of oxygen?
(*a*) H_2O_2 (*b*) O_2F_2
(*c*) KO_2 (*d*) IF_2

10. The reaction occurring at cathode is
(*a*) Oxidation (*b*) Reduction
(*c*) Hydrolysis (*d*) None of these

Directions Qs. 11-20: *In each of the following questions there are two statements Assertion (A) is given followed by a corresponding statement of Reason (R) just below it. Of the statements, mark the correct answer:*

(*a*) If both assertion and reason are true and reason is the correct explanation of assertion.
(*b*) If both assertion and reason are true but reason is not the correct explanation of assertion
(*c*) If assertion is true but reason is false
(*d*) If assertion is false but reason is true.

11. Assertion (A): $Mg_{(s)} + F_{2(s)} \rightarrow MgF_{2(s)}$, magnesium loses electrons and acts as a reducing agent.
Reason (R): Reduction in general means acceptance of electron (s) by a reactant.

12. Assertion (A): In a reaction
$Zn(s) + CuSO_4(aq) \rightarrow ZnSO_4(aq) + Cu(s)$,
Zn is a reductant but itself get oxidized.
Reason (R): In a redox reaction, oxidant is reduced by accepting electrons and reductant is oxidized by losing electrons.

13. Assertion (A): Both, oxygen atom in O_2 or O_3 has an oxidation number equal to zero.
Reason (R): Oxygen is assigned an oxidation number –2 in almost all their compounds.

14. Assertion (A): Oxidation number of magnesium in Mg^{2+} is +2.
Reason (R): Magnesium belongs to 2nd group of the periodic table.

15. Assertion (A): Oxidation state of hydrogen is + 1 in H_2O while –1 in CaH_2.
Reason (R): CaH_2 is a metal hydride and for hydrides, hydrogen is assigned the oxidation number of –1.

16. Assertion (A): In ClF_3, chlorine has the oxidation number –1.
Reason (R): Electron affinity of chlorine is greater than that of fluorine.

17. Assertion (A): Oxidation number of S in HSO_3^- is +3.
Reason (R): Sulphur is in different oxidation states in different compounds.

18. Assertion (A): In $HClO_4$, chlorine has the oxidation number of +7.
Reason (R): $HClO_4$ (perchloric acid) has two peroxide linkages.

19. Assertion (A): Oxidation number of carbon in CH_2O is zero.
Reason (R): CH_2O (formaldehyde), is a covalent compound.

20. Assertion (A): White P reacts with caustic soda, the products are PH_3 and NaH_2PO_2. This reaction is an example of disproportionation reaction.
Reason (R): In disproportionation reaction one and the same substance may act simultaneously as an oxidising agent and as a reducing agent.

ANSWERS

1	2	3	4	5	6	7	8	9	10
(*b*)	(*a*)	(*d*)	(*a*)	(*b*)	(*a*)	(*c*)	(*d*)	(*c*)	(*b*)
11	**12**	**13**	**14**	**15**	**16**	**17**	**18**	**19**	**20**
(*b*)	(*a*)	(*c*)	(*b*)	(*a*)	(*d*)	(*d*)	(*c*)	(*b*)	(*a*)

10 NATURAL AND ARTIFICIAL RADIOACTIVITY

Radioactivity : Henry Becquerel in 1895 accidentally found that uranium and its compounds emit certain mysterious radiations.

The substances like uranium and its compounds which emit such active rays are known to be radioactive and the phenomenon radioactivity.

If a substance emits radiations by itself, it is known as **natural radioactivity.** Some of the compounds become radioactive when they are kept in contact with radioactive substances. This means that they emit radiations upon exposure to rays from radioactive substances and the phenomenon of this type is called **artificial or induced radioactivity.**

Presently about 40 natural radioactive elements are known.

Nature of Radiations

Upon careful investigations of the radioactive emanations it was found that three types of rays, designated as *alpha* (α) *beta* (β) and *gamma* (γ) are emitted. These rays can be recognised by their behaviour as outlined in the table.

(a) Alpha (α) Rays

(i) Nature. Alpha particles are — helium nuclei: (He^{++}), that is, helium atom which have lost their two orbital electrons.

They are shown in symbols as ${}^{4}_{2}He$.

(ii) Penetrating power. They can penetrate through thin layers metal foil and a few centimetres of air. They can be stopped by 0.1 mm. thick aluminium foil.

(iii) Ionisation of gases. Their ionisation power is the maximum due to their massive nature.

(iv) Effect on zinc sulphide plate. Due to their high kinetic energy they excite luminosity when they strike a zinc sulphide plate.

(v) Effect on photographic plate. Their effect on the photographic plate is quite pronounced.

(b) Beta (β) Rays

(i) Nature. Beta rays are particles having same charge and mass as electrons. They are shown in symbols as ${}_{-1}{}^{\circ}\beta$ or ${}_{-1}e^{\circ}$

(ii) Penetrating power. They can stopped by 1 cm. thick aluminium sheet.

(iii) Ionization of gases. They are poor ionizers than alpha particles.

(iv) Effect on zinc sulphide plate. Normally due to low kinetic energy, their effect on the photographic plate should be less but actually it is more than the alpha particles.

(c) Gamma (γ) rays

(i) Nature. They do not consist of particles but are electromagnetic waves. They are not deflected in a magnetic field as its has no charge.

(ii) Penetrating power. They can penetrate up to 8 cm of thick lead sheet or 25 cm of thick iron sheet.

(iii) Ionizing power. They have still poor ionizing power as compared to alpha and beta rays.

(iv) Effect on zinc sulphide plate. They have very little effect.

(v) Effect on photographic plate. They have very little effect.

TABLE : Properties of Radioactive Radiations

Rays	*Charge*	*Mass (amu)*	*Velocity**
α	2^+	4.0026	About 0.1 times the speed of light
β	1^-	0.000548	Upto 0.9 times the speed of light
γ	0	0	Speed of light

*Speed of light in vacuum is 3.00×10^8 ms^{-1}.

Transmutation of Elements

The process of conversion of one element into another is called transmutation. There are two types of transmutation processes as:

(a) ***Natural transmutation :*** Elements possessing natural radioactivity undergo a series of natural transmutation until a non-radioactive element is formed.

(b) ***Artificial transmutation :*** The natural transmutation of elements suggested the possibility of artificial transmutation or conversion of one element into another by the chemists. Rutherford, an English Physicist, was the first person to transform one element into another by artificial means. The reaction is :

$$^{14}_{7}N + ^{4}_{2}He \longrightarrow ^{17}_{8}O + ^{1}_{1}H$$

Nuclear Fission Reactions

The heavy nuclei are unstable and when bombarded with neutrons, they split up into two major fragments along with more neutrons. Thus, when ^{235}U is bombarded with neutrons, the following reaction occurs:

$$^{235}U + ^{1}n \rightarrow ^{140}Ba + ^{93}Kr + 2\tfrac{1}{2}\ ^{1}n$$

The splitting of a nucleus into two heavy nuclei is called **nuclear fission.** The neutrons formed in the initial fission, attack other uranium atoms releasing even more neutrons so that a chain of fission reaction is soon built up.

Some other heavy nuclei which undergo fission under slow neutron bombardment are ^{239}Pu and ^{223}U and those fissionable under fast neutron bombardment are ^{238}U and ^{232}Th.

Nuclear Fusion Reactions : Just as the fission of large nuclei is accompanied by large energy changes, the fusion of light nuclei is accompanied by mass losses and the evolution of large quantities of energy. Some examples of nuclear fusion reactions are shown below:

$$4\ ^{1}H \rightarrow ^{4}He + 2\,^{\circ}e \text{ (positrons)} \quad ...(1)$$

$$^{2}D + ^{2}D \rightarrow ^{4}He + \gamma \quad ...(2)$$

$$^{2}D + ^{3}T \rightarrow ^{4}He + ^{1}n \quad ...(3)$$

The preparation of "hydrogen bomb" is based on the fusion reaction in which two deuterium atoms unite to form helium atom with the release of a vast amount of energy.

The energy of the sun and stars are believed to originate from nuclear fusion reactions.

A great deal of efforts is now being made to make peaceful uses of this vast source of energy.

Applications of Radioisotopes

(i) Medicinal Uses : C^{60} is used in treatment of cancerous growth.

Radioactive I^{131}, is used as medicine to determine thyroid disorders.

(ii) Thickness Measurement : Mostly γ-rays due to their high penetrating power are often used for this purpose.

(iii) Sterilisation : Gamma rays can destroy bacteria present in materials without raising their temperature much. Used to preserve perishable food stuffs, sterilise cloth etc.

(iv) Luminescence : When radiations are made to fall on some materials, they emit light. This finds application in illuminating dials of clocks and watches.

MULTIPLE CHOICE QUESTIONS

1. The α-particle is

(a) Negatively charged particle

(b) Hydrogen nucleus

(c) Helium nucleus

(d) Proton

2. An element with atomic number 84 and mass number 218 loses one α-partices in three succesive stages, the resulting element will have

(a) At no. 74 and mass number 218

(b) At no. 82 and mass number 214
(c) At no. 84 and mass number 114
(d) At no. 84 and mass number 214

3. When $_{86}Ra^{326}$ emits an α-partilce, a new element is formed. In which of the group in periodic table does it fall?
(a) V group (b) Zero group
(c) I group (d) II group

4. In melting lattice structure of solid
(a) Remains unchanged
(b) Becomes compact
(c) Changes
(d) None of the above

5. The atomic wt. of thorium is 232 and atomic number 90 at the end of disintegration we obtain an isotope of lead (At. wt. 208, At. no. 82). The no. of emitted (a) and (b) particles are
(a) $\alpha = 6, \beta = 4$ (b) $\alpha = 2, \beta = 3$
(c) $\alpha = 6, \beta = 3$ (d) $\alpha = 4, \beta = 6$

6. Radio-activity was discovered by
(a) J.J. Thomson
(b) Madame Curie
(c) Henry Becquerel
(d) Rutherford

7. Gamma rays are
(a) Low energy waves
(b) High energy protons
(c) High energy electron
(d) High energy electromagnetic waves

8. On bombarding $_7N^{14}$ with α-particles the nuclei of the product formed, after the release of a proton, will be
(a) $_8O^{18}$ (b) $_8F^{15}$
(c) $_8O^{17}$ (d) $_9F^{18}$

9. On bombarding $_8O^{16}$ with deutrons, the nuclei of the product formed will be
(a) $_9F^{18}$ (b) $_6P^{15}$
(c) $_8O^{17}$ (d) $_7N^{14}$

10. The atomic wt. of uranium is 238 and its atomic number 92. In course of its radioactive disintegration 8α and 6β particles are emitted, the atomic wt. and atomic no. of the resultant atoms are
(a) At. wt. 224 At. no. 80
(b) At no. 230 At. no. 76
(c) At wt. 206 At. no. 76
(d) At wt. 206 At. no. 82

11. Out of the following the one which has no charge is
(a) Gamma rays (b) Beta rays
(c) Alpha rays (d) Cathode rays

12. If a radioactive isotope with atomic number A and mass number M emits an alpha particle, the atomic number and mass number of that new isotope will become
(a) A; M — 4 (b) A — 4; M — 2
(c) A — 2; M — 4 (d) A — 2; M — 3

13. Which of the following is used in dating archaeological findings?
(a) $_{92}U^{235}$ (b) $_{53}Ba^{105}$
(c) $_6C^{14}$ (d) $_1H^1$

14. Hydrogen bomb is based on the phenomenon of
(a) Nuclear explosion
(b) Chemical reaction
(c) Nuclear fusion
(d) Nuclear fission

15. If uranium (mass no. 238, atomic number 92) emits α-particle the product has mass no. and atomic number
(a) 234 + 90 (b) 238 + 92
(c) 236 + 92 (d) 238 + 90

16. The age of most ancient geological formation is estimated by
(a) Uranium lead method
(b) Radium-silicon method
(c) Carbon 14 dating method
(d) Potassium-Argon method

17. When the quantity of a radioactive substance is increased two times, the number of atoms disintegrating per unit time is
(a) increased by square of two
(b) increased, but not a great extent
(c) doubled
(d) trippled

18. The half-life of a radioactive element is 50 days. How long will it take far its activity reduce to 1/4 of it original value?

(*a*) 100 days (*b*) 75 days
(*c*) 50 days (*d*) 25 days

19. Who observed that when the nucleus of uranium atom was bombarded with fast moving neutrons, it becomes so very unstable that it is immediately broken into two nuclei of nearly equal mass besides other fragments?

(*a*) Chadwick (*b*) Einstein
(*c*) Rutherford (*d*) J. J. Thomson

20. When a radioactive substance is subjected to a vacuum, the rate of disintegration per second

(*a*) increases only if the products are gaseous
(*b*) increases considerably
(*c*) decreases
(*d*) is not affected

Directions Qs. 21-26: *In each of the following questions there are two statements Assertion (A) is given followed by a corresponding statement of Reason (R) just below it. Of the statements, mark the correct answer:*

(*a*) If both assertion and reason are true and reason is the correct explanation of assertion.
(*b*) If both assertion and reason are true but reason is not the correct explanation of assertion
(*c*) If assertion is true but reason is false
(*d*) If assertion is false but reason is true.

21. Assertion (A): Cathode rays get deflected by electric or magnetic fields but light rays remain unaffected by these fields.
Reason (R): Light rays are electromagnetic waves and not charged particles.

22. Assertion (A): An electron is not deflected on passing through certain region of space. This observation confirms that there is no magnetic field in that region.
Reason (R): The deflection of electron depends on angle between velocity of electron and direction of magnetic field.

23. Assertion (A): All the radioactive elements are ultimately converted to lead.
Reason (R): All the elements above lead are unstable.

24. Assertion (A): Amongst alpha, beta and gamma rays, α-particle has maximum penetrating power.
Reason (R): The alpha particle is heavier than beta and gamma rays.

25. Assertion (A): The ionising power of β-particle is less compared to α-particles but their penetrating power is more.
Reason (R): The mass of β–particle is less than the mass of α–particle.

26. Assertion (A): The mass of β-particles when they are emitted is higher than the mass of electrons obtained by other means.
Reason (R): β-particle and electron, both are similar particles.

ANSWERS

1	2	3	4	5	6	7	8	9	10
(*c*)	(*d*)	(*b*)	(*c*)	(*a*)	(*c*)	(*d*)	(*c*)	(*a*)	(*d*)
11	**12**	**13**	**14**	**15**	**16**	**17**	**18**	**19**	**20**
(*a*)	(*c*)	(*c*)	(*c*)	(*a*)	(*a*)	(*c*)	(*a*)	(*a*)	(*d*)
21	**22**	**23**	**24**	**25**	**26**				
(*a*)	(*d*)	(*b*)	(*d*)	(*b*)	(*b*)				

B INORGANIC CHEMISTRY

11 HYDROGEN

Position of hydrogen in Periodic table

Hydrogen and helium together make up period 1 of the periodic table. Position of helium is at the head of the noble gases in group O. Hydrogen being only one other element of period 1, it would be placed at the head of any of the other groups without interfering with the positioning of the subsequent element. Hydrogen is found to be resembled with alkali metals as well as halogens. This makes its position anomalous in the periodic table. This dual behaviour can be explained by the fact that it has an equal tendency to gain or loose one electron.

Resemblance with Alkali Metals

***(i) Electropositive Character and the Valency*:** Hydrogen as well as the alkali metals exhibit an electropositive character and are monovalent.

***(ii) Formation of Cations*:** The halides of both hydrogen and alkali metals, when dissolved in water give rise to the positive ions, *i.e.*, H^+, Na^+ etc.

***(iii) Electronic Configuration*:** The alkali metals and hydrogen both have one electron in their outermost orbit which they have the tendency to loose.

***(iv) Affinity for Electronegative Elements (or non-metallic elements)* :** Hydrogen as well as alkali metals have a strong affinity for non-metals and 'little affinity for the metals.

Resemblance with Halogens

***(i) Non-metallic Character*:** Hydrogen has a non-metallic character like halogens. It is a gas and is a bad conductor of heat and electricity.

***(ii) Electronic Configuration*:** Hydrogen has one electron in the outermost shell. It needs one electron to acquire the stable configuration of helium.

(iii) Ionization Energy: The ionization energy of hydrogen is of the same order as that of halogens and much different from that of alkali metals.

***(iv) Atomicity*:** Hydrogen molecule is diatomic like the halogen molecule. The alkali metals are monoatomic.

(*v*) Replacement of Hydrogen by Halogens: Hydrogen can be replaced by the halogens in such compounds like hydrocarbons, alkyl radicals etc. which indicates the similar nature of hydrogen and the halogens.

Isotopes of Hydrogen

1. *Protium* : This isotope is most common. Its nucleus which consist of a solitary proton has no neutron at all. There is a solitary electron outside the nucleus. It has the same mass number 1, as well as the same atomic number 1. Its symbol is ${}_1H^1$.

2. *Deuterium:* Deuterium occurs in water (D_2O). Its nucleus contain a proton and a neutron thus have the mass number equal to 2. There is however a solitary electron outside the nucleus as expected thus have the Atomic number 1. It is generally represented by the symbol ${}_1H^2$ or ${}_1D^2$ and referred to as heavy hydrogen.

3. *Tritium:* Its nucleus consist of one proton and two neutrons thus having the mass number equal to 3 and a solitary electron outside the nucleus makes its atomic number 1. It is radioactive in character.

Heavy Water

It was discovered by Urey who confirmed that 6×10^3 parts of ordinary water contain 1 part of heavy water. Lewis and Denials (1933) prepared a few milli-litres of heavy water. When ordinary water is electrolysed hydrogen is liberated more readily than deuterium, hence the residual water becomes richer in heavy water. By repeated electrolysis of water, pure D_2O is produced.

Deuterium oxide is a colourless liquid resembling water in general. It has higher molecular mass and therefore, has higher density, m.p. and b.p. Since heavy water has less dielectric constant than H_2O, ionic compounds are less soluble in it than in ordinary water. In India, heavy water is manufactured at BARC, Mumbai.

It is used *(i)* in the study of exchange reactions by isotopic method; *(ii)* in the preparation of heavy hydrogen and its other compounds; *(iii)* as a moderator in nuclear reactions for slowing down the fast neutrons and absorbing some of them.

Soft and Hard Water

Water which easily forms lather with soap is known as soft water, and water which does not form lather with soap easily is known as hard water. Hardness is of two types: *(i)* temporary hardness due to the presence of bicarbonates of Ca^{2+} and Mg^{2+}, and *(ii)* the permanent hardness due to the presence of chlorides and sulphates of Ca^{2+} and Mg^{2+}.

Soaps are sodium and potassium salts of fatty acids, such as stearic, oleic and palmitic. When they are dissolved in water, they lower its surface tension and facilitate cleansing. Hard water cannot be used for washing clothes, as much soap is wasted in precipitating insoluble Ca^{2+} and Mg^{2+} salts. Moreover, these insoluble precipitates stick to the fibres of clothes and harm them.

Hard water cannot be used in boilers, as hard scales are deposited on inside the boiler and steam pipes. Due to this insulating layer, more fuel is required to generate steam. Moreover, hydrolysis of cations, Ca^{2+} or Mg^{2+}, would produce H_3O^+ ions which corrode the boiler and shorten its life.

Removal of Hardness: Temporary hard water is softened by *(i)* boiling or *(ii)* by treating with calculated quantity of lime in the form of lime water or milk of lime.

(i) $Ca(HCO_3)_2 \rightarrow CaCO_3$

(ii) $Ca(HCO_3)_2 + Ca(OH)_2 \rightarrow 2CaCO_3 + 2H_2O$

Permanent hardness is removed by one of the following methods:

1. ***Addition of sodium carbonate:***

   $CaCl_2 + Na_2CO_3 \rightarrow CaCO_3 + 2NaCl$

2. ***Permutit or ion-exchange method:*** Water is passed over permutit, a complex of sodium aluminium silicates $Na_2H_4Al_2\ Si_2O_2$ abbreviated as Na_2Z (Sodium zeolite). Ca^{2+} and Mg^{2+} ions of water are replaced by Na^+

   $$\underset{\text{in hard water}}{Ca^{2+}} + \underset{\text{in permutit}}{2Na^+} \rightarrow \underset{\text{in hard water}}{2Na^+} + \underset{\text{trapped in permutit}}{Ca^{2+}}$$

3. ***Calgon process:*** Water is passed through Calgon—a complex of $Na_2(Na_4P_6O_{16})$. It removes Ca^{2+} and Mg^{2+} from water, probably by absorption upon its colloidal anions.

   $(Na_4P_6O_{16})^{2-} + Ca^{2+} \rightarrow (Na_2CaP_6O_{16})^{2-} + 2Na^+$

MULTIPLE CHOICE QUESTIONS

1. Electronic configuration of hydrogen is
   (a) $1s^1$ *(b)* $2s^2$
   (c) $2s^1$ *(d)* $1s^2$

2. The radioactive isotope of hydrogen is
   (a) tritium *(b)* deuterium
   (c) helium *(d)* para hydrogen

3. The most abundant isotope of hydrogen is
(*a*) deuterium (*b*) tritium
(*c*) para hydrogen (*d*) protium

4. Deuterium and Tritium differ w.r.t
(*a*) neutrons (*b*) electrons
(*c*) mesons (*d*) Both (*b*) and (*c*)

5. H_2 is not adsorbed by
(*a*) palladium (*b*) silver
(*c*) cobalt (*d*) gold

6. Hardness in water is due to the presence of
(*a*) soluble salt of Mg
(*b*) soluble salt of K
(*c*) bicarbonates of Ca and K
(*d*) soluble salts of Ca and Mg

7. Water has maximum density at
(*a*) 277 K (*b*) 173 K
(*c*) 377 K (*d*) 267 K

8. For military purposes hydrogen is obtained from
(*a*) lithium aluminium hydride
(*p*) potassium hydride
(*c*) calcium hydride
(*d*) sodium hydride

9. Metal hydrides on treatment with water give
(*a*) hydrogen peroxide
(*b*) basic oxide
(*c*) hydrogen
(*d*) alkali

10. Permanent hardness of water can be removed by
(*a*) adding sodium bicarbonate
(*b*) adding sodium carbonate
(*c*) adding soda-lime
(*d*) boiling with NH_3

11. Heavy water is
(*a*) an oxide of para hydrogen
(*b*) an oxide of ortho hydrogen
(*c*) contains salts dissolved in it
(*d*) an oxide of deuterium

Directions Qs. 12-16: *In each of the following questions there are two statements Assertion (A) is given followed by a corresponding statement of Reason (R) just below it. Of the statements, mark the correct answer:*

(*a*) If both assertion and reason are true and reason is the correct explanation of assertion.
(*b*) If both assertion and reason are true but reason is not the correct explanation of assertion
(*c*) If assertion is true but reason is false
(*d*) If assertion is false but reason is true.

12. Assertion (A): Hydrogen is a diatomic molecule.
Reason (R): Hydrogen atom contains one electron, one proton and one neutron.

13. Assertion (A): Hydrogen shows resemblance with alkali metals as well as halogens.
Reason (R): Hydrogen exists in atomic form only at high temperature.

14. Assertion (A): Hydrogen has only three isotopes namely protium, deuterium and tritium.
Reason (R): Protium is radioactive in nature.

15. Assertion (A): Hydrogen cannot be prepared in laboratory.
Reason (R): Hydrogen of high purity is obtained by electrolysing warm aqueous barium hydroxide between nickel electrodes.

16. Assertion (A): Deuterium is also known as heavy hydrogen.
Reason (R): Deuterium is a good conductor of heat and electricity.

ANSWERS

1	2	3	4	5	6	7	8	9	10
(*a*)	(*a*)	(*d*)	(*a*)	(*c*)	(*d*)	(*a*)	(*c*)	(*a*)	(*b*)
11	**12**	**13**	**14**	**15**	**16**				
(*d*)	(*c*)	(*b*)	(*c*)	(*d*)	(*c*)				

12 GROUP-I ELEMENTS

Sodium Hydroxide (Caustic Soda)

Preparation

(i) Causticizing Process (Gossage Process): It is an old process involving heating of a 10% solution of Na_2CO_3 with a little excess of milk of lime, $Ca(OH)_2$

$$Na_2CO_3 + Ca(OH)_2 \rightarrow CaCO_3\downarrow + 2NaOH$$

(ii) Electrolytic Method : Caustic soda is manufactured by the electrolysis of a concentrated solution of sodium chloride. Since chlorine is one of the by-products, it may react with NaOH forming NaCl and sodium hypochlorite. To avoid reaction between NaOH and Cl_2 specially designed electrolytic cells are used.

Physical Properties

It is white, translucent, hygroscopic solid. It absorbs moisture and CO_2 from the atmosphere forming sodium carbonate.

It is highly soluble in water and the solution is bitter to taste, corrosive and soapy to touch. It is only sparingly soluble in alcohol.

Chemical Properties

(i) Reaction with Acids : It is a strong alkali and neutralized all acids or acidic oxides forming salts

$$NaOH + HCl \rightarrow NaCl + H_2O$$

(ii) Action of Non-Metals : Non-metals like B, Si, P (white), S, and halogens react with NaOH forming sod. salt of the corresponding oxy-acid.

$$2B + 6NaOH \rightarrow \underset{\text{Sod.borate}}{2Na_3BO_3} + 3H_2$$

(iii) Action on Metals : Weakly electropositive metals like Zn, Al and Sn, dissolve in NaOH solution to liberate hydrogen gas.

$$Zn + 2NaOH \rightarrow Na_2ZnO_2 + H_2$$

Uses

Sodium hydroxide is used in the manufacture of sodium metal, soap (from oils and fats), rayon, paper, dyes and drugs. It is used in petroleum refining, for mercerizing cotton to make cloth unshrinkable, and as a reagent in the laboratory.

Sodium Carbonate

It exists in various forms, namely anhydrous sodium carbonate Na_2CO_3 (soda-ash), monohydrate, $Na_2CO_3.H_2O$ (crystal carbonate), heptahydrate, $Na_2CO_3.7H_2O$ and decahydrate, $Na_2CO_3.10H_2O$ (washing soda or sal soda).

Preparation

Solvay Process (Ammonia Soda Process)

$$NH_3 + CO_2 + H_2O \rightarrow NH_4HCO_3$$

$$NH_4HCO_3 + NaCl \xrightarrow{30°C} \underset{\text{Sod. bicarbonate}}{NaHCO_3\downarrow} + NH_4Cl$$

$$2NaHCO_3 \xrightarrow{250°C} Na_2CO_3 + H_2O + \underset{\text{(used again)}}{CO_2}$$

$$\underset{\text{Amm. chloride}}{2NH_4Cl} + \underset{\text{Slaked lime}}{Ca(OH)_2} \rightarrow CaCl_2 + 2H_2O + \underset{\text{(used again)}}{2NH_3}$$

Chemical Properties

On heating, the monohydrate changes to anhydrous salt (m.p. 853°C) which does not decompose on further heating even to redness.

$$Na_2CO_3 + 2H_2O \rightleftharpoons \underset{\text{(Weak acid)}}{H_2CO_3} + \underset{\text{(Strong alkali)}}{2NaOH}$$

$$CaCl_2 + Na_2CO_3 \rightarrow CaCO_3\downarrow + 2NaCl$$

It reacts with carbonates of metals like Al, Fe, Sn etc. When formed are immediately hydrolysed to hydroxides.

$$Fe_2(SO_4)_3 + 3Na_2CO_3 \rightarrow Fe_2(CO_3)_3 \xrightarrow{3H_2O} 2Fe(OH)_3 + 3CO_2$$

$$Na_2CO_3 + H_2O + CO_2 \rightarrow \underset{\text{Sod. bicarbonate}}{2NaHCO_3}$$

Sodium Bicarbonate or Baking Soda

It is Prepared CO_2 into Na_2CO_3 solution, also in Solvay process for sodium carbonate. It is sparingly soluble in water. Decomposes to Na_2CO_3 on heating used in Na_2CO_3 manufacture, as an ingredient of baking powders, pharmaceutically, and in fire extinguishers, in the preparation of effervescent drinks and fruit salts, as antacid.

Its aqueous solution is alkaline in nature gives yellow colour with methyl orange but no colour with phenopthalein.

It precipitates out certain metal salts as normal carbonates which are precipitated by Na_2CO_3 as basic carbonates, *e.g.*, Zn^{2+} and Mg^{2+} salts.

$$ZnCl_2 + 2NaHCO_3 \rightarrow ZnCO_3\downarrow + 2NaCl + H_2O + CO_2$$

When equimolecular amounts of Na_2CO_3 and $NaHCO_3$ are dissolved in water and the resulting solution cooled to 35°C, crystals of sodium sesquicarbonate, $Na_2CO_3 . NaHCO_3 . 2H_2O$ separate out. It is neither deliquescent nor efflorescent and is used for wool washing.

Sodium Chloride

Sodium Chloride has M.P. 801°C, b.p. 1439°C, density 2.17. Occurs naturally as rock salt which is widely distributed and occurs extensively in thick beds and to about 3% in sea water. Obtained commercially by solar evaporation of sea water, by mining or usually as brine by passing water into underground deposits. The brine contains Ca^{2+}, Fe^{2+}, Mg^{2+} as impurities, these are removed by precipitation with Na_2CO_3 and NaOH. Pure NaCl is obtained by passing HCl gas into a saturated aqueous solution of NaCl. Sodium chloride is used for the formation of NaOH and Na_2CO_3 (for the alkali and glass industries) and chlorine (for the chlorocarbon industry) and is one of the chemical industries most important raw materials. Solid NaCl is used for the removal of ice and snow and as a preservative. The sodium chloride structure is cubic with octahedral co-ordination about Na` and Cl. Many other MX compounds crystallizes with this structure.

MULTIPLE CHOICE QUESTIONS

1. The raw material used in Solvay process for manufacture of sodium carbonate comprises
(*a*) Sodium chloride and carbon dioxide
(*b*) Ammonia and carbon dioxide
(*c*) Sodium chloride, limestone and ammonia
(*d*) Sodium chloride, limestone and carbon dioxide

2. Soda ash is
(*a*) $Na_2CO_3.H_2O$ (*b*) $NaHCO_3$
(*c*) Na_2CO_3 (*d*) NaOH

3. Sodium thiosulphate is used in photography to
(*a*) Convert metallic silver to silver salt
(*b*) Reduce silver bromide to silver salt
(*c*) Remove unreduced silver
(*d*) Remove undecomposed AgBr as a soluble silver thiosulphate complex

4. Sodium carbonate reacts with SO_2 in an aqueous medium to give
(*a*) Na_2SO_4 (*b*) $NaHSO_4$
(*c*) Na_2SO_3 (*d*) $NaHSO_3$

5. Sodium thiosulphate is prepared by
(*a*) Boiling an Na_2SO_3 solution with S in an acidic medium

(*b*) Boiling an NaOH solution with S
(*c*) Neutralizing an H_2SO_3 solution with NaOH
(*d*) Both (*b*) and (*d*)

6. The fluxes that can be used for the removal of basic impurities like CaO and FeO from an ore are
(*a*) SiO_2 (*b*) $Na_2B_4O_7.10H_2O$
(*c*) Both (*a*) and (*b*) (*d*) $CaCO_3$

7. Which one of the following is formed in Solvay process?
(*a*) Na_2CO_3 (*b*) $NaHCO_3$
(*c*) $CaCl_2$ (*d*) All

8. Mortar is a mixture of
(*a*) $CaCO_3$ & CaO
(*b*) Slaked lime and water
(*c*) Slaked line, sand and water
(*d*) None

9. Calcium is extracted by the electrolysis of
(*a*) Fused mixture of $CaCl_2$ and CaF_2
(*b*) $CaCl_2$ solution
(*c*) Fused mixture of $CaCl_2$ and NaF
(*d*) $Ca_3(PO_4)_2$ solution

10. Which of the following represent calcium chlorite
(*a*) $Ca(ClO_2)_2$ (*b*) $CaClO_2$
(*c*) $Ca(ClO_3)_2$ (*d*) $Ca(ClO_4)_2$

Directions Qs. 11-19: *In each of the following questions there are two statements Assertion (A) is given followed by a corresponding statement of Reason (R) just below it. Of the statements, mark the correct answer:*

(*a*) If both assertion and reason are true and reason is the correct explanation of assertion.
(*b*) If both assertion and reason are true but reason is not the correct explanation of assertion
(*c*) If assertion is true but reason is false
(*d*) If assertion is false but reason is true.

11. Assertion (A): Na is a *s*-block element.
Reason (R): Na receive its outermost electrons in *s*-orbital.

12. Assertion (A): Group 1(1*s*) elements are known as the alkali metals.
Reason (R): *s*-orbital can accommodate only two electrons.

13. Assertion (A): Radium is most abundant *s*-block element.
Reason (R): *s*-block elements are non-radioactive nature.

14. Assertion (A): *s*-block elements are highly electropositive.
Reason (R): The valence electrons present in *s*-orbital are loosely held.

15. Assertion (A): Sodium cannot be obtained by chemical reduction of its ore.
Reason (R): Sodium is one of the strongest reducing agents.

16. Assertion (A): NaOH is a stronger base than KOH.
Reason (R): KOH is more soluble in water than NaOH.

17. Assertion (A): Aqueous solution of Na_2CO_3 is alkaline in nature.
Reason (R): When dissolved in water, Na_2CO_3 undergoes anionic hydrolysis.

18. Assertion (A): Sodium reacts with oxygen to form Na_2O_2 whereas potassium reacts with oxygen to form KO_2.
Reason (R): Potassium is more reactive than sodium.

19. Assertion (A): Potassium is not obtained by the electrolysis of fused KCl.
Reason (R): Potassium vapourises at the melting point of KCl.

ANSWERS

1	2	3	4	5	6	7	8	9	10
(*c*)	(*c*)	(*d*)	(*d*)	(*d*)	(*c*)	(*d*)	(*c*)	(*a*)	(*a*)
11	**12**	**13**	**14**	**15**	**16**	**17**	**18**	**19**	
(*a*)	(*b*)	(*d*)	(*a*)	(*a*)	(*d*)	(*a*)	(*b*)	(*a*)	

13 GROUP-II ELEMENTS

Magnesia, MgO

It occurs naturally as periclare. Prepared by combustion of Mg in O_2 or ignition of $Mg(OH)_2$, $MgCO_3$, or $Mg(NO_3)_2$ sulphate or carbonate. It has NaCl structure. Hydrated to $Mg(OH)_2$.

Oxide prepared by heating magnesite ($MgCO_3$) is called calcined magnesia.

$$2Mg(NO_3)_2 \rightarrow 2MgO + 4NO_2 + O_2$$

$$2MgSO_4 \rightarrow 2MgO + 2SO_2 + O_2$$

It is a basic oxide and hence reactors with acids to form salts.

$$MgO + 2HCl \rightarrow MgCl_2 + H_2O$$

It is reduced by carbon at very high-temperature to form magnesium carbide.

$$MgO + 3C \rightarrow MgC_2 + CO$$

It dissolves in aqueous solution of magnesium chloride or bromide forming basic salts such as $MgCl_2.4Mg(OH)_2.5H_2O$.

It is used as a refractory lining in electric furnace because it has high melting point. It is also used as a basic lining in smelting process of metallurgy. It is used as an insulator (when mixed with asbestos) for lagging steam pipes and boilers. It is also used as rubber filler in the flocculent form, and magnesia, when mixed with a saturated solution of magnesium chloride, sets to a firm hard mass like cement known as Sorel's cement or magnesia cement, $MgCl_2.5MgO.xH_2O$. This is used in plaster casting, repairing porcelain ware, in filling teeth and as a substitute for tiles.

Magnesium Hydroxide, $Mg(OH)_2$

Occurs naturally as bracite, and $Mg(OH)_2$ layers occur in silicates *e.g.*, the chondrodites.

$$MgCl_2 + Ca(OH)_2 \rightarrow Mg(OH)_2 + CaCl_2$$

It is a white powder, very slightly soluble in water. It is basic in nature and hence dissolves in acids like HCl. It dissolves in NH_4Cl solution, and is therefore, not precipitated in group III of qualitative analysis. A suspension of magnesium hydroxide in water is used in medicine as an antacid under the name of *milk of magnesia.*

Magnesium Sulphate, Epsom Salt, $MgSO_4.7H_2O$

It is obtained from $MgCO_3$ and H_2SO_4, occurs naturally as kieserite or reichardite, $MgSO_4.H_2O$. Hydrates give $MgSO_4$ at 200°*(c)* Epsom salts are used as purgative as dressing for cotton goods and in dyesing.

Preparation

$$\underset{\text{Magnesite}}{MgCO_3} + H_2SO_4 \rightarrow MgSO_4 + H_2O + CO_2$$

$$\underset{\text{Dolomite}}{MgCO_3.CaCO_3} + 2H_2SO_4 \rightarrow$$

$$MgSO_4 + CaSO_4 \downarrow + 2H_2O + 2CO_2$$

Commercial Method : By boiling Kieserite in water and cooling the resulting solution when crystals of epson salt separate out.

$$MgSO_4.H_2O + 6H_2O \rightarrow MgSO_4.7H_2O$$

Properties **:** It is a colourless, efflorescent crystal-line substance with bitter taste. It forms a number of hydrates. When heated to 150°C, it changes to monohydrate which on further heating changes to anhydrous state at 200°*(c)*

Uses : It is used as purgative in medicine, in dyeing and tanning process. It is also used in making fire-proof fabrics, as a filler in paper industry, in making platinized $MgSO_4$ used as catalyst in the Grillo's process for the manufacture of sulphuric acid, and in the manufacture of ceramics, cement and match boxes.

Calcium Oxide, Lime, quicklimes, burnt lime, CaO

A white solid with the NaCl structure. Much lime is used in building and agriculture, the main use is in the chemical process industry. Large quantities are used in the manufacture of Na_2CO_3 and NaOH, SO_2 absorption, steel (45%), refractories, CaC_2, glass, pulp and paper and sugar. It is also employed in the treatment of water and sewage and in one concentration and refining and soil stabilization. When heated in oxy-hydrogen flame, it gives an intense white light called **Lime Light**. On adding water, it gives a hissing sound and forms calcium hydroxide, commonly known as *slaked lime.* The reaction is exothermic and known as *slaking of lime.*

Paste of lime in water is called milk of lime while its clear filtrate is known as lime water. Chemically, either of these is calcium hydroxide.

Preparation

$$\underset{\text{line stone}}{CaCO_3} \xrightleftharpoons{804°C} CaO + CO_2$$

The reaction is reversible and thus in order to assure the complex decomposition of $CaCO_3$, CO_2 formed must be swept away by a current of air. Temperature should not be very high because at high temperature, clay (present as impurity in lime stone) will react with lime to form fusible silicates.

Commercially lime is prepared by heating $CaCO_3$ to a temperature between 900°C and 1200°*(c)*

Uses: It is used in manufacture of bleaching powder, slaked lime and lime colours. It is also used in manufacture of calcium carbide, cement, glass, mortar (slake lime + sand + water), etc. It is also used in the conversion of Na_2CO_3 into caustic soda, in the purification of sugar, in the purification of coal gas and softening of water, as basic lining in furnace and for producing ammonia gas in the laboratory.

Calcium Hydroxide (Slaked Lime), $Ca(OH)_2$

The powder from the action of an equivalent quantity of water on CaO is slaked lime, a suspension with excess water is milk of lime, a clear solution of $Ca(OH)_2$ in water is lime water. Used as an industrial alkali and in the preparation of mortar which sets to a solid by reconversion of the $Ca(OH)_2$ to $CaCO_3$.

Preparation

(i) $CaO + H_2O \rightarrow Ca(OH)_2$

(ii) Caustic alkali + a soluble calcium salt → $Ca(OH)_2$

Properties : It is a white amorphous powder, only sparingly soluble in water. Its solubility decreases with the increase in temperature. When dried and heated to redness, it loses a molecule of water and converted into calcium oxide (lime). Lime water [Lime + $H_2O \rightarrow Ca(OH)_2$] is frequently used for the detection of CO_2 gas as CO_2 gas turns lime water milky due to formation of $CaCO_3$.

$$Ca(OH)_2 + CO_2 \rightarrow CaCO_3\downarrow + H_2O.$$

Uses : Its main uses are manufacture of bleaching powder, caustic soda, white washing, for softening of water and in preparation of *soda lime* (mixture of cal. hydroxide and caustic soda) which is a good absorbent for a number of gases. Soda lime is also used in decarboxylation of sodium salts of fatty acids.

Gypsum ($CaSO_4.2H_2O$)

Preparation **:** Prepared by adding dilute sulphuric acid or soluble sulphate (*e.g.*, Na_2SO_4) to the solution of a calcium salt.

$$CaCl_2 + H_2SO_4 \rightarrow CaSO_4 + 2HCl$$

Properties : Natural anhydrite and $CaSO_4$ prepared by dehydrating the dihydrate above 650°C is only slowly soluble in water and is used as a filler for paper and other materials. $CaSO_4$ prepared by dehydration of the dihydrate at 60 – 90°C is used as drying agent.

Gypsum and anhydrite are used for H_2SO_4 production and gypsum is used as a soil additive and as an inert additive to pharmaceuticals and insecticides.

Gypsum is used for the manufacture of Plaster of Paris and cement, in the manufacture of ammonia (at Sindri) and for filling or glazing paper. It is also used to make the soil fertile effected by sea water due to flood.

Plaster of Paris, $CaSO_4 \cdot \frac{1}{2}H_2O$

$CaSO_4 \cdot \frac{1}{2}H_2O$, prepared by heating gypsum at 130°C, is used as Plaster of Paris.

$$CaSO_4.2H_2O \xrightarrow{120°C} CaSO_4.\frac{1}{2}H_2O + 1\frac{1}{2}H_2O$$

Plaster of Paris is a white powder which when mixed with water takes up the water of crystallisation again, thus converted back into the dihydrate and sets to a hard mass with slight expansion. Setting takes about 10-15 minutes and may be catalysed by the addition of common salt or delayed by the use of alum or borax.

$$\underset{\text{Plaster of Paris}}{CaSO_4.\ 1/2\ H_2O} \xrightarrow[\text{setting}]{H_2O} \underset{\text{Orthorhombic}}{CaSO_4.2H_2O} \xrightarrow{\text{Hardening}} \underset{\text{Monoclinic (gypsum)}}{CaSO_4.2H_2O}$$

Plaster of Paris or gypsum when heated to about 200°C is converted into anhydrous calcium sulphate. The anhydrous form (anhydrite) is known as *dead burnt* because it does not set like Plaster of Paris when moistened with water.

Plaster of Paris is used in making moulds for casting and also for preparing black board chalks. Anhydrite is used for the manufacture of H_2SO_4 and ammonium sulphate. Anhydrous calcium sulphate may be used as a drying agent.

MULTIPLE CHOICE QUESTIONS

1. Plaster of Paris is
(*a*) $CaSO_4.2H_2O$ (*b*) $(CaSO_4)_2.H_2O$
(*c*) $CaSO_4.MgO$ (*d*) $CaSO_4.5H_2O$

2. Epsom salt is
(*a*) $MgSO_4.7H_2O$ (*b*) $Na_2SO_4.10H_2O$
(*c*) $MgCO_3.CaCO_3$ (*d*) $Al_2O_3.H_2O$

3. The main constituent of bones is
(*a*) CaF_2 (*b*) $CaSO_4$
(*c*) $Ca_3(PO_4)_2$ (*d*) $CaSO_4$

4. A mixture of MgO and $MgCl_2$ is called
(*a*) Sorel cement
(*b*) Portland cement
(*c*) Double salt
(*d*) None

5. Calcium is obtained by the
(*a*) Roasting of limestone
(*b*) Reduction of calcium chloride with carbon
(*c*) Electrolysis of an aqueous solution of calcium chloride
(*d*) Electrolysis of molten calcium chloride

6. Which of the following are commonly used in making toothpaste?
(*a*) $Mg(OH)_2$ (*b*) $MgCO_3$
(*c*) $CaCl_2$ (*d*) Both (*a*) and (*b*)

7. Gypsum is
(*a*) $MgSO_4.\ 2H_2O$
(*b*) $CaSO_4.\ 2H_2O$
(*c*) $2CaSO_4$
(*d*) $CaSO_4.\ 3H_2O$

8. Which is quick lime?
(*a*) $Ca(OH)_2$ (*b*) CaO
(*c*) $CaCO_3$ (*d*) $Ca(OH)_2 + H_2O$

9. Plaster of Paris hardens by
(*a*) Giving off CO_2
(*b*) Changing into $CaCO_3$
(*c*) Uniting with water
(*d*) Giving out water

10. Magnesium metal is prepared by
(*a*) Reduction of MgO by coke
(*b*) Electrolysis of aqueous solution of $Mg(NO_3)_2$

(*c*) Displacement of Mg by iron from magnesium sulphate solution

(*d*) Electrolysis of molten magnesium chloride

11. Mortar is a mixture of

(*a*) $CaCO_3$ & CaO

(*b*) Slaked lime and water

(*c*) Slaked line, sand and water

(*d*) None

12. Gypsum on heating at 120 – 130 °C gives

(*a*) Hemihydrate

(*b*) Monohydrate

(*c*) Dihydrates

(*d*) Anhydrous salt

13. Gypsum is added to clinker during cement manufacture to

(*a*) Decrease the rate of setting of cement

(*b*) Make the cement impervious

(*c*) Bind the particles of calcium silicate

(*d*) To facilitate the formation of colloidal gel

14. Slaked lime is used in the manufacture of

(*a*) Cement (*b*) Fire bricks

(*c*) Pigment (*d*) White-wash

15. Which removes temporary hardness of water and is used in the manufacture of bleaching powder?

(*a*) Slaked lime (*b*) Plaster of Paris

(*c*) Epsom (*d*) Hydrolith

Directions Qs. 16-22: *In each of the following questions there are two statements Assertion (A) is given followed by a corresponding statement of Reason (R) just below it. Of the statements, mark the correct answer:*

(*a*) If both assertion and reason are true and reason is the correct explanation of assertion.

(*b*) If both assertion and reason are true but reason is not the correct explanation of assertion

(*c*) If assertion is true but reason is false

(*d*) If assertion is false but reason is true.

16. Assertion (A): Be resemble Al.

Reason (R): Be^{2+} has almost same charge density as Al^{3+}.

17. Assertion (A): Be does not impar any characteristic colour to the bunsen flame.

Reason (R): Due to its very high ionization energy, beryllium requires a large amount of energy for excitation of the electrons.

18. Assertion (A): Magnesium is extracted by the electrolysis of fused mixture of $MgCl_2$, NaCl and $CaCl_2$.

Reason (R): Calcium chloride acts as a reducing agent.

19. Assertion (A): Magnesium keeps on burning when heated in CO_2 atmosphere.

Reason (R): Magnesium has a strong affinity for oxygen.

20. Assertion (A): In fused state, calcium chloride cannot be used to dry alcohol or NH_3.

Reason (R): $CaCl_2$ is not a good dessicant.

21. Assertion (A): Beryllium halides dissolve in organic solvents.

Reason (R): Beryllium halides are ionic in character.

22. Assertion (A): $BeCl_2$ fumes in moist air.

Reason (R): $BeCl_2$ reacts with moisture to form HCl gas.

ANSWERS

1	**2**	**3**	**4**	**5**	**6**	**7**	**8**	**9**	**10**
(*b*)	(*a*)	(*c*)	(*a*)	(*d*)	(*d*)	(*b*)	(*b*)	(*c*)	(*b*)
11	**12**	**13**	**14**	**15**	**16**	**17**	**18**	**19**	**20**
(*c*)	(*a*)	(*a*)	(*d*)	(*a*)	(*a*)	(*a*)	(*c*)	(*a*)	(*c*)
21	**22**								
(*c*)	(*a*)								

14 GROUP-III ELEMENTS

Alumina (Aluminium Oxide), Al_2O_3

The oxide of aluminium formed by heating aluminium hydroxide or most aluminium salts of oxy acids. Occurs naturally as corundum, obtained by heating α-AlO(OH), m.p. 2045°C, b.p. 2980°C, almost insoluble in water.

It is the most stable compound of aluminium. It occurs in nature as colourless *corundum* and several coloured oxides like *ruby* (red), topaz (yellow), sapphire (blue), amethyst (violet) and emerald (green). These coloured oxides are used as precious stones. Hydrated alumina occurs as bauxite.

Preparation : In laboratory it can be prepared by igniting aluminium hydroxide, aluminium sulphate, aluminium nitrate or ammonium alum.

$$Al_2(SO_4)_3 \rightarrow Al_2O_3 + 3SO_3$$

$$(NH_4)_2SO_4.Al_2(SO_4)_3.24H_2O \rightarrow Al_2O_3 + 2NH_3 + 4SO_3 + 25H_2O$$

Properties : α-Al_2O_3 is rhombic lattice and γ-Al_2O_3 is cubic lattice. α-Al_2O_3 is dense, hard and resistant to chemical attack. The great strength of Al-O-Al bond and dense crystal structure of aluminium oxide are responsible for its hardness and its use as an abrasive. α-Al_2O_3 occurs in the earth's crust as corundum, Al_2O_3. γ-Al_2O_3 is less dense, soft and has a high surface area. It is readily soluble in aqueous acids and alkalies and used as an excellent selective absorbent in dehydration, decolorisation and chromato-graphy.

It is used for the manufacture of aluminium, for making heat resistant bricks and lining for muffles, in chromatography for separation of mixtures, as a bauxite cement which sets quickly and is resistant to sea water. Bauxite cement is made by fusing bauxite and lime. It is also used as abrasive in the form of fused bauxite (alundum). It is also used for preparing artificial *precious gems.*

Alums

In general the terms is given to double sulphates of the type M_2SO_4, $M'_2(SO_4)_3.24H_2O$ where M is a univalent cation like Na^+, K^+ and NH_4^+ and M′ is a trivalent cation like Al^{3+}, Fe^{3+} and Cr^{3+}. In alums each metal ion is surrounded by six water molecules. In solution the alums behave chemically as mixture of the two constituent sulphates. Sodium and ammonium derivatives of aluminium sulphates are also commercial material. These are also prepared in the same manner as alum.

Types of Alum

(i) Potash alum – $K_2SO_4.Al_2(SO_4)_3.24H_2O$

(ii) Sodium alum – $Na_2SO_4Al_2(SO_4)_3.24H_2O$

(iii) Ammonium alum – $(NH_4)_2SO_4.Al_2(SO_4)_3.24H_2O$

(iv) Ferric alum (ferric amm-alum) – $(NH_4)_2SO_4.Fe(SO_4)_3.24H_2O$

(v) Chrome alum (Pot. chromium alum)– $K_2SO_4.Cr_2(SO_4)_3.24H_2O$

Of all these, potash alum is the most important and is commonly known as alum.

Alum is used in the dyeing industries for the production of mordants and pigments, in dressing leather, sizing paper, in water proofing fabrics, in fireproofing, in gelatin hardening and medicinally as a styptic and astringent, and as a coagulant for water purification.

Preparation :

(i) From Bauxite or aluminium sulphate: Bauxite is boiled with H_2SO_4 to form aluminium sulphate. To this solution, calculated amount of potassium sulphate is added. The solution is evaporated and then cooled to give crystals of potash alum.

(ii) From Pyrites Shale (Alum Shale) : (mixture of aluminium silicate with iron pyrites) is roasted when iron sulphide is converted into ferrous sulphate and H_2SO_2. The latter reacts with a alumina or aluminium silicate to form aluminium sulphate leaving behind silica.

$$2FeS_2 + 2H_2O + 7O_2 \rightarrow 2FeSO_4 + 2H_2SO_4$$

$$Al_2O_3.xSiO_2 + 3H_2SO_4 \rightarrow Al_2(SO_4)_3 + xSiO_2\downarrow + 3H_2O$$

Silica is removed by evaporation

Sodium Tetraborate (Borax)

It occurs naturally as tincal in certain inland lakes in India. It is obtained with tincal by extracting it with water and concentrating the solution till crystals of borax $Na_2B_4O_7 10H_2O$ separates out.

Preparation : Sodium tetraborate can also be prepared from certain minerals such as boracite, colemanite. Firstly the mineral is powdered and boiled with sodium carbonate.

$$Ca_2B_6O_{11} + 2Na_2CO_3 \longrightarrow 2CaCO_3 + Na_2B_4O_7 + 2\ NaBO_2$$

Borax is crystallised from filtrate. Sodium metaborate present can be converted into boron by passing carbon dioxide through it.

$$4NaBO_2 + CO_2 \longrightarrow Na_2B_4O_7 + Na_2CO_3$$

Uses : Borax is used in the manufacture of enamels, glazes and optical glass, soap and drying oils. It is also used as an antiseptic. It is largely used in water softening.

MULTIPLE CHOICE QUESTIONS

1. Borax is
(*a*) $Na_2B_4O_7$ (*b*) $Na_2B_4O_7.4H_2O$
(*c*) $Na_2B_4O_7.7H_2O$ (*d*) $Na_2B_4O_7.10H_2O$

2. Sapphire is a mineral of
(*a*) Cu (*b*) Zn
(*c*) Al (*d*) Hg

3. Borax on heating with cobalt oxide forms a blue bead of
(*a*) $CO(BO_2)_2$ (*b*) $CoBO_2$
(*c*) $CO_3(BO_3)_2$ (*d*) $Na_3CO(BO_3)_2$

4. Aluminium becomes passive in nitric acid because it
(*a*) Is a noble metal
(*b*) Forms a thin film of oxide
(*c*) Has positive reduction potential
(*d*) None

5. Aluminium reduces most of the metallic oxides due to its greater affinity for
(*a*) Oxygen (*b*) Metals
(*c*) Electrons (*d*) Protons

6. Aluminium is mainly extracted from
(*a*) Magnetite (*b*) Bauxite
(*c*) Alumina (*d*) Haematite

7. Which is not a mineral of aluminium?
(*a*) Anhydrite (*b*) Bauxite
(*c*) Corundum (*d*) Diaspore

8. Al and Ga have the same covalent radii because of
(*a*) Greater shielding power of *s*-electrons of Ga atoms
(*b*) Poor shielding power of *s*-electrons of Ga atoms
(*c*) Poor shielding power of *d*-electrons of Ga atoms
(*d*) Greater shielding power of *d*-electrons of Ga atoms.

9. Which is used as a mordant?
(*a*) $AlCl_3$ (*b*) $Al_2(SO_4)_3$
(*c*) Alum (*d*) Al_2O_3

10. Corundum is
(*a*) SiO_2 (*b*) Al_2O_3
(*c*) CaF_2 (*d*) Cr_2O_3

11. Aqueous solution of potash alum is
(*a*) Alkaline (*b*) Acidic
(*c*) Neutral (*d*) Soppy

12. All alums contain
(*a*) One monovalent and one trivalent metal
(*b*) Both monovalent metals
(*c*) One divalent and one monovalent metal
(*d*) Both divalent metals

13. The blue coloured mineral 'Lapis Lazuli' used as semi precious stone is
(*a*) Sodium alumino silicate
(*b*) Zinc cobaltate
(*c*) Prussian blue
(*d*) Basic copper carbonate

14. The purification method used for mineral $Al_2O_3.2H_2O$ is
(*a*) Froth floatation
(*b*) Leaching
(*c*) Liquation
(*d*) Magnetic separation

15. Purification of alumina is essential because
(*a*) Impure alumina has a very poor conductor of electricity
(*b*) Impure alumina has a very high melting point
(*c*) Impure alumina cannot react with the oxidizing agent
(*d*) It is difficult to purify aluminium metal

16. Which of the following is alum?
(*a*) $Na_2SO_4.Al_2(SO_4)_3.24H_2O$
(*b*) $(NH_4)_2SO_4.FeSO_4.6H_2O$
(*c*) $KCl.MgCl_2.6H_2O$
(*d*) $FeSO_4.(NH_4)_2SO_4.6H_2O$

17. Alum is used by dyer of cloth
(*a*) For fire-proofing fabrics
(*b*) As first-aid for cuts
(*c*) For softening hard water
(*d*) As mordant

Directions Qs. 18-22: *In each of the following questions there are two statements Assertion (A) is given followed by a corresponding statement of Reason (R) just below it. Of the statements, mark the correct answer:*

(*a*) If both assertion and reason are true and reason is the correct explanation of assertion.
(*b*) If both assertion and reason are true but reason is not the correct explanation of assertion
(*c*) If assertion is true but reason is false
(*d*) If assertion is false but reason is true.

18. Assertion (A): Boron differs from aluminium and other members of group 13 in a number of properties.
Reason (R): Boron shows anomalous behaviour.

19. Assertion (A): B_2H_6 contains two three centre electron pair bonds.
Reason (R): B_2H_6 is electron deficient molecule.

20. Assertion (A): Oxidising acids such as HNO_3 convert boron to H_3BO_3.
Reason (R): Boron reacts with all types of acids.

21. Assertion (A): Boric acid acts as a strong acid in presence of catechol.
Reason (R): Boric acid does not act as a proton donor but behaves as a Lewis acid.

22. Assertion (A): $Al(OH)_3$ is amphoteric in nature.
Reason (R): Al – O and O – H bonds can be broken with equal ease in $Al(OH)_3$.

ANSWERS

1	2	3	4	5	6	7	8	9	10
(*d*)	(*c*)	(*a*)	(*b*)	(*a*)	(*b*)	(*a*)	(*c*)	(*c*)	(*b*)
11	**12**	**13**	**14**	**15**	**16**	**17**	**18**	**19**	**20**
(*b*)	(*d*)	(*a*)	(*b*)	(*d*)	(*a*)	(*d*)	(*b*)	(*a*)	(*c*)
21	**22**								

15 GROUP-IV ELEMENTS

Coal

Coal is a readily combustible black or brownish-black sedimentary rock. Coal is a fossil fuel. Coal is extracted from the ground by mining, either underground or in open pits.

Formation of coal

Coal is believed to have been formed about 300 million years ago under the Earth by a process called carbonization.

Types of Coal

Coal is classified into four main types, depending on the amount of carbon, oxygen, and hydrogen present. The higher the carbon content, the more energy the coal contains.

Lignite is the lowest rank of coal, with a heating value of 4,000-8,300 British thermal units (Btu) per pound. Lignite is crumbly and has high moisture content. Lignite is mainly used to produce electricity. It contains 25-35 percent carbon. Sometimes called brown coal.

Subbituminous coal typically contains less heating value than bituminous coal (8,300-13,000 Btu per pound) and more moisture. It contains 35-45 percent carbon. this coal generally has a lower sulfur content than other types.

Bituminous coal was formed by added heat and pressure on lignite. It has two to three times the heating value of lignite. Bituminous coal contains 11,000-15,500 Btu per pound. Bituminous coal is used to generate electricity and is an important fuel for the steel and iron industries. It contains 45-86 percent carbon.

Anthracite was created where additional pressure combined with very high temperature inside the earth. Like bituminous coal, anthracite coal is a big energy producer, containing nearly 15,000 Btu per pound. It contains 86-97 percent carbon.

Uses of coal

1. On destructive distillation, coal yields useful products like coal gas, coal tar, ammonia and coke.
2. Generation of electricity
3. Domestic fuel for cooking and heating
4. Manufacture of synthetic petroleum and natural gas
5. To heat water in steam engines and boilers

Coke

Another major use of coal is in iron and steelmaking. The iron industry uses coke ovens to melt iron ore. Coke, an almost pure carbon residue of coal, is used as a fuel in smelting metals. Coke is the solid carbonaceous material derived from destructive distillation of low-ash, low-sulfur bituminous coal. Cokes from coal are grey, hard, and porous.

Uses: Coke is used as a fuel and as a reducing agent in smelting iron ore in a blast furnace.

Coke forms a desirable fuel for stoves and furnaces. Coke may be burned with little or no smoke under combustion conditions.

Coke was one of the materials used in the heat shielding on NASA's Apollo program space vehicles and Mars Pathfinder vehicle.

Coke may be used to make synthesis gas, a mixture of carbon monoxide and hydrogen.

Silicates

Silicates occurs in the earth's crust in abundance in the form of silicate mineral and alumino silicate clays. The sodium silicate Na_2SiO_3 is the only common silicate which is soluble in water. It can be easily prepared by fusing sodium carbonate with silica (SiO_2) at 1500°*(c)* Most of the silicates are insoluble in water. This is due to the great strength of Si-O bond. This bond can only be broken by treating it with hydrofluoric acid.

Structure of Silicates : All the silicates are comprised of SiO_4^{4-} units which are formed by sp^3 hybridization. There are four unpaired electrons in the excited silicon atom. These form four bonds with oxygen atoms yielding SiO_4 species. Since it involves sp^3 hybridization, it has a tetrahedral structure.

Complex Silicates

In some cases, the oxygen atom of SiO_4 species tends to complete their octet by sharing electrons with other silicon atom. The number of oxygen atoms which in this way form bridges to other silicon atom may be one, two, three or all the four. The resulting silicate chains are negatively charged anions because any oxygen atom which is not a bridge atom has to pick up an electron from some metals to complete octet. There are different types of complex silicates can be formed. These are pyrosilicates, chain silicates, cyclic silicates, sheet silicates etc.

Zeolites

Zeolites are microporous, aluminosilicate minerals commonly used as commercial adsorbents. The term zeolite was originally coined in 1756 by Swedish mineralogist Axel Fredrik Cronstedt, who observed that upon rapidly heating the material stilbite, it produced large amounts of steam from water that had been adsorbed by the material.

Properties: Zeolites have a porous structure that can accommodate a wide variety of cations, such as Na^+, K^+, Ca^{2+}, Mg^{2+} and others. These positive ions are rather loosely held and can readily be exchanged for others in a contact solution. Some of the more common mineral zeolites are analcime, chabazite, clinoptilolite, heulandite, natrolite, phillipsite, and stilbite. An example mineral formula is: $Na_2Al_2Si_3O_{10}2H_2O$, the formula for natrolite.

Zeolites are the aluminosilicate members of the family of microporous solids known as "molecular sieves."

Uses : 1. Zeolites are widely used as ion-exchange beds in domestic and commercial water purification, softening, and other applications.

2. Synthetic zeolites are widely used as catalysts in the petrochemical industry, for instance in fluid catalytic cracking and hydro-cracking.

3. Zeolites have uses in advanced reprocessing methods, where their micro-porous ability to capture some ions while allowing others to pass freely allow many fission products to be efficiently removed from nuclear waste and permanently trapped.

4. Zeolites can be used as solar thermal collectors and for adsorption refrigeration. In these applications, their high heat of adsorption and ability to hydrate and dehydrate while maintaining structural stability is exploited.

5. The largest single use for zeolite is the global laundry detergent market.

6. Synthetic zeolite is also being used as an additive in the production process of warm mix asphalt concrete. When added to Portland cement as a pozzolan, it can reduce chloride permeability and improve workability. It reduces weight and helps moderate water content while allowing for slower drying which improves break strength.

7. Thomsonites, one of the rarer zeolite minerals, have been collected as gemstones. When polished by a lapidary the thomsonites sometimes display chatoyancy.

8. Zeolite-based oxygen concentrator systems are widely used to produce medical-grade oxygen. The zeolite is used as a molecular sieve to create purified oxygen from air using its ability to trap impurities.

Glass

The name glass is given to any amorphous and transparent solid which is produced on the solidification of a liquid. The word is generally used for the transparent substance obtained when white sand is fused with oxides and carbonates of alkaline earth metals and the molten mixture is cooled.

1. *The ordinary glass (soda glass) also known as window glass* has approximately the composition.

$$Na_2O.CaO.6SiO_2$$

[SiO_2 75%, sodium oxide 15%, CaO 8% aluminium oxide 20% (as an impurity)]

Types of Glass & its Uses

1. Soda Glass : It is the cheapest & most common glass. It is prepared by fusing soda ash, sand, limestone. It is also called soft glass. It fuses at comparatively low temperatures. The major disadvantage of using this glass is that it is brittle & breaks easily. It cracks when subjected to sudden changes of temperature. Soda glass is used for the manufacture of window glass, mirrors, common glassware etc. It is easily attacked by chemcials.

2. Hard Glass : It is obtained by fusing potassium carbonate & limestone. It is used for making hard glass apparatus. It is more resistant to the action of acids.

3. Lead Crystal Glass : It is made from potassium carbonate, lead oxide & sand. Lead glass has high refractive index. It, therefore, sparkles & is used for making expensive glass ware. The surface of lead glass objects is often cut into decorative patterns to reflect light. cut glass show extraordinary sparkle.

4. Pyrex Glass : It is made by fusing a mixture of sand, lime, borax ($Na_2B_4O_7.10H_2O$) & alkali carbonates. It has good chemical laboratory apparatus, pharmaceutical containers, etc. In home, it is familiar with oven ware.

5. Optical Glass : It is specially made so as to be free of strains & defects. It is used for making lenses for spectacles, microscopes, cameras, telescopes & other optical instruments.

6. Photochromic Glass : It is a special variety of glass that temporarily darkens when exposed to bright light. It is, therefore, very useful as a sun-shield. This automatic darking property of photochromatic glass is because of the presence of silver bromide.

7. Safety Glass : It is prepared by placing a layer of transparent plastic between the two layers of glass by means of suitable adhesive. The 3 layers are joined together by the action of heat & pressure. The glass does not break easily under ordinary impact, & that is why it is known as safety glass. It is used in making wind screens of aeroplanes, automobiles, bulletproof glass etc.

8. Glass Fibres : It is form of fibres, has varied applications in most of the industries glass wool is a bundle of loose glass fibres, which is an excellant heat insulator. It is used as insulating material in refrigerators, ovens etc. another important class of glass fibre is optical fibres which are extensively used in telecommunication, surgical operation etc.

9. Coloured Glass : These are obtained by adding certain colouring material, such as metallic oxides, to the molten mass. Different additions may produce differenct coloured glasses.

MULTIPLE CHOICE QUESTIONS

1. Carbon atoms in diamond ore bonded with each other in following configuration
(*a*) Linear (*b*) Tetrahedral
(*c*) Planar (*d*) Octahedral

2. Which of the following is a conductor of electricity and heat?
(*a*) Graphite (*b*) Diamond
(*c*) CO_2 (*d*) SiO_2

3. Most impure form of carbon is
(*a*) Wood charcoal (*b*) Animal charcoal
(*c*) Lamp black (*d*) Graphite

4. Lead pencil contains
(*a*) Pb (*b*) FeS
(*c*) Graphite (*d*) P

5. Activated charcoal is employed to remove colouring matter from pure substances, it works by

(*a*) Bleaching (*b*) Oxidation
(*c*) Reduction (*d*) Adsorption

6. The most reactive and inert form of carbon are respectively
(*a*) Diamond and Graphite
(*b*) Charcoal and Diamond
(*c*) Coal and Charcoal
(*d*) Lamp black and Gas black

7. Water gas is an important industrial fuel. It is a mixture of
(*a*) H_2O + air (*b*) $CO + H_2$
(*c*) $CO + CO_2$ (*d*) $H_2O + CO$

8. When a mixture of air and steam is passed over red hot coke, the outgoing gas contains
(*a*) Producer gas (*b*) Water gas
(*c*) Coal gas (*d*) None

9. Producer gas is a mixture of
(*a*) CO and N_2 (*b*) CO_2 and H_2
(*c*) CO and H_2 (*d*) CO_2 and N_2

10. Synthesis gas is a mixture of
(*a*) Steam and carbon monoxide
(*b*) Carbon monoxide and nitrogen
(*c*) Hydrogen and carbon monoxide
(*d*) Hydrogen and methane

11. Ordinary glass is
(*a*) Sodium silicate
(*b*) Calcium silicate
(*c*) Calcium and sodium silicates
(*d*) Copper silicate

12. The principal constituent of pyrex glass is
(*a*) Zn (*b*) B
(*c*) Pb (*d*) Cl

13. Which glass can withstand high temperatures of the order of 1000 °C?
(*a*) Quartz (*b*) Pyrex
(*c*) Potash (*d*) Potash lead

14. Glass reacts with
(*a*) Oleum (*b*) HF
(*c*) HNO_3 (*d*) $K_2Cr_2O_7$

15. Ultra violet rays can be checked by
(*a*) Flint glass (*b*) Crooke's glass
(*c*) Soda glass (*d*) Pyrex glass

Directions Qs. 16-18: *In each of the following questions there are two statements Assertion (A) is given followed by a corresponding statement of Reason (R) just below it. Of the statements, mark the correct answer:*

(*a*) If both assertion and reason are true and reason is the correct explanation of assertion.
(*b*) If both assertion and reason are true but reason is not the correct explanation of assertion
(*c*) If assertion is true but reason is false
(*d*) If assertion is false but reason is true.

16. **Assertion (A):** Between $SiCl_4$ and CCl_4, only $SiCl_4$ reacts with water.
Reason (R): $SiCl_4$ is ionic and CCl_4 is covalent.

17. **Assertion (A):** Al forms $[AlF_6]^{3-}$ but B does not form $[BF_6]^{3-}$.
Reason (R): B does not react with F_2.

18. **Assertion (A):** Si-Si bonds are much stronger then Si-O bonds.
Reason (R): Silicon forms double bonds with itself.

ANSWERS

1	2	3	4	5	6	7	8	9	10
(*b*)	(*a*)	(*a*)	(*c*)	(*d*)	(*b*)	(*b*)	(*d*)	(*a*)	(*c*)
11	**12**	**13**	**14**	**15**	**16**	**17**	**18**		
(*c*)	(*b*)	(*a*)	(*b*)	(*b*)	(*c*)	(*c*)	(*d*)		

16 GROUP-V ELEMENTS

Nitric Acid, HNO_3

It was first called aqua *fortis*, meaning strong water, because, it attacks nearly all the metals.

The structure of nitric acid can be best explained by assuming that nitric acid is a resonance hybrid of the following two structures.

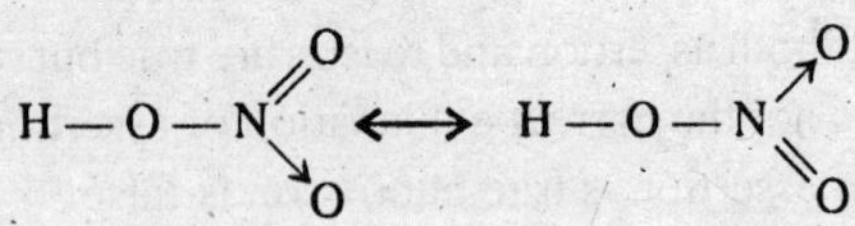

The nitrate ion, NO_3^- has a planar and symmetrical structure with the sp^2 hybridized nitrogen atom at the centre and the oxygen atoms at the corners of an equilateral triangle.

Preparation

(i) By distilling sodium nitrate or potassium nitrate (nitre) with conc. sulphuric acid (Laboratory Method)

$$NaNO_3 + H_2SO_4 \xrightarrow{200°C} NaHSO_4 + HNO_3$$

$$NaHSO_4 + NaNO_3 \rightarrow Na_2SO_4 + HNO_3$$

(ii) From air (Birkeland Eyde electric arc process)

$$N_2 + O_2 \rightleftharpoons 2NO - 43.2 \text{ kcal}$$

$$2NO + O_2 \xrightarrow{50°C} 2NO_2$$

$$2NO_2 + H_2O \rightarrow HNO_2 + HNO_3$$

Properties : Pure nitric acid is a colourless fuming liquid having a pungent smell. The yellow colour of the usual nitric acid is due to the presence of nitrogen dioxide, NO_2 (formed by decomposition of HNO_3 in sunlight) in it.

$$4HNO_3 \xrightarrow[\text{or heat}]{\text{Sunlight}} 4NO_2 + 2H_2O + O_2$$

However, the yellow colour of the acid may be decolourised by warming it to 60-80 °C and bubbling dry air through it. It has extremely corrosive action on the skin.

Nitric acid is used in the manufacture of explosives (T.N.T., picric acid, nitroglycerine, dynamite, and gun cotton), fertilizers like ammonium nitrate and calcium nitrate, artificial silk dyes, drugs, perfumes, etc. It is an important laboratory reagent. It is used as a solvent for metals for etching designs or names upon copper, brass, and bronze article. It is also used in the purification of silver and gold.

Ammonia, NH_3

It is the most important of the nitrogen hydrides. It is a colourless gas with a characteristic smell. Readily liquefied by cooling or compression. It is very soluble in water, NH_3 occurs naturally in some gases from hot springs.

It is used to produce HNO_3 urea, NH_4NO_3, Nylon the ultimate products being fertilizers, explosives and plastics, foams and films. It just burns in oxygen. It is generally a better solvent for organic compounds (unless NH_3 complexes formed. Metals frequently dissolve to give blue reducing solutions containing solvated electrons. It is detected by smell; blackening of $Hg_2(NO_3)_2$ paper and Nessler's reagent.

Nitrogen of ammonia involves sp^3 hybridisation. Since one pair of electrons on nitrogen remains free; the ammonia molecule gets distorted and acquires, a pyramidal structure instead of a regular tetrahedron. The bond angle H—N—H is $\approx 107°$.

Preparations

(i) By heating ammonium salt with a base, *e.g.*, slaked lime (lab. method).

$$2NH_4Cl + Ca(OH)_2 \rightarrow CaCl_2 + 2H_2O + 2NH_3\uparrow$$

The gas so formed is dried with quick lime, CaO.

(ii) Cyanamide Process (Commercial Method)

$$CaO + 3C \xrightarrow{800-1000°C} CaC_2 + CO\uparrow$$

$$CaC_2 + N_2 \rightarrow \underset{\substack{\text{Cal.Cyanamide}\\ \text{Nitrolim}}}{CaCN_2} + C$$

$$CaCN_2 + 3H_2O\,(\text{steam}) \xrightarrow{3\text{ atm.}} CaCO_3 + 2NH_3$$

(iii) Haber's Process (from air) Commercial Method.

$$N_2 + 3H_2 \xrightarrow[300\text{atm}]{500°C,\text{catalyst FetMO}} 2NH_3$$

(The H_2 is generally from steam-reforming process.)

***Reactions** : (i) Basic Nature :* It behaves as a strong Lewis base since nitrogen of ammonia has a lone pair of electrons. Its solution in water is best described as *aqueous ammonia* or NH_3 (aq.) Its ionisation in water is represented as

$$NH_3(aq.) + H_2O \rightleftharpoons NH_4^+ + OH^-$$

Because of its ability to give hydroxyl ions, aqueous ammonia has basic character and thus turns red litmus blue and phenoph-thalein pink. It reacts with HCl to give white fumes of ammonium chloride.

(ii) Combustion : Ordinarily it is neither combustible or supporter of combustion. However, a jet of NH_3 can be made to burn in oxygen with greenish yellow flame.

$$4NH_3 + 3O_2 \rightarrow 2N_2 + 6H_2O$$

(iii) Decomposition :

$$2NH_3 \xrightarrow[\text{spark}]{\text{red hot or electric}} N_2 + 3H_2$$

(iv) Formation of amides

$$2Na + 2NH_3 \xrightarrow{300°C} 2NaNH_2 + H_2$$

(v) Formation of complex ions :

Ammonia combines with certain cations like Ag^+, Ca^{2+}, Hg^{2+}, etc., to form very stable soluble complex cations.

(a) $AgCl + 2NH_4OH\,(\text{excess}) \rightarrow$

$$\underset{\substack{\text{Silver ammine}\\ \text{chloride}}}{[Ag(NH_3)_2]Cl} + H_2O$$

(b) $HgCl_2 + 2NH_4OH \rightarrow$

$$\underset{\substack{\text{Mercuriammine}\\ \text{chloride}}}{[Hg(NH_2)]Cl}\, NH_4Cl + 2H_2O$$

(vi) Reaction with Nessler's Reagent

$$HgCl_2 + 2KI \rightarrow 2KCl + HgI_2 \text{ (red ppt)}$$

$$HgI_2 + 2KI \rightarrow K_2HgI_4 \text{ (Nessler's reagent)}$$

$$2K_2HgI_4 + NH_3 + 3KOH \rightarrow$$

$$H_2N.HgO.\,HgI\downarrow + 7KI + 2H_2O$$

It is used as a laboratory reagent and in preparation of nitric acid (Ostwald process), sodium bicarbonate (Solvay process), ammonium sulphate and ammonium calcium phosphate which are used as fertilizers and also used in preparation of ammonium nitrate which is used as explosive.

Rock Phosphate

Phosphorite, phosphate rock or rock phosphate is a non-detrital sedimentary rock which contains high amounts of phosphate bearing minerals. The phosphate content of phosphorite is at least 20% which is a large enrichment over the typical sedimentary rock content of less than 0.2%. The phosphate is present as fluorapatite typically in cryptocrystalline masses referred to as collophane. The dark brown to black beds range from a few centimetres to several metres in thickness. The layers contain the same textures and structures as fine grained limestones and may represent diagenetic replacements of carbonate minerals by phosphates.

Production : The two main sources for phosphate are guano, formed from bird droppings, and rocks containing concentra-tions of the calcium phosphate mineral, apatite.

Phosphate rock is mined, beneficiated, and either solubilized to produce wet-process phosphoric acid, or smelted to produce elemental phosphorus. Phosphoric acid is reacted with phosphate rock to produce the fertilizer triple superphosphate or with anhydrous ammonia to produce the ammonium phosphate fertilizers.

Use : For general use in the fertilizer industry, phosphate rock or its concentrates preferably have levels of 30% phosphorus pentoxide (P_2O_5), reasonable amounts of calcium carbonate (5%), and <4% combined iron and aluminium oxides.

Rock phosphate is also used in animal feed supplements, food preservatives, anti-corrosion agents, cosmetics, fungicides, ceramics, water treatment and metallurgy.

Safety Matches

The paste applied to the side of a match box is a mixture of Phosphorous Trisulphide, glass powder and sand to cause frictions and glue. The match head contains a readily combustible substance like atimony trisulphide, potassium chlorate which maintain the combustion and a binding material like glue.

MULTIPLE CHOICE QUESTIONS

1. Which of the following is acidic?
(*a*) SO_3 (*b*) N_2O
(*c*) BeO (*d*) HgO

2. The starting material in Birkeland and Eyde process for the manufacture of HNO_3 is
(*a*) NH_3 (*b*) NO_2
(*c*) Air (*d*) Chile salt petre

3. The molecular formula of phosphorus is
(*a*) P_1 (*b*) P_4
(*c*) P_2 (*d*) P_5

4. The chemical used for cooling in refrigeration or in manufacture of ice is
(*a*) CS_2 (*b*) NH_4OH
(*c*) NH_4Cl (*d*) Liquid NH_3

5. In the treatment of leukemia which one is used :
(*a*) White phosphorus
(*b*) Red phosphorus
(*c*) Scarlet-phosphorus
(*d*) P_{32} isotope

6. One mole of calcium phosphide on reaction with excess water gives
(*a*) One mole of phosphine
(*b*) Two moles of phosphoric acid
(*c*) Two moles of phosphine
(*d*) One mole of phosphorus pentoxide

7. Nitric acid may be kept in a bottle of
(*a*) Ag (*b*) Sn
(*c*) Pb (*d*) Al

8. The catalyst used in Haber's process of NH_3 is
(*a*) Pt (*b*) V_2O_5
(*c*) Fe (*d*) Mo

9. Coloured oxide of nitrogen is
(*a*) N_2O (*b*) NO
(*c*) N_2O_4 (*d*) NO_2

10. Phosphorus is present in bone as
(*a*) $Ca_3(PO_4)_2$ (*b*) $FePO_4$
(*c*) Ca_3P_2 (*d*) Cu_3P_2

11. Superphosphate of lime is used in
(*a*) Cement industry (*b*) Glass industry
(*c*) Agriculture (*d*) Metallurgy

12. NH_3 molecule can enter into employer formation through
(*a*) Ionic bond
(*b*) Covalent bond
(*c*) Co-ordinate bond
(*d*) Electron deficient bond

13. Nitric acid (conc.) oxidizes phosphorus to
(*a*) H_3PO_4 (*b*) P_2O_5
(*c*) H_3PO_3 (*d*) $H_4P_2O_7$

14. Conc. HNO_3 is heated with P_2O_5 to form
(*a*) N_2O (*b*) NO
(*c*) NO_2 (*d*) N_2O_5

15. Phosphorus pentoxide cannot be used to dry
(*a*) Nitrogen
(*b*) Ammonia
(*c*) Hydrogen sulphide
(*d*) Sulphur dioxide

16. The metal which forms amide on passing NH_3 on it at 300 °C is .

(*a*) Magnesium (*b*) Lead
(*c*) Aluminium (*d*) Sodium.

17. Red phosphorus is chemically unreactive because

(*a*) It does not contain P-P bonds
(*b*) It does not contain tetrahedral P_4 molecules
(*c*) It does not catch fire in air even upto 400°C
(*d*) It has a polymeric structure

18. Red and yellow phosphorus are :

(*a*) Allotropes (*b*) Isobars
(*c*) Isomers (*d*) Isotopes

19. White phosphorus reacts with caustic soda to give PH_3 and NaH_2PO_2. This reaction is an example of

(*a*) Oxidation
(*b*) Reduction
(*c*) Neutralisation
(*d*) Oxidation and Reduction

20. Which sulphide is used in the manufacture of "strike anywhere" matches :

(*a*) P_2S_5 (*b*) P_2S_3
(*c*) P_4S_3 (*d*) None

Directions Qs. 21-27: *In each of the following questions there are two statements Assertion (A) is given followed by a corresponding statement of Reason (R) just below it. Of the statements, mark the correct answer:*

(*a*) If both assertion and reason are true and reason is the correct explanation of assertion.
(*b*) If both assertion and reason are true but reason is not the correct explanation of assertion
(*c*) If assertion is true but reason is false
(*d*) If assertion is false but reason is true.

21. Assertion (A): Nitrogen has higher ionization energy than that of oxygen.
Reason (R): Nitrogen has smaller atomic size than that of oxygen.

22. Assertion (A): HNO_3 is a stronger acid than HNO_2.
Reason (R): In HNO_3 there are two nitrogen - oxygen bonds while in HNO_2 there is only one.

23. Assertion (A): Although PF_5, PCl_5 and PBr_3 are known, the pentahalides of nitrogen have not been observed.
Reason (R): Phosphorus has lower electronegativity than nitrogen.

24. Assertion (A): White phosphorus is more reactive than red phosphorus.
Reason (R): Red phosphorus consists of P_4 tetrahedral units linked to one-another to form linear chains.

25. Assertion (A): SO_3 acts as a Lewis base.
Reason (R): S atom in SO_3 contains no lone pair of electrons.

26. Assertion (A): SO_3 has a planar structure.
Reason (R): S atom in SO_3 is sp^2d hybridized.

27. Assertion (A): PCl_5 is covalent in gaseous and liquid states but ionic in solid state.
Reason (R): PCl_5 in solid state consists of tetrahedral PCl_4^+ cation and octahedral PCl–

ANSWERS

1	2	3	4	5	6	7	8	9	10
(*a*)	(*c*)	(*b*)	(*d*)	(*d*)	(*c*)	(*d*)	(*c*)	(*d*)	(*a*)
11	**12**	**13**	**14**	**15**	**16**	**17**	**18**	**19**	**20**
(*c*)	(*c*)	(*a*)	(*d*)	(*b*)	(*d*)	(*d*)	(*a*)	(*d*)	(*c*)
21	**22**	**23**	**24**	**25**	**26**	**27**			
(*c*)	(*c*)	(*b*)	(*b*)	(*d*)	(*c*)	(*b*)			

17 GROUP-VI ELEMENTS

Oxides of Sulphur

Sulphur forms a number of oxides, as S_2O—sulphur suboxide, SO—sulphur monoxide, S_2O_2—sulphur sesquioxide, SO_2—sulphur dioxide, SO_3—sulphur trioxide, S_2O_7—sulphur heptoxide, SO_4—sulphur tetroxide.

Sulphur Dioxide

Sulphur dioxide occurs in volcanic gases. It is also found in the air near industrial centres formed during combustion of coal.

Preparation : *(i)* Sulphur dioxide is prepared in the laboratory usually by the action of hot concentrated sulphuric acid on copper turnings.

$$Cu + 2H_2SO_4 \longrightarrow CuSO_4 + SO_2 + 2H_2O$$

(ii) It can also be prepared by dropping concentrated hydrochloric acid on crystals of sodium bisulphite.

$$NaHSO_3 + HCl \longrightarrow NaCl + H_2O + SO_2$$

Physical Properties : It is a colourless gas with a pungent and suffocating odour. It is about twice as heavy as air. It is largely soluble in water.

Chemical Properties : *(i)* Sulphur dioxide decomposes to an appreciable extent when heated at 1200°C producing the sulphur trioxide and sulphur.

$$3SO_2 \longrightarrow 2SO_3 + S$$

(ii) Sulphur dioxide combines directly with the halogen forming sulphuryl halides. As

$$SO_2 + Cl_2 \longrightarrow SO_2Cl_2$$

The reaction takes place in presence of camphor.

(iii) Sulphur dioxide reacts with hydrogen sulphide when the two gases are brought in contact with each other, as

$$SO_2 + 2H_2S \longrightarrow 2H_2O + 3S$$

Uses : Sulphur dioxide is largely used for manufacture of sulphuric acid and as a bleaching agent. Also used as a disinfectant for killing, disease, germs.

Sulphur Trioxide

Preparation : Sulphur trioxide is obtained in small amounts mixed with sulphur dioxide, when sulphur is burnt in air.

(i) It can be prepared by heating ferric sulphate strongly as

$$Fe_2(SO_4)_3 \longrightarrow Fe_2O_3 + 3SO_3$$

(ii) It can also be prepared by dehydrating sulphuric acid with the phosphorus pentoxide.

$$6H_2SO_4 + P_4O_{10} \longrightarrow 4H_3PO_4 + 6SO_3$$

Physical Properties : Sulphur trioxide exist in three forms : *(i)* α-Sulphur trioxide-This is a colourless liquid which solidifies in the form of prismatic needle like crystals. *(ii)* β-sulphur trioxide. The α-form, if allowed to stand at 16° for some time changes to β-form. Thus, this solidifies giving asbestos like needle. *(iii)* γ-sulphur trioxide—the γ form is obtained from β-form on complete drying.

Chemical Properties : *(i)* Chemically the three forms are almost identical, although α-form is comparatively active than the rest. Any of these dissolves in water forming sulphuric acid as

$$H_2O + SO_3 \longrightarrow H_2SO_4$$

(ii) It reacts with basic oxide to form salts known as sulphates.

$$CaO + SO_3 \longrightarrow CaSO_4$$

(iii) It dissolves in con*(c)* sulphuric acid giving fuming sulphuric acid.

$$SO_3 + H_2\,SO_4 \longrightarrow H_2S_2O_7 \text{ (oleum)}$$

Sulphuric Acid

Free sulphuric does not occur in nature. It is found in some mineral springs, where it is supposed to be formed by the oxidation of iron pyrites in presence of air and moisture.

$$2FeS_2 + 2H_2O + 7O_2 \longrightarrow 2H_2SO_4 + 2FeSO_4$$

Sulphuric acid can be easily obtained by the contact process. Firstly this process involves the oxidation of sulphur dioxide by air in presence of a catalyst to give sulphur trioxide,

$$2SO_2 + O_2 \rightleftharpoons 2SO_3$$

which is dissolved in 98% sulphuric acid to give oleum.

$$SO_3 + H_2SO_4 \longrightarrow H_2S_2O_7 \text{ (Oleum)}$$

The sulphuric acid of any desired concentration can be obtained by diluting oleum with water.

$$H_2S_2O_7 + H_2O \longrightarrow 2H_2SO_4$$

Physical Properties : Pure sulphuric acid is a colourless, viscous heavy liquid. At 444°C, the vapour of sulphuric acid dissociates into steam and sulphur trioxide. Concentrated sulphuric acid has a strong affinity for water and a large amount of heat is produced on mixing the acid with water.

Chemical Properties :

(i) Acidic properties : It is quite probable that pure anydrous sulphuric acid is completely covalent. However, in actual practice, this condition is hardly tenable. Dilute sulphuric acid is diabasic and ionises as

$$H_2\,SO_4 \rightleftharpoons H^+ + HSO_4^-$$

$$HSO_4^- \rightleftharpoons H^+ + SO_4^{2-}$$

(ii) It reacts with more electronegative metals with the evolution of hydrogen.

$$Zn + H_2SO_4 \longrightarrow ZnSO_4 + H_2$$

$$Fe + H_2SO_4 \longrightarrow FeSO_4 + H_2$$

(iii) Dehydrating property : Due to the large affinity for the water it easily removes water. As ethyl alcohol is easily dehydrated to give ether.

$$2C_2H_5OH \xrightarrow[-H_2O]{H_2SO_4} C_2H_5-O-C_2H_5$$

(iv) Oxidising property : Hot concentrated sulphuric acid act as a strong oxidising agent. As

$$Cu + 2H_2SO_4 \longrightarrow CuSO_4 + 2H_2O + SO_2$$

(v) Aromatic compounds react with sulphuric acid formic sulphonic acids.

$$C_6H_6 + H_2SO_4 \longrightarrow C_6H_5\,SO_3H + H_2O$$

Uses : Sulphuric acid has extensive uses in industry, and laboratory. It is used for the preparation of fertilizers, manufacture of the important chemicals, in the preparation of dyes, drugs disinfectants, as a dehydrating and a laboratory reagent.

Hydrogen Peroxide

Since hydrogen peroxide decomposes into water and oxygen,Hydrogen peroxide is a week dibasic acid and gives two series of salts, *e.g.*, NaO—OH and NaO—ONa. This suggests the presence of two hydroxyl groups in the molecule.

Preparation:

1. *From sodium peroxide :* By adding calculated quantity of sodium peroxide to ice-cold 20% sulphuric acid solution :

$$Na_2O_2 + H_2\,SO_4 \rightarrow Na_2\,SO_4 + H_2O_2$$

Due to low temperature, $Na_2SO_4 . 10H_2O$ crystallises out. The clear solution is decanted and distilled in vacuum then a 3% solution of hydrogen peroxide is obtained.

2. *Electrolytic process*: By the electrolysis of 50% solution of H_2SO_4 at 10ºC using a current of high density and platinum anode and graphite cathode. The product obtained at the anode is hydrolysed or distilled under reduced pressure.

$$2H_2SO_4 \rightleftharpoons 2H^+ + 2HSO_4^-$$

At cathode: $2H^+ + 2e \rightarrow 2H \rightarrow H_2$

At anode: $2HSO_4^- - 2e \rightarrow 2HSO_4 \rightarrow H_2S_2O_6$

$$H_2S_2O_6 + 2H_2O \xrightarrow{\text{distilled}} 2H_2SO_4 + H_2O_2$$

3. ***Auto-oxidation method*:** By oxidising 2-ethyl-anthraquinol dissolved in a mixture of benzene and cyclo-hexanol, 20% solution of H_2O_2 is produced.

Anthraquinol derivative is regenerated.

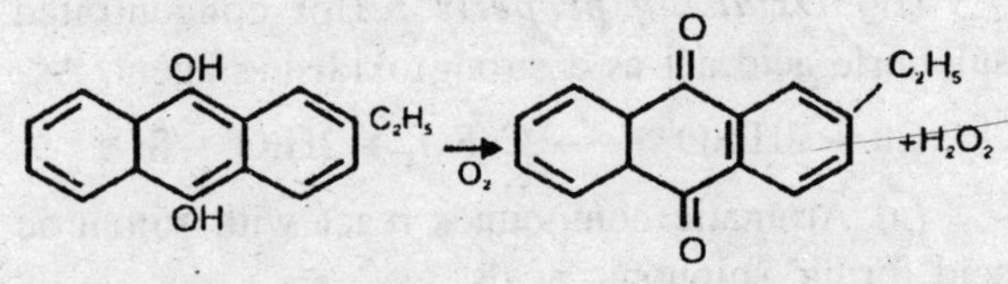

2-ethylanthraquinol   2-ethylanthraquinone

Anthraquinol derivative is regenerated.

Properties: Pure H_2O_2 is pale blue oily liquid. Pure H_2O_2 is acidic but dilute aqueous solution is neutral.

1. It decomposes explosively at about 143°C
   $$2H_2O_2 \rightarrow 2H_2O + O_2$$
2. It acts as an oxidising agent, both in acidic and alkaline medium.
3. It also acts as a reducing agent, both in acidic and alkaline medium.
   $$2KMnO_4 + 3H_2SO_4 + 5H_2O_2 \rightarrow K_2SO_4 + 2MnSO_4 + 8H_2O + 5O_2$$

Uses:

1. It is a non-poisonous antiseptic and its dilute solution is widely used for washing wounds, teeth, ear, nose, et*(c)*, under the name **perhydrol**.
2. It is also a mild bleaching agent and is used to bleach delicate materials like silk, wool, ivory, hair and straw, which are likely to be damaged by chlorine.
3. As a source of oxygen in rockets.
4. As an oxidising agent in laboratory.
5. For restoring the colour of oil paintings.

Allotropes of Sulphur

The structures of sulphur in solid, liquid and gaseous phases are complicated. Rhombic sulphur is the solid allotrope stable at room temperature. It is yellow, readily soluble in carbon disulphide, from which it can be crystallised. Above 369 K, the transition temperature, rhombic sulphur is no longer stable, slowly changing to monoclinic sulphur, and if rhombic sulphur is melted, allowed to partly solidify, and the remaining molten sulphur is poured off, there remain long needle-like crystals (almost colourless) of monoclinic sulphur. On standing at room temperature, monoclinic sulphur slowly changes to the rhombic form. Both these allotropes contain S_8 molecules with rings of eight sulphur atoms.

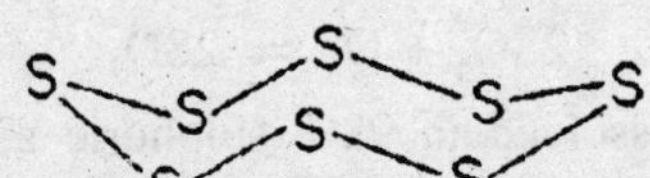

The polymeric nature of molten sulphur can be recognised if molten sulphur is poured in a thin stream into cold water, when a plastic rubbery mass known as plastic sulphur is obtained. This is only slightly soluble in carbon disulphide, but on standing it loses its plasticity and reverts to the soluble rhombic form. If certain substances, for example iodine or oxides of arsenic, are incorporated into the plastic sulphur, the rubbery character can be preserved.

MULTIPLE CHOICE QUESTIONS

1. The highest bond strength is shown by
(a) O–O bond *(b)* S–S bond
(c) Se–Se bond *(d)* Te–Te bond

2. Which one has the lowest boiling point?
(a) H_2O *(b)* H_2S
(c) H_2Se *(d)* H_2Te

3. Bond angle is minimum for
(a) H_2O *(b)* H_2S
(c) H_2Se *(d)* H_2Te

4. The sulphur molecule (S_8) possesses
(a) Cubical structure
(b) Spherical structure

(*c*) Tetrahedral structure
(*d*) W-shaped ring structure

5. SO_2 is dried by
(*a*) CuO (*b*) HNO_3
(*c*) P_2O_5 (*d*) Anhyd. $CaCl_2$

6. Which of the following is used to absorb sulphur dioxide?
(*a*) Conc. H_2SO_4
(*b*) KOH solution
(*c*) Water
(*d*) Anhydrous $CaCl_2$

7. Bleaching action of SO_2 is due to
(*a*) Reduction (*b*) Oxidation
(*c*) Hydrolysis (*d*) Its acidic nature

8. Which of the following is acidic?
(*a*) SO_3 (*b*) N_2O
(*c*) BeO (*d*) HgO

9. Sulphuric acid cannot be used
(*a*) As a picking agent
(*b*) In lead storage batteries
(*c*) In white paints
(*d*) In manufacture of dyes

10. When conc. H_2SO_4 comes in contact with sugar, it becomes black due to
(*a*) Hydrolysis
(*b*) Hydration
(*c*) Decolourisation
(*d*) Dehydration

Directions Qs. 11-18: *In each of the following questions there are two statements Assertion (A) is given followed by a corresponding statement of Reason (R) just below it. Of the statements, mark the correct answer:*

(*a*) If both assertion and reason are true and reason is the correct explanation of assertion.
(*b*) If both assertion and reason are true but reason is not the correct explanation of assertion
(*c*) If assertion is true but reason is false
(*d*) If assertion is false but reason is true.

11. Assertion (A): The electronic sturcture of O_3 is

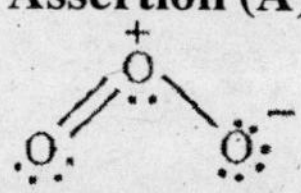

Reason (R): 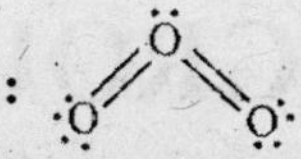 structure is not allowed because octet around O cannot be expanded.

12. Assertion (A): Sulphuric acid is more viscous than water.
Reason (R): Concentrated sulphuric acid has a great affinity for water.

13. Assertion (A): SO_3 acts as a Lewis base.
Reason (R): S atom in SO_3 contains no lone pair of electrons.

14. Assertion (A): SO_3 has a planar structure.
Reason (R): S atom in SO_3 is sp^2d hybridized.

15. Assertion (A): Among chalcogens, tendency of catenation is maximum for sulphur.
Reason (R): S-S bond dissociation energy is higher than O-O bond dissociation energy.

16. Assertion (A): In S_8 molecules, each S atom is bonded to two S atoms.
Reason (R): Each S atom in S_8 molecules is sp^3 hybridized, having two lone pairs of electrons and bonded to two other S-atoms.

17. Assertion (A): In SO_2, the bond angles is 119° whereas in SO_3, the bond angle is 120°.
Reason (R): S atom in both SO_2, and SO_3 is sp^2-hybridized.

18. Assertion (A): Both SO_2 and SO_3 are reducing agents.
Reason (R): SO_3 bleach the articles by reduction.

ANSWERS

1	2	3	4	5	6	7	8	9	10
(*b*)	(*b*)	(*d*)	(*d*)	(*c*)	(*b*)	(*a*)	(*a*)	(*c*)	(*d*)
11	**12**	**13**	**14**	**15**	**16**	**17**	**18**		
(*a*)	(*b*)	(*d*)	(*c*)	(*b*)	(*a*)	(*b*)	(*d*)		

18 GROUP-VII ELEMENTS

Fluorine

Fluorine is found only in the combined state. It is widely distributed. The main ores are *fluorspar* (CaF_2), cryolite ($AlF_3.3NaF$) and *apatite* [$CaF_2.3Ca_3(PO_4)_2$]. Traces of fluorides occur in sea water, bones, teeth, blood, milk etc.

Now a days, fluorine is obtained commercially by the electrolysis of an equimolar mixture of anhydrous HF and KF at about 250°*(c)* Carbon is used as the anode and a steel vessel serves as the cathode.

Uses : It is used in preparation of

(i) Difluorodichloromethane (Freon 12), CF_2Cl_2, which is used in refrigeration and air conditioning in place of NH_3 and SO_2. It is non-toxic and non-inflammable. It is also used as a solvent for *(d)(d)*T. and other insecticide.

(ii) Polytetrafluoroethylene (Teflon), $(-F_2C-CF_2)_n$. It is a new plastic. It is resistant to acids alkalies and oxidising agents. Sodium fluoride and cryolite are used as insecticides.

(iii) Sulphur hexafluoride is used in X-ray and high voltage machines. Its use is due to its high dielectric constant and insulating properties.

(iv) Fluorine, itself, has been used as an oxidising agent in rockets.

Chlorine

It is not found in free state. Main source is NaCl from sea water, other sources are rock salts and other chlorine minerals like *carnallite,* $MgCl_2.KCl.6H_2O$, horn silver AgCl, sylvine, KCl etc.

Manufacture

By the electrolysis brine (aq. solution of NaCl) in Nelson Cell. This is the cheapest method and gives purest chlorine.

$$NaCl \rightleftharpoons Na^+ + Cl^-$$

$$H_2O \rightleftharpoons H^+ + OH^-$$

At cathode: $H^+ + e^- \rightarrow H$

$$H + H \rightarrow H_2\uparrow$$

At anode: $Cl^- \rightarrow Cl + e^-$

$$Cl + Cl \rightarrow Cl_2\uparrow$$

Deacon's Process (Obsolete Method)

$$\underset{(4\,vol.)}{4HCl} + \underset{(1\,vol)}{O_2}\ (\text{from air}) \xrightarrow[CuCl_2\,(\text{catalyst})]{400-500°C} 2H_2O + 2Cl_2\uparrow$$

The temperature should not be below 350°C because at this temperature oxygen would not decompose hydrochloric acid at all.

Chlorine gas, this obtained, contains N_2 and O_2 and is used for the manufacture of bleaching powder by Hasenclever's process (dilute chlorine process).

Uses : Chlorine is used as a bleaching agent, in sterilisation of drinking water, in swimming pools as germicide and disinfectant. It is also used in preparation of domestic antiseptic solutions (NaOCl), in the manufacture of bleaching powder, chlorates, bromine, HCl, $CHCl_3$, CCl_4, CH_2Cl_2, DDT, synthetic plastics, rubbers etc. In the extraction of gold and platinum also chlorine is used. Its one more important use is in warfare for the manufacture of poisonous gases like phosgene ($COCl_2$), tear gas ($CCl_3.NO_2$) and mustard gas ($ClC_2H_4 - S - C_2H_4Cl$)

Bromine

It is found only in combined state. Bromides of Na, K and Mg are found in sea water and certain mineral springs. Carnallite, $KCl.MgCl_2.6H_2O$, contains about 0.01 to 0.1 bromine as $MgBr_2$.

Commercial Preparation

(i) From Carnallite : The mineral carnallite is dissolved in water and KCl and $MgCl_2$ are crystallised out. The mother liquor contains about 0.25% of bromine as $MgBr_2$ and is known as bittern. Bittern is treated with chlorine gas

$$MgBr_2 + Cl_2 \rightarrow MgCl_2 + Br_2$$

Purification : Bromine contaminated with H_2O, Cl_2 and I_2 is distilled over a mixture of KBr and zinc oxide to remove Cl_2 and I_2 respectively.

$$2KBr + Cl_2 \rightarrow 2KCl + Br_2$$

$$2ZnO + 2I_2 \rightarrow 2ZnI_2 + O_2$$

Water is removed by distilling over conc. H_2SO_4.

(ii) From Sea Water: Sea water contains about 0.006% by weight of bromine as $MgBr_2$. It is liberated by passing chlorine gas.

$$MgBr_2 + Cl_2 \rightarrow MgCl_2 + Br_2$$

Bromine vapours led by current of air are absorbed in Na_2CO_3 to form bromide-bromate mixture.

$$3Na_2CO_3 + 3Br_2 \rightarrow 5NaBr + NaBrO_3 + 3CO_2\uparrow$$

Bromine is now liberated from this mixture by means of dil. HCl or dil. H_2SO_4.

$$5NaBr + NaBrO_3 + 6HCl \rightarrow 6NaCl + 3H_2O + 3Br_2$$

$$5NaBr + NaBrO_3 + 3H_2SO_4 \rightarrow 3Na_2SO_4 + 3H_2O + 3Br_2$$

Purification : Method is same as is done in preparation from carnallite.

Uses : It is used in the preparation of ethyl bromide which is used in the manufacture of tetraethyl lead (TEL), an important anti-knock compound in the petroleum industry

It is used in the preparation of ethylene bromide (used to increase the efficiency of TEL), dyestuffs, NaBr and KBr (used in medicine as sedatives), AgBr (used in photography), etc. It is also used as an oxidising agent.

Iodine

It is the rarest of all the halogens. It occurs only in the combined state. Its main source is Relp (Varee), the ash obtained on burning sea-weeds, which contains about 1% iodine as iodides of alkali metals. Caliche (crude chile saltpetre, $NaNO_3$) contains iodine as sodium iodate.

Commercial Preparation

(i) From Sea Weeds : The ash of sea-weeds is extracted with water, sulphates and chlorides of Na and K present in it are removed by crystallisation. The mother liquor is now heated with MnO_2 and conc. H_2SO_4 in a cast iron retort and iodine vapours are collected in earthen jars called aludels or udells.

$$2NaI + 3H_2SO_4 + MnO_2 \rightarrow 2NaHSO_4 + MnSO_4 + 2H_2O + I_2$$

(ii) From Caliche (Crude Salt Petre) : Its hot solution is cooled to remove crystals of $NaNO_3$. The mother liquor is treated with a calculated amount of sodium bisulphite when iodine is precipitated out due to reduction of sodium iodate ($NaIO_3$) present in caliche.

$$2NaIO_3 + 5NaHSO_3 \rightarrow 2Na_2SO_4 + 3NaHSO_4 + H_2O + I_2\downarrow$$

$NaHSO_3$ (if present in excess) will react with iodine to form HI.

$$NaHSO_3 + I_2 + H_2O \rightarrow NaHSO_4 + 2HI$$

Purification : The precipitated iodine is contaminated with iodine chloride (ICl), iodine bromide (IBr) and water. It is distilled over potassium iodide to remove ICl and IBr.

$$KI + ICl \rightarrow KCl + I_2;$$

$$KI + IBr \rightarrow KBr + I_2$$

Water is removed by distilling over conc. H_2SO_4. Further purification may be done by sublimation.

Uses : It is used in tincture of iodine, iodex, iodoform, iodole and certain iodides. Solutions of iodine in KI are used in the treatment of goitre. It is used as a reagent in the laboratory. It is also used to increase production of eggs.

Hydrogen Chloride or Hydrochloric Acid

In aqueous solution, hydrogen chloride forms hydrochloric acid. The concentrated acid contains about 40% hydrogen chloride (about 12 M). Hydrochloric acid is a strong monobasic acid, dissolving metals to form salt and evolving hydrogen. The reaction may be slow if the chloride formed is insoluble (for example lead and silver are attacked very slowly). The rate of attack on a metal also depends on concentration ; thus aluminium is attacked most rapidly by 9 M hydrochloric acid, while with other metals such as zinc or iron, more dilute acid is best.

Hydrogen chloride is formed:

1. By the direct union of hydrogen and chlorine. Very pure hydrogen chloride is made by direct union of pure hydrogen and chlorine in a quartz vessel.

2. As the product of the hydrolysis of many substances in which chlorine is covalently bound, for example:

$$SOCl_2 + 2H_2O \rightarrow H_2SO_3 + 2HCl$$

$$PCl_3 4\text{-} 3H_2O \rightarrow H_3PO_3 + 3HCl$$

On the large scale, hydrogen chloride can be produced by the same reaction, which is usually carried a stage further by stronger heating, i.e.

$$NaCl + NaHSO_4 \rightarrow Na_2SO_4 + HCl\uparrow$$

Uses of hydrogen chloride

Hydrogen chloride is sometimes used in the preparation of an ester, for example ethyl benzoate, where it acts as both an acid catalyst and a dehydrating agent. Hydrochloric acid is used primarily to produce chlorides, for example ammonium chloride. It is extensively used in the manufacture oi aniline dyes, and for cleaning iron before galvanising and tin-plating.

Bleaching Powder

Calcium hypochlorite is a chemical compound with formula $Ca(ClO)_2$. It is widely used for water treatment and as a bleaching agent (bleaching powder). This chemical is considered to be relatively stable and has greater available chlorine than sodium hypochlorite (liquid bleach).

Preparation

It is manufactured by the calcium process:

$$2Ca(OH)_2 + 2Cl_2 \rightarrow Ca(ClO)_2 + CaCl_2 + 2H_2O$$

Bleaching powder is actually a mixture of calcium hypochlorite ($Ca(ClO)_2$) and the basic chloride $CaCl_2$, $Ca(OH)_2$, H_2O with some slaked lime, $Ca(OH)_2$.

Properties

Calcium hypochlorite is a yellow white solid which has a strong smell of chlorine. It is not highly soluble in water, and is more preferably used in soft to medium-hard water. It has two forms: a dry form and a hydrated form. The hydrated form is safer to handle.

Calcium hypochlorite reacts with carbon dioxide to form calcium carbonate and release chlorine:

$$2Ca(ClO)_2 + 2CO_2 \rightarrow 2CaCO_3 + 2Cl_2 + O_2$$

Calcium hypochlorite reacts with hydrochloric acid to form calcium chloride:

$$Ca(ClO)_2 + 4HCl \rightarrow CaCl_2 + 2H_2O + 2Cl_2$$

Uses : Calcium hypochlorite is used for the disinfection of drinking water or swimming pool water. It is used as a sanitizer in outdoor swimming pools in combination with a cyanuric acid stabilizer, which reduces the loss of chlorine due to ultraviolet radiation. The calcium content hardens the water and tends to clog up some filters; hence, some products containing calcium hypochlorite also contain anti-scaling agents.

Calcium hypochlorite is also an ingredient in bleaching powder, used for bleaching cotton and linen. It is also used in bathroom cleaners, household disinfectant sprays, moss and algae removers, and weedkillers.

Bleaching powder is used also in sugar industry for bleaching sugar cane juice before its crystallization.

MULTIPLE CHOICE QUESTIONS

1. Fluorine reacts with water to form
(*a*) Fluorine water
(*b*) Oxygen
(*c*) Ozone
(*d*) B & C

2. Bromine water is decolourised by
(*a*) SO_2 (*b*) C_2H_4
(*c*) C_2H_2 (*d*) All

3. Which statement is correct about halogens?
(*a*) They are all diatomic and form univalent ions
(*b*) They are all capable of exhibiting several oxidation states
(*c*) They are all diatomic and form divalent ions
(*d*) They can mutually displace each other from the solution of their compounds with metals

4. Conc. HNO_3 reacts with I_2 to give
(*a*) HI (*b*) HOI
(*c*) $HOIO_2$ (*d*) $HOIO_3$

5. The main source of fluorine is
(*a*) Fluorspar
(*b*) Feldspar
(*c*) Chlorofluoro carbon
(*d*) Liquid air

6. Chlorine is produced by the
(*a*) Electrolysis of aqueous NaCl
(*b*) Action of HCl on MnO_2
(*c*) Action of conc. H_2SO_4 on NaCl in the presence of MnO_2
(*d*) All of the above

7. The bond energies of the halogens follow the order
(*a*) Cl > F > I > Br
(*b*) Cl > Br > F > I
(*c*) F > Cl > Br > I
(*d*) I > Br > Cl > F

8. Some metals, *e.g.*, Cu & Fe, acquire a protective coating of
(*a*) Fluoride (*b*) Chloride
(*c*) Bromide (*d*) Iodide

9. The increasing order of the dipole moments of halogen acids is
(*a*) HI > HBr < HCl < HF
(*b*) HF < HCl < HBr < HI
(*c*) HBr < HI < HCl < HF
(*d*) HCl < HF < HBr < HF

10. Bleaching powder slowly loses its activity when it stands in air. This is due to :
(*a*) Reaction with moisture to liberate O_2
(*b*) Autooxidation
(*c*) Loss of $CaCl_2$
(*d*) Formation of $Ca(OH)_2$

11. Chlorine bleaches only in the
(*a*) Absence of acid
(*b*) Presence of alkali
(*c*) Absence of moisture
(*d*) Presence of moisture

12. Which halogen gives a brown colour with starch?
(*a*) F_2 (*b*) Cl_2
(*c*) Br_2 (*d*) I_2

13. In which of the following reactions is bromine liberated?
(a) $KBr(aq) + I_2 \xrightarrow{\Delta}$
(b) $HI(aq) + KBr(aq) \xrightarrow{\Delta}$
(c) $KBr(aq) + F_2 \xrightarrow{\Delta}$
(d) $KBr(aq) + SO_2 \xrightarrow{\Delta}$

14. Antichlor is a compound
(a) Which absorbs chlorine
(b) Which removes Cl_2 from a material
(c) Which liberates Cl_2 from bleaching powder
(d) Which acts as a catalyst in the manufacture of Cl_2

15. Bleaching powder is an example of
(a) An acidic salt
(b) A complex salt
(c) A double salt
(d) A mixed salt

Directions Qs. 16-22: *In each of the following questions there are two statements Assertion (A) is given followed by a corresponding statement of Reason (R) just below it. Of the statements, mark the correct answer:*
(a) If both assertion and reason are true and reason is the correct explanation of assertion.
(b) If both assertion and reason are true but reason is not the correct explanation of assertion
(c) If assertion is true but reason is false
(d) If assertion is false but reason is true.

16. **Assertion (A):** Among nitrogen halides NH_3, the dipole moment is highest for NI_3 and lowest for NF_3.
Reason (R): Nitrogen halides NH_3, have trigonal pyramidal structure.

17. **Assertion (A):** HClO is stronger acid than HBrO.
Reason (R): Greater is the electronegativity of the halogen, greater will be attration of electron pair towards it and hence more easily the H^+ ion will be released.

18. **Assertion (A):** HOF bond angle in HFO is lower than HOCl bond angle in HClO.
Reason (R): Oxygen is more electronegative than halogens.

19. **Assertion (A):** Oxidising power of perhalates are in the order $BrO_4^- > IO_4^- > ClO_4^-$.
Reason (R): The standard reduction potential for BrO_4^- is more positive and decreases in the order $BrO_4^- > IO_4^- > ClO_4^-$.

20. **Assertion (A):** F atom has less negative electron affinity than Cl atom.
Reason (R): Additional electron are repelled more effectively by 3*p*– electrons in Cl than any 2*p*– electrons in F atom.

21. **Assertion (A):** HBr is a stronger acid than HI.
Reason (R): Bond energy of HI is smaller than that of HBr.

22. **Assertion (A):** Cl_2 gas bleaches the articles permanently.
Reason (R): Cl_2 is a strong reducing agent.

ANSWERS

1	2	3	4	5	6	7	8	9	10
(d)	(b)	(b)	(d)	(c)	(c)	(a)	(b)	(c)	(c)
11	12	13	14	15	16	17	18	19	20
(b)	(d)	(a)	(d)	(d)	(b)	(a)	(c)	(a)	(c)
21	22								
(d)	(c)								

19 GROUP-0 (NOBLE GAS)

General Properties of Noble gases

1. The increase in atomic radius (in this group, the actual radius of the free atom).
2. The increase in melting point and boiling point, and the very narrow liquid range.
3. The large ionisation energies, as expected for atoms with complete quantum levels.
4. The small negative electron affinities of helium and neon. The simple fact that the noble gases exist as separate atoms-a unique property at ordinary temperatures-is sufficient indication of their chemical inactivity.

Helium

Helium is the chemical element with atomic number 2 and an atomic weight of 4.002602, which is represented by the symbol He. It is a colorless, odorless, tasteless, non-toxic, inert monatomic gas that heads the noble gas group in the periodic table. Its boiling and melting points are the lowest among the elements and it exists only as a gas except in extreme conditions. Helium is the second lightest element and is the second most abundant in the observable universe, being present in the universe in masses more than 12 times those of all the heavier elements combined. Its abundance is also similar to this in our own Sun and Jupiter. There are eight known isotopes of helium, but only helium-3 and helium-4 are stable. Helium has a valence of zero and is chemically unreactive under all normal conditions. It is an electrical insulator unless ionized.

Uses :

1. Helium has been used in quantity as a substitute for hydrogen in filling airships.
2. A mixture of 80% helium and 20% oxygen is used instead of air in diving apparatus because helium, unlike nitrogen, is not appreciably soluble in blood even under pressure
3. A similar helium-oxygen mixture has been used to assist breathing in cases of asthma and other respiratory diseases.
4. Helium has two important scientific uses. First, liquid helium is used to realize very low temperatures, in order to study peculiar phenomena which occur near the absolute zero-cryogenics.
5. Some metals attain enormously high electrical conductivity when cooled down to near absolute zero, and hence powerful electro-magnets can be made using very small coils cooled in liquid helium.
6. Secondly, it is used in gas thermometers for low temperature measurement.

MULTIPLE CHOICE QUESTIONS

1. Arrange H, F, Li and K in decreasing order of ionisation energy.
   (*a*) H > F > Li > K
   (*b*) F > H > Li > K
   (*c*) K > H > F > Li
   (*d*) Li > K > H > F
2. Which gas is filled in electric bulbs/tubes?
   (*a*) O_2 (*b*) N_2
   (*c*) Ar (*d*) He
3. Which one of the following is used in colour discharge tube?
   (*a*) Ne (*b*) Ar
   (*c*) Kr (*d*) He
4. The gas used for inflating the tyres of aeroplane is
   (*a*) H_2 (*b*) He
   (*c*) N_2 (*d*) Ar
5. The principle source of helium is
   (*a*) Air (*b*) Monazite sand
   (*c*) Radium (*d*) All
6. The gaseous mixture used by divers for respiration is
   (*a*) $N_2 + O_2$ mixture
   (*b*) He + O_2 mixture
   (*c*) Ar + O_2 mixture
   (*d*) Ne + O_2 mixture
7. Noble gases can be separated by
   (*a*) Passing them through same solutions
   (*b*) Electrolysis of their compounds
   (*c*) Absorption and desorption on coconut charcoal
   (*d*) None of the above
8. Coconut charcoal at –100 °C absorbs a mixture of
   (*a*) He and Kr (*b*) Ar, Kr and Xe
   (*c*) Kr and Xe (*d*) He and Ne
9. Asthma patients use a mixture of _____ for respiration.
   (*a*) O_2 and H_2 (*b*) O_2 and He
   (*c*) O_2 and Ar (*d*) O_2 and Ne
10. Helium gives a characteristic spectrum with
    (*a*) Orange and red lines
    (*b*) Orange lines
    (*c*) Yellow lines
    (*d*) Green lines
11. The gas used in gas thermometer is
    (*a*) He (*b*) O_2
    (*c*) Xe (*d*) Ne

Directions Qs. 12-13: *In each of the following questions there are two statements Assertion (A) is given followed by a corresponding statement of Reason (R) just below it. Of the statements, mark the correct answer:*

(*a*) If both assertion and reason are true and reason is the correct explanation of assertion.
(*b*) If both assertion and reason are true but reason is not the correct explanation of assertion
(*c*) If assertion is true but reason is false
(*d*) If assertion is false but reason is true.

12. **Assertion (A):** Deep sea divers use helium-oxygen mixture for respiration.
    Reason (R): Helium is inert in nature.
13. **Assertion (A):** Xenon forms fluorides.
    Reason (R): Because 5 *d*-orbitals are available for valence shell expansion.

ANSWERS

1	2	3	4	5	6	7	8	9	10
(*b*)	(*c*)	(*b*)	(*b*)	(*b*)	(*b*)	(*c*)	(*b*)	(*b*)	(*c*)
11	**12**	**13**							
(*a*)	(*b*)	(*b*)							

20 METALLURGICAL PROCESS

Occurrence of Metals

1. ***Native or Free State*** **:** Only those metals which are not attacked by the oxygen or any other non-metals can occur in the native or free state, for example, silver, gold, copper etc.

2. ***Combined State*** **:** Since most of the metals have a tendency to react with oxygen, sulphur, halogens, therefore these metals does not occur in free state. They occur in the combined forms as their compounds. The various compounds of metals which occur in the earth's crust are known as the minerals. For example, Cu_2O, ZnS, NaCl etc.

A mineral which can be used for the extraction of metals conveniently and economically is called an **ore.**

Metallurgy

The process of extracting a metal in the free state from its ore is known as Metallurgy. The various steps involved in the metallurgical processes are :

1. ***Concentration of ore*** **:** Metallic ores are often found mixed with, non-metallic or rocky materials, *i.e.*, quartz, feldspar and other silicates. These impurities are known as the gangue. Before the ore is subjected to metallurgical process, it is necessary to remove these impurities. This operation is called **concentration of ore.** Various methods are as the follows.

Froth floatation process : This process is only used for the concentration of sulphide ores such as galena (PbS), cinnabar (HgS) etc.

The crushed and finally ground powdered ore is mixed with water and in small quantity with pine oil in a large tank. Air is bubbled through the solution to agitate the contents. Consequently the froath (foam) which carries the ore particles with it to the surface leaving behind the gangue particles is formed. The ore particles rise to the surface with foam while the gangue particles are wetted with water and settle down. The foam formed on the top is transferred into another tank while they are washed with water to recover ore particles.

2. ***Calcination*** **:** Calcinations is the process of heating the concentrated ore in a limited supply of air at a temperature insufficient to melt it. The ore is subjected to the heat in order to expel the organic matter and moisture present in the ore. Calcination is also done to remove water from hydrated oxide ore or carbon dioxide from a carbonate ore. For example, when limestone is heated, carbon dioxide is given off and limestone is said to be calcinated.

$$CaCO_3 \rightarrow CaO + CO_2$$

3. ***Roasting*** **:** The process of roasting is generally of two types :

(i) Reductive roasting : The calcinated ore which contains metallic oxides is converted into metal by the reduction process. In this process the ore is mixed with suitable quantities of powdered coke and the flux. The flux is the substance which has a tendency to react with impurities to form fusible silicates called **slag.** The mixture of ore + powdered coke + flux is then heated strongly at high temperature sufficient to melt it in the furnace, generally in blast furnace. The selection of flux depends upon the impurities. For example,

$SiO_2 + CaO \rightarrow Ca\ SiO_3$ (slag)
(Acidic impurity) (A Calium silicate)

$FeO \rightarrow SiO_2\ FeSiO_3$
(Basic impurity) (Acidic flux) Iron silicate (slag)

(ii) Oxidative roasting : In oxidative roasting the calcinated ore is generally converted into oxide. It is the process of heating the concentrated ore in a free supply of air at temperature insufficient to melt in order to bring oxidation. It is generally carried in case of sulphide ores and mostly in the reverberatory furnace. For example,

$$CuS + 2O_2 \longrightarrow CuSO_4$$
$$ZnS + 2O_2 \longrightarrow ZnSO_4$$

4. *Smelting* : By the process of smelting the roasted or the calcinated ore is reduced to the free metal. The roasted ore is mixed with carbon (coal or coke) and heated in a furnace generally in the reverberatory furnace in controlled supply of air. Carbon and carbon monoxide reduce the oxide to the free metal. For example,

$$Fe_2O_3 + 3C \longrightarrow 2Fe + 3CO$$
$$Fe_2O_3 + 3CO \longrightarrow 2Fe + 3CO_2$$

Electrolytic Process for Extraction of Metals: In case of highly electropositive and chemically active elements such as alkali metals, alkaline earth metals and aluminium, it is very difficult to reduce their oxide with carbon. Even if the metal is formed by the reduction of the metallic oxide at high temperature, the metal thus liberated combines with carbon producing metallic carbides. For example, sodium is obtained by the electrolysis of fused sodium chloride containing calcium chloride using iron cathode and graphite anode in accordance with the following reactions :

$$7NaCl \rightleftharpoons Na^+ + Cl^-$$

At the cathode, $Na^+ + e^- \longrightarrow Na$

At the anode; $2Cl^- - 2e^- \longrightarrow Cl_2$

Likewise, aluminium is extracted by the electrolysis of alumina dissolved in molten cryolite ($3NaF\ .\ AlF_3$) in an iron tank lined inside with a thick layer of carbon which serves as cathode.

At the cathode, $Al^{3+} + 3e^- \longrightarrow Al$

At the anode, $2O_2^- \longrightarrow O^{2-} + 4e^-$

Reduction by Aluminium (Goldschmit thermite process) : In such cases, aluminium is used as reducing agent because aluminium is more electropositive than magnesium or chromium. This process is known as **aluminothermy.**

$$Cr_2O_3 + 2Al \longrightarrow Al_2O_3 + 2Cr$$
$$3Mn_3O_4 + 8Al \longrightarrow 4Al_2O_3 + 9Mn$$

These reactions are highly exothermic. Consequently the metals are obtained in the molten state.

Hydrometallurgy or Precipitation by a more Electropositive Metal : Hydro-metallurgy involves the dissolution of the ore using a suitable chemical reagent followed by precipitation of the metal by a more electropositive metal. The extraction of silver and gold is based, on this process.

5. *Purification or Refining :* The metal extracted form the ore by any method are usually impure and are known as crude metals. These crude metals generally contain the following impurities :

(i) Other metals originally present in the ore.

(ii) Non-metals formed by the reduction in the furnace

(iii) Substances taken up in the furnace, as slag, flux etc.

The process of removal of impurities from a crude metal is known as refining. The following methods are generally employed for refining.

Liquation Process : This process is employed for the refining of the metals having low melting points while the impurities present in them have high melting points. The impure metal is placed on the slopping floor of the earth and is gently heated in an inert atmosphere of carbon monoxide or nitrogen to prevent oxidation of the metal by air. The metal melts and flows down while non-fusible impurities are left behind.

Oxidation Process : Oxidation process is employed for the refining of metals when impurities

TABLE : Summary of Extraction of Some Important Metals

Metal	*Important minerals*	*Main method of Extraction*	*Equations of Extraction*
Ag	Silver glance—Ag_2S* Horn silver—AgCl Native silver	Dissolution with NaCN, followed by precipitation with zinc.	$Ag_2S + 4NaCN_2 \rightarrow Na[Ag(CN)_2] + Na_2S$ $2Na[Ag(CN)_2] + Zn \rightarrow Na_2[Zn(CN)_4] + 2Ag$
Al	Bauxite—$Al_2O_3 . 2H_2O$ Cryolite—$3NaF . AlF_3$	Electrolysis of Al_2O_3 dissolved in molten cryolite.	$Al^{3+} + 3e^- \rightarrow Al$
Au	Native gold	Dissolution with KCN in air, followed by precipitation with zinc.	$4Au + 8KCN + 2H_2O + O_2 \rightarrow 4K[Au(CN)_2] + 3KOH$ $2K[Au(CN)_2] + Zn \rightarrow K_2[Zn(CN)_4] + 2Au$
Cu	Copper pyrites—$Cu_2S . Fe_2S_3$ Copper glance—Cu_2S Malachite—$CuCO_3 . Cu(OH)_2$	Oxidation of Cu_2S to Cu_2O followed by reduction of Cu_2O with Cu_2S.	$2Cu_2S + 2O_2 \longrightarrow 2Cu_2O + 2SO_2$ $2Cu_2O + Cu_2S \longrightarrow 6Cu + SO_2$
Fe	Haematite—Fe_2O_3 Magnetite—Fe_3O_4 Iron pyrites—FeS_2	Reduction of oxide with carbon monoxide.	$Fe_2O_3 + 3CO \rightarrow 2Fe + 3CO_2$
Pb	Galena—PbS Tin stone—SnO_2	Reduction of SnO_2 with carbon.	$SnO_2 + 2C \rightarrow Sn + 2CO$
Zn	Calamine—$ZnCO_3$	carbon.	

consisting of other metals gets oxidised more readily than the metal itself. The impure metal is exposed in molten state to the oxidizing influence of the air in a suitable furnace. The oxides of other metals formed can be removed from the surface by skimming. For example, copper oxide is employed to impure molten copper to provide oxygen required for the oxidation process.

Electrolytic Refining : A large number of metals are refined by this method which is carried out by electrolysis.

The impure metal is cast on the thick rod and made as anode while the thick sheets of pure metal act as the cathode. A solution of soluble salt of the metal is used as electrolyte. On passing electric current pure metal from the electrolyte deposits on the cathode and an equivalent amount dissolves from the anode. The impurities present in the anode either get dissolved in the electrolyte or fall at the bottom as **"anode mud".**

Poling : This process is employed to remove reducible oxides from the metals. For example, blister copper containing cuprous oxide as impurity is purified by poling.

Cupellation : This process is specifically employed to purify silver containing lead as an impurity. Impure silver is heated in bone-ash crucibles (called *cupels)* on the floor of a reverberatory furnace under a blast of air. Lead is oxidised to lead oxide (litharge), most of which is swept away by the blast of air while last *traces* of it are absorbed by the porous material of the cupel. As the heating continues, a bright mass of molten silver is left in the cupel.

Common Alloys

It is a mixture of metallic species and usually prepared by adding other metals or non metals to solvent metals in the liquid state. Alloys are stronger than their constituent metals.

Alloy	Composition	Uses
Alloys of Aluminium		
1. Alluminium/ Bronze	Al 10%, Cu 90%	non-corrodible vessels
2. Duralumin	Al 95%, Cu 3% Mn 1%, Cu 3%	Airplanes and Automobile
3. Magnalium	Al 90%, Mg 10%	Balance beams
4. Y-Alloy	Al 92.5%, Cu 4% Mg 1.4%, Ni 2%	non-corrodible vessels
Lead Alloys		
1. Types Metals	Pb 82%, Sb 15%, Sn 3%	Printing
2. Solder	Pb 50%, Sn 50%	soldering
3. Pewter	Pb 20%, Sn 80%	making

Alloy	Composition	Uses
Special Alloy Steels		
1. Manganese steel	Mn 10-18%	Very hard and resistant to wear
2. Nickel steel	Ni 2-4%	Resists corrosion, hard and elastic.
Alloys of Copper		
1. Brass	Cu 80%, Zn 20%	Wires, parts of machinery utensils
2. Bronze	Cu 80%, Zn 10% Sn 10%	Statues, coins, cooking utensils
3. Gun metal	Cu 90%, Sn 10%	Gun barrels
4. Bell metals	Cu 80%, Sn 20%	Bell, gangs
5. Germam silver	Cu 60%, Zn 20% Ni 20%	Silver wire, plating

MULTIPLE CHOICE QUESTIONS

1. Bauxite has the formula
(*a*) $Al_2O_3.H_2O$ (*b*) $Al_2O_3.2H_2O$
(*c*) $Al_2O_3.3H_2O$ (*d*) $KAlSi_3O_8$

2. Carborundum is
(*a*) SiC (*b*) Al_2O_3
(*c*) $Al_2O_3.2H_2O$ (*d*) $KAlSi_3O_8$

3. Silver is refined by
(*a*) Cupellation
(*b*) Poling
(*c*) Van-Arkel method
(*d*) Liquation

4. Smelting involves the reduction of a metal oxide with
(*a*) Carbon monoxide
(*b*) Magnesium
(*c*) Carbon
(*d*) Aluminium

5. The ores of Ag and Au are concentrated using their solubility in
(*a*) NH_3 (*b*) HNO_3
(*c*) KCN (*d*) HCl

6. The most common ore of copper is
(*a*) Cu_2S
(*b*) $CuFeS_2$
(*c*) $CuCO_3.Cu(OH)_2$
(*d*) Cu_5FeS_4

7. The impurities present in a mineral are called
(*a*) Gangue (*b*) Alloy
(*c*) Slag (*d*) Flux

8. The concentration of chromite ($FeO.Cr_2O_3$) is carried out by
(*a*) Gravity separation
(*b*) Froth flotation
(*c*) Magnetic separation
(*d*) Roasting

9. Froth floatation is used in the metallurgy of
(*a*) Oxide ores
(*b*) Chloride ores
(*c*) Carbonate ores
(*d*) Sulphide ores

10. In froth floatation, the oil used is
(*a*) Olive oil (*b*) Pine oil
(*c*) Coconut oil (*d*) Mustard oil

11. In a blast furnace, iron oxide is reduced by
(*a*) Silica
(*b*) Carbon
(*c*) Carbon monoxide
(*d*) Lime

12. Roasting is carried out in
    (*a*) A blast furnace
    (*b*) A reverberatory furnace
    (*c*) A muffle furnace
    (*d*) An electric furnace

13. Silver is extracted from its ore using NaCN, air and on active metal. The method is known as the
    (*a*) Pattinson process
    (*b*) Parkes process
    (*c*) Amalgamation method
    (*d*) Cyanide process

14. The roasting of an ore is carried out in
    (*a*) An excess of air
    (*b*) The presence of O_2 only
    (*c*) The absence of air
    (*d*) A limited supply of air

15. When a sulphide ore is roasted, the product obtained is usually
    (*a*) A metal (*b*) A sulphite
    (*c*) An oxide (*d*) A nitride

Directions Qs. 16-24: *In each of the following questions there are two statements Assertion (A) is given followed by a corresponding statement of Reason (R) just below it. Of the statements, mark the correct answer:*

(*a*) If both assertion and reason are true and reason is the correct explanation of assertion.
(*b*) If both assertion and reason are true but reason is not the correct explanation of assertion
(*c*) If assertion is true but reason is false
(*d*) If assertion is false but reason is true.

16. **Assertion (A):** Zinc and not copper is used in the recovery of Ag from the complex $[Ag(CN)_2]^-$.
    Reason (R): Zinc is a powerful reducing agent than copper.

17. **Assertion (A):** Chalcocite is roasted and not calcined during recovery of copper.
    Reason (R): Copper pyrite is an ore of copper.

18. **Assertion (A):** Aluminium metal is used as a reducing agent in the extraction of chromium.
    Reason (R): Reduction by aluminium is known as Gold Schmidt aluminothermic process.

19. **Assertion (A):** Magnesium oxide is used for the lining in steel making furnace.
    Reason (R): Magnesium oxide acts as a flux to remove impurities of Si, P and S through slag formation.

20. **Assertion (A):** Graphite is used as anode.
    Reason (R): Graphite helps in electron conduction.

21. **Assertion (A):** Coke and flux are used in smelting.
    Reason (R): The phenomenon in which ore is mixed with suitable flux and coke and is heated to fusion is known as smelting.

22. **Assertion (A):** Sodium is not prepared by the electrolysis of its aqueous solution.
    Reason (R): On electrolysis of an aqueous solution of NaCl, the Na^+ ions are not discharged at cathode, on the contrary H^+ ions get discharged from H2.

23. **Assertion (A):** Elements of alkali metal group are strong reducing agent.
    Reason (R): Alkali metals are mainly in the form of chlorides, carbonates, sulphates and nitrates.

24. **Assertion (A):** SiO_2 is used to remove impurities of CaO in an ore.
    Reason (R): SiO_2 is an acidic flux.

ANSWERS

1	2	3	4	5	6	7	8	9	10
(*b*)	(*a*)	(*a*)	(*c*)	(*c*)	(*b*)	(*a*)	(*c*)	(*d*)	(*b*)
11	**12**	**13**	**14**	**15**	**16**	**17**	**18**	**19**	**20**
(*c*)	(*b*)	(*d*)	(*a*)	(*c*)	(*a*)	(*b*)	(*b*)	(*a*)	(*a*)
21	**22**	**23**	**24**						
(*b*)	(*a*)	(*b*)	(*b*)						

C ORGANIC CHEMISTRY

21 ORGANIC CHEMISTRY : SOME BASIC PRINCIPLES

Tetrahedral Neture of Carbon

The atomic number of carbon is 6 and its atomic weight is 12. It has 4 electrons in the last orbit and tends to gain four more electrons by forming four covalent bonds with other hydrogen atoms. Structural formula of the simplest hydrocarbon methane (CH_4) can be written as

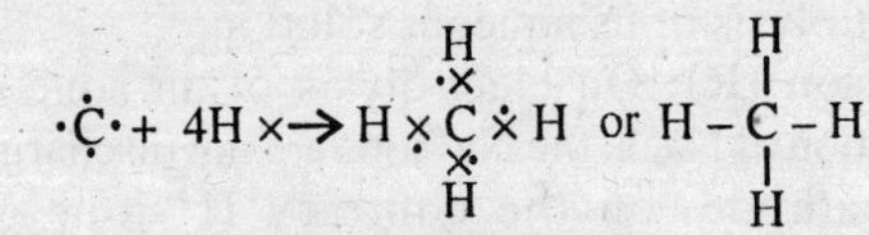

Carbon atom is tetravalent. That is, it has a valence of 4. According to Le Bel and Van't Hoff, the four valencies of carbon do not lie in one plane. They are directed towards the corners of a regular tetrahedron so that the angle between any two valencies is 109°28'

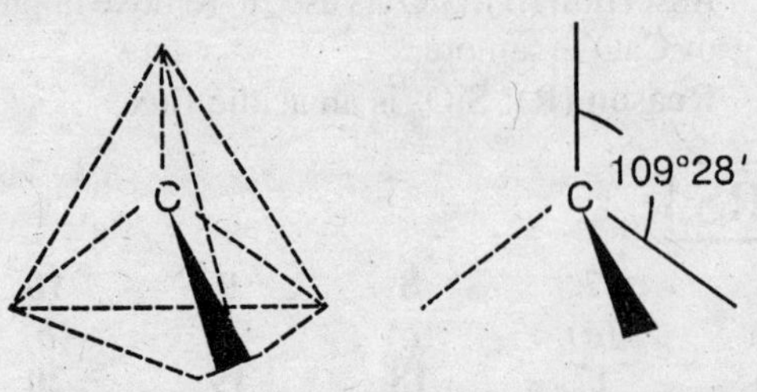

Fig. : Four valencies of carbon are directed towards the corners of a regular tetrahedron. The angle between any two valencies is 109°28'.

Carbon-Carbon Single Bond : The molecules of hydrocarbons, ethane and propane contain two and three carbon atoms respectively linked by singh covalent bonds.

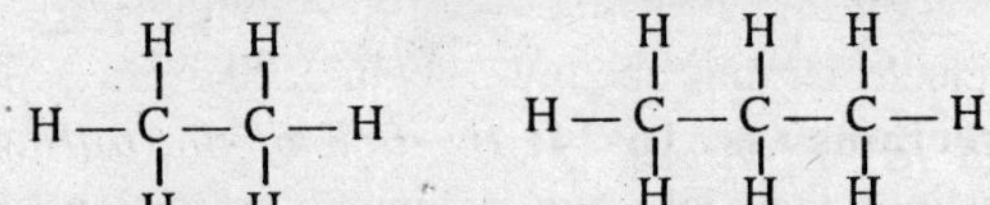

Ethane Propane

Carbon-Carbon Double Bond : In some compounds, two of the valencies of a carbon atom may be satisfied by union with the two valencies of another carbon atom. Thus, in ethylene the two carbons are united by two covalent-bonds.

Ethylene Carbon – Carbon Double bond

Such a union involving two covalent-bonds between adjacent carbon atoms is called Double Bond.

Carbon-Carbon Triple Bond : Sometimes two adjacent carbons are linked together by three covalent bonds. Thus acetylene molecule is represented as

H—C≡C—H —C≡C—

Acetylene Triple bond

Such a union involving three covalent-bonds between adjacent carbon atoms is called Triple Bond.

Sigma Bonds

A sigma bond is formed by the linear or end-to-end overlap of orbitals. They may be obtained.

(*a*) By the overlap of two *s*-orbitals :

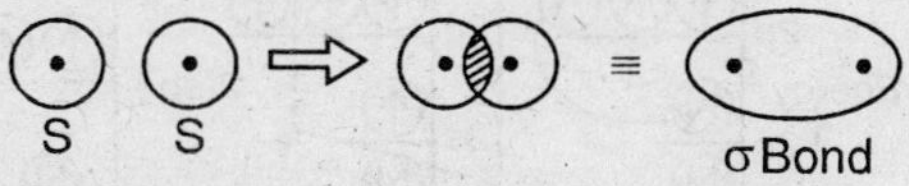

(*b*) By the overlap of p_x and an *s*-orbital:

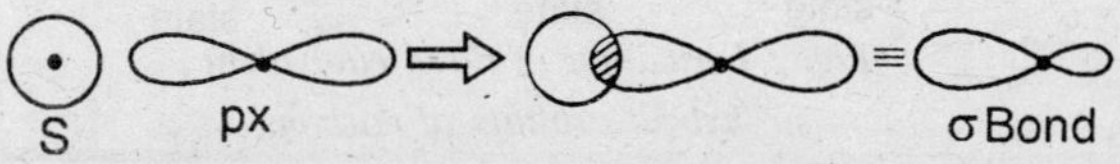

(*c*) By the overlap of two *p*-orbitals:

Fig.

Sigma bonds are symmetrical around the line drawn between the two nuclei. They are represented by a single line drawn between the atomic symbols. The electrons that occupy a sigma bond are called *Sigma (σ) Electrons.*

Pi-Bond

A pi bond is formed by parallel or side to side overlap *p*-orbitals for example :

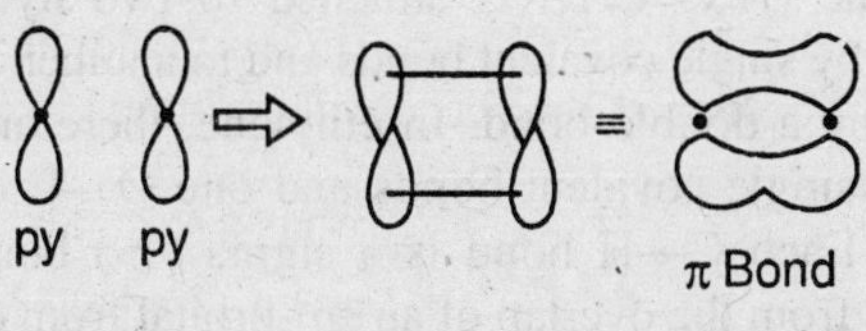

Fig.

The electrons that occupy a pi bond are known as the *Pi (π) Electrons*. There are two important differences between sigma and pi bonds.

(*i*) Pi electrons are loosely held than a pair of electrons in a σ-bond. They are of higher energy. As a result, "π-bonds are more easily broken and are more reactive than σ-bonds".

(*ii*) Rotation of atoms is not possible around a π-bond. This restriction in rotation around a π bond is responsible for Cis and Trans isomerism in alkenes.

Hybridization

Hybridization is the process whereby atomic orbitals of differing type but similar energies are combined to yield a set of equivalent hybrid orbitals.

The types of hybridization encountered in organic compounds are sp^3 (involved in saturated organic compounds containing only single covalent bond), sp^2 (involved in organic compounds having carbon linked by double bonds, *e.g.* ethylenes) and *sp* (involved in organic compounds having carbon linked by a triple bond, *e.g.* acetylenes). The bond angles and geometry associated with the three types of hybridization are summarised below.

Type	*Bond Angle*	*Geometry*	*Number of Unused p-orbital (s)*	*% of s-orbital Character*
sp^3	109.28'	Tetrahedral	0	25
sp^2	120°	Trigonal Planar	1	33.3
sp	180°	Linear	2	50

1. sp^3 *Hybridization Of Carbon*

Electronic con-figuration of carbon in its ground or atomic state is

$$\text{C (Ground State)} \doteq \frac{1s^2\,2s^2\,2px^1\,2py^1\,2pz^0}{\text{valence shell}}$$

In terms of energy level diagram, this electronic configuration may be represented as in following figure: Formation of four equivalent sp^3-Hybrid orbitals of Carbon.

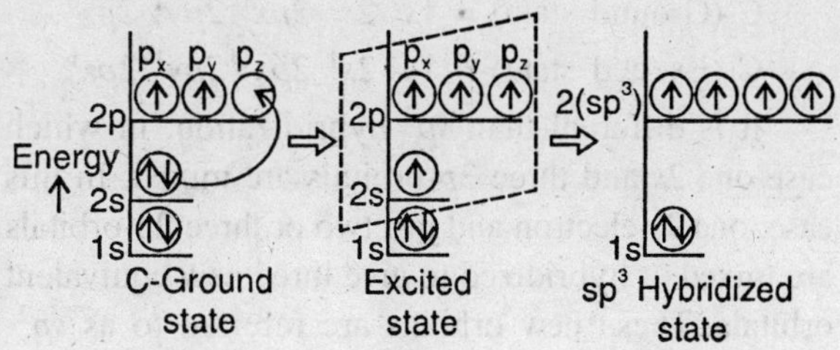

Fig.

Since there are only two unpaired electrons, (half-filled orbitals), it might be expected that only two single covalent bonds will be formed.

Each sp^3-orbital contains one electron. Since each sp^3-orbital is obtained from one *s* and three *p*-orbitals, it has 25% *s*-character and 75% *p*-

character. Each sp^3-orbital has a large lobe and a small lobe as shown in the following figure:

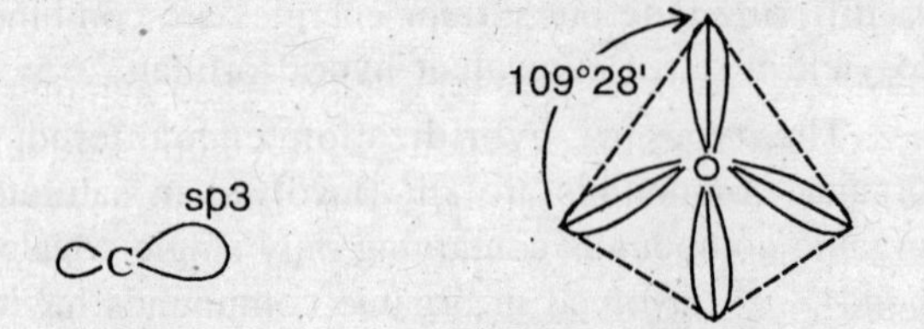

Shape of an sp^3 hybrid orbital — Orientation of four sp^3 hybrid orbitals

Fig.

Bonding in methane : In methane, carbon forms single covalent bond with four hydrogen atoms. Since the carbon atom is attached to four carbon atoms it was sp^3-orbitals to form these bonds. Following figure shows how the bonds in methane are formed.

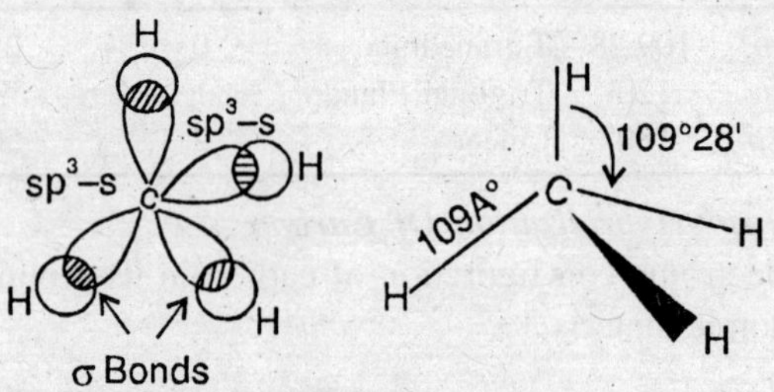

Fig. : Bonding in methane

Each C—H-covalent bond is the result of the overlap of an sp^3-orbital from carbon and 1*s*-orbital from hydrogen.

2. *sp²-Hybridization of Carbon :*

C (Ground state) = $1s^2\ 2s^2\ 2px^1\ 2py^1\ 2pz^{\circ}$

C (Excited state) = $1s^2\ 2s^1\ 2px^1\ 2py^1\ 2pz^1$

It is different than sp^3 hybridization, in which case one 2*s* and three 2*p*-orbitals are mixed. In this case, one 2*s*-electron and just two or three 2*p*-orbitals are mixed or hybridized to give three new equivalent orbitals. These new orbitals are referred to as sp^2-hybrid orbitals or simple sp^2-orbitals, because they are formed by the interaction of one *s* and two *p*-orbitals. Third $2p_z$-orbital is left unhybridized. The electron configuration of the carbon atom in its sp^2-hybridized state is

C (Hybridized state) = $1s^2\ (sp^2)^1\ 2(sp^2)^1\ 2(sp^2)^1 2p_z{}^1$

The processes which lead to the formation of three equivalent sp^3-hybrid orbitals are summarised in following figure :

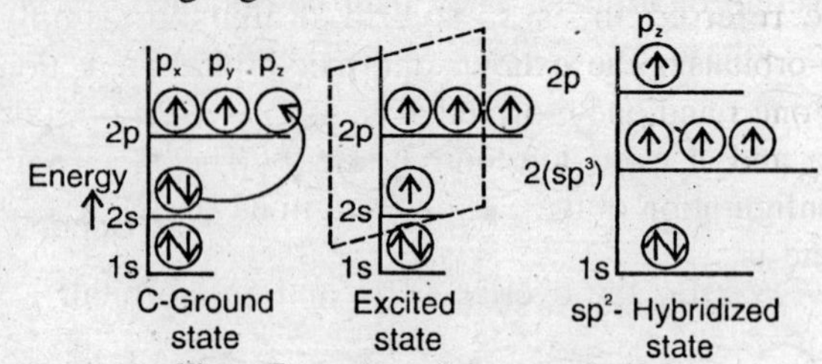

Fig. : Formation of three equivalent sp^2-hybrid orbitals of carbon.

Each sp^2-orbital contains an unpaired electron. The shape of an sp^2-orbital is similar to that of an sp^3-orbital.

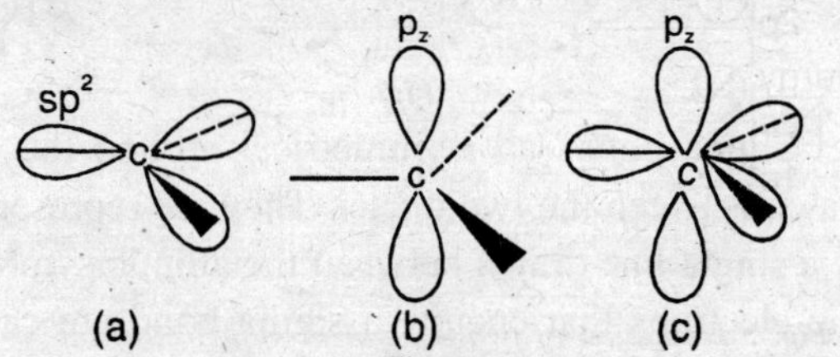

Fig.

(*a*) Orientation of three sp^2-orbitals.

(*b*) The unhybridized p_z-orbital.

(*c*) Orientation the p_z-orbital in relation to the sp^2-orbitals.

Bonding in ethylene : Each carbon atom in ethylene ($H_2C{=}CH_2$) is attached to two hydrogen atoms by single covalent bonds and to another carbon atom by a double bond. In ethylene, there are four C—H single covalent bonds and one O—C double bond. Each C—H bond is a sigma (σ) bond and results from the overlap of an sp^2-orbital from carbon and 1*s*-orbital from hydrogen.

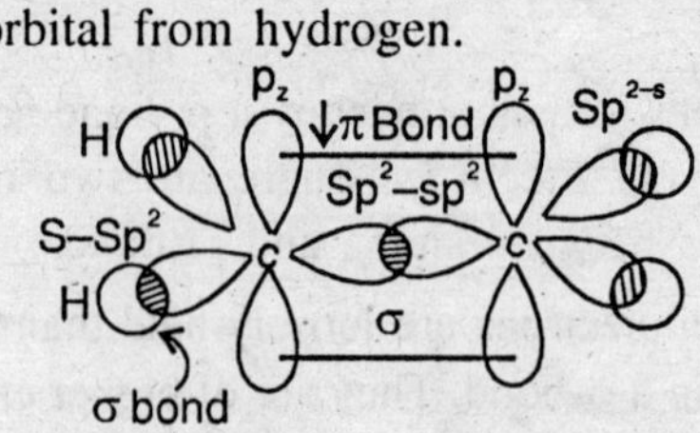

Fig. : Bonding in ethylene.

3. *sp-Hybridization of Carbon*

C (Ground state) = $1s^2\ 2s^2\ 2px^1\ 2py^1\ 2pz^0$

C (Excited state) = $\dfrac{1s^2\ 2s^1\ 2px^1\ 2py^1\ 2pz^1}{\text{Valence shell}}$

In this case, the 2*s*-electron and just one of the three 2*p*-orbitals are mixed or hybridized to give two new equivalent orbitals. These two new orbitals are referred to as *sp* hybrid orbitals or simply *sp*-orbitals because they are formed by the interaction of one *s* and one *p*-orbital. The other two 2*p*-orbitals (p_y and p_z) are left unhybridized. The electronic configuration of the carbon atom in its *sp*-hybridized state is :

C (Hybridized state)

$$= \frac{1s^2\,2(sp)^1\,2(sp)^1\,2py^1\,2pz^1}{\text{Valence shell}}$$

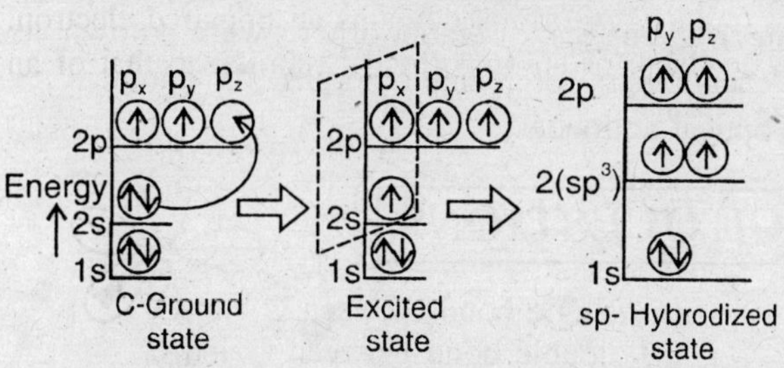

Fig. : Formation of two equivalent sp-hybrid orbitals.

Each *sp*-orbital contains an unpaired electron. The shape of an *sp*-orbital is similar to that of on sp^2-orbital.

Bonding in acetylene : Each carbon atom is attached to one hydrogen atom by a single covalent bond and to another carbon atom by a triple bond. Since each carbon is attached to two other atoms, it uses *sp* hybrid plus two unhybridized 2*p*-orbitals (p_y and p_z) to form its bonds.

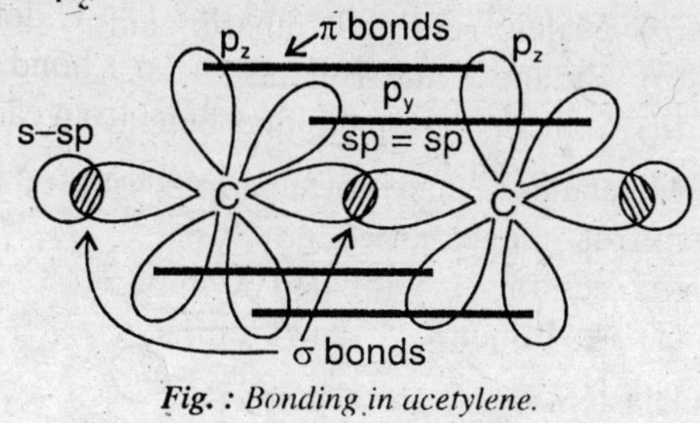

Fig. : Bonding in acetylene.

The decreasing order of the size of the three hybrid orbitals is opposite to that of the decreasing

$$sp^3 > sp^2 > sp$$

order of the *s*-orbital character in the three hybrid orbitals.

The relative energy of the *s*, *p* and different *sp*-hybrid orbitals follows the following order.

$p > sp^3 > sp^2 > sp > s$ (relative energies of orbitals)

Geometrical Isomerism

When two carbon atoms of a compound, containing two different atoms or groups are linked by a double bond the independent rotation of the carbon atoms is not possible and position of groups or atoms becomes fixed in the space.

This type of isomerism arises due to the presence of a double bond about which there is no free rotation, is called geometrical isomerism and the compounds showing this type of isomerism are known as *cis*–trans isomers.

This also, results from a restriction in rotation about single bonds in cyclic compounds.

Examples :

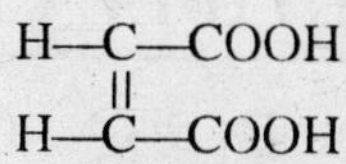

Maleic acid
(*cis*-form)

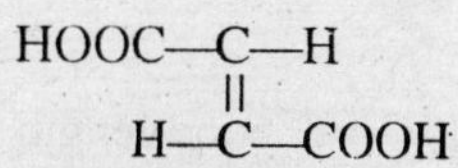

Fumatic acid
(*trans*-form)

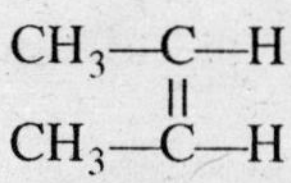

Butene-2
(*cis*-form)

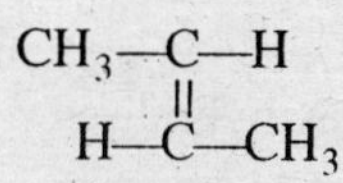

Butene-2
(*trans*-form)

The dipole moment of *cis* form is always higher than that of *trans* form. The trans isomer of some alkene derivatives have zero dipole moment. *Cis*-isomers have low melting point and high boiling point and they are soluble in water due to their dipole moments. The trans isomers have high melting points and low boiling points and they are sparingly soluble in water.

Optical isomerism

The light whose vibrations occur only in one direction or in one plane is called 'plane polarised light'. The device (Nicol prism) that brings about the polarisation of light is known as 'polariser'

The substance which rotates the plane of polarised light are said to be *optically active* and this property is known as *optical activity*. The phenomenon of optical activity was discovered by the French physicist Biot in 1815. The substance which rotate the plane of polarised light to the right (clockwise) are known as *dextro-rotatory* while those which rotate to the left (anticlockwise) are known as *laevorotatory* and they are known as *d* and *l* forms or (+) & (–) forms respectively.

1. *Specific rotation :* "Specific rotation is defined as the rotation produced by a solution of length 10 centimeters and unit concentration (1g/ml) for the given wavelength of light at the given temperature."

2. *Asymmetric carbon atom and molecular chirality :* A plane which divides an object into two symmetrical halves, is said to be 'Plane of Symmetry'. For example, a person or a hat has a plane of symmetry.

An object lacking a plane of symmetry is called *Dissymmetric* or *Chiral* (pronounced as *Ki–ral*). A symmetric object is referred to as *Achiral.*

3. *Racemic mixture or racemic modifications :* The optical rotatory powers of two isomers are equal in magnitude but opposite in sign. An equimolar mixture of the two isomers *Dextrorotatory isomer* or (+) isomer. *Laevorotatory Isomer* or (–) isomer, therefore will not rotate the plane of polarised light at all and is said to be *Racemic Mixture.*

4. *Discovery of enantiomerism :* Pairs of compounds whose structures differed only in being mirror images or each other, were known as 'enantiomers' or 'enantiomorphs' or 'antimers' or 'optical antipodes'.

MULTIPLE CHOICE QUESTIONS

1. C—C bond length is minimum when carbon is hybridised

(*a*) *sp* (*b*) *dsp*^{2}
(*c*) *sp*^{3} (*d*) All the above

2. The central carbon atom of the carbonium ion is hybridised

(*a*) *sp* (*b*) *dsp*^{2}
(*c*) *sp*^{2} (*d*) *sp*^{3}

3. The tetrahedral nature of the carbon was given by

(*a*) Van't Hoff
(*b*) Lewis
(*c*) Le-Bel and Van't Hoff
(*d*) None of the above

4. In neopentane the number of primary carbon atoms are

(*a*) 4 (*b*) 5
(*c*) 2 (*d*) 1

5. In the case of HCHO carbon is hybridised

(*a*) *sp* (*b*) *n*(N–1) *sp*^{2}
(*c*) *sp*^{2} (*d*) *dsp*^{3}

6. The hybrid orbitals involved in the formation of Butyne—1 are hybridised

(*a*) *sp* only (*b*) *sp* and *sp*^{3}
(*c*) *sp* and *sp*^{2} (*d*) *dsp* and *sp*^{2}

7. Geometrical isomerism exist due to the presence of

(*a*) single bond between C and C
(*b*) double bond between C and C
(*c*) triple bond between C and C
(*d*) All of the above

8. The process of separating a racemic mixture into optical isomers is known as

(*a*) dissolution (*b*) isomerisation
(*c*) resolution (*d*) pasteurisation

9. Which of the following is a pair of diastereomers?

(*a*) Cl and 1 tartaric acid
(*b*) Maleic acid and fumaric acid
(*c*) Butane-2 and Butene-3
(*d*) 2 methyl propane and butene-1

10. Molecules that are not superimposable on their mirror images are

(*a*) achiral (*b*) symmetric
(*c*) asymmetric (*d*) chiral

11. Diastereomers are a pair

(*a*) stereoisomers that cannot be separated from each other
(*b*) stereoisomers that are not minor image of each other
(*c*) stereoisomers that have identical chemical properties
(*d*) stereoisomers that are minor image of each other

12. Optical activity in a compound can be detected by means of
(*a*) voltmeter (*b*) potentiometer
(*c*) ammeter (*d*) polarimeter

13. Which of the following compounds would exhibit geometrical isomerism?
(*a*) 2-pentane
(*b*) 2-butene
(*c*) 1-butene
(*d*) 1, 2 dichloroethane

14. Isomers that are mirror image of each other are called
(*a*) enantiomers (*b*) diastereomers
(*c*) optical isomers (*d*) stereomers

15. A mixture of equal parts of enantiomers is known as
(*a*) diastereomers (*b*) optical antipodes
(*c*) enantiomers (*d*) racemate

Directions Qs. 16-23: *In each of the following questions there are two statements Assertion (A) is given followed by a corresponding statement of Reason (R) just below it. Of the statements, mark the correct answer:*

(*a*) If both assertion and reason are true and reason is the correct explanation of assertion.
(*b*) If both assertion and reason are true but reason is not the correct explanation of assertion
(*c*) If assertion is true but reason is false
(*d*) If assertion is false but reason is true.

16. Assertion (A): Methylene has sextet of electrons around carbon but bears no formal charge.
Reason (R): Methylene behaves as a nucleophile.

17. Assertion (A): Tertiary carbonium ions are generally formed more easily than primary carbonium ions.
Reason (R): Hyperconjugative as well as inductive effect, due to additional alkyl groups, stabilize tertiary carbonium ions.

18. Assertion (A): The greater the *s* character of the hybrid orbitals, the greater is the electronegativity.
Reason (R): The change in hybridisation is associated with a change in electronegativity.

19. Assertion (A): The *sp* hybrid orbital contains more *s* character and hence it is closer to its nucleus.
Reason (R): Hybridisation has no effect on the bond length and bond enthalpy in organic compounds.

20. Assertion (A): Polarisation of one σ bond caused by the polarisation of adjacent σ bond is referred to as the inductive effect.
Reason (R): The substituents can be classified as electron withdrawing or electron donating groups relative to hydrogen.

21. Assertion (A): Heterolytic fission involves the breaking of a covalent bond in such a way that both the electrons of the shared pair are carried away by one of the atoms.
Reason (R): Heterolytic fission occurs readily in polar covalent bonds.

22. Assertion (A): The stability of carbonium ion is influenced by both resonance and inductive effect.
Reason (R): More the number of alkyl groups, the greater the dispersal of positive charge and therefore, more the stability of carbonium ion is observed.

23. Assertion (A): The order of reactivity of carbonium ions is $2° > 3° > 1°$.
Reason (R): Carbon atom in carbonium ions is in sp^2 state of hybridisation.

ANSWERS

1	2	3	4	5	6	7	8	9	10
(*a*)	(*c*)	(*c*)	(*a*)	(*c*)	(*c*)	(*b*)	(*c*)	(*b*)	(*d*)
11	**12**	**13**	**14**	**15**	**16**	**17**	**18**	**19**	**20**
(*b*)	(*d*)	(*b*)	(*a*)	(*d*)	(*c*)	(*a*)	(*b*)	(*c*)	(*b*)
21	**22**	**23**							
(*b*)	(*a*)	(*d*)							

22 ALIPHATIC AND AROMATIC HYDROCARBONS

Hydrocarbons are the compounds which contains only carbon and hydrogen. They are of the following two types:

(i) Aliphatic Hydrocarbons

(ii) Aromatic Hydrocarbons.

Aliphatic Hydrocarbons

1. ***Bergius Process :*** To obtain a mixture of hydrocarbons powdered coal is hydrogenated at about 750 K and 200-250 atm. pressure in presence of a catalyst, as

$$C + H_2 \xrightarrow[200-250\text{ atm}]{\text{Iron oxide } 750\text{K}} \text{Mixture of Hydrocarbons}$$

2. ***Fischer process :*** When we mix water gas and hydrogen gas and heat the mixture to 480 K under 1 – 10 atm. pressure, a mixture of hydrocarbons with water is obtained.

Aromatic Hydrocarbons

From coal : The aromatic hydrocarbons can be obtained from coal by its destructive distillation or pyrolysis. When coal is heated in absence of air, the various products are obtained after destructive distillation. The products at different temperature range are given below.

Fraction	*Temperature (K)*	*Main Constituent*
Light oil fraction	up to 443 K	Benzene, toluene, xylene
Heavy oil fraction	500 – 550 K	Naphthalene Cl (reso)
Green oil	550 – 650 K	Anthracene
Residue	Non-volatile	Carbon

Alkanes

The hydrocarbons in which all the carbon atoms are linked in a straight chain are called normal alkanes. They have general formula ($C_n H_{2n+2}$). The branched chain alkanes are called *iso* or *neo* depending upon terminal group as

$$\underset{n\text{-pentane}}{CH_3CH_2CH_2CH_3}, \quad \underset{iso\text{-pentane}}{CH_3-CH(CH_3)\ CH_2-CH_3}$$

Preparations of Alkanes : Alkanes can be prepared by the following methods:

1. When we heat sodium salt of carboxylic acid with soda lime, we obtain alkanes, as

$$CH_3COONa + NaOH \rightarrow CH_4 + Na_2CO_3$$

2. **Catalytic reduction :** When alkene or alkynes are hydrogenated in presence of a catalyst, say platinum or nickel, the corresponding alkanes are formed.

$$CH_2 = CH_2 + H_2 \xrightarrow[250°C]{Ni} C_2H_6$$

$$CH_3\ C \equiv CCH_3 + 2H_2 \xrightarrow[250°C]{Ni} CH_3CH_2CH_2CH_3$$

3. **From alkyl halides :** The two alkyl radicals combines to form parent alkane by reacting with metallic sodium in presence of dry ether. This reaction is known as Wurtz reaction.

$$CH_3Br + 2Na + CH_3Br \rightarrow CH_3 - CH_3 + 2NaBr$$

4. **Kolbe's reaction :** The electrolysis of solution of a salt of organic acid yields alkanes, as

$$2CH_3COOK + 2H_2O \rightarrow$$
$$CH_3 - CH_3 + 2CO_2 + 2KOH + H_2$$

Chemical Properties

1. **Halogenation :** Alkanes react with halogens in presence of ultra-violet light, as

$$CH_4 + Cl_2 \rightarrow CH_3Cl + HCl$$

$$CH_3Cl + Cl_2 \rightarrow CH_2Cl_2 \xrightarrow{Cl_2} CHCl_3 \xrightarrow{Cl_2} CCl_4$$

2. **Oxidation Reaction :** In limited supply of air, alkanes form aldehydres or carboxylic acids, as

$$CH_4 + \frac{1}{2}O_2 \xrightarrow{200°C, 100\ atm.} CH_3OH$$

$$C_2H_6 + O_2 \rightarrow CH_3CHO + H_2O$$
$$2C_2H_6 + 3O_2 \rightarrow + 2C_2H_5\ COOH + 2H_2O$$

3. **Nitration:** When made to react with nitric acid, they yield nitro derivative as

$$R - H + HONO_2 \xrightarrow{400°C} RNO_2 + H_2O$$

Nitration can occur only for higher alkanes in which a hydrogen atom is replaced by $-NO_2$ group.

Alkenes

In IUPAC system alkenes are named by replacing *ane* of parent alkane by *ene.* The name of the compound is given by the longest chain containing double bond, as,

$$\underset{\text{2-Butene}}{CH_3CH = CHCH_3}$$

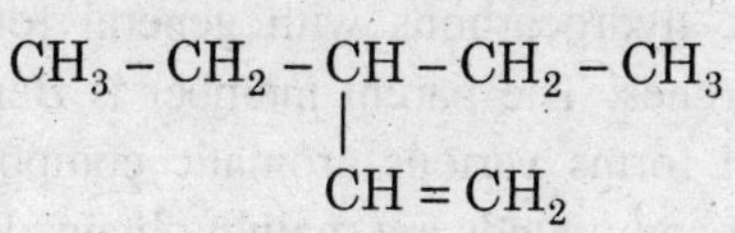

3 ethyl-1-hexane

Preparation : Alkenes can be prepared by the following methods:

1. **From alcohols :** Alcohols can be dehydrated in presence of concentrated sulphuric acid to yield alkenes, as

$$C_2H_5OH \xrightarrow[Heat]{Conc.-H_2SO_4} CH_2 = CH_2 + H_2O$$

2. **From alkyl halides :** Alkyl halides can be dehydrohalogenated in presence of alcoholic potassium hydroxide/zinc to yield alkenes

$$C_2H_5Br + KOH\ (alcoholic) \rightarrow CH_2 = CH_2 + KBr$$
$$CH_2Br - CH_2Br + Zn \xrightarrow{Heat} CH_2 = CH_2 + ZnBr_2$$

Cracking : On large scale alkene can be obtained by cracking the natural gas and high boiling fraction of petroleum.

$$C_8H_{18} \rightarrow C_6H_{14} + C_2H_4$$

Chemical Reactions

1. **Oxidation :** Alkenes can be oxidised by alkaline potassium permanganate solution/ silver to yield alcohols/alkanes as,

$$CH_2 = CH_2 \xrightarrow[\text{or Bayer's Reagent}]{Alk.\ KMNO_4} \underset{\text{glycol}}{CH_2OH - CH_2OH}$$

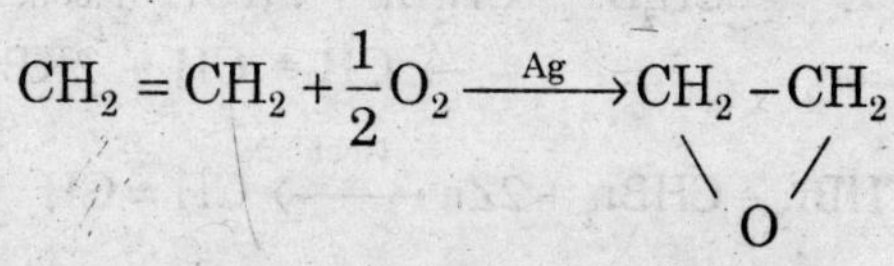

Epoxyethane

2. **Ozonolyis :** Ozone adds up to alkenes to form ozonide which can be hydrolysed to aldehyde or ketone.

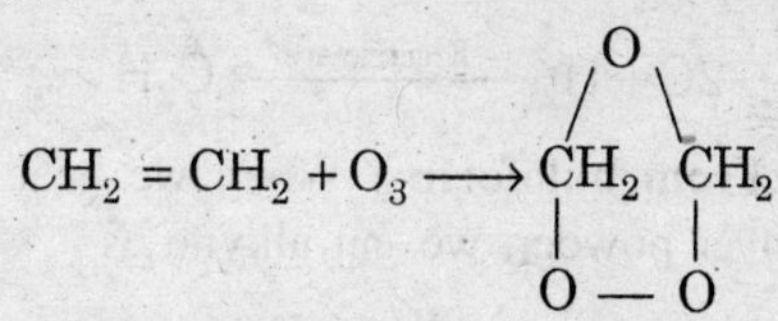

Ozonide

$$+H_2O \xrightarrow{Zn} 2HCHO + H_2O_2$$

3. **Polymerisation :** Alkenes can be polymerised to yield polycompounds as

$$n\ CH_2 = CH_2 \xrightarrow[pressure]{high\ temp.} \underset{\text{Polyethylene}}{(-CH_2 - CH_2 -)n}$$

4. **Addition reactions :** Alkenes are reactive since the π-bond in between carbons is very weak. They take part in electrophilic addition reactions, as,

$$CH_2 = CH_2 + Br_2 \rightarrow CH_2Br - CH_2Br$$

Markownikov Rule (For unsymmetrical double bond) : It states that during the addition to unsymmetrical double bond, the negative part of attracting reagent joins the carbon atom which carries smaller number of hydrogen atoms.

Alkynes

In IUPAC system the *ane* of alkanes is changed to *yne* as.

$$\underset{\text{propyne}}{CH_3CH \equiv CH} \quad \underset{\text{3-methyle-1-butyne}}{CH_3 - \underset{|}{\overset{CH_3}{CH}} - C \equiv CH}$$

Preparation : Alkynes can be prepared by following methods :

1. **From alkyl halides :** Dihalides can be dehydrohalogenated with alcoholic KOH/Zinc to yield alkynes, as

$$CH_2Br - CH_2Br + 2KOH \text{ (alcoholic)} \rightarrow CH \equiv CH + 2KBr + 2H_2O$$

$$CHBr_2 - CHBr_2 + 2Zn \xrightarrow{Heat} CH = CH + 2Zn\ Br_2$$

2. **Berthelot's synthesis :** Carbon can be made to directly combine with hydrogen to yield alkynes, as

$$2C + H_2 \xrightarrow{\text{Electric arc}} C_2H_2$$

3. **From haloform :** When we heat haloform with silver powder, we get alkyne as

$$2CHI_3 + 6Ag \xrightarrow{Heat} CH \equiv CH + 6AgI$$

Chemical Properties : Alkynes undergoes following chemical reactions:

1. **Halogenation :** The addition reaction takes place as

$$CH \equiv CH + Cl_2 \rightarrow CHCl = CHCl \rightarrow CHCl_2 - CHCl_2$$

$$CHCl_2 - CHCl_2 \xrightarrow{Ca(OH)_2} CHCl = CCl_2 + HCl$$

2. **Hydrolysis :** Alkynes can be hydrolysed in presence of heavy metal salts to yield alcohol and to aldehydes.

$$CH \equiv CH + H_2O \xrightarrow{H^+} CH_2 = \underset{\underset{OH}{|}}{CH} \rightarrow CH_3\,CHO$$

3. **Acidic property :** Alkynes react with metals forming acetylides evolving hydrogen.

$$CH \equiv CH + 2Na \rightarrow NaC \equiv CNa + H_2$$

4. **Ozonolysis :** Alkynes undergoes ozonolysis as

$$CH \equiv CH + O_3 \longrightarrow \begin{array}{c} O \\ / \quad \backslash \\ CH - CH \\ | \qquad | \\ O - O \end{array} + H_2O$$

$$\xrightarrow{-} H_2O_3 + \underset{\text{(glyonol)}}{\begin{array}{c} CHO \\ | \\ CHO \end{array}} \xrightarrow{[O]} \underset{\text{Oxalic acid}}{\begin{array}{c} COOH \\ | \\ COOH \end{array}}$$

5. **Polymerisation :** The polymerisation of alkynes takes place as

$$CH \equiv CH + HC \equiv CH \xrightarrow{CuCl} H - C \equiv C - CH = CH_2$$

$$\xrightarrow{C_2H_2} CH_2 = CH - C \equiv C - CH = CH_2$$

$$3CH \equiv CH \xrightarrow{Heat} C_6H_6$$

6. **Alcoholysis :** It reacts with alcohols to form ether which can be further hydrolysed to alcohol and aldehyde as

$$CH \equiv CH + CH_3OH \xrightarrow{KOH} \underset{\text{Methyl vinyl ether}}{CH_3OCH = CH_2}$$

$$CH_3OCH = CH_2 + H_2O \xrightarrow{H^+} CH_3OH + CH_3CHO$$

Aromatic Hydrocarbon

The aromatic hydrocarbons with general formula C_nH_{2n-6} are arenes. The parent member is Benzene (C_6H_6) which forms various aromatic compounds. The aromatic compounds are mainly obtained from distillation of coal and petroleum as by-products.

Structure of Benzene : Benzene contain 3π-bonds which can be shown at alternate position as

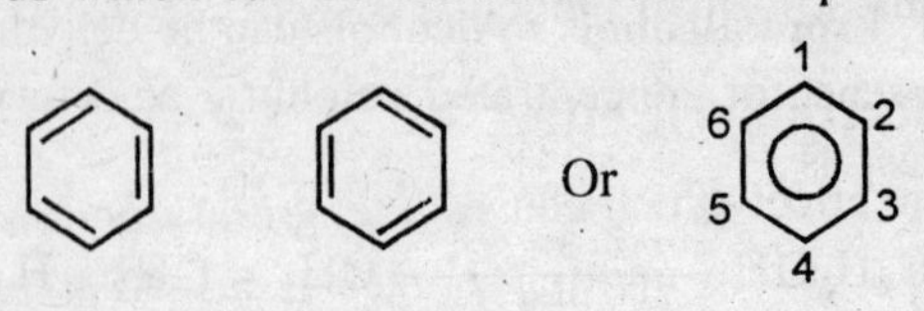

Orientation : The replacement of one hydrogen atom in benzene with other group produces only one derivative as all the six hydrogen atoms are equivalent. But replacement of two hydrogen atoms with other two groups produces three isomers known as ortho (1, 2), meta (1, 3) and para (1, 4).

The determination of the position of new substitutent under the influence of already present substitute is known as *orientation.*

Ortho- and Para-directing groups : –OH, –OR, $–NH_2$, –Cl, –Br –I and $–CH_2Cl$, etc.

Meta-directing groups : $–NO_2$, –CHO, –COOH, –COOR, $–COCH_3$, $–SO_3H$ and –CN.

Nature of Reagents

1. ***Electrophiles*** are electron loving reagents. They are electron deficient and so act as a Lewis acid and seek electrons.

Examples:

H^+; H_3O^+; NO_{2+} (from HNO_3); SO_2OH^+; NO^+; Br^+; BF_3 = $C^+ = O$ and CO_2.

2. ***Nucleophiles*** are nucleus loving reagents. They are electron rich and so, act as a Lewis base and donate electrons.

Examples :

Cl^-; Br^-; I^-; OH^-; CN^-; RO^- (alkoxide ion); ≡ N; NH_3 and $RCOO^-$.

Inductive effect

The minor displacement of bonding electron-pair towards more electronegative atom or group is called inductive effect or I-effect. The electron attracting atom or group *e.g.*, $NO_2 > F > Cl > I > OH > C_6H_5$ causes a negative inductive effect (–I-effect), while the electron releasing or electron repelling substituent *e.g.*, $(CH_3)_3C - > (CH_3)_2\ CH^- > CH_3\ CH_2 - CH_3 - >$ causes positive inductive effect (+I-effect).

Electromeric effect

The effect which causes a temporary polarisation at the seat of a multiple bond by shift of an electron-pair from one atom (or part) to the other under the influence of an electrophilic reagent is called electromeric effect.

$$A = B \xrightarrow{\text{electrophile}} A^+ = \ddot{\bar{B}}$$

Mesomeric effect

The effect which causes extreme polarity on atoms joined by multiple bond owing to a complete transfer of *pi* electron-pair to one or the other atom under the influence of an electrophilic reagent is called mesomeric effect.

$$C = C \xrightarrow{\text{electrophile}} C^+ - \ddot{C}$$

Methods of Preparation : Benzene and its homologues can be synthesized from the following methods:

1. **Polymerization of acetylenes :** Benzene and many of its homologues are produced by passing the vapour of corresponding acetylynic hydrocarbon through a red hot tube (500°C). Thus,

$$\underset{\text{Acetylene}}{3C_2H_2} \longrightarrow \underset{\text{Benzene}}{C_6H_6}$$

2. **Reduction of diazonium salts :** The diazonium salts when treated with alcohol yield the corresponding hydrocarbon.

$$C_6H_5N_2Cl + CH_3\ CH_2OH \longrightarrow C_6H_6 + HCl + CH_3CHO$$

Chemical Properties

1. **Halogenation :** Benzene and its homologues undergoes halogenation with chlorine, bromine either in the benzene nucleus. or in the side chain.

(i) In dark and in presence of a halogen carrier, they react with chlorine to form nuclear halogen derivative :

$$C_6H_6 + Cl_2 \longrightarrow \underset{\text{Chlorobenzene}}{C_6H_5Cl} + HCl$$

$$\underset{\text{Toluene}}{H_3CC_6H_5} + Cl_2 \longrightarrow \underset{\text{Chlorotoluene}}{H_3CC_6H_4Cl} + HCl$$

(ii) When chlorine is passed into boiling toluene, it attacks the side chain and the hydrogen atoms of the methyl group are successively replaced.

$$C_6H_5CH_3 \xrightarrow{Cl_2} C_6H_5CHCl_2 \xrightarrow{Cl_2}$$
Toluene — Benzyl Chloride

$$C_6H_5CHCl_2 \xrightarrow{Cl_2} C_6H_5CCl_3$$
Benzyl chloride — Benzotrichloride

2. **Nitration :** When benzene is treated with the concentrated nitric acid, it undergoes substitution in the benzene ring yielding the nitro compounds.

$$C_6H_5\boxed{H + HO}NO_2 \longrightarrow C_6H_5NO_2 + H_2O$$
Nitrobenzene

$$H_3C_6H_4\boxed{H + HO}NO_2 \rightarrow H_3\ CC_6H_4NO_2 + H_2O$$
Nitrotoluene

3. **Sulphonation :** The benzene hydrocar-bons can be easily sulphonated in the benzene nucleus by heating with the concentrated sulphuric acid.

$$C_6H_5\boxed{H + HO}SO_3H \longrightarrow C_6H_5SO_3H + H_2O$$
Benzene Sulphonic acid

4. **Friedel Craft's reaction :** When benzene and other aromatic hydrocarbons react with alkyl halide in presence of anhydrous aluminium chloride, they form alkylated hydrocarbons.

$$C_6H_6 + RCl \xrightarrow{AlCl_3} C_6H_5R + HCl$$

$$C_6H_6 + CH_3Cl \longrightarrow C_6H_5CH_3 + HCl$$
Toluene

$$C_6H_6 + ClCOCH_3 \longrightarrow C_6H_5COCH_3 + HCl$$
Acetyl Chloride — Acetophenone

$$2C_6H_6 + O_2 \longrightarrow 2C_6H_5OH$$
Phenol

5. **Oxidation :** Benzene is very stable to oxidising agents and is not attacked by hot chromic acid. The homologues of benzene having an aliphatic side chain are oxidised by the oxidising agents.

$$C_6H_5CH_3 + 3O \longrightarrow C_6H_5COOH + H_2O$$

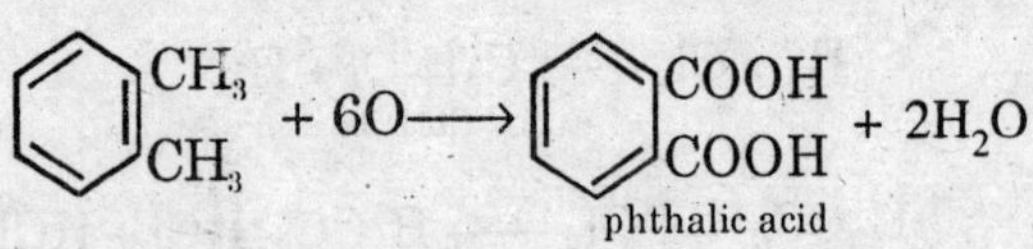

phthalic acid

6. **Addition reaction**

(i) Hydrogenation : On catalytic hydrogenation with nickel, benzene and homologues add three molecules af hydrogen to form hexahydro compound.

$$C_6H_6 + 3H_2 \longrightarrow C_6H_{12}$$
Cyclohexane

(ii) Addition of halogens : In bright sunlight and in the absence of a catalyst, benzene and its homologues add three molecules of chlorine or bromine to form hexahalocyclo hexanes.

$$C_6H_6 + 3Cl_2 \longrightarrow C_6H_6Cl_6$$
Benzene hexachloride (BHC)

Natural Petroleum

It is dark greenish-brown, viscous liquid found deep in the earth's crust. Petroleum has got its name from Latin words, *petra* = rock and *oleum* = oil.

Petroleum is classified as *paraffinic base* if, after the removal of volatile hydrocarbons, the petroleum contains mainly alkanes, and as *asphaltic base* if it contains cycloalkanes.

Knocking : Violent explosion in the cylinder of an internal combustion engine, often due to over-compression of the mixture of air and gasoline, before sparking is called *knocking.* The compound which are added to the gasoline to eliminate knocking are known as *antiknock compounds.* Lead tetraethyl, $Pb(C_2H_5)_4$ is very good antiknock.

Octane Number : The octane number of a fuel tells about the antiknock characteristic of a fuel. High octane number of a fuel means higher antiknock characteristic and vice-versa. *n*-heptane, which has poor antiknock property is assigned at octane number of zero, while 2, 2, 4-trimethyl pentane (or iso-octane) with high antiknock properties is given a value of 100.

TABLE : Typical Fractionation of Petroleum

Sl.No.	*Fraction*	*Boiling point*	*Composition*	*Uses*
1.	Uncondensed gases	0 to 6°C	C_1 to C_4	As fuel, for making carbon for rubber industry, manufacture of gasoline
2.	Crude naphtha:	6 to 150°C	C_5 to C_{11}	
	(*a*) Petroleum ether	6 to 70°C	C_5 to C_7	As a solvent in varnish and rubber industries, for dry cleaning.
	(*b*) Gasoline	70 to 120°C	C_7 to C_{11} and Cycloalkanes	Motor fuel, dry cleaning, for making petrol air gas. In dry cleaning
	(*c*) Solvent naphtha or benzene	120 to 150°C		solvent, as a solvent for varnishes, paints fats and oils.
3.	Kerosene oil	161 to 300°C	C_{11} to C_{16} and aromatics	For illumination, for making oil gas and as domestic fuel.
4.	Gas oil, fuel oil or diesel oil	300 to 330°C	C_{11} to C_{16}	Fuel for diesel engines, for conversion to gasoline by cracking.
5.	Heavy oil:	300°C and above		
	(*a*) Lubricating oils	——	C_{17} to C_{30}	For lubricating machines for making toilet creams, ointments and medicines.
	(*b*) Vaseline (petroleum jelly)	——		
	(*c*) Greases	——		As lubricant.
	(*d*) Paraffin wax		C_{20} to C_{34}	For making candles, water proofing fabrics
6.	Petroleum coke		Essentially graphite	Artificial asphalt fuel, electrodes

Processes in Petroleum Industry

1. ***Cracking :*** Thermal decomposition of higher paraffins to give lower hydrocarbons of low boiling point is termed as cracking, *e.g.*,

$$\underset{\substack{\text{decane}\\(\text{bp} = 200^\circ \text{C})}}{C_{10}H_{22}} \xrightarrow[\text{high temperature}]{\text{high pressure, catalyst}} \underset{\substack{n\text{-pentane}\\(\text{bp}=27^\circ\text{C})}}{C_5H_{12}} + \underset{\text{pentene}}{C_5H_{10}}$$

(i) Liquid phase cracking : It is the cracking in liquid phase. Heavy oil is cracked by heating at a suitable temperature (450 – 550°C) under 7 to 68 atmospheric pressure. High pressure is necessary to maintain the cracker material in liquid phase.

(ii) Vapour phase cracking : Kerosene oil or any other oil with similar bp range is heated to 600°C under 3.5 to 10 atmospheric pressure. Cracking occurs in the vapour phase.

2. ***Polymerisation :*** It is the chemical union of two or more molecules of the same compound to form larger molecules resulting in the formation of a new compound of the same empirical formula of greater molecular mass.

3. ***Alkylantion :*** In this process an alkane adds to an alkene to give a larger molecule of higher octane value.

4. ***Isomerisation :*** In this process *n*-butane available in natural and thermally cracked gas is changed into iso-butane.

5. ***Reforming :*** Sometimes the octane number of a gasoline is raised by catalytic cracking. This is achieved by

(i) Dehydrogenation :

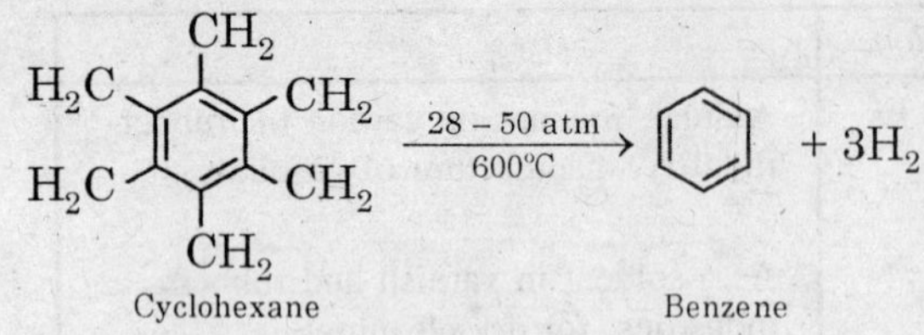

(ii) Cyclisation of alkanes :

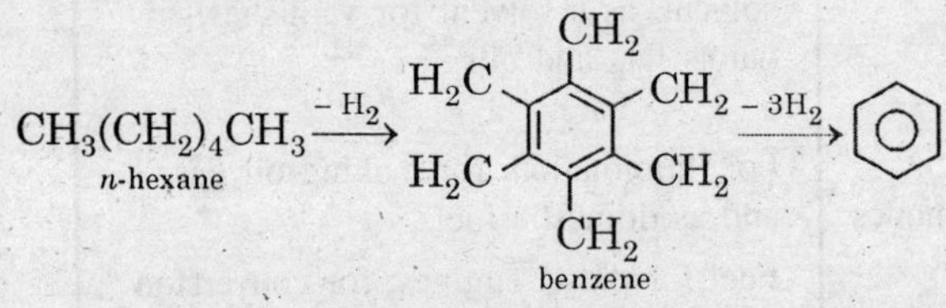

6. ***Refining of gasoline :*** The gasoline obtained by the fractionation of natural petroleum is known as straight run gasoline. This gasoline may have octane values ranging from 20 to 73. The octane value of the gasoline can be improved by adding antiknock compound but the presence of colour, odour and sulphur compounds lowers its response to antiknock compounds. To remove sulphur compounds gasolines is washed with concentrated sulphuric acid or sodium hydroxide. Washed gasoline is further treated with alkaline sodium plumbite to oxidise sulphur compounds to disulphides.

Resonance Energy of Benzene

Benzene's special stability is due to the formation of the delocalized π-molecular orbital.

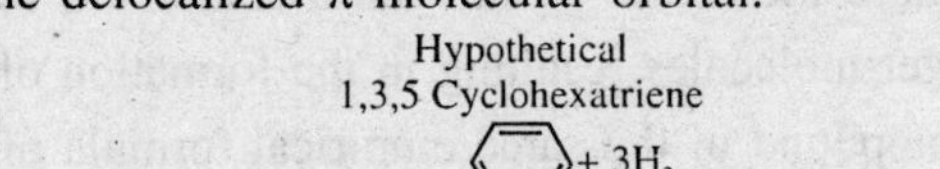

Fig. Heats evolved on hydrogenation of one mole of some cyclic compounds

The hydrogenation of a compound add stability. This stability is sometimes called *Resonance Energy*. "Resonance energy is a measure of how much more stable a resonance hybrid structure is than its extreme resonance structures."

Aromaticity

The aromatic compounds apparently contain alternate double and single bonds in a cyclic structure, and resemble benzene in chemical behaviour. They undergo substitution rather addition reactions. This characteristic behaviour is called *Aromatic character* or *aromaticity*. Aromaticity is, in fact, a property of the sp^2-hybridized planar rings in which the *p*-orbitals (one on each atom) allow cyclic delocalisation of π-electrons.

Criteria for Aromaticity

1. An aromatic compound is cyclic and planar.

2. Each atom in an aromatic ring has a *p*-orbital. These *p*-orbitals must be parallel so that a continuous overlap is possible around the ring.

3. The cyclic π-molecular orbital (electron cloud) formed by overlap of *p*-orbitals must contain $(4n + 2)$ π-electrons where n = integer 0, 1, 2, 3, etc. This is known as Huckel Rule.

Electrophilic Substitution Reactions

Benzene undergoes electrophilic substitution reactions. The benzene ring with delocalised π-electrons is an electron-rich system. It is attacked by electrophiles, giving substitution products. These reactions can be represented as

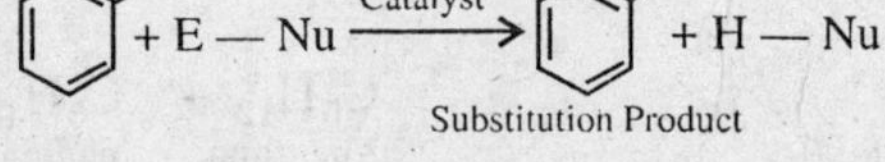

where E^+ is any electrophile and Nu^-: is a nucleophile. Such reactions in which hydrogen atom of the aromatic ring is replaced by an electrophile are called electrophilic aromatic substitution reactions.

General Mechanism : All electrophilic aromatic substitution reactions follow the same three-step mechanism :

Step 1 : Formation of an Electrophile :

E — Nu + Catalyst → E^+ + Nu — catalyst

Step 2 : The electrophile attacks the aromatic ring to form a carbonium ion.

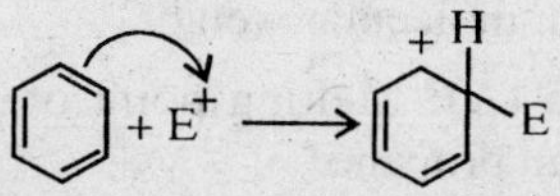

Benzene

The intermediate ion is resonance stabilised. It is a hybrid of the following three structures:

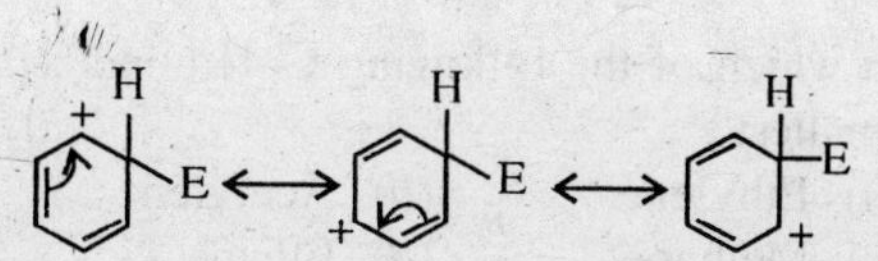

Step 3 : Loss of proton gives the substitution product.

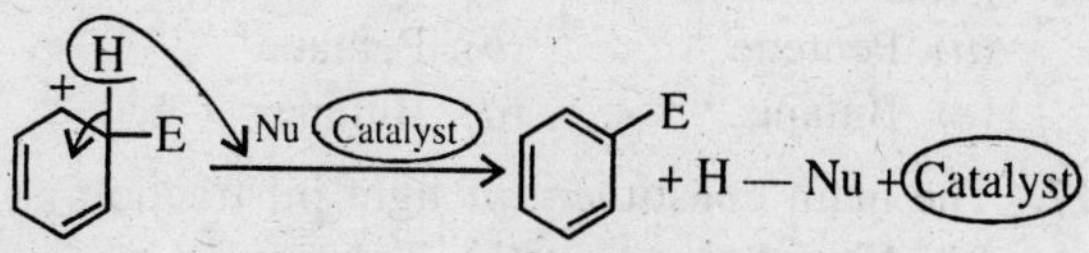

Carbonium ion

Substitution Product

MULTIPLE CHOICE QUESTIONS

1. The major component of natural gas is
(*a*) C_4H_8 (*b*) C_2H_5
(*c*) C_4H_{10} (*d*) CH_3

2. Tetraethyl lead is used as
(*a*) petroleum additive
(*b*) ethylating agent
(*c*) antrepeller
(*d*) catalyst

3. The higher fraction of crude oil burn with
(*a*) smokiest flame (*b*) pinkish flame
(*c*) blue flame (*d*) red flame

4. Cestane number of a fuel having 40% methyl naphthalene and 60% *n*- hexadexane is
(*a*) 50% (*b*) 30%
(*c*) 60% (*d*) fact

5. To which temp. the crude oil is initially heated before its fractional distillation can be carrieid out?
(*a*) 545°C (*b*) 350°C
(*c*) 645°K (*d*) 650°K

6. The compound that is used as an antiknock compound is
(*a*) R_2Pb (*b*) R_2PbBr
(*c*) $(CH_3)_4$ Pb (*d*) $(C_2H_5)_4$ Pb

7. Gasoline is a liquid mixture of hydrocarbons containing .... compounds.
(*a*) C_1 to C_5 (*b*) C_5 to C_{10}
(*c*) $C_8 — C_{15}$ (*d*) $C_{18} — C_{22}$

8. Aromatisation is the conversion of
(*a*) higher hydrocarbon to lower ones
(*b*) mixing of aromatic hydrocarbon
(*c*) a hydrocarbon to benzene
(*d*) None of the above

9. Gas oil is the same as
(*a*) diesel oil (*b*) petrol
(*c*) kerosene oil (*d*) lubricating oil

10. Crude oil is a complex mixture of
(*a*) acids (*b*) alcohols
(*c*) hydrocarbons (*d*) esters

11. A fuel having high octane number containing mainly
(*a*) branched hydrocarbons
(*b*) straight chain alkane
(*c*) alkenes
(*d*) All of the above

12. In which of the following the alkanes are insoluble?
(*a*) Benzene (*b*) CCl_4
(*c*) C_2H_5OH (*d*) CS_2

13. Vapours of *n*-hexane when passed over special catalysts under high pressure and temperautre given benzene. This process is called
(*a*) cyclisation (*b*) aromatisation
(*c*) isomerisation (*d*) crystalization

14. Alkenes are isomeric with
(*a*) cycloalkenes (*b*) cycloalkanes
(*c*) alkanes (*d*) alkenes

15. In which of the following C–H bond will be smallest?
(*a*) Ethylene (*b*) Acetylene
(*c*) Methane (*d*) Ethane

16. Which of the following will show maximum possible isomers?
(*a*) Pentene (*b*) Pentane
(*c*) Butane (*d*) Butene

17. The main constituent of light oil fraction is:
(*a*) Naphthalene (*b*) Toluene
(*c*) Phenol (*d*) Benzene

18. Ethylene when heated with Cl_2 at 723K gives
(*a*) ethylene oxide
(*b*) vinyl chloride
(*c*) ethylene chloride
(*d*) polyethene

19. The compound with maximum boiling point is
(*a*) neopentane (*b*) hexane
(*c*) *n*-pentane (*d*) isopentane

20. Benzene on ozonolysis gives
(*a*) glycol (*b*) glycolic acid
(*c*) acetic acid (*d*) glyoxal

Directions Qs. 21-28: *In each of the following questions there are two statements Assertion (A) is given followed by a corresponding statement of Reason (R) just below it. Of the statements, mark the correct answer:*

(*a*) If both assertion and reason are true and reason is the correct explanation of assertion.
(*b*) If both assertion and reason are true but reason is not the correct explanation of assertion
(*c*) If assertion is true but reason is false
(*d*) If assertion is false but reason is true.

21. Assertion (A): Boiling point of alkanes increases with increase in molecular weight.
Reason (R): van der Waal's forces increase with increase in molecular weight.

22. Assertion (A): Melting point of n-butane is higher than propane.
Reason (R): It is called oscillation effect.

23. Assertion (A): Chlorination of methane gives a mixture of products.
Reason (R): Chlorination proceeds through the formation of carbanion.

24. Assertion (A): Iodination of alkanes is reversible.
Reason (R): Iodination is carried out in presence of iodic acid.

25. Assertion (A): Replacement of carbon atom of alkane by NO_2 group is known as nitration.
Reason (R): Nitration of alkanes follows free radical mechanism.

26. Assertion (A): Oxidation of alkanes generates large heat.
Reason (R): Heat generated by burning coal is used to generate electricity.

27. Assertion (A): Isobutane on oxidation with $KMnO_4$ gives tert-butyl alcohol.
Reason (R): Oxidising agents have no effect on alkanes.

28. Assertion (A): Halogenation of alkanes is catalysed by tetraethyl lead.
Reason (R): Halogenation proceeds through free radical mechanism.

ANSWERS

1	2	3	4	5	6	7	8	9	10
(*d*)	(*a*)	(*a*)	(*c*)	(*d*)	(*d*)	(*b*)	(*c*)	(*a*)	(*c*)
11	**12**	**13**	**14**	**15**	**16**	**17**	**18**	**19**	**20**
(*a*)	(*c*)	(*b*)	(*a*)	(*b*)	(*a*)	(*d*)	(*b*)	(*c*)	(*d*)
21	**22**	**23**	**24**	**25**	**26**	**27**	**28**		
(*a*)	(*b*)	(*c*)	(*c*)	(*d*)	(*b*)	(*c*)	(*d*)		

23 HALOGEN DERIVATIVES

Chloroform

Chloroform is a colourless liquid with a sweat sickly smell and taste. It produces unconsiousness when its vapours are inhaled.

Preparation : Chloroform is prepared on industrial scale by treating ethyl alcohol with a suspension of bleaching powder. The reactions takes place in the following steps:

1. Bleaching powder produces both chlorine and calcium hydroxide

$$CaOCl_2 + H_2O \rightarrow Ca(OH)_2 + Cl_2$$

2. Ethyl alcohol is oxidised by chlorine and acetaldehyde is formed.

$$CH_3CH_2OH + Cl_2 \rightarrow CH_3CHO + 2HCl$$

3. Acetaldehyde react more of chlorine to form chloral.

$$CH_3CHO + 3Cl_2 \rightarrow \underset{\text{Chloral}}{CCl_3CHO} + 3HCl$$

4. Chloral is then hydrolysed by calcium hydroxide to give chloroform.

$$2CCl_3CHO + Ca(OH)_2 \rightarrow 2CHCl_3 + Ca\begin{matrix} HCOO \\ HCOO \end{matrix}$$

Pure chloroform required for use in medicines can also be prepared by heating chloral hydrate with caustic soda solution.

$$CCl_3CH(OH)_2 + NaOH \rightarrow CHCl_3 + HCOONa + H_2O$$

Uses : Chloroform has the following uses:

1. It is used in medicine.
2. It is used as laboratory reagent.
3. It is used as a preservative.
4. It is used as a solvent for fats, waxes.

DDT (Dichlorodiphenyltrichloroethane)

Preparation : When chloroben-zene is treated with chloral in presence of conc. sulphuric acid, the D.D.T. is formed, as

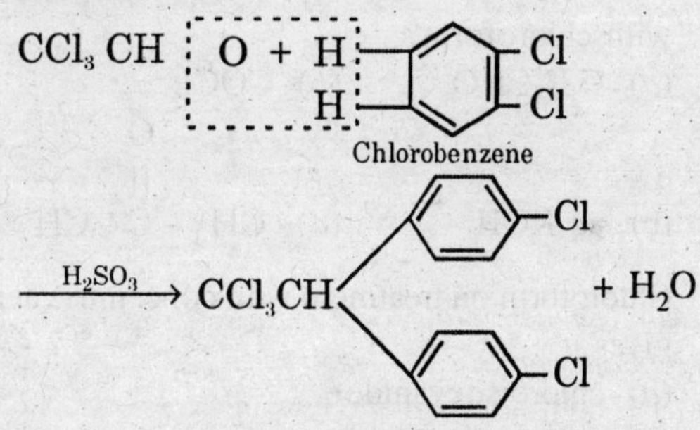

D.D.T.

DDT is a white amorphous powder having a mild smell. It is insoluble in water but it dissolves readily in most of the organic solvents. It is a contact poison and kills the insects primarily by contact in several hours. DDT persists in the living organism and in the plants and is also found to have adverse effect on some species of wild animals.

Tetrachloromethane, CCl_4 or Carbon Tetrachloride

Preparation : (1) By the exhaustive chlorination of Methane

$$CH_4 + 4Cl_2 \xrightarrow{\text{Diffused Sunlight}} CCl_4 + 4HCl$$

(2) By the action of chloroform

$$CHCl_3 + Cl_2 \rightarrow CCl_4 + HCl$$

(3) By chlorinolysis (Cleavage of C—C bond by chlorine) of hydrocarbons at about 400°C and a pressure of 74-100 atmospheres.

$$\underset{\text{Propane}}{CH_3.CH_2.CH_3} + 9Cl_2 \rightarrow \underset{\substack{\text{Carbon}\\\text{tetrachloride}\\\text{(liquid)}}}{CCl_4} + \underset{\substack{\text{Hexachloro}\\\text{ethene}\\\text{(Solid)}}}{C_2Cl_6} + 8HCl\uparrow$$

Properties : Carbon tetrachloride is less reactive than other halogen derivatives.

(1) It is a colourless liquid.

(2) It is non-inflammable and poisonous.

(3) It is insoluble in water but soluble in ethanol and ether.

(4) It is a good solvent for oils, fats and greases.

(5) Action of Hydrogen fluoride in the presence of SbF_5

$$CCl_4 + 2HF \xrightarrow{SbF_5} \underset{\text{Freon-12}}{CCl_2F_2} + 2HCl$$

Freon-12 (dichlorodifluoromethane) is widely used as a refrigerant and propellant in aerosol sprays of all kinds.

Uses : Carbon tetrachloride is used.

(*i*) As fire extinguisher under the name of pyrene. It gives dense, incombustible vapours.

(*ii*) As a solvent for oils, fats, resins and in dry-cleaning.

(*iii*) In medicine for the elimination of hookworms as helminthicide.

(*iv*) As a fumigant.

MULTIPLE CHOICE QUESTIONS

1. Which of the following will give chloretone with chloroform?

(*a*) C_2H_5CHO (*b*) $COCl_2$

(*c*) aq KOH (*d*) $CH_3-\overset{\overset{\displaystyle O}{||}}{C}-CH_3$

2. Chloroform on treatment with conc. nitric acid gives

(*a*) chloroisocyanide

(*b*) CCl_4

(*c*) CCl_3NO_2

(*d*) CCl_3NO

3. Tear gas is

(*a*) ethyl carbonate

(*b*) methylene chloride

(*c*) chloropicrin

(*d*) gasoline

4. A bottle of chloroform when left in open for a few days results

(*a*) formation of phosgene

(*b*) viscous polymer

(*c*) halogenation

(*d*) no change

5. Chloroform when heated with silver powder gives

(*a*) acetylene (*b*) CH_3Cl

(*c*) $AgC \equiv CAg$ (*d*) $C_2H_4Cl_2$

6. Chloroform when heated with aniline and alcoholic caustic alkali gives

(*a*) sweet smelling substance

(*b*) benzyl chloride

(*c*) carbylamine

(*d*) None of the above

7. Chlorobenzene when heated with ammonia in the presence of Cu_2O under pressure and temperature gave

(*a*) aniline (*b*) toluene

(*c*) 2 chloroaniline (*d*) benzyl chloride

8. Chlorobenzene is lesser reactive towards nucleophilic substitution reactions as compared to ethyl chloride because

(*a*) C–Cl bond length is lesser than in ethyl chloride

(*b*) the benzene ring has +I effect

(*c*) the molecular is planar in nature

(*d*) None of these

9. D.D.T. is obtained by the interaction of (*a*) chloral and chlorobenzene

(*b*) chlorine and toluene
(*c*) chlorine and benzene
(*d*) chloral and bromobenzene

10. If chloroform is left exposed to air in the presence of sunlight
(*a*) polymerisation takes place
(*b*) a poisonous gas is formed
(*c*) explosion takes place
(*d*) No reaction takes place

11. When ethyl amine is heated with chloroform and alcoholic KOH a compound with offensive smell is obtained. The compound is
(*a*) a secondary amine
(*b*) an isocyanide
(*c*) a cyanide
(*d*) None of these

12. A compound used as hypnotic is
(*a*) phenyl isocyanide
(*b*) chloroform
(*c*) chloropicrin
(*d*) chloretone

13. The hydrogen atom in chloroform is
(*a*) Acidic (*b*) Basic
(*c*) Neutral (*d*) None of these

14. The halogen atom is least reactive in
(*a*) Chlorobenzene (*b*) Chloroethane
(*c*) 2-Chloropropane (*d*) 3-Chloropropene

15. Which of the following is a camphor substitute?
(*a*) $CHCl_3$ (*b*) CF_2Cl_2
(*c*) $CF_3CHClBr$ (*d*) C_2Cl_6

Directions Qs. 16-23: *In each of the following questions there are two statements Assertion (A) is given followed by a corresponding statement of Reason (R) just below it. Of the statements, mark the correct answer:*

(*a*) If both assertion and reason are true and reason is the correct explanation of assertion.
(*b*) If both assertion and reason are true but reason is not the correct explanation of assertion
(*c*) If assertion is true but reason is false
(*d*) If assertion is false but reason is true.

16. Assertion (A): Boiling point of alkyl halides increases with increase in molecular weight.
Reason (R): Boiling point of alkyl halides are in the order RI > RBr > RCI > RF.

17. Assertion (A): Alkyl halides are soluble in organic solvents.
Reason (R): *p*-dichlorobenzene possesses low melting point.

18. Assertion (A): Chloroform is stored in dark coloured bottles.
Reason (R): Chloroform is used as a fumigant.

19. Assertion (A): Alkyl iodides darken on standing.
Reason (R): Alkyl iodides are prepared by Finkelstein reaction.

20. Assertion (A): H atoms of chloroform is basic.
Reason (R): Chloroform is heavier than water.

21. Assertion (A): Vinyl chloride is reactive than alkyl chloride.
Reason (R): Stability of alkyl halide decreases as the strength of C-X bond decreases.

22. Assertion (A): CCl_4 is not a fire extinguisher.
Reason (R): CCl_4 is insoluble in water.

23. Assertion (A): Iodoform gives precipitate with $AgNO_3$.
Reason (R): Iodoform test is used for confirming the presence of acetaldehyde, methyl ketones and alcohols.

ANSWERS

1	2	3	4	5	6	7	8	9	10
(*d*)	(*c*)	(*c*)	(*a*)	(*a*)	(*c*)	(*a*)	(*d*)	(*a*)	(*b*)
11	**12**	**13**	**14**	**15**	**16**	**17**	**18**	**19**	**20**
(*b*)	(*d*)	(*a*)	(*a*)	(*d*)	(*b*)	(*c*)	(*b*)	(*b*)	(*d*)
21	**22**	**23**							
(*d*)	(*d*)	(*b*)							

24 HYDROXY COMPOUNDS

The class of compounds which are regarded as hydroxy derivatives of hydrocarbons are called alcohols. They are classified *as mono-*, *di-* and *trihydric* alcohols according to the number of hydroxyl groups present in their molecules, which is one, two or three respectively. Alcohols containing four or more hydroxyl groups are known as *polyhydric* alcohols. Thus, we have:

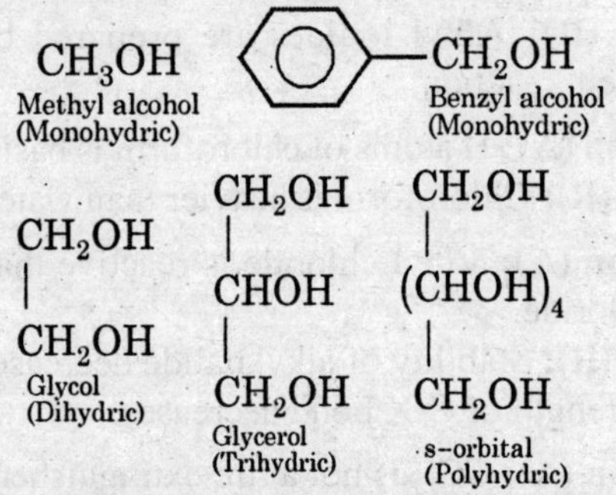

Preparation of Alcohols

1. Grignard Synthesis : It is the most important method for the preparation of primary, secondary and tertiary alcohols. The general reaction may be depicted as follows:

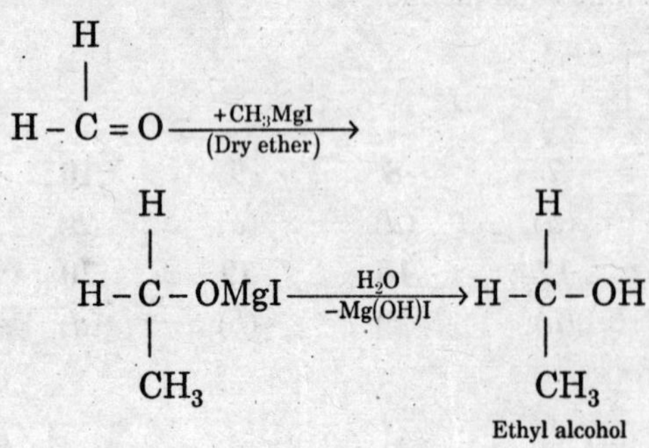

(i) When the carbonyl compound used is *aldehyde other than formaldehyde,* the final product is a *secondary alcohol.*

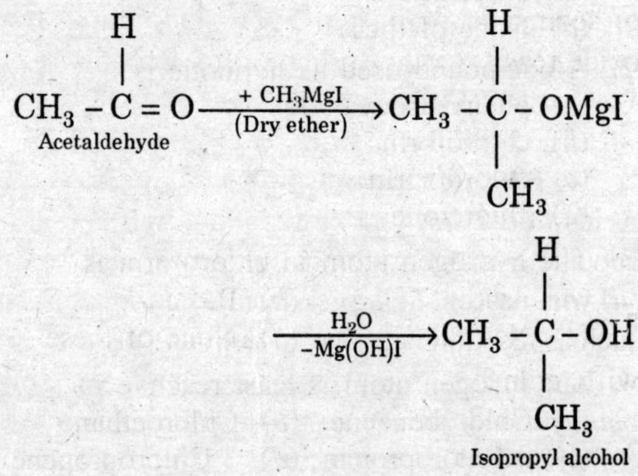

(ii) Use of a *Ketone* to react with Grignard reagent leads to the ultimate formation of a *tertiary alcohol.*

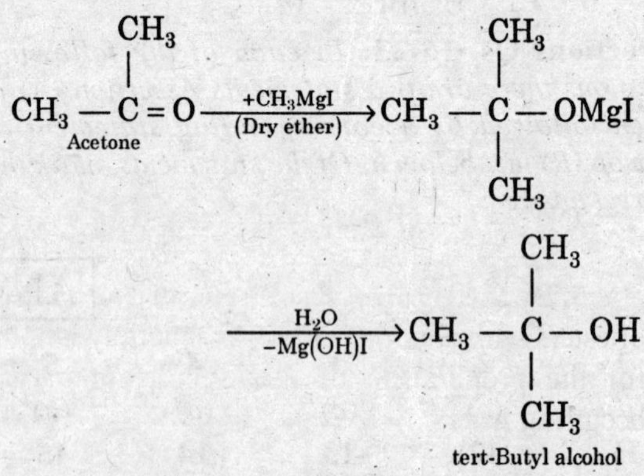

2. By the Hydration of Alkenes : It is now possible to carry out *direct hydration of alkenes* by treating more reactive alkenes with water at low temperature and high pressure in the presence of a catalyst such as $Al_2O_3/Mn_3(PO_4)_2$. For example,

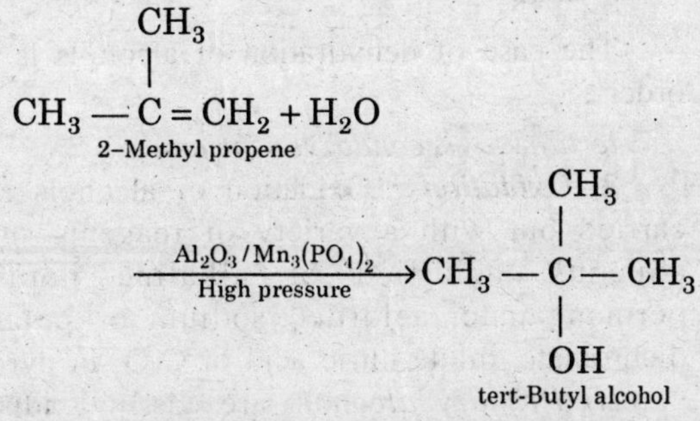

3. By the Hydrolysis of Alkyl Halides : The hydrolysis of alkyl halides with an aqueous solution of sodium or potassium hydroxide or moist silver oxide results in the formation of alcohols.

$$\underset{\text{Ethyl bromide}}{C_2H_5Br} + \underset{\text{aq.}}{KOH} \longrightarrow \underset{\text{Ethyl alcohol}}{C_2H_5OH} + KBr$$

4. By the Reduction of Aldehydes, Ketones, Acids and Esters : The reduction may be brought about *(i)* with hydrogen in the presence of a catalyst, *(ii)* with nascent hydrogen from some suitable source such as reaction between sodium and ethanol or *(iii)* with complex metal hydrides such as sodium borohydride ($NaBH_4$) and lithium aluminium hydride ($LiAlH_4$). For example,

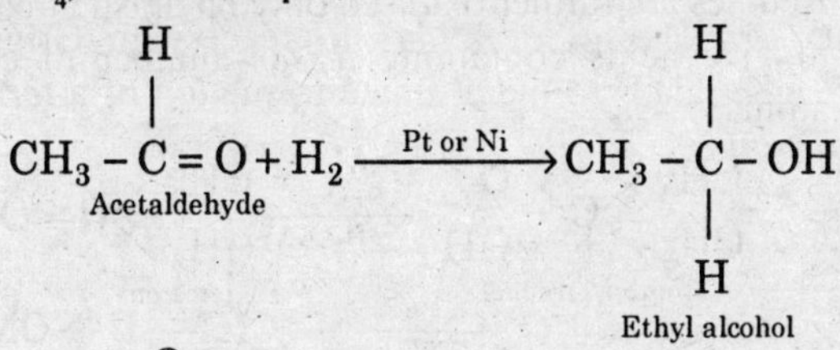

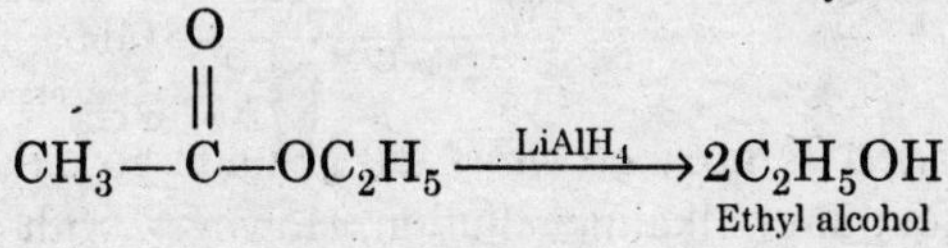

5. By the Hydrolysis of Esters : The hydrolysis of esters with alkali solutions is generally employed for the preparation of alcohols from naturally occurring esters.

$$\underset{\text{Cetyl Palmitate}}{C_{15}H_{31}-COOC_{16}H_{33}} + \underset{\text{aq.}}{NaOH} \longrightarrow \underset{\text{Sodium palmitate}}{C_{15}H_{31}-COONa} + \underset{\text{Cetyl alcohol}}{C_{16}H_{33}OH}$$

Physical Properties of Alcohols : The lower alcohols are colourless liquids having a distinct alcoholic smell and a burning taste. The higher members are colourless, tasteless waxy solids.

The boiling points of isomeric alcohols are in the order *primary > secondary > tertiary.* With the increase in molecular mass, the boiling points increase regularly.

The lower alcohols are highly soluble in water but the solubility decreases with the increase in molecular masses.

Reactions Involving the Cleavage of O—H bond

The general order of reactivity in this type of reaction is : *primary alcohol > secondary alcohol > tertiary alcohol.*

1. *Reaction with metals :* Alcohols react with active metals like sodium, potassium, magnesium and aluminium to liberate hydrogen and form metal *alkoxides.*

$$2\underset{\text{Ethyl alcohol}}{CH_3CH_2OH} + 2Na \longrightarrow \underset{\text{Sodium alkoxide}}{2CH_3CH_2ONa} + H_2$$

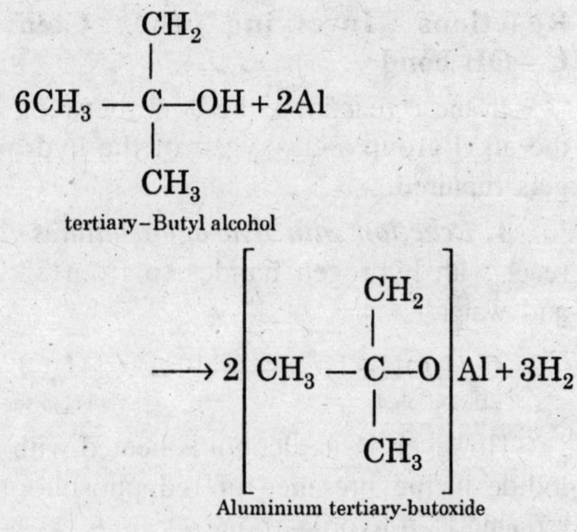

the order of acidic strength alcohols is *primary > secondary > tertiary.*

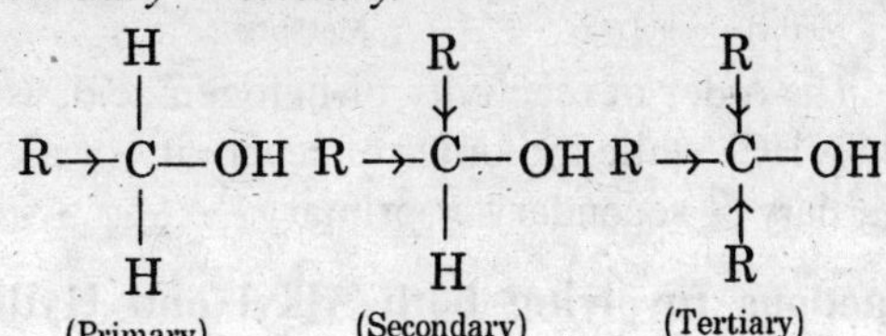

2. *Reaction with Grignard reagents :* Alcohols react with Grignard reagents to produce hydrocarbons as shown below:

$$\underset{\text{Methyl alcohol}}{CH_3O}\,\underset{\text{Ethyl magnesium iodide}}{\boxed{H + C_2H_5}\,MgI} \rightarrow \underset{\text{Ethane}}{C_2H_6} + CH_3OMgI$$

3. ***Reaction with organic acids :*** Alcohols react with organic acids to form esters. This is a slow reversible reaction termed as *esterification.* It is generally carried out in the presence of concentrated sulphuric acid or dry hydrogen chloride.

$$\underset{\text{Acetic acid}}{CH_3O}\boxed{OH+H}\underset{\text{Ethyl alcohol}}{OC_2H_5} \overset{H^+}{\rightleftharpoons} \underset{\text{Ethyl acetate}}{CH_3COOC_2H_5} + H_2O$$

4. ***Reaction with acid halides or acid anhydrides :*** When alcohols are heated with an acid halide or acid anhydride, the hydrogen of –OH group is replaced by acyl group (RCO–) and, once again, an ester is formed.

$$\underset{\text{Acetyl chloride}}{CH_3COCl} + HOC_2H_5 \longrightarrow \underset{\text{Ethyl acetate}}{CH_3COOC_2H_5} + HCl$$

Reactions Involving the Cleavage of C—OH bond

In these reactions, the bond between carbon of the alkyl group and oxygen of the hydroxyl group gets ruptured.

1. ***Reaction with hydrogen halides :*** Alcohols react with hydrogen halides to from alkyl halides and water.

$$\underset{\text{Ethyl alcohol}}{C_2H_5OH} + HCl \xrightarrow{ZnCl_2} \underset{\text{Ethyl chloride}}{C_2H_5Cl} + H_2O$$

However, if an alcohol is heated with hydrogen iodide in the presence of red phosphorus, it gets reduced to a hydrocarbon :

$$\underset{\text{Methyl alcohol}}{CH_3OH} + 2HI \xrightarrow{P} \underset{\text{Methane}}{CH_4} + H_2O + I_2$$

The order of reactivity of halogen acids is HI > HBr > HCl while the order of reactivity of alcohols is tertiary > secondary > primary.

Reactions Involving both Alkyl and Hydroxyl Groups

1. ***Dehydration :*** When alcohols are heated with sulphuric acid, phosphoric acid or boric acid or when their vaporous are passed over heated aluminium oxide, alcohols get dehydrated to form alkenes. For example,

$$\underset{\text{Ethyl alcohol}}{CH_3CH_2OH} \xrightarrow[\text{or } Al_2O_3\ 623\text{ K}]{H_2SO_4(433\text{–}443\text{ K})} \underset{\text{Ethylene}}{CH_2 = CH_2} + H_2O$$

The ease of dehydration of alcohols is in the order :

tertiary > secondary > primary

2. ***Oxidation :*** Oxidation of alcohols can be carried out with a variety of reagents such as aqueous, acidified or alkaline potassium permanganate, acidified sodium or potassium dichromate, dilute nitric acid or CrO_3 in pyridine.

(i) Primary alcohols are easily oxidised to aldehydes and then to acids containing the *same* number of carbon atoms as the original alcohols.

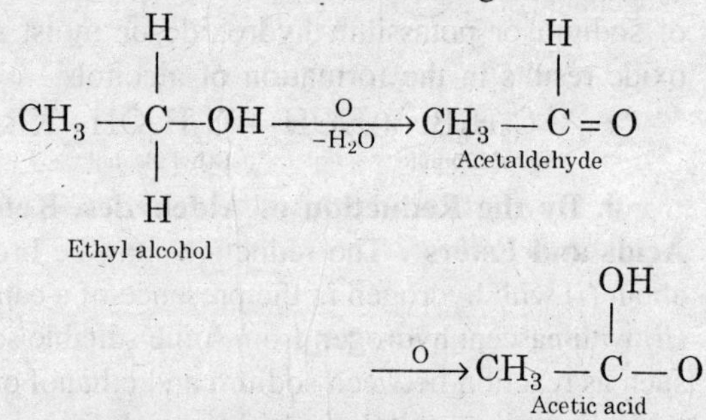

(ii) Secondary alcohols on oxidation first give ketones with the same number of carbon atoms. Ketones are further oxidised only on drastic treatment to give acids containing *lesser* number of carbon atoms.

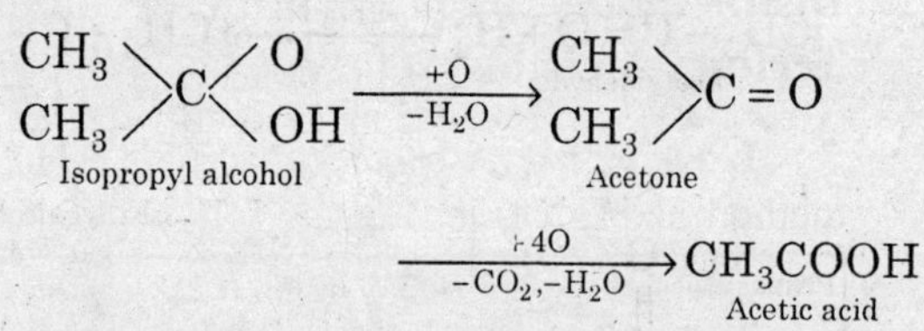

(iii) Tertiary alcohols fail to undergo oxidation in neutral or alkaline solution. However, with acidic oxidising agents they are oxidised to give ketones and acids containing lesser number of carbon atoms than the original alcohols.

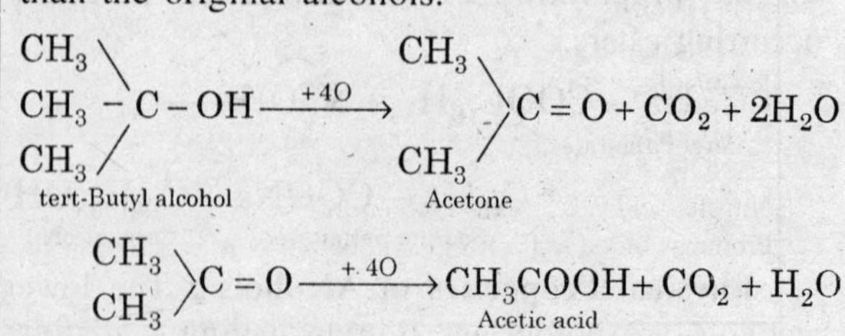

3. ***Action of hot reduced copper :*** Different classes of alcohols furnish different products when their vapours are passed over reduced copper at high temperature (475-575 K).

(i) Primary alcohols are dehydrogenated to form aldehydes.

$$\underset{\text{Ethyl alcohol}}{CH_3-\overset{\overset{H}{|}}{\underset{\underset{H}{|}}{C}}-OH} \xrightarrow{Cu,575\ K} \underset{\text{Acetaldehyde}}{CH_3-\overset{\overset{H}{|}}{C}=O}+H_2$$

(ii) Secondary alcohols are also dehydrogenated to give ketones.

$$\underset{\text{Isopropyl alcohol}}{(CH_3)_2C(H)OH} \longrightarrow \underset{\text{Acetone}}{(CH_3)_2C=O}+H_2$$

(iii) Tertiary alcohols are dehydrated to give alkenes.

$$\underset{\text{tert-Butyl alcohol}}{CH_3-\overset{\overset{CH_3}{|}}{\underset{\underset{CH_3}{|}}{C}}-OH} \longrightarrow \underset{\text{Iso-butylene}}{CH_3-\overset{\overset{CH_3}{|}}{C}=CH_2}+H_2O$$

Distinction between Primary, Secondary and Tertiary Alcohols

1. ***Victor Meyer's method :*** This is an indirect method and is carried out as follows:

Primary alcohol	Secondary alcohol	Tertiary alcohol
RCH_2OH	R_2CHOH	$R_3C\text{–}OH$
↓ $P + I_2$	↓ $P + I_2$	↓ $P + I_2$
RCH_2I	R_2CHI	R_3CI
↓ $AgNO_2$	↓ $AgNO_2$	↓ $AgNO_2$
RCH_2NO_2	R_2CHNO_2	R_3CNO_2
↓ ONOH Nitrous acid	↓ HONO Nitrous acid	↓ HNO_2 No action
$R\text{-}C\text{–}NO_2$ ‖ NOH	$R_2C\text{–}NO_2$ | NO	
Nitrolic acid Produces blood red colour with alkali	Pseudo nitrol blue colour	Colourless solution

2. ***Reaction with hot reduced water :*** When the vapours of alcohols are passed over hot reduced copper at 475-575 K, primary and secondary alcohols lose hydrogen to form aldehydes and ketones respectively, while tertiary alcohols get dehydrated to form alkenes.

Methyl alcohol

(i) As a solvent for paints, varnishes, celluloid, etc.

(ii) In the production of formaldehyde which is itself used in the manufacture of plastics.

(iii) For denaturing ethyl alcohol, *i.e.* rendering it unfit for drinking purposes.

(iv) As an antifreeze for automobile radiators.

(v) As a component of motor-spirit blends.

Ethyl Alcohol (Ethanol or Methyl Carbinol C_2H_5OH)

Ethyl alcohol is the alcohol of wine, beer, whiskey, and similar beverages. It is often referred to as simply alcohol. Ethyl alcohol is also known as *Grain Alcohol*, since it can be prepared from starchy grains.

Preparation : Molasses is the dark brown coloured mother liquor left after the crystallisation of sucrose from con-centrated sugar-cane juice. It contains about 50% uncrystallisable cane sugar, glucose and fructose.

$$\text{Ethanol} \xrightarrow{\text{distillation}} 90\%\ C_2H_5OH \xrightarrow{\text{rectification}} \underset{\text{(Rectified spirit)}}{93-95\%\ C_2H_5OH}$$

Chemical reactions involving addition of yeast

$$\underset{\text{Sucrose}}{C_{12}H_{22}O_{11}}+H_2O \xrightarrow{\text{invertase}} \underset{\text{Glucose}}{C_6H_{12}O_6}+\underset{\text{Fructose}}{C_6H_{12}O_6}$$

$$\underset{\text{Glucose or Fructose}}{C_6H_{12}O_6} \xrightarrow{\text{Zymase}} \underset{\text{Ethyl alcohol}}{2C_2H_5OH}+2CO_2\uparrow$$

(2) Manufacture from Starch : Wheat, barley, rice, maize and potatoes are the important starchy materials used for this purpose

$$2(C_6H_{10}O_5)_n + nH_2O \xrightarrow[50-60°C]{\text{Malt extract (diastase)}} 2C_{12}H_{22}O_{11}$$

Mash (Starch) — Maltose

$$C_{12}H_{22}O_{11} + H_2O \xrightarrow{\text{Maltase}} 2C_6H_{12}O_6 \xrightarrow{\text{Zymase}} 2C_2H_5OH + 2CO_2\uparrow$$

Maltose — Glucose

Uses:

(i) As a solvent for dyes, drugs, tinctures paints, varnishes, etc.

(ii) As a preservative for biological specimens.

(iii) In scientific apparatus like thermometers and spirit levels.

(iv) As a starting material for the preparation of ether, chloroform, iodoform, acetaldehyde and ethyl chloride.

(v) As an important reaction medium and extracting solvent in organic chemistry.

Absolute Alcohol (100% pure ethyl alcohol) Rectified spirit (95.87% ethyl alcohol).

Preparation

(i) Laboratory method : Rectified spirit is kept in contact with quicklime (CaO) for 24 hours and then distilled over it. Final traces of water are removed by adding anhydrous $CuSO_4$ (white in colour) till it ceases to turn blue. It is then distilled to get absolute alcohol.

(ii) Industrial Method : By mixing rectified spirit with excess of benzene followed by distillation when first ternary mixture (74.1% benzene + 18.4% ethanol + 7.4% water) is distilled at 64.8°C, then binary mixture (67.6% benzene + 32.4% ethanol) at 68.2°C and finally absolute alcohol at 78.3°C.

Glycerol

The glycerol is the only one simple and most important trihydric alcohol found mainly in all the animals and vegetable oils and fats.

Methods of Preparation : It can be prepared from the following methods :

1. **From fats or oils :** Generally all the fats and oils contains the triesters of glycerol with the organic acids, thus on hydrolysis with sodium hydroxide they yield glycerol, as

$$\begin{array}{l}CH_2OOCR\\|\\CHOOCR\\|\\CH_2OOCR\end{array} + 3NaOH \longrightarrow \begin{array}{l}CH_2OH\\|\\CHOH\\|\\CH_2OH\end{array} + RCOONa$$

Glycerol

2. **From sugar :** The glycerol can also be prepared from the fermentation of sugar (molasses) in presence of sodium sulphate.

$$C_6H_{12}O_6 \xrightarrow[NaSO_4]{} C_3H_5(OH)_3 + CH_3CHO + CO_2$$

Glycerol

Physical Properties : Glycerol is a colourless syrupy liquid having the sweet taste. It is hygroscopic and mixes with water and in- soluble in ether. It also mixes with alcohols.

Chemical Properties

1. **With sodium :** Glycerol react with sodium to form the mono sodium glycerolate, as

$$\left\{\begin{array}{l}CH_2OH\\|\\CHOH\\|\\CH_2OH\end{array} + Na \longrightarrow \begin{array}{l}CH_2ONa\\|\\CHOH\\|\\CH_2OH\end{array}\right\}$$

Mono Sodium Glycerolate

2. **With acids :**

(i) With HCl : When it is made to react with concentrated hydrochloric acid then it yields the glycerin dichlorohydrin.

$$\begin{array}{l}CH_2OH\\|\\CHOH\\|\\CH_2OH\end{array} + HCl \longrightarrow \begin{array}{l}CH_2Cl\\|\\CHOH\\|\\CH_2OH\end{array} \xrightarrow{HCl} \begin{array}{l}CH_2Cl\\|\\CHOH\\|\\CH_2Cl\end{array}$$

G. Monochlorohydrin G. Dichlorohydrin

(ii) With nitric acid: When it is made to react with nitric acid it form glycerol trinitrate.

$$\begin{array}{l}CH_2OH\\|\\CHOH\\|\\CH_2OH\end{array} + 3HONO_2 \longrightarrow \begin{array}{l}CH_2ONO_2\\|\\CHONO_2\\|\\CH_2ONO_2\end{array} + 3H_2O$$

G. Trinitrate

3. **Dehydration :** Glycerol looses two molecules of water when heated with potassium hydrogen sulphate to form acrolein.

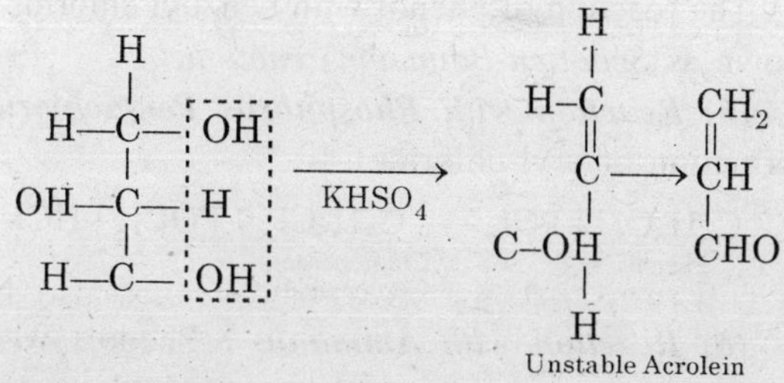

Phenols

"Phenols are compounds containing an —OH group attached directly to an aromatic ring". Like alcohols, they may be monohydric or polyhydric according to the number of —OH groups that they contain. Phenols are usually named by common system or as derivatives of the parent phenol (C_6H_5OH). Examples are:

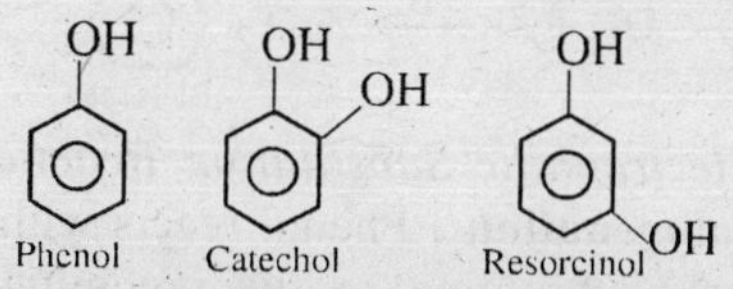

Acidic nature of phenol : The phenoles differs from aliphatic alcohols as they are acidic in nature. This peculiar behaviour of phenole is attributed to the fact that the OH group is attached to double bonded carbon atom instead of single bonded in the case of alcohols. Due to resonance, oxygen atom acquires positive charge. This helps in the separation of hydrogen of OH bond in H^+ ion. And therefore phenoles are Acidic in Nature.

Physical properties : The phenoles are colourless crystalline solids or liquids having a characteristics odour. They are steam volatile. Phenoles and cresols are sparingly soluble in water while *di*-trihydric phenoles are fairly soluble and they dissolved readily in organic solvents.

Preparation :

From chlorobenzene (Dow's Process): This involves the hydrolysis of chlorobenzene with aqueous NaOH at high temperature and pressure followed by treatment with dilute HCl.

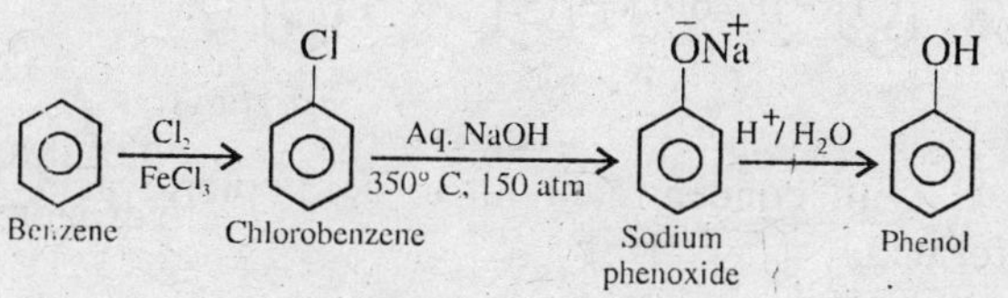

From Cumene Process : This involves air-oxidation of cumene (Isopropyl-benzene) followed by treatment with dilute HCl

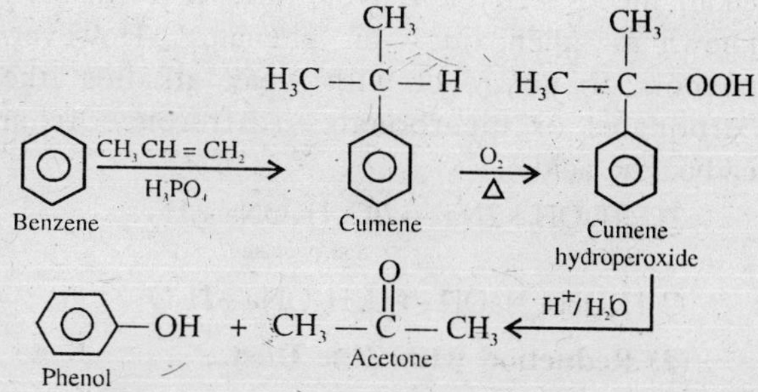

This process accounts for 80% of the total world production of phenol.

From Sodium Benzene Sulphonate : This involves fusion of sodium benzene-sulphonate with solid NaOH at 300°C followed by treatment with dil. HCl.

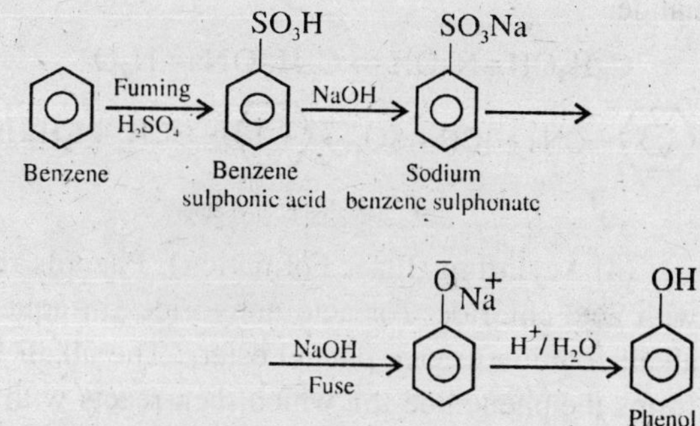

From Benzene Diazonium Salts : This reaction can be performed easily in the laboratory and simply requires warming a solution of benzene diazonium chloride, prepared from aniline, on a water bath at 50°C.

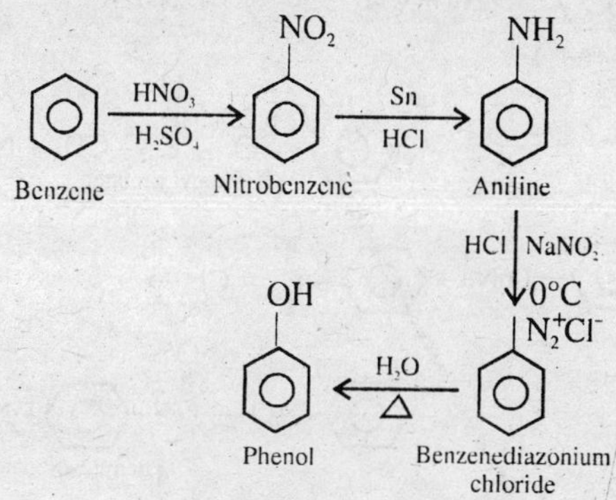

Chemical Properties

(1) Acidic Character (Salt formation) : Unlike alcohols, phenols are acidic in nature and react with alkali metals and their hydroxides to form salts known as 'phenoxides' or 'phenates'. However, phenols do not react with weak alkalies like carbonates or bicarbonates (difference from carboxylic acids).

$$\underset{\text{Phenol}}{2C_6H_5OH} + 2Na \rightarrow \underset{\text{Sod. phenate}}{2C_6H_5ONa} + H_2$$

$$C_6H_5OH + NaOH \rightarrow C_6H_5ONa + H_2O$$

(2) Reduction with Zinc Dust :

$$\underset{\text{Phenol}}{C_6H_5OH} + Zn \xrightarrow{\text{distill.}} \underset{\text{Benzene}}{C_6H_6} + ZnO$$

(3) Alkylation : (Ether formation).

(a) Phenols react with alkyl halides in alkali solution to form phenyl ethers. The alkali first forms the phenoxide ion which then reacts with the alkyl halide.

$$C_6H_5OH + NaOH \rightarrow C_6H_5\overset{-}{O}\overset{+}{Na} + H_2O$$

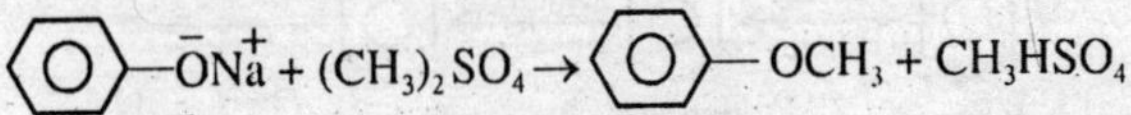

Anisole
(Methyl phenyl ether)

(4) Acylation (Ester Formation): Phenols react with acid chlorides (or acid anhydrides in aqueous alkali solution to give phenyl esters. The alkali first forms the phenoxide ion which then reacts with the acid chloride.

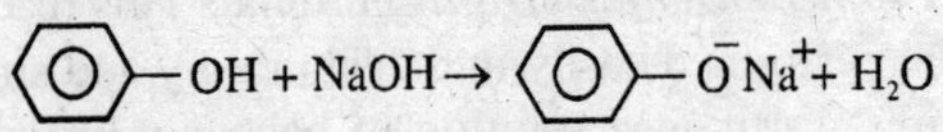

Phenol — Sod. Phenoxide

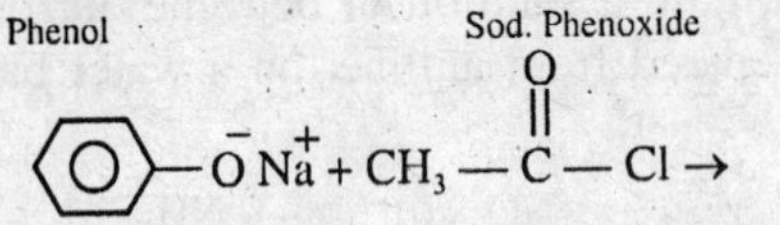

Sod. Phenoxide

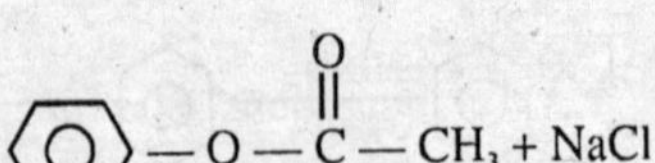

Phenyl acetate

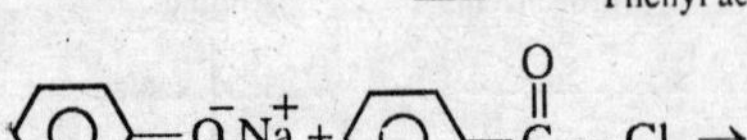

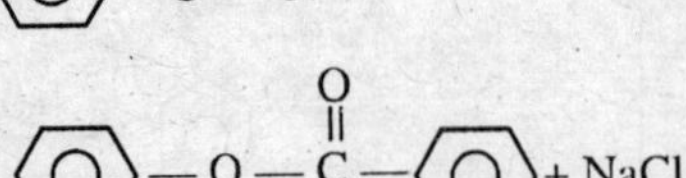

Phenyl benzoate

The reaction of phenol with benzoyl chloride is known as *Schotten-Baumann reaction.*

(5) Reaction with Phosphorus Pentachloride (formation of aryl chloride).

$$\underset{\text{Phenol}}{C_6H_5OH} + PCl_5 \rightarrow \underset{\substack{\text{Chlorobenzene} \\ \text{(Poor yield)}}}{C_6H_5Cl} + POCl_3 + HCl$$

(6) Reaction with Ammonia : Phenols when heated with ammonia in presence of anhydrous zinc or calcium chloride give the corresponding aniline.

$$\underset{\text{Phenol}}{C_6H_5OH} + HNH_2 \xrightarrow{ZnCl_2} \underset{\text{Aniline}}{C_6H_5NH_2} + H_2O$$

(7) Reaction with Zinc Dust : When phenol is distilled with zinc dust, low yield of benzene is obtained.

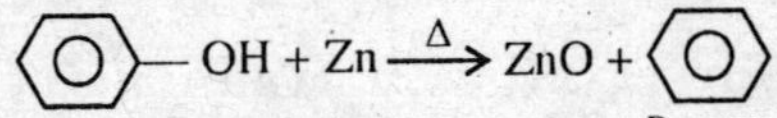

Benzene

(8) Electrophilic Substitution Reactions :

(i) Halogenation : Phenol reacts with bromine water (aqueous bromine) to give precipitate of 2, 4, 6-tribromophenol. Chlorine reacts in the same way.

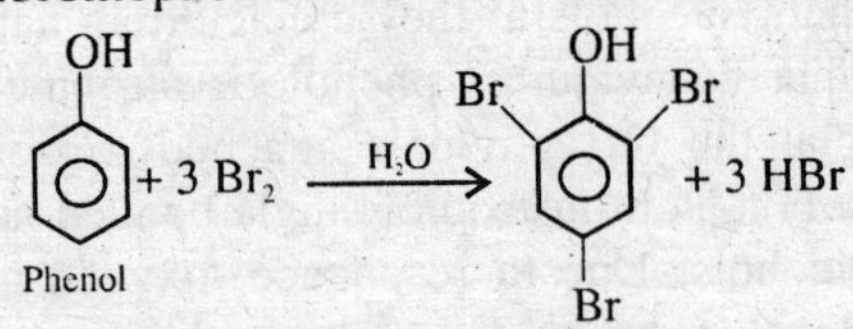

Phenol — 2,4,6-Tribromophenol

In fact, phenols are so reactive towards bromine water that even substituent like —SO_3H, —NO_2,—COOH, etc. if present, in *o*- and *p*-positions, may be replaced by halogens.

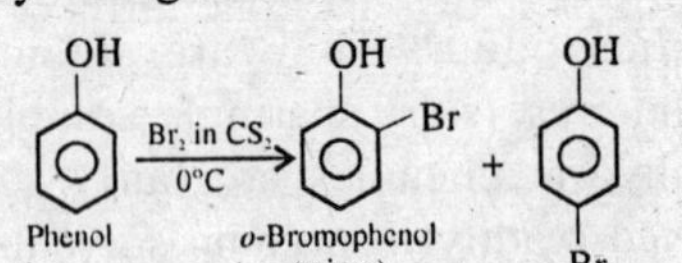

Phenol — *o*-Bromophenol (minor) — *p*-bromophenol (major)

(ii) Nitration : Phenol reacts with dilute nitric acid to give a mixture of *o*- and *p*-nitrophenol.

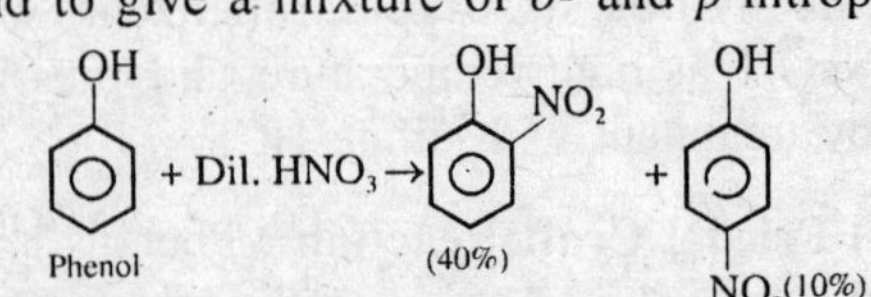

Phenol — (40%) — (10%)
o- and *p*- nitrophenol

with concentrated nitric acid, picric acid is formed.

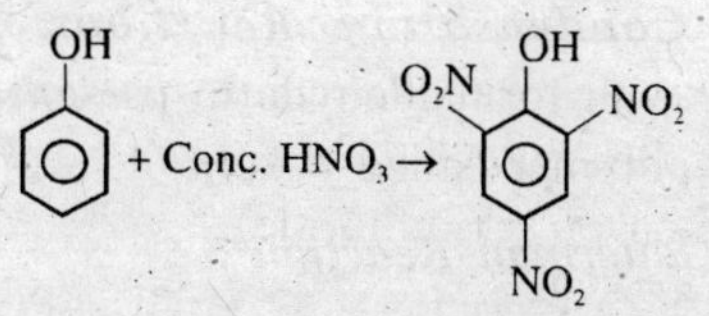

Picric acid
(2,4,6-trinitrophenol)

The yield is low because of excessive oxidation of phenol and hence picric acid is always prepared indirectly

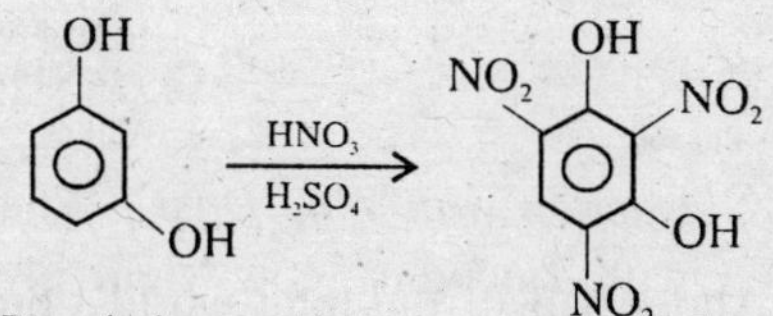

Resorcinol

2,4,6-trinitro resorcinol
(Styphinic acid)

(iii) Sulphonation : When phenol is treated with concentrated sulphuric acid at 20°C, *o*-phenolsulphonic acid is the main products. At 100°C, *p*-phenolsulphonic acid is the main product.

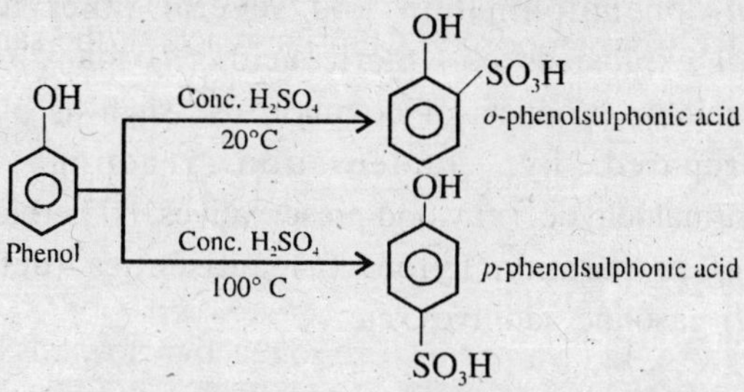

(iv) Alkylation : When phenol is warmed with an alkene or alcohol in the presence of H_2SO_4, a mixture of *o*- and *p*-alkylphenol is formed.

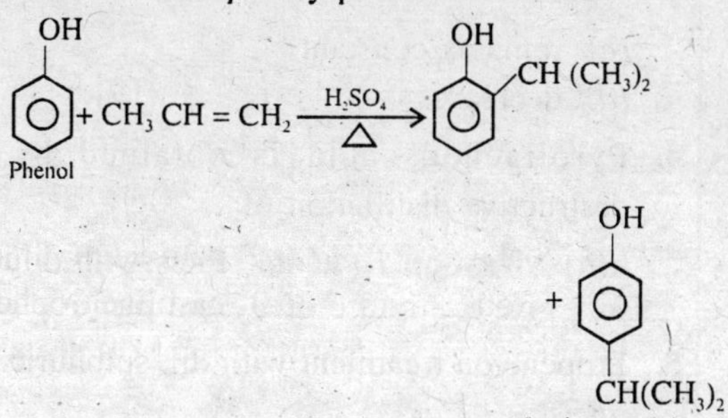

o & *p* isopropylphenol

(v) Friedel Craft Reaction : Phenols undergo Friedel-Craft alkylation and acylation forming predominantly *para*-isomers. However, the yields are poor presumably due to overactivity of phenols.

(a) Alkylation :

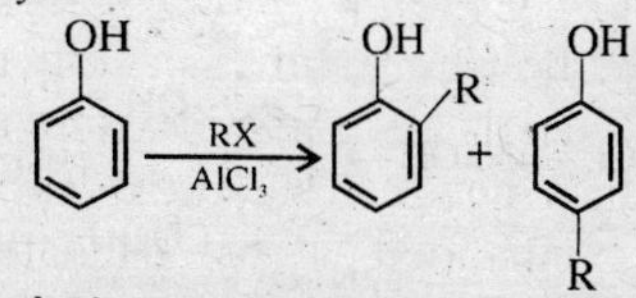

(b) Acylation :

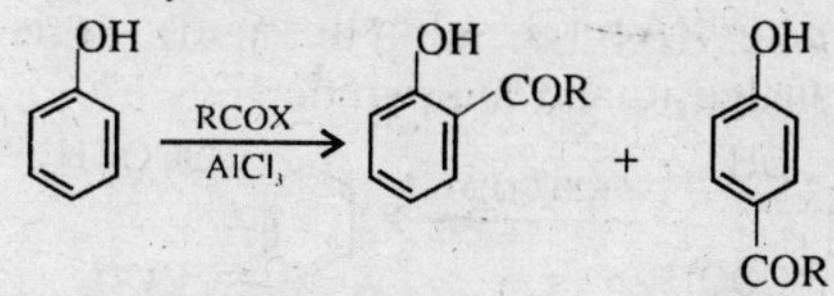

(9) Hydrogenation : Phenols when hydrogenated in presence of nickel catalyst at about 150-200°C give the corresponding cyclohexanol

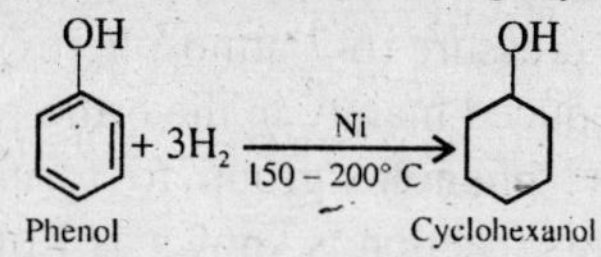

(10) Fries Rearrangement :

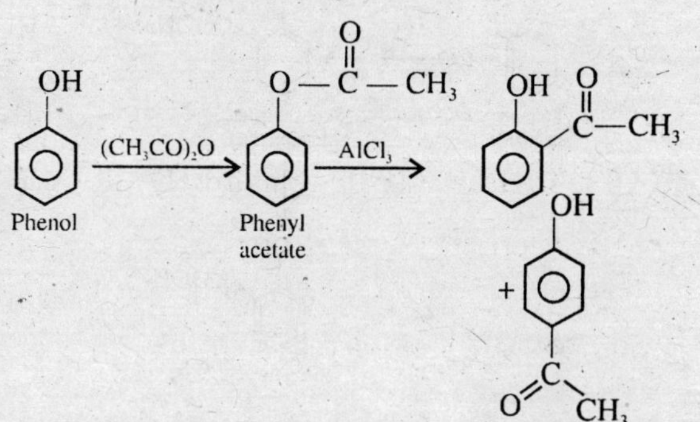

o- and *p*-hydroxyacetophenone

The Fries rearrangement is also given by other phenols and introduces —COR group in the *ortho*- and/or *para*-position.

(11) Reimer-Tiemann Reaction : This involves the treatment of phenol with chloroform in aqueous sodium hydroxide solution followed by acid-hydrolysis.

Salicyaldehyde is formed. If carbon tetrachloride is used in place of chloroform, salicylic acid is formed.

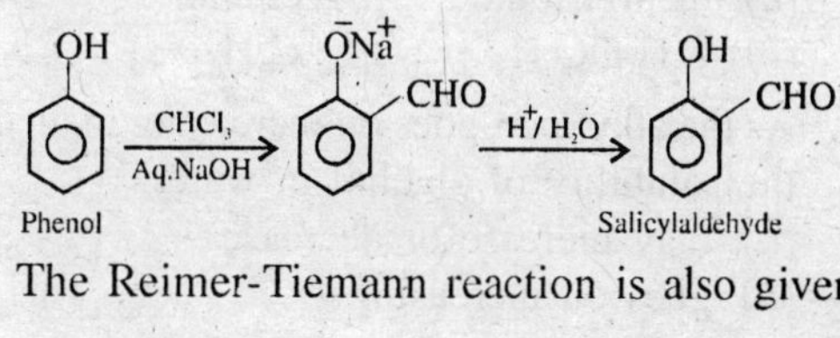

The Reimer-Tiemann reaction is also given by

other phenols and introduces —CHO group in the ortho position.

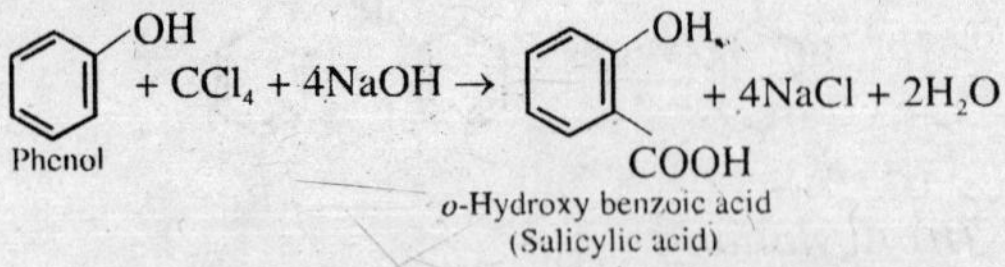

(i) Aspirin (Acetyl salicylic acid): It is an important analgesic and antipyretic

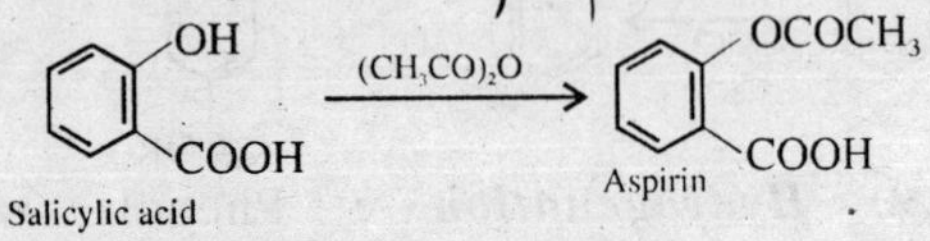

(12) Kolbe Reaction : When sodium salt of a phenol is heated with carbon dioxide gas at 120-140°C under pressure (6-7 atmospheres), a carboxyl group is introduced mainly in the ortho position with respect to the phenolic group to form *o*-hydroxy benzoates. This reaction is known as Kolbe reaction.

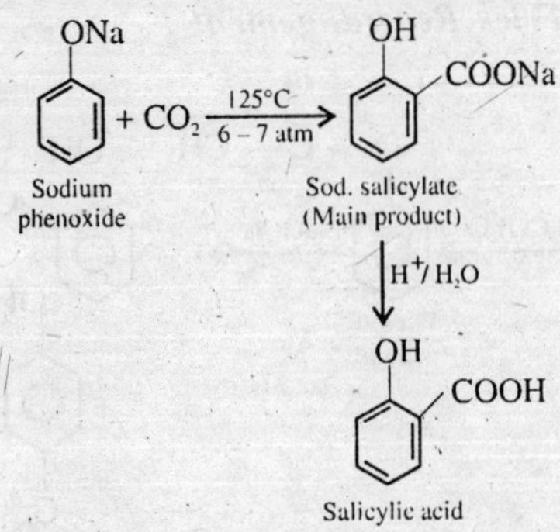

(13) Condensation Reactions : Phenol condenses with formaldehyde in presence of acids to form a polymer called bakelite.

(14) Gatterman Reaction :

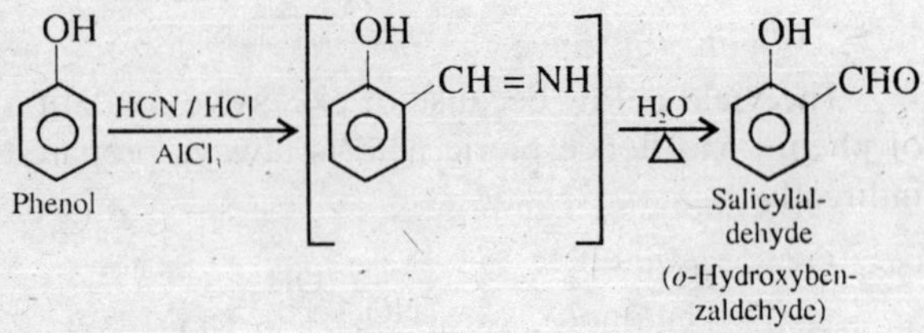

Uses :

(i) About half of the total world's production of phenol (two billion pounds) is used for making phenol-formaldehyde resins e.g. Bakelite.

(ii) Phenol is used as starting material for *(a)* pharmaceutical products such as salol, aspirin (analgesic and antipyretic compound), *(b)* phenolphthalein and several other dyes, *(c)* explosives e.g, picric acid, *(d)* many other valuable products of common use such as plastic prepared by condensation reaction with formaldehyde, *(e)* wood preservatives, *(f)* herbicides *(g)* preparation nylon, *(h)* antiseptics (dettol), *(i)* gasoline additives, etc.

MULTIPLE CHOICE QUESTIONS

1. Secondary alcohol can be prepared from Grignard reagent by reacting Grignard reagent with
(a) ethoxyethane *(b)* methanal
(c) propanal *(d)* facetaldehyde

2. Ethyl alcohol when treated with red P and hydrogen iodide gives
(a) methyl iodide *(b)* ethane
(c) $C_2H_5OC_3H_4$ *(d)* C_2H_4

3. As the alkyl part goes on increasing in alcohol the solubility of alcohol in water
(a) may increase or decrease
(b) goes on increasing
(c) remains constant
(d) decreases

4. Pyroligenous acid is obtained by the destructive distillation of
(a) coal *(b)* petroleum
(c) wood *(d)* gasoline

5. Propene on treatment with dil. sulphuric acid gives
(a) 1-propanol
(b) 2-propanol
(c) Both *(a)* and *(b)*
(d) None of the above

6. The only primary alcohol that can be prepared by the indirect hydration of alkenes is
   (a) methyl alcohol (b) isobutyl alcohol
   (c) propyl alcohol (d) ethyl alcohol

7. Which will have the maximum solubility in water?
   (a) Tertiary butyl alcohol
   (b) Isobutyl alcohol
   (c) Methanol
   (d) 1 butanol

8. Alcohols are isomeric with
   (a) ethers
   (b) esters
   (c) carbonylic acids
   (d) aldehydes

9. The main constituents of pyroligenous acids are
   (a) CH_3COOH, CH_3OH and CH_3COCH_3
   (b) CH_3OH, C_2H_5OH and acetone
   (c) C_2H_5OH, CH_3COCH_3 and CH_3COOH
   (d) None of the above

10. Primary alcohols can be prepared from alkenes by
    (a) mercuration and demercuration of alkenes
    (b) direct hydration of alkenes
    (c) diboration of alkenes
    (d) All of the above

11. Reduction of which of the following will give only ethyl alcohol?
    (a) propyl methanoate
    (b) methyl methanoate
    (c) formic acid
    (d) ethyl acetate

12. Which of the following alcohols will not undergo acetylation with acetic anhydride directly in the presence of pyridine?
    (a) isobutyl alcohol
    (b) secondary butyl alcohol
    (c) tertiary butyl alcohol
    (d) None of the above

13. Tertiary alcohols are resistant to oxidation due to
    (a) not having α-hydrogen atom
    (b) greater stearic hindrance
    (c) large +1 effect of allyl group
    (d) All of the above

14. The reaction

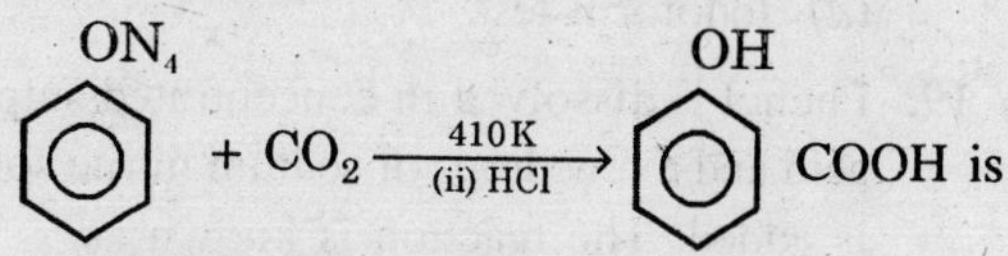

    (a) Reimer Timann reaction
    (b) Wurtz reaction
    (c) Van't Hoff reaction
    (d) Kolbe's reaction

15. Which of the following is the most acidic?

(a) 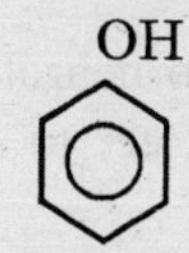

(b) 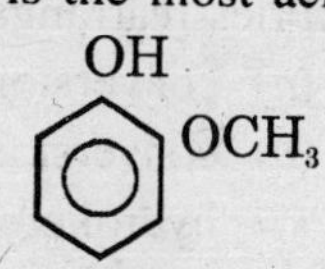

(c) 

(d) 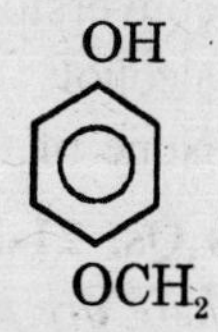

16. Which of the following is a weaker acid than phenol?

(a) 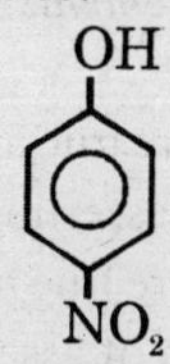

(b) 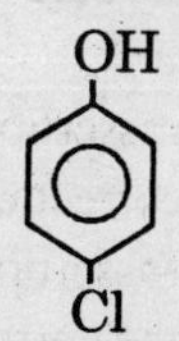

(c) 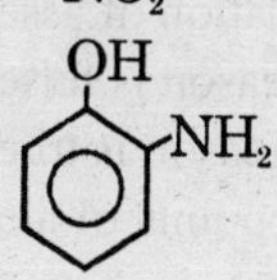

(d) 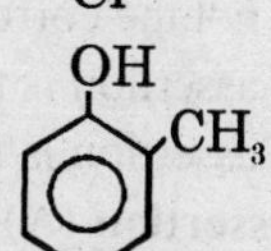

17. Which of the following is the least acidic?

(a) 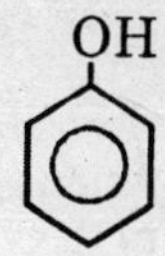

(b) 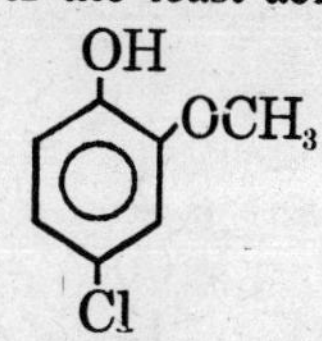

(c) 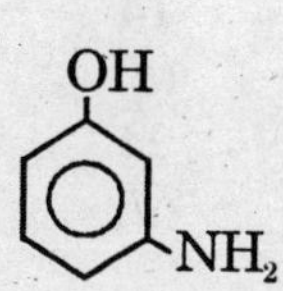

(d) 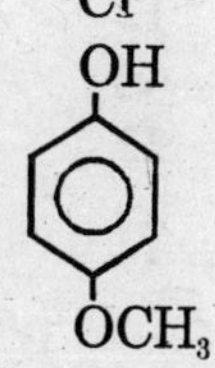

18. Which of the following tests is referred to as red, blue and white test?
(*a*) Victor Meyer's test
(*d*) Hofmann test
(*d* Lucas test
(*d*) Iodoform test

19. Phenol is dissolved in concentrated sulphuric acid and a few drops of sodium nitrite solution is added. The reaction is known as
(*a*) Libermann reaction
(*b*) Victor Meyer test
(*c*) Lucas reaction
(*d*) Precipitation

20. Fries rearrangament is
(*a*) conversion of phenol to benzaldehyde
(*b*) conversion of phenyl ester to *o* and *p*-hydroketone
(*c*) conversion of *p*-alcohol to secondary alcohol
(*d*) None of the above

Directions Qs. 21-28: *In each of the following questions there are two statements Assertion (A) is given followed by a corresponding statement of Reason (R) just below it. Of the statements, mark the correct answer:*

(*a*) If both assertion and reason are true and reason is the correct explanation of assertion.
(*b*) If both assertion and reason are true but reason is not the correct explanation of assertion
(*c*) If assertion is true but reason is false
(*d*) If assertion is false but reason is true.

21. Assertion (A): Boiling points of alcohols are lower than hydrocarbons.
Reason (R): Among isomeric alcohols, boiling point decreases in the order: 1 ° > 2° > 3°.

22. Assertion (A): The solubility of alcohols increases with increase in branching.
Reason (R): Ethanol and methanol are immiscible in water.

23. Assertion (A): The oxygen of OH group in alcohols is attached to sp^3 hybridised carbon.
Reason (R): The bond angle in alcohols C–O–H is 109°28'.

24. Assertion (A): The boiling point of alcohols are in the order *t*-butyl alcohol > sec. butyl alcohol > *n*-butyl alcohol.
Reason (R): The difference between boiling points of two consecutive members of alcohols is about 18 - 20°C.

25. Assertion (A): Alcohols are soluble in water.
Reason (R): Alcohols show H-bonding.

26. Assertion (A): Alcoholic fermentation involves conversion of sugar into ethyl alcohol by yeast.
Reason (R): Fermentation involves the slow decomposition of complex organic compounds into simpler substances through the agency of complex nitrogenous compounds called enzymes.

27. Assertion (A): Grignard reagents are alkyl or aryl magnesium halides.
Reason (R): Alcohols can be used as solvent for Grignard reagent.

28. Assertion (A): The water solubility of the alcohols follow the order t-butyl alcohol>*s*-butyl alcohol>*n*-butyl alcohol.
Reason (R): Alcohols form H-bonding with water to show soluble nature.

ANSWERS

1	2	3	4	5	6	7	8	9	10
(*c*)	(*b*)	(*d*,	(*c*)	(*b*)	(*d*)	(*a*)	(*a*)	(*a*)	(c)
11	**12**	**13**	**14**	**15**	**16**	**17**	**18**	**19**	**20**
(*d*)	(*c*)	(*a*)	(*d*)	(*c*)	(*d*)	(*d*)	(*a*)	(*a*)	(*b*)
21	**22**	**23**	**24**	**25**	**26**	**27**	**28**		
(*d*)	(*c*)	(*c*)	(*d*)	(*a*)	(*a*)	(*c*)	(*b*)		

25 ETHERS

Ethers are a class of compounds having the general formula R—O—R. The two R groups attached to oxygen may be identical or different. Moreover, both these groups may be alkyl or both may be aryl or one may be alkyl and the other aryl.

Thus, we have :

$$\underset{\text{Dimethyl ether (Symmetrical)}}{CH_3-O-CH_3} \qquad \underset{\text{Ethyl methyl ether (Unsymmetrical)}}{CH_3-O-C_2H_5}$$

Preparation of Ethers

Industrial preparation of diethyl ether : The manufacture of diethyl ether is carried out by heating excess of ethyl alcohol with concentrated sulphuric acid at 413 K.

$$\underset{\text{Ethyl alcohol}}{2C_2H_5OH} \xrightarrow{H_2SO_4,413K} \underset{\text{Diethyl ether}}{C_2H_5-O-C_2H_5} + H_2O$$

A mixture of ethyl alcohol and sulphuric acid in equimolecular proportion is heated by superheated steam. The temperature is kept at 413 K and more of ethyl alcohol is added at the same rate at which ether distills over.

Absolute ether : Diethyl ether required for use in many laboratory processes must be free of traces of water and alcohol. The moisture and alcohol-free ether, known as *absolute ether,* is obtained by distillation of ordinary ether from concentrated sulphuric acid, followed by storage over metallic sodium.

Physical Properties : Except methyl ether and methyl ethyl ether which are gases at ordinary temperature, all other ethers are colourless, volatile and pleasant smelling liquids. They are highly inflammable.

The weak polarity of ether does not, however affect their boiling points significantly.

Ethers are freely soluble in organic solvents although they dissolve in water also to an appreciable extent. The solubility in water is probably due to the formation of hydrogen bonds between water and ether molecules. They are all lighter than water.

Chemical properties

1. ***Formation of oxonium salts :***

$$CH_3-\ddot{O}-CH_3 + HCl \longrightarrow \underset{\text{Dimethyl oxonium chloride}}{\left[CH_3-\overset{H}{\dot{O}^+}-CH_3\right]Cl^-}$$

$$C_2H_5-\ddot{O}-C_2H_5 + H_2SO_4 \longrightarrow \underset{\text{(Diethyl oxonium hydrogen sulphate)}}{\left[C_2H_5-\overset{H}{\dot{O}^+}-C_2H_5\right]HSO_4^-}$$

2. ***Formation of peroxides :***

$$\underset{\text{Ethyl ether}}{CH_3-CH_2-O-CH_2-CH_3} + O_2 \longrightarrow \underset{\text{Ethyl ether peroxide}}{CH_3-CH_2-O-\underset{O-O-H}{\underset{|}{CH}}-CH_3}$$

3. *Action of acids :*

$$C_2H_5-O-C_2H_5 + HI \rightarrow C_2H_5I + C_2H_5OH$$

Ethyl ether — Ethyl iodide Ethyl alcohol

The order of reactivity of halogen acids in the reaction is

HI > HBr > HCl

Uses of Ethers : Diethyl ether is by far the most useful of all ethers. It is used:

(*i*) As an industrial solvent for oils, fats, gums, *etc.*, and as an extracting solvent.

(*ii*) As a refrigerant.

(*iii*) As a reaction medium in lithium aluminium hydride reductions.

MULTIPLE CHOICE QUESTIONS

1. Methyl chloride reacts with sodium ethoxide to form
(*a*) dimethyl ether (*b*) methyl alcohol
(*c*) ethyl iodide (*d*) ethyl methyl ether

2. Which of the following is not an isomer of ethyl ethanoate?
(*a*) Methyl propanoate
(*b*) Propyl methanoate
(*c*) Butanoic acid
(*d*) Propanone

3. When diethyl ether is heated with concentrated HI, it produces
(*a*) ethanol (*b*) iodoform
(*c*) ethyl iodide (*d*) methyl iodide

4. When diethyl ether is heated with an excess of PCl_5 it yields
(*a*) ethyl chloride
(*b*) diethyl ether peroxide
(*c*) ethanoyl chloride
(*d*) perchlorodiethyl ether

5. The reactivity of halogen acids with ether follows the order
(*a*) HI > HBr > HCl
(*b*) HCl > HBr > HI
(*c*) HBr > HI > HCl
(*d*) HCl > HI > HBr

6. Diethyl ether can be distinguished from ethyl alcohol by its reaction with
(*a*) Na
(*b*) PCl_5
(*c*) 2, 4-dinitrophenylhydrazone
(*d*) None of these

7. $CH_3\underset{\displaystyle Br}{\underset{|}{C}H}CH_3 \xrightarrow[\text{kOH}]{\text{alc.}} A \xrightarrow[\text{Peroxide}]{\text{HBr}} B \xrightarrow{CH_3ONa} C$

In the above reaction sequence, the final product is
(*a*) diethyl ether
(*b*) I-methoxypropane
(*c*) isopropyl alcohol
(*d*) propylene glycol

Directions Qs. 8-9: *In each of the following questions there are two statements Assertion (A) is given followed by a corresponding statement of Reason (R) just below it. Of the statements, mark the correct answer:*

(*a*) If both assertion and reason are true and reason is the correct explanation of assertion.
(*b*) If both assertion and reason are true but reason is not the correct explanation of assertion
(*c*) If assertion is true but reason is false
(*d*) If assertion is false but reason is true.

8. Assertion (A): Ethers behave as bases in the presence of mineral acids.
Reason (R): This is due to the presence of lone pairs of electrons on oxygen.

9. Assertion (A): Etherates are coordination complexes of ethers with Lewis acids.
Reason (R): Ethers are easily cleaved by mineral acids such as HCl and H_2SO_4 at 373 K.

ANSWERS

1	2	3	4	5	6	7	8	9	10
(*d*)	(*d*)	(*c*)	(*a*)	(*a*)	(*a*)	(*b*)	(*b*)	(*a*)	(*c*)

26 ALDEHYDES AND KETONES

Aldehydes and ketones are collectively called Carbonyl Compounds. In aldehydes, the carbonyl carbon is bonded to one hydrogen and one alkyl group. Formaldehyde, HCHO, in which the carbonyl carbon is bonded to two hydrogen atoms is an exception. In ketones, the carbonyl carbon is always bonded to two alkyl groups. These alkyl groups may be same or different.

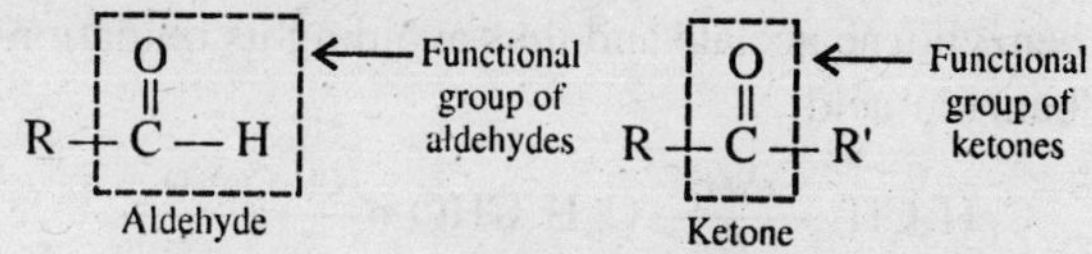

Most of the properties of aldehydes and ketones are a result of the presence of the carbonyl group.

Formula	*Common name*	*IUPAC name*
HCHO	Formaldehyde	Methanal
CH_3CHO	Acetaldehyde	Ethanal
CH_3CH_2CHO	Propionaldehyde	Propanal
CH_3COCH_3	Dimethyl ketone or Acetone	Propane
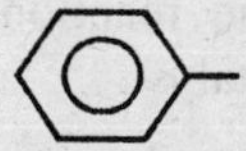–CO–CH_3	Methyl Phenyl ketone or Acetophenone	—

Preparation

1. Oxidation of Alcohols : Primary alcohols form aldehydes while secondary alcohols form ketones.

$$\underset{\text{Ethyl alcohol}}{CH_3CH_2OH} \xrightarrow[\text{[O]}]{K_2Cr_2O_7 + \text{dil. } H_2SO_4} \underset{\text{Acetaldehyde}}{CH_3CHO} + H_2O$$

$$\underset{\text{Isopropyl alcohol}}{CH_3CHOHCH_3} \xrightarrow[\text{[O]}]{K_2Cr_2O_7 + \text{dil. } H_2SO_4} \underset{\text{Acetone}}{CH_3.CO.CH_3}$$

Ketones are not easily oxidised further and can be obtained in high yield by this method.

2. Catalytic Dehydrogenation of Alcohols :

(*a*) $\underset{\text{Methanol}}{CH_3OH} \xrightarrow[300°C]{Cu} \underset{\text{Methanal}}{HCHO} + H_2$

(*b*) $\underset{\text{Iso-propanol}}{(CH_3)_2CHOH} \xrightarrow[300°C]{Cu} \underset{\text{Acetone}}{(CH_3)_2CO} + H_2$

Aldehydes are prepared from primary and ketones from secondary alcohols.

3. By the Hydration of Alkynes :

$$\underset{\text{Acetylene}}{CH \equiv CH} + H_2O \xrightarrow[HgSO_4,\ 60°C]{\text{Dil. } H_2SO_4} \underset{\text{Unstable}}{[CH_2{=}CHOH]} \rightarrow \underset{\text{Acetaldehyde}}{CH_3CHO}$$

$$CH_3C \equiv CH + H_2O \xrightarrow{\text{dil. } H_2SO_4} [CH_3\overset{OH}{\overset{|}{C}} = CH_2] \rightarrow CH_3.\overset{O}{\overset{||}{C}}.CH_3$$

Propyne — Unstable — Acetone

4. Aromatic Aldehydes and Ketones

Preparation : Benzaldehyde can be prepared by

(i) Oxidising toluene with chromium trioxide in presence of acetic anhydride to trap benzaldehyde as benzylidene acetate and thus avoiding its oxidation to benzoic acid

$$\underset{\text{Toluene}}{C_6H_5CH_3} \xrightarrow{CrO_3} \underset{\text{Benzaldehyde}}{C_6H_5CHO} \xrightarrow{(CH_3CO)_2O}$$

$$\underset{\text{Benzylidene acetate}}{C_6H_5CH(OCOCH_3)_2} \xrightarrow[\text{hydrolysis}]{\text{alkaline}} \underset{\text{Benzaldehyde}}{C_6H_5CHO}$$

(ii) By treating Benzene (aromatic compound) with a mixture of carbon monoxide and dry HCl gas under pressure and in the presence of anhydrous $AlCl_3$(Gattermann-Koch synthesis).

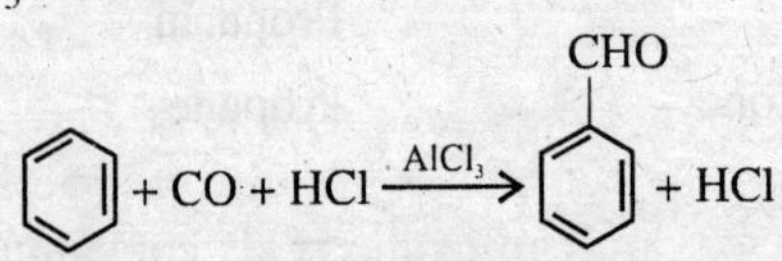

Benzene

Aromatic ketones are prepared by Friedel-Craft acylation :

(a) $$\underset{\text{Benzene}}{C_6H_6} + CH_3COCl \xrightarrow[AlCl_3]{\text{anhydrous}} \underset{\text{Acetophenone}}{C_6H_5COCH_3} + HCl$$

Aromatic ketones (as well as aliphatic ketones) can be prepared by treating benzoyl chloride (acetyl chloride) with dimethyl cadmium.

$$\underset{\text{Benzylchloride}}{2C_6H_5COCl} + \underset{\text{Dimethyl cadmium}}{(CH_3)_2Cd} \rightarrow \underset{\text{Acetophenone}}{2C_6H_5.CO.CH_3} + CdCl_2$$

Physical Properties

(i) Formaldehyde is a gas at room temperature. Acetaldehyde boils at 20°C. Other lower aldehydes and ketones are colourless liquids.

(ii) Lower aldehydes possess rather unpleasant, pungent smells, whereas the ketones have pleasant, sweet odours.

(iii) Density of aldehydes and ketones is less than that of water.

(iv) Aldehydes and ketones are polár compounds.

(v) Lower aldehydes and ketones are soluble in water. This is because of their ability to form hydrogen bonds with water molecules. Higher members containing more than five carbons are virtually insoluble in water.

Chemical Properties

(1) Addition of Sodium Bisulphite, HCN, Grignard Reagent, etc. :

$$\underset{\text{Acetaldehyde}}{CH_3-\overset{O}{\overset{||}{C}}-H} + NaHSO_3 \rightarrow \underset{\text{Acetaldehyde bisulphite}}{CH_3-\overset{OH}{\overset{|}{\underset{SO_3Na}{\underset{|}{C}}}}-H}$$

$$\underset{\text{Acetone}}{CH_3-\overset{O}{\overset{||}{C}}-CH_3} + NaHSO_3 \rightarrow \underset{\text{Acetone bisulphite}}{CH_3-\overset{OH}{\overset{|}{\underset{SO_3Na}{\underset{|}{C}}}}-CH_3}$$

Aldehydes and ketones react with hydrogen cyanide to form Cyanohydrins. The reaction is carried in the presence of a basic catalyst.

$$\underset{\text{Acetaldehyde}}{CH_3-\overset{O}{\overset{||}{C}}-H} + HCN \rightarrow \underset{\text{Acetaldehyde cyanohydrin}}{CH_3-\overset{OH}{\overset{|}{\underset{CN}{\underset{|}{C}}}}-H}$$

$$\underset{\text{Acetone}}{CH_3-\overset{O}{\overset{||}{C}}-CH_3} + HCN \rightarrow \underset{\text{Acetone Cyanohydrin}}{CH_3-\overset{OH}{\overset{|}{\underset{CN}{\underset{|}{C}}}}-CH_3}$$

HCN is a very poisonous gas. It is produced *in situ* by the action of dilute sulphuric acid on potassium cyanide.

Aldehydes and ketones react with Grignard reagents to give an addition product which can be hydrolysed with dilute acid to yield an alcohol.

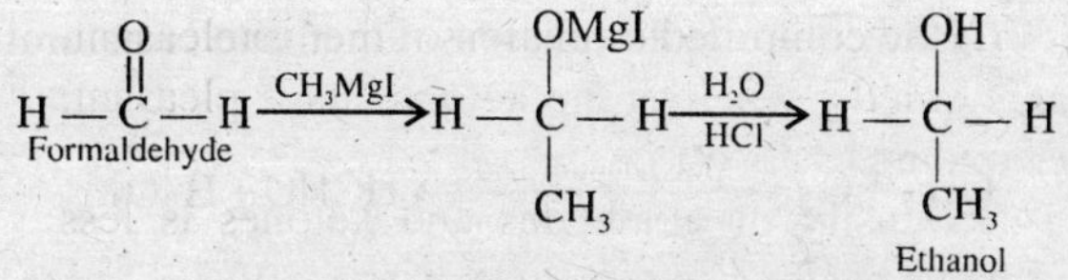

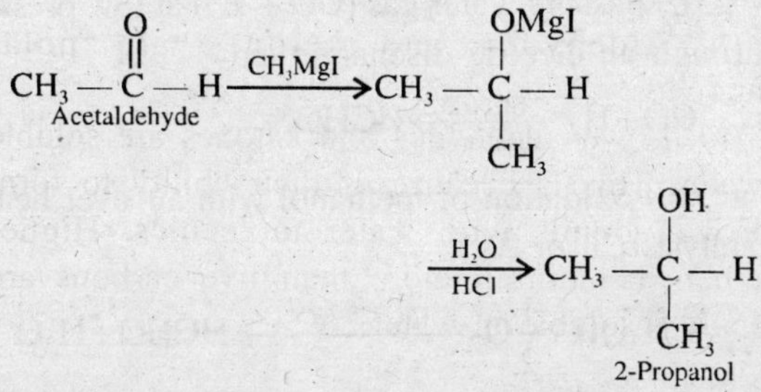

(2) Addition of Ammonia : Aldehydes (except formaldehyde) react with ammonia to form solid aldehyde ammonias. For example,

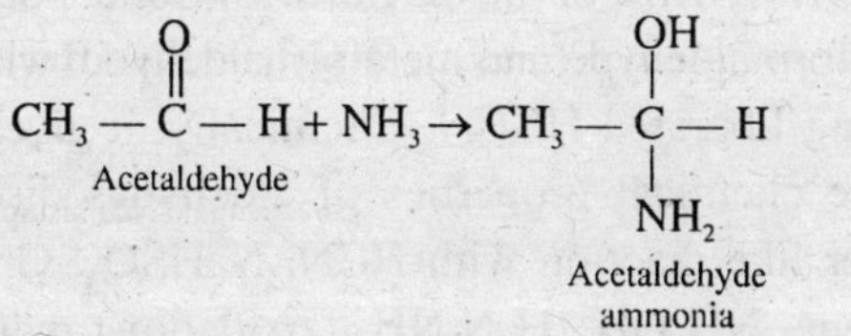

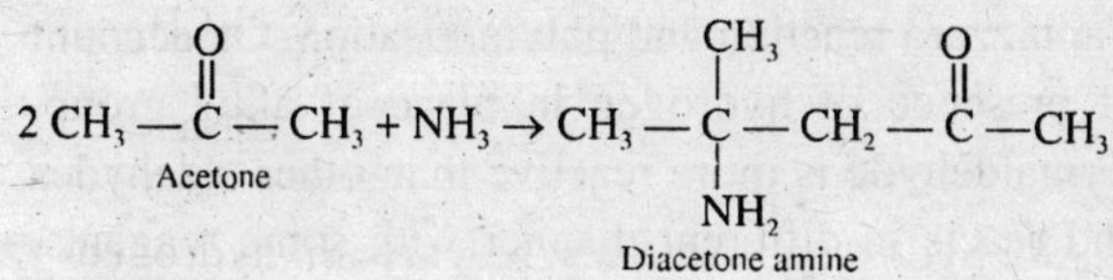

(3) Addition of Alcohols :

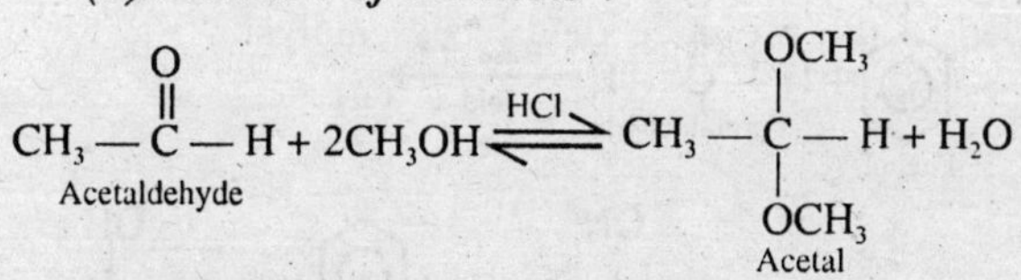

Ketones do not react with alcohols to form the corresponding hemiketals and ketals.

(4) Reduction to Alcohols : Aldehydes and ketones can be reducted to alcohols by treatment with hydrogen and Ni or Pt catalyst. Aldehydes give primary alcohols. Ketones give secondary alcohols.

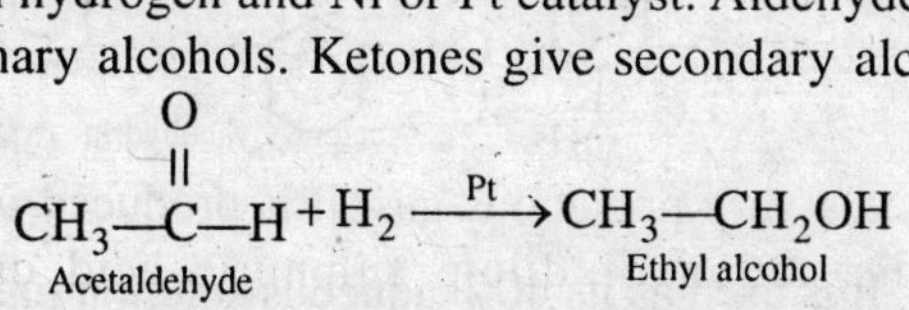

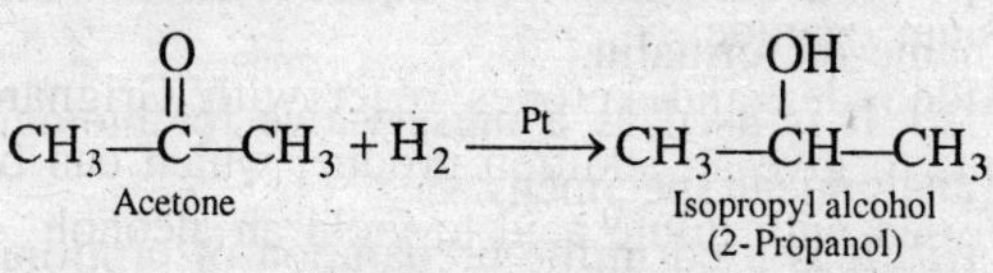

(5) Clemmensen Reduction : This involves use of zinc-mercury amalgam in hydrochloric acid as the reducing agent.

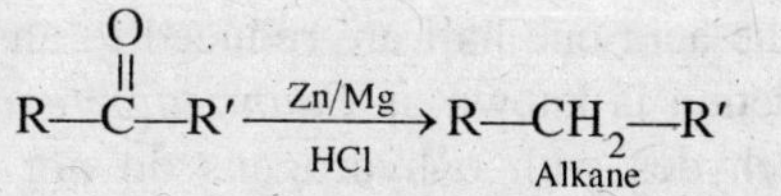

(6) Wolf-Kishner Reduction : This involves the use of a basic solution of hydrazine as the reducing agent.

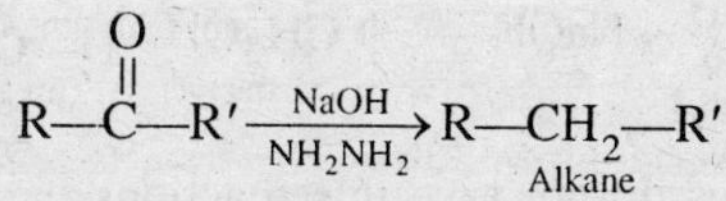

(7) Reduction to Pinacols : Ketones when reduced in neutral or alkaline medium, form pinacols (symmetrical 1, 2-diols). For example, acetone undergoes reduction with magnesium amalgam to form 2, 3-dimethylbutane-2, 3 diol.

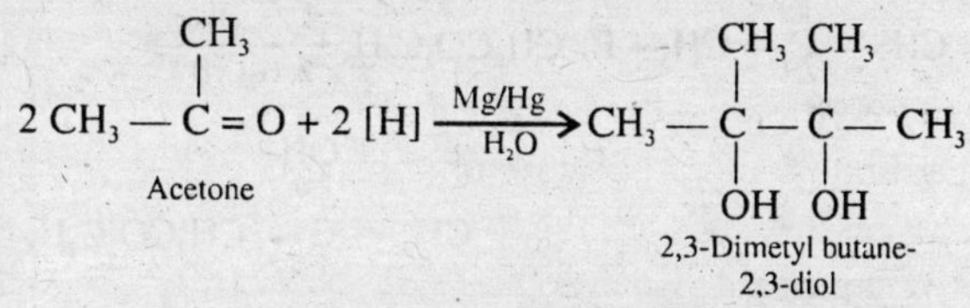

Aldehydes do not give this reaction.

(8) Oxidation of Aldehydes : Aldehydes can be oxidised with sodium (or potassium) dichromate in acidic medium to form carboxylic acids containing the same number of carbon atoms.

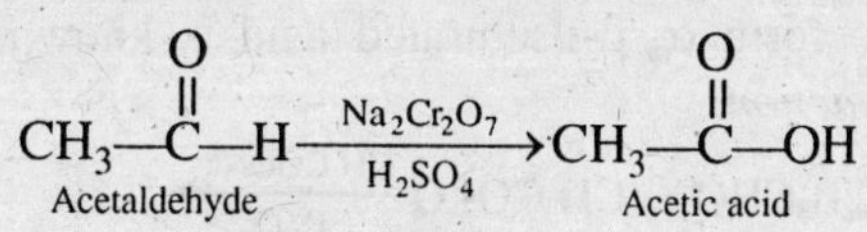

(9) Oxidation of Ketones :

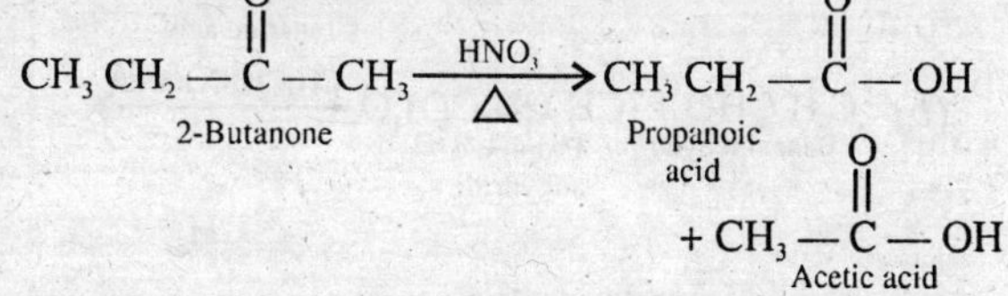

Ketones do not react with Tollen's reagent, Fehling's solution or Benedict's solution.

(10) Haloform Reaction : Acetaldehyde and methyl ketones react rapidly with halogens (Cl_2, Br_2, or I_2) in the presence of alkali to form haloform.

(11) Cannizzaro Reaction : Aldehydes which lack an α-hydrogen when heated with concentrated NaOH, undergoes a disproportonation reaction. One half of the aldehyde molecules are oxidised to a carboxylic acid one half are reduced to an alcohol. This reaction is known as *Cannizzaro reaction.*

Aldehydes with α-hydrogens do not undergo this reaction, under these conditions, they undergo an Aldol condensation reaction.

$$\underset{\substack{\text{Formaldehyde}\\ \text{(No }\alpha\text{-Hydrogen)}}}{2H{-}\overset{\overset{\displaystyle O}{\|}}{C}{-}H} + NaOH \xrightarrow{\Delta} \underset{\text{Methanol}}{CH_3OH} + \underset{\substack{\text{Sodium}\\ \text{formate}}}{H{-}\overset{\overset{\displaystyle O}{\|}}{C}{-}\overset{-}{O}\overset{+}{Na}}$$

Ketones do not give this reactions :

(12) Reformatsky Reaction : This involves the treatment of an aldehyde or ketone with α-bromo ester in the presence of zinc. The product (after acid-hydrolysis) is a hydroxy ester. For example,

$$\underset{\text{Acetone}}{CH_3-\overset{\overset{\displaystyle O}{\|}}{C}-CH_3} + Br\,CH_2CO_2C_2H_5 \xrightarrow[2,\ H_2O/H^+]{1.Zn,\ ether} \underset{\substack{\text{Ethyl 3 – hydroxy –}\\ \text{3 – methyl butanoate}}}{CH_3-\overset{\overset{\displaystyle OH}{|}}{\underset{\underset{\displaystyle CH_3}{|}}{C}}-CH_2CO_2C_2H_5}$$

(13) Perkin Reaction : Condensation of an aromatic aldehyde with acid anhydride in presence of sodium salt of the acid from which anhydride is derived to form α, β-usaturated acid is known as *Perkin Reaction.*

$$(a)\ \underset{\text{Benzaldehyde}}{C_6H_5CHO} + \underset{\substack{\text{Acetic}\\ \text{anhydride}}}{(CH_3CO)_2O} \xrightarrow[180^\circ C]{CH_3COONa} \underset{\text{Cinnamic acid}}{C_6H_5CH = CHCOOH}$$

$$(b)\ \underset{\text{Benzaldehyde}}{C_6H_5CHO} + \underset{\substack{\text{Propionic}\\ \text{anhydride}}}{(\overset{\beta}{C}H_3\overset{\alpha}{C}H_2CO)_2O} \xrightarrow{CH_3CH_2COONa} \underset{\alpha\text{-Methyl cinnamic acid}}{C_6H_5CH = \overset{\overset{\displaystyle CH_3}{|}}{C}-COOH}$$

Formaldehyde, Methanal, HCHO

It can be prepared by general methods. It can be manufactured by

(i) the controlled oxidation of methane or natural gas.

$$CH_4 + O_2 \xrightarrow[\text{heat, under pressure}]{\text{Molybdenum oxide}} HCHO + H_2O$$

(ii) passing water gas ($CO + H_2$) at low pressure through an electric discharge and

$$CO + H_2 \xrightarrow[\text{discharge}]{\text{Electric}} HCHO$$

(iii) oxidation of methanol with air over heated catalyst (Cu or Ag).

$$CH_3OH + \frac{1}{2}O_2 \xrightarrow{Cu,\ 525\text{-}625^\circ C} HCHO + H_2O$$

Properties : It is a colourless, pungent smelling gas, extremely soluble in water. Since it is a gas, it is marketed as 40% aqueous solution under the name of FORMALIN or in the form of solid polymers, 'paraformaldehyde and metaformaldehyde' which on heating liberated HCHO. Chemically, it gives most of the chemical properties of aldehydes discussed earlier like reaction with HCN, $NaHSO_3$, Grignard reagent, NH_2OH, $H_2N.NH_2$, oxidation, reduction, Cannizzaro reaction and polymerisation. On account of presence of hydrogen in place of alkyl group, formaldehyde is more reactive than other aldehydes and reacts in different manner with some reagents.

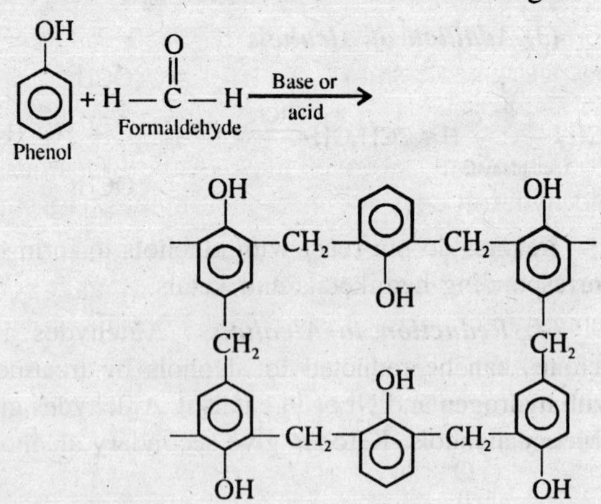

Uses :

(i) It is used as its 40% aqueous solution under the name of formalin.

(ii) It is used as a preservative for biological and anatomical specimens.

(iii) It is used in the preparation of urotropine.

(iv) It is used in the preparation of bakelite, a synthetic plastic.

(v) It is used in silvering of mirror.

Acetaldehyde, Ethanal, CH_3CHO

Acetaldehyde is the most important member of the aldehyde family. It can be manufactured in the following ways :

(i) By hydration of acetylene

$$CH \equiv CH + H_2O \xrightarrow[1\% HgSO_4, 60°C]{40\% H_2SO_4} CH_3CHO$$

(ii) By the catalytic dehydrogenation of ethanol in the presence of heated copper (300°C).

(iii) From ethylene : (Wacker process). This involves the treatment of ethylene with an acidified aqueous solution of palladium chloride and cupric chloride

$$\underset{\text{Ethylene}}{CH_2{=}CH_2} + PdCl_2 + H_2O \xrightarrow[H^+]{CuCl_2} \underset{\text{Acetaldehyde}}{CH_3{-}\overset{\overset{\large O}{\|}}{C}{-}H} + Pd + 2HCl$$

Properties : It is a colourless volatile liquid, b.p. 21°C. It has a characteristic pungent smell. Acetaldehyde is soluble in water, chloroform, ethanol and diethyl ether.

In water it is hydrated to the extent of 58% forming ethylidene hydroxide. The aqueous solution has an agreeble smell.

$$CH_2CHO + H_2O \rightleftharpoons CH_3CH(OH)_2$$

Chemically, it gives most of the properties of aldehydes. It does not undergo Cannizzaro reaction.

Polymerisation : *(i)* When acetaldehyde is treated with a small amount of conc. H_2SO_4 at room temperature, a cyclic trimer 'Paraldehyde' is formed.

(ii) When acetaldehyde is treated with a small amount of conc. H_2SO_4 at 0°C a cyclic tetramer 'Metaldehyde' is formed.

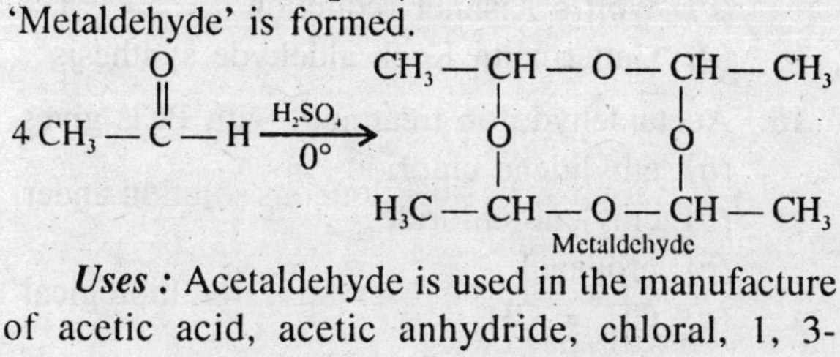

Metaldehyde

Uses : Acetaldehyde is used in the manufacture of acetic acid, acetic anhydride, chloral, 1, 3-Butadiene (used in rubber) and dyes. It is also used as an antiseptic inhalant in nose troubles.

Acetone, Dimethyl Ketone, Propanone, $CH_3.CO.CH_3$

Manufacture

(i) By the oxidation of isopropyl alcohol with oxygen at 500°C.

$$CH_3.CHOH.CH_3 + O_2 \xrightarrow{500°C} CH_3.CO.CH_3$$

(ii) By the catalytic dehydrogenation of isopropyl alcohol.

$$CH_3.CHOH.CH_3 \xrightarrow{Cu, 500°C} CH_3.CO.CH_3 + H_2$$

(iii) By Wacker Process (from propene).

$$CH_3CH{=}CH_2 + PdCl_2 + H_2O \rightarrow CH_3.CO.CH_3 + Pd + 2HCl$$

Properties : It is a colourless, inflammable liquid with characteristic pleasant smell. It responds all the general reactions of ketones discussed earlier. Some reactions deserving special attention are mentioned below :

(i) Reaction with Ammonia :

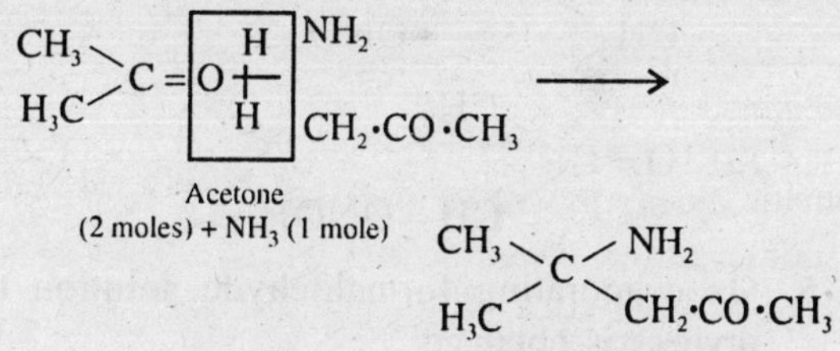

Acetone (2 moles) + NH_3 (1 mole)

Diacetoneamine

(ii) Condensation (aldol condensation) :

$$\underset{\text{Acetone (2 moles)}}{(CH_3)_2\,CO + HCH_2.CO.CH_3} \xrightarrow{Ba(OH)_2} \underset{\text{Diacetone alcohol}}{(CH_3)_2\overset{\overset{\large OH}{|}}{C}{-}CH_2{-}COCH_3}$$

Uses :

(i) It is used for storing acetylene.

(ii) It is very frequently used as solvent.

(iii) It is used in the preparation of chloroform (anaesthetic), iodoform. (antiseptic), chloretone (hypnotic and sedative), etc.

MULTIPLE CHOICE QUESTIONS

1. Oxidation of secondary alcohol to ketone with aluminium tertiary butanide is known as
   (a) Meerwein Pondorff Verley reaction
   (b) Oppenauer oxidation
   (c) Kolbe's reaction
   (d) Stephen's reaction

2. On passing a mixture of propylene and O_2 under pressure into an aqueous solution of palladium and cupric chloride....... is obtained.
   (a) acetaldehyde (b) benzaldehyde
   (c) acetone (d) propanal

3. Reduction of acid chloride with H_2 in the presence of Pb and $BaSO_4$ is known as
   (a) Meerwein Pondorff Verely reduction
   (b) Rosenmund reduction
   (c) Wolff's Kishner reduction
   (d) Kolbe's reduction

4. Acetone when treated with nitrous acid gives
   (a) CH_3COCH_2. NO
   (b) $CH_3COCH = NOH$
   (c) 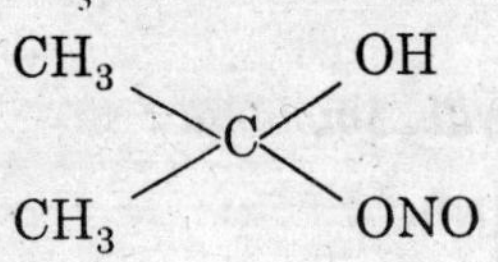
   (d) 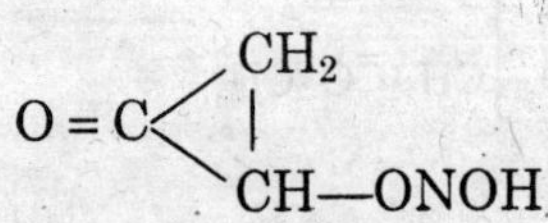

5. On evaporating formaldehyde solution to dryness is obtained
   (a) troxane (b) formic acid
   (c) CO and H_2 (d) paraformaldehyde

6. Trimer of formaldehyde is known as
   (a) metaformaldehyde
   (b) trioxymethylene
   (c) trioxane
   (d) All of the above

7. Which of the following is liquid at room temperature?
   (a) Metaldehyde
   (b) Trioxane
   (c) Polyoxymethylenes
   (d) Paraldehyde

8. Methyl cyanide is dissolved in ethyl formate and reduced with stannous chloride and hydrochloric acid. The product is then steam distilled. The distillate will be
   (a) ethyl amine (b) formaldehyde
   (c) methyl alcohol (d) acetaldehyde

9. Acetone when distilled with conc. sulphuric acid gives
   (a) diacetone alcohol
   (b) mesitylene
   (c) mesityl oxide
   (d) toluene

10. Acetaldehyde when treated with ethyl alcohol in the presence of dil HCl gives
    (a) formaldehyde (b) crotonaldehyde
    (c) propanal (d) acetal

11. Which of the following is the most reactive?
    (a) CH_3CHO (b) HCHO
    (c) CH_3COCH_2 (d) $CH_3CH_2COCH_3$

12. Which of the following will react with acetaldehyde to give white crystalline solid soluble in water?
    (a) K_2CO_3 (b) K_2HSO_3
    (c) $NaHSO_3$ (d) $NaHSO_4$

13. Which of the following will give ketone with Grignard reagent?
    (a) KCN (b) RCN
    (c) RCdCl (d) R - C - Cl

14. The reaction
    $CH_3CN \xrightarrow[HCl]{SnCl_2} CH_3CN = NH \xrightarrow{H_3/O^+} CH_3CHO$ is known as
    (a) Kolbe's reduction
    (b) Stephen's reduction
    (c) Wolf's Kishner reduction
    (d) Gattermann Koch aldehyde synthesis

15. Acetaldehyde on treatment with PCls gives
    (a) ethylidene chloride
    (b) ethylene chloride
    (c) propanal
    (d) $CH_2 = CH_2$

16. Fries rearrangement is

(*a*) $C_6H_5OCOCH_3 \xrightarrow[AlCl_3]{Anhyd} pHOC_6H_4COCH_3$

(*b*) $C_6H_5OH + CHCl_3 + NaOH \rightarrow HOC_6H_4CHO + NaCl + H_2O$

(*c*) $C_6H_5OH + HCHO \xrightarrow{NaOH} o$ and $pHOC_6H_4CH_2OH$

(*d*) $C_6H_5\ OH + NaNO_2 + HCl \xrightarrow{273K} pNO.C_6H_4\ OH + H_2O$

17. Which of the following is a sedative?
(*a*) Paraformaldehyde
(*b*) Benzaldehyde
(*c*) Metaldehyde
(*d*) Paraldehyde

18. Which of the following does not reduce Fehling's solution?
(*a*) Formaldehyde (*b*) Paraldehyde
(*c*) Propanal (*d*) Acetaldehyde

19. At room temperature, formaldeyde is
(*a*) gas (*b*) liquid
(*c*) solid (*d*) None of these

20. A water-soluble compound 'A' (C_3H_6O) does not evolve hydrogen gas with sodium but gives a pink colour with Schiff's reagent. The compound A is
(*a*) ether (*b*) acid
(*c*) aldehyde (*d*) ester

Directions Qs. 21-27: *In each of the following questions there are two statements Assertion (A) is given followed by a corresponding statement of Reason (R) just below it. Of the statements, mark the correct answer:*

(*a*) If both assertion and reason are true and reason is the correct explanation of assertion.
(*b*) If both assertion and reason are true but reason is not the correct explanation of assertion
(*c*) If assertion is true but reason is false
(*d*) If assertion is false but reason is true.

21. **Assertion (A):** Lower aldehydes and ketones are soluble in water but the solubility decreases as the molecular weight increases.
Reason (R): Distinction between aldehydes and ketones can be made by Tollen's reagent.

22. **Assertion (A):** In formaldehyde all the four atoms are in same plane.
Reason (R): The carbon atom in formaldehyde is sp^2 hybridised.

23. **Assertion (A):** Benzaldehyde is more reactive than ethanol towards nucleophilic attack.
Reason (R): The overall effect of –I and +R effect of phenyl group decreases the electron density on the carbon atom of $>C = O$ group in benzaldehyde.

24. **Assertion (A):** α-Hydrogen atoms in aldehydes and ketones are acidic.
Reason (R): The anion left after the removal of α-hydrogen is stabilised by inductive effect.

25. **Assertion (A):** Acetaldehyde undergoes aldol condensation with dilute NaOH.
Reason (R): Aldehydes which do not contain α-hydrogen undergo aldol condensation.

26. **Assertion (A):** Fehling's solution can be used to distinguish benzaldehyde from acetaldehyde.
Reason (R): The C – H bond of CHO group in benzaldehyde is stronger than C – H bond of CHO group in acetaldehyde.

27. **Assertion (A):** Aldol condensation can be catalysed both by acids and bases.
Reason (R): β-Hydroxy aldehydes or ketones readily undergo acid catalysed dehydration.

ANSWERS

1	2	3	4	5	6	7	8	9	10
(*b*)	(*c*)	(*b*)	(*b*)	(*d*)	(*d*)	(*d*)	(*d*)	(*b*)	(*d*)
11	12	13	14	15	16	17	18	19	20
(*b*)	(*c*)	(*b*)	(*b*)	(*a*)	(*a*)	(*d*)	(*b*)	(*a*)	(*c*)
21	22	23	24	25	26	27			
(*b*)	(*a*)	(*a*)	(*c*)	(*c*)	(*a*)	(*b*)			

27 NITRO COMPOUNDS AMINES

The more important amongst such classes of compounds include nitriles or cyanides (—C ≡ N group), nitro compounds (—NO_2 group), amines (—NH_2, > NH or —N) and diazonium salts (—N_2X group).

Structures of Nitro Group : The nitro group (–NO_2) which constitutes the functional group of nitro compounds has been assigned the structure

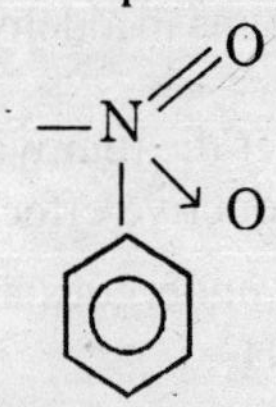

Nitrobenzene

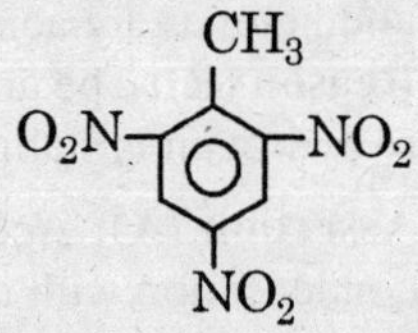

2,4,6, Trinitrotolune (T.N.T.)

Nitrobenzene

Aromatic nitro compounds are almost invariably prepared by direct nitration. The nitrating agents generally used are :

1. *Mixture of concentrated or fuming nitric acid and concentrated sulphuric acid.* This mixture is the most commonly used nitrating agent.

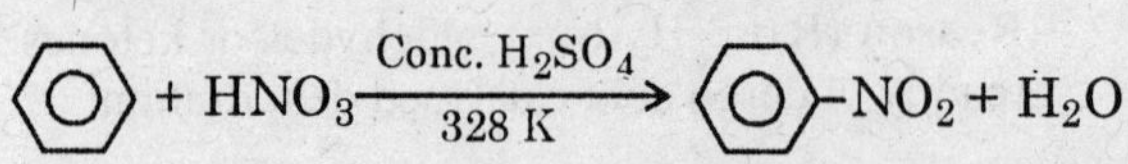

Nitrobenzene

In case of aromatic nitro compounds, nitrobenzene and some other compounds are yellow oily liquids but most of them are yellow crystalline solids. Some of them have typical smells as for instance nitrobenzene has a smell of bitter almonds. They are heavier than water and insoluble in it but dissolve readily in organic solvents.

The melting and boiling points of nitro compounds gradually increase with the number of nitro groups.

Chemical Properties

1. **Reduction :** Nitro compounds form a variety of products on reduction under different conditions. We would illustrate their behaviour on reduction by considering the case of nitrobenzene.

The reduction of nitro benzene is known to proceed in stages involving a series of intermediate compounds as shown below:

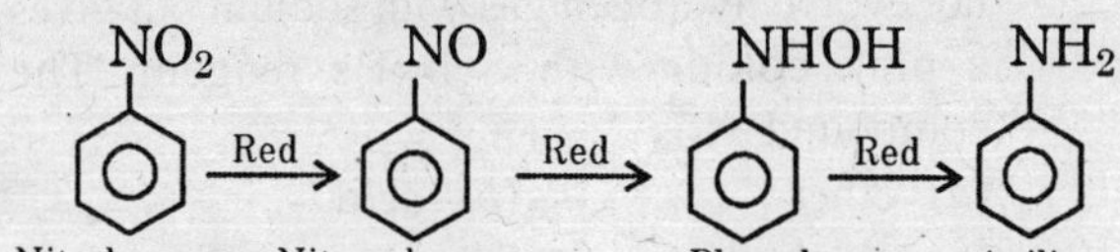

Nitrobenzene Nitrosobenzene Phenyl hydroxylamine Aniline

(i) Reduction in neutral medium: When reduced in a neutral medium, *e.g.*, with zinc dust and ammonium chloride solution, nitrobenzene furnishes phenylhydroxylamine

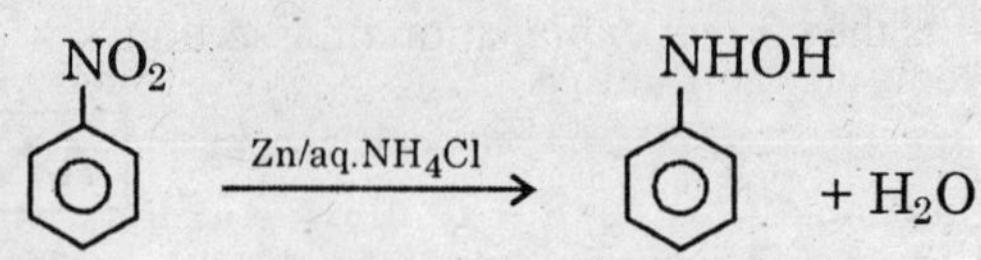

Phenylhydroxyl amine

(ii) Reduction* in *alkaline medium : On reduction in alkaline medium, nitrobenzene yields different products depending upon the nature of the reducing agent employed.

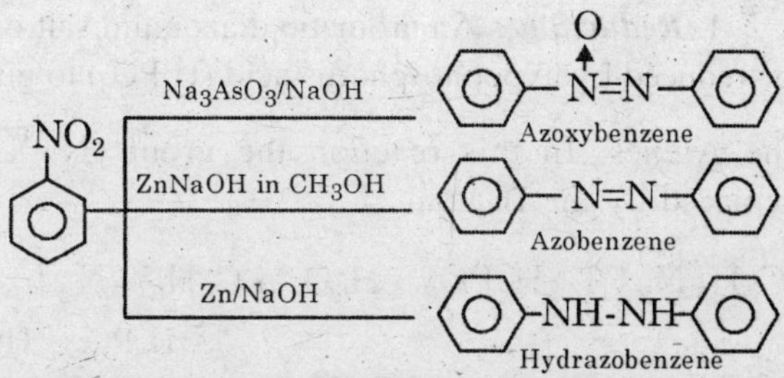

(iii) Catalytic reduction : On reduction with hydrogen in the presence of nickel as a catalyst, nitrobenzene gives aniline.

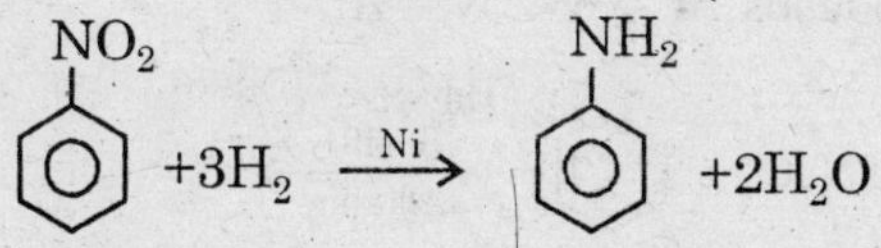

Uses : Nitroalkanes are used in the manufacture of hydroxylamine, intermediates in the production of explosives, detergents and in drugs and as solvents in plastic industry.

Trinitrotoluene

Trinitrotoluene (abbreviated as TNT), or more specifically, 2,4,6-trinitrotoluene, is a chemical compound with the formula $C_6H_2(NO_2)_3CH_3$. This yellow-colored solid is sometimes used as a reagent in chemical synthesis, but it is best known as a useful explosive material with convenient handling properties. The explosive yield of TNT is considered to be the standard measure of strength of bombs and other explosives. In chemistry, TNT is used to generate charge transfer salts.

Preparation: Industrially, TNT is synthesized in a three-step process. First, toluene is nitrated with a mixture of sulfuric and nitric acid to produce mono-nitrotoluene or MNT. The MNT is separated and then renitrated to dinitrotoluene or DNT. In the final step, the DNT is nitrated to trinitrotoluene or TNT using an anhydrous mixture of nitric acid and oleum.

Uses : TNT is one of the most commonly used explosives for military and industrial applications. It is valued because of its insensitivity to shock and friction, which reduces the risk of accidental detonation. TNT melts at 80 °C (176 °F), far below the temperature at which it will spontaneously detonate, allowing it to be poured as well as safely combined with other explosives. TNT neither absorbs nor dissolves in water, which allows it to be used effectively in wet environments. Additionally, it is stable compared to other high explosives.

Azo Compound

Azo compounds are compounds bearing the functional group R-N=N-R', in which R and R' can be either aryl or alkyl. The N=N group is called an azo group. The name azo comes from azote, the French name for nitrogen that is derived from the Greek a (not) + zoe (to live).

Azo pigments are colorless particles, which have been colored using an azo compound. Azo pigments are important in a variety of paints including artist's paints. They have excellent coloring properties, again mainly in the yellow to red range, as well as lightfastness. The lightfastness depends not only on the properties of the organic azo compound, but also on the way they have been adsorbed on the pigment carrier. Many azo pigments are non-toxic, although some, such as dinitroaniline orange, orthonitroaniline orange, or pigment orange 1, 2, and 5 have been found to be mutagenic.

Aromatic Diazonium Salts

This important class of the compounds are characterized by the presence of functional group– $\overset{+}{N} \equiv N$ (diazonium ion) directly attached to the aryl group Ar. The aryldiazoniumion combines with the anions such as Cl^-, Br^-, NO_2^-, BF_4^-, HSO_4^- etc. and form salts which are known as aryldiazonium salts.

$$Ar - \boxed{\overset{+}{N} \equiv N} \quad \text{functional group} \qquad \{Ar - \overset{+}{N} \equiv N\}X^-$$

Aryldiazonium salt

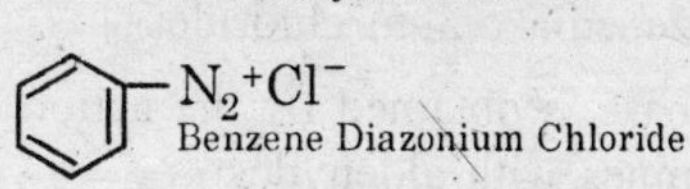

Fig. Structure

Methods of Preparation

1. ***Diazotisation*** **:** Aryl diazonium salts can be prepared by the action of primary amine with nitrous acids at about 0° – 5°C. The nitrous acid used in this reaction can be obtained by taking sodium nitrite and a mineral acid in the reaction mixture.

$$ArNH_2 + HNO_2 + HCl \xrightarrow{0^o-5^oC} Ar\overset{+}{N_2}\overset{-}{Cl} + 2H_2O$$

$$NaNO_2 + HCl \rightarrow HNO_2 + NaCl$$

The process of converting an amine to the diazonium salts is called "Diazotisation." Thus, from this procedure benzene diazonium chloride can be prepared from aniline as such

$$\underset{\text{(Aniline)}}{C_6H_5NH_2} + NaNO_2 + 2HCl \longrightarrow \underset{\text{Benzene diazonium chloride}}{C_6H_5\overset{+}{N_2}\overset{-}{Cl}} + NaCl + 2H_2O$$

Physical Properties : Aryl diazonium salts are colourless crystalline solids which turn brown when exposed to air. They are unstable and freely soluble in water, sparingly soluble in ethanol and insoluble in ether. The aqueous solutions of these salts are neutral and conduct electric current through them and showing presence of ions.

Chemical Properties : Aromatic diazonium salts are highly reactive, compounds and undergo many useful reactions. As the synthetic reagents, these salts compete with the Grignard reagent.

1. ***Reduction :*** An aromatic diazonium salt can be reduced by hypophosphorus acid (H_3PO_2) to give the Arenes. In this reaction the group $\overset{+}{N_2}\overset{-}{X}$ is replaced by the H atom.

$$C_6H_5\overset{+}{N_2}\overset{-}{Cl} + H_3PO_2 + H_2O \rightarrow C_6H_6 + N_2 + H_3PO_3 + HCl$$

2. ***Diazo coupling reactions :*** The aromatic diazonium salts react readily with phenol to form highly coloured derivatives known as Azo compounds $Ar - N \equiv N - Ar$.

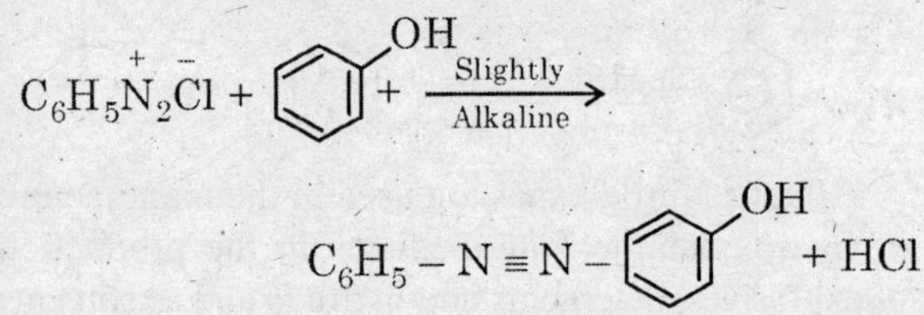

p-Hydroxy Azobenzene

Such reactions in which the diazonium ion is coupled to the ring of phenol are called as *"Diazo coupling reactions"*.

MULTIPLE CHOICE QUESTIONS

1. Aniline is purified by
(*a*) smelting
(*b*) distillation with pressure
(*c*) sublimation
(*d*) steam distillation

2. Which of the following is a carcinogen?
(*a*) Ethyl amine
(*b*) N-nitroso amine
(*c*) Aniline
(*d*) Benzenediazonium chloride

3. Schiff's base is obtained by the action of
(*a*) *p*-amines with aldehydes
(*b*) *p*-amines with acid anhydrides
(*c*) *p*-amines with acid chloride
(*d*) All of the above

4. Which of the following will give diazonium salt with sodium nitrite and HCl?
(*a*) $C_6H_5NHC_6H_5$ (*b*) $C_6H_5CONH_2$
(*c*) $C_6H_5NH_2$ (*d*) $C_2H_5NH_2$

5. Which of the following will give white precipitate with an aqueous solution of ethyl amine?
(*a*) Cuprous sulphate
(*b*) Zinc sulphate
(*c*) Ferrous sulphate
(*d*) $[Ag(NH_3)_2]^+$

6. The compound $C_5H_{13}N$ is optically active and reacts with nitrous acid to give $C_5H_{11}OH$. The compound is
(*a*) 1, N dimethyl propanamine
(*b*) N methyl butanamine
(*c*) 1-amino pentane
(*d*) 2-amino pentane

7. Which of the following pair of isomers are metamers?
(*a*) Ammonium cyanate and urea
(*b*) Ethano nitrile and ethanoisonitrile
(*c*) Propyl amino and isopropyl amine
(*d*) $(C_2H_5)_2$ NH and $CH_3NHC_3H_7$

8. *p*-bromo aniline can be obtained from aniline by reacting it

(*a*) first with acetyl chloride and then with bromine

(*b*) with water bromine water

(*c*) with HCl and then with Br_2 in CS_2

(*d*) with chlorine in CS_2 at 273 K

9. Aniline condenses with benzaldehyde to give

(*a*) N-benzoyl aniline

(*b*) benzoyl aniline

(*c*) benzal aniline

(*d*) N-benzal aniline

10. Diazonium salts are used in the solutions in which they are prepared because they are

(*a*) not obtained in the solid state

(*b*) explosive

(*c*) poisonous

(*d*) corrosive

11. Which compound is obtained when ethyl isocyanate is hydrolysed?

(*a*) Propionamide (*b*) Ethyl cyanide

(*c*) Ethyl acetate (*d*) Ethyl amine

12. Nitrobenzene when reduced with Sn and HCl gives

(*a*) hydrazobenzene (*b*) azoxy benzene

(*c*) aniline (*d*) None of these

Directions Qs. 13-20: *In each of the following questions there are two statements Assertion (A) is given followed by a corresponding statement of Reason (R) just below it. Of the statements, mark the correct answer:*

(*a*) If both assertion and reason are true and reason is the correct explanation of assertion.

(*b*) If both assertion and reason are true but reason is not the correct explanation of assertion

(*c*) If assertion is true but reason is false

(*d*) If assertion is false but reason is true.

13. **Assertion (A):** Methyl isocyanide reacts with ozone to form methyl isocyanate.

Reason (R): Methyl isocyanate was responsible for Bhopal tragedy.

14. **Assertion (A):** Alkyl cyanides on reduction with H_2/Ni give 2° amine.

Reason (R): Cyanides and isocyanides are functional isomers of each other.

15. **Assertion (A):** Isocyanides on reduction with halogens, sulphur, mercuric oxide, ozone etc. to form addition products.

Reason (R): Isocyanides on reduction with H_2/Ni gives 2° amine.

16. **Assertion (A):** Alkyl isocyanides have lower boiling point than alkyl cyanides.

Reason (R): Alkyl isocyanides are highly reactive because of the presence of lone pair of electrons on carbon atom.

17. **Assertion (A):** Treatment of alkyl halide with alcoholic solution of potassium cyanide gives isocyanide.

Reason (R): Cyanides are used for preparation of amines and carboxylic acids.

18. **Assertion (A):** Alkyl cyanide can be prepared by carbylamine reaction.

Reason (R): Ethyl amine when heated with chloroform in presence of alcoholic KOH, ethylisocyanide is formed.

19. **Assertion (A):** Alkyl isocyanides are hydrolysed by an aqueous solution of an alkali.

Reason (R): Alkyl isocyanides are less stable than isomeric cyanides.

20. **Assertion (A):** CN^- ion is an ambident nucleophile.

Reason (R): Nucleophiles are electron rich nucleophile.

ANSWERS

1	2	3	4	5	6	7	8	9	10
(*d*)	(*b*)	(*a*)	(*c*)	(*b*)	(*d*)	(*d*)	(*a*)	(*c*)	(*b*)
11	**12**	**13**	**14**	**15**	**16**	**17**	**18**	**19**	**20**
(*a*)	(*c*)	(*b*)	(*d*)	(*b*)	(*b*)	(*d*)	(*d*)	(*d*)	(*b*)

28 CARBOXYLIC ACIDS

Organic compounds which contain the carboxyl functional group (—COOH) are called the *Carboxylic acids*. Their general formula is

$$R-\overset{\overset{\large O}{\|}}{C}-OH \text{ or } R-COOH \text{ or } R-CO_2H$$

where R is an alkyl group.

Formula	***Common Name***
HCOOH	Formic acid
CH_3COOH	Acetic acid
CH_3CH_2COOH	Propionic acid
$CH_3CH_2CH_2COOH$	Butyric acid

Preparation

(*i*) *By the oxidation of alcohols, aldehydes and ketones:*

$$\underset{\text{Ethanol}}{CH_3CH_2OH} \xrightarrow[H_2SO_4]{K_2Cr_2O_7} \underset{\text{Acetaldehyde}}{CH_3CHO} \xrightarrow{[O]} \underset{\text{Acetic acid}}{CH_3COOH}$$

$$\underset{\text{iso-Propanol}}{(CH_3)_2CHOH} \xrightarrow[H_2SO_4]{K_2Cr_2O_7} \underset{\text{Acetone}}{(CH_3)_2CO} \xrightarrow{[O]} \underset{\text{Acetic acid}}{CH_3COOH}$$

Aryl nitrites may easily be prepared by fusing the alkali salt of sulphonic acid with KCN or by *Sandmeyer reaction.*

$$\underset{\text{Pot. Benzene sulphonate}}{C_6H_5SO_3K} + KCN \rightarrow \underset{\text{Phenyl cyanide}}{C_6H_5CN} \xleftarrow[KCN]{Cu_2(CN)_2} \underset{\text{Benzene diazonium chloride}}{C_6H_5N_2Cl}$$

(*ii*) *By the hydrolysis of cyanides (nitriles):*

$$\underset{\text{Methyl cyanide}}{CH_3.C \equiv N} \xrightarrow{\text{acid or alkali}} \underset{\text{Acetamide}}{CH_3CONH_2} \xrightarrow{H_2O} \underset{\text{Acetic acid}}{CH_3COOH} + NH_3$$

(*iii*) *From Grignard reagent and carbon dioxide (dry ice) :*

$$\underset{\text{Carbon dioxide}}{\overset{\overset{\large O}{\|}}{C}=O} \xrightarrow{CH_3MgI} CH_3-\overset{\overset{\large O}{\|}}{C}-OMgI \xrightarrow{\text{hydrolysis}} \underset{\text{Acetic acid}}{CH_3-\overset{\overset{\large O}{\|}}{C}-OH}$$

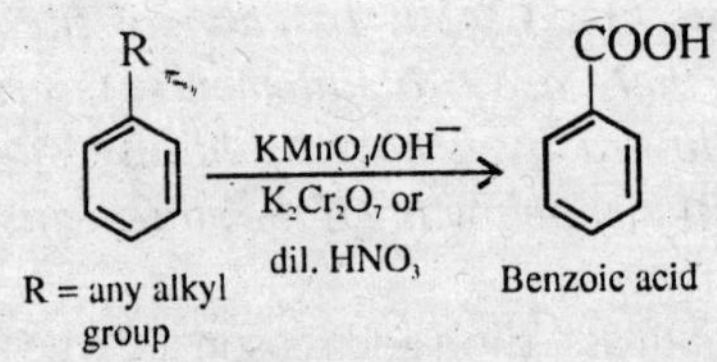

(*a*) $C_6H_6 + \underset{\text{Carbonyl chloride}}{COCl_2} \xrightarrow{\text{Anhy. AlCl}_3} \underset{\text{Benzoyl chloride}}{C_6H_5COCl} \xrightarrow{H_2O} C_6H_5COOH$

$$C_6H_6 + COCl_2 + C_6H_6 \xrightarrow{\text{Anhy. AlCl}_3} \underset{\text{Benzophenone}}{C_6H_5.CO.C_6H_5} + 2HCl$$

(*iv*) *Carboxylation of Alkenes :*

$$\underset{\text{Ethylene}}{CH_2{=}CH_2} + CO + H_2O \xrightarrow[400^\circ]{H_3PO_4} \underset{\text{Propionic acid}}{CH_3CH_2COOH}$$

$$CH_3—CH{=}CH_2 + CO + H_2O \xrightarrow[400°]{H_3PO_4} CH_3—\underset{|}{\overset{CH_3}{CH}}—COOH$$

Propene   Isobutyric acid

Formic Acid (Methane Acid) HCOOH

Formic acid occurs in the sting of bees, wasps, red ants and stinging nettles. In traces, it is also present in perspiration, urine, blood. It was first obtained in 1670 by the distillation of red ants and hence the name (Latin Formica = ant)

Preparation : It is prepared

(i) By the oxidation of formaldehyde

$$HCHO \xrightarrow{K_2Cr_2O_7—H_2SO_4} HCOOH$$

(ii) By the hydrolysis of hydrogen cyanide, HCN.

(iii) **In laboratory** it is prepared by heating glycerol with oxalic acid at 100-110°C. Since formic acid boils at 100.8°C, anhydrous formic acid can't be obtained by fractional distillation. Anhydrous formic acid can be obtained by distilling a mixture of sodium formate and sodium hydrogen sulphate.

$$HCOONa + NaHSO_4 \rightarrow HCOOH + Na_2SO_4$$

(iv) ***Manufacture :*** By heating NaOH and CO at 200°C under 6-10 atmospheric pressure.

$$NaOH + CO \xrightarrow[6\text{-}10\,atm]{200°C} HCOONa$$

$$\underset{\text{Sod. formate}}{HCOONa} + NaHSO_4 \xrightarrow{distillation} HCOOH + Na_2SO_4$$

Properties Physical : Formic acid is a colourless, sharp smelling, irritating liquid, b.p. 100.8°C. It is miscible with water, ethanol and ether. It is very corrosive and causes burns and blisters on the skin.

Chemical : Formic acid is unique because of having both an aldehyde group and a carboxyl group.

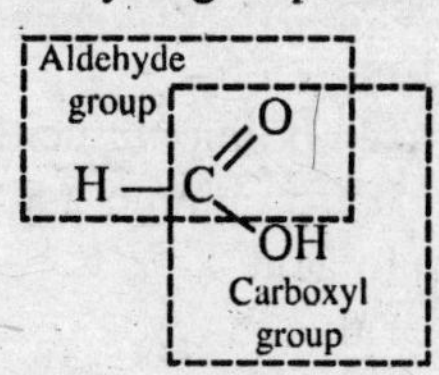

The reactions of formic acid are in keeping with its above structural feature.

In addition to properties discussed earlier, formic acid gives the following reactions.

(1) Action of conc. H_2SO_4 :

$$HCOOH \xrightarrow{H_2SO_4,\,heat} H_2O + CO$$

(2) Action of Heat : $HCOOH \xrightarrow{160°C} H_2 + CO_2$

$$HCOOH + \underset{\text{Mercuric chloride}}{2HgCl_2} \rightarrow \underset{\text{Mercurous chloride}}{Hg_2Cl_2}\downarrow + 2HCl + CO_2\uparrow$$

(3) Salt formation : Formic acid is the strongest acid among all the members of the homologous series. It forms salts with alkalis, carbonates and bicarbonates.

$$HCOOH + NaOH \rightarrow \underset{\text{Sal. formate}}{HCOO^-Na^+} + H_2O$$

$$HCOOH + NaHCO_3 \rightarrow HCOO^-Na^+ + CO_2\uparrow + H_2O$$

(4) Esterification : Formic acid reacts with alcohols to form esters. It is not necessary to add a mineral acid to catalyse the reaction since the formic acid acts as its own catalyst.

$$HCOOH + CH_3OH \overset{\Delta}{\rightleftharpoons} \underset{\text{methyl formate}}{HCOOCH_3} + H_2O$$

Uses: (1) It is used in the textile industry for dyeing processes.

(2) It is used in tanning for removing lime from the hides.

(3) It is used as a coagulant for rubber latex.

(4) It is used in fruit preservation.

(5) It is also used as an antiseptic, in the treatment of gout and in the laboratory preparation of carbon monoxide.

Ethanoic Acid, CH_3COOH (Acetic acid)

Acetic acid occurs in fruit juices which have become sour as a result of fermentation. It is a chief constituent of vinegar.

Manufacture

Quick Vinegar Process : This process is limited to the production of vinegar for food industry. Vinegar is 4 to 6 per cent acetic acid and is obtained from fermented liquors containing 12 to 15 per cent ethanol. The aqueous solution of ethanol in contact with air and under the influence of *Bacterium aceti* produces acetic acid.

$$2CH_3CH_2OH + O_2 \xrightarrow{B.aceti} 3CH_3CHO + 2H_2O$$
Ethanol — Acetaldehyde

$$2CH_3CHO + O_2 \xrightarrow{B.aceti} 2CH_3COOH$$
Acetic acid

Properties : At ordinary temperature, acetic acid is a colourless, corrosive liquid with a sharp vinegar odour and sour taste (b.p. 118°C, m.p. 16.6°C). Its vapours are suffocating and cause damage to lungs. Below 16.6°C pure acetic acid is converted into ice-like solid, hence it is called as Glacial Acetic Acid. It gives all the general reactions discussed earlier.

Uses : Acetic acid is used :

(*i*) in the manufacture of dyes, perfumes and rayon.

(*ii*) as a coagulant in the manufacture of rubber from latex and casein from milk.

(*iii*) as vinegar for table purpose and for manufacturing pickles.

(*iv*) in the manufacture of chemicals like acetone, acetic anhydride, etc.

Aspirin

Salicylic acid derivative; most important synthetic analgesic and antipyretic and antinflammatory agent.

Synthesis :

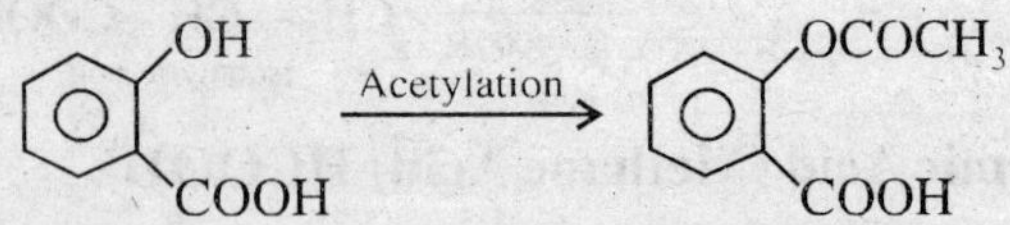

Salicylic acid — Acetyl salicylic acid (Aspirin)

It is common remedy for relieving pain. However, it is not recommended for children. The toxic side-effect is gastric irritation leading to ulcers. Therefore, it should not be taken empty stomach. It is used widely in the treatment of acute state of rheumatism.

The calcium salt of aspirin is better than simple aspirin which is water soluble. It has fewer undesirable side effects and induces analgesia faster than aspirin. A familar preparation of soluble asprin is *Disprin.* Disprin contains calcium carbonate and anhydrous citric acid besides asprin and this reduces pain faster than aspirin.

MULTIPLE CHOICE QUESTIONS

1. Weakest acid among the following is:
(*a*) CH_3COOH (*b*) C_6H_5OH
(*c*) H_2O (*d*) C_2H_2

2. Which of the following is least acidic?
(*a*) C_2H_5OH (*b*) CH_3COOH
(*c*) C_6H_5OH (*d*) $ClCH_2COOH$

3. An organic acid when heated strongly with P_2O_5, gave rise to a colourless gas which burns with a pale blue flame. The acid is:
(*a*) Acetic acid (*b*) Formic acid
(*c*) Formalin (*d*) Benzoic acid

4. Vinegar contains:
(*a*) 10 to 20% acetic acid
(*b*) 10% acetic acid
(*c*) 6 to 10% acetic acid
(*d*) 100% acetic acid

5. When acetyl chloride reacts with any amine, the reaction is known as:
(*a*) Saponification (*b*) Esterification
(*c*) Acetylation (*d*) Condensation

6. Salicylic acid treated with Zn dust, will give the product
(*a*) benzoic acid (*b*) diphenyl
(*c*) benzene (*d*) toluene

7. Formic acid is not representative member of carboxylic acids because
(*a*) it does not contain alkyl group
(*b*) it contains an aldehydic group while the other acids do not
(*c*) it is the first member of the series
(*d*) it is a gas

8. Lactic acid is
(*a*) β-hydroxy propionic acid
(*b*) α-hydroxy propionic acid
(*c*) propionic acid
(*d*) None of the above

9. Which of the following methods will give a carboxylic acid?
   (*a*) Oxidation of primary alcohol with acidified $KMnO_4$
   (*b*) Oxidation of toluene with acidified $K_2Cr_2O_7$
   (*c*) Hydrolysis of cyanide in acidic medium
   (*d*) All of the above

10. Which of the following reduces ammoniacal silver nitrate?
   (*a*) Formic acid (*b*) Propionic acid
   (*c*) Benzoic acid (*d*) Oxalic acid

11. When methyl magnesium iodide reacts with CO_2 and subsequent hydrolysis takes places it gives
   (*a*) propionic acid (*b*) ethyl alcohol
   (*c*) formic acid (*d*) acetic acid

12. Which of the following will be most highly ionised in water?
   (*a*) $RCH_2Cl\ CH_2COOH$
   (*b*) $CH_3CH_2CH_2COOH$
   (*c*) $CH_3CH_2CH_2ClCOOH$
   (*d*) $CH_3CH_2ClCH_2COOH$

13. The acid that makes lemon sour is
   (*a*) citric acid (*b*) nitric acid
   (*c*) oxalic acid (*d*) tartaric acid

14. Secondary alcohols on heating with reduced copper to 573 K give
   (*a*) aldehydes (*b*) ketones
   (*c*) alkenes (*d*) primary alcohols

Directions Qs. 15-22: *In each of the following questions there are two statements Assertion (A) is given followed by a corresponding statement of Reason (R) just below it. Of the statements, mark the correct answer:*

(*a*) If both assertion and reason are true and reason is the correct explanation of assertion.
(*b*) If both assertion and reason are true but reason is not the correct explanation of assertion
(*c*) If assertion is true but reason is false
(*d*) If assertion is false but reason is true.

15. **Assertion (A):** Boiling points of acids are lower than the corresponding alcohols.
   Reason (R): Esters are hydrolysed with dilute mineral acids to give carboxylic acid.

16. **Assertion (A):** Carboxylic acids are stabilised by resonance.
   Reason (R): Chloroacetic acid is weaker than acetic acid.

17. **Assertion (A):** Benzoic acid does not give Friedel Craft's reaction.
   Reason (R): Benzoic acid is obtained by catalytic oxidation of toluene with air in presence of Co-Mn as catalyst.

18. **Assertion (A):** First four aliphatic monocarboxylic acids are colourless.
   Reason (R): Carboxylic acids with more than five carbon atoms are insoluble in water.

19. **Assertion (A):** Acetic acid in vapour state shows a molecular mass of 120.
   Reason (R): It undergoes intermolecular hydrogen bonding.

20. **Assertion (A):** Highly branched carboxylic acids are more acidic than unbranched acids.
   Reason (R): Hydrogen bonding in carboxylic acid is stronger than alcohols.

21. **Assertion (A):** More electropositive element has greater electron donating effect.
   Reason (R): Me_3SiCH_2COOH is more acidic than Me_3CCH_2COOH.

22. **Assertion (A):** Carboxylic acids do not give characteristic reactions of carbonyl group.
   Reason (R): Carboxylic acids exist as cyclic dimers in solid, liquid and even in vapour state.

ANSWERS

1	2	3	4	5	6	7	8	9	10
(*d*)	(*a*)	(*b*)	(*c*)	(*c*)	(*a*)	(*b*)	(*a*)	(*d*)	(*a*)
11	**12**	**13**	**14**	**15**	**16**	**17**	**18**	**19**	**20**
(*d*)	(*c*)	(*a*)	(*c*)	(*d*)	(*c*)	(*b*)	(*c*)	(*a*)	(*d*)
21	**22**								
(*c*)	(*b*)								

29 ESTERS

Esters are derivatives of carboxylic acids in which the —OH group has been replaced by —OR (alkoxy) group.

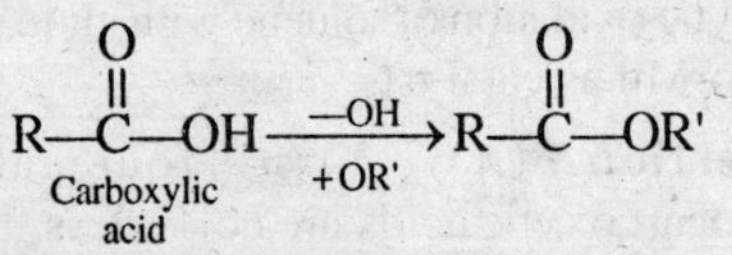

The esters of carboxylic acids are often referred to as 'Carboxylic Esters'. The type formula of this class of acid derivatives is

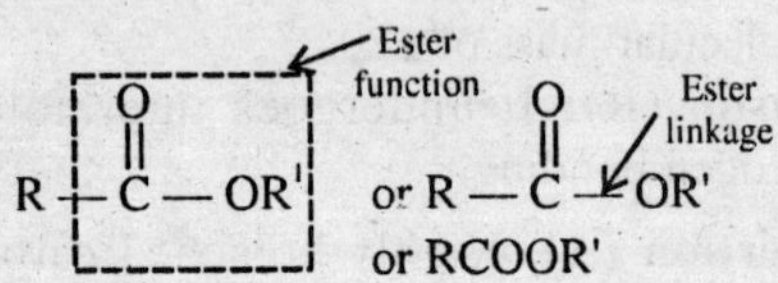

The ester functional group (or function) may be represented as —CO—OR' or —COOR'.

Formula	*Common Name*	*IUPAC Name*
$H{-}\overset{O}{\overset{\|}{C}}{-}OCH_3$	Methyl formate	Methyl methanoate
$CH_3{-}\overset{O}{\overset{\|}{C}}{-}OCH_2CH_3$	Ethylacetate	Ethylethanoate
$CH_3(CH_2)_2{-}\overset{O}{\overset{\|}{C}}{-}OCH_3$	Methyl butyrate	Methyl butanoate

Ethyl Acetate

Ethyl acetate (systematically, ethyl ethanoate, commonly abbreviated EtOAc or EA) is the organic compound with the formula $CH_3COOCH_2CH_3$. This colorless liquid has a characteristic sweet smell (similar to pear drops) like certain glues or nail polish removers, in which it is used. Ethyl acetate is the ester of ethanol and acetic acid; it is manufactured on a large scale for use as a solvent.

Production : Ethyl acetate is synthesized industrially mainly via the classic Fischer esterification reaction of ethanol and acetic acid. This mixture converts to the ester in about 65% yield at room temperature:

$$CH_3CH_2OH + CH_3COOH \rightarrow CH_3COOCH_2CH_3 + H_2O$$

The reaction can be accelerated by acid catalysis and the equilibrium can be shifted to the right by removal of water. It is also prepared industrially using the Tishchenko reaction, by combining two equivalents of acetaldehyde in the presence of an alkoxide catalyst:

$$2CH_3CHO \rightarrow CH_3COOCH_2CH_3$$

Reactions : Ethyl acetate can be hydrolyzed in acidic or basic conditions to regain acetic acid and ethanol. The use of an acid catalyst accelerates the hydrolysis, which is subject to the Fischer equilibrium mentioned above. In the laboratory, and usually for illustrative purposes only, ethyl esters are typically hydrolyzed in a two step process starting with a stoichiometric amount of strong base, such as sodium hydroxide. This reaction gives ethanol and sodium acetate, which is unreactive toward ethanol:

$$CH_3CO_2C_2H_5 + NaOH \rightarrow C_2H_5OH + CH_3CO_2Na$$

Uses : Ethyl acetate is primarily used as a solvent and diluent, being favored because of its low cost, low toxicity, and agreeable odor. For example, it is commonly used to clean circuit boards and in some nail varnish removers (acetone and acetonitrile are also used). Coffee beans and tea leaves are decaffeinated with this solvent. It is also used in paints as an activator or hardener. Ethyl acetate is present in confectionery, perfumes, and fruits. In perfumes, it evaporates quickly, leaving only the scent of the perfume on the skin.

Ethyl acetate is the most common ester in wine, being the product of the most common volatile organic acid - acetic acid, and the ethyl alcohol generated during the fermentation.

In the field of entomology, ethyl acetate is an effective asphyxiant for use in insect collecting and study.

Methyl Salicylate

Methyl salicylate (oil of wintergreen or wintergreen oil) is a natural product of many species of plants. Some of the plants which produce it are called wintergreens, hence the common name. Plants containing methyl salicylate produce this organic ester (a combination of an organic acid with an alcohol) most likely as an anti-herbivore defense. If the plant is infested with herbivorous insects, the release of methyl salicylate may function as an aid in the recruitment of beneficial insects to kill the herbivorous insects. Aside from its toxicity, methyl salicylate may also be used by plants as a pheromone to warn other plants of pathogens such as tobacco mosaic virus. Numerous plants produce methyl salicylate in very small amounts.

Commercial production : Methyl salicylate can be produced by esterifying salicylic acid with methanol. Commercial methyl salicylate is now synthesized, but in the past, it was commonly distilled from the twigs of Sweet Birch (Betula lenta) and Eastern Teaberry (Gaultheria procumbens).

Uses : 1. Methyl salicylate is used as a rubefacient in deep heating liniments, and in small amounts as a flavoring agent at no more than 0.04%. It is also used to provide fragrance to various products and as an odor-masking agent for some organophosphate pesticides. If applied in too high quantities it can cause stomach and kidney problems.

2. Methyl salicylate is among the compounds that attract male orchid bees, who apparently gather the chemical to synthesize pheromones; it is commonly used as bait to attract and collect these bees for study.

3. Methyl salicylate can be used as a transfer agent, to produce a manual copy of an image on a surface.

4. Methyl salicylate is added in small amounts to glacial acetic acid to lower its freezing point while transport in cold countries.

5. Methyl salicylate is used as a simulant or surrogate for the research of chemical warfare agent sulfur mustard, due to its similar chemical and physical properties.

6. Methyl salicylate is one of several antiseptic ingredients in Listerine mouthwash produced by the Johnson & Johnson company.

Ethyl benzoate

Ethyl benzoate is the ester formed by the condensation of benzoic acid and ethanol. It is a colorless liquid that is almost insoluble in water, but miscible with most organic solvents. As with many volatile esters, ethyl benzoate has a pleasant odor. It is a component of some artificial fruit flavors.

MULTIPLE CHOICE QUESTIONS

1. Which of the following has most acidic hydrogen?
   (*a*) 3-hexanone
   (*b*) 2, 4-hexanedione
   (*c*) 2, 5-hexanedione
   (*d*) 2, 3-hexanedione

2. Hydrolysis of an ester gives acid A and alcohol B. The acid reduces Fehling's solution.

Oxidation of alcohol B gives acid A. The ester is:
(*a*) Methyl formate (*b*) Ethyl formate
(*c*) Methyl acetate (*d*) Ethyl acetate

3. Waxes are long chain compounds belonging to the class:
(*a*) Acids (*b*) Alcohals
(*c*) Esters (*d*) Ethers

4. Which of the following is a flavouring agent called 'oil of winter green'?
(*a*) Olive oil (*b*) Vinegar
(*c*) Methyl acetate (*d*) Methyl salicylate

5. Which of the following esters is used for artificial flavour of pine apple?
(*a*) Octyl acetate (*b*) Isobutyl formate
(*c*) Ethyl butyrate (*d*) Isoamyl valerate

6. IUPAC name of $CH_3COOC_2H_5$ is:
(*a*) Methyl propanoate
(*b*) Ethyl ethanoate
(*c*) Aceto ethene
(*d*) Ethoxy ethane

7. Preparation of b-hydroxy ester is favoured by:
(*a*) Cannizzaro's reaction
(*b*) Reformatsky reaction
(*c*) Claisen condensation
(*d*) Witting reaction

8. Ammonalysis of an ester gives:
(*a*) Amine (*b*) Amide
(*c*) Uride (*d*) None of these

9. Organic esters on hydrolysis yields:
(*a*) Acids (*b*) Alcohols
(*c*) Both (*a*) and (*b*) (*d*) None of these

10. Acetic anhydride reacts with diethyl ether in the presence of anhydrous $AlCl_3$ to give:
(*a*) Ethyl acetate
(*b*) Methyl propionate
(*c*) Methyl acetate
(*d*) Propionic acid

11. Aceto acetic ester behaves as:
(*a*) An unsaturated hydroxyl compound
(*b*) A keto compound
(*c*) Both of these ways
(*d*) None of these

12. Which of the following are constituents of oil of wintergreen?
(*a*) Aspirin (*b*) Salol
(*c*) Methyl salicylate (*d*) None of these

13. When diethyl ether is heated with concentrated HI, it produces
(*a*) ethanol (*b*) iodoform
(*c*) ethyl iodide (*d*) methyl iodide

Directions Qs. 14-15: *In each of the following questions there are two statements Assertion (A) is given followed by a corresponding statement of Reason (R) just below it. Of the statements, mark the correct answer:*
(*a*) If both assertion and reason are true and reason is the correct explanation of assertion.
(*b*) If both assertion and reason are true but reason is not the correct explanation of assertion
(*c*) If assertion is true but reason is false
(*d*) If assertion is false but reason is true.

14. **Assertion (A):** Boiling points of acids are higher than the corresponding alcohols.
Reason (R): Esters are reduced to aldehyde by DIBAL-H.

15. **Assertion (A):** Esters which contain α-hydrogens undergo Claisen condensation.
Reason (R): $LiAlH_4$ reduction of esters gives acids.

ANSWERS

1	2	3	4	5	6	7	8	9	10
(*b*)	(*c*)	(*c*)	(*d*)	(*b*)	(*c*)	(*a*)	(*b*)	(*a*)	(*a*)
11	**12**	**13**	**14**	**15**					
(*c*)	(*c*)	(*c*)	(*b*)	(*c*)					

30 POLYMERS

A substance consisting of a large number of repeating molecular units in a linear sequence is known as a polymer. These repeating molecular units, *i.e.*, monomers add in a sequential manner to yield a polymer, and this process of combination of monomers to form the giant molecules is known as *"Polymerisation"*.

$$\underset{\text{Monomer}}{nP} \longrightarrow \underset{\text{Polymerisation}}{P - P - P - P} \ldots\ldots \underset{\text{Polymer}}{Pn}.$$

Some polymers contain only one kind of monomers. They are known as "Homopolymers" and some polymers are such that the no monomer unit is same and these are known as "Co-polymers". Polymerisation is of two types:

(*i*) **Additional Polymerisation :** Here the monomer unit contain multiple bond. They undergo addition reaction with one another to yield a single giant molecule (addition polymer) and the process is known as Additional Polymerisation. Thus, the polymerisation of ethylene in presence of oxygen yields the additional polymer (polyethyl)

$$nCH_2 = CH_2 \xrightarrow{O_2 Heat} \underset{\text{Polyethylene}}{-(CH_2 - CH_2)_n}$$

(*ii*) **Condensation Polymerisation :** This type of polymerisation involves the elimination of a small molecule such as water, in the condensation of any two monomer units. Thus, when a dicarboxylic acid reacts with diamine, it forms a polyamide, *e.g.*, condensation of adipic acid with hexaethylene diamine yields the polyamide, *i.e.*, Nylon 66, as

$$\underset{\text{Adipic acid}}{HOOC\,(CH_2)_4COOH} - \underset{\text{Hexaethylene diamine}}{H_2N\,(CH_2)_6NH_2}$$

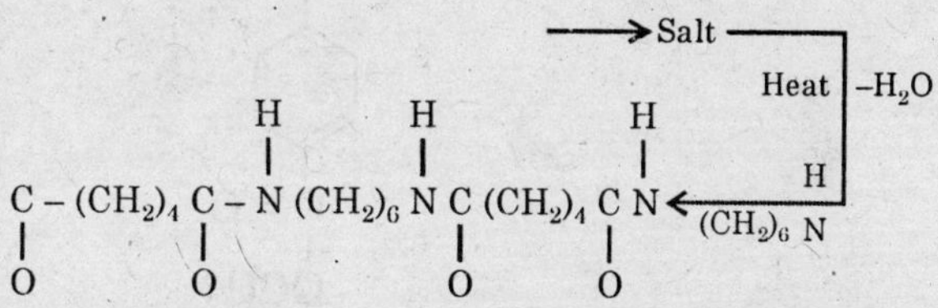

Nylon-66

The process of condensation polymerisation is largely used in the manufacture of synthetic fibre.

Polyethylene

Polymerisation of ethylene at high temperature and pressure in presence of oxygen as a catalyst yield polyethylene. Ethylene is first compressed to a pressure of 1000 atm and then in presence of oxygen.

$$nCH_2 = CH_2 \longrightarrow \underset{\text{Polyethylene}}{(..CH_2 - CH_2...)_n}$$

Dacron (Polyester)

It is a condensation polymer to terephathalic acid and ethylene glycol and is used for the manufacture of synthetic fibre. The polymer is made by the condensation of ethylene glycol and methyl terephallate. And methyl terephallate is itself obtained by subjecting a mixture of P-Xylene and methanol to oxidation with nitric acid. Here, first terephallate acid formed is udergoes esterification with the methanol to yield dimethyl ester which when treated with ethylene glycol give dacron.

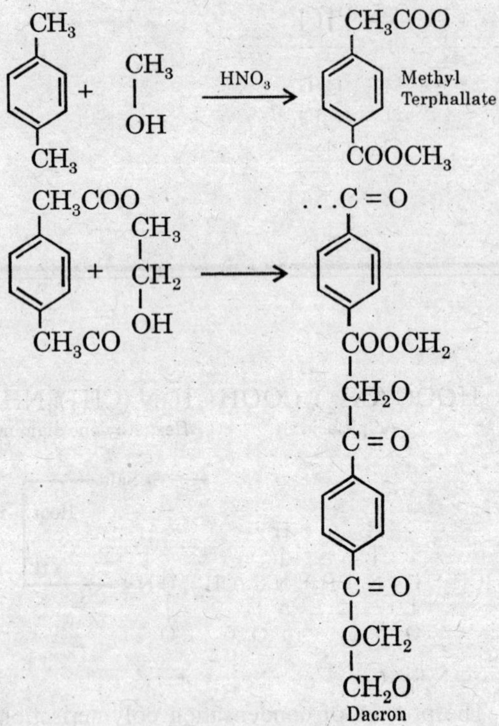

Nylon-66

Condensation of dibasic acids with diamines yields the plastic polyamides—the nylon. When adipic acid is allowed to copolymerise with hexametyl diamine it yields nylon - 66.

$$\underset{\text{Adipic Acid}}{HOOC-(CH_2)_4-COOH} - \underset{\text{Hexamethyl diamine}}{H_2N-(CH_2)_6\,NH_2}$$

$$\downarrow$$

$$HN-(CH_2)_6-NH-\underset{\underset{O}{\|}}{C}-(CH_2)_4CO\;NH-(CH_2)_6$$

$$\ldots\underset{\underset{O}{\|}}{C}-(CH_6)_4-\underset{\underset{O}{\|}}{C}-NH$$

Adipic acid required for the formation of nylon may be obtained by the oxidation of cyclo hexanol or bezene derived cyclohexane by nitric acid, as

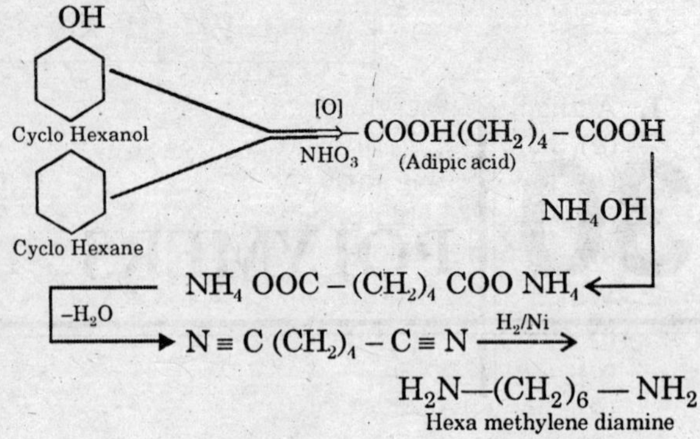

Rubber

Chemically, natural rubber is a polymer of isoprene. This is proved by the fact that destructive distillation of rubber yields isoprene.

$$\underset{\text{Isoprene}}{CH_2=\underset{\underset{CH_3}{|}}{C}-CH=CH_2} \quad \underset{\text{Natural rubber}}{(-CH_2-\underset{\underset{CH_3}{|}}{C}-CH-CH_2)_n}$$

Vulcanization of Rubber : Natural rubber can be made elastic by heating with about 5 per cent of sulphur. This process is known as the vulcanization of rubber.

A number of synthetic substitutes of rubber have been developed. They are not identical with natural rubber but they show many properties similar to it. Copolymeri-sation of 1, 3 - butadiene with styrene yields a grade of rubber known as "Buna-S" and copolymerisation of butadiene and acrylo-nitrile yields a synthetic rubber "Buna-N" which is highly resistant to abrasion. Chloroprene undergoes free radical polymerisation to yield a rubber like material known as neoprene.

$$\underset{\text{Chloroprene}}{CH_2=\underset{\underset{Cl}{|}}{C}-CH=CH_2} \rightarrow \underset{\text{Neoprene}}{(-CH_2-\underset{\underset{Cl_3}{|}}{C}=CH-CH_2^-)_n}$$

Butyl Rubber : The polymerisation of isobutylene in presence of boron trifluoride as a catalyst at low temperature yield butyl rubber.

$$\text{Isobutelene} \xrightarrow[+H_2O]{BF_3} \text{Butyl Rubber}$$

MULTIPLE CHOICE QUESTIONS

1. Acrilan (Polyacrylonitrile) is a
(*a*) hard and thorny solid
(*b*) low melting material
(*c*) water soluble polymer
(*d*) a condensation polymer

2. Which of the following polymer is developed by step growth polymeri-zation?
(*a*) Polyisoprene
(*b*) Polyvinyl chloride
(*c*) Nylon-66
(*d*) Polybutadiene

3. An example of chain growth polymer is :
(*a*) Nylon-6 (*b*) Bakelite
(*c*) Terylene (*d*) Teflon

4. Which of the following polymers is formed by condensation of adipic acid and hexamethylene diamine?
(*a*) Bakelite (*b*) Nylon-6
(*c*) Nylon-66 (*d*) Glyptal

5. Identify the synthetic polymer among the following polymers
(*a*) PVC (*b*) Protein
(*c*) Nucleic acids (*d*) Cellulose

6. Adipic acid and hexa methylenediamine on polymerization form
(*a*) Nylon-6 (*b*) Nylon-66
(*c*) Terylene (*d*) Melamine

7. In the sample in question number 16; mass average molecular mass of the sample is
(*a*) 50,000 (*b*) 51,740
(*c*) 55,750 (*d*) 58,000

8. Tetrafluoroethylene on polymerization gives
(*a*) Teflon (*b*) PCTFE
(*c*) PMMA (*d*) None of these

9. Ethylene glycol and terepthalic acid on polymerization form
(*a*) Stretchlon (*b*) Nylon
(*c*) Terylene (*d*) None of these

10. Polystyrene is a/an
(*a*) addition polymer
(*b*) condensation polymer
(*c*) dimer
(*d*) repeat unit

Directions Qs. 11-15: *In each of the following questions there are two statements Assertion (A) is given followed by a corresponding statement of Reason (R) just below it. Of the statements, mark the correct answer:*
(*a*) If both assertion and reason are true and reason is the correct explanation of assertion.
(*b*) If both assertion and reason are true but reason is not the correct explanation of assertion
(*c*) If assertion is true but reason is false
(*d*) If assertion is false but reason is true.

11. Assertion (A): Thermoplastics become hard on heating and soft on cooling.
Reason (R): Thermoplastic are linear polymers which are soluble in many organic solvents.

12. Assertion (A): Bakelite is a thermosetting plastic.
Reason (R): In thermosetting plastics, the polymeric chains are held together by strong covalent bonds.

13. Assertion (A): Natural rubber is a polymer of isoperne.
Reason (R): Isoprene is a pentene.

14. Assertion (A): Vulcanisation depends upon the amount of sulphur used.
Reason (R): Vulcanisation involves cooling of rubbers.

15. Assertion (A): Buna-S is a condensation polymer.
Reason (R): In condensation polymerisation, condensation takes place between two or more functional groups to form large polyfunctional molecules.

ANSWERS

1	2	3	4	5	6	7	8	9	10
(*a*)	(*c*)	(*d*)	(*c*)	(*a*)	(*b*)	(*b*)	(*a*)	(*c*)	(*a*)
11	**12**	**13**	**14**	**15**					
(*d*)	(*b*)	(*c*)	(*c*)	(*d*)					

31 BIO-CHEMISTRY

Carbohydrates

The term carbohydrate was originally coined for the group of carbon compounds which contain hydrogen and oxygen known as "Hydrates of Carbon" and they can be represented by general formal "$C_x (H_2O)_y$." *e.g.* glucose $C_6H_{12}O_6$ or $C_6 (H_2O)_6$ and sucrose $C_{12}H_{22}O_{11}$.

Chemistry of D (+) Glucose

D (+) Glucose : Naturally occurring glucose is dextrotatory and known as dextrose. Generally it occurs in ripe grapes and honey and is a normal constituent of blood. It is optically active and has a specific rotation of + 52.5°C.

Preparation

Hydrolysis of starch : Glucose is obtained by the hydrolysis of strach with dilute acids.

$$(C_6 H_{10} O_5)_n + {}_n H_2O \xrightarrow[\text{Press. 3–5 atm}]{\text{dil } H_2SO_4} {}_nC_6H_{12}O_6 \ (\text{Glucose})$$

Chemical Reactions

1. *Acetylation :* The acetylation of glucose with acetic anhydride yield penta acetyl derivative, as

$$\begin{array}{c} CHO \\ | \\ (CHOH)_4 \\ | \\ CH_2OH \end{array} + (CH_3CO)_2O \rightarrow \begin{array}{c} CHO \\ | \\ (CHOCOCH_3)_4 \\ | \\ CH_2OCOCH_3 \end{array}$$

2. *Condensation with hydroxyl amine :* Glucose forms an oxime when made to react with hydroxylamine.

3. *Oxidation :* Mild oxidising agent like bromine-water oxidise the aldehydic group to the corresponding acid as the glucose gives gluconic acid.

$$\begin{array}{c} CHO \\ | \\ (CHOH)_4 \\ | \\ CH_2OH \end{array} \xrightarrow[Br_2/H_2O]{[O]} \begin{array}{c} COOH \\ | \\ (CHOH)_4 \\ | \\ CH_2OH \end{array} \text{ Gluconic acid}$$

4. *Reduction :* When glucose is reduced with sodium amalgam in aqueous solution, the aldehydic group is reduced to alcoholic group and sorbitol is formed.

5. *Formation of osazone :* When glucose is treated with excess of phenyl hydrazine it forms an osazone.

Amino Acids

Amino acids are the monomer units which make up all the peptides and proteins. Hydrolysis of proteins, or peptides yield a mixture of amino acids, which have been found off 26 kinds and 10 of them are essential amino acids. Acid hydrolysis is the most satisfactory for the proteins which do not contain acid libile amino acids.

All the amino acids exist as in the internal salt, in which the acidic hydrogen of the carboxyl group has been transferred to the basic NH_2 group.

$$R{-}\underset{\underset{NH_2}{|}}{CH}{-}COOH \rightleftharpoons R{-}\underset{\underset{NH_3^+}{|}}{CH}{-}COO^-$$

Such a doubly charged ion is known is Zwitterion or dipolar ion.

Chemical Characteristics : Amino acids are neutral in water, it is basic in alkaline medium and acidic in acidic medium. Therefore, the characteristic reactions of amino group must be carried out in alkaline medium and that of carboxyl group in acidic medium.

1. **Alkylation :** Amino acids undergoes alkylation as follows :

$$\underset{\displaystyle NH_2}{R-\underset{|}{CH}-COO^-} + R'X \rightarrow \underset{\displaystyle NHR'}{R-\underset{|}{CH}-COO^-}$$

2. **With acid chlorides :** Amino acids reacts with acids chlorides in the acidic medium as

$$\underset{\displaystyle NH_2}{R-\underset{|}{CH}-COOH} + SOCl_2 \xrightarrow{H^+} \underset{\displaystyle NH_2}{R-\underset{|}{CH}-COOH}$$

Proteins

Proteins are the principal substances which make up a large part of skin, muscle, blood, harmones etc. They derive their name from the Greek word proteins which means Primary. They occur in protoplasm of all living cells. Their composition varies from the source from which they are isolated and they contain.

Characteristics : (*i*) Proteins neither have a characteristics melting point nor characteristics decomposition temperature.

(*ii*) They form a colloidal solution in water.

(*iii*) Proteins can be precipitated either reversibly or irreversibly. The irreversible precipitation can be carried out with the strong inorganic acids and they are said to be denatured. The reversible precipitation may be carried out by the addition of salts such as $(NH_4)_2$ SO_4 or $HgSO_4$ etc.

(*iv*) Proteins can behave as anions or cations at different pH and have isoelectric point.

2. *Classification on the basis of chemical complexity :*

(*a*) ***Simple Proteins :*** *Albumins, Globulins, Prolamins etc.*

(*b*) ***Conjugated Proteins :*** *Nucleo proteins, Glycoproteins, Phosphoproteins, Lipo proteins etc.*

(*c*) ***Derived proteins :*** *Denatured proteins, Primary proteins, Secondary proteins, Peptides and peptones etc.*

MULTIPLE CHOICE QUESTIONS

1. Vitamin 'A' deficiency leads to the disease known as
(*a*) blindness (*b*) beri-beri
(*c*) scurvy (*d*) T.B.

2. Ascorbic acid is a
(*a*) vitamin (*b*) enzyme
(*c*) protein (*d*) None of these

3. Fats and oils are mixture of
(*a*) acids (*b*) alcohols
(*c*) esters (*d*) aldehydes

4. A certain compound gives negative test with ninhydrin and positive test with Benedict's solution. The compound is
(*a*) a protein
(*b*) a monosaccharide
(*c*) a liquid
(*d*) an amino acid

5. The sugar part of nucleic acid is
(*a*) ribose (*b*) glucose
(*c*) formose (*d*) sorbose

6. Detergent has a common group
(*a*) $R\text{-}OSO_3Na$ (*b*) R-ONa
(*c*) R-COONa (*d*) NaOH

7. Chemically considering, digestion is basically
(*a*) anabolism (*b*) hydrogenation
(c) hydrolysis (*d*) None of these

8. Genes are composed principally of
(*a*) DNA (*b*) ATP
(c) RNA (*d*) Hormones

9. Which of the following contains iron?
(*a*) Chlorophyll (*b*) Ascorbic acid
(*c*) Haemoglobin (*d*) Vitamin K

10. Proteins are hydrolysed by enzymes into
(*a*) dicarboxylic acid
(*b*) hydroxy acid
(*c*) amino acid
(*d*) aromatic acid

11. Cane sugar is
(*a*) Lactose (*b*) Maltose
(*c*) Fructose (*d*) Sucrose

12. Which one of the following is a final waste product of protein metabolism in human beings?
(*a*) Amino acid (*b*) Urea
(*c*) Ammonia (*d*) Uric acid

13. Which of the following is commonly called polyamide?
(*a*) Nylon (*b*) Rayon
(*c*) Terylene (*d*) Orlon

Directions Qs. 14-22: *In each of the following questions there are two statements Assertion (A) is given followed by a corresponding statement of Reason (R) just below it. Of the statements, mark the correct answer:*

(*a*) If both assertion and reason are true and reason is the correct explanation of assertion.
(*b*) If both assertion and reason are true but reason is not the correct explanation of assertion
(*c*) If assertion is true but reason is false
(*d*) If assertion is false but reason is true.

14. Assertion (A): Glucose and fructose are reducing sugars.
Reason (R): Glucose and fructose contain a free aldehydic and ketonic group adjacent to a $>$ CHOH group respectively.

15. Assertion (A): Sucrose is a disaccharide.
Reason (R): Sucrose is dextro rotatory.

16. Assertion (A): Tollen's reagent, Benedict's solution and Fehling's solution are reducing agents.
Reason (R): Mild oxidising agents like chlorine or bromine water convert glucose into gluconic acid.

17. Assertion (A): Fructose reduces Fehling's solution and Tollen's reagent.
Reason (R): Fructose does not contain any aldehyde group.

18. Assertion (A): Starch and cellulose reduce Tollen's reagent.
Reason (R): Starch and cellulose on hydrolysis give hundreds to thousands of monosaccharides molecules.

19. Assertion (A): Acid or enzymatic hydrolysis of sucrose to give an equimoiar mixture of glucose and fructose is called inversion.
Reason (R): Sucrose is the only naturally occurring disaccharide which is reducing.

20. Assertion (A): The specific rotation of a freshly prepared solution of α-glucose decreases from $+112°$ to $52.7°$ while that of β-glucose increases from $+19°$ tc $52.7°$.
Reason (R): The change in specific rotation of an optically active compound with time to an equilibrium value is called mutarotation.

21. Assertion (A): Cellulose can be easily converted into fibres.
Reason (R): Cotton contains about 90% cellulose.

22. Assertion (A): Glucose is used in silvering of mirrors.
Reason (R): Glucose is less acidic than a monohydric alcohol.

ANSWERS

1	2	3	4	5	6	7	8	9	10
(*a*)	(*a*)	(*c*)	(*b*)	(*a*)	(*a*)	(*c*)	(*a*)	(*c*)	(*c*)
11	**12**	**13**	**14**	**15**	**16**	**17**	**18**	**19**	**20**
(*d*)	(*b*)	(*a*)	(*a*)	(*b*)	(*d*)	(*b*)	(*d*)	(*b*)	(*b*)
21	**22**								
(*b*)	(*c*)								

BIOLOGY

1

Nature and Scope of Biology

Biology is the Study of life and living organism. The term biology was coined by Jean Baptiste Lamarck and G. Treviranus independently in 1802.

Zoology (from ancient Greek *zoon*, animal and *logos*, knowledge). The branch of biology which deals with study of structure, function and evolution animals. Father of Zoology : Aristotle.

The study of plants is called Botany (botane = herb). Theophrastus (370-287 B.C.) prepared a list of 500 plants and is called 'Father of Botany'.

BRANCHES OF BIOLOGY

- Anatomy: Study of internal structure.
- Bacteriology: Study of structure, reproduction and function of bacteria.
- Biogeography: Study of geographical distribution of plants and animals.
- Cardiology: Study of heart.
- Cytology: Study of cell structure and function.
- Dermatology: Study of skin.
- Ecology: Study of plant and animals in relation to their environment.
- Embryology: Study of formation, growth and development of embryo and gametes.
- Entomology: Study of insects.
- Epidemiology: Study of epidemic disease.
- Etiology: Study of causes of disease.
- Evolution: Study of change in the inherited traits of a population through successive generations.
- Genetics: Study of heredity and variations.
- Gynaecology: Study of health of female reproductive organs.
- Haematology: Study of blood.
- Helminthology: Study of parasitic worms.
- Herpetology: Study of lizards and snakes.
- Histology: Study of structures of tissues.
- Ichthyology: Study of fishes.
- Immunology: Study of immunity or resistance to disease.
- Kalology: Study of human beauty.
- Karyology: Study of whole set of chromosomes.
- Limnology: Study of fresh water ecology.
- Malacology: Study of molluscs.
- Mammalogy: Study of mammals.
- Microbiology: Study of micro organisms.
- Molecular Biology: Study of biology at molecular level.
- Molecular Genetics: Study of genetics at molecular level.
- Morphology: Study of form and structure of organisms.
- Mycology: Study of fungi.
- Nematology: Study of nematodes.
- Neonatology: Study of medical care of new born infant.

- ❍ Nephrology: Study of function and disease of kidneys.
- ❍ Neurology: Study of function and disease of nervous system.
- ❍ Odontology: Study of teeth and gums.
- ❍ Oncology: Study of cancers and tumours.
- ❍ Ophthalmology: Study of eyes.
- ❍ Ornithology: Study of birds.
- ❍ Orthopaedics: Study of diagnosis and repair of disorders of locomotory system (knees, joints etc).
- ❍ Otology: Study of eggs particularly those of birds.
- ❍ Paediatrics: Study of medical care of infants, children and adolescents.
- ❍ Palaeobotany: Study of fossil plants.
- ❍ Parasitology: Study of parasites their lost and the relationship.
- ❍ Pathology: Study of disease pathogens, their transmission and effects.
- ❍ Pharmacognosy: It is a branch of pharmacology that deals with natural drugs and their constituents.
- ❍ Phycology: Study of algae.
- ❍ Physiology: Study of process and functioning of an organism.
- ❍ Pomology: Study of fruits and fruit yielding plants.
- ❍ Radiology: Study of x-rays and other imaging techniques for medical diagnosis.
- ❍ Space Biology: Study of living organism and their surrounding environment in the outer spaces.
- ❍ Taxonomy: Study of classification, nomenclature and identification of organisms.
- ❍ Toxicology: Study of harmful effects of drug and other substances.
- ❍ Urology: Study of structure, action, functions and disorders of urinary tract (urinogenital tracts in males).
- ❍ Venereology: Study and treatment of sexually transmitted disease.
- ❍ Virology: Study of viruses.
- ❍ Xylology: Study of anatomy of wood/xylem.
- ❍ Xylotomy: Preparation of small silver of wood using microtome.
- ❍ Zymology: Study of fermentation processes.

FAMOUS INDIAN BIOLOGISTS

Bryology: S.R. Kashyap (Father of Indian Bryology), P.N. Mehra

Cytogenetics: R.P. Roy, M.S. Swaminathan

DNA Finger Printing: Lalji Singh

Ecology: R. Mishra (Father of Indian Ecology)

Embryology: P. Maheshwari (Father of Indian Embryology)

***In-vitro* Fertilization:** Indira Hinduja

Morphology: V. Puri

Mycology: T.S. Sadasivam

Ornithology: Salim Ali

Palaeobotany: Birbal Sahni (Father of Indian Palaeobotany)

Plant Pathology: K.C. Mehta

Plant Breeding: B.P. Pal, M.S. Swaminathan

Plant Physiology: J.C. Bose, J.J. Chinoy

Phycology: M.O.P. Iyengar (Father of Indian Algology)

Pteridology: P.N. Mehta

Xylotomy: K.A. Chaudhari

SOME IMPORTANT BOOKS

***Anatomia Plantarum* by** Marcello Malpighi

Classification of Flowering Plants *by* Rendle

***Die Naturlichen Pflanzen Familien* by** Engler and Prantl

***Elora Laponica* by** Linnaeus

Essay on Population *by* Malthus

Flora of India *by* Roxburgh

***Genera Plantarum* by** Bentham and Hooker

Genera Plantarum secundum ordines naturales deposita by de Jussieu
Historia Animalism by Aristotle
Historia Generalis Plantarum by John Ray
On the Structure of Human Body *by* Vasalius
Origin of Life on Earth *by* Oparin
Philosophic Zoologique *by* Lamarck
Secret Life of Plants *by* Webster
Species Plantarum by Linnaeus
System Naturae by Linnaeus
The Double Helix *by* Watson
The Families of Flowering Plants *by* Hutchinson

FATHER OF VARIOUS DISCIPLINES IN BIOLOGY

Discipline	*Father*
Anatomy (Modern)	Andreas Vesalius
Antibiotics	Alexander Fleming
Antiseptic Surgery	Joseph Lister
Ayurveda	Charak
ATP Cycle	Lipman
Bacteriology	Robert Koch
Biochemical Genetics	Archibald Garrod
Biology	Aristotle
Blood Circulation	William Harvey
Blood Group	Landsteiner
Botany	Theophrastus
Bryology	Hedwig
Chemotherapy	Paul Ehrlich
Chromatography	Michael Tswett
Comparative Anatomy	Bason George Cuvier
Cytology	Robert Hooke
Cytology (Modern)	Carl P. Swanson
DNA Finger Printing	Allec Jaffreys
Electrocardiography	Einthoven
Ethology	Konard Larentz
Embryology	Aristotle
Endocrinology	Thoman Addison
Epidemiology	John Snow
Eugenics	Francis Galton
Evolutionary ideas	Empedodes
Experimental Genetics	T.H. Morgan
Experimental Physiology	Galen

Discipline	*Father*
Genetics	Gregor Johann Mendel
Genetic Engineering	Paulberg
Gerentology	Korenchevsk
Green Revolution	Norman E. Borloug
Green Revolution of India	M.S. Swaminathan
Histology	Francois Bichat
Immunology	Edward Jenner
Indian Ecology	R. Mishra
Indian Mycology	E.J. Butler
Medicine	Hippocrates
Microbiology	Louis Pasteur
Microscopy	A.V. Leeuwenhoek
Mycology	Micheli
Modern Genetics	Bateson
Modern Embryology	K.V. Baer
Mutation	Hugo de Vries
Palaeobotany (Indian)	Birbal Sahni
Palaeontology	Leonardo da Vinci
Parasitology	Platter
Plant Pathology	Anton de Bary
Plant Anatomy	Nehemiah Grew
Phycology (Indian)	O.P. Iyengar
Plant Tissue Culture	Haberlandt
Plant Physiology	Stephen Hales
Radiation Genetics	H.J. Muller
Stress Physiology	Hans Selye
Taxonomy	Carolus Linnaeus
Virology	Stanley

IMPORTANT DAYS

Dates	*Days*
January 30	World Leprosy Eradication Day
February 28	National Science Day (Vigyan Divas)
March 15	World Disabled Day
March 21	World Forestry Day
March 23	World Meteorological Day
March 24	World Tuberculosis Day
April 7	World Health Day

Dates	Days
April 17	World Haemophilia Day
April 22	Earth Day
May 7	World Asthma Day
May 8	World Red Cross Day
May 8	World Thalassaemia Day
May 11	International Migratory Bird Day
May 31	Anti-tabacco Day
May 25	World Thyroid Day
June 1	World Milk Day
June 5	World Environment Day
June 14	World Blood Donor Day
July 1	Doctor's Day
July 11	World Population Day
August 20	World Mosquito Day
September 2	Coconut Day
September 16	World Ozone Day
September 21	Alzheimer's Day
October 1	Blood Donation Day
October 4	World Animal Welfare Day
October 16	World Food Day
October 17	World Trauma Day
October 20	World Osteoporosis Day
October 24	World Polio Day
November 14	World Diabetes Days
November 17	National Epilepsy Day
November 20	World Piles Day
December 1	World AIDS Day
December 2	National Pollution Prevention Day
December 3	World Conservation Day
December 23	Kishan Divas (National Farmer's Day)
December 29	International Biodiversity Day

IMPORTANT BOTANICAL GARDENS OF INDIA

- Garden of the Forest Research Institute, Dehradun.
- Garden of the Indian Agriculture Research Institute, New Delhi.
- Indian Botanic Garden, Sibpur, Kolkata.
- Lloyd Botanic Garden, Darjeeling.
- National Botanic Garden, Lucknow.

IMPORTANT BOTANICAL GARDENS OF THE WORLD

- Arnold Arboretum, Boston, USA.
- Bondes Garden, Viena, Austria.
- Botanical Garden, Berlin, Germany.
- Kebun Raya (Botanic Garden), Bogor, Java.
- Main Botanic Garden, Moscow.
- Missouri Botanic Garden, USA.
- Nikitisky Botanic Garden, Yalta, Ukraine.
- Royal Botanic Garden, Kew, London.
- Royal Botanic Garden, Edinburg.

LIFE SPAN OF SOME ORGANISMS

Organism	*Period*
Banyan Tree	200 years to 4000 years
Cat	35 years to 40 years
Dog	20 years to 30 years
Eagle	90 years
Elephant	65 years
Ficus religosa	2000 years to 3000 years
Horse	60 years
May fly	1 day
Monkey	26 years
Mouse	4 years
Man	100 years
Parrot	140 years
Sequoia tree	300 years to 4000 years
Tortoise	200 years
Tiger	22 years
White Oak Tree	500 years

SOME EXCEPTIONS IN BIOLOGY

- TMV (Plant virus) have RNA as genetic material instead of DNA.
- *Cuscuta* (total stem parasite) is a dicot without any cotyledons.
- *Rhizophora* and *Ceripos* roots are negatively geotropic unlike others plants.
- Desert Kangaro Rat never drinks water, they used only metabolic water.
- Fishes have gills for respiration (lung fishes have alveolar sacs)
- Alligators and Crocodiles have four chambered heart like mammals.
- Like birds, Duckbilled Platypus and Spiny Ant-eaters lay eggs but they are mammals.
- Generally birds do not have urinary bladder. Ostrich are exceptions to this and they have one urinary bladder.
- Mammals have seven cervical vertebrae. Sea cows have eight vertebrae.
- Salamanders unlike others can reproduce in their larval stages i.e. neoteny.

MULTIPLE CHOICE QUESTIONS

1. Animals like plants are respire:
(*a*) during day only
(*b*) during night only
(*c*) during day and night
(*d*) when required only

2. Study of animal behaviour comes under:
(a) Ecology (*b*) Ethology
(*c*) Euthenics (*d*) Eugenics

3. Area which cannot be subjected to scientific analysis is :
(*a*) human intelligence (*b*) human faith
(*c*) animal behaviour (*d*) Origin of life

4. Study of injuries and wounds is known as:
(*a*) Tricology (*b*) Tumorology
(*c*) Traumatology (*d*) Torpedology

5. The study of chemical aspect of living organisms:
(*a*) Microbiology (*b*) Biochemistry
(*c*) Cryobiology (*d*) Ethology

6. Dendrology is the study of:
(*a*) Grasses (*b*) Flowers
(*c*) Trees and shrubs (*d*) Trees

7. Agrostology is cultivation of:
(a) oil seeds (*b*) crops
(*c*) grasses (*d*) fruit trees

8. The primitive vascular plants are:
(A) Ferns (*b*) brown algae
(*c*) Cycas (*d*) Vernalisation

9. The father of taxonomy is:
(a) Charles Darwin (*b*) John Ray
(*c*) Julian Huxley (*d*) Carolus Linnaeus

10. The branch of zoology dealing with study of fishes is called:
(A) Herpetology (*b*) Ichthyology
(*c*) Pisciculture (*d*) Saurology

11. Mobile animals according to '*Susruta Samhita*' include
(*a*) Jan Gama (*b*) Matsya
(*c*) Kulacara (*d*) all above

12. Who demonstrated that RNA is the genetic material of Tobacco Mosaic Virus?
(*a*) Hershey and Chase
(*b*) Gierer and Schramm
(*c*) Jacob and Monod
(*d*) Hans Fraenkel-Conrat *et. al.*

13. Professor Salim Ali is associated with
(*a*) ornithology (*b*) serpentology
(*c*) ichthyology (*d*) saurology

14. Who for the first time developed antiseptic surgery?
(*a*) Pasteur (*b*) Susruta
(*c*) J. Lister (*d*) Caraka

15. Biochemical nature of the genes was established by
(*a*) Avery, Machleod and McCarty
(*b*) Johanssen
(*c*) Jacob and Monod
(*d*) H. Khoranna and Ochoa

16. Chance discoveries are referred to as
(*a*) serendipity (*b*) homeostasis
(*c*) phagocytosis (*d*) entropy

17. The genetic material TMV is
(*a*) RNA (*b*) DNA
(*c*) both above (*d*) none above

18. The scientists who are responsible to make first cloned animal, a sheep, are
(*a*) Lewis and others
(*b*) Ian Wilmut and co-workers
(*c*) Yanofsky and others
(*d*) Harris and Watkins

19. While solving a problem a scientific method is employed. Which of the following is incorrect?
(*a*) The observations are made directly or indirectly
(*b*) A possible hypothesis is formulated to explain the phenomenon
(*c*) Hypothesis is subjected to experiment
(*d*) No control is kept with an experiment

20. The term 'sarcode' was given to living substance of cell by
(*a*) Robert Brown (*b*) Robert Hooke
(*c*) Dujardin (*d*) Purkinje

Directions Qs. 21-25: *In the following questions there are two statements:* ***Assertion (A)*** *and* ***Reason (R)****. Consider them accordingly and answer:*

(*a*) If (A) and (R) both are true and (R) is correct explanation of (A).
(*b*) If (A) and (R) both are true and (R) is not correct explanation of (A).
(*c*) If (A) is true and (R) is wrong.
(*d*) If (A) and (R) are wrong.

21. Assertion (A): Human genome project involves studies on hundreds of animal and plant species along with human.
Reason (R): It help us to trace our own evolutionary history.

22. Assertion (A): The formation of a superbug by Dr. Chakravarty has blessed our aquatic biome.
Reason (R): It can remove oil spills by process of bioremediation.

23. Assertion (A): None of us should use clothing made out of Shahtoosh.
Reason (R): One shawl is made by killing hundreds of endangered young ones of chiru.

24. Assertion (A): A person living in hilly area has more active bone marrow.
Reason (R): At High altitude less oxygen is present.

25. Assertion (A): Penguins live in icy cold regions.
Reason (R): They can not be eaten by polar bear.

ANSWERS

1	2	3	4	5	6	7	8	9	10
(*c*)	(*b*)	(*b*)	(*c*)	(*b*)	(*c*)	(*c*)	(*a*)	(*d*)	(*b*)
11	**12**	**13**	**14**	**15**	**16**	**17**	**18**	**19**	**20**
(*d*)	(*d*)	(*a*)	(*c*)	(*a*)	(*a*)	(*a*)	(*b*)	(*d*)	(*c*)
21	**22**	**23**	**24**	**25**					
(*a*)	(*a*)	(*a*)	(*a*)	(*b*)					

2

Biological Systematics and Classification

PRINCIPLES AND METHODS OF TAXONOMY

Diversity is the occurrence of different varieties/ forms, develops due to organic evolution and their adaptation to overcome competition. About 1.5 million species of animals have been identified, classified and assigned their respective phyletic status.

Systematics (Gk. systema–order or sequence): It is the scientific study of types and diversity of organisms and their comparative and evolutionary relationships based on comparative anatomy, physiology, ecology, biochemistry and other fields. It is based on following fundamentals.

Taxonomy (Caroleus Linnaeus-Father of Taxonomy) (Gk. Taxis–arrangement, nomos-law) : The term Taxonomy coinod by De Candolle. It is the study of identification, nomenclature and classification of an organism.

- **Identification :** It is to determine the exact place or position of an organism in the set plan of classification.
- **Classification :** It is the arrangement of organisms in hierarchical series of groups and subgroups on the basis of similarities and differences in their traits so as to bring out their relationships.
- **Nomenclature :** It is the process of providing scientific names to the organism.
- **Key :** It provides a convenient way for easy identification of an organism by using diagnostic or distinguishing characters.

The taxonomic groups are arranged in definite order from higher to lower categories. Total seven obligate categories are :

Kingdom

Division/Phylum

Class

Order

Family

Genus

Species

Biological Classification

Biological classification, or scientific classification in biology, is a method by which biologists group and categorize organisms by biological type, such as genus or species. Biological classification is a form of scientific taxonomy.

Linnaeus (1735)	Haeckel (1866)	Chatton (1925)	Copeland (1938)	Whittaker (1969)
2 kingdoms	**3 kingdoms**	**2 empires**	**4 kingdoms**	**5 kingdoms**
(not treated)	Protista	Prokaryota	Monera	Monera
		Eukaryota	Protoctista	Protista
Vegetabilia	Plantae			Fungi
			Plantae	Plantae
Animalia	Animalia		Animalia	Animalia

COMPARATIVE ACCOUNT OF CLASSIFICATIONS OF ANGIOSPERMS

Taxonomy began with man's purely utilitarian interest in the flora of the earth. It grew with apparent observations of the various types of plants and their relationship which were, of course, based on the studies made by unaided eyes. The word taxonomy was introduced by A.P. de Candolle to designate the theory of plant classification. It has summoned innumerable people of the world since times gone by. It led to the systematic grouping of plants.

Classification is the inclusion of a plant or plants in categories based on a definite system and also according to the rules of International Code of Botanical Nomenclature (ICBN).

SYSTEMS OF CLASSIFICATION

(A) Artificial System

According to this system, plants are classified on the basis of one or few convenient characteristics, e.g.

(1) ***Theophrastus (370 BC-285 BC),*** Father of Botany, ancient ages) in his book *'Historia Plantarum'* classified about 480 plants into Trees, Shrubs, Undershrubs and Herbs.

(2) ***Linnaeus (1707-1778, Middle ages) :*** Carlous Linnaeus called *'father of plant taxonomy'* described a sexual system of plant classification, details of which were published in *'Species Plantarum'* (1753). He took into consideration, the nature and number of stamens and carpels. He divided plants into 24 classes.

(B) Natural System (Modern Classification)

According to this system, overall resemblance (all important characters of plants are taken into consideration.

Bentham & Hooker (1862-1883) gave most important system of classification of angiosperms and published in three volumes of *'Genera Plantarum'*. They described 202 families.

Bentham & Hooker divided plant kingdom into two subkingdoms.

Subkingdom 1 : Cryptogamia—it includes non-flowering plants.

Subkingdom 2 : Phanerogamia—it includes flowering plants. It includes :

Class-1 Dicotyledonae
Class-2 Gymnospermae
Class-3 Monocotyledonae

Class-1 Dicotyledonae

It is further divided into three subclasses

(i) Subclass Polypetalae Petals free.

Series (A) Thalamiflorae **:** Petals & stamens are hypogynous.

Order 1 : *Ranales* : Families 8
Order 2 : *Perietales* : Families 9
Order 3 : *Polygalinae* : Families 4
Order 4 : *Caryophyllineae* : Families 4
Order 5 : *Guttiferales* : Families 6
Order 6 : *Malvales* : Families 3

Series (B) Disciflorae **:** The sepals are free or fused. These sepals are fused with ovary or free. A fleshy disc is found below ovary. Sometimes this disc is divided into some small glands. Here ovary is superior.

Order 7 : *Geraniales* : Families 10
Order 8 : *Olacales* : Families 3
Order 9 : *Celastrales* : Families 4
Order 10 : *Sapindales* : Families 5

Series (C) Calyciflorae : Sepals are fused or free (sometimes) They are fused with ovary. Sepals are arranged in one or two whorls. Ovary is usually inferior.

Order 11 : *Rosales* : Families 9
Order 12 : *Myrtales* : Families 6
Order 13 : *Passiflorales* : Families 7
Order 14 : *Ficoidales* : Families 2
Order 15 : *Umbellales* : Families 3

(ii) Subclass Gamepetalae : The petals are fused. Stamens are usually epipetalous. Carpel may be syncarpour or apocarpous.

Series (A) Inferae : Overy is inferior and stamen usually equal to petal lobes.

Order 1 : *Rubiales* : Families 2
Order 2 : *Asterales* : Families 4
Order 3 : *Campanales* : Families 3

Series (B) Heteromerae :

Order 4 : *Ericales* : Families 6
Order 5 : *Primulales* : Families 3
Order 6 : *Ebenales* : Families 3

Series (C) Bicarpellate : Ovary is superior. Stamens are usually equal to corolla lobes or lesser in number. Stamens are alternate to petals. Carpels are usually two, sometimes they may be lesser or more than two.

Order 7 : *Gentianales* : Families 6
Order 8 : *Polemonials* : Families 5
Order 9 : *Personales* : Families 8
Order 10 : *Lamiales* : Families 5

(iii) Subclass Monochlamydae : Incomplete— In flowers perianth is in one whorl. Sometimes perianth may be absent. This subclass is divided into 8 series.

Series (A) Curvembrayae : Families 6
Series (B) Multiovulate Aquaticae : Family 1
Series (C) Multiovulate Terrestres : Families 3
Series (D) Microembryeae : Families 4
Series (E) Damphnales : Families 5
Series (F) Achlamydosporeae : Families 3
Series (G) Unisexuales : Families 9
Series (H) Ordines Anomali : Families 4

Class-2 Gymnospermae

It includes 3 families.

Class-3 Monocotyledonae

Series (A) Microspermae : Families 3
Series (B) Epigynae : Families 7
Series (C) Coronaireae : Families 8
Series (D) Calycineae : Families 3
Series (E) Nudiflorae : Families 5
Series (F) Apocarpae : Families 3
Series (G) Glumaceae : Families 5

(C) Phylogenetic System

These systems are based on evolutionary and genetical tendancies of plants. Some of the important phylogenetic systems are as follows :

(1) ***Engler & Prantl (1884-1930) :*** They formulated a detailed classification and published in 23 volumes of *"Die Nuturlichen Pflanzen Families"*.

(2) ***Hutchinson (1959) :*** He published a book *'Families of flowering plants'* in two volumes. (Vol. I. Dicotyledons, Vol. II. Monocotyledons.)

He divided angiosperms into 411families.

(*i*) Dicotyledonae Groups
Division 1 : Lignosae-246 families
Division 2 : Herbacae-96 families

(*ii*) Monocotyledonae Groups
(*a*) Calyciflorae
(*b*) Corolliflorae
(*c*) Glumiflorae.

(3) ***A.L. Takhtajan (1967) :*** In his book *"A System of Phylogeny of Flowering Plants"* classified Angiosperms (Magnoliophyta). The outline of his classification is as follows :

Division: Magnoliophyta or Angiospermae
Class: Magnoliatae (or Dicotyledons)
Divided into 7 subclasses; 15 super orders and 20 orders.
Class: Liliatae or Monocotyledons
Divided into 4 subclasses; 5 super orders and 20 orders.

MULTIPLE CHOICE QUESTIONS

1. Linnaeus is known as
   (a) Father of Taxonomy
   (b) Father of Zoology
   (c) Father of Botany
   (d) Father of Genetics
2. Plant and animal classification was given by
   (a) Gregor John Mendel
   (b) Charles Darwin
   (c) Carl Van Linnaeus
   (d) Aristotle
3. Three kingdom system was proposed by
   (a) Aristottle (b) Linnaeus
   (c) Whittaker (d) Haeckel
4. Binomial system refers to
   (a) One name
   (b) One generic and one specific name
   (c) Common name
   (d) One international name
5. Ernst Meyer gave the
   (a) Classification of organism into five kingdom
   (b) Term ecological niche
   (c) Trinomial nomenclature
   (d) Biological concept of species
6. The term systematics was coined by
   (a) John Ray (b) Theophrastus
   (c) Carolus Linnaeus (d) A.P. de Candolle
7. Hierarchy of categories was introduced by
   (a) John Ray (b) Bauhin
   (c) Linnaeus (d) A.P.de Candolle
8. The standard size of herbarium sheet is
   (a) 25 × 38 cm (b) 21 × 36 cm
   (c) 15 × 30 cm (d) 29 × 41 cm
9. The basis of karyotaxonomy is
   (a) Chromosome number
   (b) Sedimentation rate of ribosomes
   (c) Chromosome banding
   (d) Number of nucleoli
10. A convenient way of easy identification of an organism by applying diagnostic characters is called
    (a) Classification (b) Key
    (c) Systematics (d) None of these
11. Number of families in Bentham and Hooker's classification is
    (a) 196 (b) 202
    (c) 206 (d) 186
12. Adansonian taxonomy brings out relationships with the help of
    (a) Evolutionary tendencies
    (b) Statistics of all known traits
    (c) Similarities and dissimilarities
    (d) All of the above
13. Species having many sub-species are
    (a) Polytypic/macrospecies
    (b) Sibling
    (c) Allopatric
    (d) Monotypic/Microspecies
14. ICBN is connected with
    (a) Correct nomenclature of biochemicals
    (b) Correct nomenclature of animals
    (c) Correct nomenclature of plants
    (d) All of the above
15. Specimen on which original publication of author was based is
    (a) Holotype (b) Lactotype
    (c) Syntype (d) Isotype
16. Binomial nomenclature was mentioned for the first time in the book
    (a) Philosphie Zoologique
    (b) Systema Naturae
    (c) Historia Plantarum
    (d) None of these
17. In Five-kingdom classification, the kingdom that includes the blue-green algae, nitrogen bacteria and methanogenic bacteria is
    (a) Monera (b) Protista
    (c) Plantae (d) Fungi
18. The largest herbarium of the world is located at
    (a) New York (b) Berlin
    (c) Geneva (d) Kew
19. The forest research institute is located at
    (a) Delhi (b) Pune
    (c) Jammu (d) Dehradun

20. The foundation of Indian Botanical Garden, Kolkata was laid in the year
(*a*) 1757 (*b*) 1797
(*c*) 1787 (*d*) 1807

21. Holotype is the specimen
(*a*) Synonym for lectotype
(*b*) Synonym for neotype
(*c*) Synonym for paratype
(*d*) Used by author of the name designated by him as the nomenclatural type

22. Scientific name of plants are mostly derived from which language ?
(*a*) Greek (*b*) Latin
(*c*) French (*d*) Sanskrit

23. The Royal Botanical Garden is situated in
(*a*) Kew (*b*) New Delhi
(*c*) Paris (*d*) Washington

24. According to Whittaker, one of the following is most primitive evolutionary group
(*a*) Fungi
(*b*) Monera
(*c*) Plantae
(*d*) Protista

25. The classification of Engler and Prantl is one which is based on
(*a*) Overall similarities
(*b*) Plant habit
(*c*) Utilization system
(*d*) Common evolutionary descent

Directions Qs. 26-30: *In each of the following questions, a statement of* ***Assertion*** ***(A)*** *is given followed by a corresponding statement of* ***Reason*** ***(R)*** *just below it. Of the statements, mark the correct answer:*
(*a*) If both assertion and reason are true and reason is the correct explanation of assertion.
(*b*) If both assertion and reason are true but reason is not the correct explanation of assertion
(*c*) If assertion is true but reason is false
(*d*) If assertion is false but reason is true.

26. **Assertion (A):** Vernacular names are chosen for nomenclature.
Reason (R): Vernacular names are familiar to local residents.

27. **Assertion (A):** Organisms are named according to binomial nomenclature.
Reason (R): Names consist of two words-one scientific and other vernacular.

28. **Assertion (A):** The same specific name can be given to organisms belonging to different genera.
Reason (R): Two organisms of different species may have same specific names.

29. **Assertion (A):** Names of organisms are generally given in Latin.
Reason (R): Latin language has names for all organisms.

30. **Assertion (A):** The hierachy includes seven obligate categories.
Reason (R): Intermediate categories are used to make taxonomic position more informative.

ANSWERS

1	2	3	4	5	6	7	8	9	10
(*a*)	(*c*)	(*d*)	(*b*)	(*d*)	(*c*)	(*c*)	(*d*)	(*a*)	(*c*)
11	**12**	**13**	**14**	**15**	**16**	**17**	**18**	**19**	**20**
(*b*)	(*d*)	(*a*)	(*c*)	(*a*)	(*b*)	(*a*)	(*d*)	(*d*)	(*c*)
21	**22**	**23**	**24**	**25**	**26**	**27**	**28**	**29**	**30**
(*d*)	(*b*)	(*a*)	(*b*)	(*d*)	(*d*)	(*c*)	(*c*)	(*c*)	(*b*)

3

Virus

The word 'virus' in Latin means a poison. The viruses were first discovered by D.J. Ivanowski (1891) while studying a plant disease known as tobacco mosaic. They are the most perfect obligate parasites. They show the following living characters:

1. They multiplies and metabolises inside the living cells.
2. The synthesise nucleic acids and proteins specific to their structural organisation.
3. They carry DNA or RNA which means mutations are inheritable in them.
4. They exhibit recombinations if two or more viruses are seen in the same cell.

The viruses also exhibit non-living characters:

1. In 1935 W.M. Stanley isolated pure crystals of Tobacco Mosaic Virus and concluded the virus are made of protein and nucleic acid. He obtained Nobel Prize in Chemistry in the year 1946.
2. They have no locomotory organs and do not react to stimuli.
3. They do not consist of enzymes therefore they cannot multiply outside the host cells.
4. The length does not increase in a complete lifespan of virus.
5. When the protein and nucleic acids of viruses are separated, they are incapable to infecting host cells.

Carolous Causius firstly recorded a viral disease in tulips in 1576. In 1886 Mayer described virus disease in tobacco. Twort and Helle discovered bacteriophages. Schelsinger showed that a bacteriophage consists of protein and DNA. Viruses are chemically nucleoproteins *i.e.,* a molecule of RNA or DNA enclosed in protein coat. Plant viruses and RNA viruses whereas animal viruses and bacteriophages contains DNA.

The two phases of virus are—extracellular virion and intracellular consisting of nucleic acid. The shape of viruses is variable. Viruses are supposed to be modified plasmids. Virion is a structually complete virus which is also called nucleocapsid.

CLASSIFICATION

According to Holmes (1948), viruses are classified on the basis of their hosts into three kinds :

(*a*) **Animal viruses (Zoophagineae) :** They infect man, fowl, pigeon, parrot, dog, cow, etc., and arthropods insects. They usually have DNA or may also have RNA (e.g. retroviruses).

(*b*) **Plant viruses (Phytophagineae) :** They infect angiosperms, like potato, tobacco, sugarcane, cucurbits, beans and higher plants. They have RNA and lack lipoprotein coat over capsid.

(*c*) **Bacterial viruses (Phagineae) :** They are commonly called bacteriophages or phages i.e., eaters of bacteria. They have DNA.

VIRUSES AND NUCLEIC ACIDS

With reference to the number of strands, four types of nucleic acids have been found in viruses :

(1) Double stranded DNA

(2) Single stranded DNA

(3) Double stranded RNA

(4) Single stranded RNA

MULTIPLE CHOICE QUESTIONS

1. Plant viruses usually contain :
(*a*) RNA (*b*) DNA
(*c*) DNA and RNA (*d*) DNA or RNA

2. Viruses were discovered by :
(*a*) Stanley (*b*) Ivanowski
(*c*) Twort (*d*) Herelle

3. Viruses are defined as :
(*a*) a bunch of genomes which are looking for a suitable host for replication
(*b*) micro-organisms which contains DNA only
(*c*) smallest micro-organisms causing disease
(*d*) micro-organisms which contains RNA only

4. RNA particles causing symptoms like those of viral diseases is known as :
(*a*) virion (*b*) vira
(*c*) viroid (*d*) None of these

5. Coliphage ϕX 174 contains :
(*a*) double stranded RNA
(*b*) single stranded DNA
(*c*) double stranded DNA
(*d*) single stranded RNA

6. Viruses differ from living organisms in that, they :
(*a*) can grow inside a living host
(*b*) can multiply inside a living host
(*c*) can undergo mutation inside a living host
(*d*) are non-cellular

7. Virus which attacks *E.coli* is called :
(*a*) bacteriophage (*b*) coliphage
(*c*) phytophage (*d*) zoophage

8. Virus is a.............................stage between living and non-living.
(*a*) balanced
(*b*) threshold
(*c*) transitional
(*d*) None of these

9. The virus DNA which is incorporated into bacterial DNA is termed as :
(*a*) episome (*b*) mesosome
(*c*) free DNA (*d*) T-phage

10. Virus was crystallized and isolated first by :
(*a*) J.D. Ivanowski (*b*) F.C. Bawden
(*c*) W.M. Stanley (*d*) K.M. Smith

11. Little leaf disease of brinjal is caused by :
(*a*) viruses (*b*) bacteria
(*c*) fungi (*d*) None of these

12. The smallest virus has a size of :
(*a*) 1 nm (*b*) 3 nm
(*c*) 6 nm (*d*) 2 nm

13. Which of the following is devoid of protein coat, with a constitution of RNA only?
(*a*) Tobacco mosaic virus
(*b*) Poxvirus
(*c*) Potato spindle tuber virus
(*d*) None of the above

14. Genetic modification brought about by a virus in a bacteria is known as :
(*a*) transformation (*b*) transduction
(*c*) conjugation (*d*) candidiasis

15. Viruses are believed to be :
(*a*) living organisms
(*b*) non-living organisms
(*c*) a transitional group between living and non-living world
(*d*) living organisms which have lost the power of multiplication

16. Influenza virus contains :
(*a*) DNA (*b*) RNA
(*c*) Both (*a*) and (*b*) (*d*) None of these

17. Edward Jenner discovered :
(*a*) vaccination against small pox
(*b*) vaccination against chicken pox
(*c*) immunization against small pox
(*d*) immunization against chicken pox

18. Roux Sarcoma virus contains :
(*a*) RNA (*b*) DNA
(*c*) DNA and RNA (*d*) None of these

19. Which of the following is a viral disease of potato ?
(*a*) Early blight of potato
(*b*) Late blight of potato
(*c*) Leaf roll of potato
(*d*) None of the above

20. The term 'Virus' has been derived from a Latin word meaning :

(*a*) poison (*b*) secretion
(*c*) cell (*d*) venom

21. Which of the following is an example of DNA containing plant virus?

(*a*) Tobacco mosaic virus
(*b*) Potato mosaic virus
(*c*) Cauliflower mosaic virus
(*d*) Tomato mosaic virus

22. Rabies is caused by :

(*a*) viruses
(*b*) slime moulds
(*c*) fungi
(*d*) bacterium

23. A clear zone formed in a bacterial colony due to viral action is called :

(*a*) plague (*b*) phage
(*c*) streak (*d*) kranz

24. An antiviral substance produced by many vertebrates in response to viral infection for resisting the multiplication of viruses is known as :

(*a*) antivirin (*b*) antigen
(*c*) virion (*d*) interferon

25. Viruses with two types of symmetry, *i.e.*, icosahedral head and helical tail are called:

(*a*) picornaviruses (*b*) binal viruses
(*c*) riboviruses (*d*) none of these

Directions Qs. 26-30: *In each of the following questions, a statement of* **Assertion (A)** *is given followed by a corresponding statement of* **Reason (R)** *just below it. Of the statements, mark the correct answer:*

(*a*) If both assertion and reason are true and reason is the correct explanation of assertion.
(*b*) If both assertion and reason are true but reason is not the correct explanation of assertion
(*c*) If assertion is true but reason is false
(*d*) If assertion is false but reason is true.

26. Assertion (A): Viruses are living organisms.

Reason (R): They donot possess their own biochemical machinery.

27. Assertion (A): Lytic cycle is shown by a virulent phage.

Reason (R): The host cell undergoes lysis.

28. Assertion (A): All RNA viruses cannot directly multiply in host cells.

Reason (R): RNA viruses need reverse transcriptase to synthesize DNA.

29. Assertion (A): Some viruses cause cancer.

Reason (R): Cancer viruses trigger uncontrolled growth of cell.

30. Assertion (A): In lysogenic cycle, the host cell does not get lysed.

Reason (R): Viral genome is integrated with the host DNA.

ANSWERS

1	2	3	4	5	6	7	8	9	10
(*a*)	(*b*)	(*c*)	(*a*)	(*b*)	(*d*)	(*b*)	(*c*)	(*a*)	(*c*)
11	**12**	**13**	**14**	**15**	**16**	**17**	**18**	**19**	**20**
(*a*)	(*d*)	(*c*)	(*b*)	(*c*)	(*c*)	(*a*)	(*a*)	(*c*)	(*a*)
21	**22**	**23**	**24**	**25**	**26**	**27**	**28**	**29**	**30**
(*c*)	(*a*)	(*a*)	(*d*)	(*b*)	(*d*)	(*b*)	(*d*)	(*a*)	(*a*)

4

Bacteria

BACTERIA

Anton Van Leeuwenhoek (1683) of Holland observed sewage water, saliva and excretion peeled off from the teeth, under the crude microscope prepared by him. In these he found numerous tiny animalcules. These were bacteria. Linnaeus (1758) placed them in genus vermes. Ehrenberg (1829) first of all called them *Bacteria*. Louis Pasteur (1812-95) worked on fermentation and reported that it takes place by bacteria. Anton Van Leeuwenhoek is called **Father of Bacteriology**.

Bacteria are prokaryotic and single celled microganisms that differ from all other organisms (the eukaryotes) in lacking a true nucleus and organelles such as mitochondria, chloroplasts, and lysosomes. Their genetic material consists of a single loop of double stranded DNA, whereas, the genetic material of eukaryotes consists of multiple chromosomes, which are complex structures of DNA and protein.

Forms of Bacteria

(1) Spherical or cocci : The word cocci is derived from the Greek word kokkos which means a particle. They are the smallest bacteria. They may be of following types :

(*a*) *Micrococci :* When they live alone e.g. Micrococcus

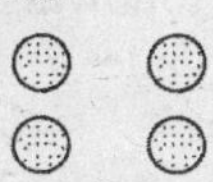

(*b*) *Diplococci :* When they are found in pairs, e.g. *Diplococcus pneumoniae*

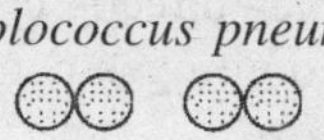

(*c*) *Streptococci :* When they are found in chains, e.g. *Streptococcus lactis*

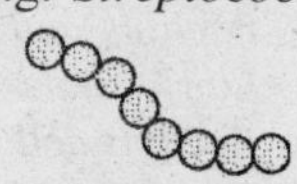

(*d*) *Tetrads :* When they line in tetrahedral pakets

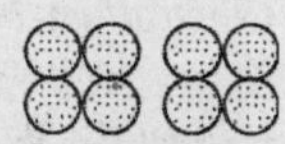

(2) Bacilus = rod-shaped : When they are rod shaped. This is derived from Greek word, Bacillum which means a stick. They are of the following types.

(*a*) *Single Bacillus :* When only one rod like structure represents the bacterium. This is most common.

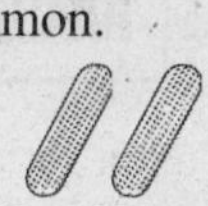

(*b*) *Diplobacillus :* When they are found in pairs.

(*c*) *Streptobacillus :* When they are found in chains e.g. *Bacillus anthracis*.

(3) Spirilli = Spiralor Helical : The word spirilli is derived from the Greek word spira which means coil e.g. *Spirillum ruprem*.

(4) Comma : These are comma (,) like bacteria e.g. *Vibrio*.

Motility

(*i*) *Atrichous :* When flagellum is absent e.g. *Pasteurella*. These are non-motile.

(*ii*) *Monotrichous :* When there is only one flagellum found on one end of the bacterium. e.g. *Pseudomonas*, *Thiobacillus*.

(*iii*) *Lophotrichous :* When there is a bunch of flagellum found on one end or both the ends of the bacteria, e.g. *Vibrio*.

(*iv*) *Amphitrichous :* When there is a bunch of flagella or single flagella on each ends of the bacterium, e.g. *Nitrosomonas*.

(*v*) *Peritrichous :* When the flagella are on whole body of the bacterium. e.g. *Escherichia*, *Clostridium*.

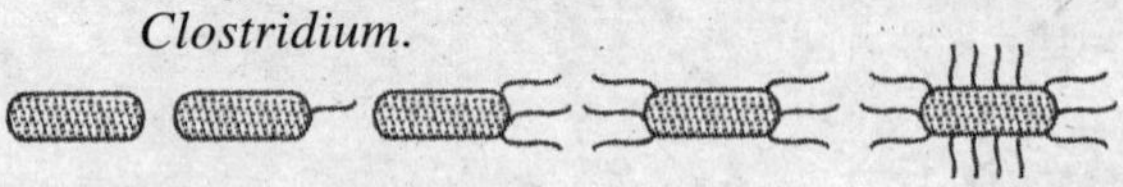

(i) Atrichous (ii) Monotrichous (iii) Lophotrichous (iv) Amphitrichous (v) Peritrichous

Parasitic Bacteria

Common human diseases caused by bacteria are:

Disease	Causal bacteria
Anthrax	*Bacillus anthracis*
Cholera	*Vibrio cholerae*
Diarrhoea	*Bacillus coli*
Diphtheria	*Corynebacterium diphteriae*
Dysentry	*Bacillus dysentrae*
Food poisoning	*Clostridium botulinum*
Gonorrhoea	*Neisseria gonorrhoeae*
Jaundice	*Leptospira icterohaemorrhagiae*
Leprosy	*Mycobacterium leprae*
Pneumonia	*Streptococcus pneumoniae*
Plague	*Pasteurella pestis*
Rheumatic fever	*Streptococcus* spp.
Syphilis	*Treponema pallidum*
Tetanus	*Clostridium tetani*
Tuberculosis	*Mycobacterium tuberculosis*
Typhoid fever	*Salmonella typhi*
Whooping cough	*Bordetella pertussis*

Common plant diseases caused by bacteria are:

Disease	***Host Plant***	***Causal bacteria***
Red rot of sugarcane	*Saccharum officinarum*	*Pseudomonas rubrilineans*
Angular	*Gossipium spp.*	*Xanthomonas malvacearum*
Leaf spot	*Capsicum annum*	*Xanthomonas vesicatora*
Blight of Paddy	*Oryza sativa*	*Xanthomonas oryzae*
Citrus canker	*Lemon/orange*	*Phytomonas citri*
Bean blight	*Phaseolus spp.*	*Xanthomonas phaseoli indicus*
Fire blight	*Members of family Rosaceae*	*Erwinia amylovora*
Tundu or bacterial rot	*Triticum aestivum*	*Coryne bacterium tritici*
Soft rot	Carrots, trunips	*Erwinia carotovora*

MULTIPLE CHOICE QUESTIONS

1. Bacteria are usually classified as plants because they :
(*a*) produce spores (*b*) have cell walls
(*c*) cannot move (*d*) produce enzymes

2. Which of the following is the best source of bacteria?
(*a*) Water (*b*) Soil
(*c*) Air (*d*) Human body

3. Germ theory of disease was proposed by :
(*a*) Robert Koch (*b*) L. Pasteur
(*c*) T. Boveri (*d*) F. Jacob

4. *Escherichia coli* is a bacterium which is a common inhabitant of :
(*a*) fertile soil (*b*) water
(*c*) human intestine (*d*) All of these

5. True nucleus is absent in :
(*a*) fungi (*b*) algae
(*c*) lichens (*d*) bacteria

6. The smallest bacteria are :
(*b*) vibrio (*b*) cocci
(*c*) spirilla (*d*) bacilli

7. The nuclear material in a bacterium is made up of :
(*a*) DNA (*b*) RNA
(*c*) Chromosome (*d*) None of these

8. Food poisoning is generally caused by :
(*a*) *Escherichia coli*
(b) *Clostridium botulinum*
(*c*) Both (*a*) and (*b*)
(*d*) None of the above

9. Which of the following bacteria can oxidise hydrogen sulphide to sulphur?
(*a*) *Rhizobium* (*b*) *Nitrosomonas*
(*c*) *Beggiatoa* (*d*) *Nitrobacter*

10. Bacteria which directly converts atmospheric nitrogen into nitrogen compounds are called:

(*a*) Nitrifying bacteria
(*b*) Nitrogen fixing bacteria
(*c*) Denitrifying bacteria
(*d*) Putrefying bacteria

11. The pigment similar to chlorophyll found in green sulphur bacteria is :
(*a*) bacterioviridim
(*b*) bacteriochlorophyll
(*c*) chlorophyll
(*d*) both (*a*) and (*c*)

12. Bacteria reproduce :
(*a*) asexually only (*b*) sexually only
(*c*) mostly asexually (*d*) mostly sexually

13. Tetanus is caused by :
(a) Bacillus coli
(*b*) *Streptococcus* spp.
(*c*) *Pasteurella pestis*
(*d*) *Clostridium tetani*

14. Which of the following is the electron donor in bacterial photosynthesis?
(*a*) H_2O_2 (*b*) H_2O
(*c*) O_2 (*d*) H_2S

15. Of the following, which is not a bacterial action?
(*a*) Sewage disposal
(*b*) Nitrogen fixation
(*c*) Emulsification of fat
(*d*) None of the above

16. Which of the following statement is correct?
(*a*) All bacteria are autotrophic
(*b*) All bacteria are heterotrophic
(*c*) All bacteria are photosynthetic
(*d*) Most of the bacteria are heterotrophic but some are autotrophic

17. A free living bacterium capable of fixing atmospheric nitrogen is :
(*a*) *Staphylococcus* (*b*) *Streptococcus*
(*c*) *Azotobacter* (*d*) None of these

18. Bacteria reproduce asexually by means of :
(*a*) fission (*b*) endospores
(*c*) cyst (*d*) All of these

19. The bacteria which do not bear flagella are termed as :
(*a*) atrichous (*b*) lophotrichous
(*c*) amphitrichous (*d*) peritrichous

20. Bacterial cell-wall is composed of :
(*a*) chitin (*b*) mucopeptide
(*c*) lignin (*d*) suberin

Directions Qs. 21-25: *In each of the following questions, a statement of* **Assertion (A)** *is given followed by a corresponding statement of* **Reason (R)** *just below it. Of the statements, mark the correct answer:*

(*a*) If both assertion and reason are true and reason is the correct explanation of assertion.
(*b*) If both assertion and reason are true but reason is not the correct explanation of assertion
(*c*) If assertion is true but reason is false
(*d*) If assertion is false but reason is true.

21. **Assertion (A):** Archaebacteria are called ancient bacteria.
Reason (R): Archaebacteria are presently living.

22. **Assertion (A):** *E. coli* is the most studied prokaryote.
Reason (R): *E. coli* is Gram +ve bacteria.

23. **Assertion (A):** Pili are motile appendages of bacteria.
Reason (R): Pili participate in conjugation.

24. **Assertion (A):** Bacteria and cyanobacteria do not possess a nucleus.
Reason (R): Prokaryotes lack membrane bound organelles.

25. **Assertion (A):** Bacterial cell walls are not like the plant cell.
Reason (R): Bacterial cell wall is not made up of cellulose.

ANSWERS

1	2	3	4	5	6	7	8	9	10
(*b*)	(*b*)	(*b*)	(*c*)	(*d*)	(*b*)	(*a*)	(*b*)	(*c*)	(*b*)
11	**12**	**13**	**14**	**15**	**16**	**17**	**18**	**19**	**20**
(*a*)	(*c*)	(*d*)	(*d*)	(*c*)	(*d*)	(c)	(*d*)	(*a*)	(*b*)
21	**22**	**23**	**24**	**25**					
(*b*)	(*c*)	(*d*)	(*a*)	(*a*)					

5

Algae

ALGAE

- **Linnaeus** (1753) introduced the term **algae.**
- **Phycology :** Study of algae.
- **Father of Indian Phycology :** M.O.P. Iyengar.
- Chlorophyll bearing, nonvascular, aquatic forms having accessory spores for asexual multiplication and nonjacketed gametangia for sexual reproduction.
- Absence of embryo stage.
- According to **Whittaker,** algae included under kingdom plantae are **chlorophyceae, phaeophyceae** and **rhodophyceae.**
- **F.E. Fritsch** (1935) divided algae into the following 11 classes on the basis of number and mode of attachment of flagella in motile cells, thallus structure, chemical nature of pigments, reserve food material, method of reproduction and varieties in life cycle– Chlorophyceae (green algae), Xanthophyceae (yellow-green algae), Chrysophyceae, Bacillariophyceae (Diatoms), Cryptophyceae, Dinophyceae (Dinoflagellates), Chloromonadineae, Euglenieae (Euglenoids), Phaeophyceae (brown algae), Rhodophyceae (red algae) and Myxophyceae or Cyanophyceae (blue-green algae).

Green Algae – Chlorophyta

- Cosmopolitan in distribution. They may be– **Terrestrial** (*Ulothrix, Zygogonium*), **Epiphytes** (*Protococcus, Trenteopohlia*), **Endophytes** (*Coleochaete nitellum*), **Epizoic** (*Cladophora, Characium*), **Endozoic** (*Zoochlorella, Chlorella*), **Parasites** (*Cephaleuros*), **Thermophilic** (*Chlorella*), **Cryophytes** (*Chlamydomonas, Scottiela*) and **Lichens**.
- Thallus organization ranges from–**Unicellular flagellate forms** (*e.g., Chlamydomonas*), **unicellular nonflagellate forms** (*Chlorella*), **colonial flagellate form** (*Volvox*) **colonial nonflagellate coenobia** (*Hydrodictyon*), **coenocytic** (*Caulerpa*), **filamentous unbranched** (*Oedogonium*), **filamentous branched** (*Cladophora*), **pseudoparenchymatous** (*Ulva*), **Siphenaceous** (*Caulerpa*). *Aetabularia* (Umbrella like) considered as largest unicellular plant, has a holdfast (with nucleus), stalk and an umbrella like cap.
- Cell possess one or more chloroplasts which may be **cup shaped** (*Chlamydomonas*), girdle shaped (*Ulothrix*), star shaped (*Zygnema*), spiral (*Spirogyra*), reticulate (*Chlamydomonas*), parietal reticulate (*Oedogonium*), parietal band shaped (*Hydrodictyon*) or disc-shaped (*Chara*). Chloroplasts have 2-20 thylakoid lamellae. **Pyrenoids** are present to store starch.
- Asexual reproduction by formation of zoospores (*Ulothrix*), daughter coenobia (*Volvox*), aplanospores (*Oedogonium*), akinetes (*Pithophora*), hypnospores and autospores.
- Sexual reproduction by isogamous (*Ulothrix*), anisogamous (*Chlamydomonas*) and oogamous (*Volvox*) methods.

Comparison of Red, Brown and Green Algae

Basis	Red Algae	Brown Algae	Green Algae
(1) Habitat	Mostly marine	Mostly marine	Mostly fresh water
(2) Organisation	Few unicellular forms	Unicellular forms absent	Many unicellular forms present
(3) Chloroplast (Thylakoids or lamellae,	Not stacked	Stacked in group of three	Stacked in group of 2-20
(4) Chlorophyll	Chl. *a* and chl *d* present	Chl *a* and chl *c* present	Chl *a* and chl *b* present
(5) Fucoxanthin	Present	Present	Absent
(6) Phycobilins	Present	Absent	Absent
(7) Reserve food	Floridean starch	Laminarian	Starch
(8) Motility	Absent	Present	Present

MULTIPLE CHOICE QUESTIONS

1. In *Spirogyra,* meiosis takes place during :
(*a*) zoospore formation
(*b*) zygospore formation
(*c*) gamete formation
(*d*) None of these

2. *Chlorella* - generally uses the greater amount of energy in :
(*a*) locomotion
(*b*) chemical reaction
(*c*) mitosis
(*d*) xanthophyll

3. On germination, zygospore of each *Spirogyra* gives rise to :
(*a*) four plants (*b*) three plants
(*c*) one plant (*d*) two plants

4. A thallus of *Volvox* is known as :
(*a*) colony (*b*) filament
(*c*) coenobium (*d*) coenocyte

5. Auxospores occur in :
(*a*) *Ulothrix* (*b*) *Oedogonium*
(*c*) Diatoms (*d*) *Chara*

6. A multicellular fragment of an algal filament capable of growth into a new plant is termed :
(*a*) Hormogonium
(*b*) Trichogyne
(*c*) Trichome
(*d*) Hormocyst

7. An example of Agarophyte is :
(*a*) *Nostoc* (*b*) *Geliarum*
(*c*) *Fucus* (*d*) Diatom

8. Fusion of mature individuals which directly act as gametes is called :
(*a*) anisogamy (*b*) isogamy
(*c*) hologamy (*d*) autogamy

9. Cell-wall of *Spirogyra* is composed up of :
(*a*) chitin (*b*) pectin
(*c*) cellulose (*d*) Both (*b*) and (*c*)

10. Major light absorption bands of phycoerythrin and phycocyanin are :
(*a*) 480, 622 nm (*b*) 610, 625 nm
(*c*) 670, 700 nm (*d*) 555, 565 nm

11. Algae are characterised by :
(*a*) thalloid plant body
(*b*) unicelled sex organs
(*c*) aquatic habitat
(*d*) pyrenoids

12. *Oscillatoria* belongs to :
(*a*) Rhodophyta (*b*) Chlorophyta
(*c*) Cyanophyta (*d*) Phaeophyta

13. A holdfast cell is present in the plant of :
(*a*) *Spirogyra* (*b*) *Oedogonium*
(c) Ulothrix (*d*) *Chlamydomonas*

14. Photosynthetic organelle of blue green algae is :
(*a*) photosynthetic lamellae

(*b*) chloroplast
(*c*) chromatophore
(*d*) None of these

15. The chloroplast in the cells of *Ulothrix* are :
(*a*) discoid (*b*) collar shaped
(*c*) rod shaped (*d*) spiral

16. Sometimes the two filaments of Spirogyra exhibit a ladder like structure. It is characteristic of :
(*a*) lateral conjugation
(*b*) scalariform conjugation
(*c*) conjugation
(*d*) asexual reproduction

17. In spirogyra
(*a*) Asexual reproduction occurs by zoospores
(*b*) filaments in which scalariform conjugation occurs are homothallic
(*c*) filaments in lateral conjugation are heterothallic
(*d*) filaments in which lateral conjugation occurs are homothallic

18. Gametes of *Ulothrix* are :
(*a*) biflagellate (*b*) non-flagellate
(*c*) quadriflagellate (*d*) None of these

19. Thermal algae are those which grow :
(*a*) in deserts where temperature is around 70°C
(*b*) in tropical regions of world where temperatures are usually high
(*c*) in hot water springs where tempera-ture is around 70°C
(*d*) on rocks exposed to high temperature

20. The chloroplasts of *Spirogyra* have :
(*a*) smooth margin
(*b*) wavy margin
(*c*) smooth or wavy margin
(*d*) None of these

Directions Qs. 21-22: *In each of the following questions, a statement of* **Assertion (A)** *is given followed by a corresponding statement of* **Reason (R)** *just below it. Of the statements, mark the correct answer:*
(*a*) If both assertion and reason are true and reason is the correct explanation of assertion.
(*b*) If both assertion and reason are true but reason is not the correct explanation of assertion
(*c*) If assertion is true but reason is false
(*d*) If assertion is false but reason is true.

21. **Assertion (A):** *Euglena* forms palmella.
Reason (R): It occurs when conditions are favourable.

22. **Assertion (A):** *Euglena* is studied as an animal as well as a plant.
Reason (R): *Euglena* is more an animal than a plant.

ANSWERS

1	2	3	4	5	6	7	8	9	10
(*b*)	(*b*)	(*c*)	(*c*)	(*c*)	(*a*)	(*b*)	(*c*)	(*c*)	(*a*)
11	**12**	**13**	**14**	**15**	**16**	**17**	**18**	**19**	**20**
(*d*)	(*c*)	(*c*)	(*a*)	(*b*)	(*b*)	(*d*)	(*a*)	(*c*)	(*c*)
21	**22**								
(*c*)	(*b*)								

6

Fungi and Lichens

FUNGI

- **Mycology** (Gk. *mykes*–mushroom) Study of fungi.
- **Father of plant pathology :** de Bary
- **Father of Indian Mycology and Indian Plant Pathology :** E.J. Butler
- **Father of mycology :** Pier Antonio Micheli (1729)
- **Achlorophyllous** (absence of chlorophyll), **Thallophytic** (not differentiated into stem, axis and leaves) and **heterophytic** organisms.
- Based on their habitat, fungi are **coprophilous** (dung inhabiting), **keratinophilous** (on hair, horns), **corticolous** (bark), **epixylic/ xylophilous/lignocolous** (on wood) etc.
- Body is made up of fine tubular, colourless filaments called **hyphae** which form complex network called **mycelium.**
- Nutrition is **heterotrophic** and **absorptive,** may be parasites or saprophytes. They possess special absorbing organs **(haustoria)** to obtain nourishment from host.
- Food reserve is glycogen (animal starch) and oil globules.
- Eukaryotic cell possess all cell organelles except **plastids.** Cell wall is made up of **fungus cellulose** or **chitin.**
- Hyphae can be **septate** or **aseptate (multinucleate** and **coenocytic).**
- Vegetative reproduction by fragmentation, fission, budding, gonidia, rhizomorphs, sclerotia etc.
- Asexual reproduction through conidia, oidia, ascospores, endospores, chlamydospores, pycniospores, basidiospores, teleutospores, uredospores etc.
- Sexual reproduction may be isogamous, anisogamous to oogamous.
- Perenating structures are **rhizomorphs** and **sclerotia.**
- Life cycle may be **haploid** (*e.g.,* phycomycetes and ascomycetes), **haploid-dikaryotic** cycle (*e.g.,* basidiomycetes), **haploid-diploid** cycle (*e.g.,* yeast) and **diploid** cycle (*e.g.,* oomycetes of phycomycetes).

SOME ANTIBIOTICS

	ANTIBIOTICS	FUNGUS
1.	Penicillin	*Penicillium notatum*
2.	Citrinin	*Penicillium citrinum*
3.	Griseofulvin	*Penicillium griseofulvum*
4.	Patulin	*Aspergillus clavatus*
5.	Fumigatin	*Aspergillus fumigatus*
6.	Proliferin	*Aspergillus proliferans*
7.	Jawaharin	*Aspergillus niger*
8.	Trichothecin	*Trichothecium roseum*
9.	Cephalosporin	*Emericellopsis minimum*
10.	Viridin	*Gliocladium virens*
11.	Ramycin	*Mucor ramannianus*
12.	Campestrin	*Agaricus campestris*
13.	Chaetomin	*Chaetomium cochliodes*

FUNGAL ORGANIC ACIDS

	ORGANIC ACIDS	FUNGUS
1.	Citric acid	*Aspergillus niger*
2.	Oxalic acid	*Aspergillus niger*

3. Itaconic acid	*Aspergillus terreus*
4. Gluconic acid	*Aspergillus niger*
5. Fumaric acid	*Rhizopus stolonifer*
6. Kojic acid	*Aspergillus flavus*
7. Gallic acid	*Aspergillus gallomyces*
8. Lactic acid	*Rhizopus oryzae*

FUNGAL MYCOTOXIN

MYCOTOXIN	FUNGUS
1. Amatoxins	*Amanita phalloides*
2. Phallotoxins	*Amanita phalloides*
3. Coprine	*Coprinus atramentanius*
4. Ochratoxins	*Aspergillus ochraceus*
5. Aflatoxins	*Aspergillus flavus*
6. Zearalenone	*Fusarium*
7. Trichothecenes	*Fusarium, Cephalosporium*

LICHEN

- Term **Lichen** was coined by **Theophrastus** (371-284 BC).
- Murison (1699) called such plants as **Musco-fungus**.
- **Composite** or **dual organism** which are formed by a fungus partner or **mycobiont** generally **ascomycete** and an algal partner or **photobiont** or **phycobiont** mostly blue-green alga (*Gloeocapsa*, *Nostoc*, *Scytonema*) or green algae (*Trebouxia* etc.)
- **Lichenology :** Study of lichens.
- Worldwide in distribution, pioneer colonizers of barren rocks and mountains.
- Constitute dominant vegetation of **Tundra region** forming food of animals like **reindeer.**
- Lichens may be–**Corticolous** (bark of trees *e.g., Usnea*), **Lignicolous** (wood directly *e.g., Cyphelium*), **Saxicolous** (rocky substrata *e.g.,Porina*), **Terricolous** (on the ground e.g., *Lecidea, Cladonia*), **Marine** (siliceous rocky shore of sea *e.g., Caloplaca, Verrucaria*) and **Fresh water** (hard siliceous rocks in fresh water *e.g., Hymenelia*).

MULTIPLE CHOICE QUESTIONS

1. Ringworms in man are caused by :
(*a*) *Trichosporon* (*b*) *Aspergillus*
(*c*) Sporoboromyces (*d*) Microsporum

2. Plants can be made disease resistant by :
(*a*) inducing mutation
(*b*) treatment with fungicides
(*c*) heat treatment of seeds
(*d*) breeding with their wild relatives

3. Which one of the following is used in preparation of cheese?
(*a*) *Aspergillus* (*b*) *Penicillium*
(*c*) *Rhizopus* (*d*) *Mucor*

4. An open ascocarp is called :
(*a*) Apothecium (*b*) Hysterothecium
(*c*) Perithecium (*d*) Cleistothecium

5. Which of these is used in the baking of bread?
(*a*) *Saccharomyces octosporus*
(*b*) *Saccharomyces cerevisae*
(*c*) *Saccharomyces ludwigii*
(*d*) *Rhizopus stolonifer*

6. The mycelium of *Mucor* breaks up into a chain of bead like structures called oidia if it is kept in a medium rich in sucrose solution. The phenomenon is known as :
(*a*) Palmella stage (*b*) Torula stage
(*c*) Scerotia (*d*) Trama

7. Green mould is the common name of :
(*a*) *Rhizopus* (*b*) *Neurospora*
(*c*) *Mucor* (*d*) *Penicillium*

8. A group of spores constitute a :
(*a*) pustule (*b*) sorus
(*c*) sterigma (*d*) sclerptium

9. Reproduction in *Mucor* takes place by :
(*a*) isogametes (*b*) anisogametes
(*c*) agametes (*d*) None of these

10. Basidiospores are characteristic of :

(*a*) Bread mould (*b*) Mushroom
(*c*) *Aspergillus* (*d*) Yeast

11. The most common group of algae in lichens is :
(*a*) green algae (*b*) brown algae
(*c*) red algae (*d*) blue-green algae

12. The number of nucleus/nuclei in a conidiosporangium are :
(*a*) 2-4 (*b*) 1
(*c*) 4-6 (*d*) 3-8

13. Lichens are known as pioneers of vegetation in a :
(*a*) Hydrosere (*b*) Halosere
(*c*) Psammosere (*d*) Lithosere

14. A fungus which shows haplontic, haplo-diplontic and diplontic life cycle is :
(*a*) Yeast (*b*) Mushroom
(*c*) *Mucor* (*d*) *Aspergillus*

15. The branch of biology which deals with bacteria, viruses and fungi is known as :
(*a*) Bacteriology (*b*) Microbiology
(*c*) Mycology (*d*) Virology

16. Stinking smut is caused by :
(a) Ustilago nuda
(b) Ustilago tritici
(c) Tilletia carries
(*d*) *Ustilago maydis*

17. The Irish famine occurred in 1845 in North Europe due to the destruction of potato plants by :
(*a*) a saprophytic fungus
(*b*) a parasitic fungus
(*c*) a saprophytic bacterium
(*d*) a parasitic bacterium

18. *Usnea,* a fructicose lichen is commonly known as :
(*a*) Irish moss
(*b*) Old man's beard
(*c*) Reindeer moss
(*d*) None of these

19. The fructification of lichens is usually :
(*a*) apothecium
(*b*) cleistothecium
(*c*) carpogonium
(*d*) carposporagium

20. Which of the following is a facultative parasite?
(*a*) *Fusarium* (*b*) *Puccinia*
(*c*) *Mucor* (*d*) None of these

21. The carps of plectomycetes are also termed as :
(*a*) Perithecium (*b*) Cleistothecium
(*c*) Apothceium (*d*) Hymenium

22. In *Puccinia* diploidisation occurs in the :
(*a*) Uredospore (*b*) Telentospore
(*c*) Pycniospore (*d*) Aeciospore

23. Fungal hyphae arranged in a parallel manner constitute :
(*a*) fruiting body (*b*) fructification
(*c*) trama (*d*) hymenium

24. Which of the following groups contain coenocytic mycelium?
(*a*) Phycomycetes (*b*) Deuteromycetes
(*c*) Basidiomycetes (*d*) Lichens

25. Mycorrhiza is a relationship between fungi and roots of higher plants. It is a :
(*a*) parasitic relationship
(*b*) saprophytic relationship
(*c*) symbiotic relationship
(*d*) epiphytic relationship

Directions Qs. 26-30: *In each of the following questions, a statement of* **Assertion** *(A) is given followed by a corresponding statement of* **Reason** *(R) just below it. Of the statements, mark the correct answer:*

(*a*) If both assertion and reason are true and reason is the correct explanation of assertion.
(*b*) If both assertion and reason are true but reason is not the correct explanation of assertion
(*c*) If assertion is true but reason is false
(*d*) If assertion is false but reason is true.

26. **Assertion (A):** Fungi are heterotrophic.
Reason (R): They lack chlorophyll.

27. **Assertion (A):** Hypha is a structural unit of mycelium.

Reason (R): Hypha constitutes the fungal thallus.

28. **Assertion (A):** Yeasts are holocarpic.

Reason (R): They form pseudomycelium.

29. **Assertion (A):** Fungal hyphae form a tissue known as plectenchyma.

Reason (R): In pseudoparenchyma, the hypha are loosely held together.

30. **Assertion (A):** Aseptate mycelium is called coenocytic.

Reason (R): Basidiomycetres show such type of mycelium.

ANSWERS

1	2	3	4	5	6	7	8	9	10
(*d*)	(*d*)	(*b*)	(*a*)	(*b*)	(*b*)	(*d*)	(*b*)	(*a*)	(*b*)
11	**12**	**13**	**14**	**15**	**16**	**17**	**18**	**19**	**20**
(*a*)	(*d*)	(*d*)	(*a*)	(*b*)	(*c*)	(*b*)	(*b*)	(*a*)	(*a*)
21	**22**	**23**	**24**	**25**	**26**	**27**	**28**	**29**	**30**
(*b*)	(*b*)	(*d*)	(*a*)	(*c*)	(*a*)	(*a*)	(*b*)	(*c*)	(*c*)

7

Bryophytes

BRYOPHYTES

- **Bryophyta** (Gk. *Bryon* : moss ; *phyton* : plants.
- Term **Bryophyta** was first used by **Robert Braun** (1864). He included algae, fungi, lichen and mosses under it.
- **Bryology :** Study of bryophytes.
- Found in tropical and temperate habitats of the world. Very commonly grow on damp soils, rocks, walls, bark of trees etc.
- The **amphibians of the plant kingdom** because also require water for dehiscence of antheridia, liberation of antherozoids, transfer of antherozoids from antheridia to archegonia, opening of archegonial neck and movement of antherozoids into the archegonial neck.
- They may be aquatic, (*Riccia, Riella, Ricciocarpus*), epiphytes (*Dendroceros, Radula*), saprophytes (*Buxbaunia, Cryptothallus*) and in dry habitats *i.e.,* (*Polytrichum juniperinum* (dry heaths), *Tortula desertorum* (deserts) and *Porella platyphylla* (dry rocks).
- Vascular tissues and roots are absent, rhizoids present which may be unicellular or multicellular.
- Life cycle with two phases :
  (*i*) **Gametophytic** (haploid) phase – Dominant, long-lived, green and independent.
  (*ii*) **Sporophytic** (diploid) phase – Short lived, dependent upon the gametophyte (parasitic over gametophyte).
- Gametophytes may be **thalloid** or **leafy shoot** (stem like central axis and leaf like appendages).
- Xylem-like **hydroids** (conduct water) and phloem like **leptoids** (conduct the assimilates) present.
- Accessory spores (mitospores) not formed.
- Asexual (vegetative) reproduction by fragmentation, tubers, adventitious branches, gemmae, buds, formation of innovations and secondary protonema.
- **Oogamous** sexual reproduction. Sex organs jacketed (with sterile jacket) and multicellular.
- The male reproductive organs **(antheridia)** and female reproductive organs **(archegonia)** present.
- Antheridia produce biflagellated male gametes (sperms or antherozoids) for which an external layer of water is essential for the swimming.
- Internal fertilization, produces an embryo which grows into a **sporophyte** or **sporogonium.**
- Sporogonium is protected by **calyptra** (produced by wall of venter). It is responsible for the production (asexual) of haploid spores (meiospores).
- The spores are first cells of gametophytic generation. It germinate directly or through a juvenile filamentous stage called **protonema**.
- **Heterologous** or heteromorphic alternation of generation.

Classification of Bryophytes

- According to Rothmaler (1951) - 3 classes
  (1) **Hepaticopsida –** Popularly called as **liver worts.**
  e.g.–Riccia, Marchantia, Pellia, Porella etc.
  (2) **Anthoceratopsida –** Commonly called as **horn worts.**
  e.g.–Anthoceros, Megaceros, Notothylos etc.
  (3) **Bryopsida –** Commonly known as **musci**, largest class of bryophyta.
  e.g. Funaria, Sphagnum, Andreaea etc.

ECONOMIC IMPORTANCE

- Bryophytes help in **soil conservation** as they bind soil particles and prevent soil erosion by running water.
- Mosses along with lichen play an important role in the formation of soil and ecological succession.
- *Sphagnum* used in nursery because it can retain water two hundred times more than their own weight.
- *Sphagnune* is also used as fodder and fuel.
- *Sphagnum* plants form a compact dark coloured **peat** (rich in carbon,. superior absorptive power and antiseptic properties). Alcohol, peat tar, paraffin, nitrates, tanning material, ammonium sulphate, brown dye etc. are obtained from peat.
- The tea prepared from *Polytrichum commune* is used to dissolve kidney and gall bladder stones.
- Many bryophytes used as **indicator plants** and **antibiotic** substances.

MULTIPLE CHOICE QUESTIONS

1. Moss plants occur in moist regions :
(*a*) Water is essential for their fertilization process
(*b*) They are free floating plants
(*c*) They have high rates of transpiration
(*d*) None of these

2. Two types of cells - hyaline and chlorophyllose green are characteristic of bryophytes in :
(*a*) *Funaria hygrometrica*
(*b*) *Polytrichum commune*
(*c*) *Sphagnum papiolossum*
(*d*) *Porella heterophylla*

3. One of the common characters between *Funaria* and *Dryopteris* is :
(*a*) monoecious gametophyte
(*b*) dioecious gametophyte
(*c*) heterosporous sporophyte
(*d*) stele is protostele

4. Dehiscence of moss capsule is achieved by:
(*a*) hygroscopic nature of peristome teeth
(*b*) breaking down of annulus
(*c*) disintegration of calyptra
(*d*) swelling of columella

5. In *Funaria,* calyptra is derived from :
(*a*) antheridium (*b*) archegonium
(*c*) capsule (*d*) columella

6. The thalloid bryophytes are known as:
(*a*) mosses (*b*) hornworts
(*c*) liverworts (*d*) ferns

7. Dioecious gametophytes are seen in:
(*a*) *Funaria* (*b*) Fern
(*c*) *Anthoceros* (*d*) *Marchantia*

8. Pseudoelaters and elaters are meant for :
(*a*) dispersal of spores
(*b*) mechanical strength
(*c*) conduction of sap
(*d*) nutrition

9. Moss capsule represents :
(*a*) Sporophyte
(*b*) Gametophyte
(*c*) A part of sorus
(*d*) A part of protonema

10. Chloroplasts are present in spores of :
(*a*) *Rhizopus* (*b*) *Funaria*
(*c*) Yeast (*d*) *Dryopteris*

11. Which of the following has largest gametophyte ?
(*a*) *Funaria* (*b*) *Selaginella*
(*c*) *Oryza* (*d*) *Pinus*

12. Vegetative reproduction in *Funaria* takes place by :
(*a*) sporangiophores (*b*) fragmentation
(*c*) protonema (*d*) buds

13. Which of these is least likely to grow along sea coast?
(*a*) Green algae (*b*) Brown algae
(*c*) Lichens (*d*) Mosses

14. Oblique septae are found in the rhizoids of :
(*a*) *Funaria*
(*b*) *Marchantia*
(*c*) *Anthoceros*
(*d*) *Riccia*

15. Nurse cells are found in the sporogonium of:
(*a*) *Riccia*
(*b*) *Anthoceros*
(*c*) *Marchantia*
(*d*) *Sphagnum*

16. In *Marchantia,* antheridia are produced on structures which are stalked known as :
(*a*) archegoniophores
(*b*) antheridiophores
(*c*) gynophore
(*d*) None of these

17. Censor mechanism during spore disposal is characteristic of :
(*a*) Bog moss (*b*) *Funaria*
(*c*) Liverworts (*d*) *Anthoceros*

18. Sporophyte in mosses is attached to gametophyte with the help of :
(*a*) seta (*b*) theca
(*c*) apophysis (*d*) foot

19. "Amphibians of plant kingdom" is/are :
(*a*) Bryophytes
(*b*) Gymnosperms
(*c*) Angiosperms
(*d*) Both (*b*) and (*c*)

20. If protonema develops from seta of a moss-plant, it will be :
(*a*) haploid (*b*) triploid
(*c*) diploid (*d*) polyploid

Directions Qs. 21-25: *In each of the following questions, a statement of* ***Assertion*** *(**A**) is given followed by a corresponding statement of* ***Reason*** *(**R**) just below it. Of the statements, mark the correct answer:*
(*a*) If both assertion and reason are true and reason is the correct explanation of assertion.
(*b*) If both assertion and reason are true but reason is not the correct explanation of assertion
(*c*) If assertion is true but reason is false
(*d*) If assertion is false but reason is true.

21. **Assertion (A):** Bryophytes are ecologically the most important plant groups.
Reason (R): They help in completing ecological succession on rocks.

22. **Assertion (A):** Bryophytes show distinct alternation of generation.
Reason (R): The sporophyte and gametophyte are independent of each other.

23. **Assertion (A):** The peristome teeth helps in dispersal of spores in *Funaria.*
Reason (R): The hygroscopic movement of the peristome teeth assists in liberating spores from spore sac.

24. **Assertion (A):** In bryophytes there is progressive sterlization of sporogenous tissue as we move from lower to higher forms.
Reason (R): The sporogenous tissue is present inside the capsule that includes columella.

25. **Assertion (A):** Bryophytes bear close resemblance with the pteridophytes.
Reason (R): Vascular tissue is present in both bryophytes and pteridophytes.

ANSWERS

1	2	3	4	5	6	7	8	9	0
(*a*)	(*c*)	(*a*)	(*a*)	(*b*)	(*c*)	(*d*)	(*a*)	(*a*)	(*b*)
11	**12**	**13**	**14**	**15**	**16**	**17**	**18**	**19**	**20**
(*a*)	(*c*)	(*d*)	(*a*)	(*a*)	(*b*)	(*b*)	(*d*)	(*a*)	(*c*)
21	**22**	**23**	**24**	**25**					
(*a*)	(*c*)	(*a*)	(*b*)	(*c*)					

8

Pteridophytes

PTERIDOPHYTES

- Pteridophytai (Gk. *Pteron* : feather; *phyta* : plants) : Plants with feather like fronds or **ferns**.
- Term **Pteridophyta** was first introduced by **Haeckel** in 1866.
- Most primitive group which flourished well during **Devonian** and **Carboniferous periods** of palaeozoic era.
- Flowerless, seedless, spore producing vascular plants which have successfully invaded the land.
- **Sporophyte** is **main independent** plant body, differentiated into **true roots, true stem** and **true leaves.**
- Leaves may be **megaphyllous** (large pinnatified leaves *e.g.,* ferns) or **microphyllous** (small simple, uninerved leaves *e.g., Equisetum* and *Selaginella*). Leaves are **scaly** (*e.g., Equisetum*) or small and sessile (*e.g., Lycopodium, Selaginella*) or large, petiolate and compound (e.g., Ferns).
- Vascular system (xylem + phloem) well developed. Cambium absent. True vessels absent in xylem while in phloem, companion cells are absent and sieve cells are present.
- The stele (central cylinder–xylem, phloem, pericycle and sometimes medullary rays as well as pith) is **protostele** (*e.g., Lycopodium, Selaginella*) or **Siphonostele** (*e.g., Equisetum*) or **Dictyostele** (*e.g. Pteris, Dryopteris*).
- Reproduction by **spores** produced inside the **sporangia** (diploid) borne on leaves **(sporophylls)** usually on the ventral surface. In some cases, they are borne in the axil of stem and **leaves** (*e.g., Selaginella*), or in some specialised structure as **fertile spike** (e.g., *Ophioglossum*), **sporangiophores** (*e.g., Equisetum*), **sporocarps** (*e.g., Marsilea*).
- They may be **homosporous** (one type of spores *e.g., Lycopodium, Pteris, Equisetum*) or **heterosporous** (two different types of spores *i.e.,* smaller **microspores** and larger **macrospores** *e.g., Marsilea, Selaginella*).
- A sporangium originates from **a group of superficial** cells *i.e.,* **eusporangiate development** *e.g., Selaginella, Equisetum, Lycopodium* or from a **single initial cell** (*e.g., Pteris, Marsilea*).
- Sporophylls may be **uniformly distributed** (*e.g., Pteris, Adiantum*) or aggregated to **form compact mass** (strobili) at the stem apices (*e.g., Selaginella, Equisetum*).
- The sporangia present in **sporocarps** in aquatic pteridophytes *e.g., Azolla, Marsilea* and *Salvinia*.
- In higher ferns sporangia are either borne singly or in groups called **sori** which may be—**Simple** (all sporangia mature together *e.g., Botrychium*), **gradate** (mature sporangia in centre and immature sporangia on periphery. *e.g.,* Marsilea) or **mixed** (*e.g., Pteris*).
- Spores are produced by **meiosis** so these are **haploid.** Spore germinate to produce **gametophyte** which may be—
  (*i*) In **homosporous** pteridophytes, **exosporic** (outside the spore wall) gametophyte development, prothalli are usually **monoecious.**
  (*ii*) In **heterosporous** pteridophytes, **endosporic** gametophyte development, and gametophytes are **dioecious.**

- Sex organs *i.e.,* male sex organ **(antheridia)** and female sex organ **(archegonia)** are either embedded in the prothallus (*e.g., Lycopodium, Selaginella*) or are projected (*e.g., Pteridium*).
- Antheridium is surrounded by a single **antheridial jacket** and **antherozoids** are unicellular, uninucleate and biflagellate.
- Archegonium is differentiated into **neck** and **venter** (embedded in prothallus). Neck consists of 4 vertical rows of cells and possess **two** neck canal cells which may be more than two (*e.g. Lycopodium*) or only one (*e.g.. Pteridium*).
- Water is required for fertilization, antherozoids are chemically attracted towards archegonium due to **malic** or **fumaric acid.**
- Zygote (diploid) develop into a well developed **sporophyte** with roots, stem and leaves.
- In some pteridophytes gametophyte arise directly from the cells of the sporophyte without meiosis and spore formation *i.e.,* **apospory** while in **apogamy,** sporophytes arise directly from gametophytes without gametes and sexual fusion.
- Extinct group of pteridophytes are—**Club mosses, Horse tails** and **Ferns.**
- The alternation of generation is heterologous **(heteromorphic)** type.

MULTIPLE CHOICE QUESTIONS

1. Spore bearing organ of a fern is called :
(*a*) sporophyll (*b*) sorus
(*c*) ramenta (d) None of these

2. The 13-celled male gametophyte of *Selaginella* has :
(*a*) 12 cells of antheridium + 1 prothallial cell
(*b*) 10 cells of antheridium + 3 prothallial cells
(*c*) 9 cells of antheridium + 4 prothallial cells
(*d*) 8 cells of antheridium + 5 prothallial cells

3. Sporophyte of *Pteris* :
(*a*) is independent of gametophyte from the beginning
(*b*) is dependent on the gametophyte only in the beginning
(*c*) supplies water and inorganic salt to the gametophyte
(*d*) None of these

4. The stele in fern is :
(*a*) eustele (*b*) siphonostele
(*c*) dictyostele (*d*) None of these

5. The number of megaspores in *Selaginella rupestris* is :
(*a*) one (*b*) two
(*c*) three (*d*) four

6. The following is the most primitive vascular plant :
(*a*) Ferns (*b*) *Cycas*
(*c*) *Sphagnum* (*d*) *Psilotum*

7. Stem is polystelic in :
(*a*) *Selaginella* (*b*) *Cycas*
(*c*) *Pinus* (*d*) All of these

8. Circinate venation is found in :
(*a*) *Cycas*
(*b*) Fern
(*c*) Both (*a*) and (*b*)
(*d*) None of these

9. Gametangia of fern are produced on the :
(*a*) leaves (*b*) sorus
(*c*) prothallus (*d*) ramenta

10. Fern differs from moss because it has :
(*a*) swimming antherozoids
(*b*) independent gametophyte
(*c*) multiciliate sperms
(*d*) biciliate sperms

11. Spores in fern are produced through :
(*a*) mitosis (*b*) mitosis and meiosis
(*c*) meiosis (*d*) None of these

12. In the archegonium of fern there are/is :
(*a*) four neck canal cells

(*b*) three neck canal cells
(*c*) one neck canal cell having one nucleus
(*d*) one neck canal cell having two nuclei

13. In *Pteris,* the petiole has a groove and its stele is :
(*a*) Horse shoe shaped
(*b*) W-shaped
(*c*) M-shaped
(*d*) Without any definite shape

14. Pteridophytes are so-called :
(*a*) phanerogams
(*b*) spermatophytes
(*c*) vascular cryptogams
(*d*) amphibians of plant kingdom

15. Rhizophore of *Selaginella* is :
(*a*) stem (*b*) root
(*c*) leaf (*d*) none of these

16. Companion cells are usually not found in :
(*a*) stem of maize
(*b*) rhizome of *Pteris*
(*c*) stem of wheat plant
(*d*) stem of sunflower

17. *Selaginella* and fern share the following character :
(*a*) Protostelic stele
(*b*) Leptosporangiate sporangium
(*c*) Heterospory
(*d*) Absence of ovules

18. Leptosporangiate sporangium is :
(*a*) single-celled
(*b*) multicelled and formed from many cells
(*c*) multicellular
(*d*) multicelled and formed from a single cell

19. The venation of fern leaf is :
(*a*) reticulate (*b*) parallel
(*c*) furcate (*d*) none of these

20. Fern prothallus is :
(*a*) homothallic (*b*) heterothallic
(*c*) heteromorphic (*d*) heterotrophic

21. ***Marsilea*** are rich in
(*a*) Starch (*b*) Protein
(*c*) Gulucose (*d*) None of these

22. Which one of the following is walking fern?
(*a*) *Cyathaea* (*b*) *Adiatum*
(*c*) *Alsophila* (*d*) None of these

Directions Qs. 23-25: *In each of the following questions, a statement of* **Assertion (A)** *is given followed by a corresponding statement of* ***Reason*** **(R)** *just below it. Of the statements, mark the correct answer:*
(*a*) If both assertion and reason are true and reason is the correct explanation of assertion.
(*b*) If both assertion and reason are true but reason is not the correct explanation of assertion
(*c*) If assertion is true but reason is false
(*d*) If assertion is false but reason is true.

23. **Assertion (A):** The gametophytic prothallus of *Lycopodium* is monoecious.
Reason (R): *Lycopodium* is a heterosporous plant.

24. **Assertion (A):** In *Dryopteris,* annulus is responsible for the dehiscence of sporangium.
Reason (R): Development of sporangium is of eusporangiate type.

25. **Assertion (A):** Protostele is the simplest and most primitive type of stele found in pteridophytes.
Reason (R): Siphonostele is thought to have been evolved from a protostele.

ANSWERS

1	2	3	4	5	6	7	8	9	10
(*a*)	(*a*)	(*a*)	(*c*)	(*a*)	(*d*)	(*a*)	(*c*)	(*c*)	(*c*)
11	**12**	**13**	**14**	**15**	**16**	**17**	**18**	**19**	**20**
(*c*)	(*d*)	(*a*)	(*c*)	(*a*)	(*b*)	(*d*)	(*d*)	(*a*)	(*a*)
21	**22**	**23**	**24**	**25**					
(*a*)	(*b*)	(*c*)	(*c*)	(*b*)					

9

Gymnosperms

GYMNOSPERMS

- Gymnosperm (*gymnos* = naked; *sperma* = seed).
- Term **"Gymnospermae"** was coined by Theophrastus.
- Commonly known as **naked seed plant** because their ovules (later becomes seed) are **not covered** and lie naked on the **megasporophylls** or **ovuliferous scales** or **carpels** which are arranged into cones.
- Leaves **dimorphic** (2 types) (*a*) green **foliage** leaves (needle shaped or pinnately compound) (*b*) brown **scaly** leaves (minute, deciduous). Leaves with sunken stomata and covered with thick cuticle.
- **Lateral veins** are absent in the leaves of most gymnosperms. Instead **transfusion tissue (hydrostereom)** present for lateral translocation of nutrient.
- Vascular bundles in stem are **conjoint, collateral** and **open** arranged in a ring.
- Xylem with **tracheids and parenchyma only.** Vessels absent (present only in members of **Gnetales).** Phloem lacks companion cells.
- **Monoxylic** (*e.g., Cycas*) or **pycnoxylic** (*e.g., Pinus*) wood.
- **Heterosporous** (2 types of spores)—microspores and megaspores which are produced in microsporangia (microsporophylls) and megasporangia (megasporophylls), respectively. These sporophylls produce compact **cones** or **strobili.** Flowers are absent.
- Generally **monosporangiate** (unisexual) cones and development of microsporangia is **eusporangiate.**
- Ovules are **orthrotropous** and **unitegmic (bitegmic** in *Gnetum*) and lies naked on megasporophyll. Each ovule has a mass of tissue **(nucellus).** Megasporophylls aggregate to form female cones which persist for several years.
- Female gametophyte develops **archegonia** with a single egg and a venter canal cell. **Neck canal cells are absent.** Male gametes are **ciliated** and **motile** (*e.g.,* Cycas, Ginkgo) or non-ciliated and non-motile (*e.g., Pinus, Gnetum*).
- Pollination by wind **(anemophily)** which bring pollen grains over the micropylar end of ovules *i.e.,* direct pollination. **Siphonogamic** (through pollen tube) fertilization. **Meroblastic** embryo development. Sometimes exhibit **polyembryony**.
- The number of cotyledons in seed is **one** (*e.g., Cycas*) or **two** (*e.g., Pinus*).

CLASSIFICATION OF GYMNOSPERMS

(1) **Cycads** *e.g., Cycas, Zamia*

(2) **Conifers** *e.g., Pinus*

(3) **Ginkophytes** *e.g.,* Ginkgo or Maiden Hair tree

(4) **Gnetophytes** *e.g., Gnetum, Ephedra*

ECONOMIC IMPORTANCE

- Seeds of *Pinus gerardiana*, commonly called **chilgoza** (dry fruit).
- An alkaloid **ephedrine** obtained from *Ephedra* is for the preparation of medicines in the treatment of asthma, cough and bronchitis.

  Copal — *Agathis australis*

Manila copal—*Agathis alba*

Rosin—As residue after the distillation of pine terpentine

Canada balsam—*Abies balsamea*

Amber—Fossil resin from extinct species of *Pinus*.

- Many gymnospermous plants (*e.g.*, *Cycas revoluta*, *Cryptomeria*, *Araucaria*, *Thuja* etc.) are cultivated as ornamental plants in gardens. *Araucaria* (monkey puzzle) is form Australia and South America.
- **Tannins** (complex phenolic substance) from the bark of ***Tsugo candensis.***

MULTIPLE CHOICE QUESTIONS

1. Mycorrhiza is found in :
(*a*) *Pinus* (*b*) *Cycas*
(*c*) *Funaria* (*d*) None of these

2. The edible part of *Pinus* seed is :
(*a*) perisperm (*b*) endosperm
(*c*) epispore (*d*) tegma

3. The turpentine oil used in varnish industry as a thinning agent is obtained from :
(*a*) *Pinus* (*b*) *Cycas*
(*c*) *Gnetum* (*d*) None of these

4. Which one of the following is a living fossil?
(*a*) *Pinus* (*b*) *Dryopteris*
(*c*) *Selaginella* (*d*) *Ginkgo*

5. *Pinus* and *Cycas* possess :
(*a*) non-motile sperms
(*b*) non-motile and motile sperms respectively
(*c*) motile sperms
(*d*) motile and non-motiole sperms respectively

6. The smallest gymnosperm among the following is :
(*a*) *Sequoiadendron giganteum*
(*b*) *Zania pygmea*
(*c*) *Pinus roxburghii*
(*d*) *Cycas revoluta*

7. *Pinus* differs from mango in having :
(*a*) tree habit
(*b*) presence of wood
(*c*) tree habit
(*d*) ovules not enclosed within ovary

8. The number of archegonia in *Pinus* ovule is:
(*a*) 1 to 4 (*b*) 2 to 4
(*c*) 2 to 6 (*d*) 2 to 8

9. In gymnosperms, pollination is exclusively by :
(*a*) animals (*b*) wind
(*c*) water (*d*) insects

10. The ovuliferous and carpellary scales are seen in .... of Pinus.
(*a*) ovule
(*b*) microsporophyll
(*c*) megasporophyll
(*d*) megasporangium

11. The longest cones are seen in :
(*a*) *Cycas revoluta*
(*b*) *Cycas pectinata*
(*c*) *Cycas siamensis*
(*d*) *Cycas rumpii*

12. Trifoliar spur consisting of three needle-like green leaves is seen in :
(*a*) *Pinus sylvestris*
(*b*) *Cycas circinalis*
(*c*) *Pinus quadrifolia*
(*d*) *Pinus longifolia*

13. Red wood tree is :
(*a*) *Cedrus* (*b*) *Pinus*
(*c*) *Dalbergia* (*d*) *Sequoia*

14. Sunken stomata are the characteristics of :
(*a*) *Wheat* (*b*) Mango
(c) Dalbergia (d) Cycas

15. Which is false with regard to *Pinus*?
(*a*) Resin canals are present
(*b*) Transfusion tissue is present
(*c*) Tracheids have bordered pits
(*d*) Plants are dioecious

16. Pollen grains in *Cycas* at the time of shedding are :
(*a*) 5-celled (*b*) 2-celled
(*c*) 3-celled (*d*) 4-celled

17. Each stamen in *Cycas* is represented by a :
(*a*) megasporophyll
(*b*) microsporophyll
(*c*) male cone
(*d*) megasporangium

18. Diploxylic vascular bundles are found in :
(*a*) *Cycas* (*b*) *Sequoia*
(*c*) *Pinus* (*d*) *Ginkgo*

19. Dimorphic roots are found in :
(*a*) *Pinus* (*b*) *Ephedra*
(*c*) *Gnetum* (*d*) *Cycas*

20. In the rachis or leaflet of *Cycas* the exarch bundles :
(*a*) have more of centripetal xylem
(*b*) have more of centrifugal xylem
(*c*) have equal amount of both centripetal and centrifugal xylem
(*d*) variable

Directions Qs. 21-25: *In each of the following questions, a statement of* **Assertion (A)** *is given followed by a corresponding statement of* **Reason (R)** *just below it. Of the statements, mark the correct answer:*

(*a*) If both assertion and reason are true and reason is the correct explanation of assertion.
(*b*) If both assertion and reason are true but reason is not the correct explanation of assertion
(*c*) If assertion is true but reason is false
(*d*) If assertion is false but reason is true.

21. Assertion (A): Conifers show monopodial growth.
Reason (R): In conifers the main axis of trunk rises straight from base and reaches upto the tip.

22. Assertion (A): Gymnosperms are classified as spermatophyta.
Reason (R): Their sporophytic plant body is differentiated into true stem, leaves and roots.

23. Assertion (A): Gymnosperms resemble the pteridopytes in some features.
Reason (R): Both exhibit similar external and internal characters of the sporophytes.

24. Assertion (A): Coralloid roots of *Cycas* become swollen at their tips.
Reason (R): This swelling is caused due to the entry of certain bacteria and blue green algae in the roots.

25. Assertion (A): The needles of *Pinus roxburghii* show xerophytic characters.
Reason (R): Epidermis of needle contain numerous stomata and a thick parenchymatous layer is present just below the epidermis.

ANSWERS

1	2	3	4	5	6	7	8	9	10
(*a*)	(*b*)	(*a*)	(*d*)	(*b*)	(*d*)	(*d*)	(*d*)	(*b*)	(*c*)
11	**12**	**13**	**14**	**15**	**16**	**17**	**18**	**19**	**20**
(*a*)	(*d*)	(*c*)	(*d*)	(*d*)	(*c*)	(*b*)	(*a*)	(*d*)	(*a*)
21	**22**	**23**	**24**	**25**					
(*a*)	(*b*)	(*c*)	(*a*)	(*c*)					

10

Angiosperms

ANGIOSPERMS

- Term **angiosperm** means **enclosed seed** because the ovules or potential seeds are enclosed within a hollow ovary.
- More than 2,50,000 different species which occupy a very wide range of ecological habitats.
- They include trees, shrubs, vines and herbs adapted to almost every kind of environment. Some **hydrophytes** (*Hydrilla* – water lili etc.) have taken water secondarily and even in shallow seas (*e.g.* eel grass–*Zostera*). Some are **epiphytes** (*e.g.* Vanda), many are xerophytes.
- Leaves exhibit reticulate or parallel venation forming **areoles.**
- The libriform fibre and true vessels in xylem and companion cells in phloem are present.
- Sporophytic plant bears **flowers** (site of sexual reproduction) which is a modified shoot. Flowers consists of 4 whorl of appendages – sepals, petals, stamen and carpels which arise from **thalamus. Sepals (calyx)** and **petals (corolla)** are two outer accessory reproductive structures while, **Stamen (androecium)** and **carpel (gynoecium)** constitute the two inner essential parts of a flower.
- Each stamen (microsporophylls) consists of stalk like **filament** and flattened globoid top called **anther**. Anther produces tectate pollen grains with exine differentiated into rod-like **columellae** covered by a **tectum**. Stamens are **bilaterally symmetrical.**
- Each carpels (megasporophylls) exhibit upper terminal **stigma,** middle elongated **style** and basal swollen **ovary**. Ovules borne on placentae are present in ovary.
- Female gametophyte is highly reduced and consists of **single egg cell, two synergids, three antipodals** and **two polar nuclei.** Archegonia absent.
- Pollination takes place by wind, insects, birds, water, human and other animals. Pollination may be self or cross. **Entomophilus** (insect pollinated plants) are generally colourful and showy, possess nectar and scent etc.
- The pollen grain germinates on stigma and form a pollen tube which carries tube nucleus.
- The most characteristic feature of angiosperms is **double fertilization** *i.e.,* one of the male gamete fuses with the egg and produce diploid zygote which develops into embryo or new sporophyte and the other male gamete fuses with **two polar nuclei (triple fusion)** and form triploid **primary endosperm nucleus.**
- The sepals, petals, stamens, style, stigma wither at the time of fruit formation. The ovules ripen into **seeds** and ovary ripens into **fruits.**
- The seed possess **dormant embryo along with food material** in endosperm or cotyledons.
- They have wide nutritional modes *e.g.* **autotrophs** (most of the green plants), **saprophytes**. (*Monotropa*, *Corallorhiza*), **parasites** (*Cuscuta, Rafflesia*) and **insectivorous plants** (*Nepenthes, Drosera*) etc.
- According to **Warming** (1895), plants are divided into 5 major groups *i.e.,* **hydrophytes** (*Hydrilla*, *Wolffia* etc.), **mesophytes** (*Brassica*, sunflower etc.), **xerophytes** (*Cynodon, Asparagus* etc.), **halophytes** (*Spartina, Portulaca* etc.) and **epiphytes** (*Vanda*).

MULTIPLE CHOICE QUESTIONS

1. Which of the following families belong to inferae under Gamopetalae in Bentham and Hooker's system of classification?
(*a*) Cucurbitaceae (*b*) Brassicaceae
(*c*) Liliaceae (*d*) Fabaceae

2. First step in Taxonomy is :
(*a*) identification (*b*) naming
(*c*) description (*d*) classification

3. Phylogenetic system of classification was given by :
(*a*) Bentham, Hooker and Hutchinson
(*b*) Bentham, Hooker and Linnaeus
(*c*) Engler, Prantle and Hutchinson
(*d*) Hutchinson and Aristotle

4. Emasculation means :
(*a*) removal of anthers
(*b*) removal of stigma
(*c*) removal of gynoecium
(*d*) removal of calyx and corolla

5. Who is known as the father of Taxonomy?
(*a*) Aristotle (*b*) Linnaeus
(*c*) Eicher (*d*) Engler

6. The basic taxonomic unit upon which the systems of classification are constructed is :
(*a*) species (*b*) order
(*c*) family (*d*) class

7. Tapioca plant, which stores food in large fleshy storage roots, belongs to genus :
(*a*) *Ipomoea* (*b*) *Manihot*
(*c*) *Solanum* (*d*) *Asparagus*

8. "Mustard" is a common name for :
(*a*) *Brassica oleracea*
(*b*) *Brassica nigra*
(*c*) *Raphanus sativus*
(*d*) *Cedrus*

9. Scientific name of mango is *Mangifera indica* L. here represents :
(*a*) botanist Linnaeus who gave the name.
(*b*) that name is latin.
(*c*) that name is not written properly, first letter of indica should be capital.
(*d*) letter L. has nothing to do with the name of tree.

10. Food is stored in underground parts of plants in members of family :
(*a*) Solanaceae, Fabaceae and Malvaceae
(*b*) Solanaceae, Brassicaceae and Liliaceae
(*c*) Solanaceae, Malvaceae and Liliaceae
(*d*) Liliaceae, Brassicaceae and Malvaceae

11. The term "New systematics" was given by :
(*a*) Karl Prantle (*b*) George Bentham
(*c*) Julian Huxley (*d*) J.B. Amici

12. What is common in cyathium and hypanthodium inflorescence?
(*a*) Nectaries (*b*) Unisexual flowers
(*c*) Ostiole (*d*) Fleshy receptacle

13. In the system of classification by Bentham and Hooker, dicotyledonous plants have been classified as :
(*a*) Polypetalae, Gamopetalae, Mono chlamydeae
(*b*) Polypetalae, Gamopetalae, Dichla-mydeae
(*c*) Archechlamydeae, Monochlamydeae, Glumiflorae
(*d*) None of these

14. Which taxonomic term may be substituted for any other?
(*a*) Oder (*b*) Species
(*c*) Taxon (*d*) Class

15. Potato belongs to family :
(*a*) Asteraceae (*b*) Brassicaceae
(*c*) Poaceae (*d*) Solanaceae

16. In ovary only one ovule is present in :
(*a*) Asteraceae (*b*) Lamiaceae
(*c*) Solanaceae (*d*) None of these

17. Fleshy fruits with persistent calyx are found in:
(*a*) Solanaceae (*b*) Asteraceae
(*c*) Lamiaceae (*d*) Both (*b*) and (*c*)

18. The floral characters similar to Malvaceae and Liliaceae are :
(*a*) superior syncarpous ovary with axile placentation
(*b*) inferior syncarpous ovary with axile placentation
(*c*) monodelphous androecium and superior syncarpous ovary
(*d*) monodelphous androecium and capsular fruits

19. Characters from flowers are used as a basis for classification because :
(*a*) flowers show variety of colours
(*b*) flowers can be preserved easily
(*c*) reproductive parts are more conservative than vegetative parts
(*d*) None of these

20. The floral formula of Solanaceae is :
(*a*) $\oplus P_2 A_3 G_1$
(*b*) $\oplus K_{4-5}\ A_{10}\ \underline{G}_{(2)}$
(*c*) $\oplus K_{2+2}\ C_4\ A_{2+4}\ \underline{G}_{(1)}$
(*d*) $\oplus K_{(5)}\ C_{(5)}\ A_5\ \underline{G}_{(2)}$

21. Syngamy was discovered by :
(a) Hofmeister
(*b*) Nawaschan
(*c*) Straburger
(*d*) Seargant

22. Entamophily is pollination by :
(*a*) air (*b*) water
(*c*) insects (*d*) animals

23. The flowers in Solanaceae are :
(*a*) actinomorphic, bisexual and epigynous
(*b*) zygomorphic, unisexual and hypo-gynous
(*c*) actinomorphic, bisexual and hypo-gynous
(*d*) actinomorphic, bisexual and perigynous

24. National Botanical garden is situated at :
(*a*) Calcutta (Kolkata) (*b*) Shimla
(*c*) Lucknow (*d*) Delhi

25. The maize grain is a :
(*a*) seed (*b*) fruit
(*c*) embryo (*d*) None of these

Directions Qs. 26-30: *In each of the following questions, a statement of* ***Assertion*** *(A) is given followed by a corresponding statement of* ***Reason*** *(R) just below it. Of the statements, mark the correct answer:*
(*a*) If both assertion and reason are true and reason is the correct explanation of assertion.
(*b*) If both assertion and reason are true but reason is not the correct explanation of assertion
(*c*) If assertion is true but reason is false
(*d*) If assertion is false but reason is true.

26. Assertion (A): Sclerenchyma fibres constitute the major mechanical tissue of the plants.
Reason (R): These are non lignified cells which possess pectocellulose thickening in specific areas of their walls.

27. Assertion (A): Death of seive tube member results in the death of its adjacent companion cells.
Reason (R): Both are derived from the same mother cell.

28. Assertion (A): In stem, secondary vascular growth is by both intra and inter fascicular cambium.
Reason (R): In root, secondary vascular growth arises from conjuctive parenchyma and pericycle.

29. Assertion (A): Glandular trichomes are hollow hairs that contain silicious tips filled with poision.
Reason (R): Not all the stinging hairs produce essential oils.

30. Assertion (A): Endodermis in dicot roots functions as biological check post.
Reason (R): The endodermal cells possess casparian strips which prevent wall to wall movement of substances.

ANSWERS

1	2	3	4	5	6	7	8	9	10
(*b*)	(*a*)	(*c*)	(*a*)	(*b*)	(*a*)	(*b*)	(*b*)	(*a*)	(*b*)
11	**12**	**13**	**14**	**15**	**16**	**17**	**18**	**19**	**20**
(*c*)	(*b*)	(*a*)	(*c*)	(*d*)	(*a*)	(*a*)	(*a*)	(*c*)	(*d*)
21	**22**	**23**	**24**	**25**	**26**	**27**	**28**	**29**	**30**
(*b*)	(*c*)	(*c*)	(*c*)	(*b*)	(*c*)	(*a*)	(*b*)	(*d*)	(*a*)

11

Cell as a Unit of Life

UNITY AND DIVERSITY OF LIFE

It looks strange but all life forms, whether plants or animals, have the same basic living strucure called cell and yet all of them are different as they exist in nature. *Cells are basic unit of life and structural unit of an organism.*

Cell: Cell is a unit of stucture and biological activity made of an organised mass of protoplasm surrounded by a protective and selectively permeable covering.

Cytology: It is the study of form, structure and contents of cells as observed under the microscope. The term was coined by Hertwig (1893) who also wrote a book on 'Cells and Tissues'.

Units of Measurement

1 mm = 0.1 cm = 0.001 m

1 micrometre/micron/μ m/μ = 0.001 mm
= 10^{-3} mm = 10^{-6} m (10^{-4} cm)

1 nanometre/millimicron/nm/μ m = 0.001 μ m
= 10^{-3} μ m = 10^{-6} mm = 10^{-9} m = (10^{-7} cm)

1 angstrom/Å = 0.1 nm = 10^{-4} mm
= 10^{-7} mm 10^{-10} m (=10^{-8} cm)

Smaller units are **picometre** = 10^{-12} m, **femtometre** = 10^{-15} m and **attometre** = 10^{-18} m.

Microscopy: It is the practice of using microscopes for the study of finer details of small objects including cells and tissues.

$$\text{Magnification} = \frac{\text{Size of Image with the Instrument at retina}}{\text{Size of Image with Unaided Eye at retina}}$$

Magnification of a microscope is roughly equal to the multiple of magnifying power of objective lens and ocular lens eye (eye piece), e.g., 45 × objective and 105 eye piece = 450 × magnification.

Resolving power or **limit of resolution** is the ability to distinguish two close object is as distinct. Its values is calculated by **abbe equation**

$$L_m = \frac{0.61\lambda}{NA}$$

λ is wavelength of light used in illumination of object. NA or numerical aperture is product of refractive index of medium (*n*) and sine of angle subtended by optical axis and outer ray covered by objective (sine α). Refractive index for lense of optical instrument is 1.6, oil immersion 1.4 and air is 1.0. Sine α is less than 1. The value for best objective is sin 70° = 0.94. In air the value of numerical aperture is less than one while for immersion objective it is 1.0 - 1.35.

Microtomy: Microscopic examination requires sectioning and staining. Sectioning is done by hand or machine known as **microtome** (developed by W.His). Extremely thin sections (20-100 nm) are obtained by **ultramicrotome** using glass or diamond knife.

X-ray Crystallography (Bragg, 1913): X-ray microscope (developed by Kirkpatrick) is useful in studying detailed structures of chemicals present in solid state (e.g., haemoglobin, insulin, DNA, RNA) due to diffraction of X-rays (wavelength 1 – 10 Å) by different atoms of the substance. Watson and Crick (1953) found out the structure of DNA from X-ray diffraction obtained by Astbury and Franklin (1953).

Cytochemistry: A number of **dyes** or **stains** are known to colour specific parts. Certain dyes can be used even in case of living materials. They are called **vital stains**, e.g., neutral red, methylene blue. Fuelgen or Schiff's reaction was developed by Fuelgen and

Rossenbeck (1924). It employs basic fuchsin after acid hydrolysis.

Autoradiography: Radioactive isotopes (^3H, ^{14}C, ^{32}P, ^{35}S) are incorporated in intermediate or raw materials (e.g., CO_2, uridine, thymidine, amino acids) for studying metabolic pathways including DNA → RNA → Protein pathway.

Cell Fractionation: It is the technique of separating and rupturing cells, and separating their components for study. It has two steps, homogenisation and differential centrifugation.

(i) **Homogenisation:** It is the rupturing of cells in order to release various constitutuents and then keeping them in their natural state by means of a special 0.25 M sucrose solution kept at 0° – 4° C. It is carried out by (a) Grinding in pestle and mortar or potter homogeniser. In potter homogeniser two cylinders separated by narrow gap are made to rotate. It produces a shearing force that causes rupturing of cells (b) Pressure (c) Osmotic shock (d) Ultrasonic vibrations (produced by piezo-electric crystal) transmitted to suspension by means of steel rod. The slurry contains homogenate as well as intact cells and other debris. They are separated by low speed centrifugation. The supernatant is homogenate. It is liquid containing cellular constituents.

(ii) **Differential Centrifugation:** It is separation of different components of homogenate by subjecting it to successively higher centrifugal force that brings about **velocity sedimentation**.

Svedberg Unit (S): It is a unit to measure sedimentation rate or **sedimentation coefficient** of a particle in an **analytical ultracentrifuge**. It is related to weight and volume of a material, e.g. 4S tRNA has a molecular weight of about 25000 daltons.

Microspectrophotometry: Solutions, gases and solids containing chemicals are subjected to different radiations. Absorption spectra are compared to know the chemicals.

MULTIPLE CHOICE QUESTIONS

1. Electron microscope employs
(*a*) fluorochromes
(*b*) polarizer and analyser filters
(*c*) objective and occular lenses
(*d*) electromagnetic lenses

2. Resolution power of a compound microscope is
(*a*) 100 μ*m* (*b*) 10 μ*m*
(*c*) 1 μ*n* (*d*) 0.25 μ*n*

3. Wavelength of radiations used in electron microscope is
(*a*) less than visible light
(*b*) more than visible light
(*c*) equal to visible light
(*d*) like the X-rays

4. Resolution power is the capacity to
(*a*) distinguish two close points as distinct
(*b*) distinguish two object
(*c*) distinguish between two organelles
(*d*) magnify

5. Which of the following is not used in autorandiography ?
(*a*) C^{14} (*b*) C^{12}
(*c*) S^{35} (*d*) P^{32}

6. Cell fractionation technique is used for
(*a*) separation of liquids
(*b*) separation of cellular components
(*c*) analysis of cell components
(*d*) all above

7. The term 'protoplasm' was coined by
(*a*) Purkinje (*b*) Vort-Mobed
(*c*) Sydney Fox (*d*) Huxley

8. Protoplasm is a
(*a*) heterogenous system
(*b*) colloidal system
(*c*) polyphasis colloidal system
(*d*) homogenous system

9. The exact composition and structure of protoplasm has been described by .... theory
(*a*) fibrillar (*b*) alveolar
(c) colloidal (*d*) granular

10. Which is absent in prokaryotes ?
(*a*) Ribosome (*b*) Nucleoid
(*c*) Cell wall (*d*) Dictyosome

11. In bacterial cell wall plasma membrane folds inside to form
(*a*) mesosomes (*b*) mitochondria
(*c*) nucleoid (*d*) lomasomes

12. Wall of young cells in plants in mainly composed of
(*a*) cellulose (*b*) starch
(*c*) glycogen (*d*) proteins

13. The mature cells in the mammalian body *i.e.,* RBC loses their capacity for
(*a*) aerobic respiration (*b*) DNA replication
(*c*) RNA synthesis (*d*) all above

14. Chondrioids are found in
(*a*) bacteria (*b*) aerobic bacteria
(*c*) anaerobic bacteria (*d*) cynobacteria

15. A cell can be declared dead when it
(*a*) loses its capacity to divide but otherwise physiologically active due to intact nucleus
(*b*) has lost its nucleus
(*c*) is metabolically dead
(*d*) both (*a*) and (*b*)

16. One of the following is high in smaller cells
(*a*) surface area per unit volume
(*b*) nucleocytoplasmic ratio
(*c*) metabolic activity
(*d*) all above

17. The cell size is mainly dependent upon
(*a*) its physiological state
(*b*) oxygen requirement
(*c*) basic metabolism
(*d*) all the above

18. Who applied the cell theory to plants ?
(*a*) Schwann (*b*) Schleiden
(*c*) Dutrochet (*d*) Virchow

19. The basis of life is
(*a*) nucleic acid (*b*) nucleoprotein
(*c*) lipid (*d*) proteins

20. Prokaryotic cells occur in
(*a*) PPLO (*b*) bacterial
(*c*) cynobacteria (*d*) all of these

21. Cell theory was given by
(*a*) Robert Hooke
(*b*) Robert Brown
(*c*) Schleiden & Schwann
(*d*) Leeuwenhoek

Directions Qs. 22-25: *In the following questions, statement of* **Assertion (A)** *is given followed by a corresponding statement of* **Reason (R)** *just below it. Of the statements, mark the correct answer—*

(*a*) If both assertion and reason are true and reason is the correct explanation of assertion
(*b*) If both assertion and reason are true but reason is not the correct explanation of assertion
(*c*) If assertion is true but reason is false
(*d*) If assertion is false but reason is true.

22. Assertion (A): Resolving power is the ability to distinguish between two close points.
Reason (R): Microscope reduces resolving power.

23. Assertion (A): Ribosomes came to our knowledge after the discovery of electron microscope.
Reason (R): Electron microscope has a high magnification and resolving power.

24. Assertion (A): Radioactive tracers are used to detect tumors.
Reason (R): Tracers locate such tissues.

25. Assertion (A): Cell like aggregates of complex organic compounds are called coacervates.
Reason (R): Coacervates had power to grow and divide.

ANSWERS

1	2	3	4	5	6	7	8	9	10
(*d*)	(*d*)	(*a*)	(*a*)	(*b*)	(*b*)	(*a*)	(*c*)	(*c*)	(*d*)
11	**12**	**13**	**14**	**15**	**16**	**17**	**18**	**19**	**20**
(*a*)	(*a*)	(*d*)	(*b*)	(*d*)	(*d*)	(*d*)	(*b*)	(*a*)	(*d*)
21	**22**	**23**	**24**	**25**					
(*c*)	(*c*)	(*d*)	(*a*)	(*b*)					

12

Structural Organisation of Cell

Cell is a unit of biological activity made of an organised mass of protoplasm surrounded by a protective and selectively permeable covering.

Living beings or organisms are made of one or more cells. Cell is a basic unit of structure and function of organisms. It was discovered by Robert Hooke (1665). He wrote a book 'Micrographia' and coined the term *cell*. Leeuwenhoek (1672) was the first to see *free cells*. He observed bacteria, protozoa, red blood cells, sperms etc. under his microscope. In 1809, Lamarck came to the conclusion that all living beings are formed of cells. Dutrochet (1824) believed in individuality of cells. Living substances was first observed by Corti (1772) and was named *sarcode* by Dujardin (1836). Robert Brown (1831) discovered nucleus while Von Mohl (1838, 1846) & Purkinje (1839) named the jelly-like living substance or sarcode of the cell as *protoplasm*.

CELL THEORY

Schleiden and Schwann (1839) compared their findings and formulated *cell theory* which states that :

(*i*) Living beings are made of one or more cells

(*ii*) A cell is a mass of protoplasm having a nucleus.

(*iii*) Cells are similar in basic structure and metabolism.

(*iv*) The functions of an organism are due to activities and interactions of its cells.

Cell theory was first modified in the light of Virchow's findings (1855, 1858) that cells develop from pre-existing cell *Omnis cellula-e-cellula*. It is known as law of other modifications were carried out in the cell theory. The modern cell theory is called 'cell principle'.

SURFACE VOLUME RATIO

Metabolically active cells are small since small cells have higher nucleo-cytoplasmic ratio for better control and higher surface-volume ratio for quicker exchange of materials. Larger cells can be efficient only if they are elongated, branched or possess membrane extensions like microvilli. Large sized animals do not possess large sized cells but more numerous small cells.

Comparison of Cell Organisation in Prokaryotic and Eukaryotic cells

	Cellular Component	*Prokaryotic Cells*	*Eukaryotic cells*
1.	Nuclear Envelope	Absent	Present
2.	DNA	naked	combined with proteins
3.	Chromosome	Single	Multiple
4.	Division	amitosis	mitosis or meiosis
5.	Ribosomes	70 s (50 s + 30 s)	80 s (60 s + 40 s)
6.	Endomembranes	Absent	Present
7.	Mitochondria	Absent	Present
8.	Chloroplast	Absent	Present in plant cells

CELL WALL

It is the outer rigid protective covering of plant cells and some prokaryotes. Cell wall has a matrix of hemicellulose, pectin, glycoprotein and water in which are embedded microfibrils.

(i) Middle lamella

It is the outermost region which functions as a cementing layer between two cells. It is absent on the outer free surface.

(ii) Primary wall

It is thin (0.1 - 3.0 μm), elastic, first wall layer. It contains microfibrils oriented variously, commonly forming a loose network.

(iii) Secondary wall

It is laid inner to primary wall. It is less hydrated i.e. less hydrophilic in nature (30-40%) as compared to primary wall. Cellulose content is high with long closely arranged straight and parallel microfibrils. Hemicellulose content is 25%.

In gymnosperms tension wood, the innermost layer contains xylan and is called 'tertiary wall'.

Hemicellulose is made of polymerised arabinogalactans, xylans, mannans, etc. It binds microfibrils with matrix.

Lignin is formed from coniferyl and coumaryl alcohols and aldehydes through dehydrogenation and polymerisation. Lignin reduces hydration and increases hardness of wall.

Cell wall possesses minute cytoplasmic bridges between adjacent cells. They are called 'plasmodesmata' (Strassburger, 1901). Plasmodesmata produce a continuum of living material called *symplasm* cell walls and intercellular spaces constitute a nonliving component of plant body known as *apoplasm.* Certain thick cell walls possess unthickened areas called *pits*.

CELL MEMBRANES

Cell membranes or biomembranes are quasifluid film like partitions of about 75 Å thickness. They are selectively permeable. Plasmalemma or plasma membrane lies on the outside of protoplast and separates it from external environment. *Tonoplast* occurs around vacuoles. Most of the eukaryotic cell organells have their own membrane coverings. Under electron microscope, a cell membrane appears *trilaminar* or *tripartite* with a middle electron transport layer and electrons dense layer on either side. Chemically, it has proteins (60-70%), lipids (20-40%) and carbohydrates (1-5%). Proteins are both structural and enzymatic. Lipids are generally phospholipids with polar hydrophilic heads and nonpolar hydrophilic tails.

Deniell & Davson Model (1935)

The model states that a double phospholipid layer is surrounded on either side by a layer of hydrated globular proteins or P-L-L-P. The hydrophobic or nonpolar tails of the two lipid layers are towards the centre.

Robertson Model (1959)

Robertson also gave Unit Membrane Concept, which proposed that all biomembranes have a trilaminar structure with an electron transparent lipid bilayer (35 Å) lying sandwitched between electrons dense protein layers (20 Å each).

Fluid Mosaic Model (Singer & Nicolson, 1972)

Cell membranes are quasifluid with viscous lipid bilayer having proteins at places (mosaic) both on the surface and inside. Singer & Nicolson have described cell membranes as protein icebergs in sea of lipids. External proteins are called Extrinsic Or Peripheral (30% of Total) While Internal Proteins Are Named Intrinsic Or Integral (70% of total). Membrane proteins may function as enzymes, permeases, carriers, receptors etc. Proteins extending from outside to inside are called 'tunnel proteins' or 'transmembrane proteins'. They function as channels. At places the external proteins and lipid molecules bear oligosaccharides forming glycoproteins and glycolipids for recognition; attachment and antigenic properties. They make the cell membrane asymmetric.

CYTOPLASM (Strasburger, 1882)

It is the bulk of protoplasm excluding nucleus. Cytoplasm is differentiated into cytoplasmic matrix, cell organelles and cell inclusions. Cytoplasmic matrix, or cytosol is jelly-like semifluid that exist in both sol and gel states. Cytoplasmic matrix is often differentiated into outer gel part called 'ectoplast' (plasmagel) and inner sol part called 'endoplast' (plasmasol). Endoplast is usually in perpetual motion called **cyclosis,** cytoplasmic or protoplsamic streaming (Amici, 1818). In **rotation** type of cyclosis, cytoplasmic matrix continuously flows in one direction. (e.g. young cells of *Hydrilla* leaf). In *Circulation* type of cyclosis, matrix flow in different directions in the same cell (e.g. cell of staminal hair of Tradescantia = Rhoea).

Nucleus

1. It discovered by a English botanist, Robert Brown in 1831, and is designated as the '*Director of the cell.*'
2. It is the largest organelle within eukaryotic cells, ranges from 5 to 25 mm in size.
3. Its nucleus is called karyology.
4. Some cells like those of **sieve tubes of vascular plants** and **RBCs of mammals** do not posses nuclei at maturity. As they can no longer divide, their life span is limited.
5. It consists of nuclear envelope or **karyotheca**
   (*i*) nucleoplasm or nuclear sap or karyolymph or **karyoplasm**
   (*ii*) nucleolus
   (*iii*) nuclear reticulum or **chromatin reticulum**.
6. The term **'chromosome'** was coined by **Waldeyer** is 1888.
7. Their number is constant in the nucleus for a particular species.
8. The lowest diploid number known among plants is 4, found in *Haplopappus gracillis* whereas *Ophioglossum reticulatum* has as high as 1,260 chromosomes.
9. Among animals *Parascaris univalens* has only 2 chromosomes while *Aulacantha* has 1,600 chromosomes per cell.

Nucleosomes

1. **R.D. Kornberg** and **Thomas (1974)** proposed a model to explain the structure of nucleosome.
2. The term **'nucleosome'** for chromatin subunits was suggested by **Oudet** *et al*, in 1975.
3. Histones are of *five types* and they are designated as $\mathbf{H_1}$, $\mathbf{H_2A}$, $\mathbf{H_2B}$, $\mathbf{H_3}$ and $\mathbf{H_4}$.

Endoplasmic Reticulum

1. They were first discovered by Claude (1951) under electron microscope.
2. The term "endoplasmic reticulum' was first coined by **Porter** *et. al* in 1945.
3. E.R. is abundant in liver, pancreas and other actively synthesizing cells. E.R. is absent in **RBC**, ova, **embryonic cells** and **prokaryotes**.
4. There are two types of endoplasmic reticulum : **Rough endoplasmic reticulum and Smooth endoplamic reticulum**
   (*i*) Tubular forms with ribosomes. Tubular forms without ribosomes.
   (*ii*) It is more developed in cells actively **engaged in protein synthesis.** It is abundant in the cells engaged in glycogen and **lipid metabolism**.
5. However, rough ER may be converted into smooth ER and vice versa, depending upon the metabolic requirements of the cell.

Functions

(*i*) It provides mechanical support to the cytoplasm.
(*ii*) It plays an important role in the intracellular transport and storage of proteins and sometimes lipids.
(*iii*) It synthesizes lipids and lipoproteins.
(*iv*) Smooth ER is involved in glycogenolysis.
(*v*) It forms other cellular organelles such as Golgi bodies, peroxisomes, and glyoxysomes.
(*vi*) It forms nuclear envelope at the end of the telophase stage of the cell division.

Golgi Complex Structure

1. An Italian physician **Camillo Golgi** (1898) first recognised this structure in nerve cells of Barn owl and hence named after the discoverer.
2. It is found in all the eukaryotic cells except **red blood cells** and **sieve tubes** of angiosperms:
3. It is also absent in prokaryotes, certain fungi and sperm cells of bryophytes and pteridophytes.

4. They are very well formed and larger in **nerve cells** and **gland cells,** however poorly developed in muscle cells.
5. The Golgi complex is morphologically very similar in both plant and animal cells.
6. In plant cells and in invertebrate tissues, dictyosomes are dispersed throughout the cytoplasm.

Functions

(*i*) It packs and transports certain materials like proteins and polysaccharides out of the cell (cell secretion).

(*ii*) It is the seat for formation of glycolipids and glycoproteins.

(*iii*) It synthesises hemicellulose, cellulose and pectin compounds during cell division and hence involved in cell wall formation in plants.

(*iv*) It helps in storage of secretory products.

(*v*) It gives rise to lysosomes and vacuoles.

(*vi*) It forms acrosome in sperms.

Centrosome

1. The first description of centrosome was given by **Van Beneden** (1880).
2. The term 'centrosome' was coined by **Boveri** (1881).
3. It is found in cytoplasm near the nucleus in animal cells and certain lower plants, however, it is absent in seed plants, prokaryotes and Amoeba.
4. The radiating structure comprises, **astral-rays,** the **centrosphere** and two small bodies called **centrioles** or **diplosomes.**
5. Centrioles are independent, cylindrical bodies (0.2 × 0.5 µm) lying perpendicular to each other at a distance 0.8 µm.
6. The wall of each centriole has **nine triplets** of **microtubules** arranged in a circle.
7. The microtubules of centrioles are madeup of tubulin and some lipids. The interior of centriole is filled with a homogeneous cytoplasm of low density.
8. RNA and DNA have also been observed in them.
9. Both centrioles may give rise to a flagellum but in most cases only one does and the other centriole remains at right angle to the **kinetosome** or basal body.
10. They help in spindle formation during cell division of animal cells.

Cilia and Flagella

1. **Cilia** and **flagella** are composed of three major parts:
   (*i*) a central **axoneme** or **shaft.**
   (*ii*) the surrounding plasma membrane. and
   (*iii*) the interposed cytoplasmic matrix.
2. The axonemal elements of nearly all cilia and flagella (as well as the tails of sperm cells) contain the same '9 + 2' arrangement of microtubules.
3. In the centre of axoneme are two **singlet microtubules** that run the length of the cilium.
4. Projections from the central microtubules occur periodically along their length to form an enclosing **sheath.**
5. Each of the central microtubules is composed of 13 **protofilaments.**
6. Nine **doublet microtubules** surround the central sheath. One microtubule of each doublet (*i.e.*, A subfibre) is composed of 13 protofilaments. The ajoining B subfibre is 'incomplete', consisting of 11 protofilaments.
7. **Radial spokes** having a periodicity of 24, 32 and 40 nm extend from each A subfibre inward to the central sheath.
8. Adjacent doublets are joined by **nexin** or **interdoublet links.**
9. Extending from each A subfibre are two **dynein arms** – an outer and inner arm.
10. Each beat of cilium and flagellum involves the same pattern of microtubule movement involving power (**effective**) **stroke** and the **recovery stroke**. Power stroke occurs in a single plane but the recovery may not occur in the same plane as the power stroke.

Functions

(*i*) Locomotion – Creation of water current for food ingestion

(*ii*) Filteration – internal circulation of fluids.

(*iii*) Sensory functions – cleaning of respiratory tract.

Lysosomes

1. These organelles were first observed and coined by **de Duve**.
2. They are variable in size ranging from 0.2 to 0.8 μm but may be as large as 5 μm in **mammalian kidney cells** and in **phagocytes**.
3. Their shape is often spherical but may by irregular and are surrounded by a **single unit membrane**.
4. They are found mostly in animal cells and in some plant cell *e.g. Neurospora*, cauliflower and bean.
5. They contain numerous hydrolytic (digestive) enzymes in latent form, *i.e.,* the enzymes are enclosed by a membrane and are not readily available to the substrate.
6. The lysosomal membrane is resistant to the enzymes that it encloses and the entire process of digestion is carried out within the lysosome. The membrane protects the rest of the cell from the possible destructive effect of the enzymes.
7. Most of the lysosomal enzymes act in an acidic medium.
8. It has been observed that if lysosomes are injured by a toxic substance, the enzymes thus released can digest the whole cell, hence lysosomes are called **'suicidal bags'**.
9. Lysosomes show considerable **polymorphism** and their four types are recognised.
   (*i*) Primary lysosomes or Storage granules
   (*ii*) Secondary lysosomes or hytero-phagosomes or digestive vacuoles
   (*iii*) Residual bodies
   (*iv*) Autophagic vacuoles or cytolysosomes or autophagosomes

Glyoxysomes

1. **Beevers** (1961) was the first person to discover these organelles and were described later by R.W. Briedenbach (1967).
2. They are about 0.5 to 1 μm in size and are surrounded by a **single unit membrane**.
3. They are found in plant cells, particularly, in germinating fatty seeds *e.g. Ricinus* (castor) and groundnut where fat is being converted into carbohydrates by a process called **glyoxylate cycle**.
4. Glyoxysomes contain two important enzymes, isocitrate lyases and maltase synthetase alongwith several others.
5. Structure of Glyoxysomes are similar to peroxisomes.

Plastids

1. One of the main distinguishing characteristic of most plant cells is the presence of **plastids**.
2. The credit for the discovery of these organelles is generally given to **Schimper** (1880).
3. The term 'plastid' was coined by **E. Haeckel** (1866).
4. Plastids generally contain pigments and may synthesise and accumulate various substances. However, some plastids are devoid of pigments and they serve primarily for the storage of various substances.
5. The plastids are of three types:
   (*a*) Leucoplasts
   (*b*) Chromoplasts and
   (*c*) Chloroplasts.

Ribosomes

1. Ribosomes were first observed by Claude (1941) who named them 'microsomes'.
2. **Palade** (1955) described them in animal cells under the electron microscope as dense particles or granules and Robert (1958) coined the term **'ribosome'**.
3. They are one of the **smallest organelles** of the cell.
4. In prokaryotes, the ribosomes are free in the cytoplasm.
5. In eukaryotes, they are either free in the cytoplasmic matrix or attached to the membranes of endoplasmic reticulum by glycoproteins called **ribophorins.** In eukaryotes, ribosomes are also found in mitochondria and chloroplasts.
6. They are approximately spherioid in shape with a diameter of 23 Å.
7. Each ribosome consists of a large and a small subunits.
8. Ribosomes are generally measured on the basis of their sedimentation coefficient which is expressed in **Svedberg (S)** units, as given under:

Functions

(*i*) Ribosomes are the site for protein synthesis in the cell.

(*ii*) It is the assembly shop or protein factory where amino acids are linkled to form proteins.

(*iii*) It gives protection of mRNA and newly synthesized polypeptide chains from enzymes, nucleases and proteases respectively.

Mitochondria

1. They were first observed by **Kolliker** (1880) in the muscle cells.
2. Flemming called them 'fila' (1882).
3. Altmann in 1894 described them as 'biplasts'.
4. Later, they were called **'mitochondria'** by **Benda** in 1897.
5. The presence of mitochondria in **plant cells** was first detected by **F. Meves** in 1904 in *Nymphaea.*
6. **Kingsbury** (1912) suggested that mitochondria are seat for cellular respiration.
7. Later, **Hogeboomet** (1948) demonstrated that mitochondria are the seat for cellular respiration.
8. **Structure of a mitochondrion**

   (*i*) Structurally, a mitochondrion consists of two membranes, an outer limiting membrane and an inner membrane.

   (*ii*) The inner membrane is most frequently convoluted to form a large number of infoldings called **cristae** or **mitochondrial crests.**

   (*iii*) The membranes of the mitochondrion separate it into two distinct compartments.

   (*iv*) An outer compartment sometimes called intermembranal space is formed between the outer and inner membrane.

   (*v*) The inner compartment, also called mitochondrial chamber, is bound by the inner membrane.

   (*vi*) This mitochondrial chamber is filled with dense proteinaceous material called the **matrix,**

   (*vii*) The mitochondrial matrix contains most of the enzymes of Kreb's cycle.

   (*viii*) The inner membrane in the crest region is covered by the particles of 8.5 nm size. These particles are called **elementary** or **F_1 particles** or **oxysomes** or **respiratory assemblies.**

Function

They are seat for cellular respiration and produce energy in the form of ATP (*i.e.*, cellular currency) needed for cellular functions. Thus they are called **power houses.**

MULTIPLE CHOICE QUESTIONS

1. Robert Hooke while observing thin sections of cork cells actually observed
(*a*) cell wall (*b*) cellulose
(*c*) nuclei (*d*) protoplasm

2. The 'cell doctrine' was given by
(*a*) Robert Hooke
(*b*) Schwann, Schleiden and Virchow
(*c*) Dutrochet
(*d*) Huxley

3. Peptidoglycan compound murein is present in cell wall of
(*a*) plant (*b*) animal
(*c*) bacteria (*d*) algae

4. One of the following cell is connecting link between prokaryotes and eukaryotes, therefore, called as mesokaryote
(*a*) protozoan (*b*) *Acetabularia*
(*c*) dinoflagellate (*d*) diatom

5. The molecular model of plasma membrane was first proposed by
(*a*) Watson and Crick
(*b*) Daniell and Davson
(*c*) Skoog and Miller
(*d*) Flemming and Jenson.

6. The model given by Wolpers was
(*a*) trilaminar model (*b*) unit membrane
(*c*) lattice model (*d*) fluid mosaic.

7. The term plasma membrane was given by
(*a*) Robertson
(*b*) Nageli
(*c*) N. Grew
(*d*) Danielli and Davson.

8. One of the following model of plasma membrane suggested that molecules show globular arrangement
(*a*) Lattice model
(*b*) Fluid mosaic
(*c*) bimolecular leaflet
(*d*) Micellar model.

9. The typical biomembranes of all the cell types are
(*a*) unilaminar (*b*) bilaminar
(*c*) trilaminar (*d*) quadrilaminar

10. Which of the following structure is not bounded by cell membrane?
(*a*) Spherosomes
(*b*) Mitochondria
(*c*) Ribosomes
(*d*) Lysosomes

11. The phospholipid bilayer
(*a*) is readily permeable to large, polar molecules
(*b*) is entirely hydrophobic
(*c*) is entirely hydrophilic
(*d*) has different lipids in the two layers

12. The most abundant types of lipids comprising the plasma membrane are
(*a*) cholesterol (*b*) triglycerides
(*c*) steroid (*d*) phospholipids

13. The structure of a eukaryotic cell which is characterized by selective permeability is
(*a*) chemical composition of cell wall
(*b*) cell membrane
(*c*) permeability of cell wall
(*d*) all above.

14. Tonoplast is a differentially permeable membrane which surrounds the
(*a*) cytoplasm of prokaryotes
(*b*) mitochondria
(*c*) nucleus of eukarytoes
(*d*) vacuole.

15. Pinocytosis and phagocytosis are concerned with
(*a*) cell wall (*b*) plasmalemma
(*c*) nucleus (*d*) none of these

16. The major driving force for formation of membrane lipid bilayer is
(*a*) vander Waal forces
(*b*) electrostatic forces
(*c*) hydrophobic interactions
(*d*) hydrogen bonding.

17. Ions enter the cell through plasma memberane by
(*a*) some carriers present in the membrane
(*b*) cytochrome pump across the membrane
(*c*) phosphatidic acid cycle
(*d*) all the above

Directions Qs. 18-20: *In each of the following questions, a statement of* **Assertion (A)** *is given followed by a corresponding statement of* **Reason (R)** *just below it. Of the statements, mark the correct answer—*
(*a*) If both assertion and reason are true and reason is the correct explanation of assertion
(*b*) If both assertion and reason are true but reason is not the correct explanation of assertion
(*c*) If assertion is true but reason is false
(*d*) If assertion is false but reason is true.

18. **Assertion (A):** Primary wall is first formed wall of the cell.
Reason (R): Some cells possess only primary wall.

19. **Assertion (A):** Lignin is a polymer.
Reason (R): Lignin reduces the water content of the wall.

20. **Assertion (A):** Mitochondria is an important cell organelle of both eukaryotes and prokaryotes.
Reason (R): They are called power houses.

ANSWERS

1	2	3	4	5	6	7	8	9	10
(*a*)	(*b*)	(*c*)	(*c*)	(*b*)	(*c*)	(*b*)	(*d*)	(*c*)	(*c*)
11	**12**	**13**	**14**	**15**	**16**	**17**	**18**	**19**	**20**
(*d*)	(*d*)	(*b*)	(*d*)	(*b*)	(*b*)	(*d*)	(*b*)	(*b*)	(*d*)

13

Cell Reproduction

CELL DIVISION AND CELL CYCLE

All living cells grow and divide. Splitting of cells is called **cell division**. The period between the end of one division and the end of next is termed as the **cell cycle** which is divided into four periods. The first three periods (**G_1, S, G_2**) constitute **interphase** stage.

(i) **G_1-phase.** It is the time 'gap' between the end of mitosis and the start of DNA synthesis. It is the pre-DNA synthesis phase in which chromosomes are fully extended and some proteins and RNA are synthesised. It can vary from 3–4 hours to several days, months or even years.

(ii) **S-Phase.** it is the DNA synthesis phase in which DNA and **histones** are synthesised. This stage lasts for 7-8 hours.

(iii) **G_2-Phase**. It is the time 'gap' between the end of DNA synthesis and the beginning of mitosis. This stage lasts for 2-5 hours. During this stage, diploid cells contain 4X times the amount of DNA. It is followed by mitosis.

(iv) **Mitosis** or **M-Phase**. It represents only a small part of life cycle of a cell. It's duration is about an hour in most cells.

MITOSIS

- The process of mitosis was first discovered in animal cells by **W. Flemming** (1882) in Salamander (*Tritus maculosa*) and in plant cells by **Strasburger**.
- It occurs in the vegetative as well as reproductive cells of eukaryotes. It results in the formation of two daughter cells which receive exactly the same number and kind of chromosomes which the parent cells had. In strict etymological sense, mitosis refers to the nuclear division (**karyokinesis**) in eukaryotic cells. It has nothing to do with the division of cytoplasm (**cytokinesis**).
- Mitosis is a continuous process though for understanding, it may be divided into four phases (stages).

1. Prophase

(*i*) In the beginning of this stage chromosomes appear as thin threads and they start to condense.

(*ii*) The two identical chromatids of a single chromosome are called **sister chromatids** which are attached firmly to each other at a specific point called **centromere** or **kinetochore**.

(*iii*) An important change that occurs during late prophase is the **disintegration of the nucleous** with nucleoplasm.

(*iv*) At the end of prophase, nuciear envolope disappears, with the result nucleoplasm and cytoplasm mix together, a process which appears to play a role in the development of spindle material.

2. Metaphase

(*i*) In the beginning of this stage, spindle apparatus develops.

(*ii*) The spindle consists of spindle fibres which are formed by the aggregation of microtubules extending between the poles.

(*iii*) Now the chromosomes move to an imaginary plane and lie perpendicular to the middle of the spindle apparatus.

(*iv*) This plane is called **equatorial plane** or **metaphase plate**. The chromosomes become attached by their centromeres to some spindle fibres. Many spindle fibres do not make connection with chromosome remaining extended from pole to pole. The function of these spindle fibres is not known.

3. Anaphase

(*i*) The centromere divides in the beginning of anaphase and this process is carried out simultaneously in all the chromosomes resulting in the separation of sister chromatids which now are called **daughter chromosomes**.

(*ii*) The centromeres of these chromosomes move apart and the daughter chromosomes begin their migration towards the poles of the spindle apparatus due to shortening of microtubules of the spindle fibres.

D. Telophase

(*i*) During telophase, the chromosomes seggregate at the poles; the spindle apparatus in disassembled and a nuclear envelope is formed from endoplasmic reticulum around each chromosome resulting in the formation of two daughter nuclei.

(*ii*) finally, chromosomes uncoil and a nucleolus in reorganised in each daughter nucleus at the chromosomal organizer site.

4. Cytokinesis

(*i*) It is the process of segmentation and separation of cytoplasm into two parts each having a nucleus.

(*ii*) In plants, cytokinesis occurs by cell plate formation.

5. Significance of mitosis

(*i*) Mitosis provides the opportunity for the growth and development to the organs and the organisms.

(*ii*) It maintains the same genetic constitution (both quantitatively and qualitatively) in the daughter cells as in the parent cell.

MEIOSIS

- Meiosis was first worked out by **Strasburger** in 1888. Later in more detail by **Farmer** and **Moore** in 1905 who also coined the term, **'meiosis'**.
- Meiosis is a complicated cell division which occur in reproductive cells (germ cells) of sexually reproducing organisms and **reduces the number of chromosomes to half**.
- Meiosis comprises two successive nuclear divisions, of which first division is a **reduction division** whereby the chromosome number **(2n) is reduced to half (n)** and the second division is **mitotic** in nature.
- These two division of meiosis are :

A. **Meiosis I** or heterotypic division or reduction division

B. **Meiosis II** or homotypic division or mitotic division.

A. Meiosis I or reduction division

It consists of four stages, namely Prophase I, Metaphase I, Anaphase I and Telophase I.

1. Prophase I: It is the longest and very complex stage and can be divided into five substages: Leptotene, Zygotene, Pachytene, Diplotene and Diakinesis.

(*i*) **Leptotene** or **leptonema** or **thin-thread stage** (Leptos = thin; nema = thread) :

(*a*) Nucleus increases in size and chromosomes become more apparent as **thin threads**.

(*b*) Although DNA synthesis had already occured in interphase, no longitudinal doubleness of chromosomes is visibly evident owing to contraction. They show bead-like thickenings called chromomeres.

(*c*) The tips (**telomeres**) of leptotene chromosomes are generally associated with the nuclear envelope, this peculiar arrangement is called **'bouquet'**.

(*ii*) **Zygotene or zygonema** (Zygon = adjoining)

(*a*) During zygotene, homologous chromosomes pair up and the phenomenon of pairing is called **synapsis**. Since each parent contributes a haploid set of chromosomes to a diploid offspring, all the diploid cells possess pairs of morphologically and genetically similar chromosomes, often called **homologous chromosomes**.

(*b*) These chromosomes pair lengthwise with each other and pairing may be initiated at one or more sites along the length of the chromosomes.

(*iii*) **Pachytene** or **thick-thread stage** (phacus = thick)

(*a*) It is a longest stage of prophase 1.

(*b*) At this stage, chromosomes contract longitudinally and they become shortened, thickened and coiled. Now, the pairs of homologous chromo-somes are called **bivalents**.

(*c*) Each chromosome is a bivalent has **two chromatids** by the middle of the pachytene stage, as a result of which a bivalent really consists of **four chromatids** and is called **'tetrad'**.

(*d*) Electron micrographic studies have shown that each chromatid has its own centromere. Thus, in a tetrad, there are **four centromeres, two homologous chromosomes** and **four chromatids**.

(*e*) During crossing over, at one point, only two chromatids, one from each of the two homologous chromosomes, take part. As the chromosomes continue to contract, the chiasmata have tendency to move towards the ends of the paired homologous chromosomes, a process called **terminalization**.

(*iv*) **Diplotene** or **diplonema** or **double-thread stage**

(*a*) This stage is generally recognised by the tendency of homologous chromosomes of the bivalents (tetrads) to pull away somewhat from each other due to repulsion.

(*b*) The seperation is generally not complete and they are held together by their chromatids at one or more specific points called chiasmata which represent the places of exchange of chromosome.

(*c*) During crossing over, at one point, only two chromatids, one from each of the two homologous chromosomes, take part. As the chromosomes continue to contract, the chiasmata have tendency to move towards the ends of the paired homologous chromosomes, a process called **terminalization.**

(*v*) **Diakinesis.** During this phase, bivalents totally separate, nuclear membrane disappear and terminalization is completed.

2. Metaphase I:

(*i*) On the disappearance of the nuclear envelope, spindle fibres appear.

(*ii*) The homologous chromosomes (tetrads) become arranged at the equatorial plane (metaphase plate) and a bundle of spindle microtubules attaches to each centromere.

(*iii*) This attachment occurs in such a way that both chromatids of a single chromosome are attached to one pole and both chromatids of the other homologous chromosome are attached to the other pole.

3. Anaphase I:

(*i*) The homologous chromosomes separate from one another *i.e.,* one chromosome with its two sister chromatids or each homologous pair move to opposite poles.

(*ii*) Thus, each pole has a **haploid** number of chromosomes (instead of a diploid set as in mitosis) *i.e.,* a reduction in chromosome number is achieved at this stage.

4. Telophase I:

(*i*) Chromosomes with their chromatids reach the poles.

(*ii*) Nuclear envelope and nucleolus reappear, resulting in the formation of two haploid daughter nuclei with each chromosome having two chromatids.

B. Meiosis II or homotypic division

It also consists of four stages :

1. **Prophase II** : It is very short stage. Chromosomes are double, each with two

chromatids joined together by a centromere. At late prophase II nuclear envolpe disintegrates and spindle fibres are formed.

2. **Metaphase II :** Chromosomes become arranged on the equatorial plane. Spindle microtubules attach to each chromosome so that sister chromatids are connected to opposite poles.
3. **Anaphase II :** The sister chromatids of each chromosome are separated from each other and move to opposite poles, as each of the members of a pair of sister centromeres split from each other. The separated chromatids are now called daughter chromosomes.
4. **Telophase II :** The daughter chromosomes reach the poles and thus each pole receives a haploid set of chromosomes. Nuclear envolpe is formed around each set of chromosomes and nucleus in organised. Thus, four haploid nuclie are formed.

Cytokinesis : It occurs either by furrowing as in animal cells or by cell plate formation in plant cell.

Significance of meiosis

(*i*) Meiosis causes restoration of the original number of chromosomes in an individual.

(*ii*) It provides an opportunity to produce variations in the individuals as a result of crossing over while the **variations are the root cause of evolution**.

(*iii*) Failure of meiosis will lead to the formation of diploid gametes which on fusion form polypoids.

MULTIPLE CHOICE QUESTIONS

1. In prokaryotes
(*a*) there are no meiotic divisions
(*b*) mitosis proceeds as in eukaryotes
(*c*) the genetic information is not carried in chromosomes
(*d*) the chromosomes are identical to those of eukaryotes.

2. The chromosomal set is
(*a*) constant for a given species
(*b*) is same for all the plants
(*c*) varies with the age for a given organism
(*d*) is same for all the organisms.

3. 'U' or 'V' or 'J' shaped chromosomes are seen in
(*a*) prophase (*b*) metaphase
(*c*) telophase (*d*) anaphase

4. In the anaphase of mitosis
(*a*) chromosomes are in the middle of the cell
(*b*) chromosomes are at one pole
(*c*) chromosomes move away towards the poles from the middle
(*d*) there are no chromosomes

5. Which of the statement is incorrect regarding meiosis ?
(*a*) No duplication of centromeres occur in meiosis I
(*b*) chromosomes do not duplicate in meiosis II
(*c*) none
(*d*) both (*a*) and (*b*).

6. In which of the following nucleolus is not visible?
(*a*) Leptonema (*b*) Zygonema
(*c*) Pachynema (*d*) Diplonema.

7. In meiosis
(*a*) meiosis II reduces the chromosome number from diploid to haploid
(*b*) DNA replicates between meiosis I and II
(*c*) the chromatids that make up a chromosome in meiosis II are identical
(*d*) homologous chromosomes separate from one another in anaphase I.

8. Synapsis is characteristic of
(*a*) leptotene (*b*) pachytene
(*c*) zygotene (*d*) diplotene.

9. Meiosis is also called as haplosis as haploid cell
(*a*) meiosis occurs in haploid cell
(*b*) daughter cells of the meiosis are haploid
(*c*) meiosis reduce the number of chromosome to half
(*d*) meiosis nullify the effect of fertilization.

10. Centrioles are
(*a*) microbodies (*b*) microtubules
(*c*) protein bodies (*d*) glycogen bodies

11. Which statement about the cell cycle is not true?
(*a*) It consists of mitosis and interphase
(*b*) The cell's DNA replicates during G1
(*c*) A cell can remain in G1 for weeks or much longer
(*d*) Most proteins are formed throughout all subphases of interphase.

12. Spindle fibres are made of
(*a*) actin proteins and RNA
(*b*) cellulose and RNA
(*c*) tubulin protein and RNA
(*d*) both (*a*) and (*c*)

13. Types of coiling found during mitosis is
(*a*) plectonemic
(*b*) paranemic
(*c*) both of the above
(*d*) none of the above.

14. One of the following cells can divide by meiosis
(*a*) Sporocyte
(*b*) Meiocytes
(*c*) Primary spermatocyte and primary oocytes
(*d*) All the above.

15. In which region of the metaphase plate the larger chromosomes get arranged ?
(*a*) Peripheral region
(*b*) Variable from plant to plant
(*c*) Central region
(*d*) Not constant.

Directions Qs. 16-20: *In the following questions, statement of **Assertion (A)** is given followed by a corresponding statement of **Reason (R)** just below it. Of the statements, mark the correct answer—*
(*a*) If both assertion and reason are true and reason is the correct explanation of assertion
(*b*) If both assertion and reason are true but reason is not the correct explanation of assertion
(*c*) If assertion is true but reason is false
(*d*) If assertion is false but reason is true.

16. **Assertion (A):** Cytokinin promotes cell division.
Reason (R): Cytokinin is a mitogen.

17. **Assertion (A):** Interphase is also called intermitosis.
Reason (R): Interphase occurs between two dividing M-phase.

18. **Assertion (A):** Interphase is resting stage.
Reason (R): The interphase cell is metabolically inactive.

19. **Assertion (A):** Interphase-phase is longer than M-phase of cell cycle.
Reason (R): Interphase-phase occupies 75-90% of the total generation time.

20. **Assertion (A):** Interphase is divisible into three stages–G_1, S and G_2.
Reason (R): S-phase is also called invisible M-phase.

ANSWERS

1	2	3	4	5	6	7	8	9	10
(*a*)	(*a*)	(*d*)	(*c*)	(*c*)	(*d*)	(*d*)	(*c*)	(*c*)	(*b*)
11	**12**	**13**	**14**	**15**	**16**	**17**	**18**	**19**	**20**
(*b*)	(*c*)	(*a*)	(*d*)	(*a*)	(*a*)	(*a*)	(*c*)	(*a*)	(*b*)

14

Macromolecules of Cell

BIOMOLECULES

About 5000 chemicals occur in a cell. About 1000 chemical reactions are being carried out at any time in a cell. All molecules or chemicals functional in living organisms, especially the organic ones, are known as **biomolecules**. In animals, the cells are also bathed in an extracellular fluid of chemicals. The collection of different types of chemicals present in cells is called **cellular pool.**

Macromolecules are large sized complex molecules of high molecular weight which are commonly formed through condensation or polymerisation of repeated units of micromolecules, *e.g.,* polysaccharides, lipids, proteins, nucleic acids.

WATER

1. Life evolved in water and living things contain 70-90% of water.
2. On an average human body contains 66% of water whereas some marine organisms have high concentration of 96%. It is highest in human embryo of about 95%.
3. Water concentration is directly related to **age** and **metabolic activity** of an organism.

Uniqueness of water molecule

- Molecule is polar in nature.
- With weaker hydrogen bond than covalent bonds they help molecules of water to cling together.
- High boiling point (100°C) and low freezing point of 10°C (without hydrogen bonds, water would have boiled at –80° and frozen at – 100°).
- Temperature of water (liquid) rises and falls slowly than other liquids.
- Universal solvent for all chemical reactions.
- High heat of vaporization (1g hot water needs 540 calories to get converted to steam).

CARBOHYDRATES

1. Carbohydrates consist of **carbon, hydrogen** and **oxygen in** molar ratio of **1 : 2 : 1**. They are the energy sources for cells and form structural elements too.
2. Simplest carbohydrates include **simple sugars** (monosaccharides) having six carbon atom and serve **C-H bonds**-the energy storing bonds.
3. Some simple sugars are *glucose*, *fructose* and *galactose* too.
   (*i*) **Glucose** (grape sugars or dextrans): most widespread of all sugars found in blood of animals and produced by photosynthesis in plants.
   (*ii*) **Fructose** (fruit sugar): most common in fruits and is sweetest of all natural sugars
   (*iii*) **Galactose**: a constituent of agar agar.

LIPIDS

1. **Lipids** are not soluble in water because of lack of any polar group. Common forms of lipids are:
   (i) **fats** and (ii) **oils**. (iii) Other forms include *phospholipids,* ***steroids*** and *waxes*
2. Uses of Lipids (*i*) **Energy reserves** by organisms (*ii*) **Insulation** (in form of blubber in skin of whales)
3. Fats and oils are **triglycerides** (Abbreviated as TG) (one glycerol + three fatty acids). These are considered macromolecules of lpids.

Type	*Location*	*Molecular structure*	*Functions*
Cellulose	Cell wall of all plants	Linear long chains 6,000 or more units of β glucose	For tensile strength, mechanical support, permebale to water and solutes; food for ruminants and microorganisms
Starch	Storage food of plants	Occurs in two forms: **amylose** (straight chain) and **amylopectin** (branched chain)	Energy source of plants (from photosynthesis)
Glycogen	In liver and muscles of animals and also in fungi	Highly branched chain	Source of energy in metabolism and respiratory fuel.
Chitin	Exoskeleton of insects and crustaceans; cell wall of bacteria	Linear long chain like cellulose (N-acetyl glucosamine)	Gives shape, strength and retains water.

$$\textbf{3 Fatty Acids + Glycerol} \underset{\text{hydrolysis}}{\overset{\text{condensation}}{\rightleftharpoons}} \textbf{fats + 3 molecules of water}$$

4. Lipids are used as 'long term energy storage compounds' in animals and plants because of high concentration of C–H bonds.
5. Fatty acids are either saturated (with nil double bonds between carbon atoms) or unsaturated (with double bonds)
6. Lipid bilayers are the basic framework of all **biological membranes.**
7. On an average, fats yield **9 kcal** of chemical energy per gram of fat as compared to **4 kcal**/g from carbohydrates
8. Animal fats contain more calories than vegetable fats
9. Waxes, such as bees-wax are esters of long chain **alcohols**. They do not contain **glycerols**.
10. Examples of waxes are
    *Animals : ear wax (for trapping dust and dirt) and honeybee wax (for making honeycomb)
    * Plants : cutin (forming cuticle)

PROTEINS

1. Discovered by dutch chemist, **G.J. Mulder,** (1939) while investigating coagulating substances in milk and egg. Later J.J. Berzelius coined the term 'proteins'.
2. Protein molecules are very large with **high molecular weight** ranging from 5000-25,000,00.
3. Form the frame work of protoplasm.
4. Amino acids are building blocks of proteins, perhaps **first molecules** formed in the atmosphere of the earth.
5. All of these groups are bonded to a central carbon atom as shown.

$$\begin{array}{c} H \\ | \\ H_2N\text{–}C\text{–}COOH \\ | \\ R \end{array}$$

6. A polypeptide is a long chain of amino acid joined by a **covalent-peptide bond.**
7. It is formed in presence of an enzyme by the **condensation of an amino group** of one molecule with the **carboxylic acid** group of a second molecule.
8. The 20 most common amino acids and their side groups (R) grouped into five groups (based on the side groups)

   Nonpolar amino acids – alanine, valine, leucine, isoleucine

   Polar uncharged amino acids – glycine, serine, threonine, asparagine

   Ionizable amino acids – glutamic acid, aspartic acid, histidine, lysine, arginine

   Aromatic amino acids – phenylalanine, tryptophan, tyrosine

   Special function amino acids –proline, methionine, cysteine

9. Amino acids are classified into three groups as per their reaction in a solution :
   (*a*) **Neutral:** e.g. asparagine
   (*b*) **Acidic:** e.g. glutamic acid, glutamine and aspartic acid
   (*c*) **Basic:** e.g. arginine, lysine and hydroxylysine
10. Neutral amino acids can be further classified into:
   (i) **Aliphatic:** e.g. Glycine, valine, leucine, isoleucine, serine and threonine
   (ii) **Aromatic:** e.g. phenylamine, tyrosine
   (iii) **Heterocyclic:** e.g. tryptophan, histidine and proline
   (iv) **S-containing:** e.g. methionine, cysteine and cystine
11. Aminoacids are classified into three types, as per their synthesis in the body of organisms :

Essential amino acid

e.g. valine, leucine isoleucine, phenylalanine, threonine, lysine, tryptophan, methionine

Non-essential amino acid

e.g. serine, tyrosine, aspartic acid, cystine, cysteine, glutamic acid, proline, alanine, hydroxyproline, hydroxylysine

Semi-essential amino acid

e.g. histidine, arginine

MULTIPLE CHOICE QUESTIONS

1. The phospholipid bilayer
(*a*) is readily permeable to large, polar molecules
(*b*) is entirely hydrophobic
(*c*) is entirely hydrophilic
(*d*) has different lipids in the two layers

2. The most abundant types of lipids comprising the plasma membrane are
(*a*) cholesterol (*b*) triglycerides
(*c*) steroid (*d*) phospholipids

3. Commonly used table sugar is
(*a*) glucose (*b*) lactose
(*c*) sucrose (*d*) maltose

4. Carbohydrates are stored in animal body as
(*a*) glucose (*b*) starch
(*c*) glycogen (*d*) cellulose

5. All of the following are ketoses except
(*a*) xylulose (*b*) ribulose
(*c*) erythrose (*d*) fructose

6. One of them is a water soluble plant pigment
(*a*) carotene (*b*) phycocyanin
(*c*) chlorophyll (*d*) anthocyanin

7. One of the following aminoacids out of 20 contributes hydrophobic properties to the protein
(*a*) Serine (*b*) Histidine
(*c*) Tyrosine (*d*) Leucine

8. The example of chromoprotein is
(*a*) cytochromes (*b*) haemoglobin
(*c*) both above (*d*) none above

9. Pick up the fibrous protein
(*a*) collagen (*b*) elastin
(*c*) keratin (*d*) all above

10. Which one is a common plant terpene?
(*a*) camphor (*b*) peppermint
(*c*) both (*a*) and (*b*) (*d*) none above

11. Regarding the primary structure of proteins the true statement is
(*a*) each peptide bond is wholly double bond in character
(*b*) glysidic bonds are common
(*c*) the hydrogen bonding is common
(*d*) the only kind of linkage involved is covalent

12. Interferon is
(*a*) a protein
(*b*) a potential used on the treatment of cancer
(*c*) a carbohydrate
(*d*) a steroid

13. Pickup the example of esters
(*a*) fats (*b*) oils
(*c*) starch (*d*) both (*a*) and (*b*)

14. The functional group of fats is a
(*a*) glycerol (*b*) amine
(*c*) ketone (*d*) both (*b*) and (*c*)

15. Glycerides of fatty acids are
(*a*) lecithin
(*b*) phospholipid of plasma membrane
(*c*) vegetable oils
(*d*) all above

16. Lipids are soluble in all of the following except
(*a*) ether (*b*) water
(*c*) chloroform (*d*) benzene

17. One of the following is considered universal solvent
(*a*) alcohol (*b*) water
(*c*) ether (*d*) all the above

18. Inulin is a polymer of
(*a*) glucose (*b*) galactose
(*c*) fructose (*d*) arabinose

Directions Qs. 19-25: *In each of the following questions, statement of* ***Assertion (A)*** *is given followed by a corresponding statement of Reason (R) just below it. Of the statements, mark the correct answer—*

(*a*) If both assertion and reason are true and reason is the correct explanation of assertion
(*b*) If both assertion and reason are true but reason is not the correct explanation of assertion
(*c*) If assertion is true but reason is false
(*d*) If assertion is false but reason is true.

19. Assertion (A): Disaccharides show optical activity.

Reason (R): Cellobiose is an example of disaccharide.

20. Assertion (A): Starch gives blue colouration with iodine.

Reason (R): Starch is a storage polysaccharide of plant.

21. Assertion (A): Sucrose is a reducing sugar.

Reason (R): All disaccharides are reducing sugar.

22. Assertion (A): Grape contains fructose.

Reason (R): Fructose is called fruit sugar.

23. Assertion (A): Essential fatty acids must be present in the animal diet.

Reason (R): Deficiency of essential fatty acid produces phyenoderma.

24. Assertion (A): Oil can be converted into fat through the process of hydrogenation.

Reason (R): Hard fats are called drying oils.

25. Assertion (A): Waxy coatings are present on the surfaces of xerophytic leaves.

Reason (R): Bees are used to extract wax from plants.

ANSWERS

1	2	3	4	5	6	7	8	9	10
(*d*)	(*d*)	(*c*)	(*c*)	(*c*)	(*d*)	(*d*)	(*c*)	(*d*)	(*c*)
11	**12**	**13**	**14**	**15**	**16**	**17**	**18**	**19**	**20**
(*d*)	(*a*)	(*d*)	(*a*)	(*d*)	(*b*)	(*b*)	(*c*)	(*d*)	(*b*)
21	**22**	**23**	**24**	**25**					
(*d*)	(*d*)	(*a*)	(*c*)	(*c*)					

15

Enzymes

- The term 'enzyme' was first coined by **W. Kuhne** (1878)
- The first enzyme to be purified and crystallized was **urease**. It was achieved by **J. B. Sumner** in 1926, who had suggested the proteinous nature of enzymes.
- Buchner (1897) got Nobel Prize for first isolating an enzyme zymase from yeast.
- An enzyme has to be a protein, though it can have non-protein part also attached to it, called as **holoenzyme** or **conjugated** enzyme.

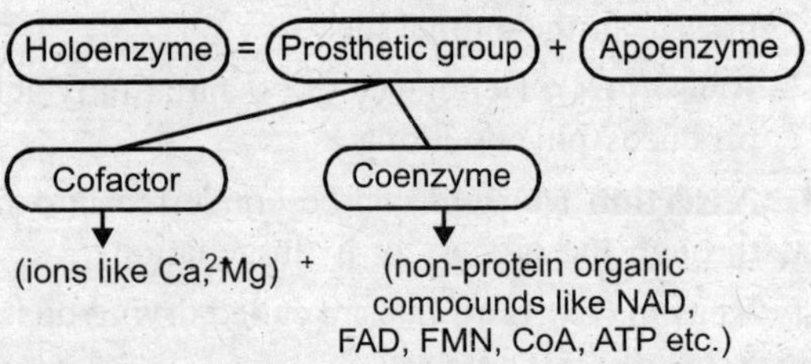

Properties of enzymes

1. **Proteinaceous and colloidal nature:** Enzymes are giant size protein molecular (molecular weight 1-1200 million) so they have low rate of diffusion and form colloidal system.
2. **Catalytic nature:** Enzymes are catalysts catalytic power of an enzyme is measured by *turn over number*, *i.e.*, the number of substrate molecules converted into products per unit time when the enzyme is fully saturated with substrate.
3. **Specificity :** Enzymes are very specific to substrate molecules of a reaction.

Factors affecting enzyme activity

1. **Temperature:** Enzymes being thermolabile, their activity increases repidly from 0°C-35°C. Above this temperature the rate starts declining. They get denatured at 60°C. For most of animals the optimum temperature is between 25°C-40°C.
2. **pH:** Each enzyme acts at a specific pH *e.g.* lipase between pH 4.0 – 5.0; arginase at 10.0; pepsin at 1.5-1.6 and ribonuclease at 7.0 –7.5.
3. **Substrate concentration:** For a known quantity of enzyme the reaction is directly proportional to substrate concentration, upto certain concentration after which the increasing concentration of a substrate does not increase the velocity of reaction.
4. **Enzyme concentration:** The velocity of enzymatic reactions in directly proportional to enzyme concentration.

Enzyme action

Two major theories are proposed to explain enzyme action, namely

1. **Fischer's lock and key model:** Proposed by **Email Fisher** in 1898. An enzyme and a substrate unite at the active site more or less like a key fits a lock thus forming an enzyme substrate complex.
2. **Koshland's induced fit model:** Koshland proposed in 1958 this is now being accepted. This theory explains the inducible change that is brought about at the active site of an enzyme by a substrate so as to form a suitable ES Complex.

Competitive and non-competitive inhibitions

When the active site or catalytic site of an enzyme is occupied by a substance other than the substrate of that enzyme, its activity gets inhibited. This is a type of reversible inhibition. In such inhibition both enzyme substrate (ES) and enzyme inhibitor (EI) complexes are formed. The actual amounts of ES and EI depends on :

(i) affinity between enzyme and substrate/ inhibitor
(ii) actual concentration of substrate and inhibitor present and
(iii) time of preincubation of enzyme with the substrate or inhibitor.

Competitive Inhibition	*Non-competitive inhibition*
1. Reversible	Irreversible or reversible
2. Inhibitor and substrate resemble each	It does not resemble other in structure.
3. Inhibitor binds the active site	Inhibitor does not bind the active site
4. V_{max} is same	V_{max} lowered
5. K_m increased	K_m altered
6. Inhibitor cannot bind with ES complex	Inhibitor can bind with ES complex
7. Lowers the substrate affinity to enzyme	It does not change the substrate affinity for the enzyme
8. Complex is E-I	Complex is E-S-I or E-I

Allosteric inhibition

1. There is a mixed kind of inhibition when the inhibitor binds to the enzyme at a site other than the active site but on a different region in the enzyme molecule called allosteric site.

2. Allosteric inhibition does not follow the Michaelis-Menten hyperbolic kinetics, instead it gives sigmoid kinetics.

3. The regulatory substance regulate the enzyme activity through allosteric transformation of the enzyme.

4. They attach with the allosteric sites of the enzyme molecule so that the shape of enzyme is changed and as such it fails to join with the substrate or shows more affinity towards it.

5. In the former case the enzyme action is inhibited while in the latter it is activated.

MULTIPLE CHOICE QUESTIONS

1. Enzymes increase the rate of reactions by
(*a*) increasing the free energy of activation
(*b*) decreasing the energy of activation
(*c*) changing the equilibrium constant of the reaction
(*d*) increasing the free energy change of the reaction.

2. If a man is exposed to an environment of 100% oxygen, the concentration of which enzyme is increased in lungs
(*a*) superoxide dismutase
(*b*) catalase
(*c*) carbonic anhydrase
(*d*) all of the above

3. Compounds structurally similar to substrate and inducing enzyme synthesis are
(*a*) constitutive enzymes
(*b*) gratuitous inducer
(*c*) coordinated inducers
(*d*) none of the above

4. One of the following is an example of coenzyme
(*a*) Zn^{2+} (*b*) Lipase
(*c*) Vitamin B (*d*) Lysine

5. In the case of competitive inhibition of an enzyme
(*a*) V_{max} is increased
(*b*) K_m is increased
(*c*) the extent of inhibition remains the same in high substrate concentrations
(*d*) none of the above

6. An example of irreversible inhibition is the action of nerve gas poisons on
(*a*) lipase
(*b*) acethylcholinesterase
(*c*) urease
(*d*) amylase

7. Cytochrome oxidase is an example of
(*a*) exoenzyme (*b*) endoenzyme
(*c*) coenzyme (*d*) proenzyme

8. Velocity of the reaction almost doubles with 10°C rise in temperature called as .... of an enzyme
(*a*) V_{10} (*b*) T_{10}
(*c*) Q_{10} (*d*) P_{10}

9. The active site of an enzyme is formed by
(*a*) R groups of amino acids
(*b*) aminogroups of amino acids
(*c*) COOH groups of amino acids
(*d*) exposed sulfur bonds

10. Hydrogen cyanide is an example of
(*a*) coenzyme
(*b*) cofactor
(*c*) non competitive inhibitor
(*d*) allosteric modulor

11. Endopeptidase of plant kingdom is
(*a*) trypsin (*b*) pepsin
(*c*) papain (*d*) urease

12. In a chemical reaction, the
(*a*) rate depends on the value of ΔG
(*b*) rate depends on the activation energy
(*c*) entropy change depends on the activation energy
(*d*) activation energy depends on the value of ΔG

13. In 1938, Max Perutz a political refugee at Cambridge University published X-ray photograph crystal data for two proteins
(*a*) haemoglobin and pepsin
(*b*) pepsin and chymotrypsin
(*c*) myoglobin and haemoglobin
(*d*) immunoglobin and haemoglobin.

14. Cheese is prepared by which enzyme?
(*a*) lactase (*b*) renin
(*c*) lipase (*d*) trypsin

Directions Qs. 15-20: *In the following questions, statement of* ***Assertion (A)*** *is given followed by a corresponding statement of* ***Reason (R)*** *just below it. Of the statements, mark the correct answer—*
(*a*) If both assertion and reason are true and reason is the correct explanation of assertion
(*b*) If both assertion and reason are true but reason is not the correct explanation of assertion
(*c*) If assertion is true but reason is false
(*d*) If assertion is false but reason is true.

15. Assertion (A): Lysozyme of tears is an example of endoenzymes.
Reason (R): Enzymes functional inside living cells are known as endoenzymes.

16. Assertion (A): Pectinase is used for cleaning fruit juices.
Reason (R): Pectinase works against middle lamella.

17. Assertion (A): Enzymes are generally globular proteins.
Reason (R): Ribozyme and ribonuclease-P are proteinaceous enzymes.

18. Assertion (A): Desmolysing enzymes are those which catalyse the reactions by hydrolysis.
Reason (R): Digestive enzymes are hydrolysing in nature.

19. Assertion (A): Enzymes lower the activation energy.
Reason (R): A substrate molecule can be acted upon by a particular enzyme.

20. Assertion (A): Catalysts and enzymes initiate the reactions.
Reason (R): They form short lived complexes with the substrate molecules.

ANSWERS

1	2	3	4	5	6	7	8	9	10
(*b*)	(*a*)	(*b*)	(*c*)	(*b*)	(*b*)	(*c*)	(*c*)	(*a*)	(*c*)
11	**12**	**13**	**14**	**15**	**16**	**17**	**18**	**19**	**20**
(*c*)	(*b*)	(*c*)	(*b*)	(*d*)	(*a*)	(*c*)	(*d*)	(*b*)	(*d*)

16

Plant Growth and Development

- **Plant growth** : A vital process which brings about a slow, permanent and irreversible increase in size or volume or mass accompanied by an increase in dry weight.
- **Development** : A process in which there is sequence of qualitative changes, towards a higher or more complex state. Plants exhibit a regular sequence of seed germination, growth, differentiation, maturation, seed formation and senescence.
- The growth of a multicellular organism can be divided into 3 phases *i.e.,* cell division, cell expansion and cell differentiation.

Dormancy and seed germination

- Internal inhibition of germination of an otherwise viable seed even when it is placed in most favourable external conditions is called as **dormancy.**
- The suspension of growth of a viable seed may be due to exogenous control such as a change in environmental conditions and is called as **quiescence.**
- **Dormancy of seed** may be due to : Tough seed coats (*e.g., Capsella, Lepidium*), rudimentary embryos (*Eranthis hiemalis*), impermeable seed coat (*Chenopodium, Brassica, Xanthium*), mechanically resistant seed coats, excess salts (*Atriplex*), after ripening (Oat, Barley, Wheat/and even by the presence of germination inhibitors *e.g.,* abscisic acid, phenolic acids, short-chain fatty acids and coumarin).
- **Natural breaking of seed dormancy**: Leaching of inhibitors and salts; development and after-ripening of embryo; development of growth hormones; destruction of inhibitors by heat, cold, light and oxidation; weakening of impermeable and tough seed coats by mechanical abrasions, microbial activity and digestive eyzymes of the alimentary canal of animals.
- **Artificial breaking of seed dormancy** : By mechanical scarification (machine threshing, chipping or filing), chemical scarification (hot water, fat solvents, mineral acid etc.), vigrous shaking, pressure (about 2000 kg for 5-20 min), chilling treatment, light (red light breaks domancy and induces seed germi-nation in Lettuce and Tobacco), counter acting inhibitors (soaking seeds in KNO_3, thiourea, ethylene chlorohydrin, gibberellin etc.).
- **Seed germination** : This is the first step towards plant growth. After seed dormancy and necessary conditions for growth are present, the dormant embryo resumes metabolic activities and growth. This process is called **seed germination.**
- **Water, oxygen and temperature** are the essential condition for seed germination. **Red region** of visible spectrum is most effective for the onset of seed germination.
- **Vivipary** : Phenomenon of producing young ones in advanced state of development. Seed germinate inside the fruit, while still attached to the parent plant. It is found in mangrove plants *e.g., Rhizophora, Sonneratia* and *Heritiera*, such plants grow in marshy land. The seedling of such plant is heavy and falls down into salt rich marshy water with plumule remaining outside.

Phases of growth

- In plants, the growth in length is due to enlargement and elongation of cells at the apical regions and in thickness due to the activity of lateral and intercalary meristems.
- In unicellular organisms growth consists of a single step while in higher plants, growth involves three steps or phases *i.e.,* **phase of cell division** (formative phase), **phase of cell enlargement** (elongation phase) and **phase of cell maturation** (maturation phase). The phases are also known as **regions.**
- **Grand period of growth** : time internal from the formative phase to maturation phase.
- **Growth curve** : The mathematical curve which represents the variation in growth rate, it is somewhat flattened S-shaped curve (sigmoid growth curve). It is differentiated into three phases :

  (*a*) Lag phase (slow phase)

  (*b*) Log phase (exponential phase)

  (*c*) Stationary phase (steady state)

Measurement of growth

- **Horizontal microscope** : Used to measure growth in length of a potted plant and herb/ seedling.
- **Auxanometer** : Devised to measure the growth in length of a plant.

Growth regulators

In all plant, minute quantities of some chemical substances (plant growth regulators or phytohormones) are present which regulate growth and differentiation. These are organic compounds and are capable of influencing physiological activities leading to promotion, inhibition and modification of growth.

Auxin

- Weak organic acids with the acidic group.
- Indole-3-acetic acid (IAA) and other natural or synthetic growth regulating substances having structures and functions similar to IAA are called **auxins.**
- **First plant growth hormone** to be discovered.
- First indication of their existence came from the work of **Darwin** (1880).
- **F.W. Went** (1928) proved that the coleoptile tips contained a substance **auxin** and found that the degree of curvature of coleoptile is proportional to the amount of active substance in the agar blocks. This test is known as **avena curvature test.**
- Maximum concentration of auxins are found in the growing tips of the plant. From tip, it is transported **towards the base.**
- Same concentration of auxin exhibit different responses in different parts of the plant.

Functions

- **Cell elongation** (stimulate the cell walls to stretch in more than one directions) ; **cell division** (in the vascular cambium that results in secondary growth) ; **phototropism ; geotropism** (high concentrations of auxins are stimulatory for stem *i.e.,* negatively geotropic, on the other hand, higher concentrations of auxins are inhibitory for roots *i.e.,* positively geotropic) ; **apical dominance** (auxins in apical bud inhibit the development of lateral buds) ; **prevention of abscission layer** (leaves and fruits have to produce auxin continuously to prevent the formation of abscission layer) and **root initiation** (very low concentration of auxins promote the growth of root).
- **Auxin precursors** : Compounds which can be converted into auxins.
- **Anti-auxins** : Compounds which inhibit the actions of auxins.
- **Free-auxins** : Auxins which can easily be extracted. Active form.
- **Auxin bioassays** : The *Avena* curvature test and root growth inhibition test are the bioassays for examining auxin activity.

Gibberellins (GA)

- The discovery of gibberellin (GA) was based upon the observations made in Japan in early 1800s on the **bakane** or **foolish seedling disease of rice.**

- Widely distributed in nature particularly in angiosperms.
- Chemically, all GA are **terpenes** and weak acids.
- Found in abundance in young, expanding organs; being synthesised especially in embryos, young apical leaves, buds seeds and root tips.

Functions

- **Overcoming genetic dwarfism** (help in stem elongation) ; **bolting (elongation of intermodes to floweing in hosette plants) and flowering** (*e.g.,* Cabbage) ; **breaking of dormancy and germination of seeds** ; **increases fruit yield** etc.
- Gibberellin bioassay is performed through **dwarf maize/pea test** and **cereal endosperm test.**
- **Yabuta** (1935) coined the term gibberellin.
- About 100 different types of gibberellins have been isolated. They are denoted as GA_1, GA_2, GA_3 and so on.
- Antigibberellins like **maleic hydrazide, phosphon D** and **chorocholins chloride** (CCC) are also called **antiretardants.**

Cytokinins

- **Skoog** (1954 – 1956) found that coconut milk contained a substance which stimulated cell division in **tobacco pith cultures.**
- Miller and his colleagues isolated and purified the substance in crystalline form from **herring sperm DNA.** This substance was identified as **6-furfuryl aminopurine.** They named this compound **kinetin** because of its property to activate cell division (cytokinesis).
- First naturally occurring cytokinin to be chemically identified was from young maize (*Zea mays*) grains in 1963, and was called **zeatin.**

Functions

- **Cell division and differentiation** (in the presence of auxins, cytokinin promote cell division even in non-meristematic tissues) ; **control of apical dominance** (cytokinin stimulate the growth of lateral buds) ; **delaying senescence** (by controlling protein synthesis and mobilisation of resources) ; **induce flowering ; breaking the dormancy of seeds** and **initiation of interfascicular cambium.**

Ethylene

- Crocker *et. al.* (1935) identified ethylene as natural plant hormone.
- Ethylene is formed from amino acid **methionine.**
- **Auxin IAA** is known to stimulate ethylene formation.
- **Only gaseous** natural plant growth regulator.
- It does not generally move through the air spaces in the plants. Rather, it escapes from the plants surface.

Functions

- **Apical dominance** (in pea seedlings and suppresses the growth of apical buds) ; **stem swelling** (transverse expansion of stem) ; **leaf abscission** (formation of abscission zone is leaves, flowers and fruits) ; **epinasty** (leaf bending) ; **fruit growth and ripening** (a rise in the rate of respiration) ; **accelerates flowering ; sex expression** (increases the number of female flowers and fruits in cucumber plants) and **graviperception** (reduces sensitivity to gravity, seeddlings develop tight epicotyl hook).
- Commercial compound **"Ethaphone** (2-chloroethyl phosphoric acid) breaks down to release ethylene in plants.

Abscisic acid (ABA)

- Fredericle Addicott (1963) and P.F. Wareing (1964) discovered this hormone.
- Liu and Crans (1961) isolated ABA from mature cotton fruit.
- Chemically, ABA is a **Terpenoid**.
- ABA synthesized in the leaves, stems, fruits and seeds and distributed through the vascular system especially phloem.

- Isolated chloroplasts can synthesize it either from mevalonic acid or xanthophyll like violoxanthin.

Functions

- **Growth inhibitor** (antagonistic to all the three growth promoters) ; **bud dormancy** (by inhibiting growth processes) ; **senescence** (of leaves) ; **abscission** (of leaves, flowers and fruits in plants) ; **stomatal closure ; hardiness** (promotes cold hardiness and inhibits growth of pathogens) ; **promotes flowering** (in some short day plants) etc.
- It is a **stress hormone** *i.e.,* helps the plants to cope with adverse environmental conditions.
- ABA is sprayed on tree crops to regulate fruit drop at the end of the season.

Photoperiodism

- The response of plants to the relative lengths and alternations of light and dark periods with regard to the initiation of flowering is called photoperiodism.
- The effect of photoperiods on flowering was discovered by **Garner** and **Allard** (1920) in case of Maryland Mammoth variety of Tobacco (an SDP).

Vernalization

- Term **Vernalization** was coined by **Lysenko** (1928) to refer to the method of accelerating the flowering ability of biennials or winter annuals, by exposing their soaked seeds to low temperature for a few weeks.
- The site of vernalization is believed to be **shoot apical meristem.**
- Vernalisation effect is reversible and the reverse process is called **devernalization.**
- The vernalisation stimulus can be transmitted between plants by grafting hence the stimulus is believed to be a hormone called **vernalin.**

Senescence

- The sum of deteriorative processes which naturally terminate that functional life of an organism.
- Ageing is a sum total of changes in the total plant or its constituents while senescence represents degenerative and irreversible changes in a plant.
- Senescense can be delayed by removing flowers from the plants.

Abscission

- Shedding of leaves, foliage branches, floral parts, fruits etc. without causing any injury is called abscission.
- It generally occurs due to change in the hormonal balance.
- Abscisic acid is known to play an important role in abscission.

MULTIPLE CHOICE QUESTIONS

1. Cyclosis in a plant cell is
(*a*) induced movement
(*b*) spontaneous movement
(*c*) growth movement
(*d*) none of the above

2. Touch me not plant which show seis monastic movement is a member of family
(*a*) mimosaceae (*b*) myrtaceae
(*c*) annonace (*d*) cruciferae

3. The pigment responsible for light sensitive plant is
(*a*) chlorophyll (*b*) carotene
(*c*) phytochrome (*d*) cytochrome

4. Vernalization of seeds....their development
(*a*) inhibits (*b*) accelerates
(*c*) both A and B (*d*) none of these

5. Wheat and potato are
(*a*) SDP
(*b*) LDP
(*c*) day-neutral plants
(*d*) none of these

6. The length of time that is required by a plant to manifest its response to a stimulus is termed
(*a*) presentation time
(*b*) relaxation time
(*c*) reaction time
(*d*) none of the above

7. A hypothetical hormone is
(*a*) vernalin (*b*) florigen
(*c*) both A and B (*d*) dormin

8. Short-day plants (SDP) are
(*a*) short-night plants
(*b*) long-night plants
(*c*) long-day plants
(*d*) day-neutral plants

9. Low temperature is lethal because it has
(*a*) desication effect (*b*) chilling effect
(*c*) freezing effect (*d*) all of these

10. The growth rate is highest in
(*a*) red light (*b*) white light
(*c*) green light (*d*) blue light

11. The term 'phytochrome' was given by
(*a*) Mothes
(*b*) Borthwick and Hendrick
(*c*) Sorokin et al
(*d*) Wickson and Thimman

12. The rate of growth is highest in
(*a*) lag phase (*b*) log phase
(*c*) steady state (*d*) none above

13. Pickup the long day plant
(*a*) Wheat (*b*) potato
(*c*) *Xanthium* (*d*) both A and B

14. Movements in tendrils,lianas and twiners brought about by growth is known as
(*a*) epistrophe (*b*) circumnutation
(*c*) nutation (*d*) none of the above

15. For short day plants, the critical period is
(*a*) light (*b*) ultraviolet rays
(*c*) auxin (*d*) none above

16. Plants get protection from fungi and bacteria by
(*a*) abscisic acid (*b*) auxin
(*c*) phenolics (*d*) ethylene

17. Effect of gravity on geotropism can be known by
(*a*) auxanometer (*b*) galvanometer
(*c*) porometer (*d*) clinostat

18. Movement exhibited by dorsiventral leaf is
(*a*) plageotropic (*b*) diageotropic
(*c*) autonomus (*d*) positive geotropic

19. Went demonstrated the role of auxins using
(*a*) root tips (*b*) coleoptiles
(*c*) callus tissue (*d*) lateral meristems

20. Indole-3 Acetic Acid is a
(*a*) cytokinin
(*b*) auxin
(*c*) gibberellin
(*d*) secondary metabloite

Directions Qs. 21-25: *In each of the following questions, a statement of* ***Assertion*** *(A) is given followed by a corresponding statement of* ***Reason*** *(R) just below it. Of the statements, mark the correct answer:*

(*a*) If both assertion and reason are true and reason is the correct explanation of assertion.
(*b*) If both assertion and reason are true but reason is not the correct explanation of assertion
(*c*) If assertion is true but reason is false
(*d*) If assertion is false but reason is true.

21. **Assertion (A):** Cytokinins are produced in apical meristems region only.
Reason (R): Cytokinin show delay senescence.

22. **Assertion (A):** Sigmoid growth curve consists of four parts.
Reason (R): Lag phase is called as grand phase of growth.

23. **Assertion (A):** Phototropism in plants is caused by auxin.
Reason (R): The plant showing bending has more elongated cells on the illuminated side.

24. **Assertion (A):** Auxin and ethylene regulate leaf abscission.
Reason (R): The plant showing bending has more elongated cells on the illuminated side.

25. **Assertion (A):** ABA causes stomatal closing.
Reason (R): ABA causes reduction of osmotically active solutes from guard cells.

ANSWERS

1	2	3	4	5	6	7	8	9	10
(*d*)	(*a*)	(*c*)	(*b*)	(*b*)	(*a*)	(*c*)	(*b*)	(*d*)	(*b*)
11	**12**	**13**	**14**	**15**	**16**	**17**	**18**	**19**	**20**
(*b*)	(*b*)	(*d*)	(*a*)	(*a*)	(*c*)	(*d*)	(*c*)	(*b*)	(*b*)
21	**22**	**23**	**24**	**25**					
(*d*)	(*c*)	(*c*)	(*b*)	(*a*)					

17

Photosynthesis

Synthesis of organic compounds (carbohydrates) from inorganic raw materials (CO_2 and Hydrogen donor) by green plants in the presence of light (solar energy)

$$CO_2 + 2H_2O \xrightarrow[\text{Green plants}]{\text{Light energy}} (CH_2O)_n + H_2O + O_2$$

This equation was given by **C.B. Van Niel** for photosynthesis.

- Only 0.2% of the light energy falling on earth is utilised by photosynthetic organisms.
- **Cyanobacteria:** Where photosynthesis was first appeared.

SITE FOR PHOTOSYNTHESIS

- Mainly leaves, though all green parts containing chloroplasts have the potential to carry on photosynthesis.
- Chlorophyll (green pigment) is present in chloroplasts *i.e.,* localised.

PHOTOSYNTHETIC PIGMENTS

(1) Chlorophyll

- Total 9 types; **chlorophyll 'a'** and **chlorophyll 'b'** are important types of chlorophyll.
- **Chlorophyll :** Molecule consist of a porphyrin ring (hydrophilic) and a **phytol** tail (lipophilic).
- **Porphyrin ring :** Flat, square structure having smaller **pyrrole** rings with a magnessium atom at the centre. Chiorophyll ***a*** ($C_{55}H_{72}O_5N_4Mg$) possess a **CH_3 (methyl)** group and chlorophyll ***b*** ($C_{55}H_{70}O_6N_4Mg$) possess a **- CHO (aldehyde)** group. The head is joined to a long' hydrocarbon tail *i.e.,* **Phytol**.
- **Chlorophyll *a* :** most abundant, only pigment in all photosynthetic plants.
- Chlorophyll synthesis in flowering plants require **light** and **iron**.
- Algae, bryophytes, pteridophytes and gymnosperms develop chlorophyll even in absence of light.
- Raw materials for chlorophyll synthesis are glycine (amino acid) and succinyl co-A.

(2) Carotenoids

- Yellow to orange pigments, absorb light strongly in the blue-violet range.
- Called shield pigments because protect chlorophyll from photo oxidation by light of high intensity and also from oxygen produced during photosynthesis.
- Absorb light and transfer' it to chlorophyll for use in photosynthesis.
- Two types: ***(a)*** **Carotenes :** (Orange coloured) - hydrocarbons (tetrapenes)
  (b) **Xanthophylls :** (Yellow) - contain oxygen

(3) Phycobilin

- Water soluble, open tetrapyrrole pigments found in red algae and blue green algae.
- Found in phycobilisomes attached to thylakoids and transform light energy to chlorophyll. 3 types *i.e.,* **Phycocyanin** (blue), Allophycocyanin (blue) and Phycoerythrin (red).

Quantasome

- Present as small units on the membranes of thylakoids.

- This term coined by **Park** and **Biggins** (1964) for a group of pigment molecules required for carrying out a photochemical reaction.
- Each quantasome consists of about **250-300 chlorophyll** molecules, carotenoids, quinone compounds, sulpholipids, phospholipids, proteins etc.

MECHANISM OF PHOTOSYNTHESIS

Two-phases *i.e.,* light reaction and dark reaction.

(A) Photochemical phase (light reaction or Hill reaction)

- Takes place in thylakoids of grana.
- Solar energy is trapped by chlorophyll and is stored in form of chemical energy (ATP) and as reducing power (NADPH)
- **Photosystems :** Groupings of photosynthetic units which possess photocentres and are surrounded by specific light harvesting pigment molecules. 2 types.

***(a)* Photosystem-I (PS I) :** Present in stroma thylakoid and non-appressed parts of granal thylakoid.

- Photocentre P_{700} along with accessory pigments which absorb at or below 700 nm constitute the PS-I.

***(b)* Photosystem-II (PS II) :** Present in appressed part of granal thylakoids.

- Possess chlorophyll ***a,*** chlorophyll ***b*** and carotenoids.
- The cluster of pigment molecules which transfer their energy to P_{680} absorb light at or below the wavelegnth of 680 nm.

Light reaction : In which light energy is transformed by photophosphorylation into chemical energy in the form of the energy source ATP and $NADPH_2$ which is used to reduce CO_2, Takes place in following steps.

1. Excitation of pigments by light so that excited pigments lose electrons and become oxidised.

$$\text{Chlorophyll} \xrightarrow{\text{Light energy}} \underset{\text{oxidised}}{\text{Chlorophyll}^+} + \underset{\text{electron}}{e^-}$$

2. The electron is captured by an electron acceptor. This represent the conversion of light energy into chemical energy. Depending upon whether the electron lost by chlorophyll molecule is returned to it or not, the further events may be cyclic or non-cyclic.

Non-cyclic photophosphorylation (Hill and Bendall proposed Z-scheme for electron transport)

- Light energised ATP synthesis in which the electron expelled by excited photocentres does not return to them.
- Both PS I and PS II are involved.
- Coordinate their activities to produce assimilatory power (ATP and NADPH).
- Oxygen is released during photolysis of water.
- Takes place at one place only.

Cyclic photophosphorylation

- Light energised ATP synthesis in which the electron expelled by excited photo centre is returned to it after passing over a chain of electron carriers.
- It is related to photo act I (PS I) only.
- Electron released from chlorophyll return back.
- There is no photolysis of water and release of oxygen.
- NADP is not oxidised.
- Takes place at two places.
- It occurs under low light conditions, wavelength longer than 680 nm and when CO_2 fixation is being inhabited.

(B) Biosynthetic phase (Dark phase reaction)

Dark reaction takes place in stroma. It comprises a series of reactions controlled by enzymes. The sequence of these reactions was determined in *Chlorella* and *Scendesmus* by **Calvin, Benson** and **Bassham** using radioactive carbon ^{14}C techniques. This is also known as **Calvin Cycle** or **Calvin-Benson Cycle** or **C_3 Cycle or Reductive Pentose Pathway or Photosynthetic Carbon Reduction or PCR.**

(a) **Carboxylation** *i.e.,* acceptance of CO_2 by RuBP - CO_2 receptor, carboxylation of RuBP is catalysed by the enzyme **RuBP carboxylase** or **Rubisco** (most abundant protein on earth).

(b) **Glycolytic reversal :** The 6 molecules of PGA (3 phosphoglyceric acid) utilises 6 ATP molecules to form 6 molecules of 1,3-diphosphoglyceric acid. These, in turn, utilizes 6 NADPH supplied by the light reactions and get converted to a 3-C glyceraldehyde phosphate.

(c) **Regeneration of RuBP :** RuBP is regenerated through the formation of 4-C, 5-C and 7-C intermediates. Five molecules of triose phosphate (TP) and three molecules of ATP are involved in the regeneration of RuBP.

PHOTORESPIRATION (DICKER AND TIO, 1959)

- Respiration which is initiated in chloroplasts and takes place in light only is called **Photo respiration.**
- The enzyme Rubisco catalyses the carboxylation reaction where CO_2 combines with RuBP, called **oxygenation**.
- Also called **Photosynthetic Carbon Oxidation Cycle** and involves an interaction of three organelles *i.e.,* **chloroplast, peroxisome** and **mitochondria**.
- The function of photorespiration is to recover some of the carbon from the excess glycolate. However, there is a **wasteful loss of carbon** as **CO_2** and **loss of energy**.
- It is stimulated by high light intensity, high temperature, ageing of leaf and high oxygen concentration.
- Two molecules of phosphoglycolate (2C) produced by oxygenation of RuBP are changed into one molecule of phosphoglycerate (3C) and one molecule of CO_2.
- 75% of carbon lost to oxygenation is recovered by PCO cycle.

HATCH-SLACK PATHWAY OR C_4-PATHWAY OR DICARBOXYLIC ACID PATHWAY

- C_4 plants possess a characteristic leaf anatomy (**Kranz anatomy**). In which vascular bundles are surrounded by two rings of cells *i.e.* **inner ring (bundle sheath cells** having starch-rich chloroplasts lacking grana) and **outer ring (mesophyll cells)**.
- It C_4 plants CO_2 is picked up by phosphoenol pyruvate (PEP - 3 Compound) in the **cytoplasm of mesophyll cells** and form **oxaloacetate** (4C) which is converted to malic acid or aspartic acid (4–C) for transport to bundle sheath cells.
- In bundle sheath cells, CO_2 is released and is assimilated through Calvin cycle. Pyruvate is released which passes back into mesophyll cells where PEP is regenerated.
- The C_4 Pathway is more eftlcient than the C_3 pathway due to the absence of photorespiration.

CRASSULACEAN ACID METABOLISM (CAM)

- Takes place in succulents (*e.g.,* Cactus, *Bryophyllum, Kalanchoe, Sedum* etc.) and other plants that normally grow in dry conditions. Such plants have a photosynthetic pathway similar to C_4 plants.
- CO_2 taken up by the leaves on green stems through open stomata during night and fixed as malic acid with the help of **phosphoenol pyruvate**.
- Calvin cycle takes place during the day when stomata are closed. Fomation of malic acid during dark is called acidification (Phase I) and the release of CO_2 for actual photosynthesis during day is called deacidification (Phase II).
- CAM is a type of adaptation that allows certain plants (*e.g.,* pineapple) to carry out photosynthesis without much loss of water which is inevitable in plants with C_3 and C_4 mechanisms.

FACTORS AFFECTING PHOTOSYNTHESIS

1. Light

Provides energy for photosynthesis.

(i) **Light quality:** Photosynthesis successfully takes place only in the visible light (3800 Å - 7600 Å wave length). Rate of photosynthesis is maximum in red light, average in blue light and minimum in green light.

(ii) **Light intensity :** It has a direct relationship with the rate of photosynthesis. As the intensity of light (μ mol. $m^{-2}s^{-1}$) increases, the rate of photosynthesis increases. At higher light intensities, the rate of photosynthesis decreases.

(iii) **Light duration:** Long continuous periods of light increase the rate of photosynthesis.

2. Carbon dioxide

As the CO_2 concentration increases upto 0.9%, the rate of photosynthesis also increases. At the same time very high concentration of CO_2 becomes toxic to plants and inhibit photosynthesis.

3. Temperature

Generally, the rate of photosynthesis increases by increase in temperature upto 40°C. High temperature results into the deactivation of enzymes and so, the dark reaction is affected. The effect of temperature varies from plant to plant. C_4 plants have higher optimum temperature than C_3 plants. 25-30°C is the optimum temperature for photosynthesis.

4. Chlorophyll

Emerson (1929) found the direct relationship between the chlorophyll content and the rate of photosynthesis.

Willstatter studied **assimilatory number** *i.e.,* **Photosynthesis/chlorophyll unit** in leaves of different ages of plants and the effect of amount of chlorophyll on the rate of photosynthesis. Assimilatory number decreases with the age of the leaf.

CHEMOSYNTHESIS

Such reactions in which reduction of carbon dioxide to carbohydrates does not require light and takes place in presence of energy obtained from oxidations of inorganic substances is called **chemosynthesis** *e.g.,* **Nitrifying bacteria** *(Nitrosomonas, Nitrosococcus)* obtain energy by oxidising ammonia to nitrites), sulphur bacteria *(e.g., Beggiatoa* convert H_2S to sulphur), Carbon monooxide bacteria *(e.g., Carboxydomonas* convert CO to CO_2), **methane bacteria** *(e.g., Methanococcus* oxidise methane into CO_2 and HzO) etc.

MULTIPLE CHOICE QUESTIONS

1. Most of the experiment work in photosynthesis has been done on
(*a*) *Chlamydomonas*
(*b*) *Chlorella*
(*c*) *Hydrilla*
(*d*) Spinach leaf

2. Green plants contain
(*a*) chlorobium chlorophyll
(*b*) chlorophyll a and b
(*c*) bacteriochlorophyll a
(*d*) phycoerythrin

3. One of the following is not the part of chlorophyll
(*a*) carbon (*b*) iron
(*c*) hydrogen (*d*) nitrogen

4. In a chlorophyll molecule, the magnesium is situated in the
(*a*) centre of porphyrin
(*b*) corner of phytol
(*c*) phytol chamber
(*d*) isocyclic ring

5. The principle of limiting factors in photosynthesis was given by
(*a*) Hill (*b*) Arnon
(*c*) Blackman (*d*) Calvin

6. Plastocyanin is rich in
(*a*) Cu (*b*) Fe
(*c*) Zn (*d*) Mg

7. The correct molecular formula for chlorophyll 'A' is

(*a*) $C_{55}H_{72}O_5N_4Mg$
(*b*) $C_{55}H_{72}O_6N_4Mg$
(*c*) $C_{55}H_{70}O_6N_4Mg$
(*d*) $C_{55}H_{72}O_4N_3Mg$

8. Maximum absorption of light by Chlorophyll *a* and *b* takes place in the following range of the spectrum
(*a*) green (*b*) yellow
(*c*) violet-blue (*d*) orange-red

9. The most efficient converter of sunlight is
(*a*) potato (*b*) tomato
(*c*) sugar cane (*d*) papaya

10. The site of light reaction is
(*a*) granum (*b*) stroma
(*c*) unit membrane (*d*) stroma lamellae

11. Light energy is trapped in P_{700} pigment in
(*a*) Photosystem I (*b*) Photosystem II
(*c*) both (*d*) none

12. Light trapping reaction takes place in
(*a*) chloroplast (*b*) grana
(*c*) lamellae (*d*) stroma

13. Photochemical reaction of photosynthesis occurs in
(*a*) thylakoids (*b*) forest
(*c*) stroma (*d*) all above

14. First visible product of photosynthesis is
(*a*) glucose (*b*) sucrose
(*c*) cellulose (*d*) starch

15. Rate limiting step in photosynthesis is
(*a*) light reaction (*b*) dark reaction
(*c*) both A and B (*d*) none above

Directions Qs. 16-20: *In each of the following questions, a statement of **Assertion (A)** is given followed by a corresponding statement of **Reason (R)** just below it. Of the statements, mark the correct answer:*
(*a*) If both assertion and reason are true and reason is the correct explanation of assertion.
(*b*) If both assertion and reason are true but reason is not the correct explanation of assertion
(*c*) If assertion is true but reason is false
(*d*) If assertion is false but reason is true.

16. **Assertion (A):** Besides chlorophyll *a*, blue-green and red algae show the presence of phycobiliproteins.
Reason (R): Chlorophyll and phycobiliproteins are integrated into thylakoids.

17. **Assertion (A):** The phycobiliproteins form chromophores.
Reason (R): Phycobiliproteins are not associated with thylakoids.

18. **Assertion (A):** All prokaryotes show the presence of only one photosystem, *i.e.* PS I.
Reason (R): No photolysis of water takes place in the prokaryotes lacking PS II.

19. **Assertion (A):** Blue-green bacteria perform oxygenic photosynthesis.
Reason (R): Purple bacteria and green bacteria perform anoxygenic photosynthesis.

20. **Assertion (A):** Photosynthetic bacteria contain bacteriochlorophylls.
Reason (R): The cellular absorption spectra of each prokaryotes are distinctive.

ANSWERS

1	2	3	4	5	6	7	8	9	10
(*b*)	(*b*)	(*b*)	(*b*)	(*c*)	(*a*)	(*a*)	(*c*)	(*c*)	(*a*)
11	**12**	**13**	**14**	**15**	**16**	**17**	**18**	**19**	**20**
(*a*)	(*b*)	(*a*)	(*a*)	(*b*)	(*c*)	(*c*)	(*d*)	(*b*)	(*b*)

18

Cellular Respiration

Respiration is an energy liberating enzymatically controlled multistep catabolic process of stepwise breakdown of organic substances inside the living cells.

Type of respiration

1. **Aerobic :** Takes place inside the mitochondria in the presence of molecular oxygen in most of plants and animals called aerobes. Divided into 4 phases-Glycolysis, oxidative decarboxylation, Krebs or TCA cycle and Electron Transport system.
2. **Anaerobic (Fermentation) :** Involves the production of energy from the food nutrients in the absence of oxygen.

Respiratory Ratio or quotient (RQ)

The ratio of the volume of carbon dioxide released to the volume of oxygen taken in respiration in the given period of time at standard temperature and pressure.

$$RQ = \frac{\text{Moles of } CO_2 \text{ produced}}{\text{Moles of } O_2 \text{ absorbed}}$$

It provides the nature of the substrate being respired by a particular tissue.

(1) Carbohydrate RQ = 1 or unity

(2) Fats RQ = 0.7 *i.e.*, less than one

(3) Proteins RQ = less than unity (usually 0.5 to 0.9)

MECHANISM OF RESPIRATION

Glycolysis (splitting of sugar) results in the break down of glucose molecules into pyruvic acid. It Completed in the **cytoplasm** of the living cells. It is also known as **Embden-Meyerhof-Parnas** (EMP) pathway. It has 3 steps.

1. **Phosphorylation of Glucose :** Initial, 6-carbon sugar is transformed into 6-carbon sugar diphosphate. This addition of phosphate groups is called phosphory-lation, it activates the sugar and preventing it from getting out of the cell.
2. **Splitting of fructose-1, 6-diphosphate :** Fructose-l, 6-diphosphate is broken down into two molecules of 3-carbon sugar phosphates *i.e.,* dihydroxy acetone phosphate and glyceraldehyde-3-phosphate (undergoes further glycolysis). Both the molecules are interconvertible.
3. Formation of Pyruvic acid-3-phosphoglyceraldehyde in combination with one inorganic phosphate and coenzyme NAD is converted into 1, 3-diphosphoglyceric acid which is dephosphorylated to 3-phosphoglyceric acid and further changed to 2-phosphoglyceric acid. On dehydration 2-phosphoglyceric acid forms phosphoenol pyruvate or PEP which transformed to pyruvic acid.

- In glycolysis, a molecule of glucose is broken down to two molecules of pyruvic acid. There is gain of 2NADH and 2ATP or equivalent to 8 ATP.

Fate of pyruvic acid during anaerobic respiration

- When oxygen is limiting, NADH and pyruvic acid begin to accumulate.
- Under anaerobic conditon plants carry out **fermentation** and leading to the formation of CO_2 and either ethanol or lactic acid.
- Pyruvic acid releases one molecule of CO_2 to produce acetaldehyde which reoxideses NADH and itself reduced to ethanol.

- These reaction are catalysed by **pyruvic acid decarboxylase and alcohol dehydrogenase enzymes.**
- Net yield of ATP in anaerobic respiration is of only 2 molecules of ATP that too during glycolysis is anaerobic respiration.

CITRIC ACID CYCLE

- Sir Hans Krebs (1937) discovered the second phase in the aerobic oxidation of pyruvic acid. Got Nobel prize in 1953.
- It takes place in the matrix of mitochondria.
- Acetyl Co-A is a common intermediate of carbohydrate and fat metabolism, it is a substrate entrant of Kreb's cycle or connecting link between glycolysis and Krebs' cycle.
- Acetyl Co-A produced not only from pyruvic acid but also that produced as a result of β-oxidation of fatty acids undergoes TCA cycle.
- In total there are five oxidations and three decarboxylations (one decarboxylation and one oxidation outside Kreb's cycle)
- The net products from two pyruvates are 8NADH, $2FADH_2$ and 2ATP *i.e.,* 30 ATP/15 per pyruvic acid or in Krebs' cycle 24 ATP/12 per acetyl Co-A.

ELECTRONS TRANSPORT SYSTEM (ETS) AND OXIDATIVE PHOSPHORYLATION

- Terminal oxidation takes place is two steps *i.e.,* electron transport and oxidative phosphory-lation.
- **Electron transport system (ETS) :** Through which the electron passes from one carrier to another and takes place in the inner mitochondrial membrane. ETS consists of flavins, FeS complex, quinones (Ubiquinone or Co-Q) and cytochromes (cyt b, cyt c, cyt c] cyt a and cyt a). Cytochromes possess iron which can undergoes $Fe^{2+} \rightarrow Fe^{3+} \rightarrow Fe^{2+}$ change during passage of electrons. cyt a_3 additionally contains copper which helps in transferring electrons to oxygen.
- **Complex I : NADH dehydrogenase** which oxidises electrons from NADH produced in the mitochondrial matrix, during Kreb's cycle and electrons are then transferred to ubiquinone in the inner membrane.
- **Complex II : Succinate dehydrogenase** – Ubiquinone receives electrons from NADH and also by reducing equivalents via $FADH_2$ which is generated during oxidation of succinate, through the activity of the enzyme succinate dehydrogenase.
- **Complex III : Dihydroubiquinone, cytochrome** oxidoreductase and two forms of **cytochrome b and cytochrome c_1** The reduced ubiquinone (ubiquinol) is oxidised with the transfer of electrons to cytochrome c via this complex.
- **Complex IV : Cytochrome c, Cytochrome oxidase, cytochrome a, cytochrome a_3 and two copper centres :** Electrons are handed over to cyt $C_1 \rightarrow$ cyt c $\rightarrow$ cyt a $\rightarrow$ cyt a_3. Both cyt a and cyt a_3 function as cytochrome oxidase and handover electrons to oxygen. Reduced coenzyme NADH + H^+ helps is pushing out 3 pairs of H^+ to outer chamber while $FADH_2$ sends two pairs of H^+ to outer chamber.
- Cytochrome c acts as mobile carrier for transfer of electrons between complex III and IV.
- **Complex V : ATP synthase–**Electrons pass from one carrier to another via complex I to IV in the electron transport chain, they are coupled to ATP synthase for the production of ATP from ADP and inorganic phosphate.
- 3 pairs of protons ($6H^+$) are transported from mitochondrial matrix to outer side of the inner mitochondrial membrane.
- **Oxidative phosphorylation:** The accumulation of protons outside the inner membrane results in their higher concentration outside the inner membrane than in the matrix creating a **proton gradient** (Δ.pH) and **electric potential** (Δ.ψ) across the membrane.
- The process of synthesis of ATP at the expense of proton motive force is called chemiosmosis which is explained by Mitchell (l961, Nobel Prize 1978).
- ATPase is a multienzyme complex containing two parts F_0 and $F_1.F_1$ (acting as ATP sythethase) catalyses the synthesis of

ATP from ADP. For each pair of protons flowing back into the matrix, one molecule of ATP is synthesized.

- F_0 is an integral membrane protein complex that forms the channel through which protons cross the inner membrane.
- $FADH_2$ donates its electrons to Co-Q and not to FMN, only 4 protons are transported outside the membrane. Therefore, the back flow of these 4 protons through $F_0 - F_1$ complex results in the formation of only 2 ATP molecules.
- Total 38 ATP molecules are yielded during aerobic respiration of glucose. In most eukaryotic cells, 2 ATP molecules are expended in transporting the NADH produced in glycolysis into the mitochondrion for further oxidation in such cells, the net gain of **36 ATP** molecules instead of 38.

FACTORS AFFECTING RESPIRATION

1. **Temperature :** Affecting the activity of enzymes. Q_{10} = 2 or 3. Optimum temperature is 20 – 30°C. At high temperature the rate of respiration declines with time and at very low temperature, the respiration rate is insignificant.
2. **CO_2 :** Increase in CO_2 concentration and absence of O_2 adversely affect the rate of aerobic respiration.
3. **Light :** Control respiration by raising the temperature of an organism.
4. **O_2-Extinction point :** *i.e.*, minimum oxygen concentration at which aerobic respiration takes place which is 3 - 10% of oxygen.
5. **Respiratory substrates :** Higher availability of respiratory substrates increases the rate of respiration upto a certain limit.
6. **Climacteric fruits :** Such fruits have high rate of respiration during their ripening *e.g.*, Banana, Apple.
7. **Inhibitors :** A number of chemicals inhibit respiration *e.g.*, azide, cyanide, malonate, carbon monoxide etc.

MULTIPLE CHOICE QUESTIONS

1. Fermentation is carried out by
(*a*) all bacteria (*b*) all fungi
(*c*) some bacteria and some fungi
(*d*) all micro-organisms

2. During anaerobic respiration in plants
(*a*) CO_2 is takes in
(*b*) CO_2 is given out
(*c*) O_2 is given out
(*d*) O_2 is taken in and CO_2 is given out

3. Which of the following is widely accepted to respire in absence of oxygen?
(*a*) Yeast (*b*) *Chlorella*
(*c*) Potato (*d*) Grass

4. When a yeast cell respires,ATP formed are
(*a*) 4 B 2
(*c*) 6 (*d*) 8

5. During respiration pyruvic acid is formed by
(*a*) glycolysis (*b*) Kreb's cycle
(*c*) HMP pathway (*d*) all above

6. The hexose monophosphate shunt is also called
(*a*) EMP pathway (*b*) HMP pathway
(*c*) TCA pathway (*d*) citric acid cycle

7. Embden-Meyerhof pathway deals with
(*a*) citric acid cycle
(*b*) HMP shunt pathway
(*c*) glucoronic acid pathway
(*d*) glycolysis

8. Metal that can inhibit glycolysis is
(*a*) arsenic (*b*) manganese
(*c*) magnesium (*d*) copper

9. HMP shunt in the eye lens provides a transparent protein called
(*a*) eledin (*b*) keratin
(*c*) glutathione (*d*) both (*a*) and (*b*)

10. Pyruvic acid before entering the Kerb's cycle is changed into
(*a*) acetyl CoA (*b*) malic acid
(*c*) succinic acid (*d*) citric acid

11. Oxidation of succinic acid to fumaric acid is accompained by reduction of
(*a*) FAD (*b*) NAD
(*c*) FMN (*d*) NADP

12. The product formed by malic dehydrogenase is
(*a*) maleic acid
(*b*) oxaloacetic acid
(*c*) fumaric acid
(*d*) succinc acid

13. The form in which carbon enters the citric acid cycle is
(*a*) citric acid (*b*) pyruvate
(*c*) acetyl coenzyme (*d*) oxaloacetate

14. The carbon atoms in citric acid molecule are
(*a*) two (*b*) six
(*c*) four (*d*) three

15. During aerobic cycle how many metabolic steps release O_2?
(*a*) one (*b*) none
(*c*) Six (*d*) twelve

16. Maximum molecules of ATP produced under aerobic conditions in eukaryotes is
(*a*) 0 (*b*) 36
(*c*) 38 (*d*) 46

17. Cytochromes are concerned in
(*a*) photosynthesis
(*b*) cellular digestion
(*c*) cellular respiration
(*d*) both (*a*) and (*c*)

18. Protein rich pulses have R.Q. equal to
(*a*) one (*b*) more than one
(*c*) less than one (*d*) none above

19. In CAM the R.Q. will be .... at night
(*a*) zero (*b*) 1
(*c*) more than one (*d*) less than one

20. Respiration and photosynthesis both require
(*a*) magnesium (*b*) cytochrome
(*c*) energy (*d*) all above

21. All of the following are prosthetic groups in pyruvate to acetyl Co-A reaction, except
(*a*) Mn
(*b*) thiamine pyrophosphate & coenzyme A
(*c*) lipoic acid
(*d*) NAD

Directions Qs. 22-25: *In each of the following questions, a statement of* **Assertion (A)** *is given followed by a corresponding statement of* **Reason (R)** *just below it. Of the statements, mark the correct answer:*

(*a*) If both assertion and reason are true and reason is the correct explanation of assertion.
(*b*) If both assertion and reason are true but reason is not the correct explanation of assertion
(*c*) If assertion is true but reason is false
(*d*) If assertion is false but reason is true.

22. Assertion (A): Aerobic respiration consists of two steps.
Reason (R): Glycolysis is called as common pathway.

23. Assertion (A): Both hexokinase and glucokinase require divalent cation Mg^{++} or Mn^{++}.
Reason (R): The divalent cations act as catalysts.

24. Assertion (A): Glucokinase is present in liver and muscle cells.
Reason (R): Glucokinase phosphorylates all types of hexoses.

25. Assertion (A): Hexokinase and glucokinase both differ in their sugar specificity.
Reason (R): Both hexokinase and glucokinase cause phosphorylation of D-glucose.

ANSWERS

1	2	3	4	5	6	7	8	9	10
(*c*)	(*b*)	(*a*)	(*b*)	(*a*)	(*b*)	(*d*)	(*a*)	(*c*)	(*a*)
11	**12**	**13**	**14**	**15**	**16**	**17**	**18**	**19**	**20**
(*a*)	(*b*)	(*c*)	(*b*)	(*b*)	(*b*)	(*d*)	(*c*)	(*a*)	(*d*)
21	**22**	**23**	**24**	**25**					
(*a*)	(*d*)	(*c*)	(*d*)	(*b*)					

19

Heredity and Variation

MENDEL'S LAWS OF INHERITANCE

G. J. Mendel is called as father of genetics. He worked on garden pea and formulates laws which explain the manner of inheritance of characters. Mendel described his result in 1866, but his work was recognized in 1900, when Mendel's laws were rediscovered by Hugo-de Vries, Carl Correus and E. von Tschemark.

Mendel's Experiments

Mendel choose garden pea as the plant material for his experiment, due to following advantages :

(1) Well defined characters
(2) Bisexual flowers
(3) Predominantly self-fertilization
(4) Easy hybridization

Besides these features, garden pea had pure lines due to natural self-fertilization for a number of years.

Mendel's success was mainly based on the fact that he considered a single character at one time. Seven pairs of contrasting characters were chosen for the study. These characters are :

Seed character

1. Shape — Round, wrinkled
2. Cotyledon colour — Yellow, green
3. Coat colour — Grey, white

Pod character

4. Shape — Inflated, constricted
5. Colour — Green, yellow

Stem character

6. Position of pod — Axial, terminal
7. Plant height — Tall, dwarf

Results of Mendel's experiments : Mendel presented the data and conclusion derived from his experiments in a paper entitled "Experiments in Plant Hybridization" which was read before the Brunn Natural History Society in 1865 and was published in the proceeding of the society in 1886.

Parents

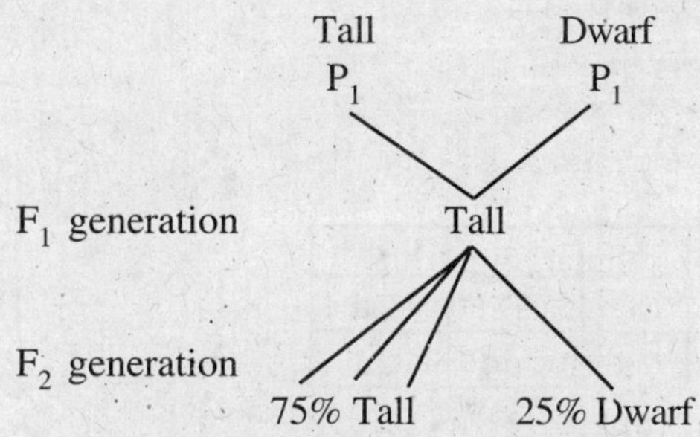

Inheritance of a Mendelian character in F_1 and F_2 generation

Mendel's results can be explained using a specific example when tall plants were crossed with dwarf plants, all plants in the F_1 generation were tall. The plants used in the initial cross are referred to as P_1 or parents. When the F_1 plants were self fertilized, both tall and drawf plants were obtained in the F_2 generation.

Backcross and Test cross : The F_1 individuals obtained in a cross are usually selfed to get the F_2 progeny. They can also be crossed into one of the two parents from which they were derived. Such a cross of F_1 individual with either of the two parents is known as a 'backcross', on the other hand, when it is crossed with recessive parent, both phenotypes appear in the progeny. While both these crosses are "backcrosses" only the cross with the recessive parent is known as a "Testcross". If is called a testcross, because it is used to test whether an individual is homozygous (pure) or heterozygous (hybrid).

This can be illustrated with the help of the example of tall (DD) and dwarf (dd) plants. In the F_1 generation only tall plants (Dd) appear. These plants can be backcrossed with either of the two parents as shown below :

F_1 Dd × DD P_1

D d   D D

↓

	D	D
D	DD	DD
d	Dd	Dd

DD = Pure tall (50%)

Dd = Hybrid tall (50%)

(Back cross)

F_1 Dd × dd P_2

D d   d d

↓

	d	d
D	Dd	Dd
d	dd	dd

Dd = Hybrid tall (50%)

dd = Dwarf (50%)

(Test cross)

Principle of segregation (Law of purity of gametes)

Mendel's principle of segregation is inherent in the results of Mendel. It is obvious that though in F_1 the dominant phenotype appears, the recessive phenotype is not lost but reappears in F_2. This suggested that there is no blending of mendelian factors in F_2, but they stay together and only one is expressed. At the time of the formation of gametes, these two factors obviously separate or segregate, otherwise recessive type will not appear in F_2. A gamete may carry either the dominant or the recessive factor but not both as we find in F_1 individuals. This concept of segregation is often called Mendel's first principle.

While a 3 : 1 ratio in F_2 generation of a monohybrid cross suggested that segregation of alleles does take place.

Principle of independent assortment

Mendel tried to find out whether one pair of characters is completely independent of another pair or else if there was any deviation.

P_1 Yellow round seeds (Pure) × Green wrinkled seeds (Pure) P_2

↓

F_1 Yellow round seeds

↓

F_2 Yellow round (9/16) | Yellow wrinkled (3/16) | Green round (3/16) | Green wrinkled (1/16)

9 : 3 : 1 : 1

When the F_1 seeds which were yellow and round (hybrid) were grown F_2 seeds were obtained which showed all the four possible combinations *i.e.*

(1) yellow and round seeds

(2) yellow and wrinkled seeds

(3) green and round seeds

(4) green and wrinkled seeds

These four kinds of seeds were obtained in 9 : 3 : 3 : 1 ratio.

The above results would be expected if the two pairs of characters are believed to behave independent of each other. If the individual characters segregate in 3 : 1 ratio, the law of probability can be applied.

P_1 GGWW Yellow round seeds   ggww green wrinkled seeds P_2

↓   ↓

gametes GW × gw

↓

F_1 Gg Ww Yellow round seeds

↓

GW   Gw   gw   gw gametes

	GW	Gw	gw	gw
GW	GWGW Yellow round	GWGw Yellow round	GWgw Yellow round	GWgw Yellow round
Gw	GwGW Yellow round	GwGw Yellow wrinkled	Gwgw Yellow round	Gwgw Yellow wrinkled
gw	GWgw yellow round	Gwgw yellow round	gwgw green round	gwgw green round
gw	GWgw yellow round	Gwgw yellow wrinkled	gwgw green round	gwgw green wrinkled

Yellow round seeds — 9

Yellow wrinkled seeds — 3

Green round seeds — 3

Green wrinkled seeds — 1

Therefore, it can be concluded that the two characters under consideration are assorting in an independent manner. This principle is known as Mendel principle of "Independent assortment".

Dihybrid cross : A dihybrid cross of round and yellow seeds with those having wrinkled and green seeds are 9 : 3 : 3 : 1.

CHROMOSOMAL BASIS OF INHERITANCE

Chromosome

Chromosome were first seen by Hofmeister in 1848 and were called chromosomes by W. Waldeyer in 1888. In mitosis chromatin condenses to become visible as chromosomes.

- The number of chromosome in a gamete is called **genome** or haploid chromosome number and is represented by '*n*'.
- Somatic cells of most organisms contain two set of chromosomes *i.e.,* diploid (2*n*) *e.g.,* every normal human being has 46 or 23 pairs (2*n*) chromosomes in each body cell.
- Plants with lowest number of chromosome are *Haplopappus gracilis* and *Brachycome lineariloba* (2*n* = 4) and plant with highest number of chromosomes is *Ophioglossum* (2n = 1262 chromosome).
- Chromosome number is used in the identification of species and in tracing the relationship within the species.

Numerical Changes in Chromosomes (Heteroploidy)

(*i*) **Aneuploidy : (*a*) Hypoploidy** (loss of one or more chromosomes) *e.g.* **Monosomy** (2n : 1), **Nullisomy** (2*n* : 2) etc.

(*b*) **Hyperploidy** (addition of one or more chromosomes to complete chromosomal complement) *e.g.* **Trisomy** (2n + 1), **Tetrasomy** (2n + 2) etc.

(*ii*) **Euploidy** : Chromosome number is exact multiple of the haploid number *e.g.* monoploidy (X), diploidy (2X), triploidy (3X), tetraploidy (4X), pentaploidy (5X), hexaploidy (6X) etc.

- **Autopolyploidy:** Same basic set of chromosomes is multiplied *e.g.* common doob grass (*Cyanodon*) is natural autotriploid. Other examples are seedless water melons, sugar beet, tomato, banana are autotriploids.
- **Allopolyploidy** (Amphidiploidy): Increase in chromosome number involve the sets from different species. Common wheat (*Triticum aestivum*) is natural hexaploid amphidiploid.

- Polyploid has important applications in crop breeding and horticulture.

Size of chromosomes

- In general plants (monocots as compared to dicots) have larger chromosomes as compared to animals.
- **Smallest chromosome** ≈ 0.25 μm in fungi and birds.
- **Longest chromosome** ≈ 30 μm in *Trillium.*

GROSS MORPHOLOGY OF CHROMOSOMES

- At metaphase each chromosome appears as a **cylindrical** double structure made up of two longitudinal halves **(sister chromatids)**

- **Centromere** (or **Primary constriction**) : Single point where sister chromatids are held together. Each centromere bears plate like **kinetochore** from where microtubules join the chromosomes at the time of cell division.
- There are four types of chromosomes with respect to position of centromere and the resulting relative length of the arms *i.e.,* **Metacentric** (centromere middle in position, equal arms and chromosome looks V-shaped in anaphase), **Submetacentric** (centromere near the centre of the chromosome, slightly unequal arms and chromosome appears J or L-shaped in anaphase), **Acrocentric** (centromere near one end of the chromosome and very unequal arms) and **Telocentric** (centromere at one end of the chromosome, only one sided arms and chromosome remains rod-shaped in anaphase also).
- **Secondary constriction:** Besides the primary constriction, a chromosome may possess an additional one termed as **secondary constriction**. Part of the chromosome beyond the secondary constriction is called **satellite.** Such chromosome bearing a satellite is called **sat chromosome**.
- **Nucleolar organisers:** Secondary constrictions involved in the organisation of nucleolus are called **nucleolar organisers** and the chromosomes as **nucleolar chromosomes.**

MOLECULAR ORGANISATION OF CHROMOSOMES

- **Nucleosome:** In interphase, chromosomes change into long, thin **chromatin fibres** as a chain of similar subunits called **nucleosomes**. Each nucleosome is connected to the next by a short **DNA linker**. A nucleosome and a linker are together referred to as a **chromatosome**.
- **Solenoid:** Nucleosome formation gives a 2nm thick chromatin fibre, a thickness of about 10nm which further folds to form a 30nm thick chromatic fibre *i.e.,* solenoid. All the folded loops of chromatin are held by a nuclear scaffold made of non-histone proteins.
- Nucleosome is made up of core particle of histone proteins (two molecules each of H_2A, H_2B, H_3 and H_4 histones) wrapped by DNA strand. Fifth histone molecule *i.e.,* H_1, attaches to the DNA near the bead.

Histones proteins: Low molecular weight, rich in basic amino acid **lysine** and **arginine**. Five types: **H_1, H_2A, H_2B, H_3 and H_4.**

H_1 : rich in lysine, holds together a chromosome fibre.

H_2A and H_2B : slightly lysine rich.

H_3A and H_4 : arginine rich, take part in organisation.

DNA and histones are loosely bound together in about 1:1 ratio to form **deoxyribo-nucleoprotein** (DNA) also called **chromatin.**

Non-histone proteins : High molecular weight, rich in **tyrosine** and **tryptophan**. Three types : **structural proteins** (from axis or scaffold of chromosome), **enzymatic proteins** (*e.g.,* DNA polymerase, RNA polymerase which helps in replication and transcription) and **regulatory proteins** (control gene expression).

LINKAGE

- The inheritance of genes of the same chromosome together and capacity of these genes to retain their parental combination in subsequent generation is called **linkage**. The genes located in the same chromosome are called **linked genes** and these present in different chromosomes are called **unlinked genes**. **Bateson** and **Punnett** (1906) discovered the principle of linkage in sweet pea *(Lathyrus odoratus)* which was put forwarded by **Morgan** (1910) from his work on fruit fly *(Drosophila).* Linkage is of two types : complete and incomplete linkage.

CROSSING OVER (GENETIC RECOMBINATION)

- The process of exchange of genetic material or segments between non-sister chromatids of two homologous chromosomes.

- Janssen (1909) observed **chiasmata** formation during the prophase of meiosis-I.
- Morgan (1910) observed that chiasmata formation leads to crossing over of alleles by breakage and reunion of homologous chromosomes.
- Crossing over takes place by **synapsis** (pairing of homologous chromosomes), followed by **duplication of chromosomes** (bivalent nature of chromosome changes to four stranded stage or tetravalent) and **crossing over** (chiasmata formation and terminalization).
- There are two theories with respect to relationship between crossing over and chiasmata formation *i.e.*, **chiasma type** or **one plane** theory (Janssen, 1909) and '**classical** or **two plane** theory (Sharp, 1934).

SEX DETERMINATION

It is determined at the time of fertilization when the male and female gametes fuse together.

1. **XX-XY type :** *e.g.* human being (total chromosomes 46).
    - 22 pairs of homologous autosomes + 1 pair of sex chromosome
    - **Male** : sperm = 22 autosomes + X or Y chromosome
    - **Female** : ovum = 22 autosomes + X chromosome
2. **XX-XO type :** Female with two homomorphic sex chromosomes XX. Male is heterogametic *i.e.,* 2 types of sperms : **gymnosperms** (with X) and **androsperms** (without X)
    - Fertilization with gymnosperms yield female offspring and with androsperms yield male offspring.
    - The sex ratio in the offspring is 1:1.
3. **ZW-ZZ type :** Male with two homomorphic sex chromosome (ZZ) Female with two heteromorphic sex chromosome (ZW)
    - Egg (Z) + Sperm (Z) → Zygote (ZZ) → Male
    - Egg (W) + Sperm (Z) → Zygote (ZW) → Female
    - Offspring have a sex ratio of 1:1.
4. **ZO-ZZ type :** Female is heterogametic, produces two types of eggs : half with Z and half without Z chromosomes.
    - Male is homomorphic and homogametic, produces one type of sperms *i.e.,* each with Z chromosome.
    - Sperm (Z chromosome) + Egg (Z containing) → Male offspring (ZZ)
    - Sperm (Z chromosome) + Egg (without Z) → Female offspring (ZO)
    - In some animals environment determine the sex *e.g. Bonellia* (chemical) and American alligator, Snapping turtle and some reptiles (temperature).
    - Non-allosomic genic determination of sex is found in bacteria. **Fertility gene** or factor located in **plasmid** determines the sex.

GENE MUTATION

"Mutation" term introduced by **Hugo de Vries** in 1901.

- Mutation is defined as a rare, random, discontinuous, inheritable variation in the amount or the structure of the genetic material (DNA) of a cell or organism.
- **Germinal Mutation** : mutation in germ cells.
- **Somatic Mutation** : mutation in somatic cells.
- **Spontaneous Mutation** : mutation which arise suddenly in nature.
- **Induced mutation** : mutagens (radiation, some chemical etc.) applied to create mutations.

Chromosomal aberrations or Mutations

Visible changes in the phenotype.

1. **Deletion (Deficiency) :** A part of chromosome is lost, may be **terminal** or **intercalary** *e.g.* **Cri-du-chat syndrome** due to deletion of half of short arm of chromosome 5 in humans. Infant's cry resembles the sound of a cat in pain due to this syndrome.
2. **Inversion :** A segment of a chromosome separates and rejoins it in an inverted

position, may be **terminal** or **intercalary** (may be **pericentric** *i.e.,* inverted segment includes the centromere or **paracentric** *i.e.,* inverted segment is on one side of centromere). Inversion can be detected by G-banded chromosome karyotyping.

3. **Translocation :** A segment of a chromosome breaks off and joins a non-homologous chromosome, may be **simple** (interstitial) or **reciprocal** (two non-homologous chromosomes mutually exchange segments). *e.g.* **chronic myloid leukemia** (CML) carried by chromosome 22 of malignant cells shortened through translocation of a segment of its long arm.
4. **Duplication :** Occurrence of a chromosome segment more than once in a chromosome, may be **tandem sequence** or in **reverse order**. Acquistion of additional genetic material can lead to new functions or phenotypic or serious deleterious effects. *e.g.,* the development of bar eye in *Drosophila.*

SEX LINKED INHERITANCE IN MAN

1. **Red green colour blindness :** Wilson (1911) detected the specific gene for **daltonism** (colour blindness) in the X-chromosome. This disease is due to sex chromosomal gene mutation.
2. **Haemophilia :** This disease is inherited through an X-linked recessive gene due to which blood lacks the capacity to coagulate. This disease is due to sex chromosomal gene mutation.

MULTIPLE CHOICE QUESTIONS

1. Identical twins are born when :
   (*a*) Two ova crosses with two sperms
   (*b*) One ovum crosses with one sperm
   (*c*) Two sperms crosses with one ovum
   (*d*) None of the above
2. Sex linked inheritance was discovered by :
   (*a*) Mendel (*b*) Johonsen
   (*c*) Morgan (*d*) Darwin
3. Germplasm theory was proposed by :
   (*a*) Weismann (*b*) Morgan
   (*c*) Darwin (*d*) Lamarck
4. Muton is :
   (*a*) a part of chromosome
   (*b*) a type of ribosome
   (*c*) a small part of gene
   (*d*) None of the above
5. Animals of similar genetic constitution are called :
   (*a*) Heterozygous (*b*) Homozygous
   (*c*) Genotypes (*d*) Phenotypes
6. The breakage in linkage association with the consequent formation of a new linkage is termed as :
   (*a*) Inheritance (*b*) Crossing over
   (*c*) Dominance (*d*) None of these
7. Crossing over generally takes place during :
   (*a*) Amitosis (*b*) Mitosis
   (*c*) Meiosis (*d*) None of the above
8. Haemophilia is :
   (*a*) Enzyme
   (*b*) Disease
   (*c*) Sex linked disease
   (*d*) None of the above
9. Polydipsia and Polyuria is related to :
   (*a*) Haemophilia
   (*b*) Colour blindness
   (*c*) Nephrogenic diabetes insipidus
   (*d*) None of the above
10. Factors which can induce genetic mutations in organisms are called :
    (*a*) Mutants (*b*) Mutagens
    (*c*) Muton (*d*) None of these
11. William Shockley is related to :
    (*a*) Eugenics (*b*) Mutation
    (*c*) Haemoglobin (*d*) Blood groups
12. What is the genotypic ratio of the F_2 generation of a dihybrid cross?
    (*a*) 9 : 3 : 3 : 1
    (*b*) 4 : 4 : 6 : 2
    (*c*) 1 : 2 : 2 : 4 : 1 : 2 : 1 : 2 : 1
    (*d*) None of the above

13. The first person who induce mutations by using X-rays was?
(*a*) Hugo de Vries (*b*) Muller
(*c*) Mendel (*d*) None of these

14. The chromosome number in plants can be increased by :
(*a*) Hormone treatment
(*b*) Colchicine treatment
(*c*) Heat treatment
(*d*) None of the above

15. Operon concept was proposed by :
(*a*) Jacob and Monod
(*b*) Watson and Crick
(*c*) Beadle and Tautam
(*d*) None of the above

16. Colchicine can induce polyploidy in :
(*a*) Animals only (*b*) Plants only
(*c*) Both (*a*) and (*b*) (*d*) None of these

17. To get a triploid plant, a tetraploid plant should by crossed with a :
(*a*) Diploid plant (*b*) Haploid plant
(*c*) Tetraploid plant (*d*) Triploid plant

18. The human chromosomes number i.e. 46 were discovered by :
(*a*) Mendel (*b*) H.J. Mullar
(*c*) Tijo and Laren (*d*) Watson and Crick

19. The basis of Klinefelter's syndrome and supermale are the following characteristic of sex chromosomes :
(*a*) XXY and XYY (*b*) XO and XXX
(*c*) XXX and YYY (*d*) None of these

20. The basis of Turner's syndrome and super-females are the following characteristic of sex chromosomes :
(*a*) XO and XXX (*b*) XXX and XYY
(*c*) XXXY and XXYY (*d*) None of these

Directions Qs. 21-25: *In each of the following questions, a statement of* **Assertion (A)** *is given followed by a corresponding statement of* **Reason (R)** *just below it. Of the statements, mark the correct answer:*
(*a*) If both assertion and reason are true and reason is the correct explanation of assertion.
(*b*) If both assertion and reason are true but reason is not the correct explanation of assertion
(*c*) If assertion is true but reason is false
(*d*) If assertion is false but reason is true.

21. Assertion (A): Mendel was succesful in his hybridization experiments.
Reason (R): Garden pea proved ideal experimental material.

22. Assertion (A): Variations do appear even in the clones.
Reason (R): Sexually reproducing organism have abundant genetic variations.

23. Assertion (A): Mutations are discontinuous variations.
Reason (R): Mutations occur suddenly.

24. Assertion (A): Blastogenic variations are inheritable.
Reason (R): Blastogenic variations appear in germ cells.

25. Assertion (A): Every somatic cells and tissue of the body produces gemmule.
Reason (R): Gemmules unite in gametes.

ANSWERS

1	2	3	4	5	6	7	8	9	10
(*a*)	(*c*)	(*a*)	(*c*)	(*c*)	(*b*)	(*c*)	(*c*)	(*c*)	(*b*)
11	**12**	**13**	**14**	**15**	**16**	**17**	**18**	**19**	**20**
(*a*)	(*c*)	(*b*)	(*b*)	(*a*)	(*b*)	(*a*)	(*c*)	(*a*)	(*a*)
21	**22**	**23**	**24**	**25**					
(*b*)	(*b*)	(*a*)	(*a*)	(*b*)					

20

Animal Nutrition

Nutrition is Sum of processes involved in taking of all the nutrients required by the body, their internment, distribution and utilization for obtaining energy, repair and growth materials and body protection.

Nutrients : Materials required by the living organisms to sustain their life, they include oxygen, water and food.

Dietetics : Science of study and regulation of diet for specific requirements.

Dietetist or dietition : Specialist associated with the needs and problems of nutrition.

On the basis of mode of nutrition, the organisms are classified as **autotrophs** and **heterotrophs.** Autotrophs synthesise their own organic food from inorganic raw material (autotrophic nutrition). Heterotrophs are unable to manufacture their own food (heterotrophic nutrition).

- **Macroelements (about 21 minerals) :** Essential for human nutrition, required in more than 100 mg per day *e.g.* Na, Ca, S, K, Mg, Cl.
- **Microelements** or **Trace elements :** Required in very small amounts *i.e.*, less than 100 mg per day e.g. Fe, I, Zn, Co, Cu, Mn, Mo etc.
- About 20 vitamins are thought to be required in small amount in human nutrition.

MODES OF NUTRITION IN ANIMALS

1. **Holozoic nutrition :** Engulfment of the whole or part of a plant or an animal, either in solid or in liquid state. *e.g.*, most of the free-living protozoans and animals.
2. **Saprozoic nutrition :** Organisms obtain food from dead and decaying organic food by first pouring digestive juices over the same and then sucking the digested food. *e.g.* spiders, horse-fly etc.
3. **Parasitic nutrition :** A smaller organism (parasite) draws food from another organism (host). *e.g. Plasmodium, Taenia,* Trypanosoma, Ascaris etc.
4. **Symbiotic nutrition or mutualism :** Two organisms or animals live in association and derive nutrition from each other. *e.g. Escherichia coli* and human. *E.coli* synthesize vitamin B_{12} for humans which human provide food to bacterium.

TYPES OF HOLOZOIC NUTRITION

1. **Herbivores :** Consume only algae or plant material e.g. cow, horse and rabbit.
2. **Carnivores :** Flesh eating animals *e.g.* lion, tiger and dog.
3. **Omnivores :** Feed both plants and animals *e.g.* cockroach, rat, crows, man etc.
4. **Cannibals :** Eat upon the members of their own species *e.g.* some fishes, cockroach, certain snakes etc.
5. **Predators :** When the larger animals feed upon the smaller animal species *e.g.* birds like eagle, kite etc.
6. **Piscivores :** Feed on fishes *eg.* crocodile, penguins, kingfisher etc.
7. **Insectivores :** Feed upon insects *e.g.* frog, lizards, small bats etc.
8. **Scavengers or Carrian eaters :** Feed up other dead and decaying material *e.g.* vultures, crow, hyena etc.
9. **Grainivorous :** Feed upon grains *e.g.* pigeon.

10. **Coprophagous :** Feed upon faeces *e.g.* dung beetle, pigs etc.
11. **Detritivorous :** Feed chiefly upon dead organic matters present in the mud *e.g.* earthworms.
12. **Myrmecophagous :** Take ants as prey *e.g.* spiny ant eater, scaly ant eater.
13. **Larvaevorous :** Feed on larvae *e.g. Gambusia* fish.
14. **Sanguinivores :** Suck blood of other animals *e.g.* leeches, mosquitoes, fleas etc.
15. **Folivorous :** Eat leaves *e.g.* sheep, goat etc.
16. **Frugivores :** Eat fruits *e.g.* birds, monkey etc.
17. **Fluid feeders :** Feeding on plant fluids *e.g.* butterflies, male mosquitoes etc.

DIGESTIVE SYSTEM OF HUMAN

Consists of alimentary canal or gut and some accessary digestive organs.

- **Alimentary canal :** Coiled, muscular tube, extending from mouth to the anus, complete, about 6–9 metres long, consists of many specialised sections.
- Arranged sequentially, these are mouth, oral or buccal cavity, pharynx, oesophagus, stomach, small intestine, large intestine, rectum and anus.
- **Buccal or oral cavity :** Has two lateral walls (**cheeks**), an anterior roof (hard palate), a posterior roof (soft palate) and a pair of jaws the upper maxilla and lower mandible.
- **Tongue or lingua :** A thick, musculo-sensory organ, present on the floor of buccal cavity, greatly protrusible, posteriosly attached to the floor of buccal cavity by a soft ligamentous fold (*frenulum*). *Dorsum* is lined by a **straitified squamous epithelium** posses a median groove and three types of lingual papillae
  - (*i*) **Circumvallate or Vallate papillae :** 8 – 12 in number, largest in size, in a Λ – shaped manner, at the the back of the tongue.
  - (*ii*) **Fungiform papillae :** Numerous, mushroom-shaped, red coloured, present on the tip and sides of the tongue.
  - (*iii*) **Filiform papillae :** Most abundant, smallest size, conical-shaped, present on anterior two-third part of the tongue.
  - (*iv*) **Foliate papillae :** 4–5 transverse fold, at each border of the tongue near the base.

Teeth (Dentenes)

Each half of the jaw holds four types of teeth.

(i) Incisors or cutting teeth

Chiesel shaped, present in anterior part of buccal cavity, used for cutting, chopping or gnawing.

(ii) Canine or tearing teeth

More pointed, dagger shaped, used for tearing or ripping or shredding.

(iii) Premolars and molars (cheek or grinding teeth)

Broad, have 2 – 4 cusped crown, located in the posterior part of the buccal cavity, used for shearing, crushing and grinding.

Dental formula *i.e.,* arrangement of teeth, I^{12}, $C^{1/1}$, $PM^{2/2}$, $M^{3/3}$ or 2123/2123 = 32

(iv) Wisdom teeth

Third pair of molars of permanent set erupt between the eighteenth and twenty-fifth year and sometimes never erupt.

Human teeth are **the codont** (embeded injaw sockets *i.e.,* alveoli), diphyodont (milk teeth are replaced by permanent teeth in the adult), heterodont (four types of teeth).

Structure of tooth

Made up of three parts *i.e.,* **crown** (exposed part of tooth, made up of shining, whitish and hard substance *i.e.,* **enamel** which is mainly formed of calcium phosphate, forms chewing surface of tooth), **neck** (covered by fleshy skin *i.e.,* **gum** or **gingiva**) and **root** (embedded in the socket of jaw bone and is fixed by a bone like cementum and a thick fibrous peridontal membrane).

Internally, a tooth is made up of a hard, bone-like substance *i.e.*, **dentine** which encloses the **pulp cavity** (contain soft, gelatinous and vascular connective tissue *i.e.,* **pulp** that enters via **pulp canal** at the base. The pulp cavity is lined by dentine – forming cells *i.e.*, **odontoblasts.**

Stomach

Most distensible part, thick, muscular, J-shaped organ present on the left side of upper part of abdominal cavity, suspended by a mesentery *i.e.,* **mesogaster.** Four parts *i.e.,*

(*i*) **Cardiac stomach :** Broad upper part, receives oesophagous through **cardiac aperture** guarded by cardiac sphincter which prevents **regurgitation** of food.
(*ii*) **Fundus :** Apex of stomach, contains gas or air.
(*iii*) **Body :** Main part, dome shaped.
(*iv*) **Pyloric entrum :** Lower, narrow part, opens in the duodenum by pyloric aperture or **pylorus** which is guarded by ***pyloric*** sphincter, it regulates passage of food into intestine.

Stomach helps in mechanical churning and chemical digestion of food. It acts as food reservoir.

Small Intestine

Longest part, about 20ft or 6 metres, narrow, tubular, divided into three parts.

(*i*) **Duodenum :** Widest, shortest (about 25 cm), most flexed part, forms a C-shaped curve, receives bile pancreatic duct.
(*ii*) **Jejunum :** Middle part, about 8 ft long.
(*iii*) **Ileum :** Longest part (12 ft), opens into the large intestine, the opening is guarded by an **ileo-caecal valve**.

Small intestine is the main region where digestion and absorption of food takes place.

Large intestine or Colon

About 5 feet (1.5 metres) long, shorter but wider than small intestine, forms an arc around the small intestine. Three parts.

(*i*) **Caecum :** Small, pouch like, ends into a tubular structure (13 cm long) called **vermiform appendix.** Caecum and vermiform appendix are **vestigial** as they are not involved in cellulose digestion.
(*ii*) **Colon :** Long sized, sacculated structure, differentiated into four regions *i.e.,* ascending colon (extending upto liver on the right side); transverse colon (crosses the abdominal cavity below the pancreas); descending colon (running downward on the left side and sigmoid or pelvic colon (S-shaped), enters the pelvis and joins the rectum). It is related with conservation of water, sodium or other minerals and formation of faeces.
(*iii*) **Rectum :** Slightly dilated part, about 13 cm long and related to temporary storage of faeces, leads into and canal (3-8 cm long), opens out by anus and guarded by two anal sphincter muscles.

THE ACCESSORY DIGESTIVE ORGANS

A. Tongue

Gustatory receptors, located in the buccal cavity, muscular organ with bony attachments (styloid process and hyoid bone).

B. Digestive glands

1. Salivary glands

(*i*) **Parotid glands :** Largest, present just below the external ear, saliva is carried by the parotid ducts to the second upper molar tooth.
(*ii*) **Submaxillary glands :** Lie beneath the jaw angles, secretion is carried by sub-maxillary or Wharton's ducts which open below the tongue.
(*iii*) **Sublingual glands :** Smallest sized, lie beneath the tongue and open at the floor of buccal cavity by a number of small **sublingual** or **Bartholin's ducts** or *ducts* of Rivinus.
Saliva = 95% water + 0.2% minerals + 0.3% organic compounds like mucin and a starch splitting enzyme **ptyalin** (salivary amylase).

2. Liver

Largest sized, weighs about 1.5 kg, reddish brown gland, present in the posterior concavity of the diaphragm in the right upper part of abdomen, multilobulated gland, two main (larger **right** and smaller **left lobe**) and two smaller lobes (**quadrate** and **caudate** lobes).

Gall bladder : Present on the lower surface of right liver lobe, thin walled, pear shaped sac, stores bile secreted by the liver. Bile is drained from the liver by a bile duct which is made by the joining of a cystic duct from the gall bladder and a common hepatic duct from different liver lobes.

Bile duct and pancreatic duct (from pancreas) join to form **hepatopancreatic** duct just near the duodenum. This duct opens in duodenum which is guarded by a **sphincter** of **oddi.**

3. Pancreas

Second largest gland, yellow coloured, 12–15 cm long, **compound racemose gland,** present in the loop of duodenum, made up of **head, body** and **tail, heterocrine gland,** glandular cells secrete the **pancretatic juice** which is drained by a **pancreatic duct** which joins bile duct. Liver and pancreas act as **endocrine** and **exocrine** glands, whereas the gall bladder acts as a **storing organ.**

GASTROINTESTINAL SECRETIONS AND DIGESTION

Exocrine tissues : Salivary glands, secretory cells in the stomach and intestinal epithelium, and the secretory cells of the liver and pancreas.

Exocrine secretion : Consists of aquedus mixtures of substances *i.e.*, water, ions, mucus and enzymes.

Role of digestive enzymes in digestion in mammals (Humans)

Gland	Enzyme	pH	Site of action	Substrates	Products of action
Salivary Gland (Saliva)	Salivary amylase or ptyalin	6.7	Buccal cavity, Oesophagous,	Starch	'Limit' dextrins, maltose, isomaltose
Gastric Gland (Gastric Juice)	(*i*) Pepsinogen → Pepsin	2 – 8	Stomach	Protein casein (milk)	Peptones paracasein (curd)
	(*ii*) Proprennin → Rennin (calf)		Stomach	Casein	Paracasein
	(*iii*) Hydrochloric acid		Stomach	Pepsinogen	pepsin
	(*iv*) Gastric lipase (infants)		Stomach	Milk fat	Fatty acids and glycerol
Pancreas (Pancreatic juice)	(*i*) Pancreatic Amylase	8.8	Small intestine	Starch glycogen	'Limit' dextrins, maltose, Isomaltose
	(*ii*) Trypsinogen $\xrightarrow{\text{enterokinase}}$ (Trypsin)		Protein	Peptides	
				Chymotryp-sinogen	Chymotrypsin
				Percarboxy Peptidases	Carboxypeptidases
	(*iii*) Elastase			Elastin	Large peptides
	(iv) Lipase			Fat	Fatty acids, monoglycerides, Glycerol
	(*v*) Chymotrypsinogen → Chymotrypsin			casein (milk)	Paracasein (curd)
	(*vi*) Carboxypeptidases			Larger peptides	Smaller peptides
	(*vii*) Amino peptidases			Larger	Oligopeptides
	(*viii*) DNAase			DNA	Deoxyribonucleotides
	(*xi*) RNAase			RNA	Deoxyribonucleotides
Small Intestine (Intestinal) (juice)	(*i*) Enteropeptidase or Enterokinase	8.3	Small Intestine	Trypsiogen	Trypsin
	(*ii*) Aminopeptidases			Peptides	Smaller peptides, amino acids
	(*iii*) Dipeptidases			Dipeptides	Amino acids
	(*iv*) Nucleotidases			Nucleotides Phosphoric acid	Nucleosidases, inorganic phosphate
	(*v*) Nucleosidases			Nucleosides	Sugars, purines
	(*vi*) Isomaltase			'Limit' dextrin Isomaltose	2 Glucose
	(*vii*) Maltase			Maltose	2 Glucose
	(*viii*) Sucrase (Invertase)			Sucrose	Glucose, fructose
	(*ix*) Lactase			Lactose	Glucose, galactose
	(*x*) Lipase			Triglycerides	Monoglycerides, fatty acids
Liver Bile	Bile Salts, no enzymes	8.0	Duodenum	Fats	Fat droplets

HORMONAL CONTROL OF DIGESTION IN HUMAN

The activities of **gastrointestinal (GI) tract** are coordinated by **nervous** and **endocrine** systems of the body. **Gastrointestinal peptide hormones** are the hormones secreted by the endocrine cells found scattered in the mucosa of stomach and intestine, These hromones regulate the digestive activities. Most of the gastroinestinal hormones are "**local hormones**" as act upon the same part of gut which secrete them.

Hormone	Source	Stimulus	Target/Action
1. Gastrin	G-cells or Argentaffin cells of Pyloric stomach	Vagus nerve activitiy; peptides and proteins stomach	(*i*) Stimulates secretion of gastric juice. (*ii*) Constricts cardiac sphincter and duodenum
2. Secretin	S–cells of mucosa of duodenum and jejunum	Acidic chyme entry into duodenum	(*i*) Release of sodium bicarbonates in pancreatic juice (*ii*) Stops up secretion of bile (*iii*) Inhibits secretion of gastrin
3. Cholecystokinin (CCK)	I-cells of mucosa of	Presence of fats in duodenum	(*i*) Release of enzymes in pancreatic juice (*ii*) Release of bile from gall bladder
4. Gastric Inhibitory peptide (GIP)	Upper part of small intestine *i.e.*, duodenum	Monosaccharides and fats (fatty) chyme in duodenum	(*i*) Inhibits gastric mobility and and secretion.
5. Enterogastrone	Duodenum	Chyme entry into duodenum	(*ii*) Slows gastric contractions to delay its emptying (*iii*) Stops secretion of gastric juice
6. Duocrinin	Duodenum	Acidic chyme in intestine	Activates Brunner's glands for producing mucus and HCO_3
7. Enterocrinin	Duodenum	Acidic chyme in intestine	Release of enzymes from Lieberkuhn's crypts
8. Motilin	Duodenum	Food/acidity	Mobility of stomach and intestine.
9. Villikinin	Intestinal epithelium	Food in duodenum	Accelerates movement of villi

ABSORPTION AND ASSIMILATION OF DIGESTED PRODUCTS IN HUMAN

Absorption

The digested food passes from the alimentary canal into the blood and lymph through its mucous membrane. This is called absorption. Absorption takes place by two types of processes *i.e*, **passive** and **active** absorption.

(*A*) *Passive absorption*

Takes place by physical processes like simple diffusion, osmosis and facilitated diffusion.

(*i*) **Simple diffusion :** Small and water soluble molecules takes place via concentration gradient across the apical membrane of the brush border of the cells lining the intestinal villi. They include fatty acids, monoglycerides, cholesterol and other fat soluble substances.

(*ii*) **Osmosis :** Water is absorbed by osmosis, Absorption of any solute raises the osmotic pressure of the blood and brings about absorption of an equivalent amount of water.

Water is absorbed partly in the small intestine and mostly in the large intestine.

(*B*) *Active absorption*

The nutrients are absorbed through the intestinal mucosa against concentration gradient. By active absorption, the nutrients can be absorbed completely from the intestinal lumen. There are two processes.

(*i*) **Active transport :** Involves the carrier molecules called permeases or translocases which are generally proteinous in nature. Glucose, galactose, amino acids, Na^+ etc. are absorbed by active transport. For active absorption of Na^+, a sodium pump operates in the cell membrane,

(*ii*) **Endocytosis :** Large sized liquid or solid nutrients are taken in some vesicles through the plasma membrane.

The end products of fats are a mixture of mono glycerides, fatty acids and glycerol. In the intestinal lumen, fatty acids, monoglycerides and the bile salts aggregate to form water-soluble molecules *i.e.*, **mixed micelles** which are taken by the intestinal mucosal cells generally by the process of pinocytosis.

The intestinal absorptive cells then, collect the reconstructed triglycerides, combine them with phospholipids and cholesterol, and release into the lymph in the form of protein-coated water soluble fat globules or droplets *i.e.,* **chylomicrons.**

Chylomicrons are transported to blood via the **central lacteal** of the lymphatic system for distribution throughout the body.

EGESTION OR DEFAECATION

- The waste material discharged from the alimentary canal is called **faeces** *or* **stool.**
- Faeces is made up of 75% water and 25% solid matter which is further made up of undigested roughage (30%), fats (20%)), inorganic matter (15%), proteins (3%) and bacteria (3%).
- Brown colour of faeces is due to **stercobilinogen** and **stercobilin** (formed from bile pigments by the fermentive action of colon bacteria).
- Odour of faeces is due to skatole, indole, mercaptans, H_2S and CH_4.
- Faeces are given out through the anus by the process of **defaecation or egestion or evacuation**.
- Egestion is partly voluntary and partly involuntary process.

Some Important Minerals and their Functions

Mineral	Functions
1. Calcium	Component of bones and teeth; required for blood clotting, muscle contraction, ATP-ase, nerve impulse transmission, heart functioning etc. Deficiency causes **tetany** and **rickets**.
2. Chlorine	Chief anion of extracellular fluid, involved in HCl synthesis and acid-base balance.
3. Copper	Synthesis of haemoglobin; co-factor for enzymes *e.g.* oxidases, component of **haemocyanin**. Deficiency causes causes **anaemia**.
4. Iodine	Metabolic control as part of thyroid hormone thyroxine. Deficiency produces *goitre.*
5. Iron	Formation of Hb and myoglobin, respiratory enzymes *e.g.* cytochromes and, O_2 transport enzymes. Deficiency causes **anaemia**.
6. Phosphorus	Bone and teeth formation, part of nucleic acids, energy metabolism, phosphorylation, maintain normal blood pH (buffer action). Deficiency reduces growth, metabolism and causes **rickets** in children.
7. Potassium	Osmotic balance; muscle contraction; nerve-impulse conduction. Deficiency causes rickets among children.
8. Sodium	Osmotic balance; acid-base balance; muscle contraction; nerve impulse conduction, bile salts.
9. Sulphur	Components of hormones (*e.g.* insulin), necessary for normal metabolism. Deficiency causes skin patches.

10. Zinc — Component of at least 70 enzymes *e.g.* carbonic anhydrase and some peptidases.
11. Flourine — Maintain enamel and prevents dental caries. Deficiency causes dental caries.
12. Magnesium — Co-factor of enzymes *e.g.* Hexokinase. Causes heart and vascular irregularities.
13. Manganese — Normal reproductive function, co-factor for enzymes. Deficiency causes **reproductive failure.**

Vitamins are organic compounds, always taken from outside the body along with food in small amounts (mg or mg/day) so called **micro-nutrients** and are vital for certain metabolic activities of life.

- Absence or deficiency causes **avitaminosis or deficiency disease** while excess causes **hypervitaminosis** and metabolic disorders.
- Term **vitamin** coined by C. Funk (1911) who discovered vitamin B_1.
- On the basis of solubility vitamins are of two types :

(*i*) **Fat soluble vitamins :** Derived from 5-carbon hydrocarbon – **Isoprene** soluble in organic solvents. *e.g.* vitamin A, D, E, and K.

(*ii*) **Water soluble vitamins :** Vitamin B and C.

Some Important Vitamins

Vitamins	Functions	Deficiency Symptoms
(A) Fat Soluble Vitamins		
1. Vitamin A (Retinol) epithelial tissue.	Body growth, principal component of retinal pigments; promotes normal vision, growth of bones and teeth, and maintenance of Night blindness, Keratomalacia, Retarded growth	Xerophthalmia,
2. Vitamin D (Calciferol)	Antirachitic, absorption and retention of Ca and P, growth and maintenance of bone	Rickets in children, Osteomalacia in women
3. Vitamin E (Tocopherol)	Prevents oxidative breakdown of vitamins, also for normal functions of gonads.	Reversible sterility in rats, Erythrocyte breakdown.
4. Vitamin K (Phylloquinone)	Prothrombin formation in liver	Haemorrhage
(B) Water Soluble Vitamins		
1. Vitamin C (Ascorbic acid)	Absorption of iron, formation of dentine collagen and bone matrix, amino acid oxidation, anti-oxidant, recovery from toxicity and burns	Scurvy
2. Vitamin B complex		Beri-Beri
(*i*) Vitamin B_1 (Thiamine)	Formation of cocarboxylase enzyme involved in decarboxylation (citric acid cycle)	
(*ii*) Vitamin B_2 (Riboflavin)	Maintenance of skin and oral mucosa, component of FMN and FAD	Cheilosis, glossitis
(*iii*) Vitamin B_3 (Pantothenic acid)	Participates in metabolism as coenzyme A	Dermatitis in chicken, Early ageing in man
(*iv*) Vitamin B_5 (Niacin)	In oxidation-reduction reactions as coenzyme NAD or NADP	Pellagra
(*v*) Vitamin B_6 (Pyridoxine)	Component of coenzyme pyridoxal phosphate	Anaemia, convulsions, impairment of antibody synthesis
(*vi*) Vitamin H or B_4 (Biotin)	Involved in carboxylation and decarboxylation	Dermatitis, muscular pain, growth failure
(*vii*) Folic acid	Formation and nucleoturation of RBCs	Macrocytic anaemia
(*viii*) Vitamin B_{12} (Cyanoco-balamine)	Important for nucleoprotein synthesis and formation of RBCs, Myelin formation	Pernicious anaemia

NUTRITIONAL DEFICIENCIES AND DISORDERS

- **Malnutrition :** Impairment of health due to improper intake of food or nutrients. It covers problems of undernutrition and overnutrition.
- **Undernutrition :** Taking a diet deficient in calories. It may also result from inability to digest, assimilate or utilize the necessary nutrients.
- **Overnutrition :** Intake of excess food and nutrients which are stored as increased body mass, may produce overweight and obesity (excessive accumulation of fat in tissues).
- **Hypercholesterolemia :** Blood cholesterol content becomes abnormally high, ultimately leading to cardiac disorder, mainly due to excess intake of saturated fats like butter, ghee, vegetable oils, red meat, eggs etc.

Nutritional Disorder

Nutrition	Disorders	Deficiency symptoms
1. Protein	Kwashiorkor (Protein energy malnutrition, PEM)	Stunted growth in the children; loss of appetite; anaemia; bulging eyes; protruded belly; darkening of the skin; repeated diarrhoea; atrophy of muscles, oedema of hands, feet and face
2. Protein and Calorie	Marasmus	Shrieveled appearence, subcutaneous fat used, ribs prominent, skin dry and wrinkled, stunted growth and repeated diarrhoea, extreme thinning of limbs (emaciation).
3. Iron	Anaemia, failure to mature RBC	Reduced number and size of RBC, reduced haemoglobin content, reduced learning ability, increased risk of infection and even death during child birth.
4. Iodine	Goitre	Enlargement of thyroid gland, cretinism in childhood causes retarded physical, mental and sexual growth.
5. Flouride	Dental caries	Appearence of cavities in the teeth due to acids produced during anaerobic respiration of bacteria.

MULTIPLE CHOICE QUESTIONS

1. Trypsin, amylase and enterokinase are the enzyme in the :
(*a*) Digestive system
(*b*) Respiratory system
(*c*) Urinogenital system
(*d*) None of the above

2. Which one of the following vitamins is take part in blood clotting?
(*a*) A (*b*) B
(*c*) E (*d*) K

3. Enzymes present in the pancreatic juice are :
(*a*) Pepsin, peptidases and lipase
(*b*) Trypsin, amylase and lipase
(*c*) Lipase, sugar and peptidases
(*d*) None of the above

4. Enzymes present in bile are :
(*a*) Sugar enzymes
(*b*) Peptidases
(*c*) Trypsin and Pepsin
(*d*) None of the above

5. At what degree enzymes of man are inactivated?
(*a*) 37°C (*b*) 35°C
(*c*) –10°C (*d*) –80°C

6. The largest gland in the coelom is :
(*a*) Pancreas (*b*) Liver
(*c*) Thyroid (*d*) Parathyroid

7. Coprophagy is associated with :
(*a*) Frog (*b*) Dog
(*c*) Rabbit (*d*) Scorpion

8. When a man starves, he starts consuming his own :
(*a*) Protein (*b*) Fat
(*c*) Glycogen (*d*) None of these

9. Deficiency of ascorbic acid causes :
(*a*) Haemophilia (*b*) Scurvy
(*c*) Beri-Beri (*d*) None of these

10. Hormone secretin stimulates the secretion of :
(*a*) Bile (*b*) Salivary juice
(*c*) Pancreatic juice (*d*) All of the above

11. Liver of rabbit is made up of :
(*a*) 6 lobes (*b*) 4 lobes
(*c*) 5 lobes (*d*) 3 lobes

12. To prevent entry of food into trachea is guarded by :
(*a*) Soft plate (*b*) Epiglottis
(*c*) Glottis (*d*) None of these

13. Which one of the following enzymes does not function after food passes from stomach to the intestine?
(*a*) Trypsin (*b*) Pepsin
(*c*) Steapsin (*d*) None of these

14. Formation of protein is :
(*a*) Dehydration analysis
(*b*) Dehydration synthesis
(*c*) Both (*a*) and (*b*)
(*d*) None of the above

15. Liver is said to be a digestive gland although it secretes :
(*a*) Many enzymes (*b*) No enzymes
(*c*) Amylase (*d*) Pepsin and trypsin

16. Vermiform appendix of man is a :
(*a*) Process in the rectum
(*b*) Process of the blood coagulation
(*c*) Diverticulum from the caecum
(*d*) None of the above

17. Increased excretion of sodium and ammonium salts by the kidneys is called :
(*a*) Alkaline tide (*b*) Diabetes
(*c*) Haemophilia (*d*) None of these

18. Night blindness is caused due to deficiency of:
(*a*) Vitamin E (*b*) Vitamin B_{12}
(*c*) Vitamin K (*d*) Vitamin A

19. Deficiency of thiamine caused :
(*a*) Beri-Beri
(*b*) Polyneuritis
(*c*) Loss of appetite and weight
(*d*) All the above

20. The enzyme invertase converts :
(*a*) Cellulose into starch
(*b*) Starch into glucose
(*c*) Glucose into sucrose
(*d*) Sucrose into glucose and fructose

Directions Qs. 21-25: *In each of the following questions, statement of* ***Assertion (A)*** *is given followed by a corresponding statement of* ***Reason (R)*** *just below it. Of the statements, mark the correct answer—*
(*a*) If both assertion and reason are true and reason is the correct explanation of assertion
(*b*) If both assertion and reason are true but reason is not the correct explanation of assertion
(*c*) If assertion is true but reason is false
(*d*) If assertion is false but reason is true.

21. **Assertion (A):** Digestion is necessary for the absorption of all macroelements.
Reason (R): Digestion breaks large complex molecules to simple smaller molecules which can be easily absorbed.

22. **Assertion (A):** Animals are said to derive energy ultimately from the sun.
Reason (R): Animals depend upon autotrophs for nutrition.

23. **Assertion (A):** Hepatopancreas of prawn is equivalent to the liver and pancreas of higher animals.
Reason (R): Hepatopancreas stores the absorbed nutrients from stomach and midgut.

24. **Assertion (A):** Digestion in *Amoeba* is said to be intracellular.
Reason (R): In *Amoeba* digestion takes place outside the body and products are engulfed by endocytosis.

25. **Assertion (A):** Rumen of alimentary canal of ruminant animals harbour numerous bacteria and protozoa.
Reason (R): Bacteria and protozoa help in secretion of gastric juice in the rumen.

ANSWERS

1	**2**	**3**	**4**	**5**	**6**	**7**	**8**	**9**	**10**
(*a*)	(*d*)	(*b*)	(*d*)	(*c*)	(*b*)	(*c*)	(*b*)	(*b*)	(*b*)
11	**12**	**13**	**14**	**15**	**16**	**17**	**18**	**19**	**20**
(*c*)	(*b*)	(*b*)	(*b*)	(*b*)	(*c*)	(*a*)	(*d*)	(*d*)	(*d*)
21	**22**	**23**	**24**	**25**					
(*d*)	(*a*)	(*b*)	(*c*)	(*c*)					

21

Gas Exchange and Transport

- Malpighi (1661) : Discovered blood capillaries.
- Halls (1732) : Measured blood pressure for the first time.
- **Stokes-Adams syndrome :** Delayed pick up of heart.
- **Ectopic pacemaker :** Shifting of pacemaker to some *other* part of heart.
- **Arrythmia :** Irregular heart beat due to damage to S.A. node.
- Antarctic icefish, *Chaenocephalus,* has colourless blood and is without RBCs.

RESPIRATORY SYSTEM

Respiration is Catabolic process which involves exchange of environmental oxygen and body's carbon dioxide through a liquid medium to utilize the oxygen for the oxidation of glucose in the mitocondria to produce the energy.

- Respiration is a **physico-chemical process.**
- Respiration may be direct or indirect.

Direct respiration

Exchange of gases takes place on the principle of diffusion through **moist body surface.** Found in unicellular organisms like aerobic bacteria protists and metazoans like sponges, coelenterates, flatworms, round worms etc.

Indirect respiration

A respiratory surface is present for exchange of gases. In this the transportation of gases between the respiratory organs and the body cells is brought about by the blood of the circulatory system *e.g.,* through skin (annelids and amphibians), gills (most of crustaceans, molluscs, all fishes and some amphibians); lungs (snails, some amphibians, all reptiles, birds and mammals).

Aerobic respiration

Takes place in the presence of molecular oxygen. It involves intake of oxygen and liberation of carbon dioxide.

Anaerobic respiration

Takes place in the absence of molecular oxygen, also called as fermentation. In muscles, lactic acid is formed as a product which is later oxidised aerobically in muscles and liver. Mammalian RBCs exhibit anaerobic respiration as these lack the mitochondria.

Internal respiration

The uptake of oxygen by tissue cells, oxidation of respiratory substrate (cellular respiration), liberation of energy and elimination of carbon dioxide evolved.

External respiration

Intake of oxygen by the blood from water or air in the respiratory organs and elimination of carbon dioxide.

Breathing or ventilation

Refers to the bringing in of fresh air and taking out of foul air. Rate of breathing is 12 – 14 (14 – 18 in adolescents) times per minute in human adults and 18 – 22 times/min in children (40 – 60/min in neonates).

VARIOUS RESPIRATORY PIGMENTS

- **Haemoglobin :** Oxygenated blood is red in colour; found in REC of vertebrates and blood plasma of invertebrates; metal base is ferrous ions.
- **Haemoerythrin :** Colourless; oxygenated red or brown in colour; found in annelida; metal base is iron.
- **Haemocyanin :** Colourless; oxygenated blue in colour; metal base is copper; found in phylum arthropoda (except insecta) and in mollusca.
- **Echinodrome :** Red in colour; metal base is iron; found in echinodermata.
- **Vanadium :** Metal base is vanadium; found in coelomic fluid of protochordates.
- **Chlorocruorin :** Green in colour; oxygenated blood red in colour, metal base is iron; formed in annelida *e.g., Sebella.*

RESPIRATORY DISORDERS

1. **Bronchitis :** Inflammation of the bronchi.
   - Characterised by **hypertrophy and hyperplasia** of the sero-mucous gland and goblet cells lining of the bronchi.
   - **Symptoms :** Regular coughing with thick greenish yellow sputum indicating infections and excessive secretion of mucus.
   - Caused due to cigarette smoking and exposue to air pollutants like carbon monooxide.
   - **Prevention and cure :** By avoiding smoking, irritating chemicals and pollutants. Bronchodilator drugs provide symptomatic relief. Intake of suitable antibiotics cure infection.
2. **Bronchial asthma :** Due to narrowing of bronchi and spasms in bronchial muscles (smooth).
   - Generally caused due to the hypersensitivity of the bronchiole to the foreign substances present in the air passing through it.
   - **Symptoms :** Coughing, or difficulty in breathing, mainly during expiration. Mucous glands become overactive producing a lot of mucus that clogs bronchioles and bronchi.
   - **Prevention and cure :** Avoiding exposure to the foreign substance or allergens; hypersensitisation; treatment includes antibiotic therapy, bronchodilator drugs as well as inhalers for symptomatic relief.
3. **Emphysema :** Inflammation or abnormal distension of the bronchiole or alveolar sac which results into the loss of elasticity of these parts.
   - Caused due to cigarette smoking and chronic bronchitis.
   - Lung size increases but ventilation is poor.
   - **Prevention and cure :** By avoiding exposure to cigarette smoke and pollutants. its a chronic lung disease resulting in irreversible distension and loss of elasticity of alveoli which can not be cured permanently.
   - Symptomatic treatment involves use of branchiodilators, antibiotics and oxygen therapy.
4. **Pneumonia :** Acute infection due to inflammation of the alveoli of the lung. Caused mainly due to infection of the bacteria, *Streptococcus pneumoniae.*
   - **Prevention and cure :** By proper and timely vaccination, can be cured by the use of antibiotics, symptomatic treatment like bronchodilator drugs.

MULTIPLE CHOICE QUESTIONS

1. The rate of breathing is lowest when we are :
   (*a*) Eating (*b*) Sleeping
   (*c*) Playing (*d*) Running
2. Chloride shift is essential for :
   (*a*) CO_2 transport (*b*) O_2 transport
   (*c*) N_2 transport (*d*) All the above
3. A iron-containing protein myoglobin is present in the :
   (*a*) Spleen (*b*) Lungs
   (*c*) Muscles (*d*) None of these
4. Amount of air exchanged in breathing can be measured with a :

(a) Thermometer (b) Barometer
(c) Potometer (d) Spirometer

5. Which animal has a double trachea?
(a) Snakes (b) Whales
(c) Penguins (d) None of these

6. The organs of sound production in birds is :
(a) Larynx (b) Syrinx
(c) Pharynx (d) Vocal sac

7. Excess of carbon dioxide in the body fluids is known as:
(a) Dysponea (b) Hypercapnia
(c) Apnoea (d) None of these

8. Lack of oxygen in body fluids is known as :
(a) Hyperpnoea (b) Anoxemia
(c) Anoxia (d) Hypoxia

9. In which form CO_2 is carried in the blood?
(a) Potassium carbonate
(b) Sodium carbonate
(c) Magnesium carbonate
(d) None of the above

10. In anaerobic respiration the pyruvic acid in muscles will form :
(a) Acetaldehyde (b) Lactic acid
(c) Acetic acid (d) Alcohol

11. Oxygen dissociation curve of haemoglobin is:
(a) Linear (b) Sigmoid
(c) Hyperblue (d) None of these

12. Pyruvic acid, ATP and energy are the end product of :
(a) Kerbs cycle (b) Citric acid cycle
(c) Glycolysis (d) None of these

13. Krebs cycle begins with :
(a) Citric acid (b) Pyruvic acid
(c) Malic acid (d) Hydrochloric acid

14. The end product of Krebs cycle is :
(a) Pyruvic acid (b) Citric acid
(c) Malic acid (d) Lactic acid

15. Whales respire through :
(a) Gills (b) Lungs
(c) Skin (d) Buccal cavity

Directions Qs. 16-20: *In each of the following questions, statement of **Assertion (A)** is given followed by a corresponding statement of **Reason (R)** just below it. Of the statements, mark the correct answer—*

(a) If both assertion and reason are true and reason is the correct explanation of assertion
(b) If both assertion and reason are true but reason is not the correct explanation of assertion
(c) If assertion is true but reason is false
(d) If assertion is false but reason is true.

16. Assertion (A): Oxidation of nutrients releases bond energy.
Reason (R): Oxidation of nutrients is done by using molecular oxygen.

17. Assertion (A): Aerobic animals are not truely aerobic
Reason (R): They produce lactic acid anaerobically.

18. Assertion (A): Aerobic respiration involves the exchange of respiratory gases twice.
Reason (R): Exchange occurs from lung to heart and then heart to lung.

19. Assertion (A): Respiratory gas exchange occurs through osmosis.
Reason (R): Respiratory gas goes from higher partial pressure region to the region of lower partial pressure.

20. Assertion (A): First step of gas exchange occurs through body surface in some animals.
Reason (R): Body surface or membrane of amphibia is thick in nature.

ANSWERS

1	2	3	4	5	6	7	8	9	10
(a)	(a)	(c)	(d)	(c)	(b)	(b)	(b)	(a)	(b)
11	**12**	**13**	**14**	**15**	**16**	**17**	**18**	**19**	**20**
(b)	(c)	(b)	(b)	(b)	(b)	(a)	(c)	(d)	(c)

22

Circulation of Body Fluid

CIRCULATORY SYSTEM OR INTERNAL TRANSPORT

It includes blood vascular system (blood, heart and blood vessels) and lymphatic system (lymph and lymph vessels). It helps in transportation of water, heat, hormones, nitrogenous waste, nutrients, respiratory gases, intermediate metabolites etc.

- Circulation of body fluids can be of following types :
  1. **Intracellular circulation** : Takes place inside the individual cells through cyclosis or cytoplasmic streaming *e.g., Amoeba, Paramecium.*
  2. **Extracellular circulation** : Fluid circulation outside the cells. It can be extra organismic circulation (outside water circulates in the body of an organism) or intra organismic circulation (involves circulation of body fluid).

Type of blood vascular system

A. Open Circulatory System

Found in arthropods and most molluscs (except cephalopod and tunicates). The main blood vessels arising from the heart pour the blood into tissue spaces called sinuses. In arthropods and molluscs, these sinuses join to form a body cavity with blood called the haemocoel.

Blood flows at a very low velocity and at low pressure due to absence of smooth muscles; direct contact *i.e.,* direct exchange of materials: respiratory pigment (if present) is dissolved in the blood plasma etc.

B. Closed Circulatory System

Formed in some annelids *e.g., Pheretima, Sepia* and all vertebrates. The blood flows around the blood through the specific blood vessels (arteries, veins and capillaries).

- Blood flows at high pressure due to the presence of smooth circular muscles in the walls of arteries and arterioles; there is no direct contact between the blood tissues and blood.
- Annelids are the simplest animals to have closed circulatory system.

BLOOD VASCULAR SYSTEM OF HUMANS

Closed type with a muscular heart, blood (fluid plasma + blood corpuscles) and blood vessels (arteries + veins).

Human Blood

Forms about 6–10% of the body weight, volume of blood in an adult about 6.8 litres complex connective tissue, consists of two components: 1. Plasma 2. Blood corpuscles

1. Plasma

Faint yellow coloured non-living fluid, forms about 55 - 60% of the blood volume, alkaline in nature, contain 91-92%, water, 7% proteins, 0.9% inorganic constituents, 0.1 % glucose and rest includes various other organic and inorganic substances. Plasma proteins mainly include albumins (4%, provides the colloid osmotic pressure in plasma), globulins (2.5%), fibrinogen (0.3%), properdin and prothrombin (0.03%). These proteins act as enzymes, coagulatin factors and antibodies. The disselved mineral ions are Na^+, K^+, Ca^{++}, Mg^{2+}, HPO_4^{2-}, PO_4^{3-}, Cl^-, HCO^{3-} etc. which maintain the pH of blood by buffering action. It helps in transportation of food components, metabolic wastes, hormones, nutrition, excretion, respiration and homeostasis; helps in homeothermy pH constancy; blood clotting; immunity; osmotic tension; as a middleman; properdin; erythropoiesis etc.

2. Blood Corpuscles

Cell which float freely in the plasma, form 40 – 45% of blood plasma. 3 types:

(i) ***Erythrocytes or red blood corpuscles (RBCs):*** Biconcave disc shaped, enucleate, reddish coloured cells, 7-8 mm is diameter, most abundant of blood corpuscles, 5.5 ± 1. 0 million/mm^3 in men, 4.8 ± 10 million/mm^3 in women, lifespan 120 days, contain haemoglobin (helps in transport of O_2) and enzyme carbonic anhydrase (required for formation of bicarbonate from CO_2), Moving RBCs tend to adhere with one another to form rows called *rouleaux.*

(ii) ***Leucocytes or white blood corpuscles (WBCs) :*** Rounded or amoeboid, nucleated, non-pigmented cells, wandering cells, larger than RBCs, 8–15 mm (may be 20 mm) in size, much less in number than RBCs (1 : 600), colourless, average WBC count is **7,000 per cubic millimetre.**

Leucocytosis : Rise in WBC count.

Leucopenia : Fall of WBC count below 4000.

Leukaemia : Pathological increase in number and denotes one type of neoplastic disease (blood cancer).

Leucopoiesis : Formaton of leucocytes takes place in bone marrow, Peyer's patches, lymph nodes, thymus, spleen etc.

On the basis of **cytoplasmic granules** leucocytes are of 2 types *i.e.,* **granulocytes** (with granules, polymorphic nucleus, produced in myeloblasts of red bone marrow, form 65% of total leucocytes) and **agranulocytes** (non-granular, non-lobulated nuclei, forms 35% of total leucocytes).

Granulocytes

(*a*) **Neutrophils :** 62% of total number of white cells, most abundant subtype, 4900 per mm^3 of blood, 2 – 7 lobulated nucleus, vary in size from 10 – 12 mm, life span 10 – 12 hours, soldiers of body (engulf microbes by phagocytosis).

(*b*) **Eosinophils (Acidophils) :** About 2 – 4% of the total leucocyte cells, coarse granules that get stained with acidic dyes (bright red with eosine), 10 – 15 μm in size, bilobed nucleus, 14 hours life span, increase in number in allergic diseases like **asthma** or **hay fever.**

(*c*) **Basophils :** About 0.5 – 2% of the total leucocyte cells, least in number 8-10 μm in size, coarse cytoplasmic granules, stained with basic dyes *e.g.,* methylene blue, nucleus S-shaped and 3 lobed, life span 8 – 12 hours, secrete heparin and histamine and helps in local anticoagulation and formation of ground substance.

Agranulocytes

(*a*) **Lymphocytes :** About 30% of total leucocyte count, 1,500 – 2,700 per cubic mm of blood, 8 – 12 μm in size, large and rounded nucleus, 2 types-small (7 μm) and large (20 μm) lymphocyte, formed in the thymus and lymphoid tissues; non-motile and non phagocytic, produces antibodies and opsonins.

(*b*) **Monocytes** : Largest sized leucocytes, 12 – 15 μm in diameter (may be 20 μm), forms about 5.3% of total leucocyte count, 100-400 mm^3 of blood, oval excentric and horse shoe-shaped formed in the lymph-nodes and spleen from the monoblasts highly motile and phagocytic in action, second line of defence (engulf the bacteria), act as macrophages or scavenger cells (remove the damaged and dead cells).

3. Blood platelets (*Throbocytes*)

Smallest sized blood corpuscles, 2-3 mm in diameter, found only in the blood of mammals, colourless, oval shaped discoidal cytoplasmic fragment formed from giant cells *i.e.,* megakaryocytes of bone-marrow, non-nucleated. 2.5 – 3.5 × 10^5 per cubic mm of blood: secrete a vasoconst or (serotonin) which causes construction of blood vessels to reduce the blood loss; release a number of platelet factors and an enzyme thromboplastin which causes the coagulation of blood and clot formation to prevent excessive bleeding.

CARDIOVASCULAR SYSTEM

Human Heart

Mesodermal in origin; hollow; muscular structure; situated behind the sternum and between the lungs in the thoracic cavity, with its apex resting on the diaphragm; looks like a blunt cone, which is 12 cm in length and 9 cm in width at the broadest point; automatic pumping organ; heavier in man (average 280-340 g) than in woman (average 230 – 280 g; form about 0.45% of body weight in males and about 0.4% of body weight in females.

Pericardium

Thin, transparent, two layared serous sac around the heart, outer non-distensible fibrous layer (parietal pericardium) and inner serous layer (visceral pericardium). A narrow cavity (pericardial cavity) between two layers, is filled with fluid (pericardial fluid) which allows frictionless movements of heart and protects the heart from mechanical shocks.

Atria or Auricles

Thin walled, push blood only upto ventricles, situated at the broader end, both auricle separated by a thin vertical **interatrial or interauricular/ septum** (prevents mixing of oxygenated and deoxygenated blood); right atrium is larger in size. Both auricles are smooth walled except a low network of ridges called **musculi pectinati**. Interatrial septum has an oval-shaped depression *i.e.,* **fossa ovalis** which is a vestige of an oval aperture *i.e.,* **foramen ovale**, present in the foetus and connecting two auricles.

Right auricle (RA) receives deoxygenated blood by 3 large veins *i.e.,* **superior vena cava (pre-caval)** from head. fore-limbs and upper part of chest; inferior vena cava (post-caval) from lower part of trunk and hind limbs and coronary sinus from wall of heart. Opening of inferior vena cava and coronary sinus are guarded by *a* **valve of Eustachus** and **valve of Thebesius**, respectively. Right auricle open in right ventricle (RV) by **right auriculoventricular aperture** guarded by **right auriculoventicular valve** which is formed of three flaps called **cusps (tricuspid valves)**.

Left Auricle (LA): Receives oxygenated blood from lungs by four pulmonary veins (two from each lung), has no valve. LA opens in left ventricle (LV) by **left auriculoventricular aperture,** guarded by left auriculoventricular valve, formed of two flaps or cusps (**bicuspid or mitral valve**).

Ventricles

Thick walled, LV is much thicker than RV because LV has to push the blood to all the body parts while RV has to push the blood to closely lying lungs only. Both ventricles are rough walled and have two ridges *i.e.,* **columnae or trabeculae carneae** and **papillary muscles or musculi papillaes** (to which chordae tendineae are attached) and grooves. **Interventricular septum** separates the both ventricles and prevent mixing of deoxygenated and oxygenated blood in ventricular part of heart.

Right ventricle (RV): From left anterior corner of right ventricle, there originates the **pulmonary arch** or **pulmonary aorta** which carries deoxygenated blood to the lungs.

Left ventricle (LV): From right anterior corner of left ventricle, there originates aortic (systemic) arch which carries oxygenated blood to various body parts.

PUMPING ACTION OF HUMAN HEART

Initiation of heart beat in under three special bundles of cardiac muscles (nodal tissues) which are :

1 .SA node-Sinu-auricular or sino-atrial node

Lies in the right wall of RA, below the opening of superior vena cava, small, flattened and ellipsoid strip of muscle fibre of 0.3 mm size, also called **right sided structure.** It has a rest potential of only– **55 to 60 m V.** Also called pacemaker as it is first to originate the cardiac impulses and determines the rate of heart beat.

2. AV Node : Atrio-ventricular node

Also called Pace setter lies in the right atrium near the junction of interauricular and interventricular septum close to opening of coronary sinus. It is stimulated by the waves of contraction initiated by SA node. It generates the cardiac impulses which are conducted to the muscles of the ventricles through **bundle of His** and **Purkinje fibres** at the rate of 1.5-4 metres second.

3. Bundle of His

Also called **AV bundle,** arises from AV node, descends in the inter ventricular septum and divides into two branches which descend along two sides of interventricular septum and supply the wall of ventricle of their own side by Purkinje fibres which bring about synchronous contraction of the ventricles form the apex of heart which force the blood into the pulmonary arch and aortic arch.

CORONARY CIRCULATION

Involves the blood supply to the heart walls and also drainge of the heart wall. It is made up of coronary arteries (one pair, arise from aortic arch just above the semilunar valves, supply oxygenated blood to the heart wall) and *coronary veins* (collects blood from heart wall. most of the veins join to form a coronary sinus which open in the right auricle. Opening is guarded by a valve of Thebesius.

PORTAL SYSTEM

A portal vein (special type of vein which collects blood from some organs by a system of capillaries, but divides into a second capillary system in some intermediate organ before the blood finally returns to the heart) with its branches is called a portal system. The flow of blood through the portal system is called portal circulation..2 types :

1. **Hepatic portal system :** Hepatic portal veins carries food ladden blood from alimentary canal and associated glands like pancreas, spleen etc. to the liver which acts as an intermediate organ.
2. **Hypophyseal portal system :** Hypophyseal portal vein collects the blood from hypothalamus of the brain and forms second set of capillaries in anterior lobe of pituitary gland.

LYMPHATIC SYSTEM

1. **Lymph :** Transparent fluid derived from blood and other tissues which accumulates in the interstitial spaces as the interstitial fluid. It contains leucocytes mainly lymphocytes, water content 94%, protein content is less (3 - 4.5% instead of blood plasma with 6 - 8%), Albumin globulin ratio is 1 : 5 instead of 1 : 1 in blood. It maintains balance of blood and tissue fluid; picks up hormones from endocrine glands; adds lymphocytes etc.
2. **Lymph capillaries :** Small, thin walled, present in nearly all the tissue spaces, lined by an **endothelium** of thin and flat cells. Special milky coloured lymph capillaries called **lacteals** (present in villi of the small intestine).
3. **Lymphatic vessels :** Thinner walls, numerous valves which give them a beaded appearence, join and rejoin and finally form two main lymphatic ducts.
   (*a*) **Thoracic duct**
   (*b*) **Right lymphatic duct**
4. **Lymph nodes :** Small, oval or bean like bodies placed in the course of the lymphatic vessels, acts as filters and are the sites of formation of the lymphocytes, mainlly present in the neck, axilla, thorax, abdomen and groin. Lymphnodes are maximum in armpit and groin.

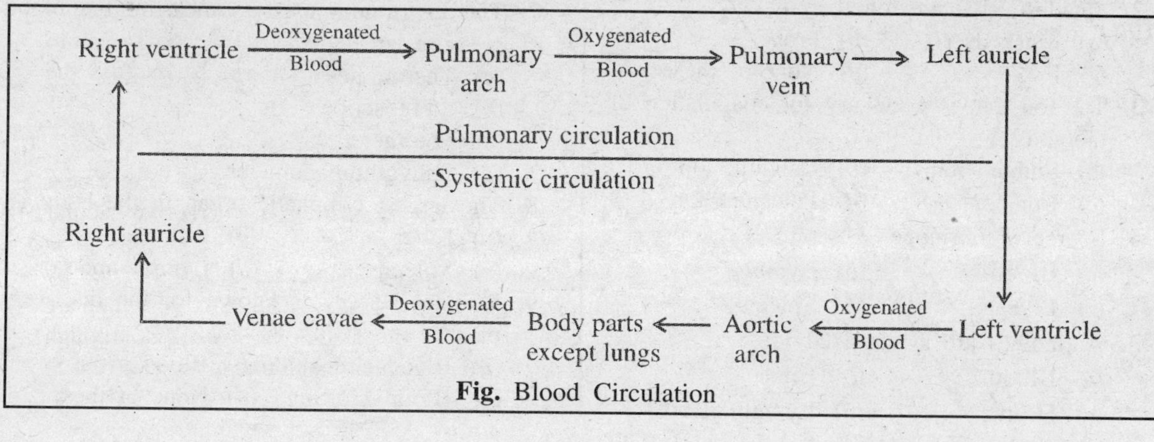

Fig. Blood Circulation

SOME IMPORTANT FACTS

- **Myogenic heart :** Found in vertebrates. The cardiac impulse originate in the cardiac muscle fibres and not brought to the heart by the nerve fibres.
- **Neurogenic heart :** Found in most crustaceans, some insects (*e.g.,* cockroach) and annelids.
- **Coronary angiography :** Radio-autograph of coronary arteries of heart.
- **Pericarditis :** Inflammation of pericardium of heart.
- **Endocarditis :** Inflammation of endocardium of heart.
- **Rheumatic heart disease (RHD) :** Defects in heart valve due to toxins produced by throat infection caused by *Streptococcus* leading to rheumatic fever.
- **Marley's law of heart** : An increase relationship between the rate of heart beat and blood pressure.
- Syphilis and Rheumatic fever causes **murmur.**
- **Aplastic anaemia :** Due to fall in rate of erythropoeisis in bone-marrow.
- **Ramanowski stain :** Used to differentiate three types of granulocytes.
- **Critical count of platelets :** If the platelet counts falls below 40,600/μl.
- Mammals have **left systemic** while birds have **right systemic.**
- **William Harvey :** First to study blood circulation in human heart.
- **Angina :** Heart pain for a short period usually in the front of the chest.
- **Myocardial infarction :** An acute heart attack in which a part of heart muscle becomes dead due to cessation of blood supply to it.
- **Ischaemia :** An inadequate flow of blood to a part of the heart due to obstruction of its blood supply.
- **Largest artery :** Systemic aorta in human body.
- **Larget vein :** Inferior vena cava in human body.
- **Negative blood pressure** is recorded in great veins.
- Malpighi (1661) : Discovered blood capillaries.
- Halls (1732) : Measured blood pressure for the first time.
- **Stokes-Adams syndrome :** Delayed pick up of heart.
- **Ectopic pacemaker :** Shifting of pacemaker to some *other* part of heart.
- **Arrythmia :** Irregular heart beat due to damage to S.A. node.

MULTIPLE CHOICE QUESTIONS

1. Defibrinated blood in:
(*a*) Blood proteins (*b*) Blood plasma
(*c*) Serum (*d*) All the above

2. Smallest RBC are found in:
(*a*) Musk deer (*b*) Frog
(*c*) Rat (*d*) None of these

3. The ion that is necessary for coagulation of blood is:
(*a*) Sodium ion (*b*) Calcium ion
(*c*) Iron ion (*d*) Potasium ion

4. Source of fibrinogen is:
(*a*) Blood (*b*) Lymph
(*c*) Liver (*d*) Spleen

5. Antibodies are synthesized in:
(*a*) Blood (*b*) Liver
(*c*) Heart (*d*) Lymph glands

6. Antibodies are produced by:
(*a*) RBC (*b*) Basophils
(*c*) Lymphocytes (*d*) Eosinophils

7. The instrument used to measure the blood pressure is:
(*a*) Thermometer
(*b*) Stethoscope
(*c*) Potometer
(*d*) Sphygmomanometer

8. The largest lymphatic organ in the body is:
(*a*) Liver (*b*) Pancreas
(*c*) Spleen (*d*) Lymph node

9. William Harvey is known for the discovery of:
(*a*) Blood circulation (*b*) Blood groups
(*c*) Blood clotting (*d*) None of these

10. The blood leaving the lungs is richer in:
(*a*) Carbon dioxide (*b*) Water
(*c*) Oxygen (*d*) Nitrogen

11. The renal portal system is found in:
(*a*) Man (*b*) Whale
(*c*) Rabbit (*d*) Frog

12. ANF (Atrial Natruretic Factor) is:
(*a*) Hormone of the heart
(*b*) An antibody
(*c*) An enzyme
(*d*) None of the above

13. The electrodes of the "pacemaker" are inserted into the:
(*a*) Ventricle
(*b*) Auricle
(*c*) Conducting system
(*d*) None of the above

14. Heart is completely enclosed by a membrane known as:
(*a*) Tympenic (*b*) Venticular septum
(*c*) Pericardium (*d*) None of these

15. Heart of rabbit is:
(*a*) Two chambered (*b*) Four chambered
(*c*) Three chambered (*d*) None of these

16. The difference between systolic and diastolic pressure is called:
(*a*) Blood pressure (*b*) Pulse pressure
(*c*) Systolic pressure (*d*) None of these

17. Blood pressure is measured usually from the:
(*a*) Branchial vein (*b*) Branchial artery
(*c*) Coronary artery (*d*) None of these

18. The first tap like sound heard at the time of measuring blood pressure represents:
(*a*) Systolic pressure (*b*) Diastolic pressure
(*c*) Both (*a*) and (*b*) (*d*) None of these

19. Double circulation is found in:
(*a*) Amphibians (*b*) Mammals
(*c*) Arthropods (*d*) None of these

20. The life of erythrocytes in mammalian blood is:
(*a*) About 120 days (*b*) About 150 days
(*c*) About 180 days (*d*) About 30 days

Directions Qs. 21-25: *In each of the following questions, statement of* ***Assertion (A)*** *is given followed by a corresponding statement of* ***Reason (R)*** *just below it. Of the statements, mark the correct answer—*

(*a*) If both assertion and reason are true and reason is the correct explanation of assertion
(*b*) If both assertion and reason are true but reason is not the correct explanation of assertion
(*c*) If assertion is true but reason is false
(*d*) If assertion is false but reason is true.

21. **Assertion (A):** A chronic vasoconstriction of arterioles may cause hypertension.
Reason (R): In vasoconstriction abnormal rise in the arterial blood pressure occurs.

22. **Assertion (A):** Fainting is a homestatic mechanism.
Reason (R): Prolonged standing leads to fainting.

23. **Assertion (A):** The increased permeability of the lymph capillaries is easily altered.
Reason (R): The increased permeability of the capillary walls leads to oedema or swelling.

24. **Assertion (A):** Arterioles possess smooth muscles on their walls.
Reason (R): These smooth muscles help in regulating blood volume flowing through a tissue or organ.

25. **Assertion (A):** Blood flows at a very slow velocity in the lacunae and sinuses of prawn.
Reason (R): This happens because of the absence of heart in the prawn.

ANSWERS

1	2	3	4	5	6	7	8	9	10
(*c*)	(*a*)	(*b*)	(*c*)	(*d*)	(*c*)	(*d*)	(*c*)	(*a*)	(*c*)
11	**12**	**13**	**14**	**15**	**16**	**17**	**18**	**19**	**20**
(*d*)	(*a*)	(*c*)	(*c*)	(*b*)	(*b*)	(*b*)	(*a*)	(*b*)	(*a*)
21	**22**	**23**	**24**	**25**					
(*a*)	(*b*)	(*b*)	(*a*)	(*c*)					

23

Excretion and Osmoregulation

Elimination of Nitrogenous Wastes

Nitrogenous waste materials are formed as the end product (mainly) of protein metabolism and small amounts by metabolism of nucleic acids.

Nitrogenous wastes are produced in two ways :

(*i*) Deamination of excess and unwanted amino acids present in the food taken (**exogenous source**) and

(*ii*) Breakdown of body's own proteins and nucleic acids **(endogenous source).**

On the basis of nature of nitrogenous end products, there are three types of excretion.

1. Ammonotelism

Main nitrogenous end product is **ammonia**; animals with such excretion are called ammonotelic animals.

Readily soluble in water and highly toxic (because of high pH). *e.g.,* many aquatic animal groups like protozoans (*e.g. Amoeba. Paramecium*); sponges (*e.g., Scypha*); Coelenteratas (*e.g., Hydra*); Crustaceans (*e.g., Palaemon*); echinoderms (*e.g., Asterias*); bony fishes (*e.g.,* Teleosts); Tadpole larva and some aquatic amphibians (*e.g.,* salamanders).

2. Ureotelism

Main nitrogenous waste is **urea**, animals are called **ureotelic** animals; 1,00000 times less toxic than ammonia, water soluble but can be concentrated to some extent (100 times in human, several hundred times in camels, sharks and kangaroo rat). *e.g.,* cartilagenous fishes, adult amphibians, aquatic reptiles (like turtles), mammals etc. Urea is produced in liver from ammonia and CO_2 through ornithene or urea cycle.

3. Uricotelism

Main nitrogenous waste is **uric acid** animals are called **uricotetlic** animals, least soluble nitrogenous waste, least toxic adaptation to terrestrial mode of life. *e.g.,* in birds, land reptiles (*e.g.,* lizards and snakes); insects (*e.g.,* Cockroach); land snails (*e.g., Helix*) and some land crustaceans (*e.g., Oniscus*). In human being main sources of uric acid are purines. Uricotelism is particularly advantageous to the land vertebrates those lay shelled eggs.

HUMAN EXCRETORY SYSTEM

One pair of kidneys, one pair of ureters, a urinary bladder and a urethra.

1. Kidney

Primary or major excretory organ, one pair, large sized, 10 cm long × 5 cm broad x 3 cm thick, bean shaped, dark red 125-170 gm in weight in adults, slightly flattened structure, present in upper part of abdominal cavity on either side of vertebral column. Concavity of kidney **(hilum or hilus renalis)** is always inward directed. Each kidney is made up of about **one million uriniferous or renal tubules** or nephrons.

2. Ureters

One pair, long, 25 – 30 mm, muscular tubular, narrow, arising from the **hilum** of kidney by its upper broader part **(renal pelvis)**, open in dorsal wall of urinary bladder, lined by **transitional epithelium.**

3. Urinary bladder

Large, thin walled distensible, pear shaped sac, present in pelvis region of abdominal cavity, wall lined with smooth muscles (*detrusor muscles*), temporarily stores urine, lined by **transitional epithelium.** Small triangular area in the wall of bladder, formed of 3 openings (2 openings of ureters and 1 internal urethral orifice at the lower most end of trigone).

4. Urethra

Muscular, tubular, extends from neck of bladder to outside. In female about 4 cm long and opens out directly by external urethral orifice above the vaginal orifice in vulva. In male urethra (20 cm long) joins the ejaculatory duct to form urinogenital canal which passes through the penis and opens by urinogenital aperture at the tip of penis and is differentiated into 3 parts – **prostatic** (about 2.5 cm through prostate gland and receiving sperm ducts), *membranous* (about 2.5 cm, free) and **penile** (about 15 cm and inside penis).

STRUCTURE OF A NEPHRON

One million per kidney, more convoluted in peripheral part *i.e., renal cortex* and is central part there tubules run straight *i.e., renal medulla. 2 types of nephrons.*

(*a*) **Cortical nephron** (about 80 – 90%) in human kidney but less in other placentals.

(*b*) **Juxta medullary** (10 – 20%).

Cortical nephron

About 3 cm long, 20 – 60 mm is diameter, formed of two parts :

1. Bowman's capsule with glamerulus

Double walled cup present in the renal cortex, lined by a thin and semipermeable squamous epithelium formed of flat cells. In the cavity of Bowman's capsule, about 50 capillaries called *glomerulus present. The glomerlus is* made up of branches of afferent arteriole which interconnect to form tuft of capillaries capped by Bowman's capsule. Glomerular capillary endothelium has numerous minute pores called **fenestrae.**

A basement membrane (between visceral epithelium and glomerular capillary endothelium) acts as **dialysing membrane.** The visceral layer, basement membrane and glomerular endothelium collectively form a filtering membrane which acts as **ultrafilter** and responsible for **ultrafiltration.**

Bowman's capsule and glomeralus, both collectively called **malpighian** body (first reportered by M. Malpighi, 1966) or renal corpuscle of Pygmalian corpuscle and is about 150 mm is diameter.

2. Nephric tubule

(*i*) **Proximal convoluted tubule** (PCT) : Continuous with Bowman's capsule, lined by brush bordered **cuboidal epithelial cells** with numerous microvilli, cells have numerous mitochondria for active transport and selective reabsorption of sodium, calcium, glucose, amino acids etc. from nephric filtrate. Some waste products present in the blood of peritubular capillaries pass into lumen of PCT through tubular secretion across the intestitial fluid. The phenomenon is called **tubular secretion** *e.g.*, H^+, NH_3 drugs, toxic substances.

(*ii*) **Loop of Henle :** Runs straight in renal medulla, U-shaped, formed of thin descending (lined by **squamous epithelium**) and thick ascending limb (lined by cuboidal epithelium). Descending limb is permeable to water but not to NaCl, while ascending limb is absolutely impermeable to water but permeable to NaCl.

(*iii*) **Distal convoluted tubule (DCT) :** Convoluted part, present in cortex, opens in the collecting tubule, lined by cuboidal epithelum with a few microvilli. DCT is permeable to both Na^+ and K^+.

(*iv*) **Collecting tubule (CT) :** Drains a number of ducts of DCT is the cortex, descends to the tip of medulla to form the collecting ducts of Bellini which empty via calyces into renal pelvis. The highly **columnar cells** of CT respond to circulating titers of ADH (antidiuretic hormone) which afrects their permeability and size of intercellular space.

Juxtamedullary Nephrons

Glomeruli are placed close to the inner margins of the cortex, and long loops of Henle are placed deep into the medulla and are associated with vasa recta. The blood first passes through the capillaries of glomerulus, and then, flows through the hair pin loops of the vasa recta of the loop of Henle. It operates when there is shortage of water and more concentrated urine is to be formed.

WATER CONSERVATION AND PRODUCTION OF CONCENTRATED URINE

Loop of Henle (LH) and Capillary loop (vasa recta) play an important role in this mechanism. Glomerular fluid in the LH and blood in the vasa recta flow in opposite directions in the two limbs of the loops, toward the renal medulla in the descending limb and toward the renal cortex in the ascending limb. These two counter-current systems help in concentrating the urine.

1. Loop of Henle

When filtrate passes through the ascending limb it loses to the interstitial fluid in the renal medulla NaCl by diffusion in its narrow region and Na^+ and Cl^- ions by active transport in its wide region of the limb. It draws out water by osmosis from the narrow region of the descending limb and also from the collecting duct (both are permeable to water). The urine leaving the collected duct is nearly as concentrated as the interstitial fluid deep in the medulla. It is 4 times as concentrated at the blood plasma in humans.

2. Vas Rectum

Its endothelial cells are freely permeable to ions, water and urea. In its increase descending capillary toward the renal medulla, water is drawn out from the blood plasma by osmosis due to a progressive increase in the concentration of the interstitial fluid and Na^+ and Cl^- ions and urea enter the plasma by diffusion. In the ascending capillary toward the renal cortex, the reverse takes place, water reenters the plasma and Na^+ and Cl^- and urea leave it due to progressive decrease in the concentration of the interstitial fluid.

Some urea diffuses from the last (urea permeable) part of the collecting duct into the interstitial fluid to further raise the latter's density for increasing water reabsorption from the collecting duct as well as from the bottom of the loop of Henle.

Counter current mechanisms operating in the Henle's loops and vasa recta concentrate the urine by maintaining high salt concentration around the nephrons and collecting duct. Production of hypertonic urine conserves body's water.

REGULATION OF KIDNEY FUNCTION BY FEEDBACK CIRCUITS

Two main hormonal controls of the kidney function by negative feed back circuits are:

1. Control by Antidiuretic Hormone (ADH)

In DCT, collecting tubules and connecting tubules, membrane permeability of water depends upon **antidiuretic hormone** (**ADH**) or **vasopressin**. ADH is produced in the hypothalamus of the brain and released into the blood stream from the pituitary gland, enhances fluid retention by making the kidneys reabsorb more water.

2. Control by Juxta glomerular apparatus (JGA)

It operates **renin-angiotensin aldosterone system** (RAAS). It responds to a decrease in blood pressure of blood volume in the afferent arteriole of the glomerulus and releases an enzyme, **renin** into the blood stream. In blood renin converts plasma protein **angiotensinogen** to a peptide called **angiotensin** II. **Angiotensin**-II. increases blood pressure by causing arterioles to constrict, increases blood volume through causing reabsorption of NaCl and water by stimulating the release of adrenal hormone **aldosterone**. Aldosterone induces the DCT to reabsorb more Na^+ and water. This leads to an increase in blood volume and pressure, completing the feedback circuit by supporting the release of renin.

MICTURITION (URINATION)

The urge for micturition takes place when 400 – 500 ml of urine has collected in urinary bladder. The release of urine is accomplished by the simulataneous contraction of the smooth muscle of the urinary bladder wall and the relaxation of the skeletal muscle sphincter around the opening of the bladder. Micturition can also be performed voluntarily.

CONSTITUENTS OF URINE

Pale yellow coloured (due to **urochrome** pigment), slightly acidic (pH = 6.0, range 4.5 – 8.0), 1.03 –

1.04 specific gravity, faint aromatic odour (due to bad smelling substances **urinod**), daily urine output in a normal adult person is 1.5 – 1.8 litres. Chemically, 95 – 96% water and 2% urea.

ABNORMAL URINE

- **Glycosuria :** Urine with sugar *e.g.*, glucose is diabetes mellitus.
- **Ketonuria :** Urine with ketonic bodies like acetoacetic acid, β-hydroxybutyric acid etc.
- **Hematuria :** Urine with blood cells in disorders like renal tuberculosis.
- **Haemoglobinuria :** Presence of haemoglobin in urine.
- **Uraemia :** Presence of excess urea in urine.
- **Albuminuria :** Urine with albumin proteins in conditions like nephritis.
- **Proteinuria :** Excess protein level in urine.
- **Jaundice :** Presence of bile pigments in urine in conditions like obstructive bileduct, hepatitis etc.
- **Pyuria :** Presence of pus cells in urine.
- **Alkaptonuria :** Presence of alkapton in urine which turns black when comes in contact with air (Black urine disease).

MULTIPLE CHOICE QUESTIONS

1. The renal organs for excretion and the genital organs for reproduction commonly called as :
(*a*) Excretory organs
(*b*) Reproductive organs
(*c*) Urogenital organs
(*d*) None of the above

2. Excess of urea in blood resulting from kidney failure is known as :
(*a*) Urochrome
(*b*) Uremia
(*c*) Urestelic
(*d*) None of these above

3. Kidneys, skin and lungs are :
(*a*) Excretory organs
(*b*) Digestive organs
(*c*) Reproductive organs
(*d*) None of the above

4. The normal range of pH of urine is :
(*a*) 7.0 (*b*) 4.8-7.5
(*c*) 3.5 - 8.0 (*d*) 5.0 - 6.0

5. Extraction of sodium is done by :
(*a*) Distal tubule (*b*) Proximal tubule
(*c*) Loop of Henle (*d*) None of the above

6. Ornithine cycle is concerned with :
(*a*) Formation of urine (*b*) Formation of urea
(*c*) Formation of blood (*d*) None of the above

7. Which one of the following have ureotelic excretion?
(*a*) Frog (*b*) Snake
(*c*) Pigeon (*d*) None of the above

8. Most toxic nitrogenous end product is :
(*a*) Uric acid (*b*) Urea
(*c*) Ammonia (*d*) None of the above

9. Which of the following is secreted by kidney?
(*a*) Vitamin K
(*b*) Erythropoietin hormone
(*c*) Erythrocytes
(*d*) Apoenzymes

10. Urea is produced by the action of :
(*a*) Uricase enzyme (*b*) Arginase enzyme
(*c*) Urease enzyme (*d*) None of the above

11. Reabsorption of water in nephrons is controlled by:
(*a*) Vasopressin (*b*) ACTH
(*c*) STH (*d*) None of the above

12. Green glands as excretory organs are found in:
(*a*) Scorpions (*b*) Spiders
(*c*) Cray fishes (*d*) None of the above

13. The structural and functional unit of kidney is:
(*a*) Medulla (*b*) Neuron
(*c*) Nephron (*d*) Tubules

14. In prawn the excretary organs are :
(*a*) Book lungs (*b*) Malpighian tubules
(*c*) Nephridia (*d*) Green glands

15. Malpighian tubules are the excretory organs in:
(*a*) Arthropods (*d*) Molluscs
(*c*) Amphibians (*d*) None of the above

16. Urea in human urine is derived from the breakdown of :
(*a*) Uric acid (*b*) Amino acids
(*c*) Glucose (*d*) All of the above

17. Waste products of adenine and guanine metabolism are excreted by man as :

(*a*) Urea (*b*) Uric acid
(*c*) Ammonia (*d*) None of the above

18. Kidney in mammals are formed from :

(*a*) Mesonephros (*b*) Pronephros
(*c*) Metanephros (*d*) None of the above

19. The yellow colour of urine of vertebrates is due to:

(*a*) Uric acid (*b*) Urochrome
(*c*) Melanin (*d*) None of the above

20. Excretory products of foetus comes out through:

(*a*) Umbilical cord (*b*) Placenta
(*c*) Both (*a*) and (*b*) (*d*) Ammiotic fluid

Directions Qs. 21-25: *In each of the following questions, statement of* ***Assertion (A)*** *is given followed by a corresponding statement of* ***Reason (R)*** *just below it. Of the statements, mark the correct answer—*

(*a*) If both assertion and reason are true and reason is the correct explanation of assertion
(*b*) If both assertion and reason are true but reason is not the correct explanation of assertion
(*c*) If assertion is true but reason is false
(*d*) If assertion is false but reason is true.

21. Assertion (A): When water has to be conserved in the body, ADH is secreted from the posterior pituitary.

Reason (R): ADH renders the walls of DCT, collecting duct and tubule impermeable to water.

22. Assertion (A): Renal threshold of glucose is said to be 180 mg per 100 ml.

Reason (R): Glucose starts appearing in the urine when its blood level exceed 180 mg per 100 ml of blood.

23. Assertion (A): Camels can go without water for long periods.

Reason (R): Camel stores water in the pouches of their rumen and fat in their hump.

24. Assertion (A): Tissue fluid contains much less protein than the blood plasma.

Reason (R): Blood capillary wall is impermeable to plasma.

25. Assertion (A): Death ensues if waste products are not removed from the body.

Reason (R): Excretory system removes the nitrogeneous waste products from the body.

ANSWERS

1	**2**	**3**	**4**	**5**	**6**	**7**	**8**	**9**	**10**
(*c*)	(*b*)	(*a*)	(*b*)	(*a*)	(*b*)	(*a*)	(*a*)	(*b*)	(*b*)
11	**12**	**13**	**14**	**15**	**16**	**17**	**18**	**19**	**20**
(*b*)	(*a*)	(*c*)	(*d*)	(*a*)	(*b*)	(*a*)	(*a*)	(*b*)	(*a*)
21	**22**	**23**	**24**	**25**					
(*c*)	(*a*)	(*c*)	(*a*)	(*b*)					

24

Movement and Locomotion

- **Movement** is a of changing place or position by the entire body or by one or more of its parts is called movement.
- **Kinesiology :** Study of movements.

BASIC TYPES OF MOVEMENT

1. Amoeboid/Pseudopodial movement

By means of temporary protoplasmic outgrowths called pseudopodia; involves change in shape of the cell body and streaming movement of cytoplasm helps in food capture and change of place.

2. Ciliary Movements

Paramecium and other ciliates swim by means of cilia. Paramecium use the cilia of the cytopharynx to drive water and food in their gullet. Cilia set up water currents that pass over the gills of bivalve molluscs (e.g., Lamellidens) for feeding; drive water through the water vascular system of echinoderms (e.g., Asterias) for locomotion.

3. Flagellar movements

Perform locomotion is euglenoids and other flagellate protists, Chlamydomonas and other unicellular colonial motile green algae, sperms etc. Flagellate choanocytes bring about circulation of water in the canal system of sponges.

4. Protoplasmic or cytoplasmic streaming (cyclosis)

It has been seen in most of the cells such as leukocytes, Amoeba and other unicellular organisms. Cyclosis helps in circulation of materials in the cells.

5. Muscular movements

Found in majority of vertebrates; including humans; brought about by sliding of myofilaments past each other in the muscle fibres. This is the ability to exert a force by alternate contraction and relaxation. Most multicellular animals have muscle fibres for movement of different organismic attaining locomotion .

HUMAN SKELETON

206 bones in adult man and some cartilage, categorised into **Axial skeleton and Appendicular skeleton.**

(A) Axial Skeleton

Skull + vertebral column + ribs + sternum; present along the longitudinal axis of the body. Made up of 80 bones (87 bones in children).

1. Skull

29 bones, endoskeleton of head, heaviest part of body.

(*i*) **Cranial bones :** 8 bones (1 frontal + 2 parietals + 1 occipital + 2 temporal + 1 sphenoid + 1 ethmoid) large, hollow part, encloses and protects the brain in its cranial cavity, forms helmet for protection of brain, cranial cavity size ranges from 1450-1600 cm in adult man. Occipital possess foramen magnum for passage of spinal cord and two *occipital condyles* for fixation with atlas vertebra. Upper jaw fused with cranium while lower jaw remains connected with the cranium by muscles.

(*ii*) **Facial bones :** 14 bones (2 nasals + 2 inferior turbines or chonchae + 1 vomer + 2 larymalse + 2 zygomatics or molars + 2 palatines + 2 maxillae + 1 mandible or lower jaw).

- **Craniostylic jaw suspension** : Mandible of lower jaw is suspended from squamosal bone of upper jaw.
- **Alveoli :** Sockets in maxilla and mandibles bone in which teeth embedded.

(*iii*) **Hyoid bone :** Bone placed at the floor of the buccal cavity.

(*iv*) **Bones of middle ear :** 6 bones (2 malleus + 2 incus + 2 stapes). Stapes is smallest bone of human body.

2. Vertebral column or Backbone

Long (70 cm), curved bony rod present on dorsal side of neck and and trunk; articulates with skull, pectoral girdle, pelvic girdle and the ribs; centrally hollow so makes the vertebral column a hollow tube through which the spinal cord coming out of the foramen magnum of the cranium possess and thus remain protected.

(*a*) Cervical (7)–curve directed forward.
(*b*) Thoracic (l2)–curve directed backward.
(*c*) Lumbar (5)–curve directed forward.
(*d*) Sacral (5)–curve directed backward,
(*e*) Coccyx (4). Sacral and coccyx are two triangular fused bones.

3. Sternum or breast bone

Thorax supported by sternum on the ventral side and the thoracic vertebrae on dorsal side, flat, narrow bone about 15 cm in length, dagger like, formed on parts-Manubrium or prosternum (upper handle like), Body or mesosternum (middle largest part) and xiphoid cartilage or xiphisternum or metasternum (lower small cartilaginous plate).

4. Ribs

12 pairs of semi-circular bones present in the thorax. Each rib is made up of posterior bony vertebral part and anterior cartilaginous sternal part. The bony part of each rib is double headed (Tubercular and capitular head)-3 types on the basis of their sternal parts.

(*i*) **True ribs** : First seven pairs whose sternal parts are directly attached to the sternum through small sized costal cartilages

(ii) **False ribs or Vertebro-chondral ribs** : Three pairs of ribs (8th, 9th and 10th pair) whose bony vertebral parts are attached to vertebrae while sternal part are attached on that of the 7th pair of ribs and not to the sternum directly.

(*ii*) **Floating ribs :** Last two pairs (11th and 12th pair) whose sternal parts remain free and do not even reach the sternum.

- **Rib cage :** Formed of thoracic vertebrae, sternum and ribs.
- **Gorilla rib** : An extra rib in some individuals, it is normally present in gorillas.

(B) Appendicular Skeleton

List along the transverse axes of body and includes two types of girdles and limb-bones. Total 126 bones.

1. Limb Bones

2 pair (1 pair forelimb or arm bones + 1 pair hindlimb or leg bones).

(*a*) **Forelimb (arm) 30** : bones (1 Humerus + 2 Radiusulna + 8 carpals + 5 metacarpals + 14 phalanges).

- **Humerus (upper arm or Brachium):** 3 parts (head, shaft and trochlea). Head articulates with glenoid cavity of pectoral girdle by a ball and socket joint has greater and lesser tuberosity. Shaft with V-shaped deltoid ridge. Trochlea with two epicondyles and a capitulum, has coronoid fossa on anterior side and olecranon fossa on posterior side.
- **Radius ulna (fore arm or Antebrachium):** 2 bones lower arm inner (radius) and outer (ulna). Ulna has an olecranon process for forming elbow joint by a hinge joint (uniaxialjoint).
- **Hand or Manus :** 3 parts (8 wrist or carpus) + 5 palm or metacarpus + 14 fingers or digits, phalangeal (14 bone) formula of hand of man is 2,3,3,3,3. **Thumb** (great finger of hand) = **pollex,** opposable to the fingers for grasping. Second index of hand = **index.**

(*b*) **Hindlimb (leg)** : Longer than arms, strengthen the legs to support body weight and help in locomotion. 30 bone (1 femur +

2 tibia fibula + 7 tarsals + 5 metatarsal + 14 phalanges + 1 patella).

- **Femur (thigh bone)** : Largest bone of human body, cylindrical with head, shaft and condyle. Head-rounded; articulates into acetabulum of pelvic girdle by a freely movable ball and socket joint; with greater and lesser trochanter. Lower end of thigh with two *condyles* and an intercondylar groove for the articulation of a sesamoid bone called knee cap or patella.
- **Tibia and fibula** (shank or skin bones). Tibia-inner, longer, thicker with tibial tuberosity at upper end and medial malleolus at distal end.
  Fibula : Thinnest bone of the man body, outer, lower end with lateral melleolus, articulates with lower end of tibia. Knee joint lies at the junction of tibial tuberosity and condyles of femur.
- **Foot bones :** 3 parts (7 ankle or tarsus + 5 instep or metatarsus + 14 toes or digits or phalanges). Phalangel formula of foot of man is 2, 3, 3, 3, 3.
- Thumb (great toe) of foot - Hallux.

2. Girdles

(*a*) **Pectoral or shoulder girdle** : 2 in number. Each half is conists of **clavicle** or **collar bone** (narrow, rod-like, S-shaped, extending between sternum and acromion process of scapula bone) and scapula or shoulder bone (flat, triangular bone present at the back of shoulder as shoulder blade). At the junction of these bones, there is a cup-like depression called **glenoid cavity**, in which the head of the humerus fits in. Scapula is produced into a scapular spine which ends into a triangular acromion process.

(*b*) **Pelvic or Hip girdle** : Present in lower part of trunk and are two in number. Each half is formed of single and semicircular innominatum or hip bone formed of fusion of ilium upper side, broadest part), ischium (*below the pubis*) and pubis (inner side). Two pubic bones are joined by a cartilagenous slightly movable joint called pubic symphysis to enclose an arch like pelvis.

- At the juncion of these three bones, there is a cup-like depression, the *acetabulum.* in which the head of femur bone of thigh fits in.
- Between the pubis and ischium bones, there is a large oval opening *obturator foramen.*

JOINTS

Meeting place of two bones. 3 main types :

1. Immovable or Fixed or Fibrous Joints

No space between bones, union of bony elements by continuous intervening fibrous tissue. 3 types :

(*a*) **Sutures** : Present amongst skull bones and innominate bones.

(*b*) **Syndesmosis** : Ligament joints as betwen radio-ulna, tibio-fibula.

(*c*) **Gomphosis** : Peg and socket joint as between tooth and jaw bone.

2. Slightly movable or cartilaginous joints (imperfect)

Adjoining bones are joined by thick elastic pads of fibro-cartilage, permits only slight movements between the bones 2 types : *Synchondrosis* and *symphysis e.g.,* between central of two vertebrae public symphysis, ribs and sternum.

3. Freely movable or synovial joints (perfect)

One bone forms a knob-like swelling while other bone forms a depression. The ends of both bones are covered by smooth *articular-cartilages.* In such joints,a space between the bones is present, called **synovial** cavity which remains filled with a viscous and slippery **synovial fluid.** These can be :

(*a*) **Ball and Socket joint *e.g.,* shoulder joint** (head of humerus and glenoid cavity of pectoral girdle) and hip joint (head of femur and acetabulum of pelvic girdle).

(*b*) **Hinge joints** : One bone is movaple or another bone in only one plane *e.g,* inter phalangeal joints, elbow joint, ankle joint, knee joint

(*c*) **Pivot Joints** : One bone exhibits the rotatory movement on another bone which remains fixed. *e.g.,* atlas and axis in humans.

(*d*) **Gliding joints :** One bone can slide upon each other *e.g.*. carpels in wrist, tarsals in ankle, zygapophyses of adjoining vertebrae and between clavicle and sternum.

(e) **Ellipsoidal joints or angular joints** : One articular end is oval and convex while the other end is elliptical and concave. *e.g.,* wrist or radio-carpal joint of humans, toes and sole.

(*f*) **Saddle joints** : Imperfectly developed ball and socket joint in which one bone is movable on another fixed bone in many direction *e.g.,* carpo metacarpal joint of human.

DISORDERS OF BONES

1. Arthritis (Arching joints)

Inflammation of joints, characterised by pain and stiffness in the joints. 3 types :

(*a*) **Rheumatoid arthritis** : Painful inflammaion of synovial membranes. It is followed by its thickening and secreting of extra-synovial fluid resulting in increased pressure and pain. The abnormal granules or pannus is produced which accumulating on the surface of the cartilage, causes erosion of cartilage, ossification of fibrous tissues, stiffening and nonmovement of joint.

(*b*) **Osteoarthritis (degenerative arthritis) :** Characterised by the degeneration of the articular cartilage and proliferation of new bones usually affected joints are of spine, knee and hands.

(*c*) **Gouty arthritis or gout :** Caused either due to excessive formation of uric acid or inability to excrete it. It generally attacks the **great toe.**

2. Osteoporosis

An age dependent systemic disorder characterised by low bone mass, micro architectural deterioration of the bone, increased fragility and proneness or susceptibility to fracture. It is caused due to deficiency of calcium or vitamin D in diet, lack of vitamin C, lack of physical stress, deficiency of estrogens, old age or cushing's syndrome. The elderly men and women are most susceptible.

3. Bursitis

Bursae or fluid filled spaces near joints swell up which may be due to deposition of calcium in the underlying tendons. *e.g.,* tennis elbow, miner's elbow, baker's cyst, ischiogluteal bursitis, house maid's knee.

4. Osteomyelitis

Inflammation of bony due to pus producing or pyogenic organism; may remain localised or spread inside involving various parts including bone marrow.

MULTIPLE CHOICE QUESTIONS

1. All the following are the properties of skeletal muscles except
(*a*) excitibility (*b*) contractility
(*c*) rhythmicity (*d*) tonicity

2. One of the following is the set of antagonistic muscles; One of the sets is wrong
(*a*) flexor; relaxor (*b*) abductor; adductor
(*c*) pronator; supinator (*d*) levator; depressor

3. the excitatory effect on skeletal muscles is produced by......salts
(*a*) Na (*b*) Mn
(*c*) Ca (*d*) Mg

4. In birds 6 - 10 caudal vertebrae are usualy fused to foim
(*a*) sesamoid bone (*b*) odonfoid process
(*c*) plow-share bone (*d*) synsacrum

5. The protein content of skeletal muscles is........%
(*a*) 10 (*b*) 20
(*c*) 40 (*d*) 30

6. In actomyosin complex for one unit of actin the number of units of myosin is
(*a*) 1 (*b*) variable
(*c*) 3 (*d*) 4

7. Which of the following is not true about rigor mortis?
(*a*) It is irreversible
(*b*) It is permanent contraction
(*c*) ATP increases in muscles
(*d*) Time of onset is not same in all muscles

8. Astragalus and calcaneum are found in
(*a*) hind limbs of rabbit

(*b*) hind limbs of frog
(*c*) fore limbs of rabbit
(*d*) fore limbs of frog

9. EDTA injected into muscles combines with Ca^{2+} and
(*a*) stops contraction
(*b*) causes contraction
(*c*) slow down contraction
(*d*) none of these

10. Half of pelvic girdles is known as
(*a*) os-innominatus (*b*) ilium
(*c*) synsacrum (*d*) suspension

11. Sharpey's fibres occur inside
(*a*) collagen (*b*) muscle
(*c*) bone (*d*) skin

12. Collar bone is known as
(*a*) scapula (*b*) clavicle
(*c*) pelvic girdle (*d*) chevron bone

13. The accumulation of uric acid crystals in the region of joints resulting in painful movement causes
(*a*) lordosis (*b*) gout
(*c*) arthritis (*d*) rheumatoid arthitis

14. Which of the following connects a muscle to a bone?
(*a*) cartilage (*b*) ligament
(*c*) tendon (*d*) disc

15. An oxygen debt develop during
(*a*) tetanus
(*b*) sarcoplasmic release
(*c*) anaerobic work
(*d*) aerobic work

Directions Qs. 16-20: *In each of the following questions, statement of* ***Assertion (A)*** *is given followed by a corresponding statement of* ***Reason (R)*** *just below it. Of the statements, mark the correct answer—*
(*a*) If both assertion and reason are true and reason is the correct explanation of assertion
(*b*) If both assertion and reason are true but reason is not the correct explanation of assertion
(*c*) If assertion is true but reason is false
(*d*) If assertion is false but reason is true.

16. Assertion (A): Locomotion enables the animal to find its partner.
Reason (R): Locomotion enables the animal to shift its entire body from place to place.

17. Assertion (A): Movement of body parts serves to change the body posture.
Reason (R): Body parts move in relation to body axis.

18. Assertion (A): There are similarities between the locomotion of unicellular organisms and multicellular animal.
Reason (R): Ciliary, flagellar and amoeboid movement occur in unicellular organiss.

19. Assertion (A): Locomotion in *Hydra* is carried out by two types of contractile cells.
Reason (R): Muscle fibres are lacking in *Hydra*.

20. Assertion (A): Starfish moves through crawling.
Reason (R): Starfish utilise a flow of water over the surface for movement.

ANSWERS

1	2	3	4	5	6	7	8	9	10
(*c*)	(*a*)	(*a*)	(*c*)	(*b*)	(*a*)	(*c*)	(*b*)	(*a*)	(*a*)
11	**12**	**13**	**14**	**15**	**16**	**17**	**18**	**19**	**20**
(*c*)	(*b*)	(*b*)	(*c*)	(*c*)	(*b*)	(*b*)	(*b*)	(*a*)	(*b*)

25

Nervous Coordination and Receptors

NERVOUS SYSTEM OF HUMANS

A nerve consists of several bundles of nerve fibres commonly called fasciculi. A fascicle contains connective tissue (**endoneurium**). surrounded by connective tissue sheath (**perineurium**). Each nerve is surrounded a dense layer of connective tissue called **epineurium**. Each nerve has its own supply of small blood vessels.

The functional unit of nervous system is nerve cell or neuron, which have a special structure but vary greatly in size and shape. Each neuron has a cell body (encloses cytoplasm and a nucleus, usually a single axon a variable number of dendrites. 3 types of neurons :

1. **Multipolar neuron** : One long axon with many dendrites, most common type *e.g.*, pyramidal cells is cerebral cortex.
2. **Bipolar neuron :** Neuron has two processes, an axon at one end and a dendrite at the other end. *e.g.,* bipolar neurons in the retina of eye.
3. **Paurounipolar or Unipolor neuron :** The cell body on a side branch of the main axon *e.g.* cells of dorsal root ganglion.

- **Neuroendocrine System** : A network of endocrine glands whose hormone production is controlled by commands form the CNS.
- **Nuclei :** Mass of neurons clustered inside the CNS.
- **Ganglia :** Mass of neurons clustered inside the PNS.
- **Tracts :** Bundles of nerve fibres within the CNS.
- **Nerves :** Bundles of nerve fibres within the PNS.

Neuroglia or Glial cells or Glia

Packing and supporting cells found in the brain and spinal cord. They perform many house keeping functions, provide nutritional support to neurons and consume waste products. They may be astrocytes, oligodendrocytes and **microglia.**

Schwann cells

In the PNS, they wrap around the axons of neurons, covering the axon with concentric layers of insulating plasma membrane.

A. CENTRAL NERVOUS SYSTEM

Brain (Encephalon)

Situated in the cranial cavity of the skull, 98% of the weight of whole CNS, average human brain weight from 1200-1400 gm, possess about 100 billion neurons, protected by meninges.

In lower vertebrates (*e.g.,* amphibians) there are two meninges, **durameter** and **piaarachnoid membrane**.

In mammals three meninges :

1. **Piameter :** Innermost layer, close to the brain, highly vascularised and nutritive, at two places it is fused with roof of brain to form *choroid plexus* for secreting *cerebrospinal fluid* (CSF).
2. **Durameter :** Outer layer, thick double layer of fibrous tissue, can resist the movement of the brain that may stretch and break the blood vessels.

Adult brain is pinkish grey, wet and wrinkled like a walnut. It has three parts :

1. **Forebrain or Prosencephalon :** Rhinencephalon (olfactory lobes), Telencephalon (cerebrum) and diencephalon.

2. **Mid-brain or Mesencephalon :** Cerebral peduncles (crura cerebri), corpora quadrigemina.
3. **Hind-brain or Rhombencephalon :** Metencephalon (cerebellum and pons) and myelencephalon (medulla).

Fore brain

Forms anterior two third to brain. It has three parts:

1. Olfactory lobe

One pair, completely covered by cerebral hemisphere dorsally. Each is differentiated into **olfactory bulb** and **olfactory tract**, control the smell.

2. Cerebral Hemisphere

One pair, 80% of weight of brain, demarcated by a median or cerebral fissure. Two cerebral hemispheres are collectively called cerebrum, which are interconnected by thick band of transverse nerve fibres called corpus callosum.

- Left half of brain controls the right side of body and viceversa.
- **Pallium :** Roof of each cerebral hemisphere and highly folded to form ridges (**gyri**) and depressions (**sulci**).
- **Corpora striata :** Ventrolateral walls of cerebrum, thick.
- **Cerebral cortex :** Peripheral portion of each cerebral hemisphere, formed of grey matter.
- **Cerebral medulla** deeper part of each cerebral hemisphere, formed of white matter.
- **Corpus striatum :** Largest nucleus, regulates planning and execution of stereotyped movements.
- Some sulci (depressions) are well developed and form three deep and wide fissures which divide each cerebral hemisphere into four **lobes**:
  (1) Anterior **frontal lobe**
  (2) Middle **parietal lobe**
  (3) Posterior **occipital lobe**
  (4) Lateral **temporal lobe**
- **Lateral ventricle or paracoel** : Fluid filled cavity in each cerebral hemisphere.
- **Frontal lobe : Each has motor area** (controls voluntary movements of muscles), premotor area (centre of involuntary) movements of muscles and for autonomic nervous system; association area (controls sensation and movements; motor speech area or Broca's area (controls speech), expressions of emotions will power, personality and intelligence.
- **Parietal lobes :** Somaesthetic area (general sensations like pains, touch, pressure, temperature), sensory speech area, taste area (gustation), to follow direction, of map, reading a clock or dressing yourself.
- **Temporal lobe :** Auditory or Acoustic area (hearing), Wernicke's area (understanding speech, decoding and interpretation of sound etc.
- **Occipital lobe** *: Visual area* (visual sensation), *i.e.,* decoding and interpretation of visual information etc.

3. Diencephalon

Completely covered by cerebral hemisphere, forms less than 1% of brain volume.

(*i*) **Thalami :** One pair, masses of grey matter, present in cerebral medulla of cerebral hemisphere, act as relay centres, relays sensory impulses except those of smell and regulates smooth muscle activity.

(*ii*) **Hypothalamus :** Floor of diencephalon, formed of one pair of masses of neurons present in cerebral medulla below thalami inner to basalganglia, controls the involuntary functions, secretes neurohormones (releasing hormones) which control secretion of trophic hormones from anterior lobe of pituitary gland *i.e.,* hypothalamus is an important link between *nervous* and endocrine systems. On the inferior surface hypothalamus possesses :

  (*a*) **Optic chiasma** (cross of two optic nerves).
  (*b*) **Pituitary gland** or hypophysis with a base called *infundibulum* and

- **Amygdala :** Almond shaped nucleus, attached to the internal tips of fork of limbic system, controls emotional behaviour like aggresion and fear.

- **Hippocampus area :** Associated with swollen lower tip of limbic system, converts information from short term to long term memory especially in learning.
- **Septum :** Linked to the hypothalamus, medial part on either cerebral hemisphere which has olfactory, hippocampal and hypothalamic connections. It is a emotion centre for sexual arousal.

Mid-brain

Completely covered by cerebral hemisphere. It has 2 parts :

1. Optic lobes

One pair, present on dorsal side, each divided transversely into upper and larger **superior coliculus** and lower and smaller **inferior coliculus** *i.e.,* there are 4 optic lobes (*Optic quadrigemina* only in mammals). Controls visual (superior coliculi) and auditory (inferior coliculi) reflexes.

2. Cerebral peduncles (Crura cerebri)

A pair of thick bands of longitudinal nerve fibres present on the floor of mid brain act as coordination centres between hind and forebrain.

Iter (Cerebral aqueduct) : Narrow canal that connects third and fourth ventricles.

Hind-brain

1. Cerebellum (Metencephalon)

Present below the occipital part of cerebral hemisphere, three lobes :

(*i*) **Cerebellar hemispheres :** One pair, large sized lobes present on lateral sides.

(*ii*) **Vermis** small, worm like median lobe between two hemispheres.

- **Cerebellar cortex** : Peripheral part formed of grey matter.
- **Cerebellar medulla :** Central part formed of white matter.
- **Cerebellar peduncle** : 3 pairs, connects cerebellum to other parts of brain, act as communication pathways. These are :

(*a*) Superior cerebellar peduncle–between cerebellum and midbrain.

(*b*) Middle cerebellar peduncle–between cerebellum and pons.

(*c*) Inferior cerebellar peduncle–between cerebellum and medulla oblongata.

- Cerebellum coordinates voluntary movements, controls equilibrium, controls rapid musclar activities etc.

2. Medulla Oblongata (Myelencephalon)

Posterior part of brain, located beneath the cerebellum.

- Fourth ventricle or Myelocoel : A cavity of medulla oblongata, continuous with central canal of spinal cord.
- Posterior choroid plexus-thin, non-nervous, roof of medulla oblongata.
- It controls involuntary functions, have a number of centres like cardiac centres (heart beat); respiratory centres (rate of respiration); vasomotor centres (contraction of blood vessels); salivary centres (secretion of saliva) etc. Also controls coughing, sneezing, vomiting, urinating, blood pressure, swallowing of food, defaecation, gut peristalsis etc.

3. Pons varolii

Thick band of transverse nerve fibres present at floor of upper part of medulla oblongata, coordinates two cerebellar hemispheres so coordinates muscles of two sides of body, Acts as neuronal link between cerebrum and cerebellum.

- **Brain stem** : Formed collectively by diencephalon, midbrain, pons varolii and medulla oblongata.
- **Histologically :** Brain is made up of 2 types of matter *i.e.,* **grey matter** (cytons and non-medullated nerve fibres) and **white matter** (medullated nerve fibre and large amount of myelin).

In medulla oblongata and pons varolii-Grey matter is internal and white matter is external, while in other parts of brain, grey matter is external and white matter is internal.

SPINAL CORD (MYELON)

Long, soft, whitish, slightly flattened, cylindrical rod present in the neural canal of vertebral column, about 45 cm long, 2 cm thick, about 35 gm in weight, exists from the skull (atlas) through formen magnum and extends as a taper to a point (conus medullaris) upto first lumbar vertebra. Conus medullaris is connected to coccygeal end of vertebral column by fibrous connective tissue derived from piameter and called filum terminale.

It has two longitudinal fissures *i.e.,* **dorsal sulcus** (deep) and **ventral sulcus** (shallow).

The diameter is not uniform, has two swellings *i.e.,* **upper cervical** and **lower lumbar** swelling.

Neurocoel or central canal : Narrow longitudinal canal of spinal cord, connected anteriorly with fourth ventricle of **medulla oblongata** but is **closed posteriorly.** It is lined by a single layered cuboidal epithelium called ependyma.

Central canal is tilled with cerebrospinal fluid. **Dorsal septum** is a thin sheet of connective tissue which extends from dorsal fissure to inward.

Spinal cord conducts sensory and motor impulses to and from the brain *i.e.,* acts as link between brain and the peripheral nerves and also controls reflex actions of the body.

CEREBROSPINAL FLUID (CSF)

Clear, colourless fluid, similar to blood plasma except that it has a much lower proportion of protein and cholesterol. The entire CNS possess 80-150 ml of CSF, produced by membranes in the roof of the ventricles. It contributes to homeostasis, protection of the delicate brains and spinal cord, medium of transport of hormones in the brain, providing buoyancy to the brain, excretion of waste products, reduction in net weight of brain from 1.4 kg to about 0.18 kg which reduces pressure at the base.

PERIPHERAL NERVOUS SYSTEM

Formed of a number of long, thin, whitish threads called nerves extending between CNS and body tissues. 2 types of nerves :

A. Cranial or Cerebral Nerves

Either arise from or end into brain; 12 pairs; 3 pairs of sensory (I, II and VII), 5 pairs of motor (III, IV, VI, XI and XIII) and 4 pairs of mixed (V, VII, IX and X) nerves.

Cranial nerves

Name	Nature	Origin	Innervation	Function
1. Olfactory	Sensory	Olfactory mucous membrane or nasal chamber	Olfactory lobe	Smell
2. Optic	Sensory	Retina of eyes	Optic chiasma	Sight
3. Occulomotor	Motor	Cerebral peduncles of mid brain	Four eye muscles, muscles of ciliary body, iris and tear glands	Movement of eye ball, pupil size, accomodation
4. Pathetic or trochlear	Motor	Near cerebral peduncle	Superior oblique eye muscle	Rotation of eye ball
5. Trigeminal (Dentist's nerve)	Mixed	From anterolateral side of medulla oblongata swells and forms *Gasserian ganglion* and divides into 3 branches		
(*a*) Ophthalmic movements	Sensory		Skin receptors of nose,	Tactile and lower jaw forehead, eyelids
(*b*) Maxillary	Sensory		Lower eyelid, cheek, upperlip upper Jaw	Sensations
(*c*) Mandibular (largest nerve)	Sensory		Lowerlip, buccal cavity, earpinna, lower teeth Muscles of lower jaw	Sensations, tongue movement Mastication
6. Abducens	Motor	Midventral side or medulla oblongata	External rectus eye muscle	Rotation of eye ball

7. Facial	Mixed	Anterolateral side of medulla oblongata behind Vth, swells and forms geniculate ganglion	Sensory fibres to taste bends on posterior 2/3 part of tongue Motor fibres to muscles of face, salivary gland, neck etc.	Tasting of food, facial expression, saliva secretion Movement of neck.
8. Auditory	Sensory	Internal ear	Anteriolateral side of medulla oblongata behind VIIth	Hearing and equilibrium,
9. Glossopharyngeal	Mixed	Lateral side of medulla oblongata	Sensory fibres to taste budson anterior 1/3 part of tongue motor fibres to parotial salivary gland	Tasting of food, salivation, swallowing muscles of pharynx and
10. Vagus or pneumogastric (*a*) Superior laryngeal (*b*) Recurrent laryngeal (*c*) Cardiac (*d*) Pneumogastric (*e*) Depressor	Mixed	Lateral side of medulla oblongata behind IXth	Sensory fibres to receptors present in wall of visceral organs. Motor fibres to muscles in the wall of visceral organs like alimentary canal, heart, blood vessels, trachea, lungs kidneys, genital tracts etc.	Sound production breathing, cardiac slowing, vasomotor and visceral reflexes gastric and pancreatic secretions; gastrointestinal movements
11. Accessory spinal	Motor	Lateral side of medulla oblongata behind Xth	Muscles of larynx, pharynx, neck and shoulder	Movement of larynx, pharynx, neck and shoulder
12. Hypoglossal	Motor	Ventral side of posterior part of medulla oblongata Hypoglossal muscles below	tongue	Movement of tongue

(B) Spinal nerves

Arise from spinal cord, 31 pairs in man (37 pairs in rabbit), mixed, divided into 5 groups-

(1) Cervical : 5 pairs-In neck region

(2) Thoracic : 12 pairs-In thoracic region

(3) Lumbar : 5 pairs-Upper part of abdomen

(4) Sacral : 5 pairs-Lower part of abdomen

(5) Coccygeal : 1 pair-Represents the tail nerves

AUTONOMOUS NERVOUS SYSTEM

Auto functioning or self governed, maintains homeostasis, controlling the smooth muscles, cardiac muscles and certain glands by both feedback loops and antagonistic control. Formed of two systems.

A. Sympathetic nervous system

B. Parasympathetic nervous system.

ANS regulates the function of involuntary visceral organs so also called visceral nervous system. Auonomic ganglia act as relay centers. Preganglionic nerve fibres of both sympathetic and paraympathetic nervous system are cholinergic.

Sympathetic and parasympathetic systems function in an antagonistic manner and facilitates in maintaining homeostasis.

Reflex Action : Rapid and Automatic Responses

An involuntary response to a stimulus, a stereotype reaction which has conscious control *e.g.,* knee jerk, withdrawl of hand on touching a hot object, blinking of eyes, yawning, sneezing, coughing etc.

Reflex pathway or arc has the following components.

1. **Sensory receptor :** Present in dorsal ganglion which lies outside the spinal cord, tissue or organ or cell which receives an external or internal stimulus *e.g.,* skin, ear, eye.
2. **Sensory neuron :** Conducts sensory impulse from receptor organ to spinal cord.
3. **Spinal cord :** Acts as modulator and changes sensory impulse into the motor impulse.

4. **Motor/Efferent nerve fibre :** Conducts motor nerve impulse from spinal cord to the effectors *e.g.,* leg muscles.
5. **Effector organ :** May be organ/muscle/gland which on being activated by a motor impulse produces work or substance to suitably deal with the stimulus.

There are more than 200 reflexes "wired" into our nervous system. Reflexes are of two types :

1. **Unconditioned or Inborn or Inherited reflexes :** Specific, predictable, purposeful, have survival value, shown by all individuals of a species without training. *e.g.,* simple reflex, compound reflex, chain reflex etc.
2. **Conditioned or acquired reflexes :** Not present at birth but develop later in life through learning,habit, experience or regular association of an indifferent stimulus with unconditioned stimulus. According to Pavlov (1929) conditioned reflex is only acquired and noninheritable property.

Reflexes may be **protective reflex, antigravity reflex** and **righting reflex** on the basis of anatomy and physiology.

On the basis of control, reflexes may be spinal reflexes (controlled by spinal cord) and cerebral reflexes (controlled by brain).

MULTIPLE CHOICE QUESTIONS

1. Nervous tissue is made up of :
   (*a*) Neurons (*b*) Nerve fibres
   (*c*) Neuroglia (*d*) All the above
2. Central nervous system is consists of :
   (*a*) Brain
   (*b*) Spinal cord
   (*c*) Nerves and ganglia
   (*d*) Both (*a*) and (*b*)
3. The centre for the control of all activities of an animal is :
   (*a*) Heart (*b*) Brain
   (*c*) Liver (*d*) Stomach
4. Diencephalon has centres for the control of :
   (*a*) Metabolism of fats and carbohydrates
   (*b*) Regulation of genital function
   (*c*) Sleep
   (*d*) All the above
5. A nerve is composed of :
   (*a*) Muscular tissue
   (*b*) Nerve fibre bundles
   (*c*) Bundle sheaths
   (*d*) None of the above
6. The system which consists of nerves arising from the brain the spinal cord is called :
   (*a*) Peripheral nervous system
   (*b*) Central nervous system
   (*c*) Autonomous nervous system
   (*d*) None of the above
7. The number of cranial nerves in a frog are:
   (*a*) 10 pairs (*b*) 15 pairs
   (*c*) 12 pairs (*d*) 4 pairs
8. In frog the auditory nerve arises from :
   (*a*) Medulla (*b*) Pharynx
   (*c*) Internal ear (*d*) Tongue
9. Which one of the following is a motor nerve?
   (*a*) Facial (*b*) Abducens
   (*c*) Vagus (*d*) Olfactory
10. The secretion of adrenal glands called :
    (*a*) Aldrin (*b*) Adrenalin
    (*c*) Sympathin (*d*) None of these
11. Which structure helps a person to maintain his equilibrium ?
    (*a*) Cochlea
    (*b*) Hammer
    (*c*) Semicircular canals
    (*d*) None of these
12. Receptors for hearing and balance are called :
    (*a*) Photoreceptors
    (*b*) Tangoreceptors
    (*c*) Statoacoustic receptors
    (*d*) Olfactoreceptors
13. Sclerotic, cloroid and retina are the layers of :
    (*a*) Intestine (*b*) Eyeball
    (*c*) Stomach (*d*) Oesophagus
14. Which of the following have solid optic lobes?
    (*a*) Frog (*b*) Snake
    (*c*) Man (*d*) Bird

15. Temperature control centre is located in the :
(*a*) Hypophysis (*b*) Hypothalamus
(*c*) Diencephalon (*d*) None of these

16. Human temperature is set at :
(*a*) 47.5°C (*b*) 37°C
(*c*) 37°F (*d*) 98.4°C

17. The sensory ganglion concerned with spinal reflex is located in :
(*a*) Dorsal root of spinal nerve
(*b*) Ventral root of spinal nerve
(*c*) Both (*a*) and (*b*)
(*d*) None of the above

18. Number of spinal nerves in rabbit are :
(*a*) 32 pairs (*b*) 23 pairs
(*c*) 73 pairs (*d*) 37 pairs

19. Constriction of the pupil is function of :
(*a*) Parasympathetic system
(*b*) Hypothalamus
(*c*) Thalamus
(*d*) None of the above

20. Brain is produced by:
(*a*) mesoderm
(*b*) ectoderm
(*c*) endoderm
(*d*) both mesoderm and ectoderm

Directions Qs. 21-25: *In each of the following questions, statement of* ***Assertion (A)*** *is given followed by a corresponding statement of* ***Reason (R)*** *just below it. Of the statements, mark the correct answer—*

(*a*) If both assertion and reason are true and reason is the correct explanation of assertion
(*b*) If both assertion and reason are true but reason is not the correct explanation of assertion
(*c*) If assertion is true but reason is false
(*d*) If assertion is false but reason is true.

21. **Assertion (A):** Repeated elicitation of a reflex may suspend the relfex response for some time.
Reason (R): Impulse can flow only in a single direction in a reflex arc.

22. **Assertion (A):** No taste sensation is evoked when distilled water is applied on human tongue.
Reason (R): Man does not possess taste buds for tasting water.

23. **Assertion (A):** Motor neuron terminates on a motor end plate at the neuromuscular junction.
Reason (R): Motor end plate acts as receptor for detecting changes in the muscle fibres.

24. **Assertion (A):** Giant squids posses very thick nerve fibres.
Reason (R): The nerve impulse travels faster in a thicker nerve fibre than in a thinner one.

25. **Assertion (A):** The chemicals stored in the synaptic vesicles are termed as neurotransmitters.
Reason (R): Synaptic vesicles release these chemicals in the synaptic cleft.

ANSWERS

1	2	3	4	5	6	7	8	9	10
(*d*)	(*d*)	(*b*)	(*d*)	(*b*)	(*a*)	(*a*)	(*a*)	(*b*)	(*b*)
11	**12**	**13**	**14**	**15**	**16**	**17**	**18**	**19**	**20**
(*c*)	(*c*)	B	(*c*)	(*b*)	(*b*)	(*a*)	(*d*)	(*a*)	(*c*)
21	**22**	**23**	**24**	**25**					
(*b*)	(*a*)	(*c*)	(*b*)	(*b*)					

26

Hormonal Coordination

ENDOCRINE SYSTEM

1. Hormones are chemical substance required in small quantities for the rational functioning of the body. Both plants and animals secrete hormones of diffenent kinds for different physiological body functions.
2. The hormone, **'Secretin'** was first discovered by **Bayliss and Starling** (1902). The term 'hormone' was coined by **Starling** (1906) for some chemical substance important for animals.
3. In animals these are secreted by endocrine glands and through blood are passed to different parts of the body. There they control and regulate the normal physiological functions of the body.
4. The word 'endocrine' is a Greek word for internal. An endocrine glands do not have ducts, they are also called **ductless glands**. As hormones are secreted within the body they are called **internal secretions**. They are also called **chemical messengers** or **chemical coordinators**.

PITUITARY GLAND (HYPOPHYSIS)

1. It is called **the master gland** as it does not controls the body functions alone. It also controls the functions of other endocrine glands. However, it is itself under the control of hypothalamic neurosecretory cells.
2. It is a small (in adult 0.5-0.6 gm) oval shaped gland situated at the base of the brain, in **sella turcica** in the sphenoid bone and is attached by a stalk with the floor of third ventricle of diencephalon.
3. Usually it is divided into **anterior lobe** and **posterior-lobe** but may be divided into six pars:
   (i) Pars distalis (Pars anterior)
   (ii) Pars tuberalis
   (iii) Pars intermedia
   (iv) Pars nervosa (lobus nervous) or pars posterior
   (v) Median eminence of tuber cinereum and
   (vi) Infundibulum or pituitary stalk.

The first three are collectively called **adenohypophysis** while the last three as **neurohypophysis.**

Adenohypophysis

The cells of **adenohypophysis** produce seven different types of hormones, like:

(i) Somatotrophic hormone (STH) or growth hormone (GH),
(ii) Thyroid stimulating hormones (TSH) or thyrotropin
(iii) Adrenocorticotropic hormone (ACTH)
(iv) Follicle stimulating hormone (FSH)
(v) Luteinizing hormone (LH) or interstitial cell stimulating hormone (ICSH)
(vi) Luteotropic hormone (LTH) or lactogenic hormone or prolactin.

The FSH and LH are together called **gonadotropic hormone** (GTH) and TSH, ACTH and LH are called together **pituitary tropins.**

Pituitary Hormones and their functions

1. **Somatotropic:** It controls growth especially that of the bones and muscles of the body.
2. **TSH:** It is proteinaceous and stimulates the functions of thyroid gland. Chemical feed back system by TSH is to maintain normal thyroid hormone in blood.

3. **ACTH:** This controls the secretion of adrenal cortex and is a polypeptide hormone.
4. **GTH:** It controls the function of gonads (testis in male and ovary in female). It includes the following hormones which are glycoprotein in nature.
   (a) **FSH:** It stimulates spermatogenesis and also takes part in the growth of testis. It stimulates the growth of follicles in the ovary and stimulate the theca cells of mature ovarian follicle to secrete female sex hormones, oestrogens.
   (b) **LH:** In male LH activates the interstitial cells to secrete testosterone; influences the growth of male accessory sex organs like vas deferens, prostate glands etc. In female LH along with FSH help in the maturation of ova as well as in ovulation. After ovulation, it helps in luteinization or granulosa and theca interna forming corpus luteum.
5. **Prolactin:** It sitmulates corpus luteum to secrete progesterone hormone but the main function of this hormone is to initiate secretion of milk from mammary gland. It also helps in the maintenance of corpus luteum and its secretion (progesterone).
6. **ADH:** It is an octapeptide. It constricts the wall of arterioles resulting in rise of blood pressure. It helps in osmoregulation of the body by increasing tubular reabsorption of water. So it is called antidiuretic hormone.
7. **Oxytocin:** It helps in the constriction of uterine wall at the time of child birth and stimutes lactation so that milk secretion increases after birth of the child.

THYROID GLAND

This endocrine gland is situated in the neck region extending from thyroid cartilage of larynx of fourth tracheal cartilage. The weight is between 5 to 9 grams. It is a bilobed structure and the two lobes are connected by a straight structure called **isthmus** from which sometimes arises a **pyramidal lobe.**

The gland is formed by many different shaped follicles. Each follicle is lined by a single layer of **cuboidal epithelial cells.** The cavity of the follicle is filled with a **colloidal substance**. It is made of iodine and iodinated amino acids. This is secreted by cells of the follicle and is called **iodothyroglobulin.**

Thyroid Hormones and their Functions

1. **Thyroxine:** It is secreted by the follicular cells and takes major part in the metabolism of carbohydrates. In addition:
   (i) It increases the intake of O_2 by the tissues thereby increasing rate of cell restoration.
   (ii) It increase the oxidation of glucose in tissues.
   (iii) In increase basal metabolic rate and R.Q.
   (iv) It reduces the formation of ketone in blood.
   (v) It removes calcium and phosphates from the bones (if the decrease exceeds the limit, the bones become porous).
   (vi) It acts on sinu-auricular node and maintains the normal heart beat.
   (vii) It also increases the activity of nervous system.
   (viii) In increases the rate of metamorphosis in amphibians.
   (ix) In controls the secetion of other endocrine glands as under:
      (a) stimulates the adrenal cortex to secrete its hormones
      (b) normal gonadal activity depends upon thyroid function
      (c) It co-operates with antidiuretic hormone (ADH) in regulation of water
2. **Calcitonin:** Calcitonin decreases the serum Ca level and acts more rapidly opposite to parathormone.

PARATHYROID GLANDS

These are four small oval glands they lie closely with the thyroid.

Parathyroid Hormones (PTH)

The main function of **PTH** is to maintain the concentration of ionized calcium in the plasma. It also increases the rate of calcium absorption from intestine in children.

THYMUS

It is a **biloed** gland and is situated just above the heart. After sexual maturity it regresses and its place is taken by fatty and fibrous connective tissue. The inner surface is provided with trabecuale of connective tissue. Each lobule is covered by capsule of connective tissue and consists of many lobules. Each lobule is with a cortical and medullary region. The cortical region is with thymocyte cells or T-cells and medullary part is with few **thymocytes** and many rounded concentric layered structures called **Hassall's corpuscles** or **thymic corpuscles.**

Thymus hormone, thymosin increases the rate of growth, helps in the maturation of genital organs and produces **lymphocytes** and antiobodies.

PINEAL BODY

It is situated on the roof of third ventricle of brain. The gland secretes **melatonin** which helps in melanin distribution in children and suppresses secondary sexual characters.

ADRENAL GLAND

Adreanal gland is an endocrine gland and is attached on the upperside of each kidney. So it also known as supra-renal gland. It is yellow in colour and the size is larger in female than male. It is covered by a fibrous capsule. The protion below the capsule is divided into two zones, an outer **cortex** and inner **medulla.** The cortex is further subdivided into three zones. Starting from outer side these are:

(i) outer, *zona glomerulosa*
(ii) middle, *zona fasciculata* and
(iii) inner, *zona reticularis.*

Adrenal gland secretes the following hormones:

Hormones of adrenal cortex

1. **Mineralocorticoids (aldosterone):** The main function of this hormone is to control the reabsorption of sodium and chloride salts by the uriniferous tubules.
2. **Glucocorticoids:** It controls the metabolism of carbohydrates, proteins and fats. It stimulates the fats and amino acids and to form glucose.
3. **Sex hormones:** These influence changes in muscles, bones and external sex organs. These sex hormones are androgens and oestrogens.

Hormones of adrenal medulla

1. **Epinephrine:** It is also called **adrenaline** (emergency hormone). It increases the sensitivity and contractility of cardic muscles resulting in increase of blood flow. It increases oxygen consumption.
2. **Norepinephrine:** It is also formed by amino acids and is also called **noradrenaline.** It is vasoconstrictor and increases the blood pressure. Thus, it acts opposite to adrenaline.

ISLETS OF LANGERHANS

Pancreas acts both as digestive and endocrine gland. This endocrine glandular part of pancreas was discovered by **Langerhans** and so called **Islets of Langerhans.** These islects are not connected with pancreatic ducts and so thus pour their secretions directly in blood vessels.

1. **Insulin:** It is secreted by β-cells and was discovered by **Banting** (1921). Insulin helps in carbohydrate metabolism as follows:
   (a) increase the rate of glucose metabolism.
   (b) decrease blood sugar concentration by converting **excess** sugar to glycogen in liver and
   (c) increases glycogen store in tissues.
2. **Glucagon:** It is proteinous in nature and is secreted by α-cells and has the following functions:
   (a) increases blood sugar level by covering glycogen into glucose (glycogenolysis) and
   (b) it is responsible for lower blood calcium level and higher excretory level of calcium.

HORMONES SECRETED BY GONADS

1. **Ovary.** It secretes the following hormones.
   (i) **Oestrogen:** It is a steroid hormone and is secreted and is secreted by Graafin follicle of ovary. It stimulate the growth of uterine

endometrium. In addition it helps in differentiation of mammary glands.

(ii) **Progesterone:** It is also a steroid hormone and is secreted by cells of corpus luteum. It is responsible for the secretory phase of uterine endometrium. It is also called pregnancy hormone.

(iii) **Relaxin:** This hormone is also secreted during pregnancy. It helps to soften the public symphysis checks the contraction of uterus and broadens the uterine cervix for easy birth of the child.

2. **Testis.** The interstitial cells (Leydig) release testosterone which controls overall growth in males and secondary sexual characters.
3. **Placenta.** During pregnancy placenta also acts as an endocrine gland. This secretes estrogen, progesterone and relaxin. These function is association with ovarian hormones. It stops ovualtion and stimulates ovarian secretions.

MULTIPLE CHOICE QUESTIONS

1. The ductless glands are also known as:
   (*a*) Endocrine glands
   (*b*) Muscular glands
   (*c*) Secretory glands
   (*d*) Digestive glands
2. The secretions of endocrine glands are called:
   (*a*) Enzymes (*b*) Vitamins
   (*c*) Hormones (*d*) All the above
3. Message of the endocrine system is conveyed in the form of chemical substances through the:
   (*a*) Blood system (*b*) Nervous system
   (*c*) Excretory system (*d*) None of these
4. Which hormone is helpful in control of blood sugar level :
   (*a*) Insulin (*b*) Calcitonin
   (*c*) Thyroxin (*d*) None of these
5. The estrogen hormone develops:
   (*a*) Secondary sexual characters in human female
   (*b*) Secondary sexual characters in human male
   (*c*) Blood potassium level in human blood
   (*d*) None of the above
6. Pituitary gland is under the control of :
   (*a*) Thyroid (*b*) Hypothalamus
   (*c*) Adrenal (*d*) None of these
7. In woman gonadotropins is detected in the urine for confirmation of :
   (*a*) Pregnancy
   (*b*) Secondary sexual characters
   (*c*) Sexual diseases
   (*d*) AIDS
8. Pigeons milk is produced in the :
   (*a*) Salivary glands (*b*) Mammary glands
   (*c*) Both (*a*) and (*b*) (*d*) Crop glands
9. The term hormone was coined by :
   (*a*) W.M. Bayiis (*b*) G.W. Harris
   (*c*) H.G. Khorana (*d*) Robert Brown
10. Blood calcium level is controlled by :
    (*a*) Parathormone (*b*) Insulin
    (*c*) Glucagon (*d*) None of these
11. Regulation of volume of urine is caused by :
    (*a*) Aldosterone (*b*) ADH
    (*c*) Both (*a*) and (*b*) (*d*) None of these
12. In the adult the deficiency of thyroxine causes:
    (*a*) Diabetes (*b*) Scurvy
    (*c*) Goitre (*d*) Myxoedema
13. Islets of langerhans secrete a hormone known as:
    (*a*) Renin (*b*) Insulin
    (*c*) Ptyalin (*d*) None of these
14. Selective accumulation of iodine occurs in :
    (*a*) Spleen (*b*) Thyroid
    (*c*) Thymus (*d*) None of these
15. Which one of the following hormones is a steroid?
    (*a*) Glucocorticoid (*b*) Glucagon
    (*c*) Insulin (*d*) None of these
16. Metamorphosis can be accelerated by :
    (*a*) Phosphorus (*b*) Calcium
    (*c*) Potassium (*d*) Iodine
17. Pituitary gland is located in :
    (*a*) Brain (*b*) Pancreas
    (*c*) Gonad (*d*) None of these

18. Endocrine glands put their secretions directly into:
(*a*) Blood (*b*) Ducts
(*c*) Liver (*d*) Nerves

19. Blood pressure is controlled by :
(*a*) Corpus luteum (*b*) Thymus
(*c*) Adrenal (*d*) Thyroid

20. The process of aging is controlled by :
(*a*) Pineal (*b*) Pituitary
(*c*) Thymus (*d*) Hypothalamus

Directions Qs. 21-25: *In each of the following questions, statement of* ***Assertion (A)*** *is given followed by a corresponding statement of* ***Reason (R)*** *just below it. Of the statements, mark the correct answer—*

(*a*) If both assertion and reason are true and reason is the correct explanation of assertion
(*b*) If both assertion and reason are true but reason is not the correct explanation of assertion
(*c*) If assertion is true but reason is false
(*d*) If assertion is false but reason is true.

21. Assertion (A): Excessive amounts of calcitonin hormone in the blood causes softening of bones.
Reason (R): Calcitonin decreases calcium of the blood plasma.

22. Assertion (A): Failure of secretion of somatotropin from an early age causes drawfism in the patient.
Reason (R): Somatotropin hormone stimulates the body growth and elongation of long bones.

23. Assertion (A): The dietary deficiency of iodine causes exophthalmic goitre.
Reason (R): Iodine is required for the formation of thyroid hormones.

24. Assertion (A): Oxytocin is also known as Anti Diuretic hormone (ADH).
Reason (R): Oxytocin can cause constriction of uterine wall of the time of child birth.

25. Assertion (A): Failure of secretion of thyroid gland hormones from childhood causes cretinism disease.
Reason (R): Thyrotropin hormone maintains BMR and promote growth of body tissues.

ANSWERS

1	2	3	4	5	6	7	8	9	10
(*a*)	(*c*)	(*a*)	(*a*)	(*a*)	(*b*)	(*a*)	(*d*)	(*a*)	(*a*)
11	**12**	**13**	**14**	**15**	**16**	**17**	**18**	**19**	**20**
(*c*)	(*d*)	(*b*)	(*b*)	(*a*)	(*d*)	(*a*)	(*a*)	(*c*)	(*c*)
21	**22**	**23**	**24**	**25**					
(*d*)	(*a*)	(*d*)	(*d*)	(*c*)					

27

Animal Reproduction and Development

GAMETOGENESIS

- Gametogenesis is the process of formation of haploid gametes (sperms or ovum) from diploid gonads (Testis or ovaries).
- Gametes are formed by specialized division *i.e.,* meiosis/reductional. Although sperm and ova are very dissimilar yet both comprise similar phases of sequential changes.
- Gametogenesis is of two types as: Spermatogenesis and Oogenesis

SPERMATOGENESIS

- It is the process of formation of haploid, microscopic and functional sperm/ spermatozoa from diploid spermatogonia. It is under the control of FSH and androgens. Sperm formation occurs in different steps as:
  (*a*) **Multiplicative phase:** At sexual maturity the undifferentiated, diploid, germ cells divide several times mitotically to produce *spermatogonia* A and B type. Type A cells act as stem cell and type B serve as precursor of sperms *i.e.,* which enter the growth phase.
  (*b*) **Growth phase:** During this phase spermatogonia increase in size by obtaining nourishment from the nursing cells. These cells are also spermatocyte and the process is called *spermatocytogenesis.* Nurse cells are also called sertoli cells.
  (*c*) **Maturation phase:** Now each primary spermatocyte Undergoes two successive divisions as maturation divisions. The first division is meiosis and *i.e.,* reductional division which yields two haploid daughter cells called *secondary spermatocytes.* Now secondary spermatocytes undergo second maturation division *i.e.,* meiosis II (equational division). This is produces four haploid spermatids.
  (*d*) **Differentiation:** The differentiation of non motile, rounded spermatids into functional, motile spermatozoa is called *spermiogenesis / spermioteliosis.*

Structure of sperm/spermatozoa: Sperm consists of head, neck, middle piece and tail. The details of different contents are given under heading spermiotoleosis.

OOGENESIS

It involves the formation of haploid ova/ovum from the diploid egg mother cells, oogonia of the ovary. It occurs in various organisms so it support the evidence of basic relationship of the organisms. It involves two biological processes as genetical programming and packaging. Period of oogenesis is different in different animals. It usually starts before birth and completed only after the onset of puberty. Only one out of 500 cells is stimulated by FSH to mature. So oogenesis is a discontinuous process. Oogenesis occurs in the follicles of ovaries as follows:

(a) Multiplication phase

The primary germ cells of germinal epithelium of ovary undergo repeated mitotic divisions to form oogonia (2N). These oogonia perfect into stroma as a cord and later on form egg nest. This phase lasts only till the time of birth. Only one cell of egg nest will grow and form primary oocyte and other cells form regular layer around it.

(b) Growth phase

Oogonia grows into a large primary oocyte by taking food from the surrounding follicle cells (usually happens after puberty). Vitellogenesis occurs in this phase whose formation is controlled by mass of mitochondria and golgi body near the nucleus called balbiani body/ yolk nucleus.

(c) Maturation phase

Each primary oocyte undergoes two maturation divisions i.e., meiosis I and II. The primary oocyte forms unequal haploid daughter cells after meiosis I i.e., secondary oocyte and first polar body/ polocyte. In the second meiotic division, oocyte again divide into unequal daughter cells as ootid and second polar body. First polar body also divide in two polar bodies.

(d) Differentiation phase

Ootid undergo differentiation to form functional ovum. Thus only one ovum and three polar bodies are formed from one oogonium. So it is a highly wasteful process. Ovum is much larger than oogonium. Ootid undergoes following changes as:

- Nucleus of the oocyte becomes enlarged due to the production of a large amount of nuclear sap.
- Nucleolus increases greatly in size.
- The number of mitochondria increases and form mitochondria clouds.
- Golgi complex disappears completely.
- The membranes offer is usually perforated.
- In mature oocytes, formation of cortical granules with acid mucopolysaccharide takes place.
- Vitellogenesis occurs in the primary oocyte either in mitochondria or in the liver cells and then transported to oocyte.
- Ribosomes become very numerous as they actively participate in the synthesis of protein required for yolk formation.

Ovum/Egg

Ovum are always larger than all the other cells of the body. They are nearly spherical and non motile. The large size of ova is mainly due to their high food reserve in the cytoplasm. Ova are without centrioles during the second maturation division, the centrioles are taken away by the second polar body.

(A) Types of eggs on the basis of amount of yolk

- **Alecithal:** These are the eggs withcut yolk *e.g.* Eutheria.
- **Microlecithal/Oligolecithal:** These are small sized eggs with very small amount of yolk *e.g. Hydra.* tunicates sea urchin, *Amphioxus,* marsupials etc.
- **Mesolecithal:** These eggs contain moderate amount of yolk *e.g.* Annelids, molluscs, *petromyzon,* lung fishes (Dipnoi) and amphibians.
- **Macrolecithal/polylecithal:** These contain very large amount of yolk or reserve food e.g. Insects, elasmobranchs, reptiles, birds and prototherians.

(B) Types of eggs on the basis of distribution of yolk

- **Homolecital/Isolecithal:** In these eggs, small amount of yolk is uniformly distributed in the cytoplasm *e.g.* all animals with microlecithal eggs.
- **Telolecithal:** In these eggs, the moderate amount of yolk is distributed in one hemisphere *i.e.,* vegetal pole and the opposite side of egg is called animal pole. *e.g.,* It is found in mesolecithal, polylecithal and meiolecithal eggs. In meiolecithal eggs, cytoplasm is confined in a disc like area at animal pole called blastodiscs.
- **Centrolecithal:** In these eggs yolk is confined at the centre of the egg surrounded by a peripheral cytoplasm. e.g. Insects.

FERTILIZATION / SYNGAMY

- Fertilization is the fusion of haploid male and female gametes to form diploid zygote.

The type of fertilization varies in different animals but the basic facts remains the same.

- Fertilization is highly species specific *i.e.,* eggs produce a chemical, fertilizin (Glycoprotein + amino acid + Monosaccharide) and sperms has on its surface antifertilizin (acidic amino acids). This interaction of fertilizin with antifertilizin makes the egg and sperm stick *i.e.,* it results in agglutination.
- The fertilization may be :
- **External:** It takes place outside the body of the parents since gametes can remain alive in water for a short period; fertilization begins immediately after amplexus generally shortly before swelling of albumen into jelly occurs. So the purpose of amplexus in such organisms is to ensure simultaneous discharge of male and female gametes at the same place in water. e.g. All aquatic animals.
- **Internal fertilization:** In this case male deposit sperms during copulation either in the oviduct (veretebrates) or into spermathecal off female (earthworm, insects, spiders) so that fertilization takes place inside the body of female. The sperms are made capable of fertilizing the egg by secretions of the female genital tract and the phenomenon is called *capacitation.* The Various Physical And Chemical Events Of

EARLY EMBRYONIC DEVELOPMENT

- Cleavage is a series of rapid mitotic divisions of the zygote which convert the single celled zygote into multicellular structure called *blastula.*
- Cleavage results in decrease in cell size and unlike mitosis it is not followed by growth of blastomere so is called fractionating process. There is more consumption of oxygen and more synthesis of DNA during cleavage. Therefore nuclear to cytoplasmic ratio goes on increasing as cleavage progresses.
- Blastomeres show synchronous division in early stages but later this synchrony is lost.

MORULA

Cleavage produces a solid ball of small blastomeres and the embryo now looks like a mulberry and is called morula. Blastomeres remain in contact with one another of compression. True morula is absent in vertebrates of immediate appearance of blastocoel.

BLASTULATION

As the cleavage proceeds, the number of blastomeres undergo rearrangement. It results in hollow mass of cells with a cavity inside called blastocoel segmentation cavity. The single layer of cells is called blastoderm. The hollow spherical embryo at this stage is called *blastula.* The amount of yolk present in the egg affects the way blastula is formed.

Blastula in different types of eggs is as

- **Isolecithal eggs:** Blastomeres tend to anange in a single layer on the periphery and have central cavity blastocoel. The cells towards vegetal pole are of large size *e.g. Amphioxus,* echinoderms. This type of blastula is called coeloblastula.
- **Moderately Telolecithal eggs** (yolk is unequally distributed): The distribution of yolk causes fast cleavege in animal pole and slow cleavage in vegetal pole which results in macromere and micromere formation in the respective poles. So the blastocoel in such cases is eccentric and blastomeres form 2-3 layers *e.g.* Frog. This kind of blastula is *amphiblastula.*
- **Heavily Telolecithal eggs/Megalecithal eggs:** In such kind of heavily yolky eggs, the cleavage is restricted to the germinal disc as 'a result multilayered central part separates from yolk and those at the margin of the yolk form a continuous layer. So they develop as clear subgerminal cavity whose roof has *epiblast layer* and those forming floor is *hypoblast layer.* This kind of blastula is called *Discoblastula. e.g.* fishes, reptiles, birds.

- In mammals, a peripheral layer of blastomere form trophoderm/trophoblast and consists of an innercell mass attached to one side by the blastocoel. This blastula is called *blastocyst.*

FATE MAP

In all organ grade metazoans, various parts of the body are derived from three basic embryonic cell layers called primary germ layers viz. ectoderm, mesoderm and endoderm.

These area are referred to as presumptive or prospective areas forming a sort offate map on the surface of blastula. It is done by using certain vital stains like neutral red, nile blue sulphate, bismarck brown etc. It was first prepared by W. Vogt (1929).

GASTRULA

It is the process by which blastocyst is changed into three layered gastrula and the new cavity formed in it is called *archenteron* which opens out through *blastopore.* The most important features of gastrulation are:

- It is a complex, orderly, irreversible, morphogenetic movement of mass of cells.
- The rhythm of cellular division is slowed down.
- Growth if any is insignificant, nuclei become more active.
- Archenteron formed is surrounded by endoderm.
- Chemodifferentiation starts by synthesis of new proteins.
- Cell movements involve processes like emboly, epiboly, delamination, invagination, involution, ingression etc.

ORGANOGENESIS

It is the final phase of embryonic development where cell changes their form and give rise to morphologically recognised tissue and organs of new individuals.

Epiboly

The rapidly dividing micromeres of the animal hemisphere grow downwards towards vegetal pole. Cells from deeper layers migrate from their original position and spread over macromeres of the entire surface of the embryo except near gray crescent where blastopore is to be formed. Through the blastopore, a tiny white spot of yolky endodermal cells can be seen, called *yolk plug.*

Invagination

The epiboly of marginal cell zone excerts an inward pressure on macromeres which are pushed towards blastocoel. Due to invagination a small groove appears at the marginal zone it is the begining of archenteron and its opening is *blastopore.*

INVOLUTION/EMBOLY

Involution involves the morphogenetic movement of the cells as rolling in of superficial cells over the rim of blastopore into the interior and also the flow of new portions of blastoderm towards the rim of blastopore. As a result, the following type of structure is formed where blastocoel is completely reduced and displaced towards vegetal pole. The slit at which involution starts is dorsal side and its form dorsal lip of blastopore.

- The complete obliteration of blastocoel marks the end of gastrulation.
- Now all the three primary germ layers form in their embryo *i.e.,* ectoderm, mesoderm and endoderm.

EXTRA EMBRYONIC MEMBRANES IN MAMMALS

The growing mammalian embryo develops four extra embryonic membranes which perform different functions. These foetal membranes are:

1. **Amnion :** Inner side - Trophoblast
   Outer side - Mesoderm

   The space between the embryo and the amnion is called amniotic cavity filled with amniotic fluid. Embryo is suspended in amniotic cavity by means of umbilical cord. The amniotic fluid prevents dessication of the embryo and act as protective cushion against infections and shocks.

2. **Chorion:** Inner side - Somatopleuric mesoderm
   Outer side - Trophoblast
   Chorion takes part in the formation of placenta and helps in metabolic exchange between mother and foetus. It acts as an extra embryonic lung.
3. **Yolk sac:** Inner side - Endoderm
   Outer side - Splanchnopleuric mesoderm
   As human ovum is devoid of yolk, the yolk sac develps an evolutionary process *i.e.,* it is a vestigial organ. Initially it act as a gut and a site of blood cell formation till liver becomes functional (6th week)
4. **Allantois:** Inner side - Endoderm
   Outer side - Mesoderm splanchnopleuric
   It is a sac like structure pushed out from gut during third week. It is reduced and non-functional in humans except the formation of umbilical artery vein and earlier it acted as extra embryonic kidney.

PLACENTATION

Placentation is invariably associated with the viviparity of organism. It is the only, organ in animals formed of temporary connection of the tissues derived from two individuals *i.e.* mother and foetus. The part of placenta derived from foetus is called foetal placenta while that derived from mother is maternal placenta. In mammals the placenta is fully formed in about 10 weeks and lasts throughout pregnancy.

TYPES OF PLACENTA

Histologically there are following tissues which form barrier to separate foetal blood from maternal blood.

Foetal placental tissue

1. Endothelium of blood vessels
2. Chorionic connective tissue
3. Trophoblast

Maternal placental tissues

1. Uterine epithelium
2. Endometrial connective tissue
3. Endothelium of blood vessels

(A) On the basis of intimacy of embryonic and maternal tissue

Epitheliochorial	Syndesmochorial	Endotheliochorial	Haemochorial	Haemoendothelia
All the six tissue barriers present	Uterine epithelium is eroded	Uterine epithelium and uterine connective tissue are broken	Endothelium of maternal blood vessel disappear	Chorionic epithelium and mesoderm also disappears.
e.g. Horse, donkey, pig	e.g. cattle, sheep, goat	e.g. Dog, cat, shrew mole.	e.g. Apes, monkeys, humans.	e.g., rodents, lagomo-rph.
The degree of intimacy is so strong that eventually the blood vessels of the villi are eventually bathed in the blood.				

(B) On the basis foetal membranes involved in the formation of placenta

Yolk sac	Allantoic	Corionic
Derived from yolk sac and chorion	Formed of allantois and chorion	Made up of only chorion
e.g. *Opossum, Macropus* (Kangroo)	e.g. *Perameles* (Bandicoot) most of the eutherians.	e.g. Primates including Humans

(C) On the basis of injury caused to uterine placenta during parturition

Non deciduate	Contradeciduate	Deciduate
No part of placenta is shed in after birth It does not show any injury Their villi are simple apposed without contact with foetus and uterine wall. *e.g.* Cattle, horse, pig	Placenta is indeciduate and even foetal placenta is absorbed *e.g. Perameles* (Bandicoot) *Talpa* (Mole)	Some part of uterine tissue is injured and passed out as decidua in after birth *e.g.* rabbits, dogs, humans

FUNCTION OF PLACENTA

- It helps in nutrition of the foetus as the food contents passes from mother to the baby through placenta.
- It acts as a respiratory organ as oxygen diffuses from mother to foetus and CO_2 from foetus to mother's baby.
- Nitrogen wastes are transported from foetus of mother through placenta.
- It stores the food contents as glycogen, fat and minerals as iron before liver becomes functional.
- Trophoblast of placenta break the proteins before passing them to foetal blood.
- Placenta also acts as a endocrine gland as it secretes certain hormones as estrogen, progesterone, human chorionic gonadotrophin (HCG) and relaxin.
- Three to eight weeks embryo is very sensitive to certain chemicals (Teratogens) as nicotine, heroin, cocain, etc. which may cause malformation in the developing foetus. Some infections e.g. rubella, syphilis, HIV etc. may also pass through the placenta and infect the developing foetus.

MULTIPLE CHOICE QUESTIONS

1. Semen consists of an enzyme
(*a*) Ligase (*b*) Seminase
(*c*) Vasciulase (*d*) Hyaluronidase

2. In mammals the fertilization and entire embryonic development takes place within the:
(*a*) Uterus of female
(*b*) Uterus of male
(*c*) Fellopian tube of female
(*d*) Ovary of female

3. Sustentacular cells is another name of
(*a*) Oocytes (*b*) Sertolicells
(*c*) Oogonia (*d*) Leydig cells

4. Gynogenesis is shown by
(*a*) Nematodes (*b*) Few annelids
(*c*) Both above (*d*) None of these

5. The testis of a mammal is enclosed in a capsule known as :
(*a*) Tunica albuginea
(*b*) Tunica vaginalis
(*c*) Tunica propria
(*d*) None of the above

6. Mature ova are released by :
(*a*) Uterus (*b*) Ovary
(*c*) Fallopian tube (*d*) Oviduct

7. Normal menstrual cycle is completed in :
(*a*) 32 days (*b*) 82 days
(*c*) 28 days (*d*) 38 days

8. Which surgical operation is undertaken in a women for planned family?
(*a*) Vasectomy (*b*) Prostatectomy
(*c*) Tubectomy (*d*) None of the above

9. SAHELI is a :
(*a*) Contraceptive pills
(*b*) Famous actress
(*c*) Both (*a*) and (*b*)
(*d*) None of the above

10. Premenstrual syndrome is related to :
(*a*) Women (*b*) Men
(*c*) Microorganisms (*d*) None of the above

11. Cells of sertoli are found in :
(*a*) Testis of frog
(*b*) Testis of cockroach
(*c*) Testis of rabbit
(*d*) None of the above

12. In which animal the fertilization occurs externally?
(*a*) Frog (*b*) Woman
(*c*) Earthworm (*d*) None of the above

13. The gestation period of elephant is about :
(*a*) 15 months (*b*) 10 months
(*c*) 22 months (*d*) 30 months

14. Gestation period in man is about :
(*a*) 28 weeks (*b*) 35 weeks
(*c*) 32 weeks (*d*) None of the above

15. In the human menstrual cycle, ovulation occurs around :
(*a*) 8th Day (*b*) 10th Day
(*c*) 14th Day (*d*) 15th Day

16. The part of reproductive system in male which neither produces hormones nor sperms is :
(*a*) Prostrate glands (*b*) Cowpers glands
(*c*) Seminal vesicles (*d*) All the above

17. In the honey bee the eggs develop parthenogenetically as :
(*a*) Queen bees (*b*) Drones
(*c*) Worker bees (*d*) None of the above

18. The head cap of sperm is called :
(*a*) Acrosome (*b*) Centrosome
(*c*) Lysosome (*d*) Ribosome

19. In mammals, the fertilization is :
a) External (*b*) Internal
(*c*) Self (*d*) None of the above

20. The embryonic membranse which forms placenta is :
(*a*) Chorion (*b*) Allantois
(*c*) Amnion (*d*) Umbilicus

Directions Qs. 21-25: *In each of the following questions, statement of* ***Assertion (A)*** *is given followed by a corresponding statement of* ***Reason (R)*** *just below it. Of the statements, mark the correct answer—*
(*a*) If both assertion and reason are true and reason is the correct explanation of assertion
(*b*) If both assertion and reason are true but reason is not the correct explanation of assertion
(*c*) If assertion is true but reason is false
(*d*) If assertion is false but reason is true.

21. Assertion (A): In human male, testis are extra abdominal and lie in scrotal sacs.
Reason (R): Scrotum acts as thermoregulator and keeps testicular temperature lower by 2°C for normal spermatogenesis.

22. Assertion (A): At puberty human male develops secondary sexual characters.
Reason (R): At puberty there is decreased secretion of testosterone in male.

23. Assertion (A): Fallopian tubes is lined by ciliated epithelium, whose cilia beats towards uterus.
Reason (R): Cilia lining the funnel helps to pick up and push the released ovum into the oviduct.

24. Assertion (A): Clitoris of female is homologous to tip of penis.
Reason (R): Both have same origin and richly supplied with nerves and blood vessels.

25. Assertion (A): Corpus luteum secretes the female hormone progesterone.
Reason (R): After ovulation, a ruptured follicle turned into yellowish solid mass of cells called corpus albicans.

ANSWERS

1	2	3	4	5	6	7	8	9	10
(*d*)	(*a*)	(*b*)	(*c*)	(*a*)	(*b*)	(*c*)	(*c*)	(*a*)	(*a*)
11	**12**	**13**	**14**	**15**	**16**	**17**	**18**	**19**	**20**
(*c*)	(*a*)	(*c*)	(*b*)	(*c*)	(*d*)	(*b*)	(*a*)	(*c*)	(*b*)
21	**22**	**23**	**24**	**25**					
(*a*)	(*c*)	(*a*)	(*a*)	(*c*)					

28

Ecosystem

A complex system where biotic and abiotic factors interact in an environment is called an **ecosystem.** The term was coined by **AG Tansley** in 1935. Examples: lake, pond, an estuary or a forest.

The structure of any ecosystem is a result of two components, viz-

(A) ABIOTIC FACTORS

The abiotic factors, all together, constitute the environment part of ecosystem and are grouped into three main types.

1. Climate factors

These factors constitute the aerial environment such as light, wind, rainfall, temperautre etc.

I. Light

(i) **Solar energy.** or sunlight **is the source of energy** in an ecosystem. The solar radiations reaching earth consist of visible light and small proportion of ultraviolet and infrared rèadiation.

(ii) The **visible light** lies in the range of 400-750mμ.

(iii) The wave length of UV light is below 390mμ and of infrared above 750mμ.

(iv) **Ultraviolet** rays are most harmful as they **kill protoplast.**

(v) Light is measured by lightmeter or photometer or **luxmeter.**

Effect of light on plants

(i) Light is necessary for chlorophyll production and brings about photosynthesis.

(ii) Light increases temperature, regulates the rate of transpiration and also effects the opening and closing of stomata.

(iii) Depending on the preference of light, plants are classified as

(a) **Heliophytes.** Sun loving plants require full sunlight. e.g. Sunflower.

(b) **Sciophytes.** Shade loving plants which require low level of sunlight.e.g. *Abies, Picea.*

(iv) Light affects the movements of plants. This phenomena is called **Heliotropic effect.**

(v) Stems are usually positively and roots negatively phototropic.

Effect of light in ocean/lake water

More than 10% of incident light is lost by reflection. **Penetration** gradually **decreases as depth increases.** Depending on light penetration, the water bodies can be divided into three zones:

(*i*) **Euphotic Zone:** The upper most zone extends from surface to roughly 80 meters, maximum penetration of light. Most **autotrophs are present** in this zone.

(*ii*) **Diphotic Zone: Middle region** extends from 80 to upto 200 mts. Light penetration decreases and light is diffused. Primary comsumers are found in this region.

(*iii*) **Aphotic Zone:** Last zone found below 200 metres, no light penetration. Predators and decomposers are found in this zone.

II. Temperature

(*i*) Solar energy get converted to heat on the earth surface (measured in **calories**).

(*ii*) Reradiated energy heats the earth's atmosphere, therefore, layers near the earth are warmer.

(*iii*) Temperature at which metabolic activity is favourable is called **activity limit.**

(*iv*) Temperature at which organisms dies is called **lethal limit.**

(*v*) Temprature is maximum at equator but decreases gradually towards the poles.

(*vi*) Generally, temperature decreases at 10^0F for every degree of latitude north or south of equator and for every 300 ft. of altitude.

III. Atmosphere and Air

Atmosphere is multilayered gaseous envelope surrounding the earth. It extneds to 1000 km above earth's surface and serves as reservoir various elements. It is made up of 5 layers.

Composition of Air

1. 99% of air is made up of mixture of two gases: N_2 (78%) and O_2 (21%).
2. 1% of air is made up of Argon (0.93%), CO_2 (0.03%), Hydrogen, Helium (.0005%) and Neon (.0018%).
3. 0.4% of air is made up of water vapour.

Effect of Air

1. Air is the medium in which living beings survive.
2. Wind causes water loss by transpiration rain and it is also responsible for soil eroson.
3. Brings about dispersal of seeds, pollen grains and microorganisms.
4. Wind can be measured by **anemometer.**

IV. Rainfall or Precipitation

The atmosphere moisture falling on the earth is termed as precipitation. It may be in form of rain, dew, frost, sleet and snow.

Plant adaptations

On the basis of water relationship, plants are classified into 4 major groups:

- **Hydrophytes** grow in abundance of water.
- **Xerophytes** grow in scarcity of water.
- **Halophytes** grow in saline conditions.
- **Mesophytes** grow in moderate conditions of water.

2. Edaphic factors

Study of soil is called **Pedology.** It is defined as weathered surface of earth's crust mixed with organic materials, generally made up of sand, silt and clay. It forms the substratum for plants and animals.

Depending of agencies of transport , soil are of following types:

(a) **Alluvial** soil are transported by running water.

(b) **Colluvial** soil are transported by force of gravity.

(c) **Eolian** soil are transported by winds.

(d) **Glacial** soil are transported by glaciers.

Types of soil

- **Sand:** They contain less % of silt (15%) and more sand.
- **Loam:** They contain 20-30% clay, 30-45% silt and rest sand.
- **Sandy loam:** It has 70-80% sand.
- **Loamy sand:** It has 21-51-% clay or silt and rest is sand.
- **Clayey loam:** It contains 20-30% clay.
- **Clay:** It has more than 31% clay.

3. Physiographic or Topographic Factors

These factors concerned with physical geography, *i.e.*, they bring about variations in climate. The main geographic features are:

- **Height of mountains**
- **Direction of mountains and valleys**
- **Steepness of slope**
- **Exposure of slope to sun, wind etc.**

(B) BIOTIC FACTORS

Include the living organisms of the environment like plant, animals, bacteria and viruses. They are classified into the following groups:

1. **Producers** carry out photosynthesis, e.g. plants and algae.
2. **Consumers** are the organisms which eat or devour other organisms. Further they are divides into primary, secondary and tertiary consumers.
    - **Primary consumers** (herbivores) eat the producers like plants and algae.
    - **Secondary consumers** kill and eat the herbivores. They are also termed as carnivores. As these carnivores directly depend on herbivores, they are specially called as primary carnivores. Foxes and wolves are the secondary consumers in a terrestrial ecosystem.
    - **Tertiary consumers** kill and eat the secondary consumers, thus also called secondary carnivores. e.g. Lions and tigers.

3. **Reducers** or **decomposers** are organisms which breakup the dead bodies of plants and their waste products. They include fungi and certain bacteria.

Dynamics of ecosystem

Energy. The main source of energy is the radiation energy derived from the sun.

Primary production is the amount and the rate of energy produced by autotrophs. the total organic material actually present (biomass) in plants is called **net primary production.**

Net primary Production = Gross primary production - Respiration

Secondary production. The energy trapped by the producers (primary production) is utilized by consumers. The producers are consumed by the herbivores that are eaten by the primary carnivores that in turn are consumed by the secondary carnivores.

Trophic levels. Each food chain contains many steps like producers, herbivores, primary carni vores and so on. Each step of the food chain is called trophic level. The number of trophic levels in a food chain is always restricted to 4 or 5. Quite often the chains are very complicated with many trophic levels.

Food Chain : It is the sequence of populations/ organisms through which food and its contained energy pass in an ecosystem with members at one step becoming food of later step of the sequence. Common food chains are predator food chains.

Detritus food chain is the *shortest*. Parasitic food chain is the third type.

(*a*) *Aquatic Food Chain :*

Phytoplankton → Zooplankton → Small Crustaceans → Water Insects → Small Fish → Larger Fish → King Fishes.

(*b*) *Land Food Chain :*

● Grass → Grasshopper → Frog → Snake → Peacock → Falcon → Eagle.

● Herbs → Rabbit → Fox → Wolf → Tiger

Food Web : It is a network of food chains which are interconnected at different trophic levels so as to form a number of feeding connections amongst members of a biotic community. e.g. Rabbits preyed upon by Wild dog, Fox, Wild cat and Jackal. These carnivores can also prey upon mice, shrews, and squarrels. Wolf preys upon Jackal, Fox, Wild cat and Sheep/goat. It is eaten by Tiger. Tiger preys upon Deer, Antelope and a number of other animals. In Sunderbans, Tiger does not have its natural preys. It feeds there on fish and crabs.

Flow of Energy

Ecosystem is continuously obtaining solar energy which is trapped by producers—1.15% in grasslands, 0.9% in Savannah, 0.81% in mixed forest, 5% in crops and 10-12% in sugarcane. The rest is dissipated as heat. A part of energy stored in food is released during respiration. Some of the released energy is used in carrying out body activities while the rest is dissipated. Herbivores feed on producers. Part of the food energy is wasted in digestion and assimilation. Some of the assimilated food is broken down to release energy for performing body activities.

Some of the energy is lost as heat. A very small portion becomes part of the body of herbivore. Herbivores are eaten by primary carnivores, the latter by secondary carnivores and so on. At every step, a lot of energy is wasted. Flow of energy in an ecosystem is unidirectional.

ECOLOGICAL PYRAMIDS/ELTONIAN PYRAMIDS (ELTON, 1927)

Ecological pyramid is the graphic representation of an ecological parameter (number, biomass, energy) sequence-wise in various trophic levels of a food chain. It can be upright, inverted or spindle-shaped.

(a) Pyramid of Numbers : It is an ecological/ eltorian pyramid which employs the numbers of individuals per unit area at various trophic levels sequence-wise with producers at the base and various consumers at successively higher levels. It is generally *upright* as maximum number of individuals occur at the producer level, followed by smaller number of herbivores, fewer primary carnivores and still fewer higher levels of carnivores. However, a single large-sized tree can provide food to several herbivores (e.g. birds, squirrels) which support a few carnivores. Such a pyramid of number is *spindle shaped*. When the herbivores support a large number. of ecotoparasites, which are further parasitised by still larger number of hyperparasites, the pyramid of numbers is *inverted.*

(b) Pyramid of Biomass : It is the amount of living matter present in the individual/population/ community/trophic level/ecosystem. Biomass is measured on both fresh and dry weight basis. It is ecological/electronian pyramid which represents the biomass in various trophic levels. It is more real than the numbers at various trophic levels. Only 10-20% biomass is transferred from one trophic level to the next – 10% law. 1000 kg (one tonne) vegetation shall form only 100 kg of herbivore, 10 kg of primary carnivore, 1 kg of secondary carnivore and only 0.1 kg of tertiary carnivore. Therefore, with increasing dependence on animal diet (nonvegetarian), the human population will require higher amount of farm produces—total nonvegetarian 10 times, 50% non vegetarian 5 times, vegetarian with milk 1.5 times. Pyramid of biomass is upright, exception equatic food chain involving short lived plankton (biomass at one time will be less than consumers at higher levels.

(c) *Pyramid of Energy :* It is ecological/eltorian pyramid that gives graphic representation of amount of energy trapped by different traphic levels per unit area. It is always upright because during transfer of energy from one trophic level to the next, a lot of wastage occurs in feeding, digestion, assimilation and respiration.

HOMOESTASIS/BIOLOGICAL EQUILIBRIUM

It is the maintenance of a functional dynamic balance or state of equilibrium amongst various components of an ecosystem. Homeostasis is maintained through a number of controls/cybernetics/limitations like :

(i) Carrying Capacity : It is the maximum number of individuals the environment can support. Near it, the rate of population growth begins to decline. Beyond it, the population decreases rapidly.

(ii) Recycling of Wastes : It controls the availability of raw materials and space.

(iii) Self regulation : Mortality decreases and natility increases in case of small population while reverse is true in case of larger population.

(iv) Feedback System : It is of two types, positive and negative. In *positive feedback* the increased availability of a resource increase number of utilisers. For example, predator population will increase with the increase in population of prey. In *negative feedback*, the increase in number of members of a higher trophic level will decrease the number of individuals at the lower level which in turn shall reduce their number as well.

Habitat : It is a specific place/locality delimited by a combination of factors and barriers where an individual/population/community resides, e.g. pond, desert, river, bark of tree, burrow, intestine.

Microhabitat : It is part of the habitat having a specific property, e.g. forest floor, tree canopy, tree trunk, edge of a pond.

Niche/Ecological Niche : It is specific/part of habitat occupied by individuals of a species which is circumscribed by its range of tolerance, range of movement, microclimate, type of food and its availability, shelter, type of predator, timing of activity. Tadpole and adult frog occupy different ecological niches as the former is herbivorous aquatic while the latter is carnivorous amphibian. Water Bug and water Boatman live in shallow edges of ponds but occupy different niches as the former is predator while the latter scavenger. Both owl and cat feed on shrews and mice. They occupy the same niche because of being *ecological equivalents* though their habitats are different.

Life Forms

Raunkiaer (1934) has distinguished plants into five forms on the basis of size, shape, branching, crown, life span and perennation.

(i) Therophytes : Annual plants which perennate in the form of seeds.

(ii) Cryptophytes : Perennial plants with underground storage parts,

(a) Geophytes : Subterranean perennating structures (root, root tuber, bulb, stem tuber, rhizome, corm).

(*b*) ***Helophytes :*** (Marshy plants). Perennating structures are embedded in mud.

(*c*) ***Hydrophytes :*** Aquatic plants.

(*iii*) ***Hemicryptophytes :*** Peremating structures occur at ground level. Aerial shoots die on the onset of winter, e.g. rosette plants.

(*iv*) ***Chemaephytes :*** Small plants of cold areas where perennating buds or shoot apices lie at or above the ground level.

(*v*) ***Phanerophytes :*** Pernnial herbs, shrubs, and trees, epiphytes, succulenty lianas etc. where perennating buds occur at 10 cm or more heigh above ground level.

MULTIPLE CHOICE QUESTIONS

1. The plant, *Zizpyphus rotundifolia* is an indicator of
(*a*) sandy loam soil
(*b*) high lime soil
(*c*) soil formation
(*d*) soil erosion

2. Mangrove vegetation is found in
(*a*) Sundarbans (*b*) Kullu valley
(*c*) Western Ghats (*d*) Dehradun Valley

3. Organisms which can tolerate narrow range of temperature are called
(*a*) polythermal (*b*) stenothermal
(*c*) oligothermal (*d*) eurythermal

4. Plants flourishing in acidic soils are
(*a*) psammophytes (*b*) oxalophytes
(*c*) sciophytes (*d*) heliophytes

5. The efficiency of any ecosystem is best depicted by pyramid of
(*a*) energy (*b*) number
(*c*) biomass (*d*) volume

6. Biotic components include
(*a*) producers only
(*b*) consumers only
(*c*) producers and consumers only
(*d*) producers, consumers and decomposers

7. The world 'ecosystem' was coined by
(*a*) Elton (*b*) Tansley
(*c*) Odum (*d*) Billing

8. The primary consumers in a pond ecosystem is
(*a*) phytoplankton (*b*) zooplankton
(*c*) fishes (*d*) bacteria

9. The type of food chain in which organic matter decomposed is converted into energy rich compounds is called
(*a*) detritus food chain
(*b*) grazzing food chain
(*c*) cybernatics
(*d*) none above

10. The study of fresh water ecology is
(*a*) hydrology (*b*) ecology
(*c*) limnology (*d*) geology

11. The energy which is utilized by the living world is primarily
(*a*) heat (*b*) electricity
(*c*) light (*d*) kinetic

12. Alluvial soils are mostly found in which region
(*a*) Eastern Ghats (*b*) Southern India
(*c*) Upper Gangetic plains
(*d*) Eastern Himalaya

13. The amount of energy, utilised by herbivores from the plants is
(*a*) 5% (*b*) 10%
(*c*) 50% (*d*) 90%

14. An ecosystem resist change because it is in a state of
(*a*) imbalance
(*b*) homeostasis
(*c*) shortage of components
(*d*) deficiency of light

15. The spread of prickly pear (*Opuntia*) in India can be controlled by
(*a*) Letting loose cattles to consume them
(*b*) Encourage the birds for seed dispersal
(*c*) Spraying hormones
(*d*) Introducing cochineal insects

16. In ecological sense desert animals are termed as
(*a*) arboreal (*b*) benthos
(*c*) cursorial (*d*) xeric

17. Many angiosperms and ferns synthesize a molecule similar to ecdysone. It causes the insect herbivores that eat it to
(*a*) become tiny adults and leave few offsprings
(*b*) die
(*c*) stop deeding
(*d*) fail to reproduce

18. The vegetation found between tree line and snow line is called
(*a*) Canyon effect (*b*) Savanaha
(*c*) Alpine meadow
(*d*) Open deciduous forest

19. Sargassum and Laminaria are the members of which of the following ecosystem?
(*a*) Terrestrial biome
(*b*) Tropical rain forest
(*c*) Fresh water biome
(*d*) Marine biome

20. The pyramid of energy in any ecosystem is
(*a*) always upright or inverted
(*b*) may be upright
(*c*) always inverted
(*d*) none of the above

Directions Qs. 21-25: *In each of the following questions, a statement of **Assertion (A)** is given followed by a corresponding statement of **Reason (R)** just below it. Of the statements, mark the correct answer:*

(*a*) If both assertion and reason are true and reason is the correct explanation of assertion.
(*b*) If both assertion and reason are true but reason is not the correct explanation of assertion
(*c*) If assertion is true but reason is false
(*d*) If assertion is false but reason is true.

21. Assertion (A): Producers are also known as transducers.
Reason (R): Herbivores are called key industry animals.

22. Assertion (A): Food web consists of several food chains.
Reason (R): Food web decreases the stability of an ecosystem.

23. Assertion (A): Ecological pyramids are also called Eltonian pyramids.
Reason (R): An ecological pyramid is always upright.

24. Assertion (A): Tree and its dependents form upright pyramid of number.
Reason (R): Pond ecosystem show upright pyramid of number.

25. Assertion (A): Pyramid of biomass may be upright or inverted.
Reason (R): Pyramid showing total biomass produced per unit time is always upright.

ANSWERS

1	2	3	4	5	6	7	8	9	10
(*a*)	(*a*)	(*b*)	(*b*)	(*a*)	(*d*)	(*b*)	(*b*)	(*a*)	(*c*)
11	**12**	**13**	**14**	**15**	**16**	**17**	**18**	**19**	**20**
(*c*)	(*c*)	(*b*)	(*b*)	(*d*)	(*d*)	(*a*)	(*c*)	(*d*)	(*d*)
21	**22**	**23**	**24**	**25**					
(*b*)	(*c*)	(*c*)	(*d*)	(*b*)					

29

Application of Biology

DRUGS

They are chemical substances used in the treatment, cure, prevention or diagnosis of disease or used otherwise to enhance someone's physical or mental well being.

Drug are usually divided into two categories:

A. **In physical dependency** the drug gives a satisfaction and makes a person addict to it. Later larger doses are needed and the body develops tolerance. Subsequently the process of stopping the continuous use of the drug causes severe and unpleasant physical symptoms, (withdrawal symptoms).

B. **Psychological dependency** means that the addict becomes dependent on the drug and stopping the pleasure would be unbearable to them.

Drug are classified on the basis of their action as following:

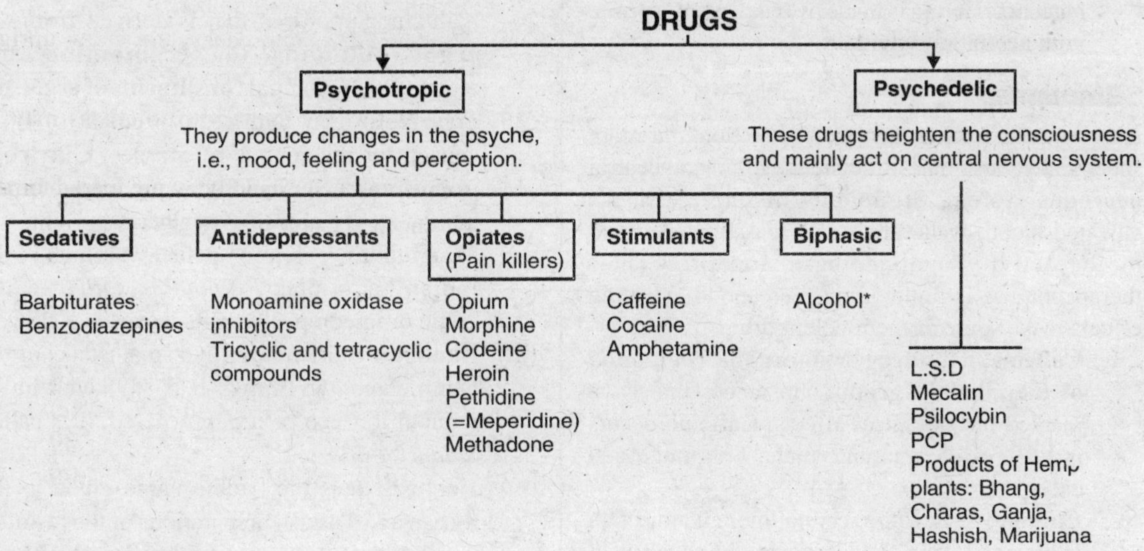

Antidepressants

They are essential for the treatment of severe cases of depression and are prescribed in most cases of clinical depression. Four types are in use as given

Opiates

These narcotic drugs produce analgesia (pain relief), narcosis (state of stupor or sleep) and addiction (physical dependence on the drug). In some people narcotics also produce euphoria (a feeling of great elation). Medically speaking, narcotics are the most powerful painkillers available, but they are used with great caution because of their addictive properties. Some opiates (pain killers) with their effects are summerized:

(i) **Opium:** It is narcotic drug obtained from the immature fruits of the opium poppy (*Papaver somniferum*), of family Papaveraceae, by collecting and drying the juice from slightly incised fruits.

(ii) **Morphine:** It is narcotic **analgesic drug** used in

Antidepressants

Monoamine Oxidase Inhibitors (MAO)	Selective Serotonin reuptake Inhibitor (SSRI)	Tricyclic antidepressants (TCA)	Newer Antidepressants
Phenelzine (Nardil) Tranylcypromine (Parnate)	Fluoxetine (Prozac) Sertraline (Zoloft) Paroxetine (Paxil) Fluvoxamine maleate (Luvox)	Clomipramine (Anafranil) Amoxapine (Asendin) Amitriptyline (Elavil, Endep) Doxepin (Sinequan)	Citalopram hydrobromide (Celexa)

medicine in the form of its hydrochloride, sulfate, acetate and tartrate salts. In its power to reduce the level of physical distress, morphine is among the most important caused by **cancer**.

(iii) **Heroin:** It is also called **Diacetylmorphine**, a highly addictive morphine derivative that makes up a large portion of the illicit traffic in narcotics. Heroin is made by **treating morphine** with **acetic anhydride**.

Stimulants

Any drug that excites any bodily function, but more specifically those that stimulate the brain and central nerveous system. Stimulants induce alertness, elevated mood, wakefulness, increased speech and motor activity and decrease appetite. Their therapeutic use is limited, but their mood-elevating effects make some of them potent drugs of abuse.

1. **Caffeine:** It is **nitrogenous organic compound of the alkaloid group,** substances that have marked physiological effects. Caffeine occurs in tea, coffee, guarana, mate, kola nuts, and cacao.
2. **Cocaine:** It is white, crystalline, alkaloid that is obtained from the leaves of the cocal plant (*Erythroxylum coca*), a bush commonly found growing wild in Peru, Bolvia and Ecuador.

Hallucinogenic Drug (Psychedelic)

A substance that produces psychological effects that are normally associated only with dreams, schizophenia, or religious exaltation. It produces changes in perception, thought and feeling, ranging from distortions of what is sensed (illustions) to sensing objects where there are none to be sensed (hallucinations).

(i) **LSD:** It is also called **Lysergic acid diethylamide** or **Lysergide**, potent synthetic hallucinogenic drug that is derived from the **ergot alkaloids** (as **ergotamine** and ergonovine, **principal constituents of ergot**, the grain deformity and toxic infectant of flour caused by the fungus of grasses, **Claviceps purpurea**) LSD usually is preapared in the laboratory by chemical synthesis.

(ii) **Marijuana:** It is also spelled Marihuana, the Indian hemp plant, *Cannabis sativa* or the crude drug composed of its leaves and flower.

(iii) **Bhang:** It is dried laves and small stalks of the hemp, *Cannabis sativa*. It is smoked with or without tobacco or it is powdered. It is called *Siddhi* or *Sabzi*.

(iv) **Hashish:** It is the Arabic name given to the Syrian and Turkish preparations of the resinous hemp leaves, prepared by heating the bhang with water and butter.

(v) **Charas:** It is the resin that exudes naturally from the hemp plant.

(vi) **Ganja:** It consists of flower and fruits of female hemp plant. It is brownish green resembling bhang.

TOXICITY OF CHEMICALS AND SELF-MEDICATION

Steroids and their effects

Steroid drug treatment is now widely used to fight various diseases and inflammation (allergies). But unwanted side effects of steriod occur because they are given in much higher amounts than would normally be found in the body.

Anabolic steroids

The anabolic steroids used by atheletes are almost exclusively those with structures similar to the chief male sex hormone, testosterone. Anabolic steroid promote protein accumulation by the muscles.

Steroid abuse and health hazard: Major side effects from abusing anabolic steroid include:

1. Liver and kidney tumors and cancer.
2. Jaundice (yellowish pigmentation of skin, tissues and body fluid).
3. Fluid retention
4. High blood pressure
5. Increases LDL (bad cholesterol) and decreases HDL (good cholesterol).
6. Severe ache and trembling.

Asprin

It is mild, non-narcotic analgesic useful in the relief of headache, muscular and joint aches. It is also effective in reducing fever, inflammatin and swelling and thus has been used for treatment of rheumatoid arthritis, rheumatic fever and mild infection.

Tobacco: Effects And Control

Tobacco is in legal use eveywhere in the world yet it causes **far more deaths** than all other **psychoactive substances** combined. About 3 million premature deaths a year (6 per cent of the world total) are already attributed to tobacco smoking. Tobacco is responsible for about 30 percent of all cancer deaths in developed countries.

Components of Tobacco Smoke

Smoke contains over **4,000 chemicals**; 43 of them are **carcinogenic** (cancer promoting). These include chemicals as polycyclic hydrocarbons, beta naphtylamines and also nitrosamines, which have been long recognized as carcinogenic in nature. Tobacco is produced from plant called 'tobacco plant' (**Nicotiana tabecum**).

BIOTECHNOLOGY

Originally, it meant the branch of technology connected with actions and requirements of human beings. It is now the technological exploitation or biological process including genetic engineering for obtaining useful materials for agriculture, industry, domestic, medical and other uses.

Alcoholic Fermentation

Louis Pasteur found for the first time that beer and butter milk are produced due to activity of Yeast and Yeast-like microorganisms. Yeast species used in alcoholic fermentation are *Saccharomyces cerevisiae* (Brewer's Yeast), S.*ellipsoidens* (Wine Yeast) S.sake (Sake yeast) and S.*pireformis* (Giger Beer/alc Yeast).

Food Yeast

Common food yeast is *Torulopsis utilis.* It is cultured over molasses and starchy materials like Potato at 5pH. Ammonium phosphate is added for stimulating growth. Waste sulphite liquor of paper industry is also used.

Baker's Yeast

Selected strains of *Saccharomyces* are used as Baker's yeast. It is available as powder or cake.

Cheese

It is nutritive product of curdling and fermentation of milk that contains proteins (20-35%), fat (20-30%), minerals and vitamins. **Raw cheese.** Milk is curdled with the help of cheese culture *Streptococus lactis, S. cremoris, Leuconsotoc citrovourm* etc. Curd is heated gently to separated cheese from liquid called whey. Any liquid left in cheese is allowed to drain by hanging it in cloth.

Yoghurt (Yogurt, yoghurt)

It is produced by curdling of milk with the help of *Streptococcus thermophilus* and *Lactobacillus bulgariscus at* 40–46°C for four hours.

Butter Milk

(i) It is acidulated product which is formed by inoculating skimmed milk with sarter culture of *Streptococcus cremoris, S. lactis, Lactobacillus acidophilus, Leuconostoc citrovorum* or *L. dextranicum* at 22°C for 18 hours. (ii) Acidulated liquid left after churning of butter from curd.

Curd

Indian curds is prepared by inoculating cream milk (also skimmed milk) with *Lactobacillus acidophilus* at less than 40°C. The curd is eaten as such, salted or sweatened, churned to prepare lassi or obtain butter and butter milk.

Curd, yoghurt and butter milk have more easily digestible proteins, low lactose but high vitamin content and presence of other useful biochemicals, some of which are antimutagenic and anticancer.

Enzymes

Rene Reamur (1752) demonstrated digestive power of gastric juice. Dufrunfaut (1830) found that malt extract has the potential to change into sugar. Payen and Persoz (1883) prepared alchoholic extract of malt and named it diastase. Kuhne (1878) named the biocatalyst as **enzyme**. **Immobilisation of enzymes.** It is the process of fixing enzyme to or enclosing it in a solid support so as to protect it.

Dextran

It is a plasma extract having 6–10% solution of dextrins which is given in case of haemorrhage, shock and dehydration. Dextrins are soluble polyglucans or polymers of D-glucose.

Steroids

They are complex cyrstallisable lipids having a tetracyclic hydrocarbon core (one 5-carbon and three 6-carbon rings) and a long side chain. They are constituents of cell membranes, hormones and some important biochemicals–cholestrol, progesterone, oestrogen, testosternone, cortisterone, cortisone. Steroids are used medicinally in correcting hormonal imbalance, as anabolic stimulants, birth control pills (progesterone ± estrogen), antifertility drugs (e.g. diosgenin), anti-inflammation, for relieving pain and suppressing immune responses.

Antibiotics

The term was coined by Waksman (1942). Clinically, anitbiotic is a substance produced by a microorganisms without harming the host. Antibiosis was first shown by Babes (1885). Pasteur and Joubert (1877) found some bacteria inhibited growth of anthrax causing bacterium, Bacillus anthracis. Vuillemin (1899) proposed concept of antiobiosis. Rene Dubois was first to produce an antibiotic named tyrothricin from Bacillus brevis. However, first antibiotic is generally associated with the name of Alexander Fleming (1928) when he discovered penicillin from Penicillium natatum (did not allow growth of Staphylococus aureus). Waksman and Woodruff isolated actinomycin in 1941 and streptothricin in 1942. Waksman and Albert (1943) and Waksman (1944) discovered streptomycin Burkholder (1947) isolated chloromycetin.

Monoclonal Antibodies (Magic antibodies)

They are pure antibodies against a particualr antigen which are obtained through clonal culture of hybridomas. The technique was developed by Kohler and Milstein (1974). A rat/mouse is injected with a**n antigen.**

Insulin. Sharpy-Shafer (1916) proposed that diabetes is due to failure of pancreas to secrete a chemical, he named insulin. Banting and Best

(1921) isolated insulin from pancreas of Dog and demonstrated its efficacy in curing diabetes in humans. Later insulin was extracted from pancreas of slaughtered cattle and pigs.

DNA Fingerprinting

In 1985, Jeffreys *et al* found that DNA of each individual has some noncistrinic hyper-variable **repeat minisatellite sequences.** They have core units of 11–60 bp and are flanked by conserved restriction sites. These repeats minisatellite sequences restrictions sites are now commonly called **VNTRs** or **variable number of tandem repeats.** They are similar only in monozygotic twins.

VNTRs become different due to small deletions, inserts and mutation. The same are passed down in inheritance. However, during sexual reproduction there is reshuffling of DNA molecules. Therefore, each individual has distinct combination of VNTRs which is peculiar to that perosn only.

Application. (i) DNA finger printing is highly reliable method of indetification of individuals involved in rape, murder and other crimes. Through this technique perosns indulged in killings of a former prime-minister and a former chief-minister were identified. (ii) DNA typing or finger printing is a sure method in solving paternity and maternity disputes. (iii) It is being used to study relationships amongst various groups of people for knowing historical migrations and invasions. (iv) The method provides information about human lineage and relationship with great apes. (v) It can help identify individuals and races with specific genes carrying inherited diseases, proneness to various diseases as well as resistance to them. (vi) It is useful in guiding breeding programmes for endangered animlas.

PRINCIPLES OF DISEASE AND BODY'S DEFENCE MECHANISM

Immunology is the study of the physiological responses by which the body destroys or neutralizes matter, both living and non-living as well as cells of its own which become altered in certain ways.

Cells mediating immune responses. The cells that carry out immune responses are collectively termed the immune system, but they are a diverse collection of the cells both in the blood and in organs and tissues throughout the body. The major cells types are listed alongwith their sites of production:

Cell Types	Production site
Leucocytes (WBC)	
Neutrophills	Bone marrow
Basophills	Bone marrow
Eosinophills	Bone marrow
Monocytes	Bone marrow
Lymphocytes	
B-Cells	Bone marrow, thymus
T-Cells	
Cytotoxic T-Cells	
Helper T-Cells	
Suppressor T-Cells	
Third population = NK cells	
Plasma cells	Peripheral lymphoid organs, developed from B-lymphocytes during immune reponse.
Macrophages	Almost all tissues and organs; differentiate from monocytes
Mast Cells	Almost tissues and organs; differentiate from basophills.

Allergy

1. **Allergy** is a **disorder of immune system.**
2. The body mistakes a harmless substance as an antigen (**allergen**) and reacts violently. This state of increased reactivity is known as **hypersensitivity** and may be due to cell mediated immune response or humoral response.
3. When body mistakes one of its own components (cell or chemical) as allergen, it results in **auto immune disorder.**

Allergy Types

There are two major types of allergy:

A. Immediate hypersensitivity

1. ***Anaphylactic***

1. Exposure to allergen.
2. Production of IgE.
3. Sensitization of basophils by attachment to IgE.
4. Re-exposure to allergen.
5. Exposion of basophils to release histamine or serotonin.
6. Results in swelling (edema), itching, burning, rash, coughing etc. and in extreme cases death occurs due to asphyxiation (anaphylactic shock).

2. ***Cytotoxic***

1. Exposure to allergen.
2. Production of IgG.
3. IgG attaches to antigen.
4. Destruction of antigen.
5. Observed in transfusion reaction, Rh incompatibility.
6. Takes hours to days.

3. ***Immune comlex hypersensitivity***

1. Exposure to allergen.
2. Formation of antigen antibody complex.
3. Attraction of neutrophills.
4. Release of lysosomal enzymes by neutrophills and tissue destruction.
5. Usually seen in serum sickness, farmers lung etc.
6. Takes hours to days.

B. Delayed or cellular hypersensitivity

1. Exposure to allergen.
2. Contact of allergen with T-cells.
3. Destruction of allergen.
4. Thickening and drying of tissue, redness etc.
5. Reaction occurs after 24 to 72 hours of exposure.
6. Response to microbial allergen is called infection allergy.

AIDS

The most striking example of the lack of a basic resistance mechanism is the disease called **acquired immune deficiency syndrome** (AIDS). This disease is caused by the **human immuno deficiency virus** (**HIV**) which incapacitates the immune system.

- HIV belongs to the **retro-virus** family which are unusual in that their nucleic acid core is RNA rather than DNA.
- Retroviruses possess an enzyme that, once the virus is inside a host cell, transcribes the virus's RNA into DNA, which is then integrated into the host cell's chromosomes. The association is permanent; every time the host cell reproduces, it also reproduces the retrovirus DNA.
- The virus may remain dormant inside the cell for years. Thus individuals, known to be infected with HIV, based on the finding of antibodies against HIV in their blood, will show no symptoms until the virus becomes active and undergoes rapid replication inside the cell, resulting in the cells's death. What causes the virus to become active is not known.

TECHNOLOGY FOR MEDICAL APPLICATIONS

A number of technical devices are available to mitigate human sufferings and protection from infection both diagnostic and therapeutic purposes.

Diagnostic Instruments

1. ECG

Electrocardiograph is a noninvasive instrument employed to record myo-electrical change during cardiac cycle. **Electrocardiogram** is the recording of electrical potential produced during cardiac cycle from initiation and spread of cardiac impulse from SA node to AV node AV bundle and purkinje fibres.

2. EEG

Satton (1875) found electrical activity from exposed animal brain. Hand Berger (1929) obtained electrical activity when a string galvanometer was connected to human scalp. It gave birth to electroencephalography. **Electro encephalograph** in an electricity operated instrument having electrodes which when attached to scalp gives information about the electric waves emanating from different parts of the brain. It is a noninvasive technique.

3. Evoked Potentials/ Evoked Responses (EPs/ETs)

It is record of electrical activity following repeated stimulation (100-1000) of a particular type (e.g/light, sound, touch) so that evoked response or potentials become larger than the back ground EEG waves.

4. SQUID

Super-conducting quantum interference device e. g. magnetoencephalograph. The device has a super-conducting ring containing one or more Joephsoris junctions where a change in the magnetic flux linkage of the ring by one flux quantum produces a sharp change in ring's impedance.

5. MET

It is magnetoencephalorgraphic techinque which employs SQUID for picking up weak magnetic waves emanating from brain. The technique gives information about the health of various parts of the brain.

6. Autoanalyser

It is a fully automatic, computerised instrument which can analyse qualitatively and quantitatively various biochemicals present in body fluids like urea, uric acid, ketones, cholesterol, glucose, proteins, enzymes, etc.

Imaging Instruments

1. Tomography

It is a technique of development of three-dimensional impression of an internal area through imaging or different layers. Tomography can indicate cysts, tubercular foci, calculi, cancer, etc.

2. CT scanning (Computerised/Computed Tomographic Scanning)

CT scanning is an invasive radiographic technique which passes short X-rays through the patient's body from various angles and the images picked up by special X-ray detectors. X-rays were discovered by William Roentgen (1895). **Radiology** is the scientific study of X-rays and their use in diagnosis of diseases. X-ray imaging is carried out to detect diseases of lung, heart, fractures, joints, health of hollow oragans including kidneys through formulations having barium and iodine. However, direct X-rays cannot differentiate soft tissues.

3. CAT Scanner

Computerised axial tomography (CAT) is specialised radiological technique (X-ray) for study of various parts inside the skull and location of any patholigical condition. Now replaced by CT scanning.

4. PET

Positron emission tomography or position tomographic scanning is an investigative techinque in which positron emitting radio-isotopes (generated by cyclotron) of common elements like ^{11}C, ^{13}N, ^{15}O and ^{18}F are incorporated in biochemicals like fluorodeoxy glucose (PDG), ammonia, amino acids, oxygen, carbon dioxide, etc. The biochemicals are injected or inhaled in small with electrons and form photons.

5. MRI/Magnetic Resonance Imaging

Magnetic resonance imaging is a noninvasive technique which uses strong magnetic field for generating resonance and low radio frequency in protons present in the body.

6. Sonography/Ultrasound Imaging (Olaf von Ramm; Donald, 1958)

It is a **noninvasive** technique which uses ultrasound for producing images of internal body parts.

7. Dopplar Ultarsound Scanning.

It is used in scanning blood flow in vessels, blood clots and heart abnormalities.

8. Hall Effect Imaging (HEI)

It uses a strong magnetic field and ultrasonic emissions to pinpoint diseased tissues, especially cancerous ones.

THERAPEUTIC INSTRUMENTS

1. **Pace-maker**

   It is an electric device first developed by Greatbach and Chardack (1960) which is connected to heart for covering up any deficiency of myogenic functioning so as make it beat normally.

2. **Laser**

   Laser (LASER) or light amplified by stimulated emission of radiations is a beam of very high energy particles.

3. **Intra-aortic Balloon Pump**

   It is meant for improving blood supply to heart walls and other organs.

4. **Angioplasty**

   It is a technique of opening a blocked coronary artery through **ballooning.**

5. **Angiography/Arteriography**

   It is radiography in which a radio-opaque contrast medium or fluoroscopic chemical is passed through various parts of the heart to study health of walls, valves, atria, venticles, coronary arteries, etc.

DIAGNOSTIC LABORATORY TESTS

1. Haemoglobin Estimation–Sahli's Acid Haematin Method

Haemoglobin is changed into brown coloured acid haematin with the help of hydrochloric acid (HCl). The colour is matched against brown glass standards in a comparator.

Normal range of Haemoglobin

Adult male	15.5 ± 2.5 g/dl
Adult female	14.0 ± 2.5 g/dl
New Borns	16.5 – 19.5 g/dl
Infants (3 months)	11.5 ± 1.5 g/dl
Child (1 year)	12.5 ± 1.0 g/dl
Child (Post 1 year)	13 ± 1.5 g/dl

Diurnal Variations. 10% variations occur with lowest in the morning (haemoglobin contents highest) and highest in the evening (hameoglobin contents lowest).

2. Blood Sugar Estimation

Two common methods of blood sugar estimation are (i) Folin-Wu method and glucose oxidase method. Care is taken to stop glycolytic activity of RBCs by sodium fluoride. Otherwise, RBCs consume roughly 5% of glucose every hour. Blood sugar is estimated from fasting venous blood as well as after meals (post prandial).

Normal Values

Whole Blood sugar (Fasting). 90-120 mg/100 ml. It includes 20-30 ml/dl non-glucose saccharoid fraction.

Plasma/Serum Sugar (Fasting). 70-110 mg/dl.

Plasma/Serum Post Prandial Sugar. 100-140 mg/dl

Whole Blood Post Prandial Sugar. 140-180 mg/dl.

3. Determination of Total Leucocyte Count (T.L.C.)

Normal Value of TLC

Adults	$7.5 \pm 3.5 \times 10^3$ /mm^3
Full term pregnant woman	$15 \pm 3 \times 10^3$ /mm^3
Children (4-7 years)	$11 \pm 5 \times 10^3$ /mm^3
New born	$18 \pm 8 \times 10^3$ /mm^3
Infant (1 year)	$12 \pm 6 \times 10^3$ /mm^3
Children (8-12 years)	$9 \pm 4.5 \times 10^3$ /mm^3

4. Determination of DLC (differential leucocyte count)

It is required to know certain disorders like gout, TB, allergy, worms,etc,. Buffer phosphate solution or water is required. Cells are stained with Giemsa or Leishman's stain.

5. Determination of ESR (Erythrocytic Sedimentation Rate)

If a blood mixed with anticoagulant is allowed to stand, red blood corpuscles begin to settle down slowly at the bottom. It occurs in three stages (i) formation of rouleaux or stacking or RBC. (ii) Sinking or rouleaux. (iii) Compaction of sedimented rouleaux.

There are two methods of ESR determination, wintrobe an westergren method:

Normal value of ESR by Westergren's Method

Males. 0–5 mm after first hour, average 3 mm/hour.
Females. 0–7 mm after first hour, average 5 mm/hour.

6. Lipid Profile

Lipid profile is quantitative listing of various types of lipids present in circulating blood. Pure lipids, being insoluble in water, cannot be transported through blood plasma. They get complexed with proteins to form small particles of **lipoproteins.** The latter are 'soluble' in blood plasma. Lipoproteins are formed in intestine as well as liver. They are of five types-Chylomicrons, VLDL, IDL, LDL, and HDL. **Chylomicrons** (density 0.93 g/ml).
LDL= Total cholestrol– (HDL+ Triglycerides)/2

7. Determination of Blood Urea

Urea is formed in liver from deamination product (ammonia) of amino acids. It is one of the main nitrogenous waste product excreted in urine. Concentrtion of blood urea indicates the health of excretory system.

Normal Value

30–40 mg/dl. Adults over 60 years and premature infants may have a little higher value while lower value may occur during early pregnancy due to physiological hydremia (excess water).

8. ELISA Test (Enzyme Linked Immuno-sorbent Assay Test)

ELISA is very quick diagnostic technique for detection of a wide variety of pathogens, disorders, allergens and hormones e.g. (i) Pregnancy test through presence of human chorionic gonadotropin in urine or blood. (ii) AIDS or HIV infection. (iii) Hepatitis. (iv) STD or sexually transmitted disease. (v) Rubella virus. (vi) Thyroid disorder.

9. Widal Test

It is a serological test for **enteric fever** which is caused by bacterium *Salmonella.* Several types of serological forms are known but common types are **typhoid** (*Salmonella typhi*) and **paratyphoid** (*S. paratyphi A, S. paratyphi B,* occasionlly *S. typhimurium*).

DOMESTICATION OF PLANTS AND CROP IMPROVEMENT

Plant breeders uses numerous methods to develop new varieties, but their primary techniques of development are:

1. **Selection:** Plant breeders choose plants with desirable traits for further propagation and discard plants that are inferior for those traits.
2. **Hybridization:** It involves crossing plants of different strains or types to combine in the progeny the desirable traits of both parents. Undesirable traits also enter the combination, however, so hybridization is usually followed by several generations of selection.
3. **Mutation:** Most mutaions are deleterous, but occasionally one has a distinct advantage. The plant showing the mutation may be used directly as a variety, a common practice in apples and other fruits, or the new trait may be added to existing varieties through hybridization and back crossing.
4. **Tissue culture:** This newest and most technical method of vegetative propagation involves cloning. Any plant tissue with cells that can divide can be used for tissue culture. Although cultures have been started from fruit, endosperm, pollen and embryos, tissues taken from the vascular area of stems or roots have been the most successful.

AGRICULTURAL CHEMICALS

(Pesticides)

The toxic substance used to kill animals or plants that cause economic damage to crop or ornamental plants or are hazardous to the health of domestic animals and humans are called pesticides. All pesticides interfere with normal metabolic process in the pest organism and often are classified as:

1. **Herbicides:** These agents, usually chemicals, for killing or inhibiting the growth of unwanted plants *i.e.*, weeds.
2. **Insecticides:** These toxic substance are used to kill insects. Such substances are used primarily

to control pests that infest cultivated plants or to eliminate disease-carrying insects in specific areas.

Insecticides can be classified in any of several ways, on the basis of their chemistry, their toxicological action, or their mode of penetration. They are also classified according to whether they take effect upon ingestion (stomach poisons), inhalation (fumigants), or upon penetration of the body covering (contact poisons).

3. **Fumigants:** These are the volatile, poisonous substance used to kill insects, nematodes and other animals or plants that damage stored foods or seeds, human dwelling, clothing and nursery stock.
4. **Fungicide:** They are also called **antimycotic** which kill or inhibit the growth of fungi that either cause economic damage to crop or ornamental plants or endanger the health of domestic animals or humans. Most fungicides are applied as sprays or dusts.

BIOFERTILIZERS AND BIOLOGICAL PEST CONTROL

The use of fertilizers has caused short term increases in agricultural productivity which is, however, an essential component of modern agricultural system. Indian soils are usually very poor in organic matter as well as in nitrogen contents.

BIOPESTICIDE

It is the biological control of weeds. Several biopesticides are enumerated below:

A. **Biochemical Active Ingredients**

Floral attractants/Plant volatiles e.g. Allyl isothicyanate (Mustard oil), Bergamont oil, Citronella oil, Eucalyptus oil, Geranium oil and Lemon Grass oil.

Natural insect growth regulators e.g., Azadirachtin, Dihydrozadirachtin

Pheromones e.g., German cockroach pheromone, Lauryl alcohol, Xanthine

Plant Growth regulators and herbicides e.g. Cytokinin, Ethylene, Gamma aminobutyric acid (GABA), Gibberellic Acid, Indole-3-acetic Acid

Repellents e.g., Allium sativum (*Garlic*), Cederwood oil, Cloves (crushed), Rosemary (herb), Thyme (herb).

Miscellaneous biochemicals e.g. Jojoba oil, Neem oil, Soybean oil

B. **Microbial active ingredients**

Bacteria e.g., *Agrobacterium radiobacter, Bacillus cereus, Bacillus subtills, Bacillus thuringiensis, Pseudomonas aureofaciens*

Fungi e.g., *Ampelomyces quisqualis, Beauveria bassiana, Candida oleophilla, colletotrichum gloeosporioides, Puccinia canaliculate, Trichoderma harzianum*

Viurses e.g., *Nucleopolyhedrosis Virus* (NPV) controls *Anagrapha falcifera,* Douglas fir tussock moth Gypsy moth, *Helicoverpa zea*

Virus coat proteins e.g. Papaya Ringspot Virus Coat Protein, Potato leafroll virus (PLRV) coat protein as produced in potato and Watermelon Mosaic Virus Coat Protein in Squash.

C. **Non-viable microbial pesticides e.g.** *Bacillus thuringiensis* releases delta-endotoxin in killed *Pseudomonas fluorescens*

D. **Miscellaneous microbial pesticides e.g.** *Nosema locustae*

E. **Plant pesticides e.g.** *Bacillus thuringiensis* Cry IA (b) in corn from PV CIB 4431, Potato Leafroll Virus (PLRV) replicase protein as produced in potato plants.

Bioinsecticides: Natural Pest Control

Insects are among the most successful organisms on the planet. Infact, they outweigh animals and people on the planet by about seven to one.

Researchers have developed several kinds of bioinsecticides based on viruses, fungi and bacteria. Some of them are given below e.g. *Bacillus thuringiensis,* Fungi-based Bioinsecticides, Virus-based Bio-insecticides, Baculovirus and grassphoppres.

Bioherbicide

The Weed feeding insects and microorganisms form bioherbicides.

Only those organisms should be used which will feed selectively and exclusively on weeds and not on crops. e.g. **Cochneal insect** (*Cactoblastis cactorum*), *Phytophthora palmivora, Chrysolina* beetles, *Zygograme bicolorata,* **Devine** and **Collego, Transgenic plants,** *Cercospora rodmanii, Viruses* are pathogenic, *Colletotrichum, Neochentina olchhorniae, Plusia virticillata* caterpillars.

GENETIC CONSERVATION AND NEW CROPS

Some Wild Edible Plants

	Common name	Botanical name
1.	Watercress	*Nasturtium officinale*
2.	Water hyacinth	*Eichhornia crassipes*
3.	Water or purple	*Geum rivale*
4.	Water sheild	*Brassenia schreberi*
5.	White clover	*Trifolium repens*
6.	White sweet clover	*Melilotus alba*
7.	White trilium	*Trillium grandiflonum*
8.	Whorled loosesterife	*Lysimachia quadrifolia*
9.	Wild bean	*Phaseolus polystachios*
10.	Wild black mustard	*Brassica nigra*
11.	Wild carrot	*Daucus carota*
12.	Wild common straberries	*Fragaria virginiana*
13.	Wild garlic	*Allium canadense*
14.	Wild mint	*Mentha arvensis*

Few Improved Varities of Important Plants

Wheat: Kalyan Sona, Sonalika, Safed Lerma, Chhoti Lerma, Sharbati Sonara etc.

Rice: Type 1, Type 3, Type A64, P500-28, Varangal 487, Basmati 307, Patnai S. 1092 etc.

Potato: Darjeeling Red Round, Kufri red, Kufri Kuber, Kufri Kisan, President, etc.

Cotton: *G. arboreum*: Virnar, Jarila, Bhog, Daulat etc.
G. herbaceum: Kalyan, Vijay, Digvijay, Jayawant etc.
G. hirsutum: Laxmi, L.SS. Buri 0394, Co. 4, MCU 2. etc.
G. barbadense. Andrews, P.R.S.-72.

Sugarcane: CO 1082, CO 312, CO 8152, CO 8153 etc. (CO = Coimbatore)

BIOENERGY

Energy Resources

1. **Fossil fuels or coal, oil and natural gas**
   Coal, oil and nantural gas represent the photosynthetic output of green plants which occurred millions of years ago.
2. **Biogas** or **Klargas** or **Marsh gas** is produced by a process known as **methanogenesis.** It was discovered by **Volta** in 1776 who demonstrated the presence of methane in marsh gas.

Energy Plantations

These are the plantations where fuel wood trees are grown. Such plantations would ensure the supply of firewood to nearby areas. Energy plantations have following advantages.

- To uitlize land properly, the land for growing fueldwood should be:
  (a) Farmer's land (b) Village common lands (c) Along both sides of road, canals and railway tracks (d) On degraded forests and (e) On wastelands.
- The crops which can be used for the production of alcohol are called **energy crops.** Some energy crops are:
  (i) Potato (ii) Maize (iii) Sugarcane (iv) Sugarbeet (v) Tapioca (vi) Molasses

MULTIPLE CHOICE QUESTIONS

1. Morphine is got from
(*a*) *Cajanus cajan*
(*b*) *Cannabis sativa*
(*c*) *Papaver somniferum*
(*d*) *Rauwolfia serpentina*

2. An opiate narcotic is
(*a*) Morphine (*b*) LSD
(*c*) Amphetamine (*d*) Barbiturate

3. LSD is
(*a*) Narcotic

(*b*) Stimulant
(*c*) Antidepressant
(*d*) Hallucinogen

4. Which one is correctly matched
(*a*) Cocaine – Opiate narcotic
(*b*) Bhang – Analgesic
(*c*) Reserpine – Tranquilizer
(*d*) Morphine – Hallucinogen

5. Drugs that cause malformations in embryo during pregnancy are called
(*a*) Tranquilizers
(*b*) Teratogens
(*c*) Alcoholic beverages
(*d*) Nicotine

6. Caffeine is a stimulant present in
(*a*) Coffee (*b*) Tea
(*c*) Cold drinks (*d*) All the above

7. Opium is obtained from
(*a*) *Oryza sativa*
(*b*) *Coffea arabica*
(*c*) *Papaver somniferum*
(*d*) None of these

8. Cassava is widely grown in Kerala and Tamil Nadu and second to rice as a staple food is member of family
(*a*) Asclepiadiaceae
(*b*) Asteraceae
(*c*) Euphorbiaceae
(*d*) None above

9. Pulses are obtained from the family
(*a*) fungi (*b*) liliaceae
(*c*) cycadaceae (*d*) leguminosae

10. The largest industry and our main occupation is
(*a*) oil
(*b*) agriculture
(*c*) transport and communication
(*d*) steel

11. Vegetables, drugs and narcotic plants are commonly found in the family
(*a*) malvaceae (*b*) solanaceae
(*c*) salikaceae (*d*) rubiaceae

12. The correct match for wheat variety is
(*a*) True Emmer – *Triticum dicoccum*
(*b*) Wild Emmer – *Triticum dicoccoides*
(*c*) Durum – *Triticum durum*
(*d*) all above

13. Cochineal insects are used to check the growth of
(*a*) *Eichhornia* (*b*) *Parthenium*
(*c*) *Opuntia* (*d*) *Lantana*

14. The first pesticide used in large scale is
(*a*) Bordeaux mixture
(*b*) carbofuran
(*c*) cultural
(*d*) chemical

15. One of these is nematicide
(*a*) warfarine
(*b*) zinc sulphate
(*c*) sodium fluoroacetate
(*d*) methyl bromide

16. Most herbicides interfere with
(*a*) absorption of water
(*b*) CO_2 absorption
(*c*) photolysis of water
(*d*) translocation of sugar

17. Household pesticides are
(*a*) flit (*b*) DDT
(*c*) flit and baygon (*d*) all of these

18. Straight fertlizer contains
(*a*) one primary nutrients
(*b*) two primary nutrients
(*c*) all primary nutrients
(*d*) all micronutrients

19. Green manuring increases the crop yield by
(*a*) 5-10% (*b*) 15-25%
(*c*) 30-40% (*d*) 80-90%

20. In honey the main constituent is
(*a*) Calcium (*b*) Sugar
(*c*) Protein (*d*) none of these

21. The largest industry and our main occupation is
(*a*) oil
(*b*) agriculture
(*c*) transport and communication
(*d*) steel

22. Vegetables, drugs and narcotic plants are commonly found in the family

(*a*) malvaceae (*b*) solanaceae
(*c*) salikaceae (*d*) rubiaceae

23. The first pesticide used in large scale is
(*a*) Bordeaux mixture
(*b*) carbofuran
(*c*) cultural
(*d*) chemical

24. Most herbicides interfere with
(*a*) absorption of water
(*b*) CO_2 absorption
(*c*) photolysis of water
(*d*) translocation of sugar

25. Household pesticides are
(*a*) flit (*b*) DDT
(*c*) flit and baygon (*d*) all of these

Directions Qs. 26-30: *In each of the following questions, a statement of* **Assertion (A)** *is given followed by a corresponding statement of* **Reason (R)** *just below it. Of the statements, mark the correct answer:*

(*a*) If both assertion and reason are true and reason is the correct explanation of assertion.
(*b*) If both assertion and reason are true but reason is not the correct explanation of assertion
(*c*) If assertion is true but reason is false
(*d*) If assertion is false but reason is true.

26. Assertion (A): Tranquillizers are called antipsychosis drug.
Reason (R): Tranquillizers are used to treat psychosis.

27. Assertion (A): Sedatives induce sleep.
Reason (R): Sedative reduces excitement.

28. Assertion (A): Hypnotic drugs induce sleep.
Reason (R): Central nervous system is sensitive to hypnotic drugs.

29. Assertion (A): *Azotobacter* fixes nitrogen in symbiotic form.
Reason (R): They form root nodules in the roots of leguminous plants.

30. Assertion (A): *Azospirillum lipoferum* shows associative mutualism.
Reason (R): It lives around the roots of higher plants.

ANSWERS

1	2	3	4	5	6	7	8	9	10
(*c*)	(*a*)	(*d*)	(*c*)	(*b*)	(*d*)	(*c*)	(*c*)	(*d*)	(*b*)
11	**12**	**13**	**14**	**15**	**16**	**17**	**18**	**19**	**20**
(*b*)	(*d*)	(*c*)	(*c*)	(*d*)	(*c*)	(*d*)	(*a*)	(*c*)	(*b*)
21	**22**	**23**	**24**	**25**	**26**	**27**	**28**	**29**	**30**
(*b*)	(*b*)	(*c*)	(*c*)	(*d*)	(*a*)	(*b*)	(*a*)	(*d*)	(*b*)

30

Environmental Pollution

POLLUTION AND ITS CONTROL

The word "pollutionem" (means to make dirty), is the act of polluting the environment. *Environmental pollution* is defined as the unfavourable alteration of our surroundings, wholly as a by product of man's activities through direct or indirect efforts to changes in the physical, chemical and biological characteristics of land, air or water that harmfully effect human life or any desirable living thing.

(a) Pollutant : It is a substance, chemical or factor which has the potentiality to adversesly affect natural characteristics of environment, natural resources and assets. Pollution may be a substance in wrong place, e.g. fertilizer causes water pollution.

(b) Primary Pollutant : Pollutant persisting in the environment in the form it is produced, e.g. carbon monoxide.

(c) Secondary Pollutant : Pollutant formed from a primary one through change or reaction. Nitrogen oxides and hydrocarbons react photochemically to produce peroxyacyl nitrate and ozone. The secondary pollutants may be more toxic than the primary ones. The phenomenon is called *synergism*.

AIR POLLUTION

It is the addition of particulate matter, gases and other ingredients into air which can produce an adverse effect on human beings, animals, vegetation, human assets and resources.

Major sources of atomospheric pollution are

(*i*) Combustion of fossil fuels in homes, factories, thermal plants, automobiles, aircrafts, railways etc.

(*ii*) Mining and processing

(*iii*) Chemical industries

(*iv*) Processing industries etc.

Particulate Matter

It consists of soot, flyash, dusts of various types, fur, hair, spores, pollen grains etc. Particulate matter is differentiated into settleable (larger than 10 μm) and suspended (less than 10 μm). SPM (suspended particulate matter) is maximum in Kolkata. Tokyo is the most polluted city of the world. It is differentiated into aerosol (less than 1 μm), dust (more than 1 μm) and *mist* (liquid, more than 1 μm). Particles of 2 - 4 μm size are mostly deposited in the respiratory tract.

Carbon Monoxide

It is produced due to incomplete combustion and naturally by plants as well as animals. 70% emissions are from automobiles. Carbon monoxide combines with haemoglobin, produces carboxyhaemoglobin or COHb. At 50 ppm, CO converts 7.5% of haemoglobin into carboxy-haemoglobin with 8 hours. It impairs oxygen transport resulting in giddiness, headache, decreased vision, cardiovascular malfunction and asphyxia.

Carbon Dioxide

It is a green house gas the concentration of which is constantly rising (0.033 from 0.029 and expected to be doubled by 2020). In excess it causes headache and nausea.

Sulphur Dioxide

It is produced during combustion of fossil fuels and smelting of sulphur containing ores. Sulphur dioxide produces *smog*. Maximum SO_2 pollution is found in calcutta. It causes membrane damage, destruction of chlorophyll (change to phaeophytin), necrosis and water soaked appearance. Lichens are most sensitive to SO_2 pollution. Mossess and garden pea are also destroyed. SO_2 corrodes metals, equipment, damages buildings, kmarble, paper and textiles. It produces eye irritation, damages respiratory tract, produces *asthama, bronchitis* and *emphysaema*. SO_2 alongwith nitrogen oxides produces acid rain that destroys vegetation and degrades various articles (like SO_2). Threat to Taj Mahal from Mathura refinery is due to pollutant gases comprising SO_2, H_2S and nitrogen oxides. They would convert $CaCO_3$ (marble) into calcium sulphate and calcium nitrate.

Nitrogen Oxides

They are formed electro-photo-chemically in air, burning of fossils fuels and denitrifying bacteria. Maximum NO_2 pollution is recorded from Baroda. Nitrogen oxides causes *necrosis*, *defoliation*, dieback and death of plants. Like SO_2, they corrode metals and deteriorate paints as well as various articles. At less than 0.1 ppm, nitrogen oxides impair sensation of smell. The oxides produce eye irritation, lung oedema, dilation of arteries, injury to liver and kidneys. Carcinogenic, form Secondary pollutants.

Chlorofluorocarbons/Chlorofluoromethane/ Freon/Aerosols

They are chemicals used as refrigerants, propellants and solid plastic foams. The chemicals are release as aerosols by jets flying at high altitudes. Alongwith nitrogen oxides, chlorofluoro-carbons react with ozone of ozonosphere and depletes the same. Hole in the ozone shield over Antarctic region has widened from 129 to 133 dobson units in 1994 alone. This can increase the amount of ultraviolet radiations reaching the earth.

Photochemical Oxidants

They are secondary pollutants (ozone, peroxyacyl nitrates, aldehydes and phenols) produced due to unsaturated hydrocarbons.

Peroxy-Acyl Nitrates : (PAN/PBN/PPN). Young and spongy parenchyma cells are destroyed resulting in silvering, glazing, bronzing and necrosis in leaves. In human beings they cause respiratory distress and eye irritation.

CONTROL OF AIR POLLUTION

Automobiles

(*i*) Two stroke engines fitted in two wheelers (waste fuel 20-30%) be changed to either four-stroke engines or fitted with catalytic converters specially designed for them. A low cost-catalytic converter has been developed at BHU.

(*ii*) Leaded petrol be replaced with unleaded one and diesel with low sulphur diesel.

(*iii*) Periodic checkup of pollution control for all vehicles.

Fly Ash

About 38% fly ash is produced by coal based thermal plants. It should be removed through wet method and used in building material.

Industrial Pollution : Control measures

(*i*) *Tall Chimneys :* They disperse smoke more throughly.

(*ii*) *Gravity Setting Chambers :* Particles larger than 50 μm settle down.

(*iii*) *Wet Scrubbers :* A fine spray of water or alkaline fluid is used to remove soluble gases and particles.

(*iv*) *Bag Filters :* Porousbags of teflon or polyester filterout particulate matter.

(*v*) *Cyclone Collectors :* They cause settling down of particulate matter through centrifugation.

(*vi*) *Electrostatic Precipitators :* (ESPs). They are electrically charged plates or electrodes which remove most of the particles present in chimney exhausts.

WATER POLLUTION

It is the degradation of quality of water due to addition of substances (inorganic, organic, biological, radiological) or factors (e.g. heat).

Organic Wastes/Sewage

It is waste water having food residue, animal and human excreta, detergents, discharges from commercial and industrial establishments.

Pesticides

Pesticides spread over crops also pass into water bodies due to surface run-off. In excess they cause immediate and mass scale deaths of aquatic animals. Persistent-perticides (e.g. organochlorine or chlorinated hydrocarbons like DDT) pass into food chain and increase in amount per unit weight of organisms with the rise in trophic level due to their accumulation in fat. The phenomenon is called *biomagnification / bioconcentration/biological amplification*. e.g. 0.01-0.05 parts per billion in water, 10 parts per billion or 0.001 ppm in phytoplankton, 0.04 ppm in zooplankton, 0.4 ppm in clams, 3-8 ppm in fish and 18-75 ppm in water birds like Sea-Gulls. Higher amounts of pesticides cause cerebral haemorrhage, softening of brain, liver cirrhosis, hypertension, thinning of egg shell, malfunctioning of sex hormones. Population of Bald Eagle has declined due to it.

Industrial Effluents

(*a*) **Mercury :** It is changed to water soluble dimethyl mercury, which undergoes biomagnification. Eating poisoned animals causes deformity known as *mina mata* disease which is characterized by diarrhoea, Lemolysis, impairment of various senses, meningitis and death.

(*b*) **Copper :** Hypertensions, uremia, occassional fever and coma.

(*c*) **Lead :** (also common from automobile exhausts) Anaemia, vomiting, convulsion, loss of appetite, damage to liver, kidneys and brain.

(*d*) **Zinc :** Vomiting, cramps, renal damage.

(*e*) **Cobalt :** Diarrohoea, hypotension, bone defects and paralysis.

(*f*) **Chromium :** Gastro-intestinal ulcers, nephritis, nervous system disorders.

(*g*) **Cadmium :** Anaemia, hypertension, testicular atrophy, damage to liver and kidneys, diarrhoea and skeletal deformities called *itai-itai* (ouch-ouch, first reported in 1947 in Toyoma city of Japan).

SOIL POLLUTION

It is alteration in soil caused by removal or addition of substances and factors which decreases its productivity, quality of plants and ground water. *Negative soil pollution* is reduction in soil productivity due to erosion and over use. *Positive soil pollution* is reduction is soil productivity due to addition of undesirable substances (e.g. perticides, fertilizers, industrial wastes, air pollutants washed down by rain, faulty sanition). *Landscape/third pollution* is converting fertilie land into barren one by dumping wastes (e.g. ash, sludge, garbage, rubbish, industrial wastes, broken cans, bottles etc.) over it.

RADIOACTIVE POLLUTION

It is degradation of environment due to release of radioactivity (emission of α-particles, β-particles and gamma rays) by changes in nuclides (atomic nucleus) of sunstable/radioactive elements causing short-range and long range harmful effects on living beings including humans. Radioactivity is measured is units called ROENTGENS.

NOISE POLLUTION

It is a physical form of pollution that affects the receiver directly. Noise pollution in due to release of unwanted sound generally with the value of 80 dB and above. Frequency of sound is measured in Hz (Hertz) while unit of sound in dB (deci Bel).

Range of human hearing is 20 Hz to 2000 Hz. Infrasonic vibrations (below 20 Hz) can be felt because some body parts resonate at this frequency. Ultrasonic vibrations (above 20,000 Hz) are used in imaging, cleaning, drilling cutting, welding and sealing. Moderate conservative produces 60 dB sound, loud conversation 70 dB, plying of trunk/bus 90 dB, jet aeroplane–150 dB, rocket 180 dB. Noise becomes uncomfortable above 100 dB. Sound is produced by loud speakers/public address system, transport vehicles, defence equipment, industries, domestic gudgets and outdoor equipment.

Effects of Noise

(*i*) Damage to ear drum and impairment of hearing (a ten year exposure to 80 dB impairs hearing by 15 dB) (*ii*) Interference in conversation and hearing.

(*iii*) Emotional disturbance, development of anxiety and stress (first effect).

(*iv*) Damage to eye sight, colour perception, night vision, etc.

(*v*) Hypertension, decreased heart output and gastric problems.

(*vi*) Startle reaction.

Control Measures

(*i*) Delimination of acoustic soning

(*ii*) Use of cotton pluges or ear muffs in occupational exposure

(*iii*) Development of quieter machines.

(*iv*) Restricted use of loud speakers.

(*v*) Acoustic furnishing and low voice radio/TV.

(*vi*) Regulation of noise on road.

(*vii*) Green muffler.

MULTIPLE CHOICE QUESTIONS

1. Match the following diseases with their cause

Column I	Column II
(A) Minnimata disease	(I) As
(B) Itai-Itai disease	(II) Cd
(C) Black foot disease	(III) Hg
	(IV) Nitrates

Codes :

	A	B	C
(*a*)	III	II	I
(*b*)	III	I	II
(*c*)	IV	I	II
(*d*)	II	I	IV

2. Frequency of sound is measured in

(*a*) Decibel (*b*) Hertz

(*c*) Watts (*d*) Fathoms

3. Range of human hearing is

(*a*) 2 – 20 Hz

(*b*) 20 – 200 Hz

(*c*) 200 – 2000 Hz

(*d*) 20 – 200000 Hz

4. Green house gases are

(*a*) SO_2, NO_2, CO_2

(*b*) N_2O, CH_4, SO_2

(*c*) CO_2, CH_4, CFC, NO_2

(*d*) CFC, NO_2, H_2S

5. Increase in CO_2 concentration result in following effects. Mark the correect statement suitable to the effect

(I) Rate of photosynthesis in C_3 plants increases.

(II) Rate of transpiration increases.

(III) Rate of stomatal conductance will decrease.

(IV) Plants will go more successfully in water scarcity and nutrient poor soil.

Codes :

(*a*) I, III (*b*) II, IV

(*c*) I, II, IV (*d*) I, III, IV

6. The most toxic metal pollutant of automobile exhaust is

(*a*) Hg (*b*) Cd

(*c*) Cu (*d*) Pb

7. Which of the following is non biodegradable pollutant

(*a*) Leaves (*b*) Fruits, vegetables

(*c*) Mud (*d*) Plastic

8. Which one is not a fossil fuel—

(*a*) Coal (*b*) Petrol

(*c*) Uranium (*d*) Natural gas

9. Which of the following measures can be taken to reduce the wastage of energy

(*a*) Preparation of smokeless stoves

(*b*) Use of biogas

(*c*) Growing fuel wood trees

(*d*) All of the above

10. The major pollution causing agent is

(*a*) Microbes

(*b*) Animals

(*c*) Man

(*d*) Hydrocarbon gases

11. Melting of glaciers may result from

(*a*) Depletion of ozone layer

(*b*) Increase in U.V. rays

(*c*) Increase in oxygen in the atmosphere

(*d*) Increase in CO_2 in the atmosphere

12. In the coming years the skin diseases will be more common due to
(*a*) Increase in atmospheric pressure
(*b*) Increase in air pollution
(*c*) Excess use of detergent
(*d*) Depletion of ozone

13. The algal bloom due to the additon of washed off fertilizers in aquatic reservoir is
(*a*) Biomagnification (*b*) Biobloom
(*c*) Eutrophication (*d*) Nitrification

14. Environment planning organisation is
(*a*) NEERI (*b*) WHO
(*c*) CSIR (*d*) ICMR

15. Mark the wrong statement/s
(I) Lead is a cause of soil pollution
(II) Green house effect is actually global warming
(III) Large particles are removed by cyclone reactors
(IV) CFC are used as refrigerant

Codes:
(*a*) I, III (*b*) I, III, IV
(*c*) IV only (*d*) II, III

16. Biological magnification is
(*a*) Increased accumulation of toxic substances in the food pyramid.
(*b*) Increase in reproductive potential of an organism
(*c*) Increase in biomass of different trophic levels in the pyramid of biomass.
(*d*) All of the above.

17. An indicator of fluorine pollution is
(*a*) Garden pea (*b*) Maize
(*c*) Beans (*d*) Rice

18. Units of intensity of sound is
(*a*) Decibel (*b*) Murtz
(*c*) Kalvin (*d*) Morgan

19. If a lake is contaminated with insecticide DDT, its concentration would be highest in
(*a*) Algae (*b*) Protozoa
(*c*) Small fish (*d*) Fish eating birds

20. Environmental pollution that can cause birth defects is
(*a*) SO_2 (*b*) CO
(*c*) smog (*d*) radioactivity

Directions Qs. 21-25: *In each of the following questions, a statement of* ***Assertion (A)*** *is given followed by a corresponding statement of* ***Reason (R)*** *just below it. Of the statements, mark the correct answer:*

(*a*) If both assertion and reason are true and reason is the correct explanation of assertion.
(*b*) If both assertion and reason are true but reason is not the correct explanation of assertion
(*c*) If assertion is true but reason is false
(*d*) If assertion is false but reason is true.

21. Assertion (A): Pollution is always caused by human activities.
Reason (R): Pollution is not different from contamination.

22. Assertion (A): DDT is an example of secondary pollutant.
Reason (R): DDT persists in nature.

23. Assertion (A): PAN is a secondary pollutant.
Reason (R): Formation of PAN occurs in sunlight.

24. Assertion (A): DDT persists in nature.
Reason (R): DDT is a qualitative pollutant.

25. Assertion (A): The content of CO_2 is steadily rising.
Reason (R): CO_2 is a qualitative pollutants.

ANSWERS

1	2	3	4	5	6	7	8	9	10
(*a*)	(*b*)	(*d*)	(*c*)	(*d*)	(*d*)	(*d*)	(*c*)	(*d*)	(*c*)
11	**12**	**13**	**14**	**15**	**16**	**17**	**18**	**19**	**20**
(*d*)	(*d*)	(*c*)	(*a*)	(*a*)	(*a*)	(*b*)	(*a*)	(*d*)	(*d*)
21	**22**	**23**	**24**	**25**					
(*d*)	(*d*)	(*a*)	(*b*)	(*c*)					

BIOTECHNOLOGY

1

Biotechnology : An Overview

BIOTECHNOLOGY

The term biotechnology is derived from a fusion of biology and technology. Thus biotechnology is defined as the industrial application of living organisms and their biological processes such as biochemistry, microbiology, genetic engineering, etc. in order to make best use of the microorganisms for the benefit of mankind.

The European Federation of Biotechnology (EFB) has given a definition of biotechnology as "The integration of natural science and organisms, cells, parts thereof, and molecular analogues for products and services."

Biotechnology is "the controlled use of biological agents, such as micro-organisms or cellular components, for beneficial use" (US National Science Foundation).

Table 1. : ***Chronology of some important developments in biotechnology***

Activity	Year
Yeasts used to make wine and beer.	Before 6,000 BC
Yeasts used for making leavened bread.	About 4,000 BC
Sewage treatment systems using microbes developed/established.	About 1910 AD
Large scale production of acetone, butanol and glycerol using bacteria.	1912-14
Large scale production of penicillin.	1944
Mining of uranium with the aid of microbes (Canada).	1962
First successful genetic engineering experiments.	1973
Creation of the first hybridomas.	1975
The first biotech company.	1976
World's first 'test-tube baby', Louise Brown was born through *in vitro* fertilization.	1978
Marketing of human food of fungal origin (U.K.).	1980
The use of monoclonal antibodies for diagnosis approved in U.S.A.	1981
The first gene was synthesized. The first DNA synthesizer was developed.	1981
The first genetically engineered drug, human insulin produced by bacteria was manufactured and marketed by a US company. Production of the first monoclonal antibodies for diagnostics.	1982
Approval for the use of insulin produced by genetically engineered microbes (GEMs) in U.S.A. and U.K.	1983
The first transgenic plant was created—petunia plant was genetically engineered to be resistant to kanamycin, an antibiotic.	1983

Activity	Year
The chromosomal location of the gene responsible for the genetic disorder, Huntington's disease was discovered leading to the development of genetic sçreening test.	1983
Animal interferons, produced by GEMs approved for the protection of cattle against diseases.	1984
DNA fingerprinting is first used in a criminal investigation.	1985
The first field tests of genetically engineered plants (a tobacco) were conducted.	1986
Chymosin, an enzyme used in cheese making, becomes the first product of genetic engineering to be introduced into the food supply.	1990
Human genome project is launched.	1990
The first human gene therapy trial is performed on a four year old girl with an immune disorder.	1990
The gene implicated in the inherited form of breast cancer is discovered.	1991
Techniques for testing embryos for inherited disease are developed.	1992
First commercial approval for transgenic plant by the US government.	1994
First successful xenotransplantation trial is conducted, transplanting a heart from a genetically engineered pig into a baboon.	1995
Commercial cultivation of transgenic (genetically engineered) crop varieties.	1996
First commercial introduction of a 'gene chip' designed to rapidly detect variances in the HIV virus and select the best drug treatment for patience.	1996
Dolly, the sheep is cloned from a cell of an adult sheep.	1996
Embryonic stem cells are grown successfully, opening new doors to cell or tissue based therapies.	1998
A US company announced the successful cloning of human embryonic cells from an adult skin cell.	1999
Chinese scientists clone a giant panda embryo.	1999
Indian scientists and companies start producing recombinant vaccines, hormone and other drugs.	1999
The draft of human genome sequence has been published.	2002
Dolly, the cloned sheep that made headlines in 1997, is euthanized after developing progressive lung disease. Dolly was the first successful clone of a mammal.	2003
The FDA approves the first anti-angiogenic drug for cancer, AVASTIN® (bevacizumab).	2004
Researchers at the University of Georgia successfully produce a cow cloned from the cells of a carcass.	2005
On May 7, the one billionth acre of biotech seed is planted.	2005
The National Institutes of Health begins a 10-year, 10,000-patient study using a genetic test that predicts breast-cancer recurrence and guides treatment. Patients whose cancer is deemed unlikely to recur will be spared chemotherapy. The genetic test, Oncotype DXTM was developed by the biotech company Genomic Health and is already commercially available.	2006
Researchers at the J. Craig Venter Institute create the first synthetic cell.	2010

IMPORTANCE AND SCOPE OF BIOTECHNOLOGY

Biotechnology is rapidly emerging as area of activity which has potential to impact all domains of human welfare, ranging from food processing, environmental protection to human health.

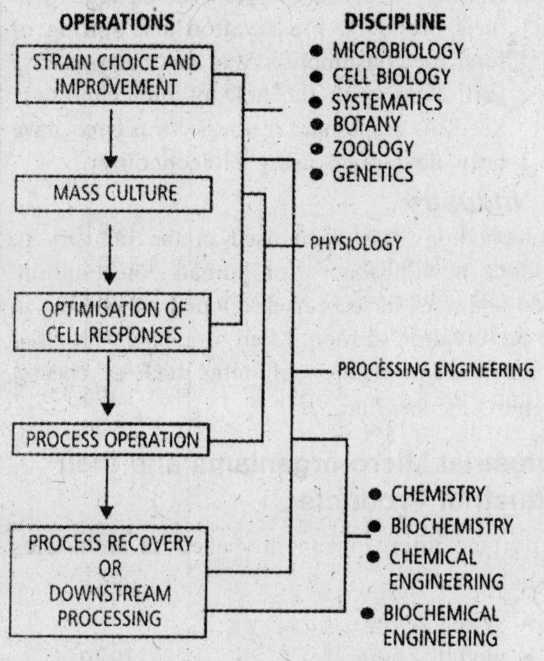

***Flow chart:** Biotechnological operations*

PRINCIPLES OF BIOTECHNOLOGY

- The two main techniques that gave birth to modern biotechnology are:
  - ❖ Genetic engineering refers to artificial synthesis, isolation, modification, combination, addition and repair of the genetic material (DNA) by man to alter the phenotype of the host organism to suit human needs.
  - ❖ Biochemical engineering involves processes that help the growth of desired microbes/eukaryotic cells in large quantities in a sterile medium (tissue culture technique) for the manufacture and multiplication of biotechnological product (antibiotics, vaccines, enzymes, medicines, hormones, etc.)

Conceptual development of the principles of genetic engineering

- The technique of genetic engineering includes formation of 'recombinant DNA (rDNA)', use of gene cloning and gene transfer.
- It checks this limitation and permits to isolate only one or a set of desirable genes without introducing undesirable genes into the target organism.
- The piece of DNA is not be able to multiply itself in the organism and can multiply only when it gets integrated into the genetic material of the recipient organism.
- The alien (foreign) piece of DNA is added to the recipient chromosome which possesses the ability to replicate.
- A specific DNA sequence present in the chromosomal DNA, called the origin of replication, is responsible for initiation of replication. An alien DNA linked with the origin of replication, can replicate and multiply itself in the host organism.
- It is also called the cloning, *i.e.*, forming multiple identical copies of any template DNA. Thus genetic engineering is alternately called recombinant DNA technology or gene cloning.
- If this sequence is present, the DNA will replicate. If it is absent, the DNA will not replicate.
- Therefore, the alien piece of DNA fails to replicate until it gets integrated with the chromosomal DNA having origin of replication.

Applications of Biotechnology

1. *Health and medicine*

- **Fighting infectious diseases :** Biotechnology is used extensively in the study of infectious diseases such as SARS (Severe Acute Respiratory Syndrome), influenza, etc. As a result more effective pharmaceuticals have been developed.

- **Development of vaccines and antibiotics :** Using technology, micro-organisms are used to develop antibiotics and vaccines to cure diseases. For example, bacteria *Bacillus polymysea* is used to produce polymyxin B antibiotic (used to cure urinary tract infections), fungus *Penicillium notatum* is used to produce penicillin (used to cure fever, pneumonia, etc.)
- **Treating genetic disorders :** Disease can occur when genes become defective due to mutations. With advance in biotechnology it will in the near future be possible to use gene therapy to replace an abnormal or faulty gene with a normal copy of the same gene. It may be used to treat ailments such as heart disease, inherited diseases such as SCID, Thallesemia.
- **In forensic science :** With the help of new techniques such as **DNA fingerprinting**, it has now become easy to identify criminals and have many other applications.

2. Environment

- **Cleaning up and managing the environment:** Cleaning up the environment using living organisms is called bioremediation. Naturally occurring, as well as genetically modified micro-organisms, such as bacteria and fungi, and enzymes are used to break down toxic and hazardous substances present in the environment.

3. Agriculture

Biotechnology has helped in production of crops with improved disease resistance; herbicide-tolerance and insecticide-resistance. Plants with improved nutritional value for livestock etc. have also been bred through biotechnology.

- **Control of pests :** One application of biotechnology is in the control of insect pests. The genetic make-up of the pest is changed by causing some mutations. These pests become sterile and cannot produce next generation.
- **Manufacturing and bio-processing :** With the help of new biological techniques it has become possible to grow on large scale, the plants that produce compounds for use in detergents, paints, lubricants and plastics etc.
- **Food and drinks :** With biotechnology, it has now become easy to process foods and their products. Preservation and storing of food for consumption later has become easy and cheap with the help of biotechnology. Seedless grapes and seedless citrus fruits have been developed using biotechnology.

4. Industry

Biotechnology has been used in the industry to produce new products for human consumption. Food additives have been developed which help in the preservation of food. Micro-organisms are used in the mass production of items such as cheese, yoghurt, alcohol, etc.

Industrial Micro-organisms and their Industrial Products

Important micro-organisms used in industries include:

- yeasts (fungi)
- moulds (fungi)
- bacteria
- filamentous bacteria (*actinomycetes*)

Microbes are used in the manufacture of several products. Some of these are:

- alcohol-containing beverages
- yoghurt (curd)
- proteins
- antibiotics and monoclonal antibodies
- vitamins, steriods and enzymes
- biogas

> The progress in gene manipulation and genetic engineering has introduced the use of cultured mammalian cells and 'hybridomas' in the industries. Hybridomas are created by fusion of cells belonging to organisms of different species.

Production of Alcohol—Containing Beverages

> **Fermentation**
>
> Fermentation is a process by which carbohydrates such as sugar are converted into alcohol.
>
> $$\text{Glucose} \xrightarrow[\text{enzyme}]{\text{yeast}} \text{Ethyl alcohol} + \text{Carbon-dioxide} + \text{ATP}$$
>
> Yeast and bacteria are capable of fermenting sugar to alcohol. Fermentation is an energy yielding process.

In the mid nineteenth century, Louis Pasteur showed that fermentation by the yeast *Saccharomyces cerevisiae* yields beer and buttermilk. Presently, yeast is being used on a large scale for fermentation.

Bakers use yeast to leaven (raise) dough to make bread. Yeast is grown on molasses and is also packed and sold. Yeast is used to raise cakes and bread while baking.

Alcoholic beverages are manufactured by fermentation of sugars by the yeast, *Saccharomyces cerevisiae*. It is called Brewer's yeast. The source of carbohydrate fermented by yeast gives the beverage its specific flavour. For example:

- Wine is obtained by fermentation of grapes. Grapes are fermented by *S. cerevisiae* and its soluble sugars (glucose and fructose) are converted into CO_2 and ethyl alcohol.
- Fermentation is carried out in large tanks called bioreactors.
- Barley malt is fermented to yield beer.

Steps taken for fermentation

(*i*) Fermenter or tank and the nutrient medium are sterilised by steam under pressure (autoclave).

(*ii*) The correct strain of yeast is selected.

(*iii*) The yeast is inoculated into the medium. Inoculation can be done in two ways:

- (*a*) Yeast can be grown as a layer on the surface of nutrient medium. This is called support growth system.
- (*b*) Cells or mycelia are suspended in a liquid medium. This is called suspended growth system.

(*iv*) Care is taken to maintain the right temperature, pH, oxygen and carbondioxide concentration.

(*v*) The medium is stirred and left to ferment.

(*vi*) The sugar in the medium gets fermented by enzymes released by yeast.

(*vii*) The fermented product is taken out (Fig. 1).

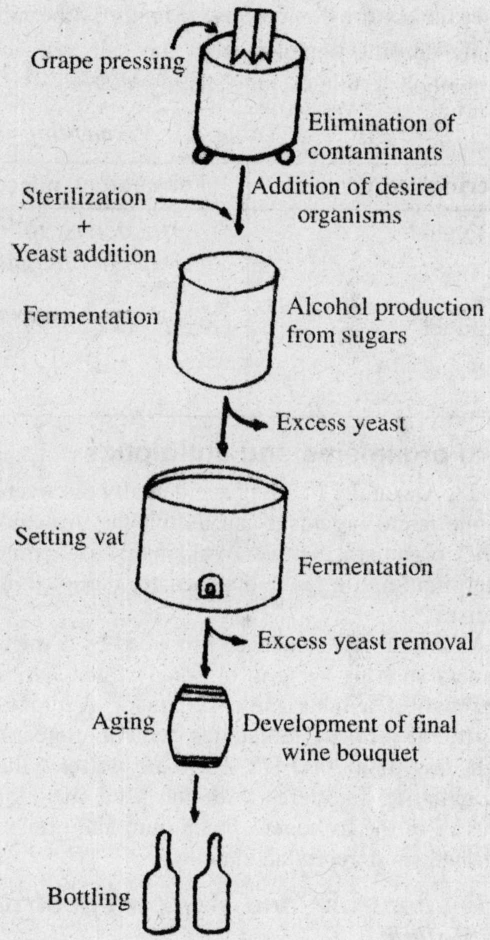

***Fig. 1. :** Obtaining wine from grapes by fermentation by Yeast*

Some alcohols manufactured by yeast fermentation are : Ethyl alcohol, butanol and glycerol. The same method also yields lactic acid and acetic acid (vinegar).

The yeast extract left after removal of the beverage can be used as animal feed. It is also a rich source of vitamins.

YOGHURT AND CHEESE MAKING

At home we add a bit of yoghurt (starter) to milk and it sets. The milk becomes yoghurt or curd due to the milk curdling enzymes released by the increasing population of bacteria, *Lactobacillus* present in the starter. (Table 2) On a commercial scale, making of yoghurt as well as cheese utilises Rennet tablets for this purpose. Rennin is the milk curdling enzyme obtained from the calf stomach. This method is not so popular any more.

Whether by bacteria or by "rennin" when milk is 'curdled', milk protein casein, separates from the liquid which is called whey. Lactic acid bacteria convert lactose in the milk into lactic acid which lowers the pH. Lowered pH causes souring which is essential for preservation.

Butter can be made by violently shaking (churning) sour milk. The fat globules separate and form butter. A starter culture of *Streptococcus cremosis*, *Leuconostoc* is added to the milk when butter, yoghurt or cheese are made.

Table 2. : ***Fermenting microbes used for dairy products***

Fermented product	Fermenting micro-organism	Description
Yoghurt	*Streptococcus thermophilus* and *Lactobacillus bulgarians*	Product made from low or non-fat milk and stabilisers like gelatin added.
Butter	*Lactococcus lactis*	Cream is incubated till desired acidity achieved followed by churning, washing and salting.

Micro-organisms and antibiotics

In 1928, Alexander Fleming accidentally discovered that one micro-organism can inhibit the growth of another organism. Selman Waksman 1942 coined the term antibiotic (anti: opposed to, biotic: living organism).

Antibiotic is a substance produced by a micro-organism such as bacteria or fungi which inhibits the growth of another micro-organism. Antibiotics are generally small molecules with a molecular weight less than 2000 Da. They are not enzymes. The antibiotic interferes with the vital metabolic processes of the pathogenic bacterium and prevents its growth and reproduction.

Wide-spectrum and narrow-spectrum antibiotics

Modern medicines have found a specific antibiotic for almost every different pathogen. *Streptomyces* bacterium yields some of the most widely used antibiotics like Chloramphenicol, Erythromycin, Tetracycline etc. These are called 'broad spectrum antibiotics' and can be used against more than one kind of pathogenic bacterium. Streptomycin and Penicillin are narrow spectrum antibiotics used against few pathogenic bacteria.

Drawbacks of antibiotics

Use of antibiotics was a big step in curing infectious diseases—a safe, sure and relatively inexpensive cure. But even now we find many people suffering from bacterial diseases. The reasons for this are:

1. Some people are allergic to a particular antibiotic.
2. Some disease causing bacteria undergo mutation and become resistant to a particular antibiotic to which they were earlier sensitive.

Sources of antibiotics

Some of the common antibiotics and their source organisms are given in table 3.

Table 3. : ***Major antibiotics and their sources***

Antibiotic group	Source
Tetracyclin	*Streptomyces sp*
Chlorotetracycline	*Streptomyces auriefaciens*
Chloramphenicol	*S. venezuelae*
Cycloheximide	*S. griseus*
Streptomycine	*S. griseus*
Cephalosporin	*Cephalosporium acremonium*
Penicillin	*Penicillium chrysogenum*

VACCINATION

In 1790, Edward Jenner observed that milkmaids did not get smallpox as they were exposed to a milder disease cowpox. Jenner infected a boy with cowpox germs and after two months with small pox germs. The boy did not get small pox. Jenner proposed that if mild or attenuated (weakened) germs were introduced into the body, they would not cause the disease. He gave the term vaccine (latin *vacca* : cow) or vaccination, for the weakened germ and its protective inoculation.

Today, the principle of vaccination has been extended to prevent attack of many diseases. When vaccines are made from attenuated disease causing bacteria, they are termed "first generation vaccines". The "second generation vaccines" have been produced by genetic engineering or recombinant DNA technology about which you shall study in the next section. Second generation vaccines for Hepatitis B virus and *Herpes* virus are already in use. Vaccines synthesised from chemicals are termed "third generation vaccines".

PRODUCTION OF VITAMINS

Vitamins are nutrients required in very small amounts for essential metabolic reactions in the body. Vitamins are produced using biotechnology. Vitamin C was the first vitamin to be produced during a fermentation process by using bacteria. B_{12} or cyanocobalamin and B_2 or Riboflavin were obtained from liver extract. The production of B_{12} involved fermentation by propionic bacteria. In nature B_2 is found in cereals, vegetables and yeast but the yield of B_2 can be enhanced hundred to three hundred fold by using microbes.

PRODUCTION OF BIOGAS

Biogas is a new conventional source of fuel. Its use can save fossil fuel (coal, kerosene, petrol etc.) which are fast getting depleted.

Biogas is made from organic waste including faecal matter. Cowdung or faeces have lignocellulose. The energy used as fuel comes from methane (CH_4). Cowdung forms the primary source of biogas. In India, cowdung is available in plenty in villages and small scale methane generating plants have been designed.

Any biodegradable (which can be decomposed by bacteria) substance can be fermented anaerobically (in the absence of oxygen) by methane-producing (methanogenic) bacteria. Cowdung or faeces are collected and put in a biogas digester or fermenter (a large vessel in which fermentation can take place). A series of chemical reactions occur in the presence of methanogenic bacteria (CH_4 generating bacteria) leading to the production of CH_4 and CO_2.

While generating biogas, few fermentation parameters have to be maintained. These are as follows:

1. Fermentation should be in an anaerobic environment and no free Oxygen should be present.
2. pH in the fermenter should be close to neutral, around 6.8 to 7.6.
3. Methanogenic bacteria are to be used for fermentation.

Several kinds of reactors have been designed. One side of the reactor is for input, that is, for introducing cowdung or faecal matter into the reactor. Other side of the reactor has an outlet for removal of biogas: What is left behind is called slurry. Gas gets stored above the slurry level. Slurry forms excellent manure.

Advantages of biogas

1. Biogas is a fuel used to cook food, light lamps etc.
2. Slurry left after biogas production forms a soil conditioner (manure).
3. Biogas is much cheaper than LPG (Liquefied Petroleum Gas) which we commonly use these days in our houses.

GENETIC ENGINEERING

An engineer fixes a machine to make it work efficiently. Body is like a machine and genes, the nucleotide sequences in DNA have the information for products to run this machine. With progress in molecular biology, techniques have been developed by which a scientist can now manipulate genetic material, replace genes or replace gene products in the body, make identical copies of these genes and store them in a gene library. This is called genetic engineering.

Importance of genetic engineering

You know that diabetes mellitus is a genetic disorder. A diabetic patient lacks a gene which has the information for synthesis of insulin, so such a person cannot secrete insulin. Take another example. A person suffering from Thalassemia lacks the gene for haemoglobin and can survive only through frequent blood transfusions. A person suffering from sickle cell anemia has an altered gene whose product makes the red blood corpuscles abnormal on exposure to oxygen because they contain faulty haemoglobin.

Human suffering from genetic disorders such as those cited above have now hope in genetic engineering. Genetically engineered copies of DNA can be produced and stored in gene libraries to be used when required.

In the previous sections of this lesson you have studied about the use of microbes to produce various products on a commercial scale. Currently bacteria are being genetically manipulated to act as biological factories to produce various kinds of proteins such as enzymes, hormones, antibodies through genetic engineering. Researchers have isolated genes which can be used to produce effective vaccines.

Researchers have also developed bacterial strains, through genetic manipulation, which can degrade harmful environmental pollutants.

MULTIPLE CHOICE QUESTIONS

1. Who discovered the first antibiotic?
A. A. Fleming B. W. Flemming
C. S. Waksman D. Louis Pasteur

2. The first antibiotic discovered was
A. penicirin B. chloromycetin
C. cephalosporin D. streptomycin

3. Diastase was identified for the first time by
A. Hansen B. Pasteur
C. Payen & Persoz D. Kuhne

4. Biosensors are
A. enzymes used in industries
B. analytic devices used to monitor the product of biological process
C. used to detect abnormal toxins in body
D. B and C

5. Terramycin is obtained from
A. *Streptomyces aurefaciens*
B. *Streptomyces venezuelas*
C. *Streptomyces griseus*
D. *Streptomyces ramosus*

6. What are important products of yeast fermentation?
A. Ethanol and acetic acid
B. Butanol, glycerol and pyruvic acid
C. Phenylethanol, ethylalcohol and caproiacid
D. All of the above

7. Following is used in the production of leavened bread
A. *Aspergillus* B. *Rhizopus*
C. *Saccharomyces* D. None of the above

8. Which one is the most important accessory chemical processes?
A. Petroleum B. Rubber
C. Ether D. None of the above

9. Which one of the following acids was produced by fermentation?
A. Citric acid B. Lactic acid
C. Acetic acid D. All of the above

10. Which one of the following enzyme is secreted by yeast for fermentation?
A. Enolase B. Invertase
C. Zymase D. Dehydrogenase

11. Which one of the following is not an antibiotic?
A. *Streptomycin* B. Citric acid
C. *Griseofulvin* D. Cephalosporin

12. In genetic engineering enzyme used is
A. topoisomerase
B. ligase
C. replicase
D. restriction endonuclease

13. To obtain virus free plants, which among the following is preferred?
A. micropropagation
B. shoot tip culture
C. root tip culture
D. grafting

14. The organism used in alcoholic fermentation is
A. *Penicillium* B. *Pseudomonas*
C. *Aspergillus* D. *Saccharomyces*

15. Restriction enzymes are used in genetic engineering because
A. they can cut DNA at specific base sequence
B. they are proteolytic enzymes which can degrade harmful proteins
C. they are nucleases that cut DNA at variable sites
D. they can join different DNA fragments

16. Barley malt is used as raw materials for the preparation of
A. beer B. amylase
C. cheese D. None of these

17. An enzyme useful in dissolving blood clots in heart attack patients is
A. thrombin
B. tissue plasminogen activate
C. progesterone
D. cholesterol

18. Insulin mainly obtained from
A. calf stomach
B. pancreas of cattle and pigs
C. steroid conversion
D. enzymatic reaction

19. The first organic acid to be produced industrially by fermentation was
A. citric acid B. butyric acid
C. lactic acid D. acetic acid

20. One of the following vaccines is produced through genetic engineering
A. typhoid B. tetanus
C. small pox D. hepatitis B

21. The micro-organism grown on molasses and sold as a food flavouring substance is
A. *Saccharomyces*
B. *Rhizopus*
C. *Acetobacter*
D. *Lactobacillus*

22. The following are the product of fermentation except
A. bread B. beer
C. dosa D. butter

23. Small pox vaccine was developed by
A. Jenner B. Pasteur
C. Flemming D. Funk

24. When a body itself develops antibodies against a particular disease, it is called
A. passive immunity
B. active immunity
C. acquired immunity
D. induced immunity

25. Vaccines produced by recombinant DNA technology are known as
A. Ist generation B. 2nd generation
C. 3rd generation D. 4th generation

ANSWERS

1	2	3	4	5	6	7	8	9	10
A	C	B	D	B	D	C	A	D	C
11	**12**	**13**	**14**	**15**	**16**	**17**	**18**	**19**	**20**
B	D	B	D	A	A	B	B	C	D
21	**22**	**23**	**24**	**25**					
A	D	A	C	B					

2

Biomolecules

BIOMOLECULES

About 5000 chemicals occur in a cell. About 1000 chemical reactions are being carried out at any time in a cell. All molecules or chemicals functional in living organisms, especially the organic ones, are known as biomolecules. In animals, the cells are also bathed in an extracellular fluid of chemicals. The collection of different types of chemicals present in cells is called cellular pool. It has two categories of chemicals, inorganic and organic. Inorganic substances are salts, mineral ions and water. The common organic substances are carbohydrates, lipids, amino acids, proteins, nucleotides, hormones and vitamins. Cellular molecules exist in two phases, aqueous (true and colloidal solution) and non-aqueous (chemicals as constituents of cellular structure like membrane, cell wall) phases. The cells maintain their specific cellular pool with the help of selectively permeable plasma membrane (selective intake and elimination of specific molecules).

Micromolecules and Macromolecules

Micromolecules, they are small sized, simple molecules of low molecular weight, *e.g.*, water, minerals, sugars, amino acids and nucleotides. Macromolecules are large sized complex molecules of high molecular weight which are commonly formed through condensation or polymerisation of repeated units of micromolecules, *e.g.*, polysaccharides, lipids, proteins, nucleic acids.

Inorganic Substances

They include water, salts and mineral ions.

Special Points to Note

Ions in Cellular pool

Elements	***Functions***
Molybdenum	Cofactor for certain enzymes in important metabolic reactions; Nitrogen fixation, aldehyde oxidation, and nucleic acid metabolism
Cobalt	Constituent of Vit. B_{12} and required for synthesis of phytohormone and auxin
Copper	Constituent of plastocyanin and cofactor of respiratory enzymes
Iodine	Constituent of thyroxine
Boron	Required by plants for sugar translocation
Zinc	Required for synthesis of phytohormone and auxin and acts as cofactor for certain enzymes
Manganese	Cofactor of certain enzymes
Iron	Constituent of haemoglobin, cytochromes and myoglobin of muscle cell; Required for maturation of chlorophyll
Magnesium	Constituent of chlorophyll; Cofactor of enzymes for glycolysis; Activates enzyme, ATPase
Sulphur	Constituent of CoA; biotin, thiamine (vitamins) and amino acids (methionine, cystine, cysteine)
Phosphorus	Constituent of lipids, nucleic acids and proteins

Calcium	Constituent of middle lamella of plant cells (calcium pectate); forms matrix of animal bones; Cofactor for blood coagulation enzymes; Forms exoskeleton of invertebrates – $CaCO_3$ in molluscan shells
Potassium	Cofactor for ATPase enzyme

Note:

1. Only six elements (carbon, hydrogen, nitrogen, oxygen, phosphorus and sulphur) make up 98% of the body weight of an organism. They are called protoplasmic elements. You can remember the acryonym CHNOPS. However, CHNO constitute 95% of human body weight. CHO in cellular materials is about 93%; in plants these form the framework *i.e.*, the cells wall, membrane.
2. Small organic molecules in living things like sugar, amino acids, fatty acids and nucleotides have a 'carbon backbone'. Carbon is most common in all living things.
3. Major elements of cellular pool (other than CHO) are as follows :

   (*a*) *Metals*

Calcium	
Potassium	Less than 5%
Sodium	
Magnesium	

   (*b*) *Non-metals*

Nitrogen	
Phosphorus	2%
Chlorine	
Sulphur	

   Trace elements of cellular pool include Copper, Zinc, Cobalt, Manganese, Molybdenum and Chromium.

Macromolecules, The giant polymers

1. Macromolecules are constructed by the formation of covalent bonds between smaller molecules called monomers. Macromolecules include polysaccharides, proteins and nucleic acids.
2. Macromolecules have specific, characteristic three-dimensional shapes that depend on the structure, properties and sequence of their monomers. Different functional groups give local sites on macromolecules specific properties that are important for their biological functioning and interactions with other macromolecules.
3. Monomers join by condensation reactions, which lose a molecule of water for each bond formed. Hydrolysis reactions use water to digest polymers into monomers.
4. Both covalent and non-covalent linkages are found between the various classes of macromolecules.
5. Glycoproteins contain an oligosaccharide 'label' that directs the protein to the proper cell destination. The carbohydrate groups of glycolipids are displayed on the cell's outer surface, where they serve as recognition signals.
6. Weaker forces bind a catalytic protein (enzyme) to a reactant. Hydrophobic interactions bind cholesterol to the protein that transports it in the blood.

WATER

1. Life evolved in water and living things contain 70-90% of water.
2. On an average human body contains 66% of water whereas some marine organisms have high concentration of 96%. It is highest in human embryo of about 95%.
3. Water concentration is directly related to age and metabolic activity of an organism.
4. **Uniquensess of water molecule**
   * Molecule is polar in nature
   * With weaker hydrogen bond than covalent bonds they help molecules of water to cling together
   * High boiling point (100°C) and low freezing point of 10°C (without hydrogen bonds, water would have boiled at –80° and frozen at –100°)
   * Temperature of water (liquid) rises and falls slowly than other liquids
   * Universal solvent for all chemical reactions

* High heat of vaporization (1g hot water needs 540 calories to get converted to steam).

CARBOHYDRATES

1. Carbohydrates consist of carbon, hydrogen and oxygen in molar ratio of 1 : 2 : 1. They are the energy sources for cells and form structural elements too.
2. Simplest carbohydrates include simple sugars (monosaccharides) having six carbon atom and serve C-H bonds-the energy storing bonds.
3. Some simple sugars are *glucose*, *fructose* and *galactose* too.
   (*i*) **Glucose (grape sugars or dextrans):** most widespread of all sugars found in blood of animals and produced by photosynthesis in plants.
   (*ii*) **Fructose (fruit sugar):** most common in fruits and is sweetest of all natural sugars.
   (*iii*) **Galactose**: a constituent of agar agar.
4. **Some important carbohydrates**
   Ascorbic acid: a sugar acid (vitamin C), naturally occurs in tomato and citrus fruits, gets easily destroyed by cooking and exposure to air and light.
   Raffinose: a trisaccharide of glucose, fructose and galactose residues.
   Insulin: a polymer of fructose found in the roots of *Dahlia.*
   Mannitol: an alcohol of mannose found in seeds and ivory nut.
   Xylans and arobans: form hemicellulose in cell wall of plants.
   Lactose: milk sugar (galactose + glucose)
   Sucrose: nonreducing commercial or canesugar.
   Arabinose: a component in wheatbran.
   Maltose: a reducing sugar (glucose + glucose) also known as malt sugar.
   Ribulose: an important intermediate in photosynthesis.
   Galactose: a constituent of gum arabic and brain glycolipids.
   Fucose: found in marine algae (sea weeds).
5. Monosaccharides link to form disaccharides, join together by dehydration and form long polymers called polysaccharides.
6. The most common polysaccharides in the living things are starch (in plants), glycogen (= animal starch) and cellulose (structural compound in plants). Each of these is a polymer of glucose.
7. Cellulose is a structural polysaccharide having 6000 units of β glucose.
8. Like cellulose, chitin (with additional nitrogen group) is a polymer of glucose with similar linkages undigestible by human but digestible by certain other animals like ruminants, snails, termites because of their symbiotic associations with certain micro-organisms present in their gut.
9. The Starch with the simplest structure is amylose (having 1000 or more glucose units).
10. Carbohydrates are easier to oxidize than fats, thus get used up first in diet.
11. Disaccharides include :
    * Maltose...(glucose + glucose)
    * Sucrose (cane sugar or table sugar)...(glucose + fructose)
    * Lactose (milk sugar)...(glucose + galactose)
    * Ribose and deoxyribose are two 5-carbon sugars (pentoses) found respectively, RNA and DNA.

12. Characteristic properties of polysaccharides :

Type	*Location*	*Molecular structure*	*Functions*
Cellulose	Cell wall of all plants	Linear long chains 6,000 or more units of β glucose	For tensile strength, mechanical support, permable to water and solutes; food for ruminants and micro-organisms.
Starch	Storage food of plants	Occurs in two forms: amylose (straight chain) and amylopec-tin (branched chain)	Energy source of plants (from photosynthesis).

Type	*Location*	*Molecular structure*	*Functions*
Glycogen	In liver and muscles of animals and also in fungi	Highly branched chain	Source of energy in metabolism and respiratory fuel.
Chitin	Exoskeleton of insects and crustaceans; cell wall of bacteria	Linear long chain like cellulose (N-acetyl glucosamine)	Gives shape, strength and retains water.

LIPIDS

1. Lipids are not soluble in water because of lack of any polar group. Common forms of lipids are:
   (*i*) fats and (*ii*) oils. (*iii*) Other forms include *phospholipids, steroids* and *waxes*
2. Uses of Lipids (*i*) Energy reserves by organisms (*ii*) Insulation (in form of blubber in skin of whales)
3. Fats and oils are triglycerides (Abbreviated as TG) (one glycerol + three fatty acids). These are considered macromolecules of lipids.

$$\textbf{3 Fatty Acids + Glycerol} \underset{\text{hydrolysis}}{\overset{\text{condensation}}{\rightleftharpoons}} \textbf{fats + 3 molecules of water}$$

4. Lipids are used as 'long term energy storage compounds' in animals and plants because of high concentration of C–H bonds.
5. Fatty acids are either saturated (with nil double bonds between carbon atoms) or unsaturated (with double bonds).
6. Lipid bilayers are the basic framework of all biological membranes.
7. On an average, fats yield 9 kcal of chemical energy per gram of fat as compared to 4 kcal/g from carbohydrates.
8. Animal fats contain more calories than vegetable fats.
9. Waxes, such as bees-wax are esters of long chain alcohols. They do not contain glycerols.
10. Examples of waxes are
    * *Animals:* ear wax (for trapping dust and dirt) and honeybee wax (for making honeycomb)
    * *Plants:* cutin (forming cuticle)
11. **Types of fatty acids:**

Common Name	Number of Carbon Atoms	Molecular Formula
SATURATED		
Butyric	4	C_3H_7 COOH
Caproic	6	C_5H_{11}COOH
Caprylic	8	C_7H_{15}COOH
Capric	10	C_9H_{19}COOH
Lauric	12	$C_{11}H_{23}$COOH
Myristic	14	$C_{13}H_{27}$COOH
Plamitic	16	$C_{15}H_{31}$COOH
Stearic	18	$C_{17}H_{35}$COOH
Arachidic	20	$C_{19}H_{39}$COOH
Behenic	22	$C_{21}H_{43}$COOH
UNSATURATED		
One double bond		
Myristoleic	14	$CH_3(CH_2)_3CH{=}CH(CH_2)_7COOH$
Plamitoleic	16	$CH_3(CH_2)_5CH{=}CH(CH_2)_7COOH$
Oleic	18	$CH_3(CH_2)_7CH{=}CH(CH_2)_7COOH$

Common Name	Number of Carbon Atoms	Molecular Formula
Two double bonds		
Linoleic	18	$CH_3(CH_2)_4CH{=}CHCH_2CH{=}CH(CH_2)_7COOH$
Three double bonds		
Linolenic	18	$CH_3CH_2CH{=}CHCH_2CH{=}CHCH_2CH{=}CH(CH_2)_7COOH$
Eleostearic	18	$CH_3(CH_2)_3CH{=}CHCH{=}CHCH{=}CH(CH_2)_7COOH$
Four double bonds		
Arachidonic	20	$CH_3(CH_2)_3(CH_2CH{=}CH)_4(CH_2)_3COOH$

12. Phospholipids have polar heads but non-polar tails.
13. Plasma membrane is essentially a phospholipid bilayer.
14. Steroids are lipids that consists of four fused carbon rings. Cholesterol is the precursor of other steroids like aldosterone (regulate Na^+ content of the blood) and sex hormones of males and females.
15. Biologically important pigments of lipid nature are:
    (*a*) Chlorophyll in plants and
    (*b*) Retinol pigments
16. The pigments of long chain lipids are called terpenes *e.g.* rubber:
17. Prostaglandins are modifed fatty acids with two non-polar tails attached to 5-carbon ring. They act as local chemical messengers and stimulate smooth muscle to contract and relax.
18. **Common lipids**

(i) Diosgenin	extracted from *Dioscorea* which is useful for its antifertility properties.
(ii) Terpenses	form an important constituent of camphor, methanol, gibberellins, ABA and carotenes.
(iii) Sterols	an important constituent of many hormones like testosterones, progesterone and cortisones. They form most of the anabolic stimulants.
(iv) Paraffin wax	a mixture of saturated hydrocarbon obtained from crude petroleum and is useful in preparing candles, polishes and consmetics.
(v) Suberin	a component in cell wall of plants which reduces water loss.
(vi) Cutin	a waxy substance that cover the epidermis of xerophytes.
(vii) Lecithin	ester of fatty acid found in egg yolk and in brain *Lecithin is an important constituent of protoplasm.*
(viii) Cerebrosides	an important constituent of nervous system.
(ix) Gangliosides	a special class of lipids occuring in the lysosomes of brain cells.
(x) Myelin	a glycolipid that ensheaths nerve fibre and chloroplast.
(xi) Lanolin	a grease extracted from raw sheep wool and is used as lubricant, leather preservative and in cosmetics.

PROTEINS

1. Discovered by dutch chemist, G.J. Mulder, (1939) while investigating coagulating substances in milk and egg. Later J.J. Berzelius coined the term 'proteins'.
2. Nitrogenous macromolecules of amino acids.
3. Nitrogen content around 16% of the molecular weight of proteins.
4. Protein molecules are very large with high molecular weight ranging from 5000-25,000,00.
5. Form the frame work of protoplasm.
6. Amino acids are building blocks of proteins, perhaps first molecules formed in the atmosphere of the earth.

7. An amino acid contains
   (i) An amino group ... $-NH_2$
   (ii) A carboxyl group ... $-COOH$
   (iii) A hydrogen atom ... $-H$
8. All of these groups are bonded to a central carbon atom as shown.

```
      H
      |
  H2N–C–COOH
      |
      R
```

9. The chemical nature and identity of each amino acid are denatured by the radical, R.
10. The functions and shape of a protein are affected by sequence of 20 types of amino acids with an amino group – NH_2, a carboxylic acid group – COOH, a hydrogen atom each attached to carbon located next to – COOH group and a side chain R which actually varies from one amino acid to other with either hydrogen or aliphatic, aromatic or heterocyclic group.
11. A polypeptide is a long chain of amino acid joined by a covalent-peptide bond.
12. It is formed in presence of an enzyme by the condensation of an amino group of one molecule with the carboxylic acid group of a second molecule.
13. The 20 most common amino acids and their side groups (R) grouped into five groups (based on the side groups)

Amino acids	*Examples*
Non-polar	– alanine, valine, leucine, isoleucine
Polar uncharged	– glycine, serine, threonine, asparagine
Ionizable	– glutamic acid, aspartic acid, histidine, lysine, arginine
Aromatic	– phenylalanine, tryptophan, tyrosine
Special function	– proline, methionine, cysteine

14. Amino acids are classified into three groups as per their reaction in a solution :
   (*a*) **Neutral:**
   Largest group divided into many types e.g. asparagine
   (*b*) **Acidic:**
   Monoamino dicarboxylic acid, e.g. glutamic acid, glutamine and aspartic acid
   (*c*) **Basic:**
   Diamino monocarboxylic acid, e.g. arginine, lysine and hydroxylysine
15. Neutral amino acids can be further classified into:
   (i) Aliphatic: e.g. Glycine, valine, leucine, isoleucine, serine and threonine
   (ii) Aromatic: e.g. phenylamine, tyrosine
   (iii) Heterocyclic: e.g. tryptophan, histidine and proline
   (iv) S-containing: e.g. methionine, cysteine and cystine
16. Some amino acids do not occur as structural components of proteins :

Types	*Examples*
1. β-alanine	coenzyme
2. tri-iodo and tetra- iodo-thyronine	thyroid
3. taurine	bile salts
4. ornithine and citrulline	metabolities of urea cycle
5. DOPA	precursor of melanin

17. Aminoacids are classified into three types, as per their synthesis in the body of organisms :
   Essential
   Cannot be synthesized by organisms in the body and are obtained from plants e.g. valine, leucine isoleucine, phenylalanine, threonine, lysine, tryptophan, methionine.
   Non-essential
   Can be synthesized by the organism and may not be the requisite components of diet *e.g.* serine, tyrosine, aspartic acid, cystine, cysteine, glutamic acid, proline, alanine, hydroxyproline, hydroxylysine.
   Semi-essential
   Required by growing children and lactating, pregnant mothers *e.g.* histidine, arginine.

18. Sulphur containing amino acids are: (*a*) cystine (*b*) cysteine and (*c*) methionine.
19. Alcoholic amino acids are: (*a*) serine and (*b*) threonine.
20. Aromatic amino acids include: (*a*) tyrosine (*b*) tryptophan and (*c*) phenylalanine.
21. Heterocyclic amino acids are: (*a*) proline (*b*) histidine (*c*) hydroxyproline and (*d*) tryptophan.
22. Semi-indispensable amino acids (slow formation in humans) are: (*a*) Arginine and (*b*) Histidine.
23. Four levels of structure of protein are :
    (i) The primary structure is a linear sequence of amino acids, e.g. ribonuclease enzyme
    (ii) The secondary structure contain α (alpha) helices and β (beta) sheets held in place by hydrogen bonding between peptide bonds e.g. silk fibroin (major protein in silk); proteins of tendons and muscles
    (iii) The tertiary structure is the final folding and twisting of polypeptide and held by hydrophobic (interactions resulting in globular structures e.g. myoglobin (consists of globin with 153 amino acid residues and one heme group)
    (iv) The quaternary structure results from the association of polypeptides in aggregates held together by non-covalent bonds e.g. haemoglobin (four globins attached to a heme group)

NUCLEIC ACIDS

1. The nucleotides (containing C, H, O, N and P) are monomers of nucleic acids. They consist of three unit molecules
   (i) a phosphate (Phosphoric acid)
   (ii) a pentose sugar and
   (iii) an organic nitrogen containing base.
2. Nucleotides are linked together by phosphate groups. The phosphate group of one nucleotide binds to hydroxyl group of another by what is called Phosphodiester bond.
3. ATP (adenosine triphosphate) is a nucleotide used in cells. It gives energy for synthesis

   ATP $\longleftrightarrow$ ADP + P + Energy.
4. Nucleic acids include the largest molecule formed by organisms.
5. Nucleic acids were first isolated by a Swiss scientist, F. Miescher (1869) from the nuclei of pus cells (dead white blood cells or leukocytes) who named it nuclein.
6. Thereafter, it was found that nuclein had acidic properties and then it was renamed as nucleic acid by Altmann in 1899.
7. Unique among the macromolecules, the nucleic acids can produce precise copies of them (replication).
8. Nucleic acids consist of a number of monomer units called nucleotides.
9. Thus a nucleic acid is a chain of ribose or deoxyribose sugars linked together by phosphodiester bonds (the phosphate group of one nucleotide binds to the hydroxyl group of another) and with an organic base.
10. Each nucleotide is formed from one molecule of pentose sugar, one phosphoric acid and one of the nitrogenous bases.
11. Within the nucleotide the combination of base with the pentose sugar constitutes a nucleoside For example,

    Nucleoside : Adenosine = Adenine + ribose
    Nucleotide : Adenosine = Adenine + ribose + phosphate
12. Some important other nucleotides are:

    AMP : Adenosine monophosphate
    CMP : Cytidine monophosphate
    CDP : Cytidine diphosphate
    TMP : Thymidine monophosphate
    GMP : Guanosine monophosphate
    GTP : Guanosine triphosphate
    UMP : Uridine monophosphate
    NAD : Nicotinamide adenine dinucleotide
    NADP : Nicotinamide adenine dinucleotide phosphate
    FMN : Flavin mononucleotide
    FAD : Flavin adenine dinucleotide
13. Nucleotides are linked together by means of phosphate-diester bridges with the pentose moiety. These bonds link the 3' carbon of pentose in one nucleotide to the 5' carbon in the pentose of the adjacent nucleotide.

14. Levine (1935) pointed out two classes of nucleic acids containing two different pentose sugars: (i) ribonucleic acid and (ii) deoxyribonuceic acid

15. Pentose sugars are of two types : ribose in RNA and deoxyribose in DNA. The only difference in that oxygen at 2' carbon is lacking in deoxyribose in DNA.

16. Purines and pyrimidines are the bases found in nucleic acids.

Bases	***Structure***		***Types***
Purines	Double ring	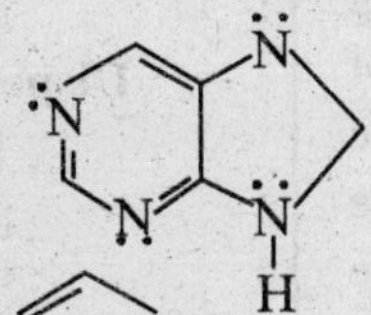	Adenine Guanine
Pyrimidines	Single ring	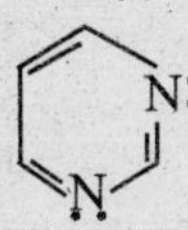	Cytosine Thymine Uracil*

* *In RNA, Uracil is found instead of thymine*

DNA

1. DNA of mitochondria and chloroplasts is not associated with histones.
2. In a deoxyribonucleic acid molecule, Amount of Purines = Amount of pyrimidines
   $$A + T = G + C$$
   This is called Chargaff's equivalence Rule (1950).
3. The ratio of $\frac{A+T}{G+C}$ is constant in any species.
4. DNA is made up of two long chains (strands) which are right-handed coiled, antiparallel (i.e., one strand runs in 5' to 3' direction and the other in the 3' to 5' direction) and complimentary to each other. They are held together by hydrogen bonds established between the bases of pairs.
5. Base pairing occurs among adenine (A), thymine (T), cytosine (C) and guanine (G).
6. Two hydrogen bonds are formed between A and T while three between C and G.
7. Watson and Crick Model of DNA
   (i) each nucleotide pair needs 3.4 Å space of double helix
   (ii) one complete turn of helix has of 10 nucleotide pairs, occupying 34 Å space.
   (iii) the diameter of DNA molecule = 20 Å.
   (iv) there is a fixed distance of 10.08 Å between the two sugar moeties in opposite strands thus only certain base pairs can fit into the structure.
   (v) although, DNA is double helical in all the organisms but single helical DNA has been found f X 174, M^{13} and G^1 phages of *Escherichia coli*. Thus, DNA is rarely single helical.
   (vi) If any long base-pair sequence exists at many different places throughout the DNA, it is called repetitive or repetitious DNA. The highly repetitive DNA is also known as satellite DNA. The repetitive DNA is not found in prokaryotes.
8. **Various forms of DNA**

   A-DNA: It contains 11 base pairs/turn base pairs are not perpendicular to axis but tilted

   B-DNA: It is usually right handed DNA containing 10 base pairs/turns

   C-DNA: It contains 9 base pairs/turns

   Z-DNA: It is a left-handed DNA

   Single-stranded DNA: Found in $\phi \times 174$ and and S-13 viruses
9. **Replication of DNA**
   (i) DNA replication is a complex process and is carried out by a multi-enzyme complex. It is often called replication apparatus or the replisome.

(ii) Before duplication, H-bonds are broken resulting in unwinding and separation of two strands of DNA molecules by enzymes known as helicases. Once the two strands are free, the base sequences of DNA are exposed to the surrounding nuclear fluid which contains all kinds of building blocks *i.e.,* deoxyribonucleoside triphosphates; deoxyadenosine triphosphate (d-ATP), deoxyguanosine triphosphate (d-GTP), deoxycytidine triphosphate (d-CTP) and thymidine triphosphate (TTP).

(iii) The bases of separated strands now attract their complementary bases towards them. The cellular enzyme, DNA polymerase that catalyses the synthesis of DNA, acts only on nucleoside triphosphates and it removes the two of their three phosphates, releasing tremendous amount of energy and forming nucleotides.

(iv) Nucleotides thus formed, join together to form a long chain on the original DNA strand. However, new DNA strands are synthesized in short segments, always in 5' to 3' direction.

(v) These segments then join together by the action of another enzyme called polynucleotide ligase or DNA ligase; discovered by Khorana *et al.* 1967.

(vi) The short DNA piece are called Okazaki fragments after their discoverer.

10. **Meselson and Stahl (1958):** Alternate hypothesis for DNA replication:

Conservative

Two strands of the parent DNA molecule act as template for new double stranded molecule.

The parent DNA molecule is preserved intact and goes into one daughter cell.

Semi-conservative

Two strands of parental molecule separate like the two side of a zipper.

Each act as the template for formation of new strand which then become bound to it to form a complete molecule.

Each daughter cell inherits a DNA molecule that is a hybrid, consisting of one new and one parental strand. It is the most acceptable theory.

Dispersive

The parental DNA is broken up into short segments and act as template for the formation of two new double helices which are thereafter somehow joined together.

RNA

1. Ribonucleic acid (RNA) is found in the nucleus as well as in the cytoplasm and its main function is protein synthesis but in many viruses, it functions as genetic material.
2. RNA molecules are single stranded, however, double stranded RNA is found in wound tumor plant virus, rice dwarf virus and reovirus.
3. A single strand of RNA folds back, due to which pairing of bases can occur easily.
4. The primary structure of RNA is similar to that of DNA except for the following differences:
   (i) Pentose sugar in RNA is ribose while in DNA, it is deoxyribose.
   (ii) RNA contains uracil as one of the pyrimidine bases while DNA contains thymine.
   (iii) RNA is generally single stranded while DNA is double stranded.
5. Cellular organisms have three types of RNA: rRNA, tRNA and mRNA.
   (A) Ribosomal RNA (rRNA) (Discovered by Jacob and Monad in 1961).
   (i) occurs in ribosomes and forms 80% of the total RNA in the cell.
   (ii) largest amongst RNAs.
   (iii) In eukaryotic cells, ribosomes contain four different types of rRNA namely 5S, 5.8S, 18S and 28S.
   (iv) In prokaryotic cells, the large sub-unit of ribosomes contains 23S and molecules

while their small sub-unit contains 16 S rRNA molecule.

(v) Ribosomal RNA gets synthesized in the nucleus by an enzyme called RNA polymerase I, although its 5 S component is synthesized outside the nucleus.

(B) Transfer RNA (tRNA)

(i) It is also known as S-RNA (soluble) or adaptive RNA.

(ii) It is about 10-15% of the total RNA of the cell.

(iii) Its molecule are the smallest ones with 75-85 nucleotides and function as adaptors for carrying a molecule to site of protein synthesis.

(iv) tRNA is folded to form a clover leaf like structure.

(v) There are at least 20 types of tRNA molecule in every cell, at least one corresponding to each of the 20 amino acids required for protein synthesis.

(vi) The tRNA is synthesized in the nucleus by an enzyme called RNA polymerase III presumably it might be synthesized in the nucleus.

(C) Messenger RNA (mRNA)

(i) It constitutes 5 to 10% of the total cellular RNA.

(ii) It transcribes the genetic information coded in the structure of DNA molecule and translates into protein.

(iii) It is heterogeneous because the size of its molecule varies greatly. On an average mRNA contains 900-1500 nucleotides in *E. coli.*

(iv) The average life span of some mRNAs is about two minutes after which the molecules are broken down by ribonuclease. However, it may survive for many hours or even days in eukaryotes.

(v) The mRNA is synthesized on DNA template in the nucleus by an enzyme called RNA polymerase II.

(vi) The process through which RNA is synthesized along the DNA template is called transcription. It was discovered earlier by Mulder.

Special Points to Note

1. 80% of dry weight of plants is carbohydrates.
2. Pentose and hexose are most common sugars in monosaccharides.
3. Glucose (= dextrose grape sugar) is universal sugar.
4. Ascorbic acid (Vit. C) is one of the most important acid sugars.
5. Most common sugar happens take be fructose (= levulose) found in fruits.
6. The most abundant structural polysaccharide: cellulose.
7. The most abundant nutrient polysaccharides: glycogen (animals) and starch (plants).
8. Natural silk is a polyamide while artificial silk is polysaccharide.
9. The term 'protein' was first coined by J.J. Berzelius (*Proteios* - Means primary). It was discovered earlier by Mulder.
10. Plants are able to synthesize all amino acids.
11. Simplest amino acid is glycine.
12. The term 'lipid' was first coined by Bloor (1943).
13. Oxidation of 1 gm of fats yields 9300 cals while 1 gm of protein or carbohydrate release 4100 calories.
14. **Abundance of some important elements (%):**

Elements	Earth's crust	Human body
Oxygen	46.60	65.0
Carbon	0.03	18.5
Hydrogen	0.14	9.5
Nitrogen	0.01	3.3
Calcium	3.63	1.5
Phosphorus	0.12	1.0

15. **Inorganic and organic compounds in active protoplasm:**

Substance	*Percentage*
Water	70-90
Protein	10.0
DNA	0.4
RNA	0.7
Lipid	2.5
Other organic compounds	0.4
Inorganic compounds	1.5

16. Griffith discovered the 'hereditary role of DNA' in 1928, with his experiment performed on *Diplococcus pneumoniae*.
17. Thereafter, Avery, Macleod and Mc carty (1944) received Noble prize when they provided the first evidence that DNA is the genetic material by repeating Griffith's transformation experiment.
18. Hershey and Chase (1952) further confirmed that 'DNA is the genetic material' while studying T_2 bacteriophage of *E. *coli.*
19. It is well established now that all the genetic information of living organisms is stored in the linear sequence of the four bases of DNA molecules.
20. Based on X-ray diffraction data of Wilkins and Franklin (1950); Watson and Crick (1953) proposed a double helical model of DNA.
21. Rapid cooling (quenching) of denatured DNA fixes it in a permanently denatured state.
22. Total amount of DNA/genome of an organism is called C value. It is characteristic of a species. It is expressed in picograms (pg) or base pairs (bp), 1 pg = 0.965×10^9 bp.
23. Important glycoproteins and their functions:

Glycoprotein		Function	% of carbohydrates
1.	Ceruloplasmin	Copper transport protein of blood plasma	7
2.	Erythropoietin	Hormone	33
3.	Fibrinogen	Blood clotting protein	5
4.	FSH	Hormone	4
5.	Glucose-oxidase	Enzyme	15
6.	Heptoglobin	Haemoglobin-binding protein of blood plasma	23
7.	Luteinizing hormone	Hormone	20
8.	Mucin	Mucus secretion	50-60
9.	Ovalbumin	Albumin in hen's egg	1
10.	Peroxidase	Enzyme	18
11.	Transferrin	Iron transport protein of blood plasma	6

24. Some important proteins and their amino acids :

Protein	Molecular weight	Amino acids
ACTH	4500	39
Insulin	5700	51
Cytochrome C	15,600	140
Trypsin	20.000	180
Haemoglobin	64,500	574

ENZYMES, VITAMINS AND HORMONES

Enzymes

1. With cells, a very large variety of chemical reactions take place. They are termed as metabolism. Every reaction in a metabolic pathway requires an enzyme generally globular proteins speeds up chemical reactions without getting permanently changed.
2. Till date more than 2000 enzymes have been recorded in living organisms.
3. About 70% of the enzymes of a eukaryotic cells are present in their mitochondria.
4. Viruses do not at all have any enzymes.
5. Enzymes are synthesized inside the living cells; some function inside the cells called endoenzymes and others outside the cell called exoenzymes or extracellular enzymes.
6. Enzymes start acting by
   (i) bringing two substrates together in correct orientation or
   (ii) by stressing particular chemical bonds of a substrate together in correct orientation and
   (iii) lowering the activation energy required for new bonds to form.
7. The term 'enzyme' was first coined by W. Kuhne (1878).

8. The first enzyme to be purified and crystallized was urease. It was achieved by J. B. Sumner in 1926, who had suggested the proteinous nature of enzymes.
9. Buchner (1897) got Nobel Prize for first isolating an enzyme zymase from yeast.
10. An enzyme has to be a protein, though it can have non-protein part also attached to it, called as holoenzyme or conjugated enzyme.

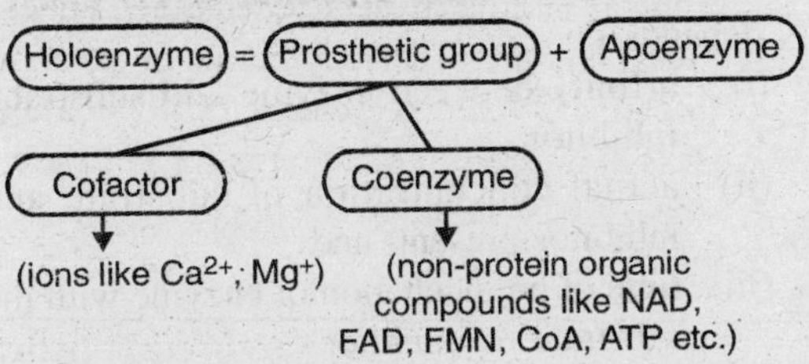

11. **Properties of enzymes**
   (*i*) ***Proteinaceous and colloidal nature:*** Enzymes are giant size protein molecular (molecular weight 1-1200 million) so they have low rate of diffusion and form colloidal system.
   (*ii*) ***Catalytic nature:*** Enzymes are catalysts. Catalytic power of an enzyme is measured by *turn over number*, *i.e.*, the number of substrate molecules converted into products per unit time when the enzyme is fully saturated with substrate.
   Turn over number of some enzymes:
   Lysozymes ...
   30
   Sucrase ...
   10,000
   Catalase ...
   5,000,000
   Carbonic anhydrase ...
   36,000,000
   (*iii*) ***Specificity:*** Enzymes are very specific to substrate molecules of a reaction.
   (*iv*) ***Thermolability:*** Enzymes are sensitive to heat being proteinaceous.
   (*v*) ***Reversibility of a reaction:*** Enzymes are capable of reversal in chemical reactions.
   (*vi*) ***Feedback inhibition:*** Some enzymes would catalyze the synthesis of small molecules (amino acids) in a number of steps. The enzyme catalysing the first step in this biosynthetic pathway is inhibited by the end product of the reaction (it accumulates in the cell and stops further synthesis of that substance). This regulatory mechanism is called feed back inhibition.
12. **Factors affecting enzyme activity**
   (i) **Temperature:** Enzymes being thermolabile, their activity increases repidly from 0°C-35°C. Above this temperature the rate starts declining. They get denatured at 60°C. For most of animals the optimum temperature is between 25°C-40°C.
   (ii) **pH:** Each enzyme acts at a specific pH *e.g.* lipase between pH 4.0–5.0; arginase at 10.0; pepsin at 1.5–1.6 and ribonuclease at 7.0 –7.5.
   (iii) **Substrate concentration:** For a known quantity of enzyme the reaction is directly proportional to substrate concentration, upto certain concentration after which the increasing concentration of a substrate does not increase the velocity of reaction.
   (iv) **Enzyme concentration:** The velocity of enzymatic reactions is directly proportional to enzyme concentration.
13. **Enzyme action:** Two major theories are proposed to explain enzyme action, namely
   (i) **Fischer's lock and key model:** Proposed by **Email Fisher** in 1898. An enzyme and a substrate unite at the active site more or less like a key fits a lock thus forming an enzyme substrate complex.
   (ii) **Koshland's induced fit model:** Koshland proposed in 1958 this is now being accepted. This theory explains the inducible change that is brought about at the active site of an enzyme by a substrate so as to form a suitable ES Complex.
14. **Enzyme Kinetics:** Rates of reaction under different experimental conditions
   (i) Donor Michaelis and Maud Menten (1913) demonstrated a mathematical

relationship also between velocity of enzyme reaction and concentration of substrate in a given time period.

(ii) The enzyme catalysed reactions consists of three steps:

$$E + S \xrightarrow{\text{Binding}} ES \xrightarrow{\text{Catalysis}} EP \xrightarrow{\text{Release}} P + E$$

E = Enzyme
S = Substrate
P = Product

(iii) Rate of reaction μ ES.

$$\frac{dP}{dt} \propto ES \quad \text{or} \quad \frac{dP}{dt} = K_{cal}[ES]$$

Here $\frac{dP}{dt}$ = rate of formation of product in given time interval, K_{cal} = Catalytic constant.

(iv) The rate of formation (V) of product can be calculated by Michaelis-Menten equation:

$$V = V_{max}\frac{[S]}{[S] + K_m}$$

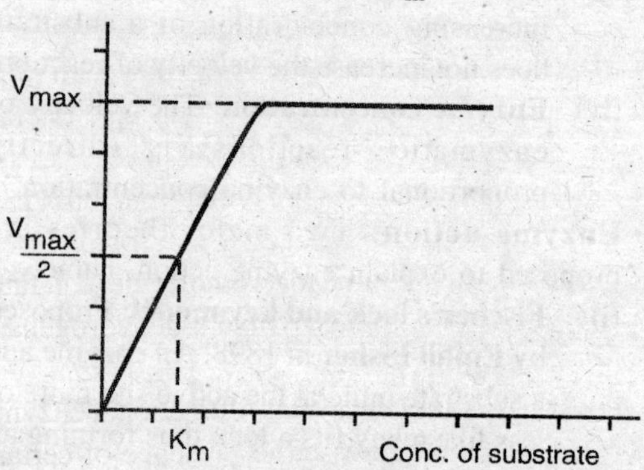

Where V_{max} = rate of reaction where enzyme is fully saturated with substrate.

K_m = Michaelis – Menten constant

(v) Michaelis-Menten Constant (K_m) is that concentration of substrate at which the rate of reaction is half maximal.

(vi) Maximal velocity (V_{max}) is that velocity at which the enzyme reaches a state of saturation due to presence of greater concentration of substrate.

15. **Enzyme Inhibitions**

A. Competitive and non-competitive inhibitions:

When the active site or catalytic site of an enzyme is occupied by a substance other than the substrate of that enzyme, its activity gets inhibited. This is a type of reversible inhibition. In such inhibition both enzyme substrate (ES) and enzyme inhibitor (EI) complexes are formed. The actual amounts of ES and EI depends on:

(i) affinity between enzyme and substrate/inhibitor,

(ii) actual concentration of substrate and inhibitor present, and

(iii) time of preincubation of enzyme with the substrate or inhibitor.

Competitive Inhibition	*Non-competitive inhibition*
1. Reversible	Irreversible or reversible
2. Inhibitor and substrate resemble each	It does not resemble other in structure
3. Inhibitor binds the active site	Inhibitor does not bind the active site
4. V_{max} is same	V_{max} lowered
5. K_m increased	K_m altered
6. Inhibitor cannot bind with ES complex	Inhibitor can bind with ES complex
7. Lowers the substrate affinity to enzyme	It does not change the substrate affinity for the enzyme
8. Complex is E-I	Complex is E-S-I or E-I

B. Allosteric inhibition:

(i) There is a mixed kind of inhibition when the inhibitor binds to the enzyme at a site other than the active site but on a different region in the enzyme molecule called allosteric site.

(ii) Allosteric inhibition does not follow the Michaelis-Menten hyperbolic kinetics, instead it gives sigmoid kinetics.

(iii) The regulatory substance regulate the enzyme activity through allosteric transformation of the enzyme.

(iv) They attach with the allosteric sites of the enzyme molecule so that the shape of enzyme is changed and as such it fails to join with the substrate or shows more affinity towards it.

(v) In the former case the enzyme action is inhibited while in the latter it is activated.

16. **The Operon Hypothesis:**

(i) Jacob and Monod (1961) suggested that protein synthesis is controlled by DNA segments and gave inducible operon model. They illustrated this in E. coli for catabolism of lactose.

(ii) Three enzymes are required for lactose catabolism. These are produced by three structural genes : cistron-Z, cistron-Y and cistron-a.

(iii) These genes are associated with control genes. These include:

(a) Regulator gene that produces some specific enzyme, the repressor which binds to the operator gene and suppresses or shuts off the synthesis of mRNA from operator gene.

(b) Promoter gene initiates the transcription of structure genes and controls the rate of mRNA synthesis.

(c) Operator gene receives the enzyme RNA polymerase.

(iv) Operon repressible model was given by Charles Yanofsky and coworkers while working on tryptophan metabolism in *E. coli*.

17. **Some Very Important Enzymes**

(i) *Proenzyme (zymogen) :* an inactive proteolytic enzyme (e.g. pepsinogen).

(ii) *Zymase:* An active form of an enzyme is called zymase (e.g. pepsin).

(iii) *Exoenzyme:* That acts elsewhere from their origins.

(iv) *Endoenzyme:* That acts within the cell (e.g. cytochrome P_{450}).

(v) *Conjugate enzyme (Holoenzyme):* An enzyme containing non-protein part alongwith protein part.

(vi) *Coenzyme:* an organic group that is required for certain enzymes that acts as a donor or acceptor of electrons or functional group in a reaction e.g. coenzyme A functions in transfer of acetyl groups in lipid and carbohydrate matabolism.

(vii) *Apoenzyme:* The protein part of holoenzyme.

(viii) *Inducible enzyme:* An enzyme which is synthesized in presence of a particular substrate.

(ix) *Constitutive enzyme:* The enzymes which are synthesized always by the cell irrespective of the presence of substrate in a cell.

(x) *Allosteric enzyme:* Enzymes which do not follow Michaelis-Menten hyperbolic kinetic but instead give sigmod kinetics (Monod *et. al.* 1975) e.g. aspartate carbamylase.

(xi) *Monomeric enzyme:* Certain enzymes with only one polypeptide in their structure e.g. ribonuclease.

(xii) *Oligomeric enzymes:* Enzymes with more than one polypeptide chain e.g. Lactate dehydrogenase and hexokinase.

(xiii) *Multienzyme:* Many different enzymes that catalyse a reaction site or different sites of the same macromolecule, e.g. fatty acid synthetase, pyruvate dehydrogenase and others.

(xv) *Repressible enzyme:* A functional enzyme which is repressed in presence of certain chemical *e.g.* glucokinase.

(xvi) *Transferase:* Enzymes that catalyse transfer of a funtional group from one substrate to another *e.g.* aspartate and alanine transaminase.

(xvii) *Ligase (synthetase)*: Enzymes involved in joining together substrates *e.g.* RNA synthetases.

(xviii) *Hydrolase:* Enzymes that bring about hydrolysis *e.g.* glucose 6 phosphatase, pepsin and trypsin.

(xix) *Isomerase:* Enzyme that catalyzes the conversion of one isomer to another *e.g.* epimerase.

(xx) *Lyase:* An enzyme that facilitate removal of small molecules from large substrates *e.g.* fumerase.

(xxi) *Oxireductase (oxidoreductase):* An enzyme which brings oxidation-reduction reaction *e.g.* glutamic dehydrogenase and cytochrome oxidase.

(xxii) *Proteolytic enzyme:* An enzyme that hydrolyses the peptide bonds between aminoacids in a protein chain.

(xxiii) *Sarcolytic enzyme:* Enzyme that digests carbohydrates *e.g.* sucrase. (*carbohydrase*)

(xxiv) *Lipolytic enzyme* (*lipase*): An enzyme that hydrolyzes the ester linkages between glycerol and fatty acids of triglycerides.

(xxv) *Restriction enzyme:* Bacteria contain nucleases that recognize short nucleotide sequences within duplex DNA and cleave the backbone at highly specific sites on both strands of duplex. These enzymes are called restriction endonucleases or restriction enzymes.

Vitamins

1. Vitamins are an essential accessory to food factors, necessary for complete normal metabolism of the energy yielding constituents of the diet, they are a must for normal body growth and development. Normal balanced diet usually contains adequate amounts of various vitamins for normal body functions. Excess of vitamins are needed for growing children, pregnant and lactating females and in conditions where metabolic activities are increased.
   (i) Vitamins were called Essential accessory food factors in 1880 by F. Crowl and Hophkins.
   (ii) The term 'Vitamin' was coined by Funk in 1912.
2. **Characteristics**
   (i) They are the organic compounds synthesized by plants
   (ii) Their daily requirement is low
   (iii) Few of them can be stored in the body e.g. Vit. A, D and K (in liver)
   (iv) Few can be synthesized in the body e.g. Vit. D in skin and Vit. K and B in intestine
   (v) They are of two types, as under–

Fat Soluble	*Water Soluble*
They cause diseases in excess.	Also cause disease in deficiency
They cause diseases only in deficiency	They can be excreted only by liver through bile
They can be excreted by kidney with urine	
Examples Vit. A, D, E, K	Examples : Vit. B. and C

3. **MODELS OF ACTION OF VITAMINS**
   (i) The vitamins act as coeyzmes in enzymes reactions in the body and act as electron acceptors of donors, e.g., in tissue respiration, the oxidases and dehydrogenases, are needed where the protein part (apoenzyme) is enzyme specific, while electron which can last for acceptor or donor part is co-enzyme. Vitamins concerned in such like reactions are vitamins A, C, E, K, B and niacin etc.
   (ii) Certain vitamins (Vit. B, being most important) act as CO, splitters e.g. oxidation of pyruvic acid bicarboxylase and Vit. B_1 as coenzyme).
   (iii) Still others (e.g. Pantothenic acid) act as activators e.g. activation of acetates to form acetyl CoA.

Hormones

1. Those organic compounds which are synthesized away from place of their actions are called hormones. First hormone discovered was 'Secretin', from duodenal mucous membrane. It was discovered by W.M. Bayliss and Starling (1903).
2. The word "hormone" was coined by Starling (1905).
3. In 1889 Brown Sequard injected himself testicular extracts of dog to improve the sexual capacity at the age of 72.
4. J.F. Von Mering and Oscar Minkowski (1889) discovered *Diabetes mellitus* in dogs.

A. Animals Hormones

(Classification, chemically)

Amino Acid	Derivative	Peptides	Proteins (polypeptides)	Steroids
Epinephrine	Catecholamines	Oxytocin	Gonodotropic	Adrenal cortex
Norepinephrine	of Adrenal medulla	Vasopressin	Somatotropic	Pituitary
Thyroxine	(Thyroid)	MSH	Thyrotropic	Testes
		Insulin		Ovaries
		Glucagon		
		Calcitonin		Placenta
		Parathormone		

5. Difference between hormones and enzymes

	Hormones	Enzymes
(i)	produced in endocrine glands	produced anywhere
(ii)	aminoacids, proteins, peptides, steroids	complex proteins
(iii)	low molecular weight (easily diffusible)	high molecular weight (not diffusible)
(iv)	produced away from site of action	on some cell or cavity
(v)	used up in reaction	remain unaffected in reaction
(vi)	Their action is reversible	Their action is reversible
(vii)	Effective in low concentration	The rate of reaction increases with concentration
(viii)	They can act slow/fast	They act fast

6. Some Endocrine Glands and Their Hormones

Endocrine Gland and Hormone	*Target Tissue*	*Principal Actions*
(i) **Hypothalamus**		
* Releasing and inhibiting hormones	Anterior lobe of pituitary gland	Stimulates or inhibits secretion of specific hormones
(ii) **Posterior lobe of pituitary** (storage and release)		
* Oxytocin	Uterus	Stimulates contraction
	Mammary glands	Stimulates ejection of milk into ducts
* Antidiuretic hormone	Kidneys (collecting ducts)	Stimulates reabsorption of water; conserves water
(iii) **Anterior lobe of pituitary**		
* Growth hormone	General	Stimulates growth by promoting protein synthesis
* Prolactin	Mammary glands	Stimulates milk production
* Thyroid-stimulating hormone (TSH)	Thyroid gland	Stimulates secretion of thyroid hormones; stimulates increase in size of thyroid gland
* Adrenocorticotropic hormone (ACTH)	Adrenal cortex	Secretion of adrenal cortical hormones
* Gonadotropic hormones (follicle-stimulating hormone [FSH] and lutenizing hormone [LH])	Gonads	Stimulate gonad function and growth

(iv) **Thyroid gland**		
* Thyroxine (T_4) and triodothyronine (T_3)	General	Stimulate metabolic rate; essential to normal growth and development
* Calcitonin	Bone	Lowers blood calcium level by inhibiting bone breakdown by osteoclasts
(v) **Parathyroid glands**		
* Parathyroid hormone	Bone, kidneys digestive tract	Increases blood calcium level by stimulating bone breakdown; stimulates calcium reabsorption by kidneys; activates vitamin D
(vi) **Islets of Langerhans of pancreas**		
* Insulin	General	Lowers glucose concentration in the blood by facilitating glucose uptake and utilization by cells; stimulates glycogenesis; stimulates fat storage and protein synthesis
* Glucagon	Liver, adipose tissue	Raises glucose concentration in the blood by stimulating glycogenolysis and gluconeogenesis; mobilizes fat
(vii) **Adrenal medulla**		
* Epinephrine and norepinephrine	Muscle, cardiac-muscle, blood vessels, liver, adipose tissue	Help body cope with stress; increase heart rate, blood pressure, metabolic rate; re-route blood; mobilize fat; raise blood sugar level
(viii) **Adrenal cortex**		
* Mineralocorticoids (aldosterone)	Kidney tubules	Maintain sodium and phosphate balance
* Glucocorticoids (cortisol)	General	Help body adapt to long-term stress; raise blood glucose level; mobilize fat
(ix) **Pineal gland**		
* Melatonin	Gonads, pigment cells, other tissues	Influences reproductive processes in hamsters and other animals; pigmentation in some vertebrates; may control biorhythms in some animals; may help control onset of puberty in humans
(x) **Ovary**		
* Oestrogens	General; uterus	Develop and maintain sex characteristics in female; stimulate growth of uterine lining
* Progesterone	Uterus; breast	Stimulates developement of uterine lining
(xi) **Testis**		
* Testosterone	General; reproductive structures	Develops and maintains sex characteristic of male; promotes spermatogenesis; responsible for adolescent growth spurt.

B. Plant Hormones (Phytohormones)

Five classes of hormones regulate plant growth and development : Only three of these—the auxins, cytokinins, and gibberellins—are, in fact, hormone *classes.*

Plants produce hormones in small concentrations, but even a minute amount of any of these chemicals has profound effect on target cells. Just a few molecules of a hormone can redirect the metabolism and development of a plant's cells. Plant hormones do this

(i) by altering the expression of genes,
(ii) by activating or inhibiting enzymes and other proteins, or
(iii) by changing the properties of membranes.

	Hormone	Major Functions	Production and Location
1.	Auxins	Stimulate stem elongation, root growth, differentiation and branching, developemnt of fruit; apical dominance; phototropism. and gravitropism (respone to gravity)	Meristems of apical buds and young leaves; embryo within seed
2.	Cytokinins	Affect root growth and differentiation; stimulate cell division and growth, germination, and flowering; delay again.	Synthesized in roots and transported to other organ
3.	Gibberellins	Promote seed and bud germination, stem elongation, leaf growth; stimulate flowering and development of fruit; affect root growth and differentiation.	Meristems of apical buds, roots and young leaves; embryo
4.	Abscisic acid (ABA)	Inhibits growth; closes stomata during water loss; helps maintain dormancy.	Leaves, stems, green fruit
5.	Ethylene	Promotes fruit ripening; opposes some auxin effects; promotes or inhibits growth and development of roots, leaves and flowers, depending on species.	Tissues of ripening fruits, nodes of stems, dying leaves

Special Points to Note

1. Carbonic anhydrase—the fastest acting enzyme can speed up a reaction 10 million times!
2. RNA molecules can act as catalysts also, greatly accelerating the reactions thus called ribozymes
3. The activity of almost every enzyme in a cell is regulated by feedback Inhibition, (also called negative feedback). For example, the amino acid asparate becomes the amino acid threonine by sequence of five enzymatic reactions. Threonine, the end product, when present in excess, binds to an allsteric site of the first enzyme and thus inhibits the substrate binding.
4. In a human body 200 molecules of carbonic acid are formed per hour which is very slow for our requirement. However, in presence of an enzyme, carbonic anhydrase, over 600,000 molecules of carbonic acid as estimated, from per second!

MULTIPLE CHOICE QUESTIONS

1. Number of essential amino acids in man is :
A. 7 B. 8
C. 10 D. 14

2. In which of the following groups all are poly-saccharides?
A. glycogen, sucrose and maltose
B. glycogen, cellulose and starch
C. sucrose, glucose and fructose
D. maltose, lactose and fructose

3. Cell theory states that :
A. cells are reproduced by mitosis and meiosis
B. cells are structural units of plants and animals
C. all cells are living
D. all cells have nuclei

4. The proteins are synthesized at :
A. ribosomes B. mitochondria
C. centrosomes D. golgi bodies

5. Vitamin D is synthesized in the skin by the action of sunlight on :

A. 7—hydro cholesterol
B. Cephlano-cholesterol
C. Cholesterol
D. All of the above

6. The calcium pectate and magnesium pectate are found in :

A. secondary wall of a plant cell
B. primary wall of a plant cell
C. the middle lamella of a plant cell
D. the tertiary wall of a plant cell

7. Phospholipids are :

A. amphipathic B. amphibolic
C. hydrophobic D. All of the above

8. Linoleic acid is unsaturated fatty acid and its content is highest in :

A. Cotton oil B. Groundnut oil
C. Sunflower oil D. Coconut oil

9. The source of turpentine oil is :

A. Fern B. Sunflower
C. Pinus D. Moss plant

10. The cell wall of plants is made up of cellulose, which is a type of:

A. protein B. lipid
C. polysaccharide D. amino acid

11. The number of amino acids used in protein synthesis is:

A. 20 B. 35
C. more than 50 D. more than 100

12. Which of the following is related with vitamin B_2?

A. FMN/FAD B. NAD
C. NADH D. $NADH_2$

13. Steroid is:

A. Thyroxine derivative
B. Vitamin A
C. Thyroid acid
D. Ester and fatty acid

14. How many base pairs are present in one turn of ADNA?

A. Ten B. Nine
C. Eleven D. Twelve

15. On hydrolysis a nucleoside would not yield:

A. Purine B. Pyrimidine
C. Pentose sugar D. Phosphoric acid

16. An ATP represents:

A. The bonds with high energy
B. Two moles of phosphorus
C. Three atoms of high energy phosphate
D. None of the above

17. Contents of nucleic acids in protoplasm is:

A. 35% B. 29%
C. 10% D. 2%

18. Nucleic acids occur in:

A. Bacteria only B. Viruses only
C. Mammals only D. All forms of life

19. DNA resembles RNA as both have:

A. Polymers of nucleotide
B. Similar sugars
C. Similar pyrimidine bases
D. Ability to replicate

20. Protein conjugated to carbohydrates is:

A. Lecithoprotein
B. Glycoprotein
C. Lipoprotein
D. Metalloprotein

ANSWERS

1	2	3	4	5	6	7	8	9	10
B	B	B	A	A	C	A	C	C	C
11	**12**	**13**	**14**	**15**	**16**	**17**	**18**	**19**	**20**
A	A	C	C	D	A	D	D	A	B

3

Cell and Development

INTRODUCTION

Unity and Diversity of Life

It looks strange but all life forms, whether plants or animals, have the same basic living strucure called cell and yet all of them are different as they exist in nature. *Cells are basic unit of life and structural unit of an organism.*

Cell: Cell is a unit of stucture and biological activity made of an organised mass of protoplasm surrounded by a protective and selectively permeable covering.

Cytology: It is the study of form, structure and contents of cells as observed under the microscope. The term was coined by Hertwig (1893) who also wrote a book on 'Cells and Tissues'.

Cell Biology: Cell biology is the study of all aspects of cells and their components including their structure, biochemistry, development and physiology. It is a unifying subject that is akin to studying life because of the fundamental processes of cells are not only similar but are actually life processes that make the cells living entities.

Tools and Techniques for Cell Study: Tools for cell study are devices and instruments called microscopes. Techniques are skills required to aid study of cells and their components. They are histological preparations, cytochemistry, autoradiography, cell fractionation, biochemical techniques and tissue culture.

Units of Measurement

1 mm = 0.1 cm = 0.001 m

1 micrometre/micron/μ m/μ = 0.001 mm
= 10^{-3} mm = 10^{-6} m (10^{-4} cm)

1 nanometre/millimicron/nm/μ m = 0.001 μ m
= 10^{-3} μ m = 10^{-6} mm = 10^{-9} m = (10^{-7} cm)

1 angstrom/Å = 0.1 nm = 10^{-4} mm
= 10^{-7} mm 10^{-10} m (=10^{-8} cm)

Smaller units are picometre = 10^{-12} m, femtometre = 10^{-15} m and attometre = 10^{-18} m.

Common unit of measurement in microscopy (as well cytology) and photometry is nm (nanometre) while unit for measurement of cells in µm (micrometre, previously called micron or µ).

Microscopy: It is the practice of using microscopes for the study of finer details of small objects including cells and tissues.

$$\text{Magnification} = \frac{\text{Size of Image with the Instrument at retina}}{\text{Size of Image with Unaided Eye at retina}}$$

Magnification of a microscope is roughly equal to the multiple of magnifying power of objective lens and ocular lens eye (eye piece), e.g., 45 × objective and 105 eye piece = 450 × magnification.

Resolving power or limit of resolution is the ability to distinguish two close object is as distinct. Its values is calculated by Abbe equation

$$L_m = \frac{0.61\lambda}{NA}$$

λ is wavelength of light used in illumination of object. NA or numerical aperture is product of refractive index of medium (*n*) and sine of angle subtended by optical axis and outer ray covered by objective (sine α). Refractive index for lense of optical instrument is 1.6, oil immersion 1.4 and air is 1.0. Sine α is less than 1. The value for best objective is sin 70° = 0.94. In air the value of numerical aperture is less than one while for immersion objective it is 1.0 – 1.35. Therefore, $L_m = \frac{0.61\lambda}{1.3}$ or 0.5λ or one half of the wavelength of illuminating light. The average wavelength of light used in microscope is 250 –300 nm or 0.25–0.3 μm for optical microscope. Use of blue light or blue filters helps in achieving a resolution limit of 200 nm or 0.2 μm. It is 100 μm (= 0.1 mm) in case of naked eye and 2–10 Å for electron microscope.

Depending upon the type of illumination, there are four types of microscopes—light microscope, electron microscope, ultraviolet microscope and X-ray microscope. Dark-field, phase-contrast, intereference and polarising microscopes are varied forms of light microscope.

Microtomy: Microscopic examination requires sectioning and staining. Sectioning is done by hand or machine known as microtome (developed by W.His). Extremely thin sections (20–100 nm) are obtained by ultramicrotome using glass or diamond knife.

X-ray Crystallography (Bragg, 1913): X-ray microscope (developed by Kirkpatrick) is useful in studying detailed structures of chemicals present in solid state (e.g., haemoglobin, insulin, DNA, RNA) due to diffraction of X-rays (wavelength 1–10 Å) by different atoms of the substance. Watson and Crick (1953) found out the structure of DNA from X-ray diffraction obtained by Astbury and Franklin (1953).

Cytochemistry: A number of dyes or stains are known to colour specific parts. Certain dyes can be used even in case of living materials. They are called vital stains, e.g., neutral red, methylene blue. Fuelgen or Schiff's reaction was developed by Fuelgen and Rossenbeck (1924). It employs basic fuchsin after acid hydrolysis.

Tracer Isotopes: Radioactive isotopes function like normal elements but emit radiations. Thus they can be located by Geiger counter or scintillation counter and autoradiography, e.g., ^{3}H, ^{14}C, ^{32}P, ^{35}S, ^{42}K. Instead of radioactive isotopes, heavy isotopes like ^{15}N and ^{18}O can be used and are analysed by mass spectroscopy (photosynthetic evolution of oxygen from water by Ruben *et al*, 1941) and density gradients centrifugation.

Autoradiography: Radioactive isotopes (^{3}H, ^{14}C, ^{32}P, ^{35}S) are incorporated in intermediate or raw materials (e.g., CO_2, uridine, thymidine, amino acids) for studying metabolic pathways including DNA → RNA → Protein pathway.

Cell Franctionation: It is the technique of separating and rupturing cells, and separating their components for study. It has two steps, homogenisation and differential centrifugation.

(i) Homogenisation: It is the rupturing of cells in order to release various constituents and then keeping them in their natural state by means of a special 0.25 M sucrose solution kept at 0° – 4° C. It is carried out by (a) Grinding in pestle and mortar or potter homogeniser. In potter homogeniser two cylinders separated by narrow gap are made to rotate. It produces a shearing force that causes rupturing of cells (b) Pressure (c) Osmotic shock (d) Ultrasonic vibrations (produced by piezo-electric crystal) transmitted to suspension by means of steel rod. The slurry contains homogenate as well as intact cells and other debris. They are separated by low speed centrifugation. The supernatant is homogenate. It is liquid containing cellular constituents.

(ii) Differential Centrifugation: It is separation of different components of homogenate by subjecting it to successively higher centrifugal force that brings about velocity sedimentation.

Density Gradient Centrifugation: A density gradient of sucrose (or caesium chloride) is established in a tube with denser solution at the bottom and a gradual decrease in density towards the top. Homogenate is poured over the top of sucrose solution and the tube is centrifuged at high speed of 50,000 – 5,00,000 × g for 1–3 hours. Different components of homogenate separate and form bands at different positions as per their size and density or sedimentation coefficients called Svedberg unit. This method of separation of homogenate fractions as per their Svedberg value and density of the sucrose solution is called isopyknic or equal density centrifugation.

Buoyant Desnity Centrifugation: Homogenate components are separated according to their buoyant density independent of size, shape or Svedberg unit.

A high concentration of sucrose or caesium chloride (about 50%) is mixed with cell extract. By this method Mathew Meselson and Francklin Stahl (1958) could separate heavy DNA fraction having $^{15}N/^{15}N$ from intermediate DNA fraction having $^{15}N/^{14}N$ and light DNA fraction with $^{14}N/^{14}N$ proving semiconservative replication of DNA.

Svedberg Unit (S): It is a unit to measure sedimentation rate or sedimentation coefficient of a particle in an analytical ultracentrifuge. It is related to weight and volume of a material, e.g. 4S tRNA has a molecular weight of about 25000 daltons.

Microspectrophotometry: Solutions, gases and solids containing chemicals are subjected to different radiations. Absorption spectra are compared to know the chemicals.

Other Methods of Franctionation of Macromolecules of Cells: A number of methods have been developed to separate macromolecules of the cells. Two of the more common and important methods are chromatography and electrophoresis.

Chromatography (Michael Tswett, 1906): It is a technique of separating molecular components of a mixture of a mobile phase by using the principle of partition coefficient due to which different molecules move at different rates through a stationary phase and get separated.

Electrophoresis: Particles of different sizes and charges are separated due to their movement to different distances under the influence of an electric field, e.g., nucleic acids, proteins, amino acids, nucleotides, etc. A base material is used for the passage of molecules of the mobile phase. In PAGE (polyacrylamide gel electrophoresis) the base material is polyacrylamide or polymer of acrylamide and methylene bisacarylamide). In agarose gel electrophoresis, the base material is agarose.

CELL—THE BASIC UNIT OF LIFE

1. Robert Hooke coined the term 'Cell' for small structures in a piece of cork under a microscope (1665). His observations were published in his book, *Micrographia*.
2. Leeuwenhoek (1674) observed one celled structures like bacteria, protozoans and blood cells. Dutrochet later confirmed that plants were also made up of cells.
3. Cell is now defined as structural and functional unit of the body. There are two general classes of cells: Prokaryotic cells and Eukaryotic cells.
4. **Theories in cytology**
   (i) **Colloidal theory:** Protoplasm is polyphasic colloidal solution that generates energy of life.
   (ii) **Protoplasm theory:** It states that an organism is made up of protoplasm and cells are simply an accumulation of protoplasm.
   (iii) **Organismal theory:** According to it an organism is a continuous mass of protoplasm segmented into cells. Idea of cell theory was first given by H.J. Dutrochet (1824).
   (iv) **Cell theory:** The credit of forming the cell theory goes to German botanist, M.J. Schleiden (1838) and German zoologist, T. Schwann (1839).
   Outlines of cell theory include: (a) A cell is the basic unit of life and (b) cells arise *de novo*.
   (iv) **Modern cell theory** or **cell principle** or **cell doctrine**. It states that
      (a) all organism are made up of cells.
      (b) cells arise from pre-existing cells (*Omnis cellulae cellula*).
      (c) all organisms start their life from a single cell.
      (d) multiplication of cells and their growth leads to the growth of an organism.
5. **Cell size**
   Two important factors contribute to the cell size:
   (i) Nucleocytoplasmic ratio, and
   (ii) Surface area of a cell.
6. **Cellular totipotency**
   (i) The capacity of a cell to form whole of the organism is called *totipotency*, as called by *Morgan*.
   (ii) Pluripotency means the ability to dedifferentiate completely after cell maturity.

(iii) German botanist, Gottlieb Haberlandt was the first to culture plant tissues '*in vitro*'.

(iv) Callus is the mass of undifferentiated cells formed by growth of a totipotent cell in culture.

(v) Callus was first time diffentiated to whole plant of tobacco by Skoog and Miller.

7. Comparison of sizes

Organs	> 1 mm
Tissues	10-100 μm
Cells	0.3-10 μm
Cell components	1-200 nm
Molecules and Atoms	< 1 nm

8. (i) Smallest cell is *Mycoplasma gallisepticum* (0.1 μm) which is also referred to as PPLO (Pleuro-pneumonia like organism) or Jokers of plant kingdom.

(ii) Largest prokaryote is *Spirillum volutans* (15 μm).

(iii) Largest plant cell is an alga, *Acetabularia* (8-10 cm).

(iv) Longest animal cell is egg of Ostrich (175 mm × 158 mm).

(v) Longest cell fibre is Ramie fibre of *Boehmeria nivea* (550 mm)

(vi) Longest animal cell is a nerve cell (1 m long).

9. Prokaryotes: These cells are small in size and are exhibited by bacteria and cyanobacteria (Blue-green algae).

They have the following characteristic features:

(i) The cell wall is made of murein derived from amino acids (glutamic acid, alanine, glycine and lysine). Sometimes instead of lysine similar substance like N-acetyl glucosamine and N-acetylmuramic acid are present.

(ii) Plasma membrane forms infoldings called mesosomes or chondrioids which control DNA replication, cell division and respiration as they store respiratory enzymes.

(iii) Membrane bound organelles such as mitochondria, plastids, ER and Golgi bodies are absent.

(iv) Vacuoles are absent, however, protosynthetic prokaryotes including both bacteria and blue-green algae possess gas bubbles.

(v) Flagella (if present) lack 9 + 2 fibrillar structure but have a single fibril made up of flagellin proteins.

(vi) Cytoplasmic streaming (cyclosis) is absent.

(vii) These cells divide by simple fission.

(viii) Organized nucleus is absent but a simple structure called nucleoid or chromatin body or genophore is present. The nucleoid lacks nuclear envelope and nucleolus but contains a single circular molecule of double-stranded DNA which is also called a ring-chromosome. The DNA does not have histone proteins associated with it, but contain acidic proteins.

10. Eukaryotes: These cells contain a membrane-bound nucleus having nucleolus and chromosomes, and membrane bound organelles. They are of two types: plant cells and animals cells.

11. The plant cell: A typical plant cell can be differentiated into three parts:

(a) Cell Wall (b) Protoplasm and (c) Vacuole.

12. Protoplasm: The term 'protoplasm' was first given by Purkinje (1839) for the living substance of the cell. In 1846, Hugo Von Mohl introduced a distinction between the protoplasm and cell sap. Protoplasm is a viscous colourless fluid and is the seat for all the physiological functions. Later in 1868, Huxley suggested that 'protoplasm is the physical basis of life'.

13. Properties of protoplasm

Physical

(i) Protoplasm is a polyphasic colloidal system.

(ii) Its specific gravity is slightly above that of water.

(iii) Its viscosity has been found to be like that of glycerine.

(iv) It has power of responding to external stimuli, like heat, electric shocks, application of chemicals, etc. This property of protoplasm is called irritability.

(v) It exhibits streaming movement e.g., rotatory movements in the leaves of aquatic plants like *Hydrilla* and *Vallisneria*.

(vi) It exhibits circulatory movements called cyclosis as seen in the staminal hair of *Rhoeo discolour*.

(vii) Amoeboid movement of the protoplasm can also be noticed in myxomycetes and *Amoeba*.

(viii) In general, the pH of cytoplasm is slightly acidic *i.e.*, 6.8, however, pH of the nuceoplasm is 7.6 to 7.8.

Chemical

(i) Generally in the active stage protoplasm is composed of more than 75 per cent water and less than 25 per cent other materials representing the dry weight.

(ii) The dry matter is rougly 90 per cent organic and 10 percent inorganic.

(iii) In aquatic plants, the amount of water may go upto 95 per cent whereas the water content of the protoplasm of seeds and spores is as low as 10-15 per cent.

(iv) Among organic substances, proteins, carbohydrates, fats and nucleic acids are the most important ones.

(v) Among the inorganic compounds, phosphates, chlorides, sulphates and carbonates of alkaline and alkaline earth metals are present out of which Ca, Mg, K, Na and Fe the most common elements.

14. The protoplasm of a cell can be distinguished into two parts:

(i) Nucleoplasm and

(ii) Cytoplasm.

STRUCTURE OF A CELL

Every cell is made up of a small mass of *protoplasm* having a *nucleus*, a number of organelles and a covering membrane.

Cell
(on the basis of DNA)

Prokaryotic cell
[Primitive type, genetic material is *not* organised in the form of nucleus]
(e.g. in bacteria, archae bacteria, mycoplasma or PPLO = pleuropreumonina like organisms spirochaetes, rickettsiac, chlamydiae and cyanobacteria or blue green algae.)

Eukaryotic cell
[Organised nucleus and several mebrane covered cell organelles.]
(Except monera, cells of all other kingdoms art eukaryotic.)

BIOMEMBRANES

1. All cells and most of the organelles are enveloped by a thin membrane called plasma membrane. This was discovered by Nageli and Crammer in 1855. It was called 'plasmalemma' by J.Q. Plowe (1931).

2. Feature of plasma membrane

Occurrence	:	Found in animals, plants, fungi and bacterial cells
Thickness	:	75 Å
Composition Lipids	:	They form the main bulk and exist in form of phospholipids (30 to 40%). They mostly occur in form of lecithin and have hydrophilic heads and hydrophobic tails.
Proteins	:	They form the backbone with 40-60% of the plasma membrane and

		form structural components, carrier molecules and enzymes. They exist in intrinsic and extrinsic forms.
Carbohydrates	:	They are about 1 to 6%.
Salts and Water	:	They are present in small amounts.

3. Structure: Various models can be as under

(i) **Danielli and Davson model:** It is also called bimolecular leaflet model or sandwich model or lamellar model which was proposed by Danielli and Davson (1934). According to this model, the two globular proteins layers were thought to sandwich a phospholipid bilayer.

(ii) **Lattice model:** It was proposed by Wolpers (1941). According to him lipids were thought to be distributed in the meshes of protein frame work.

(iii) **Robertson unit membrane model:** This concept of Robertson (1959) explains that a phospholipid bilayer is surrounded by two continuous protein layers.

(iv) **Fluid mosaic model:** Finally in 1972 this model was introduced by S. Singer and G. Nicholson. According to it the protein molecules were put within the phospholipid bilayer and two types of globular proteins, extrinsic and intrinsic were introduced in this model. This is the most acceptable theory.

(v) **Micellar model:** This was proposed by Hillier and Hoffman (1953). The model explains that lipid micelles (globular subunits) form the building block and the protein globules are arranged on either sides.

(vi) **Greater membrane model:** It was given by Lehminger. According to him. The inner surface of the membrane is covered by unconjugated proteins and outer surface by glycoproteins to which oligosaccharide side chains are attached.

4. The plasma membrane is permeable but differentially (selectively) and performs following functions.

(i) Transport

(a) Plasma membrane acts as a barrier but permits few substances into and out of the cell.

(b) The nutrient materials are taken in while the waste products are expelled out.

(ii) Cell recognition and Adhesion

(a) Cells like mammalian leucocytes are able to differentiate between foreign cells and other cell of blood by adhering to the foreign particles and engulfing them by phagocytosis.

(b) Sialic acid (an amino-sugar) present on the surface of plasma membranne is possible involved in cell recognition and adhesion.

(iii) Antigen Specificity

(a) The specificity of antigens of the cells is determined by the glycoproteins present on the cell surface.

(b) The different blood group system is based on the relationship between antigens on RBC and antibodies in blood serum.

(IV) Hormone Receptors

(a) Hormone receptors are located on the outer surface in the plasma membrane.

(b) These recognize the specific hormones and relay the information of the hormone to the anterior of the cell.

(v) Secretion

(a) Secretion occurs in the ribosome at the endoplasmic reticulum and the secretory products are condensed in Golgi complex.

(b) The secretory vesicles, from Golgi complex, carrying the products, move to cell periphery and fuse with the plasma membrane.

(vi) Oxidative Phosphorylation

(a) The inner membrane of the plasma membrane has the electron transport chain that can transfer electrons.

(b) Electrons from substrate pass along this chain and cause reduction and oxidation of each component.

(c) During the process free energy is decreased, half of which is utilized in synthesis of ATP.

(d) This is called oxidative phosphorylation.

(vii) Endocytosis

(a) It is the process of transport of substances to the interior of the cell by formation of vesicles.

(b) It includes phagocytosis (cell eating) and pinocytosis (cell drinking).

(c) Phagocytosis involves ingestion of solid particles whereas pinocytosis refers to ingestion of fluid material.

(viii) Exocytosis

(a) It is reverse of endocytosis and also called as cell vomiting.

(b) Membrane of vesicles (formed in cell interior) reach the periphery and fuse with plasmamembrane thereby releasing their contents to the outside of cell.

(xi) Chemoreception

(a) The chemoreceptors present in membranes of bacteria and protozoa are proteins.

(b) In mammals, the binding proteins associated with the cell membranes act as chemoreceptors for taste and smell.

(c) In man, specific receptor proteins for four distinct tastes sour, sweet, bitter and salty, have been found to occur.

(x) Transmission

(a) The surface membrane of the nerve cells is found to be involved in the transmission of nerve impulses.

Separation and Protection

(b) Plasma membrane separates and protects the living protoplasm from its non-living environment.

(xi) Excitation

Membrane of excitable cells (*e.g.* muscle and nerve cells) is capable of series of reversible structural and functional changes.

(xii) Enzymatic Action of the External Environment

Several enzymes are integral part of plasma membrane and by virtue of which the membranes can act chemically on their immediate environment.

(xiii) Division and Fusion

Cell membranes can divide to form new cells or vesicles within a plasma membrane, both processes are intimately related.

(xiv) Regeneration and repair

(a) Plasma membrane is able to :

* reform rapidly over exposed regions of cytoplasm.
* fuse with one another if ruptured.
* expand and contract during cell movement or changes in cell shape.

(b) These abilities suggest fluid and dynamic nature of membrane.

5. Plasma membranes are modified to perform certain specialised functions as under :

Modifications		**Characteristics and Functions**
Microvilli	...	Found in intestinal epithelial cells; increase area of absorption
Desmosomes	...	bears filaments called tonofibrils for firm adhereing of cells
Inter-digitations	...	finger like foldings
Plasmodesmata	...	also called tight junctions; barriers for diffusion
Basal foldings	...	infoldings contain mitochondria and common in nephrons

Special Points to Note

1. In epithelial cells, water moves in and out by endocytosis and exocytosis processes respectively called cytopemphis.
2. Mammalian leucocytes recognized foreign cells like bacteira and engulf them by '*phagocytosis*'.
3. The term 'phagocytosis' was coined by Metchnikoff.
4. '*Pinocytosis*' (cell drinking) was first coined by Lewis in tissue cultured cells (1931).
5. In hypertonic solution, an animal cell shrinks called 'crenation' whereas if the plasma membrane shrinks away from the cell wall in plant cell, it is defined as plasmolysis.
6. Plasma membranes are involved in the biogenesis of cell organelles.
7. Many cell organelles like ribosomes, centrioles, basal bodies and nucleoli are not bounded by plasma membrane.
8. Cell surface receptors in membranes bind specifically to chemicals like hormones and transmit informations to the cell.
9. The movement of sodium ions from the axoplasm into the interstitial fluid across the plasma membrane of axon by active transport is called sodium pump.
10. The antigenic determinants present on the plasma membranes are glycoproteins.
11. The amino acids involved in cell recognition located on the surface of the membranes in sialic acid.
12. Cell recognition proteins the MHC (major histocompatibility complex) glycoproteins are different for each person and thus serve as 'fingerprints' of the cell. Thus, organ transplants are difficult to achieve. Cells with foreign MHC glycoprotein are attached by blood cells responsible for immunity.
13. In addition, animal membranes contain glycolipids and cholesterol.
14. The cell membrane was defined as 'protein icebergs in a sea of lipids', by Singer and Nicholson.
15. The cell membrane contains about 20% of water.
16. Biologically useful molecules enter and exit out of the cell across the membrane where carrier proteins or Permeases are required for facilitated and active transport.
17. The molecules that can pass through the plasma membrane is between 1 to 15 Å.

CELL ORGANELLES

I. Cell Wall

1. Monerans, plant protistans, fungi and plants have characteristic cell wall around their plasma membrane which is lacking in animals cells.
2. The cell wall is non-living which is secreted and maintained by the protoplasm. It constitutes a frame-work that provides mechanical support to the organisms.
3. In plants it is composed principally of cellulose.
4. Adjacent walls are cemented by calcium pectate which forms middle lamella.
5. On the basis of development and structure, the plant cell wall can be divided into three layers
   (a) Primary wall,
   (b) Secondary wall, and
   (c) Tertiary wall.
6. **Plasmodesmata:** In most of the plant cells, there are bridges of cytoplasmic material called plasmodesmata which establish the continuity between the adjacent cells through the pores present in the cell wall. They have a thin plasma membrane and may contain microtubules.

II. Nucleus

1. It discovered by a English botanist, Robert Brown in 1831, and is designated as the '*Director of the cell.*'
2. It is the largest organelle within eukaryotic cells, ranges from 5 to 25 mm in size.
3. Its nucleus is called karyology.
4. Some cells like those of sieve tubes of vascular plants and RBCs of mammals do not posses nuclei at maturity. As they can no longer divide, their life span is limited.
5. It consists of nuclear envelope or karyotheca
   (i) nucleoplasm or nuclear sap or karyolymph or karyoplasm

(ii) nucleolus
(iii) nuclear reticulum or chromatin reticulum.

6. The term 'chromosome' was coined by Waldeyer in 1888.
7. Their number is constant in the nucleus for a particular species.
8. The lowest diploid number known among plants is 4, found in *Haplopappus gracillis* whereas *Ophioglossum reticulatum* has as high as 1,260 chromosomes.
9. Among animals *Parascaris univalens* has only 2 chromosomes while *Aulacantha* has 1,600 chromosomes per cell.
10. According to centromeric position chromosomes are of five types :
    (i) **Telocentric chromosomes:** centromere situated on one end.
    (ii) **Acrocentric chromosomes:** with centromere little away from the tip and have one very small arm.
    (iii) **Sub-metacentric chromosomes:** These are L-shaped chromosomes with unequal arms.
    (iv) **Metacentric chromosomes:** centromere in the middle with equal arms. These are V-shaped chromosomes.
    (v) **Acentric :** without centromere.
11. **Structure of chromosomes**
    (i) The chromosome consists of proteinaceous matrix and two spirally coiled chromonemata, each one of which contains a single DNA molecule.
    (ii) When the coils of chromonemata are easily separable, they are referred to as paranemic coils.
    (iii) When chromonemata remain so intertwined that their coils are not easily separable then such coils are called plectonemic.
    (iv) Each chromonema (singular) possesses bead-like thickenings called chromomeres.
    (v) However, recent studies have shown that each individual chromosome in the eukaryotic cell contains a single chromonema which represents a single DNA molecule.
    (vi) Each chromosome has one (sometimes two or more) clear zone, called centromere or kinetochore or Primary constriction.
12. **Chemical composition of chromosomes**
    (i) DNA
    (ii) RNA
    (iii) Proteins : two major classes are found
        (a) basic proteins called **histones** and
        (b) a heterogenous, largely acidic group of proteins collectively referred to as **non-histone** chromosomal proteins.
13. **Nucleosomes**
    (i) **R.D. Kornberg** and **Thomas (1974)** proposed a model to explain the structure of nucleosome.
    (ii) The term **'nucleosome'** for chromatin subunits was suggested by **Oudet** *et al*, in 1975.
    (iii) Histones are of *five types* and they are designated as $\mathbf{H_1}$, $\mathbf{H_2A}$, $\mathbf{H_2B}$, $\mathbf{H_3}$ and $\mathbf{H_4}$.
14. **Special chromosomes**
    A. **Polytene chromosomes** or **Salivary gland chromosomes**
    (i) These chromosomes were first observed by **Balbiani** in 1881.
    (ii) The name **'polytene chromosome'** was assigned to them by **Killar**.
    (iii) They are found in the tissues of dipteran larvae, such as salivary glands, gut, trachea, fat body cells and malpighian tubules.
    (iv) Polytene chromosomes are of rare occurrence in plants, although observed by **W. Nageli** (1962) in the embryo of *Phaseolus coccineus*.
    (v) In *Drosophila melanogaster* the volume of polytene chromosomes is about 1,000 times greater than that of somatic chromosomes.
    (vi) The total length is about 2,000 μm as compared to 7.5 μm in other somatic chromosomes.

B. Lampbrush chromosomes

(i) These chromosomes were first observed by Flemming in 1882.

(ii) They were described in detail in shark oocytes by Ruckert (1892) who also coined the name 'lampbrush chromosomes' to them.

(iii) They occur at the diplotene stage of meiotic prophase in oocytes of amphibians and also in the giant nucleus of the unicellular alga, *Acetabularia*.

(iv) Each lampbrush chromosome has many fine lateral projections or loops giving them a characteristic 'hairy' appearance.

(v) They are best visualised in Salamander (5200 μm) about three times longer than the polytene chromosomes of Drosophila.

III. Endoplasmic Reticulum (Filamental Structure)

1. Garneir for the first time had described them as filamentous structure, ergastoplasm.
2. They were first discovered by Claude (1951) under electron microscope.
3. The term "endoplasmic reticulum' was first coined by Porter *et. al*. in 1945.
4. E.R. is abundant in liver, pancreas and other actively synthesizing cells. E.R. is absent in RBC, ova, embryonic cells and prokaryotes.
5. There are two types of endoplasmic reticulum : Rough endoplasmic reticulum and Smooth endo-plamic recticulum

   (i) Tubular forms with ribosomes.
   Tubular forms without ribosomes.

   (ii) It is more developed in cells actively engaged in protein synthesis. It is abundant in the cells engaged in glycogen and lipid metabolism.
6. However, rough ER may be converted into smooth ER and vice versa, depending upon the metabolic requirements of the cell.
7. **Characteristics of endoplasmic reticulum**

   (i) E.R. occurs in three forms cisternae, vesicles and tubules.

   (ii) E.R. originates from the ground substance (hyaloplasm) or infoldings of plasma membrane or evaginations of nuclear membrane. However, exact nature of origin is unknown.

   (iii) E.R. present in retinal cells is called myeloid bodies.

   (iv) Those present in muscle cells are called as sarcoplasmic reticulum.

   (v) Microsomes are the fragments of endoplasmic reticulum containing ribosomes.

   (vi) They contain large amount of RNA and phospholipids.

   (vii) High rate of protein synthesis is seen here and are also involved in the synthesis of triglycerides and steroids.
8. **Functions of endoplasmic reticulum**

   (i) It provides mechanical support to the cytoplasm.

   (ii) It plays an important role in the intracellular transport and storage of proteins and sometimes lipids.

   (iii) It synthesizes lipids and lipoproteins.

   (iv) Smooth ER is involved in glycogenolysis.

   (v) It forms other cellular organelles such as Golgi bodies, peroxisomes, and glyoxysomes.

   (vi) It forms nuclear envelope at the end of the telophase stage of the cell division.

IV. Golgi Complex Structure

1. An Italian physician Camillo Golgi (1898) first recognised this structure in nerve cells of Barn owl and hence named after the discoverer.
2. These are also known as lipochondrion, Golgi body or Golgi apparatus or Idiosome.
3. **Characteristics**

   (i) It is found in all the eukaryotic cells except red blood cells and sieve tubes of angiosperms:

   (ii) It is also absent in prokaryotes, certain fungi and sperm cells of bryophytes and pteridophytes.

   (iii) They are very well formed and larger in nerve cells and gland cells, however poorly developed in muscle cells.

   (iv) The Golgi complex is morphologically very similar in both plant and animal cells.

(v) In plant cells and in invertebrate tissues, dictyosomes are dispersed throughout the cytoplasm.

(vi) Their number varies from 3 to 7 in most plant and animal cells.

(vii) Each complex is formed by stack of flattened disc-shaped cisternae and associated with secretory vesicles.

(viii) In general, each Golgi apparatus has three components :

(a) a stack of flattened sacs called cisternae with dilated ends.

(b) clusters of tubules and vesicles of about 60 μm size.

(c) large vesicles or vacuoles filled with amorphous or granular content.

(ix) The golgi cisternae in a dictyosome are arranged in parallel and are separated by a space of 20 to 30 μm.

(x) However, in certain protista *e.g. Euglena*, there may be as many as 10 or 20 cisternae.

(xi) Each stack of cisternae forming a dictyosome is a polarised structure having a proximal or forming face generally convex and closer to the nuclear envelope or the ER and a Distal or maturing face of concave shape which encloses a region containing large secretory vacuoles.

4. Origin: Elements of Golgi complex may arise from:

(a) ER (b) Nuclear envelope (c) Pre-existing Golgi complex.

5. Composition: The Golgi complex is composed of about 60% protein and about 40% lipids and polysaccharides.

6. Functions:

(i) It packs and transports certain materials like proteins and polysaccharides out of the cell (cell secretion).

(ii) It is the seat for formation of glycolipids and glycoproteins.

(iii) It synthesises hemicellulose, cellulose and pectin compounds during cell division and hence involved in cell wall formation in plants.

(iv) It helps in storage of secretory products.

(v) It gives rise to lysosomes and vacuoles.

(vi) It forms acrosome in sperms.

V. Microtubules

1. They were first observed by de Robertis and Franchi (1953) in nerve fibres.

2. The term 'microtubule' was coined by Slautter Back (1953).

3. They are found in all eukaryotic cells either free in cytoplasm or forming part of centrioles, cilia and flagella. They are about 250 Å in diameter, several micrometer in length and with a wall of 60 Å thickness consisting of 13 subunits.

4. The main component of microtubules is **tubulin** protein.

5. Functions:

(i) Microtubules together with microfilaments functions as cytoskeleton.

(ii) Microtubules maintain the shape of some cell processes such as axons and dendrities.

(iii) They transport molecules and granules within the cells and thus they function as microcirculatory system.

(iv) They help in spindle fibre formation during cells division.

(v) They provide frame work for cyclosis.

VI. Centrosome

1. The first description of centrosome was given by Van Beneden (1880).

2. The term 'centrosome' was coined by Boveri (1881).

3. It is found in cytoplasm near the nucleus in animal cells and certain lower plants, however, it is absent in seed plants, prokaryotes and Amoeba.

4. The radiating structure comprises, astral-rays, the centrosphere and two small bodies called centrioles or diplosomes.

5. Centrioles are independent, cylindrical bodies (0.2 × 0.5 μm) lying perpendicular to each other at a distance 0.8 μm.

6. The wall of each centriole has nine triplets of microtubules arranged in a circle.
7. The microtubules of centrioles are made up of tubulin and some lipids. The interior of centriole is filled with a homogeneous cytoplasm of low density.
8. RNA and DNA have also been observed in them.
9. Both centrioles may give rise to a flagellum but in most cases only one does and the other centriole remains at right angle to the kinetosome or basal body.
10. They help in spindle formation during cell division of animal cells.

VII. Cilia and Flagella

1. Cilia and flagella are composed of three major parts:
   (i) a central axoneme or shaft,
   (ii) the surrounding plasma membrane, and
   (iii) the interposed cytoplasmic matrix.
2. The axonemal elements of nearly all cilia and flagella (as well as the tails of sperm cells) contain the same '9 + 2' arrangement of microtubules.
3. In the centre of axoneme are two singlet microtubules that run the length of the cilium.
4. Projections from the central microtubules occur periodically along their length to form an enclosing sheath.
5. Each of the central microtubules is composed of 13 protofilaments.
6. Nine doublet microtubules surround the central sheath. One microtubule of each doublet (*i.e.*, A subfibre) is composed of 13 protofilaments. The ajoining B subfibre is 'incomplete', consisting of 11 protofilaments.
7. Radial spokes having a periodicity of 24, 32 and 40 nm extend from each A subfibre inward to the central sheath.
8. Adjacent doublets are joined by nexin or interdoublet links.
9. Extending from each A subfibre are two dynein arms—an outer and inner arm.
10. Each beat of cilium and flagellum involves the same pattern of microtubule movement involving power (effective) stroke and the recovery stroke. Power stroke occurs in a single plane but the recovery may not occur in the same plane as the power stroke.
11. *Difference between cilia and flagella* :

Flagellum	Cilium
Long whip like about 150 μm	short
structure ranging from 5–10 μm,	
Less in number per cell (1–4)	more in
number, about few hundred per cell.	
Present near one end of the cell	
present all over the surface	
Exhibit undulating movements	exhibit
sweeping rhythm.	
Move independently	show
coordinating movements.	

12. **Functions.**
   1. Locomotion – Creation of water current for food ingestion
   2. Filteration – internal circulation of fluids.
   3. Sensory functions – cleaning of respiratory tract.

VIII. Microbodies

They are minute spherical bodies in a cell bounded by single unit membrane. The following kinds are recognized namely,

(A) Lysosomes
(B) Spherosomes
(C) Peroxisomes
(D) Glyoxysomes
(E) Lomasomes.

(A) Lysosomes

1. These organelles were first observed and coined by de Duve.
2. They are variable in size ranging from 0.2 to 0.8 μm but may be as large as 5 μm in mammalian kidney cells and in phagocytes.
3. Their shape is often spherical but may by irregular and are surrounded by a single unit membrane.

4. They are found mostly in animal cells and in some plant cell *e.g. Neurospora*, cauliflower and bean.
5. They contain numerous hydrolytic (digestive) enzymes in latent form, *i.e.*, the enzymes are enclosed by a membrane and are not readily available to the substrate.
6. The lysosomal membrane is resistant to the enzymes that it encloses and the entire process of digestion is carried out within the lysosome. The membrane protects the rest of the cell from the possible destructive effect of the enzymes.
7. Most of the lysosomal enzymes act in an acidic medium.
8. It has been observed that if lysosomes are injured by a toxic substance, the enzymes thus released can digest the whole cell, hence lysosomes are called 'suicidal bags'.
9. Lysosomes show considerable polymorphism and their four types are recognised.
   (a) **Primary lysosomes or Storage granules:** They are small about 0.4 μm in size. Their enzymatic content is synthesised by ribosomes on ER and from there it penetrates into Golgi region and then into the lysosomes.
   (b) **Secondary lysosomes or hetero-phagosomes or digestive vacuoles:** They are formed by the fusion of primary lysosomes with vacuoles containing engulfed material.
   (c) **Residual bodies :** They are formed if the digestion is incomplete and thus, they contain undigested material.
   (d) **Autophagic vacuoles or cytolysosomes or autophagosomes:** They digest the various parts of the cell. Autophagy plays an important role in the metamorphosis of amphibians.

(B) Spherosomes

1. They were first discovered by Hanstein (1880) in plant cells.
2. They are characteristic of plant cells.
3. They term 'spherosome' was coined by Dangeard.
4. They are about 0.5 to 1 μm in diameter and have a single limiting membrane.
5. They contain hydrolytic enzymes like protease, ribonuclease, phosphatase, esterase, etc. and thus are somewhat similar to lysosomes of animals.
6. However, the primary function of spherosomes is storage and transport of lipids.
7. Lipid synthesising enzymes were reported in spherosomes by Semadeni (1967).
8. They are formed by budding of endoplasmic reticulum.

(C) Poroxisomes

1. These structures where first described from liver and kidney cells by Rodhin (1954).
2. In plant cells, they were first observed in germinating seeds by Tolbert (1969).
3. The term 'peroxisome' was first used by de Duve and are also called as uricosomes.
4. Peroxisomes are ovoid or granular structures, limited by a single unit membrane and have a diameter of 0.5 to 1.0 μm.
5. In green leaves of C_3 plants, peroxisomes carry out photorespiration.
6. In animal cells they carry out lipid metabolism.
7. They contain important enzymes as oxidases (peroxide producing enzyme), peroxidases and catalases (which break down toxic peroxides to water and oxygen).

(D) Glyoxysomes

1. Beevers (1961) was the first person to discover these organelles and were described later by R.W. Briedenbach (1967).
2. They are about 0.5 to 1 μm in size and are surrounded by a single unit membrane.
3. They are found in plant cells, particularly, in germinating fatty seeds *e.g. Ricinus* (castor) and groundnut where fat is being converted into carbohydrates by a process called glyoxylate cycle.
4. Glyoxysomes contain two important enzymes, isocitrate lyases and maltase synthetase alongwith several others.
5. Structure of Glyoxysomes are similar to peroxisomes.

(E) Lomasomes

1. They were discovered by Moore and McAllister (1961) in fungi.

2. These membranous and vesicle like structures are present between plasmalemma and hyphal cell wall.
3. They seem to help in cell wall formations.
4. They can be compared with mesosomes of prokaryotes.

IX. Plastids

1. One of the main distinguishing characteristic of most plant cells is the presence of plastids.
2. The credit for the discovery of these organelles is generally given to Schimper (1880).
3. The term 'plastid' was coined by E. Haeckel (1866).
4. Plastids generally contain pigments and may synthesise and accumulate various substances. However, some plastids are devoid of pigments and they serve primarily for the storage of various substances.
5. The plastids are of three types:
   (a) Leucoplasts
   (b) Chromoplasts and
   (c) Chloroplasts.

X. Ribosomes

1. Ribosomes were first observed by Claude (1941) who named them 'microsomes'.
2. Palade (1955) described them in animal cells under the electron microscope as dense particles or granules and Robert (1958) coined the term 'ribosome'.
3. They are one of the smallest organelles of the cell.
4. In prokaryotes, the ribosomes are free in the cytoplasm.
5. In eukaryotes, they are either free in the cytoplasmic matrix or attached to the membranes of endoplasmic reticulum by glycoproteins called ribophorins. In eukaryotes, ribosomes are also found in mitochondria and chloroplasts.
6. During protein synthesis, many ribosomes become attached to the mRNA molecule, forming a polyribosome or polysome which is the site for polypeptide synthesis.
7. They are approximately spheroid in shape with a diameter of 23 Å.
8. Each ribosome consists of a large and a small subunits.
9. The association and dissociation of these subunits depend upon the Mg^{++} concentration in the cytoplasm. The two subunits unite to form a complete ribosome when Mg^{++} concentration is 0.001 M and when Mg^{++} concentration is lowered, the subunits of each ribosome redissociate. If Mg^{++} concentration is increased ten folds (0.01 M), ribosomes combine in pairs to form dimer each with a sedimentation coefficient of 120 S.
10. Ribosomes are generally measured on the basis of their sedimentation coefficient which is expressed in Svedberg (S) units, as given under:

Examples	Ribosomes	Subunits
1. Eukaryotes	80 S	60 S and 40 S
2. Prokaryotes	70 S	50 S and 30 S
3. Fungi (Yeast and *Neurospora*)	70 *S*	50 *S* and 30 *S*
4. Mitochondria	50–60 *S* (Variable)	40 *S* and 30 *S*
5. Chloroplast	70 *S*	50 *S* and 30 *S*

11. **Composition:** Ribosomes are composed of RNA (rRNA) and protein present almost in equal amounts. In addition, they have enzymes, lipids and metallic ions (Mg, Ca, Mn and Fe).
12. **Important functions of ribosomes**
    (i) Ribosomes are the site for protein synthesis in the cell.
    (ii) It is the assembly shop or protein factory where amino acids are linked to form proteins.
    (iii) It gives protection of mRNA and newly synthesized polypeptide chains from enzymes, nucleases and proteases respectively.

XI. Mitochondria

1. They were first observed by Kolliker (1880) in the muscle cells.
2. Flemming called them 'fila' (1882).

3. Altmann in 1894 described them as 'biplasts'.
4. Later, they were called 'mitochondria' by Benda in 1897.
5. The presence of mitochondria in plant cells was first detected by F. Meves in 1904 in *Nymphaea*.
6. Kingsbury (1912) suggested that mitochondria are seat for cellular respiration.
7. Later, Hogeboomet (1948) demonstrated that mitochondria are the seat for cellular respiration.
8. The shape, size and number of mitochondria in a cell are extremely variable. Mitochondria are generally rod-shaped (filamentous) about 0.2-2 μm long. They are in general uniformly distributed throughout the cytoplasm, however, in some cases, they accumulate preferentially around the nucleus or in the peripheral cytoplasm. There are fewer mitochondria in green plant cells than in animal cells, since some of their function is taken over by chloroplast in plant cells. Some unicellular algae *e.g. Microsterias* and *Micromonas* have a single mitochondrion. In a normal liver cell, there are about 1000 to 1600 mitochondria per cell.
9. **Structure of a mitochondrion**
   (i) Structurally, a mitochondrion consists of two membranes, an outer limiting membrane and an inner membrane.
   (ii) The inner membrane is most frequently convoluted to form a large number of infoldings called cristae or mitochondrial crests.
   (iii) The membranes of the mitochondrion separate it into two distinct compartments.
   (iv) An outer compartment, sometimes called intermembranal space, is formed between the outer and inner membrane.
   (v) The inner compartment, also called mitochondrial chamber, is bound by the inner membrane.
   (vi) This mitochondrial chamber is filled with dense proteinaceous material called the matrix,
   (vii) The mitochondrial matrix contains most of the enzymes of Kreb's cycle.
   (viii) The ribosomes in mitochondria are smaller than cytoplasmic eukaryotic ribosomes and there is considerable variation in the sizes of the subunits and the monomers formed from them as given below

Animal mitochondrial ribosome

Type	Sedimentation coefficient
Monomer	50 – 60 *S*
Large Subunit	40 – 45 *S*
Small Subunit	30 – 35 *S*

   (ix) The inner membrane in the crest region is covered by the particles of 8.5 nm size. These particles are called elementary or F_1 particles or o..ysomes or respiratory assemblies.
   (x) There are about 10^4 to 10^5 elementary particles per mitochondrion. These particles contains a special ATPase enzyme involved in oxidative phosphorylation.
10. **Composition:** Mitochondria consists of approx. 65-70% proteins, 30-35% lipids and small amounts of DNA and RNA.
11. **Function of mitochondrion:** They are seat for cellular respiration and produce energy in the form of ATP (*i.e.*, cellular currency) needed for cellular functions. Thus they are called power houses.

XII. Vacuole

1. Vacuoles are liquid-filled cavities, the presence of which is characteristic features of plant cells.
2. In young dividing cells, vacuoles are small and numerous and remain scattered throughout the cytoplasm while in mature cells there is a single large central vacuole in each cell and it is surrounded by a thin membrane called tonoplast.
3. It is filled with a fluid called cell sap which is the solution of sugars, salts, organic acids, glucosides and alkaloids.
4. Many vacuoules contain a pigment called anthocyanin which gives red colour in very acidic sap and blue to purple colour in alkaline medium.

5. Due to the change of the colour of anthocyanin with the change in pH, anthocyanin has been called 'vegetable chameleon'.
6. Functions of vacuoles-
   (a) The vacuole acts as a sub-cellular storage organ of useful metabolities.
   (b) It also maintains turgidity of the cell which is very useful for herbaceous plants.

XIII. Hyaloplasm

1. Hyaloplasm or matrix of the cytoplasm contains some nonprotoplasmic materials called ergastic substances produced during metabolic activities of the protoplasm.
2. They are characterised by their inactivity in protoplasm. They are present in the form of solutions in vacuolar sap or may remain dispersed in colloidal condition even in the cytoplasm.
3. The ergastic matters can be placed into three groups:

   (i) **Reserve materials:** They are food materials of the plants and they are concerned with their nutrition and maintenance of turgidity of cells, particularly in herbaceous plants.

   (ii) **Excretory materials:** These substances regain accumulated in laticiferous tissues and ultimately pass out through leaf, fruit etc. They are secretory products of the protoplasm *e.g.* enzymes, nectar and pigments such as anthocyamin.

   (iii) **Excretory materials:** These substances remain accumlated in laticiferous tissues and ultimately pass out through leaf, fruit etc. They are alkaloids (*e.g.*, 'Quinine' in the bark of '*Cinchona*', 'Caffeine' in coffee beans, 'Theine' in tea leaves, 'Nicotine' in tobacoo leaves, 'Reserpine' in the root of *Rauwolffia serpentina*, 'Morphine' in poppy fruits); glucosides in Crucifers; fats and essential oils in *Eucalyptus* and resins and gums in Conifers.

Mojor Functions of Cell Organelles

Cell Organelles	Functions
Cell wall (plant cells)	It forms a thick cellulose wall around the plasma membrane giving strength and rigidity to the cell.
Plasma membrane	It is diffrentially permeable membrane through which extracellular substances may be selectively sampled and cell products may be liberated.
Cytoplasm	It contains machinery for carrying out the instructions sent from the nucleus.
Hyaloplasm or Matrix	It contains enzymes for glycolysis and structural materials such as sugars, amino acids, water, vitamins, nucleotides etc.
Microtubules	They transport water, ions and small molecules of various substances, form structural unit of centrioles, basal granules, cilia, flagella and mitotic spindle and play active role in the cyclosis of cytoplasm and in the movement of chromosomes during cell division.
Microfilaments	They maintain the structure of cell and form contractile components of muscle cell.
Centrioles	They form the spindle and have significant role in the movement of chromosomes during cell division.
Basal granules	They have important role in ciliary and flagellar movements.
Cillia and Flagella	These are extracellular projections of the cell which provide movement to the cells and serve various purpose of cells.

Endoplasmic reticulum	It provides greatly expanded surfaces area for biochemical reactions which normally occur at or across membrane surfaces and provide skeletal support to the colloidal complex of hyaloplasm. It also has transporting business of extracellular and intracellular chemical molecules.
Ribosomes	These are the sites for synthesis of protein molecules.
Golgi complex	It stores the synthetic proteins and enzymes of cells and forms secretory vesicles and lysosomes.
Lysosomes (animal cell only)	It contains intracellular digestive enzymes which aid in disposal of bacteria and other foreign bodies.
Peroxisomes	They contain enzymes for hydrogen peroxide metabolism purine metabolism, gluconeogenesis and photorespiration.
Sphaerosomes	They store fats.
Vacuoles	They are storage depots for excess water, waste products and soluble pigments.
Mitochondria	They are the organelles for the production of energy key Kreb's cycle, electron transport chain and beta-oxidation of fatty acids.
Plastids	They are the structures for storage of starch, pigments and other cellular products. Photosynthesis occurs in chloroplasts.
Nucleus	It regulates growth and reproduction of cell.
Nuclear membrane	It provides selective continuity between nuclear and cytoplasmic materials.
Nucleoplasm	It contains materials for building DNA and different kinds of RNA molecules which act as intermediates between nucleus and cytoplasm.
Chromosomes	They are the bearers of hereditary instructions and regulations of cellular processes.
Nucleolus	It disappears during replication and synthesizes the ribosomes and proteins for the formation of ribosomes.

Special Points to Note

1. The terms 'thylakoid' was first coined by Menke (1961).
2. RNA and DNA were first seen by Ris and Plant in 1962 in chloroplast.
3. The ribosomes in chloroplast were first seen by Jacobson *et. al.* in 1963.
4. The quatasome was discovered by Park and Biggins.
5. The plastid system carrying the genetic information is collectively called plastidome.
6. Ribosomes of mitochondria were first observed by Kislevet *et. al* (1965).
7. A circular DNA molecule was first observed in mitochondria by Nass and Nass (1963).
8. The elementary particles of F_1 particles were first observed in mammalian mitochondria by Fernandez Moran (1962).
9. Since mitochondria and chloroplast are able to replicate and synthesize some of their own proteins, they are thought to be semiautonomous cellular inclusions.
10. About 40% of the proteins in mitochondria are enzymes.
11. Hyaloplasm contain raphides and sphaeroraphides (crystals of calcium oxalate); cystolith (calcium carbonate crystals) and organic acids like tartaric acid (in tamarind), citric acid (in citrus fruits) and malic acid (as in apple).
12. Histones contain high amount of either lysine or arginine and small amount of aspartic acid and glutamic acid. Only amino acids which never form histones are sulphur containing ones and tryptophan.

13. Histone proteins are basic and non histone proteins are acidic in nature. Non histone proteins are also known as High Mobility Group Proteins (HMGS).
14. The nucleus of average human cell is only 6 μm (6×10^{-6} m) in diameter, yet it contains 1.8 m of DNA which is distributed amongst 46 chromosomes each consisting of single DNA molecule about 40 mm.
15. Approximate chemical composition of a typical mammalian cell is

Components	*% of total cell weight*
H_2O	70
Inorganic ions (Na^+, K^+, Mg^{++}, Ca^{++}, Cl^-)	1.0
Miscellaneous small metabolites	3.0
Proteins	18
RNA	1.1
DNA	0.25
Phospholipids and other lipids	5.0
Polysaccharides	2.0

16. The Relative Volumes of Organelles in Plant Cells

Organelle	*Young Cell*	*Mature Cell*
Nucleus	32.4	0.23
Vacuole	4.93	83.3
Mitochondria	5.35	1.01
Plastids	3.72 (proplastids)	3.16 (chloroplasts)
Dictyosomes	0.40	0.04
Hyaloplasm	52.9	12.1
Lipid	0.23	0.00

17. Young cells tend to be small with large nuclei; as they grow and differentiate, the cell volume increases and certain organelles become more prominent, depending on the specialization of the cell. The relative volumes of organelles (the volume of the organelle expressed as a percentage of the volume of the whole cell) are given here for a young cell from a shoot tip and for a mature, photosynthetic cell from the outer part of the stem.

MULTIPLE CHOICE QUESTIONS

1. Which of the following criteria can be used to distinguish between prokaryotic and eukaryotic cells?
   A. whether or not they have plasma membrane
   B. whether or not they have a nucleus (membrane-enclosed nucleoplasm)
   C. comparison of the sequences of homologous proteins
   D. comparison of the signature sequences of 16S rRNA

2. Which one of the following arrangement of the metal ions Na^+, K^+, Mg^{++} and Ca^{++} in the order of decreasing concentrations is correct with respect to a quiescent mammalian cell?
   A. K^+, Na^+, Mg^{++}, Ca^{++}
   B. Na^+, K^+, Mg^{++}, Ca^{++}
   C. Mg^{++}, K^+, Na^+, Ca^{++}
   D. Ca^{++}, Mg^{++}, K^+, Na^+

3. The primary thermodynamic factor that favours the formation of a lipid bilayer in aqueous surroundings is
   A. hydrogen bonds
   B. hydrophobic force
   C. ionic interactions
   D. covalent bonds

4. What property of biomembranes is responsible for their self-sealing nature?
   A. hydrophilicity of the phospholipid head group
   B. presence of protein in biomembranes
   C. presence of cholesterol in biomembranes
   D. hydrophobicity of the fatty acid side chains of phospholipids

5. Most abundant lipid in plasma membrane is
   A. cholesterol   B. sterol
   C. glycolipid   D. phospholipids

6. A poikilothermic organism living in the Arctic would have, compared to that organism living in the temperate climate zone, a plasma membrane richer in
   A. cholesterol
   B. long chain fatty acids
   C. protein
   D. unsaturated fatty acids
7. Proteins that are embedded in or reach across a membrane are characterized by
   P. a stretch of at least 16 to 21 hydrophobic amino acids in their primary structure
   Q. a high lysine, arginine and histidine content
   R. a high glutamate and aspartate content
   S. at least one α-helical or β-sheet domain
   A. P, Q B. P, R
   C. P, S D. All of these
8. Diffusion across the plasma membrane is more rapid if a substance is
   A. a protein
   B. hydrophilic
   C. high in its oil : water partition coefficient
   D. large and globular in shape
9. When an ion or solute is moved against a concentration gradient using energy, the process is called
   A. diffusion
   B. transport
   C. active transport
   D. regulated diffusion
10. Which one of the following is a *correct* statement for Na-K ATPase?
   A. it gives out 3 Na-ions and takes in 2 K-ions
   B. it gives out 2 Na-ions and takes in 3 K-ions
   C. it gives out 3 Ca-ions and takes in 2 K-ions
   D. it gives out 3 Na-ions and takes in 2 Ca-ions
11. Which of the following are *correct* about transport ATPase
   P. P, F and V type transport ATPase transport ATPase
   Q. Na^+/K^+ pump is an example of P type transport ATPase
   R. all members of the ABC superfamily contain two cytosolic ATP-binding domains
   S. F and V-class ATPase do not form phosphoprotein intermediates and transport only protons
   A. P, R B. Q, R, S
   C. All of these D. P, Q, R
12. The process by which a cell secretes macromolecules by fusing a vesicle to the plasma membrane is called
   A. endocytosis B. exocytosis
   C. pinocytosis D. phagocytosis
13. Receptor mediated endocytosis from plasma membrane requires which one of the following coat proteins
   A. clathrin B. SNARE
   C. arrestin D. glycophorin
14. Which one of the following receptors is associated with an ion channel
   A. acetylcholine receptors
   B. EGF receptor
   C. corticosteroid receptor
   D. growth hormone receptor
15. Which of the following statement is *correct* for the endoplasmic reticulum (ER)?
   A. the smooth ER is the site for the synthesis, modifications and sorting of proteins
   B. the rough ER is the site for lipid biosynthesis
   C. the rough ER is the site for the modifications and sorting of proteins
   D. the rough ER is the site for protein synthesis and smooth ER is involved in modifications and sorting of proteins
16. Membranes of the following two organelles are contiguous
   A. ER and golgi
   B. nucleus and ER
   C. golgi and plasma membrane
   D. golgi and lysosomes
17. Which of the following statements is *true* for smooth endoplasmic reticulum
   A. it is made up of a single lipid layer membrane structure
   B. it is involved in protein biosynthesis
   C. it is involved in lipid biosynthesis
   D. it is a site for interleukin-2 biosynthesis

18. Marker enzyme of golgi apparatus is
A. acetyl CoA synthetase
B. pyruvate kinase
C. galactosyl transferase
D. cytochrome oxidase

19. A distinctive features of the lysosome is that it has
A. a lower pH than the cytoplasm
B. a reduced hydrolase activity
C. DNA
D. ribosome

20. In autophagy
A. a single lysosome surrounds a single cell organelle
B. cellular constituents are recycled
C. internal membranes can be synthesized
D. mitochondria are assembled

21. During the synthesis of secretory proteins the following events happen
1. the signal recognition particle (SRP) binds to the signal sequence
2. the two parts of a cytosolic ribosome binds to the mRNA
3. protein synthesis stops
4. the signal sequence is translated
5. the ribosome and protein are transferred to a translocon
6. the secretory protein folds
7. the signal sequence is cleaved by a signal peptidase
8. the SRP binds to its receptor in the ER membrane

The *correct* order for these events is
A. 4, 1, 2, 3, 6, 8, 7, 5
B. 2, 4, 1, 3, 8, 5, 7, 6
C. 1, 4, 2, 3, 6, 5, 8, 7
D. 2, 4, 1, 3, 5, 8, 7, 6

22. Which of the following show *correct* order of the secretory pathway?
A. RER → Golgi → Secretory vesicle → Cell exterior
B. SER → Golgi → Secretory vesicle → Cell exterior
C. Golgi → SER → Secretory vesicle → Cell exterior
D. Golgi → Lysosome → SER → Secretory vesicle → Cell exterior

23. Which of the following cell components is *not* part of the cytoskeleton of eukaryotic cells?
A. microfilaments
B. mitochondria
C. intermediate filaments
D. microtubules

24. A monoclonal antibody binds to G-actin but not to F-actin, what does this tell you about epitope recognized by the antibody?
A. the epitope in F-actin induces an enzymatic activity inactivating mAb
B. the epitope in F-actin is surrounded by Myosin thus no mAb interaction is observed
C. the epitope is likely to be a structure that is buried when G-actin polymerizes to form F-actin
D. a peptide fragment released from F-actin blocks the mAb interaction with the epitope

25. Which of the following protein targeting mechanisms is common for mitochondria and chloroplast organelles?
A. co-translational
B. post-translational
C. co-translational and post-translational
D. *De novo* protein synthesis

ANSWERS

1	2	3	4	5	6	7	8	9	10
B	A	B	D	D	D	C	C	C	A
11	**12**	**13**	**14**	**15**	**16**	**17**	**18**	**19**	**20**
C	B	A	A	C	B	C	C	A	B
21	**22**	**23**	**24**	**25**					
B	A	B	C	B					

4

Genetics and Molecular Biology

HEREDITY AND VARIATION

This topic concerns the science of transmission of characters from one generation to the next and the mechanism of action of hereditary units as they bring about the characteristics which they controls. It is genetics.

Pre-Mendelian history

Even before this science got founded, man had some vague idea regarding heredity. Several different theories were forwarded by different persons as below:

Vapour and fluid theory (by Pythogoras)

He suggested that some moist vapour is given out from the brain, nerves and other parts of the body of male during coitus. One account of these vapours, the off-springs exhibit similarities with the male parent.

Semen Theory (by Empedocles)

He suggested that both parents produce semen which arises directly from its various body parts. The semen from both parents mixes and produces a new individual.

Aristotle's idea (by Aristotle)

He suggested that the semen of male is highly purified blood and possesses life-giving power. The semen of female was its menstrual fluid which furnishes nourishing material to the developing off-spring.

Preformation theory (by Swammerdam)

According to this theory the gametes had the miniature copy of adults and the development of embryo was actually only the enlargement of parts that were already present in the sperm or egg. He got support from Hartsoeker (1694) and Dalemponus (1699) who called the miniature human being, *Homunculus*.

Particulate theory (by Maupertuis)

According to this theory the body of each parent gives rise to minute particles. In sexual reproduction, these particles from both the parents unite together to form the daughter individuals.

Encasement theory (by Charles Bonnet)

According to this theory every female contains within her body miniature prototypes of all the creatures.

Theory of epigenesis (by C.F. Wolff)

In this theory the germ cells contain definite but undifferentiated substances which, after fertilization, become organized into various complex body organs which form the adult. This idea was referred to as epigenesis.

Pangenesis theory (by Charles Darwin)

According to this theory every cell, tissue and organ of animal body produces many minutes particles, known as *gemmules* or *pangenes*. These gemmules get deposited in the reproductive organs and by blending of two kinds of pangenes, an individual is formed.

Weismann theory (by A. Weismann)

He called reproductive cells as *germplasm* and rest of the body as *somatoplasm*. The germplasm forms the bridge of life and passes from one generation to the next.

Mendelian Genetics

G. J. Mendel conducted hybridization experiments with garden pea *Pisum sativum*. Though first published in 1866, such valuable work of Mendel remained unnoticed, as it was published in an obscure Journal by an unknown monk. His work was rediscovered after 35 years, in 1900, when three botanists, Hugo de Vries from Holland, C. Correns from Germany and E.V. Tschermak from Austria independently discovered Mendel's paper and reviewed Mendel's laws of inheritance.

Variations

Variations are differences found amongst individuals of the same species, race and family.

Sources of variation

Replication of DNA is so nearly perfect that there is almost nil possibility of variations occuring in the genotypes of asexually reproducing organisms, except, may be as the result of environmental influences. In case of sexually reproducing oraganisms there however is ample opportunity for genetic variation to arise. Two processes occuring during meiosis and the fusion of gametes during fertilisation provide for introducing unlimited genetic variation into the population.

Classification of variation

Variations are classified into the following types on many basis:

I. Basis: *Affected trait*

1. **Morphological variations:** They are the differences that are found in form and structure.
2. **Physiological variations:** They are the variations observed in various functions of the body like basal metabolic rate etc.
3. **Cytological variations:** They are those variations which occur in the number of cells, cell constituents and their products.
4. **Behaviouristic variations:** The variations are connected with behaviour of the organisms towards different conditions.

II. Basis: *Impact*

1. **Beneficial variations:** They are variations which help the organism in adapting to a particular environment.
2. **Harmful variation:** These variations make organism unfit to its environment.
3. **Neutral variations:** They do not affect organism in any way.

III. Basis: *Parts*

1. **Meristic variations:** These are the variations which involve changes in the number of parts, segments etc, *e.g.*, in human beings hands and feet can have extra digits (6 instead of 5). The phenomenon is known as polydactyly. The number of digits can be less due to fusion called as syndactyly.
2. **Substanctive variations:** They are variations which affect whole of the organism. *e.g.* In human beings, substantive variations are found in height and weight of individuals belonging to same age.

IV. Basis: Cells affected

1. **Blastogenic or Germinal variations:** The variations which affect the germ cells of an organism and are inheritable. They are trasmitted from one generation to the next. They are caused because of number of factors as:
   (a) Chance separation of chromosomes
   (b) Chance combination of chromosomes
   (c) Crossing over
   (d) Chromosome aberrations
   (e) Change in chromosome number
   (f) Gene mutations
2. **Somantogenic or Somatic variations:** They are the variations which affect only the somatic cells of the organisms. They are neither inherited nor transmitted to the next generation, *e.g.*, well developed body muscles of an athlete, boring of pinna etc. Somatogenic variations are caused by three types of factors:
   (*a*) ***Environmental factors***: They include temperature, light, nutrition, water supply, humidity, medium or habitat, wind, topography, enemies, etc. Such variations are called as environmental variations.
   (*b*) ***Use and disuse of organs***: *e.g.* vocal organs are more developed in singer than a nonsinger.
   (*c*) ***Conscious efforts***: Variations produced in human beings due to conscious efforts include receiving of education, learning of an art, development of some habits, boring of pinnae etc.

Heredity

Mendel's experiments

Mendel studied the inheritance of seven different pairs of contrasting characters in garden pea (*Pisum sativum*) plant but considered only one pair at a time. The pea plant has a number of contrasting characters. Out of these Mendel selected only seven characters:

Characters	Dominant	Recessive
1. The length of the stem	Tall	Dwarf
2. The position of the flower Terminal	Axial	
3. The colour of the pod	Green	Yellow
4. The shape of the pod constrict	Inflated	
5. The shape of the seed Wrinkled	Round	
6. The colour of the seedcoat	Coloured	White
7. The colour of the cotyledon	Yellow	Green

Now it is known that the genes controlling the seven pea characters are located on four chromosome, namely chromosomes 1, 4, 5 and 7. However the genes for the characters are not linked.

Law of Dominance

Each organism is formed of a bundle of characters. Each character is controlled by a pair of factors of genes (T or t). Each factor of the paired factors (Tt) is responsible for the expression of a particular variety *e.g.* (tall or dwarf) of a characters height). As per the law of dominance one factor in a pair, may mask or prevent the expression of the other. The variety that shows in the F_1 generation of monohybrid cross is known as dominant while those which did not appear in the F_1 generation is called recessive. A recessive factor freely expresses itself in the absence of its dominant allele. This law is formulated based on the monohybrid experiment.

Law of segregation

Mendel's first law: The two genes of a particular character remain uncontaminated till they are inside the organism. During gamete formation the paired genes segregate and enter different gametes. Thus each gamete contains only one of the paired genes that are responsible for any particular character. During gamete formation, the genes of the particular character separate and enter different gametes. This is the law of segregation. It is also formulated based on monohybrid experiment. For example, during gamete formation in F_1 hybrid tall plants the paired factors (Tt) present in the plants segregate independently and enter different gametes. So each gamete receives either T or t from the paired factors Tt that are responsible for the expression of a single character.

Law of independent assortment

Mendel's second law: This law is based on dihybrid experiment. As per this law, the genes for each pair of characters separate independently from those or other characters during gamete formation. For example, during gamete formation of a dihybrid cross, the factors, for yellow colour assort out independently of the factors for round shape. The gene-Y may combine with the dominant gene-R or the recessive gene r of the other character and enter a gamete. Similarly the gene-y may combine with the dominant gene-R or the recessive gene-r and enter a gamete. Thus, the F_1 hybrid plants produce four types of gametes and they are YR, Yr, yR and yr.

Genetic Interaction: Factor Hypothesis

1. As per Mendel each character is controlled by a pair of factors or genes. But, later, it was proved that in many cases the expression of a single character is controlled by the interaction of more than one pair of genes. This is known as *interaction of genes* or *factor hypothesis*, proposed by *Bateson* and *Punnet*. It states that certain characters are controlled by the interaction of two or more genes.
2. **Non allelic and allelic gene interaction:** The genic interaction may occur in between genes located in the same chromosome or different chromosomes. This type of genic interaction is known as *non allelic gene interactions.* The genic interaction may also occur, between the two alleles of a single type of gene. This type of genic interaction is called *allelic gene interaction*.

Polygenic (Quantitative) Inheritance

Mendel stated that each character is controlled by two genes located at the same locus, but certain characters are controlled by more than two genes located at different loci, called *multiple genes* or *polygenic* or *cumulative genes*. The characters controlled by cumulative genes are not sharply marked from one individual to other. But they are continuous with many grades of intermediates. Thus these characters are quantitative rather qualitative. Thus, this is also called *quantitative inheritance*, e.g. (i) colour of wheat kernel, (ii) skin colour in man (iii) eye colour in *Drosophilla* (iv) body size in chicken (v) coat colour in cattle and (vi) height in man.

GENES AND CHROMOSOMES

Gene

Gene is a unit of hereditary material located on a chromosome that, by itself or with other genes, determines a characteristic in an organism. Genes may exist in a number of forms, termed alleles. For example, a gene controlling the characteristic 'height' in peas, may have two alleles, one for 'tall' and another for 'short'.

Genes can occasionally undergo changes, called mutations, to new allelic forms.

A gene can be defined as the smallest hereditary unit capable

(i) either of recombination
(ii) or of mutation
(iii) or of controlling a specific function.

Chromosome

It is one of a group of thread like structures of different lengths and shapes in nuclei of eukaryotic cells. During nuclear reactions they are tightly coiled and are easily visible through the light microscope. After division, they uncoil and become difficult to see. They controll cell activity because they carry genes. They consist of DNA with RNA and protein. The number of chromosomes per nucleus is characteristic of the species for example man has 46.

Discovery of Chromosomes

Hofmeister (1848): First observed chromosomes in microsporocytes of *Tradescantia*.

Flemming (1879): Observed splitting of chromosomes during cell division and coined the term, chromatin.

Benedin (1883): Observed doubling of chromosomes number in the fertilized egg or threadworm.

Roux (1883): He believed the chromosomes take part in inheritance.

W. Waldeyer (1888): He coined the term '*chromosome*'.

Benedin and Boveri (1887): They found a *fixed number* of *chromosomes* in each species.

Sutton and Boveri (1902): They proposed *chromosome theory of inheritance*.

Kinds of Chromosome

1. **Viral chromosomes:** In viruses and bacteriophages a single molecule of DNA or RNA represents the viral chromosome. The DNA viral chromosomes occur mostly as linear complex and is found in animal viruses and bacteriophages while RNA containing viral chromosomes are single stranded and occur in plant viruses.
2. **Bacterial chromosomes:** In bacteria and cyanobacteria, the hereditary matter is oragnised into a single large, circular molecule of double stranded DNA, which is loosely packed in the nuclear zone. It is known as bacterial chromosome or nucleoid. It has DNA and some RNA but no protein. It remains attached either to the plasma membrane or to its invagination. In addition to main chromosome, a few small ring-shaped DNA molecules also occur in the cytoplasm. These are known as plasmids. The contour length of circular DNA of *E. coli* is about 1100 μm, as the cell diameter is only 1.2 μm, the chromosome is highly coiled or folded. It shows about 50 folds or domains.
3. **Eukaryotic chromosomes:** Chromosomes of eukaryotic cells are specific individualized bodies, formed of deoxyribonucleo-proteins (DNA + Proteins). Their number, shape and size varies in organisms of different species.

Chromosome Number

The number of chromosomes remains throughout their lives individual and in all the individuals of a species. Therefore, chromosome number is a perfect identification of species and in tracing the relationship within the species. The number of chromosomes in the somatic cells is always just the double of what is present in the gametes. Therefore, gametes are described as haploid (n) and somatic cells are diploid (2n). For example, in man sperm and ova possess 23 chromosomes (n), while the normal body cells have 46 or 23 pairs (2n) of chromosomes.

Chromosome Cycle

Chromosomes show cyclic change in their shapes and sizes during a cell cycle. In the nondividing interphase nucleus the chromosomes form an interwoven network of fine twisted but uncoiled threads of chromatin and are invisible. During cell division the chromatin threads condense into compact structures by helical coiling. In prophase of cell division the chromosomes appear as distinct threads and by metaphase and also in anaphase these become short compact bodies having difinite shapes and sizes. In anaphase these appear as rod-shaped, V-shaped L-shaped or J-shaped. The chromosomes are studied and described at metaphase or at anaphase stage. In telophase they again uncoil to form the chromatin net.

Classification of Chromosomes

Depending upon the position of kinetochores, the chromosomes are classified in four types:

1. Acrocentric are rod-shaped chromosomes with terminal centromere.
2. Telocentric are chromosomes with a terminal centromere.
3. Submetacentric are L-shaped chromosomes with centromere slightly away from the mid-point so that the two arms of the chromosome are unequal.
4. Metacentric chromosomes have centromere in the middle so that the two arms of the chromosome are almost equal.

Giant Chromosomes

These are exceptionally larger chromosomes, described as unusual chromosomes by A.M. Winchester. There are two types of giant chromosomes, namely *polytene chromosomes* and *lampbrush chromosomes*.

FURTHER ABOUT GENE

Linkage

It is defined as the tendency of two or more genes to remain together in the original combination in the same chromosome during the process of inheritance for number of generations. This theory of linkage was propounded by T.H. Morgan, *e.g.*, the genes in man for haemophilia are linked with colour blindness.

Coupling and repulsion

The terms coupling and repulsion were, coined by, Bateson and Punnet in 1906. These are two aspects of linkage. The two dominant alleles or recessive alleles of a character coming from the same parent tend to enter the same gamete and inherit together for many generations. This is called coupling. Similarly, the two dominant alleles or recessive alleles of a character coming from two parents separate from each other and enter different gametes called repulsion.

Types of Linkage

There are two types of linkage. They are complete linkage and incomplete linkage:

1. **Complete linkage:** It is the phenomenon in which two or more genes or characters are inherited together for a number of generations. Complete linkage is due to the fact that there occurs no break in the chromosomes. As a result of complete linkage, the young ones inherit only the parental characters. New characters do not appear among the young ones. So complete linkage produces only parental combination and new combinations do not arise. This phenomenon is very rare. It is found only in male *Drosophila*.
2. **Incomplete linkage:** Here, the linked genes on certain occasions separate. This leads to the formation of new combinations among the young ones. Incomplete linkage is due to the breakage of chromosomes, during gametogenesis. Incomplete linkage is found in female *Drosophilla*. The new genic combination of the offspring produced in the F_2 generation of different from their parent in their phenotype and genotype. So the

incomplete linkage involves the accidental breakage of chromosomal segments or linked genes, resulting in new combination of genes.

Linkage groups

All the genes which are linked with one another form a linkage group. Since linked genes are present in the same chromosome, the number of linkage group in an animal or plant is equal to the haploid number of chromosomes present in its cells. This hypothesis was given by Sutton and was proved by experiments on *Drosophila* by T.H. Morgan.

Strength of linkage

The strength of linkage between any two pairs of linked genes of a chromosome depend upon the distance between them. Closely located genes show strong linkage, while genes widely located show weak linkages.

Theories of linkage

There are two theories on linkage, as following:

1. **Differential multiplication theory:** Bateson (1930) assumed that linkage is due to the differential multiplication of cells containing different combinations of genes. After the segregation of genes during gamete formation the set of gametes possessing parental combinations multiplies rapidly than the set having non parental combination. This theory was disproved because after segregation of genes, there is only one division.
2. **Chromosome theory of linkage:** This theory was proposed by Morgan and Castle. This theory states that: Genes located in the same chromosome are inherited together and are said to be linked. The linked genes are arranged in a linear fashion on the chromosome. The degree of linkage is determined by the distance between genes

Linkage is affected by the distance, age, temperature, X-rays.

SEX DETERMINATION

Sex Determination in Plants

1. **Bryophytes (by Allen):** Male gametophyte has seven autosomes and Y chromosome. The female has seven autosomes and an X chromosome. After union of egg and sperm a diploid zygote is formed with 14 autosomes and one X and one chromosome.
2. **Higher plants (by Correns):** In higher plants either male or female is heterogametic.
   (i) Angiosperms: Female is homogametic (XX) and male is heterogametic (XY).
   (ii) Dioscorea: Female is homogametic (XX) and male is heterogametic (XO).
   (iii) Flagellaria: Female is heterogametic (ZW) and male is homogametic (ZZ).
   (iv) Melandrium album: The diploid female plants have 22 autosomes and 2X chromosomes. The diploid male plants have 22 autosomes and XY chromosome.

Sex Determination in Animals

It is done by following theories:

1. **Theory of heterogamesis:** (By correns) They reported two types of chromosomes autosomes and allosomes (sex chromosomes). Examples:
   (i) In man, *Drosophila*: Female has XX and male has XY.
   (ii) In birds, butterflies and few fishes: Male is XX (ZZ) and female is XY(ZW).
   (iii) In grasshopper: Male is XO and female is XX.
2. **Genic balance theory:** (by Bridges) According to this theory sex is determined by the relative number of X chromosomes and autosomes. It is actually the ratio between the X chromosomes and autosomes which determines the sex.
3. **Quantitative theory of sex determination:** (by Goldschmidt) He proposed this theory to explain the existance of intersexes in gypsy moth, *Lymantria dispar*.
4. **Haploidy and diploidy mechanism of sex determination:** Development of an ovum into young one without fertilization is parthenogenesis. *e.g.* in honeybees wasps and ants females are diploid and male (drones) are haploid.
5. **Gynandromorphs:** They are individuals which show male characters on some parts of the body and female characters on other parts of the body. They are also called gynanders. They are sterile.

6. **Cytological basis of sex determination:** (by Barr and Bertram) The mammalian cells of certain sexes contain a darkly staining body in the nucleus called as sex chromatin or Barr body. A normal female has one X chromosome and one Barr body. Male has none.
7. **Environmental sex determination:** (by Baltzar) *Bonellia* is marine worm which exhibits sexual dimorphism. The male lives as a parasite in the uterus of the adult female.
8. **Hormonal theory of sex determination, Free-martin:** (by Lillie) If the twins of opposite sex are born to cattle, male will be normal but the female will be sterile. This is called as Free-Martin. The male develops gonads little earlier and placenta pass sex hormones into the female twin causing sterility.
9. **Determination of sex by parasites:** In case of crabs, if a crustacean parasite, *Sacculina* is present in male, it is changed to female in course of time.

INHERITANCE

Sex Linked Inheritance

Chromosomes are the carriers of genes. Genes are present both on autosomes and sex chromosomes. The genes present on the autosomes control the somatic characters of the organism. But the genes present on the sex chromosomes control the sex-characters. However, certain genes present in the sex chromosomes control the somatic characters. The body characters whose genes are located on the sex chromosomes are called sex linked characters because they follow sex during inheritance.

Sailent feature of sex linked inheritance

1. The body characters controlled by genes located on sex chromosomes are called sex linked characters.
2. The sex linked characters are transmitted along with the sex of the animals.
3. The sex linked genes are located on X or Y or both XY chromosomes.
4. Most of the sex linked genes are located on the X chromosomes and their alleles are absent from Y chromosomes.
5. Most of the sex linked characters are recessive.
6. They are more common in man than in woman.
7. They follow criss-cross inheritance or zig-zag inheritance. That is, the sex linked character is transmitted from the father to the grandson through his daughter.

The sex linked inheritance was first discovered by Morgan in 1910 in *Drosophila melanogaster*.

There are five types of sex linked inheritance. They are as follows:

1. **X linked inheritance:** Certain sex linked genes are located only on X chromosomes and their alleles are absent from Y chromosome. These genes are called X linked genes *e.g.* colourblindness, haemophilia in man and eyecolour in *Drosophila*.
2. **Y linked inheritance:** The sex linked genes located on Y chromosomes only are called Y linked genes. The Y linked genes are confined to males only. Hence they are also called holandric genes (Holos = whole; andros = male). These genes are transmitted directly from the father to the son. Their mode of inheritance is called Y linked inheritance.

   e.g. hypertrichosis (hairs in the pinna), ichthyosis hystrix (hystrix is characterised by the presence of scales on the body.
3. **XY linked inheritance:** Certain sex linked genes are located on both X and Y chromosomes. They are called XY linked genes and their mode of inheritance is called XY linked inheritance. *e.g.* Xeroderma pigmentosum, nephritis, retinitis pigmentosa etc.
4. **Completely sex linked inheritance:** The X and Y chromosomes are not similar. The X chromosome is larger and straight and the Y chromosome is smaller and is curved. The similar parts of X and Y are called homologous regions. They have the same types of genes. The remaining parts of the X and Y chromosomes which are not similar are called non-homologous regions or differential regions. They do not contain similar type of genes. The genes located on non homologous regions are inherited together because crossing

over does not occur in these regions. So the genes located on non homologous regions are called completely sex linked genes, and their inheritance is called completely sex linked inheritance. *e.g.* Haemophilia, colour blindness etc.

5. **Incompletely sex linked inheritance:** The genes located on homologous regions of sex chromosomes do not inherit together because of crossing over in these regions. So these genes are called incompletely sex linked. *e.g.* retinitis pigmentosa, nephritis etc.

Crossing Over (i.e. Recombination of Genes)

1. Morgan termed the word 'crossing over'.
2. Crossing over may be defined as the interchange of chromosomal parts between non-sister chromatids of a homologous pair resulting in recombination of genes.
3. According to Bridges and Darlington the crossing over is completely absent from male *Drosophila*.
4. In rats also the amount of crossing over in males is lower than that of females.
5. In maize, mice, poultry and man, crossing over (recombination) of linked genes takes place both in females and males.

Salient Features of Crossing Over

1. Crossing over is the interchange of chromosomal segments between the homologous chromosomes.
2. It occurs during meiosis or gametogenesis.
3. The cross over occurs only between non sister chromatids of the homologous chromosomes.
4. The number of crossing over depends upon the length of the chromosome. The longer the length, the higher the percentage of crossing over.
5. When the genes are located apart from one another the chance of crossing over is higher.

The Chromosome Map

1. The chromosome map can be defined as a line on which the genes are represented by points, separated by distances proportional to the amount of crossing over.
2. These chromosome maps are also referred to as cross over maps since they are sketched by the amount of crossing over.
3. In mapping the genes a unit of distance is used and it is called as map unit or Morgan unit.
4. The first chromosome map was made in 1911 by Sturtevant and soon after additional maps were made by Bridges and others.
5. *Drosophila* was the earliest material used by the scientists for constructing maps.

Mutations

Mutation is sudden change of a gene or chromosome from one to another. It results in an alternation in the character under its control. Dobzhansky stated that mutation is a mistake or misprint in cell division. The term, 'mutation' was introduced by de Vries.

Types of Mutations

Mutations can be classified into several types as per nature, rate of efficiency stage of development, as follows:

1. **Somatic mutation:** It is the mutation occurring in the somatic cells of organisms. It is not inherited, disappears at the death of the animals.
2. **Germinal mutation:** It occurs in the germinal cells. It is a heritable mutation.
3. **Zygotic mutation:** When mutation occurs in the zygote, it is known as zygotic mutation.
4. **Dominant mutation:** When mutation produces a dominant gene, it is known as dominant mutation that expresses the character immediately.
5. **Recessive mutation:** If the mutation of a gene produces recessive gene, it is called recessive mutation. Most of the natural mutations are recessive in nature. When mutation is recessive, the character is not expressed immediately; but it is expressed only when the mutant gene occurs in double doses.
6. **Back mutation:** It is the reversion of mutation. It is rare when a mutant gene changes into the original gene, thus called reverse mutation.
7. **Lethal mutation:** Sometimes mutation affects the vital functions of an animal; as a result the mutant dies, hence, it is called lethal mutation.
8. **Spontaneous mutation:** If mutation occurs in the absence of any obvious cause, it is called

spontaneous mutation. Most of the mutations occuring in nature are of this type.

9. **Induced mutation:** They are generally external factors called mutagenic agents or mutagens.
10. **Macromutations:** Some mutations are large with prominent phenotypic effects. These mutations are called macromutations.
11. **Micromutations:** Mutations which bring about slight phenotypic effects are called micromutations.
12. **Gene mutations:** The sudden change of gene from one form into another is gene mutation. It is also called point mutation as it affects only a particular locus in the chromosomes.
13. **Chromosomal mutations:** The sudden changes occurring in the number of genes in the arrangement of genes and in the number of chromosomes is termed chromosomal mutation or chromosomal aberration.
14. **Biochemical mutation:** Mutation causing changes in the metabolities or their end-products is known as biochemical mutation. Most of the enzymes are proteins and are synthesized by genes. Each enzyme is controlled by different gene. Any defect in any gene shall thus, result in metabolic disorders, *e.g.* phenylketonurea and alkaptonurea.

Causes of Mutation

1. Ionizing radiations (X-rays; gamma rays; alpha rays, beta rays, ultraviolet rays, infra-red rays).
2. Chemical mutagens (mustard gas, nitrous acid, nitrogen mustard, colchicine, formaldehyde, peroxides, caffiene, phenol etc.)
3. Temperature
4. Deamination
5. Tautomerization
6. Base analogs

Cytoplasmic Inheritance

1. Most of the phenotypic characters are controlled by genes present in the chromosomes, but some characters are expressed by factors present in the cytoplasm. These factors lying in the cytoplasm are called plasmagenes. They also transmit characters from one generation to another.
2. The transmission of characters controlled by plasmagenes is called cytoplasmic inheritance or extrachromosomal inheritance.
3. This was first described by Correns in 1908.

Characteristics of Cytoplasmic Inheritance

1. Plasmagenes are situated outside the chromosomes.
2. Like chromosomal genes, plasmagenes are also self-replicating.
3. Plasmagenes are transmitted by means of cytoplasm only.
4. Like chromosomal genes, plasmagenes are also capable of mutation.
5. The offspring receive cytoplasm only from female gamete and not from male gamete. Hence the plasmagenes of female parent alone are contributed to the offspring. As a result the cytoplasmic inheritance is said to be a maternal inheritance.
6. The results of reciprocal crosses are not the same.

HUMAN GENETICS

1. The basic principles or inheritance are same for all living organisms including human. Genetics which deals with man is termed as human genetics, pioneered by Sir Archibald Garrod, as british physician. In 1901, he opined that certain inborn errors of metabolism like phenylketonurea and alkaptonurea are controlled by genes and are inherited in simple Mendelian fashion.
2. In recent years, the development of newer techniques is used in understanding the mechanism of inheritance of a large number of characters in man. Two of them are as follows:
   (i) **Pedigree analysis:** It defined as a chart made by using internationally accepted symbols by taking family history from the patient known as index or proband case. The females are represented by circles and males by squares.
   (ii) The drawback of limited progeny has been overcome by studying population genetics.

(iii) **Biochemical genetics:** has helped in metabolic studies in human populations.
(iv) **Somatic hybridization:** has added to the knowledge of biochemical basis of habitable features.

Aneuploidy

It is the condition resulting from non-disjunction of homologous chromosomes at meiosis, in which one or more chromosomes are missing from or added to the normal somatic chromosome number.

Barr Body

1. Barr and Bertram (1949) observed a deeply stained chromatin body in the nerve cells of female cat which was absent in the male.
2. This chromatin body is called sex-chromatin or *Barr body*.
3. Such barr body has been observed in the skin, oral epithelium and blood cells of man and other mammals. Human males are called **sex** chromatin negative.
4. The sex chromatin appears in the interphase nucleus as a small chromocentre, heavily stained with basic dyes. It exists in four positions:
   (i) Attached to the nucleus as in nerve cells of certain species.
   (ii) Attached to the nuclear membrane as in cells of epidermis or of the oral mucosa.
   (iii) Free in the nucleoplasm as in neurons after electric stimulation, and
   (iv) as a nuclear expansion. The best example is the nucleophil leukocyte of female in which the sex chromatin (Barr body) appears as a small rod called the drumstick.

Normal Human Chromosomes

Two plant pathologists Tijo and Levan reported the correct number of human chromosomes as 2n = 46 in 1956. Human chromosomes have been classified on the basis of length, the position of centromere, banding pattern, identification by the presence of satellite bodies and morphological feature. By the Denver-London system all the autosomal chromosomes have been numbered from 1 to 22 in decreasing order of length or size.

GENETIC MATERIAL

Replications, Transcription and Translation

DNA is the genetic material. Chromosomes carry and transmit the hereditary characters. It is formed mainly by two types of chemical substance: nucleic acids and proteins. Nucleic acids are responsible for transmission of hereditary character. The nucleic acid which carries the genetic information is DNA.
Nucleic acids: The nucleic acid was first isolated in 1868 by Miescher from the nuclei of pus cells. He called it 'nuclein'. The term 'nucleic acid' was coined by Altman in 1889. There are two types of nucleic acids: DNA (deoxyribonucleic acid) and RNA (ribonucleic acid).

Deoxyribonucleic acid (DNA)

1. DNA is a nucleic acid which contains deoxyribose sugars.
2. It is present in all cells except plant viruses. In eukaryotic cells, DNA is present in the chromosomes of nucleus. Further, the mitochondria and plastids also contain DNA.
3. In eukaryotic nucleus, the DNA is in the form of a double helix. In bacteria, mitochondria and plastids the DNA molecules are circular. In viruses and bacteriophages they are coiled.
4. The nuclear DNA is bound to basic proteins called histones. DNA is present in every nucleated cell and it carries the genetic information. It is conveniently isolated from viruses, thymus gland, spleen, leucocytes etc.
5. Chromosomal DNA is made up of very long DNA molecules. It is a polymer of about 10^{10} deoxy-ribonucleotides.
6. Generally there are only four types of deoxyribonucleotides found in DNA molecule, viz.:
   (i) adenine deoxyribonucleotide (dA)
   (ii) thymine deoxyribonucleotide (dT)
   (iii) guanine deoxyribonucleotide (dG)
   (iv) cytosine deoxyribonucleotide (dC)
   Nucleotides of each of the two helical bands are bound to each other by covalent 3′–5′ phospho-diester linkage. This double helix

model got Nobel prize to Watson, Crick and M. Wilkins.

7. The length of hydrogen bond is approx. 3.0 Å. the ratio of purine to pyrimidine bases in the DNA molecule is always around 1 (*i.e.*, G + A/T + C = 1). This is known as Chargaff's rule.
8. DNA exist, in several conformations depending on the base composition and under different physical conditions. The same base pairing rules apply always and changes do not alter the information content of the DNA. The conformations of DNA have been determined by X-rays crystallography. By far the most common conformation is B-DNA., though there are other types also like A-DNA and Z-DNA.

Ribonucleic acid (RNA)

1. Ribonucleic acid is a polymer of ribonucleotides of Adenine, Uracil, Guanine and Cytosine, joined together by 3′–5′ phosphodiester bonds. Thymine is absent in RNA. RNA is found in the nucleolus, Nissl grannules, ribosomes, mitochondria and cytoplasm. The pentose sugar of the nucleotide is D-ribose.
2. The secondary structure of RNA has many coil formations of the polyribonucleotide chain. These are coiled structures which are stabilized by hydrophobic interactions between the purine and pyrimidine bases.
3. There are intra-chain hydrogen bonds between G-C and A-U.
4. The tertiary structure of RNA shows the folding of the molecule into 3-D structure. Cross-linking too occurs at various sites stabilized by hydrophobic and hydrogen bonds producing a compactly coiled globular structure.
5. There are mainly three types of RNA in plants and animals. They are:
   (i) messenger RNA or m-RNA
   (ii) transfer or soluble RNA or t-RNA
   (iii) ribosomal RNA or r-RNA.
6. The main function of each of these RNA is protein synthesis.

Kinds of RNA

1. **Messenger RNA or m-RNA:** This is the most heterogeneous class of RNA with respect to its size and stability. It carries mainly adenine, guanine, cytosine and uracil as the major bases and methylpurines and methylpyrimidines as minor bases.
2. **Transfer RNA or t-RNA:** These are also called as soluble or s-RNA. They remain largely in cytoplasm. The t-RNAs are relatively small, single-stranded globular molecules. There are at least 20 different t-RNA molecules.
3. **Ribosomal or r-RNA:** A ribosome is present in the cytoplasm and is a nucleoprotein. It is one the ribosome that the m-RNA and t-RNA interact during the process of protein biosynthesis.

Replication of DNA

The synthesis of new DNA molecules from pre-existing DNA is called replication. Watson and Crick suggested that there is self duplication of DNA molecule. There are three types of DNA replication:

(i) Semiconservative method
(ii) Conservative and
(iii) Dispersive method

Major enzymes involved in DNA replication

1. DNA polymerase I is the major enzyme though DNA polymerase II and DNA polymerase III are also found in prokaryotes while eukaryotic DNA polymerases are called alpha, beta, gamma and sigma.
2. DNA-ligase brings about the process of connecting two DNA chains.
3. A specific RNA polymerase which synthesises short strands of RNA called as primase.
4. In addition to the enzymes roughly 10 other proteins carry out DNA replication.

Genetic Code

1. It is defined as the structure of nitrogen bases (nucleotides) in mRNA molecule which contains the information for the synthesis of protein molecules.

2. A codon is the sequence of nitrogen bases (nucleotides) in mRNA which codes for a single amino acid.
3. **Triplet code:** *Nirenberg* and *Mathaei* (1961) experimentally proved that a single amino acid is determined by a sequenc of three nitrogen bases. The sequence of the nitrogen bases determining a single amino acid is called a triplet code. In 1964, H.G. Khorana synthesized polyribonucleotides of mRNA of known sequences. He synthesized an alternating co-polymer such as UGU-GUG-UGU-GUG. Use of this particular alternating polynucleotide as a mRNA in the *in vitro* protein synthesizing systems gave rise to the formation of the alternating polypeptide (Valine-Cysteine-Valine-Cysteine). Thus Khorana determined the exact sequence of nucleotides in various genetic codons.

Salient Features of a Genetic Code

1. **Triplet:** A single amino acid is specified by the sequence of three nucleotides in the mRNA. This is called a codon. Thus a codon is a triplet.
2. **Universal:** Genetic code is universal in nature in viruses, bacteria, plants and animals.
3. **Linear arrangement:** The codons in DNA, mRNA and the amino acids in polypeptide chains have linear arrangement.
4. **Commaless:** There is no comma or punctuation or spaces between the adjacent codons.
5. **Non-overlapping:** Codons are non-overlapping.
6. **Degeneracy:** A single amino acid may be specified by many codons, called degeneracy. Such codons are called degenerate codons.
7. **Initiation codon:** The synthesis of a polypeptide chain is initiated by a codon called initiation codon, located in the beginning of a cistron. It is constant in most cases. It is AUG and it codes for the amino acid, methionine.
8. **Termination codon:** The synthesis of a polypeptide chain is completed by a codon called termination codon, located at the end of a cistron. It may be UAA, UGA or UAG. The termination codon does not code for any amino acid. So, it is called non-sense codon.
9. Then the genetic informatin is expressed by a two step process:
   (i) Transcription generates a single stranded RNA identical in sequence with one of the strands of duplex DNA.
   (ii) Translation converts the nucleotide sequence of the RNA into the sequence of amino acids making a protein. RNA can be converted to DNA by an enzyme, reverse transcriptase the phenomenon is called as teminism.

Mechanism of protein synthesis

It is involved stepwise- Transcription, Activation of amino acids, Attachment of amino acids with tRNA, Translation, Elongation of polypeptide chain, and Termination of polypeptide chain.

GENE

Its Expression, Regulation and Manipulation

Genetic expression in prokaryotes

1. Gene expression gets regulated at several different levels as at transcription, mRNA processing, mRNA turnover, translation and enzyme function. The regulation of transcription, though, is the most important.
2. The first category of regulatory mechanisms involves rapid turn-on and turn-off of gene expression in response to environmental changes. This is very important in micro-organisms because of the frequent exposure of these organisms to sudden changes in environment. It seems to be less important in higher eukaryotes.
3. The second category includes preprogrammed circuits of gene expression. In these cases, certain events as infection by a virus triggers the expression of certain genes. The product of one or more of these genes functions by turning off the transcription of the first set of genes and/or turning on the transcription of a second set of genes.
4. Gene expression as such refers to the molecular mechanism phenotype by which a gene

expresses. The gene expresses a phenotype by synthesizing a protein or an enzyme which determines the character. The gene contains the blue print or the information for the protein or enzyme. The information is transferred to mRNA by the synthesis of mRNA called transcription. The amino acids are linked in a particular sequence as per the sequence of nucleotides in the mRNA. It is called translation.

5. Jacob and Monod proposed in 1961 mechanism of the expression of genes in *Escherichia coli*, an intestinal bacterium:
   A. Each metabolic reaction is controlled by a set of genes. All the genes controlling reaction constitute as operon.
   B. Catabolism of lactose in *E. Coli* involves a set of enzymes produced by a set of genes. All these genes in catabolism of lactose are called lac operon.
   C. The operon model is formed of two types of genes namely:
      (i) Structural genes have the codes for the synthesis of enzymes (proteins) necessary for the metabolic reactions.
      (ii) Control genes control the activity of structural genes. The control genes are the three types, namely regulator, promotor and operator genes.

Genetic expression in eukaryotes

1. Here, the chromosomal proteins play important roles in gene expression. The chromosomal proteins are of two types. They are histones and non-histones. The regulation of gene expression involves an interaction between these two types of protein. Histones inhibit protein synthesis and non-histones induce RNA synthesis.
2. There are four main steps in the expression of genes. One or more of these steps may be involved in regulation. They are:
   (i) Replication (ii) Transcription (iii) Processing and (iv) Translation.

Genetic Engineering (gene manipulation)

It may be defined as the extracellular (*i.e., in vitro*) creation of new forms or arrangements of DNA in such a way as to allow the incorporation or continued propagation of the altered genetic condition in nature. Genetic engineering, in short, mean manipulation of genes under controlled laboratory conditions.

Gene Cloning

The importance of genetic engineering as a modern technique comes from two chief reasons, namely that (1) it allows the isolation and study of single genes in large amounts and complete purity (gene splicing) and (2) these genes can now be expressed and re-introduced into cells from the same or different species (recombinant DNA technique). These two techniques together are called gene cloning.

Recombinant DNA technology: It provides methods by which genes or other segments of large chromosomes can be isolated, replicated and studied by nucleic acid sequencing (which is otherwise very laborious if one has to search for a gene in the main DNA molecule at random). Electron microscopy is very handy in this technology. The various steps involved in preparation of recombinant DNA are:

1. Selection of DNA of interest referred to as the foreign or target or passenger DNA.
2. A cloning vector or vehicle to carry the inserted piece of target DNA.
3. Restriction endonucleases that makes internal cuts at specific sites on DNA.
4. DNA ligase—the enzyme that joins pieces of nucleotides/DNA together.
5. A host to replicate the vector containing the foreign DNA. This can be a prokaryotic or eukaryotic cell.
6. Screening test for recombinants created by insertion of DNA.

Plasmids

Basic tools for genetic engineering are the plasmids. Important characteristics of naturally-occurring plasmids are:

1. They influence characteristics of their bacterial hosts.

2. They replicate independently of the main chromosome.
3. They are usually specific to one or a new species of bacteria.
4. Some may undergo reversible integration into bacterial chromosome.
5. Some can pick up and transfer chromosomal genes.
6. They are transferred by conjugation.
7. They usually contain upto 40 genes varying from bacteria to bacteria.
8. They do not occur free in nature.

It may be noted that except for the last three properties of plasmids, the other characters are shared with viruses.

Plasmids produce effects on the host's phenotype; a few of these effects are listed below:

1. F-factors increase fertility.
2. R-factors cause resistance to antibiotics, heavy metals, UV radiation or virus infection.
3. They are usually involved in the production of antibiotics, bacteriocins (certain toxins), enzymes or antigens.
4. They can metabolize sugars and hydrocarbons.
5. Ti plasmids of *Agrobacterium* can induce tomours in susceptible plants.

The utility of plasmids arises from the fact that new genes can be tailored into plasmids and then multiplied repeatedly. This manipulation of DNA *in vitro* is recombinant DNA technology. In this technology, certain important enzymes are used. A nuclease hydrolyses nucleic acids into nucleotides. An exonuclease chews the DNA from one end whereas an endonuclease cuts DNA in an intercalary position in between two nucleotides. Restriction endonucleases cut DNA at specific sequences.

Important Enzymes of Genetic Engineering

1. Exonuclease
2. Endonuclease
3. Restriction Enzyme
4. DNA ligase
5. Reverse transcriptase
6. DNA polymerase
7. Alkaline phosphatase

CLONING

Twining of Genome

A clone is a collection of genetically identical individuals. The word 'clone' comes from the Greek word 'Klan', meaning 'twig'. Cloning provides a short cut method to amplify the number of plants/ animals with a desirable genetic constitution. No sex is involved in the production of clones. Clones have no choice but to have the same genes as their single parent. Members of a clone are genetically identical. The arrival of worlds' first clone of a adult animal named Dolly created a cascade in this technology and thus provided huge boot in the modern biotechnological work. Dolly was created at Edinburghs' Roslin Institute from a single cell taken from the udder of a six year old sheep, turned into an embryo and then implanted in a surrogate mother.

Significance of Cloning

(i) New organisms with high and desired characters can now be obtained in pure form.

(ii) Specific compounds for medicines which are made available by the specific organism can be obtained in abundant amount by cloning.

(iii) Organs could be formed by cloning method for transplantation, such that it get accepted by human system without any rejection.

(iv) Main organs of transplantation are liver, cornea, heart, kidney and skin etc.

(v) The success in animal cloning also has some adverse effects. Man has thoughts of producing human clones which is thought to be unethical. We are afraid of the time when man can clone a criminal.

World of Clones

The aim of research is to employ cloned breeding animals in the mass production of sought-after proteins which could lead to significant progress in the manufacture of pharmaceutical drugs. Dr. Xiangzhong Yang *et. al.* at the University of Connecticut cloned four healthy calves from the skin cells taken from the ear of an ageing Japanese breeding bull.

The number of mammalian species that scientists have cloned from adult cells is eight including sheep, mice, rabbits, goats, cats, pigs and cattle, *viz.*:

1. **Dolly:** Scottish biological research team headed by lan Wilmut cloned the first mammal - a sheep (lamb) (early 1997).
2. **Polly:** A lamb with human protein gene. She is the first genetically engineered cloned sheep.
3. **Molly:** Another cloned sheep introduced with human genes (July 1997).
4. **George and Charlie:** Two bulls from the genetically tailored cows have been born on a cattle ranch, Texas. It is estimated that such cows will be able to produce large amount of albumin.
5. **Tetra:** The cloned monkey was made by splitting a very early embryo into four pieces, (January 2000).
6. **ANDi:** This first cloned monkey by embryo splitting derived its name from a backward abbreviation for "inserted DNA'. ANDi is the worlds first genetically altered primate produced by inserting a jellyfish gene into the embryo of a rhesus monkey.
7. **Cc:** Cc (carbon copy cat) has (14th Feb. 2002) been created at Texas by transplanting DNA from a female three-coloured (tortoise shell or calico) cat into an egg cell whose nucleus had been removed. It is only one of 87 implanted cloned embryos to survive.
8. **Millie, Christa, Alexis, Carrel and Dotcom:** They are five piglets successfully cloned by PPL therapeutics with collaboration of Roslin Institute which had cloned dolly.
9. **Prometea:** First cloned member of equine family (horses) (born on Aug. 7, 2003 at Itlay) (Cremona). It is the first to have been gestated in the womb of the same female that donated the skin cell from which the clone was grown.

MULTIPLE CHOICE QUESTIONS

1. Characteristic unique to DNA is
   A. denaturation and renaturation
   B. polymer complex
   C. replication
   D. resistance to temperature change
2. In the Meselson and Stahl experiment, *E. coli* cells grown on heavy nitrogen were transferred to light nitrogen. What % of DNA can be expected to be constituted of light nitrogen after 3 generation of multiplication?
   A. 25 B. 50
   C. 75 D. 100
3. Any DNA molecule that has the ability to replicate autonomously is called
   A. plasmid B. chromosome
   C. genome D. replicon
4. In genetics, an origin is
   A. a protein that binds to DNA at the point where replication begins
   B. a protein that binds to RNA at the point where translation begins
   C. the point in DNA where transcription starts
   D. the point in DNA where replication starts
5. Replication of DNA in eukaryotes commences from
   A. one end of chromatid extending to the other end
   B. both ends of the chromatid simultaneously
   C. the centromere to either of the ends of chromatids
   D. several sites of the DNA of the chromatid simultaneously
6. The fact that DNA polymerase I from *E. coli* has a 5′ – 3′ exonuclease activity
   A. implies that the enzyme has multiple subunits
   B. implies that DNA polymerase I can use both DNA and RNA as primers
   C. makes the enzyme able to detect thymine dimers in double - stranded DNA
   D. enables the enzyme to play an important role in DNA replication

7. The essential initiator protein at the *E. coli* origin of replication is

A. DNA *A* B. DNA *B*
C. DNA *C* D. DNA *E*

8. Okazaki fragments

A. require the activity of only a DNA polymerase for synthesis
B. require only RNA polymerase activity for synthesis
C. are made when DNA is exposed to UV radiation
D. are composed of both DNA and RNA

9. Approximately how many Okazaki fragments are synthesized during one round of replication of the *E. coli* genome?

A. 5000 to 10000 B. 4×10^6
C. 2500 to 5000 D. 2

10. DNA replication undergoing in an *E. coli* cell resulted into two circular DNAs interlocked with each other; this may be due to a defective gene encoding

A. primase
B. DNA topoisomerase I
C. DNA polymerase
D. DNA topoisomerase II

11. What is the role of gyrase in DNA replication?

A. to relax or prevent supercoiling of the double helix
B. to separate the two template strands of DNA at each replication fork
C. to synthesize a short RNA primer for each of the new DNA strand
D. to protect single-stranded regions of DNA from damage

12. When the human genome draft sequence was released, which of the following observation was least expected?

A. the large amount of repetitive DNA
B. the size of the total genome
C. the size of individual chromosomes
D. the small number of protein-coding genes

13. Which of the following are genes found in all retroviruses?

P. *gag* Q. *pol*
R. *env* S. *onc*

A. P, Q B. P, S
C. R, S D. P, Q, S

14. Human genomic DNA is digested into fragments approximately 1 kb in size, denatured and then renatured. Which of the following statements is *true*

A. all fragments will renature at the same rate
B. fragments composed largely of repetitive DNA sequences will renature fastest
C. fragments composed largely of non-repetitive DNA sequences will renature fastest
D. fragments with high A : T content will renature fastest

15. If a segment of an mRNA molecule has the sequence 5′ GUACCGAUCG 3′, which of the following could have been the template DNA molecule?

A. 5′ GCUAGCCAUG 3′
B. 5′ GUACCGAUCG 3′
C. 5′ CATGGCTAGC 3′
D. 5′ CGATCGGTAC 3′

16. Which of these substrates is best suited for measuring RNA synthesis by RNA polymerase

A. γP^{32} ATP B. αP^{32} d ATP
C. γP^{32} d ATP D. αP^{32} UTP

17. The prokaryotic RNA polymerase holoenzyme has the subunit structure

A. $\alpha\beta\beta'\sigma$ B. $\alpha\beta\sigma$
C. $\alpha\beta$ D. $\alpha\beta\beta$

18. Mode of action of rifampicin in *E. coli* is through

A. cell division
B. initiation of transcription
C. RNA polymerase binding to DNA template
D. inhibition of the oxidation potential

19. There are three kinds of RNA polymerases (I, II, III) in eukaryotic cells, each specific for one class of RNA molecule. Which of the following is a correct match?

A. RNA pol I–rRNA; RNA pol II–tRNA
B. RNA pol II–mRNA; RNA pol III–rRNA
C. RNA pol I–rRNA; RNA pol II–mRNA
D. RNA pol I–tRNA; RNA pol III–rRNA

20. A new antibiotic was discovered which strongly inhibited mRNA precursor transcripts and snRNA transcripts. This antibiotic was predicted to be an inhibitor of
A. RNA polymerase I
B. RNA polymerase II
C. RNA polymerase III
D. helicase

21. Transcription termination of mRNA genes in eukaryotes occurs
A. at polyadenylation sites by the action of a termination factor
B. by the formation of a strong hairpin structure in the vicinity of polyadenylation site
C. termination factor bound to the termination site in the vicinity of polyadenylation site
D. at pause sites following the polyadenylation sites

22. Which one of the following regions in the mature mRNA is not found in the gene for the mRNA?
A. start codon B. leader sequence
C. poly A tail D. exon

23. The poly A end of eukaryotic mRNAs originates from
A. transcription of corresponding poly T region of the respective gene
B. addition of a pre-synthesized poly-A-tail to the 3′ end of the primary transcript
C. sequential addition of A residues at the 3′ end of the primary transcript
D. by as yet an unknown mechanism

24. Introns are released as lariats during splicing of
A. pre-mRNA transcripts and group II intron containing transcripts
B. pre-mRNA transcripts only
C. group I intron containing transcripts
D. tRNA precursors

25. Operons
A. are of approximately uniform in size
B. do not bind proteins
C. are found in some eukaryotic genes
D. are shorter and smaller in lower eukaryotes than higher eukaryotes

ANSWERS

1	2	3	4	5	6	7	8	9	10
C	C	D	D	D	D	A	D	C	D
11	**12**	**13**	**14**	**15**	**16**	**17**	**18**	**19**	**20**
A	D	D	B	D	A	A	B	C	B
21	**22**	**23**	**24**	**25**					
C	C	C	A	C					

5

Proteins and Gene Manipulation

INTRODUCTION

Proteins are essential to maintain the structure and function of all life forms. The word 'protein' itself is derived from the Greek word protos, meaning "primary" or "first". Proteins are vital for the growth, repair, and maintenance of muscles, blood, internal organs, skin, hair, and nails; and its functions are endless. Each and every property that characterizes a living organism is affected by proteins, whether it is a bacteria or human body. Nucleic acids, another major biological macromolecule, are also essential for life that encodes genetic information—mostly specific for the structure of proteins—and the expression of that information depends almost entirely on proteins. The fertilization of an egg with a sperm and the development and differentiation of the resultant zygote into a fully developed organism, and its growth and maintenance of life activities up to its death is controlled and programmed by a large number of proteins.

3-D Shape of Proteins

A single cell develops into a multicellular embryo through a large number of complicated biochemical reactions mediated and controlled by various types of proteins expressed during the course of its development. Every function in the living cell depends on proteins. They make us who we are and make our cells operate properly. A cell cannot function without proteins. The shape of a protein determines its biological activity. A single protein may have varying structure and more than one function. Proteins have many different biological functions. Proteins are even classified according to their biological roles. The key to appreciating how different proteins function in these different ways lies in an understanding of protein structure and its three dimensional shape. Proteins interact with other molecules like small molecules, other proteins, nucleic acids, lipids, etc. and these interactions form the basis of its biological roles. Structural complementarity is the means of molecular recognition that allows molecules to interact. The structure of one molecule is complementary to that of its partner(s) in the interaction, like pieces in a puzzle, or a lock and its key. Proteins, by virtue of their architectural diversity, are ideal for such complementary interactions. In short, the structure or the molecular shape of the protein determines its function.

The detailed study on the structure of proteins requires protein extraction and its purification to its homogeneity. The purified protein has to be analyzed by various biochemical and instrumental methods to get the details about its chemical composition. The pure protein obtained has to be crystallized to study its 3-D shape by X-ray crystallography (X-ray diffraction technique). Another equally powerful technique to elucidate the 3-D structure of protein is NMR spectroscopy. The structural studies of protein, thus has two parts—the first part is determination of amino acid sequence and the second part is the elucidation of 3-D shape of the protein formed by the specific folding of the polypeptide chain controlled by a number of molecular forces.

Structure-Function Relationship in Proteins

Proteins are the workhorses of the cells and in every activity there is the involvement of one or more proteins in different ways. Now it is very clear that each protein has a specific three-dimensional shape determined by the amino acid sequence and various other intermolecular interactions. This 3-D shape has got a great influence on its biological function that it performs in the cells. We consider two proteins as example to understand the importance of 3-D structure on its specific function: a proteolytic enzyme Chymotrypsin and the oxygen carrying protein haemoglobin.

Characterization of Proteins

The protein purified has to be identified and characterized. There are various methods to accomplish this. The purity of the protein can be determined by the SDS-PAGE or by the isoelectric focusing (IEF) or by 2-D Gel Electrophoresis.

Identification or characterization of protein can be carried out with the help of any one of the following techniques or by a combination of these:

1. 2-D Gel Electrophoresis
2. Peptide finger printing
3. Protein Sequencing and
4. Mass Finger printing by Mass Spectrometry

PROTEIN BASED PRODUCTS

Proteins, as a biological macromolecule is very important for the existence of normal activity in a living system. The biological roles played by various types of proteins have been discussed elsewhere. Because of the multifaceted role of proteins, it is important commercially also. Proteins may be classified as follows based on its commercial importance:

1. Blood derived proteins and vaccines
2. Therapeutic antibodies and enzymes
3. Hormones and growth factors as therapeutic agents
4. Regulatory factors
5. Proteins and enzymes used in analytical applications
6. Industrial enzymes and proteins
7. Non-catalytic Functional Proteins
8. Nutraceutical proteins.

PROTEOMICS

A large number of sequencing projects initiated in the late eighties and early nineties have now been completed. The entire genome sequences of numerous prokaryotes, the eukaryotes like *Saccharomyces cerevisiae* and the first multicellular organism *Caenorabditis elegans* and *Homo sapience* have been deposited into public databases (GenBank, http://www.ncbi.nlm.nih.gov/Entrez/Genome/org.html). Significantly, analysis of the sequence data has revealed a myriad of uncharacterized genes having homologues in many different organisms, yet their functions have not been deciphered in any of these organisms. Illustrating this statistically, 24% of *Haemophilus influenzae,* 40% of *Escherichia coli* and 43% of *Saccharomyces cerevisiae* open reading frames (ORFs) remain completely uncharacterized.

The Key Technologies in Proteomics

- Reproducible 2D Gel Technology
- Staining and Scanning Technology
- Mass Spectrometry for Identification
- Databases (protein and genome)
- Database Searching Algorithms

Genes and Proteins

Owing to concerted effort by numerous state and commercial establishments, the human genome has completely been sequenced in 2000. All 24 human chromosomes are mapped, the defects of hundreds of genes responsible for the development of hereditary diseases have been revealed. A new medical concept has been advanced whereby all diseases can be divided into two major groups: (a) hereditary diseases—a consequence of transmission of defective genes from parents to their children and (b) nonheritable or so called socially-significant diseases, which make up more than 95% of all human diseases and result from disturbances in normal gene's expression regulation. Thus, all human diseases, this way or another, are associated with the genome; the only difference is that the diseases of the former group are due to a defect(s) in gene's structure while those of the latter group are caused by disturbances in gene expression's regulation. The cataloguing of human genes is an achievement, which can hardly be overestimated: for many years to come this catalogization will serve as a basis for the development of basic biochemistry, molecular biology and genome-associated applied sciences. The future of this area of research called 'genomics' is even more brilliant. Currently the emphasis has swung from sequencing the human genome to sequencing the genomes of animals and micro-organisms, particularly pathogenic, and plant genomes. These researches promise enormous achievements in medicine, especially in the struggle against infectious diseases and in agriculture.

How the informational structure (gene) is connected with the actually working molecular machine (protein)? To answer this question, we have to consider the results those few works, where the expression map of mRNA was compared with the proteinous map in the same cellular system. It failed to reveal any strict correlation between the two maps. Thus, the informational knowledge cannot be directly converted into the knowledge of actually operating protein molecules. As a result, a new area of research has appeared which is called 'proteomics' and which deals with inventory of proteins. At first glance this is an utterly impossible task. While the human genome map is the same for all human cells (24 chromosomes), in the proteomic map each cell is individual. Although the cell may have only 33,000 of functional genes, the numerous modification reactions may increase the number of proteins in it up to several millions. There are two definitions for proteomics: a narrow one (the so-called structural proteomics) and a broader one, encompassing both the structural and functional proteomics. In a narrow sense of the word, 'proteomics' is a science dealing with cataloguing of proteins based on a combination of several methods: 2-D-electrophoresis, mass-spectrometric analysis of molecular mass, sequencing of electrophoretically separated proteinaceous biological material with subsequent analysis of the results obtained, with the help of bioinformatical and computational methods.

In a broader sense, the terms 'proteome analysis' or 'proteomics' can be used not only for cataloguing proteins of a biological subject but also for the monitoring of reversible post-translational modification of proteins by specific enzymes, *i.e.*, phosphorylation, glycosylation, acylation, phrenylation, sulphurization, etc. To date, more than 300 different types of post-translational modification have been characterized with the aid of proteomics. Another aspect of functional proteomics is to clarify the composition of functionally active complexes that constitute different metabolic chains and also, to determination of the interactions between various proteins or subunits of oligomeric complexes by a combination of methods to isolate these complexes and subsequent mass-spectrometric analysis. Lately, structural proteomics is often called expressional proteomics while functional proteomics is also designated as cell-mapping proteomics, since it elucidates the interactions of proteins within metabolic pathways. Therefore, in short the number of proteins and the number of genes are not equal and they are non-linear. The number of protein easily outnumbers the number of genes.

Types of Proteomics

Even though proteomics can be defined as the study of the protein complement of the genome, the dynamic nature of proteome made it difficult to study and understand. The total protein expression profile always changes with time, and micro and macro environmental conditions. Only a small percentage of the total gene is expressing at a particular given time. Information about the genome and gene structure can predict the structure and function of its protein complement to a limited extent because of the post-transcriptional modifications that the protein undergoes. These modifications are not represented in the respective gene. This is what that makes proteomics difficult to study in comparison to genomics. Following are the major types of proteomics identified:

1. **Expression Proteomics:** This is the qualitative and quantitative study of the expression of total proteins under two different conditions. For example, expression proteomics of normal cells and diseased cells can be compared to understand the protein which is responsible for the diseased state or the protein which is expressed due to the disease. Using this

method the disease specific protein can be identified and characterized by comparing the protein expression profile of the entire proteome or of the subproteome between the two samples.

2. **Structural Proteomics:** Structural proteomics, as the name indicates is about the structural aspects including the 3-D shape and structural complexities of functional proteins. This includes the structural prediction of a protein when its amino acid sequence is determined directly by sequencing or from the gene by a method called homology modelling. This can be carried out by doing a homology search and computational methods of protein structural studies and predictions. Apart from this, structural proteomics can map out the structure and function of protein complexes present in a specific cellular organelle.
3. **Functional Proteomics:** This is an assembly of proteomic method to analyze and understand the properties of macromolecular networks involved in the life activities of a cell. By these mothods it will be possible to identify the specific protein molecules and their role in the individual metabolic activities and their contribution to the metabolic network that operates in the system. This forms one of the major objectives of functional proteomics. For example, the recent elucidation of the protein network that involved in the functioning of nuclear pore complex has led to the identification of novel proteins involved in the translocation of macromolecules between cytoplasm and nucleus through these complex pores.

MULTIPLE CHOICE QUESTIONS

1. A genetic unit that codes for amino acid sequence of a complete functional polypeptide could be termed as
   A. recon B. intron
   C. exon D. cistron
2. An aminoacyl t-RNA synthetase is responsible for
   A. formation of a peptide bond
   B. binding of m-RNA to ribosomes
   C. attaching an amino group to an organic acid
   D. joining an amino acid to t-RNA
3. GTP is required by which of the following steps in protein synthesis?
   A. aminoacyl-tRNA synthetase activation of amino acids
   B. attachment of ribosomes to endoplasmic reticulum
   C. translocation of tRNA-nascent protein complex from A to P sites
   D. attachment of mRNA to ribosomes
4. Formylmethionine tRNA is present in
   A. yeast B. bacteria
   C. insects D. mammals
5. All of the following factors are required to initiate translation of mRNA in prokaryotes *except*
   A. formylmethionine-tRNAf
   B. 30S ribosomal subunits
   C. guanosine triphosphate
   D. EF-Tu
6. GUG codes for valine in both prokaryotes and eukaryotes but when GUG is initiation codon, this codes
   A. methionine B. valine
   C. tryptophan D. None of these
7. Synthesis of peptide bond is catalyzed by
   A. A site of ribosome B. P site of ribosome
   C. 23S rRNA D. tRNA
8. The translation of mRNA into the amino acid sequence of a polypeptide in prokaryotes is terminated at the end of the message by one of three chain termination codons in the mRNA. The stop codon is recognized by
   A. a specific uncharged tRNA
   B. a specific aminoacyl-tRNA
   C. a specific ribosomal RNA
   D. a release factor

9. *E.coli* release factor1 (RF1) recognizes which codons?
   A. UAA only B. UGA only
   C. UGA and UAA D. UAG and UAA

10. Which of the following occurs when the eukaryotic translational machinery encounters the UAG codon
   A. the ribosome can recognize the UAG codon as a translational stop codon and does not allow any tRNA to bind the mRNA
   B. the termination factors recognize this codon and translation ends
   C. this codon is recognized by the corresponding anti-codon of an empty tRNA molecule that is not linked to an amino acid
   D. the non-recognition of this codon by any tRNA ultimately causes the translational machinery to stop

11. What is the E-site of the ribosome?
   A. the site where eukaryotic mRNA is processed
   B. an exciting term made up by your instructor
   C. the site where tRNA exits the prokaryotic ribosome
   D. the site where the endonuclease *Eco*RI restricts the ribosome

12. What is the *correct* order of the following steps in protein synthesis?
   1. a peptide bond is formed
   2. the small ribosomal subunit is loaded with initiation factors, m-RNA, and initiation aminoacyl-tRNA
   3. the intact ribosome slides forward three bases to read a new codon
   4. the primed small ribosomal subunit binds with the large ribosomal subunit
   5. elongation factors deliver aminoacyl-tRNA to bind to the A site

   A. 1, 2, 5, 4, 3 B. 2, 3, 4, 5, 1
   C. 4, 5, 1, 2, 3 D. 2, 4, 5, 1, 3

13. Which of the following is *not* characteristic of eukaryotes?
   A. ribosomes are larger than those of prokaryotes
   B. all mRNAs specify a single protein or a polyprotein
   C. no specific ribosome binding sites exist in mRNA comparable to those of the Shine-Dalgarno sequences in prokaryotes
   D. N-formylmethionine initiates all nascent polypeptide chains

14. Which one of the following statements *correctly* describes initiation of protein synthesis in *E. coli*?
   A. the initiator tRNA binds to the Shine-Delgarno sequence
   B. three initiation factors are involved and IF2 binds to GTP
   C. the intermediate containing IF1, IF2, IF3 initiator tRNA and mRNA is called the 30S initiation complex
   D. binding of the 50S subunit releases IF1, IF2, GMP and PPi

15. How many energy bonds are expended in the formation of a peptide bond?
   A. 2 B. 4
   C. 3 D. 6

16. Which of the following is not an antibacterial antibiotic?
   A. tetracycline B. streptomycin
   C. nystatin D. nalidixic acid

17. An antibiotic which inhibits translation in both prokaryotes and eukaryotes is
   A. chloromycetin B. puromycin
   C. actinomycin–D D. tetracycline

18. All aminoglycoside antibiotics inhibit protein synthesis by
   A. binding to small ribosomal subunit
   B. binding to large ribosomal subunit
   C. binding to both small and large ribosomal subunits
   D. inactivating eEF-2

19. Chloramphenicol inhibits
   A. cell wall synthesis in bacteria
   B. protein synthesis on 70S ribosome
   C. protein synthesis on 80S ribosome
   D. DNA replication

20. An antibiotic that resembles the 3′ end of a charged tRNA molecule is
A. streptomycin B. penicillin
C. tetracycline D. puromycin

21. Diptheria toxin inhibits protein synthesis by
A. causing formation of ADP-EF2 complex
B. release of peptidyl-t-RNA from the P site
C. binding to factor eEF-2
D. inhibiting peptide bond formation

22. Chaperone proteins help in
A. protein folding and assembly
B. protein stability
C. Both of the above
D. None of the above

23. Which of the following statements is *not* true for ubiquitin mediated proteolysis?
A. ubiquitinylated proteins are degraded in the lysosome
B. ubiquitin is polypeptide of 76 amino acids
C. ubiquitination occurs at a specific amino acid of the target proteins
D. following ubiquitination, proteins are degraded by proteasomes

24. Which of the following is involved in the majority of the ATP-dependent cytosolic degradation of proteins in eukaryotes?
A. cathepsins B. calpains
C. lysosome D. 26S proteosomes

25. The term protein splicing refers to
A. removal of intervening sequences between the genes
B. splicing out of introns from RNA
C. removal of intervening protein sequences from the translated protein
D. joining (splicing) of two different gene products to generate a novel protein

ANSWERS

1	2	3	4	5	6	7	8	9	10
D	D	C	B	D	A	C	D	D	B
11	**12**	**13**	**14**	**15**	**16**	**17**	**18**	**19**	**20**
C	D	D	B	B	C	B	A	B	D
21	**22**	**23**	**24**	**25**					
A	C	A	D	C					

6

Recombinant DNA Technology

RECOMBINANT DNA TECHNOLOGY

Genetic engineering may be defined as construction and utilisation of new DNA molecules that have been engineered by recombinant DNA techniques. The technique of genetic engineering is in the production of recombinant DNA. Recombinant DNA, as the name suggests, involves cutting a piece of original DNA and inserting in its place a different segment of DNA. The recombined or recomposed DNA is then copied multifold inside bacterial cells and stored in a gene library for use when required. The multiple copies of the gene are termed cloned DNA or cloned genes.

> Causing genetic change by artificially manipulating DNA is genetic engineering.
>
> Clone is a group of genetically identical cells. Such cells are descendents of a single cell. When a bacterium with recombinant DNA divides several times, it provides a clone containing a specific segment of DNA from another species.
>
> The production of genetically identical individuals or genetic material from a single cell is called cloning.

Recombinant DNA technology resulted from the two discoveries made while experimenting with bacteria:

(*i*) presence of plasmids or extra chromosomal DNA fragments in the bacterial cell which replicate along with bacterial DNA and can be used as a vector for carrying foreign DNA.

(*ii*) presence of specific enzymes which attack and cut DNA at specific sites.

Tools and Steps in Recombinant DNA Technology

Recombinant DNA technology is a "cut and paste" technology. Specific nucleotide sequences are cut from the DNA of humans, other animals or plants and "pasted" into plasmids. DNA of the plasmid carrying nucleotide sequence of another organism is the recombinant DNA. It is then inserted into bacteria. Bacteria divide repeatedly and a clone of bacteria with the recombinant DNA is obtained.

Five requirements for recombinant DNA technology are:

(*i*) Cell culture
(*ii*) Restriction endonuclease enzyme
(*iii*) Plasmids
(*iv*) Ligases
(*v*) Host bacteria

(*i*) **Cell culture:** Cultured cells of an animal or plant (or even a bacterium) carrying the required gene (nucleotide sequence of DNA) in its nucleus.

(*ii*) **The enzyme Restriction endonuclease:** Restriction endonucleases cut short specific DNA sequences. There are many different restriction endonucleases found in bacteria. Each of these enzymes very specifically recognises a particular DNA sequence (usually 4 to 6 bases) and cuts it. These enzymes are the "molecular scissors". Either they cut both the strands at the same place or at different places so that the two DNA strands hang out at the two ends. Two cuts at the two ends of a DNA segment releases the cut part as the restriction fragment. The ends are single stranded and called sticky ends.

Thus, a piece of DNA containing a particular gene can be obtained by selecting a particular restriction endonuclease.

(*iii*) **Plasmids:** Plasmids are extra chromosomal DNA molecules in a bacterial cell which have sequences matching those of the required gene and can be similarly cut by the same restriction enzymes. Plasmids can readily enter bacteria, yeast or other speedily reproducing cells.

(*iv*) **DNA ligase:** It is an enzyme which can seal one DNA fragment with another DNA segment, both having sticky ends. Ligase is the "molecular glue".

(*v*) **Host Bacteria:** Host bacteria are the bacteria whose plasmid is used for carrying foreign DNA.

Sequences of Steps in Recombinant DNA Technology

1. Specific restriction enzyme is selected.
2. Cell culture with required gene in the cells is obtained.
3. Restriction enzyme cuts the DNA at two ends of the specific gene and a restriction fragment is obtained (fig.).
4. Same restriction enzyme cuts a matching DNA sequence from a plasmid (fig.).
5. Ligase joins the restriction fragment in the place vacated by the cut DNA segment of the plasmid. The plasmid becomes a recombinant plasmid containing a foreign DNA fragment (fig.). Its DNA is the recombinant DNA. Since plasmids can carry foreign DNA, they are called clonal vectors. Bacteriophages (viruses) can also function as clonal vectors.
6. The recombinant plasmids then enter the bacteria.
7. Bacteria divide. Recombinant plasmids replicate along with bacterial DNA.
8. A large population of bacteria (more than a million) containing recombinant DNA can be obtained in less than ten hours.
9. Multiple identical copies of DNA fragments inserted into plasmids or bacteriophage (bacterial virus) then obtained and preserved in a DNA library.
10. These DNA fragments are the cloned DNA.

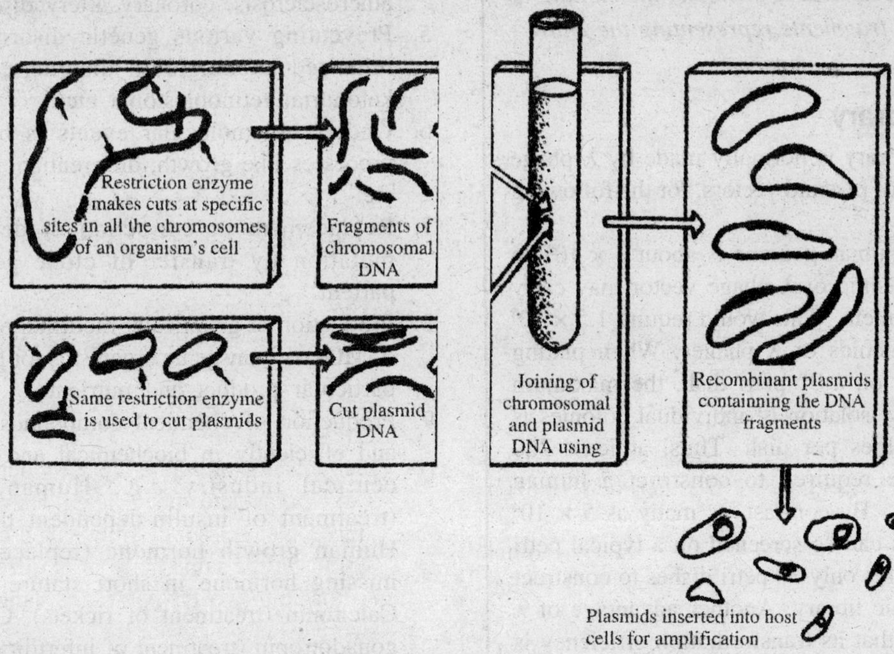

Fig. 1 : Steps in formation of multiple copies of recombinant DNA for DNA library.

Preparation of a DNA Library

DNA library is a collection of cloned DNA fragments. There are two types of DNA library:

The genomic library contains DNA fragments representing the entire genome of an organism.

The *c*DNA library contains only complementary DNA molecules synthesized from *m*RNA molecules in a cell.

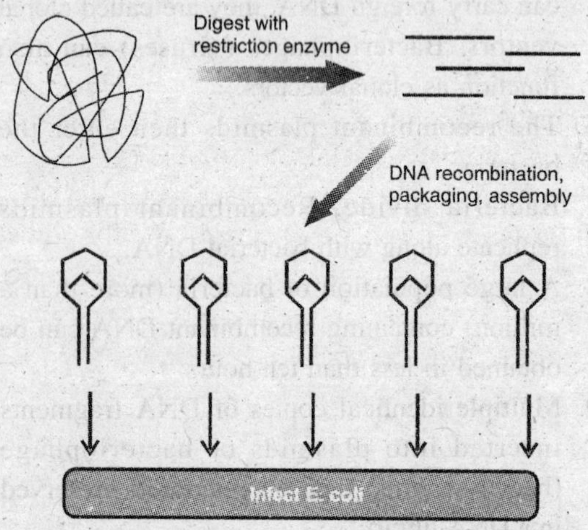

Fig. 2 : Preparation of the genomic library using λ phage vectors. It is basically the cloning of all DNA fragments representing the entire genome.

Genomic Library

The genomic library is normally made by λ phage vectors, instead of plasmid vectors, for the following reasons:

The entire human genome is about 3×10^9 bp long while a plasmid or λ phage vector may carry up to 20 kb fragment. This would require 1.5×10^5 recombinant plasmids or λ phages. When plating *E. coli* colonies on a 3" petri dish, the maximum number to allow isolation of individual colonies is about 200 colonies per dish. Thus, at least 700 petri dishes are required to construct a human genomic library. By contrast, as many as 5×10^4 λ phage plagues can be screened on a typical petri dish. This requires only 30 petri dishes to construct a human genomic library. Another advantage of λ phage vector is that its transformation efficiency is about 1000 times higher than the plasmid vector.

cDNA Library

The advantage of cDNA library is that it contains only the coding region of a genome. To prepare a cDNA library, the first step is to isolate the total mRNA from the cell type of interest. Because eukaryotic mRNAs consist of a poly-A tail, they can easily be separated. Then the enzyme reverse transcriptase is used to synthesize a DNA strand complementary to each mRNA molecule. After the single-stranded DNA molecules are converted into double-stranded DNA molecules by DNA polymerase, they are inserted into vectors and cloned.

Application of Recombinant DNA Technology

1. Preparation of gene maps.
2. In revealing details of various infections, diseases such as "inborn errors of metabolism."
3. Finding out the complete nucleotide sequence of genome of an organism and identification of genes.
4. Detecting cytogenetic abnormalities *e.g.*, Down's syndrome, multifactorial disorders, atherosclerosis, coronary artery disease etc.
5. Preventing various genetic disorders *e.g.*, inherited haemoglobin disorders, phenylketonuria, retinoblastoma etc.
6. Understand molecular events is biological processes like growth, differentiation, ageing etc.
7. Replacement or correction of deleterious mutation by transfer of clone gene in a patient.
8. Production of genetically modified organisms (GMOs) or transgenic organisms for providing particular product and nutrient.
9. Production of chemical compounds cheaply and efficiently in biochemical and pharmaceutical industry *e.g.*, Human insulin (treatment of insulin-dependent diabetes), Human growth hormone (replacement of missing hormone in short stature people), Calcitonin (treatment of rickets), Chorionic gonadotropin (treatment of infertility), Blood clotting factor VIII/IX (treatment of

haemophilia A/B), Tissue plasminogen activator (dissolving blood clots in heart attacks and strokes), Erythropoietin (stimulation of the formation of erythrocytes *i.e.,* RBCs in patients suffering from anaemia during repeated kidney dialysis or AIDS treatment), platelet derived growth factor (stimulation of wound healing), Interferon (treatment of viral infection, cancer), Interleukins (enhanced immune system's activity), Vaccines (for hepatitis B, herpes, influenza, pertussis, meningitis) etc..

10. **Gene Therapy:** Removal and replacement of defective genes with normal healthy functional genes is known as gene therapy *e.g.,* Sickle cell anaemia, Severe Combined Immuno-Deficiency (SCID). SCID is due to a defect in the gene for the enzyme adenosine deaminase (ADA) in 25 per cent of the cases.
11. It has several negative features as well: extensive erosion and genetic destruction of plant germplasm; ecological imbalance; production of monsters; production of dangerous toxic chemicals, production of highly lethal microbes and their use in microbiological warfare to kill humans, animals and plants.

POLYMERASE CHAIN REACTION

PCR is a rapid and versatile *in vitro* method for amplifying defined target DNA sequences present within a source of DNA. This technique was formulated in 1985 by *Kerry Mullis.* Usually, the method is designed to permit *selective amplification* of a specific target DNA sequence(s) within a heterogeneous collection of DNA sequences (e.g. total genomic DNA or a complex cDNA population). To permit such selective amplification, some prior DNA sequence information from the target sequences is required. This information is used to design two oligonucleotide primers (amplimers) which are specific for the target sequence and which are often about 15-25 nucleotides long. After the primers are added to denatured template DNA, they bind specifically to complementary DNA sequences at the target site. In the presence of a suitably heat-stable DNA polymerase and DNA precursors (the four deoxynucleoside triphosphates, dATP, dCTP, dGTP and dTTP), they initiate the synthesis of new DNA strands which are complementary to the individual DNA strands of the target DNA segment, and which will overlap each other.

Primer design: To permit selective amplification, some prior DNA sequence information from lhe target DNA is required. The information is used to design two primers (amplimers), which are specific for sequences flanking the target DNA sequence. So, for most PCR reactions, it is very important to reduce the chance of the primers binding to othe. locations in the DNA than the desired one. Hence certain rules for primer design that are important to consider. These rules follow:

1. **Primer length:** Length of the primers should not be very short or long. If the primers are too short they might hybridize to non-target sites and give undesired amplification products. Similarly, long primer influences the rate at which it hybridizes to the template DNA; long primers hybridize at a slower rate.
2. **Base composition:** The GC content should be betweem 40 and 60% with an even distribution of all four nucleotides.
3. **Nature of sequences:** Inverted repeats or any self-complementary sequences >3bp in length are to be avoided.
4. **Melting temperature:** Tm values for two primers used together should not differ by >5°C and the Tm of the amplification product should not differ from those of the primers by > 10°C.

Reaction cycle

The PCR is a chain reaction because newly synthesized DNA strands will act as templates for further DNA synthesis in subsequent cycles. After about 25 cycles of DNA synthesis, the products of the PCR will include, in addition to the starting DNA, about 10^5 copies of the specific target sequence. PCR consists of a series of cycles of three successive reactions:

- **Denaturation,** typically at about 93-95°C for human genomic DNA.
- **Primer annealing** at temperatures usually from about 50°C to 70°C depending on the T m of the expected duplex (the annealing temperature is typically about 5°C below the calculated T m). Annealing temperature must be low enough to enable hybridization between primer and template, but high enough to prevent mismatched hybrids from forming. This temperature can be calculated by determining the melting temperature or Tm of the primer-template hybrid. The T m can be determined experimentally or calculated from the simple formula:

$$Tm = (4 \times [G + C]) + (2 \times [A + T])\ °C$$

- **DNA synthesis,** typically at about 70-75°C.

These three steps **denaturation, primer annealing** and **DNA synthesis** constitute one cycle of the PCR amplification and can be carried out repetitively just by changing the temperature of the reaction mixture. The thermostability of the polymerase makes it feasible to carry out PCR. Suitable heat-stable DNA polymerases have been obtained from microorganisms whose natural habitat is hot springs. For example, the widely used *Taq* DNA polymerase is obtained from *Thermus aquaticus* and is thermostable up to 94°C, with an optimum working temperature of 80°C. Other thermostable DNA polymerase currently used in PCR are *pfu (Pyrococcus furiosus), Bst E (Bacillus stearothermophilus), Tth (Thermus thermophilus).*

After each cycle, the number of templates doubles, so that if one starts with a single double-stranded DNA molecule, after 20 cycles the number of molecules synthesized by the PCR is 1 x 10^6, and after 30 cycles the number of molecules increases to 1 x 10^9. This number is obtained by applying the following formula:

$$M_f = M_1 \times 2^n$$

Where M_1 is the initial number of molecules (template), M_f is the final number of DNA segments produced by the PCR, and n is the number of cycles performed.

MODIFICATIONS TO THE PCR TECHNIQUE

Nested PCR

Nested PCR increases the specificity of DNA amplification, by reducing non-specific amplification of DNA. Two sets of primers are being used in two successive PCR reactions. In the first reaction, one pair primers is used to generate DNA products, which besides the intended target, may still consist non-specifically amplified DNA fragments. The product(s) are then used in a second PCR reaction with a set of primers whose binding sites are just downstream of the first primer, or *nested* between the original set of primers. Binding sites are completely or partially different from the primer pair used in the first reaction, but are completely within the DNA target fragment.

Asymmetric PCR

Asymmetric PCR is used to preferentially amplify one-strand of the original DNA more than the other. It finds use in some types of sequencing and hybridization probing where having only one of the two complementary stands is required. PCR is carried out as usual, but with a great excess of the primers for the chosen strand.

RT-PCR

RT-PCR (Reverse Transcription PCR) is a method used to amplify, isolate or identify a known sequence from a cellular or tissue RNA. The PCR reaction is preceded by a reaction using reverse transcriptase to convert RNA to cDNA.

Hot-start PCR

Mixing of all PCR reagents prior to an initial heat denaturation step allows more opportunity for non-specific binding of primer sequences. Hot-start PCR is a technique that reduces non-specific amplification during the initial set up stages of the PCR. The technique may be performed manually by simply heating reaction components briefly at the melting temperature (e.g., 95°C) before adding the polymerase.

Quantitative PCR

Quantitative PCR is used to measure the quantity of a PCR product. It is the method of choice to quantitatively measure starting amounts of DNA, cDNA or RNA. Quantitative PCR is commonly used to determine whether a DNA sequence is present in a sample and the number of its copies in the sample. The method with currently the highest level of accuracy is *Quantitative real-time PCR*. It is often confusingly known as **RT-PCR** (Real Time PCR). Real Time PCR is used to quantify gene expression and also to screen for mutations and single nucleotide polymorphisms.

Touchdown PCR

Touchdown PCR is a method used to increase specificity without compromising yield. The principle is to initiate synthesis at very high annealing temperatures which permit only perfectly matched primer-template hybrids to form. The annealing temperature is dropped in a stepwise fashion with each cycle. Once copies of the target sequence have begun to accumulate over the first few cycles, high temperature annealing becomcs much less critical for specificity, as it is previous products which form the major template. These products are unlikely to have any sites for mispriming. The benefit of decreasing the annealing temperature is to increase the probability of stable primer-target interaction.

RACE (Rapid amplification of cDNA ends)

RACE is a PCR based method for locating precise start and end points of gene transcripts. First step in this method is conversion of mRNA into cDNA with reverse transcriptase. One of the primer used during reverse transcription is specific for an internal region close to beginning (5'-end) of the gene under study. Since only a small segment of mRNA is copied, the cDNA so generated will correspond exactly with the start of the mRNA. After synthesis of above cDNA, a short poly(A) tail is attached to its 3' end and it is subjected to normal PCR. The second primer will anneal to this poly(A) sequence and converts the single stranded cDNA into a double stranded molecule, which is subsequently amplified as the PCR proceeds. The sequence of the final PCR product will reveal precise position of the start or 5'-end of the transcript. This is known as 5′-RACE, so called because its results in amplification of the 5'-end of the starting RNA (Fig. 3).

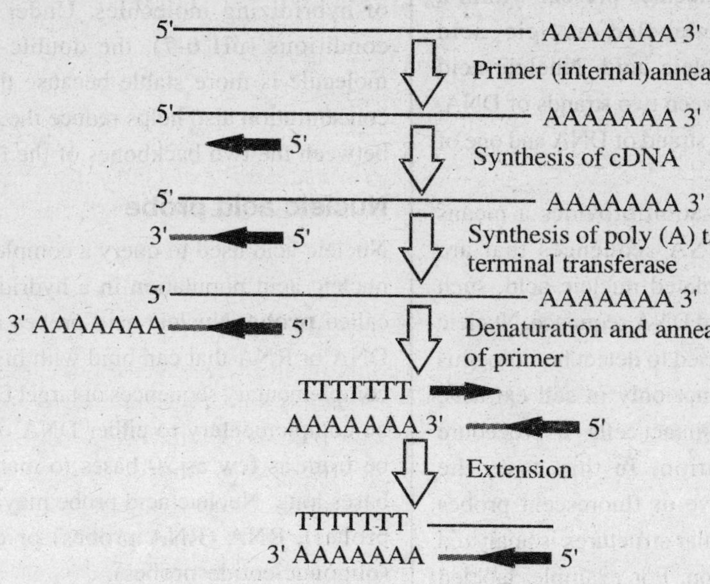

Fig. 3: *5' RACE. An internal primer is used to primer synthesis from a mRNA template. A poly(A) is added to the 3' end of the cDNA using terminal transferase. Second strand synthesis is prime using second primer.*

In 3′-RACE for isolation of 3′-end of an RNA, mRNA is first reverse transcribed to cDNA. PCR amplification of this cDNA is then carried out using a primer specific to known sequence in the mRNA under study (close to its 3'-end) and an oligo U primer. Nested PCR is then performed to improve specificity. These 5' and 3' RACE products can than be used to isolate full length cDNAs in a few days compared to weeks required for screening cDNA libraries.

NUCLEIC ACID HYBRIDIZATION

Nucleic acid hybridization is a fundamental tool in molecular genetics which takes advantage of the ability of individual single-stranded nucleic acid molecules to form double-stranded molecules (that is, to hybridize to each other). For this to happen, the interacting single-stranded molecules must have a sufficiently high degree of base complementarity. Nucleic acid hybridization is used to identify related or identical molecules. Nucleic acid hybridization assays involve a labeled nucleic acid **probe** to identify related DNA or RNA molecules present within a complex mixture of unlabeled nucleic acid molecules, the target nucleic acid. Nucleic acid hybrids can be formed between two strands of DNA, two strands of RNA, or one strand of DNA and one of RNA.

Nucleic acid hybridization provides a means for detecting DNA or RNA sequences that are complementary to any isolated nucleic acid, such as a viral genome or a cloned DNA sequence. Nucleic acid hybridization can be used to detect homologous DNA or RNA sequences not only in cell extracts, but also in chromosomes or intact cells - a- procedure called *in situ* hybridization. In this case, the hybridization of radioactive or fluorescent probes to specific cells or subcellular structures is analyzed by microscopic examination. For example, labeled probes can be hybridized to intact chromosomes in order to identify the chromosomal regions that contain a gene of interest.

Stringency of a hybridization reaction

Stringency is a term that describes the number of mismatched bases that are allowed in a hybridization reaction and still have a double-stranded hybrid. The more stringent the hybridization conditions are, the fewer the mismatched bases that are allowed before the two strands come apart. Stringency is influenced by the *temperature,* the *salt* concentration of the hybridization mixture and the pH of the hybridization reaction. All three of these conditions affect the stability of the double-stranded molecule. High stringency conditions are usually achieved with high temperatures (approaching the melting temperature of the DNA-DNA hybrid) and low salt concentrations (< 0.1 M). High salt concentrations tend to stabilize the double-stranded molecule and allow more mismatches to take place before the two strands come apart. The pH of the solution in which hybridization takes place also effects the stability of hybridizing molecules. Under slightly acidic conditions (pH 6-7), the double-stranded DNA molecule is more stable because the increased H^+ concentration also helps reduce the repulsive forces between the two backbones of the DNA molecule.

Nucleic acid probe

Nucleic acid used to query a complex heterogenous nucleic acid population in a hydridization assay is called **probe.** Nucleic acid probes are segments of DNA or RNA that can bind with high specificity to complementary sequences of target DNA. Probes can be complementary to either DNA or RNA and can be from as few as 20 bases to many thousands of bases long. Nucleic acid probe may be DNA (DNA probes), RNA (RNA probes) or oligonucleotide (oligonucleotide probes).

MULTIPLE CHOICE QUESTIONS

1. Specific fragment of DNA can be amplified by
   A. PCR
   B. molecular cloning
   C. both (A) and (B)
   D. Southern hybridization
2. A representative collection of all the DNA fragments from a given source cloned in vectors is described as
   A. DNA library B. cDNA library
   C. subcloning D. cloning
3. Molecular cloning is an ______ technique.
   A. *in vivo* B. *in vitro*
   C. both (A) and (B) D. none of the above
4. Source of DNA for cloning experiments is
   A. genomic DNA
   B. cDNA
   C. PCR product or previously isolated clone or chemically synthesized oligonucleotides
   D. all of the above
5. Random cleavage of DNA can be achieved by
   A. mechanical shearing (sonication, vortexing)
   B. chemical treatment (acid/alkali hydrolysis)
   C. non-specific endonuclease
   D. all of the above
6. Which is involved in joining of two DNA fragments?
   A. ligase
   B. homopolymer tailing
   C. both (A) and (B)
   D. polymerases
7. Double stranded oligonucleotides which contain restriction site used in DNA end modification are
   A. linkers
   B. adopters (predigested linkers with restriction enzymes)
   C. oligonucleotides
   D. primers
8. Predicting the orientation of insert DNA in the vector is called
   A. directional cloning (*e.g.* use of pair of restriction enzymes)
   B. bidirectional cloning
   C. both (A) and (B)
   D. none of the above
9. Cloning vector is usually a derivative of
   A. plasmid B. virus
   C. chromosome D. all of the above
10. Bacteriophage promoters or *E. coli* promoters flanking the multiple cloning site are referred as
    A. transcription vectors
    B. translation vectors
    C. expression vectors
    D. all of the above
11. Selection of recombinant DNA can be made easy by using
    A. insertion vectors
    B. replacement vectors
    C. both A and B
    D. none of the above
12. DNA can be transferred in to bacterial cloning host by
    A. chemical treatment with divalent cations
    B. heat shock
    C. electrophoration
    D. all of the above
13. Usually the scorable or selectable markers used in vectors are
    A. dominant
    B. recessive
    C. epistatic expression
    D. all of the above
14. cDNA is complementary to
    A. DNA B. mRNA
    C. rRNA D. ds mRNA
15. First strand cDNA is reverse transcribed from mRNA using reverse transcriptase and the primer with
    A. oligo dA
    B. oligo dT (or random hexanucleotide)
    C. oligo dC
    D. oligo dG

16. Second strand of cDNA can be synthesized using
- A. hairpin priming
- B. homopolymer tailing
- C. both A and B
- D. none of the above

17. The limitation of hairpin priming in cDNA synthesis is
- A. loss of 5′ end sequences
- B. loss of 3′ ends
- C. both A and B
- D. none of the above

18. Techniques used for identification of gene structure (promoter, intron/exon boundaries etc.,) are
- A. Nuclease mapping
- B. *in situ* hybridization
- C. inverse PCR and 5′ or 3′ RACE
- D. all of the above

19. The major disadvantage of expression of eukaryotic protein in *E. coli* is
- A. absence of post translational modification
- B. genes cannot be expressed
- C. poor translational efficiency
- D. all of the above

20. Yeast expression vector are usually based on
- A. 2μ plasmid
- B. 2m plasmid
- C. 2n plasmid
- D. 2new plasmids

ANSWERS

1	2	3	4	5	6	7	8	9	10
C	A	C	D	D	C	A	A	D	A
11	**12**	**13**	**14**	**15**	**16**	**17**	**18**	**19**	**20**
C	D	A	B	B	C	A	D	A	A

7

Genomics and Bioinformatics

INTRODUCTION

The term 'bioinformatics' is the short form of 'biological informatics', just as biotechnology is the short form of 'biological technology'. Anthony Keriavage, of the Celera Genomics, defined bioinformatics as 'Any application of computation to the field of biology, including data management, algorithm development, and data mining'. Clearly, a number of divergent areas, many of them outside biotechnology, come under bioinformatics.

Bioinformatics is the field of science in which biology, computer science, and information technology merge to form a single discipline. The ultimate goal of the field is to enable the discovery of new biological insights as well as to create a global perspective from which unifying principles in biology can be discerned.

Initial interest in Bioinformatics was propelled by the necessity to create databases of biological sequences. The first database was created within a short period after the Insulin protein sequence was made available in 1956.

The sequence information generated by the human genome research, initiated in 1988 has now been stored as a primary information source for future applications in medicine. The available data is so huge that if compiled in books, the data would run into 200 volumes of 1000 pages each and reading alone (ignoring understanding factor) would require 26 years working around the clock. For the population of about 5 billion human beings with two individuals differing in three million bases, the genomic sequence difference database would have about 15,000,000 billion entries. The present challenge to handle such a huge volume of data is to improve database design, develop software for database access and manipulation, and device data-entry procedures to compensate for the varied computer procedures and systems used in different laboratories. A single experiment can now yield data on the transcription level of 100,000 different mRNA species from a given tissue.

Functional Genomics

Functional genomics analyses the emerging knowledge about genomes to understand the gene and their product functions and interactions. Two exciting new developments are now enabling scientists to get a wealth of clues to this complicated story. The new technique, microarray technology and proteomics provide snapshots of all the genes expressed in a cell or tissue under different environmental conditions. The DNA microarray technology is used for analysing the expression of thousands of messenger RNA molecules.

Multiple Functions of the Genome

It is tempting to reduce genome function to encoding cellular proteins and RNAs. However, molecular genetics has shown that achieving this task requires cells to possess a number of additional capabilities also encoded in the genome:

1. Regulating timing and extent of coding sequence expression. 2. Organizing coordinated expression of protein and RNA molecules that function together. 3. Packaging DNA appropriately within the cell. 4. Replicating the genome in synchrony with the cell division cycle.

5. Transmitting replicated DNA accurately to progeny cells at cell division. 6. Detecting and repairing errors and damage to the genome. 7. Restructuring the genome when necessary (as part of the normal life cycle or in response to a critical selective challenge). These additional capabilities involve specific kinds of interactions between DNA and other cellular molecules. The construction of highly precise transcription complexes in RNA and protein synthesis is one example of such interactions (Ptashne, 1986). Formation of a kinetochore structure at the centromere for attachment of microtubules to ensure chromosome distribution at mitosis is another example (Volpe *et al.*, 2003).

Ever since the elaboration in the early 1960s of the operon model (Jacob & Monod, 1961) and the replicon hypothesis (Jacob, Brenner & Cuzin, 1963), we have understood specific interactions of cellular proteins with the genome to depend upon the existence of recognition signals in the DNA distinct from coding sequences. Signals like the operator or the origin of chromosome replication are completely different from any classical definition of a 'gene' as a basic unit. These recognition signals are themselves repetitive sequences in the genome, often carried by larger repeat elements. The idea that repetitive DNA is 'junk' without functional significance in the genome is simply not consistent with an extensive and growing literature, only a minor part of which is cited here.

APPLICATIONS OF BIOINFORMATICS

Genomics

Estimating the number of genes in an organism basing on the number of nucleotide base pairs was not reliable, due to the presence of high numbers of redundant copies of many genes. Genomics has corrected this situation. Useful genes can be selected from a gene library thus constructed and inserted into other organisms for improvement or harmful genes can be silenced. In the areas of structural genomics, functional genomics and nutritional genomics, bioinformatics plays a vital role.

Proteomics

Proteomics involves the sequencing of amino acids in a protein, determining its three-dimensional structure and relating it to the function of the protein. Before computer processing comes into the picture, extensive data, particularly through crystallography and NMR, are required for this kind of study. With such data on known proteins, the structure and its relationship to function of newly discovered proteins can be understood in a very short time. In such areas, bioinformatics has an enormous analytical and predictive potential. Protein folding alone of the most significant and fundamental problem in biological science realizing this, IBM in December 1999, had built a supercomputer, which is 2 million times faster than the today's fastest desktop PC. This new computer nicknamed "Blue Gene" by IBM researchers will be capable of performing more than one quadrillion operations per second. Better understanding of how proteins fold will give scientists and doctors better insight into diseases and ways to combat them.

Cheminformatics and drug design

It is now possible, through computer algorithm based bioinformatic procedures, to identify and structurally modify a natural product, to design a drug with the desired properties and to assess its therapeutic effects, theoretically. Such procedures, similar to an architect's on board plan before construction, are described as *in silico* (in the computer, based on silicon chip technology), as opposed to the earlier *in vitro* (in experimental models) and *in vivo* (in clinical trials) methods. The risk involved in the earlier random processes of drug discovery methods is largely removed by bioinformatics. Cheminformatics involves organisation of chemical data in a logical form to facilitate the process of understanding chemical properties, their relationship to structures and making inferences.

Molecular Phylogenies

Phylogeny is the origin and evolution of organisms. Biologists have constructed very elegant systems of classifications for the known organisms, though problems persist. Extensive work was carried out this way, comparing a very large number of organisms of plants and animals. Amino acid sequences and characteristics of proteins are also used in systematics.

Drug Modification

Several synthetic products are quite useful but cannot be used by one and all for certain side effects in some people. For example, aspartame (marketed under different trade names) is a dipeptide of aspartic acid and phenylalanine, and is 300 times sweeter than cane sugar. Aspartame is widely used as an alternate sweetener by diabetics and others who cannot take sweeteners loaded with calories. Unfortunately, pregnant women and people suffering from phenylketonuria, a disorder due to an impaired metabolism of phenylalanine, should not use aspartame. It would be useful if phenylalanine were substituted by some other amino acid without affecting its sweetness, to remove the restriction on its use.

PARTNERSHIP IN BIOINFORMATICS

(*a*) **Data Gatherers:** Enormous amounts of basic data from biomolecular chemistry and related areas, very painstakingly gathered over long years by experimental and analytical scientists, are the body and substance of bioinformatics.

(*b*) **Data Processors:** The second party use skills of complex software, to serve the needs of the data gathered.

(c) **Process Product Users:** End users of products.

Today it is possible to perform (using heuristic algorithms) 80% accurate searches perhaps 90-95% accuracy from the leading software systems. Sensitive algorithms which improve the search accuracy, such as hidden Markov models and Smith-Waterman algorithm, are also available but take more time to execute the search. Now, to handle these demanding needs, computers are being designed around the biologists.

Leading bioinformatics companies are developing software systems which permit research scientists to integrate their diverse data and tools under Common Graphical User Interfaces (GUIs). It also permits scientists to share information and provides powerful solution to archive data. The whole area of biology can immensely benefit from the bioinformatic approach. Bioinformatics tools for efficient research will have significant implications in life sciences and betterment of human lives.

MULTIPLE CHOICE QUESTIONS

1. Action of topoisomerase leads to change in
   A. linking number of single-stranded linear DNA
   B. linking number of double-stranded linear DNA
   C. linking number of closed circular single stranded DNA
   •D. linking number of closed circular double stranded DNA

2. Which of the following statement about topoisomerases are *correct*?
   P. type I topoisomerases act by making a transient break in one strand of DNA
   Q. type II topoisomerases act by introducing a transient double-strand break Topoisomerases
   R. DNA gyrase is a type II topoisomerase
   S. DNA gyrase removes the positive supercoiling and introduce negative supercoiling
   A. P and Q B. P, Q and R
   C. P, R and S D. All of the above

3. All of the following statements about telomerase are correct *except*
   A. the RNA component acts as a template for the synthesis of a segment of DNA
   B. it adds telomeres to the 5′-ends of the DNA strands
   C. it provides a mechanism for replicating the ends of linear chromosomes
   D. it recognizes a G-rich single strand of DNA

4. All of the following cell types contain the enzyme telomerase which protects the length of telomeres at the end of chromosomes, *except*
A. germinal B. somatic
C. haemopoetic D. tumor

5. Reverse transcriptases are encoded by retroviruses. The only enzyme from the mammalian cells with reverse transcriptase like activity is
A. telomerase B. ribonuclease
C. ribozyme D. caspase

6. Which of the following statement is *incorrect*
A. the protein that binds to the 9-mers in the *oriC* site of *E. coli* is DNA *A*
B. the enzyme that replicates DNA in mitochondria is called DNA pol γ
C. the origin of replication site in the *E.coli* chromosome is about 245 bp long
D. None of the above

7. Mitochondrial DNA is replicated from
A. a single *ori* site bidirectionally
B. two different *ori* sites in the same direction
C. two different *ori* sites at different times in opposite directions
D. many sites bidirectionally, like nuclear chromosomes

8. Chloroplast DNA is replicated from
A. a single *ori* site bidirectionally
B. two different *ori* sites in the same direction
C. two different *ori* sites simultaneously and in opposite directions
D. many sites bidirectionally, like nuclear chromosomes

9. Which of the following genome feature increases as the complexity of an organism increases?
P. genome size
Q. number of genes
R. density of genes in the genome
S. average size of individual genes
A. P, Q B. P, S
C. P, R, S D. R, S

10. Approximately what percentage of human genome contains protein coding genes?
A. ~2% B. ~20%
C. ~50% D. ~80%

11. Non-coding DNA in eukaryotic cells may include
A. introns B. pseudogenes
C. spacer DNA D. All of the above

12. Scientists estimate that greater numbers of proteins are produced in humans than the number of predicted genes. This is because
A. the relationship between genes and proteins is not known
B. primary RNA transcripts can be alternatively spliced to make different mRNA
C. DNA replication occurs throughout life, so that genes are produced as a person gets older
D. both the coding and the template strand of a single gene is transcribed

13. Satellite DNA consists of
A. extra chromosomal DNA
B. short repetitive nucleotide sequences
C. ribosomal RNA genes
D. single gene regions

14. Highly repetitive sequences are normally located in chromosome near
A. centromere B. telomere
C. chromocenter D. All of the above

15. Simple sequence DNA
P. reassociates with the intermediate component of a $Cot_{1/2}$ curve
Q. consists of individual short repeated sequences that are dispersed throughout the genome
R. accounts for approximately 10% of mammalian genome
S. has a characteristic buoyant density based on its nucleotide composition
A. P, Q B. P, S
C. R, S D. P, Q, R

16. Highly repetitive sequences in DNA are characterized by
A. 2 to 10 base pairs repeated in tandem 10^5 to 10^7 times in a genome
B. a slowness to reanneal as compared to unique sequences
C. unique sequences transcribed together
D. their absence in human DNA

17. Which of the following is the most likely mechanism for the origin of multigene families
A. endosymbiosis
B. gene duplication
C. convergent evolution of dissimilar genes
D. horizontal gene transfer

18. Gene duplication
P. has occurred many times in eukaryotic genomes
Q. is critical for generating new genes
R. creates gene families
S. occurs only within gene clusters
A. P, Q B. P, S
C. P, Q, S D. P, Q, R

19. Amplification of genes involves
A. removal of histones from the DNA to allow transcription of the gene
B. multiple duplications of the gene via replication
C. multiplication of extra chromosomal elements only
D. invertebrate genomes only

20. Globin genes
A. are derived from a common ancestral gene
B. are organized into a single cluster
C. are organized differently in all higher eukaryotes
D. are all expressed simultaneously

21. How are the eight human globin genes organized?
A. they are randomly distributed on the human chromosomes
B. they are in two clusters, one on chromosome 11 and the other on chromosome 16
C. they are in a single cluster on the X chromosome
D. they are in four clusters, alpha, beta, gamma and delta

22. DNA sequence rearrangements are involved in all of the following processes *except*
A. immunoglobulin gene expression in mammals
B. Mating-type switching in yeast
C. antigen switching in trypanosomes
D. intron splicing in ciliates

23. Similarities between RNA polymerase and DNA polymerase activities in prokaryotes include
P. the requirement for a template
Q. the requirement for a primer
R. synthesis of the nascent chain in the 5′ to 3′ direction
S. the 3′ to 5′ exonuclease editing function
A. P, Q B. P, R
C. P, S D. P, Q, R

24. Which one of the following statements about *E.coli* RNA polymerase is *false?*
A. the holoenzyme includes the sigma factor
B. the core enzyme includes the sigma factor
C. it requires Mg^{2+} for its activity
D. it requires Zn^{2+} for its activity

25. How many different types of polypeptide subunits are present in RNA polymerase of *E.coli*?
A. two B. five
C. three D. four

ANSWERS

1	2	3	4	5	6	7	8	9	10
D	D	B	B	A	D	C	C	C	A
11	**12**	**13**	**14**	**15**	**16**	**17**	**18**	**19**	**20**
D	B	B	D	C	A	B	D	B	A
21	**22**	**23**	**24**	**25**					
B	D	B	B	D					

8

Microbial Culture

MICROBIAL CULTURE

As stated previously microbial cultures should be provided with required chemical and physical environment for proper multiplication and physiological state, so that the cells can carry out the required bioconversion satisfactorily. The chemical environment of a microbial cell is its nutritional conditions in which it is growing. It also includes the correct pH and temperature.

Nutrients for Microbial Culture

Like any other living system, micro-organisms also require a source of energy, carbon, nitrogen, oxygen, iron and other minerals, micronutrients and water for growth and multiplication. All these nutrients that are essential for the growth and multiplication of microbial organisms are supplied in the form of nutrient media. For the laboratory scale cultivation we may use certain costly media components, but for industrial purposes it should be economical and readily available. The media that we use for growing microbes may be synthetic, semi synthetic or completely natural. If the nutritive components of the media are not of natural origin, such nutritive media are known as synthetic media. Such media we can synthesize in the laboratory following certain recipes, mixing the required salts, minerals and carbon source. There are a large number of commercially available nutrient media, which contain both natural as well as salts and minerals. Such nutrient media are known as semi synthetic media. For example, commercially available nutrient broth, trypticase soya broth (TSB), brain heart infusion (BHI) broth, yeast extract, potato dextrose agar, casein digest etc. are some examples for semi synthetic media. For laboratory scale cultivation of bacteria and other micro-organisms we can prefer these synthetic or semi synthetic media, but for industrial scale cultivation these media are not recommended on economical point of view. For commercial purpose the media that we recommend should be cheap and available round the year. The following are the minimum components required in a microbial medium for cultivation of microbes in the laboratory.

Carbon source: A simple carbon source, which is simple to use and easily available can be used as a carbon source. Sugars like glucose, lactose, sucrose, and complex polysaccharides like starch, glycogen cellulose, mixture of various carbohydrates and other compounds like cereal grain powders, cane molasses etc. are usually used as carbon source in microbial culture media. The main purpose of carbon source is to provide energy and carbon skeleton for the synthesis of various other biological compounds.

Nitrogen sources: The major types of nitrogen sources used in culture media are ammonium salts, urea, animal tissue extracts, amino acid mixtures and plant tissue extracts.

Micro-elements or Trace elements: Elements required in small amounts or in traces are to be added into the medium as salts in required amounts. The elements such as Cu, Cobalt, Iron, Zinc, Mn, Mg etc. are the micro-elements.

Growth Factors: Growth factors are certain organic compounds that are essential for the growth and multiplication of cells, but cannot be synthesised by the cells. Such compounds should be supplemented in the medium. Certain amino acids and vitamins are also included in this category.

Antifoams: This is not a nutritive component of the media. The media rich in nutritive components like starch, protein and other organic material and also the proteins and other products secreted by the growing cells can result in excessive foaming, while the culture media is agitated for aeration. To prevent the formation of foam some antifoaming agents are included in the media. Certain types of fatty acids such as olive oil and sunflower oil, and silicones are commonly used in the cell cultures as antifoam agents.

Energy sources: The carbon sources used in the culture media such as carbohydrates, sugars, proteins, lipids etc./can work as energy source for the growth and metabolism of the microbial cells.

Water: Water is the base of any culture media, whether it is liquid or solid. In solid culture media like the media of Solid State Fermentation or agar media the quantity of water is comparatively less than liquid media. In laboratory experiments usually single distilled water or double distilled water is used. But in large-scale microbial cultivation for industrial purposes, the pH and the dissolved salts present should considered when formulating the media requirements and its concentration. Water is also required for a large number of other services in the laboratory such as cooling, heating, steaming etc. Therefore, any laboratory should be provided with a source of clean water of consistent quality.

Culture Procedures

1. **Sterilization:** The media and the culture vessel have to be sterilized to prevent the growth of unwanted micro-organisms and thus contamination. If the laboratory scale experiments are carried out in 100-1000 ml flasks, or in lesser volumes like 50 ml or 10 ml, the media along with the culture flasks or vials can be steam sterilized with an autoclave. Depending upon the quantity of the materials autoclaved, the sterilization can be carried out alternatively in a pressure cooker of convenient size. Steam sterilization with an autoclave or pressure cooker is carried out at 120° C for 15-20 minutes under 15 psi pressure.

   When microbes are to be cultivated in a fermentor for large-scale operation, it is convenient to sterilize the fermentor as a whole with media or without media. Media may be sterilized separately or *in situ* in the fermentor itself. Steam is used for the sterilization of the media and fermentor, by passing the steam through the sterilization jacket or the coil around the fermentor. When the fermentor is sterilized without media in it, steam can be sparged into the vessel through all openings and allowing it to exit very slowly. Sparging is a process by which sterile air or steam allowed to pass through the medium in the vessel with the help of a sparging device placed at the bottom of the fermentor. The steam pressure is held at 15 psi for 20-30 minutes while circulating or holding the steam within the vessel or in the jacket.

2. **Environment for Microbial Growth:** The nutrient composition of the medium, the ionic concentration of salts, pH and temperature influence the growth of micro-organisms in culture and its metabolic state. Most of the bacteria grow at neutral pH, where as yeast and fungi prefer acidic pH. Similarly different organisms prefer different optimum temperature for active growth and multiplication. The optimum temperature has to be maintained in the culture with the help of an incubator in the case of small-scale cultures and circulating water of the appropriate temperature through the jacket of the fermentor.

3. **Aeration and Mixing:** Mixing of the broth is essential for the uniform distribution of the nutrients and the microbial population in the culture. Aeration is needed for the easy gas exchange between the medium and the environment. Aerated medium will be rich in oxygen. Aeration and mixing can be easily achieved by shaking the medium on a shaker in the case of small-scale cultures (shake flasks cultures). In large-scale cultivation in bioreactors the transfer of oxygen to organisms is very

difficult because it requires proper mixing. In fermentors, the proper mixing of cells, media components and oxygen is achieved by stirring the medium with the help of mechanical stirrer and baffles attached to it. Baffles help in maintaining turbulence. Microbial free air passed through the media ensures proper aeration and this forced aeration also help in the mixing of media, cells and oxygen.

Types of Microbial Cultures

The culturing of microbial system can be achieved in different ways. The type of culture method sometimes depends on the type of the microbial system or on the type of the product that we expect. For example, one can get two entirely different products from the same organism by changing the nutritional and other parameters or even culturing vessels.

1. Batch Culture,
2. Fed-batch Culture,
3. Continuous Culture.

APPLICATIONS OF MICROBIAL CULTURE TECHNOLOGY

As we have discussed earlier, the microbial cultures have a large number of applications. In our daily life they play an important role in different ways. Microbial cells can be used for the production of various substances depending on the metabolic activities of the cells. The following are the major applications of microbial cultures for basic research and industiral applications:

1. Production of whole microbial cells as food and vaccines.
2. Production of primary metabolites like acids, alocohols, enzymes and microbial polysaccharides.
3. Production of secondary metabolites like antibiotics and biodegradable plastics.
4. Microbial leaching of metals, effluent and waste treatment.
5. Microbial cells as agents of biotransformation of organic compounds.
6. As host cells for the production of recombinant proteins, gene cloning and many other molecular biology researches.

The oldest of the uses of microbial cultures is to produce fermented foods like curds, cheese and wines. Here, the whole cells are used. Whole cells are also used for the production of single cell proteins and certain vaccines like tetanus vaccine, typhoid vaccines and tuberculosis vaccines. For the production of single cell proteins selected micro-organisms like yeast and spirulina are allowed to grow in the culture (may be starch factory effluents) and the whole microbial cells are dried and used as food material both for human consumption and cattle feed. Production of organic acids, alocohols, various types of enzymes and polysaccharides are examples for primary metabolites. Antibiotics and other types of organic molecules like hydrocarbons are examples for secondary metabolites. Microbial extraction of metals like copper and iron, treatment of liquid waste or effluents using microbial systems are examples where microbial metabolism is used to convert the waste into useful products and thereby the disposal of the waste without creating any environmental pollution. In modern biology particularly in molecular biology and recombinant techniques, certain microbial cells such as *E.coli* are used as host cells for the cloning and expression of certain genes and the production of recombinant proteins. Production of human insulin in *E.coli* and hepatitis B vaccine in yeast are good examples. Microbial production of vitamin C is another example for exploiting the primary metabolism for the production of useful compounds.

Table 1 : Some microbial products and the organism

Micro-organisms	Products
Saccharomyces cerevisiae	Ethanol and production recombinant proteins like Hepatitis B surface antigens
Aspergillus niger	Citric acid
Aspergillus oryzae	Amylase
Penicillium chrysogenum	Penicillin
Streptomyces griseus	Streptomycin
Propionibacterium shormani	Vitamin B_{12}
Escherichia coli	Recombinant proteins
Alcaligenes eutrophus	Poly 3-hydroxybutyrate
Leuconostoc mesenteroides	Dextran

MULTIPLE CHOICE QUESTIONS

1. The cell walls of gram positive bacteria contain two modified sugars, viz., N-acetylglucosamine (NAG) and N-acetylmuramic acid (NAM). They are covalently linked by
A. alpha-1,4-glycosidic bond
B. beta-1,6-glycosidic bond
C. alpha-1,6-glycosidic bond
D. beta-1,4-glycosidic bond

2. Teichoic acid present in the cell wall of gram positive bacteria makes a covalent bond with
A. N-acetyl muramic acid
B. N-acetyl glucosamine
C. cell membrane lipid
D. Both A and C

3. Which of the following statements about an *E.coli* ribosomes are correct?
P. it is composed of two spherically symmetrical subunits
Q. it has a large subunit comprising 34 kinds of proteins and two different rRNA molecules
R. it has a sedimentation coefficient of 70S
S. it has two subunits, one housing the A site and the other the P site
T it has an average diameter of approximately 200 Å
U. it has a mass of approximately 270kd, one-third of which is RNA
A. Q, R, T B. P, Q, R, S, T
C. P, Q, R, T, U D. P, R, S, T, U

4. *Bacillus anthracis* capsules consist of which one of the following?
A. γ-glutamyl polypeptide
B. hyaluronic acid
C. repeating polysaccharide capsule of glucose and glucuronic acid
D. sialic acid polymers

5. A bacterial culture had an initial cell density of 10^3 cells/ml. In 6 hours the cell density reached 10^6 cells/ml. Given the formula for the number of generations,

$n = (\log_{10} N_t - \log_{10} N_0)/0.301$

The number of generations (n) the cells have undergone is
A. 3 B. 10
C. 15 D. 20

6. Choose the correct pair
A. Tryptic soy agar – supportive media
B. Blood agar – enriched media
C. MacConkey agar – selective as well as differential media
D. Beef extract – defined media

7. In the penicillin-enrichment method, auxotrophic mutants are isolated based on the following principle
A. both the auxotroph and prototroph grow but at different rates in the minimal medium
B. only the prototroph grows and gets killed while the auxotroph remains unaffected
C. while the prototroph grows readily, auxotroph grows with a lag phase
D. only the auxotroph growth is supported by minimal medium

8. What properties is/are *not* expected to be significantly different between hot-spring bacteria (that live at 120 degrees Celsius) and regular intestinal *E.coli?*
P. Number of cysteines in the proteins
Q. Number of methionines in the protein
R. Molecular weight of the protein
S. GC richness of the genomic DNA
T. Richness of saturated fatty acids in plasma membranes
A. P and T B. Q and R
C. R and S D. P and R

9. The group of bacteria that don't have cell walls are the
A. archaebacteria B. mycobacteria
C. mycoplasma D. Both A and B

10. Mycoplasmas are different from other prokaryotes by
A. presence of chitin in cell walls
B. presence of murein in cell walls
C. presence of proteins in cell walls
D. absence of cell wall itself

11. Actinomycetes
- A. a gram negative bacteria
- B. bears spores
- C. intracellular obligate parasite
- D. lack peptidoglycan cell wall

12. Clostridia are anaerobes and they form ATP by 'stickland' reaction. For this they use
- A. palmitic acid
- B. pyruvic acid
- C. ethanol
- D. amino acids

13. Which of the following statement is *correct*?
- A. bacteria chromosome has many origin of replication
- B. bacterial chromosome is present mostly in positively supercoiled structure
- C. bacterial chromosome is a circular *ss*DNA
- D. bacterial chromosome has no centromere and telomere

14. A plasmid is
- A. a fragment of a chromosome that is transferred during bacterial conjugation
- B. a piece of DNA carrying a multiple cloning site
- C. a circular *ds*DNA which can replicate independently
- D. a piece of DNA carrying gene essential for bacteria

15. Which of the following statements regarding plasmid is *incorrect?*
- A. mostly circular *ds*DNA, with multiple copies of pseudogenes
- B. a bacterial cell can harbor more than one type of plasmids
- C. replication of plasmid may be unidirectional or bidirectional
- D. θ-replication in plasmid is more common form of replication than rolling circle replication

16. The copy number of a plasmid is the
- A. number of copies of the plasmid per immediately after cell division
- B. number of plasmids that take part in conjugation
- C. number of plasmids containing *Ori*V
- D. number of identical genes on plasmid

17. F′ plasmids
- A. are responsible for high frequency recombination
- B. carry some chromosomal genes
- C. are those plasmids that have never been incorporated into a bacterial chromosome
- D. None of these

18. Plasmids do which of the following?
- A. direct synthesis of conjugation pili
- B. provide resistance to certain antibiotics
- C. induce the formation of tumors in plants
- D. All of these

19. RNA as a genetic material is found in which of the following organisms
- A. *Plasmodium*
- B. *Staphylococcus aureus*
- C. *Schizosaccharomyces cerevisiae*
- D. Polio virus

20. Which one of the following viruses replicate in the cytoplasm?
- A. SV40
- B. adenovirus
- C. vaccinia virus
- D. herpex simplex virus

ANSWERS

1	2	3	4	5	6	7	8	9	10
D	D	A	A	B	D	B	B	C	D
11	**12**	**13**	**14**	**15**	**16**	**17**	**18**	**19**	**20**
B	D	D	C	A	A	B	D	D	C

9

Plant Tissue Culture and its Application

PLANT TISSUE CULTURE

Growing the plant cells, tissues and organs on a artificial, synthetic medium under controlled conditions is called plant tissue culture. Plant tissue culture has become a major thrust area in plant biotechnology.

Concept

The basic concept of plant tissue culture is totipotency, differentiation, dedifferentiation and redifferentiation.

Totipotency

The inherent potential of any living plant cell to develop into entire organism is called totipotency. This is unique to plant cells.

Differentiation

The meristematic tissue is differentiated into simple or complex tissues.

Dedifferentiation

Reversion of mature tissue into meristematic state leading to the formation of callus is called dedifferentiation.

Redifferentiation

The ability of the callus to develop into shoot or root or embryoid.

The origin and development of plant tissue culture

The beginning of plant tissue culture was made as early as 1898, when a German Botanist G. Haberlandt successfully cultured individual plant cells, isolated from different tissues. But only during 1934 to 1939, a foundation of plant tissue culture was laid down by three scientists (Gauthret, White and Nobecourt) due to discovery of plant growth regulators such as auxins and vitamins.

During next twenty years (1940 to 1960) a variety of growth regulator such as cytokinins were identified for their effect on cell division, growth and differentiation.

After 1960, *in vitro* culture of plant cells, tissues and organs was reasonably well developed. Research in this area was initiated in early 1960s by Prof. P. Maheshwari and Prof. S. Narayanaswamy at the Department of Botany, University of Delhi in India. Consequently, media and culture techniques for a variety of plant materials became known, which are now extensively utilised in all areas of plant improvement programmes.

Basic techniques of plant tissue culture

1. Culture vessels

The culture vessels used for plant tissue studies includes Erylenmayer flask (conical flask), petri plates and culture tubes (25 × 150 mm).

2. Culture medium

The important media used for all purpose experiment are Murashige and Skoog medium (MS medium), Gamborg medium (B5 medium), White medium (W medium) and Nitsch medium. The culture medium is closed with cotton plug/ or aluminium foil sheet. The pH of the medium is adjusted to 5.8 (acidic range).

3. Sterilization

Sterilization is the technique employed to get rid of the microbes such as bacteria and fungi in the

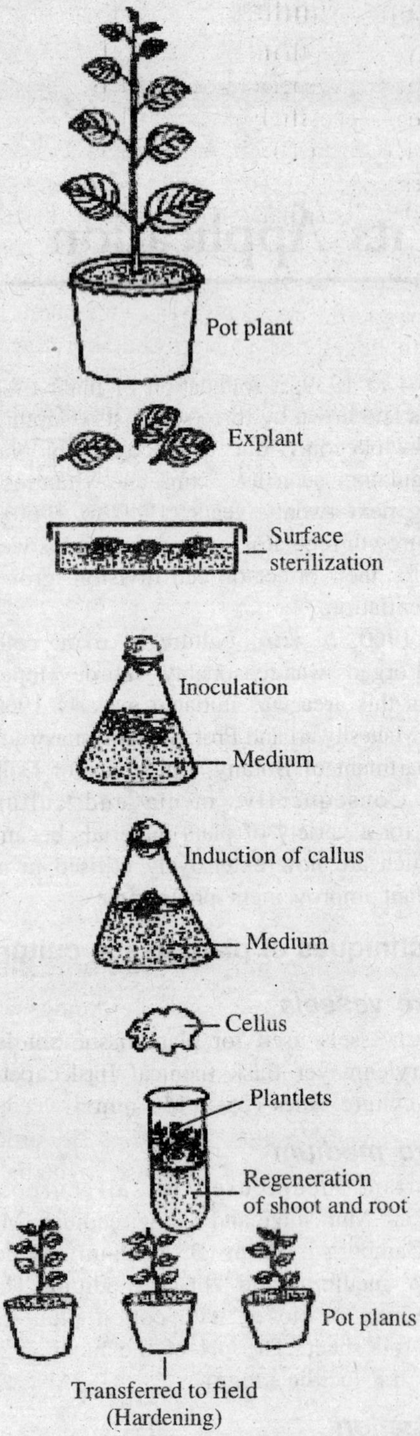

Fig 1. : Techniques of tissue culture

culture medium and plant tissues. So, it is important to sterilize the culture medium and plant tissue. The culture medium can be sterilised by keeping it in an autoclave and maintaining the temperature of 121°C for 15 minutes. The plant tissue (inoculum) is to be surface sterilised.

Chemical sterilization

By treating the inoculum in any one of the chemical sterilizant such as Sodium hypochlorite, Calcium hypochlorite, Mercury chloride for 15 to 20 minutes followed by repeated washing in sterile water upto 3 to 5 times.

4. Inoculation

Transfer of explant (root, stem, leaf, etc.) on to a culture medium is called inoculation. The inoculation is carried out under aseptic condition for which an apparatus called laminar air flow chamber is used. Flamed and cooled forceps are used for transfer of plant materials to different culture media kept in glasswares.

5. Incubation

The culture medium with the inoculum is incubated at 26 ± 2ºC with the light intensity at 2000 to 4000 lux (unit of intensity of light) and allowing photoperiod of 16 hours of light and 8 hours of darkness.

6. Induction of callus

Due to activity of auxins and cytokinins, the explant is induced to form callus. The callus is an unorganized mass of undifferentiated tissue. The mechanism of callus formation is that auxin induce cell elongation and cytokinin induces cell division as a result of which masses of cells are formed.

7. Morphogenesis

Formation of new organs from the callus under the influence of auxin and cytokinin is called morphogenesis. Roots and shoots are differentiated from the callus. Such embryos are called somatic embryos result in the formation of young plantlet.

There are two types of morphogenesis

(*a*) **Organogenesis:** Formation of new organs such as shoot and root is known as organogenesis. The development of shoot from the callus is called caulogenesis and formation of root is called rhizogenesis respectively.

(*b*) **Embryogenesis:** Formation of embryos (*i.e.* bipolar structure having shoot and root) from the callus is called embryogenesis. These embryos arise from somatic callus tissue and are called somatic embryos or embryoids or somaclonal embryos.

8. Hardening

Exposing the plantlets to the natural environment in a stepwise manner is known as hardening. Finally the plantlets are gradually transferred to the soil.

TRANSGENIC PLANTS

Introduction of foreign gene

In genetically engineered plant cells, a bacterium *Agrobacterium* is mainly involved in transfer of foreign gene. However, *Agrobacterium* cannot infect all plants since it has a narrow range of host specificity. Therefore other techniques have been developed to introduce foreign DNA into plant cells. Novel methods of ensuring DNA uptake into cells include electroporation and mechanical delivery or biolistics.

Electroporation is a process of creating temporary pores in the cell membrane by application of electric field. Creation of such pores in a membrane allows introduction of foreign molecules such as DNA, RNA, antibodies, drugs, etc. into cytoplasm. The development of this technique is due to contribution of biophysics, bioengineering, cell and molecular biology. While the technique is now used widely to create transgenic micro-organisms, plants and animals, it is used increasingly for application of gene therapy.

The mechanical particle delivery or gene gun methods to deliver DNA on microscopic particles into target tissue or cells. The process is increasingly used to introduce new genes into a range of bacterial, fungal plant and mammalian species. It is the main method of choice for genetic engineering of many plant species including rice, corn, wheat, cotton and soyabean.

Transgenic (modified) plants

Presently, more than 50 types of genetically engineered plant species, called transgenic plants have been successfully developed. These plants were made to resist insect, pests, viruses or herbicides through incorporation of foreign gene into DNA of host plant cells. Initially transgenic plants were developed more in dicotyledons, but now extended to several monocotyledons like wheat, maize, rice and oats. Transgenic plants have also been developed and are suitable for food industries (delaying ripening in tomato).

Gene pharming, the use of transgenic plants as bioreactors or factories for production of speciality chemicals and pharmaceuticals is being pursued by a number of firms. Plants have been engineered to produce human proteins, such as hormones, in their seeds. A weed called mouse-eared cress has been engineered to produce a biodegradable plastic (polyhydroxybutyrate, or PHB) in tissue granules.

Transgenic dicotyledonous plants

1. *Nicotiana tabacum*	4. *Helianthus annuus*
2. *Beta vulgaris*	5. *Solanum tuberosum*
3. *Glycine max*	6. *Gossypium hirsutum*

Transgenic monocotyledonous plants

1. *Asparagus sp.*	3. *Zea mays*
2. *Oryza sativa*	4. *Avena sativa*

Herbicide resistance in transgenic plants

Under normal circumstances, herbicides affect photosynthesis or biosynthesis of essential amino acids. Under field conditions, application of herbicides not only kills the unwanted weeds but also greatly affects the field crops. In order to protect the crops against exposure to herbicides, scientists after intensive research isolated a gene from Streptomyces *hygroscopicus* which encodes an enzyme, capable of inactivating the herbicide 'Basta'. Transgenic plants with this gene have been developed, demonstrating effectiveness of this gene for protection against herbicide 'Basta'. Thus, herbicide-tolerant crop plants have now been developed by genetically manipulating plant genomes resistant to specific herbicides.

Improved resistance to insect pests and microbial diseases

Genes from *Bacillus thuringiensis* (Bt_2) have been introduced into several crops, including tomato and cotton, and field-testing has demonstrated impressive results against many pests. Spore preparation of this bacterium is used as a biological insecticide during the last 20 years. Insecticidal activity depends on a toxic protein called delta endotoxins. The toxin gene (Bt_2) from *Bacillus thuringinesis* has been isolated and used for *Agrobacterium*. Ti plasmid mediated transformation of tobacco, cotton and tomato plants. The transgenic plants were resistant to the *Manducta sexta*, a pest of tobacco. India had acquired technology from U.S.A. for introducing Bt toxin gene in cotton for the development of resistance against pests in this major cash crop of India. Widespread use of insecticides, fungicides and pesticides for crop protection undoubtedly has damaging effects on the environment and hence it is important to improve the control of pests and diseases by genetic means. Genetic modification of plants is an attempt for ècofriendly measures against environmental degradation.

Through genetic modification, the oil-producing soya bean was tailored to produce a wide range of industrial lubricants, cosmetic compounds and detergents that are biodegradable. A whole new area of biotechnology has been opened up and plants are made to synthesize many novel substances including functional human antibody fragments.

Practical application of genetic transformation

- By genetic manipulation, it is possible to obtain plants with insecticidal property. Thus, application of chemical pesticides to the crop plants is reduced.
- Genetic manipulation is also carried out in crops for making desirable storage proteins, vitamins, amino acids and for the suppression of antinutritional protein synthesis.
- Plants are made to produce large amount of secondary metabolites having high commercial value.
- Losses during storage and transport of some crops can be as high as 80 per cent. This is mainly due to biological activities—bruising, heat and cold damage in soft fruits and vegetables. In tomato, the enzyme *polygalactronase* breaks down cell wall constituents, thus leading to softening of the fruit during ripening. By inhibiting the *polygalactronase* by antisense genes the tomato can remain dormant fresh until mature and be transported in a firm solid state. Antisense RNA is a RNA molecule capable of controlling and expression of particular enzymes which are involved in ripening processes.
- Genetic manipulation of flower and leaf colour, abundance of flowers, perfume and shape are now the major targets for decorative plant industries.

Transgenic microbes

The genetically engineered micro-organisms are being used for the commercial production of some non-microbial products such as insulin, interferon, human growth hormone and viral vaccines.

Use of genetically engineered bacterial strain

In 1979, for the first time Anand Mohan Chakrabarty, an Indian born American scientist developed a strain of *Pseudomonas putida* that contained a hybrid plasmid derived by combining parts of CAM and OCT. (CAM and OCT are the plasmids which contain the genes responsible for the decomposition of the hydrocarbons like camphor and octane respectively present in the oil.) This strain could grew rapidly on crude oil because it was capable of metabolizing hydrocarbons more efficiently. The bacterial strain called the superbug was produced on a large scale in laboratory, mixed with straw and dried. When the straw was spread over oil slicks, the straw soaked up the oil and bacteria broke up the oil into non-polluting and harmless products. In this way, pollution of land and water due to the oil slicks can be remedied and the phenomenon is called bioremediation. It is defined as the use of living micro-organisms to degrade environmental pollutants or prevent pollution. The contaminated sites are restored and future pollution is prevented.

Benefits from release of genetically modified micro-organisms into the environment

- Protection of environment—Bioremediation of polluted environment.
- Micro-organisms producing enzymes for food industry.
- Micro-organisms with improved efficiency of fermentation.
- Improved micro-organisms for milk industry.
- Micro-organisms as live attenuated vaccines for health care.
- Increasing efficiency of plant nutrition, pest control (safe biopesticides), protection of plants from climatic stress and protection of plants from tumour formation and disease.

Tissue culture technology in India

India is reported to have one of the largest groups of tissue culture scientists in the world. Most of the research is directed towards the development of improved plants for agriculture, horticulture and forestry using tissue culture methods.

The Department of Biotechnology (DBT), Government of India, New Delhi is playing a vital role in promoting research in the area of plant tissue culture. Several laboratories are being supported by providing funds for development of tissue culture technology for the improvement of crop plants.

The important Biotechnology centres are:

- Indian Agricultural Research Institute (IARI), New Delhi.
- Central Institute of Medicinal and Aromatic plants (CIMAP), Lucknow.
- Dr.M.S. Swaminathan Research Institute (MSSRI), Chennai.
- Bhaba Atomic Research Centre (BARC), Mumbai.

Applications of plant tissue culture

- Embryo culture technique is applied to overcome embryo abortion, seed dormancy and self-sterility in seeds.
- By plant tissue culture techniques, a plant cell of potato and tomato were brought together through protoplasmic fusion and the hybrid cell was made to develop into a pomato plant. In pomato, the stem bears the tubers and the branches produced tomatoes.
- Plant tissue culture is applied in the area of plant physiological and biochemical research to study the cell cycle, metabolism in cells, nutritional, morphogenetical and developmental studies in plants.
- In recent years, plant tissue culture methods are employed in plants for the introduction of foreign gene into plant cells through DNA coated microparticles and delivering these particles into a host cell by using a gene gun.
- Plant tissue culture is an important technique for the production of secondary metabolites in large quantities.
- Protoplasmic fusion encourages genomes of incompatible crops to come together to form somatic hybrids.
- Tissue culture helps in induction of haploidy in anther culture *i.e.* useful for mutation breeding, triploidy through endosperm culture for inducing parthenocarpic fruits and polyploidy for increase in biomass or yield.
- Several commercial establishments now routinely use micropropagation for different foliage and ornamental plants.
- Through tissue culture methods using bud proliferation and multiple shoot formation, ornamental plants are produced in large numbers.
- Virus free germplasm are produced through apical meristem culture *e.g.* banana.
- Artificial synthetic seeds are produced through somatic embryogenesis.

MULTIPLE CHOICE QUESTIONS

1. The process of *de novo* organ formation referred as
   A. Organogenesis B. Morphogenesis
   C. Both A and B D. Rhizogenesis
2. Totipotency is not found in
   A. terminal leaf
   B. terminal bud
   C. terminally differentiated structures such as sieve tubes/tracheids
   D. all of the above
3. Production of haploid plants from female egg nucleus or ovum is called as
   A. androgenesis B. gynogenesis
   C. microgenesis D. embryogenesis
4. An individual plant selected for propagation is referred as
   A. mother plant B. stock plant
   C. ortet D. all of the above
5. A group of asexually produced plants are called as
   A. ramets B. plantlets
   C. seedlings D. saplings
6. 'Undefined' part of the tissue culture medium comprises
   A. tender coconut water or coconut milk
   B. fruit juices
   C. yeast extract
   D. all of the above
7. ''Solidifying agent'' which is commonly used in tissue culture medium is
   A. agar B. gellan gum
   C. gelrite D. All of the above
8. *In vivo* nutrient flow is achieved by water flow resulting from root pressure and water potential. How it is being done in *in vitro?*
   A. very humid atmosphere
   B. continuously opened stomata
   C. callus
   D. Both A and B
9. Plant tissue culture media is usually prepared with a pH range of
   A. 5.4-5.8 B. 4-6.8
   C. 7.0 D. 7.4-7.8
10. Higher, toxic concentration of phosphate are usually prevented by converting them into
    A. phytic acid B. phosphoric acid
    C. uric acid D. phosphorous
11. Examples of undefined supplements used in plant tissue culture media are
    A. yeast and potato extract
    B. malt extract
    C. banana homogenate and coconut milk
    D. All of the above
12. The role of organic acids such as succinate, malate and fumarate in plant tissue culture media
    A. act as chelating agent and hence improve availability of other nutrients
    B. buffer the medium against pH range
    C. act as essential nutrients to plant cell
    D. All of the above
13. Name the substance added to the culture medium to induce callus induction
    A. auxins B. cytokinins
    C. GA_3 D. Ethylene
14. Name the bacteria known as natural genetic engineer of plants
    A. *Agrobacterium tumefaciens*
    B. *E.coli*
    C. *Pseudomonas*
    D. All of the above
15. Golden rice is rich in ______
    A. Vitamin K
    B. Pro-Vitamin A
    C. Vitamin E
    D. Vitamin A derivatives
16. Breeding associated with molecular markers (nucleic acid) is known as
    A. molecular genetics
    B. molecular breeding
    C. molecular cloning
    D. All of the above

17. NCEDs are involved in
A. ABA biosynthesis
B. GA biosynthesis
C. Auxin biosynthesis
D. Ethylene biosynthesis

18. ABA catabolism is mediated by
A. ABA-8' carboxylase
B. ABA-8' hydroxylase
C. ABA-8' aminotransferase
D. ABA-8' oxygenase

19. Ethyl Methyl Sulphonate (EMS) is a
A. Alkylating agent B. Base analogue
C. Acid D. All of the above

20. Identify the steroidal plant growth regulator among the following
A. Brassinosteroid B. Abscissic acid
C. Gibberellic acid D. Auxin

21. Gossypol a toxic substance in cotton is normally contained in
A. Seed B. Leaf
C. Stem D. Fruit wall

22. Doubled haploid lines can be generated by
A. Protoplast fusion B. Transformation
C. Anther culture D. RNAi technology

23. Glyphosate resistant gene *gox* is isolated from
A. Arthrobacter B. Achromobacter
C. Bacillus D. *E.coli*

24. EUI trait in rice is related to
A. GA metabolism B. ABA metabolism
C. Auxin metabolism D. All of the above

25. C4 rice has been developed by transforming rice with
A. *PEPC* B. *PPDK*
C. Both A and B D. None of the above

ANSWERS

1	2	3	4	5	6	7	8	9	10
C	C	B	D	A	D	D	D	A	A
11	**12**	**13**	**14**	**15**	**16**	**17**	**18**	**19**	**20**
D	D	A	A	B	B	A	B	A	A
21	**22**	**23**	**24**	**25**					
A	C	B	A	C					

10

Animal Cell Culture

ANIMAL CELL CULTURE TECHNIQUES

Eukaryotic cells are much more difficult to culture than most prokaryotes. They demand complex media and are very susceptible to contamination and overgrowth by microbes such as bacteria, yeasts and fungi. Before dealing with the details let us discuss some of the important characteristics of animal cells and its growth in culture.

Features of Animal Cell Growth in Culture

The animal tissue culture can be divided into organ culture, tissue culture and cell culture. In organ culture, the entire embryos, organs or tissues are excised from the body either by vivisection or shortly after brain death. In such cases the normal physiological functions of organs or tissues are maintained. The cells remain fully differentiated and scale-up is not recommended in organ cultures. It is characterized by slow growth and fresh explantation is required for every experiment.

When fragments of excised tissue are grown in a defined medium it is called the tissue culture in true sense. In these cases most of the normal functions of the cultured tissues may be maintained even though the original organization of the tissue is lost. Compared to organ cultures the tissue cultures are better for scale-up but not ideal. When the excised tissues or tissue out growths are dispersed, mostly enzymatically, into a cell suspension, which may then be cultured as a monolayer or suspension culture, it is termed as animal cell cultures. The cell cultures can be maintained for a very long time as a particular type of cell line and are very suitable for scale-up studies. But these cells may lose some of its differentiated characteristics in course of time.

Animal cells, unlike plant and microbial cell cultures show limited growth even if the best of nutritive and physical conditions are maintained. In plant and microbial cell cultures, the cells continue to grow if the adequate physical and chemical environments are maintained in the medium. The behaviour of animal cells in culture depends on the source organs or the tissue from which the cells are excised. In most cases, it grows and divide only to a limited number of generations and then it dies or stops its growth, even if all the appropriate physical and nutritive factors are provided. In some cases the cells will not divide in culture as ir the case of neurons. But the cells like fibroblasts or skin cells can undergo cells division for some generations and then it comes to a static phase and eventually dies. Thus mortality is associated with all animal cell cultures unless it is an immortal cell line or cancer cell line. Cells derived from tissue such as kidneys, liver, muscles, and embryos grow and spread as a monolayer and fill the surface of the container in which they are growing. Once the cells spread to the entire surface and come in contact with the walls of the plate or the container, the cells stop further growth and this phenomenon is termed as 'contact inhibition'.

TRANSGENIC ANIMALS

A *transgenic animal* contains in its genome, a gene or genes introduced by one or the other technique of tranfection. The gene introduced by transfection is called *transgene*. In animals, *transfection* specifies the introduction of a DNA segment, either naked or integrated into a vector, into an animal cell. It may be pointed out that the same phenomenon is known as *transformation* in all other organisms. However, in case of animals, transformation has long been used to described the change of normal, *i.e.*, nontumourous, cells in culture to tumour like cells. Therefore, it because necessary to use the term transfection to avoid confusion.

Transfection may be transient or permanent (stable). In *transient tranfection*, the introduced gene(s) is(are) gradually lost from the daughter cells of transfected cells. But in case of *stable transfection*, the introduced gene(s) is(are) retained and expressed in all the cells derived from the transfected cells. Since most of the animal vectors are unstable, *i.e.*, are gradually lost, in the extrachromosomal state, stable transfections are ordinarily due to the integration of introduced gene(s) into the cell genome.

1. Objectives of Gene Transfer

The various objectives for which transgenic cells/animals are produced may be summarised into the following five categories.

1. Genes have bee transferred and expressed into cultured cell lines to obtain the proteins encoded by them.
2. Genetic modification of animals may be aimed at improving their milk, meat, wool, etc. production.
3. Genes have been transferred into animals with a view to obtain a large scale production of the proteins encoded by these genes in the milk, urine or blood of such animals; such animals are often called *bioreactors* and the approach is referred to as *molecular farming* or even as *gene farming*.
4. A special case of gene transfer aims at alleviating or even eliminating the symptoms and consequent miseries of genetic diseases. In this approach, normal and functional copies of the defective gene (that produced the genetic diseases) are introduced into the patient (*gene therapy*).
5. Finally, specific transgenic animals strains or lines are created to fulfil specialized experimental and/or biomedical needs. A good example of such animals are "knock out" mice strains in which specific genes have been replaced or knocked out by their disrupted counterparts through a process of homologous recombination. In simple words, the strategy is as follows. The gene to be knocked out say, *tk* (thymidine kinase) gene, is cloned, and disrupted by insertion with its coding region of another DNA sequence, *e.g.*, the *neo* gene of *E. coli*. The disrupted gene is used to transfect embryonic stem cells (ES cells, Section 7.1) where it could replace the normal *tk* gene present in the genome of ES cells by homologous recombination. The ES cells containing such recombination can be selected by a suitable experimental design.

2. Proliferation of Axillary Buds

This approach of micropropagation utilizes cultures of either shoot tip or nodal explants on a medium supplemented with usually a cytokinin (*e.g.*, 1-3 mg/l BAP) and often an auxin. The cytokinin stimulates pre-existing shoot buds present in the explants (apical meristems in shoot tips, and axillary buds in nodal explants) to develop into shoots. Each leaf on such shoots has an axillary bud. In many species, *e.g.*, strawberry, these axillary branching in a culture reaches the maximum. The individual shoots are then excised and subcultured onto fresh medium to initiate a new cycle of multiplication by axillary branching. In most plant species, each explant would produce 5-6 shoots in 4-5 weeks; this would give upto 5^{10}-6^{12} plants in one year from a single initial explant provided all the shoots produced could be cultured and there was 100% survival.

In some species, *e.g.*, blueberry, *Dalbergia sisso*, etc., the axillary buds present on growing shoots do not develop into new shoots irrespective of the GR combination or culture condition used. In such cases, each preexisting shoot bud present in an explant develops in a single shoot. Each such shoot is excised and cut into pieces to obtain nodal explants of appropriate size which are subcultured to initiate a new cycle of micropropagation. In such cases, the rate of multiplication will depend on the number of nodal explants obtained from each shoot. This approach may give a multiplication rate of 3-4 in about 6 weeks.

Micropropagation of woody trees presents several problems, e.g, poor growth *in vitro*, deterioration on subculturing, difficulties in rooting, browing of medium, fungi and bacteria present within plant tissues (endophytic) causing contamination, etc. Often juvenile or seedling explants of woody species are quite responsive, but explants from mature trees are problematic especially in terms of rooting. Therefore, explants from elite mature/old trees are obtained as follows : some branches of a tree are cut off, and the new shoots arising on the remaining portions of such branches are used when they are about 1 year old. Alternatively, suckers, copices etc. may be used. In some cases, *e.g., Feijoa, Eucalyptus* etc., serial subculturing *in vitro* improves rooting and brings about rejuvenation of shoots.

Maintenance of Isolates/Strains

Industrial microbiology continuously uses specific micro-organism isolates/strains as research, assay, development and production cultures. These strains are highly valuable and must be preserved over long periods without any genetic and, as a result, phenotypic changes. A microbial strain discovered to produce a useful product is called *research culture* since it is now used for studies on product isolation and identification, strain improvement, etc. The micro-organisms used for assays of the biological activities of microbial products as well as for various screening techniques are known as *assay cultures*. A research culture producing a metabolite of sufficient commercial interest is used for developing the large scale fermentation procedures; it is now termed as *development culture*. These cultures become *production cultures* if their product is good enough to be marketable.

The various cultures can be maintained for reasonably prolonged period using one of the following approaches.

1. *Low temperature storage* at 2-6°C on agar slants or in liquid medium; useful for relatively shorter periods (2-6 months) for working stock cultures.
2. *Storage as frozen cultures* at – 20 to – 100°C (both agar slants and liquid cultures).
3. Storage as lyophilized cells under vacuum at low temperatures (5°C or even – 20 to – 70°C). Lyophilization removes the free water from cells and spores. Protective agents like skimmed milk or sucrose reduce the determinetal effects of lyophilization. Lyophilization is easily done by using automated freeze-drying machines in which the cells are first frozen and then their free water is withdrawn under high vacuum.
4. Storage of vegetative *cells/spores in liquid nitrogen* (– 196°C) or in the vapour phase of liquid nitrogen (– 167°C).
   Preservation of lyophilized cells in the most widely used method, but storage in liquid nitrogen is also as effective and widely used. These methods can theoretically maintain the micro-organisms indefinitely.
5. *Air dried at room temperature on sterile loam sand or on other natural substrates* like maize seed, rice bran, etc., and stored at room temperature or in a refrigerator. Bacterial cultures may remain viable upto 70-80 years. It is more suited for organisms that form spores.
6. Storage in glycerine stabs (0.85 ml of cell suspension mixed with 0.15 ml of sterile glycerol and stored at – 70 or – 75°C). This is commonly used for recombinant bacteria.

Procedure 1 for culture maintenance is used for working stock cultures, while those listed from 2 to 6 are used for the primary stock cultures. *Working stock cultures* are maintained in a vigorous and uncontained condition and are used quite frequently. They must be checked frequently for characteristic features and contamination. In contrast, *primary stock cultures* are kept in long-term storage for later use; they are used to produce new working stock culture as per need. A *stock culture* may be simply defined as a culture which serves as a source of inoculum.

APPLICATIONS OF ANIMAL CELL CULTURES

Animal cell cultures are used for various basic researches as well as for commercial purposes as the source of valuable therapeutic chemicals and agents. There is an ever-increasing list of proteins and other biologicals that can be produced by culturing mammalian and other types of cells. Some of the major services and different products obtained from mammalian cells have been summarised below. In addition to the natural compounds produced by the cultured cells, genetically altered cells can be used for the production of recombinant proteins.

- Vaccine production
- Monoclonal antibody production
- Enzymes and hormones production
- *In vitro* skin and other tissues and organs by stem culturing
- Viral cultivation

Some of the products both recombinant and natural proteins produced by the cell cultures are tabulated below:

Table: *Products of animal cell cultures available in the market or under clinical trials*

Products	Medical Application
Urokinase	Heart attacks
Tissue plasminogen activator	Heart attacks
FSH, Factor VIII	Haemophilia A
Factor IX	Haemophilia B
Human-rabies, mumps, rubella etc. Veterinary-FMD vaccine, New Castle's Disease etc.	Vaccines
Monoclonal antibodies Hybridoma technology	Various diagnostic tools
Baculovirus Bioinsecticides	Virus cultivation
Interferons and interleukins	Used as Immunoregulators
Toxicological testing	Cultured cells as experimental models
Stem cells	Stem cell culturing various basic research for clinical applications

MULTIPLE CHOICE QUESTIONS

1. An antibody which is directed against CD3 surface antigen of T–cells that removes antigen bearing cells from circulation and from the graft is

A. OKT3 (anti-CD3 Moab)
B. CD3 antigenic body
C. polyclonal antibodies
D. IgG

2. Which is the first drug to be produced by mammalian cell culture?

A. Tissue plasminogen Activator
B. IgG
C. Insulin
D. Streptovidin

3. Alpha and Beta amylase help in

A. Biosynthesis of lipids
B. Breakdown of lipids
C. Breakdown of starch
D. Breakdown of sucrose

4. Which of these disease causing bacteria was discovered by Robert Koch?

A. *Bacillus anthracis*
B. *Salmonella typhimurium*
C. *Shigella dysenteriae*
D. *Clostridium perfringens*

5. Vortexing in stirred tank reactors can be prevented by

A. Using axial flow impeller
B. Using a turbine impeller
C. Using baffles in the reactor
D. Using multiple impellers

6. Super-bug is

A. Artificial plasmid
B. Oil degrading arthropod
C. Resaltifying bacteria
D. High infecture virus

7. Red tides are often caused by blooms of
   A. Zooplankton B. *Rhodospirillum*
   C. Trichodesmium D. *Navicula*
8. Thalassemia is caused mainly by
   A. Point mutation in β globin chain
   B. Deletion mutation in β globin chain
   C. Insertional mutation
   D. None of the above
9. In general, the higher the intensity of exercise, the greater the proportional contribution of
   A. aerobic energy production
   B. anaerobic energy production
   C. the TCA cycle (Krebs' cycle) to the production of ATP
   D. the electron transfer chain to the production of ATP
10. Liver glycogen breakdown is stimulated by
   A. insulin B. glucagon
   C. adrenaline D. Both B and C
11. The sex of a child is dictated by the inheritance of
   A. the number of X chromosomes from the mother
   B. a recessive allele on the X chromosome
   C. a single Y chromosome from the mother
   D. a single Y chromosome from the father
12. Which of the following is not a dietary antioxidant?
   A. vitamin C B. lipoic acid
   C. vitamin K D. beta-carotene
13. When the value of selection coefficient is unity (S = 1), there would be
   A. No survival B. 10% survival
   C. 50% survival D. 75% survival
14. Myeloma-Spleen fusion cells are selected on
   A. LB medium B. DMEM medium
   C. HAT medium D. PDA medium
15. The p53 gene is responsible for
   A. triggering cells to grow uncontrollably
   B. damaging DNA
   C. initiating transcription of p21 which binds to cyclins
   D. Degrading damaged DNA
16. Which one of the following statements is not true about Gene Therapy
   A. Somatic gene therapy is an attempted correction of a genetic defect in an individual and his/her descendants
   B. All approved gene therapies for humans are somatic gene therapies
   C. Most current gene therapies use viruses to carry transgenes into human cells
   D. The first successful human gene therapy was for adenosine deaminase deficiency, a condition causing severe combined immune deficiency (SCID)
17. Molecules referred to as abzymes are
   A. catalytic antigens
   B. catalytic antibodies
   C. non-protein catalyst
   D. catalytic ribosomes
18. "stone leprosy" is caused by
   A. water pollution B. acid rain
   C. air pollution D. radioactive pollution
19. A fungus has been reported to possess psycho therapeutic properties
   A. *Psitocybe mexicana* B. Nux-vomica
   C. Cascara D. Ginseng
20. T-cells (lymphocytes) synthesize antigen binding proteins, which is called as
   A. antibody B. immunoglobulin
   C. T-cell receptor D. None of the above

ANSWERS

1	2	3	4	5	6	7	8	9	10
A	A	C	A	D	A	C	A	B	D
11	**12**	**13**	**14**	**15**	**16**	**17**	**18**	**19**	**20**
D	C	A	C	C	A	B	B	A	C

MATHEMATICS

1

Sets

Set: A set is a well defined collection of objects. Objects must be distinct and clearly specified. If a is an element of a set A, then we write $a \in A$ and say a belongs to A or a is in A or a is a member of A. If a does not belongs to A, then we write $a \notin A$.

Methods for Describing Sets:

(*a*) **Listing Method (Roster Method):** We list all the elements and enclose them in curly brackets *e.g.*, $A = \{1, 2, 3, 4\}$.

(*b*) **Set Builder Method (Property Form):** Let P be a property or collection of properties. If x satisfies the property, then x is in the set. This can be written symbolically as

$$A = \{x: P(x)\}.$$

e.g., $A = \{x : x \text{ is a natural number less than } 5\}$.

$$A = \{1, 2, 3, 4\}.$$

Finite Set and Infinite set: A set which has only finite number of elements is called finite set and which has infinite number of elements, is called infinite set.

Null Set: A set having no element is called a null or void or empty set and is denoted by ϕ or $\{\}$, *e.g.*, $\phi = \{x : x \text{ is an immortal man on earth}\}$ or $\phi = \{x : x \text{ is an integer such that } x = x + 5\}$.

Universal Set: A set consisting of all possible elements which occur in the discussion is called the universal set.

Subset: Let A and B be two sets in the same universal set. If every element of A is also an element of B, *i.e.*, $x \in A \Rightarrow x \in B$, then we say that A *is* a subset of B or that A is contained in B. It is denoted by $A \subseteq B$.

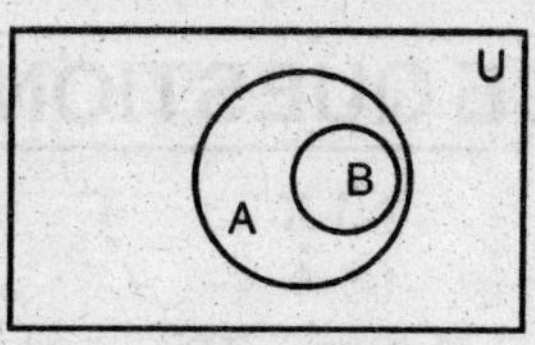

Equality of Two Sets: Two sets A and B are said to be equal iff they contain exactly the same elements, *i.e.*, $A \subseteq B$ and $B \subseteq A$.

Proper and Improper Subsets: The null set ϕ is subset of every set and every set is subset of itself, *i.e*, $\phi \subset A$ and $A \subset A$ for every set A. They are called improper subsets of A. Thus every non-empty set has two subsets. It should be noted that ϕ has only one subset ϕ which is improper.

Thus all other subsets of A are called its proper subsets.

Thus if $A \subseteq B$, $A \neq B$, $A \neq \phi$, then A is said to be proper subset of B.

e.g., Let $A = \{1, 2\}$. Then A has ϕ, $\{1\}$, $\{2\}$, $\{1, 2\}$ as its subsets out of which ϕ and $\{1, 2\}$ are improper and $\{1\}$ and $\{2\}$ are proper subsets.

Power Set: The set of all subsets of a set A is called the power set of A. It is denoted by P(A).

Thus $n(A) = p \Rightarrow n\{P(A)\} = 2^p$.

Some Important Deductions:

(*a*) $A \subseteq A, \forall A$

(*b*) $\phi \subseteq A \ \forall A$

(*c*) $A \subseteq U$, (the universal set) $\forall$ A in U

(*d*) $A = B \Leftrightarrow A \subseteq B, B \subseteq A$.

OPERATIONS ON SETS

Union of Two Sets: The union of two sets A and B is the set of all elements in set A or in set B and is written as $A \cup B$. Thus $x \in A \cup B \Leftrightarrow x \in A$ or $x \in B$ and $x \notin A \cup B \Leftrightarrow x \notin A$ and $x \notin B$.

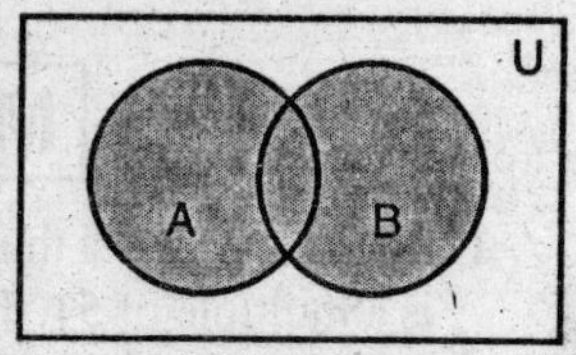

Also $A \cup B = \{x : x \in A \text{ or } x \in B\}$. *e.g.*, Let $A = \{a, b, c\}$, $B = \{c, d, e\}$ Then $A \cup B = \{a, b, c, d, e\}$.

***Important Deductions*:**

(*a*) $A \cup A = A$, $A \cup U = U$, $A \cup \phi = A$

(*b*) $A \subseteq A \cup B$; $B \subseteq A \cup B$

(*c*) $A \cup B = B \cup A$ (*commutative law*)

(*d*) Let A and B be two finite sets such that $n(A) = p$, $n(B) = q$

Then min. $n(A \cup B) = \max. (p, q)$ and max. $n(A \cup B) = p + q$

(*e*) $A \subseteq B \Rightarrow A \cup B = B$

(*f*) $(A \cup B) \cup C = A \cup (B \cup C)$ (*associative law*)

Intersection of Two Sets: The intersection of two sets A and B is the set of elements which are common to both A and B. Intersection of sets A and B is denoted by $A \cap C$.

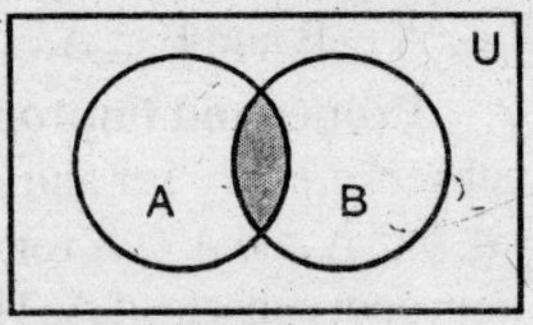

Therefore, $x \in A \cap B \Leftrightarrow x \in A$ and $x \in B$.

$\therefore A \cap B = \{x: x \in C \text{ and } x \in B\}$,

Also $x \notin A \cap B \Leftrightarrow x \notin A$ or $x \notin B$.

e.g., Let $A = \{1, 2, 3\}$ and $B = \{2, 3, 5\}$,

Then $A \cap B = \{2, 3\}$.

Disjoint Sets: If A and B have no element in common, then A and B are called disjoint sets. Thus A and B are called disjoint $\Leftrightarrow A \cap B = \phi$, *e.g.*, the sets $\{1, 2\}$, $\{3, 4\}$ are disjoint sets.

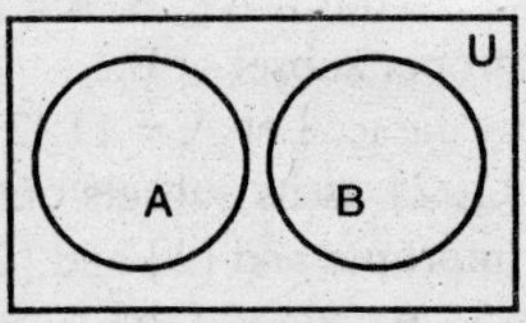

Difference of Sets: Let A and B be two sets. Then A – B (A difference B or simply A minus B) is the set of elements which belong to A but which do not belong to B.

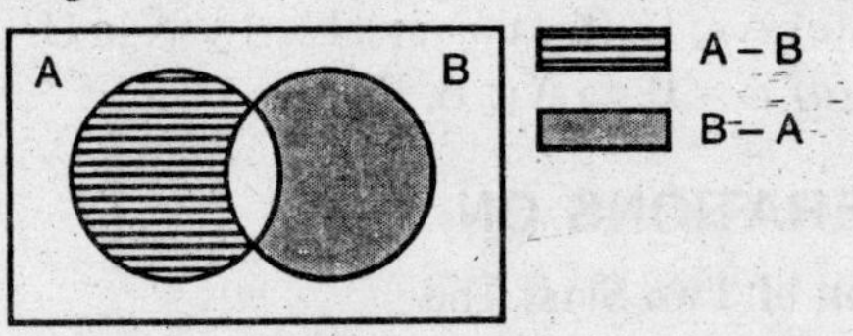

Thus $x \in A - B \Leftrightarrow x \in A$ and $x \notin B$ and so $x \notin A - B \Leftrightarrow x \notin A$ or $x \in B$

Also $A - B = \{x : x \in a, x \notin B\}$, *e.g.*, consider the sets $A = \{1, 2, 3\}$, $B = \{3, 4, 5\}$. Then $A - B = \{1, 2\}$, $B - A = \{4, 5\}$. As another example, R – Q is the set of all irrational numbers.

Symmetric Difference: $(A - B) \cup (B - A)$ is called the symmetric difference of A and B. It is usually denoted by $A \Delta B$. Also $A \Delta B = (A \cup B) - (A \cap B)$.

Complement of a Set: The complement of a set A is the set of elements which do not belong to A, that is, the difference U – A, where U is the univeral sets. It is denoted by A' or A^C or C(A). *e.g.*, consider $U = \{1, 2, ..., 10\}$ and $A = \{1, 3, 5, 7, 9\}$.

Then $A' = \{2, 4, 6, 8, 10\}$

***Some Deductions*:**

(*a*) $A \cap A' = \phi$, $A \cup A' = U$ (the universal set)

(*b*) $U' = \phi$, $\phi' = U$

(*c*) $(A')' = A$

(*d*) $A - B = A \cap B'$

De MORGAN'S LAW

De Morgan's Laws:

(*a*) $A - (B \cup C) = (A - B) \cap (A - C)$

(*b*) $A - (B \cap C) = (A - B) \cup (A - C)$

(*c*) $(A \cup B)' = A' \cap B'$

(*d*) $(A \cap B)' = A' \cup B'$

Some More Results Regarding the Order of Finite Sets: Let A, B and C be finite sets and U be the finite universal set, then

(*a*) If A and B are disjoint, then $n(A \cup B) = n(A) + n(B)$

(*b*) $n(A \cup B) = n(A) + n(B) - n(A \cap B)$

(*c*) $n(A - B) = n(A) - n(A \cap B)$, *i.e.*, $n(A) = n(A - B) + n(A \cap B)$.

(*d*) $n(A' \cup B') = n(A \cap B)' = n(U) - n(A \cap B)$

MULTIPLE CHOICE QUESTIONS

1. Let $a = \{x : x$ is a multiple of $3\}$ and $B = \{x : x$ is a multiple of $5\}$. Then $A \cap B$ is given by

(*a*) $\{3, 6, 9, ...\}$ (*b*) $\{5, 10, 15, 20, ...\}$

(*c*) $\{15, 30, 45, ...\}$ (*d*) None of these

2. If X and Y are two sets, then $X \cup (Y \cap X)'$ equals

(*a*) X (*b*) Y

(*c*) ϕ (*d*) None of these

3. Let $n(U) = 700$, $n(A) = 200$, $n(B) = 300$, $n(A \cap B) = 100$. Then $n(A' \cap B') =$

(*a*) 400 (*b*) 600

(*c*) 300 (*d*) None of these

4. If A and B be two subsets of a set U, then which of the following is false?
(*a*) $A \cap B = A$ (*b*) $A \cap U = A$
(*c*) $A \cap B \in A$ (*d*) $B \cup A \cap B$

5. It is given that *n* (P(S)) = 64, where P (S) is power set of S, then *n* (S) is:
(*a*) 2 (*b*) 4
(*c*) 8 (*d*) 6

6. If A and B are two sets, then $A \cup (A \cap B)$ is:
(*a*) $A \cap B$ (*b*) A
(*c*) B (*d*) None of these

7. If A and B are two subsets of universal set U, then A – B is equal to:
(*a*) $A \cap B$ (*b*) $A \cap B'$
(*c*) $A' \cap B$ (*d*) $A' \cap B'$

8. If $P, Q \subset U$, then $P \cap (P \cup Q)$; is equal to:
(*a*) P (*b*) Q
(*c*) f (*d*) None of these

9. In a town of 840 persons, 450 persons read Hindi, 300 read English and 200 read both. The number of persons who read neither, is:
(*a*) 210 (*b*) 290
(*c*) 180 (*d*) 260

10. Which of the following is false?
(*a*) If *n* (A) = 20, *n* (B) = 25 and *n* $(A \cap B)$ = 10, then *n* $(A \cup B)$ = 35
(*b*) If *n* (A) = 25, *n* (B) = 20, *n* $(A \cup B)$ = 30, then *n* $(A \cap B')$ = 10
(*c*) $n(A \cap B') = n(A) - n(A \cap B)$
(*d*) All of the above are true.

11. Let A and B have 3 and 6 elements respectively. What can be the maximum and minimum number of elements in $A \cup B$?
(*a*) 9 and 6 (*b*) 6 and 3
(*c*) 9 and 3 (*d*) 6 and 12

12. Which of the following is false?
(*a*) $A \subset B \Rightarrow (C - B) \subset (C - A)$
(*b*) $A \subset B \Rightarrow (A \cup C) \subset (B \cup C)$
(*c*) $A \subset B \Rightarrow (U - B) \subset (U - A)$
(*d*) $A \subset B \Rightarrow (B \cap C) \subset (A \cap C)$

13. If A = {*a, c, d, g*} and B = {*b, d, j, k*}, then which of the following is true?
(*a*) $A \cap B$ is a null set
(*b*) A and B are disjoint sets
(*c*) $A \cap B$ is a singleton set
(*d*) All of the above are true

14. If A and B are subsets of a set X, then $[A \cap (X - B)] \cup B$ is equal to:
(*a*) $A \cup B$ (*b*) $A \cap B$
(*c*) A (*d*) B

15. The number of elements in the power set *P*(*S*) of the set $S = \{2, \{1, 4\}\}$ is:
(*a*) 2 (*b*) 4
(*c*) 8 (*d*) None of these

16. Let *A* = {1, 2, 3, 4, 5, 6, 7, 8, 9}. How many subsets does *A* have which contain 4, 5 and 6?
(*a*) 8 (*b*) 16
(*c*) 64 (*d*) None of these

17. Let *A* and *B* be two sets. The correct statement among the following is :
(*a*) $(A - B) \cap (A \cup B) = \phi$
(*b*) $A \cup (A - B) = A \cap B$
(*c*) $(A \cup B) - (A \cap B) = (A - B) \cup (B - A)$
(*d*) $(A - B) \cup A = A \cup B$

18. If $(A - B) \subset (B - A)$, then
(*a*) $A - B = B - A$ (*b*) $B \subset A$
(*c*) $A \subset B$ (*d*) $A \cup B = A$

19. If *A* = {1, 2, 3}, *B* = {3, 4, 5}, the symmetric difference of *A* and *B* is given by
(*a*) {1, 2, 3, 4} (*b*) {1, 3, 4, 5}
(*c*) {2, 3, 4, 5} (*d*) {1, 2, 4, 5}

20. Which of the following is false?
(*a*) $A \cap (B - C) = (A \cap B) - (A \cap C)$
(*b*) $B' - A \neq (A \cap B)$
(*c*) $A - B' \neq (A \cap B)$
(*d*) $A - (A - B) \neq A \cup B$

21. Which of the following is false?
(*a*) $A \subset B$ and $C \subset D \Rightarrow A \times C \subset B \times D$
(*b*) $A \times (B \cap C) = (A \times B) \cup (A \times C)$
(*c*) $A \cap (B \cup C) = (A \cap B) \cup (A \cap C)$
(*d*) $(A \times B) \cap (C \times C) = (A \cap C) \times (B \cap D)$

22. Which of the following is false?
(*a*) $A \cap B = \phi \Rightarrow$ there is no common element to both
(*b*) $A \cup B = A \Rightarrow B \subset A$
(*c*) $A \subset B$ and $B \subset A \Rightarrow A = B$

(*d*) If A contains 5 elements, the power set contains 25 elements.

23. Which of the following is false?

(*a*) Every subset of finite set is finite

(*b*) A subset of an infinite set may be finite

(*c*) $\{0\} \neq 0$

(*d*) $\{\phi\}$ is a null set

24. Which of the following is false?

(*a*) If $n(A) = 20$, $n(B) = 25$ and $n(A \cap B) = 10$, then $n(A \cup B) = 35$

(*b*) If $n(A) = 15$, $n(B) = 20$, $n(A \cup B) = 30$, then $n(A \cap B') = 10$

(*c*) If $n(A) = 15$, $n(B) = 20$, $n(A \cup B) = 30$, then $n(A \cap B') = 15$

(*d*) None of these

25. Let A and B have 3 and 6 elements respectively. What can be the maximum and minimum number of elements in $A \cup B$?

(*a*) 9 and 6 (*b*) 6 and 13

(*c*) 9 and 3 (*d*) 6 and 12

ANSWERS

1	2	3	4	5	6	7	8	9	10
(*c*)	(*c*)	(*c*)	(*a*)	(*d*)	(*b*)	(*b*)	(*c*)	(*b*)	(*a*)
11	**12**	**13**	**14**	**15**	**16**	**17**	**18**	**19**	**20**
(*c*)	(*c*)	(*c*)	(*a*)	(*b*)	(*c*)	(*c*)	(*c*)	(*d*)	(*c*)
21	**22**	**23**	**24**	**25**					
(*b*)	(*d*)	(*d*)	(*d*)	(*d*)					

SOME SELECTED EXPLANATORY ANSWERS

1. $A = \{3, 6, 9, 12, 15, 18, ...\}$

$B = \{5, 10, 15, ...\}$

$\therefore A \cap B = \{15, 30, 45, ...\}$.

Since x is a multiple of 3 and 5 both iff it is a multiple of H.C.R. of 3 and 5, *i.e.*, of 15.

Note: $aN \cap bN = hN$, (h = H.C.F. of a, b)

2. $X \cap (Y \cup X)' = X \cap (Y' \cap X')$

$= (X \cap X') \cap Y' = \phi \cap X' = \phi$

3. $n(A' \cap B') = n(A \cup B)'$

$= 700 - n(A \cup B)$

$= 700 - [200 + 300 - 100] = 300$

14. See the figures

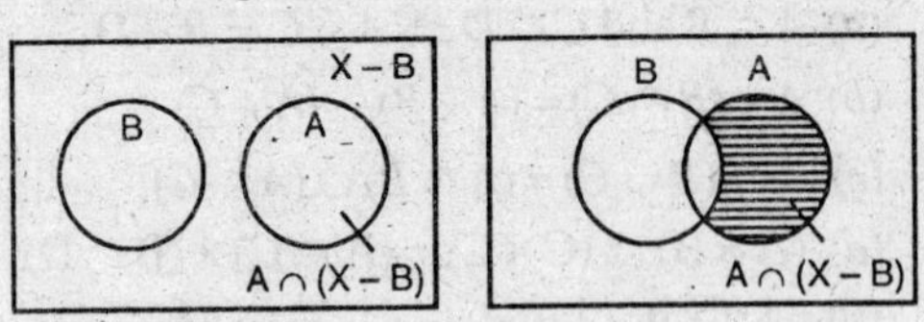

In both cases

15. Here, S has 2 elements 2 and $\{1,4\}$

$\therefore$ Total number of elements in power set $= 2^2 = 4$.

16. Consider 4, 5 and 6 as one element, then total number of subsets $= 2^7 = 128$.

The number of subsets which contain 4, 5 and 6 $= 128 - 64 = 64$

19. $A - B = \{1,2\}$ and $B - A = \{4,5\}$

$\therefore A \Delta B = (A - B) \cup (B - A)$

$= \{1,2\} \cup \{4,5\} = \{1,2,4,5\}$

22. Number of elements of power set $= 2^5 = 32$

24. Statement (A) is true

Again $n(A \cap B) = n(A) + n(B) - n(A \cup B)$

$n(A \cap B) = 15 + 20 - 30 = 5$

$\therefore \quad n(A \cap B') = n(A) - n(A \cap B)$

$= 15 - 5 = 10$

2

Relation and Function

CARTESIAN PRODUCT

Cartesian Product: Cartesian product of sets A and B denoted by A × B, is the set of all ordered pairs of which first coordinates are elements of the set A and second coordinates, the elements of set B. Symbolically,

$A \times B = \{(a, b) : a \in A, b \in B\}$.

Example: Let $A = \{a, b\}$, $B = \{1, 2, 3\}$.

Then $A \times B = \{(a, 1), (a, 2), (a, 3), (b, 1), (b, 2), (b, 3)\}$

***Some Deductions*:**

(*a*) $B \subset C \Leftrightarrow A \times B \subset A \times C$

(*b*) $n(A) = p, n(B) = q \Rightarrow n(A \times B) = pq$

THEORY OF RELATION

Relation: Let A and B be two non-empty sets. Then every subset of A × B defines a relation from A to B and every relation from A to B is a subset of A × B. Let $R \subseteq A \times B$ and $(a, b) \in R$. Then we say that *a* is related to *b* by the relation R and write it as $a\ R\ b$. If $(a, b) \notin R$, we write it as $a \not R\ b$.

Example: Let $A = \{1, 2, 3, 4, 5\}$, $B = \{1, 3\}$. We set a relation from A to B as: $a\ R\ b$ iff $a \le b$; $a \in A$, $b \in B$.

Then $R = \{(1, 1), (1, 3), (2, 3)\} \subset A \times B$.

Domain and Range of a Relation: Let R be a relation from A to B, that is, let $R \subseteq A \times B$. Then ***Domain* R** $= \{a : a \in A, (a, b) \in R$ for some $b \in B\}$, *i.e.*, dom. R in the set of all the first elements of the ordered pairs which belong to R.

Also ***Range*** R $= \{b : b \in B, (a, b) \in R$ for some $a \in A\}$, *i.e.*, range R is the set of all the second elements of the ordered pairs which belong to R. Thus domain $R \subseteq A$, range $R \subseteq B$.

Inverse Relation: Let $R \subseteq A \times B$ be a relation from A to B. Then inverse relation $R^{-1} \subseteq B \times A$ is defined by $R^{-1} = \{(b, a) : (a, b) \in R\}$. It is clear that

(*a*) $a\ R\ b \Leftrightarrow b\ R^{-1}\ a$

(*b*) domain R^{-1} = range R and range R^{-1} = domain R

(*c*) $(R^{-1})^{-1} = R$.

e.g., Let $A = \{1, 2, 3, 4\}$, $B = \{a, b, c\}$ and $R = \{(1, a) (1, c), (2, a)\}$. Then

(*i*) domain $R = \{1, 2\}$, range $R = \{a, c\}$.

(*ii*) $R^{-1} = \{(a, 1), (c, 1), (a, 2)\}$.

Reflexive Relations: R is a reflexive relation if $(a, a) \in R$, $\forall\ a \in R$. It should be noted that if $\exists$ any $a \in A$ such that $a \not R a$, then R is not reflexive.

Example: Let $A = \{1, 2, 3\}$ and $R = \{(1, 1), (1, 3)\}$. Then R is not reflexive since $3 \in A$ but $(3, 3) \notin R$.

Symmetric Relations: R is called a symmetric relation on A if $(x, y) \in R \Rightarrow (y, x) \in R$.

That is, if *x* is R-related to *y*, then *y* is also R-related to *x*.

It should be noted that R is symmetric iff $R^{-1} = R$.

Anti-symmetric Relations: R is called an anti-symmetric relation if $(a, b) \in R$ and $(b, a) \in R \Rightarrow a = b$. Thus, if $a \neq b$ then *a* may be related to *b* or *b* may be related to *a*, but never both.

Example: Let N be the set of natural numbers. A relation $R \subseteq N \times N$ is defined by:

$x\ R\ y$ iff *x* divides *y* (*i.e.*, *x/y*). Then $x\ R\ y$, $y\ R\ x \Rightarrow x$ divides *y*, *y* divides $x \Rightarrow x = y$.

Transitive Relations: R is called a transitive relation if $(a, b) \in R$, $(b, c) \in R \Rightarrow (a, c) \in R$. In other words, if a is related to b, b is related to c, then a is related to c. Transitivity fails only when there exists a, b, c such that a R b, b R c but $a \not R c$.

Examples: Consider the set A = {1, 2, 3} and the relations

$R_1 = \{(1, 2), (1, 3)\}$ $R_2 = \{(1, 2)\}$
$R_3 = \{(1, 1)\}$ $R_4 = \{1, 2), (2, 1) (1, 1)\}$.

Then R_1, R_2 and R_3 are transitive while R_4 is not transitive since in R_4, $(2, 1) \in R_4$, $(1, 2) \in R_4$ but $(2, 2) \notin R_4$.

Equivalence Relations

Equivalence Relations: A relation R in a set A is called an equivalence relation if

(*a*) R is reflexive, *i.e.*, $(a, a) \in R$, $\forall a \in A$,
(*b*) R is symmetric, *i.e.*, $(a, b) \in R$
$\Rightarrow (b, a) \in R$,
(*c*) R is transitive, *i.e.*, $(a, b), (b, c) \in R$
$\Rightarrow (a, c) \in R$.

THEORY OF FUNCTION

Definition of Function

Let A and B be two non-empty sets. Then a function 'f' from set A to set B is a rule which associates elements of set A to elements of set B such that

- An element of set A are associated to unique element is set B.
- All elements of set A are associated to element in set B.

If f is a function from a set A to set B, then we write $f : A \rightarrow B$ or A $\xrightarrow{f}$ B. Which is read as f is a function from A to B or f maps A to B.

Domain: Set A is called domain of f *i.e.* Set of those elements from which function is to be defined.

Co-Domain: Here set B is called co-domain of function.

Range: Set of images of each element in A, is called range of f.

Note: Range $\subseteq$ Co-domain

Equality of two function: Two function f and g are said to be equal functions, if and only if

(*i*) domain of f = domain of g
(*ii*) co-domain of f = co-domain of g
(*iii*) $f(x) = g(x)$ $\forall x \in$ their domain

Example:

If $A = \{1, 2\}$, $B = \{10, 13\}$, $f : A \rightarrow B$, $f(x) = x^2 + 9$ and $g = A \rightarrow B$, $g(x) = 3x + 7$, then $f = g$ because domains and co-domains of both f and g are same also $f(1) = 10 = g(1)$; $f(2) = 13 = g(2)$

Composite Function

Consider two function
$f : X \rightarrow Y$, $g : Y \rightarrow Z$
One can define h : $X \rightarrow Z$ such that $h(x) = g\{f(x)\}$
Domain of $gof(x)$ *i.e.*
$f\{f(x)\} = \{x : x \in \text{Dom g}, g(x) \in \text{Dom} f\}$

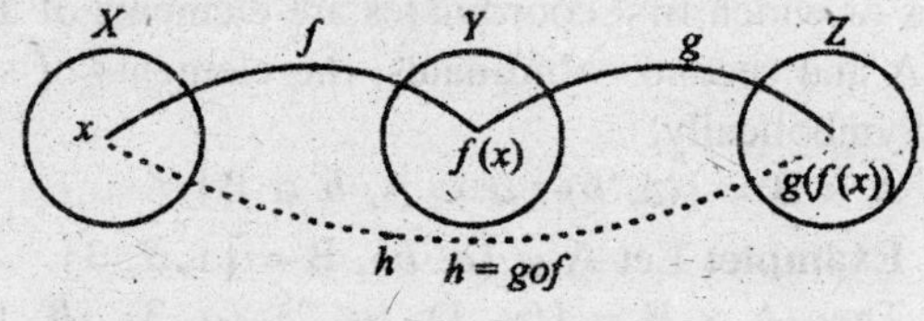

Classification of Function

The following are the kinds of function:

(*i*) One-One Function (Injective):

If each element in the domain of a function has a distinct image in the co-domain the function is said to be one-one function and is also known as Injective Function.

Example: $f : R \rightarrow R^+$ given by $y = e^x$
$g : R \rightarrow R$, $g(x) = 3x - 7$
are one-one functions or, $f : A \rightarrow B$ is one-one
$\Leftrightarrow a \neq b \Rightarrow f(a) \neq f(b)$ for all $a, b \in A$
$\Leftrightarrow f(a) = f(b) \Rightarrow a = b$ for all $a, b \in A$

(*ii*) Many-one Function

If there are two or more than two elements of domain having the same image then $f(x)$ is called Many - One function.

Example: $f : A \rightarrow B$ is a many - one function if there exist $x, y \in A$ such that $x \neq y$ but $f(x) = f(y)$.

(*iii*) Onto Function (Surjective):

Let $f : X \rightarrow Y$ be a function. If each element in the co-domain Y has atleast one pre-image in the domain X *i.e.* Range f - Co-domain, the f is called onto.

Onto function is also called surjective and if function be both one-one and onto then function is called Bijective.

or, $f : A \to B$ is a surjection iff for each $b \in B$, then there exist $a \in A$ such that $f(a) = b$.

Example: If $f : R^+ \to R$ is defined by $y = \log_2 x$, then $f(x)$ is onto function.

(*iv*) Into Function:

If there exist one or more than one element in the co-domain Y which is not an image of any element in the co-domain X. Then f is into.

In other words $f : A \to B$ is an into function, if it is not an onto function.

Example: Let $f : R \to R$ is defined by $y = x^2 + 1$, then f(x) is an into function. But when $f: R \to R^+$ is defined by $y = x^2 + 1$, then $f(x)$ is not into function.

Inverse Function

If $f : X \to Y$ be a function defined by $y = f(x)$ such that f is both one-one and onto, then ther exists a unique function $g : Y \to X$ such that for each $y \in Y$, $g(y) = x$. The function g so defined is called the inverse of f and denoted by f^{-1}. Also f is the inverse of g and the two functions f and g are said to the inverse of each other.

$f(f^{-1}(x)) = x, \forall x \in Y$ and $f^{-1}(f(x)) = x, \forall x \in X$

MULTIPLE CHOICE QUESTIONS

1. Let R be a relation defined in the set {1, 2, 3, 4, 5, 6} as R = {(*a*, *b*) : *b* = *a* + 1}, then R is not:
(*a*) reflective (*b*) transitive
(*c*) symmetric (*d*) none of these

2. If $f : R \to R$ be given by $f(x) = (3 - x^3)^{1/3}$, then $fof(x)$ is:
(*a*) $x^{1/3}$ (*b*) x^3
(*c*) x (*d*) $(3 - x^3)$

3. Let $f : R - \left\{-\frac{4}{3}\right\} \to R$ be a function defined as $f(x) = \frac{4x}{3x+4}$. The inverse of f is the map g: Range $f \to R - \left\{-\frac{4}{3}\right\}$ given by
(*a*) $g(y) = \frac{3y}{3-4y}$ (*b*) $g(y) = \frac{4y}{4-3y}$
(*c*) $g(y) = \frac{4y}{3-4y}$ (*d*) $g(y) = \frac{3y}{4-3y}$

4. Let A = {1, 2, 3} and let $f : A \to A$ defined by f = {(1, 2), (2, 3) (3, 1)}. Find f^{-1}
(*a*) {(1, 2), (2, 1), (3, 2), (3, 1)}
(*b*) {(2, 1), (3, 2), (1, 3)}
(*c*) {(1, 2), (2, 3), (3, 1)}
(*d*) None of these

5. Find the domain of the real function $f(x) = \sqrt{9 - x^2}$.
(*a*) $\{x \in R : -3 \le x \le 3\}$
(*b*) $\{x \in R : -3 \le y \le 3\}$
(*c*) $x \in R$
(*d*) $R - (-3, 3)$

6. Find range of the real function $f(x) = \frac{1}{1-x^2}$
(*a*) $R - \{-1, 1)$ (*b*) $(0, 1)$
(*c*) $(-1, 0) \cup (1, 2)$ (*d*) $\{y \in R : y \ge 1\}$

7. Let $f : R \to R$, $g : R \to R$ be two functions given by $f(x) = 2x - 3$, $g(x) = x^3 + 5$. Then $(fog)^{-1}(x)$ is equal to
(*a*) $\left(\frac{x+7}{2}\right)^{1/2}$ (*b*) $\left(x - \frac{7}{2}\right)^{1/2}$
(*c*) $\left(\frac{x-2}{7}\right)^{1/3}$ (*d*) $\left(\frac{x-7}{2}\right)^{1/3}$

8. Let $f: R \to R$ be defined by $f(x) = 7x - 3$. Find f^{-1}.
(*a*) f is not invertible (*b*) $\frac{x+3}{4}$
(*c*) $\frac{x+3}{7}$ (*d*) $\frac{7x+3}{4}$

9. If $f : R$ is defined by $f(x) = \frac{x}{x^2+1}$. Find $f(f(2))$.
(*a*) $\frac{2}{5}$ (*b*) $\frac{29}{5}$

(c) $\frac{3}{5}$ (d) $\frac{10}{29}$

10. Let $f = \{(1, 3), (2, 1), (3, 2)\}$ and $g = \{(1, 2), (2, 3), (3, 1)\}$, then find (gof) (1) and $(fog)(2)$ respectively.

(a) 1, 2 (b) 3, 1
(c) 2, 1 (d) 2, 3

11. A set A contains m elements and a Set B contain n $(n > m)$ elements. The number of injective functions that can be defined from A to B is

(a) $n!$ (b) $\frac{n!}{m!}$

(c) $\frac{n!}{(n-m)!}$ (d) $(n-m)!$

12. In the set A = {1, 2, 3, 4, 5}, a relation R is defined by

$R = \{(x, y) \mid x, y \in A \text{ and } x < y\}$. The R is

(a) reflexive (b) symmetric
(c) transitive (d) None of these

13. For real numbers x and y, we write $x \mathrel{R} y \Leftrightarrow x - y + \sqrt{2}$ is an irrational number. Then the relation R is

(a) reflexive (b) symmetric
(c) transitive (d) None of these

14. The function $f : C \to R : f(z) = |z|$ is

(a) many-one, into (b) many-one, onto
(c) one-one, into (d) one-one, onto

15. The range of $f(x) = 1 - |x|$ is

(a) $]-\infty, 1]$ (b) $[-1, 1]$

(c) $[1, \infty[$ (d) $]-\infty, -1]$

16. The domain of the function

$$f(x) = \sqrt{\log\left(\frac{1}{|\sin x|}\right)} \text{ is}$$

(a) $R - \{-\pi, \pi\}$ (b) $R - \{n\pi \mid n \in z\}$

(c) $R - \{2n\pi \mid n \in z\}$ (d) $(-\infty, \infty)$

17. The domain of definition of

$f(x) = \sqrt{\log(\log x) - \log(4 - \log x) - \log 3}$ is

(a) $(10^3, 10^4)$ (b) $[10^3, 10^4]$

(c) $[10^3, 10^4)$ (d) $(10^3, 10^4]$

18. If the function

$f(x) = \frac{x}{x-1}$, express $f(3x)$ in terms of $f(x)$

(a) $\frac{3f(x)}{3f(x)-1}$ (b) $\frac{3f(x)}{3f(x)-3}$

(c) $\frac{3f(x)}{2f(x)+1}$ (d) $3f(x) - 1$

19. If $f(x) = \tan x$, then the value of $f(x) + f(-x)$ is equal to

(a) $2 \tan x$ (b) $-2 \tan x$
(c) ∞ (d) 0

20. Given $f(x) = \log\left(\frac{1+x}{1-x}\right)$ and $g(x) = \frac{3x + x^3}{1 + 3x^2}$, then $fog(x)$ equals

(a) $-f(x)$ (b) $3f(x)$
(c) $[f(x)]^3$ (d) None of these

ANSWERS

1	2	3	4	5	6	7	8	9	10
(d)	(c)	(b)	(b)	(a)	(d)	(d)	(c)	(d)	(a)
11	**12**	**13**	**14**	**15**	**16**	**17**	**18**	**19**	**20**
(c)	(c)	(a)	(a)	(a)	(b)	(c)	(c)	(d)	(b)

SOME SELECTED EXPLANATORY ANSWERS

1. Given R = $\{(a, b) : b = a + 1\}$, $a, b \in \{1, 2, 3, 4, 5, 6\}$

$\Rightarrow$ R = {(1, 2), (2, 3), (3, 4), (4, 5), (5, 6)}

2. f : R and $f(x) = (3 - x^3)^{1/3}$

and $f[f(x)] = [3 - \{f(x)\}^3]^{1/3}$

$= (3 - 3 + x^3)^{1/3} = x$

5. It is clear that $f(x) = \sqrt{9 - x^2}$ is not defined when $(9 - x^2) < 0$, *i.e.*, when $x^2 > 9$, *i.e.*, when $x > 3$ or $x < -3$.

$\therefore$ dom (f) = $\{x \in R : -3 \le x \le 3$

6. Clearly, $\frac{1}{(1-x^2)}$ is not defined when $1 - x^2 = 0$, *i.e.*, when $x = \pm 1$.

$\therefore$ dom(f) = R – {–1, 1}.

Also, $y = \frac{1}{1-x^2} \Rightarrow (1-x^2) = \frac{1}{y}$

$\Rightarrow$ $x = \sqrt{1 - \frac{1}{y}}$

Clearly, x is not defined when $\left(1 - \frac{1}{y}\right) < 0$

or $1 < \frac{1}{y}$ or $y < 1$.

7. Since f : R $\to$ R and g : R $\to$ R, given by $f(x) - 2x - 3$ and $g(x) = x^3 + 5$ respectively, are bijections. Therefore f^{-1} and g^{-1} exists.

We have $f(x) = 2x - 3$

$\therefore$ $f(x) = y \Rightarrow 2x - 3 = y$

$\Rightarrow$ $x = \frac{y+3}{2} \Rightarrow f^{-1}(y) = \frac{y+3}{2}$.

Thus, f^{-1} is given by $f^{-1}(x) = \frac{y+3}{2}$ for all $x \in$ R.

Similarly, $g^{-1}(x) = (x - 5)^{1/3}$ for all $x \in$ R.

Now, $(fog)^{-1}(x) = (g^{-1} of^{-1})(x) = g^{-1}(f^{-1}(x))$

$= g^{-1}\left(\frac{x+3}{2}\right) = \left(\frac{x+3}{2} - 5\right)^{1/3} = \left(\frac{x-7}{2}\right)^{1/3}$

9. Given $f(x) = \frac{x}{x^2+1} \Rightarrow f(2) = \frac{2}{2^2+1} = \frac{2}{5}$

$\Rightarrow$ $f(f(2)) = f\left(\frac{2}{5}\right) = \frac{5}{\left(\frac{2}{5}\right)^2 + 1} = \frac{\frac{5}{5}}{\frac{4}{25}+1}$

$= \frac{\frac{2}{5}}{\frac{29}{25}} = \frac{10}{29}$

14. I. Given function is $f : C \to R : f(z) = |z|$.

Let $z_1 = x + iy$ and $z_2 = x - iy$.

$f(z_1) = |z_1| = \sqrt{x^2 + y^2}$ ...(*i*)

$f(z_2) = |z_2| = \sqrt{x^2 + y^2}$ ...(*ii*)

(*i*) and (*ii*) implies $f(x + iy) = f(x - iy)$.

So f is many-one function.

II. Since the modulus of a Complex number can never be negative, so range $(f) \ne$ R.

f is into.

16. $f(x) \in R$ if $\log\left\{\frac{1}{|\sin x|}\right\} \ge 0$

$\Rightarrow \frac{1}{|\sin x|} \ge 1$ and $|\sin x| \ne 0$

$\Rightarrow |\sin x| \ne 0 \left[\because \frac{1}{|\sin x|} \ge 1 \text{ for all } x\right]$

$\Rightarrow x \ne n\pi, n \in Z$.

$\therefore$ Domain of $f(x) = R - \{n\pi : n \in Z\}$.

19. Given function is $f(x) = \tan x$.

$\Rightarrow f(-x) = \tan(-x) = -\tan x$

($\because$ tan x is an odd function)

$\therefore f(x) + f(-x) = \tan x - \tan x = 0$.

3

Trigonometric Functions

TRIGONOMETRICAL RATIOS AND IDENTITIES

1. An angle is positive, if it is measured in anti-clockwise direction.
2. An angle is negative, if it is measured in clockwise direction.
3. **Measurement of angle**

   Angle in radian = $\frac{\text{Length of the arc}}{\text{Radius of the circle}}$ $\theta = \frac{l}{r}$

4. **Basic Trigonometrical Identities**

   (*i*) $\sin^2\theta + \cos^2\theta = 1$

   (*ii*) $1 + \tan^2\theta = \sec^2\theta$,

   $\theta \neq (2n+1)(\pi/2)$. Where n is integer.

   (*iii*) $1 + \cot^2\theta = \text{cosec}^2\theta, \theta \neq n\pi$

   (*iv*) $\sin^2\theta + \cos^2\theta = 1$

   $\Rightarrow |\sin\theta| \leq 1$ and $|\cos\theta| \leq 1$

   $\Rightarrow -1 \leq \sin\theta \leq 1$ and $-1 \leq \cos\theta \leq 1$

   (*v*) $1 + \tan^2\theta = \sec^2\theta \Rightarrow |\sec\theta| \geq 1$

   $\Rightarrow -\infty < \sec^2\theta \leq -1$ or $1 \leq \sec\theta < \infty$

   (*vi*) $1 + \cot^2\theta = \text{cosec}^2\theta \Rightarrow |\text{cosec}\theta| \geq 1$

   $\Rightarrow -\infty < \text{cosec}\theta \leq -1$ or $1 \leq \text{cosec}\theta < \infty$

5. **Table for Reduction Formulae:**

   S means "sin and its reciprocal cosec are positive and rest negative in II quadrant".

   T means "tan and cot are +ve and rest are –ve in III quadrant"

   C means "cos and sec are +ve and rest are –ve in IV quadrant".

   Here, if angle is $\pi \pm \theta$ or $2\pi - \theta$, then name of the ratio remains same. If angle is $\frac{\pi}{2} \pm \theta$ or $\frac{3\pi}{2} \pm \theta$, then change sine to cosine, cosine to sine, tangent to cotgent, cotgent to tangent, sec to cosec and cosec to sec.

θ	π/2 – θ	π/2 + θ	π – θ	π + θ	3π/2 – θ	3π/2 + θ	2π – θ	– θ
sinθ	cosθ	cosθ	sinθ	–sinθ	–cosθ	–cosθ	–sinθ	–sinθ
cosθ	sinθ	–sinθ	–cosθ	–cosθ	–sinθ	sinθ	cosθ	cosθ
tanθ	cotθ	–cotθ	–tanθ	tanθ	cotθ	–cotθ	–tanθ	–tanθ
cotθ	tanθ	–tanθ	–cotθ	cotθ	tanθ	–tanθ	–cotθ	–cotθ
secθ	cosecθ	–cosecθ	–secθ	–secθ	–cosecθ	cosecθ	secθ	secθ
cosecθ	secθ	secθ	cosecθ	–cosecθ	–secθ	–secθ	–cosecθ	–cosecθ

6. **Trigonometrical ratios for sum and difference of two angles**

   (*i*) $\sin(A+B) = \sin A\cos B + \cos A\sin B$

   (*ii*) $\sin(A-B) = \sin A\cos B - \cos A\sin B$

   (*iii*) $\cos(A+B) = \cos A\cos B - \sin A\sin B$

   (*iv*) $\cos(A-B) = \cos A\cos B + \sin A\sin B$

   (*v*) $\sin(A+B)\sin(A-B) = \sin^2 A - \sin^2 B = \cos^2 A - \cos^2 B$

(*vi*) $\cos(A+B)\cos(A-B)=\cos^2 A-\sin^2 B$

$=\cos^2 B-\sin^2 A$

(*vii*) $\tan(A+B)=\dfrac{\tan A+\tan B}{1-\tan A\tan B}$

(*viii*) $\tan(A-B)=\dfrac{\tan A-\tan B}{1+\tan A\tan B}$

(*ix*) $\cot(A+B)=\dfrac{\cot A\cot B-1}{\cot A+\cot B}$

$(A\neq m\pi, B\neq n\pi, A+B\neq k\pi)$

(*x*) $\cot(A-B)=\dfrac{\cot A\cot B+1}{\cot A-\cot B}$

$(A\neq n\pi, B\neq m\pi, A-B\neq k\pi)$

7. Multiple angles and half angles

(*i*) $\sin 2\theta=2\sin\theta\cos\theta$

(*ii*) $\sin\theta=2\sin\frac{\theta}{2}\cos\frac{\theta}{2}$

(*iii*) $\cos 2\theta=\cos^2\theta-\sin^2\theta$

$=2\cos^2\theta-1$

$=1-2\sin^2\theta$

(*iv*) $\cos\theta=\cos^2\frac{\theta}{2}-\sin^2\frac{\theta}{2}$

$=2\cos^2\frac{\theta}{2}-1$

$=1-2\sin^2\frac{\theta}{2}$

(*v*) $\sin 2\theta=\dfrac{2\tan\theta}{1+\tan^2\theta}$

(*vi*) $\cos 2\theta=\dfrac{1-\tan^2\theta}{1+\tan^2\theta}$

(*vii*) $\tan 2\theta=\dfrac{2\tan\theta}{1-\tan^2\theta}$

(*viii*) $\sin 3\theta=3\sin\theta-4\sin^3\theta$

(*ix*) $\cos 3\theta=4\cos^3\theta-3\cos\theta$

(*x*) $\tan 3\theta=\dfrac{3\tan\theta-\tan^3\theta}{1-3\tan^2\theta}$

8. Sum and difference into Products

(*i*) $\sin C+\sin D=2\sin\left(\dfrac{C+D}{2}\right)\cos\left(\dfrac{C-D}{2}\right)$

(*ii*) $\sin C-\sin D=2\cos\left(\dfrac{C+D}{2}\right)\sin\left(\dfrac{C-D}{2}\right)$

(*iii*) $\cos C+\cos D=2\cos\left(\dfrac{C+D}{2}\right)\cos\left(\dfrac{C-D}{2}\right)$

(*iv*) $\cos C-\cos D=-2\sin\left(\dfrac{C+D}{2}\right)\sin\left(\dfrac{C-D}{2}\right)$

9. Transformations of Products into sum or difference

(*i*) $2\sin A\cos B=\sin(A+B)+\sin(A-B)$

(*ii*) $2\cos A\sin B=\sin(A+B)-\sin(A-B)$

(*iii*) $2\cos A\cos B=\cos(A+B)+\cos(A-B)$

(*iv*) $2\sin A\sin B=\cos(A-B)-\cos(A+B)$

10. Greatest and Least Value of $a\sin\theta+b\cos\theta$

Let $a=r\cos\alpha$ and $b=r\sin\alpha$

Then, $a^2+b^2=r^2$ or $r=\sqrt{a^2+b^2}$

$\therefore a\sin\theta+b\cos\theta=r(\cos\alpha\sin\theta+\cos\theta\sin\alpha)$

$=r\sin(\theta+\alpha)$

But $-1\le\sin(\theta+\alpha)\le 1$

$\therefore\ -r\le r\sin(\theta+\alpha)\le r$

$\therefore\ -r\le a\sin\theta+b\cos\theta\le r$

$\therefore\ -\sqrt{a^2+b^2}\le a\sin\theta+b\cos\theta\le\sqrt{a^2+b^2}$

$\therefore$ Greatest value of $a\sin\theta+b\cos\theta$ is $\sqrt{a^2+b^2}$ and least value of $a\sin\theta+b\cos\theta$ is $-\sqrt{a^2+b^2}$.

TRIGONOMETRICAL EQUATIONS

Some Important Results:

1. If $\sin\theta=\sin\alpha$, then $\theta=n\pi+(-1)^n\alpha$,

   Where $\alpha\in\left[-\frac{\pi}{2},\frac{\pi}{2}\right]$ called principal value of θ and $n\in Z$.

2. If $\cos\theta=\cos\alpha$, then $\theta=2n\pi\pm\alpha$.

   Where $\alpha\in[0,\pi]$ called principal value of θ and $n\in Z$.

3. If $\tan\theta=\tan\alpha$, then $\theta=n\pi+\alpha$.

Where $\alpha \in \left(-\frac{\pi}{2}, \frac{\pi}{2}\right)$ called principal value of θ and $n \in Z$.

4. If $\sin^2\theta = \sin^2\alpha$, then $\theta = n\pi \pm \alpha$.

Where $\alpha \in \left[-\frac{\pi}{2}, \frac{\pi}{2}\right]$ and $n \in Z$.

5. If $\cos^2\theta = \cos^2\alpha$, then $\theta = n\pi \pm \alpha$.

Where $\alpha \in [0, \pi]$ and $n \in Z$.

6. If $\tan^2\theta = \tan^2\alpha$, then $\theta = n\pi \pm \alpha$.

Where $\alpha \in \left(-\frac{\pi}{2}, \frac{\pi}{2}\right)$ and $n \in Z$.

MULTIPLE CHOICE QUESTIONS

1. The minimum value of $3\cos\theta + 4\sin\theta + 8$ is
   (*a*) 5 (*b*) 9
   (*c*) 7 (*d*) 3

2. $\sec^2\theta = \dfrac{4xy}{(x+y)^2}$ is true if and only if
   (*a*) $x + y = 0$ (*b*) $x = y, x \neq 0$
   (*c*) $x = y$ (*d*) $x \neq 0, y \neq 0$

3. If $\tan\alpha = m/(m+1)$ and $\tan\beta = 1/(2m+1)$, then $\alpha + \beta$ is equal to
   (*a*) $\pi/2$ (*b*) $\pi/3$
   (*c*) $\pi/4$ (*d*) None of these

4. $\tan A + 2\tan 2A + 4\tan 4A + 8\cot 8A$ is equal to
   (*a*) $\tan 2A$ (*b*) $\cot A$
   (*c*) $\sin 3A$ (*d*) None of these

5. If A, B, C and D are angles of a cyclic quadrilateral, then $\cos A + \cos B + \cos C + \cos D$ is equal to
   (*a*) 0
   (*b*) 1
   (*c*) $\cos(A+B)\sin(C+D)$
   (*d*) None of these

6. $\tan 20° \tan 40° \tan 60° \tan 80°$ is equal to
   (*a*) 1 (*b*) 2
   (*c*) 3 (*d*) None of these

7. $\sin 12° \sin 48° \sin 54°$ is equal to
   (*a*) 1 (*b*) 1/4
   (*c*) 1/8 (*d*) 1/16

8. $\sin 36° \sin 72° \sin 108° \sin 144°$ is equal to
   (*a*) 3/8 (*b*) 7/8
   (*c*) 1/16 (*d*) None of these

9. $\left(1+\cos\frac{\pi}{8}\right)\left(1+\cos\frac{3\pi}{8}\right)\left(1+\cos\frac{5\pi}{8}\right)\left(1+\cos\frac{7\pi}{8}\right)$ is equal to
   (*a*) 1/2 (*b*) 1/8
   (*c*) 3/16 (*d*) 5/16

10. If $\cos(\alpha+\beta) = 4/5$ and $\sin(\alpha-\beta) = 5/13$ and α, β lie between 0 and $\pi/4$, then $\tan 2\alpha$ is equal to
    (*a*) 56/33 (*b*) 21/65
    (*c*) 63/22 (*d*) None of these

11. If α and β are the solutions of the equation $a\tan\theta + b\sec\theta = c$, then $\tan(\alpha+\beta)$ is equal to
    (*a*) $\dfrac{2ab}{a^2+c^2}$ (*b*) $\dfrac{2ab}{a^2-c^2}$
    (*c*) $\dfrac{2ac}{a^2-c^2}$ (*d*) None of these

12. If $a\cos 2\theta + b\sin 2\theta = c$ has α and β as its solutions, then $\tan\alpha - \tan\beta$ is equal to
    (*a*) $\dfrac{ab}{a+c}$ (*b*) $\dfrac{2b}{a+c}$
    (*c*) $\dfrac{a+b}{a+c}$ (*d*) $\dfrac{c-a}{c+a}$

13. $\dfrac{1}{\sin 10°} - \dfrac{\sqrt{3}}{\cos 10°}$ is equal to
    (*a*) 2 (*b*) 3
    (*c*) 4 (*d*) 1

14. If $\cos A = m\cos B$, then

$\cot\left(\frac{A+B}{2}\right) \Big/ \tan\left(\frac{B-A}{2}\right)$ is equal to

(*a*) $\frac{m}{m^2-1}$ (*b*) $\frac{m-1}{m+1}$

(*c*) $\frac{mn}{m-1}$ (*d*) None of these

15. The smallest value of x for which $\tan(x+100°) = \tan(x+50°)\tan(x)\tan(x-50°)$ is

(*a*) 45° (*b*) 30°

(*c*) 35° (*d*) None of these

16. $\sin^2\theta - \cos\theta = \frac{1}{4}, 0 \le \theta < 2\pi$, then θ is given by

(*a*) $2\pi/3$ (*b*) $4\pi/3$

(*c*) $5\pi/3$ (*d*) None of these

17. $2\sin^2\theta = 3\cos\theta, 0 \le \theta \le 2\pi$, then the value of θ is

(*a*) $\pi/3$ and $5\pi/3$ (*b*) $\pi/3$ and $2\pi/3$

(*c*) $2\pi/3$ and $4\pi/3$ (*d*) None of these

18. The general solution of $4\cos^2 x\sin x - 2\sin^2 x = 3\sin x$ is

(*a*) $n\pi + (-1)^n(\pi/5)$

(*b*) $2n\pi \pm (\pi/10)$

(*c*) $n\pi + (-1)^n(3\pi/5)$

(*d*) $n\pi + (-1)^n(-3\pi/10)$

19. If $(1-\tan\theta)(1+\tan\theta)\sec^2\theta + 2\sec^2\theta = 0$, then the value of θ in the interval $(-\pi/2, \pi/2)$ is

(*a*) Only $\pi/3$ (*b*) Only $-\pi/3$

(*c*) Both $\pi/3$ and $-\pi/3$ (*d*) None of these

20. The general solution of the equation $\cot\theta - \tan\theta = \sec\theta$ is

(*a*) $\theta = n\pi + (-1)^n(-\pi/2)$

(*b*) $\theta = 2n\pi \pm (\pi/2)$

(*c*) $\theta = n\pi + (-1)^n(\pi/6)$

(*d*) None of these

21. If $\sin 5x + \sin 3x + \sin x = 0$, the value of x between $0 < x < \pi/2$ is

(*a*) $\pi/6$ (*b*) $\pi/12$

(*c*) $\pi/3$ (*d*) None of these

22. The number of solution of the equation $\cos\theta\cos 2\theta\cos 3\theta = \frac{1}{4}$ is

(*a*) 2 (*b*) 4

(*c*) 6 (*d*) 8

23. The solution of $\tan\theta + \sec\theta = \sqrt{3}, 0 \le \theta \le 2\pi$ is

(*a*) $\pi/3$ (*b*) $2\pi/3$

(*c*) $5\pi/6$ (*d*) None of these.

24. One value of θ which satisfies the equation $\sin^4\theta - 2\sin^2\theta - 1 = 0$ lies between 0 and 2π is

(*a*) No root (*b*) One root

(*c*) Two roots (*d*) None of these

25. The number of solutions of the equation $\sin(e^x) = 5^x + 5^{-x}$ is

(*a*) 0 (*b*) 1

(*c*) 2 (*d*) Infinitely many

ANSWERS

1	2	3	4	5	6	7	8	9	10
(*d*)	(*b*)	(*c*)	(*b*)	(*a*)	(*c*)	(*c*)	(*d*)	(*b*)	(*a*)
11	12	13	14	15	16	17	18	19	20
(*c*)	(*d*)	(*c*)	(*d*)	(*b*)	(*c*)	(*a*)	(*d*)	(*c*)	(*c*)
21	22	23	24	25					
(*c*)	(*c*)	(*d*)	(*a*)	(*a*)					

SOME SELECTED EXPLANATORY ANSWERS

1. Let $r\cos\alpha = 3$ and $r\sin\alpha = 4$

$\therefore\ r = \sqrt{3^2+4^2} = 5$

$\therefore 3\cos\theta + 4\sin\theta + 8$

$= r\cos\alpha\cos\theta + r\sin\alpha\sin\theta + 8$

$= r\cos(\alpha-\theta)+8$

$= 5\cos(\alpha-\theta)+8$

Since, $-1 \le \cos(\alpha-\theta) \le 1$,

Hence, minimum value of $\cos(\alpha-\theta)$ is -1.

$\therefore$ Minimum value of $3\cos\theta+4\sin\theta+8$

$= 5(-1)+8 = 3$

2. If both $x=0, y=0$,
then it becomes an indeterminant form.

3. $\tan(\alpha+\beta) = \dfrac{\tan\alpha+\tan\beta}{1-\tan\alpha\tan\beta}$

$= \dfrac{\{m/(m+1)\}+\{1/(2m+1)\}}{1-\{m/(m+1)\}\{1/(2m+1)\}}$

$= \dfrac{2m^2+2m+1}{2m^2+3m+1-m} = 1$

$\Rightarrow\ \alpha+\beta = \pi/4$

7. $\sin 12°\sin 48°\sin 54°$

$= \dfrac{1}{2}\left[(2\sin 12°\sin 48°).\sin 54°\right]$

$= \dfrac{1}{2}\left[(\cos 36° - \cos 60°)\cos 36°\right]$

$= \dfrac{1}{2}\left[\left(\dfrac{\sqrt{5}+1}{4} - \dfrac{1}{2}\right)\dfrac{\sqrt{5}+1}{4}\right]$

$= \dfrac{1}{2}\left[\left(\dfrac{\sqrt{5}+1}{4}\cdot\dfrac{\sqrt{5}-1}{4}\right)\right] = \dfrac{1}{8}$

8. Given expression
$= (\sin 36°\sin 144°)(\sin 72°\sin 108°)$

$= (\sin 36°\sin 36°)(\cos 18°\cos 18°)$

$= \sin^2 36°\cos^2 18°$

$= \left(\dfrac{\sqrt{10-2\sqrt{5}}}{4}\right)^2\left(\dfrac{\sqrt{10+2\sqrt{5}}}{4}\right)^2$

$= \dfrac{10-2\sqrt{5}}{16}\cdot\dfrac{10+2\sqrt{5}}{16}$

$= \dfrac{100-20}{256} = \dfrac{5}{16}$

12. $a\cos 2\theta + b\sin 2\theta = c$

$\Rightarrow a\left(\dfrac{1-\tan^2\theta}{1+\tan^2\theta}\right) + b\left(\dfrac{2\tan\theta}{1+\tan^2\theta}\right) = c$

$\Rightarrow (a+c)\tan^2\theta - 2b\tan\theta + (c-a) = 0$

$\therefore\ \tan\alpha\tan\beta = \dfrac{c-a}{c+a}$

16. $4(1-\cos^2\theta) - 4\cos\theta - 1 = 0$

$\therefore 4\cos^2\theta + 4\cos\theta - 3 = 0$

or, $(2\cos\theta+3)(2\cos\theta-1) = 0$

$\Rightarrow \cos\theta = -3/2$ is not possible

$\therefore\ \cos\theta = 1/2 = \cos\pi/3$ or $\cos 5\pi/3$

$\therefore\ \theta = \pi/3$ or, $5\pi/3$

24. $\sin^2\theta = \dfrac{2\pm\sqrt{4+4}}{2}$

$= 1\pm\sqrt{2}$

$\sin^2\theta = 1+\sqrt{2} > 1$ Which is impossible

$\therefore\ \sin^2\theta = 1-\sqrt{2} < -1$

$\Rightarrow \sin\theta$ will become imaginary.

$\therefore$ Given equation has no root.

25. $\sin(e^x) = 5^x + 5^{-x} \ge 2\forall x$

$(\because \text{A.M} \ge \text{G.M.})$

$\Rightarrow\ e^x$ has no answer.

$\Rightarrow\ x$ has no valid value.

4

Principle of Mathematical Induction

Let $P(n)$ be a statement involving the natural number n. To prove that $P(n)$ is true for all natural number n, we proceed as follows :

1. Prove that $P(1)$ is true.
2. Assume that $P(k)$ is true.
3. Using steps 1 and 2, prove that $P(k + 1)$ is true. Sometimes step 2 is replaced by 2(*a*).

2(*a*) assumes that $P(n)$ is true for all $n \in N$, where $1 \le n \le k$

The First Principle of Mathematical Induction comprises steps 1, 2 and 3. Steps 1, 2(*a*) and 3 constitute another principle of mathematical induction, usually called the Second Principle of Mathematical Induction.

The key word in step 2 is *assume*. You are not trying to prove it's true for $n = k$, you're going to accept on faith that it is, and show it's true for the next number, $n = k + 1$. If it later turns out that you get a contradiction, then the assumption was wrong.

Annotated Example of Mathematical Induction

Prove $1 + 4 + 9 + ... + n^2 = n(n + 1)(2n + 1)/6$ for all positive integers n.

Another way to write "for every positive integer n" is $\forall n \in Z^+$. This works because Z is the set of integers, so Z^+ is the set of positive integers. The upside down A is the symbol for "for all" or "for every" or "for each" and the symbol that looks like a weird e is the "element of" symbol. So technically, the statement is saying "for every n that is an element of the positive integers", but it's easier to say "for every positive integer n".

Identify the General Term and nth Partial Sum Before Beginning the Problem

The general term, a_n, is the last term on the left hand side, $a_n = n^2$

The nth partial sum, S_n, is the right hand side.

$S_n = n(n + 1)(2n + 1)/6$

Find the next term in the general sequence and the series

The next term in the sequence is a_{k+1} and is found by replacing n with k+1 in the general term of the sequence, a_n.

$a_{k+1} = (k + 1)^2$

The next term in the series is S_{k+1} and is found by replacing n with k +1 in the nth partial sum, S_n. You may wish to simplify the next partial sum, S_{k+1}.

$S_{k+1} = (k+1)[(k+1) + 1][2(k+1) + 1]/6$

$S_{k+1} = (k + 1)(k + 2)(2k + 3)/6$ (This will be our Goal in step 3)

We will use these definitions later in the mathematical induction process. We're now ready to begin.

(*i*) Show the statement is true for n = 1, that is, show that $a_1 = S_1$.

a_1 is the first term on the left or you can find it by substituting $n = 1$ into the formula for the general term, a_n.

S_1 is found by substituting $n = 1$ into the formula for the nth partial sum, S_n.

LHS : $a_1 = 1$

RHS : $S_1 = 1(1+1)[2(1) + 1]/6 = 1(2)(3)/6 = 1$

So, you can see that the left hand side equals the right hand side for the first term, so we have established the first condition of mathematical induction.

(*ii*) Assume the statement is true for $n = k$.
The left hand side is the sum of the first k terms, so we can write that as S_k. The right hand side is found by substituting $n = k$ into the S_n formula.
Assume that $S_k = k(k + 1)(2k + 1) / 6$

(*iii*) Show the statement is true for $n = k + 1$
What we are trying to show is that
$S_{k+1} = (k + 1)(k + 2)(2k + 3) / 6$. This was our goal from earlier.
We begin with something that we know (assume) it is true and add the next term, a_{k+1}, to both sides.
$S_k + a_{k+1} = k(k + 1)(2k + 1)/6 + a_{k+1}$
On the left hand side, $S_k + a_{k+1}$ means the "sum of the first k terms" plus "the k+1 term", which gives us the sum of the first $k + 1$ terms, i.e. S_{k+1}.
This often gives students difficulties, so let us think about it this way. Assume k =10. Then S_k would be S_{10}, the sum of the first 10 terms and a_{k+1} would be a_{11}, the 11th term in the sequence. $S_{10} + a_{11}$ would be the sum of the 10 terms plus the 11th term which would be the sum of the first 11 terms.
On the right hand side, replace a_{k+1} by $(k+1)^2$, which is what you found it was before beginning the problem.

$$S_{k+1} = \frac{k(k+1)(2k+1)}{6+(k+1)^2}$$

Now, try to turn your right hand side into goal of

$$\frac{(k+1)(k+2)(2k+3)}{6}.$$

You need to get a common denominator, so multiply the last term by 6/6.

$$S_{k+1} = \frac{k(k+1)(2k+1)}{\frac{6+6(k+1)^2}{6}}$$

Now simplify. It is almost always easier to factor rather than expand when simplifying. This is especially aided by the fact that your goal is in factored form. You can use that to help you factor. You know that you want a $(k+1)(k+2)(2k+3)$ in the final form. We see right now that there is a $(k+1)$ that is common to both of those, so let's begin by factoring it out.

$S_{k+1} = (k + 1)[k(2k + 1) + 6(k + 1)] / 6$

What's left inside the brackets [ ] doesn't factor, so we expand and combine like terms.

$S_{k+1} = (k + 1)(2k^2 + k + 6k + 6) / 6$

$S_{k+1} = (k + 1)(2k^2 + 7k + 6) / 6$

Now, try to factor $2k^2 + 7k + 6$, keeping in mind that you need a $(k+2)$ and $(2k+3)$ in the goal that you don't have yet.

$S_{k+1} = (k + 1)(k + 2)(2k + 3) / 6$

Hey! That's what our goal was. That's what we were trying to show. That means we did it!

Conclusion

The conclusion is found by saying "Therefore, by the principle of mathematical induction" and restating the original claim.

Therefore, by the principle of mathematical induction,

$1 + 4 + 9 + \cdots + n^2 = n(n + 1)(2n + 1) / 6$ for all positive integers n.

Summations

Earlier in the chapter we had some summation formulas that were very melodious. In the following examples, c is a constant, and x and y are functions of the index.

You can factor a constant out of a summation

$$\sum cx = c\sum x$$

The sum of a constant times a function is the constant times the sum of a function.

The sum of a sum is the sum of the sums

$$\sum (x+y) = \sum x + \sum y$$

The summation symbol can distribute over addition.

The sum of a difference is the difference of the sums

$$\sum (x-y) = \sum x - \sum y$$

The summation symbol can distribute over subtraction.

Sum of the Powers of the Integers

Now, we're going to look at the sum of the whole number powers of the natural numbers.

Sigma Notation = Closed Form

Expanded

$1 + 1 + 1 + ... + 1$ (n times)

$1 + 2 + 3 + ... + n$

$1 + 4 + 9 + ... + n^2$

$1 + 8 + 27 + ... + n^3$

$1 + 16 + 81 + ... + n^4$

$1 + 32 + 243 + ... + n^5$

The closed form for a summation is a formula that allows you to find the sum simply by knowing the number of terms.

Finding Closed Form

Find the sum of : $1 + 8 + 22 + 42 + ... + (3n^2 - n - 2)$

The general term is $a_n = 3n^2 - 2$, so what we're trying to find is $\sum(3k^2 - k - 2)$, where the Σ is really the sum from $k = 1$ to n I'm just not writing those here to make it more accessible.

Take the original, open form of the summation, $\sum(3k^2 - k - 2)$,

Distribute the summation sign, $\sum 3k^2 - \sum k - \sum 2$

Factor out any constants, $3\sum k^2 - \sum k - 2\sum 1$

Replace each summation by the closed form given above. The closed form is a formula for a sum that doesn't include the summation sign, only n.

$$3\left[\frac{n(n+1)(2n+1)}{6}\right] - \left[\frac{n(n+1)}{2}\right] - 2[n]$$

Now get a common denominator, in this case, 2.

$$\frac{n(n+1)(2n+1)}{2} - \frac{n(n+1)}{2} - \frac{4n}{2}$$

Remember that the word factor begins with the letter F and anytime you have a choice of doing something in mathematics that starts with the letter F, that's probably where you should start. So, do not expand, factor instead.

The common factor is n so we'll factor that out of each term. The whole expression is over 2.

$n\,[(n + 1)(2n + 1) - (n + 1) - 4]\,/\,2$

Now expand inside the brackets [ ].

$n\,[\,2n^2 + 3n + 1 - n - 1 - 4\,]\,/\,2$

Simplify like terms.

$n\,(\,2n^2 + 2n - 4\,)\,/\,2$

Notice the common factor of 2 inside the parentheses, let's factor that out.

$2n\,(n^2 + n - 2\,)\,/\,2$

The 2 in the numerator and the 2 in the denominator divide out and we can factor the rest to get the closed form for the sum.

$n\,(n + 2)\,(n - 1)$

Isn't that beautiful? At this point, we could write a mathematical induction problem similar to those in the book for this problem. It would read ...

Prove:

$$\sum_{t=1}^{n}(3k^2 - k - 2) = n(n+2)(n-1), \forall n \in Z$$

We're not going to prove that statement, but that is how the book came up with many of the problems that you're asked to prove. You take an open form, find the closed form, and then put it in the text as a problem for the student to prove.

Pattern Recognition

Sometimes you have to figure out what the general term of a sequence is. Here are some guidelines.

(*i*) Calculate the first several terms of the sequence. Sometimes it helps to write the term in factored and expanded form.

(*ii*) Try to find a recognizable pattern. Here are some things to look for :

- (*a*) Linear patterns: $an+b$ (will have a common difference)
- (*b*) Quadratic pattern : $an^2 + b$ (the perfect squares plus/minus a constant)
- (*c*) Cubic pattern : $an^3 + b$ (the perfect cubes plus/minus a constant)
- (*d*) Exponential patterns: $2^n + b$, $3^n + b$ (powers of 2 or 3 plus/minus a constant)
- (e) Factorial patterns: $n!$, $(2n)!$, $(2n - 1)!$ (factoring these really helps)

(*iii*) After you have your pattern, then you can use mathematical induction to prove the conjecture is correct.

Finite Differences

Finite differences can help you find the pattern if you have a polynomial sequence.
The *first differences* are found by subtracting consecutive terms. If the first differences are all the same, then the pattern is linear.

The *second differences* are found by subtracting consecutive *first* differences. If the second differences are all the same, then the pattern is quadratic. Remember that you can find a quadratic model by taking the equation $y = ax^2+bx+c$ with three points. Then solve the system of equations that results. The analogy here is that you can find $a_n = an^2+bn+c$ by substituting in three terms in the sequence for a_n and their corresponding position in the sequence for n. Then solve the system of linear equations.

This can be extended to *third differences* by subtracting consecutive *second* differences. If the third differences are all the same, the pattern is cubic. You can fit a cubic model with four points and the model $a_n = an^3+bn^2+cn+d$.

Example for Finite Differences

Find the general term of the sequence 1, –2, –1, 4, 13, 26, 43, 64, 89, ...

The first finite differences are found by subtracting consecutive terms in the original sequence. That is, take – 2–1 = –3, –1– (–2) = 1, 4 – (–1) = 5, 13 – 4 = 9, 26 –13 = 13, etc.

The first finite differences are: –3, 1, 5, 9, 13, 17, 21....

Since these aren't all the same, your sequence is not a linear (first degree) polynomial.

Move on to the second finite differences. These are the differences in the consecutive terms of the first finite differences. 1–(–3) = 4, 5 –1= 4, 9 – 5 = 4 , 13 – 9 = 4, 17 – 13 = 4, etc.

The second finite differences are 4, 4, 4, 4, 4, 4, 4, ...

Since the second finite differences are all the same, your sequence is that of a second degree polynomial and you can write the general term as $a_n = An^2 + Bn + C$.

Plug in the values 1, 2, 3 (since there are three variables) into the expression.

(*i*) When $n = 1$, $1A + 1B + 1C = 1$ (the first term is 1)

(*ii*) When $n = 2$, $4A + 2B + 1C = -2$ (the second term is 2)

(*iii*) When $n = 3$, $9A + 3B + 1C = -1$ (the third term is -1)

Now, solve that system of linear equations (I recommend using matrix inverses $AX=B$, so $X = A^{-1}B$). In our case, we get $A = 2$, $B = -9$, and $C = 8$, so the general term of our sequence is $a_n=2n^2-9n+8$.

If you want to check it, pick any value for n and plug it into the general term. You should get the nth number in the sequence. For example, if $n = 6$, then $a_6 = 2(6)^2 - 9(6) + 8 = 26$. Check the sequence, sure enough, the 6th number is 26.

MULTIPLE CHOICE QUESTIONS

1. Using method of induction the value of $(\cos\theta + i\sin\theta)^n$ where $n \in$ N is
   (*a*) $\cos n\theta + i\sin n\theta$ (*b*) $\cos n\theta - i\sin n\theta$
   (*c*) $2\sin n\theta$ (*d*) None of these

2. If n is any odd positive integer, then $n(n^2-1)$ is divisible by
   (*a*) 10 (*b*) 9
   (*c*) 24 (*d*) None of these

3. If p is a natural number, then $3p^{n+1} + (p+1)^{2n-1}$ is divisible by
   (*a*) $p^2 + p + 1$ (*b*) $p + 1$
   (*c*) $p^2 + 1$ (*d*) None of these

4. The expression $n^3 + 3n^2 + 5n + 3$, $n \in$ N is divisible by
   (*a*) 3 (*b*) 4
   (*c*) 5 (*d*) 6

5. If P(n) be the statement "$n^2 < 10$"
then which of the following is not true.
(*a*) P(1) (*b*) P(2)
(*c*) P(3) (*d*) P(4)

6. If P(n) be the statement "n (n +1) (n + 2) is divisible by 12"
then which of the following is not true.
(*a*) P(2) (*b*) P(3)
(*c*) P(4) (*d*) P(5)

7. For all n Î N, $3^{2n+2} - 8n - 9$ is divisible by
(*a*) 3 (*b*) 8
(*c*) 5 (*d*) 9

8. The product of any three consecutive integers are always divisible by
(*a*) 8 (*b*) 6
(*c*) 5 (*d*) 9

ANSWERS

1	2	3	4	5	6	7	8
(*a*)	(*c*)	(*a*)	(*a*)	(*d*)	(*d*)	(*b*)	(*b*)

SOME SELECTED EXPLANATORY ANSWERS

4. For $n = 1$, $n^3 + 3n^2 + 5n + 3 = 12$,
which is divisible by 3 and 4
For $n = 2$, $n^3 + 3n^2 + 5n + 3 = 33$
which is divisible by 3 only.

5. P(n) : $n^2 < 10$
P(1) : 1 < 10, which is true
Now, P(4) : 16 < 10, which is not true.

6. P(n) : n (n + 1) (n + 2) is divisible by 12.
P(2) : $2 \times 3 \times 4 = 24$, which is divisible by 12.
Hence P (2) is true.
P(3) : $3 \times 4 \times 5 = 60$, which is divisible by 12.
Hence P(3) is true.
Again, P(4) : $4 \times 5 \times 6 = 120$, which is divisible by 12.
Hence P(4) is true.
Now, P(5) : $5 \times 6 \times 7 = 210$, which is not divisible by 12.
Thus P(5) is not true.

8. $P(n) = n(n + 1)(n + 2)$
for $n = 1$; $P(1) = 1(1 + 1)(1 + 2) = 6$
which is divisible by 6
for $n = 2$, $P(2) = 2(1 + 2)(2 + 2) = 24$
which is divisible by 6
hence it is true for $P(2)$.
For $n = k$, let $P(k)$ be true
$\therefore$ $P(k) = k(k + 1)(k + 2)$ is divisible by 6.
Let $k(k + 1)(k + 1) = 6l$ for some $l \in$ N.
Now consider that $P(k + 1)$ is also true.
$P(k + 1)$
$= (k + 1)(k + 2)(k + 3)$
$= k(k + 1)(k + 2) + 3(k + 1)(k + 2)$
$= 6l + 3(k + 1)(k + 2)$
$= 6l + 6t = 6(l + t)$

Note: $(k + 1)(k + 2) = 2t$ as $(k + 1)(k + 2)$ are consecutive integers so, one of then must be even, hence divisible by 2.
It means $(k + 1)(k + 2)$ is divisible by $2 = 2t$ (say)
Thus $P(k + 1)$ is divisible by 6.
Therefore $P(k + 1)$ is true.
$\therefore$ $P(k)$ is true $\Rightarrow$ $P(k + 1)$ is true.
Hence, by principle of mathematical induction. $P(n)$ is true for all natural numbers.

5

Complex Number and Quadratic Equation

1. **Complex Number:** The number $a + ib$, where a and b are real numbers, is called a complex number.

or

The ordered pair (a, b), where a and b are real numbers, is called a complex number. If $z = a + ib$, then a is called the real part denoted by $R_e(z)$ and b the imaginary part denoted by $I_m(z)$ of the complex number $z = a + ib$. A complex number $z = x + iy$, $x, y \in R$, is purely real if $y = 0$, *i.e.*, if $I_m(z) = 0$ and is purely imaginary if $x = 0$, *i.e.*, if $R_e(z) = 0$.

2. **Equal Complex Numbers:** Let $z_1 = x_1 + iy_1$ and $z_2 = x_2 + iy_2$, then $z_1 = z_2$, *i.e.*, $x_1 + iy_1 = x_2 + iy_2$, $\Leftrightarrow x_1 = x_2$, $y_1 = y_2$, A complex number $z = x + iy = 0 \Leftrightarrow x = 0, y = 0$.

3. **Modulus and Argument of a Complex Number:** Let $z = x + iy$, then the modulus of z is the positive real number $\sqrt{(x^2 + y^2)}$ and is denoted by $|z|$, *i.e.*, $|z| = \sqrt{(x^2 + y^2)}$.
The argument or amplitude of a complex number $z = x + iy \neq 0$ is any one of the numbers which are solutions of the system of equations $\cos\theta = \dfrac{x}{\sqrt{(x^2+y^2)}}$, $\sin\theta = \dfrac{y}{\sqrt{(x^2+y^2)}}$ and is denoted by arg (z) or Arg (z). The argument of a complex number is not unique. If θ is a value of the argument, then $2n\pi + \theta$, where $n \in I$, are also values of the arguments of z. Principal value of the argument of the complex numbers z is the argument θ, which satisfy the inequality $-\pi < \theta \leq \pi$.

Note that the argument of the complex number 0 is not defined.

4. **Conjugate of a Complex Number:** If $z = x + iy$ is a complex number, then the complex number $x - iy$ denoted by $\bar{z}$ is called the conjugate of the complex number $\bar{z}$, *i.e.*, $z = x - iy$.

5. **Properties of Conjugates and Moduli:**
   - (*a*) $|z| = |\bar{z}|$
   - (*b*) $|z|^2 = z.\bar{z}$
   - (*c*) $\overline{(z_1 \pm z_2)} = \bar{z}_1 \pm \bar{z}_2$
   - (*d*) $\left(\overline{z_1 z_2}\right) = \bar{z}_1.\bar{z}_2$
   - (*e*) $\left(\overline{z_1 / z_2}\right) = \bar{z}_1 / \bar{z}_2$; $z_2 \neq 0$
   - (*f*) $|z| = 0$ iff $z = 0$
   - (*g*) $|z_1 + z_2| \leq |z_1| + |z_2|$
   - (*h*) $|z_1 - z_2| \geq |z_1| - |z_2|$.

6. **Polar form of a Complex Number:**
   - (*a*) The complex number $z = x + iy$, $x, y \in R$ can be represented in polar form as $z = r(\cos\theta + i\sin\theta)$ where $x = r\cos\theta$, $y = r\sin\theta$, $r = \sqrt{(x^2 + y^2)} = |z|$ and θ, which is arg z, is the solution of the equations
   $\cos\theta = x/\sqrt{(x^2 + y^2)} = x/r$ and $\sin\theta = y/\sqrt{(x^2 + y^2)} = y/r$.
   - (*b*) If $z_1 = r_1(\cos\theta_1 + i\sin\theta_1)$ and $z_2 = r_2(\cos\theta_2 + i\sin\theta_2)$, then $|z_1| = r_1$, $|z_2| = r_2$, arg $z_1 + \theta_1$ and arg $z_2 + \theta_2$
   Now $z_1 z_2 = r_1 r_2 \cos(\theta_1 + \theta_2)$
   $\therefore$ $|z_1 z_2| = r_1 r_2$, arg $(z_1 z_2) = \theta_1 + \theta_2$
   Thus we have $|z_1 z_2| = |z_1|\,|z_2|$ and arg $(z_1\, z_2) = $ arg $z_1 -$ arg z_2

Also $(z_1/z_2) = (r_1/r_2) \cos(\theta_1 - \theta_2)$

$\therefore$ $|z_1/z_2| = (r_1/r_2)$ and $\log(z_1/z_2) = \theta_1 - \theta_2$

Thus we have $|z_1/z_2| = |z_1| / |z_2|$ and $\arg(z_1/z_2) = \arg z_1 - \arg z_2$.

7. Cube Roots of Unity:

(*a*) The cube roots of unity are given by

$$1,\ \omega = \frac{-1+\sqrt{3}i}{2},\ \omega^2 = \frac{-1-\sqrt{3}i}{2}.$$

(*b*) If ω is the complex cube root of unity, then $\omega^3 = 1$ and $\omega^2 + \omega + 1 = 0$.

Quadratic Equation: If the polynomial $f(x)$ when equated to zero is satisfied by one or more particular values of x, then $f(x) = 0$ is called an equation. The general form of the quadratic equation in x is $ax^2 + bx + c = 0$, when $a \neq 0$, *e.g.*, $f(x) = x^2 - 3x + 2 = 0$ is a quadratic equation as it is satisfied for $x = 2$ and 1 only.

Roots of an Equation $ax^2 + bx + c = 0$

We have, $ax^2 + bx + c = 0$

or $a^2x^2 + abx + ac = 0$

or $(ax + \frac{1}{2}b)^2 = \frac{1}{4}(b^2 - 4ac)$ or $ax + \frac{1}{2}b$

$= \pm \frac{1}{2}\sqrt{(b^2 - 4ac)} \Rightarrow x = \dfrac{-b \pm \sqrt{(b^2 - 4ac)}}{2a}$

If α and β be the roots of the eqn.,

and $\alpha > \beta$, then $\alpha = \dfrac{-b + \sqrt{(b^2 - 4ac)}}{2a}$

and $\beta = \dfrac{-b - \sqrt{(b^2 - 4ac)}}{2a}$

Obviously $\sigma_1 = \alpha + \beta = -b/a$,

$\sigma_2 = \alpha\beta = c/a$, $\alpha - \beta = \sqrt{(b^2 - 4ac)}/a$

The quantity $b^2 - 4ac = \Delta$ is called the discriminant of the equation.

Nature of the Roots of $ax^2 + bx + c = 0$

(*a*) Supposing $a, b, c \in R$ and $a \neq 0$, then

(*i*) roots are real and distinct iff $\Delta > 0$,

(*ii*) roots are real and equal iff $\Delta = 0$,

(*iii*) roots are imaginary iff $\Delta < 0$,

(*iv*) If $p + iq$ is a root then the other root must be $p - iq$, $q \in R$, $i = \sqrt{(-1)}$

(*b*) (*i*) If $a, b, c \in Q$ (the set of rational numbers) and Δ is a perfect square of a rational number, then roots must be rational.

(*ii*) If $a, b, c \in Q$ and $p + \sqrt{q}$ $(p, q \in Q)$ is a root, then the other root must be $p - \sqrt{q}$.

(*c*) If $a = 1$, b, c are integers and roots are rational, they must be integers.

(*d*) If a and c are of opposite sign, the roots must be of opposite sign.

(*e*) If $ax^2 + bx + c = 0$ is satisfied by more than two values, it is an identity and $a = b = c = 0$ and vice versa.

(*f*) If roots are equal in magnitude but opposite in sign, then $b = 0$, $ac > 0$.

(*g*) If roots are reciprocal of each other, then $c = a$

Some Important Propositions:

(*a*) If α, β are roots of $f(x) = ax^2 + bx + c = 0$, then $f(x) = a(x - \alpha)(x - \beta)$.

(*b*) The equation whose roots are α, β is $x^2 - (\alpha + \beta)x + \alpha\beta = 0$.

(*c*) **One Common Root:** Two quadratic equations $f(x) = a_1x^2 + b_1x + c_1 = 0$ and $g(x) = a_2x^2 + b_2x + c_2 = 0$ has only one common root α if $a_1\alpha^2 + b_1\alpha + c_1 = 0$, $a_2\alpha^2 + b_2\alpha + c_2 = 0$. So by the method of cross-multiplication, we get

$$\frac{\alpha^2}{b_1c_2 - b_2c_1} = \frac{\alpha}{c_1a_2 - c_2a_1} = \frac{1}{a_1b_2 - a_2b_1},$$

$(a_1b_2 - a_2b_1 \neq 0)$.

Thus the required condition for one common root is $(c_1a_2 - c_2a_1)^2 = (b_1c_2 - b_2c_1)(a_1b_2 - a_2b_1)$, and if $c_1, c_2 \neq 0$ (*i.e.*, $\alpha \neq 0$),

we have $\alpha = \dfrac{b_1c_2 - b_2c_1}{c_1a_2 - c_2a_1}$ or $\dfrac{c_1a_2 - c_2a_1}{a_1b_2 - a_2b_1}$.

(*d*) **Both Common Roots:** Both roots of the equations $a_1x^2 + b_1x + c_1 = 0$ and $a_2x^2 + b_2x + c_2 = 0$ are common iff

$$\frac{a_1}{a_2} = \frac{b_1}{b_2} = \frac{c_1}{c_2}$$

(*e*) If α is a repeated root of $f(x) = 0$, α is also a root of the equation $f'(x) = 0$.

(*f*) If α is repeated common root of $f(x) = 0$ and $g(x) = 0$, then α is also a common root of the equations $f'(x) = 0$ and $g'(x) = 0$.

MULTIPLE CHOICE QUESTIONS

1. The complex number $\sin x + i\cos 2x$ and $\cos x - i\sin 2x$ are conjugate to each other for

(*a*) $x = n\pi$ (*b*) $x = (n + ½)\pi$
(*c*) $x = 0$ (*d*) No value of x

2. If $\begin{vmatrix} 6i & -3i & 1 \\ 4 & 3i & -1 \\ 20 & 3 & i \end{vmatrix} = x + iy$, then

(*a*) $x = 3, y = 1$ (*b*) $x = 1, y = 3$
(*c*) $x = 0, y = 3$ (*d*) $x = 0, y = 0$

3. If $\frac{(1+i)x-2i}{3+i} + \frac{(2-3i)y+i}{3-i} = i$, then the real values of x and y are given by

(*a*) $x = -3, y = -1$ (*b*) $x = 3, y = -1$
(*c*) $x = 3, y = 1$ (*d*) $x = 1, y = -3$

4. The argument of $(1 - i\sqrt{3})/(1 + i\sqrt{3})$ is

(*a*) $\pi/3$ (*b*) $\frac{2\pi}{3}$
(*c*) $\frac{4\pi}{3}$ (*d*) $\frac{-2\pi}{3}$

5. The least positive integer n which will reduce $\left(\frac{1+i}{1-i}\right)^n$ to unity is

(*a*) 2 (*b*) 4
(*c*) 8 (*d*) 12

6. $1 + i^2 + i^4 + i^6 + ... + i^{2n}$ is

(*a*) positive
(*b*) negative
(*c*) 0
(*d*) cannot be determined

7. If ω is a cube root of unity, then the value of $(1 + \omega - \omega^2)(1 - \omega + \omega^2)$ is

(*a*) 1 (*b*) 0
(*c*) 2 (*d*) 4

8. $1, \omega, \omega^2$, are the cube roots of unity, then the value of $(1 + \omega)^3 - (1 + \omega^2)^3$ is

(*a*) 2ω (*b*) 2
(*c*) -2 (*d*) 0

9. $\left[\frac{-1+\sqrt{(-3)}}{2}\right]^{3n} + \left[\frac{-1-\sqrt{(-3)}}{2}\right]^{3n} =$

(*a*) 0 (*b*) 1
(*c*) 2 (*d*) 3

10. If $(x + iy)^{1/3} = a + ib$, then $x/a + y/b =$

(*a*) 0 (*b*) 1
(*c*) -1 (*d*) None of these

11. If α and β are imaginary cube roots of unity and $x = a + b$, $y = a\alpha + b\beta$, $z = a\beta + b\alpha$, then $xyz =$

(*a*) $(a + b)^3$ (*b*) $a^3 + b^3$
(*c*) $(a - b)^3$ (*d*) $a^3 - b^3$

12. If the cube roots of unity are $1, \omega, \omega^2$, then the roots of the equation $(x - 1)^3 + 8 = 0$ are

(*a*) $-1, 1 + 2\omega, 1 + 2\omega^2$
(*b*) $-1, 1 - 2\omega, 1 - 2\omega^2$
(*c*) $-1, -1, -1$
(*d*) None of these

13. If $x + iy = \sqrt{\left(\frac{a+ib}{c+id}\right)}$, then $(x^2 + y^2)^2 =$

(*a*) $(a^2 + b^2)/(c^2 + d^2)$
(*b*) $(a^2 - b^2)/(c^2 + d^2)$
(*c*) $(a^2 - b^2)/(c^2 - d^2)$
(*d*) $(a - b)/(c - d)/\{(a + b)(c + d)\}$

14. If $Z(3 + 4i) = 2 + 3i$ then the value of Z is:

(*a*) $\frac{18-i}{25}$ (*b*) $\frac{9+9i}{16}$
(*c*) $\frac{18+i}{25}$ (*d*) $\frac{5+2i}{9}$

15. If $x = -2 - \sqrt{3}\,i$, then the value of $2x^4 + 5x^3 + 7x^2 + 41$ is:

(*a*) $4 + \sqrt{3}\,i$ (*b*) $4 - \sqrt{3}\,i$
(*c*) $\sqrt{3} + 4i$ (*d*) $\sqrt{3} - 4i$

16. If $r(\cos\theta + i\sin\theta) = 1$, then:
(*a*) $r = 0, \theta = 0$ (*b*) $r = 0, \theta = \frac{\pi}{2}$
(*c*) $r = 1, \theta = 0$ (*d*) $r = 1, \theta = \frac{\pi}{2}$

17. The value of $\left(\frac{\cos\theta + i\sin\theta}{\cos\theta - i\sin\theta}\right)^4$ is:
(*a*) 1 (*b*) 0
(*c*) $\cos 4\theta - i\sin 4\theta$ (*d*) $\cos 8\theta + i\sin 8\theta$

18. $1 + i$ equal to:
(*a*) $\cos\frac{\pi}{2} + i\sin\frac{\pi}{2}$
(*b*) $\cos\frac{\pi}{4} + i\sin\frac{\pi}{4}$
(*c*) $\sqrt{2}\left(\cos\frac{\pi}{4} + i\sin\frac{\pi}{4}\right)$
(*d*) $\frac{1}{\sqrt{2}}\left(\cos\frac{\pi}{4} + i\sin\frac{\pi}{4}\right)$

19. If $1, \omega, \omega^2$ are the cube roots of unity, then $(x - y)(x - \omega y)(x - \omega^2 y)$ is equal to:
(*a*) $x - y$ (*b*) $x^2 - y^2$
(*c*) $x^3 - y^3$ (*d*) $x^3 + y^3$

20. If one root of $5x^2 + 13x + k = 0$ is reciprocal of the other, then k is equal to
(*a*) 0 (*b*) 5
(*c*) 1/6 (*d*) 6

21. If α and β are the roots of the equation $ax^2 + bx + c = 0$, then $(1 + \alpha + \alpha^2)(1 + \beta + \beta^2) =$
(*a*) 0 (*b*) positive
(*c*) negative (*d*) None of these

22. If α, β are the roots of $x^2 + px + q = 0$, then $(\omega\alpha + \omega^2\beta)(\omega^2\alpha + \omega\beta) =$
(*a*) $p^2 - q$ (*b*) $p^2 + q$
(*c*) $p^2 - 3q$ (*d*) $p^2 + 3q$

23. Let α, β be the roots of the equation $(x - a)(x - b) = c, c \neq 0$, then the roots of the equation $(x - \alpha)(x - \beta) + c = 0$ are
(*a*) a, c (*b*) b, c
(*c*) a, b (*d*) $a + c, b + c$

24. If one root of $x^2 - x - k = 0$ is square of the other, then $k =$
(*a*) $2 \pm \sqrt{3}$ (*b*) $3 \pm \sqrt{2}$
(*c*) $2 \pm \sqrt{5}$ (*d*) $5 \pm \sqrt{2}$

25. If the ratio of the roots of $ax^2 + 2bx + c = 0$ is same as the ratio of the roots of $px^2 + 2qx + r = 0$, then
(*a*) $b/ac = q/pr$ (*b*) $b^2/ac = q^2/pr$
(*c*) $2b/ac - q^2/pr$ (*d*) None of these

26. If 8, 2 are the roots of $x^2 + ax + \beta = 0$, and 3, 3 are the roots of $x^2 + ax + b = 0$, then the roots of $x^2 + ax + b = 0$ are
(*a*) 8, – 1 (*b*) – 9, 2
(*c*) – 8, – 2 (*d*) 9, 1

27. If $(1 + m^2)x^2 + 2mcx + (c^2 - a^2) = 0$ has equal roots, then
(*a*) $c^2 = a^2(1 - m^2)$ (*b*) $c^2 = a^2(1 - m)$
(*c*) $a^2 = c^2(m^2 + 1)$ (*d*) $c^2 = a^2(1 + m^2)$

28. Both roots of the equation $(x - b)(x - c) + (x - a)(x - c) + (x - a)(x - b) = 0$
(*a*) positive (*b*) negative
(*c*) real (*d*) None of these

29. If a, b, c are in G.P., the equations $ax^2 + 2bc + c = 0$ and $dx^2 + 2ex + f = 0$ have a common root if $d/a, e/b, f/c$ are in
(*a*) A.P. (*b*) G.P.
(*c*) H.P. (*d*) None of these

30. Let α and β be the roots of the equation $x^2 + x + 1 = 0$. The equation whose roots are α^{19}, β^7 is
(*a*) $x^2 - x - 1 = 0$ (*b*) $x^2 - x + 1 = 0$
(*c*) $x^2 + x - 1 = 0$ (*d*) $x^2 + x + 1 = 0$

31. If $3 + 4i$ is a root of the equation $x^2 + px + q = 0$, then
(*a*) $p = 6, q = 25$ (*b*) $p = 6, q = 1$
(*c*) $p = -6, q = -7$ (*d*) $p = -6, q = 25$

32. The equation whose roots are the reciprocals of the roots of $ax^2 + bx + c = 0$, when $c \neq 0$ is
(*a*) $(a/c)x^2 - bx - 1 = 0$
(*b*) $cx^2 - (b/c)x + a = 0$

(c) $bx^2 + cx + a = 0$
(d) $cx^2 + bx + a = 0$

33. The equation $x - 2/(x - 1) = 1 - 2/(x - 1)$ has
(a) no root
(b) one root
(c) two roots
(d) infinitely many roots

34. The value of p for which the difference between the roots of the equation $x^2 + px + 8 = 0$ is 2 are
(a) ± 2 (b) ± 4
(c) ± 6 (d) ± 8

35. The coefficient of x in the equation $x^2 + px + q = 0$ was taken as 17 in place of 13 and its roots were found to be -2 and -15. The roots of the original equation are
(a) 2, 15 (b) 10, 3
(c) $-10, -3$ (d) -2, 15

36. The real roots of the equation $x^2 + 5|x| + 4 = 0$ are
(a) $(-1, -4)$ (b) $(1, 4)$
(c) $(-4, 4)$ (d) None of these

37. The number of solutions of the equation $|x^2| - 3|x| + 2 = 0$ is
(a) 4 (b) 1
(c) 3 (d) 2

38. If $1, \omega, \omega^2, \ldots, \omega^{n-1}$ are the n roots of unity, then $(1 - \omega)(1 - \omega^2)\ldots(1 - \omega^{n-1})$ equals
(a) 0 (b) 1
(c) n (d) n^2

39. If α, β are the roots of $x^2 + px + q = 0$ and also of $x^{2n} + p^n x^n + q^n = 0$ and if $\frac{\alpha}{\beta}, \frac{\beta}{\alpha}$ are the roots of $x^n + 1 + (x + 1)^n = 0$, then n is
(a) an odd integer
(b) an even integer
(c) any integer
(d) None of these

40. If $x = 2 + i\sqrt{3}$, then the value of, $4x^2 + 8x + 13$ is:
(a) $33 + 24\sqrt{3}\,i$ (b) $45 - 24\sqrt{3}\,i$
(c) $45 + 24\sqrt{3}\,i$ (d) $55 - 24\sqrt{3}\,i$

ANSWERS

1	2	3	4	5	6	7	8	9	10
(d)	(d)	(b)	(d)	(b)	(d)	(d)	(d)	(c)	(d)
11	**12**	**13**	**14**	**15**	**16**	**17**	**18**	**19**	**20**
(b)	(b)	(a)	(c)	(b)	(c)	(d)	(c)	(c)	(b)
21	**22**	**23**	**24**	**25**	**26**	**27**	**28**	**29**	**30**
(b)	(c)	(c)	(c)	(b)	(d)	(d)	(c)	(a)	(d)
31	**32**	**33**	**34**	**35**	**36**	**37**	**38**	**39**	**40**
(d)	(d)	(a)	(c)	(d)	(d)	(a)	(c)	(b)	(a)

SOME SELECTED EXPLANATORY ANSWERS

1. If $\sin x + i\cos 2x$ and $\cos x - i\sin 2x$ are conjugate to each other, then

$$\overline{(\sin x + i\cos 2x)} = \cos x - i\sin 2x$$
$$\Rightarrow \sin x - i\cos 2x = \cos x - i\sin 2x$$
$$\Rightarrow \sin x = \cos x \text{ and } \cos 2x = \sin 2x$$
$$\Rightarrow \tan x = 1 \text{ and } \tan 2x = 1$$

Both these equations cannot hold simultaneously.

2. $$x + iy = \begin{vmatrix} 6i+4 & 0 & 0 \\ 4 & 3i & -1 \\ 20 & 3 & i \end{vmatrix} \text{ By } R_1 \to R_1 + R_2$$

$= (6i + 4)(3i^2 + 3) = 0 \Rightarrow x = 0, y = 0$

3. Given expression
$\Rightarrow 4x + 9y = 3, 2x - 7y = 13$
$\Rightarrow x = 3, y = -1$

5. Given expression $= \left[\frac{(1+i)^2}{(1-i)(1+i)}\right]^n = (i)^n$

which is equal to 1 for least positive integer $n = 4$

6. Given expression
$= 1 - 1 + 1 - 1 + \ldots + (-1)^n$
which cannot be determined, unless n is known.

7. $(1 + \omega - \omega^2)(1 - \omega + \omega^2)$
$= (1 + \omega + \omega^2 - 2\omega^2)(1 + \omega + \omega^2 - 2\omega)$
$= (0 - 2\omega^2)(0 - 2\omega) = 4\omega^3 = 4$

8. $(1 + \omega)^3 - (1 + \omega^2)3$
$= (-\omega^2)^3 - (-\omega)^3 = -\omega^6 + \omega^3 = 0$

10. $x + iy + (a + ib)^3$
$= a^3 + i^3b^3 + 3ai^2b^2 + 3ia^2b$
$= (a^3 - 3ab^2) + i(3a^2b - b^3)$
$\Rightarrow \quad x = a^3 - 3ab^2, y = 3a^2b - b^3$
$\Rightarrow \quad x/a + y/b = (a^2 - 3b^2) + (3a^2 - b^2)$
$= 4(a^2 - b^2).$

11. α, β are complex cube roots of unity.
$\therefore$ let $\alpha = \omega$ and $\beta = \omega^2$
$\therefore xyz = (a + b)(a\gamma + b\beta)(a\beta + b\alpha)$
$= (a + b) [(a^2 + b^2) \alpha\beta + ab(\alpha^2 + \beta^2)]$
$= (a + b) [(a^2 + b^2) \omega^3 + ab(\omega^2 + \omega^4)]$
$= (a + b) [a^2 + b^2 - ab(\omega^2 + \omega)]$
$= (a + b) [a^2 + b^2 - ab] = a^3 + b^3$

12. $(x - 1)^3 + 8 = 0$
$\Rightarrow (x - 1)$
$\Rightarrow (-8)^{1/3}$
$\Rightarrow (-2)(1)^{1/3}$
$\Rightarrow x - 1 = -2.1, -2\omega, -2\omega^2$
$\Rightarrow x = -1, 1 - 2\omega, 1 - 2\omega^2$

13. $x + iy = \sqrt{[(a + ib)/(c + id)]}$
$\therefore x - iy = \sqrt{[(a - ib)/(c - id)]}$
Multiplying and squaring, we get
$(x^2 + y^2)^2 = (a^2 + b^2)/(c^2 + d^2)$

16. Given that $r (\cos\theta + i \sin\theta) = 1 + 0i$
$\therefore \quad r\cos\theta = 1,$
$r \sin\theta = 0$
$\therefore \quad r^2\cos^2\theta + r^2\sin^2\theta = (1)^2 + (0)^2$
$\Rightarrow \quad r = 1$ and $\tan\theta = \frac{0}{1} = 0 \Rightarrow \theta = 0$

17. $\left(\frac{\cos\theta + i\sin\theta}{\cos\theta - i\sin\theta}\right)^4$
$= (\cos\theta + i\sin\theta)^4 . (\cos\theta - i\sin\theta)^{-4}$
$= (\cos\theta + i\sin\theta)^4 (\cos\theta + i\sin\theta)^4$
$= (\cos\theta + i\sin\theta)^8 = \cos 8\theta + i\sin 8\theta$

20. Let α and $1/\alpha$ be the roots.
$\therefore$ Product of roots $= \alpha(1/\alpha) = 1$
$= k/5 \Rightarrow k = 5$

22. $\alpha + \beta = -p, \alpha\beta = q$
$(\omega\alpha + \omega^2\beta)(\omega^2\alpha + \omega\beta)$
$= \alpha^2 + \beta^2 + (\omega^2 + \omega)\alpha\beta = \alpha^2 + \beta^2 - \alpha\beta$
$(\because \omega^3 = 1, 1 + \omega + \omega^2 = 1)$
$= (\alpha + \beta)^2 - 3\alpha\beta = p^2 - 3q$

27. If the roots are equal, then
$(2mc)^2 - 4. (1 + m^2)(c^2 - a^2) = 0$
$\Rightarrow c^2 = a^2 (1 + m^2)$

28. The equation can be written as
$3x^2 - 2(a + b + c)x + (bc + ca + ab) = 0$
$\therefore B^2 - 4AC = 4(a + b + c)^2 - 12 (bc + ca + ab)$
$= 2[(a - b)^2 + (b - c)^2 + (c - a)^2] \geq 0$
$\therefore$ Both the roots are real.

31. Roots are $3 + 4i$ and $3 - 4i$
$\therefore$ Sum $= 3 + 4i + 3 - 4i = -p$
and product $= (3 + 4i)(3 - 4i) = q$
$\therefore p = -6, q = 25$

32. $\alpha + \beta = -b/a, \alpha\beta = c/a.$
The equation whose roots are $1/\alpha$ and $1/\beta$ is
$x^2 - (1/\alpha + 1/\beta)x + (1/\alpha)(1/\beta) = 0$
or $\alpha\beta x^2 - (\alpha + \beta)x + 1 = 0$
or $cx^2 + bx + a = 0$

33. The equation can be written as
$x^2 - 2x + 1 = 0, x \neq 1$
or $(x - 1)^2 = 0$
which is not satisfied, $\because x \neq 1$
Hence the equation has no root

6

Linear Inequality

1. Elementary Properties of Inequalities : If a, b, c be real numbers

(*i*) $a > b$ and $b > c \Rightarrow a > c$

(*ii*) $a > b \Rightarrow a + c > b + c$

(*iii*) $a > b$ and $c > 0 \Rightarrow ac > bc$

(*iv*) $a > b$ and $c > 0 \Rightarrow ac < bc$

(*v*) $a > b$ and $c = 0 \Rightarrow ac = bc$

(*vi*) $a > b \Rightarrow \frac{1}{a} < \frac{1}{b} = (a \neq 0, b \neq 0)$.

2. If $\{a_1 + a_2 + \ldots\ldots + a_n\}$, $\{b_1 + b_2 + \ldots\ldots + b_n\}$ be two sets of real numbers such that $a_i > b_i$ for $i = 1, 2, 3, \ldots. n$, then

$a_1 + a_2 + \ldots.. + a_n > b_1 + b_2 + \ldots.. + b_n$.

3. If $\{a_1, a_2, \ldots.., a_n\}$, $\{b_1, b_2, \ldots.., b_n\}$ be two sets of positive numbers such that $a_i > b_i$ for $i = 1, 2, 3, \ldots. n$, then

$a_1\, a_2 \ldots\ldots. a_n > b_1\, b_2 \ldots\ldots. b_n$.

4. If $a > b > 0$ and n be a positive integer, then $a_n > b_n$.

5. If $a > b > 0$ and n be a negative integer, then $a_n < b_n$.

6. Weiertrass's Inequalities : If $a_1, a_2, \ldots\ldots, a_n$ are all positive numbers less than 1 and $S_n = a_1 + a_2 + \ldots\ldots. + a_n$, then

(*i*) $1 - S_n < (1 - a_1)(1 - a_2)(1 - a_3) \ldots. (1 - a_n) < \frac{1}{1 + S_n}$

(*ii*) $1 + S_n < (1 + a_1)(1 + a_2)(1 + a_3) \ldots.. (1 + a_n) < \frac{1}{1 - S_n}$.

7. Arithmetic Geometric and Harmonic Means: Let $(a) = \{a_1, b_2, \ldots., a_n\}$ be a set of n positive numbers.

The **arithmetic mean** (A.M.) of the set is defined by

$$\frac{a_1 + a_2 + \ldots. + a_n}{n}.$$

The **geometric mean** (G.M.) of the set is defined by

$$\sqrt[n]{(a_1 . a_2 \ldots. a_n)}.$$

The **harmonic mean** (H.M.) of the set is defined by

$$\frac{n}{\frac{1}{a_1} + \frac{1}{a_2} + \ldots. + \frac{1}{a_n}}$$

8. Let $\{a_1, a_2, \ldots\ldots., a_n\}$ be any finite set of positive real numbers, and let A, G and H denote their arithmetic, geometric and harmonic means then

$A \geq G \geq H$, *i.e.*,

(*i*) If $a_1, a_2, \ldots.., a_n$ be n positive numbers, then

$$\frac{a_1 + a_2 + \ldots. + a_n}{n} \geq \sqrt[n]{(a_1 . a_2 \ldots. a_n)}$$

i.e., $A \geq G$;The equality occurs when $a_1 = a_2 = \ldots. = a_n$.

(*ii*) If $a_1, a_2, \ldots.., a_n$ be n positive numbers, then

$$(a_1 a_2 \ldots. a_n)^{\frac{1}{n}} \geq \frac{n}{\frac{1}{a_1} + \frac{1}{a_2} + \ldots \frac{1}{a_n}}$$

i.e., $G \geq H$.

MULTIPLE CHOICE QUESTIONS

1. If a, b are unequal positive numbers, then

(a) $\frac{a^5+b^5}{2} < \frac{a^3+b^3}{2} \cdot \frac{a^2+b^2}{2}$

(b) $\frac{a^5+b^5}{2} = \frac{a^3+b^3}{2} \cdot \frac{a^2+b^2}{2}$

(c) $\frac{a^5+b^5}{2} > \frac{a^3+b^3}{2} \cdot \frac{a^2+b^2}{2}$

(d) None of these

2. If a, b, c, d be all greater than 1, then $(a+1)(b+1)(c+1)(d+1)$ is

(a) $< 4(ab+cd)$ (b) $< 8(ab+cd)$

(c) $< 8(abcd+1)$ (d) None of these

3. If a, b, c, d be positive numbers, not all equal,

then is $(a+b+c+d)\left(\frac{1}{a}+\frac{1}{b}+\frac{1}{c}+\frac{1}{d}\right)$

(a) > 9 (b) < 9

(c) < 16 (d) > 16

4. If $x, y, z \in R^+$ and $x+y+z=1$, then $(1-x)(1-y)(1-z)$ is

(a) $\le \frac{27}{8}$ (b) $\le \frac{8}{27}$

(c) $\le \frac{9}{4}$ (d) $\le \frac{4}{9}$

5. If $x, y, z \in R^+$ a, b, c be all positive and $abc = k^3$, then $(1+a)(1+b)(1+c)$

(a) $\ge (1-k)^3$ (b) $\ge (1+k)^3$

(c) $(1+k)^{\frac{3}{2}}$ (d) $(1-k)^{\frac{3}{2}}$

6. If and $x^3+y^3+z^3=81$, then $x+y+z$ is

(a) ≥ 3 (b) ≤ 3

(c) ≥ 9 (d) ≤ 9

7. Let $x, y, z \in R^+$ and $x^2+y^2+z^2=27$, then

(a) $x^3+y^3+z^3 \le 81$

(b) $x^3+y^3+z^3 \ge 81$

(c) $x^3+y^3+z^3 \ge 80$

(d) None of these

8. If $a, b, c \in R^+$, then $\frac{a}{b+c}+\frac{b}{c+a}+\frac{c}{a+b}$

(a) $\ge \frac{1}{2}$ (b) $\ge \frac{3}{2}$

(c) $\le \frac{3}{2}$ (d) $\le \frac{1}{2}$

9. If $a^2+b^2+c^2=1$, then $ab+bc+ca$ lies in the interval

(a) $\left[\frac{1}{2}, 1\right]$ (b) $\left[0, \frac{1}{2}\right]$

(c) $\left[-\frac{1}{2}, 1\right]$ (d) $\left[-1, \frac{1}{2}\right]$

10. If $a, b, c \in R^+$, then $\left(\frac{a+b+c}{n}\right)^{a+b+c+\ldots+k}$

(a) $> a^a.b^b.c^c.\ldots.k^k$

(b) $< a^a.b^b.c^c.\ldots.k^k$

(c) $= a^a.b^b.c^c.\ldots.k^k$

(d) None of these

11. If n quantities a, b, c, ... k are in A.P., then $\frac{1}{a}+\frac{1}{b}+\frac{1}{c}+\ldots+\frac{1}{k}$ is

(a) $< \frac{2n}{a+k}$ (b) $> \frac{2n}{a+k}$

(c) $< \frac{n}{2(a+k)}$ (d) $> \frac{n}{2(a+k)}$

12. If x is a +ve proper fraction, then $(1+x)^{1+x}.(1-x)^{1-x}$

(a) > 1 (b) < 1

(c) > -1 (d) < -1

13. Which of the following statement is true?

(a) $(n!)^2 = n^n$ (b) $(n!)^2 < n^n$

(c) $(n!)^2 > n^n$ (d) None of these

14. If $x + y + z = 1$, then $\frac{1}{x}+\frac{1}{y}+\frac{1}{z}$ is

(a) > 9 (b) < 9

(c) $> \frac{1}{9}$ (d) $< \frac{1}{9}$

15. If x, y, and z are real and different and $u = x^2 + 4y^2 + 9z^2 - 6yz - 3zx - 2xy$ then u is always

(a) non-negative

(b) zero

(c) non-positive

(d) None of these

ANSWERS

1	2	3	4	5	6	7	8	9	10
(c)	(c)	(d)	(b)	(b)	(d)	(b)	(b)	(c)	(b)
11	**12**	**13**	**14**	**15**					
(b)	(a)	(c)	(a)	(a)					

SOME SELECTED EXPLANATORY ANSWERS

1. **Case I :** Let $a > b > 0$. Then $a^3 > b^3$, $a^2 > b^2$ and $(a^3 - b^3)(a^2 - b^2) > 0$

Case II : Let $b > a > 0$. Then $a^3 < b^3$, $a^2 < b^2$ and $(a^3 - b^3)(a^2 - b^2) > 0$

$(a^3 - b^3)(a^2 - b^2) > 0$ if a and b are positive,

or $a^5 + b^5 > a^3b^2 + b^3a^2$ $a \neq b$

or $a^5 + b^5 > a^5 + b^5 + a^3b^2 + a^2b^3$

$= (a^3 + b^3)(a^2 + b^2)$

or $\frac{a^5+b^5}{2} > \frac{a^3+b^3}{2}.\frac{a^2+b^2}{2}$.

2. $(a - 1)(b - 1) > 0$ since $a - 1 > 0$, $b - 1 > 0$

or $ab + 1 > a + b$

or $2(ab + 1) > ab + 1 + a + b$

$= (a + 1)(b + 1)$

$(a + 1)(b + 1) < 2(ab + 1)$ ...(i)

Similarly $(c + 1)(d + 1) < 2(cd + 1)$ ...(ii)

Now $ab > 1$, $cd > 1$

from (i)

$(ab + 1)(cd + 1) < 2(abcd + 1)$ ...(iii)

$(a + 1)(b + 1)(c + 1)(d + 1) < 4(ab + 1)(cd + 1)$

from (i) and (ii) $< (abcd + 1)$ from (iii)

or $(a + 1)(b + 1)(c + 1)(d + 1) < 8(abcd + 1)$.

5. We have

$(1 + a)(1 + b)(1 + c) = 1 + \Sigma a + \Sigma ab + abc$

Now $\frac{\Sigma a}{3} \geq \sqrt[3]{abc}$ or $\Sigma a \geq 3k$

and $\frac{\Sigma ab}{3} \geq \sqrt[3]{a^2b^2c^2}$ or $\Sigma ab \geq 3k^2$

$\therefore$ $(1 + a)(1 + b)(1 + c) \geq 1 + 3k + 3k^2 + k^3 = (1 + k)^3$.

9. $(a + b + c) \geq 0$

$\Rightarrow a^2 + b^2 + c^2 + 2(ab + bc + ca) \geq 0$

$\Rightarrow ab + bc + ca \geq -\frac{1}{2}$ $[a^2 + b^2 + c^2 = 1]$

Again $(b - c)^2 + (c - a)^2 + (a - b)^2 \geq 0$

$\Rightarrow 2(a^2 + b^2 + c^2) - 2(ab + bc + ca) \geq 0$

$\Rightarrow ab + bc + ca \leq a^2 + b^2 + c^2 = 1$

Hence $-\frac{1}{2} \leq ab + bc + ca \leq 1$.

15. We have

$$u = \frac{1}{2}\left[2x^2 + 8y^2 + 18z^2 - 12yz - 6zx - 4xy\right]$$

$$= \frac{1}{2}\left[(x-2y)^2 + (2y-3z)^2 + (3z-x)^2\right] \geq 0.$$

7

Permutation and Combination

PERMUTATION

Each of arrangements which can be made by taking some or all of a number of things is called a permutation. If the arrangement is in a line, then it is a linear permutation, if it is in a circle, then it is a circular permutation.

Fundamental Theorem

I. If a certain thing can be done in m ways and when it has been done in any one these ways, a second thing can be done in n ways in which two things can be done in mn ways.

The number of permutations of n different thing taking r at a time

$$^nP_r = \frac{n!}{(n-r)!} = n(n-1)(n-2)...(n-r+1)$$

where $(0 \le r \le n)$, $^nP_0 = 1$, $^nP_1 = n$

Note: *In the above, repetition was not allowed, so that if we could fill the first place in n ways, then second can be filled in (n –1), third in (n –2) ....and rth in (n –r + 1) ways. Hence, total place could be filled in n(n – 1)(n – 2) ...(n – r + 1) ways.*

Now, if repetition is allowed, then each of the r places can be filled in n ways. Hence, by fundamental theorem all the r places can be filled in $n.n.n...r$ times *i.e.,* n^r ways.

Number of permutations of n different things taken all at a time :

$$^nP_n = n(n-1)(n-2)...[n-(n-1)]$$
$$= n(n-1)(n-2)...3.2.1 = n!$$

COMBINATION

Each of the groups or selections which can be made by taking some or all a number of things without reference to the order of things in each group is called combination.

Number of combinations of n different things taken r at a time :

$$^nC_r = \frac{n!}{(n-r)!r!} = \frac{^nP_r}{r!}$$

Number of combinations of n different things taken all at a time is

$$^nC_n = \frac{n!}{n!(n-n)!} = \frac{1}{0!} = 1$$

Number of permutations of n things taken all at a time when p of them are similar of one type, q of them are similar of another type, r of them are simliar of a third type and the remaining $n-(p+q+r)$ are all different is

$$\frac{n!}{p!q!r!}$$

If some or all of n things be taken at a time, then the number of combinations is given by $^nC_1 + {^nC_2} + ... + {^nC_n} = 2^n - 1$

(*i*) $^nC_r = {^nC_{n-r}}$; $^nC_0 = {^nC_n} = 1$, $^nC_1 = n$

(*ii*) $^nC_x = {^nC_y} \Leftrightarrow x = y$ or $x + y = n$

(*iii*) $^nC_r + {^nC_{r-1}} = {^{n+1}C_r}$

The number of combinations of n distinct things taken r at a time when k particular things always occur is:

$$^{n-k}C_{r-k}$$

- If k particular objects never occur, then no. of combinations of n distinct objects taken r at a time is

$$^{n-k}C_r$$

The number of permutations of n distinct things taken r at a time when k particular things always occur is

$$^{n-k}C_{r-k}.\ r!$$

If k particular things never occur, then the no. of permutations of n distinct things taken r at a time is

$$^{n-k}C_r.r!$$

Combinations or Selection Out of Identical Things

(*i*) The number of combinations of r things out of n identical things is 1.

(*ii*) The total number of selections of zero, one, two....or n things from n identical things is $n + 1$.

(*iii*) The number of selections of some or all out of $(p + q + r)$ things where p are alike of one kind, q are alike of second kind and r are alike of third kind is $\{(p + 1)(q + 1)(r + 1) - 1\}$.

Circular Permutation

(*i*) **Arrangements round a circular table**

A circular table has no head and arrangements like those in the figure are considered indentical.

Here, having fixed one thing, the remaining $(n - 1)$ things can be arranged round the table in $(n - 1)!$ ways.

(*ii*) Arrangements of heads around a circular wire.

A circular differs from a circular table because when we turn it over we see that the other side presents an arrangement of different beads different from that on the first side.

If the wire on the left is turned over we obtain the arrangement on the right. We can see that the only two different arrangements of three coloured beads on a circular wire can be found on opposite side of the same wire, *i.e.*, there is really one arrangement.

Therefore, n beads on a circular wire can be arranged in $\frac{1}{2}(n-1)!$ different ways.

Arrangements in Groups

(*i*) The number of ways in which n different things can be arranged in r different groups is

$$n!\,^{n-1}C_{r-1}$$

as blank groups are not permitted.

(*ii*) The number of ways in which n things of the same kind can be distributed into r different groups is

$$^{(n+r-1)}C_{r-1} \text{ or, } ^{n-1}C_{r-1}$$

according as blank groups are not permitted.

The greatest value of ^nC_r

(*i*) If n is even, ^nC_r is greatest when $r = n/2$

(*ii*) If n is odd, ^nC_r is greatest when

$$r = (n - 1)/2 \text{ or } (n + 1)/2$$

MULTIPLE CHOICE QUESTIONS

1. How many words can be formed out of the word LADIES so that the vowels are never together?

(*a*) 144 (*b*) 696
(*c*) 720 (*d*) 576

2. How many different words can be formed from the word DAUGHTER so that ending and beginning letters are consonants?

(*a*) 7200 (*b*) 14400
(*c*) 360 (*d*) None of these

3. How many words can be formed by taking 4 letters at a time out of the letters of the word CONDITION?

(*a*) 610 (*b*) 858
(*c*) 738 (*d*) 126

4. The number of odd integers between 1000 and 9999 with no digit repeated is

(*a*) 3333 (*b*) 2120
(*c*) 2240 (*d*) 3331

5. How many 5-letter words can be formed that contain at least one of the 5 vowels ?

(*a*) $\binom{16}{5} - 1$ (*b*) $26^5 - 21^5$

(*c*) $\sum_{r=1}^{5} \binom{21}{5-r}\binom{5}{r}$ (*d*) None of these

6. A student is allowed to select at the most n books from a collection of $(2n+1)$ books. If the total number of ways in which he can select a book is 63, then the value of n is
(a) 1 (b) 3
(c) 7 (d) 5

7. To fill 12 vacancies there are 25 candidates of which 5 are from scheduled castes. If 3 of the vacancies are reserved for scheduled caste candidates while the rest are open to all, the number of ways in which the selection can be made is
(a) 1679600 (b) 4974200
(c) 994840 (d) None of these

8. If $n_1, n_2, ..., n_p$ are p positive integers, whose sum is an even number, then the number of odd integers among them is
(a) even (b) odd
(c) even and odd (d) None of these

9. In a class of 10 students there are 3 girls A, B and C. In how many different ways can they be arranged in a row such that number two of the three girls are consecutive?
(a) 720 × 7! (b) 336 × 7!
(c) 56 × 7! (d) 120 × 7!

10. 25 points, no 3 collinear, are given in a plane. How many triangles do they determine?
(a) 2300 (b) 75
(c) 1975 (d) 242

11. There are 3 copies each of 4 different books. In how many ways can they be arranged in a shelf?
(a) $12!/(4!)^3$ (b) $12!/(3!)^4$
(c) 12! (d) None of these

12. Five balls of different colours are to be placed in three boxes of different size. Each box can hold all five balls. In how many different ways can the balls be placed so that no box remains empty?
(a) 160 (b) 150
(c) 140 (d) 190

13. The number of non-negative integral solutions of $x_1 + x_2 + x_3 + x_4 \le n$ is
(a) ${}^{n+3}C_3$ (b) ${}^{n+4}C_4$
(c) ${}^{n+5}C_5$ (d) ${}^{n+4}C_n$

14. The number of numbers that can be formed with the digits 1, 2, 3, 4, 3, 2, 1 so that the odd digits always occupy odd places is
(a) 24 (b) 18
(c) 12 (d) None of these

15. A polygon has 44 diagonals, then the number of its sides is
(a) 11 (b) 7
(c) 8 (d) None of these

16. A question paper is divided into two parts A and B. Each part contains 5 questions. The number of ways in which a candidate can answer 6 questions selecting at least two questions from each part is
(a) 50 (b) 100
(c) 200 (d) None of these

17. The total number of 3-digit number that can be composed from the digits 1, 2, 3..., 9, when the repetition of digits is not allowed, is
(a) 224 (b) 280
(c) 324 (d) 405

18. The number of combinations of n letters out of $3n$ letters ; of which n are a, n are b and rest n are alike, is given by
(a) 2^{n-1} (b) $n.2^n$
(c) $(n + 2).2^{n-1}$ (d) None of these

19. Given five different green dyes, four different blue dyes and three different red dyes, the number of combinations that can be chosen, taking at least one green and one blue dye are
(a) 31 (b) 15
(c) 8 (d) 3720

20. Five balls of different colours are to be placed in 3 boxes of different sizes. Each box can hold all the 5 balls. The number of different ways of placing these balls in these boxes so that no box is empty are
(a) 90 (b) 15
(c) 150 (d) None of these

ANSWERS

1	2	3	4	5	6	7	8	9	10
(d)	(b)	(c)	(c)	(b)	(b)	(b)	(a)	(b)	(a)
11	**12**	**13**	**14**	**15**	**16**	**17**	**18**	**19**	**20**
(b)	(b)	(b)	(b)	(a)	(c)	(a)	(c)	(d)	(c)

SOME SELECTED EXPLANATORY ANSWERS

1. In the word LADIES, there are three vowels A, E and I. If we consider all the three vowels as a single letter, then there are only 4 letters which can be arranged in 4! ways but three vowels can be arranged in 3! ways.

$\therefore$ The words in which 3 vowels come together

$= 4! \times 3!$

$\therefore$ The words in which vowels are never together = Total words – words in which vowels are together

$= 6! - 4! \times 3! = 720 - 144 = 576$

2. Here, total letters are 8; 3 vowels and 5 consonants. Here 2 consonants can be chosen in ${}^5C_2$ ways and these two consonants can be put in 2! ways. The remaining 6 letters can be arranged in 6! ways.

The words beginning and ending letters with consonant $= {}^5C_2 \times 2! \times 6! = 14400$.

4. The numbers between 1000 and 9999 must contain 4 digits. Since, the number is odd, so its last digit must be one of the five digits 1, 3, 5, 7 or 9. We can choose last digit in 5 ways. The first digit can not be 0, hence first digit of the number can be chosen in 8 ways, second digit can be chosen in 8 ways (because it can be 0), third in 7 ways.

$\therefore$ Total no. of odd numbers between 1000 and $9999 = 8 \times 8 \times 7 \times 5 = 2240$

5. Total number of 5-letters words

$= 26 \times 26 \times 26 \times 26 \times 26 = (26)^5$

The words which have no vowels

$= 21 \times 21 \times 21 \times 21 \times 21 = (21)^5$

The words which have at least one vowel

$= (26)^5 - (21)^5$

7. 3 reserved vacancies can be filled in ${}^5C_3$ ways. Now the remaining candidates are 22 and remaining vacancies are 9, which can be filled in ${}^{22}C_9$ ways.

$\therefore$ Total no. of ways in which selection can be made $= {}^5C_3 \times {}^{22}C_9 = 4974200$

8. Given the sum of given numbers is even, it means that the no. of odd numbers in the given numbers is even.

$\therefore$ The sum of odd integers in the given integers will be even.

9. Since, the no. of boys is 7. Therefore, there are 8 places among boys which can be filled by 3 girls in ${}^8P_3$ ways and the 7 boys can be arranged in 7! ways

$\therefore$ Required no. of arrangements

$= {}^8P_3 \times 7! = 336 \times 7!$

10. A triangle is formed from 3 points

$\therefore$ Total no. of triangles $= {}^{25}C_3 = 2300$

16. The candidate can select 6 questions as under

(*i*) 2 questions from part A + 4 questions from part B.

(*ii*) 3 questions from part A + 3 questions from part B.

(*iii*) 4 questions from part A + 2 questions from part B.

$\therefore$ Total number of ways

$= {}^5C_2 \times {}^5C_4 + {}^5C_3 \times {}^5C_3 + {}^5C_4 \times {}^5C_2$

$= 10 \times 5 + 10 \times 10 + 5 \times 10 = 200$

8

Binomial Theorem

BINOMIAL EXPRESSION

Any expression which contains only two terms is known as binomial expression. The formula by which any positive integral power of a binomial expression can be expanded in the form of a series is known as binomial series.

(*i*) $(x+a)^n = {}^nC_0x^n + {}^nC_1x^{n-1}a + {}^nC_2x^{n-2}a^2 + \cdots + {}^nC_r\,x^{n-r}a^r + \ldots + {}^nC_na^n$

(*ii*) $(x-a)^n = {}^nC_0x^n - {}^nC_1x^{n-1}a + {}^nC_2x^{n-2}a^2 + \ldots + (-1)^r\,{}^nC_r\,x^{n-r}a^r + \ldots + (-1)^n a^n$

I. General Term

The general term of $(x-a)^n$ is $(r+1)^{th}$ term and denoted by T_{r+1}.

$\therefore T_{r+1} = {}^nC_rx^{n-r}a^r$ for $(x+a)^n$

$T_{r+1} = {}^nC_rx^{n-r}(-a)^r$ for $(x-a)^n$

II. Middle Term (s)

Since the number of terms in the expansion of $(x+a)^n$ is $(n+1)$

(*i*) If n is **even** there will be **only one middle** term, *i.e.*, $\left(\frac{n}{2}+1\right)$th

(*ii*) If n is **odd,** there will be **two middle** terms, *i.e.*, $\left(\frac{n+1}{2}\right)$th and $\left(\frac{n+3}{2}\right)$th

III. Greatest Term in $(x+a)^n, x > 0$

$$\frac{T_{r+1}}{T_r} = \frac{{}^nC_rx^{n-r}a^r}{{}^nC_{r-1}x^{n-r+1}a^{r-1}}$$

$$= \frac{n-r+1}{r}.\frac{a}{x}$$

$T_{r+1} \geq T_r$ as $\frac{n-r+1}{r}.\frac{a}{x} \geq 1$

$$\Rightarrow r \leq \frac{n+1}{\frac{x}{a}+1} = m(\text{say})$$

IV. Independent Term

Let T_{r+1} be the term independent of x in $(x+a)^n$, equate to zero the index of x and find the value of r.

MULTIPLE CHOICE QUESTIONS

1. The coefficient of x^n in the expansion of $(1-2x+3x^2-4x^3+\ldots)^{-n}$ is

(*a*) $\frac{(2n)!}{(n!)^2}$ (*b*) $\frac{(2n)!}{n!}$

(*c*) $\frac{(2n)!}{(n!)^3}$ (*d*) None of these

2. In the expansion of $(1+x)^{50}$, the sum of the coefficients of odd powers of x is

(*a*) 0 (*b*) 2^{49}

(*c*) 2^{51} (*d*) None of these

3. Find the term independent of x in the expansion of $(2x^2-1/x)^{12}$.

(a) 7910 (b) 7920
(c) 7930 (d) 7900

4. If the coefficients of x^7 and x^8 in $\left[2+\frac{x}{3}\right]^n$ are equal, find the value of n.
(a) 45 (b) 65
(c) 75 (d) 55

5. Find the constant term in the expansion of $\left(x^3-\frac{1}{x^2}\right)^{15}$.
(a) ${}^{15}C_9$ (b) $-{}^{15}C_9$
(c) ${}^{15}C_6$ (d) ${}^{14}C_6$

6. Let n be a positive integer. If the coefficient of 2nd, 3rd, 4th terms in the expansion of $(1+x)^n$ are in A.P., then find the value of n.
(a) 7 (b) 8
(c) 4 (d) 5

7. $C_1+2C_2+3C_3+....+nC_n=$
(a) 2^{n-1} (b) $n\cdot 2^{n-1}$
(c) $n2^n$ (d) 2^n

8. $C_0+2C_1+3C_2+...+(n+1)C_n=$
(a) $(n+1)2^n$ (b) $(n+1)2^{n-1}$
(c) $(n+2)2^{n-1}$ (d) $(n+2)2^{n-2}$

9. The term independent of x in $\left(\frac{3}{2}x^2-\frac{1}{3x}\right)^9$ is
(a) $\frac{7}{18}$ (b) $\frac{5}{18}$
(c) $\frac{11}{18}$ (d) $\frac{13}{18}$

10. The given positive integers $r>1$, $n>2$ and coefficients of $(3r)$th and $(r+2)$th terms in the expansion of $(1+x)^{2n}$ are equal, then
(a) $n=2r$ (b) $n=3r$
(c) $n=2r-1$ (d) None of these

11. The number of integral terms in the expansion of $\left(\sqrt{3}+\sqrt[8]{5}\right)^{256}$ is
(a) 32 (b) 33
(c) 34 (d) 35

12. If the sum of middle terms is S in the expansion of $\left(2a-\frac{a^2}{4}\right)^9$, then the value of S is:
(a) $\left(\frac{63}{32}\right)a^{14}(8+a)$ (b) $\left(\frac{63}{32}\right)a^{13}(8+a)$
(c) $\left(\frac{63}{32}\right)a^{14}(8-a)$ (d) $\left(\frac{63}{32}\right)a^{13}(8-a)$

13. If the rth term in the expansion of $\left(\frac{x}{2}-\frac{2}{x^2}\right)^{10}$ contains x^4, then r is equal to :
(a) 4 (b) 3
(c) 2 (d) None of these

14. If the coefficients of x^2 and x^3 in the expansion of $(3+kx)^9$ are equal, then the value of k is :
(a) $-\frac{9}{7}$ (b) $\frac{9}{7}$
(c) $\frac{7}{9}$ (d) None of these

15. If the expansion of $\left(x^2+\frac{2}{x}\right)^n$ for positive integer n has 13th term independent of x, then n is:
(a) 16 (b) 18
(c) 20 (d) None of these

ANSWERS

1	2	3	4	5	6	7	8	9	10
(a)	(b)	(b)	(d)	(b)	(a)	(b)	(c)	(a)	(a)
11	12	13	14	15					
(b)	(d)	(b)	(b)	(b)					

SOME SELECTED EXPLANATORY ANSWERS

2. $(1+x)^{50} = 1 + {}^{50}C_1 x + {}^{50}C_2 x^2 + {}^{50}C_3 x^3 + ... + {}^{50}C_{50} x^{50}$

Putting $x = 1$ and -1 and adding, we get

$2^{50} = 1 + {}^{50}C_1 + {}^{50}C_2 + {}^{50}C_3 + ... + {}^{50}C_{50}$ ...(*i*)

$0 = 1 - {}^{50}C_1 + {}^{50}C_2 - {}^{50}C_3 + ... + {}^{50}C_{50}$ ...(*ii*)

Subtracting (*ii*) from (*i*), we get

$2^{50} = 2\left({}^{50}C_1 + {}^{50}C_3 + ... {}^{50}C_{49}\right)$

$\Rightarrow {}^{50}C_1 + {}^{50}C_3 + ... {}^{50}C_{49} = 2^{49}$

4. Coefficients of T_8 and T_9 are equal

$\therefore {}^nC_7 2^{n-7}\left(\frac{1}{3}\right)^7 = {}^nC_8 2^{n-8}\left(\frac{1}{3}\right)^8$

$\Rightarrow \frac{n!}{7!(n-7)!}.2 = \frac{n!}{8!(n-8)!}.\frac{1}{3}$

or, $\frac{2}{n-7} = \frac{1}{8.3} \Rightarrow n = 55$

5. Constant term means term which is independent of *x*.

$T_{r+1} = {}^{15}C_r \left(x^3\right)^{15-r}\left(-\frac{1}{x^2}\right)^r$

$= (-1)^r \, {}^{15}C_r x^{45-5r}$

The index of *x* is zero, *i.e.*, $45 - 5r = 0$

$\Rightarrow r = 9$

$\therefore T_{10} = (-1)^9 \, {}^{15}C_9 = -{}^{15}C_9$

10. $T_{3r} = {}^{2n}C_{3r-1} \cdot x^{3r-1}$

$T_{r+2} = {}^{2n}C_{r+1} \cdot x^{r+1}$

${}^{2n}C_{3r-1} = {}^{2n}C_{r+1}$ (given)

$\therefore 3r - 1 + r + 1 = 2n$

(since ${}^nC_r = {}^nC_8$ then $r + s = n$)

11. *r*th term of $\left(\sqrt{3} + \sqrt[8]{5}\right)^{256}$

i.e., $T_{r+1} = C_r(3)^{(256-r)/2}(5)^{r/8}$

$\frac{256-r}{2}$ and $\frac{r}{8}$ are both positive integers.

$r = 0, 8, 16, 24, 32, ..., 256$

Hence total number of terms are 33.

13. Here $T_r \; {}^{10}C_{r-1}\left(\frac{x}{2}\right)^{10-(r-1)} = \left(-\frac{2}{x^2}\right)^{r-1}$

$= {}^{10}C_{r-1}\left(\frac{x}{2}\right)^{11-r} (-2)^{r-1} x^{13-3r}$

Since, rth term contains x^4, so that

$13 - 3r = 4$

or $3r = 9 \Rightarrow r = 3$

15. Here, $T_{13} = {}^nC_{12} (x^2)^{n-12} \left(\frac{2}{x}\right)^{12}$

$= {}^nC_{12} (2)^{12} (x)^{2n-36}$

If 13th term is independent of *x*, then

$2n - 36 = 0$ or $n = 18$.

9

Sequences and Series

1. Sequence: A sequence is a function defined on the domain N (the set of natural numbers) with co-domain R or C. If R is the co-domain, it is called real sequence and if C is the co-domain, it is called complex sequence.

If f: N $\to$ S is a sequence, then it is usually written as $\{f(1), f(2), ...\} \equiv \{f(x)\}$.

If $t_n = f(n)$, then sequence is also written as $\{t_1, t_2, t_3 ...\} = \{t_n\}$.

2. Series: If $\{t_1, t_2, t_3, ...\}$ is a sequence, then $S = t_1 + t_2 + t_3 + ...$ is called the corresponding series.

3. Arithmetic Progression: An arithmetic progression (A.P.) is a sequence whose terms increase or decrease by a fixed number. Thus if sequence $\{t_1, t_2, t_3, ...\}$ is such that $t_n - t_{n-1}$ is a constant for all $n \in$ N, it is an A.P. This fixed number is called the common difference of the A.P.

4. *n*-th Term and Sum of *n*-terms of an A.P.: If a is the first term and d the common difference of an A.P. then its n-th term t_n is given by $t_n = a + (n - 1)\, d$, and the sum of n terms, denoted by S_n is given by

$$S_n = \frac{n}{2}[2a + (n-1)\,d] = \frac{n}{2}(a + l),$$ where a is the first term and l is the last term.

5. Arithmetic Mean (A.M.): Arithmetic mean (A.M.), A of any two numbers a and b is given by $(a + b)/2$ and so a, A, b are in A.P. If $x_1, x_2 ..., x_n$ are n numbers then their A.M. is given by $(x_1 + x_2 + ... + x_n)/n$.

6. The *n* Arithmetic Means: The numbers $A_1, A_2, ..., A_n$ are said to be n arithmetic means between a and b if $a, A_1, A_2, ..., A_n, b$ is an A.P. Obviously, $b = t_{n+2} = a + (n + 1)\, d$,

$$\therefore d = \frac{b-a}{n+1} \text{ and so } A_1 = a + \frac{b-a}{n+1}, ..., A_r = a + \frac{r(b-a)}{n+1}.$$

7. Geometric Progression: A geometric progression (G.P.) is a sequence of numbers such that the first term is non-zero and each of the succeeding term is equal to the preceding one multiplied by a certain non-zero number, that is, constant for the given sequence. This non-zero number is called the common ratio of G.P.

8. *n*-th Term and Sum of *n* Terms of a G.P.: If a is the first term and r its common ratio of a G.P., then n-th term $= t_n = ar^{n-1}$ and sum of n terms, denoted by S_n is given by

$$S_n = \frac{a(r^n - 1)}{r-1} = \frac{a(1-r^n)}{1-r} \text{ if } r \neq 1.$$

9. Sum of Infinite G.P.: If $-1 < r < 1$, *i.e.*, $|r| < 1$, then the sum of the infinite G.P.

$$a + ar + ar^2 + ... = a/(1 - r).$$

10. Geometric Mean (G.M.): If $a, b > 0$, then the geometric mean of a and $b = \sqrt{(ab)}$.

If $a_1, a_2 ..., a_n > 0$, then their geometric mean is given by $(a_1 \,.\, a_2 .... a_n)^{1/n}$.

If n geometric means $g_1, g_2 ..., g_n$ are to be inserted between two positive real numbers a and b then $a, g_1, g_2, ..., g_n, b$ are in G.P.

so $b = ar^{n-1}$, *i.e.*, $r = (b/a)^{1/(n+1)}$ and then $g_1 = ar, g_2 = ar^2, ..., g_n = ar^n$.

11. Harmonic Progression: The sequence $x_1, x_2, x_3, ..., x_n ...$, where $x_n \neq 0$ for each n, is said to be in H.P. if the sequence $1/x_1, 1/x_2, 1/x_3, ...$ is an A.P. It should be noted that no term of H.P. can be zero.

12. *n*-th Term of a H.P.: If $x_1, x_2, x_3, ...$ is a H.P., then the corresponding A.P. is $1/x_1, 1/x_2, ...$ Hence $a = a/x_1$ and $d = 1/x_2 - 1/x_1$, then x_n

$$(n\text{-th term of H.P.}) = \frac{1}{a+(n-1)d}$$

13. Harmonic Mean (H.M.): If a and b are two non-zero numbers, then the harmonic mean of a and b, denoted by H is given by

$$\frac{1}{H} = \frac{1}{2}\left(\frac{1}{a}+\frac{1}{b}\right) \text{ or } H = \frac{2ab}{a+b}.$$

If a_1, a_2, a_n, an be n non-zero numbers, then their harmonic mean H is given by

$$\frac{1}{H} = \frac{1}{2}\left(\frac{1}{a_1}+\frac{1}{a_2}...+\frac{1}{a_n}\right).$$

If $H_1, H_2, ..., H_n$ are n harmonic means between two non-zero numbers a and b, then

$$\frac{1}{a}, \frac{1}{H_1}, \frac{1}{H_2}, \frac{1}{H_3}, ..., \frac{1}{H_n}, \frac{1}{b} \text{ are in A.P.}$$

Let $x_1, x_2, x_3 ...$ be in A.P. and $y_1, y_2, y_3 ...$ in G.P. then $x_1y_1, x_2y_2, x_3y_3, ...$ is said to be an arithmetico geometric sequence (A.G.S.). An A.G.S. has the form $x, (x + d)r, (x + 2d)r^2, ...$ (1)

14. Formulae for $\Sigma n, \Sigma n^2, \Sigma n^3$:

(*a*) $\Sigma n = \sum_{r=1}^{n} r = 1, 2, 3, + ... + n = \frac{1}{2}n(n+1)$

(*b*) $\Sigma n^2 = \sum_{r=1}^{n} r^2 = \frac{1}{6}n(n+1)(2n+1)$.

(*c*) $\Sigma n^3 = \sum_{r=1}^{n} r^3 = \left[\frac{1}{2}n(n+1)\right]^2$.

15. Relation Between A.M., G.M. and H.M.: If A, G and H are respectively the A.M., G.M. and H.M. of two numbers a and b then

(*a*) $AH = G^2$ (*b*) $A \geq G \geq H$

MULTIPLE CHOICE QUESTIONS

1. If $1, \log_y x, \log_z y, -15\log_x z$ are in A.P., then

(*a*) $z = x^3$ (*b*) $y = z^{-3}$

(*c*) $x^2 = y^{-1}$ (*d*) None of these

2. The number of identical terms in two A.P's 3, 7, 11, ... 407 and 2, 9, 16, ...709 is

(*a*) 10 (*b*) 11

(*c*) 12 (*d*) 13

3. If the A.M. between a and b is twice as great as their G.M., then $a : b$ is equal to

(*a*) $(2+\sqrt{3}):(2-\sqrt{3})$

(*b*) $(2-\sqrt{3}):(2+\sqrt{3})$

(*c*) $\sqrt{3}-2:2+\sqrt{3}$

(*d*) None of these

4. The harmonic mean of two numbers is 4, their A.M., A and G.M., G satisfy the relation, $2A + G^2 = 27$. The numbers are

(*a*) 1, 3 (*b*) 1, 4

(*c*) 3, 6 (*d*) None of these

5. The value of $x+y+z$ is 15, if a, x, y, z, b are in A.P., while the value of $\frac{1}{x}+\frac{1}{y}+\frac{1}{z}$ is 5/3, if a, x, y, z, b are in H.P., then

(*a*) $a=9, b=1$ (*b*) $a=2, b=3$

(*c*) $a=5, b=7$ (*d*) None of these

6. If $a^{1/x} = b^{1/y} = c^{1/z}$ and a, b, c be in G.P. then x, y, z are in

(*a*) A.P. (*b*) G.P.

(*c*) H.P. (*d*) None of these

7. If the mth, n^{th} and p^{th} terms of an A.P. and G.P. be equal and be respectively x, y and z, then $x^{y-z}y^{z-x}z^{x-y}$ is equal to

(*a*) 0

(*b*) 1

(*c*) $(x+y+z)(x+y+z)$

(*d*) None of these

8. If a, b, c are in A.P. and a^2, b^2, c^2 in H.P., then

(*i*) a, b, c are in G.P.

(*ii*) $-a/2$, b, c are in G.P.

(*iii*) $a = b = c$

Which one among the following is true?

(*a*) Only (*i*) and (*ii*) (*b*) Only (*ii*) and (*iii*)

(*c*) Only (*i*) and (*iii*) (*d*) None of these

9. If the mth term of an H.P. is n and n^{th} term is m, then $(m+n)^{th}$ term is

(*a*) $\frac{mn}{m-n}$ (*b*) $\frac{m-n}{m+n}$

(*c*) $\frac{mn}{m+n}$ (*d*) None of these

10. If $\frac{1}{b-a}+\frac{1}{c-a}=\frac{1}{a}+\frac{1}{c}$, then a, b, c are in

(*a*) A.P. (*b*) G.P.

(*c*) H.P. (*d*) H.P. and G.P. both

11. The sum of first n terms of the series $1^2+2.2^2+3^2+2.4^2+5^2+2.6^2+\ldots$, when n is even, is

(*a*) $n(n+1)/2$ (*b*) $n^2(n+1)^2/2$

(*c*) $n(n+1)^2/2$ (*d*) None of these

12. $1+3+7+15+31+\ldots$ to n terms is equal to

(*a*) 2^n-n+2 (*b*) 2^n-2

(*c*) $2^{n+1}-n-2$ (*d*) None of these

13. The sum of the series

$$\frac{1}{5.8}+\frac{1}{8.11}+\frac{1}{11.14}+\ldots\infty, \text{ is}$$

(*a*) $\frac{1}{5}$ (*b*) $\frac{1}{8}$

(*c*) $\frac{1}{11}$ (*d*) $\frac{1}{15}$

14. If $\frac{a^n+b^n}{a^{n-1}+b^{n-1}}$ is the arithmetic mean between two numbers a and b, then value of n is

(*a*) 0 (*b*) 1

(*c*) -1 (*d*) None of these

15. If p, q, r are in A.P. then pth, qth and rth terms of any G.P. are in

(*a*) A.P. (*b*) G.P.

(*c*) H.P. (*d*) None of these

16. The sum of the series

$$1+\frac{1+2}{2!}+\frac{1+2+3}{3!}+\frac{1+2+3+4}{4!}+\ldots\infty \text{ is}$$

(*a*) e (*b*) $\frac{3e}{2}$

(*c*) $2e$ (*d*) $\frac{5e}{2}$

Note:

Remember $\sum_{n=0}^{\infty}\frac{1}{n!}=\sum_{n=0}^{\infty}\frac{1}{(n-1)!}=\sum_{n=0}^{\infty}\frac{1}{n-2}=e$

17. The value of $1+\frac{2^3}{2!}+\frac{3^3}{3!}+\frac{4^3}{4!}+\ldots\infty$ is

(*a*) e^3 (*b*) e^2

(*c*) $3e$ (*d*) $5e$

18. If $S_n=\frac{1^2.2}{1!}+\frac{2^2.3}{2!}+\frac{3^2.4}{3!}+\ldots\infty$, then S has the value

(*a*) $3e$ (*b*) $5e$

(*c*) $7e$ (*d*) $9e$

19. The sum of the series $\frac{1}{n^2}+\frac{1}{2n^4}-\frac{1}{3n^6}+\ldots$ is

(*a*) $\log\left(\frac{n^2}{n^2+1}\right)$ (*b*) $\log\left(\frac{n^2+1}{n^2}\right)$

(*c*) $\log\left(\frac{n^2}{n^2-1}\right)$ (*d*) None of these

20. Sum of the series $\frac{1}{1.2}+\frac{1}{3.4}+\frac{1}{5.6}+\ldots\infty$ is

(*a*) $\log_e 2$ (*b*) $\log_e 3$

(*c*) $2\log_e 2$ (*d*) $\log_e 5$

ANSWERS

1	2	3	4	5	6	7	8	9	10
(*b*)	(*d*)	(*a*)	(*c*)	(*a*)	(*a*)	(*b*)	(*b*)	(*c*)	(*c*)
11	**12**	**13**	**14**	**15**	**16**	**17**	**18**	**19**	**20**
(*c*)	(*c*)	(*d*)	(*b*)	(*b*)	(*b*)	(*d*)	(*c*)	(*c*)	(*a*)

SOME SELECTED EXPLANATORY ANSWERS

1. $\log_y x = 1+d \Rightarrow x = y^{1+d}$

$\log_z y = 1+2d \Rightarrow y = z^{1+2d}$

and $-15\log_x z = 1+3d \Rightarrow z = x^{-(1+3d)/15}$

$\therefore x = y^{1+d} = z^{(1+2d)(1+d)} = x^{-(1+d)(1+2d)(1+3d)/15}$

$\Rightarrow (1+d)(1+2d)(1+3d) = -15$

$\Rightarrow d = -2$

$\therefore x = y^{1+d} = y^{-1},\ y = z^{1-4} = z^{-3}$

$\therefore y = z^{-3}$ is the correct answer.

4. $A = \frac{a+b}{2}$ or $a+b = 2A$

$G = \sqrt{ab}$ or $ab = G^2$

$H = \frac{2ab}{a+b} = 4$

$\therefore G^2 - AH$ gives $G^2 = 4A$

Also $2A + G^2 = 27$ or $2A + 4A = 27$

$\therefore \quad A = \frac{27}{6} = \frac{9}{2}$

$\therefore \quad \frac{a+b}{2} = \frac{9}{2}$ or, $a+b = 9$ ...(*i*)

Also $G^2 = 4A = 4.\frac{9}{2} = 18$

$\therefore \quad ab = 18$ ...(*ii*)

From (*i*) and (*ii*)

$a = 3,\ b = 6$

6. *a, b, c* are in G.P.

$\therefore b^2 = ac$ ...(*i*)

Also $a^{1/x} = b^{1/y} = c^{1/z} = k$ (say)

$\therefore a = k^x, b = k^y, c = k^z$

Put the values of *a, b, c* in (*i*), we get

$k^{2y} = k^x.k^z = k^{x+z}$

$\therefore \quad 2y = x+z \Rightarrow y = \frac{x+z}{2}$

$\therefore$ *x, y, z* are in A.P.

7. $x = a+(m-1)d = AR^{m-1}$

$y = a+(n-1)d = AR^{n-1}$

$z = a+(p-1)d = AR^{p-1}$

$\therefore y-z = (n-p)d, z-x = (p-m)d$ and

$x-y = (m-n)d$

$\therefore \quad x^{y-z}y^{z-x}z^{x-y}$

$= \left(AR^{m-1}\right)^{(n-p)d}\left(AR^{n-1}\right)^{(p-m)d}\left(AR^{p-1}\right)^{(m-n)d}$

$= A^0R^0 = 1$

9. $T_m = \frac{1}{n}$ and $T_n = \frac{1}{m}$ (given) of an A.P.

$\therefore A + (m-1)D = \frac{1}{n}$ and $A + (n-1)D = \frac{1}{m}$

Solving these, we get $A = D = \frac{1}{mn}$

T_{m+n} of A.P. $= A + (m+n-1)D$

$= \frac{1+m+n-1}{mn} = \frac{m+n}{mn}$

$\therefore T_{m+n}$ of H.P. $= \frac{mn}{m+n}$

11. *n* is even, then

$S = 1^2 + 2.2^2 + 3^2 + 2.4^2 + ... + 2.n^2$

$$=\left(1^2+2^2+\ldots+n^2\right)+\left(2^2+4^2+\ldots+n^2\right)$$

$$=\left(1^2+2^2+\ldots+n^2\right)+2^2\left(1^2+2^2+\ldots+\left(\frac{n}{2}\right)^2\right)$$

$$=\frac{n(n+1)(2n+1)}{6}+4\frac{\frac{n}{2}\left(\frac{n}{2}+1\right)\left(2.\frac{n}{2}+1\right)}{6}$$

$$=\frac{n(n+1)(2n+1)}{6}+\frac{n(n+1)(n+2)}{6}$$

$$=\frac{n(n+1)}{6}\left[(2n+1)+(n+2)\right]$$

$$=\frac{n(n+1)^2}{2}$$

15. Here p, q, r are in A.P.

$2q = p + r$ ...(*i*)

$t_p = p$th term of G.P. $= aR^{p-1}$

$t_q = q$th term of G.P. $= aR^{q-1}$

$t_r = r$th term of G.P. $= aR^{r-1}$

Now, $t_q^2 = a^2R^{2q-2}$ (*ii*)

Also, $t_pt_r = a^2R^{p+r-2} = a^2R^{2q-2}$ ...(*iii*) [by (*i*)]

From (*ii*) and (*iii*)

$t_q^2 = t_p.t_r$

Hence, pth, qth and rth terms are in G.P.

19. Given series

$$=-\left[-\left\{\left(\frac{1}{n^2}\right)+\frac{1}{2}\left(\frac{1}{n^2}\right)^2+\frac{1}{3}\left(\frac{1}{n^2}\right)^3+\ldots\right\}\right]$$

$$=-\log\left(1-\frac{1}{n^2}\right)=-\log\left(\frac{n^2-1}{n^2}\right)$$

$$=\log\left(\frac{n^2}{n^2-1}\right)$$

20. Here, $T_1 = 1-\frac{1}{2}$

$T_2 = \frac{1}{3}-\frac{1}{4}$

$T_3 = \frac{1}{5}-\frac{1}{6}$ and so on

$\therefore T_1 + T_2 + T_3 + \ldots\ldots\infty$

$$=\left(1-\frac{1}{2}+\frac{1}{3}-\frac{1}{4}+\frac{1}{5}-\frac{1}{6}+\ldots\right)$$

$$=\log_e(1+1)=\log_e 2$$

10

Coordinates and Straight Line

RECTANGULAR CARTESIAN COORDINATE SYSTEM

Distance Formula

(*a*) The distance between two points (x_1, y_1) and $(x_2, y_2) = \sqrt{[(x_2 - x_1)^2 + (y_2 - y_1)^2]}$.

(*b*) The distance of the point (x_1, y_1) from the origin $= \sqrt{(x_1^2 + y_1^2)}$.

Section Formula

(*a*) The coordinates of the point P(x, y), dividing the line joining the points A(x_1, y_1) and B(x_2, y_2) in the ratio $m : n$ (internally and externally). If P divide AB internally, then

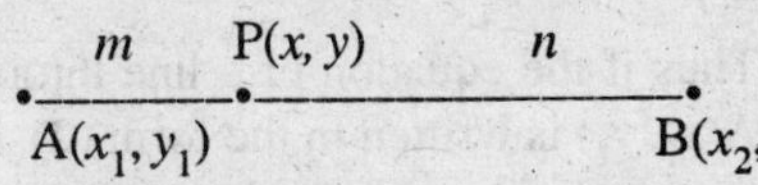

$$x = \frac{mx_2 + nx_1}{m+n}, \qquad y = \frac{my_2 + ny_1}{m+n}$$

If P divide AB externally, then $m : n$

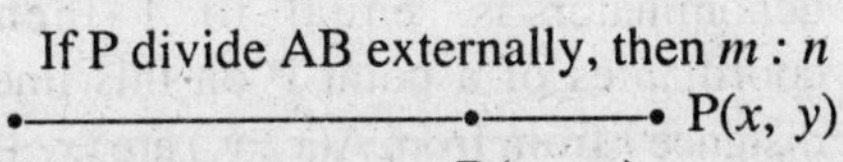

$$x = \frac{mx_2 - nx_1}{m-n}, \qquad y = \frac{my_2 - ny_1}{m-n}$$

(*b*) The coordinates of the mid-point of line joining A(x_1, y_1) and B(x_2, y_2) are $\left(\frac{x_1 + x_2}{2}, \frac{y_1 + y_2}{2}\right)$

Area of a Triangle: The area of a triangle whose vertices are A(x_1, y_1), B(x_2, y_2) and C(x_3, y_3) is

$$= \tfrac{1}{2} |(x_1y_2 + x_2y_3 + x_3y_1) - (y_1x_2 + y_2x_3 + y_3x_1)|$$

$$\Rightarrow \tfrac{1}{2} |x_1(y_2 - y_3) + x_2(y_3 - y_1) + x_3(y_1 - y_2)| \text{ or}$$

$$\frac{1}{2}\begin{vmatrix} x_1 & y_1 & 1 \\ x_2 & y_2 & 1 \\ x_3 & y_3 & 1 \end{vmatrix}$$

Area of a Quadrilateral: The area of a quadrilateral whose vertices are A(x_1, y_1), B(x_2, y_2), C(x_3, y_3) and D(x_4, y_4) is $= \tfrac{1}{2} |(x_1y_2 + x_2y_3 + x_3y_4 + x_4y_1) - (y_1x_2 + y_2x_3 + y_3x_4 + y_4x_1)|$.

Centroid of a Triangle: The centroid of a triangle is the point of intersection of its medians (the line joining a vertex to the middle point of the opposite side). The centroid divides the medians in the ratio 2 : 1. If (x_1, y_1), (x_2, y_2) and (x_3, y_3) are the coordinates of the vertices of a triangle, then the coordinates of its centroid are

$$\left(\frac{x_1 + x_2 + x_3}{3}, \frac{y_1 + y_2 + y_3}{3}\right)$$

Circumcentre of a Triangle: The point of intersection of the perpendicular bisectors of the sides of a triangle is called its circumcentre. It is the centre of the circle which passes through the vertices of the triangle and so its distance from the vertices of the triangle is the same and this distance is known as the circum-radius of the triangle.

In-centre of a Triangle: The centre of the circle which touches the sides of a triangle is called its in-centre. It is the point of intersection of the internal bisectors of the in-centre from the sides of the triangle is the same and this distance is called the in-radius of the triangle. If A(x_1, y_1), B(x_2, y_2) and C(x_3, y_3) are the coordinates of the vertices of a triangle, then the coordinates of its in-centre are

$$\left(\frac{ax_1 + bx_2 + cx_3}{a+b+c}, \frac{ay_1 + by_2 + cy_3}{a+b+c}\right), \text{ where}$$

BC $= a$, CA $= b$ and AB $= c$.

Ortho-centre of a Triangle: The ortho-centre of a triangle is the point of intersection of the lines through the vertices and perpendicular to the opposite sides of the triangle.

Collinearity of Three given Points: The three given points A, B, C are collinear, *i.e.*, lie on the same straight line, if

(*a*) area of Δ ABC is zero

(*b*) slope of AB = slope of AC

(*c*) any one of the three points (say C) lie on the straight line joining the other two points (here points A and B).

Slope of a Line: The slope of a line joining two points $A(x_1, y_1)$ and $B(x_2, y_2)$ is denoted by m and is given by

$$m = \frac{y_2 - y_1}{x_2 - x_1} = \tan\theta,$$

Where θ is the angle which the line makes with the positive direction of x-axis. This angle θ is taken positive when it is measured in the anticlockwise direction from the positive direction of the axis of x.

EQUATION OF A LINE IN VARIOUS FORMS

Straight Lines: The general form of the equation of a straight line is $ax + by + c = 0$, (the first degree equation in x and y) where a, b and c are real constants not all equal to zero simultaneously.

(*i*) ***Lines Parallel to Axes:***

(*a*) The equation of a straight line || to x-axis and at a distance b from it is given by $y = b$.

(*b*) The equation of the straight line || to y-axis and at a distance a from it is given by $x = a$.

(*c*) The equations of the straight lines || to x-axis and y-axis respectively and passing through the pt. (a, b) are $y = b, x = a$.

(*d*) Equation of x-axis is $y = 0$ and equation of y-axis is $x = 0$.

(*ii*) ***Slope Intercept Form:*** The equation of a straight line which cuts off an intercept c on y-axis and makes an angle θ (in anticlockwise direction) with the positive direction of x-axis is given by $y = mx + c$, where $m = \tan\theta$. The number $m = \tan\theta$ is called the slope or the gradient of the line. Obviously, $y = mx$ is the equation of the line through the origin.

(*iii*) ***Slope Point Form:*** The equation of a straight line passing through the point $A(x_1, y_1)$ and having slope m is given by $y - y_1 = m(x - x_1)$ ...(1)

This equation can also be written as

$$\frac{x - x_1}{\cos\theta} = \frac{y - y_1}{\sin\theta} = r \text{ (say)} \qquad ...(2)$$

$(\because m = \tan\theta)$

This form of the line is called parametric form or symmetrical form of the line. The coordinates of any point P on this line are taken as $(x_1 + r\cos\theta, y_1, + r\sin\theta)$

$\therefore$ Distance of point p from A (x_1, y_1)

$= \sqrt{[(x_1 + r\cos\theta - x_1)^2 + (y_1 + r\sin\theta - y_1)^2]}$

$= r$

Thus if the equation of a line through point $A(x_1, y_1)$ is written in the form (2), in which the coefficients of x and y are equal to 1 and the sum of squares of the terms in the denominators is equal to 1 then the coordiantes of a point P on this line at a distance r from from $A(x_1, y_1)$ are $(x_1 + r\cos\theta, y_1 + r\sin\theta)$

(*iv*) ***Two Points Form:*** The equation of a straight line passing through two points $A(x_1, y_1)$ and $B(x_2, y_2)$ is given by $y - y_1 = \frac{y_2 - y_1}{x_2 - x_1}(x - x_1)$

and its slope $= \frac{y_2 - y_1}{x_2 - x_1}$

$= \frac{\text{difference of ordinates}}{\text{difference of abscissae}}$

(*v*) ***Intercept Form:*** The equation of a straight line making intercepts a and b on the axes of x and y respectively is given by $x/a + y/b = 1$

(*vi*) ***Normal (or Perpendicular) Form:*** The equation of a straight line on which the perpendicular from origin is of length p and makes an angle α with the positive direction of x-axis is given by

$$x \cos \alpha + y \sin \alpha = p.$$

(*vii*) ***Slope and Intercepts on Coordinate Axes of the Line Ax + By + C = 0***

(*a*) *To find slope of the line:* The equation

$$Ax + By + C = 0 \qquad ...(1)$$

can be written as, $y = -(A/B)x - (C/A)$

Comparing with $y = mx + c$, we have $m = -(A/B)$

$$\text{Thus slope of line} = \frac{\text{coefficient of } x}{\text{coefficient of } y}$$

(*b*) *To find intercepts on y-axes:* The equation (1) can be written as

$$\frac{x}{(-C/A)} + \frac{y}{(-C/B)} = 1$$

Comparing with $x/a + y/b = 1$, we have

a = intercept on x-axis = $-$ C/A

b = intercept on y-axis = $-$ C/B

ANGLE BETWEEN TWO LINES

Angle Between Two Lines: If m_1 and m_2 are the slopes of the two lines, then the angle θ between them is given by

$$\theta = \tan^{-1}\left|\frac{m_1 - m_2}{1 + m_1 m_2}\right|$$

(*a*) *Condition for two lines to be parallel:* The two lines whose slopes are m_1 and m_2 are parallel if $\theta = 0°$ or $180°$,

$\therefore \tan\theta = 0 \Rightarrow m_1 = m_2$.

If the two lines are $a_1x + b_1y + c_1 = 0$ and $a_2x + b_2y + c_2 = 0$, then $m_1 = a_1/b_1$ and $m_2 = -a_2/b_2$, *i.e.*, the coefficients of x and y in the two equations are proportional.

(*b*) *Condition for two lines to be perpendicular:* The two lines whose slopes are m_1 and m_2 are perpendicular if $\theta = 90°$.

$\therefore \tan\theta = \infty \Rightarrow = 1 + m_1m_2 = 0$ or $m_1m_2 = -1$

If the two lines are $a_1x + b_1y + c_1 = 0$ and $a_2x + b_2y + c_2 = 0$, then the two lines are perpendicular if $m_1m_2 = (-a_1/b_1)(a_2/b_2) = -1$

or $a_1a_2 + b_1b_2 = 0$

i.e., the product of coeffs. of x + product of coeffs. of $y = 0$.

Equation of Line Parallel to $ax + by + c = 0$: The equation of the straight line which is parallel to the line $ax + by + c = 0$ is given by $ax + by + \lambda = 0$. Thus to write the equation of any line parallel to a given line, do not change the coefficients of x and y and change the constant term only.

Equation of Line Perpendicular to $ax + by + c = 0$: The equation of a straight line which is perpendicular to the line $ax + by + c = 0$ is given by $bx - ay + \lambda = 0$. Thus to write the equation of any line perpendicualr to a given line, interchange the coefficient of x and y in the given equation and change the sign of one of these coefficients and also change the constant term.

Point of Intersection of Two Given Lines: Let the two given lines be $a_1x + b_1y + c_1 = 0$ and $a_2x + b_2y + c_2 = 0$. Solving these two equations, the point of intersection of these two lines is given by

$$\left(\frac{b_1c_2 - b_2c_1}{a_1b_2 - a_2b_1}, \frac{c_1a_2 - c_2a_1}{a_1b_2 - a_2b_1}\right).$$

Concurrent Lines: Three or more straight lines are said to be concurrent lines if they meet in a point.

DISTANCE OF A POINT FROM A LINE

Distance of a Point from a Line: The distance of a point from a line is calculated as the length of perpendicular from a given point (x_1, y_1) to a line. If the equation of line is $ax + by + c = 0$ then distance is

$$\frac{ax_1 + by_1 + c}{\sqrt{(a^2 + b^2)}}$$

Equation of the Bisectors of the Angles Between Two Lines: Let the two given lines be $a_1x + b_1y + c_1 = 0$ and $a_2x + b_2y + c_2 = 0$. Then the equations of the bisectors of the angles between two lines are

$$\frac{a_1x + b_1y + c_1}{\sqrt{(a_1^2 + b_1^2)}} = \pm \frac{a_2x + b_2y + c_2}{\sqrt{(a_1^2 + b_1^2)}} \qquad ...(1)$$

Remember that, every point of the bisectors of angles is equidistant from the given lines.

MULTIPLE CHOICE QUESTIONS

1. If the equation of the locus of a point equidistant from the points (a_1, b_1) and (a_2, b_2) is $(a_1 - a_2)x + (b_1 - b_2)y + c = 0$, then the value of '$c$' is

(a) $\frac{1}{2}\left(a_2^2 + b_2^2 - a_1^2 - b_1^2\right)$

(b) $a_1^2 - a_2^2 + b_1^2 - b_2^2$

(c) $\frac{1}{2}\left(a_1^2 + a_2^2 + b_1^2 + b_2^2\right)$

(d) $\sqrt{a_1^2 + b_1^2 - a_2^2 - b_2^2}$

2. Locus of centroid of the triangle whose vertices are $(a \cos t, a \sin t)$, $(b \sin t, -b \cos t)$ and $(1, 0)$, where t is a parameter, is

(a) $(3x - 1)^2 + (3y)^2 = a^2 - b^2$
(b) $(3x - 1)^2 + (3y)^2 = a^2 + b^2$
(c) $(3x + 1)^2 + (3y)^2 = a^2 + b^2$
(d) $(3x + 1)^2 + (3y)^2 = a^2 - b^2$

3. A straight line through the origin O meets the parallel line $4x + 2y = 9$ and $2x + y + 6 = 0$ at points P and Q respectively. Then, the point O divides the segment PQ in the ratio :

(a) 1 : 2 (b) 3 : 4
(c) 2 : 1 (d) 4 : 3.

4. Let $P(-1, 0)$, $Q(0, 0)$ and $R\left(3, 3\sqrt{3}\right)$ be three points. Then, the equation of the bisector of the angle PQR is :

(a) $\frac{\sqrt{3}}{2}x + y = 0$ (b) $x + \sqrt{3}y = 0$

(c) $\sqrt{3}x + y = 0$ (d) $x + \frac{\sqrt{3}}{2}y = 0$

5. The equation of base of an equilateral triangle is $x + y = 2$ and vertex is $(2, -1)$, then the length of the side of the triangle is equal to :

(a) $\sqrt{\frac{2}{3}}$ (b) $\sqrt{\frac{1}{3}}$

(c) $\sqrt{\frac{3}{2}}$ (d) $\sqrt{3}$.

6. Equations of the bisectors of the lines $3x - 4y + 7 = 0$ and $12x + 5y - 2 = 0$ are given by :

(a) $21x + 77y - 101 = 0, 11x - 3y + 9 = 0$
(b) $11x - 6y + 111 = 0, 22x - 13y + 104 = 0$
(c) $15x - 9y + 67 = 0, 15x + 4y + 33 = 0$
(d) None of these.

7. The orthocentre of the triangle formed by the lines $x + y = 1, 2x + 3y = 6$ and $4x - y + 4 = 0$ lies in

(a) IIIrd quadrant (b) IInd quadrant
(c) I[st] quadrant (d) IVth quadrant

8. The reflection of $(4, -13)$ in the line $5x + y + 6 = 0$ is

(a) $(3, 4)$ (b) $(1, 2)$
(c) $(-1, -14)$ (d) $(-4, 13)$

9. The points (a, b), (c, d), (a, d) and (c, b) taken in order and $a, b, c, d, \in R$ are

(a) collinear (b) square
(c) rhombus (d) None of these

10. A line passing through P(4, 2) has intercepts $(a, 0)$ and $(0, b)$ respectively. If O is the origin, then locus of centre of circumcircle of a Δ formed by the line and the axes is

(a) $\frac{2}{x} + \frac{1}{y} = 2$ (b) $\frac{2}{x} + \frac{1}{y} = 1$

(c) $\frac{1}{x} + \frac{2}{y} = 1$ (d) $\frac{2}{x} + \frac{2}{y} = 1$

11. If a straight line passes through $R(\alpha, \beta)$ and its segment between the axes is bisected at this point, then its equation is given by

(a) $\frac{x}{\alpha} + \frac{y}{\beta} = 2$ (b) $\frac{x}{\alpha} + \frac{y}{\beta} = 1$

(c) $\frac{x}{\beta} + \frac{y}{\alpha} = 1$ (d) None of these

12. If the line $y = mx$ meet the lines $x + 2y - 1 = 0$ and $2x - y + 3 = 0$ at the same point, then m is equal to

(*a*) 1 (*b*) –1
(*c*) 2 (*d*) –2

13. If $ax + by + c = 0$, $ax + by + c_1 = 0$, $bx + ay + c = 0$ and $bx + ay + c_1 = 0$ are sides of a parallelogram, then the diagonals include an angle,

(*a*) $\tan^{-1}\left(\dfrac{a^2+b^2}{a^2-b^2}\right)$ (*b*) $\dfrac{\pi}{6}$

(*c*) $\dfrac{\pi}{2}$ (*d*) $\tan^{-1}\left(\dfrac{a^2-b^2}{a^2+b^2}\right)$

14. If $x+\alpha y+\alpha=0$, $\beta x+y+\beta=0$ and $\gamma x + \gamma y + 1 = 0$ (α, β, γ are distinct and $\neq 1$) are concurrent, then the value of $\dfrac{\alpha}{\alpha-1}+\dfrac{\beta}{\beta-1}+\dfrac{\gamma}{\gamma-1}$ is

(*a*) 0 (*b*) –1
(*c*) –2 (*d*) 1

15. The equation of bisector of that angle of line which contains (1, 1) between the lines $x+y=3$ and $2x-y=2$ is

(*a*) $3x=10$
(*b*) $3y=10$
(*c*) $\left(5+2\sqrt{2}\right)x+\left(5+2\sqrt{2}\right)y=0$
(*d*) None of these

ANSWERS

1	2	3	4	5	6	7	8	9	10
(*a*)	(*b*)	(*b*)	(*c*)	(*a*)	(*a*)	(*c*)	(*c*)	(*b*)	(*b*)
11	**12**	**13**	**14**	**15**					
(*a*)	(*b*)	(*c*)	(*d*)	(*d*)					

SOME SELECTED EXPLANATORY ANSWERS

1. (a_1, b_1) and (a_2, b_2) satisfy the equation.

So that $a_1(a_1-a_2)+b_1(b_1-b_2)+c=0$
and $a_2(a_1-a_2)+b_2(b_1-b_2)+c=0$
Adding, we get

$$(a_1+a_2)(a_1-a_2)+(b_1+b_2)(b_1-b_2)+2c=0$$

$$2c=-\left[a_1^2-a_2^2+b_1^2-b_2^2\right]$$

$$c=\frac{1}{2}\left[a_2^2+b_2^2-a_1^2-b_1^2\right]$$

6. The equations of the bisectors of the angles between given lines are

$$\frac{3x-4y+7}{\sqrt{3^2+4^2}}=\pm\frac{12x+5y-2}{\sqrt{(12)^2+5^2}}$$

$$\Rightarrow \frac{3x-4y+7}{5}=\pm\frac{12x+5y-2}{13}$$

$$\Rightarrow 13(3x-4y+7)=\pm5(12x+5y-2)$$

$$\Rightarrow 21x+77y-101=0, 99x-27y+81=0$$

$$\Rightarrow 21x+77y-101=0,\quad 11x-3y+9=0$$

9. Let the points be PQRS. Then

Slope of PS = Slope of QR = 0.

∴ PS || QR || x-axis

Also $PQ^2 = QR^2$

Slope of PR = Slope of QS = ∞

∴ PR || QS || y-axis

Also $AC^2 = BD^2$

Thus PQRS represents a square

12. The equation of lines $x + 2y - 1 = 0$ and $2x - y + 3 = 0$ intersect each other at (–1. 1). Because the lines are concurrent, so (–1, 1) lies on $y = mx$.

$\Rightarrow m=-1$

11

Conic Section

Conic Sections: The locus of a point which moves such that its distance from a fixed point bears a constant ratio to its distance from a fixed straight line is called a conic section. The fixed point is called the focus and the fixed line is called the directrix and the constant ratio (say e) is called the eccentricity of the conic section. The conic section is called

(*a*) parabola, if $e = 1$
(*b*) ellipse, if $e < 1$
(*c*) hyperbola, if $e > 1$

General Equation: The second degree linear equation.

$ax^2 + 2hxy + by^2 + 2gx + 2fy + c = 0$, represents a conic section, which is

(*a*) parabola, if $h^2 - ab = 0$ and $\Delta \neq 0$
(*b*) ellipse, if $h^2 - ab < 0$ and $\Delta \neq 0$
(*c*) hyperbola, if $h^2 - ab > 0$ and $\Delta \neq 0$

The Standard Equation: The standard equation of

(*a*) parabola is $y^2 = 4ax$
(*b*) ellipse is $x^2/a^2 + y^2/b^2 = 1$
(*c*) hyperbola is $x^2/a^2 - y^2/b^2 = 1$

Axes of a Conic Section: The line (or lines) about which the given conic is symmetrical is (or are) called the axis (or axes) of the conic.

(*a*) A parabola has one axis.
(*b*) An ellipse have two axes, called the major and minor axes.
For ellipse $x^2/a^2 + y^2/b^2 = 1$ the length of axes are $2a$ and $2b$ which are major and minor axes if $a > b$ and lie along x-axis and y-axis respectively and $b^2 = a^2(1 - e^2)$.
(*c*) A hyperbola have two axes called transverse and conjugate axes.
For hyperbola $x^2/a^2 - y^2/b^2 = 1$, the length of transverse and conjugate axes, are $2a$ and $2b$ and lie along axis of x and y respectively and $b^2 = a^2(e^2 - 1)$.

Vertices: The points of intersection of the conic and its axis are called the vertices of the conic. A parabola has one vertex $(a, 0)$ while ellipse and hyperbola each have two vertices $(a, 0)$ and $(-a, 0)$.

Foci and Directrices:

(*a*) For the parabola $y^2 = 4ax$, focus is $(a, 0)$ and the directrix is $x = -a$.
(*b*) For the ellipse $x^2/b^2 + y^2/b^2 = 1$, the foci are $(ae, 0)$ and $(-ae, 0)$ and the corresponding directrices are $x = a/e$ and $x = -a/e$ respectively.
(*c*) For the hyperbola $x^2/a^2 - y^2/b^2 = 1$, the foci are$(ae, 0)$ and $(-ae, 0)$ and the corresponding directrices are $x = a/e$ and $x = -a/e$ respectively.

Focal Property:

(*a*) If $P(x, y)$ is a point on the parabola $y^2 = 4ax$ and $S(a, 0)$ is its focus then $PS = x + a$.
(*b*) If P is a point on the ellipse whose foci are S and S' and $2a$, the length of its major axis, then $PS + PS' = 2a$.
(*c*) If P is a point on the hyperbola whose foci are S and S' and 2a the length of its transverse axis, then $|PS - PS'| = 2a$.

Focal Chord: Any chord through the focus of a conic in called its focal chord.

Latus Rectum: The cord (or chords) through the focus (or foci) and perpendicular to the major axis (or transverse axis) is called latus rectum. The length of L.R. of

(*a*) parabola is $4a$
(*b*) ellipse is $2b^2/a$
(*c*) hyperbola is $2b^2/a$

Tangents: The equation of tangent at the point (x_1, y_1) on the

(*a*) parabola is $yy_1 = 2a(x + x_1)$

(*b*) ellipse is $\frac{xx_1}{a^2}+\frac{yy_1}{b^2}=1$

(*c*) hyperbola is $\frac{xx_1}{a^2}-\frac{yy_1}{b^2}=1$

Condition for Tangency: The condition that the line $y = mx + c$ is tangent to

(*a*) parabola is $c = a/m$

(*b*) ellipse is $c^2 = a^2m^2 + b^2$

(*c*) hyperbola is $c^2 = a^2m^2 - b^2$

Tangents in Terms of *m*: If *m* is the slope of the tangent, then its equation for

(*a*) parabola is $y = mx + a/m$

(*b*) ellipse are $y = mx \pm \sqrt{(a^2m^2 + b^2)}$

(*c*) hyperbola are $y = mx \pm \sqrt{(a^2m^2 - b^2)}$

Chord of Contact: The line joining the points of contact of the tangents from an external points to a conic is called the chord of contact of the point. The equation of chord of contact of the point (x_1, y_1) for

(*a*) parabola is $yy_1 = 2a(x + x_1)$

(*b*) ellipse is $\frac{xx_1}{a^2}+\frac{yy_1}{b^2}=1$

(*c*) hyperbola is $\frac{xx_1}{a^2}-\frac{yy_1}{b^2}=1$

Polar of a Point: The polar of a point is the locus of the points of intersection of the tangents drawn to the conic at the ends of chords passing through the given point. The point is called the pole of the polar. The equation of the polar of the point (x_1, y_1) for

(*a*) parabola is $yy_1 = 2a(x + x_1)$

(*b*) ellipse is $\frac{xx_1}{a^2}+\frac{yy_1}{b^2}=1$

(*c*) hyperbola is $\frac{xx_1}{a^2}-\frac{yy_1}{b^2}=1$

Tangents Through a Point: The equation of the pair of tangents drawn from a point (x_1, y_1) to the conic $S = 0$ is $SS_1 = T^2$, where

(*a*) for parabola $y^2 = 4ax$, $S = y^2 - 4ax$, $S_1 = y_1^2 - 4ax_1$, $T = yy_1 - 2a(x + x_1)$

(*b*) for ellipse $x^2/a^2 + y^2/b^2 = 1$,
$S = x^2/a^2 + y^2/b^2 - 1$, $S_1 = x_1^2/a^2 + y_1^2/b^2 - 1$,
$T = xx_1/a^2 + yy_1/b^2 - 1$

(*c*) for hyperbola $x^2/a^2 - y^2/b^2 = 1$,
$S = x^2/a^2 - y^2/b^2 - 1$, $S_1 = x_1^2/a^2 - y_1^2/b^2 - 1$,
$T = xx_1/a^2 + yy_1/b^2 - 1$.

Parametric Equations of Conics:

(*a*) The coordinates of a point on the parabola $y^2 = 4ax$, in terms of parameter *t* are taken as $(at^2, 2at)$. The point is denoted by *t*. Thus parametric equation of parabola are $x = at^2$, $y = 2at$.

(*b*) The coordinates of any point on the ellipse $x^2/a^2 + y^2/b^2 = 1$ are taken as $(a \cos\phi, b \sin\phi)$, ϕ is called the eccentric angle of the point and the point is denoted by ϕ. Thus parametric equations of ellipse are $x = a\cos\phi$, $y = b\sin\phi$.

(*c*) The coordinates of any point on the hyperbola $x^2/a^2 - y^2/b^2 = 1$ are taken as $(a\sec\phi, b\tan\phi)$. The point is denoted by ϕ. Thus parametric equations of hyperbola are $x = a\sec\phi$, $y = b\tan\phi$.

Equation of Tangent: The equation of tangent at the point

(*a*) *t* on $y^2 = 4ax$ is $yt = x + at^2$

(*b*) ϕ on $x^2/a^2 + y^2/b^2 = 1$
is $(x/a)\cos\phi + (y/b)\sin\phi = 1$

(*c*) ϕ on $x^2/a^2 - y^2/b^2 = 1$
is $(x/a)\sec\phi + (y/b)\tan\phi = 1$.

Equation of Normal: The equation of normal at the point

(*a*) *t* on $y^2 = 4ax$ is $y = -tx + 2at + at^3$

(*b*) ϕ on $x^2/a^2 + y^2/b^2 = 1$
is $ax\sec\phi - by\,\text{cosec}\,\phi = a^2 - b^2$

(*c*) ϕ on $x^2/a^2 - y^2/b^2 = 1$
is $ax\cos\phi + by\cot\phi = a^2 + b^2$.

Focal Chord: If PQ is a focal chord of parabola $y^2 = 4ax$ and coordinates of P are $(at^2, 2at)$ then coordinates of Q will be $(a/t^2, -2a/t)$.

Point and the Conic: Let $S = 0$ be a conic section. Then the point $P(x_1, y_1)$ lie

(*a*) inside the conic if $S(x_1, y_1) < 0$

(*b*) on the conic if $S(x_1, y_1) = 0$

(*c*) outside the conic if $S(x_1, y_1) > 0$

MULTIPLE CHOICE QUESTIONS

1. The locus of the point of intersection of the lines $\sqrt{3}x - y - 4\sqrt{3}k = 0$ and $\sqrt{3}kx - ky - 4\sqrt{3} = 0$ represent a hyperbola then eccentricity equals

(a) $\sqrt{2}$ (b) 2
(c) 9 (d) none of these.

2. Consider the set of hyperbola $xy = k, k \in R$. Let e_1 and e_2 be the eccentricties when $k = 16$ and $k = 25$ respectively then the value of $e_1^2 - e_2^2$ equals

(a) –2 (b) –3
(c) 9 (d) none of these.

3. If the chords of contacrt of tangents from two points (x_1, y_1) and (x_2, y_2) to the hyperbola $\frac{x^2}{a^2} - \frac{y^2}{b^2} = 1$ are at right angle then $\frac{y_1 y_2}{x_1 x_2}$ equals

(a) $\frac{a^2}{b^2}$ (b) $\frac{-a^2}{b^2}$
(c) $\frac{-a^4}{b^4}$ (d) $\frac{-b^4}{a^4}$

4. Equation of normal to the hyperbola $xy = c^2$ at $(ct, c/t)$ is

(a) $xt^3 + yt + ct^4 + c = 0$
(b) $xt^3 - yt - ct^4 - c = 0$
(c) $xt^3 - yt + ct^4 - c = 0$
(d) none of these

5. The second degree equation $x^2 + 4y^2 + 2x + 16y + 13 = 0$ represent itself as

(a) a parabola
(b) a pair of straight line
(c) an ellipse
(d) a hyperbola

6. The equation of the chord of $y^2 = 8x$ which is bisected at (2, –3) is

(a) $4x - 3y + 1 = 0$
(b) $4x - 3y - 1 = 0$
(c) $4x + 3y + 1 = 0$
(d) $3x + 4y = 1$

7. For what value of k, the line $x + y = 1$ touches the parobola $y^2 = kx$.

(a) 4 (b) –4
(c) –2 (d) none of these

8. The length of the latus-rectum of the parabola $ay^2 + by = x - c$ is

(a) $1/a$ (b) $a/2$
(c) $1/4a$ (d) $4a$

9. If the line $2x + \sqrt{6}\,y = 2$ touches the hyperbola $x^2 - 2y^2 = 4$ then the point of contact is

(a) $(-2, -\sqrt{6})$ (b) $(-4, -\sqrt{6})$
(c) $(-4, -\sqrt{6})$ (d) none of these

10. The curve described parametrically by $x = t^2 + t + 1$ and $y = t^2 - t + 1$ represent

(a) a parabola
(b) an ellipse
(c) a hyperbola
(d) a pair of straigh lines

11. The equation of the ellipse, whose length of major axis is 20 and foci are (0, ± 5) is

(a) $\frac{x^2}{100} + \frac{y^2}{75} = 1$ (b) $\frac{x^2}{75} + \frac{y^2}{100} = 1$
(c) $\frac{x^2}{400} + \frac{y^2}{25} = 1$ (d) $\frac{x^2}{25} + \frac{y^2}{400} = 1$

12. Let E be the ellipse $\frac{x^2}{9} + \frac{y^2}{4} = 1$ and C be the circle centred at (0, 0) with radius 3. Let P and Q be the points (1, 2) and (2, 1) respectively then

(a) Q lies outside C and E
(b) P lies inside both C and E
(c) P lies inside C but outside E
(d) Q lies inside C but outside E

13. If the line $2x - 1 = 0$ is the directrix of the parabola $y^2 - kx + 6 = 0$ then one of the values of k is

(a) –6 (b) 6
(c) 1/4 (d) –1/4

14. The equation of common tangent to the curves $y^2 = 12\sqrt{3x}$ and $4xy = -1$ is

(a) $y = -3x + \sqrt{3}$ (b) $y = 3x - \sqrt{3}$

(c) $y = 3x + \sqrt{3}$ (d) none of these

15. If $x + y = k$ is normal to the parabola $y^2 = 12x$ then k equals

(a) 12 (b) −9

(c) −3 (d) 9

16. If $xy = m^2 - 9$ be a rectangular hyperbola whose brances lie only in the second and fourth quadrant, then

(a) $|m| \geq 3$ (b) $|m| < 3$

(c) $m \in R - \{|m|\}$ (d) none of these

17. There are exactly two points on the ellipse $\frac{x^2}{a^2}+\frac{y^2}{b^2}=1$ whose distance from the centre of the ellipse are equal and equal to $\sqrt{\frac{a^2+2b^2}{2}}$. Eccentricity of this ellipse is equal to

(a) $\frac{\sqrt{3}}{2}$ (b) $\frac{1}{\sqrt{3}}$

(c) $\frac{1}{\sqrt{2}}$ (d) $\frac{2}{\sqrt{3}}$

18. Which of the following is a common tangent to the ellipses $\frac{x^2}{a^2+b^2}+\frac{y^2}{b^2}=1$ and $\frac{x^2}{a^2}+\frac{y^2}{a^2+b^2}=1$.

(a) $by = ax + \sqrt{a^4 + a^2b^2 + b^4}$

(b) $ay = bx - \sqrt{a^4 + a^2b^2 + b^4}$

(c) $by = ax - \sqrt{a^4 - a^2b^2 + b^4}$

(d) $ay = bx - \sqrt{a^4 - a^2b^2 + b^4}$

19. Consider a point P on the ellipse $\frac{x^2}{a^2}+\frac{y^2}{b^2}=1$ and a corresponding point Q on its auxiliary circle. If R be the foot of perpendicular drawn from focus S_1 to the tangent draw to the auxiliary circle at Q then triangle S_1PR is necessarily

(a) equilateral (b) isosceles

(c) right angled (d) scalene

20. Consider an ellipse and a concentric circle. The circle passes through the foci of the ellipse and intersects the ellipse in four distinct points. The length of major axis of the ellipse is 15 units. If S_1 and S_2 are the foci of the ellipse, P be one of point of intersection of ellipse and circle and area of triangle PS_1S_2 is 26 sq. units, then eccentricity of the ellipse is equal to

(a) $\frac{2}{3}$ (b) $\frac{7}{15}$

(c) $\frac{13}{15}$ (d) $\frac{11}{15}$

ANSWERS

1	2	3	4	5	6	7	8	9	10
(b)	(d)	(d)	(b)	(c)	(c)	(b)	(a)	(c)	(a)
11	**12**	**13**	**14**	**15**	**16**	**17**	**18**	**19**	**20**
(b)	(c)	(a)	(c)	(d)	(b)	(c)	(a)	(b)	(d)

SOME SELECTED EXPLANATORY ANSWERS

1. From given equation we have

$$k = \frac{\sqrt{3}x - y}{4\sqrt{3}} = \frac{4\sqrt{3}}{\sqrt{3}x + y}$$

$$\Rightarrow (\sqrt{3}x - y)(\sqrt{3}x + y) = (4\sqrt{3})^2$$

$$\Rightarrow 3x^2 - y^2 = 48 \Rightarrow \frac{x^2}{16} - \frac{y^2}{48} = 1$$

For eccentricity , $b^2 = a^2(e^2 - 1)$

$\Rightarrow 48 = 16(e^2 - 1) \Rightarrow e^2 = 4 \Rightarrow e = 2$

2. From given
$xy = 16$ & $xy = 25$ both are rectangular hyperbolas and eccentricity for rectangular for rectangular hyperbola is $\sqrt{2}$

$\therefore e_1 = e_2 = \sqrt{2} \;\therefore\; e_1^2 - e_2^2 = 2 - 2 = 0$

4. Equation of normal at (x_1, y_1) to the hyperbola $xy = c^2$ is $xx_1 - yy_1 = x_1^2 - y_1^2$

$\Rightarrow x(ct) - y\left(\frac{c}{t}\right) = c^2t^2 - \frac{c^2}{t^2}$

$\Rightarrow xt^3 - yt - ct^4 + c = 0$

6. Equation of chord to the parabola $y^2 = 4ax$ at a point (x_1, y_1) is given by $T = S'$ where $T = yy_1 - 2a(x + x_1)$ and $S' = y_1^2 - 4ax_1$
Here $(x_1, y_1) = (2, -3)$, $a = 2$ as $y^2 = 4(2)x$
$\therefore T = S' \Rightarrow y(-3) - 2(2)(x + 2) = 3^2 - 4(2)(2)$
$\Rightarrow -3y - 4(x + 2) = -7 \Rightarrow 4x + 3y + 1 = 0$

7. $x + y = 1 \Rightarrow y = -x + 1$
Using it to the parabola $y^2 = kx$
$\Rightarrow x^2 - 2x + 1 = kx$
$\Rightarrow x^2 - x(2 + k) + 1 = 0$ using $D = 0$
$\Rightarrow (2 + k)^2 = 4 \Rightarrow 2 + k = \pm 2 \Rightarrow k = 0, -4$
$\Rightarrow k = 4$ is required value

9. Given hyperbola is $x^2 - 2y^2 = 4$
$\therefore$ Equation of tangent to the given hyperbola at (x_1, y_1) is $xx_1 - 2yy_1 = 4$. On comparing this equation with $2x + \sqrt{6}\, y = 2$.

$\Rightarrow \frac{x_1}{2} = \frac{-2y_1}{\sqrt{6}} = \frac{4}{2} \Rightarrow x_1 = 4,\; y_1 = -\sqrt{6}$

$\therefore$ point of contract $(x_1, y_1) = (4, -\sqrt{6})$ which is choice (*c*).

12. Given $P(1, 2)$ and $Q(2, 1)$ and equation of circle be $x^2 + y^2 = 9$ for which $C(0, 0)$ and $r = 3$
$\therefore 1^2 + 2^2 = 5 < 9$
$\Rightarrow$ P(1, 2) lies inside the circle and in the same way $Q(2, 1)$ also lies inside the circle C.

Again $\frac{x^2}{9} + \frac{y^2}{4} = 1$ using $P(1, 2)$ we have

$\frac{1^2}{9} + \frac{4}{4} > 1$, so P lies outside the ellipse E

and or $Q(2, 1)$ gives $\frac{4}{9} + \frac{1}{4} = \frac{25}{36} < 1$ which means Q lies inside E.

$\therefore$ P lies insides the circle C but outside the ellipse E.

$\therefore$ (*c*) is correct choice.

15. If the line $y = mx + k$ is normal to $y^2 = 4ax$ then
$k = -2am - am^3$
In our case $m = -1$, $a = 3$
$\therefore k = -2(-1)(3)\; 3(-1)^3$
$\Rightarrow k = 6 + 3 = 9.$

12

Three Dimensional Geometry

Coordinates of a Point in Space: Let O be a fixed point, known as origin as let OX, OY and OZ be three mutually perpendicular lines, taken as *x*-axis, *y*-axis and *z*-axis respectively, in such a way that they form a right-handed system.

The planes XOY, YOZ and ZOX are in known as *xy*-plane, *yz*-plane and *zx*-plane respectively.

In the adjoining figure, OX, OY and OZ shows the positive *x*, *y* and *z* axes.

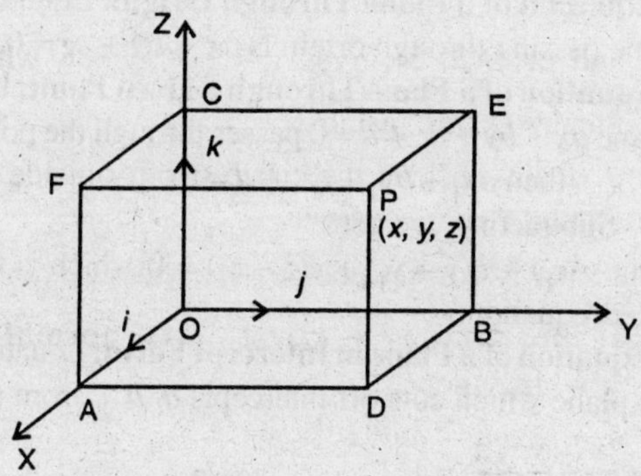

Let P be a point in space and distances of P from *y*-*z*, *z*-*x* and *x*-*y* planes by x, y, z respectively (with proper signs) then we say that coordinates of P are (x, y, z).

Also $OA = x$, $OB = y$, $OC = z$.

Position Vector of a Point: Let *i, j, k* be unit vectors along OX, OY and OZ respectively. Then position vector of a point $P(x, y, z)$ is

$\overrightarrow{OP} = xi + yj + zk$.

Distance Formula: The distance between two points $A(x_1, y_1, z_1)$ and $B(x_2, y_2, z_2)$ is given by $AB = \sqrt{[(x_2 - x_1)^2 + (y_2 - y_1)^2 + (z_2 - z_1)^2]}$

Distance from Origin: Let O be the origin andP (x, y, z) be any point, then $OP = \sqrt{(x^2 + y^2 + z^2)}$.

Distance of a Point from Coordinate Axes: Let $P(x, y, z)$ be any point in the space. Let PA, PB and PC be the perpendiculars drawn from P to the axes OX, OY and OZ respectively. Then

$$PA = \sqrt{(y^2 + z^2)},$$
$$PB = \sqrt{(z^2 + x^2)},$$
$$PA = \sqrt{(x^2 + y^2)},$$

Section Formula: Let $P(x_1, y_1, z_1)$ and $Q(x_2, y_2, z_2)$ be two points and $P(x, y, z)$ divide PQ in the ratio $m_1 : m_2$. Then R is (x, y, z)

$$= \left(\frac{m_1x_2 + m_2x_1}{m_1 + m_2}, \frac{m_1y_2 + m_2y_1}{m_1 + m_2}, \frac{m_1z_2 + m_2z_1}{m_1 + m_2}\right).$$

If (m_1/m_2) is positive, R divides PQ internally and if (m_1/m_2) is negative, then externally. In particular, coordinate of the point dividing PQ externally in the ratio $m_1 : m_2$ is given by

$$\left(\frac{m_1x_2 - m_2x_1}{m_1 - m_2}, \frac{m_1y_2 - m_2y_1}{m_1 - m_2}, \frac{m_1z_2 - m_2z_1}{m_1 - m_2}\right).$$

Mid-point: Mid-point of PQ is given by $\left(\frac{x_1 + x_2}{2}, \frac{y_1 + y_2}{2}, \frac{z_1 + z_2}{2}\right)$

Centroid of a Triangle: Let $A(x_1, y_1, z_1)$, $B(x_2, y_2, z_2)$, $C(x_3, y_3, z_3)$, be the vertices of a triangle. Then its centroid G is given by

$$G = \left(\frac{x_1 + x_2 + x_3}{3}, \frac{y_1 + y_2 + y_3}{3}, \frac{z_1 + z_2 + z_3}{3}\right).$$

Direction Cosines: Let α, β, γ be the angles which a directed line makes with the positive directions of the axes of *x*, *y* and *z* respectively then cos α, cos β, cos γ are called the direction cosines of the line.

The direction cosines are usually denoted by $< l, m, n >$. Then $l = \cos\alpha$, $m = \cos\beta$, $n = \cos\gamma$.

Direction Ratios: Let a, b, c be proportional to the d.c.'s l, m, n, then a, b, c are called the direction ratios.

Relation Between Direction Cosines: $OP^2 = [x^2 + y^2 + z^2] \Rightarrow r^2 = r^2(\cos^2\alpha + \cos^2\beta + \cos^2\gamma) = r^2(l^2 + m^2 + n^2)$ $\therefore l^2 + m^2 + n^2 = 1$

D.R.'s and D.C.'s: Let a, b, c be d.r.'s and l, m, n the d.c.'s of a line, then $\frac{a}{l} = \frac{b}{m} = \frac{c}{n} = r$ (say)

$\therefore a^2 + b^2 + c^2 = r^2(l^2 + m^2 + n^2) = r^2$

where $r = \sqrt{[a^2 + b^2 + c^2]}$

$\therefore$ Direction cosines l, m, n are given by $l = \frac{a}{r}$,

$m = \frac{b}{r}$, $n = \frac{c}{r}$, where $r = \sqrt{[(a^2 + b^2 + c^2)]}$

Direction Ratios of the Line Joining two Points: Let $A(x_1, y_1, z_1)$ and $B(x_2, y_2, z_2)$ be two points, then d.r.'s of AB are $x_2 - x_1, y_2 - y_1, z_2 - z_1$, and d.c.'s of AB are $\frac{1}{r}(x_2 - x_1)$, $\frac{1}{r}(y_2 - y_1)$, $\frac{1}{r}(z_2 - z_1)$ where $r = \sqrt{[\Sigma(x_2 - x_1)^2]}$.

Angle Between Two Lines: Let θ be the angle between the lines with d.c.'s l_1, m_1, n_1 and l_2, m_2, n_2, then

$\cos\theta = l_1l_2 + m_1m_2 + n_1n_2$. Also $\sin^2\theta = \Sigma(l_1m_1)^2$

Perpendicularity and Parallelism: Let the two lines have their d.c.'s given by l_1, m_1, n_1 and l_2, m_2, n_2 respectively then they are perpendicular if $\theta = 90°$ *i.e.*, $\cos\theta = 0$ if $l_1l_2 + m_1m_2 + n_1n_2 = 0$

Also the two lines are parallel if $\theta = 90°$,

i.e., $\phi \sin\theta = 0$. *i.e.*, $\frac{l_1}{l_2} = \frac{m_1}{m_2} = \frac{n_1}{n_2}$

Length of Projection of PQ on a given Line: Let $P(x_1, y_1, z_1)$ and $Q(x_2, y_2, z_2)$ be two points and d.c.'s (not d.r.'s) of the given line be l, m and n. Then length of projection of PQ on the given line is given by $(x_2 - x_1)l + (y_2 - y_1)m + (z_2 - z_1)n$.

The Plane: A geometrical locus is a plane if it is such that if P and Q are any two points on the locus, then every point on the line PQ is also a point on the locus.

Equation of a Plane: The equation of every plane is of the first degree, *i.e.*, of the form $ax + by + cz + d = 0$, in which a, b, c are constants, where $a^2 + b^2 + c^2 \neq 0$ (*i.e.*, $a, b, c \neq 0$ simultaneously).

Planes Parallel to the Coordinate Planes:

(*a*) Equation of y-z plane is $x = 0$

(*b*) Equation of z-x plane is $y = 0$

(*c*) Equation of x-y plane is $z = 0$

(*d*) Equation of the plane parallel to x-y plane at a distance c (if $c > 0$, towards positive z-axis) is $z = c$. Similarly, planes parallel to y-z plane and z-x plane.

Equation of a Planes Parallel to the Axes: If $a = 0$, the plane is parallel to x-axis, *i.e.*, equation of the plane parallel to x-axis is $by + cz + d = 0$

Similarly, equations of planes || to y-axis and || to z-axis are $ax + cz + d = 0$ and $ax + by + d = 0$, respectively.

Equation of a Plane Through Origin: Equation of plane passing through origin is $ax + by + cz = 0$.

Equation of a Plane Through a given Point: Let the plane $ax + by + cz + d = 0$ passes through the point (x_1, y_1, z_1), then $ax_1 + by_1 + cz_1 + d = 0$.

$\therefore$ Subtracting, we get

$a(x - x_1) + b(y - y_1) + c(z - z_1) = 0$ which is the required equation.

Equation of a Plane in Intercept Form: Equation of the plane which cuts off intercepts a, b, c from the axes is $\frac{x}{a} + \frac{y}{b} + \frac{z}{c} = 1$.

Equation of a Plane in Normal Form: If the length of the perpendicular distance of the plane from the origin is p and direction cosines of this perpendicular are (l, m, n), then the equation of the plane is $lx + my + nz = p$.

Transformation of the Equation of a Plane to the Normal Form: To reduce any equation to the normal form, first write the constant term on the right-hand side and make it positive, then divide each term by $\sqrt{(a^2 + b^2 + c^2)}$, where a, b, c are coefficients of x, y and z respectively.

e.g.,
$$\frac{ax}{\pm\sqrt{(a^2+b^2+c^2)}} + \frac{by}{\pm\sqrt{(a^2+b^2+c^2)}} + \frac{cz}{\pm\sqrt{(a^2+b^2+c^2)}} = \frac{d}{\pm\sqrt{(a^2+b^2+c^2)}}$$

where + sign is to be taken if $d > 0$ and – sign is to be taken if $d < 0$.

Equation of a Plane Through Three Points: The equation of the plane through three non-collinear points (x_1, y_1, z_1), (x_2, y_2, z_2), (x_3, y_3, z_3) is

$$\begin{vmatrix} x & y & z & 1 \\ x_1 & y_1 & z_1 & 1 \\ x_2 & y_2 & z_2 & 1 \\ x_3 & y_3 & z_3 & 1 \end{vmatrix} = 0$$

Angle Between Two Planes: Consider two planes $ax + by + cz + d = 0$ and

$a'x + b'y + c'z + d' = 0$.

Angle between these planes is the angle between their normals. Since direction ratios of their normals are a, b, c and a', b', c' respectively, θ, the angle between them is given by

$$\cos\theta = \frac{aa' + bb' + cc'}{\sqrt{(a^2 + b^2 + c^2)}\sqrt{(a'^2 + b'^2 + c'^2)}}$$

∴ Planes are perpendicular if $aa' + bb' + cc' = 0$ and they are parallel if $a/a' = b/b' = c/c'$.

Planes Parallel to a given Plane: Equation of a plane parallel to the plane $ax + by + cz + d = 0$ is $ax + by + cz + d' = 0$. d' is to be found by other given condition.

A Plane Through the Line of Intersection of Two given Planes: Consider two planes

$u = ax + by + cz + d = 0$

and $v = a'x + b'y + c'z + d' = 0$.

The equation $u + \lambda v = 0$, λ a real parameter, represents the plane passing through the line of intersection of given planes and if planes are parallel, this represents a plane parallel to them.

Perpendicular Distance of a Point from the Plane: Perpendicular distance p, of the point $A(x_1, y_1, z_1)$ from the plane $ax + by + cz + d = 0$ is

givne by $p = \dfrac{|ax_1 + by_1 + cz_1 + d|}{\sqrt{(a^2 + b^2 + c^2)}}$.

Bisectors of Angles Between Two Planes: Let the equations of the two planes be $ax + by + cz + d = 0$ and $a_1x + b_1y + c_1z + d = 0$. Then equations of bisectors of angles between them are given by

$$\frac{ax + by + cz + d}{\sqrt{(a^2 + b^2 + c^2)}} = \pm \frac{a_1x + b_1y + c_1z + d_1}{\sqrt{(a_1^2 + b_1^2 + c_1^2)}}$$

Volume of a Tetrahedron: Volume of the tetrahedron with vertices $A(x_1, y_1, z_1)$, $B(x_2, y_2, z_2)$, $C(x_3, y_3, z_3)$ and $D(x_4, y_4, z_4)$ is given by

$$V = \frac{1}{6}\begin{vmatrix} x_1 & y_1 & z_1 & 1 \\ x_2 & y_2 & z_2 & 1 \\ x_3 & y_3 & z_3 & 1 \\ x_4 & y_4 & z_4 & 1 \end{vmatrix}$$

If $ax^2 + by^2 + cz^2 + 2fyz + 2gzx + 2hxy = 0$ represents a pair of planes, then the angle between the planes is given by $\tan\theta$

$$= \frac{2\sqrt{f^2 + g^2 + h^2 - ab - bc - ca}}{a + b + c}$$

Equation of a Straight Line in Symmetrical Form:

(*a*) *One point form:* Let $A(x_1, y_1, z_1)$ be a given point on the straight line and l, m, n the d.c.'s of the line, then its equation is

$$\frac{x - x_1}{l} = \frac{y - y_1}{m} = \frac{z - z_1}{n} = r \text{ (say)}$$

It should be noted that $P(1 + lr, y_1 + mr, z_1 + nr)$ is a general point on this line at a distance r from the point $A(x_1, y_1, z_1)$, *i.e.*, $AP = r$. One should note that for $AP = r$, l, m, n must be d.c.'s not d.r.'s. If a, b, c are direction ratios of the line, then equation of the line is

$$\frac{x - x_1}{l} = \frac{y - y_1}{m} = \frac{z - z_1}{n} = r, \text{ but here } AP \neq r.$$

(*b*) Equation of the line through two points $A(x_1, y_1, z_1)$ and $B(x_2, y_2, z_2)$ is $\dfrac{x - x_1}{x_2 - x_1} = \dfrac{y - y_1}{y_2 - y_1} = \dfrac{z - z_1}{z_2 - z_1}$.

Angle Between a Line and a Plane: Let equations of the line and plane be

$\dfrac{x - x_1}{l} = \dfrac{y - y_1}{m} = \dfrac{z - z_1}{n}$, and $ax + by + cz + d = 0$

respectively and θ be the angle which line makes with the plane. Then $(\pi/2 - \theta)$ is the angle between the line and the normal to the plane. So

$$\sin\theta = \frac{al+bm+cn}{\sqrt{(a^2+b^2+c^2)}\sqrt{(l^2+m^2+n^2)}}$$

Line is parallel to plane if $\theta = 0$,

i.e., if $al + bm + cn = 0$

Line is $\perp$ to the plane if line is parallel to the normal of the plane, *i.e.*, if $\frac{a}{l} = \frac{b}{m} = \frac{c}{n}$.

Equation of any Plane through a Given Line: Equation of any plane through the line

$\frac{x-x_1}{l} = \frac{y-y_1}{m} = \frac{z-z_1}{n}$ is $A(x-x_1)+B(y-y_1)+C(z-z_1)=0$, where $Al+Bm+Cn=0$.

There can be infinitely many planes through a given line. Particular plane can be found if extra condition is given.

Condition that two given lines should intersect, *i.e.*, be coplanar, *i.e.*, testing of skewness or coplanarity of two given lines. Equation of plane containing two intersecting lines:

Let the two lines be

$$\frac{x-\alpha_1}{l_1} = \frac{y-\beta_1}{m_1} = \frac{z-\gamma_1}{n_1} \quad ...(1)$$

and $$\frac{x-\alpha_2}{l_2} = \frac{y-\beta_2}{m_2} = \frac{z-\gamma_2}{n_2} \quad ...(2)$$

These lines will intersect if they lie in one plane. Now equation of the plane containing (1) is

$a(x-\alpha_1)+b(y-\beta_1)+c(z-\gamma_1)=0$ .... (3)

where $al_1 + bm_1 + cn_1 = 0$ .... (4)

The plane (3) will contain (2) if point $(\alpha_2, \beta_2, \gamma_2)$ lies on it and the normal to plane is perpendicular to (2), *i.e.*, if $a(\alpha_2-\alpha_1)+b(\beta_2-\beta_1)+c(\gamma_2-\gamma_1)=0$...(5)

where $al_2 + bm_2 + cn_2 = 0$ .... (6)

Eliminating *a, b, c* from (4), (5) and (6), the required condition is

$$\Delta = \begin{vmatrix} \alpha_2-\alpha_1 & \beta_2-\beta_1 & \gamma_2-\gamma_1 \\ l_1 & m_1 & n_1 \\ l_2 & m_2 & n_2 \end{vmatrix} = 0 \quad .... (7)$$

If we eliminate *a, b, c* from (3), (4), (6), the plane containing the two lines is

$$\begin{vmatrix} x-\alpha_1 & y-\beta_1 & z-\gamma_1 \\ l_1 & m_1 & n_1 \\ l_2 & m_2 & n_2 \end{vmatrix} = 0 \quad .... (8)$$

Magnitude and Equation to the Shortest Distance:

Suppose the equation of the lines are

$$\frac{x-\alpha}{l} = \frac{y-\beta}{m} = \frac{z-\gamma}{n} \quad ...(1)$$

and $$\frac{x-\alpha'}{l'} = \frac{y-\beta'}{m'} = \frac{z-\gamma'}{n'} \quad ...(2)$$

Let λ, μ, ν, be d.c.'s of the S.D. Since S.D. is perpendicular to each of the given lines, therefore $l\lambda+m\mu+n\nu=0; l'\lambda+m'\mu+n'\nu=0$.

Solving these, we get

$$\frac{\lambda}{mn'-m'n} = \frac{\mu}{nl'-n'l} = \frac{\nu}{lm'-l'm} = \frac{1}{\sqrt{\Sigma(mn'-m'n)^2}}$$

$$\therefore \lambda = \frac{mn'-m'n}{\sqrt{\Sigma(mn'-m'n)^2}},\ \mu = \frac{nl'-n'l}{\sqrt{\Sigma(mn'-m'n)^2}},$$

$$\nu = \frac{lm'-l'm}{\sqrt{\Sigma(mn'-m'n)^2}}$$

$P(\alpha, \beta, \gamma)$ is a point on line (1) and $Q(\alpha', \beta', \gamma')$ is a point on line (2). S.D. will be the projection of the line PQ, joining these points on the line whose d.c.'s are λ, μ, ν.

$\therefore$ S.D. $= (\alpha-\alpha')\lambda + (\beta-\beta')\mu + (\gamma-\gamma')\nu$

$$= \frac{(\alpha-\alpha')(mn'-m'n)+(\beta-\beta')(nl'-n'l)+(\gamma-\gamma')(lm'-l'm)}{\sqrt{\{\Sigma(mn'-m'n)^2\}}}$$

$$= \begin{vmatrix} \alpha-\alpha' & \beta-\beta' & \gamma-\gamma' \\ l & m & n \\ l' & m' & n' \end{vmatrix} \div \sqrt{\{\Sigma(mn'-m'n)^2\}}$$

Equation to the plane containing line (1) and the S.D. is $\begin{vmatrix} x-\alpha & y-\beta & z-\gamma \\ l & m & n \\ \lambda & \mu & \nu \end{vmatrix} = 0$ .... (3)

Equation to the plane containing the line (2) and the S.D. is $\begin{vmatrix} x-\alpha' & y-\beta' & z-\gamma' \\ l' & m' & n' \\ \lambda & \mu & \nu \end{vmatrix} = 0$ .... (4)

Equation (3) and (4) will represent the S.D. between the given lines (1) and (2).

MULTIPLE CHOICE QUESTIONS

1. The ratio in which the line joining (2, 4, 5), (3, 5, –4) is divided by the Y-Z plane is
(*a*) 2 : 3 (*b*) 3 : 2
(*c*) –2 : 3 (*d*) 4 : –3

2. The straight lines whose direction cosines are given by $al + bm + cn = 0$, $fmn + gnl + hlm = 0$ are perpendicular if
(*a*) $\frac{f}{a}+\frac{g}{b}+\frac{h}{c} = 0$
(*b*) $\sqrt{(a/f)} + \sqrt{(b/g)} + \sqrt{(c/h)} = 0$
(*c*) $\sqrt{(af)} = \sqrt{(bg)} = \sqrt{(ch)}$
(*d*) $\sqrt{(a/f)} = \sqrt{(b/g)} = \sqrt{(c/h)}$

3. If centroid of the tetrahedron OABC, where A, B and C are given by (*a*, 2, 3), (1, *b*, 2) and (2, 1, *c*) respectively, be (1, 2, 3), then distance of P(*a*, *b*, *c*) from origin is equal to
(*a*) $\sqrt{(107)}$ (*b*) $\sqrt{(14)}$
(*c*) $\sqrt{(107/4)}$ (*d*) None of these

4. The foot of the perpendicular from (0, 2, 3) to the line $\frac{x+3}{5}=\frac{y-1}{2}=\frac{z+4}{3}$ is
(*a*) (–2, 3, 4) (*b*) (2, – 1, 3)
(*c*) (2, 3, – 1) (*d*) (3, 2, – 1)

5. A plane meets the coordinate axes in A, B, C and (α, β, γ) is the centroid of the Δ ABC. Then the equation of the plane is
(*a*) $x/\alpha + y/\beta + z/\gamma = 3$
(*b*) $x/\alpha + y/\beta + z/\gamma = 1$
(*c*) $3x/\alpha + 3y/\beta + 3z/\gamma = 1$
(*d*) $\alpha x + \beta y + \gamma z = 1$

6. Equation of the plane that passes through the point (2, –3, 1) and is perpendicular to the line joining the points (3, 4, – 1) and (2, – 1, 5) is given by
(*a*) $x + 5y - 6z + 19 = 0$
(*b*) $x - 5y + 6z - 23 = 0$
(*c*) $x + 5y - 6z = 19$
(*d*) None of these

7. If the line $x = ay + b$, $z = cy + d$ and the line $x = a'y + b'$, $z = c/y + d'$ are perpendicular, then
(*a*) $aa' + bb' = 1$ (*b*) $aa' + c' = -1$
(*c*) $aa' + bb' = 0$ (*d*) None of these

8. The equations of the projection of the line $\frac{x-1}{2}=\frac{y-2}{1}=\frac{z-3}{3}$ on the plane $x + y + z - 1 = 0$ are
(*a*) $x + y + z - 1 = 0 = 2x - y - z + 3$
(*b*) $x + y - z - 1 = 0 = x + 2y - z - 3$
(*c*) $2x - y + 3z - 1 = 0 = x + y + z + 1$
(*d*) $x + 2y - 3z = 0 = x + y + z - 1$

9. The distance of the point (3, 8, 2) from the line $\frac{x-1}{2}=\frac{y-3}{4}=\frac{z-2}{3}$ measured parallel to the plane $3x + 2y - 2x + 5 = 0$ is
(*a*) 2 (*b*) 3
(*c*) 5 (*d*) None of these

10. The shortest distance between the lines
$$\frac{x-1}{2}=\frac{y-2}{3}=\frac{z-3}{4}$$
and $\frac{x-2}{3}=\frac{y-4}{4}=\frac{z-5}{5}$ is
(*a*) 1/6 (*b*) $1/\sqrt{6}$
(*c*) $(1/\sqrt{3})$ (*d*) 1/3

11. A parallelopiped is formed by planes drawn through the points (2, 3, 5) and (5, 9, 7), parallel to the co-ordinate planes. The length of a diagonal of the parallelopiped is
(*a*) 7 (*b*) $\sqrt{38}$
(*c*) $\sqrt{155}$ (*d*) None of these

12. The equation of the plane passing through the point (–2, –2, 2) and containing the line joining the points (1, 1, 1) and (1, –1, 2) is
(*a*) $x + 2y - 3z + 4 = 0$
(*b*) $3x - 4x + 1 = 0$
(*c*) $5x + 2y - 3z - 17 = 0$
(*d*) $x - 3y - 6z + 8 = 0$

13. Equation of the plane perpendicular to the plane $x - 2y + 5z + 1 = 0$ which passes through the points (2, –3, 1) and (–1, 1,–7) is given by
(*a*) $4x - 4y + z + 7 = 0$

(*b*) $4x + 7y + 2z + 11 = 0$
(*c*) $2x + y - z = 0$
(*d*) None of these

14. The coordinates of the foot of the perpendicular from the point (1, 3, 4) upon the plane $2x - y + z = 0$ are

(*a*) (1, 1, –4) (*b*) (4, –3, 1)
(*c*) (1, –3, 4) (*d*) (–1, 4, 3)

15. The image of the point (1, 3, 4) w.r.t. the plane $2x - y + z + 3 = 0$ is

(*a*) (–1, 4, 3) (*b*) (–3, 5, 2)
(*c*) (1, 3, 4) (*d*) None of these

ANSWERS

1	2	3	4	5	6	7	8	9	10
(*c*)	(*a*)	(*c*)	(*a*)	(*a*)	(*b*)	(*a*)	(*d*)	(*b*)	(*b*)
11	**12**	**13**	**14**	**15**					
(*a*)	(*d*)	(*b*)	(*d*)	(*b*)					

SOME SELECTED EXPLANATORY ANSWERS

1. Let the required ratio be $\lambda : 1$.

Then the point is

$$\frac{3\lambda+2}{\lambda+1}, \frac{5\lambda+4}{\lambda+1}, \frac{-4\lambda+5}{\lambda}$$

It lies on YZ plane, so its x-coordinate = 0

i.e., $\lambda = -2/3$

$\therefore$ Ratio is – 2: 3

i.e., 2 : 3 externally.

3. Any point on the given line is $(5r - 3, 2r + 1, 3r - 4)$. If it is the foot of the perpendicular from (1, 2, 3), then

$$5(5r-3-0)+2(2r+1-2) + 3(3r-4-3) = 0$$

i.e., $38r = 38$, *i.e.*, $r = 1$

$\therefore$ Foot of perpendicular is (2, 3, – 1).

4. Let the equation of the plane be

$x/a + y/b + z/c = 1$ ...(1)

Plane (1) meets the axes at A(a, 0,0), B(0, b, 0) and C(0, 0, c).

If (α, β, γ) is the centroid of the Δ ABC, then

$\alpha = (a + 0 + 0)/3$, $\beta = (0 + b + 0)/3$,

$\gamma = (0 + 0 + c)/3$

i.e., $a = 3\alpha$, $b = 3\beta$, $c = 3\gamma$

Putting in (1), the equation of the required plane is

$x/\alpha + y/\beta + z/\gamma + 3$.

6. The equation of the two lines can be written as

$$\frac{x-b}{a} = \frac{y-0}{1} = \frac{z-d}{c}$$

and $$\frac{x-b'}{a'} = \frac{y-0}{1} = \frac{z-d'}{c'}$$

$\therefore$ The lines are perpendicular if

$a.a' + 1.1 + c.c' = 0$, *i.e.*, $aa' + cc' = -1$.

11. The lengths of the edges are given by

$a = 5 - 2 = 3$, $b = 9 - 3 = 6$, $c = 7 - 5 = 2$

So, length of the diagonal

$$= \sqrt{9+36+4} = 7$$

12. Equation of a plane through (–2, –2, 2) is given by $a(x+2) + b(y+2) + c(z-2) = 0$.

It contains the line joining the points (1, 1, 1) and (1,–1,2), so these points also lie in the plane.

$\therefore$ $3a + 3b - c = 0$

$3a + b + 0 \cdot c = 0$

$$\therefore \quad \frac{a}{1} = \frac{b}{-3} = \frac{c}{-6} = r(\text{say})$$

or $a = r$, $b = -3r$, $c = -6r$.

$\therefore$ Equation of the plane is $x - 3y - 6z + 8 = 0$.

13

Limit and Derivatives

1. **Limit:** The number l is said to be the limit of $f(x)$ at $x = a$ if for any arbitrarily chosen positive number $\in$, however small but not zero, there exist a corresponding number δ greater than zero such that $| f(x) - l | < \in$ for all values of x for which $0 < | x - a | < \delta$. In other words, a function $f(x)$ is said to tend to a limit l [*i.e.*, $f(x) \to l$] as x tends to a [*i.e.*, $x \to a$] if the difference between $f(x)$ and l can be made as small as we please by taking x sufficiently near to a. We write it as $\lim_{x\to 0} f(x) = l$

2. **Indeterminate Forms:** For some value of x, say $x = a$ if the function $f(x)$ takes any of the following forms, $\frac{0}{0}, \frac{\infty}{\infty}, \infty - \infty, 0 \times \infty, 0^{\circ}, 1^{\infty}, \infty^{\circ}$, then $f(x)$ is said to be indeterminate at $x = a$.

3. **L-Hospital's Rule:** Let $\phi(x)$ and $\psi(x)$ be two functions, such that $\phi(a) = 0$ and $\psi(a) = 0$, then $\lim_{x\to a} \frac{\phi(x)}{\psi(x)} = \lim_{x\to a} \frac{\phi'(x)}{\psi'(x)}$ provided $\phi'$and $\psi'(a)$ are not both zero.

4. **Some Important Limits:**

(*i*) $\lim_{\theta\to 0} \frac{\sin\theta}{\theta} = 1$, where θ is measured in radians,

(*ii*) $\lim_{n\to\infty} \left(1+\frac{1}{n}\right)^n = e$

(*iii*) $\lim_{n\to -\infty} \left(1+\frac{1}{n}\right)^n = e$

(*iv*) $\lim_{n\to\infty} \left(1+\frac{a}{n}\right)^n = e$

(*v*) $\lim_{x\to 0} (1+x)^{1/x} = e$

(*vi*) $\lim_{x\to a} \frac{x^n - a^n}{x - a} = na^{n-1}, n \neq 1$

(*vii*) $\lim_{n\to\infty} a^n = \begin{cases} \infty \text{ (so does not exist)}, & \text{if } a > 1 \\ 0 & \text{if } -1 < a < 1 \\ \text{not exist}, & \text{if } a \leq -1 \end{cases}$

5. **To Find the Limit by Finding L.H.L. and R.H.L.:** We evaluate the limit of a function $f(x)$ as $x \to a$ by finding its right-hand limit (R.H.L.) and left-hand limit (L.H.L.) in the special case when the values of $f(x)$ are given by different functions of x for $x \leq a$ and for $x \geq a$. l is the limit of the function $f(x)$ as $x \to a$, if $f(a + 0) = l = f(a - 0)$
$f(a + 0)$, known as right-hand limit at $x = a$, is given by $f(a + 0) = \lim_{h\to 0} f(a + h)$ and $f(a - 0)$ known as left-hand limit at $x = a$ is given by $f(a - 0) = \lim_{h\to 0} f(a - h)$

6. **Derivative at a Point:** If $f(x)$ is a function of x defined over the open interval $]a, b[$ and if $x = c$ is any point of this interval, then $f(x)$ is said to be differentiable at the point $x = c$ iff $\lim_{h\to 0} \frac{f(c+h) - f(c)}{h}$ exists finitely. The value of this limit is called the derivative (or differential coefficient) of $f(x)$ at the point $x = c$ and is denoted by $f'(c)$, D $f(c)$ or $\left[\frac{d}{dx} f(x)\right]_{x=c}$

7. **Definition of Differential Coefficients:** Let $y = f(x)$ be a given function. Then the limit:

$\lim_{h \to 0} \frac{f(x+h)-f(x)}{h}$, if it exists, is called the differential coefficient of y w.r.t. x and is denoted by $\frac{dy}{dx}$ or dy/dx or $f'(x)$. Thus $\frac{dy}{dx} = \frac{d}{dx} f(x) = \lim_{h \to 0} \frac{f(x+h)-f(x)}{h}$

8. Fundamental Rules for Differentition:

(*i*) The differential coefficient of any constant is zero.

(*ii*) The differential coefficient of the product of a constant and a functionis equal to to the product of the constant and the differential coefficient of the function.

(*iii*) The differential coefficient of the sum (or difference) of any number of functions is the sum of their respective differential coefficients.

(*iv*) The differential coefficient of the product of two function.

Let $y = u.v$, then $\frac{dy}{dx} = u\frac{dv}{dx} + v\frac{du}{dx}$

or (first function) × (differential of second function) + (second function) × (differential of first function).

(*v*) The differential coefficient of the quotient of two function.

Let $y = \frac{u}{v}$, then $\frac{dy}{dx} = \frac{v\frac{du}{dx} - u\frac{dv}{dx}}{v^2}$

or $\frac{\text{deno.}\times(\text{d.c.of the num.}) - \text{num.}(\text{d.c.of the deno.})}{(\text{denominator})^2}$

9. Some Standard Differentials:

(*i*) $\frac{d}{dx}$ (constant) = 0

(*ii*) $\frac{d}{dx} x^n = nx^{n-1}$, where n is a constant.

(*iii*) $\frac{d}{dx} e^x = e^x$

(*iv*) $\frac{d}{dx} a^x = a^x \log_e a,\ a > 0$

(*v*) $\frac{d}{dx}(\log_e x) = \frac{1}{x},\ x > 0$

(*vi*) $\frac{d}{dx}(\log_a x) = \frac{1}{x}\log_a e,\ x > 0$

(*vii*) $\frac{d}{dx} \sin x = \cos x$

(*viii*) $\frac{d}{dx} \cos x = -\sin x$

(*ix*) $\frac{d}{dx} \tan x = \sec^2 x$

(*x*) $\frac{d}{dx} \operatorname{cosec} x = -\operatorname{cosec} x \cot x$

(*xi*) $\frac{d}{dx} \sec x = \sec x \tan x$

(*xii*) $\frac{d}{dx} \cot x = -\operatorname{cosec}^2 x$

(*xiii*) $\frac{d}{dx} \sin^{-1} x = \frac{1}{\sqrt{(1-x^2)}}$

(*xiv*) $\frac{d}{dx} \cos^{-1} x = -\frac{1}{\sqrt{(1-x^2)}}$

(*xv*) $\frac{d}{dx} \tan^{-1} x = \frac{1}{1+x^2}$

(*xvi*) $\frac{d}{dx} \cot^{-1} x = -\frac{1}{1+x^2}$

(*xvii*) $\frac{d}{dx} \sec^{-1} x = \frac{1}{|x|\sqrt{(x^2-1)}}$

(*xviii*) $\frac{d}{dx} \operatorname{cosec}^{-1} x = -\frac{1}{|x|\sqrt{(x^2-1)}}$

10. Differentiation of a Function (Chain Rule): If y is a function of t and t is a function of x, then $\frac{dy}{dx} = \frac{dy}{dx} \times \frac{dt}{dx}$. Thus if $y = f(t)$ and $t = \phi(x)$, then, $\frac{dy}{dx} = f'(t)$ and $\frac{dt}{dx} = \phi'(x)$.

$\therefore \frac{dy}{dx} = \frac{dy}{dt} \cdot \frac{dt}{dx} = f'(t) \cdot \phi'(x)$. This rule is called chain rule. This chain rule can be extended as follows. Let $y = f(t), t = \phi(z), z = \psi(x)$, then $\frac{dy}{dx}$

$= \frac{dy}{dt} \cdot \frac{dt}{dz} \cdot \frac{dz}{dx} = f'(t) . \phi'(z) . \psi'(x)$

e.g., Let $y = \log \sin x^3 = \log t$. Putting $t = \sin x^3 = \sin z, z = x^3$,

$\frac{dy}{dx} = \frac{dy}{dt} \cdot \frac{dt}{dz} \cdot \frac{dz}{dx} = (1/t).\cos z \, 3x^2$

$= (1/\sin x^3) \cos x^3.3x^2 = 3x^2 \cot x^3$.

11. Differentiation of Parametric Equations: Let x and y be the function of t (parameter), *i.e.*, let $x = f(t)$ and $y = \phi(t)$, then $\frac{dy}{dx} = \frac{dy}{dt} \Big/ \frac{dx}{dt} = \phi'(t)/ f'(t)$.

MULTIPLE CHOICE QUESTIONS

1. The value of $\frac{d}{dx}(x^x)$ is

(*a*) $x\,x^{x-1}$ (*b*) $x^x \log ex$

(*c*) $x^x \log x$ (*d*) None of these

2. If $y = \log |x|, x \neq 0$, then $\frac{dy}{dx}$ is given by

(*a*) $\frac{1}{|x|}$ (*b*) $\frac{1}{x}$

(*c*) $-\frac{1}{x}$ (*d*) Does not exist

3. If $y = \sqrt{\left(\frac{1+\cos 2\theta}{1-\cos 2\theta}\right)}$, then $\frac{dy}{d\theta}$ at $\theta = 3\pi/4$ is

(*a*) −2 (*b*) 2

(*c*) ± 2 (*d*) None of these

4. Differential coefficient of $\sec \tan^{-1} x$ is

(*a*) $x/(1+x^2)$ (*b*) $x\sqrt{(1+x^2)}$

(*c*) $1/\sqrt{(1+x^2)}$ (*d*) $x/\sqrt{(1+x^2)}$

5. If $x^y = e^{x-y}$, then dy/dx is

(*a*) $\frac{y}{1+\log x}$ (*b*) $\frac{x-y}{(1+\log x)^2}$

(*c*) $\frac{x-y}{1+\log x}$ (*d*) $\frac{\log x}{(1+\log x)^2}$

6. If $x = a \cos^3 \theta, y = a \sin^3 \theta$, then $\sqrt{[1+(dy/dx)^2]} =$

(*a*) $\tan^2 \theta$ (*b*) $\sec^2 \theta$

(*c*) $\sec \theta$ (*d*) $|\sec \theta|$

7. If $y = \sqrt{[\sin x + \sqrt{\{\sin x + \sqrt{(\sin x + ...)}\}}]}$, then $\frac{dy}{dx} =$

(*a*) $\frac{2y-1}{\cos x}$ (*b*) $\frac{\cos x}{2y-1}$

(*c*) $\frac{2x-1}{\cos y}$ (*d*) $\frac{\cos y}{2x-1}$

8. If $y = \sin^{-1}\sqrt{(1-x)} + \cos^{-1}\sqrt{x}$, then $\frac{dy}{dx} =$

(*a*) $\frac{1}{\sqrt{[x(1-x)]}}$ (*b*) $\frac{-1}{\sqrt{[x(1-x)]}}$

(*c*) $\frac{1}{\sqrt{[x(1+x)]}}$ (*d*) None of these

9. The derivative of $\tan^{-1}\left(\frac{2x}{1-x^2}\right)$ w.r.t. $\cos^{-1}\left(\frac{1-x^2}{1+x^2}\right)$ is :

(*a*) 1/2 (*b*) 1

(*c*) 0 (*d*) 2

10. If $y = x^{x^{x...\infty}}$, then $x(1 - y \log x)dy/dx =$

(*a*) x^2 (*b*) y^2

(*c*) xy^2 (*d*) None of these

11. $\lim\limits_{x\to 0} \frac{1-\cos x}{x^2}$ is equal to :

(*a*) 0 (*b*) 1/4

(*c*) 1/2 (*d*) 1

12. $\lim\limits_{x\to \pi/4} \left(\frac{\sin x - \cos x}{x - \pi/4}\right)$ equals:

(*a*) $\sqrt{2}$ (*b*) $1/\sqrt{2}$

(*c*) 2 (*d*) 1/2

13. $\lim_{x\to\infty}\sqrt{\left(\frac{x+\sin x}{x-\cos x}\right)}=?$

(*a*) 0 (*b*) 1
(*c*) −1 (*d*) None of these

14. $\lim_{x\to 0}\left\{\tan\left(\frac{\pi}{4}+x\right)\right\}^{1/x}$ is

(*a*) 1 (*b*) −1
(*c*) e^2 (*d*) *e*

15. $\lim_{n\to\infty}(4^n+5^n)^{1/n}$ is equal to

(*a*) 4 (*b*) 5
(*c*) *e* (*d*) None of these

16. $\lim_{x\to 0}\left(\frac{1+\tan x}{1+\sin x}\right)^{\text{cosec}\,x}$ is equal to

(*a*) *e* (*b*) 1/*e*
(*c*) 1 (*d*) None of these

17. $\lim_{n\to\infty}\left[\frac{1}{n^3+1}+\frac{4}{n^3+1}+\frac{9}{n^3+1}+\ldots+\frac{n^2}{n^3+1}\right]$ is equal to

(*a*) 1 (*b*) 2/3
(*c*) 1/3 (*d*) 0

18. $\lim_{x\to\infty}(1+2/x)^x$ equals

(*a*) *e* (*b*) ∞
(*c*) e^2 (*d*) 1/*e*

19. $\lim_{x\to\infty}\left(\frac{x-2}{x+2}\right)^{x+1}$ is given by

(*a*) e^{-4} (*b*) e^{-2}
(*c*) e^2 (*d*) e^4

20. $\lim_{n\to\infty}\left(\frac{n+1}{n+2}\right)^{2n+1}$ is

(*a*) *e* (*b*) e^{-2}
(*c*) e^{-1} (*d*) 1

ANSWERS

1	2	3	4	5	6	7	8	9	10
(*b*)	(*b*)	(*b*)	(*d*)	(*d*)	(*d*)	(*b*)	(*b*)	(*b*)	(*b*)
11	**12**	**13**	**14**	**15**	**16**	**17**	**18**	**19**	**20**
(*c*)	(*a*)	(*b*)	(*c*)	(*b*)	(*c*)	(*c*)	(*c*)	(*a*)	(*b*)

SOME SELECTED EXPLANATORY ANSWERS

1. $y=x^x$

$\Rightarrow \log y = x\log x$

$\Rightarrow \frac{1}{y}\frac{dy}{dx}=\log x + x\,\frac{1}{x}$

$\frac{dy}{dx}=x^x(1+\log x)=x^x\log(ex).$

3. $y=\sqrt{\left(\frac{2\cos^2\theta}{2\sin^2\theta}\right)}=\sqrt{(\cot^2\theta)}=-\cot\theta.$

[∵ cot θ is −ve in the neighbourhood of 3π/4, *i.e.*, in Q_2]

$\left(\frac{dy}{d\theta}\right)_{\theta=3\pi/4}=(\text{cosec}^2\,\theta)_{3\pi/4}=2.$

4. $\frac{d}{dx}\sec\tan^{-1}x$

$=\sec\tan^{-1}x.\tan\tan^{-1}x.1/(1+x^2)$

$=\sec\theta.\,x/(1+x^2),\ \tan^{-1}x=\theta$

$=x\sqrt{(1+\tan^2\theta)}/(1+x^2)=x/\sqrt{(1+x^2)}.$

6. $\frac{dy}{dx}=\frac{dy/d\theta}{dx/d\theta}=\frac{3a\sin^2\theta\cos\theta}{-3a\cos^2\theta\sin\theta}=-\tan\theta$

$\therefore \sqrt{[1+(dy/dx)^2}=\sqrt{(1+\tan^2\theta)}=|\sec\theta|.$

10. $y=x^y$

$\therefore \log y = y\log x$

Diff. $x(1-y\log x)\frac{dy}{dx}=y^2.$

12. Putting $x = y + \pi/4$

Given limit

$$= \lim_{y\to 0}\left(\frac{\sin(y+\pi/4)-\cos(y+\pi/4)}{y}\right)$$

$$= \lim_{y\to 0}\left(\sqrt{2}.\sin y/y\right) = \sqrt{2}.$$

13. $$\lim_{x\to\infty}\sqrt{\left(\frac{x+\sin x}{x-\cos x}\right)}$$

$$= \lim_{x\to\infty}\sqrt{\left(\frac{1+(\sin x/x)}{1-(\cos x/x)}\right)} = \lim_{x\to\infty}\sqrt{(1)} = 1$$

[This is so as $\lim_{x\to\infty}\frac{\sin x}{x}$ and $\lim_{x\to\infty}\frac{\cos x}{x}$ both equal to 0.]

15. Given limit

$$= \lim_{n\to\infty}(4^n + 5^n)^{1/n}$$

$$= \lim_{n\to\infty} 5\left[\left\{1/(4/5)^n\right\}^{(5/4)^n}\right]^{(1/n).(4/5)^n}$$

$= 5.e^0 = 5.$ $\qquad (\because (4/5)^n \to 0 \text{ as } n\to\infty)$

16. Given limit

$$= \lim_{x\to 0}\left[(1+\tan x)^{\operatorname{cosec} x}\times\left\{1/(1+\sin x)^{\operatorname{cosec} x}\right\}\right]$$

$$= \lim_{x\to 0}\left[\left\{(1+\tan x)^{\cot x}\right\}^{\sec x}\times\left\{1/(1+\sin x)^{\operatorname{cosec} x}\right\}\right]$$

$= e^{\sec 0}.\ 1/e = e.1/e = 1.$

17. Given limit

$$= \lim_{x\to\infty}\frac{1^2+2^2+3^2+\ldots+n^2}{1+n^3} = \lim_{x\to\infty}\frac{\Sigma n^2}{1+n^3}$$

$$= \lim_{x\to\infty}\frac{1}{6}\frac{n(n+1)(2n+1)}{1+n^3}$$

$$= \lim_{x\to\infty}\frac{1}{6}(1+1/n)(2+1/n)/(1/n^3+1)$$

$= (1/6).\ 1.2/(1) = (1/3).$

18. Given limit

$$= \lim_{n\to\infty}\left[\left(1+\frac{2}{x}\right)^{x/2}\right]^2 = e^2.$$

19. Given limit

$$= \lim_{n\to\infty}\left[\left(1-\frac{4}{x+2}\right)^{-(x+2)/4}\right]^{-4(x+1)/(x+2)}$$

$$= e^{-4}\left(\because \text{As } x\to\infty, \frac{-4(x+1)}{x+2} = \frac{-4(1+1/x)}{1+2/x}\to 4\right).$$

20. $$\lim_{n\to\infty}\left(\frac{n+1}{n+2}\right)^{2n+1} = \lim_{n\to\infty}\left[1-\frac{1}{n+2}\right]^{2n+1}$$

$$= \lim_{n\to\infty}\left[1-(2n+1)\frac{1}{n+2}+\frac{(2n+1)2n}{2!}\left(\frac{1}{n+2}\right)^2 - \frac{(2n+1).(2n).(2n-1)}{3!}\left(\frac{1}{n+2}\right)^3+\ldots\right]$$

$$= \lim_{n\to\infty}\left[1-\frac{2+1/n}{1+2/n}+\frac{(2+1/n).2}{2!}.\frac{1}{(1+2/n)^2} - \frac{(2+1/n)2(2-1/n)}{3!}.\frac{1}{(1+2/n)^2}+\ldots\right]$$

$$= 1-2+\frac{2^2}{2!}-\frac{2^3}{3!}+\ldots = e^{-2}.$$

14

Inverse Trigonometrical Function

1. **Inverse Function (Definition):** A function $f: \mathrm{A} \to \mathrm{B}$ is said to be invertible, if f is one-one and onto and then inverse function

   $f^{-1}: \mathrm{B} \to \mathrm{A}$ is defined by

   $f^{-1}(y) = x$

   If $f(x) = y, x \in \mathrm{A}, y \in \mathrm{B}$.

2. (*i*) $\sin^{-1} x \in [-\pi/2, \pi/2]$

   (*ii*) $\cos^{-1} x \in [0, \pi]$

   (*iii*) $\tan^{-1} x \in [-\pi/2, \pi/2]$

   (*iv*) $\cot^{-1} x \in [-\pi/2, 0) \cup (0, \pi/2]$

   (*v*) $\sec^{-1} x \in [0, \pi/2) \cup (\pi/2, \pi]$

   (*vi*) $\operatorname{cosec}^{-1} x \in [(-\pi/2, 0) \cup (0, \pi/2)]$

3. (*i*) $\sin(\sin^{-1} x) = x$ and $\sin^{-1}(\sin x) = x$

   (*ii*) $\cos(\cos^{-1} x) = x$ and $\cos^{-1}(\cos x) = x$

   (*iii*) $\tan(\tan^{-1} x) = x$ and $\tan^{-1}(\tan x) = x$

   (*iv*) $\cot(\cot^{-1} x) = x$ and $\cot^{-1}(\cot x) = x$

   (*v*) $\sec(\sec^{-1} x) = x$ and $\sec^{-1}(\sec x) = x$

   (*vi*) $\operatorname{cosec}(\operatorname{cosec}^{-1} x) = x$ and $\operatorname{cosec}^{-1}(\operatorname{cosec} x) = x$

4. (*i*) $\sin^{-1}(\cos x) = (\pi/2) - x$

   (*ii*) $\cos^{-1}(\sin x) = (\pi/2) - x$

   (*iii*) $\tan^{-1}(\cot x) = (\pi/2) - x$

   (*iv*) $\cot^{-1}(\tan x) = (\pi/2) - x$

   (*v*) $\sec^{-1}(\operatorname{cosec} x) = (\pi/2) - x$

   (*vi*) $\operatorname{cosec}^{-1}(\sec x) = (\pi/2) - x$

5. (*i*) $\sin^{-1}(-x) = -\sin^{-1} x$

   (*ii*) $\cos^{-1}(-x) = \pi - \cos^{-1} x$

   (*iii*) $\tan^{-1}(-x) = -\tan^{-1} x$

   (*iv*) $\cot^{-1}(-x) = \pi - \cot^{-1} x$

   (*v*) $\sec^{-1}(-x) = \pi - \sec^{-1} x$

   (*vi*) $\operatorname{cosec}^{-1}(-x) = -\operatorname{cosec}^{-1} x$

6. (*i*) $\sin^{-1} x = \operatorname{cosec}^{-1}\left(\frac{1}{x}\right)$

   (*ii*) $\cos^{-1} x = \sec^{-1}\left(\frac{1}{x}\right)$

   (*iii*) $\tan^{-1} x = \cot^{-1}\left(\frac{1}{x}\right)$

   (*iv*) $\cot^{-1} x = \tan^{-1}\left(\frac{1}{x}\right)$

   (*v*) $\sec^{-1} x = \cos^{-1}\left(\frac{1}{x}\right)$

   (*vi*) $\operatorname{cosec}^{-1} x = \sin^{-1}\left(\frac{1}{x}\right)$

7. (*i*) $\sin^{-1} x + \cos^{-1} x = \frac{\pi}{2}, \ (-1 \le x \le 1)$

   (*ii*) $\tan^{-1} x + \cot^{-1} x = \frac{\pi}{2}, x \in \mathrm{R}$

   (*iii*) $\sec^{-1} x + \operatorname{cosec}^{-1} x = \frac{\pi}{2}, x \le -1$ or $x \ge 1$

8. (*i*) $\sin^{-1} x + \sin^{-1} y$

   $= \sin^{-1}\left[x\sqrt{1-y^2} + y\sqrt{1-x^2}\right]$

(*ii*) $\sin^{-1} x - \sin^{-1} y$

$= \sin^{-1}\left[x\sqrt{1-y^2} - y\sqrt{1-x^2}\right]$

(*iii*) $\cos^{-1} x + \cos^{-1} y$

$= \cos^{-1}\left[xy - \sqrt{1-x^2}\sqrt{1-y^2}\right]$

(*iv*) $\cos^{-1} x - \cos^{-1} y$

$= \cos^{-1}\left[xy + \sqrt{1-x^2}\sqrt{1-y^2}\right]$

(*v*) $\tan^{-1} x + \tan^{-1} y = \tan^{-1}\frac{x+y}{1-xy}$, if $xy < 1$

(*vi*) $\tan^{-1} x + \tan^{-1} y = \pi + \tan^{-1}\frac{x+y}{1-xy}$, if $xy > 1$

(*vii*) $\tan^{-1} x - \tan^{-1} y = \tan^{-1}\frac{x-y}{1+xy}$

(*viii*) $\tan^{-1} x + \tan^{-1} y + \tan^{-1} z$

$= \tan^{-1}\frac{x+y+z-xyz}{1-xy-yz-zx}$

9. (*i*) $2\sin^{-1} x = \sin^{-1} 2x\sqrt{1-x^2}$

(*ii*) $2\cos^{-1} x = \cos^{-1} 2x\sqrt{1-x^2}$

(*iii*) $2\tan^{-1} x = \sin^{-1}\frac{2x}{1+x^2}$

(*iv*) $2\tan^{-1} x = \cos^{-1}\frac{1-x^2}{1+x^2}$

(*v*) $2\tan^{-1} x = \tan^{-1}\frac{2x}{1-x^2}$

(*vi*) $3\sin^{-1} x = \sin^{-1}\left(3x-4x^3\right)$

(*vii*) $3\cos^{-1} x = \cos^{-1}\left(4x^3-3x\right)$

(*viii*) $3\tan^{-1} x = \tan^{-1}\frac{3x-x^3}{1-3x^2}$

MULTIPLE CHOICE QUESTIONS

1. The value of sin [2 $\sin^{-1}$ (0.8)] is equal to

(*a*) sin 1.2° (*b*) sin 1.6°
(*c*) 0.48 (*d*) 0.96

2. If $\sin^{-1} x = \pi/5$, for some $x \in (-1, 1)$, then the value of $\cos^{-1} x$ is

(*a*) $3\pi/10$ (*b*) $5\pi/10$
(*c*) $7\pi/10$ (*d*) $9\pi/10$

3. The value of $\tan^{-1} 2 + \tan^{-1} 3$ is

(*a*) $-\pi/4$ (*b*) $-3\pi/4$
(*c*) $\pi/4$ (*d*) $3\pi/4$

4. $\tan\left[\frac{1}{2}\sin^{-1}\frac{2a}{1+a^2} + \frac{1}{2}\cos^{-1}\frac{1-a^2}{1+a^2}\right] =$

(*a*) $\frac{2a}{1+a^2}$ (*b*) $\frac{a}{1+a^2}$

(*c*) $\frac{2a}{1-a^2}$ (*d*) $\frac{2}{1+a^2}$

5. The value of sin [$\sin^{-1}(\sqrt{5}/4) + \tan^{-1}\sqrt{(5/11)}$] is

(*a*) $\frac{\sqrt{5}}{4\sqrt{11}}$ (*b*) $\frac{4}{\sqrt{35}}$

(*c*) $\frac{\sqrt{55}}{8}$ (*d*) None of these

6. tan [$\sin^{-1}$ (3/5) – $\cos^{-1}$ (–4/5)] is equal to

(*a*) 24/7 (*b*) –24/7
(*c*) 7/24 (*d*) –7/24

7. The value of

$\cot^{-1}\left[\frac{\sqrt{(1-\sin x)}+\sqrt{(1+\sin x)}}{\sqrt{(1-\sin x)}-\sqrt{(1+\sin x)}}\right]$ is

(*a*) $\pi - x$ (*b*) $2\pi - x$
(*c*) $x/2$ (*d*) $\pi - x/2$

8. sin [$\cot^{-1}$ cos $\tan^{-1} x$] is equal to

(*a*) $\sqrt{\left(\frac{x^2-1}{x^2+2}\right)}$ (*b*) $\sqrt{\left(\frac{x-2}{x^2+1}\right)}$

(c) $\sqrt{\left(\dfrac{x^2+1}{x^2+2}\right)}$ (d) None of these

9. If $\cos^{-1} p + \cos^{-1} q + \cos^{-1} r = \pi$, then $p^2 + q^2 + r^2 + 2pqr =$

(a) 3 (b) 1
(c) –1 (d) None of these

10. The solution of

$\sin^{-1} \dfrac{2a}{1+a^2} - \cos^{-1} \dfrac{1-b^2}{1+b^2} = \tan^{-1} \dfrac{2x}{1-x^2}$ is

(a) $\dfrac{a-b}{1-ab}$ (b) $\dfrac{1+ab}{a-b}$

(c) $\dfrac{ab-1}{a+b}$ (d) $\dfrac{a-b}{1+ab}$

11. The x satisfying $\sin^{-1} x + \sin^{-1}(1-x) = \cos^{-1} x$ are

(a) 1, 0 (b) 1, –1
(c) 0, 1/2 (d) None of these

12. The principal value of $\sin^{-1}[\sin(2\pi/3)]$ is:

(a) $-2\pi/3$ (b) $2\pi/3$
(c) $4\pi/3$ (d) None of these

13. The value of

$\cot^{-1}\left[\dfrac{\sqrt{1-\sin x}+\sqrt{1+\sin x}}{\sqrt{(1-\sin x)}-\sqrt{(1+\sin x)}}\right]$ is:

(a) $\pi - x$ (b) $2\pi - x$

(c) $x/2$ (d) $\pi - \dfrac{1}{2}x$

14. The value of $\tan[\cos^{-1}(4/5) + \tan^{-1}(2/3)]$ or $\tan[\sin^{-1}(3/5 + \cos^{-1}(3/2)]$

(a) 6/17 (b) 7/16
(c) 17/6 (d) None of these

15. The value of $\sin \cot^{-1} x$ is:

(a) $\sqrt{1+x^2}$ (b) x

(c) $(1 + x^2) - 3/2$ (d) $(1 + x^2)^{-1/2}$

16. Find the value of

$\cot^{-1}\left\{\dfrac{\sqrt{1+\sin x}+\sqrt{1-\sin x}}{\sqrt{1+\sin x}-\sqrt{1-\sin x}}\right\}$,

when $0 < x < \dfrac{\pi}{2}$.

(a) $\dfrac{x}{2}$ (b) $\dfrac{\pi}{4}+\dfrac{x}{2}$

(c) $\dfrac{\pi}{2}$ (d) $\dfrac{3\pi}{2}$

17. Simplify $\tan^{-1}\{x+\sqrt{1+x^2}\}, x \in R$.

(a) $\dfrac{\pi}{2}$ (b) $\dfrac{\pi}{2}-\dfrac{1}{2}\cot^{-1} x$

(c) $\dfrac{\pi}{2}-\dfrac{x}{2}$ (d) $\dfrac{1}{2}\cot^{-1} x^2$

18. If $\sin^{-1} x - \cos^{-1} x = \dfrac{\pi}{6}$, then $x =$

(a) $\dfrac{1}{2}$ (b) $\dfrac{\sqrt{3}}{2}$

(c) $\dfrac{-1}{2}$ (d) $\dfrac{4}{3}$

19. If $\alpha = \tan^{-1}\left(\tan\dfrac{5\pi}{4}\right)$ and $\beta = \tan^{-1}\left(-\tan\dfrac{2\pi}{3}\right)$, then

(a) $4\alpha = 3\beta$ (b) $3\alpha = 4\beta$

(c) $\alpha - \beta = \dfrac{7\pi}{12}$ (d) none of these

20. Simplify $\tan^{-1}\left(\dfrac{a+bx}{b-ax}\right), x < \dfrac{b}{a}$.

(a) $\tan^{-1}(a/b) + \tan^{-1} x$
(b) $\tan^{-1}(a/b) - x$
(c) $\tan^{-1}(x/a)$
(d) $\tan^{-1} b - \tan^{-1} a$

ANSWERS

1	2	3	4	5	6	7	8	9	10
(d)	(a)	(d)	(c)	(c)	(a)	(d)	(c)	(b)	(d)
11	12	13	14	15	16	17	18	19	20
(c)	(d)	(d)	(c)	(d)	(a)	(b)	(b	(a)	(a)

SOME SELECTED EXPLANATORY ANSWERS

1. $\sin[2\sin^{-1}(0.8)] = \sin 2\theta$
where $\sin^{-1}(0.8) = \theta$, *i.e.*, $\sin\theta = 0.8$
$= 2\sin\theta\sqrt{(1-\sin^2\theta)}$
$= 2(0.8)\sqrt{[1-(0.8)^2]} = 0.96.$

2. $\cos^{-1}x = \pi/2 - \sin^{-1}x$
$= \pi/2 - \pi/5 = 3\pi/10.$

3. $\tan^{-1}2 + \tan^{-1}3$

$= \pi + \tan^{-1}\dfrac{2+3}{1-2.3}$ $\quad\because 2.3 > 1$

$= \pi - \tan^{-1}1 = \pi - \pi/4 = 3\pi/4.$

5. Given expression
$= \sin[\sin^{-1}(\sqrt{5}/4) + \sin^{-1}(\sqrt{5}/4)]$
$= \sin(2\sin^{-1}(\sqrt{5}/4)]$
$= \sin\sin^{-1}[2.\sqrt{5}/4.\sqrt{(1-5/16)}]$
$= \sqrt{(55)}/8.$

6. Given expression
$= \tan[\sin^{-1}(3/5) - \pi + \cos^{-1}(4/5)]$
$= \tan[-\pi + 2\sin^{-1}(3/5)]$
$= \tan[2\sin^{-1}(3/5)]$
$= \tan[2\tan^{-1}(3/4)] = \tan 2\theta,$
where $\theta = \tan^{-1}(3/4)$
$= 2\tan\theta/(1-\tan^2\theta) = 24/7.$

9. Given expression
$\Rightarrow \cos^{-1}[pq - \sqrt{\{(1-p^2)(1-q^2)\}}]$
$= \pi - \cos^{-1}r = \cos^{-1}(-r)$
$\Rightarrow pq - \sqrt{\{(1-p^2)(1-q^2)\}} = -r$
$\Rightarrow (pq+r)^2 = (1-p^2)(1-q^2)$
$\Rightarrow p^2 + q^2 + r^2 + 2pqr = 1.$

10. Given expression
$\Rightarrow 2\tan^{-1}a - 2\tan^{-1}b = 2\tan^{-1}x$

$$\Rightarrow 2\left[\tan^{-1}\frac{a-b}{1+ab}\right] = 2\tan^{-1}x$$

$$\Rightarrow x = \frac{a-b}{1+ab}.$$

12. First note that $-\pi/2 \le \sin^{-1}x \le \pi/2$ for principal value of
$\sin^{-1}x$ and for $0 \le x \le 1$,
we have $0 \le \sin^{-1}x \le \pi/2$
$\therefore \sin^{-1}[\sin(2\pi/3)] = \sin^{-1}(\sqrt{3}/2) = \pi/3$
for the principal value.

14. We have $\tan[\cos^{-1}(4/5) + \tan^{-1}(2/3)]$
$= \tan[\tan^{-1}(3/4) + \tan^{-1}(2/3)]$

$$= \tan\tan^{-1}\left[\frac{(3/4)+(2/3)}{1-(3/4)\cdot(2/3)}\right]$$

$$= \tan\tan^{-1}\left[\frac{17/12}{1/2}\right] = \tan\tan^{-1}\left(\frac{17}{6}\right)$$

$$= \frac{17}{6}$$

Also, $\cos^{-1}(4/5) = \sin^{-1}(3/5)$
and $\tan^{-1}(2/3) = \cot^{-1}(3/2)$

15. We have,

$$\sin\cot^{-1}x = \sin\sin^{-1}\left[1/\sqrt{1+x^2}\right]$$

$$= 1/\sqrt{1+x^2}$$

20. We have $\tan^{-1}\left(\dfrac{a+bx}{b-ax}\right)$

$$= \tan^{-1}\left(\frac{\frac{a}{b}+x}{1-\frac{a}{b}x}\right) = \tan^{-1}\frac{a}{b} + \tan^{-1}x$$

15

Matrices

Matrices: A system of mn numbers (real or complex) arranged in a rectangular array of m rows and n columns is called a matrix of order $m \times n$ or *an* $m \times n$ matrix (to be read as 'm by n' matrix). A $m \times n$ matrix is written as

$$\begin{bmatrix} a_{11} & a_{12} & \dots & a_{1n} \\ a_{21} & a_{22} & \dots & a_{2n} \\ \dots & \dots & \dots & \dots \\ \dots & \dots & \dots & \dots \\ a_{m1} & a_{m2} & \dots & a_{mn} \end{bmatrix}$$

Where a_{ij} represent the number in the i-th row and j-th column. The numbers a_{11}, a_{12}, etc., are called the elements or entries of the matrix.

Row Matrix: A single row matrix is called a row matrix or a row vector,

e.g., the matrix $[a_{11}, a_{12}, ..., a_{1n}]$ is a $1 \times n$ row matrix.

Column Matrix: A single column matrix is called a column matrix or a column vector.

e.g., the matrix $\begin{bmatrix} a_{1j} \\ a_{2j} \\ \dots \\ \dots \\ a_{mj} \end{bmatrix}$ is a $m \times 1$ column matrix.

Square Matrix: If $m = n$, *i.e.*, if the number of rows and columns of a matrix are equal, say n, then it is called a square matrix of order n.

Null (or zero) matrix: If all the elements of a matrix are equal to zero, then it is called a null matrix and is denoted by $O_{m \times n}$ or O.

Diagonal Matrix: A square matrix in which all its elements are zero except those in the leading diagonal, is called a diagonal matrix. Thus in a diagonal matrix, $a_{ij} = 0$ if $i \neq j$.

The diagonal matrices of order 2 and 3 are

$$\begin{bmatrix} k_1 & 0 \\ 0 & k_2 \end{bmatrix}, \begin{bmatrix} k_1 & 0 & 0 \\ 0 & k_2 & 0 \\ 0 & 0 & k_3 \end{bmatrix}.$$

Scalar Matrix: A square matrix in which all the diagonal elements are equal and all other elements equal to zero is called as a scalar matrix, *i.e.*, in a scalar matrix $a_{ij} = k$, for $i = j$ and $a_{ij} = 0$ for $i \neq j$. Thus

$\begin{bmatrix} k & 0 & 0 \\ 0 & k & 0 \\ 0 & 0 & k \end{bmatrix}$ is a scalar matrix.

Transpose of a Matrix: Let A be *a*, $m \times n$ marix. Then the matrix of order $n \times m$ obtained by changing its rows into columns and columns into rows is called the transpose of A and is denoted by A' or A^T or A^t.

***Symbolically*:** If $A = [a_{ij}]_{m \times n}$ is a matrix then the transpose of the matrix A denoted by A' or A^T or A^t is defined as $A' = [a'_{ji}]_{n \times m}$ where $a'_{ji} = a_{ij}$.

***Some Important Results of Transposed Matrices*:**

(*i*) $(A + B)' = A' + B'$

(*ii*) $(kA)' = kA'$ where k is constant

(*iii*) $(AB)' = B'A'$.

Conjugate of a Matrix: If A is a matrix having complex numbers as its elements, the matrix obtained from A by replacing each element of A by its conjugate is called conjugate of A and is denoted by $\overline{A}$.

Symmetric Matrix: A square matrix A is said to be symmetric if A' = A. Thus a square matrix $A = [a_{ij}]$ is symmetric if $a_{ij} = a_{ji}$ for all values of i and j.

Skew-symmetric Matrix: A square matrix A is said to be skew-symmetric if A' = – A. Thus a square matrix $A = [a_{ij}]$ is skew-symmetric if $a_{ij} = -a_{ji}$ for all values of i and j. In particular $a_{ii} = -\bar{a}_{ii} \Rightarrow 2a_{ii} = 0 \Rightarrow a_{ii} = 0$,

i.e., all diagonal elements of a skew-symmetric matrix must vanish.

Addition of Matrices: Let there be two matrices A and B of the same order $m \times n$. Then there sum denoted by A + B is defined to be the matrix of order $m \times n$ obtained by adding the corresponding elements of A and B. Thus if $A = [a_{ij}]_{m \times n}$ and $B = [b_{ij}]_{m \times n}$, then $A + B = [a_{ij} + b_{ij}]_{m \times n}$. Note that we can find the sum A + B only if the matrices A and B are of the same order, *i.e.*, if the number of rows and columns of matrix A are equal to those of matrix B.

Multiplication of Matrices: Let A be $m \times n$ and B be $n \times p$ matrices. Then the product of matrices A and B denoted by A.B is the matrix of order $m \times p$ whose (i, j)th element is obtained by adding the products of corresponding elements of i-*th* row of A and j-th column of B.

Symbolically: Let A $[a_{ij}]_{m \times n}$ be a matrix of order $m \times n$ and $B = [b_{jk}]_{n \times p}$ be a matrix of order $n \times p$. Then the product matrix AB is denoted as a matrix $C = [c_{ik}]_{m \times p}$ of order $m \times p$. Where

$$c_{ik} = \sum_{j=1}^{n} a_{ij}\, b_{jk}, \text{ i.e., } c_{ik} = a_{i1}\, b_{1k} + a_{i2}\, b_{2k} + \dots + a_{in}\, b_{nk}.$$

Properties of Matrix Multiplication:

(*a*) **Associative Law for Multiplication:** *i.e.*, if A, B and C be three matrices of order $m \times n$, $n \times p$ and $p \times q$ respectively, then (A B)C = A (B C).

(*b*) **Distributive Law:** If A, B, C be three matrices of order $m \times n$, $n \times p$ and $p \times q$ respectively, then A(B + C) = A.B + A.C.

(*c*) Matrix multiplication is not commutative in general, *i.e.*, A.B ≠ B.A (in general).

Minor of an Element: The **minor** M_{ij} of an element a_{ij} of a square matrix A is the determinant of the matrix that remains after deleting the i-th row and j-th column in given matrix A.

Cofactor of an Element: The **cofactor** C_{ij} of an element a_{ij} of a square matrix A is $(-1)^{i+j}$ times the determinant of the matrix that remains after deleting the i-th row and j-th column in the given matrix A.

Adjoint of a Matrix: Let $A = [a_{ij}]_{m \times n}$ be a square matrix and C_{ij} the cofactor of a_{ij} in A. Then the transpose of the matrix B obtained by replacing each element of A by its cofactor in A is known as the **adjoint of A** and is denoted by Adj A.

i.e. **Adj A = B^t**, where $B = [C_{ij}]$.

$$\text{If } A = \begin{bmatrix} a_{11} & a_{12} & a_{13} \\ a_{21} & a_{22} & a_{23} \\ a_{31} & a_{32} & a_{33} \end{bmatrix}, \text{ then}$$

$$\text{Adj } A = \begin{bmatrix} C_{11} & C_{21} & C_{31} \\ C_{12} & C_{22} & C_{32} \\ C_{13} & C_{23} & C_{33} \end{bmatrix}$$

Inverse of a Matrix (Def.): Let A be *a* $n \times n$ square matrix. If there exists a matrix B, such that $AB = I_n = BA$, where I is an $n \times n$ unit matrix, then B is called the inverse of A and is denoted by A^{-1}.

Singular and Non-Singular Matrices: A square matrix A is said to be **non-singular** matrix if its inverse, *i.e.*, A^{-1} exists. If A^{-1} does not exist then the square matrix A is said to be a singular matrix.

or

A square matrix A is called a singular matrix or a non-singular matrix according as |A| = 0 or |A| ≠ 0 respectively. Every invertible matrix possesses a unique inverse.

MULTIPLE CHOICE QUESTIONS

1. If $A = \begin{vmatrix} a_{11} & a_{12} & a_{13} \\ a_{21} & a_{22} & a_{23} \\ a_{31} & a_{32} & a_{33} \end{vmatrix}$, then which of the following is true—

(*a*) $a_{11} A_{21} + a_{12} A_{22} + a_{13} A_{23} = A$

(*b*) $a_{21} A_{21} + a_{22} A_{22} + a_{23} A_{23} = A$

(*c*) $a_{31} A_{31} + a_{32} A_{32} + a_{33} A_{33} = A$

(*d*) $a_{21} A_{31} + a_{22} A_{32} + a_{23} A_{33} = A$

2. If $A = \begin{bmatrix} \cos\alpha & -\sin\alpha \\ \sin\alpha & \cos\alpha \end{bmatrix}$, then A^{-1} is

(a) not existing (b) $\begin{bmatrix} 1 & 0 \\ 0 & 1 \end{bmatrix}$

(c) $\begin{bmatrix} \cos\alpha & \sin\alpha \\ -\sin\alpha & \cos\alpha \end{bmatrix}$ (d) None of these

3. If $A = \begin{bmatrix} a & b \\ c & d \end{bmatrix}$, then

(a) $|\text{Adj } A| = |A|$ (b) $|\text{Adj } A| = |A^{-1}|$

(c) $|\text{Adj } A^{-1}| = |A|$ (d) $|\text{Adj } A^{-1}| = (|A|)^{-1}$

4. If $A = \begin{bmatrix} x & y \\ z & w \end{bmatrix}$, then Adj (Adj A) is equal to —

(a) $\begin{bmatrix} w & -z \\ -y & x \end{bmatrix}$ (b) $\begin{bmatrix} x & y \\ z & w \end{bmatrix}$

(c) $\dfrac{1}{xw - yz}\begin{bmatrix} w & -z \\ -y & x \end{bmatrix}$ (d) None of these

5. If $A = \begin{bmatrix} 0 & 0 & 1 \\ 0 & 1 & 0 \\ 1 & 0 & 0 \end{bmatrix}$, then A^{-1} is

(a) $\begin{bmatrix} 1 & 0 & 0 \\ 0 & 1 & 0 \\ 0 & 0 & 1 \end{bmatrix}$ (b) $\begin{bmatrix} 0 & 0 & 1 \\ 0 & 1 & 0 \\ 1 & 0 & 0 \end{bmatrix}$

(c) $\begin{bmatrix} 0 & 1 & 0 \\ 0 & 0 & 1 \\ 1 & 0 & 0 \end{bmatrix}$ (d) None of these

6. If $A = \begin{bmatrix} a_1 & c_1 & b_1 \\ a_2 & b_2 & c_2 \\ a_3 & b_3 & c_3 \end{bmatrix}$, $|A|$ = determinant of A,

then $|A. \text{adj } A|$ is given by

(a) $|A|$ (b) $|A|^2$

(c) $|A|^3$ (d) $|A|^{-1}$

7. The rank of the matrix $\begin{bmatrix} 2 & 2 & 2 & 2 \\ 3 & 3 & 3 & 3 \\ 4 & 4 & 4 & 4 \\ 5 & 5 & 5 & 5 \end{bmatrix}$ is

(a) 4 (b) 3

(c) 2 (d) 1

8. The values of θ lying between $\theta = 0$ and $\theta = \dfrac{\pi}{2}$ and satisfying the equation

$$\begin{bmatrix} 1+\sin^2\theta & \cos^2\theta & 4\sin 4\theta \\ \sin^2\theta & 1+\cos^2\theta & 4\sin 4\theta \\ \sin^2\theta & \cos^2\theta & 1+4\sin 4\theta \end{bmatrix} = 0 \text{ are}$$

(a) $\dfrac{7\pi}{24}$ and $\dfrac{11\pi}{24}$ (b) $\dfrac{7\pi}{24}$ and $\dfrac{\pi}{24}$

(c) $\dfrac{5\pi}{24}$ and $\dfrac{11\pi}{24}$ (d) $\dfrac{\pi}{24}$ and $\dfrac{7\pi}{24}$

9. The rank of the matrix

$\begin{bmatrix} 1 & 1 & 1 \\ a & b & c \\ a^3 & b^3 & c^3 \end{bmatrix}$ a, b, c being all real, is 3, if

(a) a, b, c are all different and $a + b + c \neq 0$

(b) $a = b = c$

(c) a, b, c are all different but $a + b + c = 0$

(d) two of the numbers a, b, c are equal but are different from the third.

10. If $A = \begin{bmatrix} 0 & 1 & 0 & 0 \\ 0 & 0 & 1 & 0 \\ 0 & 0 & 0 & 1 \\ 0 & 0 & 0 & 0 \end{bmatrix}$, then $\rho(A^2)$ is equal to

(a) 1 (b) 2

(c) 3 (d) 4

11. Let $A = \begin{bmatrix} 5 & 0 & 2 \\ 0 & 1 & 0 \\ -4 & 0 & -1 \end{bmatrix}$ and I be 3×3 unit matrix.

If $M = I - A$, then rank of $I - A$ is —

(a) 0 (b) 1

(c) 2 (d) 3

12. If ω is the cube root of unity, then the inverse of $A = \begin{bmatrix} 1 & 1 & 1 \\ 1 & \omega & \omega^2 \\ 1 & \omega^2 & \omega \end{bmatrix}$ is

(a) $\begin{bmatrix} 1 & 1 & 1 \\ 1 & \omega^2 & \omega \\ 1 & \omega & \omega^2 \end{bmatrix}$ (b) $\dfrac{1}{2}\begin{bmatrix} 1 & 1 & 1 \\ 1 & \omega^2 & \omega \\ 1 & \omega & \omega^2 \end{bmatrix}$

(c) $\frac{1}{4}\begin{bmatrix}1 & 1 & 1\\ 1 & \omega & \omega^2\\ 1 & \omega^2 & \omega\end{bmatrix}$ (d) $\frac{1}{3}\begin{bmatrix}1 & 1 & 1\\ 1 & \omega^2 & \omega\\ 1 & \omega & \omega^2\end{bmatrix}$

13. Consider the matrix

$B = \begin{bmatrix}0 & -1\\ 1 & 0\end{bmatrix}$, then B^{1023} equals

(a) $\begin{bmatrix}-1 & 0\\ 0 & -1\end{bmatrix}$ (b) $\begin{bmatrix}0 & 1\\ -1 & 0\end{bmatrix}$

(c) $\begin{bmatrix}1 & 0\\ 0 & 1\end{bmatrix}$ (d) $\begin{bmatrix}-1 & 0\\ 0 & 1\end{bmatrix}$

14. What is the value of determinant of adj [adj {adj (A^{-1})}], where A is a 3×3 matrix and $|A| = 2$?

(a) $\frac{1}{64}$ (b) $\frac{1}{256}$

(c) $\frac{1}{128}$ (d) $\frac{1}{512}$

15. If $A = \begin{bmatrix}1 & -1\\ 2 & -1\end{bmatrix}$, $B = \begin{bmatrix}a & 1\\ b & -1\end{bmatrix}$ and $(A+B)^2 = A^2 + B^2$, then the values of a and b are respectively

(a) 2, 2 (b) 3, –2

(c) 1, 4 (d) –4, 7

ANSWERS

1	2	3	4	5	6	7	8	9	10
(b)	(c)	(a)	(b)	(b)	(c)	(d)	(a)	(a)	(b)
11	**12**	**13**	**14**	**15**					
(b)	(d)	(b)	(b)	(c)					

SOME SELECTED EXPLANATORY ANSWERS

1. If $A = \begin{vmatrix}a_{11} & a_{12} & a_{13}\\ a_{21} & a_{22} & a_{23}\\ a_{31} & a_{32} & a_{33}\end{vmatrix}$, then

$a_{21}A_{21} + a_{22}A_{22} + a_{23}A_{23} = A$ is true

6. Since $A\cdot \text{adj } A = |A|\cdot I$

$\therefore$ $|A \text{ adj } A| = |A|^n$,

if A is $n\times n$ order matrix.

Here $n = 3$, hence

$|A \text{ adj } A| = |A|^3$

11. $M = I - A$

$= \begin{bmatrix}1 & 0 & 0\\ 0 & 1 & 0\\ 0 & 0 & 1\end{bmatrix} - \begin{bmatrix}5 & 0 & 2\\ 0 & 1 & 0\\ -4 & 0 & -1\end{bmatrix}$

$M = \begin{bmatrix}-4 & 0 & -2\\ 0 & 0 & 0\\ 4 & 0 & 2\end{bmatrix} \sim \begin{bmatrix}-4 & 0 & -2\\ 0 & 0 & 0\\ 4 & 0 & 2\end{bmatrix}$,

$R_3 \to R_3 + R_1$ $\rho(M) = 1$

12. Here, $A^2 = 3I$

$\therefore$ $A = 3I\cdot A^{-1}$

$\Rightarrow$ $A^{-1} = \frac{1}{3}A = \frac{1}{3}\begin{bmatrix}1 & 1 & 1\\ 1 & \omega^2 & \omega\\ 1 & \omega & \omega^2\end{bmatrix}$

15. $A + B = \begin{bmatrix}1+a & 0\\ 2+b & -2\end{bmatrix}$

$\therefore (A+B)^2 = \begin{bmatrix}a^2+2a+1 & 0\\ ab+2a-b-2 & 4\end{bmatrix}$

$A^2 = \begin{bmatrix}-1 & 0\\ 0 & -1\end{bmatrix}$

and $B^2 = \begin{bmatrix}a^2+b & a-1\\ ab-b & b+1\end{bmatrix}$

$A^2 + B^2 = \begin{bmatrix}a^2+b-1 & a-1\\ ab-b & b\end{bmatrix}$

But given that $(A+B)^2 = A^2 + B^2$

$\Rightarrow a - 1 = 0$ $\Rightarrow a = 1$

and $b = 4$

16

Determinants

Def.: A determinant of order 3 is written as $\begin{bmatrix} a_{11} & a_{12} & a_{13} \\ a_{21} & a_{22} & a_{23} \\ a_{31} & a_{32} & a_{33} \end{bmatrix}$, where horizontal lines are called rows and vertical lines are called columns. The elements a_{11}, a_{22}, a_{33} are called diagonal elements.

Minors: The determinant obtained by deleting the i-th row and j-th column passing through the element a_{ij} is called the minor of element a_{ij} and is denoted by M_{ij}.

Cofactors: The cofactor of element a_{ij} is $(-1)^{i+j}$ times the deteminant obtained by deleting the i-th row and j-th column passing through a_{ij} and is denoted by C_{ij}. Obviously $C_{ij} = (-1)^{i+j} M_{ij}$.

Properties of Determinants:

(*a*) The value of the determinant does not change when rows and columns are interchanged. The determinant obtained by interchanging the rows and columns is called the transpose of the determinant and is denoted by Δ^T. Thus $\Delta = \Delta^T$.

(*b*) If all the elements of a row (column) are zero, then the value of the determinant is zero.

(*c*) The interchange of any two rows (columns) of the determinant changes its sign. Thus if Δ^* is the new deteminant obtained on interchanging any two rows (columns), then $\Delta = \Delta^*$. If i-th and j-th row are interchanged then this operation is denoted by $R_i \rightleftharpoons R_j$.

(*d*) If all the elements of a row (column) of a determinant are multiplied by a non-zero constant, then the determinant gets multiplied by the same constant. If we apply $R_i \to pR_i$, *i.e.*, each element of i-th row is multiplied by p, then we get $\Delta^* = p\Delta$ or $\Delta = \frac{1}{p}\Delta^*$. $(p \neq 0)$, where Δ^* is the new determinant thus obtained.

(*e*) If all the elements of a row (column) are proportional (identical) to the elements of some other row (column), then determinant is zero.

(*f*) If each element of any row (column) is sum of two numbers, the determinant can be expressed as the sum of two determinants of the same order. For example,

$$\begin{vmatrix} a_1+\alpha_1 & b_1 & c_1 \\ a_2+\alpha_2 & b_2 & c_2 \\ a_3+\alpha_3 & b_3 & c_3 \end{vmatrix} = \begin{vmatrix} a_1 & b_1 & c_1 \\ a_2 & b_2 & c_2 \\ a_3 & b_3 & c_3 \end{vmatrix} + \begin{vmatrix} \alpha_1 & b_1 & c_1 \\ \alpha_2 & b_2 & c_2 \\ \alpha_3 & b_3 & c_3 \end{vmatrix}$$

It should be noted that C_2 and C_3 remains same in both determinants of right-hand side.

Product of Two Determinant of Same Order:

Let $\Delta_1 = \begin{vmatrix} a_1 & b_1 & c_1 \\ a_2 & b_2 & c_2 \\ a_3 & b_3 & c_3 \end{vmatrix}$ and

$$\Delta_2 = \begin{vmatrix} \alpha_1 & \beta_1 & \gamma_1 \\ \alpha_2 & \beta_2 & \gamma_2 \\ \alpha_3 & \beta_3 & \gamma_3 \end{vmatrix}.$$

Then row by row multiplication of Δ_1 and Δ_2 is given by

$$\Delta_1.\Delta_2 = \begin{vmatrix} a_1\alpha_1+b_1\beta_1+c_1\gamma_1 & a_1\alpha_2+b_1\beta_2+c_1\gamma_2 & a_1\alpha_3+b_1\beta_3+c_1\gamma_3 \\ a_2\alpha_1+b_2\beta_1+c_2\gamma_1 & a_2\alpha_2+b_2\beta_2+c_2\gamma_2 & a_2\alpha_3+b_2\beta_3+c_2\gamma_3 \\ a_3\alpha_1+b_3\beta_1+c_3\gamma_1 & a_3\alpha_2+b_3\beta_2+c_3\gamma_2 & a_3\alpha_3+b_3\beta_3+c_3\gamma_3 \end{vmatrix}$$

MULTIPLE CHOICE QUESTIONS

1. If $x+y+z=\pi$, then the value of

$$\Delta = \begin{vmatrix} \sin(x+y+z) & \sin B & \cos C \\ -\sin B & 0 & \tan A \\ \cos(A+B) & -\tan A & 0 \end{vmatrix} \text{ is equal}$$

to:

(*a*) 0
(*b*) 2 sin B tan A cos C
(*c*) 1
(*d*) None of these

2. The value of determinant

$$\begin{vmatrix} x+1 & x+2 & x+4 \\ x+3 & x+5 & x+8 \\ x+7 & x+10 & x+14 \end{vmatrix} \text{ is}$$

(*a*) –2 (*b*) x^2+2
(*c*) 2 (*d*) None of these

3. $\begin{vmatrix} 1 & 1 & 1 \\ {}^mC_1 & {}^{m+2}C_1 & {}^{m+2}C_1 \\ {}^mC_2 & {}^{m+1}C_2 & {}^{m+2}C_2 \end{vmatrix}$ is equal to

(*a*) $m(m+1)$ (*b*) 0
(*c*) 1 (*d*) $m(m-1)$

4. The determinant

$$\begin{vmatrix} \cos(\theta+\phi) & -\sin(\theta+\phi) & \cos 2\phi \\ \sin\theta & \cos\theta & \sin\theta \\ -\cos\theta & \sin\theta & \cos\theta \end{vmatrix} \text{ is}$$

(*a*) 0
(*b*) independent of θ
(*c*) independent of ϕ
(*d*) independent of both θ and ϕ

5. If the system of linear equations $x+4ay+az=0$, $x+3by+bz=0$, $x+2cy+cz=0$, have a non-zero solution, then a, b, c are in

(*a*) A.P. (*b*) G.P.
(*c*) H.P. (*d*) None of these

6. If $\begin{vmatrix} x & x+y & x+y+z \\ 2x & 3x+2y & 4x+3y+2z \\ 3x & 6x+3y & 10x+6y+3z \end{vmatrix} = 64$,

then the real value of x is

(*a*) 2 (*b*) 3
(*c*) 4 (*d*) 6

7. The parameter, on which the value of the determinant

$$\begin{vmatrix} 1 & a & a^2 \\ \cos(p-d)x & \cos px & \cos(p+d)x \\ \sin(p-d)x & \sin px & \sin(p+d)x \end{vmatrix}$$

does not depend upon, is

(*a*) a (*b*) p
(*c*) d (*d*) x

8. Let $f(x) = \begin{vmatrix} x^3 & \sin x & \cos x \\ 6 & -1 & 0 \\ p & p^2 & p^3 \end{vmatrix}$ where p is constant. Then $\dfrac{d^3}{dx^3} f(x)$, at $x=0$ is

(*a*) p (*b*) $p+p^2$
(*c*) $p+p^3$ (*d*) independent of p

9. The value of $\begin{vmatrix} 1 & yz & y+z \\ 1 & zx & z+x \\ 1 & xy & x+y \end{vmatrix}$ will be:

(*a*) $(y+z)(z-x)(x+y)$
(*b*) $(x-y)(y-z)(z-x)$
(*c*) $(x+y)(y+z)(z+x)$
(*d*) $xyz(x+y+z)$

10. $\begin{vmatrix} x & x^2 & 1+x^2 \\ y & y^2 & 1+y^2 \\ z & z^2 & 1+z^2 \end{vmatrix} = 0$, then the value of xyz is:

(*a*) – 1 (*b*) 1
(*c*) – 2 (*d*) None of these

11. The value of λ for which the system of equations $x+2y+3z=\lambda x$, $3x+y+2z=\lambda y$, $2x+3y+z=\lambda x$, have non-trivial solution is given by:

(*a*) 1 (*b*) 2
(*c*) 4 (*d*) None of these

12. If $\Delta = \begin{vmatrix} a & b & c \\ b & c & a \\ c & a & b \end{vmatrix}$, then Δ^2 will be:

(*a*) $\begin{vmatrix} bc-a^2 & c^2 & b^2 \\ c^2 & ca-b^2 & a^2 \\ b^2 & a^2 & ab-c^2 \end{vmatrix}$

(*b*) $\begin{vmatrix} a^2 & b^2 & c^2 \\ b^2 & c^2 & a^2 \\ c^2 & a^2 & b^2 \end{vmatrix}$

(c) $\begin{vmatrix} bc-a & c & b \\ c & ca-b & a \\ b & a & ab-c \end{vmatrix}$

(d) $\begin{vmatrix} 2bc-a^2 & c^2 & b^2 \\ c^2 & 2ca-b^2 & a^2 \\ b^2 & a^2 & 2ab-c^2 \end{vmatrix}$

13. If $\Delta = \begin{vmatrix} a_1 & b_1 & c_1 \\ a_2 & b_2 & c_2 \\ a_3 & b_3 & c_3 \end{vmatrix}$, then Δ^2 is given by:

(a) $\begin{vmatrix} A_1 & B_1 & C_1 \\ A_2 & B_2 & C_2 \\ A_3 & B_3 & C_3 \end{vmatrix}$ (b) $\begin{vmatrix} A_1 & C_1 & B_1 \\ A_2 & C_2 & B_2 \\ A_3 & C_3 & B_3 \end{vmatrix}$

(c) $\begin{vmatrix} A_3 & B_3 & C_3 \\ A_2 & B_2 & C_2 \\ A_1 & B_1 & C_1 \end{vmatrix}$ (d) None of these.

Where $A_1, B_1, C_1, ...$ etc. are cofactors of $a_1, b_1, c_1, ...$

14. The value of the determinant

$$\begin{vmatrix} 1 & \cos(\beta-\alpha) & \cos(\gamma-\alpha) \\ \cos(\alpha-\beta) & 1 & \cos(\gamma-\beta) \\ \cos(\alpha-\gamma) & \cos(\beta-\gamma) & 1 \end{vmatrix}$$

is given by:

(a) 1 (b) 0

(c) – 1 (d) None of these

15. If $\omega^3 = 1$, then the value of the determinant $\begin{bmatrix} 1 & \omega & \omega^2 \\ \omega & \omega^2 & 1 \\ \omega^2 & 1 & \omega \end{bmatrix}$ is:

(a) 0 (b) – 1

(c) 1 (d) ω

ANSWERS

1	2	3	4	5	6	7	8	9	10
(a)	(a)	(c)	(b)	(c)	(c)	(b)	(d)	(b)	(d)
11	**12**	**13**	**14**	**15**					
(d)	(d)	(a)	(b)	(a)					

SOME SELECTED EXPLANATORY ANSWERS

1. Since $x + y + z = \pi$,

$$\Delta = \begin{vmatrix} 0 & \sin B & \sin C \\ -\sin B & 0 & \tan A \\ -\cos C & -\tan A & 0 \end{vmatrix}$$

$\therefore -\Delta' = \Delta$, *i.e.* $-\Delta = \Delta$ so $\Delta = 0$

7. $C_1 \to C_1 + C_3 - 2C_2 \cos dx$ gives

$$\Delta = \begin{vmatrix} 1+a^2-2\cos dx & a & a^2 \\ 0 & \cos px & \cos(p+d)x \\ 0 & \sin px & \sin(p+d)x \end{vmatrix}$$

$= (1 + a^2 - 2 \cos dx) \sin x$,

which is independent of p.

11. The given equations can be written as

$$(1-\lambda)x + 2y + 3z = 0$$
$$3x + (1-\lambda)y + 2z = 0$$
$$(2-\lambda)x + 3y + z = 0.$$

The equations will have non-trivial solutions if rank of coefficient matrix is less than 3, *i.e.*, the coefficient matrix will be zero.

$$\therefore \begin{vmatrix} 1-\lambda & 2 & 3 \\ 3 & 1-\lambda & 2 \\ 2-\lambda & 3 & 1 \end{vmatrix} = 0$$

$(1-\lambda)[(1-\lambda)-6] - 2[3-2(2-\lambda)]$
$+ 3[9-(1-\lambda)(2-\lambda)] = 0$

i.e., $4\lambda^2 + 9\lambda + 18 = 0$

12. If $\Delta = \begin{vmatrix} a & b & c \\ b & c & a \\ c & a & b \end{vmatrix}$

then $\Delta^2 = \begin{vmatrix} a & b & c \\ b & c & a \\ c & a & b \end{vmatrix} \times \begin{vmatrix} -a & c & b \\ -b & a & c \\ -c & b & a \end{vmatrix}$

$$= \begin{vmatrix} 2bc-a^2 & c^2 & b^2 \\ c^2 & 2ac-b^2 & a^2 \\ b^2 & a^2 & 2ab-c^2 \end{vmatrix}$$

17

Continuity and Differentiability

CONTINUITY OF FUNCTION

(*i*) A single-valued function is continuous throughout the neighbourhood of $x = a$ if and only if $\lim_{x \to a} f(x)$ exists and is equal to $f(a)$.

(*ii*) A single-valued function is continuous in an interval (a, b) or $[a, b]$ if and only if it is continuous at each point of this interval.

(*iii*) A single-valued function has a discontinuity of the first kind at the point $x = a$ if $f(a+0) \neq f(a-0)$. The difference of these two values is known as the jump (status) of $f(x)$.

(*iv*) A single-valued function is piecewise-continuous on a given interval (a, b) or $[a, b]$ if and only if $f(x)$ is continuous throughout this interval except for a finite number of discontinuities of the first kind.

(*v*) A single-valued function has a discontinuity of the second kind at the point $x = a$ if both $f(a + 0)$ and $f(a - 0)$ does not exist.

(*vi*) A single-valued function has a discontinuity of mixed kind if any one of them *i.e.*, $f(a + 0)$ or $f(a - 0)$ exist.

(*vii*) A single-valued function has a discontinuity of third kind at the point $x = a$, if $f(a + 0) = f(a - 0) \neq f(a)$. It is also known as removal of discontinuity.

Some Fundamental Results on Continuous Function

I. Let $f(x)$ and $g(x)$ be two continuous functions and let c be a real number. Then

(*i*) cf is continuous

(*ii*) $f + g$ is continuous

(*iii*) $f - g$ is continuous

(*iv*) fg is continuous

II. Listed below Some Common Type of Continuous Function

(*i*) **Polynomial Function :** A function of the form $f(x) = a_0 + a_1x + a_2x^2 + \ldots + a_nx^n$, where $a_0, a_1, a_2, \ldots, a_n \in \mathrm{R}$ is called a polynomial function.

(*ii*) **Rational Function :** If P(x) and Q(x) are two polynomials, then the function of the form

$$f(x) = \frac{P(x)}{Q(x)}, \; Q(x) \neq 0$$ is called a polynomial function.

III. Trigonometric Function : All trigonometrical function *i.e.*, $\sin x$, $\cos x$, $\tan x$, $\sec x$, $\mathrm{cosec}\, x$, $\cot x$ are continuous at each point.

Differentiability at a Point : Let $f(x)$ be a real valued function defined on an open interval (a, b) and let $c \in (a, b)$. Then $f(x)$ is said to be differentiable or derivable at $x = c$, if f

$$\lim_{x \to c} \frac{f(x) - f(c)}{x - c} \text{ exist finitely.}$$

$$\Leftrightarrow \lim_{x \to c-} \frac{f(x) - f(c)}{x - c} = \lim_{x \to c+} \frac{f(x) - f(c)}{x - c}$$

$$\Leftrightarrow \lim_{h \to 0} \frac{f(c-h) - f(c)}{-h} = \lim_{h \to 0} \frac{f(c+h) - f(c)}{h}$$

$$\Leftrightarrow \mathrm{L}f(c) = \mathrm{R}f(c)$$

If $\mathrm{L}f'(c) \neq \mathrm{R}f'(c)$, we say that $f(x)$ is not differentiable at $x = c$.

Relation Between Continuity and Differentiability : If a function is differentiable at a point, it is necessarily continuous at that point *i.e.*, $f(x)$ is differentiable at $x = c \Rightarrow f(x)$ is continuous at $x = c$.

Converse : The converse of the above theorem is not necessarily true *i.e.*, a function may be continuous at a point but may not be differentiable at that point.

MULTIPLE CHOICE QUESTIONS

1. Two functions

$$f(x) = \begin{cases} x \sin \frac{1}{x}, & x \neq 0 \\ 0, & x = 0 \end{cases}$$

$g(x) = |x|$ are given, then at $x = 0$

(*a*) both f and g are continuous

(*b*) f is continuous and g is discontinuous

(*c*) f is discontinuous and g is continuous

(*d*) both f and g are discontinuous

2. If the function f is defined by $f(x) = \frac{x}{1+|x|}$, then at what points is f differentiable?

(*a*) everywhere

(*b*) except at $x = \pm 1$

(*c*) except at $x = 0$

(*d*) except at $x = 0$ or ± 1

3. The function $f(x) = \frac{|x|}{x}$ at $x = 0$ has

(*a*) discontinuity of second kind

(*b*) discontinuity of first kind

(*c*) mixed discontinuity

(*d*) continuous

4. If $f(x) = \frac{\sin(e^{x-2} - 1)}{\log(x-1)}$, $x \neq 2$ is continuous at $x = 2$, then $f(2)$ is equal to

(*a*) 0 (*b*) 2

(*c*) 1 (*d*) -2

5. If $f(x + y) = f(x).f(y)$ and $f(5) = 2, f'(0) = 3$, then the value of $f'(5)$ is

(*a*) 0 (*b*) 6

(*c*) 3 (*d*) 5

6. Discuss the continuity of the function

$$f(x) = \begin{cases} -x^2 & \text{for } x \le 0 \\ 5x - 4 & \text{for } 0 < x \le 1 \\ 4x^2 - 3 & \text{for } 1 < x < 2 \\ 3x + 4 & \text{for } x \ge 2 \end{cases}$$

(*a*) continuous at $x = 1, 2$ and discontinuous at $x = 0$

(*b*) continuous at $x = 0, 1$ and discontinuous at $x = 2$

(*c*) continuous at $x = 0, 1, 2$

(*d*) None of these

7. The set of all points when the function

$f(x) = \frac{x}{1+|x|}$ is differentiable is

(*a*) $(-\infty, \infty)$ (*b*) $(0, \infty)$

(*c*) $(-\infty, 0) \cup (0, \infty)$ (*d*) $(-\infty, 0)$

8. If $f(x) = \begin{cases} \frac{1-|x|}{1+x}, & x \neq -1 \\ 1, & x = -1 \end{cases}$, then $f([2x])$ is

(where [ ] represents the greatest integer function)

(*a*) continuous at $x = -1$

(*b*) continuous at $x = 0$

(*c*) discontinuous at $x = \frac{1}{2}$

(*d*) All are correct

9. If $F(x) = \frac{1}{x^2} \int_4^x (4t^2 - 2F'(t)\, dt)$, then $F'(4)$ equals

(*a*) $\frac{32}{9}$ (*b*) $\frac{64}{3}$

(*c*) $\frac{64}{9}$ (*d*) None of these

10. The derivative of $\log |x|$ is

(*a*) $\frac{1}{x}, x > 0$ (*b*) $\frac{1}{|x|}, x \neq 0$

(*c*) $\frac{1}{x}, x \neq 0$ (*d*) None of these

11. If $x^a y^b = (x + y)^{a+b}$, then $\frac{dy}{dx}$ is equal to

(*a*) $\frac{a}{b}$ (*b*) $\frac{b}{a}$

(c) $\dfrac{x}{y}$ (d) $\dfrac{y}{x}$

12. If $f(x)=\sqrt{[\tan x+\sqrt{[\tan x+\ldots.\infty}}$, then $\dfrac{dy}{dx}$ is

(a) $\dfrac{\tan x}{2y-1}$ (b) $\dfrac{\sqrt{\tan x}}{2y-1}$

(c) $\dfrac{\sec x}{2y-1}$ (d) $\dfrac{\sec^2 x}{2y-1}$

13. If $y=\cfrac{x}{a+\cfrac{x}{b+\cfrac{x}{a+\cfrac{x}{b+\cfrac{x}{a}+\ldots}}}}$

then $\dfrac{dy}{dx}$ is given by

(a) $\dfrac{a}{b(a+2y)}$ (b) $\dfrac{a}{b(b+2y)}$

(c) $\dfrac{b}{b(b+2y)}$ (d) $\dfrac{b}{a(a+2y)}$

ANSWERS

1	2	3	4	5	6	70	8	9	10
(a)	(a)	(b)	(c)	(c)	(d)	(a)	(d)	(a)	(c)
11	**12**	**13**							
(d)	(c)	(d)							

SOME SELECTED EXPLANATORY ANSWERS

3. Here $f(x)=\dfrac{|x|}{x}=\begin{cases}1, \text{ for } x>0\\ -1, \text{ for } x<0\end{cases}$

(i) $\lim\limits_{x\to 0+} f(x) = \lim\limits_{x\to 0+} 1=1$

(ii) $\lim\limits_{x\to 0-} f(x) = \lim\limits_{x\to 0-} f(-1)=-1.$

$\therefore$ $\lim\limits_{x\to 0+} f(x) \neq \lim\limits_{x\to 0-} f(x)$

$\therefore$ $f(x)$ is discontinuous function of first kind at $x=0$.

7. Let $f(x)=\dfrac{x}{1+|x|}=\dfrac{g(x)}{h(x)}$, when $g(x)=x$ and $h(x)$ $=1+|x|$

(i) $f(x)$ is clearly differentiable on $(-\infty,\infty)$.

(ii) The domain of $g(x)$ is differentiable on $(-\infty,\infty)$. But as $|x|$ is not differentiable at $x=0$, therefore $h(x)$ is not differentiable at $x=0$.

Thus $f(x)$ is differentiable on $(-\infty,0)\cup(0,\infty)$.

8. Given $f(x)=\begin{cases}\dfrac{1-|x|}{1+x}, & x\neq -1\\ 1, & x=-1\end{cases}$

$=\begin{cases}\dfrac{1-x}{1+x}, & x\geq 0\\ 1, & x<0\end{cases}$

$\therefore$ $f([x])=\begin{cases}\dfrac{1-[x]}{1+[x]}, & x\geq 0\\ 1, & x<0\end{cases}$

$=\begin{cases}1, & x<0\\ 1, & 0\leq x<1/2\\ 0, & 1/2\leq x<1\\ -3/2, & 1\leq x<3/2\end{cases}$

Clearly, $f(x)$ is continuous for all $x<\dfrac{1}{2}$ and discontinuous at $x=\dfrac{1}{2},1.$

18

Applications of Derivative

EQUATION OF TANGENT

(*i*) The equation of tangent at the point P(*x*, *y*) on the curve

$y = f(x)$ is $Y - y = (dy/dx).(X - x)$

where (*x*, *y*) are the coordinates of any point of the curve.

(*ii*) The equation of tangent at the point $P(x_1, y_1)$ on the curve $y = f(x)$ is $y - y_1 = \left[\frac{dy}{dx}\right]_{(x_1,y_1)} (x, x_1)$

If the tangent at the point P(*x*, *y*) on the curve $y = f(x)$ make an angle ψ with the positive direction of the axis of *x*, in anticlockwise direction, then

$\tan \psi = dy/dx$ and is called the gradient (or slope) of the curve at P(*x*, *y*).

Tangent Parallel to the axis of *x*: At the point the tangent is parallel to the axis of *x*, $dy/dx = 0$

Tangent Perpendicular to the axis of *x*: At the point the tangent is perpendicular to the axis of *x* (*i.e.*, parallel to the axis of *y*), $dx/dy = 0$.

EQUATION OF NORMAL

(*i*) The equation of the normal at the point P(*x*, *y*) on the curve $y = f(x)$ is $(Y - y) = -\frac{dx}{dy}(X - x)$ or $(Y - y)\frac{dy}{dx} + (X - x) = 0$

(*ii*) The equation of the normal at the point $P(x_1, y_1)$ on the curve $y = f(x)$ is

$$(y - y_1)\left[\frac{dy}{dx}\right]_{(x_1+y_1)} + (x - x_1) = 0$$

Angle of Intersection of Two Curves: The angle of intersection of two curves is defined as the angle between the tangents to the two curves at their point of intersection.If θ is the angle of intersection of the two curves at the point of intersection P, then

(*i*) $\theta = | \psi_1 - \psi_2 |$ or

(*ii*) $\theta = \tan^{-1}\left|\frac{m_1 - m_2}{1 + m_1 m_2}\right|$

where m_1 and m_2 are the gradients of the two curves at their point of intersection.

i.e., $m_1 = \tan \psi_1$ and $m_2 = \tan \psi_2$

where ψ_1 and ψ_2 are the angles which the tangents to the two curves at their point of intersection make with the positive direction of the axis of *x*.

Condition of orthogonal intersection of two curves: The two curves intersect orthogonally (*i.e.*, at right angles) at their point of intersection if $m_1 m_2 = -1$

Lengths of tangent, normal, subtangent and subnormal: Let the tangent and normal at the point P(*x*, *y*) on the curve meet the axis of *x* at the points T and N respectively. Let PM be the ordinate of the point P. Then,

(*i*) Length of Tangent = PT $| y \operatorname{cosec} \psi |$

$$= \left|\frac{y\sqrt{[1+(dy/dx)^2]}}{dy/dx}\right|$$

(*ii*) Length of Normal : PN = $| y \sec \psi |$ $= | y\,[1 + (dy/dx)^2]^{1/2} |$

(*iii*) Length of Sub-tangent: TM = $| y \cot \psi |$ $= | y/(dy/dx) |$

(*iv*) Length of Sub-normal: MN = $| y \tan \psi |$ $= | y(dy/dx) |$

Increasing Function: Let *f*(*x*) be a function defined in the interval I then *f*(*x*) is said to be increasing for x_1 and x_2 whenever $x_1 < x_2$, then $f(x_1) < f(x_2)$.

$\because x_2 > x_1 \Rightarrow x_2 - x_1 > 0$ again $f(x_2) > f(x_1)$
$\Rightarrow f(x_2) - f(x_1) > 0$

$$\therefore \frac{f(x_2)-f(x_1)}{x_2-x_1} > 0$$

Decreasing Function: Let $f(x)$ be a function defined in the interval I then $f(x)$ is said to be decreasing for x_1 and x_2 whenever $x_1 < x_2$, then $f(x_1) > f(x_2)$.

$\because x_2 > x_1 \Rightarrow x_2 - x_1 > 0$ again $f(x_2) < f(x_1)$
$\Rightarrow f(x_2) - f(x_1) < 0$

$$\therefore \frac{f(x_2)-f(x_1)}{x_2-x_1} < 0$$

Maxima and Minima: Maxima and minima are local properties. A function $f(x)$ has a maximum at x_0 if $f(x_0)$ is greatest in some interval $(x_0 - h, x_0 + h)$ or simply to say in some interval containing x_0 and likewise minima exists at x_0 if in some interval containing x_0, $f(x_0)$ is least. Thus $f(x)$ has maximum at x_0 if in some interval

I = $]x_0 - h, x_0 + h[$ for all $x \in$ I, $f(x) < f(x_0)$ and minima if $f(x) > f(x_0)$.

Methods of Testing for Maxima or Minima

(*i*) **First derivate test:**

Let x_0 be a critical point then

$f'(x_0 - h)$	$f'(x_0 + h)$	Result
> 0	< 0	max. at x_0
< 0	> 0	min. at x_0
> 0	> 0	neither maxima nor minima
< 0	< 0	

(*ii*) **2nd Derivative Test:** If $f(x)$ has a derivative, at x_0 and $f'(x_0) = 0$, then

$f''(x_0) > 0 \Rightarrow f(x)$ has minima at x_0.

$f''(x_0) < 0 \Rightarrow f(x)$ has maxima at x_0.

If $f''(x_0) = 0$, we cannot say anything. Then if $f'''(x_0) \neq 0$, then $f(x)$ has neither max. nor min. at x_0 and if $f'''(x_0) = 0$, then $f^{iv}(x_0) > 0 \Rightarrow$ minima at $x_0 < 0 \Rightarrow$ maxima at x_0.

In general, let $f'(x_0) = f''(x_0) = \ldots.. = f^{(n-1)}(x_0) = 0$, $f^{(n)}(x_0) \neq 0$.

If n is odd, then there is neither max. nor min. at $x = x_0$ and if n is even then $f^{(n)}(x_0) > 0$ gives min at x_0 and $f^{(n)}(x_0) < 0$ gives max. at x_0.

ROLLE'S THEOREM

Let $f(x)$ be a function such that

(*a*) $f(x)$ is continuous in $[a, b]$

(*b*) $f'(x)$ exist for every point in (a, b)

(*c*) $f(a) = f(b)$

Then $\exists$ at least a point C $\in (a, b)$ such that $f'(c) = 0$

Lagrange's Mean Value Theorem

If $f(x)$ is a function such that

(*a*) $f(x)$ is continuous in $[a, b]$

(*b*) $f'(x)$ exist in (a, b)

then $\exists$ at least one point c such that $f'(c) = \dfrac{f(b)-f(a)}{b-a}$, $c \in (a, b)$

Cauchy's Mean Value Theorem

If $f(x)$ and $g(x)$ be two functions such that

(*a*) $f(x)$ and $g(x)$ both are continuous in $[a, b]$

(*b*) $f'(x)$ and $g'(x)$ both exist in (a, b)

(*c*) $g'(x) \neq 0 \ \forall\ x \in (a, b)$

Then $\exists$ at least one point C $\in (a, b)$ such that

$$\frac{f(b)-f(a)}{g(b)-g(a)} = \frac{f'(c)}{g'(c)}$$

Derivative as a Rate Measure: Let $y = f(x)$ be a relation between two variables x and y. If δx be a small change in x and δy be the corresponding change in y, then $\delta y/\delta x$ represents the average rate of change in y with respect to x in the interval $(x, x + \delta x)$.

Taking limit as $\delta x \to 0$, the average rate of change $\delta y/\delta x$ becomes dy/dx, called the instantaneous rate of change of y with respect to x.

Velocity and Acceleration: If s is the distance moved by a particle in time t, s.t. $s = f(t)$, then the velocity v and the acceleration a of the particle at any instant t are given by

$$v = \frac{ds}{dt} \text{ and } a = \frac{dv}{dt} = \frac{d^2s}{dt^2}$$

Velocity and acceleration both are +ve in the direction of s increasing with time t.

Angular Velocity and Acceleration: Let P be the position of a moving point on the curve at time t s.t. $\angle POX = \theta$, where OX is the initial line and O the pole. Then angular velocity = $d\theta/dt$ and angular acceleration = $d^2\theta/dt^2$. Both are positive in the direction of θ increasing with time t.

Approximation and Error

We have $\lim\limits_{\delta x \to 0} \dfrac{\delta y}{\delta x} = \dfrac{dy}{dx}$

$\therefore$ For smaller values of δx, we have $\dfrac{\delta y}{\delta x} = \dfrac{dy}{dx}$ (approximately).

i.e., $\delta y = \left(\dfrac{dy}{dx}\right)\delta x$ (appr.). Thus corresponding to a small error δx in the value of x, the approximate error in the value of y is $\dfrac{dy}{dx}\delta x$.

MULTIPLE CHOICE QUESTIONS

1. If a line $y = mx + 2$ cuts the parabola $x^2 = 2y$ at points (x_1, y_1) and (x_2, y_2) such that $x_1 < x_2$, then value of m for which $\int_{x_1}^{x_2} (mx + 2 - \frac{x^2}{2})\, dx$ is minimum, is

   (a) $\sqrt{3}$ (b) $\sqrt{2}$
   (c) $\dfrac{1}{\sqrt{3}}$ (d) none of these

2. The slope of the tangent to the curve $y = \int_x^{x^2} \sin^{-1} t^2$ at $x = \dfrac{1}{\sqrt[4]{2}}$ is

   (a) $\dfrac{\pi}{12}(2^{7/4} + 3)$ (b) $\dfrac{\pi}{12}(2^{7/4} - 3)$
   (c) $\dfrac{\pi}{12}(2^{3/4} - 3)$ (d) $\dfrac{\pi}{12}(2^{5/4} - 3)$

3. The solpe of tangent to the curve $y = \int_0^{x^2} \dfrac{1}{1+t^2} dt$ at $x = 1$ is

   (a) $\dfrac{16}{17}$ (b) $\dfrac{17}{16}$
   (c) 1 (d) none of these

4. The acute angle between the curves $y = |x^2 - 2|$ & $y = |x^2 - 8|$ at their point of intersection is

   (a) $\dfrac{\pi}{4}$ (b) $\dfrac{\pi}{6}$
   (c) $\tan^{-1}\left(\dfrac{4\sqrt{5}}{19}\right)$ (d) none of these

5. The value of a for which the function $f(x) = \begin{cases} -x^3 + \sin^{-1} a, & 0 < x < 1 \\ x, & x \geq 1 \end{cases}$ has a minimum value at $x = 1$, is

   (a) 0 (b) 1
   (c) -1 (d) none of these

6. In [0, 1] lagrange's mean value theorem is not applicable to

   (a) $f(x) = \begin{cases} \frac{1}{2} - x & \forall x < \frac{1}{2} \\ \left(\frac{1}{2} - x\right)^2 & \forall x \geq \frac{1}{2} \end{cases}$

   (b) $\begin{cases} \frac{\sin x}{b}, x \neq 0 \\ 1, x = 0 \end{cases}$

   (c) $f(x) = x\,|x|$ (d) $f(x) = |x|$

7. The equation of tangent at $x = 1$ if y is given by $y = \int_{x^2}^{x^3} \dfrac{dt}{\sqrt{1+t^2}}$

   (a) $y\sqrt{2} = x + 1$ (b) $y\sqrt{2} + 1 = x$
   (c) $y = x\sqrt{2} + 1$ (d) none of these

8. The value (s) of 'a' for which the area of the triangel included between the axes & any tangent to the curve $x^a y = \lambda^a$ is constant, is/are

   (a) 1/2 (b) 1/3
   (c) 1 (d) 2

9. Set of values of 'λ' for which locat extrema of the function $f(x)$ are positive where $f(x) = \frac{2}{3}a^2x^3 - \frac{5a}{2}x^2 + 3x + \lambda$ & maxima occurs at $x = \frac{1}{3}$ is

(a) $(-4, \infty)$ (b) $\left(\frac{-3}{8}, \infty\right)$

(c) $\left(-10, \frac{3}{8}\right)$ (d) none of these

10. The length of the part of tangent to the curve $x = a\cos^3\theta$, $y = a\sin^3\theta$ between the co-ordinate axes is

(a) $2\pi\,|a|$ (b) $\pi\,|a|$

(c) $|a|$ (d) none of these

11. The interval for which $f(x) = \sin(\log_e x) + \cos(\log_e x)$ is strictly decresing

(a) $\left[e^{2k\pi+\frac{\pi}{4}}, e^{2k\pi+\frac{5\pi}{4}}\right]$ (b) $\left[e^{2k\pi+\frac{\pi}{4}}, e^{2k\pi+\frac{7\pi}{4}}\right]$

(c) $\left[e^{2k\pi+\frac{\pi}{4}}, e^{2k\pi+\frac{9\pi}{4}}\right]$ (d) none of these

12. If from a point 'P' two perpendicular tangents PA & PB are drawn to the parabola $y^2 = 8x$, where A & B are the points of contact, then minimum area of triangle PAB is

(a) 16 Sq. units (b) 8 Sq. units

(c) 4 Sq. units (d) none of these

13. If the tangent to the curve $xy + ax + by = 0$ at $(-1, 1)$ is inclined at an angle $\tan^{-1}(2)$ with x–axis, then

(a) $a = 1, b = -2$ (b) $a = -1, b = -2$

(c) $a = -1, b = 2$ (d) $a = 1, b = 2$

14. The rate of change of the curved surface of a right circular cone of radius r & height h w.r. to chang in radius, is

(a) $\frac{\pi(2r^2 + h^2)}{\sqrt{r^2 + h^2}}$ (b) $\frac{2\pi(2r^2 + h^2)}{\sqrt{r^2 + h^2}}$

(c) $\frac{\pi(r^2 + 2h^2)}{\sqrt{r^2 + h^2}}$ (d) $\pi\sqrt{r^2 + h^2}$

15. A water tank has the shape of an inverted right circular cone with its axis vertical & vertex lowest. The semi-vertical angle is $\tan^{-1}(0.5)$, water is poured into it at a constant rate of 5 cubic meter per minute, then the rate at which the level of water is rising at the instant when the depth of water in the tank is 10 m, is

(a) $\frac{1}{2\pi}$ meter/minute

(b) $\frac{1}{5\pi}$ meter/minute

(c) $\frac{1}{\pi}$ meter/minute

(d) None of these

16. The least value of 'a' so that the function $f(x) = x^2 + ax + 1$ is strictly increasing in (1, 2) is

(a) −2 (b) 2

(c) 1 (d) −1

ANSWERS

1	2	3	4	5	6	7	8	9	10
(d)	(b)	(a)	(c)	(d)	(a)	(b)	(c)	(b)	(c)
11	12	13	14	15	16				
(d)	(a)	(c)	(a)	(b)	(a)				

SOME SELECTED EXPLANATORY ANSWERS

3. $y = \int_0^{x^2} \frac{dt}{1+t^2}$

$\Rightarrow \frac{dy}{dx} = \frac{2x}{1+x^4}$ $\therefore \left(\frac{dy}{dx}\right)_{x=(1/2)} = \frac{16}{17}$

$\therefore$ Choice (*a*) is correct.

5. let $y = f(x)$, for $x \geq 1$, the function $f(x) = y = x$ is increasing. For a minimum at $x = 1$, y should be decreasing, for $0 < x < 1$

So $-x^2 + \sin^{-1} a > f(1)$

or $-x^3 + \sin^{-1} a > 1$ for $0 < x < 1$

$\Rightarrow \sin^{-1} a > 1 + x^3$ for $0 < x < 1$

$\Rightarrow \sin^{-1} a > 1$

$\therefore$ No choice matched, so (*d*) is correct.

7. Give $x = 1$ & $y = \int_{x^2}^{x^3} \frac{dt}{\sqrt{1+t^2}} = \int_{1}^{1} \frac{dt}{\sqrt{1+t^2}}$

$\therefore$ point on the curve is (1, 1)

Again $\frac{dy}{dx} = \frac{3x^2}{\sqrt{1+x^6}} - \frac{2x}{\sqrt{1+x^4}}$

$\therefore \left(\frac{dy}{dx}\right)_{x=1} = \frac{1}{\sqrt{2}}$

$\therefore$ Equation of tangent $y - y_1 = \frac{dy}{dx}(x - x_1)$

$\therefore y - 0 = \frac{1}{\sqrt{2}}(x - 1) \Rightarrow x = y\sqrt{2} + 1$

$\therefore$ Choice (*b*) is correct.

11. $f(x) = \sin(\log_e x) + \cos(\log_e x)$

$\therefore f(x) = [\cos(\log_e x) - \sin(\log_e x)]\ \frac{1}{x}$

Now $f(x)$ to be strictly decreasing

$\therefore f(x) < 0$

$\Rightarrow \frac{\cos(\log_e x) - \sin(\log_e x)}{x} < 0$

$\Rightarrow \cos(\log_e x) - \sin(\log_e x) < 0$

($\log_e x$ to be defined only for $x > 0$)

$\Rightarrow \cos(\log_e x) < \sin(\log_e x)$

$\Rightarrow 2k\pi + \frac{\pi}{4} < \log_e x < 2k\pi + \frac{5\pi}{4}\ \forall\ k$

$= \{0, 1, 2...\}$

$\Rightarrow x \in \left(e^{2k\pi+\frac{\pi}{4}}, e^{2k\pi+\frac{5\pi}{4}}\right)$

$\therefore$ Choice (*d*) is correct.

13. Given curve is $xy + ax + by = 0...(i)$ & $(-1, -1)$ lies on it

$\therefore 1 - a - b = 0 \Rightarrow 1\ a + b = 1$ *or* $b = 1 - a$

Again from (*i*) on differentiatinf w.r. to x we have

$\frac{dy}{dx} = -\frac{a+y}{b+x}$

$\left(\frac{dy}{dx}\right)_{(-1,-1)} = -\left(\frac{a-1}{b-1}\right) = 2$ ($\because \tan\theta = m$)

$\Rightarrow \frac{a-1}{1-a-1} = -2 \qquad \Rightarrow a - 1 = 2a$

$\therefore a = -1$ then b = 2 $\therefore$ Choice (*c*) is correct

16. Given $f(x) = x^2 + ax + 1,\ x \in (1, 2)$

$\Rightarrow f'(x) = 2x + a$

Now $x \in (1, 2)$

$\Rightarrow 1 < x < 2 \qquad \Rightarrow 2 < 2x < 4$

$\Rightarrow 2 < a < 2x + a < 4 + a$

$\Rightarrow 2 < a < f'(x) < 4 + a$

For $f(x)$ to be strictly increasing on (1, 2) we have

$2 + a > 0 \Rightarrow a > -2$

$\therefore$ least value of $a = -2$

$\therefore$ Choices (*a*) is correct.

19

Integrals

INTEGRATION

If $f(x)$ and $F(x)$ are two functions of x such that $\frac{d}{dx}\{f(x)\} = F(x)$, then we say that integral of $F(x)$ is $f(x) \Rightarrow \int F(x)dx = f(x)+c$, where c is any constant, which is known as constant of integration. Because of presence of 'c' in $\int F(x)dx = f(x)+c$, it is called indefinite integral of $F(x)$.

Some Standard Formulae

(Readers are advised to remember these formulas)

1. $\int x^n dx = \frac{x^n+1}{n+1} + c,\ (n \neq -1)$
2. $\int \frac{1}{x} dx = ln|x| + c, x \neq 0$
3. $\int e^x dx = e^x + c$
4. $\int a^x dx = \frac{a^x}{ln\ a} + c$
5. $\int \sin x\ dx = -\cos x + c$
6. $\int \cos x\ dx = \sin x + c$
7. $\int \sec^2 x\ dx = \tan x + c$
8. $\int \text{cosec}^2 x\ dx = -\cot x + c$
9. $\int \sec x \tan x dx = \sec x + c$
10. $\int \text{cosec}\, x \cot x\, dx = -\text{cosec}\, x + c$
11. $\int \tan x\, dx = \ln|\sec x| + c$
12. $\int \cot x\, dx = \ln \sin x + c$
13. $\int \sec x\, dx = \log_e|\sec x + \tan x| + c = \log\tan\left(\frac{\pi}{4}+\frac{x}{2}\right) + c$
14. $\int \text{cosec}\, x\, dx = \log_e|\text{cosec}\, x - \cot x| + c = \log\tan\left(\frac{x}{2}\right) + c$
15. $\int \frac{1}{\sqrt{1-x^2}} dx = \sin^{-1} x + c$ or, $-\cos^{-1} x + c$
16. $\int \frac{1}{1+x^2} dx = \tan^{-1} x + c$ or, $-\cot^{-1} x + c$
17. $\int \frac{1}{x\sqrt{x^2-1}} dx = \sec^{-1} x + c$ or, $-\text{cosec}^{-1} x + c$
18. $\int \frac{dx}{\sqrt{a^2-x^2}} = \sin^{-1}\frac{x}{a} + c$
19. $\int \frac{dx}{\sqrt{a^2+x^2}} = \ln\left|x+\sqrt{a^2+x^2}\right| + c$ or, $\sinh^{-1}\frac{x}{a} + c$
20. $\int \frac{dx}{\sqrt{x^2-a^2}} = \ln\left|x-\sqrt{a^2-x^2}\right| + c$
21. $\int \frac{dx}{x\sqrt{x^2-a^2}} = \frac{1}{a}\sec^{-1}\frac{x}{a} + c$ or, $-\frac{1}{a}\text{cosec}^{-1}\frac{x}{a} + c$
22. $\int \frac{dx}{x^2-a^2} = \frac{1}{a}\tan^{-1}\frac{x}{a} + c$ or, $-\frac{1}{a}\cot^{-1}\frac{x}{a} + c$
23. $\int \frac{dx}{x^2-a^2} = \frac{1}{2a}\ln\left|\frac{x-a}{x+a}\right| + c$ If $x > a$
24. $\int \frac{dx}{a^2-x^2} = \frac{1}{2a}\ln\left|\frac{a+x}{a-x}\right| + c$ If $x > a$
25. $\int \sqrt{x^2+a^2}\ dx =$

$$\frac{1}{2}\left[x\sqrt{x^2+a^2} + a^2\ln\left(x+\sqrt{x^2+a^2}\right)\right]$$

$$= \frac{1}{2}\left[x\sqrt{x^2+a^2} + a^2 \sinh^{-1}\frac{x}{a}\right] + c$$

26. $\int\sqrt{x^2-a^2}\,dx =$

$$\frac{1}{2}\left[x\sqrt{x^2-a^2} - a\left(x+\sqrt{x^2+a^2}\right)\right] + c$$

27. $\int\sqrt{a^2-x^2}\,dx = \frac{1}{2}\left[x\sqrt{a^2-x^2} + a^2\sin^{-1}\frac{x}{a}\right] + c$

Integration by substitution

If the integrand is not in the integrable form, then it is changed to an integrable form by using substitution—

Let $I = \int f(x)dx$ and $x = \phi(t)$

$\therefore \quad \frac{dI}{dx} = f(x)$

Now, $\quad \frac{dI}{dt} = \frac{dI}{dt}.\frac{dx}{dt}$

$$= f(x).\frac{dx}{dt} = f\{\phi(t)\}.\phi'(t)$$

$\therefore$ Integrating both sides with respect to t

$$I = \int f\{\phi(t)\}.\phi(t).dt$$

$$\int f(x)dx = \int f\{\phi(t)\}.\phi(t).dt$$

Integration by Parts

Let $f(x)$ and $\phi(x)$ be the two functions of x, then

$$\int f(x).\phi(x)dx = f(x)\int\phi(x)dx - \int\left[\frac{d}{dx}\{f(x)\}.\int\phi(x)dx\right]dx$$

Integration of Rational Function

We know that the rational function is a quotient of two polynomial $P(x)$ and $Q(x)$.

$$\int\frac{dx}{ax^2+bx+c}$$

If ax^2+bx+c can be factorized, then the integration is easily done by the method of partial fractions.

If ax^2+bx+c cannot be factorized, then

$$\text{Let } I = \int\frac{dx}{ax^2+bx+c} =$$

$$= \frac{1}{a}\int\frac{dx}{\left(x+\frac{b}{2a}\right)^2 + \left(\frac{4ac-b^2}{4a^2}\right)}$$

Let $f(x)$ be a continuous functions defined on an intervals $[a, b]$ and let the anti-derivative of $f(x)$ be $F(x)$. Then, the definite integral of $f(x)$ over $[a, b]$, denoted by $\int_a^b f(x)dx$ is defined as

$$\int_a^b f(x)\,dx = [F(x)]_a^b = F(b) - f(a)$$

Here a and b are respectively known as the lower limit and upper limit of the integral.

General Properties of Definite Integral

I. $\int_a^b f(x)\,dx = \int_a^b f(t)\,dt = \int_a^b f(u)\,du = \ldots.$

II. $\int_a^b f(x)\,dx = -\int_b^a f(x)\,dx$

III. $\int_a^b f(x)\,dx = \int_a^c f(x)\,dx + \int_c^b f(x)\,dx,$

where $a < c < b$

The above property can be generalized into the following form

$$\int_a^b f(x)\,dx = \int_a^{c_1} f(x)\,dx + \int_{c_1}^{c_2} f(x)\,dx + \ldots + \int_{c_n}^b f(x)\,dx.$$

where $a < c_1 < c_2 < c_3 \ldots. < c_{n-1} < c_n < b$

IV. If $f(x) \geq 0$ on the interval $[a, b]$ then

$$\int_a^b f(x)\,dx \geq 0$$

V. If $f(x) \leq g(x)$ on the interval $[a, b]$, then

$$\int_a^b f(x)\,dx \leq \int_a^b g(x)\,dx$$

VI. $\left|\int_a^b f(x)\,dx\right| \leq \int_a^b |f(x)|\,dx$, the equality comes when $f(x)$ is either positive or negative in the interval $[a, b]$.

VII. $\int_0^a f(x)\,dx = \int_0^a f(a-x)\,dx$

$\int_0^a \frac{f(x)}{f(x)+f(a-x)}\,dx = \frac{a}{2}$

VIII. $\int_a^b f(x)\,dx = \int_a^b f(a+b-x)\,dx$

$\int_a^b \frac{f(x)}{f(x)+f(a+b-x)}\,dx = \frac{b-a}{2}$

IX. $\int_{-a}^a f(x)\,dx =$

$$\begin{cases} 2\int_0^a f(x)\,dx, \text{ if } f(-x) = f(x) \text{ i.e., } f(x) \text{ is even function} \\ 0, \text{ if } f(-x) = -f(x) \text{ i.e., } f(x) \text{ is odd function} \end{cases}$$

X. $\int_0^{2a} f(x)\,dx$

$$= \begin{cases} 2\int_0^a f(x)\,dx, \text{ if } f(2a-x) = f(x) \\ 0, \text{ if } f(2a-x) = -f(x) \end{cases}$$

XI. If $f(x)$ is a periodic function with period T, then $\int_a^{a+T} f(x)\,dx = \int_0^T f(x)\,dx$

XII. $\int_0^{nT} f(x)\,dx = n\int_0^T f(x)\,dx$; for $n \in N$

XIII. $\int_a^{a+nT} f(x)\,dx = n\int_0^T f(x)\,dx$

MULTIPLE CHOICE QUESTIONS

1. $\int \frac{x^2 \tan^{-1} x^3}{1+x^6}\,dx$ is equal to:

(a) $\frac{1}{6}(\tan^{-1} x^3)^2 + C$ (b) $\frac{1}{2}(\tan^{-1} x^2)^2 + C$

(c) $\frac{1}{3}(\tan^{-1} x^3)^3 + C$ (d) $\frac{1}{6}(\tan^{-1} x^6) + C$

2. $\int (\log x)^2\,dx$ is equal to:

(a) $x^2 (\log x)^2 + x \log x + 2x + C$

(b) $x^2 \log x - \log x + x + C$

(c) $x^2 (\log x)^2 - 2x \log x + 2x + C$

(d) $x^2 (\log x)^2 + 2x \log x + 2x + C$

3. $\int x \sec^2(x^2+1) \tan(x^2+1)\,dx$ is equal to:

(a) $\frac{1}{4}\tan(x^2+1)$ (b) $\frac{1}{4}\sec(x^2+1)$

(c) $\frac{1}{2}\sec(x^2+1)$ (d) None of these

4. $\int \frac{xe^x}{(x+1)^2}\,dx$ is equal to:

(a) e^x (b) $(x+1)e^x$

(c) $\frac{1}{x+1}e^x$ (d) $(x-1)^2 e^x$

5. $\int \frac{\cos x \sin x}{1+\sin^4 x}\,dx$ is equal to:

(a) $\frac{1}{2}\sin^{-1}(\tan^2 x)$ (b) $\frac{1}{2}\tan^{-1}(\sin^2 x)$

(c) $\frac{1}{2}\cos^{-1}(\cot^2 x)$ (d) $\frac{1}{2}\cot^{-1}x(\cos^2 x)$

6. $\int \frac{(1+x)e^x}{\sin^2(xe^x)}$ is equal to:

(a) $\sin(x e^x)$ (b) $-\cos(x e^x)$

(c) $\tan(x e^x)$ (d) $-\cot(x e^x)$

7. $\int \frac{\cot x}{\sqrt{\sin x}}\,dx$ is equal to:

(a) $2\sqrt{\sin x}$ (b) $-2\sqrt{\sin x}$

(c) $\frac{2}{\sqrt{\sin x}}$ (d) $\frac{-2}{\sqrt{\sin x}}$

8. $\int y\frac{d^2z}{dx^2}$ is equal to:

(*a*) $y\frac{dz}{dx}-\int\frac{dy}{dx}\cdot\frac{d^2z}{dx^2}dx$

(*b*) $y\frac{dz}{dx}+z\frac{dy}{dx}-\int z\frac{d^2y}{dx^2}dx$

(*c*) $y\frac{dz}{dx}-z\frac{dy}{dx}+\int z\frac{d^2y}{dx^2}dx$

(*d*) $y\frac{d^2z}{dx^2}-\int y\frac{d^3z}{dx^3}dx$

9. The two possible values $\frac{1}{2a}\log\frac{x-a}{x+a}$ and $\frac{1}{2a}\log\frac{a-x}{a+x}$ of the integral $\int\frac{dx}{x^2-a^2}$ differ by:

(*a*) $\frac{i\pi}{2a}$ (*b*) $\frac{\pi}{2a}$

(*c*) $\frac{i\pi}{4a}$ (*d*) None of these

10. $\int e^x \sec x\,(1+\tan x)dx$ is equal to:

(*a*) $e^x(\sec x \tan x)$ (*b*) $e^x \sec x$

(*c*) $e^x \tan x$ (*d*) None of these

11. $\int e^x \sin x \cos x\, dx$ is equal to:

(*a*) $\frac{1}{\sqrt{2}}e^x(x-\tan^{-1}x)$

(*b*) $\frac{1}{2\sqrt{5}}e^x\sin(2x-\tan^{-1}x)$

(*c*) $\frac{1}{\sqrt{2}}e^x\sin(2x-\tan^{-1}2)$

(*d*) $\frac{1}{\sqrt{5}}e^x\sin(2x-\tan^{-1}2)$

12. $\int\frac{\cos x dx}{(1+\sin x)(2+\sin x)}dx$ is equal to:

(*a*) $\log\frac{2+\sin x}{1+\sin x}$

(*b*) $\log(1+\sin x)(2+\sin x)$

(*c*) $\log\frac{1+\sin x}{2+\sin x}$

(*d*) None of these

13. $\int\frac{dx}{2x^2+x-1}$ is equal to:

(*a*) $\frac{1}{3}\log\frac{2x+1}{2(x+1)}$ (*b*) $\frac{1}{3}\log\frac{2(x+1)}{2x+1}$

(*c*) $\frac{1}{3}\log\frac{2x-1}{(x+1)}$ (*d*) $\frac{1}{3}\log\frac{2(x+1)}{2x-1}$

14. $\int\frac{dx}{1+3\sin^2 x}$ is equal to:

(*a*) $\frac{1}{2}\tan^{-1}(\tan x)$ (*b*) $2\tan^{-1}(\tan x)$

(*c*) $\frac{1}{2}\tan^{-1}(2\tan x)$ (*d*) $2\tan^{-1}\left(\frac{1}{2}\tan x\right)$

15. $\int\frac{dx}{\sqrt{2x^2+3x+4}}$ is equal to:

(*a*) $\frac{1}{\sqrt{2}}\sin^{-1}\frac{4x+3}{\sqrt{23}}$

(*b*) $\frac{1}{\sqrt{2}}\sinh^{-1}\frac{4x+3}{\sqrt{23}}$

(*c*) $\frac{1}{\sqrt{2}}\cosh^{-1}\frac{4x+3}{\sqrt{23}}$

(*d*) None of these

16. $\int_1^2\frac{1}{x^2}e^{-1/x}dx$ is equal to

(*a*) $\frac{1}{\sqrt{e}}+\frac{1}{e}$ (*b*) $\frac{1}{\sqrt{e}}-\frac{1}{e}$

(*c*) $\frac{1}{\sqrt{e}}-\frac{1}{4}$ (*d*) None of these

17. $\int_0^4\frac{1}{\sqrt{x}}\sin\sqrt{x}\,dx$ is equal to

(*a*) 2 (*b*) 2cos2

(*c*) 0 (*d*) None of these.

18. $\int_1^{\pi/2} \frac{\cos(\log x)}{x} dx$ is equal to

(*a*) $2\sin\left(\log\frac{\pi}{2}\right)$ (*b*) $\cos\left(\log\frac{\pi}{2}\right)$

(*c*) $\sin\frac{\pi}{2} - \log\frac{\pi}{2}$ (*d*) $\sin\left(\log\frac{\pi}{2}\right)$

19. Find the value of the integral $\int_0^1 \sqrt{1-x^2}\, dx$.

(*a*) π (*b*) 2π

(*c*) $\frac{2\pi}{3}$ (*d*) $\frac{\pi}{3}$

20. $\int_0^{\frac{\pi}{2}} \left(\sin\frac{x}{2}+\cos\frac{x}{2}\right) \cos x\, dx$ is equal to

(*a*) $\frac{2}{3}\left(2\sqrt{2}-1\right)$ (*b*) $\frac{2}{3}\left(2\sqrt{2}+1\right)$

(*c*) $\frac{2}{3}\left(2\sqrt{3}+1\right)$ (*d*) None of these.

21. $\int_2^3 xe^{x^2} dx$ is equal to

(*a*) $e^9 - e^4$ (*b*) $\frac{1}{2}\left(e^9 - e^4\right)$

(*c*) $\frac{1}{4}\left(e^9 - e^4\right)$ (*d*) None of these

22. If $\int_0^p f(x)\, dx = K$, then the value of

$\int_0^p f(x)\, dx + \int_0^p f(p-x)\, dx$

(*a*) $K+\frac{1}{K}$ (*b*) 2K

(*c*) $K-\frac{1}{K}$ (*d*) 0

23. Find the value of

$$\int_0^1 \left[1-\frac{x}{1!}+\frac{x^2}{2!}-\frac{x^3}{3!}+\ldots+\frac{(-1)^n x^n}{n!}+\ldots\right] e^x dx$$

(*a*) 0 (*b*) e

(*c*) 1 (*d*) $\frac{1}{2}$

24. $\int_0^a \frac{x^4}{\sqrt{a^2-x^2}} dx$ is equal to

(*a*) $\frac{\pi a^4}{16}$ (*b*) $\frac{3\pi a^4}{16}$

(*c*) $\frac{\pi a^4}{13}$ (*d*) $\frac{\pi a^4}{13}$

25. $\int_0^{\frac{\pi}{2}} \frac{\sin^m x}{\sin^m x + \cos^m x} dx$ is equal to

(*a*) π (*b*) $\frac{\pi}{2}$

(*c*) $\frac{\pi}{3}$ (*d*) $\frac{\pi}{4}$

26. Which of the following is correct if

$$\int_0^{2a} f(x)\, dx = 2\int_0^a f(2-a)\, dx$$

(*a*) $f(2a-x) = 0$ and $f(x) > 0$

(*b*) $f(2a-x) = f(x)$

(*c*) $f(2a-x) = -f(x)$

(*d*) $f(x) = f(x-2a)$

27. Find the value of the integral

$\int_0^1 2\sin\left(\frac{\pi t}{2}-\frac{\pi}{4}\right) dt$

(*a*) 1 (*b*) 2

(*c*) –1 (*d*) 0

28. $\int_0^{\frac{\pi}{3}} \sin 5x \cos x\, dx$ is equal to

(*a*) $\frac{1}{16}$ (*b*) $\frac{2}{15}$

(*c*) $\frac{3}{15}$ (*d*) $\frac{3}{16}$

29. $\int_0^1 \frac{dx}{\left(x^2+1\right)^{3/2}}$ is equal to

(*a*) $\frac{1}{2}$ (*b*) $\frac{1}{2}\sqrt{2}$

(*c*) 1 (*d*) $\sqrt{2}$

30. $\int_0^\infty e^{-a^2x^2} dx$ is equal to

(*a*) $\sqrt{\pi}$ (*b*) $\frac{\sqrt{\pi}}{2a}$

(*c*) $\frac{\sqrt{\pi}}{-4a}$ (*d*) $\frac{\pi}{2a}$

ANSWERS

1	2	3	4	5	6	7	8	9	10
(a)	(c)	(d)	(c)	(b)	(d)	(d)	(c)	(a)	(d)
11	**12**	**13**	**14**	**15**	**16**	**17**	**18**	**19**	**20**
(c)	(c)	(c)	(b)	(b)	(b)	(d)	(d)	(d)	(a)
21	**22**	**23**	**24**	**25**	**26**	**27**	**28**	**29**	**30**
(b)	(b)	(c)	(b)	(d)	(b)	(d)	(d)	(b)	(b)

SOME SELECTED EXPLANATORY ANSWERS

1. Let $I = \int \frac{x^2 \tan^{-1} x^3}{1+x^6} dx$

Put $\tan^{-1} x^3 = t \Rightarrow \frac{1}{1+x^6} .3x^2\, dx = dt$

$\therefore \quad I = \frac{1}{3} \int t\, dt = \frac{1}{6} t^2$

$= \frac{1}{6} (\tan^{-1} x^3)^2 + c$

5. Let $I = \int \frac{\cos x \sin x}{1+\sin^4 x} dx$

Put $\sin^2 x = t \Rightarrow 2 \sin x \cos x\, dx = dt$

$\therefore I = \int \frac{\frac{1}{2} dt}{(1+t^2)} = \frac{1}{2} \tan^{-1} t$

$= \frac{1}{2} \tan^{-1} (\sin^2 x)$

12. Let $I = \int \frac{\cos x\, dx}{(1+\sin x)\,(2+\sin x)}$

Put $\sin x = t \Rightarrow \cos x\, dx = dt$

$\therefore \quad I = \int \frac{dt}{(1+t)\,(2+t)}$

$= \int \left(\frac{1}{t+1} - \frac{1}{t+2}\right) dt$

$= \log (t+1) - \log (t+2)$

$= \log \frac{(t+1)}{(t+2)} = \log \left(\frac{\sin x + 1}{\sin x + 2}\right)$

16. Let $I = \int_1^2 \frac{1}{x^2} e^{-\frac{1}{x}} dx$

Put $-\frac{1}{x} = t \Rightarrow \frac{1}{x^2} dx = dt$

$\therefore \quad I = \int_{-1}^{-1/2} e^t\, dt = \left[e^t\right]_{-1}^{-1/2} = e^{-1/2} - e^{-1} = \frac{1}{\sqrt{e}} - \frac{1}{e}$

22. Use $\int_0^a f(x) dx = \int_0^a f(a-x) dx$

23. $\int_0^1 e^{-x}.e^x dx = \int_0^1 dx = (x)_0^1 = 1$

29. $I = \int_0^1 \frac{dx}{(x^2+1)^{3/2}}$

Put $x = \tan\theta \Rightarrow dx = \sec^2 \theta\, d\theta$

$\therefore \quad I = \int_0^1 \frac{dx}{(x^2+1)^{3/2}} = \int_0^{\pi/4} \cos\theta\, d\theta$

$= [\sin \theta]_0^{\pi/4} = \sin \frac{\pi}{4} = \frac{1}{\sqrt{2}} = \frac{1}{2}\sqrt{2}$

20

Applications of Integral

Rectification : The process of finding the length of an arc of a plane curve between two points is called rectification.

Length of arc of a curve in cartesian coordinates

(*i*) In the curve $y = f(x)$, the length of the curve between $x = a$ and $x = b$ is

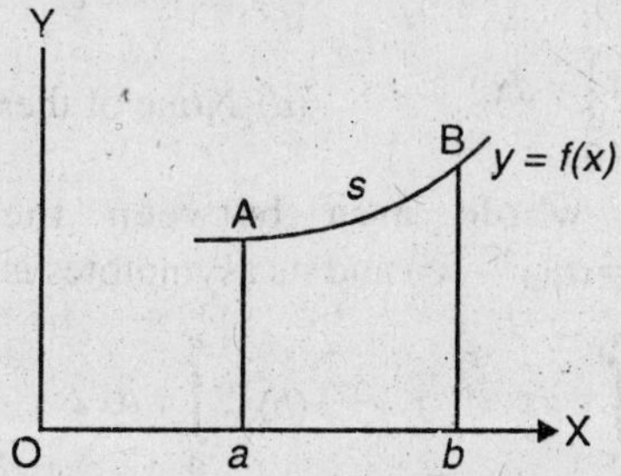

$$S = \int_{b}^{a} \sqrt{\left\{1 + \left(\frac{dy}{dx}\right)^2\right\}}\, dx$$

(*ii*) If the curve be $x = f(y)$, the length of the curve between $y = a$ and $y = b$ is

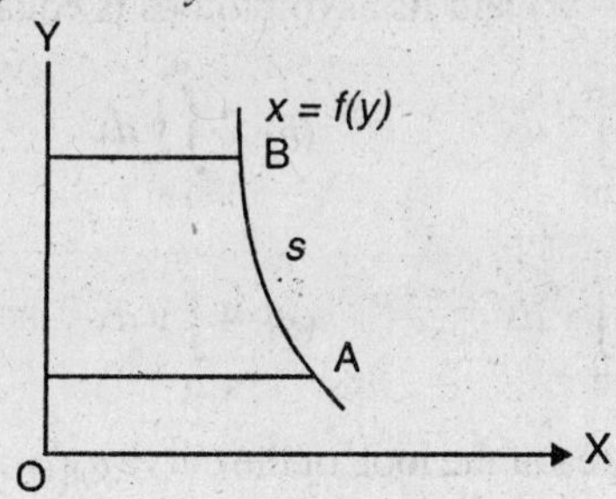

$$S = \int_{a}^{b} \sqrt{\left\{1 + \left(\frac{dx}{dy}\right)^2\right\}}\, dy$$

Length of arc of a curve in polar coordinates :

If the curve be $r = f(\theta)$, the length of the curve between the points (r_1, θ_1) and (r_2, θ_2) is

$$S = \int_{\theta_1}^{\theta_2} \sqrt{\left\{r^2 + \left(\frac{dr}{d\theta}\right)^2\right\}}\, d\theta$$

or $$S = \int_{r_1}^{r_2} \sqrt{\left\{1 + r^2\left(\frac{d\theta}{dr}\right)^2\right\}}\, dr$$

Length of arc of a curve in parametric form :

Let $x = f(t)$, $y = \phi(t)$, $\alpha \le t \le \beta$ be the equation of a curve in parametric form. Then the length of the curve between two points α and β is

$$S = \int_{\alpha}^{\beta} \sqrt{\left\{\left(\frac{dx}{dt}\right)^2 + \left(\frac{dy}{dt}\right)^2\right\}}\, dt$$

Area when the curve is given in cartesian coordinates

(*i*) If $f(x)$ is continuous and single valued function of x in the interval (a, b), then area A bounded by the curve $y = f(x)$, the axis of x and the ordinates at $x = -a$ and $x = b$ is given by

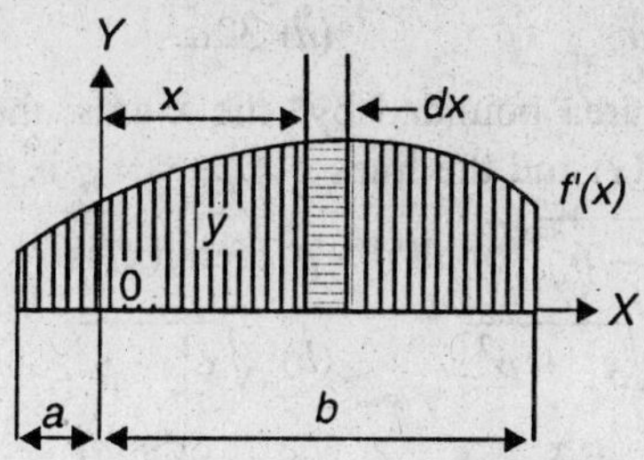

$$A = \int_{-a}^{b} f(x)\, dx$$

(*ii*) If the area is bounded by a continuous curve, the abscissae $y = -c$, $y = d$ and the axis of y, then it is given by

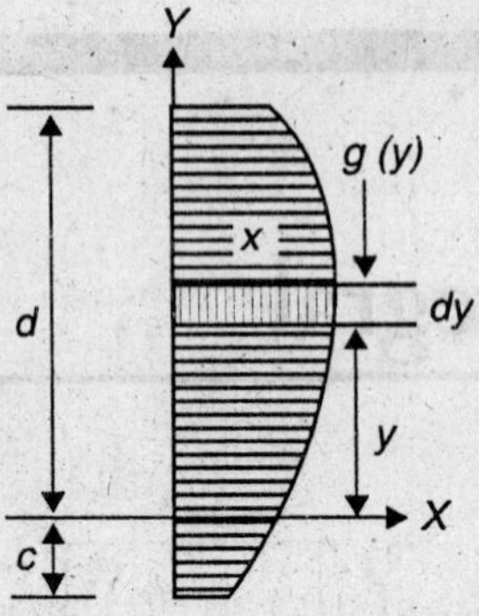

$$A = \int_{-c}^{d} g(y)\, dy$$

(*iii*) The area bounded by the curves $y = f(x)$, $y = \phi(x)$ and the ordinates $x = a, x = b$ is given by

$$A = \int_{a}^{b} \{f(x) - \phi(x)\}\, dx$$

MULTIPLE CHOICE QUESTIONS

1. Length of the arc of the parabola $y^2 = 4ax$ cut off by the line $3y = 8x$ is

(*a*) $\left(\ln 2 + \frac{15}{16}\right)$ (*b*) $a\left(\ln 2 - \frac{15}{16}\right)$

(*c*) $\left(\ln 2 - \frac{15}{16}\right)$ (*d*) $a\left(\ln 2 + \frac{15}{16}\right)$

2. The entire length of the curve $x^{2/3} + y^{2/3} = a^{2/3}$ is given by

(*a*) $8a$ (*b*) $4\sqrt{3}\, a$

(*c*) $6a$ (*d*) $\sqrt{8}\, a$

3. The length of the complete cycloid $x = a(\theta - \sin\theta), y = a(1 - \cos\theta)$ is given by

(*a*) a (*b*) $4a$

(*c*) $8a$ (*d*) $32a$

4. The area bounded by the x-axis, the curve $y = f(x)$ and the lines $x = p, x = q$ is equal to $\sqrt{q^2 - p^2}$ for all $q > p$, then $f(x)$ is

(*a*) $\sqrt{x^2 + p^2}$ (*b*) $\sqrt{x^2 - p^2}$

(*c*) $\frac{x}{\sqrt{x^2 - p^2}}$ (*d*) $\frac{(x + p)}{\sqrt{(x^2 - q^2)}}$

5. The whole area between the curve $y^2(a - x) = x^2(a + x)$ and its asymptotes is equal to

(*a*) $\int_0^a y\, dx$ (*b*) $2\int_0^a y\, dx$

(*c*) $4\int_0^a y\, dx$ (*d*) None of these

6. The whole area between the curve $x^2y^2 = a^2(y^2 - x^2)$ and its asymptotes is equal to

(*a*) $\int_0^a y\, dx$ (*b*) $2\int_0^a y\, dx$

(*c*) $4\int_0^a y\, dx$ (*d*) None of these

7. The whole area between the curve $x(x^2 + y^2) = a(x^2 - y^2)$ and its asymptotoes is equal to

(*a*) $2\int_0^a y\, dx$ (*b*) $2\int_0^{-a} y\, dx$

(*c*) $4\int_0^a y\, dx$ (*d*) $4\int_0^{-a} y\, dx$

8. The area of the loop of the curve $ay^2 = x^2(a - x)$ is

(*a*) $\frac{2}{15}a^2$ (*b*) $\frac{4}{15}a^2$

(*c*) $\frac{8}{15}a^2$ (*d*) $\frac{15}{8}a^2$

9. Area enclosed by the curve $y^2 = x(x - 1)^2$ is given by

(a) $\frac{15}{2}$ (b) $\frac{5}{2}$

(c) $\frac{2}{15}$ (d) $\frac{2}{5}$

10. The whole area enclosed by the curve $y^2 (a^2 + x^2) = x^2 (a^2 - x^2)$ is

(a) $a^2 (\pi - 2)$ (b) $a^2 (\pi + 2)$

(c) $a^2 (\pi^2 - 4)$ (d) $a^2 (\pi^2 + 4)$

11. Whole area included between the curve $x^2y^2 = a^2 (y^2 - x^2)$ and its asymptotes

(a) $2a^2$ (b) $4a^2$ (c) $6a^2$ (d) $8a^2$

12. The entire area enclosed by the curve $r = a \cos 2\theta$ is

(a) πa^2 (b) $\frac{1}{2}\pi a^2$

(c) $\frac{1}{4}\pi a^2$ (d) None of these

13. Area of the curve $r = a (1 + \sin \theta)$ is

(a) $\frac{1}{2}\pi a^2$ (b) πa^2

(c) $\frac{3}{2}\pi a^2$ (d) $2\pi a^2$

14. The curves $y^2 = 4x$ and $x^2 = 4y$ divide the square bounded by $x = 0, x = 4, y = 0, y = 4$ into

(a) two equal areas (b) three equal areas

(c) four equal areas (d) None of these

ANSWERS

1	2	3	4	5	6	7	8	9	10
(d)	(c)	(c)	(c)	(d)	(c)	(b)	(c)	(c)	(a)
11	**12**	**13**	**14**						
(b)	(b)	(c)	(b)						

SOME SELECTED EXPLANATORY ANSWERS

5. Area lying between the curve and its asymptote is

$$A_{OAB} = 2\int_0^a f(x)\,dx = 2\int_0^a y\,dx$$

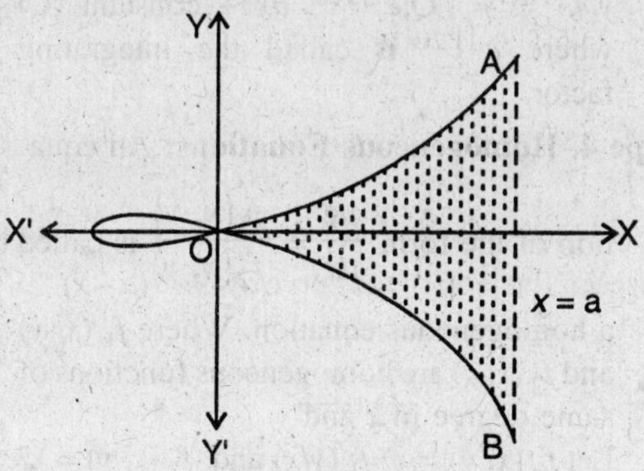

7. The curve is symmetrical about x-axis as there occur only even powers of y.

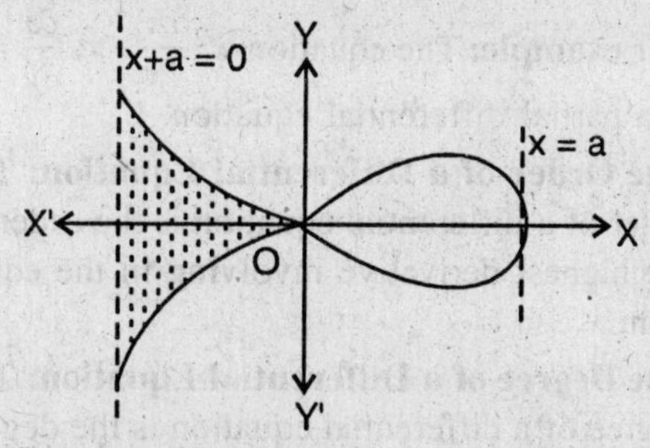

It passes through (0, 0) and $(a, 0)$. Equating to zero the coefficients of highest powers of y, the asymptote is $x + a = 0$ i.e. $x = -a$.
Here required area lying between the curve and its asymptote is

$$-2\int_{-a}^{0} f(x)\,dx = 2\int_0^{-a} f(x)\,dx = 2\int_0^{-a} y\,dx.$$

21

Differential Equations

1. **Def.:** An equation which contains independent variables, dependent variables and their derivatives w.r.t. independent variables, is called a *differential equation.*
2. **Ordinary Differential Equation:** A differential equation which involves only one independent variable is called an *ordinary differential equation.*
   For example: The equations $\frac{dy}{dx}+xy=0$ and $x\left(\frac{dy}{dx}\right)^2+y\frac{dy}{dx}+1=0$ are ordinary differential equations.
3. **Partial Differential Equation:** A differential equation which involves one or more partial derivatives is called partial differential equation.
   For example: The equation $x^2\frac{\partial z}{\partial x}+xy\frac{\partial z}{\partial y}=a$ is a partial differential equation.
4. **The Order of a Differential Equation:** The order of a differential equation is the order of the highest derivative involving in the equation.
5. **The Degree of a Differential Equation:** The degree of a differential equation is the degree of the highest order derivative involving in the equation, when the equation is free from radicals and fractional powers.
   For example:
   $\frac{dy}{dx}+xy=a$ ..... (1); $\frac{d^2y}{dx^2}+x\frac{dy}{dx}=2$ ... (2);

   $$\frac{d^2y}{dx^2}=\left[1+\left(\frac{dy}{dx}\right)^5\right]^{1/3} \quad \text{... (3)}$$

The equation (1) is of first order and first degree. Equation (2) is of second order and first degree. Equation (3) is of second order and third degree.

Type 1. Variable separation Form: The equation of this type can be put in the form.
$f(x)\,dx + g(y)\,dy = 0$
Integrating both sides, we get the solution
$\int f(x)\,dx + \int g(y)\,dy = C$

Type 2. Linear Differential Equations: A differential equation is called linear if the dependent variable and its derivative occurring in it are of the first degree and are not multiplied together. The linear differential equation is of the form $\frac{dy}{dx}+$ $Py = Q$ where P and Q are the functions of x only.
The solution of the linear differential equation is
$y.e^{\int Pdx} = \int Q.e^{\int Pdx}.\,dx$ + constant (C)
where $e^{\int Pdx}$ is called the integrating factor.

Type 4. Homogeneous Equations: An equation of the form $\frac{dy}{dx}=\frac{f_1(x,y)}{f_2(x,y)}$ is called a homogeneous equation. Where $f_1(x, y)$ and $f_2(x, y)$ are homogeneous functions of same degree in x and y.
Let $f_1(x, y) = x^n f_1(y/x)$ and $f_2(x, y) = x^n f_2(y/x)$

$$\therefore \frac{dy}{dx}=\frac{f_1(x,y)}{f_2(x,y)}=\frac{x^n f_1(y/x)}{x^n f_2(y/x)}=F(Y/X) \quad \text{... (1)}$$

Now put $y = vx \Rightarrow \frac{dy}{dx} = v + x\frac{dv}{dx}$

$\therefore$ From (1) $v + x\frac{dv}{dx} = f(v)$ ... (2)

Separating the variables, the equation (2) become

$\frac{dv}{f(v)-v} = \frac{dx}{x}$. Integrating this, we get the required solution.

Type 5: Non-homogeneous equation which can be reduced in the homogeneous form—The equation of this type is—

$$\frac{dy}{dx} = \frac{ax+by+c}{Ax+By+C} \quad ..... (1)$$

Case (I). When $\frac{a}{A} \neq \frac{b}{B}$

Now put $x = X + h$ and $y = Y + k$.

$$\therefore \frac{dy}{dx} = \frac{dY}{dX} \Rightarrow \frac{dy}{dx}$$

$$= \frac{a(X+h)+b(Y+k)+c}{A(X+h)+B(Y+k)+C}$$

$$\Rightarrow \frac{dY}{dX} = \frac{aX+bY+(ah+bk+c)}{AX+BY+(Ah+Bk+C)} \quad ... (2)$$

Now choose h and k such that $ah + bk + c = 0$, $Ay + Bk + C = 0$

$\therefore$ The equation (2) becomes $\frac{dY}{dX} = \frac{aX+bY}{AX+BY}$

Which is a homogeneous equation and can be solved by putting $Y = vX$.

Type 6: Exact Differential Equation: A differential equation which can be obtained by direct differentiation of some function of x and y is called exact differential equation, consider the equation

$M\,dx + N\,dy = 0$ is exact if $\frac{\partial M}{\partial y} = \frac{\partial N}{\partial x}$,

where M and N are the functions of x and y. Solution of a exact differential equations is $\int M dx + \int N\,dy = C$

Type 7: Change of Variable: Let the equation

$$\frac{dy}{dx} = f(ax + by + c) \quad ..... (1)$$

put $ax + by + c = v \Rightarrow a + b\frac{dy}{dx} = \frac{dv}{dx}$

Equation (1) reduced to $\frac{1}{b}\left(\frac{dv}{dx} - a\right) = f(v)$

$\frac{dv}{bf(v)+a} = dx$. Integrating we find the required solution.

MULTIPLE CHOICE QUESTIONS

1. The solution of the differential equation $(x - y^2)\,dx + 2xy\,dy = 0$ is:

(a) $ye^{y^2/x} = A$ (b) $xe^{y^2/x} = A$

(c) $ye^{x/y^2} = A$ (d) $xe^{x/y^2} = A$

2. The solution of the differential equation $\frac{dy}{dx} + y\frac{d\phi}{dx} = \phi(x)\frac{d\phi}{dx}$ is:

(a) $y = \phi(x) - 1 + ce^{-\phi}$

(b) $y = ce^{\phi}$

(c) $y = x\,\phi(x) - ce^{-\phi}$

(d) $y = [\phi(x) - 1]e^{-\phi} + c$

3. The general solution of the differential equation $\frac{dy}{dx} = \frac{y}{x} + \tan\frac{y}{x}$ is:

(a) $\sin\frac{y}{x} = c$ (b) $\cos\frac{y}{x} = cx$

(c) $\sin\frac{y}{x} = cx$ (d) $\cos\frac{y}{x} = c$

4. The differential equation $xdy - y\,dx + 2x^3dx = 0$ has the solution:

(a) $y + x^3 = c_1 x$ (b) $-y + x^3 = c_2 x$

(c) $y - x^3 = c_3 x$ (d) $y^3 - x^3 = c_4 x$

5. If $x = A\cos(mt - \alpha)$ then the differential equation satisfying this relation is:

(a) $\frac{dx}{dt} = 1 - x^2$ (b) $\frac{d^2x}{dt^2} = -\alpha^2 x$

(c) $\frac{d^2x}{dt^2} = -m^2 x$ (d) $\frac{dx}{dt} = -m^2 x$

6. The solution of the differential equation $\sec x \frac{dy}{dx} - y = \sin x$ is given by:

(a) $y + 1 + \cos x = c_3 e^{\cos x}$

(b) $y + 1 - \sin x = c_2 e^{-\sin x}$

(c) $y + 1 + \sin x = c_1 e^{\sin x}$

(d) $y - 1 - \cos x = c_4 e^{-\cos x}$

7. The integrating factor for the differential equation $(x^2y - 2xy^2)\,dx - (x^3 - 3x^2y)\,dy$ is given by:

(a) $\frac{1}{xy}$ (b) xy

(c) x^2y^2 (d) $\frac{1}{x^2y^2}$

8. The integrating factor for the differential equation $(x^3 + xy^4)dx + 2y^3\,dy = 0$ is given by:

(a) e^{-x} (b) e^{x^2}

(c) e^x (d) e^{-x^2}

9. Integrating factor of $\frac{dy}{dx} + 2y \tan x = \sin x$ is:

(a) $\cos^2 x$ (b) $\sin^2 x$

(c) $\sec^2 x$ (d) $2 \tan x$

10. The solution of $y\,(x - y)\,dy - y^2\,dx = 0$ is:

(a) $y = ce^{-y/x}$ (b) $y = ce^{y/x}$

(c) $y = ce^{-x/y}$ (d) $y = ce^{x/y}$

11. The primitive of the differential equation $(2xy^4\,e^y + 2xy^3 + y)\,dx + (x^2\,y^4\,e^y - x^2y^2 - 3x)dy = 0$, is:

(a) $x^2e^y + \frac{x^2}{y} + \frac{x}{y^3} = c$

(b) $x^2e^y - \frac{x^2}{y} + \frac{x}{y^3} = c$

(c) $x^2e^y + \frac{x^2}{y} - \frac{x}{y^3} = c$

(d) $x^2e^y - \frac{x^2}{y} - \frac{x}{y^3} = c$

12. The equation $x\,(1 + y^2)\,dx + \tan^{-1} y\,dy = 0$ is satisfied by:

(a) $y = \tan\sqrt{c - x^2}$

(b) $(\tan^{-1} y)^2 + 2x^2 = c$

(c) $\tan^{-1} y = x^2 + c$

(d) $\tan^{-1} y - 2x^2 = c$

13. The integrating factor of the equation $(x^2y^3 + xy)\frac{dy}{dx} = 1$ is:

(a) e^{y^2} (b) $e^{\frac{1}{2}y^2}$

(c) $e^{\frac{1}{2}x^2}$ (d) $e^{-\frac{1}{2}y^2}$

14. The integrating factor of the equation $y' - \frac{\tan y}{1+x} = (1 + x)e^x \sec y$ is:

(a) $\frac{1}{1+x}$ (b) $\frac{-1}{1+x}$

(c) $\frac{e^x}{1+x}$ (d) e^{x+1}

15. The solution of the equation $xp + 2y = pxy$, $\left(p = \frac{dy}{dx}\right)$ is:

(a) $xy^2 = Ae^y$ (b) $xy^2 = Ae^x$

(c) $x^2y = Ae^y$ (d) $xy^2 = Ae^y$

ANSWERS

1	2	3	4	5	6	7	8	9	10
(b)	(a)	(c)	(a)	(c)	(c)	(d)	(b)	(c)	(c)
11	**12**	**13**	**14**	**15**					
(a)	(a)	(b)	(a)	(c)					

SOME SELECTED EXPLANATORY ANSWERS

1. The given differential equation is

$$(x - y^2)\,dx + 2xy\,dy = 0$$

This is of the form M dx + N dy = 0

Here $\frac{\partial M}{\partial y} = -2y$ and $\frac{\partial N}{\partial x} = 2y$

$$\therefore \quad \frac{1}{N}\left(\frac{\partial M}{\partial y} - \frac{\partial N}{\partial x}\right) = \frac{-2y-2y}{2xy} = \frac{-2}{x}$$

$\therefore$ Integrating factor $= e^{\int \frac{-2}{x}dx} = e^{-2\log x} = \frac{1}{x^2}$

Now, multiplying the integrating factor in the given equation, we get

$$\left(\frac{1}{x} - \frac{y^2}{x^2}\right)dx + \frac{2y}{x}dy = 0$$

$\therefore$ The solution is $\int \left(\frac{1}{x} - \frac{y^2}{x^2}\right)dx = \log A$

or $\quad \log x + \frac{y^2}{x} = \log A$

or $\log \frac{A}{x} = \frac{y^2}{x}$ or $A = xe^{y^2/x}$

2. The given equation is

$$\frac{dy}{dx} + y\frac{d\phi}{dx} = \phi(x)\frac{d\phi}{dx}$$

This equation is of the form

$$\frac{dy}{dx} + Py = Q$$

where $P = \frac{d\phi}{dx}$ and $Q = \phi(x)\frac{d\phi}{dx}$

$\therefore$ I.F. $= e^{\int \frac{d\phi}{dx}dx} = e^{\int d\phi} = e^{\phi(x)}$

$\therefore$ Solution is

$$y.e^{\phi(x)} = \int \phi(x)\frac{d\phi}{dx}\,e^{\phi(x)}.\,dx + c$$

or $\quad y.e^{\phi(x)} = \int (x)\,e^{\phi(x)}\,d\phi + c$

$$= \phi(x).\,e^{\phi(x)} - e^{\phi(x)} + c$$

$$= (\phi(x) - 1)\,e^{\phi(x)} + c$$

or $\quad y = \phi(x) - 1 + ce^{-\phi(x)}$

6. $\frac{dy}{dx} - y\cos x = \sin x \cos x$

I.F. $= e^{-\int \cos x\,dx}$

$= e^{-\sin x}$

$\therefore \quad y.e.^{-\sin x} = \int \sin x \cos x e^{-\sin x}\,dx + c$

Put $\quad \sin x = t \Rightarrow \cos x\,dx = dt$

$\therefore \quad ye^{-\sin x} = \int te^{-1}\,dt + c$

$= -te^{-1} + \int 1 \cdot e^{-t}\,dt$

$= -te^{-1} + (-1)e^{-1} + c$

$= -(\sin x + 1)e^{-\sin x} + c$

or $\quad y = -\sin x - 1 + ce^{\sin x}$

$\therefore \quad y + 1 + \sin x = ce^{\sin x}$

9. This differential equation is linear,

Hence, I.F. $= e^{2\int \tan x\,dx}$

$= e^{2\log \sec x}$

$= \sec^2 x$

12. The given equation is

$$x(1 + y^2)\,dx + \tan^{-1} y\,dy = 0$$

or $\quad x\,dx + \frac{\tan^{-1} y}{1+y^2}dy = 0$

Integrating, we get

$$\int x\,dx + \int \frac{\tan^{-1} y}{1+y^2}dy = \frac{1}{2}c$$

or $\quad \frac{x^2}{2} + \frac{1}{2}(\tan^{-1} y)^2 = \frac{1}{2}c$

or $\quad x^2 + (\tan^{-1} y)^2 = c$

or $\quad y = \tan\sqrt{c - x^2}$

13. The given equation is

$$(x^2y^3 + xy)\frac{dy}{dx} = 1 \text{ or } \frac{dx}{dy} - xy = x^2y^3$$

This is a Bernaulli's equation

$$\therefore \quad \frac{1}{x^2}\frac{dx}{dy} - \frac{y}{x} = y^3 \quad \ldots (i)$$

Let $\frac{1}{x} = v \Rightarrow -\frac{1}{x^2}\frac{dx}{dy} = \frac{dv}{dy}$

From (i) $-\frac{dv}{dy} - vy = y^3$ or $\frac{dv}{dy} + vy = -y^3$

This is a linear differential equation

$\therefore \quad$ I.F. $= e^{\int y dy} = e^{\frac{1}{2}y^2}$

22

Probability

Random Experiment: If the result of an experiment is not certain and is any one of the several possible outcomes, the experiment is called a trial or a random experiment.

Sample Space: The set of all possible outcomes of an experiment is called the sample space provided no two or more of these outcomes can occur simultaneously and exactly one of these outcomes must occur whenever the experiment is conducted.

Events: The outcomes of an experiment, *i.e.*, sample points of the sample space are usually known as simple events and any subset of the sample space 'S' is called an event.

Thus throwing of a dice is an experiment, $S = \{1, 2, 3, 4, 5, 6\}$ is the sample space, $\{1\}, \ldots \{6\}$ are simple events and $\{1, 2\}$, etc., are events. The empty set ϕ is also an event as $\phi \subset S$ and it is called an impossible event. The sample space S is also a subset of S and so it is also an event. S represents the sure event, *i.e.*, certainty.

Equally Likely Events: A set of events is said to be equally likely if taking into consideration all the relevant factors there is no reason to expect one of them in preference to others.

For example, when a fair coin is tossed, the occurrence of a tail or a head are equally likely.

Exhaustive Events: A set of events is said to be exhaustive if the performance of the experiment always results in the occurrence of atleast one of them.

For example, if we throw a dice. Then the events $A_1 = \{1, 2\}$, $A_2 = \{2, 3, 3\}$ are not exhaustive as we can get A_2 as outcome of the experiment which is not the member of any of the events A_1 and A_2. If we consider the events $E_1 = \{1, 2, 3\}$ and $E_2 = \{2, 4, 5, 6\}$, then the set E_1, E_2, is exhaustive.

Mutually Exclusive Events: A set of events is said to be mutually exclusive if they have no point in common, *i.e.*, happening of one of them eliminates the happening of any of the remaining events. Thus $E_1, E_2, E_3, \ldots$ are mutually exclusive iff $E_i \cap E_j = \phi$ for $i \neq E_1$ and E_2 are mutually exclusive.

Complement of An Events: The complement of an event A, denoted by $\overline{A}$, A' or A^c, is the set of all sample points of the space other than the sample points in A.

e.g., In the experiment of throwing a fair dice, $S = \{1, 2, 3, 4, 5, 6\}$. If $A = \{1, 3, 5, 6\}$, then $\overline{A} = \{2, 4\}$

Note that $A \cap \overline{A} = S$.

Classical Definition of Probability: If there are n exhaustive mutually exclusive and equally likely outcomes of an experiment and m of them are favourable to an event A, the probability of the happening of A is defined as the ratio m/n.

Thus denoting the probability of the happening of an event A by p, we have $p = m/n$.

Clearly p is a positive number not greater than unity, so that $0 \leq p \leq 1$.

Since the number of cases in which the event A will not happen is $n - m$, the probability q that the event will not happen is given by

$$q = \frac{(n-m)}{n} = 1 - \frac{m}{n} = 1 - p. \therefore p + q = 1$$

If probability of happening of an event A is 1, then A is certain event and if probability of happening of an event A is 0, then A is impossible event.

Odds in Favour and Odds against an Event: As a result of an experiment if a of the outcomes are favourable to an event E and b of the outcomes are against it, then we say that odds are a to b in favour of E or odds are b to a against E.

Thus odds in favour of an event

$$E = \frac{\text{number of favourable cases}}{\text{number of unfavourable cases}}$$

Similarly odds against an event

$$E = \frac{\text{number of unfavourable cases}}{\text{number of favourable cases}}$$

If odds in favour of an event are $a : b$ then the probability of the occurrence of that event is $\frac{a}{a+b}$ and the probability of the non-occurrence of that event is $\frac{b}{(a+b)}$.

Addition Theorem:

(*a*) If 'A' and 'B' are any two events in a sample space S, then $P(A \cup B) = P(A) + P(B) - P(A \cap B)$.

(*b*) If 'A' and 'B' are mutually exclusive then $P(A \cap B) = 0$ so that $P(A \cup B) = P(A) + P(B)$.

(*c*) If A is any event in S, then $P(A') = 1 - P(A)$.

Total Probability Theorem: The probability that one of several mutually exclusive events $A_1, A_2, ..., A_n$ will happen, is the sum of the probabilities of the separate events. In symbol,

$$P(A_1 + A_2 + ... + A_n) = P(A_1) + P(A_2) + ... + P(A_n).$$

Conditional Probability: The probability of B under the assumption that A has occurred is called the conditional probability of B under the condition that the event A has taken place and is denoted by P(B/A). P(B/A) is read as "probability at the event B given A^B."

Conditional Probability Theorem: If A and B are any two events in the sample space S, the conditional probability of B relative to A is given by $P(B/A) = \frac{P(B \cap A)}{PA} = \frac{n(B \cap A)}{n(A)}, A \neq \phi$.

Multiplication Theorem: (Theorem of Compound Probability):

(*a*) If A and B are any events in the sample space S, then $P(A \cap B) = P(A).P(B/A)$ if $P(A) \neq 0$
$= P(B).P(A/B)$ if $P(B) \neq 0$.

(*b*) If A and B are independent events, then $P(A \cap B) = P(A).P(B)$.

Generalised Multiplication Theorem: Let $A_1, A_2, ..., A_n, A_{n+1}$ be $(n + 1)$ events such that $P(A_1 \cap A_2 \cap ... \cap A_n)$, then $P\left(\bigcap_{j=1}^{n+1} A_j\right)$
$= P(A_1)P(A_2/A_1)\,P(A_3/A_1 \cap A_2)...\,P(A_{n+1}/A_1 \cap ... \cap A_n)$.

Independent Events: Two events A and B are said to be independent if occurence of A does not depend on the occurrence or non-occurrence of the event B. Thus A and B are independent if

$P(A/B) = P(A)$ and $P(B/A) = P(B)$.

If A and B are independent events, then

(*a*) A and B' (*i.e.*, $\overline{B}$) are independent, *i.e.*, $P(A \cap \overline{B}) = P(A)\,P(\overline{B}) = P(A)[1 - P(B)]$.

(*b*) A' and B are independent, *i.e.*, $P(A' \cap B) = [1 - P(A)]P(B)$.

(*c*) A' and B' are independent so $P(A' \cap B') = P(A')\,P(B') = [1 - P(A)][1 - P(B)]$.

(*d*) $P(A'/B') = P(A'), P(A'/B) = P(A')$.

Difference Between Mutually Exclusiveness and Independence: Mutually exclusiveness is used when the events are taken from the same experiment and independence is used when the events are taken from the different experiments.

Baye's Theorem: If $H_1, H_2, ..., H_n$ are mutually exclusive and exhaustive events in a sample space S, and A is another event such that $P(A) \neq 0$, then for $j = 1, 2, ..., n$

$$P(H_j/A) = \frac{P(H_j)P(A/H_j)}{\sum_{k=1}^{n} P(H_k)P(A/H_k)}$$

Mathematical Expectation: Let X be discrete random variable which assumes the values $x_1, x_2, ..., x_n$ with the corresponding probabilities $p_1, p_2, ..., p_n$.

Then the expected value of X or the mathematical expectation of X denoted by E(X) is defined as

$$E(X) = \sum_{i-1}^{n} x_i p_i \text{ where } \sum_{i-1}^{n} p_i = 1, i.e., E(X) = x_1 p_1$$
$+ x_2 p_2 + ... + x_n x_n$, where $p_1 + ... + p_n = 1$.

Suppose a Binomial experiment has probability of success p and that of failure q, $q = 1 - p$, if there are n independent trials of the Binomial experiment then the probability of exactly x success is given by

$$P(x) = {}^nC_x\, q^{n-x} p^x.$$

MULTIPLE CHOICE QUESTIONS

1. The probability that a leap year selected at random contain 53 Sundays is
(*a*) 7/366 (*b*) 26/183
(*c*) 1/7 (*d*) 2/7

2. The probabilities of three mutually exclusive events A, B and C are given by 2/3, 1/4 and 1/6 respectively. The statement
(*a*) is true (*b*) is false
(*c*) Nothing can be said (*d*) Could be either

3. A father has 3 children with at least one boy. The probability that he has 2 boys and one girl is
(*a*) 1/4 (*b*) 3/7
(*c*) 1/3 (*d*) None of these

4. Let A and B be two events such that $P(A) = 0.4$, $P(B) = 0.3$ and $P(A \cup B) = 0.7$. Then
(*a*) A and B are independent
(*b*) A and B are exhaustive
(*c*) A and B are mutually exclusive
(*d*) None of these

5. If the letters of the word ASSASSIN are written down at random in a row, the probability that no two S's occur together is
(*a*) 3/5 (*b*) 1/28
(*c*) 1/35 (*d*) 1/14

6. A man alternately tosses a coin and throws a dice beginning with the coin. The probability that he gets a head in the coin before he gets a 5 or 6 on the dice is
(*a*) 3/4 (*b*) 1/2
(*c*) 1/3 (*d*) None of these

7. Let A be a set of 4 elements. From the set of all functions from A to A, the probability that it is an into function is
(*a*) 3/32 (*b*) 29/32
(*c*) 0 (*d*) 1

8. If A and B are two events such that $P(A \cup B) = 5/6$, $P(\overline{A}) = 1/4$, $P(B) = 1/3$, then A and B are
(*a*) mutually exclusive
(*b*) dependent
(*c*) independent
(*d*) None of these

9. In a box containing 100 bulbs, 10 are defective. The probability that out of a sample of 5 bulbs none is defective is
(*a*) $1/10^5$ (*b*) $(9/10)^5$
(*c*) ${}^{90}C_5/{}^{100}C_5$ (*d*) None of these

10. A pack of cards contains 4 aces, 4 kings, 4 queens and 4 jacks. Two cards are drawn at random. The probability that at least one of them is an ace is
(*a*) 1/5 (*b*) 3/16
(*c*) 9/20 (*d*) 1/9

11. A bag contain 10 mangoes out of which 4 are rotten, two mangoes are taken out together. If one of them is found to be good, the probability that other is also good is
(*a*) 5/18 (*b*) 8/15
(*c*) 5/13 (*d*) 2/3

12. An unbiased dice with faces 1, 2, 3, 4, 5 and 6 is round 4 times. Out of four face values obtained, the probability that the minimum face value is not less than 2 and the maximum face value is not greater than 5, is
(*a*) 16/81 (*b*) 1/81
(*c*) 80/81 (*d*) 65/81

13. Two cards are drawn successively with replacement from a well shuffled deck of 52 cards. The probability of drawing two aces is:
(*a*) $\frac{1}{13} \times \frac{1}{17}$ (*b*) $\frac{1}{13} \times \frac{1}{13}$
(*c*) $\frac{1}{13} \times \frac{4}{51}$ (*d*) $\frac{1}{52} \times \frac{1}{51}$

14. (*i*) The probability that in the toss of two dices we obtain the sum 7 or 11 is:
(*a*) 1/18 (*b*) 1/6
(*c*) 23/108 (*d*) 2/9
(*ii*) Two dices are thrown. The probability that the same of the points on two dices will be 7 is:
(*a*) 1/6 (*b*) 5/36
(*c*) 2/9 (*d*) 7/36

15. A biased dice is tossed and the respective probabilities for various faces to turn up are

Face	1	2	3	4	5	6
Probability	0.1	0.24	0.19	0.18	0.15	0.14

If an even face has turned up, then the probability that it facd 2 or face 4 is;

(*a*) 0.42 (*b*) 0.25
(*c*) 0.9 (*d*) 0.75

ANSWERS

1	2	3	4	5	6	7	8	9	10
(*a*)	(*b*)	(*b*)	(*c*)	(*d*)	(*a*)	(*b*)	(*c*)	(*c*)	(*c*)
11	**12**	**13**	**14**	**15**					
(*c*)	(*a*)	(*b*)	(*d*)	(*d*)					

SOME SELECTED EXPLANATORY ANSWERS

1. Total number of ways = $^{100}C_1 = 100$
Favourable numbers are $1^2, 2^2, ..., 10^2$.
Therefore favourable ways = 10
$\therefore$ Probability = 10/100 = 1/10.

3. Total ways = $^{80}C_2$
favourable ways = $^{20}C_2$

$$P = \frac{^{20}C_2}{^{80}C_2} = \frac{19}{316}$$

4. $P(A \cap B) = P(A) + P(B) - P(A \cup B)$
$= 0.4 + 0.3 - 0.7 = 0$
$\therefore P(A \cap B) \neq P(A)\,P(B)$ so A and B are not independent.
{Since $P(A \cap B) \neq 1$, A and B are not exhaustive.}
$P(A \cap B) = 0$ gives A and B are mutually exclusive.

5. Let A, A, I, N be arranged. Now there are 5 place for 'S' to be arranged.
$\therefore$ Ways for arrangement of 'S'

$$= {}^5C_4 . \frac{4!}{2!} = 60.$$

And total number of ways of arrangement of 8

$$\text{letters} = \frac{8!}{4!\,2!} = \frac{8.7.6.5}{2}$$

$\therefore$ P(no two S occur together)

$$= \frac{60.2}{8.7.6.5} = \frac{1}{14}.$$

7. Set A contains 4 elements.
$\therefore$ Total function that can be formed
$= 4^4 = 256$
Out of these, onto functions = 4! = 24
$\therefore$ Into functions = 256 − 24 = 232
$\therefore$ P(of into function) = 232/256 = 29/32.

9. Favourable ways of selecting 5 non-defective bulbs = $^{90}C_5$.
Total no. of ways of selection = $^{100}C_5$

$$\therefore \text{P (non-defective bulbs)} = \frac{^{90}C_5}{^{100}C_5}.$$

12. In a single throw the favourable point are 2, 3, 4 and 5 whose number is 4.
All possible outcomes are 6.
$\therefore$ P = Probability that in a single throw the minimum face values is not less than 2 and the maximum face value is not greater than 5
= 4/6 = 2/3
Since the dice is rolled four times and all the four throws are independent events, therefore the required probability = $(2/3)^4 = 16/81$.

23

Vector Algebra

1. **Scalar Quantity:** A quantity which has magnitude only and is not related to any direction is called a scalar quantity or briefly a scalar, *e.g.*, mass, volume, temperature, density, etc.

2. **Vector Quantity:** A quantity which has magnitude and also a direction in space is called a vector quantity or briefly a vector, *e.g.*, force, velocity, displacement, acceleration, etc.

3. **Notation of Vectors:** Vectors are represented by bold type letters **a, b, c,** ... or $\vec{a}, \vec{b}, \vec{c}$, ... or $\bar{a}, \bar{b}, \bar{c}$, ..., *i.e.*, to represent a vector, either we use the bold type letters or put an arrow (→) or bar (–) over the letters *a, b, c,* ... whereas scalars are written as ordinary letters without arrows or bars.

4. **Modulus of a Vector:** The positive number which is the measure of the magnitude of a vector is called its module or modulus. The modulus of a vector *a* is denoted by the corresponding italic letter *a*. The modulus *a* of a vector **a** is also written as | **a** |

   ∴ Thus $a = |\mathbf{a}|$.

5. **Null Vector:** A vector whose magnitude is zero is called a null vector or zero vector. It is denoted by 0 or $\bar{0}$ or $\vec{AA}$ or $\vec{PP}$ or $\vec{00}$. etc.

6. **Like and Unlike Vectors:** The vectors having the same direction are called the like vectors and the vectors having opposite directions are called unlike vectors.

7. **Coplanar Vectors:** The vectors which lie in the same plane or are parallel to the same plane are called coplanar vectors.

8. **Position Vector (p.v.):** Let O be a fixed point in space, called the origin and let P be any point in space, then the vector $\vec{OP}$ is called the position vector (p.v.) of the point P w.r.t. the origin O.

9. **Addition of Vectors:** Given two vectors **a** and **b**. Draw vector $\vec{OA}$ to represent **a.** Taking the terminal point A of vector **a** as the initial point of vector **b**, draw $\vec{AB}$ to represent **b.** Then the vector $\vec{OB}$ drawn from the initial point of **a** to the terminal point of **b** represents the sum of two vectors **a** and **b.** Thus if $\vec{OA} = \mathbf{a}$, $\vec{AB} = \mathbf{b}$ then $\vec{OB} = \vec{OA} + \vec{AB}$, *i.e.*, $\vec{OB} = \mathbf{a} + \mathbf{b}$. The above rule of addition of two vectors is called the triangular law of vector addition.

   In ΔOAB, $\vec{OB} = \vec{OA} + \vec{AB}$

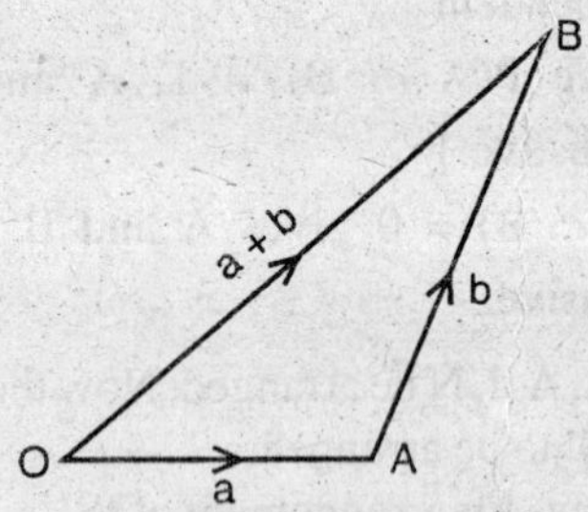

10. **Collinearly of Three Points:** The necessary and sufficient condition for three points with position vectors **a**, **b** and **c** to be collinear is that there exist three scalars *x, y, z* not all zero such that $x\mathbf{a} + y\mathbf{b} + z\mathbf{c} = 0$ and $x + y + z = 0$. Also if

    (*i*) **a**, **b**, **c**, are collinear

    (*ii*) there exists scalar *x, y, z* such the $x\mathbf{a} + y\mathbf{b} + z\mathbf{c} = 0$

(*iii*) at least one of the scalars is non-zero, then $x + y + z = 0$.

11. The Scalar (or dot) Product of Two Vectors: The scalar (or dot) product of two vectors **a** and **b** whose moduli are a and b respectively, is defined as the scalar quantity $ab \cos \theta$, where θ is the angle between the two vectors when drawn from a common origin.

The scalar product of two vectors **a** and **b** is denoted by **a·b.**

Thus $\mathbf{a}\cdot\mathbf{b} = ab \cos \theta = |\mathbf{a}||\mathbf{b}| \cos \theta$.

12. Scalar Products of Unit Vectors i, j, k:

$\mathbf{i}\cdot\mathbf{i} = 1 = \mathbf{j}\cdot\mathbf{j} = \mathbf{k}\cdot\mathbf{k}$; $\quad \mathbf{i}\cdot\mathbf{j} = \mathbf{j}\cdot\mathbf{k} = \mathbf{k}\cdot\mathbf{i} = 0$

If $\mathbf{a} = a_1\mathbf{i} + a_2\mathbf{j} + a_3\mathbf{k}$ and $\mathbf{b} = b_1\mathbf{i} + b_2\mathbf{j} + b_3\mathbf{k}$.

Then, $\mathbf{a}\cdot\mathbf{b} = a_1b_1 + a_2b_2 + a_3b_3$.

13. Condition for the Two Vectors to be Perpendicular: The two non-zero vectors

$\mathbf{a} = a_1\mathbf{i} + a_2\mathbf{j} + a_3\mathbf{k}$ and

$\mathbf{b} = b_1\mathbf{i} + b_2\mathbf{j} + b_3\mathbf{k}$ are perpendicular iff

$\mathbf{a}\cdot\mathbf{b} = 0$ *i.e.,* $a_1b_1 + a_2b_2 + a_3b_3 = 0$

14. Cosine of the Angle Between Two Vectors: If θ is the angle between two vectors

$\mathbf{a} = a_1\mathbf{i} + a_2\mathbf{j} + a_3\mathbf{k}$ and

$\mathbf{b} = b_1\mathbf{i} + b_2\mathbf{j} + b_3\mathbf{k}$ then

$$\cos \theta = \frac{a_1b_1 + a_2b_2 + a_3b_3}{\sqrt{(a_1^2 + a_2^2 + a_3^2)}.\sqrt{(b_1^2 + b_2^2 + b_3^2)}}$$

15. Projection of a Vector in the Direction of Other Vector and Components of a Vector along and Perpendicular to Other Vector:

Projection of a vector **r** along a vector $\mathbf{a} = \mathbf{r}\cdot\hat{a}$

$$= \frac{\mathbf{r}\cdot\mathbf{a}}{|\mathbf{a}|}$$

Length of projection of **r** along $\mathbf{a} = |\mathbf{r}\cdot\hat{a}|$

$$= \left|\frac{\mathbf{r}\cdot\mathbf{a}}{|\mathbf{a}|}\right|$$

Component of **r** along $\mathbf{a} = (\mathbf{r}\cdot\hat{a})\hat{a} = \dfrac{(\mathbf{r}\cdot\mathbf{a})\mathbf{a}}{|a|^2}$ and

Component of $\mathbf{r} \perp \mathbf{a} = \mathbf{r} - (\mathbf{r}\cdot\hat{a})\hat{a} = \mathbf{r} - \dfrac{(\mathbf{r}\cdot\mathbf{a})\mathbf{a}}{|\mathbf{a}|^2}$

An important result $\mathbf{a} = (\mathbf{a}\cdot\mathbf{i})\mathbf{i} + (\mathbf{a}\cdot\mathbf{j})\mathbf{j} + (\mathbf{a}\cdot\mathbf{k})\mathbf{k}$.

16. The Vector (or Cross) Product of Two Vectors: The vector (or cross) product of two vectors **a** and **b** whose moduli are **a** and **b** respectively is a vector quantity whose magnitude is **ab** $\sin \theta$, where θ is the angle between the two vectors and whose direction is that of a unit vector $\hat{n}$ perpendicular to both **a** and **b** such that **a, b** and $\hat{n}$ form a right-hand triad, *i.e.,* they are in right handed orientation.

By the right handed orientation we mean that if we turn a right handed screw from the vector **a** into the vector **b**, $\hat{n}$ will point in a direction, perpendicular to **a** and **b**, in which this right handed screw will move.

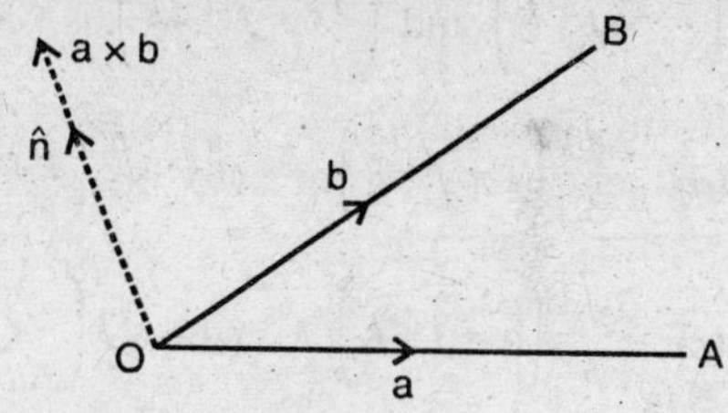

The vector product of two vectors **a** and **b** is denoted by **a × b**.

Thus $\mathbf{a} \times \mathbf{b} = ab \sin \theta.\hat{n}$.

Obviously $|\mathbf{a} \times \mathbf{b}| = ab \sin \theta = |\mathbf{a}||\mathbf{b}| \sin \theta$.

17. Properties of Vector Product:

(*i*) If **a** and **b** are parallel vectors, then $\theta = 0$ or π. Thus $\mathbf{a} \times \mathbf{b} = 0$. Thus $\mathbf{a} \times \mathbf{a} = 0$, $\mathbf{b} \times \mathbf{b} = 0$

(*ii*) If **a** and **b** are mutually perpendicular vectors, then $\theta = \pi/2$. Thus $|\mathbf{a} \times \mathbf{b}| = ab$.

(*iii*) Vector product of unit vector, **i, j, k.** If **i, j, k** is orthonormal triad of vectors

then $\quad \mathbf{i} \times \mathbf{i} = 1.1 \sin 0°.\hat{n} = 0$

Thus $\quad \mathbf{i} \times \mathbf{i} = \mathbf{j} \times \mathbf{j} = \mathbf{k} \times \mathbf{k} = 0$

$\mathbf{i} \times \mathbf{j} = 1.1 \sin 90°.\ \mathbf{k} = \mathbf{k}$.

$\because \hat{n} = \mathbf{k}$, unit vector $\perp$ to **i** and **j** according to right handed system.

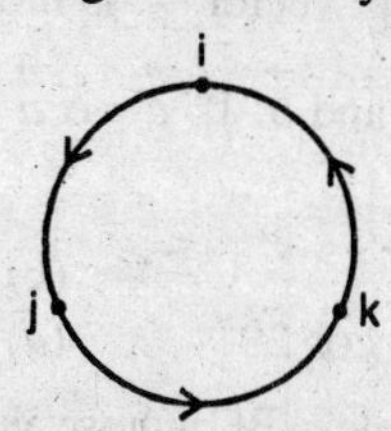

Similarly $\mathbf{j} \times \mathbf{k} = \mathbf{i}$ and $\mathbf{k} \times \mathbf{i} = \mathbf{j}$

Also $\quad \mathbf{j} \times \mathbf{i} = -\mathbf{k}$, $\mathbf{k} \times \mathbf{j} = -\mathbf{i}$, $\mathbf{i} \times \mathbf{k} = -\mathbf{j}$.

18. Area of a Triangle: Vector area of $\Delta ABC = \frac{1}{2}(\overrightarrow{AB} \times \overrightarrow{AC})$ and its magnitude $= \frac{1}{2}|\overrightarrow{AB} \times \overrightarrow{AC}|$

19. Area of a Parallelogram:

(*a*) **When adjacent sides are given:** If **a** and **b** represent the adjacent sides OA and OB of a parallelogram, then Area of the parallelogram = | **a × b** |

(*b*) **When the diagonals are given:** Area of the parallelogram OACB $= \frac{1}{2}|\overrightarrow{OC} \times \overrightarrow{AB}|$

MULTIPLE CHOICE QUESTIONS

1. A vector of magnitude 3 and perpendicular to $\left(\hat{i} - 2\hat{j} + \hat{k}\right)$ and $\left(2\hat{i} + \hat{j} - 3\hat{k}\right)$ is

(*a*) $\frac{2\sqrt{3}}{3}\left(\hat{i} + \hat{j} - \hat{k}\right)$ (*b*) $\sqrt{3}\left(\hat{i} + \hat{j} + \hat{k}\right)$

(*c*) $\frac{2\sqrt{3}}{3}\left(\hat{i} - \hat{j} + \hat{k}\right)$ (*d*) $\sqrt{3}\left(-\hat{i} + \hat{j} + \hat{k}\right)$

2. If $\vec{a}, \vec{b}, \vec{c}$ are non-coplanar unit vectors such that

$$\vec{a} \times \left(\vec{b} \times \vec{c}\right) = \frac{\left(\vec{b} + \vec{c}\right)}{\sqrt{2}}$$

where b and c are non-parallel, then the angle between $\vec{a}$ and $\vec{b}$ is

(*a*) $\frac{\pi}{4}$ (*b*) $\frac{3\pi}{4}$

(*c*) $\frac{\pi}{2}$ (*d*) None of these

3. The unit vectors $\vec{a}$ and $\vec{b}$ are perpendicular and the unit vector $\vec{c}$ is inclined at an angle θ to both $\vec{a}$ and $\vec{b}$. If $\vec{c} = \alpha\vec{a} + \beta\vec{b} + \gamma\left(\vec{a} \times \vec{b}\right)$, then

(*a*) $\alpha = \beta$ (*b*) $\gamma^2 = 1 - 2\alpha^2$

(*c*) $\gamma^2 = -\cos 2\theta$ (*d*) All of them

4. The triangle ABC whose vertices A, B and C are $7\hat{j} + 10\hat{k}$, $-\hat{i} + 6\hat{j} + \hat{k}$, $-4\hat{i} + 9\hat{j} + 6\hat{k}$ is

(*a*) Isosceles triangle

(*b*) Right angled triangle

(*c*) Isosceles right angle triangle

(*d*) None of these

5. A unit vector coplanar with $\hat{i} + \hat{j} + 2\hat{k}$ and $\hat{i} + 2\hat{j} + \hat{k}$ and perpendicular to $\hat{i} + \hat{j} + \hat{k}$ is

(*a*) $\pm\frac{1}{2}\left(\hat{j} + \hat{k}\right)$ (*b*) $\pm\frac{1}{\sqrt{2}}\left(\hat{j} + \hat{k}\right)$

(*c*) $\pm\frac{1}{\sqrt{2}}\left(\hat{j} - \hat{k}\right)$ (*d*) None of these

6. Let $\vec{a} = 2\hat{i} - 3\hat{j} - 6\hat{k}$ and $\vec{b} = -2\hat{i} - 2\hat{j} - \hat{k}$. Then the value of

$$\left[\frac{\text{The projection of } \vec{a} \text{ on } \vec{b}}{\text{The projection of } \vec{b} \text{ on } \vec{a}}\right] \text{ is equal to}$$

(*a*) $\frac{3}{7}$ (*b*) $\frac{7}{3}$

(*c*) 7 (*d*) 3

7. Vector coplanar with vectors $\hat{\imath} + \hat{\jmath}$ and $\hat{\jmath} + \hat{k}$ and parallel to the vector $2\hat{\imath} - 2\hat{\jmath} - 4\hat{k}$ is :

(*a*) $3\hat{\imath} + 3\hat{\jmath} - 6\hat{k}$ (*b*) $\hat{\imath} - \hat{\jmath} - 2\hat{k}$

(*c*) $\hat{\imath} - \hat{k}$ (*d*) $\hat{\imath} - \hat{\jmath} - \hat{k}$

8. If θ is the angle between unit vectors $\vec{a}$ and $\vec{b}$, then $\sin(\theta/2)$ is :

(*a*) $\frac{1}{2}|\vec{a} + \vec{b}|$ (*b*) $\frac{1}{2}|\vec{a} \times \vec{b}|$

(*c*) $\frac{1}{2}|\vec{a} - \vec{b}|$ (*d*) $\sqrt{\frac{1}{2}\left(1 - \vec{a}.\vec{b}\right)}$

9. If the sides of an angle are given by vectors $\vec{a} = \hat{i} - 2\hat{j} + 2\hat{k}$ and $\vec{b} = 2\hat{i} + \hat{j} + 2\hat{k}$, then the internal bisector is :

(*a*) $3\hat{i} - \hat{j} + 4\hat{k}$ (*b*) $\frac{1}{3}\left(3\hat{i} - \hat{j} - 4\hat{k}\right)$

(*c*) $3\hat{i} - \hat{j} - 4\hat{k}$ (*d*) None of these

10. Unit vector is the *xy*-plane that makes an angle of 60° with the vector $3\hat{i} - 4\hat{j}$ and an angle of 45° with the vector $\hat{i} + \hat{j}$ is :

(*a*) $\frac{\hat{i} + \hat{j}}{\sqrt{2}}$ (*b*) $\frac{\hat{i} - \hat{j}}{\sqrt{2}}$

(*c*) 0 (*d*) None of these

11. The value of p so that the unit vectors $\vec{a} = \frac{2\hat{i} + p\hat{j} + \hat{k}}{\sqrt{5 + p^2}}$ and $\vec{b} = \frac{\hat{i} - 2\hat{j} + 3\hat{k}}{\sqrt{14}}$ are orthonormal is :

(*a*) $\frac{2}{5}$ (*b*) $\frac{5}{2}$

(*c*) $\frac{3}{7}$ (*d*) $\frac{2}{7}$

12. Let $\vec{a}, \vec{b}, \vec{c}$ be three vectors such that $3\vec{a} + 4\vec{b} + 5\vec{c} = \vec{0}$. Then which of the following statement is true ?

(*a*) $\vec{a}$ is perpendicular to $\vec{b}$

(*b*) $\vec{a}$ is parallel to $\vec{b}$

(*c*) $\vec{a}$ is neither perpendicular nor parallel to $\vec{b}$

(*d*) None of these

13. If the vector $\vec{a} = 2\hat{i} - \hat{j} + \hat{k}, \vec{b} = \hat{i} + 2\hat{j} - 3\hat{k}$ and $\vec{c} = 3\hat{i} + p\hat{j} + 5\hat{k}$ are coplanar value of p is equal to :

(*a*) $-\frac{1}{4}$ (*b*) -1

(*c*) -4 (*d*) 3

14. The area of the triangle determined by the vectors $3\hat{i} + 4\hat{j}$ and $-5\hat{i} + 7\hat{j}$ is :

(*a*) $\frac{41}{2}$ sq.units (*b*) $\frac{141}{2}$ sq.units

(*c*) $\frac{-13}{2}$ sq.units (*d*) None of these

15. If $\vec{a}, \vec{b}, \vec{c}$ are mutually perpendicular vectors, each of magnitude, 3, then $|\vec{a} + \vec{b} + \vec{c}|$ is equal to:

(*a*) 3 (*b*) $3\sqrt{3}$

(*c*) 9 (*d*) None of these

ANSWERS

1	2	3	4	5	6	7	8	9	10
(*b*)	(*b*)	(*d*)	(*d*)	(*c*)	(*b*)	(*b*)	(*d*)	(*a*)	(*d*)
11	**12**	**13**	**14**	**15**					
(*b*)	(*d*)	(*c*)	(*a*)	(*b*)					

SOME SELECTED EXPLANATORY ANSWERS

1. Let $\vec{a} = \left(\hat{i} - 2\hat{j} + \hat{k}\right)$ and $\vec{b} = 2\hat{i} + \hat{j} - 3\hat{k}$.

Then $\vec{a} \times \vec{b} = \begin{vmatrix} \hat{i} & \hat{j} & \hat{k} \\ 1 & -2 & 1 \\ 2 & 1 & -3 \end{vmatrix} = 5\left(\hat{i} + \hat{j} + \hat{k}\right)$

$\therefore \quad \left|\vec{a} \times \vec{b}\right| = \sqrt{5^2 + 5^2 + 5^2} = 5\sqrt{3}$

Unit vector along $\left(\vec{a} \times \vec{b}\right) = \frac{5\left(\hat{i} + \hat{j} + \hat{k}\right)}{5\sqrt{3}}$

$$= \frac{\sqrt{3}}{3}\left(\hat{i}+\hat{j}+\hat{k}\right)$$

Required vector $= \frac{3\sqrt{3}}{3}\left(\hat{i}+\hat{j}+\hat{k}\right)$

$$= \sqrt{3}\left(\hat{i}+\hat{j}+\hat{k}\right).$$

2. $\vec{a}\times\left(\vec{b}\times\vec{c}\right) = \left(\vec{a}.\vec{c}\right)\vec{b} - \left(\vec{a}.\vec{b}\right)\vec{c}$

$\therefore \quad \vec{a}\times\left(\vec{b}\times\vec{c}\right) = \frac{\left(\vec{b}+\vec{c}\right)}{\sqrt{2}}$

$\Rightarrow \left(\vec{a}.\vec{c}\right)\vec{b} - \left(\vec{a}.\vec{b}\right)\vec{c} = \frac{\left(\vec{b}+\vec{c}\right)}{\sqrt{2}}$

$\Rightarrow \left(\vec{a}.\vec{c} - \frac{1}{\sqrt{2}}\right)\vec{b} - \left(\vec{a}.\vec{b} + \frac{1}{\sqrt{2}}\right)\vec{c} = 0$

$\Rightarrow \quad x\vec{b} - y\vec{c} = 0$

As $\vec{b}$ and $\vec{c}$ are non-parallel vectors, so $x = 0$ and $y = 0$

$\therefore \quad \vec{a}.\vec{b} = -\frac{1}{\sqrt{2}}$

or $\quad 1.1.\cos\theta = -\frac{1}{\sqrt{2}}$

or $\quad \cos\theta = \cos\left(\pi - \frac{\pi}{4}\right) = \cos\frac{3\pi}{4}$

or $\quad \theta = \frac{3\pi}{4}.$

6. $\dfrac{\text{The projection of } \vec{a} \text{ on } \vec{b}}{\text{The projection of } \vec{b} \text{ on } \vec{a}} = \dfrac{\left(\vec{a}.\vec{b}\right)/\left|\vec{b}\right|}{\left(\vec{b}.\vec{a}\right)/\left|\vec{a}\right|}$

$$= \frac{\sqrt{(2)^2+(-3)^2+(-6)^2}}{\sqrt{(-2)^2+(-2)^2+(-1)^2}}$$

$$= \frac{\sqrt{49}}{\sqrt{9}} = \frac{7}{3}.$$

11. Since $\vec{a}$ and $\vec{b}$ are orthonormal, then $\vec{a}.\vec{b} = 0$

or $\left(\frac{2\hat{i}+p\hat{j}+\hat{k}}{\sqrt{5+p^2}}\right).\left(\frac{\hat{i}-2\hat{j}+3\hat{k}}{\sqrt{14}}\right) = 0$

or $\quad 2-2p+3=0$

$\therefore \quad p = \frac{5}{2}$

14. Area of the triangle $= \frac{1}{2}\left|\vec{a}\times\vec{b}\right|$

$$= \frac{1}{2}\left|\left(3\hat{i}\times 4\hat{j}\right)\times\left(-5\hat{i}+7\hat{j}\right)\right|$$

$$= \frac{1}{2}\left|41\hat{k}\right| = \frac{41}{2}\text{sq. units}$$

15. We have, $\left|\vec{a}+\vec{b}+\vec{c}\right|^2$

$$= \left|\vec{a}\right|^2 + \left|\vec{b}\right|^2 + \left|\vec{c}\right|^2 + 2\left(\vec{a}.\vec{b}+\vec{b}.\vec{c}+\vec{c}.\vec{a}\right)$$

$$\Rightarrow \left|\vec{a}+\vec{b}+\vec{c}\right|^2 = (3)^2+(3)^2+(3)^2+2(0+0+0);$$

$$\left[\text{as } \left|\vec{a}\right| = \left|\vec{b}\right| = \left|\vec{c}\right| = 3 \text{ and } \vec{a}\perp\vec{b}, \vec{b}\perp\vec{c}, \vec{c}\perp\vec{a}\right]$$

$$\Rightarrow \left|\vec{a}+\vec{b}+\vec{c}\right|^2 = 27$$

Thus $\Rightarrow \left|\vec{a}+\vec{b}+\vec{c}\right| = 3\sqrt{3}$.

24

Statistics

Measures of Central Tendencies

In statistics, various types of averages or measures of central tendencies are studied.

They can be classified as below:

(a) Mathematical Averages:

(i) Arithmetic Mean or Mean,

(ii) Geometric Mean and

(iii) Harmonic Mean.

(b) Positional Averages: (i) Median, (ii) Mode.

Arithmetic Mean:

(a) **For a Discrete Data (Individual Series):**

(i) *Direct method:* Let $x_1, x_2, ..., x_n$ be n data-values, then A.M. x is given by

$$\bar{x} = \frac{1}{n}(x_1 + x_2 + ... + x_n) = \frac{1}{n}\sum_{i=1}^{n} x_i$$

(ii) *Short cut method:* Let A be an assumed mean, $d = x - A$, the deviation of each value of the variable from the assumed mean. Then

A.M. $\bar{x}$ is given by $\bar{x} = A + \frac{1}{n}\Sigma d$

(b) **For a Discrete Frequency Distribution:**

(i) *Direct method:* $\bar{x} = \frac{\Sigma fx}{\Sigma f} = \frac{1}{N}\Sigma fx$, where f is the frequency of the variable x.

(ii) *Short cut method:* $\bar{x} = A + \frac{1}{N}\Sigma fd$, where A = assumed mean, $d = x - A$ and $N = \Sigma f$.

(iii) *Step deviation method:* $\bar{x} = A + \frac{h}{N}\Sigma fd'$,

where $d' = \frac{1}{h}d = \frac{1}{h}(x - A)$, the step-deviations and $N = \Sigma f$.

(c) **For a Continuous Frequency Distribution (*i.e.*, data is given in the form of class intervals):** Here $x = \frac{1}{2}$(lower limit + upper limit) of the corresponding class-interval.

(d) **The Combined Mean Formula:** Let $\bar{x}_1, \bar{x}_2, ..., \bar{x}_k$, be the means of k groups of observations having repective sizes $n_1, n_2, ..., n_k$. Then $\bar{x}$ the combined mean is given by

$$\bar{x} = \frac{n_1\bar{x}_1 + n_2\bar{x}_2 + ... + n_k\bar{x}_k}{n_1 + n_2 + ... + n_k}$$

(e) **Weighted A.M.** The weighted arithmetic mean is given by

$$\bar{x}_w = \frac{\Sigma x^w}{\Sigma w}$$, where $\bar{x}_w$ is the weighted mean, w-values are the weights attached to values of the variable x.

Geometric Mean: The geometric mean of a set of n positive (> 0) observations is defined as the n-th root of their products. Thus

G.M. of $x_1, x_2, ..., x_n = (x_1.x_2. .... .x_n)^{1/n}$ or

$$\log(\text{G.M.}) = \frac{1}{n}\sum_{i=1}^{n}(\log x_i)$$

i.e. $$\text{G.M.} = \text{antilog}\left[\frac{1}{n}\sum_{i=1}^{n}\log x_i\right].$$

Harmonic Mean: Harmonic Mean (H.M.) of a set of observations is the reciprocal of the A.M. of the reciprocal of observations provided all observations are non-zero. Thus

$$\text{H.M.} = \left[\frac{1}{n}\Sigma(1/x)\right]^{-1}$$

For grouped data H.M. = $\left[\frac{1}{n}\Sigma(f/x)\right]^{-1}$, where N = Σf. It gives a better average in the computation of average speed, average price at which an article is sold.

Relationship Between A.M., G.M. and H.M.:

A.M. ≥ G.M. ≥ H.M., if all the observations are positive.

Equality sign holds only when all the observations in the series are same.

Median: Median is that value which divides the entire set of data into two equal parts.

(*a*) **Individual Series (Un-grouped Data):** First arrange the data in ascending or descending order. Then the median is

I. [½(*n* + 1)th item if *n* is odd.

II. Mean of the (½ *n*)th and (½ *n* + 1)th item if *n* is even.

$$\text{Median} = l_1 + \frac{(\frac{1}{2}N - c)}{f} \cdot (l_2 - l_1); \text{ where}$$

l_1 = lower limit of the median class

l_2 = upper limit of the median class

f = frequency of the median class

c = cumulative frequency of the class just preceding the median class.

Computation of the Quartiles, Deciles and Percentiles:

(*a*) **Individual and Discrete Series:** First arrange the data in ascending order only. Then

Quartile Q*r* = [(N + 1)*r*/4]th item in the series = 1, 2, 3.

Decile D*r* = [(N + 1)*r*/10]th item in the series, *r* = 1, 2, ..., 9.

Percentile P*r* = [(N + 1)*r*/100]th item in the series, *r* = 1, 2, ..., 99.

(*b*) **Continuous Series:**

1. *r*-th quartile: $Q_r = l_1 + \frac{(rN/4) - C}{f} \cdot (l_2 - l_1)$, $r = 1, 2, 3$

2. *r*-th decile: $D_r = l_1 + \frac{(rN/10) - C}{f} (l_2 - l_1)$, $r = 1, 2, ..., 9$.

3. *r*-th percentile: $P_r = l_1 + \frac{(rN/100) - C}{f} (l_2 - l_1)$, $r = 1, 2, ..., 99$.

Here l_1, l_2, f and c denote the usual meaning.

Mode: Mode or the modal value is the value in a series of observations which occurs with highest frequency near to the frequency of modal value, the method of grouping may be used.

$$\text{Mode} = l_1 + \frac{f_1 - f_0}{2f_1 - f_0 - f_2} \times h,$$

where l_1 = the lower limit of the modal class

f_1 = the frequency of the modal class

f_0 = the frequency of the class precedingthe modal class

f_2 = the frequency of the class succeedingthe modal class

h = the size of the modal class

Empirical Relationship Between Mean, Median, Mode:

(*a*) *In a symmetrical distribution:*
mean = median = mode.

(*b*) *In case of positive skewed curve:*
mode < median < mean.

(*c*) *In case of negative skewed curve:*
mean < median < mode.

(*d*) *For moderately asymmetrical distributions, the values of mean, median and mode have the following empirical relationship:*
Mode = Median – 2 Mean

Mean Deviation or Average Deviation:

(*a*) **Individual Series (Ungrouped Data):** Mean deviation from

(*i*) Mean $\bar{x}$ = M.D. ($\bar{x}$) = $\frac{1}{n}\Sigma|x - \bar{x}|$

(*ii*) Median M_d = M.D. (M_d) = $\frac{1}{n}\Sigma|x - M_d|$

(*iii*) Mode M_0 = M.D. (M_0) = $\frac{1}{n}\Sigma|x - M_0|$

Here $|x - \bar{x}|$, etc., are called absolute deviations.

(*b*) **Grouped Data or Frquency Distribution:**

$$\text{M.D.} = \frac{1}{n}\sum f|x - A|,$$

where A is average ($\bar{x}$ or M_d or M_0), f is the corresponding frequencies and $N = \sum f$.

Standard Deviation:

(*a*) **For Individual Series:** $\text{S.D.} = \sigma = \sqrt{\left[\frac{1}{n}\sum f(x-\bar{x})^2\right]}$.

(*b*) **For Frequency Distribution:**

$$\sigma = \sqrt{\left[\frac{1}{N}\sum f(x-\bar{x})^2\right]}\text{ ; Variance} = \sigma^2.$$

Computation of Standard Deviation:

(*a*) **Direct Method:** $\sigma = \sqrt{\left[\frac{\sum x^2}{n} - \bar{x}^2\right]}$

(*b*) **Short Cut Method:** Let A be any assumed mean and $d = x - A$. Then $\text{S.D.} = \sqrt{\left[\frac{\sum d^2}{n} - \left(\frac{\sum d}{n}\right)^2\right]}$.

For grouped data, it becomes

$$\text{S.D.} = \sigma = \sqrt{\left[\frac{\sum fd^2}{N} - \left(\frac{\sum fd}{N}\right)^2\right]},\ N = \sum f$$

MULTIPLE CHOICE QUESTIONS

1. If the mean of the set of numbers $x_1, x_2, ..., x_n$ is $\bar{x}$, then the mean of the numbers $x_i + 2_i$, $1 \le i \le n$ is

(*a*) $\bar{x} + 2n$ (*b*) $\bar{x} + n + 1$

(*c*) $\bar{x} + 2$ (*d*) $\bar{x} + n$

2. The weighted mean of first n natural numbers whose weights are equal to the squares of corresponding numbers is

(*a*) $\frac{n+1}{2}$ (*b*) $\frac{3n(n+1)}{2(2n+1)}$

(*c*) $\frac{(n+1)(2n+1)}{6}$ (*d*) $\frac{n(n+1)}{2}$

3. The arithmetic mean of 9 observations is 100 and that of 6 is 80, the combined mean of all the 15 observations will be

(*a*) 100 (*b*) 80

(*c*) 90 (*d*) 92

4. If the difference between mean and mode is 63 the difference between mean and median is

(*a*) 18 (*b*) 24

(*c*) 22 (*d*) 21

5. If a variable takes the discrete values $\alpha + 4, \alpha - \frac{7}{2}, \alpha - \frac{5}{2}, \alpha - 3, \alpha - 2, \alpha + \frac{1}{2}, \alpha - \frac{1}{2}, \alpha + 5$ ($\alpha > 0$), then the median is

(*a*) $\alpha - \frac{5}{4}$ (*b*) $\alpha - \frac{1}{2}$

(*c*) $\alpha - 2$ (*d*) $\alpha + \frac{5}{4}$

6. If the mean of the distribution

Variate x	1	2	3	4	5
Frequency of f of x	4	5	k	1	2

is 2.6, then the value of k is

(*a*) 8 (*b*) 10

(*c*) 12 (*d*) 18

7. The standard deviation of first n natural numbers is

(*a*) $\frac{n(n+1)(2n+1)}{6}$ (*b*) $\frac{n^2-1}{12}$

(*c*) $\sqrt{\left(\frac{n^2-1}{12}\right)}$ (*d*) None of these

8. Mean deviation of the series $a, a + d, a + 2d, a + 2nd$ from its mean is

(*a*) $\frac{(n+1)d}{(2n+1)}$ (*b*) $\frac{nd}{2n+1}$

(*c*) $\frac{n(n+1)d}{(2n+1)}$ (*d*) $\frac{(2n+1)d}{n(n+1)}$

9. A sample of 35 observations has the mean 80 and S.D. as 4. A second sample of 65 observations from the same population has mean 70 and S.D. 3. The S.D. of the combined sample is

(*a*) 5.85 (*b*) 5.58
(*c*) 34.2 (*d*) None of these

10. The mean and S.D. of the marks of 200 candidates were found to be 40 and 15 respectively. Later, it was discovered that a score 40 was wrongly read as 50. The correct mean and S.D. respectively are

(*a*) 14.98, 39.95 (*b*) 39.95, 14.98
(*c*) 39.95, 224.5 (*d*) None of these

ANSWERS

1	2	3	4	5	6	7	8	9	10
(*b*)	(*b*)	(*d*)	(*b*)	(*a*)	(*a*)	(*c*)	(*c*)	(*a*)	(*b*)

SOME SELECTED EXPLANATORY ANSWERS

1. Mean $= \frac{1}{n}\sum_{i=1}^{n}(x_i + 2i)$

$= \frac{1}{n}\Sigma x_i + \frac{2}{n}\sum_{i=1}^{n} i = \bar{x} + \frac{2}{n}.n(n+1)$

$= \bar{x} + n + 1$

2. $\bar{x} = \left(1\Big/\sum_{k=1}^{n} k^2\right).\sum_{1}^{n} k.k^2$

$= \frac{n^2(n+1)^2.6}{4.n(n+1)(2n+1)} = \frac{3}{2}\frac{n(n+1)}{(2n+1)}.$

5. Arranging the data as

$\alpha - 7/2, \alpha - 3, \alpha - 5/2, \alpha - 2, \alpha - 1/2,$

$\alpha + 1/2, \alpha + 4, \alpha + 5,$

median $= \frac{1}{2}(\alpha - 2 + \alpha - 1/2) = \alpha - 5/4.$

6. $\frac{1\times 4 + 2\times 5 + 3\times k + 4\times 1 + 5\times 2}{4+5+k+1+2} = 26$

$\therefore 28 + 3k = 26\,(12 + k) \Rightarrow k = 8.$

8. $\bar{x} = \frac{\frac{2n+1}{2}.[a+a+2nd]}{(2n+1)} = a + nd$

$\Sigma|x - \bar{x}| = 2d\,[1 + 2 + ... + n] = n\,(n+1)d$

$\therefore$ M.D. $= \frac{n(n+1)d}{2n+1}.$

9. Here $n_1 = 35, \bar{x}_1 = 80, \sigma_1 = 4$

$n_2 = 65, \bar{x}_2 = 70, \sigma_2 = 3$

$\therefore \quad \bar{x}_{12} = \frac{35\times 80 + 65\times 70}{35+65} = 73.5$

$\sigma_{12} = \sqrt{\left[\frac{35(16 + 42\times 25) + 65(9 + 12\times 25)}{100}\right]}$

$= \sqrt{(34.21)} = 5.85.$

10. Corrected $\Sigma x = 40\times 200 - 50 + 40 = 7990$

$\therefore$ Corrected $\bar{x}$ = 7990/200 = 39.95.

Incorrect $\Sigma x^2 = n[\sigma^2 + x^2]$

$= 200\,[40^2 + 15^2]$

$= 365000$

Corrected $\Sigma x^2 = 365000 - 2500 + 1600$

$= 364100$

$\therefore$ Corrected $\sigma = \sqrt{\left[\frac{364100}{200} - (39.95)^2\right]}$

$= \sqrt{(224.5)} = 14.98.$